Math Study:

Your Roadmap to Success in Math

The Carson Math Study System is designed to help you succeed in your math course. You will discover your own learning style, and you will use study strategies that match the way you learn best. You will also learn how to organize your course materials, manage your time efficiently, and study and review effectively.

Your Notebook

Notes (see pages xviii–xix):

Section 1.4   9/20

We can simplify an expression by combining like terms.

p. 37 def. Like terms: constant terms or variable terms that have the same variable(s) raised to the same powers.

Ex 1) 2x and 3x are like terms.
Ex 2) 5x and 7y are not like terms.

Homework (see pages xix–xxii):

Section 1.5 Homework   9/21

#1 — 15 odd

1. $5^2 + 3 \cdot 4 - 7$
   $= 25 + 3 \cdot 4 - 7$
   $= 25 + 12 - 7$
   $= 37 - 7$
   $= 30$

Quizzes/Tests (see page xxv):

Chapter 2 Quiz   10/10

For 1-4, simplify.

1. $|-8|$ $= 8$ ✓
                          4/4 = 100%
                          Nice work!
2. $|9|$ $= 9$ ✓

3. $-15 + 5$ $= -10$ ✓

4. $-8 + -6$ $= -14$ ✓

Study Materials (see pages xxiii–xxv):

Chapter 2 Study Sheets   9/10

To graph a number on a   ex) Graph -2,
number line, draw a dot on $-4 \ -3 \ -2 \ -1 \ 0 \ 1$
the mark for the number.

The absolute value of a positive
number is positive.   ex) $|7| = 7$

The absolute value of a negative
number is positive.   ex) $|-12| = 12$

Amste...
De...

Editorial Director: Christine Hoag
Editor in Chief: Maureen O'Connor
Executive Editor: Cathy Cantin
Executive Project Manager: Kari Heen
Senior Project Editor: Lauren Morse
Project Editor: Katie DePasquale
Editorial Assistant: Jonathan Wooding
Media Producers: Ceci Fleming and Vicki Dreyfus
Software Development: Mary Durnwald, Tanya Farber, and Eileen Moore
Senior Development Editor: Dawn Nuttall
Production Manager: Ron Hampton
Senior Design Supervisor: Andrea Nix
Art Director: Heather Scott
Production Services: Pre-Press PMG
Executive Marketing Manager: Michelle Renda
Marketing Manager: Adam Goldstein
Marketing Assistant: Margaret Wheeler
Prepress Services Buyer: Caroline Fell
Manufacturing Manager: Evelyn Beaton
Senior Manufacturing Buyer: Carol Melville
Senior Media Buyer: Ginny Michaud
Manager, Cover Visual Research and Permissions: Karen Sanatar

Cover photo: © Shutterstock/Hemisferic, City of Arts and Sciences, Valencia, Spain

Photo credits can be found on page App-30.

Library of Congress Cataloging-in-Publication Data

Carson, Tom, 1967–
    Elementary & intermediate algebra / Tom Carson, Bill Jordan.—3rd ed.
        p. cm.
    Includes bibliographical references and index.
    ISBN 978-0-321-62109-2
    1. Algebra—Textbooks. I. Jordan, Bill. II. Title. III. Title: Elementary and intermediate algebra.
    QA152.3.C374 2011
    512.9—dc22   2009020870

4 5 6 7 8 9—CKV—14 13 12

Addison-Wesley
is an imprint of

www.pearsonhighered.com

ISBN-10: 0-321-62109-3
ISBN-13: 978-0-321-62109-2

Your Le

To discover what ki
on page xvi (or in M
boxes and the accon
learning style.

Your L

LEARNING Strategy

Developing
good study
system and
understanding how you best
learn is essential to academic
success. Make sure you famil-
iarize yourself with the study
system outlined in the To the
Student section at the beginning of the text. Also take a m
ment to complete the Learnin
Styles Inventory found at the
end of that section to discove
your personal learning style. I
these Learning Strategy boxes,
we offer tips and suggestions
on how to connect the study
system and your learning styl
to help you be successful in t
course.

Contents

Preface

Welcome to the third edition of *Elementary and Intermediate Algebra*! Revising this series has been both exciting and rewarding, and it has given us the opportunity to respond to valuable instructor and student feedback. It is with great pride that we share with you both the improvements and additions to this edition as well as the hallmark features and style of the Carson/Jordan series.

Elementary and Intermediate Algebra, Third Edition, is the fourth book in a series that includes *Prealgebra*, Third Edition; *Elementary Algebra*, Third Edition; and *Intermediate Algebra*, Third Edition. This text is designed to be versatile enough for use in a standard lecture format, a self-paced lab, or even in an independent study format. Written in a relaxed, nonthreatening style, we have taken great care to ensure that students who have struggled with math in the past will be comfortable with the subject matter. Throughout the text, we explain *why* an algebraic process works the way it does, instead of just showing how to follow the process. In addition, problems from science, engineering, accounting, health, the arts, and everyday life link algebra to the real world. A complete Study System—including a Learning Styles Inventory, integrated Learning Strategy boxes, and a plan for developing a notebook as a personalized organization study tool—is also provided to guide students to success in the course. (See page xiii.)

Changes to the Third Edition

This revision includes refinements to the presentation of the material as well as the addition of examples, exercises, Your Turns, Instructor's Notes, and updated applications throughout the text.

The *To the Student* material within this Preface (pages xiii–xxv) has been extensively revised. In addition to revising the discussion about why mathematics is important and what it takes to succeed, a table comparing behaviors of strong and weak students has been added. Also, the explanation of the notebook system now includes sample notes, homework, and study materials.

Each Your Turn is now numbered to match its example. Additionally, examples are now correlated to corresponding groups of exercises in each exercise set. The authors have added section references to both the section Review Exercises and Cumulative Reviews. In the Annotated Instructor's Edition, the answers for the Practice Test exercises also have a reference to the corresponding section.

The *Elementary Algebra Review* has been streamlined to provide students with ample practice problems for transitional Elementary Algebra topics. Section references for each objective direct students to the relevant topics within the text itself for extra review. Two new appendixes, *Mean, Median, and Mode* and *Synthetic Division*, have also been added to the text.

To provide students with frequent opportunities to review what they have learned, Cumulative Review exercises now appear after every chapter, starting with Chapter 2.

The exercise sets have been carefully revised in response to user and reviewer feedback. Changes have been made to reflect the reorganization of topic coverage, to improve grading and pairing, to offer extra practice in problem solving, and to ensure a strong correlation between exercises and examples to help students master the math. Algebra Pyramid icons, used throughout the text to help students differentiate between constants and variables, expressions and equations and inequalities, have been added to each set of end-of-section Review Exercises in addition to the Chapter Review Exercises and the Cumulative Review exercises. They have also been added to the start of relevant sections.

Finally, new supplemental resources have been added to the Carson/Jordan series to provide additional support for students and instructors: *Worksheets for Classroom or Lab Practice, Chapter Test Prep Video* (in *Video Resources on DVD*), and enhanced topic coverage in *MyMathLab*.

Key Features

Study System   A study system is introduced in the To the Student section on pages xviii–xxv and is then reinforced throughout the text. The system recommends color codes for taking notes. The color codes are consistent in the text itself: red for definitions, blue for procedures and rules, and black for notes and examples. In addition, the study system presents strategies for succeeding in the course.

Learning Styles Inventory   A Learning Styles Inventory is presented on page xvi to help students assess their particular learning styles. Throughout the book, Learning Strategy boxes offer study tips keyed to specific learning styles.

Learning Strategy Boxes   Learning Strategy boxes with icons depicting the learning style appear where appropriate in the text to offer advice on how to use the study system effectively and how to study specific topics based on a student's individual learning style (see pages 3, 107, and 354).

The Algebra Pyramid   An Algebra Pyramid is used throughout the text to help students see how the topic they are learning relates to the big picture of algebra—particularly focusing on the relationship between constants, variables, expressions, and equations and inequalities (see pages 3, 92, and 348). In the end-of-section Review Exercises, Chapter Review Exercises, and Cumulative Review Exercises, an Algebra Pyramid icon indicates the level of the pyramid that correlates to a particular group of exercises to help students determine what actions are appropriate with these exercises, for example, whether to "simplify" or "solve" (see pages 238, 326, and 557).

Chapter Openers   Like the Algebra Pyramid, chapter openers are designed to help students see how the topics in the upcoming chapter relate to the big picture of the entire course. The chapter openers give information about the importance of the topics in each chapter and how they fit into the overall structure of the course (see pages 1, 91, and 169).

Connection Boxes   Connection boxes help students understand how math concepts are related and build on each other (see pages 92, 212, and 362).

Your Turn Practice Exercises   The Your Turn practice exercises found after most examples give students an opportunity to work problems similar to the examples they have just seen. This practice step makes the text more interactive and provides immediate feedback to help students develop confidence in their problem-solving skills (see pages 5, 97, and 294).

Real, Relevant, and Interesting Applications   The application problems in the examples and exercises reflect real situations in science, engineering, health, finance, the arts, and other areas. These real-world applications illustrate the everyday use of basic algebraic concepts and encourage students to apply mathematical concepts to solve problems (see pages 98, 211, and 302).

Thorough Explanations   Great care is taken to explain not only *how* to do the math but also *why* the math works the way it does, how concepts are related, and how the math is relevant to students' everyday lives. Knowing all of this gives students a context in which to learn and remember math concepts.

Problem-Solving Method   A five-step problem-solving method is introduced on page 94 of Section 2.1 with the following steps:

1. Understand
2. Plan
3. Execute
4. Answer
5. Check

Application examples throughout the text follow these steps, modeling the problem-solving process clearly and concisely (see pages 113, 175, and 188).

Warning Boxes   Warning boxes alert students to common mistakes and false assumptions that students often make and explain why these are incorrect (see pages 201, 309, and 353).

Of Interest Boxes   Throughout the text, Of Interest boxes offer a unique perspective on the mathematics. With historical notes and fun facts, these are designed to enhance the learning process by making math concepts interesting and memorable (see pages 20, 129, and 352).

Puzzle Problems   Mathematical brainteasers, often solved without a formulaic approach, appear at the end of selected exercise sets to encourage critical thinking (see pages 104, 378, and 403).

Collaborative Exercises   These exercises, which appear once per chapter, encourage students to work in groups to discuss mathematics and use the topics from a particular section or group of sections to solve a problem (see pages 135, 195, and 282).

Calculator Tips   The relevant functions of calculators (scientific or graphing, depending on the topic) are explained and illustrated throughout the text in the optional Calculator Tips. In addition, an occasional calculator icon in the exercise sets indicates that the problem is designed to be solved using a calculator, though one is not required (see pages 16, 44, and 277).

Review Exercises   Since continuous review is important in every mathematics course, this text includes Review Exercises at the end of each section's exercise set. These exercises review previously learned concepts not only to keep the material fresh for students, but also to serve as a foundational review for the discussion in the upcoming section (see pages 36, 302, and 358).

Chapter Summaries and Review Exercises   An extensive Summary at the end of each chapter provides a list of defined terms referenced by section and page number, a list of important formulas from the chapter, and a summary of key concepts with examples. A set of Chapter Review Exercises is also provided. Answers to all Chapter Review Exercises are given in the back of the book (see pages 156–163, 228–236, and 319–330).

Chapter Practice Tests   A Practice Test follows each set of chapter review exercises. The problem types in the practice tests correlate to the short-answer tests in the Printable Test Bank. This is especially comforting for students who have math anxiety or who experience test anxiety (see pages 89, 164, and 236). Answers to all Practice Test Exercises are provided in the back of the book. Students can also watch a video of an instructor working through each chapter test exercise solution in the Chapter Test Prep Videos, available on the *Video Resources on DVD,* in *MyMathLab,* and on **YouTube** (search "CarsonCombinedAlg" and click on "Channels").

Cumulative Reviews   Cumulative Review Exercises now appear after every chapter starting with Chapter 2. These exercises help students stay current with all the material they have learned and help prepare them for midterm and final exams (see pages 166 and 412). Answers to all Cumulative Review Exercises are provided in the back of the book.

Supplements for *Elementary and Intermediate Algebra*, Third Edition

Student Supplements

Student's Solutions Manual
- By Emily Keaton
- Contains complete solutions to the odd-numbered section exercises and solutions to all of the section-level Review Exercises, Chapter Review Exercises, Practice Tests, and Cumulative Review Exercises.

ISBN-10: 0-321-62734-2
ISBN-13: 978-0-321-62734-6

Worksheets for Classroom or Lab Practice
- By Carrie Green
- Extra practice exercises for every section of the text with ample space for students to show their work.
- These lab- and classroom-friendly workbooks also list key vocabulary terms for every text section, along with vocabulary practice problems.

ISBN-10: 0-321-62730-X
ISBN-13: 978-0-321-62730-8

Video Resources on DVD Featuring
Chapter Test Prep Videos
- Short lectures for each section of the text.
- Complete set of digitized videos on DVD for student use at home or on campus.
- NEW! Chapter Test Prep Videos, presented by Tina Levy (*Diablo Valley College*) and Jason Rosenberry (*Harrisburg Area Community College*), allow students to watch an instructor work through step-by-step solutions for all of the Chapter Test exercises from the textbook.
- Ideal for distance learning or supplemental instruction.
- All videos include optional English and Spanish subtitles.

ISBN-10: 0-321-62732-6
ISBN-13: 978-0-321-62732-2

MathXL® Tutorials on CD
This interactive tutorial CD-ROM provides algorithmically generated practice exercises that are correlated at the objective level to the exercises in the textbook. Every practice exercise is accompanied by an example and a guided solution designed to involve students in the solution process. Selected exercises may also include a video clip to help students visualize concepts. The software provides helpful feedback for incorrect answers and can generate printed summaries of students' progress.

ISBN-10: 0-321-62283-9
ISBN-13: 978-0-321-62283-9

Instructor Supplements

Annotated Instructor's Edition
- Includes answers to all exercises, including Puzzle Problems and Collaborative Exercises, printed in bright blue near the corresponding problems.
- Useful teaching tips are printed in the margin.
- A ★ icon, found in the AIE only, indicates especially challenging exercises in the exercise sets.

ISBN-10: 0-321-62133-6
ISBN-13: 978-0-321-62133-7

Instructor's Solutions Manual
- By Emily Keaton
- Contains complete solutions to all even-numbered section exercises, Puzzle Problems, and Collaborative Exercises.
- Available on the Instructor's Resource Center.

Instructor's Resource and Support Manual
- Includes resources designed to help both new and adjunct faculty with course preparation and classroom management.
- Offers helpful teaching tips specific to the sections of the text.
- Available on the Instructor's Resource Center.

Printable Test Bank
- By Laura Hoye, *Trident Technical College*
- Contains one diagnostic test per chapter, four free-response tests per chapter, one multiple-choice test per chapter, a mid-chapter check-up for each chapter, one midterm exam, and two final exams.
- Available on the Instructor's Resource Center.

TestGen®
TestGen (www.pearsonhighered.com/testgen) enables instructors to build, edit, print, and administer tests using a computerized bank of questions developed to cover all the objectives of the text. TestGen is algorithmically based, allowing instructors to create multiple, but equivalent, versions of the same question or test with the click of a button. Instructors can also modify test bank questions or add new questions. Tests can be printed or administered online. The software and testbank are available for download from Pearson Education's online catalog.

PowerPoint® Lecture Slides
- Present key concepts and definitions from the text.
- Available in MyMathLab or can be downloaded from the Instructor's Resource Center.

MyMathLab MyMathLab® Online Course (access code required)

MyMathLab is a text-specific, easily customizable online course that integrates interactive multimedia instruction with textbook content. MyMathLab gives you the tools you need to deliver all or a portion of your course online, whether your students are in a lab setting or working from home.

- **Interactive homework exercises,** correlated to your textbook at the objective level, are algorithmically generated for unlimited practice and mastery. Most exercises are free-response and provide guided solutions, sample problems, and learning aids for extra help.
- **Personalized Study Plan,** generated when students complete a test or quiz, indicates which topics have been mastered and links to tutorial exercises for topics students have not mastered. Instructors can customize the available topics in the study plan to match their course concepts.
- **Multimedia learning aids,** such as video lectures, animations, and a complete multimedia textbook, help students independently improve their understanding and performance.
- **Assessment Manager** lets you assign media resources (such as a video segment or a textbook passage), homework, quizzes, and tests. If you prefer, you can create your own online homework, quizzes, and tests that are automatically graded. Select just the right mix of questions from the MyMathLab exercise bank, instructor-created custom exercises, and/or TestGen test items.
- **Gradebook,** designed specifically for mathematics and statistics, automatically tracks students' results, lets you stay on top of student performance, and gives you control over how to calculate final grades. You can also add offline (paper-and-pencil) grades to the gradebook.
- **MathXL Exercise Builder** allows you to create static and algorithmic exercises for your online assignments. You can use the library of sample exercises as an easy starting point, or you can edit any course-related exercise.
- **Pearson Tutor Center** (www.pearsontutorservices.com) access is automatically included with MyMathLab. The Tutor Center is staffed by qualified math instructors who provide textbook-specific tutoring for students via toll-free phone, fax, email, and interactive Web sessions.

MyMathLab is powered by CourseCompass™, Pearson Education's online teaching and learning environment, and by MathXL®, our online homework, tutorial, and assessment system. MyMathLab is available to qualified adopters. For more information, visit www.mymathlab.com or contact your Pearson sales representative.

MathXL MathXL® Online Course (access code required)

MathXL® is an online homework, tutorial, and assessment system that accompanies Pearson's textbooks in mathematics or statistics.

- **Interactive homework exercises,** correlated to your textbook at the objective level, are algorithmically generated for unlimited practice and mastery. Most exercises are free-response and provide guided solutions, sample problems, and learning aids for extra help.
- **Personalized Study Plan,** generated when students complete a test or quiz, indicates which topics have been mastered and links to tutorial exercises for topics students have not mastered. Instructors can customize the available topics in the study plan to match their course concepts.
- **Multimedia learning aids,** such as video lectures and animations, help students independently improve their understanding and performance.
- **Gradebook,** designed specifically for mathematics and statistics, automatically tracks students' results, lets you stay on top of student performance, and gives you control over how to calculate final grades.
- **MathXL Exercise Builder** allows you to create static and algorithmic exercises for your online assignments. You can use the library of sample exercises as an easy starting point or use the Exercise Builder to edit any of the course-related exercises.
- **Assessment Manager** lets you create online homework, quizzes, and tests that are automatically graded. Select just the right mix of questions from the MathXL exercise bank, instructor-created custom exercises, and/or TestGen test items.

MathXL is available to qualified adopters. For more information, visit our website at www.mathxl.com or contact your Pearson sales representative.

Acknowledgments

Many people gave of themselves in so many ways during the development of this text. Mere words cannot contain the fullness of our gratitude. Though the words of thanks that follow may be few, please know that our gratitude is great.

We would like to thank the following people who have given of their time in reviewing this and previous editions of this textbook. Their thoughtful input was vital to the development of the text.

Khadija Ahmed, *Monroe County Community College*

Bruce Armbrust, *Lake Tahoe Community College*

Frank Attanucci, *Scottsdale Community College*

Daniel Bacon, *Massasoit Community College*

Kerry Bailey, *Laramie Community College*

Madelaine Bates, *Bronx Community College*

Denise Brown, *Collin County Community College*

Debra Bryant, *Tennessee Technological University*

Baruch Cahlon, *Oakland University*

Nancy Carpenter, *Johnson County Community College*

Delaine E. Cochran, *Indiana University Southeast*

Pat Cook, *Weatherford College*

Greg Cripe, *Spokane Falls Community College*

Patrick S. Cross, *University of Oklahoma*

Cheryl B. Davids, *Central Carolina Technical College*

Elias Deeba, *University of Houston–Downtown Campus*

Stephan DeLong, *Tidewater Community College–Virginia Beach Campus*

Laura Ferguson, *Weatherford College*

John Garlow, *Tarrant County Junior College*

Margret Hathaway, *Kansas City Kansas Community College*

Theresa McChesney, *Johnson County Community College*

Allen Miller, *South Plains College*

Carol Murphy, *Miramar College*

Joanne Peeples, *El Paso Community College*

Larry Pontaski, *Pueblo Community College*

Jack Sharp, *Floyd College*

Linda Shoesmith, *Scott Community College*

Mary Lou Townsend, *Wor-Wic Community College*

Beverly Vredevelt, *Spokane Falls Community College*

James Vicich, *Scottsdale Community College*

Linda J. Wagner, *Indiana University Purdue University Fort Wayne*

Walter Wang, *Baruch College*

Floyd Wouters, *University of Wisconsin Oshkosh*

We would like to extend a heartfelt thank-you to everyone at Pearson Education for giving so much to this project. We would like to offer special thanks to Cathy Cantin and Greg Tobin, who believed in us and gave us the opportunity; to Dawn Nuttall for her insightful comments during the development of this edition; to Jonathan Wooding, Katie DePasquale, Lauren Morse, and Kari Heen, for keeping us on track; and also to Marlana Voerster, Adam Goldstein, and Margaret Wheeler for their encouragement and for working so hard to get us "out there."

A very special thank-you to Ron Hampton, whose keen eyes and editorial sense were invaluable during production; and to Laura Hakala and all of the fabulous people at Pre-Press PMG for working so hard to put together the finished pages.

To Vicki Dreyfus, Peter Silvia, and all the people involved in developing the media supplements package, we are so grateful for all that you do. A special thank-you to Laura Hoye, who created the excellent Printable Test Bank, Carrie Green for her work on the Worksheets for Classroom or Lab Practice, and to Emily Keaton for her work on the solutions manuals. Thank you to Gary Williams, Laurie Delitsky of Nassau Community College, and Perian Herring, for their wonderful job of accuracy checking the manuscript and page proofs. A big thank-you goes to Elizabeth Morrison for her additional help reviewing art during production.

Finally, we'd like to thank our families for their support and encouragement during the process of developing and revising this text.

Tom Carson

Bill Jordan

To the Student

Why Do I Have to Take This Course?

Often, this is one of the first questions students ask when they find out they must take an algebra course. What a great question! But why focus on math alone? What about English, history, psychology, or science? Does anyone really use *every* topic of *every* course in the curriculum? Most jobs do not require that we write essays on Shakespeare, discuss the difference between various psychological theories, or analyze the cell structure of a frog's liver. So what's the point? The issue comes down to recognizing that general education courses are not job training. The purpose of those courses is to stretch and exercise the mind so that the educated person can better communicate, analyze situations, and solve problems, which are all valuable skills in life and any job.

Professional athletes offer a good analogy. A professional athlete usually has an exercise routine apart from their sport designed to build and improve their body. They may seek a trainer to design exercises intended to improve strength, stamina, or balance and then push them in ways they would not normally push themselves. That trainer may have absolutely no experience with their client's sport, but can still be quite effective in designing an exercise program because the trainer is focused on building basic skills useful for any athlete in any sport. Education is similar: it is exercise for the mind. A teacher's job is like that of a physical trainer. A teacher develops exercises intended to improve communication skills, critical-thinking skills, and problem-solving skills. Different courses are like different types of fitness equipment. Some courses may focus more on communication through writing papers, discussion, and debate. Other courses, such as mathematics, focus more on critical thinking and problem solving.

Another similarity is that physical exercise must be challenging for your body to improve. Similarly, mental exercise must be challenging for the mind to improve. Expect course assignments to challenge you and push you mentally in ways you wouldn't push yourself. That's the best way to grow. So as you think about the courses you are taking and the assignments in those courses, remember the bigger picture of what you are developing: your mind. When you are writing papers, responding to questions, analyzing data, and solving problems, you are developing skills important to life and any career out there.

What Do I Need to Succeed?

☑ **Adequate Time**   To succeed, you must have adequate time and be willing to use that time to perform whatever is necessary. To determine if you have adequate time, use the following guide.

Step 1. Calculate your work hours per week.
Step 2. Calculate the number of hours in class each week.
Step 3. Calculate the number of hours required for study by doubling the number of hours you spend in class.
Step 4. Add the number of hours from steps 1–3 together.

Adequate time: If the total number of hours is below 60, then you have adequate time.

Inadequate time: If the total number of hours is 60 or more, then you do not have adequate time. You may be able to hang in there for a while, but eventually, you will find yourself overwhelmed and unable to fulfill all of your obligations. Remember, the above calculations do not consider other likely elements of life such as commuting, family, recreation, and so on. The wise thing to do is cut back on work hours or drop some courses.

Assuming you have adequate time available, choosing to use that time to perform whatever is necessary depends on your attitude, commitment, and self-discipline.

☑ **Positive Attitude**   We do not always get to choose our circumstances, but we do get to choose our reaction and behavior. A positive attitude is choosing to be cheerful, hopeful, and encouraging no matter the situation. A benefit of a positive attitude is that it tends to encourage people around you. As a result, they are more likely to want to help you achieve your goal. A negative attitude, on the other hand, tends to discourage people around you. As a result, they are less likely to want to help you. The best way to maintain a positive attitude is to keep life in perspective, recognizing that difficulties and setbacks are temporary.

☑ **Commitment**   Commitment means binding yourself to a course of action. Remember, expect difficulties and setbacks, but don't give up. That's why a positive attitude is important, it helps you stay committed in the face of difficulty.

☑ **Self-Discipline**   Self-discipline is choosing to do what needs to be done—even when you don't feel like it. In pursuing a goal, it is normal to get distracted or tired. It is at those times that your positive attitude and commitment to the goal help you discipline yourself to stay on task.

Thomas Edison, inventor of the lightbulb, provides an excellent example of all of these principles. Edison tried over 2000 different combinations of materials for the filament before he found a successful combination. When asked about all his failed attempts, Edison replied, "I didn't fail once, I invented the lightbulb. It was just a 2000-step process." He also said, "Our greatest weakness lies in giving up. The most certain way to succeed is always to try just one more time." In those two quotes, we see a person who obviously had time to try 2000 experiments, had a positive attitude about the setbacks, never gave up, and had the self-discipline to keep working.

Behaviors of Strong Students and Weak Students

The four requirements for success can be translated into behaviors. The following table compares the typical behaviors of strong students with the typical behaviors of weak students.

Strong students ...	Weak students ...
are relaxed, patient, and work carefully.	are rushed, impatient, and hurry through work.
almost always arrive on time and leave the classroom only in an emergency.	often arrive late and often leave class to "take a break."
sit as close to the front as possible.	sit as far away from the front as possible.
pay attention to instruction.	ignore instruction, chit-chat, draw, fidget, etc.
use courteous and respectful language, encourage others, make positive comments, are cheerful and friendly.	use disrespectful language, discourage others, make negative comments, are grumpy and unfriendly. Examples of unacceptable language include: • "I hate this stuff!" (or even worse!) • "Are we doing anything important today?" • "Can we leave early?"
ask appropriate questions and answer instructor's questions during class.	avoid asking questions and rarely answer instructor's questions in class.

(continued)

Strong students ...	Weak students ...
take lots of notes, have organized notebooks, seek out and use study strategies	take few notes, have disorganized notebooks, do not use study strategies
begin assignments promptly and manage time wisely, and almost always complete assignments on time.	procrastinate, manage time poorly, and often complete assignments late.
label assignments properly and show all work neatly.	show little or no work and write sloppily.
read and work ahead.	rarely read or work ahead.
contact instructors outside of class for help, and use additional resources such as study guides, solutions manuals, computer aids, videos, and tutorial services.	avoid contacting instructors outside of class, and rarely use additional resources available.

Assuming you have the prerequisites for success and understand the behaviors of a good student, our next step is to develop two major tools for success:

1. **Learning Style:** complete the Learning Styles Inventory to determine how you tend to learn.
2. **The Study System:** this system describes a way to organize your notebook, take notes, and create study tools to complement your learning style.

Learning Styles Inventory

What is *your* personal learning style?

A learning style is the way in which a person processes new information. Knowing your learning style can help you make choices in the way you study and focus on new material. Below are fifteen statements that will help you assess your learning style. After reading each statement, rate your response to the statement using the scale below. There are no right or wrong answers.

3 = Often applies   **2** = Sometimes applies   **1** = Never or almost never applies

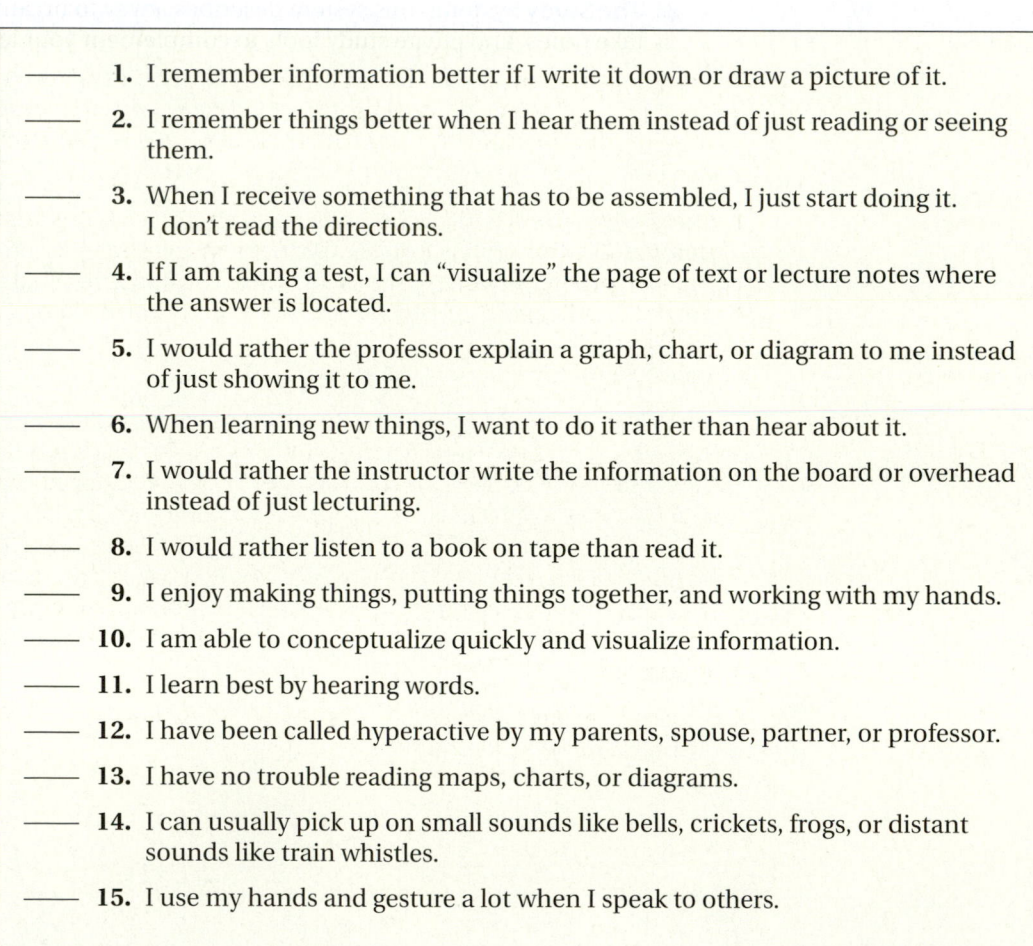

—— 1. I remember information better if I write it down or draw a picture of it.

—— 2. I remember things better when I hear them instead of just reading or seeing them.

—— 3. When I receive something that has to be assembled, I just start doing it. I don't read the directions.

—— 4. If I am taking a test, I can "visualize" the page of text or lecture notes where the answer is located.

—— 5. I would rather the professor explain a graph, chart, or diagram to me instead of just showing it to me.

—— 6. When learning new things, I want to do it rather than hear about it.

—— 7. I would rather the instructor write the information on the board or overhead instead of just lecturing.

—— 8. I would rather listen to a book on tape than read it.

—— 9. I enjoy making things, putting things together, and working with my hands.

—— 10. I am able to conceptualize quickly and visualize information.

—— 11. I learn best by hearing words.

—— 12. I have been called hyperactive by my parents, spouse, partner, or professor.

—— 13. I have no trouble reading maps, charts, or diagrams.

—— 14. I can usually pick up on small sounds like bells, crickets, frogs, or distant sounds like train whistles.

—— 15. I use my hands and gesture a lot when I speak to others.

Determining Your Learning Style   Write your score for each statement beside the appropriate statement number below. Then add the scores in each column to get a total score for that column.

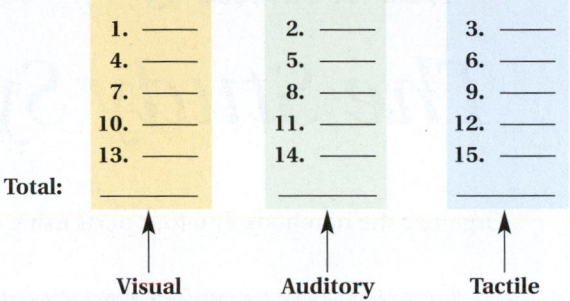

The largest total of the three columns indicates *your* dominant learning style.

Visual learners learn best by *seeing*. If this is your dominant learning style, then you should focus on learning strategies that involve seeing. The color coding in the study system (see page xviii) will be especially important. The same color coding is used in the text. Draw lots of diagrams, arrows, and pictures in your notes to help you see what is happening. Reading your notes, study sheets, and text repeatedly will be an important strategy.

Auditory learners learn best by *hearing*. If this is your dominant learning style, you should use learning strategies that involve hearing. After getting permission from your instructor, bring a tape recorder to class to record the discussion. When you study your notes, play back the tape. Also, when you learn rules, say the rule over and over. As you work problems, say the rule before you do the problem. You may also find the videotapes beneficial in that you can hear explanations of problems taken from the text.

Tactile (also known as kinesthetic) learners learn best by *touching or doing*. If this is your dominant learning style, you should use learning strategies that involve doing. Doing lots of practice problems will be important. Make use of the Your Turn exercises in the text. These are designed to give you an opportunity to do problems that are similar to the examples as soon as the topic is discussed. Writing out your study sheets and doing your practice tests repeatedly will be important strategies for you.

Note that the study system developed in this text is for all learners. Your learning style will help you decide what aspects and strategies in the study system to focus on, but being predominantly an auditory learner does not mean that you shouldn't read the textbook, do lots of practice problems, or use the color-coding system in your notes. Auditory learners can benefit from seeing and doing, and tactile learners can benefit from seeing and hearing. In other words, do not use your dominant learning style as a reason for not doing things that are beneficial to the learning process. Also, remember that the Learning Strategy boxes presented throughout the text provide tips to help you use your personal learning style to your advantage.

Index of Learning Strategies   Learning strategies keyed to visual, auditory, and tactile learners can be found on the following pages:

Visual	Auditory	Tactile
3, 6, 93, 107, 137, 156, 164, 172, 265, 276, 354, 363, 379, 417, 445, 576, 675, 861, 908, 918	3, 6, 93, 110, 156, 164, 186, 369, 418, 445, 675, 918	3, 6, 93, 110, 156, 164, 212, 244, 307, 354, 363, 445, 675, 844, 918

This Learning Styles Inventory is adapted from *Cornerstone: Creating Success through Positive Change*, Sixth Edition, by Moody/Sherfield © 2011. Reprinted by permission of Prentice Hall, Inc., Upper Saddle River, NJ.

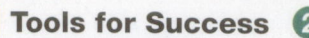

The Study System

Organize the notebook into four parts using dividers as shown:

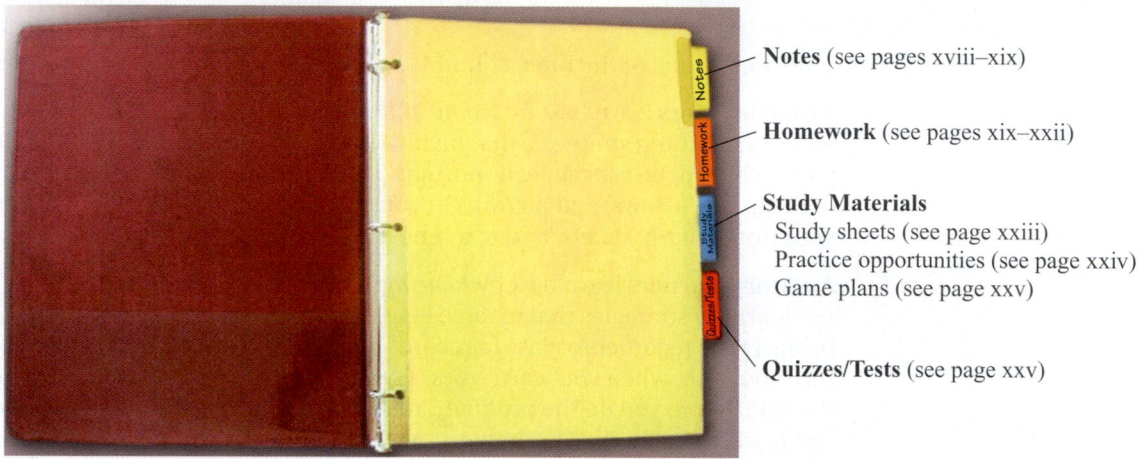

Notes (see pages xviii–xix)

Homework (see pages xix–xxii)

Study Materials
Study sheets (see page xxiii)
Practice opportunities (see page xxiv)
Game plans (see page xxv)

Quizzes/Tests (see page xxv)

Notes

- Use a color code: **red** for definitions, **blue** for rules or procedures, and **pencil** for all examples and other notes.
- Begin notes for each class on a new page (front and back for that day is okay). Include a topic title or section number and the date on each page.
- Try to write your instructor's spoken explanations along with the things he or she writes on the board.
- Mark examples your instructor emphasizes in some way to give them a higher priority. These problems often appear on quizzes and tests.
- Write warnings your instructor discusses about a particular situation.
- Include common errors that your instructor illustrates, but mark them clearly as errors so that you do not mistake them for correct.
- To speed note taking, eliminate unnecessary words like "the" and use codes for common words like $+$ for "and" and $\therefore$ for "therefore." Also, instead of writing complete definitions, rules, or procedures, write the first few words and place the page reference from the text so that you can copy from the text later.

Sample Notes with Color Code

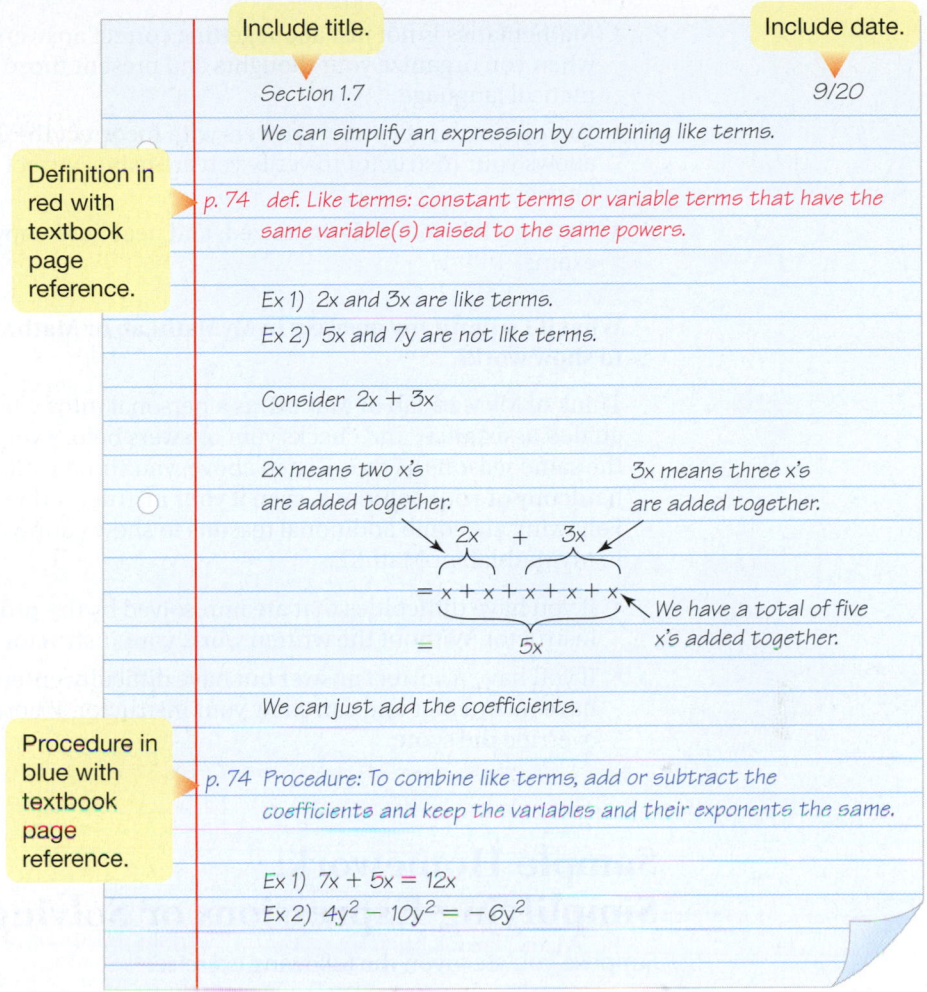

Include title.

Include date.

Section 1.7   9/20

We can simplify an expression by combining like terms.

Definition in red with textbook page reference.

p. 74   def. Like terms: constant terms or variable terms that have the same variable(s) raised to the same powers.

Ex 1) 2x and 3x are like terms.

Ex 2) 5x and 7y are not like terms.

Consider 2x + 3x

2x means two x's are added together.   3x means three x's are added together.

$$2x \ + \ 3x$$

$$= x + x + x + x + x$$ We have a total of five x's added together.

$$= \qquad 5x$$

We can just add the coefficients.

Procedure in blue with textbook page reference.

p. 74 Procedure: To combine like terms, add or subtract the coefficients and keep the variables and their exponents the same.

Ex 1) 7x + 5x = 12x

Ex 2) $4y^2 - 10y^2 = -6y^2$

Homework

This section of the notebook contains all homework. Use the following guidelines whether your assignments are from the textbook, a handout, or a computer program like MyMathLab or MathXL.

- Use pencil so that mistakes can be erased (scratching through mistakes is messy and should be avoided).

- Label according to your instructor's requirements. Usually, at least include your name, the date, and assignment title. It is also wise to write the assigned problems at the top as they were given. For example, if your instructor writes "Section 1.5 #1–15 odd," write it that way at the top. Labeling each assignment with this much detail shows that you take the assignment seriously and leaves no doubt about what you interpreted the assignment to be.

- For each problem you solve, write the problem number and show all solution steps neatly.

Why do I need to show work and write all the steps? Isn't the right answer all that's needed?

- Mathematics is not just about getting correct answers. You really learn mathematics when you organize your thoughts and present those thoughts clearly using mathematical language.

- You can arrive at correct answers with incorrect thinking. Showing your work allows your instructor to verify you are using correct procedures to arrive at your answers.

- Having a labeled, well-organized, and neat hard copy is a good study tool for exams.

What if I submit my answers in MyMathLab or MathXL? Do I still need to show work?

Think of MyMathlab or MathXL as a personal tutor who provides the exercises, offers guided assistance, and checks your answers before you submit the assignment. For the same reasons as those listed above, you should still create a neatly written hardcopy of your solutions, even if your instructor does not check the work. Following are some additional reasons to show your work when submitting answers in MyMathlab or MathXL.

- If you have difficulties that are unresolved by the program, you can show your instructor. Without the written work, your instructor cannot see your thinking.

- If you have a correct answer but have difficulty entering that correct answer, you have record of it and can show your instructor. If correct, your instructor can override the score.

Sample Homework:
Simplifying Expressions or Solving Equations

Suppose you are given the following exercise:

For Exercises 1–30 simplify.

   1. $5^2 + 3 \cdot 4 - 7$

Your homework should look something like the following:

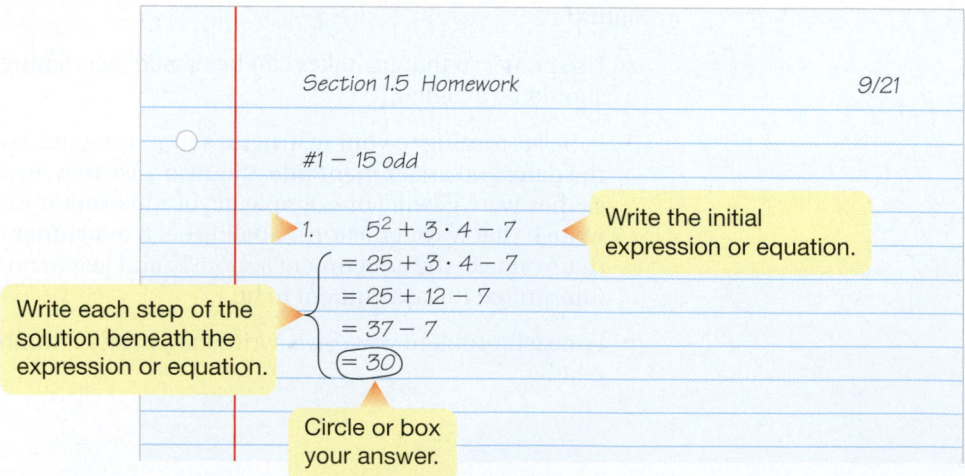

Sample Homework: Solving Application Problems

Example: Suppose you are given the following two problems.

For Exercises 1 and 2, solve.

1. Find the area of a circle with a diameter of 10 feet.
2. Two cars are traveling toward each other on the same highway. One car is traveling 65 miles per hour and the other is traveling at 60 miles per hour. If the two cars are 20 miles apart, how long will it be until they meet?

Your homework should look something like the following:

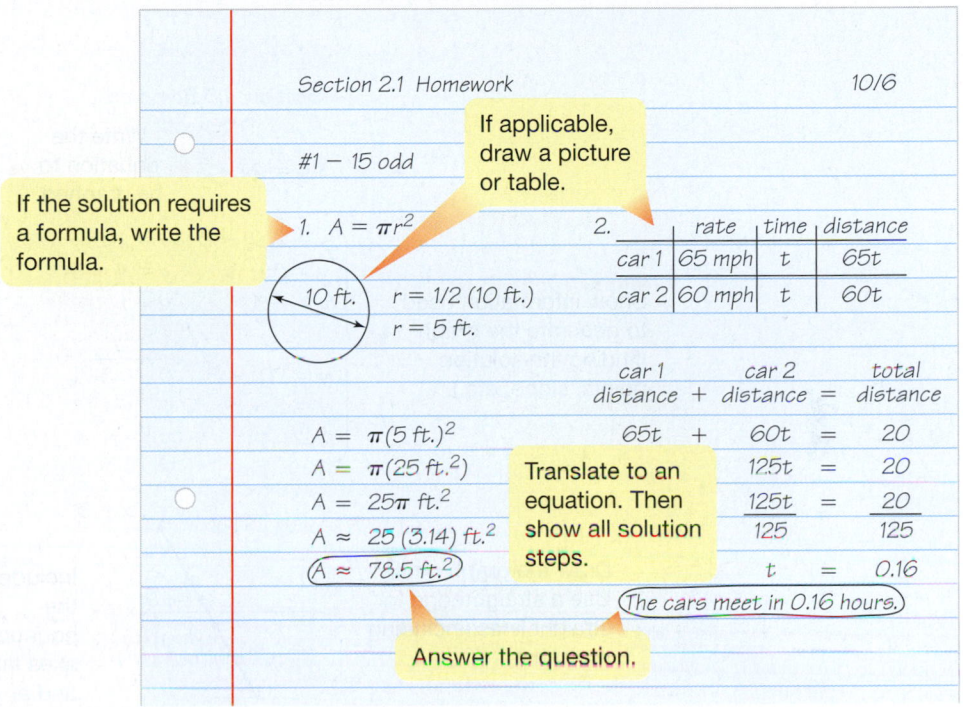

Sample Homework: Graphing

Example: Suppose you are given the following two problems.

For Exercises 1 and 2, graph the equation.

   1. $y = 2x - 3$ **2.** $y = -2x + 2$

Your homework should look something like the following:

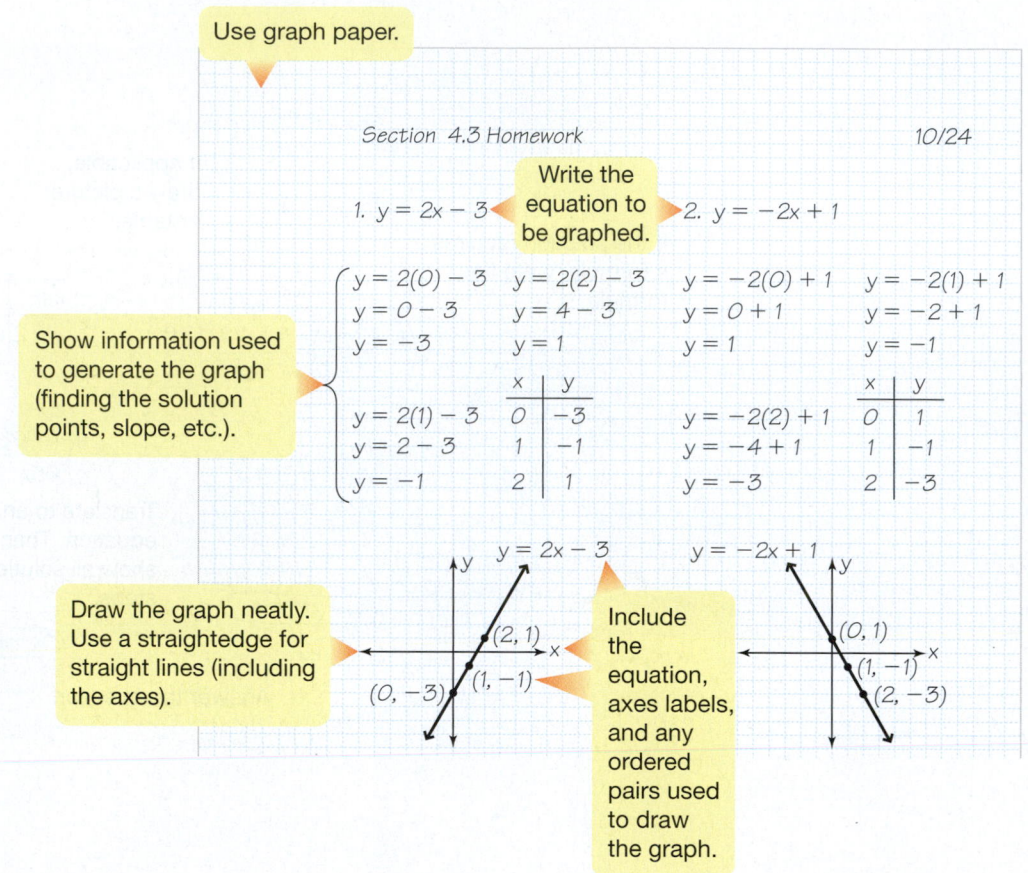

Study Materials

This section of the notebook contains three types of study materials for each chapter.

Study Material 1: The Study Sheet A study sheet contains *every* rule or procedure in the current chapter.

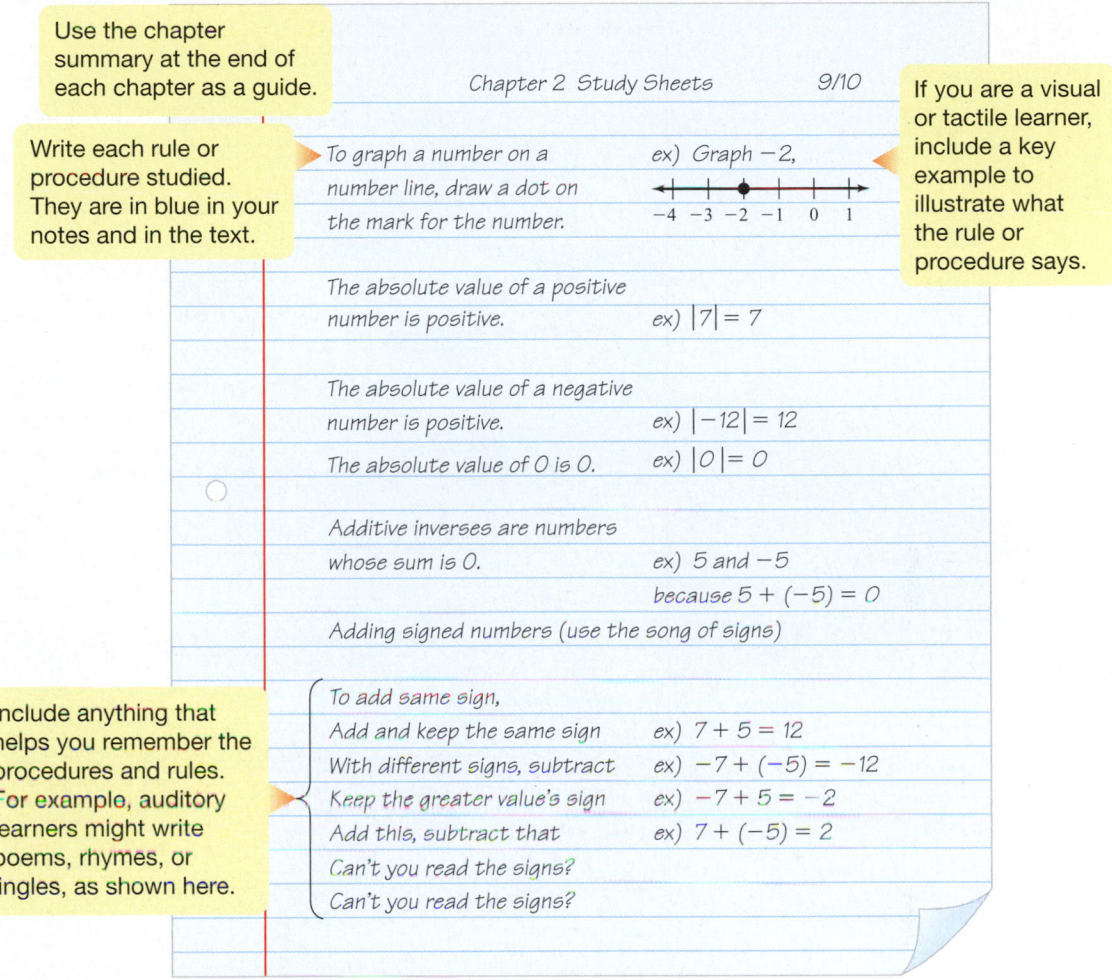

Use the chapter summary at the end of each chapter as a guide.

Write each rule or procedure studied. They are in blue in your notes and in the text.

If you are a visual or tactile learner, include a key example to illustrate what the rule or procedure says.

Include anything that helps you remember the procedures and rules. For example, auditory learners might write poems, rhymes, or jingles, as shown here.

Chapter 2 Study Sheets   9/10

To graph a number on a number line, draw a dot on the mark for the number.   ex) Graph -2,

The absolute value of a positive number is positive.   ex) $|7| = 7$

The absolute value of a negative number is positive.   ex) $|-12| = 12$

The absolute value of 0 is 0.   ex) $|0| = 0$

Additive inverses are numbers whose sum is 0.   ex) 5 and -5 because $5 + (-5) = 0$

Adding signed numbers (use the song of signs)

To add same sign,
Add and keep the same sign   ex) $7 + 5 = 12$
With different signs, subtract   ex) $-7 + (-5) = -12$
Keep the greater value's sign   ex) $-7 + 5 = -2$
Add this, subtract that   ex) $7 + (-5) = 2$
Can't you read the signs?
Can't you read the signs?

Study Material 2: The Practice Test   If your instructor gives you a practice test, proceed to the discussion of creating a game plan.

If your instructor does not give you a practice test, use your notes to create your own practice test out of the examples given in class. Include only the instructions and the problem, not the solutions. The following sample practice test was created from examples in the notes for Chapter 2 in a prealgebra course.

Chapter 2 Practice Test   10/10

For # 1 and 2, graph on a number line.

1. 4

2. -3

For #3 and 4, simplify.

3. $|-8|$

4. $|9|$

For #5 and 6, find the additive inverse of the number.

5. 13

6. -15

For #7 – 10, add.

7. $13 + (-9)$ 8. $-20 + (-6)$

9. $-15 + 8$ 10. $3 + (-24)$

> For each example in your notes, write the directions and the problem but not the solution.

> After working through the practice test, use your notes to check your solutions.

Study Material 3: The Game Plan   The game plan refines the study process further. It is your plan for the test based on the practice test. For each problem on your practice test, write the definition, rule, or procedure used to solve the problem.

The sample shown gives the rule or procedure used to solve each problem on the sample practice test on the previous page. The rules and procedures came from the sample study sheet.

Chapter 2 Game Plan   9/10

Multiple problems that use the same rule or procedure can be grouped together.

Write the rule or procedure used to solve the problems on the practice test.

#1 and 2: Draw a dot on the mark for the number.

#3 and 4: The absolute value of a positive number
          is a positive number.
          The absolute value of a negative number
          is a negative number.
          The absolute value of 0 is 0.

#5 and 6: Additive inverses are numbers whose
          sum is 0.

#7 – 10: To add same sign,
         Add and keep the same sign
         With different signs
         Subtract and keep the greater value's sign
         Add this, subtract that
         Can't you read the signs?
         Can't you read the signs?

Quizzes/Tests

Archive all returned quizzes and tests in this section of the notebook.

- Midterm and final exams questions are often taken from the quizzes and tests, so they make excellent study tools for those cumulative exams.
- Keeping all graded quizzes and tests offers a backup system in the unlikely event your instructor should lose any of your scores.
- If a dispute arises about a particular score, you have the graded test to show your instructor.

Chapter 2 Quiz   10/10

For 1-4, simplify.

4/4 = 100%
Nice
work!

1. $|-8|$ $= 8$ ✓

2. $|9|$ $= 9$ ✓

3. $-15 + 5$ $= -10$ ✓

4. $-8 + -6$ $= -14$ ✓

Foundations of Algebra

1

> " Three people were at work on a construction site. All were doing the same job, but when each was asked what the job was, the answers varied. 'Breaking rocks,' the first replied. 'Earning my living,' the second said. 'Helping to build a cathedral,' said the third. "
>
> —Peter Schultz,
> German businessman

> " Build up your weaknesses until they become your strengths. "
>
> —Knute Rockne (1888–1931)
> Notre Dame football coach

In Peter Schultz's quote, each of the three people answered the question correctly. Notice that the first two people were rather narrowly focused, whereas the third person had a greater vision and appreciation of the finished structure and his or her place in the construction process. Your curriculum and the courses in it are like a building, with the units or chapters like construction materials. Try to keep the big picture in mind as your courses and curriculum build. In the chapter openers of this book, we discuss how the mathematics in the chapter fits into the bigger picture of the entire course structure to help you see the "cathedral" we are building.

In this chapter, we review the foundation of algebra, which is arithmetic. More specifically, we review number sets, operations of arithmetic, properties of real numbers, and the evaluation and simplification of expressions. This review is by no means a complete instruction of arithmetic. Since a solid foundation is important to support the rest of our building, if you encounter a topic in this chapter that is a particular weakness for you, it is important that you consult with your instructor or other more complete sources for extra practice.

1.1 Number Sets and the Structure of Algebra

Objectives

1. Understand the structure of algebra.
2. Classify number sets.
3. Graph rational numbers on a number line.
4. Determine the absolute value of a number.
5. Compare numbers.

Objective 1 Understand the structure of algebra. Learning mathematics is like learning a language. When we learn a language, we must learn the alphabet, vocabulary, and sentence structure. Similarly, mathematics has its own alphabet, vocabulary, and sentence structure. In this section, we begin the development of the foundation of algebra with an overview of its components and structure. The basic components are **variables** and **constants**.

Definitions   *Variable:* A symbol that can vary in value.

Constant: A symbol that does not vary in value.

Variables are usually letters of the alphabet, such as x or y. Usually, constants are symbols for numbers such as 1, 2, $\frac{3}{4}$, and 6.74. However, constants can sometimes be symbols such as e or the Greek letter π, each having special numeric values. Variables and constants are used to make **expressions**, **equations**, and **inequalities**.

Definition   *Expression:* A constant, a variable, or any combination of constants, variables, and arithmetic operations that describes a calculation.

Examples of expressions:
$$2 + 6 \qquad 4x - 5 \qquad \frac{1}{3}\pi r^2 h$$

Definition   *Equation:* A mathematical relationship that contains an equal sign.

Examples of equations:
$$2 + 6 = 8 \qquad 4x - 5 = 12 \qquad V = \frac{1}{3}\pi r^2 h$$

Connection Think of expressions as phrases and equations as complete sentences. The expression $2 + 6$ is read "two plus six," which is not a complete sentence. The equation $2 + 6 = 8$ is read "two plus six is eight." Notice that the equal sign translates to the verb *is*, which makes the sentence a complete sentence.

Definition   *Inequality:* A mathematical relationship that contains an inequality symbol $(\neq, <, >, \leq, \text{ or } \geq)$.

Inequality Symbols and Their Translations

Symbolic Form	Translation
$8 \neq 3$	Eight is not equal to three.
$5 < 7$	Five is less than seven.
$7 > 5$	Seven is greater than five.
$x \leq 3$	x is less than or equal to three.
$y \geq 2$	y is greater than or equal to two.

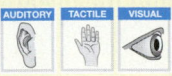
The algebra pyramid shown illustrates how variables, constants, expressions, equations, and inequalities relate. At the foundation of algebra and our pyramid are constants and variables, which are used to build expressions, which in turn are used to build equations and inequalities.

The Algebra Pyramid

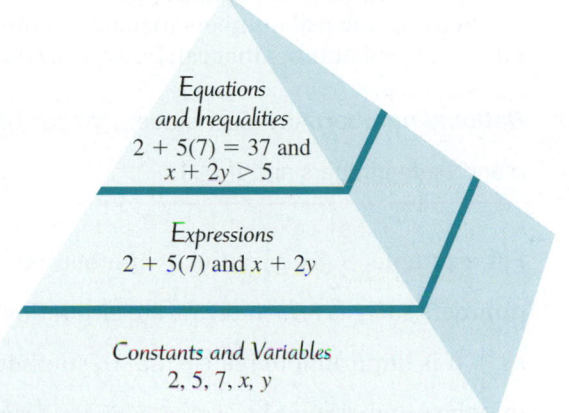

Equations and Inequalities
$2 + 5(7) = 37$ and $x + 2y > 5$

Expressions
$2 + 5(7)$ and $x + 2y$

Constants and Variables
$2, 5, 7, x, y$

Note: During this course, we move back and forth between expressions, equations, and inequalities. When we transition from one to the other, we use the algebra pyramid as a quick visual so that it's clear what we are working on.

Objective ❷ Classify number sets. Numbers can be placed into different categories using **sets**.

Definition *Set:* A collection of objects.

Braces are used to indicate a set. For example, the set containing the numbers 1, 2, 3, and 4 would be written $\{1, 2, 3, 4\}$. The numbers 1, 2, 3, and 4 are called the *members* or *elements* of this set.

Procedure **Writing Sets**

To write a set, write the members or elements of the set separated by commas within braces, $\{\ \}$.

Example 1 Write the set containing the first five letters of the alphabet.

Answer $\{A, B, C, D, E\}$

Your Turn 1 Write the set containing the first four months of the year.

Answer   Your Turn 1

$\{January, February, March, April\}$

Numbers are classified using number sets. The set of *natural numbers* contains the counting numbers 1, 2, 3, 4, ... and is written {1, 2, 3, ... }. The three dots are an *ellipsis* and indicate that the numbers continue forever in the same pattern. The set of *whole numbers* contains all of the natural numbers and 0 and is written {0, 1, 2, 3, ... }. The set of *integers* contains all of the whole numbers and the opposite (or negative) of every natural number and is written { ..., −3, −2, −1, 0, 1, 2, 3, ... }. Notice that the integers continue forever in both directions.

A number line is often useful in mathematics. The following number line is marked with the integers. Note that the positive numbers are to the right of 0 and the negative numbers are to the left.

Although we can mark and view only a portion of the number line, the arrows at the ends indicate that the line and the numbers on it continue forever in both directions. There are other types of numbers that include the integers and all numbers between them. If we traveled along the number line forever in both directions, we would encounter every number in the set of *real numbers*.

Some of the real numbers that are not integers are the **rational numbers** that contain every real number that can be expressed as a ratio of integers.

Definition   *Rational number:* Any real number that can be expressed in the form $\frac{a}{b}$, where a and b are integers and $b \neq 0$.

> **Note:** In the definition for a rational number, the notation $b \neq 0$ is important because if the denominator b were to equal 0, the fraction would be undefined. The reason it is undefined will be explained later.

For example, $\frac{3}{4}$ is a rational number because 3 and 4 are integers. Note that numbers such as 0.75 and 75% are also rational numbers because they can be written as $\frac{3}{4}$. It is important to realize that the definition of a rational number does not state that the number *must* be expressed in the form $\frac{a}{b}$. Rather, if a number *can* be expressed in the form $\frac{a}{b}$, it is a rational number. For example, the number 5 is a rational number because it can be expressed as $\frac{5}{1}$.

Example 2   Determine whether the given number is a rational number.

a. $\frac{2}{3}$

Answer   Yes, because 2 and 3 are integers.

b. 0.4

Answer   Yes, 0.4 is a rational number because it can be expressed as $\frac{4}{10}$ and 4 and 10 are integers.

c. $0.\overline{6}$

Answer   A bar written over a decimal digit indicates that the digit repeats without end. So $0.\overline{6} = 0.66666\ldots$, and we say that these decimal numbers are nonterminating decimal numbers. Usually, we encounter these numbers as quotients in certain division problems, such as when we write certain fractions as decimals. In this case, $0.\overline{6}$ is the decimal equivalent of $\frac{2}{3}$. Because $0.\overline{6}$ can be expressed as the fraction $\frac{2}{3}$, it is a rational number.

> **Note:** All nonterminating decimal numbers with repeating digits are rational numbers because they can be expressed as fractions with integers in both the numerator and denominator.

d. 3

Answer   Yes, because 3 can be thought of as $\frac{3}{1}, \frac{6}{2}, \frac{9}{3}$, etc. All integers are rational numbers.

a. $\dfrac{4}{7}$ **b.** 0.56 **c.** $0.\overline{2}$ **d.** -7

Not all numbers can be expressed as a ratio of integers. One such number is π (pronounced "pie"). Because the exact value of π cannot be expressed as a ratio of integers, it is categorized as an **irrational number**.

Definition   *Irrational number:* Any real number that is not rational.

Some other irrational numbers are $\sqrt{2}$ and $\sqrt{3}$. (Square roots are explained in more detail in Section 1.5.) Because an irrational number cannot be written as a ratio of integers, if a calculation involves an irrational number, we must leave it in symbolic form or use a rational number approximation. We can approximate π with rational numbers such as 3.14 and $\dfrac{22}{7}$. Any decimal representation of an irrational number is a nonrepeating, nonterminating decimal number. For example, $0.1010010001\ldots$ is irrational.

The rational and irrational numbers make up the **real numbers**.

Definition   *Real numbers:* The union of the rational and irrational numbers.

The following figure illustrates how the number sets relate in the real number system.

Real Numbers

Rational Numbers: Real numbers that can be expressed in the form $\dfrac{a}{b}$, where a and b are integers and $b \neq 0$, such as $-4\dfrac{3}{4}$, $-\dfrac{2}{3}$, 0.018, $0.\overline{3}$, and $\dfrac{5}{8}$.

Integers: $\ldots, -3, -2, -1, 0, 1, 2, 3, \ldots$

Whole Numbers: $0, 1, 2, 3, \ldots$

Natural Numbers: $1, 2, 3, \ldots$

Irrational Numbers: Any real number that is not rational, such as $-\sqrt{2}$, $-\sqrt{3}$, $\sqrt{0.8}$, and π.

Objective ③ Graph rational numbers on a number line.   Number lines can be useful tools when comparing numbers or solving certain arithmetic problems. Let's review how to graph a number on a number line.

Example 3   Graph on a number line.

a. $1\dfrac{3}{5}$

SOLUTION   The number $1\dfrac{3}{5}$ is located $\dfrac{3}{5}$ of the way between 1 and 2, so we divide the space between 1 and 2 into 5 equal divisions and place a dot on the 3rd mark.

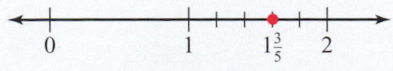

b. -2.56

SOLUTION Because -2.56 means $-2\frac{56}{100}$, we could divide the space between -2 and -3 into 100 divisions and count to the 56th mark. Because this is tedious to do, we gradually use smaller and smaller sections of the number line to graph the number.

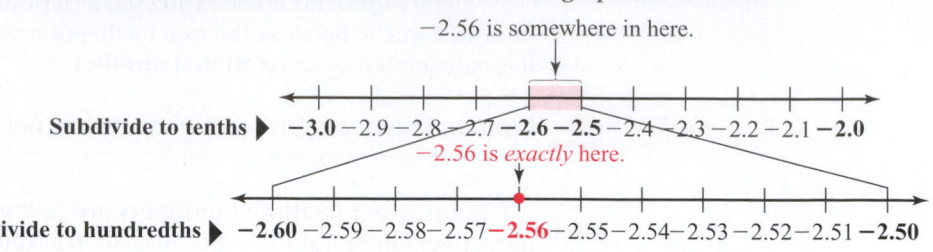

−2.56 is somewhere in here.

Subdivide to tenths ▶ **−3.0** −2.9 −2.8 −2.7 **−2.6** −2.5 −2.4 −2.3 −2.2 −2.1 **−2.0**

−2.56 is *exactly* here.

Subdivide to hundredths ▶ **−2.60** −2.59 −2.58 −2.57 **−2.56** −2.55 −2.54 −2.53 −2.52 −2.51 **−2.50**

Your Turn 3   Graph on a number line.

a. $2\frac{3}{4}$

b. -1.78

Objective ④ Determine the absolute value of a number. The word *value* indicates how much something is worth or its magnitude. In mathematics, **absolute value** indicates the magnitude of a number.

Definition   *Absolute value:* A given number's distance from 0 on a number line.

For example, the absolute value of 5 is 5 because the number 5 is 5 units from 0 on a number line. Likewise, the absolute value of -5 is 5 because -5 is also 5 units from 0 on a number line.

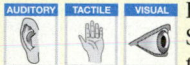
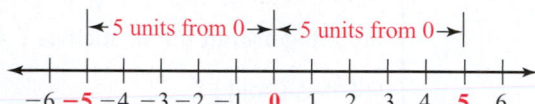

|← 5 units from 0 →|← 5 units from 0 →|

−6 **−5** −4 −3 −2 −1 **0** 1 2 3 4 **5** 6

The absolute value of a number n is written $|n|$. The examples just mentioned translate this way:

The absolute value of 5 is 5.   The absolute value of -5 is 5.

Symbolic form: $|5| = 5$ $|-5| = 5$

What about the absolute value of 0? Since there are no units between 0 and itself, the absolute value of 0 is 0.

Rule   Absolute Value

The absolute value of every real number is either positive or 0.

Answers   Your Turn 3

a.

1   2 $2\frac{3}{4}$ 3

b.

−1.8   −1.78   −1.76   −1.74   −1.72   −1.7

Example 4   Simplify.

a. $|-4.5|$

Answer $|-4.5| = 4.5$ because -4.5 is 4.5 units from 0 on a number line.

b. $\left| \dfrac{3}{8} \right|$

Answer $\left| \dfrac{3}{8} \right| = \dfrac{3}{8}$ because $\dfrac{3}{8}$ is $\dfrac{3}{8}$ of a unit from 0 on a number line.

Your Turn 4 Simplify.

a. $\left| -2\dfrac{4}{5} \right|$ **b.** $|12.8|$

Objective ⑤ Compare numbers. Number lines can also be used to determine which of two numbers is greater. Because numbers increase from left to right on a number line, the number farthest to the right will be the greater of two numbers.

Rule   Comparing Numbers

For any two real numbers a and b, a is greater than b if a is to the right of b on a number line. Equivalently, b is less than a if b is to the left of a on a number line.

Consider the following number line where we compare 2 and 8.

8 is farther right than 2.

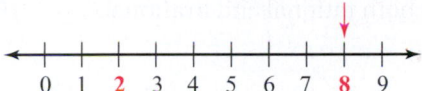

0 1 **2** 3 4 5 6 7 **8** 9

Because the number 8 is farther to the right on a number line than the number 2, we say that 8 is greater than 2, or in symbols, $8 > 2$. Or we could say that since 2 is to the left of 8, $2 < 8$.

Example 5   Use $=$, $<$, or $>$ to write a true statement.

a. 4 ☐ -4

Answer $4 > -4$ because 4 is farther to the right on a number line than -4.

b. -2.7 ☐ -2.5

Answer $-2.7 < -2.5$ because -2.7 is farther to the left on a number line than -2.5.

c. $\left| 3\dfrac{5}{6} \right|$ ☐ $3\dfrac{5}{6}$

Answer $\left| 3\dfrac{5}{6} \right| = 3\dfrac{5}{6}$ because the absolute value of $3\dfrac{5}{6}$ is equal to $3\dfrac{5}{6}$.

d. $|-0.7|$ ☐ -1.5

Answer $|-0.7| > -1.5$ because the absolute value of -0.7 is equal to 0.7, which is farther to the right on a number line than -1.5.

Your Turn 5 Use $=$, $<$, or $>$ to write a true statement.

a. -15 ☐ -21 **b.** -4.1 ☐ 0   **c.** $2\dfrac{5}{6}$ ☐ $2\dfrac{1}{4}$ **d.** $|-9.5|$ ☐ 9.5

1.1 Exercises For Extra Help *MyMathLab*

Note: Exercises marked with a ★ represent challenging exercises.

1. Define a set in your own words.

2. Define rational numbers in your own words.

3. Explain the difference between a rational number and an irrational number.

4. Explain why it is incorrect to say that absolute value is always positive.

For Exercises 5–10, answer true or false.

5. Every rational number is a real number.

6. Every real number is a rational number.

7. Every whole number is an integer.

8. Every real number is a natural number.

9. A number exists that is both rational and irrational.

10. Zero is a rational number.

For Exercises 11–20, write a set representing each description. See Example 1.

11. The days of the week

12. The last ten letters of the English alphabet

13. The vowels of the English alphabet

14. The states in the United States that do not share a border with any other state

15. The natural numbers that are multiples of 5

16. The even natural numbers

17. The odd natural numbers greater than 7

18. The even integers greater than or equal to 16

19. The integers greater than or equal to −6 but less than −2

20. The integers greater than −2.1 and less than $\frac{3}{4}$

For Exercises 21–30, determine whether each number is a rational number or an irrational number. See Example 2.

21. $-\frac{4}{5}$

22. $\frac{1}{4}$

23. 9

24. −12

25. π

26. $\dfrac{\pi}{4}$ **27.** -0.21 **28.** -0.8 **29.** $0.\overline{6}$ **30.** $0.\overline{13}$

For Exercises 31–38, graph each number on a number line. See Example 3.

31. $-\dfrac{5}{6}$ **32.** $5\dfrac{1}{2}$ **33.** $2\dfrac{3}{8}$ **34.** $-\dfrac{2}{5}$

35. -3.5 **36.** 7.4 **37.** 2.45 **38.** -7.62

For Exercises 39–48, simplify. See Example 4.

39. $|23|$ **40.** $|6|$ **41.** $|-2|$ **42.** $|-8|$ **43.** $|-5.7|$

44. $|-4.5|$ **45.** $\left|-3\dfrac{1}{8}\right|$ **46.** $\left|2\dfrac{3}{5}\right|$ **47.** $|0|$ **48.** $|-67.8|$

For Exercises 49–76, use =, <, or > to write a true statement. See Example 5.

49. $9 \quad 3$ **50.** $2 \quad 7$ **51.** $4 \quad -3$ **52.** $-6 \quad 5$ **53.** $-7 \quad -8$

54. $-19 \quad -7$ **55.** $-8 \quad 0$ **56.** $0 \quad -5$ **57.** $6.2 \quad 4.5$ **58.** $2.63 \quad 3.75$

59. $-4.3 \quad -7.6$ **60.** $-3.5 \quad -3.1$ **61.** $2\dfrac{2}{3} \quad 2\dfrac{3}{4}$ **62.** $3\dfrac{5}{6} \quad 3\dfrac{1}{4}$ **63.** $5.8 \quad |-5.8|$

64. $|-4.1| \quad 4.1$ **65.** $7.3 \quad |-8.7|$ **66.** $|-10.4| \quad 3.2$ **67.** $|4.31| \quad |-4.31|$ **68.** $|-0.59| \quad |0.59|$

69. $\left|-6\dfrac{5}{8}\right| \quad 5\dfrac{3}{8}$ **70.** $4\dfrac{2}{9} \quad \left|4\dfrac{5}{9}\right|$ **71.** $|-4| \quad |-2|$ **72.** $|-10| \quad |-8|$ **73.** $|29.5| \quad |-29.7|$

74. $|-5.36| \quad |5.76|$ **75.** $\left|-\dfrac{2}{3}\right| \quad \left|-\dfrac{4}{3}\right|$ **76.** $\left|-\dfrac{9}{11}\right| \quad \left|-\dfrac{7}{11}\right|$

★ *For Exercises 77–80, list the given numbers in order from least to greatest.*

77. $-2.56,\ 5.4,\ |8.3|,\ \left|-7\dfrac{1}{2}\right|,\ -4.7$ **78.** $2.9,\ 1,\ -12.6,\ |-1.3|,\ -9.6,\ \left|-2\dfrac{3}{4}\right|$

79. $0.4,\ -0.6,\ 0,\ 3\dfrac{1}{4},\ |-0.02|,\ -0.44,\ \left|1\dfrac{2}{3}\right|$ **80.** $1.02,\ -0.13,\ -4\dfrac{1}{8},\ -2\dfrac{1}{4},\ |-1.06|,\ -2,\ |0.1|$

1.2 Fractions

Objectives

① Write equivalent fractions.

② Write equivalent fractions with the LCD.

③ Write the prime factorization of a number.

④ Simplify a fraction to lowest terms.

In this section, we review some basic principles of **fractions**.

Definition *Fraction:* A quotient of two numbers or expressions a and b having the form $\dfrac{a}{b}$, where $b \neq 0$.

For example, $\dfrac{3}{4}$ is a fraction. The top number in a fraction is called the *numerator*, and the bottom number is called the *denominator*.

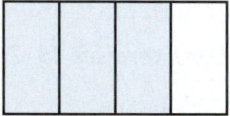

$$\dfrac{3}{4} \quad \leftarrow \quad \text{Numerator} \\ \qquad \leftarrow \quad \text{Denominator}$$

Connection We learned in Section 1.1 that if a number can be expressed in the form $\dfrac{a}{b}$, where a and b are integers, it is a rational number. So fractions that have integer numerators and denominators are rational numbers.

We can use fractions to indicate a part of a whole. For example, the rectangle to the right has been divided into four equal pieces, three of which are shaded. So the shaded region represents $\dfrac{3}{4}$ of the rectangle.

Objective ① **Write equivalent fractions.**   Now let's review how to write *equivalent fractions*, which are fractions that represent the same amount. For example, $\dfrac{1}{2}$ and $\dfrac{2}{4}$ are equivalent. We can see that they represent the same amount on a ruler.

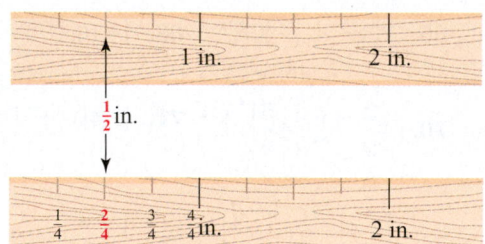

We can rewrite $\dfrac{1}{2}$ as $\dfrac{2}{4}$ by multiplying its numerator and denominator by 2.

$$\frac{1}{2} = \frac{1 \cdot 2}{2 \cdot 2} = \frac{2}{4}$$

We also can begin with $\dfrac{2}{4}$ and get back to $\dfrac{1}{2}$ by dividing both the numerator and denominator by 2.

$$\frac{2}{4} = \frac{2 \div 2}{4 \div 2} = \frac{1}{2}$$

Writing Equivalent Fractions

For any fraction, we can write an equivalent fraction by multiplying or dividing both its numerator and denominator by the same nonzero number.

This rule makes use of the fact that multiplying or dividing both the numerator and denominator by the same nonzero number is equivalent to multiplying or dividing by 1, which does not change a number.

Example 1 Find the missing number that makes the fractions equivalent.

a. $\dfrac{7}{12} = \dfrac{?}{48}$

SOLUTION Because $48 \div 12 = 4$, $12 \cdot 4 = 48$. So we multiply the numerator and denominator by 4.

$$\frac{7 \cdot 4}{12 \cdot 4} = \frac{28}{48}$$ **Multiply the numerator and denominator by 4.**

Note: When a fraction is negative, the sign can be placed to the left of the numerator, to the left of the fraction line, or to the left of the denominator.

$$-\frac{30}{42} = -\frac{5}{7} = \frac{-5}{7} = \frac{5}{-7}$$

b. $-\dfrac{30}{42} = \dfrac{?}{7}$

SOLUTION Because $42 \div 7 = 6$, $42 \div 6 = 7$. So divide the numerator and denominator by 6.

$$-\frac{30 \div 6}{42 \div 6} = -\frac{5}{7}$$ **Divide the numerator and denominator by 6.**

Your Turn 1 Find the missing number that makes the fractions equivalent.

a. $\dfrac{9}{16} = \dfrac{?}{32}$

b. $-\dfrac{54}{72} = \dfrac{3}{?}$

Objective ② Write equivalent fractions with the LCD. It is sometimes necessary to write fractions that have different denominators, such as $\dfrac{1}{2}$ and $\dfrac{1}{3}$, as equivalent fractions with a common denominator. A common denominator can be any number that is a **multiple** of the denominators.

Definition *Multiple:* A multiple of a given integer n is the product of n and an integer.

We can generate multiples of a given number by multiplying the given number by the integers. For example, to generate the positive **multiples** of 2 and 3, we multiply 2 and 3 by 1, 2, 3, 4, and so on.

Note: Every multiple of 2 is evenly divisible by 2. Similarly, every multiple of 3 is evenly divisible by 3. In general, every multiple of a number n is evenly divisible by n.

Multiples of 2	Multiples of 3
$2 \cdot 1 = 2$	$3 \cdot 1 = 3$
$2 \cdot 2 = 4$	$3 \cdot 2 = 6$
$2 \cdot 3 = 6$	$3 \cdot 3 = 9$
$2 \cdot 4 = 8$	$3 \cdot 4 = 12$
$2 \cdot 5 = 10$	$3 \cdot 5 = 15$
$2 \cdot 6 = 12$	$3 \cdot 6 = 18$

Answers Your Turn 1

a. $\dfrac{9}{16} = \dfrac{18}{32}$ b. $-\dfrac{54}{72} = -\dfrac{3}{4}$

Notice that some common multiples of 2 and 3 in our lists are 6 and 12. When we rewrite fractions with a common denominator, it is usually preferable to use the **least common multiple** of the denominators. Such a denominator is called the **least common denominator**. From our lists, we see that the least common multiple of 2 and 3 is 6. So 6 is the least common denominator of $\frac{1}{2}$ and $\frac{1}{3}$.

Definitions   *Least common multiple (LCM):* The smallest number that is a multiple of each number in a given set of numbers.

Least common denominator (LCD): The least common multiple of the denominators of a given set of fractions.

We can now write equivalent fractions for $\frac{1}{2}$ and $\frac{1}{3}$ with their LCD, 6.

$$\frac{1}{2} = \frac{1 \cdot 3}{2 \cdot 3} = \frac{3}{6} \quad \text{and} \quad \frac{1}{3} = \frac{1 \cdot 2}{3 \cdot 2} = \frac{2}{6}$$

Example 2   Write each pair as equivalent fractions with the LCD.

a. $\frac{5}{8}$ and $\frac{1}{6}$

SOLUTION   The LCD of 8 and 6 is 24.

$$\frac{5}{8} = \frac{5 \cdot 3}{8 \cdot 3} = \frac{15}{24} \quad \text{and} \quad \frac{1}{6} = \frac{1 \cdot 4}{6 \cdot 4} = \frac{4}{24}$$

b. $\frac{2}{3}$ and $\frac{5}{9}$

SOLUTION   The LCD of 3 and 9 is 9.

$\frac{2}{3} = \frac{2 \cdot 3}{3 \cdot 3} = \frac{6}{9}$, and $\frac{5}{9}$ already has the LCD as its denominator.

Your Turn 2   Write each pair as equivalent fractions with the LCD.

a. $\frac{4}{9}$ and $\frac{5}{6}$ b. $-\frac{3}{4}$ and $-\frac{5}{8}$

Objective ③ Write the prime factorization of a number.   We now consider simplifying fractions. When simplifying fractions, we work with **factors**, which are numbers or variables that are multiplied together.

Definition   *Factors:* If $a \cdot b = c$, then a and b are factors of c.

In the multiplication statement $20 = 2 \cdot 10$, the 2 and 10 are factors of 20. Notice that a factor of a given number divides the given number evenly (leaves a remainder of 0). The numbers 1, 2, 4, 5, 10, and 20 are factors of 20 because they all divide 20 evenly. When a number is expressed as a product of its factors, it is said to be in *factored form*.

Answers   **Your Turn 2**

a. $\frac{8}{18}$ and $\frac{15}{18}$ b. $-\frac{6}{8}$ and $-\frac{5}{8}$

Factored form is sometimes referred to as a *factorization*. Here are some factorizations for 20:

$$20 = 1 \cdot 20 \quad \text{or} \quad 20 = 2 \cdot 10 \quad \text{or} \quad 20 = 4 \cdot 5$$

Some natural numbers have only two distinct factors, 1 and the number itself. These numbers are called **prime numbers**.

Definition   ***Prime number:*** A natural number that has exactly two different factors, 1 and the number itself.

Note that 0 is not a prime number because it is not a natural number. (0 is a whole number.) Also, 1 is not a prime number because the only factor for 1 is itself and, by definition, a prime number must have *two different* factors.

The first prime number is 2 because it is the first natural number greater than 1 that has exactly two factors, 1 and 2. The number 4 is not a prime number because it is divisible by a number other than 1 and 4, namely 2. Here is a list of prime numbers:

$$2, 3, 5, 7, 11, 13, 17, 19, 23, 29, 31, 37, 41, 43, \ldots$$

When simplifying a fraction, it is useful to write the **prime factorization** of the numerator and denominator. For example, the prime factorization for 20 is $2 \cdot 2 \cdot 5$.

Definition   ***Prime factorization:*** A factorization that contains only prime factors.

We can find the prime factorization of a number using any factorization of the number as a starting point. Consider 20 again and suppose we factored it as $4 \cdot 5$. Because 4 is not a prime number, we can factor 4 as $2 \cdot 2$. It looks like this:

$$20 = 4 \cdot 5$$
$$= 2 \cdot 2 \cdot 5$$

We will use factor trees to find prime factorizations. The idea is to draw two branches beneath any composite number and write two factors of that number at the end of the branches. Following are two ways we can use a factor tree to find the prime factorization of 20. Notice that when a branch ends in a prime factor, we circle the prime factor. This is optional, but many people find it helpful in making the prime factors stand out.

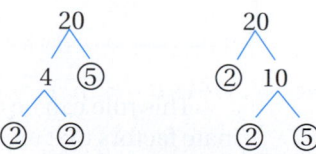

We usually write the prime numbers in a prime factorization in ascending order. So we would write $20 = 2 \cdot 2 \cdot 5$ (instead of $20 = 5 \cdot 2 \cdot 2$ or $20 = 2 \cdot 5 \cdot 2$). Note that no matter how we write the prime factorization, the product of two 2s and one 5 is always 20. No two numbers have the same prime factorization. Consequently, we say that a number's prime factorization is unique.

Example 3   Find the prime factorization of 360.

SOLUTION   Use a factor tree.

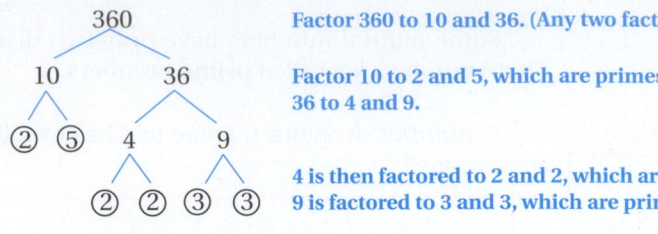

Factor 360 to 10 and 36. (Any two factors will work.)

Factor 10 to 2 and 5, which are primes. Then factor 36 to 4 and 9.

4 is then factored to 2 and 2, which are primes, and 9 is factored to 3 and 3, which are primes.

Answer $360 = 2 \cdot 2 \cdot 2 \cdot 3 \cdot 3 \cdot 5$

Your Turn 3   Find the prime factorization.

a. 84   b. 280

Objective ④ **Simplify a fraction to lowest terms.**   We can use prime factorizations to simplify fractions to **lowest terms**.

Definition   *Lowest terms:* Given a fraction $\dfrac{a}{b}$ and $b \neq 0$, if the only factor common to both a and b is 1, then the fraction is in lowest terms.

For example, $\dfrac{3}{4}$ is in lowest terms because the only factor common to both 3 and 4 is the number 1, whereas $\dfrac{6}{8}$ is not in lowest terms because 6 and 8 have a common factor of 2.

To develop the method of reducing fractions using primes, we must understand that if the numerator and denominator are identical and not both 0, then the fraction is equivalent to 1.

$$1 = \frac{1}{1} = \frac{2}{2} = \frac{3}{3} = \frac{4}{4} = \frac{5}{5} = \cdots$$

This suggests the following rule:

Rule   Simplifying a Fraction with the Same Nonzero Numerator and Denominator

$$\frac{n}{n} = \frac{1}{1} = 1, \text{ when } n \neq 0.$$

This rule can be extended to common factors within a fraction. It allows us to eliminate factors that are identical in the numerator and denominator of a fraction.

Rule   Eliminating a Common Factor in a Fraction

$$\frac{an}{bn} = \frac{a \cdot 1}{b \cdot 1} = \frac{a}{b}, \text{ when } b \neq 0 \text{ and } n \neq 0.$$

These rules allow us to write fractions in lowest terms using prime factorizations. The idea is to replace the numerator and denominator with their prime factorizations and then eliminate the prime factors that are common to both the numerator and

Answers   Your Turn 3

a. $84 = 2 \cdot 2 \cdot 3 \cdot 7$

b. $280 = 2 \cdot 2 \cdot 2 \cdot 5 \cdot 7$

denominator. Consider the fraction $\dfrac{10}{15}$. Replacing the 10 and 15 with their prime factorizations, we have the following:

$$\frac{10}{15} = \frac{2 \cdot 5}{3 \cdot 5}$$

$$= \frac{2 \cdot 1}{3 \cdot 1}$$

$$= \frac{2}{3}$$

> **Note:** The common factor of 5 is replaced with 1. In Example 4, we leave this step out.

This process of eliminating common factors is called *dividing out* the common factors and is sometimes written this way:

$$\frac{10}{15} = \frac{2 \cdot \overset{1}{\cancel{5}}}{3 \cdot \underset{1}{\cancel{5}}} = \frac{2}{3}$$

Although there are different styles of showing which common factors are eliminated, we simply highlight the common factors that are eliminated.

Procedure **Simplifying a Fraction to Lowest Terms**

To simplify a fraction to lowest terms:
1. Replace the numerator and denominator with their prime factorizations.
2. Eliminate (divide out) all prime factors common to the numerator and denominator.
3. Multiply the remaining factors.

Connection You may remember a method for simplifying fractions in which you divide out the greatest common factor. For example, to simplify $\dfrac{28}{70}$, we recognize that 14 is the greatest common factor of 28 and 70 and divide both of those numbers by 14 to get lowest terms:
$\dfrac{28}{70} = \dfrac{28 \div 14}{70 \div 14} = \dfrac{2}{5}$. This method is a perfectly good way of simplifying fractions. However, we show the prime method to prepare for simplifying rational expressions in Chapter 7.

Example 4 Simplify to lowest terms.

a. $\dfrac{28}{70}$

SOLUTION $\dfrac{28}{70} = \dfrac{2 \cdot 2 \cdot 7}{2 \cdot 5 \cdot 7} = \dfrac{2}{5}$ **Replace the numerator and denominator with their prime factorizations; then eliminate the common prime factors.**

b. $\dfrac{156}{210}$

SOLUTION $\dfrac{156}{210} = \dfrac{2 \cdot 2 \cdot 3 \cdot 13}{2 \cdot 3 \cdot 5 \cdot 7} = \dfrac{26}{35}$

Your Turn 4 Simplify to lowest terms.

a. $\dfrac{36}{48}$ b. $\dfrac{126}{315}$ c. $\dfrac{312}{378}$

Example 5 A newspaper reports that over the last year, 160 households per 1000 households were victims of a crime. Write a fraction in simplest form representing the fraction of households that were victims of crime.

Answers Your Turn 4

a. $\dfrac{3}{4}$ b. $\dfrac{2}{5}$ c. $\dfrac{52}{63}$

SOLUTION If 160 out of 1000 households were victims of crime, the fraction of households that were victims of crime is $\frac{160}{1000}$. Now we can simplify the fraction.

$$\frac{160}{1000} = \frac{2 \cdot 2 \cdot 2 \cdot 2 \cdot 2 \cdot 5}{2 \cdot 2 \cdot 2 \cdot 5 \cdot 5 \cdot 5} = \frac{4}{25}$$

Answer The fraction, in simplest form, of households that were victims of crime is $\frac{4}{25}$, which means that 4 out of 25 households were victims of crime.

Your Turn 5 Researchers gave 560 volunteers a new medication and found that 48 of them developed a headache within 30 minutes of taking the medication. Write a fraction in simplest form representing the portion of the volunteers that developed a headache.

Answer **Your Turn 5**

$\frac{3}{35}$

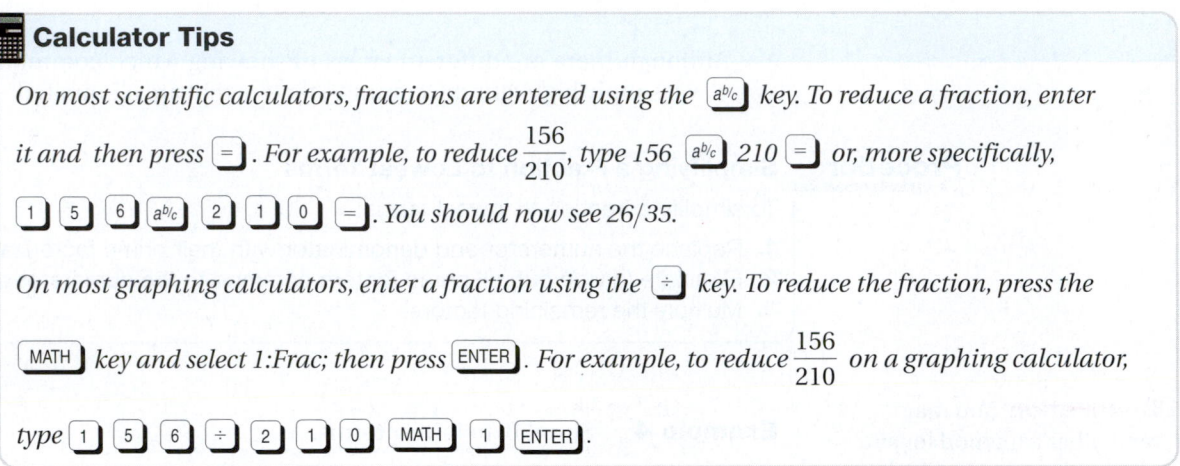

📱 Calculator Tips

On most scientific calculators, fractions are entered using the $\boxed{a^{b/c}}$ *key. To reduce a fraction, enter it and then press* $\boxed{=}$. *For example, to reduce* $\frac{156}{210}$, *type* 156 $\boxed{a^{b/c}}$ 210 $\boxed{=}$ *or, more specifically,*

$\boxed{1}$ $\boxed{5}$ $\boxed{6}$ $\boxed{a^{b/c}}$ $\boxed{2}$ $\boxed{1}$ $\boxed{0}$ $\boxed{=}$. *You should now see 26/35.*

On most graphing calculators, enter a fraction using the $\boxed{\div}$ *key. To reduce the fraction, press the*

$\boxed{\text{MATH}}$ *key and select 1:Frac; then press* $\boxed{\text{ENTER}}$. *For example, to reduce* $\frac{156}{210}$ *on a graphing calculator,*

type $\boxed{1}$ $\boxed{5}$ $\boxed{6}$ $\boxed{\div}$ $\boxed{2}$ $\boxed{1}$ $\boxed{0}$ $\boxed{\text{MATH}}$ $\boxed{1}$ $\boxed{\text{ENTER}}$.

1.2 Exercises   For Extra Help   *MyMathLab* Math XL PRACTICE WATCH DOWNLOAD

1. What are two ways to generate fractions equivalent to a given fraction?

2. Is 4 a multiple of 12? Explain.

3. Is $2 \cdot 2 \cdot 4 \cdot 7$ a prime factorization? Explain.

4. Is it possible for two different numbers to have the same prime factorization? Explain.

5. Is every whole number a composite number?

6. How do you know if a fraction is in lowest terms?

For Exercises 7–10, answer true or false; then explain.

7. Every prime number is a whole number.

8. Prime numbers are rational numbers.

9. Every composite number has at least one factor that is a prime number.

10. The number 1 is neither prime nor composite.

For Exercises 11–14, identify the fraction represented by each shaded region.

11.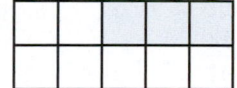

12.

13.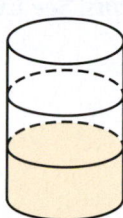

14.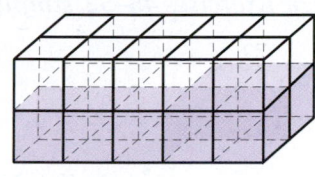

For Exercises 15–20, write the length of each line segment in lowest terms.

15.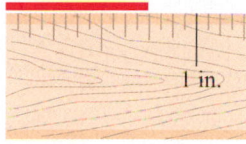

16.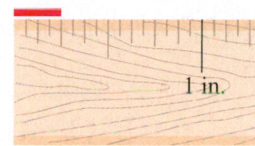

17.

18.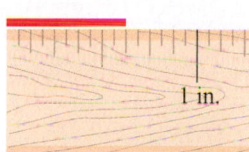

19.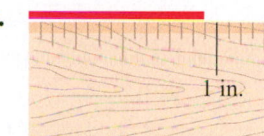

20.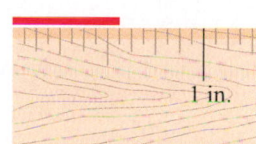

For Exercises 21–28, find the missing number that makes the fractions equivalent.

21. $\dfrac{3}{10} = \dfrac{?}{40}$

22. $\dfrac{5}{8} = \dfrac{?}{16}$

23. $-\dfrac{7}{12} = \dfrac{?}{60}$

24. $\dfrac{2}{5} = \dfrac{6}{?}$

25. $\dfrac{4}{7} = \dfrac{?}{28}$

26. $\dfrac{6}{8} = \dfrac{?}{4}$

27. $\dfrac{8}{60} = \dfrac{2}{?}$

28. $\dfrac{27}{30} = \dfrac{9}{?}$

For Exercises 29–36, write the fractions as equivalent fractions with the LCD. See Example 2.

29. $\dfrac{4}{5}$ and $\dfrac{2}{3}$

30. $\dfrac{5}{7}$ and $\dfrac{3}{11}$

31. $\dfrac{4}{9}$ and $\dfrac{7}{6}$

32. $\dfrac{5}{8}$ and $\dfrac{7}{12}$

33. $-\dfrac{11}{18}$ and $-\dfrac{17}{24}$ **34.** $-\dfrac{9}{20}$ and $-\dfrac{7}{15}$ **35.** $-\dfrac{1}{16}$ and $-\dfrac{15}{36}$ **36.** $-\dfrac{13}{21}$ and $-\dfrac{9}{14}$

For Exercises 37–44, write the prime factorization for each number. See Example 3.

37. 44   **38.** 33   **39.** 36   **40.** 42

41. 64   **42.** 48   **43.** 250   **44.** 810

For Exercises 45–52, simplify to lowest terms. See Example 4.

45. $\dfrac{72}{90}$ **46.** $\dfrac{48}{84}$ **47.** $\dfrac{63}{99}$ **48.** $\dfrac{42}{91}$

49. $-\dfrac{78}{104}$ **50.** $-\dfrac{30}{54}$ **51.** $-\dfrac{48}{90}$ **52.** $-\dfrac{24}{162}$

For Exercises 53–56, determine whether each simplification is correct. If it isn't, explain why. See Example 4.

53. $\dfrac{\overset{1}{\cancel{2}} + 3}{\underset{1}{\cancel{2}}} = \dfrac{1+3}{1} = 4$

54. $\dfrac{3 - 4 \cdot \overset{1}{\cancel{2}}}{\underset{1}{\cancel{2}}}$

55. $\dfrac{84}{240} = \dfrac{\overset{1}{\cancel{2}} \cdot \overset{1}{\cancel{2}} \cdot \overset{1}{\cancel{3}} \cdot 7}{2 \cdot \underset{1}{\cancel{2}} \cdot \underset{1}{\cancel{2}} \cdot \underset{1}{\cancel{3}} \cdot 3 \cdot 5} = \dfrac{7}{30}$

56. $\dfrac{300}{108} = \dfrac{\overset{1}{\cancel{2}} \cdot \overset{1}{\cancel{2}} \cdot \overset{1}{\cancel{3}} \cdot 5 \cdot 5}{\underset{1}{\cancel{2}} \cdot \underset{1}{\cancel{2}} \cdot \underset{1}{\cancel{3}} \cdot 3} = \dfrac{25}{3}$

For Exercises 57–60, write each fraction in lowest terms. See Example 5.

57. A student scores 294 points out of a total of 336 points in a college course. What fraction of the total points is the student's score?

58. The nutrition label on a package of frozen fish sticks indicates that a serving of fish sticks has 250 calories, with 130 of those calories coming from fat. What fraction of the calories in a serving of fish sticks comes from fat?

59. At a company, 300 of the 575 employees have optional life insurance coverage as part of their benefits package. What fraction of the employees have optional life insurance coverage?

60. Laura uses 120 square feet of a room in her home as an office for her business. The total living space of her home is 1830 square feet. For tax purposes, she needs to compute the fraction of her home that is used as an office. What is that fraction?

For Exercises 61–64, use the following bar graph, which shows the number of hours that Carla spends each week in each activity. Write all fractions in lowest terms. See Example 5.

61. What fraction of a week does Carla spend working?

62. What fraction of a week does Carla spend sleeping?

63. What fraction of a week does Carla spend in all of the listed activities combined?

64. What fraction of the week does Carla have for free time, which is time away from all of the listed activities?

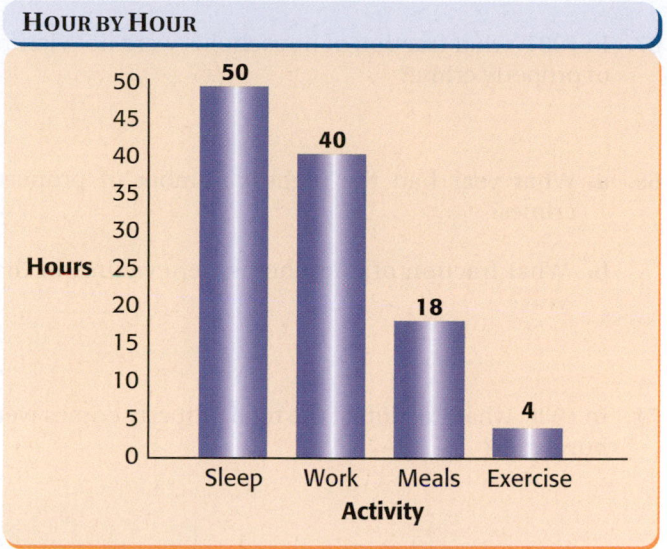

HOUR BY HOUR

For Exercises 65–72, use the following table, which shows the number of households per 1000 that were victims of property crime in each year. Write all fractions in lowest terms. See Example 5.

National Crime Victimization Survey, Property Crime Trends, 1985–2005

Year	Total Property Crime*	Burglary	Thefts	Motor Vehicle Thefts
1985	385	75	296	14
1986	373	74	284	15
1987	380	75	289	16
1988	378	74	287	18
1989	373	68	287	19
1990	349	65	264	21
1991	354	65	267	22
1992	325	59	248	19
1993	319	58	242	19
1994	310	56	235	19
1995	291	49	224	17
1996	266	47	206	14
1997	248	45	190	14
1998	217	39	168	11
1999	198	34	154	10
2000	178	32	138	9
2001	167	29	129	9
2002	159	28	122	9
2003	163	30	124	9
2004	161	30	123	9
2005	154	30	116	8

*Victimizations per 1000 households

Source: U.S. Department of Justice, Bureau of Crime Statistics.

65. What fraction of households were victims of property crime in 1987?

66. What fraction of households were victims of property crime in 1992?

67. In 1987, what fraction of households were not victims of property crime?

68. In 1994, what fraction of households were not victims of property crime?

69. a. What year had the highest number of property crimes?

   b. What fraction of households were victims in that year?

70. a. What year had the lowest number of burglaries?

   b. What fraction of households were victims of burglary in that year?

71. In 1988, what fraction of the total property crimes were burglaries?

72. In 1985, what fraction of total property crimes were motor vehicle theft?

For Exercises 73–80, write each fraction in lowest terms. See Example 5.

73. What fraction of a year is 30 days? Use 365 days for the number of days in a year.

> **Of Interest**
>
> It takes Earth one year to complete one revolution around the Sun, which is actually $365\frac{1}{4}$ days. Because it is impractical to have $\frac{1}{4}$ days on the calendar, every four years those $\frac{1}{4}$ days add up to a full day, which is why we have a leap year with 366 days.

74. What fraction of an hour is 8 minutes?

75. What fraction of a minute is 40 seconds?

76. What fraction of a foot is 4 inches?

77. In the first session of 2007, the U.S. House of Representatives had 233 Democrats and 202 Republicans. What fraction of the House of Representatives was Republican? (*Source:* Official list of United States House of Representatives.)

78. In 2007, the U.S. Senate had 49 Democrats, 49 Republicans, and 2 Independents. What fraction of the Senate was not Democrat? (*Source:* Official list of United States Senators.)

> **Of Interest**
>
> The chemical formula for a molecule of lactose is $C_{12}H_{22}O_{11}$. Notice that the subscripts in a chemical formula indicate the number of atoms of each element that are present in the molecule.

79. One molecule of lactose contains 12 carbon atoms, 22 hydrogen atoms, and 11 oxygen atoms. What fraction of the atoms that make up the molecule are carbon?

80. One molecule of glucose contains 6 carbon atoms, 12 hydrogen atoms, and 6 oxygen atoms. What fraction of the atoms that make up the molecule are not carbon?

Review Exercises

Exercises 1–3  CONSTANTS AND VARIABLES

[1.1] **1.** Write a set containing the names of the four planets closest to the Sun.

[1.1] **2.** Is 0.8 a rational or irrational number? Explain.

[1.1] **3.** Graph 3.7 on a number line.

Exercises 4–6  EXPRESSIONS

[1.1] **4.** Is $5x + 2$ an expression or an equation? Explain.

[1.1] **5.** Simplify: $|27|$

[1.1] **6.** Use $<, >$, or $=$ to make a true statement: $|-6| \quad 6$

1.3 Adding and Subtracting Real Numbers; Properties of Real Numbers

Objectives

1. Add integers.
2. Add rational numbers.
3. Find the additive inverse of a number.
4. Subtract rational numbers.

Objective 1 **Add integers.** We now turn our attention to adding and subtracting numbers. We begin with addition. First, consider the parts of an addition statement. The numbers added are called *addends*, and the answer is called a *sum*.

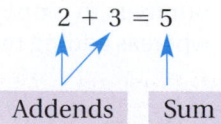

$$2 + 3 = 5$$

Addends   Sum

There are three important properties of addition that are true for all real numbers. First, the sum of a number and 0 is that number. Because adding 0 to a number does not change the identity of the number, 0 is called the *additive identity*.

$$2 + 0 = 2$$

Second, the order of addends can be changed without affecting the sum. This is known as the *commutative property of addition*.

$$2 + 3 = 5 \quad \text{or} \quad 3 + 2 = 5$$

Third, when adding three or more numbers, we can group the addends differently without affecting the sum. This is known as the *associative property of addition*.

$$2 + (3 + 4) \qquad \text{or} \qquad (2 + 3) + 4$$
$$= 2 + 7 \qquad\qquad\qquad\quad = 5 + 4$$
$$= 9 \qquad\qquad\qquad\qquad\quad = 9$$

Following is a summary of these properties.

Properties of Addition	Symbolic Form	Word Form
Additive Identity	$a + 0 = a$	The sum of a number and 0 is that number.
Commutative Property of Addition	$a + b = b + a$	Changing the order of addends does not affect the sum.
Associative Property of Addition	$a + (b + c) = (a + b) + c$	Changing the grouping of three or more addends does not affect the sum.

Example 1   Indicate whether each equation illustrates the additive identity, commutative property of addition, or associative property of addition.

a. $(4 + 7) + 2 = 4 + (7 + 2)$

Answer   Associative property of addition because the grouping is changed.

b. $0 + (-7) = -7$

Answer   Additive identity because the sum of a number and 0 is that number.

c. $(-3 + 5) + 2 = 2 + (-3 + 5)$

Answer   Commutative property of addition because the order of the addends is changed.

Your Turn 1   Indicate whether each equation illustrates the additive identity, commutative property of addition, or associative property of addition.

    a. $-6 + (3 + 5) = (-6 + 3) + 5$ **b.** $3 + 0 = 3$ **c.** $-7 + 3 = 3 + (-7)$

Adding Numbers with the Same Sign

Keeping these properties in mind, let's consider how to add signed numbers. First, we learn how to add numbers that have the same sign. It is helpful to relate adding signed numbers to money. For example, adding two positive numbers is like adding deposits, whereas adding two negative numbers is like adding debts.

$$15 + 10 = 25 \qquad\qquad\qquad -20 + (-8) = -28$$

Note: This addition is like having $15 in an account and depositing another $10 into the same account, bringing the total to $25.

Note: This addition is like having a debt balance of $20 on a credit card and charging another $8 on the same card, bringing the balance to a total debt of $28.

We can also illustrate each of these cases using a number line. The first addend is the starting point, and the second addend indicates the direction and distance to travel on the number line so that we finish at the sum.

Answers   Your Turn 1

a. associative property of addition
b. additive identity
c. commutative property of addition

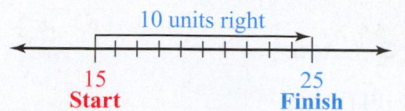
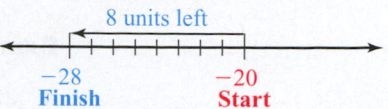

These two cases are similar and suggest the following procedure:

Procedure   **Adding Numbers with the Same Sign**

To add two numbers that have the same sign, add their absolute values and keep the same sign.

Example 2   Add.

a. $15 + 13$

SOLUTION $15 + 13 = 28$

b. $-14 + (-23)$

SOLUTION $-14 + (-23) = -37$

> **Note:** In terms of money, part a illustrates two deposits, whereas part b illustrates two debts.

Your Turn 2   Add.

a. $32 + 19$ **b.** $-62 + (-13)$

Adding Numbers with Different Signs

Now consider adding two numbers with different signs, which is like making a payment toward a debt. Consider the following cases:

$-15 + 3 = -12$ $-9 + 20 = 11$

> **Note:** This addition is like having a debt of $15 and a payment of $3 is made toward that debt. This decreases the debt to $12.

> **Note:** This addition is like having a debt of $9 and a payment of $20 is made toward that debt. Because the payment is more than the debt, we now have a credit of $11.

Using number lines:

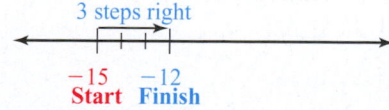
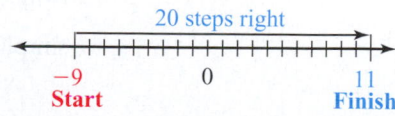

Note that the commutative property of addition indicates that in the examples, the addends can be rearranged without affecting the sum.

$3 + (-15) = -12$ $20 + (-9) = 11$

These examples suggest the following procedure:

Procedure   **Adding Numbers with Different Signs**

To add two numbers that have different signs, subtract the smaller absolute value from the greater absolute value and keep the sign of the number with the greater absolute value.

Answers **Your Turn 2**

a. 51   **b.** -75

Example 3 Add.

a. $22 + (-14)$

SOLUTION $22 + (-14) = 8$

b. $-13 + 20$

> **Note:** In terms of money, parts a and b illustrate situations in which the deposit is greater than the debt. Therefore, the result is a credit, or positive.

SOLUTION $-13 + 20 = 7$

c. $12 + (-18)$

SOLUTION $12 + (-18) = -6$

d. $-24 + 5$

> **Note:** In terms of money, parts c and d illustrate situations in which a payment toward a debt is not enough to pay off the debt. Therefore, the result is still debt, or negative.

SOLUTION $-24 + 5 = -19$

Your Turn 3 Add.

a. $15 + (-6)$ **b.** $27 + (-35)$ **c.** $-13 + 19$ **d.** $-29 + 14$

Objective ❷ Add rational numbers.

Adding Fractions with the Same Denominator

Recall that the set of rational numbers contains any number that can be expressed as a ratio of integers. This set includes the integers themselves because every integer can be expressed as a ratio with 1 as the denominator. Now let's turn our attention to the rest of the rational number set, namely fractions and decimals. Consider the intuitive fact that adding two quarters equals a half. (In money, this would be a half-dollar.) The following circle shows this sum.

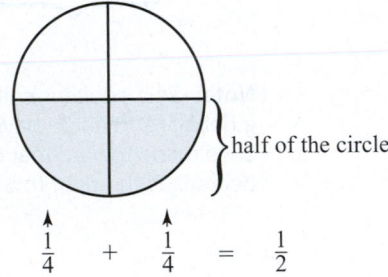

$$\frac{1}{4} \quad + \quad \frac{1}{4} \quad = \quad \frac{1}{2}$$

Notice that if the numerators are added and the denominator remains the same, the result simplifies to $\frac{1}{2}$.

$$\frac{1}{4} + \frac{1}{4} = \frac{1+1}{4} = \frac{2}{4} = \frac{1}{2}$$

Procedure To add fractions with the same denominator, add the numerators and keep the same denominator; then simplify.

Example 4 Add.

a. $\dfrac{5}{8} + \dfrac{1}{8}$

SOLUTION $\frac{5}{8} + \frac{1}{8} = \frac{5+1}{8} = \frac{6}{8}$ — Add the numerators and keep the same denominator. Because $\frac{6}{8}$ is not in lowest terms, we must simplify.

$$= \frac{2 \cdot 3}{2 \cdot 2 \cdot 2} = \frac{3}{4}$$ — Replace 6 and 8 with their prime factorizations; divide out the common factor, 2; and then multiply the remaining factors.

b. $-\frac{3}{10} + \left(-\frac{1}{10}\right)$

SOLUTION $-\frac{3}{10} + \left(-\frac{1}{10}\right) = \frac{-3 + (-1)}{10} = -\frac{4}{10}$

Note: Because the fractions have the same sign, we add and keep the same sign.

$$= -\frac{2 \cdot 2}{2 \cdot 5} = -\frac{2}{5}$$ — Simplify to lowest terms by dividing out the common factor, 2.

c. $\frac{7}{9} + \left(-\frac{4}{9}\right)$

Note: Because the two addends have different signs, we subtract the smaller absolute value from the greater absolute value and keep the sign of the number with the greater absolute value.

SOLUTION $\frac{7}{9} + \left(-\frac{4}{9}\right) = \frac{7 + (-4)}{9} = \frac{3}{9}$

$$= \frac{3}{3 \cdot 3} = \frac{1}{3}$$ — Simplify to lowest terms by dividing out the common factor, 3.

Your Turn 4 Add.

a. $\frac{3}{8} + \frac{1}{8}$ **b.** $-\frac{5}{9} + \left(-\frac{2}{9}\right)$ **c.** $-\frac{7}{12} + \frac{5}{12}$

Adding Fractions with Different Denominators

If the denominators are different, we must first find a common denominator. Recall that in Section 1.2, we wrote equivalent fractions with the LCD. Consider the addition $\frac{3}{4} + \frac{1}{6}$. The LCD of 4 and 6 is 12. We now write each fraction as an equivalent fraction with the LCD.

$$\frac{3}{4} + \frac{1}{6} = \frac{3(3)}{4(3)} + \frac{1(2)}{6(2)}$$

$$= \frac{9}{12} + \frac{2}{12}$$

$$= \frac{11}{12}$$

Note: To write $\frac{3}{4}$ as an equivalent fraction with 12 as the denominator, we multiply the numerator and denominator by 3.

Note: To write $\frac{1}{6}$ as an equivalent fraction with 12 as the denominator, we multiply the numerator and denominator by 2.

Procedure   **Adding Fractions**

To add fractions with different denominators:

1. Write each fraction as an equivalent fraction with the LCD.
2. Add the numerators and keep the LCD.
3. Simplify.

Answers   Your Turn 4

a. $\frac{1}{2}$ **b.** $-\frac{7}{9}$ **c.** $-\frac{1}{6}$

Example 5 Add.

a. $\dfrac{1}{6} + \dfrac{5}{9}$

SOLUTION The LCD of 6 and 9 is 18.

$$\dfrac{1}{6} + \dfrac{5}{9} = \dfrac{1(3)}{6(3)} + \dfrac{5(2)}{9(2)}$$ Write equivalent fractions with 18 in the denominator.

$$= \dfrac{3}{18} + \dfrac{10}{18}$$ Add numerators and keep the common denominator.

$$= \dfrac{3 + 10}{18}$$ Because the addends have the same sign, we add and keep the same sign.

$$= \dfrac{13}{18}$$

b. $-\dfrac{4}{5} + \dfrac{2}{3}$

SOLUTION The LCD of 5 and 3 is 15.

$$-\dfrac{4}{5} + \dfrac{2}{3} = -\dfrac{4(3)}{5(3)} + \dfrac{2(5)}{3(5)}$$ Write equivalent fractions with 15 in the denominator.

$$= -\dfrac{12}{15} + \dfrac{10}{15}$$ Add numerators and keep the common denominator.

$$= \dfrac{-12 + 10}{15}$$ Because the addends have different signs, we subtract and keep the sign of the number with the greater absolute value.

$$= -\dfrac{2}{15}$$

c. $-\dfrac{7}{12} + \dfrac{7}{30}$

SOLUTION The LCD of 12 and 30 is 60.

$$-\dfrac{7}{12} + \dfrac{7}{30} = -\dfrac{7(5)}{12(5)} + \dfrac{7(2)}{30(2)}$$ Write equivalent fractions with 60 as the denominator.

$$= -\dfrac{35}{60} + \dfrac{14}{60}$$ Add numerators and keep the common denominator.

$$= \dfrac{-35 + 14}{60}$$

$$= -\dfrac{21}{60}$$ Reduce to lowest terms.

$$= -\dfrac{3 \cdot 7}{2 \cdot 2 \cdot 3 \cdot 5}$$ Write the numerator and denominator in terms of prime factors.

$$= -\dfrac{3 \cdot 7}{2 \cdot 2 \cdot 3 \cdot 5}$$ Divide out the common factor of 3.

$$= -\dfrac{7}{20}$$

Your Turn 5 Add.

a. $\dfrac{1}{15} + \dfrac{5}{12}$ **b.** $\dfrac{7}{8} + \left(-\dfrac{1}{5}\right)$ **c.** $-\dfrac{7}{18} + \dfrac{5}{30}$

Answers Your Turn 5

a. $\dfrac{29}{60}$ **b.** $\dfrac{27}{40}$ **c.** $-\dfrac{2}{9}$

Adding Decimal Numbers

Now let's review adding decimal numbers, which are also a part of the set of rational numbers. Recall that to add decimal numbers, we add like place values.

Example 6   Angela has a balance of $578.26 and incurs a debt of $92.45. What is Angela's new balance?

SOLUTION   A debt of $92.45 is −$92.45. Her balance is

$$578.26 + (-92.45) = \$485.81$$

Your Turn 6   Daryl has a balance of −$452.75 on a credit card, and he makes a payment of $120. What is his new balance?

Objective ③ Find the additive inverse of a number.   What happens if we add two numbers that have the same absolute value but different signs, such as 5 + (−5)? In money terms, this is like making a $5 payment toward a debt of $5. Notice that the payment pays off the debt so that the balance is 0.

$$5 + (-5) = 0$$

Because their sum is 0, we say that 5 and −5 are **additive inverses**, or opposites.

Definition   *Additive inverses:* Two numbers whose sum is 0.

Example 7   Find the additive inverse of the given numbers.

a. 9

Answer   −9 because 9 + (−9) = 0

b. −7

Answer   7 because −7 + 7 = 0

c. 0

Answer   0 because 0 + 0 = 0

Answer   Your Turn 6

−$332.75

🖩 **Calculator Tips**

To enter a negative number or find an additive inverse on a graphing or scientific calculator, use the [(−)] *key. Note that this key is different from the subtraction key.*

To find the additive inverse of 9, type [(−)] [9] [ENTER] .

To find the additive inverse of −7, type [(−)] [(−)] [7] [ENTER] .

On older scientific calculators, the [+/−] *key is used to find the additive inverse of a number. You enter the value first, then press the* [+/−] *key. Each time you press the* [+/−] *key the sign will change.*

To find the additive inverse of 9, type [9] [+/−] .

To find the additive inverse of −7, type [7] [+/−] [+/−] .

Your Turn 7 Find the additive inverse of the given numbers.

a. -12 b. 0.67 c. $-\dfrac{3}{8}$

A minus sign can be used to indicate additive inverse.

Example 8 Simplify.

a. $-(-8)$

Answer $-(-8)$ indicates that we need to find the additive inverse of -8. So $-(-8) = 8$.

b. $-|6|$

Answer $-|6|$ indicates that we need to find the additive inverse of the absolute value of 6. So $-|6| = -6$.

c. $-|-4|$

Answer $-|-4|$ indicates that we need to find the additive inverse of the absolute value of -4. So $-|-4| = -4$.

Your Turn 8 Simplify.

a. $-(-5)$ b. $-|8|$ c. $-|-3|$

Objective ④ Subtract rational numbers. Now let's turn our attention to subtraction. The parts of subtraction are

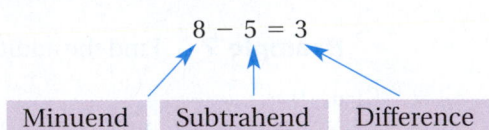

Recall that when we add two numbers that have different signs, we actually subtract. This implies that subtraction and addition are related. In fact, every subtraction statement can be written as an equivalent addition statement. Notice that the subtraction statement $8 - 5 = 3$ and the addition statement $8 + (-5) = 3$ are equivalent: They are both ways of finding the balance given an $8 deposit with a $5 debt. Note that in writing $8 - 5$ as an equivalent addition $8 + (-5)$, we change the operation sign from a minus sign to a plus sign and change the 5 to its additive inverse, -5.

Any subtraction statement can be written as an equivalent addition statement using this procedure. Consider $8 - (-5)$, which represents an $8 deposit subtracting a debt of $5. What does it mean to subtract a debt? If a bank records a debt and then realizes it made a mistake, it must reverse that debt. Subtracting, or reversing, a debt means that the bank must record a deposit to correct the mistake. In math terms, this means that $8 - (-5)$ is equivalent to $8 + 5$, which equals 13. Notice that this still follows the process we just discussed.

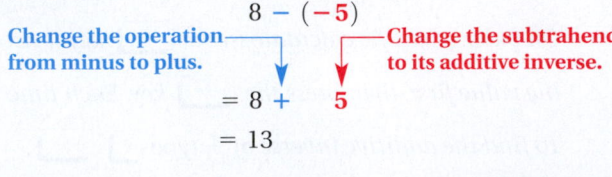

Procedure | **Rewriting Subtraction**

To write a subtraction statement as an equivalent addition statement, change the operation symbol from a minus sign to a plus sign and change the subtrahend to its additive inverse.

Example 9 Subtract.

a. $-15 - (-6)$

SOLUTION Write the subtraction as an equivalent addition.

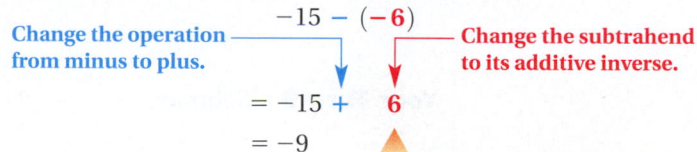

$$-15 - (-6)$$

Change the operation from minus to plus. Change the subtrahend to its additive inverse.

$$= -15 + 6$$
$$= -9$$

> **Note:** We can think of this as subtracting a debt of $6 from a debt of $15. Subtracting a debt from a debt is equivalent to making a deposit or payment against the existing debt balance. So this decreases the amount of debt.

b. $-\dfrac{2}{3} - \dfrac{3}{4}$

SOLUTION Write an equivalent addition; then find a common denominator and continue the process of adding fractions.

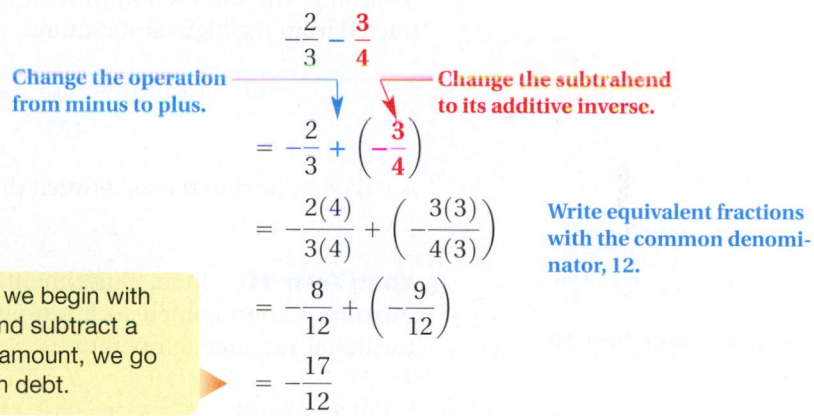

$$-\frac{2}{3} - \frac{3}{4}$$

Change the operation from minus to plus. Change the subtrahend to its additive inverse.

$$= -\frac{2}{3} + \left(-\frac{3}{4}\right)$$

$$= -\frac{2(4)}{3(4)} + \left(-\frac{3(3)}{4(3)}\right) \quad \text{Write equivalent fractions with the common denominator, 12.}$$

$$= -\frac{8}{12} + \left(-\frac{9}{12}\right)$$

$$= -\frac{17}{12}$$

> **Note:** If we begin with a debt and subtract a positive amount, we go deeper in debt.

🖩 **Calculator Tips**

To enter the problem in Example 9(b) into a scientific calculator, type

[(−)] [2] [$a^{b/c}$] [3] [−] [3] [$a^{b/c}$] [4] [ENTER] .

On a graphing calculator, type

[(−)] [2] [÷] [3] [−] [3] [÷] [4] [ENTER] .

To see the answer in fraction form, press [MATH] *and select the 1:Frac option from the menu; then press* [ENTER] .

---

c. $5.04 - 8.01$

SOLUTION Write an equivalent addition statement.

$$5.04 - \mathbf{8.01}$$
$$= 5.04 + (\mathbf{-8.01})$$
$$= -2.97$$

Note: To calculate the sum, the number with the smaller absolute value must be subtracted from the number with the greater absolute value.

$$\begin{array}{r} 8.01 \\ -5.04 \\ \hline 2.97 \end{array}$$

Because the number with the greater absolute value is negative, the result is negative.

Your Turn 9 Subtract.

a. $25 - (-17)$

b. $-\dfrac{3}{8} - \left(-\dfrac{1}{3}\right)$

c. $0.06 - 4.02$

Example 10 On Mars, the daytime high temperature can reach $-15°C$, while at night, the temperature can plummet to $-140°C$. What is the difference between the high and low temperatures?

SOLUTION *Difference* indicates that we subtract the temperatures. Because the wording in the problem is to find the "difference between the high and low temperatures," we arrange the subtraction with the low temperature subtracted from the high temperature.

$$-15 - (-140) = -15 + 140 \qquad \text{Write as an equivalent addition.}$$
$$= 125 \qquad \text{Add.}$$

Answer The difference between the high and low temperatures is $125°C$.

Answers Your Turn 9

a. 42 **b.** $-\dfrac{1}{24}$ **c.** -3.96

Answer Your Turn 10

34.4°C

Your Turn 10 In an experiment, a mixture begins at a temperature of $5.8°C$. The mixture is then cooled to a temperature of $-28.6°C$. Find the difference between the initial and final temperatures.

1.3 Exercises   For Extra Help   *MyMathLab*    PRACTICE   WATCH   DOWNLOAD

Note: Exercises marked with a ★ represent challenging exercises.

1. Explain the difference between the commutative property of addition and the associative property of addition.

2. Explain why 4 and −4 are additive inverses.

3. Why is 0 called the additive identity?

4. In your own words, explain the process of adding or subtracting two fractions with different denominators.

5. In your own words, explain how to add two numbers that have the same sign.

6. In your own words, explain how to add two numbers that have different signs.

7. Explain how to write a subtraction statement as an equivalent addition statement.

8. Explain why $6 - 5$ and $5 - 6$ have different answers, whereas $-6 - 5$ and $-5 - 6$ have the same answer. (*Hint:* Think about the properties of addition.)

For Exercises 9–20, indicate whether each equation illustrates the additive identity, commutative property of addition, associative property of addition, or additive inverse. See Examples 1 and 7.

9. $5 + 6 = 6 + 5$

10. $-3 + 7 = 7 + (-3)$

11. $0 + 8 = 8$

12. $-5 + 0 = -5$

13. $0.8 + (-0.8) = 0$

14. $-\dfrac{4}{9} + \dfrac{4}{9} = 0$

15. $(3 + 5) + 2 = 3 + (5 + 2)$

16. $6 + (2 + 3) = (6 + 2) + 3$

17. $-8 + (7 + 3) = (7 + 3) + (-8)$

18. $(5 + 3) - 4 = -4 + (5 + 3)$

19. $6.3 + (2.1 - 2.1) = 6.3 + 0$

20. $(-4.6 + 4.6) + 9.5 = 0 + 9.5$

For Exercises 21–56, add. See Examples 2–6.

21. $8 + 13$

22. $15 + 7$

23. $-6 + (-12)$

24. $-5 + (-7)$

25. $-4 + 13$

26. $-5 + 16$

27. $-14 + 5$

28. $-17 + 8$

29. $27 + (-13)$

30. $29 + (-7)$

31. $-28 + 12$

32. $-16 + 13$

33. $\dfrac{5}{8} + \dfrac{1}{8}$

34. $\dfrac{9}{16} + \dfrac{5}{16}$

35. $-\dfrac{4}{9} + \left(-\dfrac{2}{9}\right)$

36. $-\dfrac{3}{5} + \left(-\dfrac{1}{5}\right)$

37. $\dfrac{1}{6} + \left(-\dfrac{5}{6}\right)$

38. $-\dfrac{9}{14} + \dfrac{3}{14}$

39. $\dfrac{3}{4} + \dfrac{1}{6}$

40. $\dfrac{1}{4} + \dfrac{7}{8}$

41. $-\dfrac{1}{12} + \left(-\dfrac{2}{3}\right)$

42. $-\dfrac{2}{5} + \left(-\dfrac{3}{20}\right)$

43. $-\dfrac{5}{6} + \dfrac{4}{21}$

44. $-\dfrac{5}{16} + \dfrac{3}{12}$

45. $0.21 + 0.05$

46. $0.06 + 0.17$ **47.** $-0.18 + 6.7$ **48.** $-15.81 + 4.28$ **49.** $-0.28 + (-4.1)$ **50.** $-7.8 + (-9.16)$

★**51.** $-42 + |-14|$ ★**52.** $-31 + |-54|$ ★**53.** $|-2.4| + |-0.78|$ ★**54.** $|-0.6| + |-9.1|$ ★**55.** $\left|-\dfrac{3}{8}\right| + \left|\dfrac{5}{6}\right|$

★**56.** $\left|-\dfrac{4}{5}\right| + \left|\dfrac{3}{4}\right|$

For Exercises 57–70, find the additive inverse. See Example 7.

57. 5 **58.** 7 **59.** -12 **60.** -6 **61.** 0 **62.** -9 **63.** $\dfrac{5}{6}$

64. $-\dfrac{6}{17}$ **65.** -0.29 **66.** 2.8 **67.** $-x$ **68.** b **69.** $\dfrac{m}{n}$ **70.** $-\dfrac{a}{b}$

For Exercises 71–78, simplify. See Example 8.

71. $-(-2)$ **72.** $-(-15)$ ★**73.** $-(-(-4))$ ★**74.** $-(-(-1))$

75. $-|4|$ **76.** $-|10|$ **77.** $-|-12|$ **78.** $-|-5|$

For Exercises 79–102, subtract. See Example 9.

79. $6 - 15$ **80.** $8 - 20$ **81.** $-4 - 9$ **82.** $-7 - 15$ **83.** $4 - (-3)$

84. $6 - (-7)$ **85.** $-8 - (-2)$ **86.** $-13 - (-6)$ **87.** $-\dfrac{1}{5} - \left(-\dfrac{1}{5}\right)$ **88.** $-\dfrac{3}{4} - \left(-\dfrac{3}{4}\right)$

89. $\dfrac{7}{10} - \left(-\dfrac{3}{5}\right)$ **90.** $\dfrac{3}{8} - \left(-\dfrac{5}{6}\right)$ **91.** $-\dfrac{4}{5} - \left(-\dfrac{2}{7}\right)$ **92.** $-\dfrac{1}{2} - \left(-\dfrac{1}{3}\right)$ **93.** $4.01 - 3.65$

94. $8.1 - 4.76$ **95.** $0.07 - 5.82$ **96.** $0.107 - 5.802$ **97.** $-6.1 - (-4.5)$ **98.** $-7.1 - (-2.3)$

★**99.** $-|4| - |-6|$ ★**100.** $-|-9| - |-12|$ ★**101.** $|6.2| - |-7.1|$ ★**102.** $|4.6| - |-7.3|$

For Exercises 103–112, solve. See Example 10.

103. The Walt Disney Company's third-quarter financial report for 2007 contains the following information. Find the net income.

Third-quarter report for the year ending September 29, 2007

Income (in millions)	Expenditures (in millions)
Revenues = $8930	Costs and expenses = $7359
Equity in the income of investees = $96	Net interest expense = $163
	Income taxes = $521
	Minority interests = $100
	Loss from discontinued operations = $6

104. Following is a balance sheet of income and expenditures for a small business. Find the net profit or loss.

June 2010 balance sheet

Income	Expenditures
Revenue = $24,572.88	Materials = $1545.75
Dividends = $1284.56	Lease = $2700
	Utilities = $865.45
	Employee wages = $21,580.50

105. In engineering, the resultant force on an object is the sum of all forces acting on the object. The diagram shows the forces acting on a steel beam. Find the resultant force. What does the sign of the resultant force indicate?

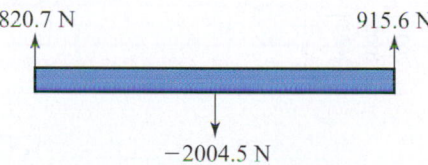

820.7 N   915.6 N

−2004.5 N

106. A family's taxable income is its income less deductions. Following is a table that shows the Brendel family's income and deductions for the 2010 tax year. What is the Brendel family's taxable income?

Income	Deductions
Mr. Brendel = $31,672.88	Mortgage interest = $6545.75
Mrs. Brendel = $32,284.56	Charitable donations = $1200
Dividends = $124.75	Medical expenses = $165.45
Miscellaneous income = $2400	Exemptions = $10,800

107. On February 13, 2008, the Dow Jones Industrial Average (DJIA) closed at 12,373.41. On February 12, 2008, the DJIA closed at 12,240.01. Find the difference in closing value from February 13 to February 12.

108. On January 28, 2008, the closing price of Dell's stock was $20.35. On January 29, 2008, the closing price was $20.56. Find the difference in closing price from January 29 to January 28.

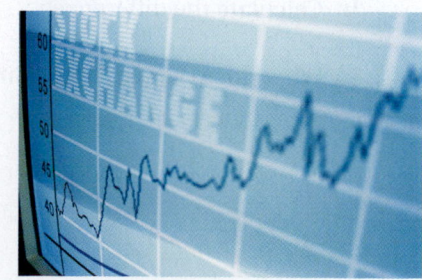

109. On February 11, 2008, the closing price of AT&T's stock was $24.78. If the price had a change of −$0.32 from the previous trading date's closing price, what was the closing price on February 8, 2008?

110. On February 14, 2008, an investor found that the NASDAQ closed at 2,320.04, which was a change of −53.89 from the previous day's closing value. What was the closing value on February 13, 2008?

111. The temperature of liquid nitrogen is −208°C. When an apple is placed in the liquid, the apple raises the temperature of the liquid to its boiling point of −196°C. Write an expression that describes the difference between the boiling point and the initial temperature; then calculate the difference.

112. Absolute zero is the temperature at which molecular motion is at a minimum. This temperature is −273.15°C. A piece of metal is cooled to −256.5°C. Write an expression that describes the difference between the piece of metal's current temperature and absolute zero; then calculate the difference.

Use the following line graph to answer Exercises 113 and 114. The graph shows the mean composite ACT score of entering college freshmen from 1985 to 2006.

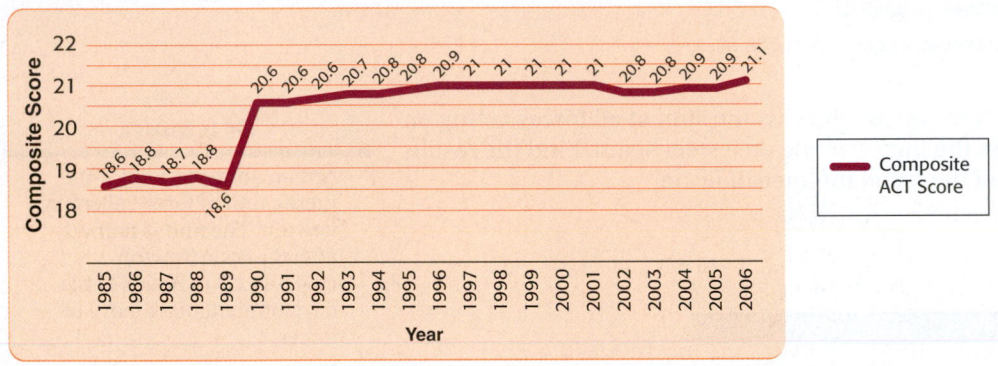

Source: U.S. Department of Education.

113. a. Write an expression that describes the difference between the mean composite scores in 1989 and 1986.

    b. Calculate the difference.

    c. What does the sign of the difference indicate about how the scores changed?

114. a. Write an expression that describes the difference between the mean composite scores in 2006 and 1986.

    b. Calculate the difference.

    c. What does the sign of the difference indicate about how the scores changed?

Example 1

each equation...

a. $4(-3) = -$...

Answer Co...
different.

b. $6(-2 \cdot 7) = $...

Answer Ass...
different.

c. $5(8 - 3) = $...

Answer Dis...

Your Turn 1

each equation...

a. $-8 \cdot 0 = 0$...

Multiplyin...

To determine...
Pay attention...

$$2 \cdot$$
$$2 \cdot$$
$$2 \cdot$$
$$2 \cdot$$
$$2 \cdot$$
$$2 \cdot (-$$
$$2 \cdot (-$$

This patte...
equals a negat...
can exchange...
itive number...

Rule Multiplying T...

When multiply...

Example 2

a. $9(-8)$

Solution $9($

b. $(-12)3$

Solution $(-$

Your Turn 2

a. $13(-7)$

Answers Your Turn 1

a. multiplicative property of 0

b. multiplicative identity

c. distributive property of multiplication over addition

Answers Your Turn 2

a. -91 **b.** -45

Use the bar graph to the right to answer Exercises 115 and 116. The graph shows the median annual income by level of education completed and gender from 1995 to 2004.

115. In what year was the difference between salaries for males and females with a bachelor's degree the greatest? How much was that difference?

116. In what year was the difference in salaries greatest between a person with a high school degree and a person with a bachelor's degree? How much was that difference?

Puzzle Problem

Fill in each square with a number 1 through 9 so that the sum of the numbers in each row, each column, and the two diagonals is the same.

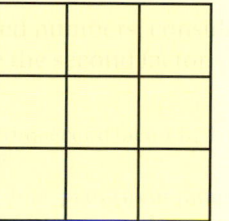

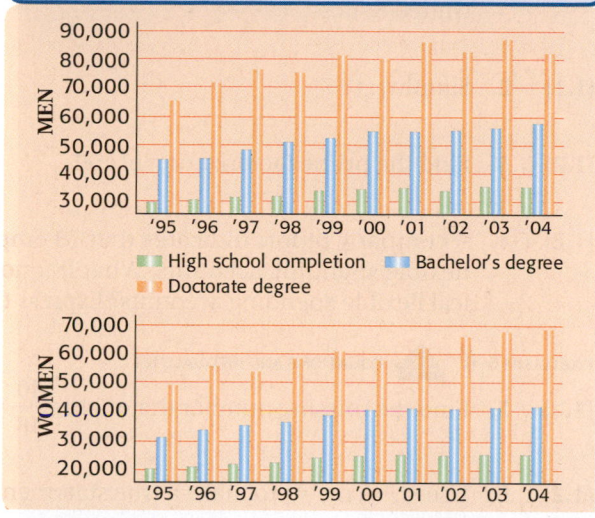

MEDIAN ANNUAL INCOME OF YEAR-ROUND FULL-TIME WORKERS

25 YEARS OLD AND OVER, BY LEVEL OF EDUCATION COMPLETED AND GENDER: 1995 TO 2004

Collaborative Exercises Dollars and Sense

Your group is to prepare a simple monthly household budget. The budget should be for a "typical" student or for a new employee in your field of study. (However, you should not use any of your personal information.) Remember to be realistic about both income and expenditures.

1. Discuss and agree upon an amount of income you think an average working student makes each month. (Or use a starting salary for a person in your field of study.)
2. Estimate average expense amounts for each of the following. If your group believes that an expense does not apply, explain why.

Taxes	Transportation/Gasoline
Housing	Child care
Electricity	Loan repayments
Water	Hair care
Telephone (home and cell)	Clothing
Cable TV	Magazine/Newspaper subscriptions
Internet service	Allowance
Food	Movies
Household supplies	Other entertainment (renting DVDs, attending concerts,
Medicine	eating out)

3. Using your estimates in Exercises 1 and 2, calculate the monthly net.
4. Discuss and list ways that a person could cut expenses to improve his or her monthly net.

Review Exercises

[1.1] **1.** Write a set containing the last names o
United States.

[1.1] **2.** Simplify: $|-7|$

[1.2] **3.** Find the prime factorization of 100.

[1.2] **4.** A company report indicates that 78 en
flexible spending accounts. What fract
ical flexible spending accounts? Express

Exercises 5–6   EQUATIONS AND INEQUALITIES

[1.2] **5.** Find the missing number that makes $\dfrac{40}{48}$

[1.2] **6.** Use $<$, $>$, or $=$ to make a true stateme

1.4 Multiply
Proper

Objectives
1. Multiply in
2. Multiply m
3. Multiply ra
4. Find the m
5. Divide rati

Objective 1 Multiply intege
multiplied to e
properties of a
ties to addition
multiplication.

Properties of I

Multiplicative pr

Multiplicative ide

Commutative pr
multiplication

Associative prop
multiplication

Distributive prop
multiplication ov

36   Chapter 1   Foundations of Algebra

Multiplying Numbers with the Same Sign

Now let's examine multiplying numbers with the same sign. Consider the following pattern to determine the rule.

$$(-2) \cdot 4 = -8$$ From the rule we already established, a negative number times a positive number is a negative product.

$$(-2) \cdot 3 = -6$$ ◄ As we decrease the positive factor by 1, the product *increases* by 2, from -8 to -6.

$$(-2) \cdot 2 = -4$$
$$(-2) \cdot 1 = -2$$ Notice that the same pattern continues with the product increasing by 2 each time we decrease the factor by 1.

$$(-2) \cdot 0 = 0$$

$$(-2) \cdot (-1) = 2$$ ◄ To continue the same pattern, when we decrease the 0 factor to -1, we must continue to increase the product by 2. This means that the product must become positive.

$$(-2) \cdot (-2) = 4$$

This pattern indicates that the product of two negative numbers is a positive number. We have already seen that the product of two positive numbers is a positive number. So we can draw the following conclusion:

Rule **Multiplying Two Numbers with the Same Sign**

When multiplying two numbers that have the same sign, the product is positive.

Example 3   Multiply.

a. $-6(-8)$

SOLUTION $-6(-8) = 48$

b. $(-9)(-7)$

SOLUTION $(-9)(-7) = 63$

Your Turn 3   Multiply.

a. $-6(-9)$ **b.** $(-3)(-13)$

Objective 2 **Multiply more than two numbers.**   Suppose we have to multiply more than two numbers, such as $(-2)(-3)(5)$. The associative property of multiplication allows us to group the factors any way we like.

$$(\mathbf{-2})(\mathbf{-3})(5) = \mathbf{6}(5)$$ Multiply the first two factors.
$$= 30$$

or

$$(-2)(\mathbf{-3})(\mathbf{5}) = (-2)(\mathbf{-15})$$ Multiply the second and third factors.
$$= 30$$

Notice that no matter which way we choose to multiply the factors, the result is the same. Also notice that the result is positive because there are two negative factors involved in the multiplication. What if we have three negative factors, as in $(-2)(-3)(-5)$?

$$(\mathbf{-2})(\mathbf{-3})(-5) = \mathbf{6}(-5)$$ Multiply the first two factors.
$$= -30$$

Answers **Your Turn 3**
a. 54   **b.** 39

38   Chapter 1   Foundations of Algebra

or

$$(-2)(-3)(-5) = (-2)(15) \quad \text{Multiply the second and third factors.}$$
$$= -30$$

With three negative factors, the result is negative. This suggests the following rule:

Rule **Multiplying with Negative Factors**

The product of an even number of negative factors is positive, whereas the product of an odd number of negative factors is negative.

Example 4 Multiply.

a. $(-1)(-1)(-2)(6)$

SOLUTION Because there are three negative factors (an odd number of negative factors), the result is negative.

$$(-1)(-1)(-2)(6) = -12$$

b. $(-1)(-3)(-7)(2)(-4)$

SOLUTION Because there are four negative factors (an even number of negative factors), the result is positive.

$$(-1)(-3)(-7)(2)(-4) = 168$$

Your Turn 4 Multiply.

a. $(-3)(4)(-5)(2)$ **b.** $(6)(-5)(-1)(2)(-1)$

Objective ③ Multiply rational numbers.

Multiplying Fractions

On a ruler, note that half of a fourth is an eighth. When preceded by a fraction, the word *of* means multiply. So the sentence would translate as follows:

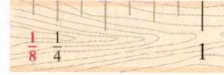

Half of a fourth is an eighth.

$$\frac{1}{2} \quad \cdot \quad \frac{1}{4} \quad = \quad \frac{1}{8}$$

Notice that the numerators multiply to equal the numerator and the denominators multiply to equal the denominator. This suggests the following rule for multiplying fractions:

Rule **Multiplying Fractions**

$$\frac{a}{b} \cdot \frac{c}{d} = \frac{ac}{bd}, \text{ where } b \neq 0 \text{ and } d \neq 0.$$

Multiplying and Simplifying

In mathematics, we are expected to simplify results to lowest terms. Consider $\frac{3}{4} \cdot \frac{5}{6}$. We can multiply first, then simplify.

$$\frac{3}{4} \cdot \frac{5}{6} = \frac{15}{24} = \frac{3 \cdot 5}{2 \cdot 2 \cdot 2 \cdot 3} = \frac{5}{8}$$

> **Note:** After multiplying, we divide out the common factor, 3.

Or as an alternative, we can divide out the common factors before multiplying the fractions. In our example, we can divide out the common factor of 3 and then multiply.

$$\frac{3}{4} \cdot \frac{5}{6} = \frac{3}{2 \cdot 2} \cdot \frac{5}{2 \cdot 3} = \frac{5}{8}$$

> **Note:** Some people prefer to divide out the common factors without breaking down to primes, like this:
> $$\frac{\overset{1}{\cancel{3}}}{4} \cdot \frac{5}{\underset{2}{\cancel{6}}} = \frac{5}{8}$$

We can divide out common factors before or after multiplying because every factor in the original numerators becomes a factor in the product's numerator and every factor in the original denominators becomes a factor in the product's denominator. Now let's put this together with the sign rules for multiplication.

Connection Using the slash marks style, Example 5(a) looks like this:

$$-\frac{\overset{1}{\cancel{2}}}{3} \cdot \frac{7}{\underset{4}{\cancel{8}}} = -\frac{7}{12}$$

Notice that we are still dividing out a factor of 2 in the 2 and 8.

Calculator Tips

On most scientific calculators, for Example 5(a), type

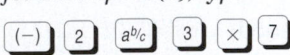 

Example 5 Multiply.

a. $-\frac{2}{3} \cdot \frac{7}{8}$

SOLUTION $\quad -\frac{2}{3} \cdot \frac{7}{8} = -\frac{2}{3} \cdot \frac{7}{2 \cdot 2 \cdot 2}$ Divide out the common factor, 2.

$$= -\frac{7}{12}$$ Because we are multiplying two numbers that have different signs, the product is negative.

b. $-\frac{4}{15} \cdot \left(-\frac{9}{16}\right) \cdot \frac{14}{15}$

SOLUTION $\quad -\frac{4}{15} \cdot \left(-\frac{9}{16}\right) \cdot \frac{14}{15} = \frac{2 \cdot 2}{3 \cdot 5} \cdot \left(-\frac{3 \cdot 3}{2 \cdot 2 \cdot 2 \cdot 2}\right) \cdot \frac{2 \cdot 7}{3 \cdot 5}$ Divide out the common factors.

$$= \frac{7}{50}$$ Because there is an even number of negative factors, the product is positive.

Your Turn 5 Multiply.

a. $\frac{5}{8} \cdot \left(-\frac{3}{10}\right)$ b. $-\frac{4}{5} \cdot \left(\frac{1}{6}\right) \cdot \left(-\frac{3}{9}\right)$

Multiplying Decimal Numbers

Recall that decimal numbers name fractions with denominators that are powers of 10. In the problem $0.3 \cdot 0.7$, we could multiply by replacing the decimal factors with their fraction equivalents.

$$0.3 \cdot 0.7 = \frac{3}{10} \cdot \frac{7}{10} = \frac{21}{100} = 0.21$$

Answers Your Turn 5

a. $-\frac{3}{16}$ b. $\frac{2}{45}$

Because we multiply the denominators, if we multiply tenths times tenths, the product will be hundredths. Notice that if we count the number of decimal places in the factors, this corresponds to the total places in the product.

$$
\begin{array}{rl}
0.7 & \longrightarrow \quad \text{1 place} \\
\times\ 0.3 & \longrightarrow \quad +\ \text{1 place} \\
\hline
0.21 & \longrightarrow \quad \text{2 places}
\end{array}
$$

Likewise, hundredths (2 places) times tenths (1 place) equals thousandths (3 places).

$$
0.03 \cdot 0.7 = \frac{3}{100} \cdot \frac{7}{10} = \frac{21}{1000} = 0.021 \quad \text{or} \quad
\begin{array}{rl}
0.7 & \longrightarrow \quad \text{1 place} \\
\times\ 0.03 & \longrightarrow \quad +\ \text{2 places} \\
\hline
0.021 & \longrightarrow \quad \text{3 places}
\end{array}
$$

Procedure   **Multiplying Decimal Numbers**

To multiply decimal numbers:

1. Multiply as if they were whole numbers.
2. Place the decimal in the product so that it has the same number of decimal places as the total number of decimal places in the factors.

Example 6   Multiply.

a. $(-5.6)(0.03)$

SOLUTION   First, we calculate the value and disregard the signs for now.

$$
\begin{array}{rl}
0.03 & \quad \text{2 places} \\
\times\quad 5.6 & \quad +\ \text{1 place} \\
\hline
0\,1\,8 & \\
+\ 0\,1\,5 & \\
\hline
0.1\,6\,8 & \longleftarrow \quad \text{3 places}
\end{array}
$$

Answer -0.168

Note: When we multiply two numbers with different signs, the product is negative.

b. $(-2)(4.3)(1.8)(-7.1)$

SOLUTION   First, we calculate the value and disregard the signs for now. Multiply from left to right.

$$
(2)(4.3) \qquad\qquad (8.6)(1.8)
$$

$$
(2)(4.3)(1.8)(7.1) = (8.6)(1.8)(7.1) = (15.48)(7.1) = 109.908
$$

Answer   109.908

Note: The product of an even number of negative factors is positive. The factors have a total of three decimal places, so the product has three decimal places.

Your Turn 6   Multiply.

a. $(-0.07)(-2.65)$ **b.** $(-1)(-0.9)(-24)(0.2)$

Objective 4 **Find the multiplicative inverse of a number.** Two numbers that multiply to equal 1 are called **multiplicative inverses**.

Definition *Multiplicative inverses:* Two numbers whose product is 1.

For example, $\frac{2}{3}$ and $\frac{3}{2}$ are multiplicative inverses because their product is 1.

$$\frac{2}{3} \cdot \frac{3}{2} = \frac{6}{6} = 1$$

Notice that to write a number's multiplicative inverse, we simply invert the numerator and denominator. Multiplicative inverses are also known as *reciprocals*.

Example 7 Find the multiplicative inverse.

a. $\frac{5}{8}$

Answer The multiplicative inverse of $\frac{5}{8}$ is $\frac{8}{5}$ because $\frac{5}{8} \cdot \frac{8}{5} = 1$.

b. $-\frac{1}{4}$

Answer The multiplicative inverse of $-\frac{1}{4}$ is -4 because $-\frac{1}{4} \cdot (-4) = -\frac{1}{4} \cdot \left(-\frac{4}{1}\right) = 1$.

c. -6

Answer The multiplicative inverse of -6 is $-\frac{1}{6}$ because $-6 \cdot \left(-\frac{1}{6}\right) = -\frac{6}{1} \cdot \left(-\frac{1}{6}\right) = 1$.

Your Turn 7 Find the multiplicative inverse.

a. $\frac{4}{5}$ b. $\frac{1}{7}$ c. -9

Objective 5 **Divide rational numbers.** Now let's turn our attention to division. The parts of a division statement are shown next.

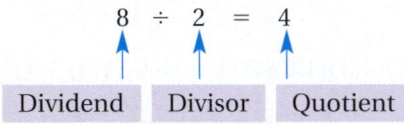

Connection Division can also be expressed in fraction form. For example, the division statement $8 \div 2 = 4$ can also be written as $\frac{8}{2} = 4$.

Sign Rules for Division

The sign rules for division are the same as for multiplication. We can see this by looking at the relationship between division and multiplication. We can use one operation to check the other. For example, if $8 \cdot 2 = 16$, then $16 \div 2 = 8$ or $16 \div 8 = 2$. In checking a multiplication, we divide the product by one of the factors to equal the other factor. If we place signs in the problem, we can say that if $8 \cdot (-2) = -16$, then $-16 \div (-2) = 8$ or $-16 \div 8 = -2$.

Answers Your Turn 7

a. $\frac{5}{4}$ b. 7   c. $-\frac{1}{9}$

When dividing two numbers that have the *same* sign, the quotient is positive.
When dividing two numbers that have *different* signs, the quotient is negative.

Example 8   Divide.

a. $48 \div (-6)$

SOLUTION $48 \div (-6) = -8$

b. $-35 \div (-7)$

SOLUTION $-35 \div (-7) = 5$

Your Turn 8   Divide.

a. $(-60) \div 12$ **b.** $-42 \div (-6)$

Division Involving 0

What if 0 is involved in division? First, consider when the *dividend* is 0 and the *divisor* is not 0, as in $0 \div 9$. The result should be $0 \div 9 = 0$. Remember that we can check division by multiplying the quotient and the divisor to equal the dividend. So we can verify that $0 \div 9 = 0$ by writing the following multiplication:

$$\text{Quotient} \cdot \text{Divisor} = \text{Dividend}$$
$$0 \quad \cdot \quad 9 \quad = \quad 0$$

Conclusion   If the dividend is 0 and the divisor is not zero, the quotient is 0.

What if the divisor is 0, as in $8 \div 0$? Again, think about the check statement.

$$\text{Quotient} \cdot \text{Divisor} = \text{Dividend}$$
$$? \quad \cdot \quad 0 \quad = \quad 8$$

There is no number that can make $? \cdot 0 = 8$ true because any number multiplied by 0 equals 0, not 8. So there also is no number that can make $8 \div 0 = ?$ true, and we say that dividing a nonzero number by 0 is *undefined*.

Conclusion   If the divisor is 0 with a nonzero dividend, the statement is undefined.

What if both the dividend and divisor are 0, as in $0 \div 0$? Again, think about the check statement.

$$\text{Quotient} \cdot \text{Divisor} = \text{Dividend}$$
$$? \quad \cdot \quad 0 \quad = \quad 0$$

Notice that any number will work for this quotient because any number times 0 equals 0. Because we cannot determine a unique quotient here, we say that $0 \div 0$ is *indeterminate*.

Conclusion   If both the dividend and divisor are 0, the statement is indeterminate.

Following is a summary of these rules.

Calculator Tips

Try dividing any number by 0 on your calculator. When you press the equal key, the screen will display "error," which is your calculator's way of indicating undefined or indeterminate.

Dividing Fractions

To determine how to divide when the divisor is a fraction, consider this problem: How many quarters are in $5? A quarter is $\frac{1}{4}$ of a dollar, so this question translates to the following division problem:

$$\text{Number of quarters} = 5 \div \frac{1}{4}$$

Now since there are 4 quarters in each of those 5 dollars, there must be 20 quarters in $5. Notice in analyzing the situation that we translated the division problem into a multiplication problem. We can say the following:

$$\text{Number of quarters} = 5 \div \frac{1}{4} = 5 \cdot 4 = 20$$

Notice that $\frac{1}{4}$ and 4 are multiplicative inverses. We can write a division problem as an equivalent multiplication problem by changing the divisor to its multiplicative inverse.

Rule **Dividing Fractions**

$$\frac{a}{b} \div \frac{c}{d} = \frac{a}{b} \cdot \frac{d}{c}, \text{ where } b \neq 0, c \neq 0, \text{ and } d \neq 0.$$

Example 9   Divide. $-\dfrac{7}{8} \div \dfrac{5}{6}$

SOLUTION $-\dfrac{7}{8} \div \dfrac{5}{6} = -\dfrac{7}{8} \cdot \dfrac{6}{5}$ **Write an equivalent multiplication.**

$$= -\frac{7}{2 \cdot 2 \cdot 2} \cdot \frac{2 \cdot 3}{5}$$ **Divide out the common factor, 2.**

$$= -\frac{21}{20}$$ **Because we are dividing two numbers that have different signs, the result is negative.**

Connection Using slash marks, Example 9 looks like this:

$$-\frac{7}{8} \div \frac{5}{6} = -\frac{7}{\underset{4}{\cancel{8}}} \cdot \frac{\overset{3}{\cancel{6}}}{5} = -\frac{21}{20}$$

Your Turn 9   Divide.

a. $\dfrac{4}{9} \div \left(-\dfrac{8}{15}\right)$

b. $-\dfrac{8}{21} \div \left(-\dfrac{6}{7}\right)$

Dividing Decimal Numbers

Division with decimal numbers can be broken into two categories: (1) if the divisor is an integer and (2) if the divisor contains a decimal point. The following procedure summarizes how to divide decimal numbers.

Answers **Your Turn 9**

a. $-\dfrac{5}{6}$ **b.** $\dfrac{4}{9}$

Dividing Decimal Numbers

To divide decimal numbers, set up a long division and consider the divisor.

Case 1: If the divisor is an integer, divide as if the dividend were a whole number and place the decimal point in the quotient directly above its position in the dividend.

Case 2: If the divisor is a decimal number,

1. Move the decimal point in the divisor to the right enough places to make the divisor an integer.
2. Move the decimal point in the dividend the same number of places.
3. Divide the divisor into the dividend as if both numbers were whole numbers. Make sure you align the digits in the quotient properly.
4. Write the decimal point in the quotient directly above its new position in the dividend.

In either case, continue the division process until you get a remainder of 0 or a repeating digit (or block of digits) in the quotient.

Example 10 Divide. $-12.8 \div (-0.09)$

SOLUTION Because the divisor is a decimal number, we move the decimal point enough places to the right to create an integer—in this case, two places. Then we move the decimal point two places to the right in the dividend. Because we are dividing two numbers with the same sign, the result is positive.

$$
\begin{array}{r}
142.22 \\
9\overline{)1280.00} \\
-9 \\
\hline
38 \\
-36 \\
\hline
20 \\
-18 \\
\hline
20
\end{array}
$$

Note: Because the 2 digit repeats without end, we write a repeat bar over the 2 just after the decimal point.

Answer $142.2\overline{2}$

Your Turn 10 Divide.

a. $32.04 \div (-12)$ **b.** $-13.12 \div 6.4$ **c.** $-0.88 \div (-0.6)$

Applications

It is often necessary to use signed numbers when solving applications problems.

Example 11 Solve.

a. For a chemistry experiment, Fred used $\frac{3}{5}$ of the contents of a bottle of hydrochloric acid that was $\frac{3}{4}$ full. What fractional part of a full bottle did he use?

SOLUTION To find $\frac{3}{5}$ of $\frac{3}{4}$, multiply $\frac{3}{5}$ and $\frac{3}{4}$.

$$\frac{3}{5} \cdot \frac{3}{4} = \frac{3 \cdot 3}{5 \cdot 4} = \frac{9}{20}$$

Answers **Your Turn 10**
a. -2.67 **b.** -2.05 **c.** $1.4\overline{6}$

Answer Fred used $\frac{9}{20}$ of a full bottle of hydrochloric acid.

b. An object's weight is a downward force due to gravity, and this force is calculated by multiplying the object's mass by the acceleration due to gravity (Force = Mass × Acceleration.) Find the force due to gravity of a truck with a mass of 1800 kilograms if the acceleration due to gravity is -9.8 m/sec^2. The force will be in newtons.

SOLUTION Force = Mass × Acceleration. **Substitute for mass and acceleration.**

$$\text{Force} = (1800)(-9.8)$$
$$= -17{,}640$$

Answer The force due to gravity of the truck is $-17{,}640$ newtons.

Your Turn 11 Solve.

a. At a company picnic, $\frac{2}{5}$ of the people attending were men and of those men, $\frac{3}{4}$ were over the age of 50. What fraction of those attending were men over the age of 50?

b. A truck is carrying a load that has a weight of 4900 newtons, which means that the downward force due to gravity is -4900 newtons. Find the mass of the load in kilograms.

Answers Your Turn 11

a. $\frac{3}{10}$ **b.** 500 kg

1.4 Exercises   For Extra Help   *MyMathLab*

 PRACTICE   WATCH   DOWNLOAD

Note: Exercises marked with a ★ represent challenging exercises.

For Exercises 1–4, state whether the result is positive or negative.

1. When multiplying or dividing two numbers that have the same sign, the sign of the result is _____.

2. When multiplying or dividing two numbers that have different signs, the sign of the result is _____.

3. When a multiplication problem contains an odd number of negative factors, the sign of the product is _____.

4. When a multiplication problem contains an even number of negative factors, the sign of the product is _____.

5. Why is $5 \div 0$ undefined?

6. Why is $0 \div 0$ indeterminate?

7. Why are 5 and $\frac{1}{5}$ multiplicative inverses?

8. Explain how to divide fractions.

9. Explain why $\frac{-a}{b} = \frac{a}{-b} = -\frac{a}{b}$, where a and b are any real numbers and $b \neq 0$.

10. Explain the difference between $6(-7)$ and $6 - 7$.

For Exercises 11–22, indicate whether each equation illustrates the multiplicative property of 0, the multiplicative identity, the commutative property of multiplication, the associative property of multiplication, or the distributive property. See Example 1.

11. $6(3 + 4) = 6 \cdot 3 + 6 \cdot 4$

12. $7(2 - 5) = 7 \cdot 2 - 7 \cdot 5$

13. $1 \cdot \left(-\dfrac{5}{6}\right) = -\dfrac{5}{6}$

14. $1 \cdot (-7) = -7$

15. $0 \cdot (-6.3) = 0$

16. $\dfrac{2}{3} \cdot 0 = 0$

17. $\dfrac{3}{5} \cdot \left(-\dfrac{1}{4}\right) = -\dfrac{1}{4} \cdot \dfrac{3}{5}$

18. $-\dfrac{2}{3} \cdot \dfrac{4}{5} = \dfrac{4}{5} \cdot \left(-\dfrac{2}{3}\right)$

19. $[4(-2)] \cdot 3 = 4(-2 \cdot 3)$

20. $-6 \cdot (5 \cdot 4) = (-6 \cdot 5) \cdot 4$

21. $(-4.5)(2 + 0.6) = (2 + 0.6)(-4.5)$

22. $-6.2(-4.3 + 7.1) = (-4.3 + 7.1)(-6.2)$

For Exercises 23–56, multiply. See Examples 2–6.

23. $6(-3)$

24. $4(-7)$

25. $(-10)5$

26. $(-8)(5)$

27. $(9)(-6)$

28. $(12)(-4)$

29. $-4(-5)$

30. $(-4)(-3)$

31. $(-7)(-6)$

32. $(-8)(-12)$

33. $\dfrac{1}{2} \cdot \left(-\dfrac{1}{2}\right)$

34. $-\dfrac{4}{5} \cdot \left(\dfrac{20}{3}\right)$

35. $\left(-\dfrac{2}{3}\right)\left(-\dfrac{3}{4}\right)$

36. $\left(-\dfrac{5}{6}\right)\left(-\dfrac{6}{5}\right)$

37. $\left(-\dfrac{5}{6}\right)\left(\dfrac{8}{15}\right)$

38. $\left(\dfrac{2}{9}\right)\left(-\dfrac{21}{26}\right)$

39. $3.8(-10)$

40. $8(-2.5)$

41. $-1.6(-4.2)$

42. $-7.1(-0.5)$

43. $-4.3(1.52)$

44. $8.1(-2.75)$

45. $6(-9)(-1)$

46. $-4(5)(-3)$

47. $6(-2)(4)$

48. $3(7)(-8)$

49. $(-4)(-6)(-7)$

50. $(-5)(-3)(-2)$

51. $12(-6)(-2)(-1)$

52. $-5(3)(-4)(-2)$

53. $(-1)(-6)(-40)(-3)$

54. $(-2)(-4)(-30)(-1)$

55. $(-2)(3)(-4)(-1)(2)$

56. $(-1)(-1)(4)(-5)(-3)$

For Exercises 57–64, find the multiplicative inverse. See Example 7.

57. $\dfrac{2}{3}$

58. $\dfrac{20}{3}$

59. $-\dfrac{5}{2}$

60. $-\dfrac{6}{7}$

61. -5

62. 17

63. 0

64. -1

For Exercises 65–92, divide. See Examples 8–10.

65. $-6 \div 2$

66. $42 \div (-7)$

67. $-56 \div (-4)$

68. $-12 \div (-4)$

69. $\dfrac{-18}{9}$

70. $\dfrac{75}{-3}$

71. $\dfrac{-63}{-7}$

72. $\dfrac{-48}{-6}$

73. $\dfrac{0}{-6}$

74. $\dfrac{0}{5}$

75. $\dfrac{-8}{0}$

76. $-21 \div 0$

77. $\dfrac{0}{0}$

78. $0 \div 0$

79. $6 \div \dfrac{2}{3}$

80. $-8 \div \dfrac{3}{4}$

81. $\dfrac{3}{5} \div \left(-\dfrac{9}{10}\right)$

82. $-\dfrac{4}{5} \div \dfrac{4}{5}$

83. $-\dfrac{2}{7} \div \left(-\dfrac{8}{21}\right)$

84. $-\dfrac{1}{3} \div \left(-\dfrac{3}{2}\right)$

85. $-\dfrac{4}{9} \div \dfrac{10}{21}$

86. $\dfrac{7}{15} \div \left(-\dfrac{35}{24}\right)$

87. $9.03 \div 4.3$

88. $8.1 \div 0.6$

89. $-36.72 \div (-0.4)$

90. $-10.65 \div (-7.1)$

91. $-14 \div 0.3$

92. $19 \div (-0.06)$

For Exercises 93–98, solve. See Example 11a.

93. In a poll where respondents can agree, disagree, or have no opinion, $\dfrac{3}{4}$ of the respondents said that they agreed and $\dfrac{2}{3}$ of those that agreed were women. What fraction of all respondents were women who agreed with the statement in the poll?

94. On a standard-size guitar, the length of the strings between the saddle and nut is $25\dfrac{1}{2}$ inches. Guitar makers must place the 12th fret at exactly half of the length of the string. How far from the saddle or nut should a guitar maker measure to place the 12th fret?

Saddle

Nut

12th fret

95. A financial planner estimates that at the rate one of his clients is increasing debt, she will have $3\dfrac{1}{2}$ times her current debt in five years. If her current debt is represented by $-\$2480$, how can her debt in five years be represented?

Of Interest

Three components affect the pitch of a string on a stringed instrument: diameter, tension, and length. Given two strings of the same diameter and same tension, if one string is half the length of the other, it will have a pitch that is one octave higher than the longer string. On a guitar, placing a finger at the 12th fret cuts the string length in half, thereby creating a tone that is an octave higher than the string's open tone.

96. Mario finds that $\frac{2}{3}$ of his credit card debt is from dining out. If his total credit card debt is represented by $-\$858$, how much of the debt is from dining out?

97. In 2000, the Garret family's only debt was credit card balance represented by $-\$258.75$. In 2010, they have a mortgage, two car loans, and three credit cards with a total debt represented by $-\$158,572.85$. How many times greater is their debt in 2010 than it was in 2000?

98. A company's stock loses value by an amount represented by $-\$\frac{3}{8}$ each day for four days. Write a representation of the total loss in value at the close of the 4th day.

For Exercises 99–102, use the fact that an object's weight is a downward force due to gravity and is calculated by multiplying the object's mass by the acceleration due to gravity, which is a constant (Force = Mass × Acceleration). The following table indicates the units. See Example 11b.

	Force	=	Mass	×	Acceleration
American measurement	pounds (lb)		slugs (s)		-32.2 ft./sec.^{2}
Metric measurement	newtons (N)		kilograms (kg)		-9.8 m/sec.^{2}

99. Find the force due to gravity on an object with a mass of 12.5 slugs. What does the sign of the weight indicate?

100. Find the force due to gravity on a person with a mass of 70.4 kilograms.

101. The blue whale is the largest animal on Earth. The largest blue whale ever caught weighed 1,658,181.8 newtons, which means that the downward force due to gravity was $-1,658,181.8$ newtons. What was the whale's mass in kilograms? (*Source: The Whale Watcher's Guide: Whale Watching Trips in North America*, Corrigan, 1999)

102. The Liberty Bell weighs 2080 pounds, which means the downward force due to gravity is -2080 pounds. What is its mass in slugs? (*Source: Ring in the Jubilee: The Epic of America's Liberty Bell*, Boland, 1973)

For Exercises 103 and 104, use the fact that in an electrical circuit, voltage (V) is equal to the product of the current, measured in amperes (A), and the resistance of the circuit, measured in ohms (Ω) (Voltage = Current × Resistance).

103. Suppose the current in a circuit is -6.4 amperes and the resistance is 8 ohms. Find the voltage.

104. An electrical technician measures the voltage in a circuit to be -15 volts and the current to be -8 amperes. What is the resistance of the circuit?

For Exercises 105 and 106, use the fact that in an electrical circuit, power, which is measured in watts (W), is the product of the voltage and the current (Power = Voltage × Current).

105. An engineer measures the voltage in a circuit to be −120 volts and the current to be −30 amperes. What is the power?

★**106.** An electrical technician determines that the power in a circuit is 400 watts. If the current is measured to be −6.5 amperes, what is the resistance in the circuit?

Review Exercises

Exercises 1–3 CONSTANTS AND VARIABLES

[1.1] **1.** Is π a rational or irrational number?

[1.1] **2.** Graph −7.2 on a number line.

[1.3] **3.** Find the additive inverse of $\frac{2}{3}$.

Exercises 4–6 EXPRESSIONS

[1.1] **4.** Simplify: $|-6.8|$

[1.3] **5.** Add: $-7 + (-15)$

[1.3] **6.** Subtract: $-\frac{5}{8} - \left(-\frac{1}{6}\right)$

1.5 Exponents, Roots, and Order of Operations

Objectives

1. Evaluate numbers in exponential form.
2. Evaluate square roots.
3. Use the order-of-operations agreement to simplify numerical expressions.
4. Find the mean of a set of data.

Objective 1 Evaluate numbers in exponential form. Sometimes, problems involve repeatedly multiplying the same number. In such problems, we can use an **exponent** to indicate that a **base** number is repeatedly multiplied.

Definitions **Exponent:** A symbol written to the upper right of a base number that indicates how many times to use the base as a factor.

Base: The number that is repeatedly multiplied.

When we write a number with an exponent, we say that the expression is in *exponential form*. The expression 2^4 is in exponential form, where the base is 2 and the exponent is 4. It is read "two to the fourth power," or simply "two to the fourth." To *evaluate* 2^4, write 2 as a factor 4 times; then multiply.

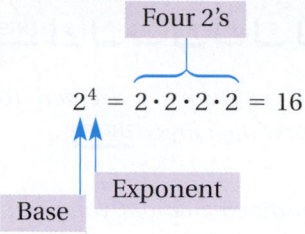

$$2^4 = 2 \cdot 2 \cdot 2 \cdot 2 = 16$$

Exponent

Base

Procedure   Evaluating an Exponential Form

To evaluate an exponential form raised to a natural number exponent, write the base as a factor the number of times indicated by the exponent; then multiply.

Example 1   Evaluate.

a. $(-7)^2$

SOLUTION   The exponent 2 indicates that we have two factors of -7. Because we multiply two negative numbers, the result is positive.

$$(-7)^2 = (-7)(-7) = 49$$

◄ **Note:** When a base is raised to the second power, we say it is **squared**. The expression $(-7)^2$ can be read as "negative seven **squared**."

b. $\left(-\dfrac{2}{3}\right)^3$

SOLUTION   The exponent 3 means that we write the base as a factor three times.

$$\left(\frac{2}{3}\right)^3 = \left(-\frac{2}{3}\right)\left(-\frac{2}{3}\right)\left(-\frac{2}{3}\right)$$

$$= -\frac{8}{27}$$

◄ **Note:** When a base is raised to the third power, we can say it is **cubed**. The expression $\left(-\dfrac{2}{3}\right)^3$ can be read as "negative two-thirds **cubed**."

Example 1 suggests the following sign rules:

Rule   Evaluating Exponential Forms with Negative Bases

- If the base of an exponential form is a negative number and the exponent is even, the product is positive.
- If the base is a negative number and the exponent is odd, the product is negative.

Your Turn 1   Evaluate.

a. $\left(-\dfrac{1}{2}\right)^5$

b. $(-0.3)^4$

Use the $\boxed{\wedge}$ *key to indicate an exponent. To evaluate* $\left(-\dfrac{2}{3}\right)^3$*, type*

The answer will be given as a decimal. To change the decimal answer to a fraction, press the $\boxed{\text{MATH}}$ *key, select 1:Frac, and then press* $\boxed{\text{ENTER}}$*.*

On some scientific calculators, you use a $\boxed{y^x}$ *key. To evaluate* $\left(-\dfrac{2}{3}\right)^3$ *on one of these calculators, type*

The answer will be given as a decimal. Consult your calculator manual to see how to change the decimal to a fraction.

There is a great deal of difference between expressions of the form $(-a)^n$ and $-a^n$. The base of the exponent in $(-a)^n$ is $-a$. So $(-a)^n$ means multiply n factors of $-a$. The base of the exponent in $-a^n$ is a. So $-a^n$ means find the additive inverse of a^n.

Example 2   Evaluate.

> **Note:** This is read as "negative 2 to the sixth power."

a. $(-2)^6$

SOLUTION $(-2)^6 = (-2)(-2)(-2)(-2)(-2)(-2) = 64$

> **Note:** This is read as "the additive inverse of two to the sixth power."

b. -2^6

SOLUTION $-2^6 = -2 \cdot 2 \cdot 2 \cdot 2 \cdot 2 \cdot 2 = -64$

c. $(-3)^3$

SOLUTION $(-3)^3 = (-3)(-3)(-3) = -27$

d. -3^3

SOLUTION $-3^3 = -3 \cdot 3 \cdot 3 = -27$

Your Turn 2   Evaluate.

a. $(-3)^4$ **b.** -3^4 **c.** $(-2)^3$ **d.** -2^3

Objective ② Evaluate square roots.   Roots are inverses of exponents. More specifically, a square root is the inverse of a square. So a square root of a given number is a number that, when squared, equals the given number. For example, 5 is a square root of 25 because $5^2 = 25$. Notice that -5 is also a square root of 25 because $(-5)^2 = 25$.

Conclusion   For every real number, there are *two* square roots, a positive root and a negative root.

For convenience, if asked to find all square roots of 25, we can write both the positive and negative answers in a compact expression like this: ± 5.

Answers   Your Turn 2
a. 81   **b.** -81 **c.** -8 **d.** -8

Example 3   Find all square roots of the given number.

a. 36

Answer   ±6

b. −25

Answer   No real-number square roots exist.

> **Note:** We do not say that this situation is undefined or indeterminate because we eventually will define the square root of a negative number. Such roots will be defined using imaginary numbers.

Explanation   Because of the sign rules for multiplication, there is no way to square a real number and get anything other than a positive result.

Our work so far suggests the following rules:

Rule   **Square Roots**

- Every positive number has two square roots, a positive root and a negative root.
- Negative numbers have no real-number square roots.

Your Turn 3   Find all square roots of the given number.

   a. 144   **b.** −81

The Principal Square Root

A symbol, $\sqrt{\ }$, called the *radical*, is used to indicate finding only the *positive* (or *principal*) square root of a given number. The given number or expression inside the radical is called the *radicand*.

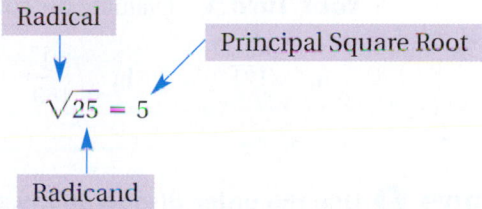

What about the square root of a fraction such as $\sqrt{\dfrac{4}{9}}$? Because $\left(\dfrac{2}{3}\right)^2 = \dfrac{2}{3} \cdot \dfrac{2}{3} = \dfrac{4}{9}$, we can say that $\sqrt{\dfrac{4}{9}} = \dfrac{2}{3}$. Notice that we can evaluate $\sqrt{\dfrac{4}{9}}$ by finding the square roots of the numerator and denominator separately, or $\sqrt{\dfrac{4}{9}} = \dfrac{\sqrt{4}}{\sqrt{9}} = \dfrac{2}{3}$. We now have the following rules for square roots involving the radical sign.

Rule   **Square Roots Involving the Radical Sign**

- The radical symbol $\sqrt{\ }$ denotes only the positive (principal) square root.

- $\sqrt{\dfrac{a}{b}} = \dfrac{\sqrt{a}}{\sqrt{b}}$, where $a \geq 0$ and $b > 0$.

Answers   **Your Turn 3**

a. ±12

b. No real-number square roots exist.

To calculate a square root, use the $\sqrt{\ }$ function, which may require pressing `2nd` and then the `x²` key. To find the square root of 144, type

`2nd` `x²` `1` `4` `4` `ENTER` .

Some graphing calculators give an open parenthesis when the $\sqrt{\ }$ function is used; so you may need to type a closed parenthesis before you press `ENTER` .

Some scientific calculators have a `√x` key. To find the square root of 144 on one of these calculators, type

`1` `4` `4` `√x` .

Example 4 Evaluate the square roots.

a. $\sqrt{144}$

b. $\sqrt{\dfrac{49}{81}}$

SOLUTION $\sqrt{144} = 12$

SOLUTION $\sqrt{\dfrac{49}{81}} = \dfrac{\sqrt{49}}{\sqrt{81}} = \dfrac{7}{9}$

c. $\sqrt{0.81}$

d. $\sqrt{-81}$

SOLUTION $\sqrt{0.81} = 0.9$

SOLUTION not a real number

Your Turn 4 Evaluate the square roots.

a. $\sqrt{121}$ **b.** $\sqrt{\dfrac{81}{169}}$ **c.** $\sqrt{0.49}$ **d.** $\sqrt{-0.09}$

Objective ③ Use the order-of-operations agreement to simplify numerical expressions. Some expressions are too complicated to evaluate using only the associative and commutative properties. If an expression contains a mixture of operations, we must be careful about the order in which we perform them. Performing the operations in a different order can change the results. Consider the expression $2 + 5 \cdot 3$.

Adding first: $\quad \mathbf{2 + 5} \cdot 3 \qquad\qquad$ Multiplying first: $\quad 2 + \mathbf{5 \cdot 3}$
$\qquad\qquad\qquad = \mathbf{7} \cdot 3 \qquad\qquad\qquad\qquad\qquad\qquad = 2 + \mathbf{15}$
$\qquad\qquad\qquad = 21 \qquad\qquad\qquad\qquad\qquad\qquad\qquad = 17$

Changing the order of the operations changes the outcome. To ensure that everyone arrives at the same answer, mathematicians developed an order of operations that everyone agrees to follow.

Procedure   Order-of-Operations Agreement

Perform operations in the following order:

1. Within grouping symbols: parentheses ( ), brackets [ ], braces { }, above/below fraction bars, absolute value | |, and radicals $\sqrt{\ }$.
2. Exponents/Roots from left to right, in order as they occur.
3. Multiplication/Division from left to right, in order as they occur.
4. Addition/Subtraction from left to right, in order as they occur.

Answers   Your Turn 4

a. 11   **b.** $\dfrac{9}{13}$ **c.** 0.7

d. not a real number

If we follow this agreement to simplify $2 + 5 \cdot 3$, the correct order is to multiply first and then add so that the answer is 17.

Embedded Parentheses

When one set of grouping symbols contains another set of grouping symbols, we say that they are *nested* or *embedded*. Work from the innermost set of grouping symbols outward.

Example 5 Simplify.

a. $-34 + 12 \div (-3) \cdot 5$

SOLUTION $-34 + 12 \div (-3) \cdot 5$

$= -34 + (-4) \cdot 5$ **Divide $12 \div (-3) = -4$.**

$= -34 + (-20)$ **Multiply $(-4) \cdot 5 = -20$.**

$= -54$ **Add $-34 + (-20) = -54$.**

> **Note:** The division and multiplication must be calculated before the addition. As we read from left to right, we see the division before the multiplication; so we divide first.

b. $-2^4 + 4|18 - 24|$

SOLUTION $-2^4 + 4|18 - 24|$

$= -2^4 + 4|-6|$ **Subtract inside the absolute value: $18 - 24 = -6$.**

$= -2^4 + 4 \cdot 6$ **Simplify the absolute value $|-6| = 6$.**

$= -16 + 4 \cdot 6$ **Evaluate the exponential form $-2^4 = -16$.**

$= -16 + 24$ **Multiply $4 \cdot 6 = 24$.**

$= 8$ **Add $-16 + 24 = 8$.**

> **Note:** Recall that the expressions $(-a)^n$ and $-a^n$ are different. The expression $(-a)^n$ means "multiply n factors of $-a$" and $-a^n$ means "find the additive inverse of a^n."

c. $(-5)^2 - 8[4 - (6 + 3)] - \sqrt{36}$

SOLUTION $(-5)^2 - 8[4 - (6 + 3)] - \sqrt{36}$

$= (-5)^2 - 8(4 - 9) - \sqrt{36}$ **Calculate within the innermost parentheses: $6 + 3 = 9$.**

$= (-5)^2 - 8(-5) - \sqrt{36}$ **Calculate within the brackets.**

$= 25 - 8(-5) - 6$ **Evaluate the exponential form and square root.**

$= 25 - (-40) - 6$ **Multiply $8(-5) = -40$.**

$= 25 + 40 - 6$ **Write $25 - (-40)$ as an equivalent addition.**

$= 65 - 6$ **Add $25 + 40 = 65$.**

$= 59$ **Subtract $65 - 6 = 59$.**

Your Turn 5 Simplify.

a. $28 - 36 \div 3(-4)$ **b.** $6|-5 - 4| + 2(-3)^3$

c. $15.8 - 0.2[(-7)^2 - (12 + 8) \div 4]$

Radical Symbols

Radical symbols can be grouping symbols. Consider $\sqrt{36 \cdot 4}$. Following the order of operations, we multiply first, then find the root of the product. However, if a radical contains multiplication, we can find the roots of the factors and multiply those roots to get the same answer.

<table>
<tr><td>Multiplying first:</td><td></td><td>Multiplying the roots:</td></tr>
<tr><td>$\sqrt{36 \cdot 4} = \sqrt{144}$</td><td>or</td><td>$\sqrt{36 \cdot 4} = \sqrt{36} \cdot \sqrt{4}$</td></tr>
<tr><td>$= 12$</td><td></td><td>$= 6 \cdot 2$</td></tr>
<tr><td></td><td></td><td>$= 12$</td></tr>
</table>

Both approaches yield the same result. The same holds true for division under a radical.

$$\sqrt{36 \div 4} = \sqrt{9} \qquad \text{or} \qquad \sqrt{36 \div 4} = \sqrt{36} \div \sqrt{4}$$
$$= 3 \qquad\qquad\qquad\qquad\qquad\qquad = 6 \div 2$$
$$\qquad\qquad\qquad\qquad\qquad\qquad\qquad\qquad = 3$$

Rule   **Square Root of a Product or Quotient**

If a square root contains multiplication or division, we can multiply or divide first, then find the square root of the result, or we can find the square roots of the individual numbers, then multiply or divide the square roots. In symbols, $\sqrt{\dfrac{a}{b}} = \dfrac{\sqrt{a}}{\sqrt{b}}$ for $a \geq 0$ and $b > 0$ and $\sqrt{ab} = \sqrt{a} \cdot \sqrt{b}$ for $a \geq 0$ and $b \geq 0$.

Note that this rule does *not* hold for addition or subtraction under a radical, such as in $\sqrt{9 + 16}$.

Adding first: Adding the roots:

$$\sqrt{9 + 16} = \sqrt{25} \qquad\qquad \sqrt{9 + 16} = 3 + 4$$
$$= 5 \qquad\qquad\qquad\qquad\qquad = 7$$

Because the results are different, we do not have a choice of how to approach the situation when the radical contains addition. The correct approach is to treat the radical as a grouping symbol and add, then find the root of the sum. Thus, the correct answer for $\sqrt{9 + 16}$ is 5. The same procedure applies when a radical contains subtraction.

Rule   **Square Root of a Sum or Difference**

When a radical contains addition or subtraction, we must add or subtract first, then find the root of the sum or difference.

Example 6   Simplify.

a. $12.4 \div 5(-3)^2 + 4\sqrt{169 - 25}$

SOLUTION $12.4 \div 5(-3)^2 + 4\sqrt{169 - 25}$

$= 12.4 \div 5(-3)^2 + 4\sqrt{144}$ **Subtract within the radical: 169 − 25 = 144.**

$= 12.4 \div 5(9) + 4(12)$ **Evaluate the exponential form and root.**

$= 2.48(9) + 4(12)$ **Divide 12.4 ÷ 5 = 2.48.**

$= 22.32 + 48$ **Multiply 2.48(9) = 22.32 and 4(12) = 48.**

$= 70.32$ **Add 22.32 + 48 = 70.32.**

Note: We divide before we multiply here because the order is to multiply or divide from left to right in the order in which those operations occur.

b. $\left(-\dfrac{1}{4}\right)^3 - \left(\dfrac{1}{2} + \dfrac{3}{8}\right) \div \sqrt{\dfrac{48}{3}}$

SOLUTION $\left(-\dfrac{1}{4}\right)^3 - \left(\dfrac{1(4)}{2(4)} + \dfrac{3}{8}\right) \div \sqrt{\dfrac{48}{3}}$

$= \left(-\dfrac{1}{4}\right)^3 - \left(\dfrac{4}{8} + \dfrac{3}{8}\right) \div \sqrt{\dfrac{48}{3}}$ **Write equivalent fractions with a common denominator in order to add within the parentheses.**

$= \left(-\dfrac{1}{4}\right)^3 - \dfrac{7}{8} \div \sqrt{16}$ **Add within the parentheses and divide within the radical.**

$$= -\frac{1}{64} - \frac{7}{8} \div 4 \qquad \text{Evaluate the exponential form and square root.}$$

$$= -\frac{1}{64} - \frac{7}{8} \cdot \frac{1}{4} \qquad \begin{array}{l}\text{Write an equivalent multiplication using the reciprocal}\\ \text{of the divisor.}\end{array}$$

$$= -\frac{1}{64} - \frac{7(2)}{32(2)} \qquad \begin{array}{l}\text{Multiply } \frac{7}{8} \cdot \frac{1}{4} \text{ to get } \frac{7}{32}; \text{ then write it as an equivalent}\\ \text{fraction with the common denominator 64 in order}\\ \text{to subtract.}\end{array}$$

$$= -\frac{1}{64} - \frac{14}{64}$$

$$= -\frac{15}{64} \qquad \text{Subtract.}$$

Your Turn 6 Simplify.

a. $-\dfrac{3}{5} \div \dfrac{1}{10} \cdot 4 + \sqrt{64 + 36}$ **b.** $5 + (0.2)^3 - 3(8 - 14)$

Fraction Lines

Sometimes fraction lines are used as grouping symbols. When they are, we simplify the numerator and denominator separately, then divide the results.

> **Connection** The expression in Example 7 is equivalent to the expression
> $[5(-6) - 2^3] \div [3(7) - 2]$.

Example 7 Simplify.

a. $\dfrac{5(-6) - 2^3}{3(7) - 2}$

SOLUTION $\dfrac{5(-6) - 2^3}{3(7) - 2}$

$$= \frac{5(-6) - 8}{21 - 2} \qquad \begin{array}{l}\text{Evaluate the exponential form in the numerator and}\\ \text{multiply in the denominator.}\end{array}$$

$$= \frac{-30 - 8}{19} \qquad \text{Multiply in the numerator and subtract in the denominator.}$$

$$= \frac{-38}{19} \qquad \text{Subtract in the numerator.}$$

$$= -2 \qquad \text{Divide.}$$

b. $\dfrac{9(3) + 15}{4^2 + 2(-8)}$

SOLUTION $\dfrac{9(3) + 15}{4^2 + 2(-8)}$

$$= \frac{27 + 15}{16 + 2(-8)} \qquad \begin{array}{l}\text{Multiply in the numerator and evaluate the}\\ \text{exponential form in the denominator.}\end{array}$$

$$= \frac{42}{16 + (-16)} \qquad \begin{array}{l}\text{Add in the numerator and multiply in the}\\ \text{denominator.}\end{array}$$

$$= \frac{42}{0} \qquad \text{Add in the denominator.}$$

Because the denominator or divisor is 0, the answer is undefined.

Answers **Your Turn 6**
a. -14 **b.** 23.008

Your Turn 7 Simplify.

a. $\dfrac{7(3-7)+1}{-1-2^3}$

b. $\dfrac{2[9-4(3+5)]}{25-(6-1)^2}$

Objective 4 **Find the mean of a set of data.** Fraction line notation can be used to indicate the proper sequence of operations in calculating an arithmetic mean, or average, of a set of data.

Procedure **Finding the Arithmetic Mean**

To find the arithmetic mean, or average, of n numbers, divide the sum of the numbers by n.

$$\text{Arithmetic mean} = \dfrac{x_1 + x_2 + \cdots + x_n}{n}$$

Note: The subscripts indicate that each x represents a different given number. So x_1 represents the first given number, x_2 represents the second, and so on, until x_n, which represents the last given number.

Example 8 Jacky has the following test scores: 86, 95, 78, 82, and 84. Find the average of her test scores.

SOLUTION $\dfrac{86 + 95 + 78 + 82 + 84}{5} = \dfrac{425}{5}$ Divide the sum of the 5 scores by 5.

$$= 85$$

Your Turn 8 This table shows the daily rainfall accumulations for the month of April in a certain city. Find the average daily accumulation for April. (Remember, April has 30 days.)

Answers **Your Turn 7**

a. 3 **b.** undefined

Answer **Your Turn 8**

0.175 in./day

Date	April 2	April 7	April 9	April 15	April 21	April 26
Accumulation (in inches)	0.5	1.25	1.0	1.25	0.25	1.0

1.5 Exercises   For Extra Help   *MyMathLab*

PRACTICE   WATCH   DOWNLOAD

Note: Exercises marked with a ★ represent challenging exercises.

1. What is another way to say "two to the third power"?

2. What is another way to say "three to the second power"?

3. Explain the difference between squaring a number and finding its square root.

4. If the base of an exponential form is a negative number and the exponent is even, will the sign of the result be positive or negative?

5. When simplifying a square root of a product of two numbers, what is the proper order of operations?

6. When simplifying a square root of a sum of two numbers, what is the proper order of operations?

For Exercises 7–12, identify the base and the exponent; then translate the expression to words. See Objective 1.

7. 7^2

8. 9^4

9. $(-5)^3$

10. $(-8)^2$

11. -2^7

12. -3^8

For Exercises 13–34, evaluate. See Examples 1 and 2.

13. 3^4

14. 2^5

15. $(-8)^2$

16. $(-2)^4$

17. -8^2

18. -2^4

19. $(-5)^3$

20. $(-3)^5$

21. -5^3

22. -3^5

23. $-(-2)^3$

24. $-(-3)^3$

25. $-(-1)^6$

26. $-(-1)^4$

27. $\left(-\dfrac{1}{5}\right)^2$

28. $\left(-\dfrac{2}{7}\right)^2$

29. $\left(-\dfrac{3}{4}\right)^3$

30. $\left(-\dfrac{1}{3}\right)^5$

31. $(0.2)^3$

32. $(0.3)^4$

33. $(-4.1)^2$

34. $(-0.2)^4$

For Exercises 35–42, find all square roots of each number. See Example 3.

35. 121

36. 49

37. -81

38. -36

39. 196

40. 169

41. 256

42. 225

For Exercises 43–54, evaluate the square roots. See Example 4.

43. $\sqrt{16}$

44. $\sqrt{36}$

45. $\sqrt{144}$

46. $\sqrt{289}$

47. $\sqrt{0.49}$

48. $\sqrt{0.01}$

49. $\sqrt{-64}$

50. $\sqrt{-25}$

51. $\sqrt{\dfrac{64}{81}}$

52. $\sqrt{\dfrac{9}{100}}$

53. $\sqrt{\dfrac{50}{2}}$

54. $\sqrt{\dfrac{48}{3}}$

For Exercises 55–102, simplify using the order of operations. See Examples 5–7.

55. $5 \cdot 3 + 4$

56. $4 \cdot 6 - 5$

57. $8 \div 2 - 4$

58. $18 \div 2 + 3$

59. $8 + 4 \div 2$

60. $9 + 6 \div 3$

61. $4 \cdot 6 - 7 \cdot 5$

62. $-3 \cdot 4 - 2 \cdot 7$

63. $12 - 2^4$

64. $8 - 3^2$

65. $8 - 3(-4)^2$

66. $16 - 5(-2)^2$

67. $4^2 - 36 \div 4(12 - 8)$

68. $3^2 - 18 \div 3(6 - 3)$

69. $-4 + 3(-1)^4 + 18 \div 3 \cdot 3$

70. $12 - 2(-2)^3 - 64 \div 4 \cdot 2$

71. $-2^3 + 8 - 7(4 - 3)$

72. $-3^3 - 16 - 5(7 - 2)$

73. $24 \div (-9 + 3)(3 + 4)$

74. $18 \div (-6 + 3)(4 + 1)$

75. $13.02 \div (-3.1) + 6^2 - \sqrt{25}$

76. $-15.54 \div 3.7 + (-2)^4 + \sqrt{49}$

77. $18.2 + 3.4[(5 + 9) \div 7 - 6^2]$

78. $16.3 + 2.8[(8 + 7) \div 5 - 4^2]$

79. $-4^3 - 3^2 + 4|7 - 12|$

80. $-2|9 - 15| + 5^2 - 3^2$

81. $-\dfrac{3}{4} \div \left(\dfrac{1}{8}\right) + \left(-\dfrac{2}{5}\right)(-3)(-4)$

82. $\dfrac{5}{6} \div \left(-\dfrac{2}{3}\right) + \left(-\dfrac{2}{7}\right)(5)(-14)$

83. $-36 \div (-2)(3) + \sqrt{169 - 25} + 12$

84. $\sqrt{100 - 64} + 18 \div (-3)(-2)$

85. $8 - 3[5 - (7 + 4)] - \sqrt{49}$

86. $4 - 8[3 - (9 + 3)] + \sqrt{64}$

87. $\sqrt{71 - 35} - 3^2[6 - (4 - 9)] + 4^3$

88. $\sqrt{83 - 58} - 2^2[9 - (3 - 8)] + 3^4$

89. $\dfrac{9}{8} \cdot \left(-\dfrac{2}{3}\right) + \left(\dfrac{1}{5} - \dfrac{2}{3}\right) \div \sqrt{\dfrac{125}{5}}$

90. $\left(\dfrac{3}{4} - \dfrac{2}{3}\right) \div \sqrt{\dfrac{9}{81}} - \left(\dfrac{16}{27}\right) \div \left(\dfrac{4}{9}\right)$

91. $\dfrac{2}{5} \div \left(-\dfrac{1}{10}\right) \cdot (-3) + \sqrt{64 + 36}$

92. $\dfrac{5}{6}(-18) \div \left(\dfrac{3}{2}\right) - \sqrt{9 + 16}$

93. $-16 \cdot \left(\dfrac{3}{4}\right) \div (-2) + |9 - 3(4 + 1)|$

94. $18 \cdot \left(-\dfrac{5}{6}\right) \div (-3) + 2|4 + 2(7 - 3)|$

95. $\dfrac{7 - |4(5) - 13|}{6^2 - 3(2 - 14)}$

96. $\dfrac{|6(-3) + 7| - 11}{5^3 - 2(6 - 12)}$

97. $\dfrac{4[5 - 8(2 + 1)]}{3 - 6 - (-4)^2}$

98. $\dfrac{3[24 - 4(6 - 2)]}{-3^3 + 4^2 + 3}$

99. $\dfrac{5^3 - 3(4^3 - 41)}{19 - (3 - 10)^2 + 38}$

100. $\dfrac{6^2 - 3(4 + 2^5)}{4 + 20 - (2 + 4)^2}$

101. $\dfrac{4(5^2 - 10) + 3}{\sqrt{25 - 16} - 3}$

102. $\dfrac{5(4 - 9) + 1}{2^3 - \sqrt{100 - 36}}$

In Exercises 103–106, a property of arithmetic was correctly used as an alternative to the order of operations. Determine what property of arithmetic was applied and explain how it is different from the order-of-operations agreement.

103. $14 - 2 \cdot 6 \cdot 3 + 8^2$
$= 14 - 2 \cdot 18 + 64$
$= 14 - 36 + 64$
$= 42$

104. $2(3 + 8) - \sqrt{81}$
$= 6 + 16 - 9$
$= 13$

105. $-6(5 + 3^2) - \sqrt{14 + 11}$
$= -6(5 + 9) - \sqrt{25}$
$= -30 + (-54) - 5$
$= -89$

106. $(-4)^3 + 2[-10 + 8 + (-3)]$
$= -64 + 2(-13 + 8)$
$= -64 + 2(-5)$
$= -64 + (-10)$
$= -74$

Find
 the mistake   **For Exercises 107–110, explain the mistakes, then simplify.**

107. $24 \div 4 \cdot 2 - 11$
$= 24 \div 8 - 11$
$= 3 - 11$
$= -8$

108. $19 - 6(10 - 8)$
$= 19 - 6(2)$
$- 13(2)$
$= 26$

109. $40 \div 2 + \sqrt{25} - 9$
$= 20 + \sqrt{25} - 9$
$= 20 + 5 - 3$
$= 22$

110. $-3^4 + 20 \div 5 - (16 - 24)$
$= 81 + 4 - (-8)$
$= 85 + 8$
$= 93$

For Exercises 111–118, solve. See Example 8.

111. Tomeka has the following test scores in a history course: 82, 76, 64, 90, and 74. What is the average of her test scores?

112. Will's math instructor gives quizzes worth 10 points each. Will has the following quiz scores: 9, 8, 4, 8, 7, 7, 6, 9, and 8. If his instructor drops the lowest quiz score, what is Will's quiz average?

★ **113.** Michael will not have to take his chemistry final if he has a test average greater than or equal to 90 on the five tests in the course. His current test scores are 96, 88, 86, and 84. Using trial and error, determine the minimum test score on the last test that will give him a test average of 90.

★ **114.** To get an A in her psychology course, Lisa must have a test average greater than or equal to 90 on four out of five tests. (The lowest test score is dropped.) Her test scores on the first four tests are 98, 68, 84, and 86. What is the minimum score on the last test that will give her a test average of 90?

115. The following chart shows the average weekly earnings for workers in the United States in a recent year. Find the average weekly earnings for the year.

AVERAGE WEEKLY EARNINGS			
Month	**Average Earnings**	**Month**	**Average Earnings**
January	$577.98	July	$589.81
February	$578.29	August	$591.50
March	$583.42	September	$592.85
April	$583.05	October	$593.87
May	$585.42	November	$596.23
June	$589.86	December	$598.60

Source: Bureau of Labor Statistics.

116. The following graphic shows the number of people 20 years and older who were unemployed in the United States for each month during a recent year. Find the average number of people unemployed per month during that year.

ALL IN A MONTH'S WORK			**YEAR 2007**
Month	**Unemployed (in thousands)**	**Month**	**Unemployed (in thousands)**
January	7043	July	7137
February	6837	August	7133
March	6738	September	7246
April	6829	October	7291
May	6863	November	7181
June	6997	December	7655

Source: Bureau of Labor Statistics, *Labor Force Statistics from Current Population Survey.*

117. The following bar graph shows the daily closing price of Walmart stock over a one-week period from February 4 to February 8, 2008. Find the average of the daily prices for that week.

CLOSING PRICE OF WALMART STOCK

February 2008

Price values: 4: 50.07, 5: 49.56, 6: 48.83, 7: 49.84, 8: 48.76 (Price in $, Date)

Source: www.marketwatch.com.

118. The following bar graph shows the Dow Jones Industrial Average (DJIA) at the close of each of the days listed. What is the average of the DJIA for that week?

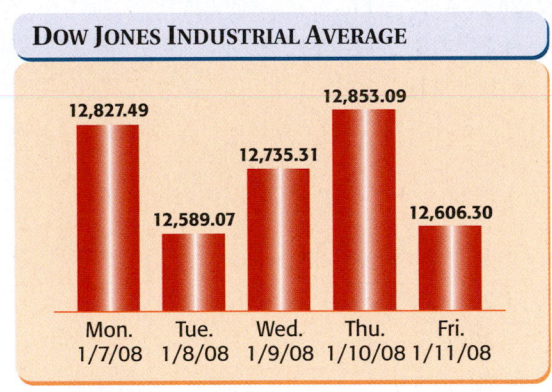

DOW JONES INDUSTRIAL AVERAGE

Mon. 1/7/08: 12,827.49; Tue. 1/8/08: 12,589.07; Wed. 1/9/08: 12,735.31; Thu. 1/10/08: 12,853.09; Fri. 1/11/08: 12,606.30

Source: www.measuringworth.com.

Review Exercises

Exercises 1–3 CONSTANTS AND VARIABLES

[1.1]   **1.** Write a set containing the whole numbers through 10.

[1.1]   **2.** Graph $-\dfrac{4}{5}$ on a number line.

[1.4]   **3.** Find the multiplicative inverse of -9.

[1.1]   **4.** Is $5x - 4y$ an equation or an expression? Explain.   [1.4]   **5.** Multiply: $-4(3)(-2)$

[1.4]   **6.** Divide: $\dfrac{-15}{-5}$

1.6 Translating Word Phrases to Expressions

Objective ① Translate word phrases to expressions.   An important technique in solving problems is to translate words in the problem to math symbols. In this section, we focus on how to translate word phrases to expressions containing variables, but we begin by translating expressions containing variables to word phrases for addition, subtraction, multiplication, and division.

Operation	Variable Expression	Word Phrases
Addition	$x + 3$	Some number plus three Three added to some number The sum of some number and three Three more than some number Some number increased by three
Subtraction	$y - 6$	Some number minus six Six subtracted from some number The difference of some number and six Six less than some number Some number decreased by six Some number less six
	$6 - y$	Six minus some number Some number subtracted from six The difference of six and some number Some number less than six Six decreased by some number Six less some number
Multiplication	$2a$	Twice some number Two times some number The product of two and some number Some number multiplied by two
Division	$\dfrac{x}{5}$	Some number divided by five The quotient of some number and five The ratio of some number and five
	$\dfrac{5}{x}$	Five divided by some number The quotient of five and some number The ratio of five and some number

The first step in translating a word phrase to math symbols is to identify the unknown amount. If a variable is not already given, select a variable to represent that unknown amount. The above translations from variable expressions to word phrases can be used as a guide.

Translating Basic Phrases

The following table contains some basic phrases and their translations.

Addition	Translation
The sum of x and three	$x + 3$
h plus k	$h + k$
t added to seven	$7 + t$
Three more than a number	$n + 3$
y increased by two	$y + 2$

Subtraction	Translation
The difference of three and x	$3 - x$
h minus k	$h - k$
Seven subtracted from t	$t - 7$
Three less than a number	$n - 3$
Two decreased by y	$2 - y$

Note: Since addition is a commutative operation, it doesn't matter in what order we write the translation.

For "the sum of x and three," we can write $x + 3$ or $3 + x$.

Note: Subtraction is not a commutative operation; therefore, the way we write the translation matters. We must translate each key phrase exactly as it was presented above. Notice that when we translate "less than" or "subtracted from," the translation is in reverse order from what we read.

Multiplication	Translation
The product of x and three	$3x$
h times k	hk
Twice a number	$2n$
Triple the number	$3n$
Two-thirds of a number	$\frac{2}{3}n$

Division	Translation
The quotient of x and three	$x \div 3$ or $\frac{x}{3}$
h divided by k	$h \div k$ or $\frac{h}{k}$
h divided into k	$k \div h$ or $\frac{k}{h}$
The ratio of a to b	$a \div b$ or $\frac{a}{b}$

Note: Like addition, multiplication is a commutative operation. This means that we can write the translation order any way we like.

h times k can be hk or kh.

Note: Division is like subtraction in that it is not a commutative operation; therefore, we must translate division phrases exactly as presented above. Notice how "divided into" is translated in reverse order of what we read.

Exponents	Translation
c squared	c^2
The square of b	b^2
k cubed	k^3
The cube of b	b^3
n to the fourth power	n^4
y raised to the fifth power	y^5

Roots	Translation
The square root of x	$\sqrt{x}$

Consider the translated phrases that follow. The key words *sum, difference, products,* and *quotient* indicate the answer for their respective operations. Also notice that all of the phrases involve the word *and*. In the translation, the word *and* separates the parts and can therefore be translated to the operation symbol indicated by the key word *sum, difference, product,* or *quotient*.

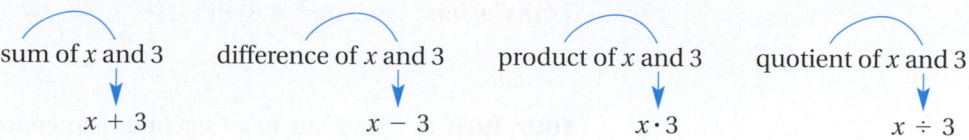

sum of x and 3   difference of x and 3   product of x and 3   quotient of x and 3

$$x + 3 \qquad x - 3 \qquad x \cdot 3 \qquad x \div 3$$

Combinations of Basic Phrases

Now that we have seen the basic phrases, let's translate phrases that involve combinations of the basic phrases.

Example 1   Translate to an algebraic expression.

a. four more than three times a number

Translation $4 + 3n$ or $3n + 4$

> **Note:** The commutative property of addition allows us to write the expression in either order.

b. six less than the square of a number

Translation $n^2 - 6$

> **Note:** Because subtraction is not commutative, we must translate the subtraction in a specific order. *Less than* indicates that *6* is the subtrahend and the *square of a number* is the minuend.

c. the sum of h raised to the fifth power and fifteen

Translation $h^5 + 15$

d. the ratio of m times n to r cubed

Translation $\dfrac{mn}{r^3}$

> **Note:** When coupled with the word *ratio*, the word *to* translates to the fraction line. The amount to the left of the word *to* goes in the numerator, and the amount to the right of the word *to* goes in the denominator.

e. one-half of v divided by the square root of t

Translation $\dfrac{1}{2}v \div \sqrt{t}$

> **Note:** When the word *of* is preceded by a fraction, it means multiply.

Your Turn 1   Translate to an algebraic expression.

a. two-thirds subtracted from the product of nine and a number

b. -6.2 increased by 9.8 times a number

Translating Phrases Involving Parentheses

Sometimes the word phrases imply an order of operations that would require us to use parentheses in the translation. These situations arise when the phrase indicates that a sum or difference is to be calculated before a higher-order operation such as multiplication, division, exponent, or root is performed.

Example 2   Translate to an algebraic expression.

a. five times the sum of x and y

Translation $5(x + y)$

> **Note:** Without the parentheses, the expression is $5x + y$, which indicates that we are to multiply 5 times x first and then add y to the result. The parentheses indicate that we add x and y first, then multiply the resulting sum by 5.

b. The sum of five times x and y

Translation $5x + y$

c. the square root of the difference of the square of x and the square of y

Translation $\sqrt{x^2 - y^2}$

> **Note:** The *square root of the difference* indicates that we are to calculate the difference before we calculate the square root. We indicate this symbolically by placing the entire subtraction under the radical.

Answers   **Your Turn 1**

a. $9n - \dfrac{2}{3}$ **b.** $-6.2 + 9.8n$

d. the product of x and y divided by the sum of x^2 and 5

Translation $xy \div (x^2 + 5)$ or $\dfrac{xy}{x^2 + 5}$

Your Turn 2   Translate to an algebraic expression.

    a. negative two times the sum of n and seven

    b. the sum of eight times a and b

    c. the difference of m and n, all raised to the fourth power

    d. the product of 5 and m divided by the difference of m and 3

Example 3   Isaac Newton developed an expression that describes the gravitational effect between two objects. The relationship is the product of the masses of the two objects divided by the square of the distance between the objects. Translate the relationship to an expression.

Translation $Mm \div d^2$ or $\dfrac{Mm}{d^2}$

> **Note:** If the same letter is to describe two different quantities, uppercase and lowercase can be used to distinguish the quantities. Subscripts can also be used, as in $m_1 m_2 \div d^2$.

Answers   **Your Turn 2**

a. $-2(n + 7)$ **b.** $8a + b$

c. $(m - n)^4$ **d.** $\dfrac{5m}{m - 3}$

Answer   **Your Turn 3**

$d \div (v_1 + v_2)$ or $\dfrac{d}{v_1 + v_2}$

Your Turn 3   Two asteroids are on a collision course with each other. We can calculate the time to impact by dividing the distance that separates the asteroids by the sum of their velocities. Translate the relationship to an expression.

1.6 Exercises   For Extra Help   *MyMathLab*

Note: Exercises marked with a ★ represent challenging exercises.

1. List three key words that indicate addition.

2. List three key words that indicate subtraction.

3. List three key words that indicate multiplication.

4. List three key words that indicate division.

5. The phrase *nine subtracted from n* translates to $n - 9$, which is in reverse order of what we read. What other key words for subtraction translate in reverse order?

6. What key words for division translate in reverse order?

7. Translate the algebraic expression $x + 7$ into five different word phrases.

8. Translate the algebraic expression $x - 5$ into six different word phrases.

9. Translate the algebraic expression $\dfrac{7}{b}$ into three different word phrases.

10. Translate the algebraic expression $3n$ into three different word phrases.

**For Exercises 11–44, translate each phrase to an algebraic expression.
See Examples 1 and 2.**

11. four times a number

12. the product of a number and four

13. the sum of four times x and sixteen

14. five more than y

15. the difference of seven times x and eight

16. six less than T

17. the quotient of negative four and the cube of y

18. the ratio of seven to the square of m

19. the product of eight and p decreased by four

20. thirteen subtracted from twice a number

21. fourteen divided into m

22. r divided by six

23. x to the fourth power increased by five

24. the sum of b cubed and seven

25. one-fifth subtracted from the product of seven and a number

26. two-thirds added to the product of four and a number

27. the product of negative three and the difference of a number and two

28. the product of three and the sum of the number and four

29. the sum of four and n, all raised to the fifth power

30. the difference of two and l, all raised to the third power

31. five less than the product of m and n

32. the product of three and a number, increased by five

33. the quotient of four and a number, decreased by two

34. seven added to the quotient of x and y

35. negative twenty-seven decreased by the sum of a and b

36. the difference of m and n subtracted from negative eight

37. six-tenths decreased by the product of four and the difference of y and two

38. eighty-one hundredths increased by the product of eight and the sum of x and three tenths

39. the difference of p and q decreased by the sum of m and n

40. the sum of a and b subtracted from the difference of c and d

41. the product of m and n subtracted from the square root of y

42. the square root of x subtracted from the product of a and b

43. a number minus the product of three and the difference of the number and six

44. the product of five and a number minus the sum of the number and two

Find

ⓧ *the mistake*   **For Exercises 45–48, explain each mistake. Then translate the phrase correctly.**

45. seventeen less than three times t

Translation: $17 - 3t$

46. four subtracted from the square of m

Translation: $4 - m^2$

47. nine times the sum of x and y

Translation: $9x + y$

48. Nineteen divided into the product of h and k

Translation: $19 \div hk$

For Exercises 49–58, translate each phrase. See Example 3.

49. The length of a rectangle is five more than the width. If the width is represented by w, write an expression that describes the length.

50. The width of a rectangle is four less than the length. If the length is represented by l, write an expression that describes the width.

51. The length of a rectangle is three times the width. If the width is represented by the variable w, write an expression that describes the length.

52. The width of a rectangle is one-fourth of the length. If the length is represented by l, write an expression that describes the width.

53. The radius of a circle is one-half of the diameter. If d represents the diameter, write an expression for the radius.

54. The diameter of a circle is twice the radius. If r represents the radius, write an expression for the diameter.

55. Lindsey has 42 coins in her change purse that are either dimes or quarters. If n represents the number of quarters she has, write an expression in terms of n that describes the number of dimes.

56. Sherice owns a total of 60 shares of stock in two companies. If n represents the number of the higher-priced stock, write an expression in terms of n for the number of shares of the lower-priced stock.

★**57.** Don passes mile marker 51 on the highway. One-fourth of an hour later, a state trooper traveling in the same direction passes the same mile marker. If t represents the amount of time it takes the trooper to catch up to Don, write an expression in terms of t that describes the amount of time Don has traveled since passing marker 51.

★**58.** Barry is jogging along a trail and passes a sign. One-third of an hour later, Dedra, who is traveling in the same direction, passes the same sign. If t represents the amount of time it takes Dedra to catch up to Barry, write an expression in terms of t that describes the amount of time Barry has traveled since passing the sign.

Exercises 59–66 contain word descriptions of expressions from mathematics and physics. Translate the descriptions to symbolic form. See Examples 1–3.

59. The perimeter of a rectangle is twice the width plus twice the length.

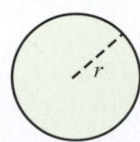

60. The area of a circle can be found by multiplying the square of the radius by π.

61. The volume of a cone is one-third of the product of π, the square of the radius, and the height of the cone.

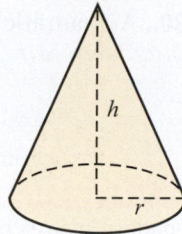

62. The volume of a sphere is four-thirds the product of π and the cube of the radius.

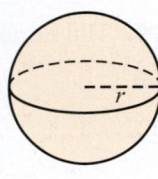

63. Albert Einstein developed an expression that describes the energy of a particle at rest, which is the product of the mass, m, of the particle and the square of the speed of light, which is represented by c.

64. The centripetal acceleration of an object traveling in a circular path is found by an expression that is the ratio of the square of the velocity, v, to the radius, r, of the orbital path.

★**65.** René Descartes developed an expression for the distance between two points in the coordinate plane. The distance between two points is the square root of the sum of the square of the difference of x_2 and x_1 and the square of the difference of y_2 and y_1.

★**66.** Albert Einstein's theory of relativity includes a mathematical expression that is the square root of the difference of one and the ratio of the square of the velocity, v, of an object in motion to the square of the speed of light, c.

For Exercises 67–72, the algebraic expressions have been translated to a word phrase that is ambiguous (unclear). Indicate why the translation is unclear and give a translation that is not ambiguous. There is more than one correct answer.

67. $3x + 4$

Translation: three times x plus four

68. $2a - 7$

Translation: two times a minus seven

69. $5(x - 1)$

Translation: five times x minus one

70. $4(y + 6)$

Translation: four times y plus six

71. $n + 5(n - 6)$

Translation: n plus five times the difference of n and six

72. $m - 3(m + 2)$

Translation: m minus three times the sum of m and two

For Exercises 73–80, translate the expression to a word phrase.

73. The area of a triangle: $\frac{1}{2}bh$

74. The area of a trapezoid: $\frac{1}{2}h(a + b)$

75. The volume of a box: lwh

76. The volume of a cylinder: $\pi r^2 h$

77. The surface area of a box: $2(lw + lh + wh)$

78. The surface area of a cylinder: $2\pi r(r + h)$

79. The slope of a line: $\dfrac{y_2 - y_1}{x_2 - x_1}$

80. A quadratic expression: $ax^2 + bx + c$

Puzzle Problem

Consecutive integers follow one another in a pattern such as 1, 2, 3, . . .
Let n represent the first integer in a set of consecutive integers.

a. Write expressions for the next two consecutive integers.

b. Suppose n is odd. Write expressions for the next two consecutive odd integers.

c. Suppose n is even. Write expressions for the next two consecutive even integers.

Review Exercises

Exercises 1–6  EXPRESSIONS

[1.4] **1.** Rewrite and simplify $2(5 + 2)$ using the distributive property.

[1.3] *For Exercises 2 and 3, simplify.*

  2. $4 - 8 + 6$

  3. $\dfrac{2}{3} - \dfrac{1}{4}$

[1.5] *For Exercises 4–6, simplify.*

  4. $-6^2 \div 3 + 4(5 - 9)$

  5. $-4|3 - 8| + 2^5$

  6. -3^2

1.7 Evaluating and Rewriting Expressions

Objectives

❶ Evaluate an expression.

❷ Determine all values that cause an expression to be undefined.

❸ Rewrite an expression using the distributive property.

❹ Rewrite an expression by combining like terms.

To be able to solve equations in Chapter 2, we must learn how to manipulate algebraic expressions. There are two actions we can perform with algebraic expressions. We can *evaluate* them and *rewrite* them.

Objective ❶ Evaluate an expression.   First, we learn how to evaluate an expression.

Procedure   **Evaluating an Algebraic Expression**

To evaluate an algebraic expression:

1. Replace the variables with their corresponding given values.

2. Calculate the numerical expression using the order of operations.

Example 1 Evaluate each expression using the given values.

a. $4m - 3(n - 4)$ when $m = 3$ and $n = 2$

SOLUTION $4\textbf{\textit{m}} - 3(\textbf{\textit{n}} - 4)$

$4(\textbf{3}) - 3(\textbf{2} - 4)$ Replace m with 3 and n with 2.

$= 4(3) - 3(-2)$ Simplify inside the parentheses first.

$= 12 + 6$ Multiply.

$= 18$ Add.

b. $y^2 - 0.2xy + 5$ when $x = 6$ and $y = -4$

SOLUTION $y^2 - 0.2xy + 5$

$(-\textbf{4})^2 - 0.2(\textbf{6})(-\textbf{4}) + 5$ Replace x with 6 and y with -4.

$= 16 - 0.2(6)(-4) + 5$ Begin calculating by simplifying the exponential form.

$= 16 - (-4.8) + 5$ Multiply.

$= 16 + 4.8 + 5$ Write the subtraction as an equivalent addition.

$= 25.8$ Add from left to right.

c. $\dfrac{5r^2}{d - 3}$ when $r = 2$ and $d = -5$

SOLUTION $\dfrac{5r^2}{d - 3}$

$\dfrac{5(\textbf{2})^2}{-\textbf{5} - 3}$ Replace r with 2 and d with -5.

$= \dfrac{5(4)}{-8}$ Calculate the top and bottom expressions separately.

$= \dfrac{20}{-8}$

$= -\dfrac{5}{2}$ or -2.5

Your Turn 1 Evaluate the expression using the given values.

a. $-5a + 3(b - 6)$ when $a = -2$ and $b = 4$

b. $m^3 - 6n^2$ when $m = -2$ and $n = 3$

c. $\dfrac{3a^2 - b}{a + 6}$ when $a = -4$ and $b = 2$

Objective ② **Determine all values that cause an expression to be undefined.** In Example 1c, what if $d = 3$? Notice that the denominator would become 0: $\dfrac{5r^2}{3 - 3} = \dfrac{5r^2}{0}$, which is undefined if $r \neq 0$ and indeterminate if $r = 0$. When asked to evaluate a division expression in which the divisor or denominator contains a variable or variables, we must be careful about what values replace the variable(s). Often we need to know what values replace the variable(s) and cause the expression to be undefined or indeterminate.

Answers **Your Turn 1**
a. 4   **b.** -62 **c.** 23

Example 2 Determine all values that cause each expression to be undefined.

a. $\dfrac{5}{x + 7}$

Answer If $x = -7$, we have $\dfrac{5}{-7 + 7} = \dfrac{5}{0}$, which is undefined because the denominator is 0.

b. $\dfrac{-y}{(y + 2)(y - 6)}$

SOLUTION Note that because $(y + 2)(y - 6)$ is a product, if either factor is 0, the entire denominator is 0.

Notice that if $y = -2$, we have
$$\dfrac{-(-2)}{(-2 + 2)(-2 - 6)}$$
$$= \dfrac{2}{(0)(-8)}$$
$$= \dfrac{2}{0}, \text{ which is undefined}$$

Also, if $y = 6$, we have
$$\dfrac{-6}{(6 + 2)(6 - 6)}$$
$$= \dfrac{-6}{(8)(0)}$$
$$= \dfrac{-6}{0}, \text{ which is undefined}$$

> **Connection** This principle—that if either factor is 0, then the entire product is 0—is known as the *zero-factor theorem*. We use this theorem extensively in Chapter 6 when solving equations such as $(y + 2)(y - 6) = 0$.

Answer Both $y = -2$ and $y = 6$ cause the expression to be undefined.

Your Turn 2 Determine all values that cause each expression to be undefined.

a. $\dfrac{9}{n + 6}$

b. $\dfrac{2m}{(m - 3)(m + 4)}$

Objective ③ Rewrite an expression using the distributive property. Recall from Section 1.4 the distributive property of multiplication over addition.

Rule **Distributive Property of Multiplication over Addition**
$$a(b + c) = ab + ac$$

This property gives us an alternative to the order of operations. Look at the numerical expression $2(5 + 6)$ and compare using the order of operations versus using the distributive property.

Following order of operations:
$$2(\mathbf{5 + 6}) = 2(\mathbf{11})$$
$$= 22$$

or

Using the distributive property:
$$\mathbf{2}(5 + 6) = \mathbf{2} \cdot 5 + \mathbf{2} \cdot 6$$
$$= 10 + 12$$
$$= 22$$

> **Note:** Using the order of operations, we add within the parentheses first, then multiply.

> **Note:** Using the distributive property, we multiply each addend by the factor outside the parentheses, then add the products.

The result is the same either way, so the distributive property allows us to say that $2(5 + 6)$ and $2 \cdot 5 + 2 \cdot 6$ are equivalent expressions.

Conclusion   We can use the distributive property to rewrite an expression in another form that is equivalent to the original form.

Example 3   Use the distributive property to write an equivalent expression and simplify.

a. $2(x + 5)$

SOLUTION $2(x + 5) = 2 \cdot x + 2 \cdot 5$
$$= 2x + 10$$

b. $-6(n - 9)$

SOLUTION $-6(n - 9) = -6 \cdot n - (-6) \cdot 9$
$$= -6n - (-54)$$
$$= -6n + 54$$

c. $\dfrac{3}{8}\left(2m + \dfrac{4}{5}\right)$

SOLUTION $\dfrac{3}{8}\left(2m + \dfrac{4}{5}\right) = \dfrac{3}{8} \cdot 2m + \dfrac{3}{8} \cdot \dfrac{4}{5}$ **Use the distributive property.**

$$= \dfrac{3}{\underset{4}{\cancel{8}}} \cdot \dfrac{\overset{1}{\cancel{2}}}{1}m + \dfrac{3}{\underset{2}{\cancel{8}}} \cdot \dfrac{\overset{1}{\cancel{4}}}{5}$$ **Divide out common factors.**

$$= \dfrac{3}{4}m + \dfrac{3}{10}$$

Connection   We can apply the distributive property to $-6(n - 9)$ even though the parentheses contain subtraction instead of addition because subtraction can be expressed as addition like this:

$$-6(n - 9) = -6[n + (-9)]$$

In fact, it is helpful to think of the expression this way because when we multiply the -6 by the -9, we get positive 54.

Your Turn 3   Use the distributive property to write an equivalent expression and simplify.

a. $3(n + 2)$ **b.** $-9(7 - 3x)$ **c.** $\dfrac{4}{5}\left(\dfrac{1}{2}y - 10\right)$

Objective 4   Rewrite an expression by combining like terms.   Many expressions are sums of expressions called **terms**. For example, the expression $5x + 3$ is a sum of the terms $5x$ and 3.

Definition   *Terms:* Expressions that are the addends in an expression that is a sum.

If an expression contains subtraction, as in $4x^2 - 6x + 8$, we can identify its terms by writing the subtraction as addition. Because $4x^2 - 6x + 8 = 4x^2 + (-6x) + 8$, its terms are $4x^2, -6x$, and 8. Notice we can see that the second term in $4x^2 - 6x + 8$ is $-6x$ without writing the equivalent addition by recognizing that the sign to the left of the term is its sign.

The numerical factor in a term is called the numerical **coefficient** or, simply, the coefficient of the term.

Definition   *Coefficient:* The numerical factor in a term.

Answers   Your Turn 3
a. $3n + 6$ **b.** $-63 + 27x$
c. $\dfrac{2}{5}y - 8$

The coefficient of $4x^2$ is 4.
The coefficient of $-6x$ is -6.
The coefficient of 8 is 8.

The coefficient of x is 1 because $1 \cdot x = x$.
The coefficient of $-x$ is -1 because $-1 \cdot x = -x$.
The coefficient of $\dfrac{2x}{3}$ is $\dfrac{2}{3}$ because $\dfrac{2}{3}x = \dfrac{2x}{3}$.

Another way to rewrite an expression is to combine **like terms**.

Definition *Like terms:* Constant terms or variable terms that have the same variable(s) raised to the same exponents.

Connection We can use the distributive property in reverse to confirm our procedure for combining like terms. Consider the expression $3x + 5x$ again. Notice that there is a common factor of x in both terms. When applying the distributive property in reverse, we write this common factor outside the parentheses and write the remaining factors as addends within the parentheses. It looks like this:

$$3x + 5x = (3 + 5)x$$

Because the parentheses contain only the addition of two numbers, we simplify by adding the numbers.

$$3x + 5x = (3 + 5)x$$
$$= 8x$$

Examples of like terms:

$3x$ and $5x$
$4y^2$ and $9y^2$
$7xy$ and $3xy$
6 and 15

Examples of unlike terms:

$2x$ and $8y$ (different variables)
$4t^2$ and $4t^3$ (different exponents)
x^2y and xy^2 (different exponents)
12 and $12x$ (different variables)

Combining Like Terms

Now consider an expression that is a sum of like terms, such as $3x + 5x$. We can rewrite this expression in a more compact form. Multiplication of 3 times x means that x is repeatedly added three times. Likewise, 5 times x means that x is repeatedly added five times. We can expand the expression out like this:

$$\underbrace{3x}\ +\ \underbrace{5x}$$
$$= \underbrace{x + x + x}\ +\ \underbrace{x + x + x + x + x}$$
$$=\qquad\qquad 8x$$

Note: After expanding, we see that there are a total of eight x's repeatedly added. We can write those eight x's as a single term, $8x$.

Notice that when we combine the like terms, we get an expression that is more compact. When we rewrite an expression in a more compact form, we say that we are *simplifying* the expression. Also notice that when combining like terms, we simply add the coefficients and keep the variable the same.

Procedure **Combining Like Terms**

To combine like terms, add or subtract the coefficients and keep the variables and their exponents the same.

Example 4 Combine like terms.

a. $9y + 7y$

SOLUTION $9y + 7y = 16y$ ◄ **Note:** We think: Nine y's plus seven y's equals sixteen y's.

b. $7x - 2x$

SOLUTION $7x - 2x = 5x$ ◄ **Note:** We think: Seven x's minus two x's leaves five x's.

c. $14n^2 - n^2$

SOLUTION $14n^2 - n^2 = 13n^2$ ◄ **Note:** The coefficient of n^2 is 1, so we think fourteen n^2's minus one n^2 leaves thirteen n^2's.

Your Turn 4 Combine like terms.

Answers Your Turn 4

a. $14r$ **b.** $8.8x$ **c.** $-\dfrac{5}{12}n^3$

a. $6r + 8r$ **b.** $6.5x + 2.3x$ **c.** $\dfrac{1}{3}n^3 - \dfrac{3}{4}n^3$

Collecting Like Terms

Sometimes expressions are more complex and contain different sets of like terms. In such cases, we combine the like terms and copy any unlike terms in the final expression. Many people like to use the commutative property of addition and rearrange the expression first so that the like terms are together, then combine them. This type of manipulation is called *collecting* the like terms.

Example 5   Combine like terms in $4y^2 + 5 + 3y^2 - 9$.

SOLUTION $4y^2 + 5 + 3y^2 - 9$

$= 4y^2 + 3y^2 + 5 - 9$ **Collect the like terms.**

$= 7y^2 - 4$ **Combine like terms.**

Collecting like terms is optional. Alternatively, many people mark through the terms as they combine them as a way of keeping track of what has been combined. Using this technique with the expression from Example 5 looks like this:

$$4y^2 + 5 + 3y^2 - 9$$

$$= 7y^2 - 4$$

Example 6   Combine like terms in $15y + 8x - y - 8x$.

SOLUTION $15y + 8x - y - 8x$ **Note:** The term $-y$ is equivalent to $-1y$.

$= 14y + 0$

$= 14y$

Sometimes collecting the like terms is beneficial, as in Example 7, where there are like terms with fraction coefficients. Some people find it easier to "see" the steps in writing the equivalent fractions when the fractions are closer together.

Example 7   Combine like terms in $\frac{1}{3}n - 9m + 2 - m + \frac{1}{4}n$.

SOLUTION $\frac{1}{3}n - 9m + 2 - m + \frac{1}{4}n$

$= \frac{1}{3}n + \frac{1}{4}n - 9m - m + 2$ **Collect the like terms.**

$= \frac{1(4)}{3(4)}n + \frac{1(3)}{4(3)}n - 9m - m + 2$ **Write fraction coefficients as equivalent fractions with their LCD, 12.**

$= \frac{4}{12}n + \frac{3}{12}n - 9m - m + 2$

$= \frac{7}{12}n - 10m + 2$ **Combine like terms.**

Your Turn 7 Combine like terms.

a. $6x^3 + 2y + 4x^3 - 6y$ **b.** $4.2x^2 - 6x + 9 - 0.7x - 12$

c. $\dfrac{3}{8}y - x + 2 - \dfrac{3}{4}y + 5x$

Connection One way to verify that two expressions are equivalent is to evaluate both using the same value(s) for the variable(s). Consider the expression in Example 5. If we evaluate the original expression $4y^2 + 5 + 3y^2 - 9$ and the simplified expression $7y^2 - 4$ using the same chosen value for y, we should get the same answer. Let's choose the number 2 for y.

Original Expression	Simplified Expression
$4y^2 + 5 + 3y^2 - 9$	$7y^2 - 4$
$4(2)^2 + 5 + 3(2)^2 - 9$	$7(2)^2 - 4$
$= 4(4) + 5 + 3(4) - 9$	$= 7(4) - 4$
$= 16 + 5 + 12 - 9$	$= 28 - 4$
$= 24$	$= 24$

Warning: Because some expressions that are not equivalent may give the same answer using this method, this "check" does not guarantee that the expressions are equivalent. It is, however, a good sign.

Notice that the simplified expression gave the same answer, yet was easier to work with. This is one of the reasons we simplify expressions. If we can simplify an expression, we have an easier time evaluating it.

1.7 Exercises For Extra Help MyMathLab

Note: Exercises marked with a ★ represent challenging exercises.

1. In your own words, explain how to evaluate an expression.

2. Explain the difference between evaluating and rewriting an expression.

3. What causes an expression to be undefined?

4. What are like terms?

5. What is the coefficient in a term?

6. Explain how to combine like terms.

For Exercises 7–28, evaluate the expressions using the given values. See Example 1.

7. $3(a + 5) - 4b; a = 2, b = 3$

8. $8n - 2(m + 1); m = 5, n = 3$

9. $4 - 0.2(x + 3); x = 1$

10. $6 - 0.4(y - 2); y = 5$

11. $2p^2 - 3p - 4; p = -3$

12. $n^2 - 8n + 1; n = -1$

13. $2y^2 - 4y + 3; y = -\dfrac{1}{2}$

14. $3r^2 - 9r + 6; r = -\dfrac{1}{3}$

15. $8 - 2(y + 4); y = -2.3$

16. $-6 - 2(l - 5); l = -0.4$

17. $-|3x| + |4y^3|; x = 5, y = -1$

18. $-|2m^2| - |4n|; m = 3, n = -2$

19. $|2r^2 - 3q|; r = -3, q = 5$

20. $|2m^2 + 2n|; m = -4, n = -5$

21. $\sqrt{c} + 4ab^2; a = -1, b = -2, c = 16$

22. $-2x^3y + \sqrt{z}$; $x = -2, y = -3, z = 4$ **23.** $-5\sqrt{a} + 2\sqrt{b}$; $a = 25, b = 4$ **24.** $-3\sqrt{h} + 3\sqrt{k}$; $h = 16, k = 9$

25. $\dfrac{6x^3}{y - 8}$; $x = 2, y = 7$ **26.** $\dfrac{4m^2}{n + 4}$; $m = 2, n = 4$ **27.** $\dfrac{15 - b^2}{2\sqrt{c} + d}$; $b = 3, c = 25, d = 144$

28. $\dfrac{5 - a^2}{3\sqrt{x} + y}$; $a = 1, x = 64, y = 36$

29. The expression $b^2 - 4ac$ is called the *discriminant* and is used to determine the types of solutions for quadratic equations. Find the value of the discriminant given the following values.

     a. $a = 2, b = 1, c = -3$ **b.** $a = 1, b = -2, c = 4$

30. The expression $ad - bc$ is used to calculate the determinant of a matrix. Find the determinant given the following values.

     a. $a = 1, b = 0.5, c = -4, d = 6$ **b.** $a = -3, b = \dfrac{4}{5}, c = 2, d = \dfrac{1}{2}$

> **Connection** You will learn all of the details about each expression presented in Exercises 29–32 in future chapters (and future courses). Notice that even without a full understanding of the context in which the expressions are used, you can still evaluate them.

31. The expression $\dfrac{y_2 - y_1}{x_2 - x_1}$ is used to calculate the slope of a line. Find the slope given each set of the following values.

     a. $x_1 = 2, y_1 = 1, x_2 = 5, y_2 = 7$ **b.** $x_1 = -1, y_1 = 2, x_2 = -7, y_2 = -2$

32. The expression $\sqrt{(x_2 - x_1)^2 + (y_2 - y_1)^2}$ is used to calculate the distance between two points in the coordinate plane. Evaluate the expression using the following values.

     a. $x_1 = 2, y_1 = 1, x_2 = 5, y_2 = 7$ **b.** $x_1 = -1, y_1 = 2, x_2 = -7, y_2 = -2$

For Exercises 33–40, determine all values that cause each expression to be undefined. See Example 2.

33. $\dfrac{-7}{5 + y}$ **34.** $\dfrac{8}{x + 3}$ **35.** $\dfrac{6m}{(m + 1)(m - 3)}$ **36.** $\dfrac{-5a}{(a - 4)(a - 2)}$

37. $\dfrac{6 + x^2}{x}$ **38.** $\dfrac{7 - y}{y}$ ★**39.** $\dfrac{x + 1}{3x - 2}$ ★**40.** $\dfrac{3y}{2y + 1}$

For Exercises 41–48, use the distributive property to write an equivalent expression and simplify. See Example 3.

41. $6(a + 2)$ **42.** $4(b - 5)$ **43.** $-8(4 - 3y)$ **44.** $-7(3 - 2m)$

45. $\dfrac{7}{8}\left(\dfrac{1}{2}c - 16\right)$ **46.** $\dfrac{4}{5}\left(-10h + \dfrac{2}{9}\right)$ **47.** $0.2(3n - 8)$ **48.** $-1.5(6x + 7)$

For Exercises 49–58, identify the coefficient of each term. See Objective 4.

49. $-6x^3$

50. $-14y$

51. m^9

52. y^5

53. $-b$

54. $-n$

55. $-\dfrac{2}{3}m$

56. $\dfrac{5}{8}a$

57. $\dfrac{y}{5}$

58. $-\dfrac{u}{3}$

For Exercises 59–84, simplify by combining like terms. See Examples 4–7.

59. $3y + 5y$

60. $6m + 7m$

61. $6a - 11a$

62. $5b - 13b$

63. $14x - 3x$

64. $-5y + 12y$

65. $-3r - 8r$

66. $-7m - 6m$

67. $6.3n - 8.2n$

68. $-5.1x^4 + 3.4x^4$

69. $\dfrac{1}{2}b^2 - \dfrac{5}{6}b^2$

70. $\dfrac{3}{4}z - \dfrac{7}{5}z$

71. $-11c - 10c - 5c$

72. $-15w - 6w - 11w$

73. $6x + 7 - x - 8$

74. $5y^2 + 6 + 3y^2 - 8$

75. $-8x + 3y - 7 - 2x + y + 6$

76. $-4a + 9b - a + 5 + 2b - 8$

77. $-10m + 7n + 1 + m - 2n - 12 + y$

78. $-3h + 7k - 5 - 8h - 7k + 19 + x$

79. $1.5x + y - 2.8x + 0.3 - y - 0.7$

80. $0.4t^2 + t - 2.8 - t^2 + 0.9t - 4$

81. $\dfrac{1}{6}c + 3d + \dfrac{2}{3}c + \dfrac{1}{7} - \dfrac{1}{2}d$

82. $\dfrac{5}{8}y + 4 - \dfrac{3}{4}x + \dfrac{2}{3} - \dfrac{1}{4}y$

83. $7a + \dfrac{4}{5}b^2 - 5 - \dfrac{3}{4}a - \dfrac{2}{7}b^2 + 9$

84. $\dfrac{1}{2}m - 3n + 14 - \dfrac{3}{8}m - \dfrac{9}{10}n - 5$

★**85. a.** Translate to an algebraic expression: fourteen plus the difference of six times a number and eight times the same number.

   b. Simplify the expression.

   c. Evaluate the expression when the number is -3.

★**86. a.** Translate to an algebraic expression: the sum of negative five times a number and eight minus two times the same number.

   b. Simplify the expression.

   c. Evaluate the expression when the number is 0.2.

Puzzle Problem

Each letter in the addition problem to the right represents a different whole number, 0–9. What number does each letter represent?

```
    F O R T Y
      T E N
  +   T E N
  ---------
  S I X T Y
```

Review Exercises

Exercise 1   CONSTANTS AND VARIABLES

[1.1]   **1.** Arrange in order from least to greatest.

$$4.2, |-6|, 4\frac{5}{8}, -2.5, -3$$

Exercises 2–6   EXPRESSIONS

[1.2]   **2.** Simplify: $\dfrac{72}{420}$

For Exercises 3–6, perform the indicated operation(s).

[1.3]   **3.** $-8 + (-14)$ **[1.4]**   **4.** $\dfrac{1}{2}(12.5)(8)$ **[1.5]**   **5.** $4(-2) + 9$ **[1.4]**   **6.** $\dfrac{3}{4}\left(5\dfrac{1}{3}\right) - 1$

Chapter 1 Summary

Defined Terms

Section 1.1
Variable (p. 2)
Constant (p. 2)
Expression (p. 2)
Equation (p. 2)
Inequality (p. 2)
Set (p. 3)
Rational number (p. 4)
Irrational number (p. 5)
Real numbers (p. 5)
Absolute value (p. 6)

Section 1.2
Fraction (p. 10)
Multiple (p. 11)
LCM (p. 12)
LCD (p. 12)
Factors (p. 12)
Prime number (p. 13)
Prime factorization
(p. 13)
Lowest terms (p. 14)

Section 1.3
Additive inverses (p. 27)

Section 1.4
Multiplicative inverses
(p. 42)

Section 1.5
Exponent (p. 50)
Base (p. 50)

Section 1.7
Terms (p. 73)
Coefficient (p. 73)
Like terms (p. 74)

The Real Number System

Real Numbers

Rational Numbers: Real numbers that can be expressed in the form $\frac{a}{b}$, where a and b are integers and $b \neq 0$, such as $-4\frac{3}{4}$, $-\frac{2}{3}$, 0.018, $0.\overline{3}$, and $\frac{5}{8}$.

Integers: $\ldots, -3, -2, -1, 0, 1, 2, 3, \ldots$

Whole Numbers: $0, 1, 2, 3, \ldots$

Natural Numbers: $1, 2, 3, \ldots$

Irrational Numbers: Any real number that is not rational, such as $-\sqrt{2}$, $-\sqrt{3}$, $\sqrt{0.8}$, and π.

Arithmetic Summary Diagram

Each operation has an inverse operation. In the diagram, the operations build from the top down. Addition leads to multiplication, which leads to exponents. Subtraction leads to division, which leads to roots.

Properties of Arithmetic

In each of the following, *a*, *b*, and *c* represent real numbers.

Additive Identity
$$a + 0 = a$$

Commutative Property of Addition
$$a + b = b + a$$

Associative Property of Addition
$$(a + b) + c = a + (b + c)$$

Multiplicative Property of 0
$$a \cdot 0 = 0$$

Multiplicative Identity
$$a \cdot 1 = a$$

Commutative Property of Multiplication
$$ab = ba$$

Associative Property of Multiplication
$$(ab)c = a(bc)$$

Distributive Property
$$a(b + c) = ab + ac$$

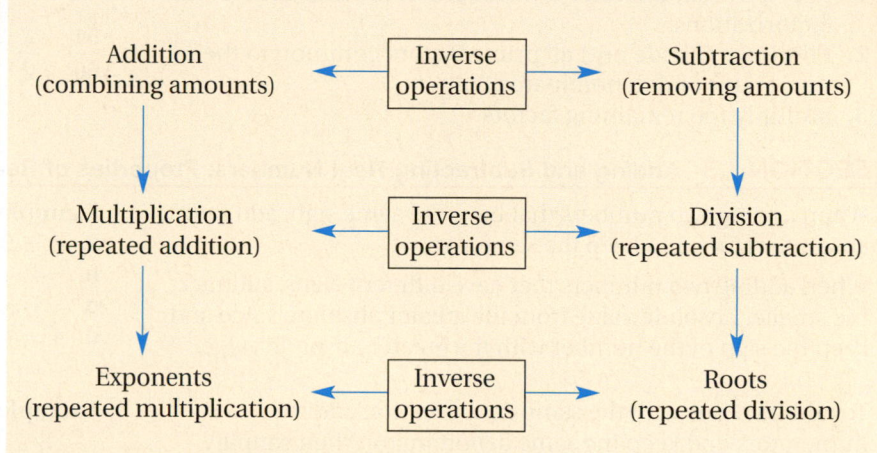

Order of Operations
1. Within grouping symbols
2. Exponents and roots from left to right
3. Multiply or divide from left to right
4. Add or subtract from left to right

Procedures, Rules, and Key Examples

Procedures/Rules	Key Examples

SECTION 1.1 Number Sets and the Structure of Algebra

To write a set, write the elements or members of the set within braces, { }.

Example 1: Write the set containing the natural numbers divisible by 3.

Answer: $\{3, 6, 9, 12, \dots\}$

The absolute value of a number is positive or 0.

Example 2: Find the absolute value.
a. $|8| = 8$
b. $|-15| = 15$
c. $|0| = 0$

For any two real numbers *a* and *b*, *a* is greater than *b* if *a* is to the right of *b* on a number line.

Example 3: Use $<$ or $>$ to make a true sentence.
a. $-8 \quad \square \quad -10$
b. $-6 \quad \square \quad 9$

Answers: a. $-8 > -10$
b. $-6 < 9$

SECTION 1.2 Fractions

For any fraction, we can write an equivalent fraction by multiplying or dividing both its numerator and denominator by the same nonzero number.

Example 1: Find the missing number in $\dfrac{5}{6} = \dfrac{?}{24}$ so that the fractions are equivalent.

Answer: $\dfrac{5}{6} = \dfrac{5 \cdot 4}{6 \cdot 4} = \dfrac{20}{24}$

(continued)

Procedures/Rules	Key Examples

SECTION 1.2 Fractions (*continued*)

To simplify a fraction to lowest terms:
1. Replace the numerator and denominator with their prime factorizations.
2. Eliminate (divide out) all prime factors common to the numerator and denominator.
3. Multiply the remaining factors.

Example 2: Reduce $\dfrac{54}{60}$ to lowest terms.

$$\frac{54}{60} = \frac{2 \cdot 3 \cdot 3 \cdot 3}{2 \cdot 2 \cdot 3 \cdot 5} = \frac{9}{10}$$

SECTION 1.3 Adding and Subtracting Real Numbers; Properties of Real Numbers

When adding two numbers that have the same sign, add their absolute values and keep the same sign.

When adding two numbers that have different signs, subtract the smaller absolute value from the greater absolute value and keep the sign of the number with the greater absolute value.

Example 1: Add.
a. $4 + 9 = 13$
b. $-4 + (-9) = -13$
c. $-4 + 9 = 5$
d. $4 + (-9) = -5$

To add fractions with the same denominator, add the numerators and keep the same denominator; then simplify.

To add fractions with different denominators:
1. Write each fraction as an equivalent fraction with the LCD.
2. Add the numerators and keep the LCD.
3. Simplify.

Example 2: Add.

a. $\dfrac{1}{10} + \dfrac{3}{10} = \dfrac{4}{10} = \dfrac{2}{5}$

b. $-\dfrac{3}{4} + \dfrac{5}{6} = -\dfrac{3(3)}{4(3)} + \dfrac{5(2)}{6(2)}$

$$= -\frac{9}{12} + \frac{10}{12}$$

$$= \frac{1}{12}$$

To write a subtraction statement as an equivalent addition statement, change the operation symbol from a minus sign to a plus sign and change the subtrahend to its additive inverse.

Example 3: Subtract.
a. $4 - 9 = 4 + (-9) = -5$
b. $-4 - 9 = -4 + (-9) = -13$
c. $4 - (-9) = 4 + 9 = 13$
d. $-4 - (-9) = -4 + 9 = 5$

SECTION 1.4 Multiplying and Dividing Real Numbers; Properties of Real Numbers

The product of an even number of negative factors is positive, whereas the product of an odd number of negative factors is negative.

Example 1: Multiply.
a. $(5)(7) = 35$
b. $(-5)(-7) = 35$
c. $(-1)(-2)(-5)(-7) = 70$
d. $(-5)(7) = -35$
e. $(5)(-7) = -35$
f. $(-2)(-5)(-7) = -70$

Rule for multiplying fractions: $\dfrac{a}{b} \cdot \dfrac{c}{d} = \dfrac{ac}{bd}$, where $b \neq 0$ and $d \neq 0$.

Example 2: Multiply.

a. $\dfrac{3}{4} \cdot \dfrac{5}{7} = \dfrac{15}{28}$

b. $\dfrac{5}{6} \cdot \dfrac{4}{9} = \dfrac{5}{2 \cdot 3} \cdot \dfrac{2 \cdot 2}{9} = \dfrac{10}{27}$

To multiply decimal numbers:
1. Multiply as if they were whole numbers.
2. Place the decimal in the product so that it has the same number of decimal places as the total number of decimal places in the factors.

Example 3: Multiply $(3.6)(2.4)$.

$$
\begin{array}{r}
2.4 \longrightarrow \quad \text{1 place} \\
\times\ 3.6 \longrightarrow \quad +\ \text{1 place} \\
\hline
144 \\
+\ 72 \\
\hline
8.64 \longleftarrow \quad \text{2 places}
\end{array}
$$

(*continued*)

Procedures/Rules	Key Examples

SECTION 1.4 Multiplying and Dividing Real Numbers; Properties of Real Numbers (*continued*)

When dividing two numbers that have the same sign, the quotient is positive.

When dividing two numbers that have different signs, the quotient is negative.

Rules for division involving 0: $0 \div n = 0$ when $n \neq 0$
$n \div 0$ is undefined when $n \neq 0$
$0 \div 0$ is indeterminate

Rule for dividing fractions: $\dfrac{a}{b} \div \dfrac{c}{d} = \dfrac{a}{b} \cdot \dfrac{d}{c}$, where $b \neq 0$, $c \neq 0$, and $d \neq 0$.

Example 4: Divide.
a. $24 \div 8 = 3$
b. $-24 \div (-8) = 3$
c. $-24 \div 8 = -3$
d. $24 \div (-8) = -3$

e. $0 \div 5 = 0$
f. $14 \div 0$ is undefined.

Example 5: Divide $\dfrac{5}{8} \div \dfrac{3}{4}$.

$$\frac{5}{8} \div \frac{3}{4} = \frac{5}{8} \cdot \frac{4}{3} = \frac{5}{2 \cdot 2 \cdot 2} \cdot \frac{2 \cdot 2}{3} = \frac{5}{6}$$

When dividing decimal numbers, set up a long division and consider the divisor.
Case 1: If the divisor is an integer, divide as if the dividend were a whole number and place the decimal point in the quotient directly above its position in the dividend.
Case 2: If the divisor is a decimal number,
1. Move the decimal point in the divisor to the right enough places to make the divisor an integer.
2. Move the decimal point in the dividend the same number of places.
3. Divide the divisor into the dividend as if both numbers were whole numbers. Make sure you align the digits in the quotient properly.
4. Write the decimal point in the quotient directly above its new position in the dividend.
In either case, continue the division process until you get a remainder of 0 or a repeating digit (or block of digits) in the quotient.

Example 6: Divide $12.88 \div 0.06$.

$$
\begin{array}{r}
214.66 \\
6\overline{)1288.00} \\
-12 \\
\hline
08 \\
-6 \\
\hline
28 \\
-24 \\
\hline
40 \\
-36 \\
\hline
40 \\
\end{array}
$$

Answers: $214.\overline{6}$

SECTION 1.5 Exponents, Roots, and Order of Operations

To evaluate an exponential form raised to a natural number exponent, write the base as a factor the number of times indicated by the exponent; then multiply.

If the base of an exponential form is a negative number and the exponent is even, the product is positive.

If the base is a negative number and the exponent is odd, the product is negative.

Every positive number has two square roots, a positive root and a negative root.

Negative numbers have no real-number square roots.

Example 1: Evaluate.
a. $3^4 = 3 \cdot 3 \cdot 3 \cdot 3 = 81$
b. $-3^4 = -3 \cdot 3 \cdot 3 \cdot 3 = -81$

c. $(-2)^4 = (-2)(-2)(-2)(-2) = 16$

d. $(-2)^5 = (-2)(-2)(-2)(-2)(-2)$
$ = -32$

Example 2: Find all square roots.
a. 49

Answer: ± 7

b. -36

Answer: No real-number square roots exist.

(*continued*)

Procedures/Rules	Key Examples

SECTION 1.5 Exponents, Roots, and Order of Operations (*continued*)

The radical symbol $\sqrt{}$ indicates to find only the positive (principal) square root.

$$\sqrt{\frac{a}{b}} = \frac{\sqrt{a}}{\sqrt{b}}, \text{ where } a \geq 0 \text{ and } b > 0.$$

Example 3: Simplify.

a. $\sqrt{81} = 9$

b. $\sqrt{-49}$ is not a real number.

c. $\sqrt{\dfrac{25}{64}} = \dfrac{\sqrt{25}}{\sqrt{64}} = \dfrac{5}{8}$

Order-of-Operations Agreement

Perform operations in the following order:
1. Within grouping symbols.
2. Exponents/Roots from left to right, in order as they occur.
3. Multiplication/Division from left to right, in order as they occur.
4. Addition/Subtraction from left to right, in order as they occur.

Example 4: Simplify.

$$35 - [29 - 2(12 + 4)] + 4^3$$
$$= 35 - [29 - 2(16)] + 4^3$$
$$= 35 - (29 - 32) + 4^3$$
$$= 35 - (-3) + 4^3$$
$$= 35 - (-3) + 64$$
$$= 35 + 3 + 64$$
$$= 102$$

To find the arithmetic mean, or average, of n numbers, divide the sum of the numbers by n.

$$\text{Arithmetic mean} = \frac{x_1 + x_2 + \cdots + x_n}{n}$$

Example 5: Carl has the following test scores: 68, 76, 82, and 78. Find the average of his test scores.

$$\frac{68 + 76 + 82 + 78}{4} = \frac{304}{4} = 76$$

SECTION 1.6 Translating Word Phrases to Expressions

To translate a word phrase to an expression, identify the unknowns, constants, and key words; then write the corresponding symbolic form.

Example 1: Translate to an algebraic expression:

a. Five more than seven times a number

Answer: $7n + 5$

b. Six times the difference of a number and nine

Answer: $6(n - 9)$

c. Four divided by a number cubed

Answer: $4 \div n^3$

SECTION 1.7 Evaluating and Rewriting Expressions

To evaluate an algebraic expression:
1. Replace the variables with their corresponding given values.
2. Calculate the numerical expression using the order of operations agreement.

Example 1: Evaluate $4x^2 - 5x$ when $x = -3$.

$$4(-3)^2 - 5(-3)$$
$$= 4(9) - 5(-3)$$
$$= 36 - (-15)$$
$$= 36 + 15$$
$$= 51$$

Distributive Property of Multiplication over Addition

$$a(b + c) = ab + ac$$

Example 2: Use the distributive property to write an equivalent expression.

$$3(x + 7) = 3 \cdot x + 3 \cdot 7$$
$$= 3x + 21$$

To combine like terms, add or subtract the coefficients and keep the variables and their exponents the same.

Example 3: Combine like terms.

$$15x^2 + 8x + 3x^2 + 7 - 9x$$
$$= 15x^2 + 3x^2 + 8x - 9x + 7$$
$$= 18x^2 - x + 7$$

Chapter 1 Review Exercises

For Exercises 1–6, answer true or false.

[1.1] **1.** π is a rational number.

[1.1] **2.** $-\sqrt{2}$ is a rational number.

[1.7] **3.** $-6x$ and $2x^3$ are like terms.

[1.7] **4.** $8xy$ and yx are like terms.

[1.5] **5.** $-6^2 = -36$

[1.5] **6.** $(-6)^2 = 36$

For Exercises 7–10, fill in the blank.

[1.1] **7.** The absolute value of every number is either ——————— or 0.

[1.2] **8.** A natural number that has exactly two different factors, 1 and the number itself, is a(n) ——————— number.

[1.3] **9.** When adding two addends that have the same sign, add their absolute values and ——————— the sign.

[1.4] **10.** If there are a(n) ——————— number of negative factors, the product is positive.

[1.1] For Exercises 11–14, write a set representing each description.

11. The months beginning with the letter M

12. The letters in the word *algebra*

13. The even natural numbers

14. The even integers

[1.1] For Exercises 15–18, graph each number on a number line.

15. $-3\dfrac{2}{5}$

16. $2\dfrac{1}{4}$

17. 8.2

18. -4.6

Exercises 19–22  **EXPRESSIONS**

[1.1] For Exercises 19–22, simplify.

19. $|-6.3|$

20. $|8.46|$

21. $\left|2\dfrac{1}{6}\right|$

22. $\left|-4\dfrac{3}{8}\right|$

Exercises 23–26 **EQUATIONS AND INEQUALITIES**

[1.1] For Exercises 23–26, use =, +, <, or > to write a true sentence.

23. $|-6.4|$ ▮ 6.4

24. -6.9 ▮ 0

25. -14 ▮ -9

26. $\left|5\dfrac{6}{7}\right|$ ▮ $\left|-5\dfrac{6}{7}\right|$

[1.2] *For Exercises 27 and 28, name the fraction represented by each shaded region.*

27.

28.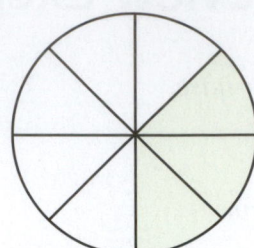

Exercises 29–32 EQUATIONS AND INEQUALITIES

[1.2] *For Exercises 29–32, find the missing number that makes the fractions equivalent.*

29. $-\dfrac{6}{7} = -\dfrac{?}{14}$

30. $\dfrac{2}{5} = \dfrac{10}{?}$

31. $\dfrac{15}{30} = \dfrac{?}{10}$

32. $\dfrac{-5}{2} = \dfrac{?}{4}$

Exercises 33–40 EXPRESSIONS

[1.2] *For Exercises 33–36, write the prime factorization for each number.*

33. 51

34. 100

35. 108

36. 84

[1.2] *For Exercises 37–40, simplify to lowest terms.*

37. $\dfrac{152}{200}$

38. $\dfrac{61}{122}$

39. $\dfrac{250}{360}$

40. $\dfrac{26}{39}$

Exercises 41–46 EQUATIONS AND INQUALITIES

[1.3–1.4] *For Exercises 41–46, indicate whether the expression illustrates the commutative property of addition, the associative property of addition, the commutative property of multiplication, the associative property of multiplication, or the distributive property of multiplication over addition.*

41. $4 + (7 + 3) = (4 + 7) + 3$

42. $3(9 + 1) = 27 + 3$

43. $\dfrac{1}{3} \cdot \dfrac{5}{6} = \dfrac{5}{6} \cdot \dfrac{1}{3}$

44. $5.6 + (11.2 + 4.3) = 5.6 + (4.3 + 11.2)$

45. $(n - 8)(-3) = -3(n - 8)$

46. $(2 \cdot 3) \cdot 7 = 2 \cdot (3 \cdot 7)$

Exercises 47–92 EXPRESSIONS

[1.3–1.5] *For Exercises 47–68, perform the indicated operation(s).*

47. $-8 - (-4.2)$

48. $6 - (-9.1)$

49. $\dfrac{2}{5} + \dfrac{1}{3}$

50. $-\dfrac{1}{4} + \dfrac{2}{3}$

51. $-7(-13)$

52. $6(-8)$

53. $-25 \div 5$

54. $-30 \div (-3)$

55. -6^2

56. $(-3)^2$

57. $7(6 + 2) - 48 \div 12$

58. $6(9 - 4) \div 3 - 1$

59. $(-2)^3(4)(-5)$

60. $3^5 \div 3^2 \div 3^2 \div 3$

61. $-\dfrac{1}{3} - \dfrac{3}{2} \div \left(\dfrac{1}{6}\right)$

62. $-6.8 + (-4.1 + 2.3)$

63. $4|-2| \div (-8)$

64. $10 \div 5 + (-5)(-5)$

65. $6 + \{3(4 - 5) + 2[6 + (-4)]\}$

66. $-1 - 3[4^2 - 3(-6)]$

67. $\dfrac{-6 + 3^2(8 - 10)}{2^3 - 16}$

68. $\dfrac{3(4 - 5) - 4 \cdot 3 + (11 - 20)}{(3 - 4)^3}$

[1.5] *For Exercises 69 and 70, find all square roots of each number.*

69. 49

70. -36

[1.6] *For Exercises 71–74, translate to an algebraic expression.*

71. twice a number subtracted from fourteen

72. the ratio of y to seven

73. a number minus twice the sum of the number and four

74. seven times the difference of m and n

[1.7] *For Exercises 75–80, evaluate the expressions using the given values.*

75. $b^2 - 4ac$; $a = -3$, $b = -1$, $c = 5$

76. $b^2 - 4ac$; $a = 2$, $b = 6$, $c = -2$

77. $-|4x| + |y^2|$; $x = -3$, $y = -2$

78. $\sqrt{m} + \sqrt{n}$; $m = 16$, $n = 9$

79. $\sqrt{m + n}$; $m = 16$, $n = 9$

80. $\dfrac{3l^2}{4 - n}$; $l = -4$, $n = 16$

[1.7] *For Exercises 81 and 82, determine all value(s) that cause each expression to be undefined.*

81. $\dfrac{n}{n + 6}$

82. $\dfrac{y}{(y - 4)(y + 3)}$

[1.7] *For Exercises 83–86, use the distributive property to write an equivalent expression and simplify.*

83. $5(x + 6)$

84. $-3(5n - 8)$

85. $\dfrac{1}{4}(8y + 3)$

86. $0.6(4.5m - 2.1)$

[1.7] *For Exercises 87–92, simplify by combining like terms.*

87. $6x + 3y - 9x - 6y - 15$

88. $5y^2 - 6y + 4y - y^2 + 9y$

89. $-6xy + 9xy - xy$

90. $8x^3 - 4x - 6x^2 - 10x^3 + 4x$

91. $-4m - 4m - 4n + 4n$

92. $14 - 6x + 4y - 8 - 10y - 8x$

[1.3] **93.** The credit card statement to the right shows when each transaction was posted to the account. Find the new balance.

4-01-09	Beginning balance	−$685.92
4-03-09	Dillard's	−$45.80
4-05-09	Payment	$250.00
4-08-09	Applebee's	−$36.45
4-16-09	CVS Pharmacy	−$12.92
4-24-09	Target credit	$32.68
4-30-09	Finance charge	−$5.18

[1.3] **94.** The bar graph to the right shows the closing price for Merck and Co., Inc. stock each day for the week beginning February 4, 2008.

    a. Find the difference between the closing price at the end of the week and the closing price at the beginning of the week.

    b. If a person bought shares at the close of 2/5 and sold at the close of 2/7, what would that person's net profit or loss be on each share?

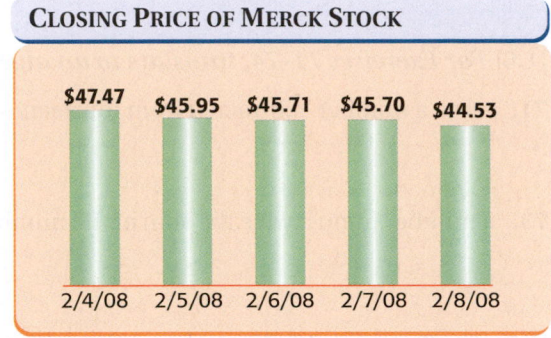

CLOSING PRICE OF MERCK STOCK

$47.47 $45.95 $45.71 $45.70 $44.53

2/4/08   2/5/08   2/6/08   2/7/08   2/8/08

[1.4] **95.** Refer to the graph in Exercise 94.

    a. If Kaye bought 200 shares at the close of 2/4, how much did she spend?

    b. If she sold all 200 shares on 2/8, what was her loss?

[1.4] **96.** The voltage in a circuit can be calculated by multiplying the current and the resistance. Suppose the current in a circuit is −9.5 amperes and the resistance is 16 ohms. Find the voltage.

[1.4] **97.** In 2000, Larry has a loan balance of −$2405.80. If he makes no payment for ten years, his balance will become −$14,896.08. How many times greater is his debt in 2010 than it was in 2000?

[1.5] **98.** A company reports the following revenues for the first six months of 2010. Find the average revenue.

Month	Revenue
January	$45,320
February	$38,250
March	$61,400
April	$42,500
May	$74,680
June	$62,800

Chapter 1 Practice Test

For Extra Help  CHAPTER Test Prep VIDEOS

Step-by-step test solutions are found on the Chapter Test Prep Videos available via the Video Resources on DVD, in *MyMathLab*, and on YouTube (search "CarsonCombinedAlg" and click on "Channels").

1. Graph $-3\frac{2}{7}$ on a number line.

2. Simplify: $|-3.67|$

3. Write the prime factorization of 100.

4. Simplify: $\dfrac{17}{51}$

5. Find the missing number: $-\dfrac{6}{5} = -\dfrac{?}{10}$

For Exercises 6 and 7, indicate whether the expression illustrates the commutative property of addition, the associative property of addition, the commutative property of multiplication, the associative property of multiplication, or the distributive property of multiplication over addition.

6. $3(2 + 5) = 6 + 15$

7. $4(7 + 1) = 4(1 + 7)$

For Exercises 8–13, calculate.

8. $8 + (-4)$

9. $\dfrac{7}{8} - \left(-\dfrac{5}{6}\right)$

10. $(-1.5)(-0.4)$

11. $-\dfrac{5}{6} \div \dfrac{2}{3}$

12. $(-4)^3$

13. -3^4

For Exercises 14–17, simplify.

14. $-12 \div 3\,(6 - 2^2)$

15. $\dfrac{6^2 + 14}{(2 - 7)^2}$

16. $-8 \div |4 - 2| + 3^2$

17. $\sqrt{25 - 16} + [(14 + 2) - 3^2]$

For Exercises 18 and 19, translate the indicated phrase.

18. twice the sum of m and n

19. five less than three times w

20. After making a payment, Karen's balance on her credit card is $-\$854.80$. If her balance prior to making the payment was $-\$1104.80$, how much did she pay toward the balance?

21. On six days over a three-month period, a city receives 6 inches, 10 inches, 4 inches, 3 inches, 2.5 inches, and 8 inches of snow. Calculate the average daily amount of snowfall over those days.

22. Evaluate $-|3x^2 + 2y|$, when $x = -1$ and $y = 4$.

23. Evaluate $\sqrt{a - b}$, when $a = 64$ and $b = -36$.

24. Use the distributive property to write an equivalent expression: $-5(4y + 9)$

25. Simplify: $3.5x - 8 + 2.1x + 9.3$

Solving Linear Equations and Inequalities

2

In Chapter 1, we reviewed numbers, arithmetic, and expressions, which form a foundation for studying equations. Remember from Chapter 1 that our most basic components are variables and constants, which make up expressions. We saw that we can perform two actions with expressions: (1) evaluate and (2) rewrite. In this chapter, we build on that foundation by creating equations and inequalities out of those expressions; then we *solve* those equations and inequalities. This development is repeated throughout the course. We study and classify new types of expressions, evaluate and rewrite those expressions, and solve equations and inequalities made from the new expressions.

Remember, as we build algebra concepts, they become more complex, just as a building becomes more complex during its construction. Try not to get discouraged if a new concept does not make sense right away. Learning new concepts is like constructing a building: During the early stages, the building is not as solid as it will be upon completion. This chapter is like the frame of the walls of a house: The frame is not as solid as the walls will be after the drywall and roof are put in place. Similarly, as we revisit equations and inequalities throughout the course, we reinforce and solidify the concepts you will learn in this chapter.

2.1 Equations, Formulas, and the Problem-Solving Process

Objectives

1. Verify solutions to equations.
2. Use formulas to solve problems.

We now move to the top tier of the Algebra Pyramid and focus on equations. An **equation** was first defined in Section 1.1. We repeat the definition, but state it a little differently.

Definition   *Equation:* Two expressions set equal.

For example, $4x + 5 = 9$ is an equation made from the expressions $4x + 5$ and 9.

Connection We can think of an equation as a complete sentence and the equal sign as the verb *is*. We can read $4x + 5 = 9$ as "Four *x* plus five is equal to nine" or "Four *x* plus five is nine."

$$4x + 5 = 9$$

"Four *x* plus five is 9."

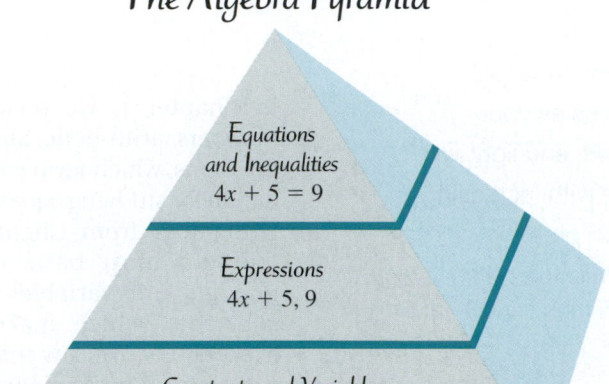

The Algebra Pyramid

Equations
and Inequalities
$4x + 5 = 9$

Expressions
$4x + 5, 9$

Constants and Variables
$4, 5, 9, x$

Objective 1   **Verify solutions to equations.** An equation can be true or false. For example, the equation $3 + 1 = 4$ is true, whereas the equation $3 + 1 = 5$ is false. When given an equation that contains a variable for a missing number, such as $x + 2 = 7$, our goal is to solve it, which means to find every number that can replace the variable and make the equation true. Such a number is called a **solution** to the equation.

Definition   *Solution:* A number that makes an equation true when it replaces the variable in the equation.

For example, the number 5 is a solution to the equation $x + 2 = 7$ because when 5 replaces *x*, it makes the equation true.

$$x + 2 = 7$$

$$\mathbf{5} + 2 = 7 \qquad \text{True}$$

By showing that the equation $x + 2 = 7$ is a true statement when *x* is replaced with 5, we verify, or *check*, that 5 is the solution to the equation.

Procedure   **Checking a Possible Solution**

To determine whether a value is a solution to a given equation, replace the variable in the equation with the value. If the resulting equation is true, the value is a solution.

Example 1 Check to see if the given value is a solution to the equation.

a. $4x + 9 = 1; x = -2$

SOLUTION $4x + 9 = 1$

$$4(-2) + 9 \overset{?}{=} 1 \qquad \text{Replace } x \text{ with } -2 \text{ and see if the equation is true.}$$
$$-8 + 9 \overset{?}{=} 1$$
$$1 = 1$$

Note: The symbol $\overset{?}{=}$ indicates that we are asking, "How does the left side compare with the right side? Are they equal or not equal?"

Because -2 makes the equation true, it is a solution to $4x + 9 = 1$.

b. $n^2 + 0.8 = 3n - 0.1; n = 0.4$

SOLUTION $n^2 + 0.8 = 3n - 0.1$

$$(0.4)^2 + 0.8 \overset{?}{=} 3(0.4) - 0.1 \qquad \text{Replace } n \text{ with 0.4 and see if the equation is true.}$$
$$0.16 + 0.8 \overset{?}{=} 1.2 - 0.1$$
$$0.96 \neq 1.1$$

Because 0.4 does not make the equation true, it is not a solution to $n^2 + 0.8 = 3n - 0.1$.

c. $\frac{3}{4}y - 1 = \frac{y}{2} + \frac{1}{3}; y = 5\frac{1}{3}$

LEARNING Strategy

AUDITORY TACTILE VISUAL As suggested in the To the Student section, set up your notebook with four sections: Notes, Homework, Study sheets, and Practice tests. Remember to use color in your notes: red for definitions and blue for rules and procedures. Organization and color coding will help you locate important concepts faster.

SOLUTION $\frac{3}{4}y - 1 = \frac{1}{2}y + \frac{1}{3}$

Note: It is convenient to write $\frac{y}{2}$ as $\frac{1}{2}y$, which we can do because $\frac{y}{2} = y \div 2 = y \cdot \frac{1}{2} = \frac{1}{2}y$.

$$\frac{3}{4}\left(5\frac{1}{3}\right) - 1 \overset{?}{=} \frac{1}{2}\left(5\frac{1}{3}\right) + \frac{1}{3} \qquad \text{Replace } y \text{ with } 5\frac{1}{3} \text{ and see if the equation is true.}$$

$$\frac{3}{4}\left(\frac{16}{3}\right) - 1 \overset{?}{=} \frac{1}{2}\left(\frac{16}{3}\right) + \frac{1}{3} \qquad \text{Rewrite } 5\frac{1}{3} \text{ as an improper fraction, divide out common factors, and then multiply.}$$

$$4 - 1 \overset{?}{=} \frac{8}{3} + \frac{1}{3}$$

$$3 = 3 \qquad \text{Therefore, } 5\frac{1}{3} \text{ is a solution to } \frac{3}{4}y - 1 = \frac{y}{2} + \frac{1}{3}.$$

Your Turn 1 Check to see if the given value is a solution to the equation.

a. $5n - 8 = 3n + 4; n = 6$ **b.** $4y - 3.5 = 2y + 3(y + 0.2); y = -1.5$

c. $\frac{2}{3}x + \frac{1}{4} = \frac{1}{5} - x^2; x = \frac{1}{6}$

Objective ❷ Use formulas to solve problems. A primary purpose of studying mathematics is to develop and improve problem-solving skills. George Polya proposed the idea that all problem solving follows a four-step outline: (1) Understand the problem, (2) devise a plan for solving the problem, (3) execute the plan, and (4) check the results. Following is an outline for problem solving based on Polya's four stages. We'll see Polya's four stages illustrated throughout the rest of the text in application problems.

Answers   Your Turn 1
a. yes **b.** no **c.** no

Problem-Solving Outline

1. **Understand** the problem.
   a. Read the question(s) (not the whole problem, just the question at the end) and write a note to yourself about what you are to find.
   b. Read the whole problem, underlining the key words.
   c. If possible or useful, draw a picture, make a list or table to organize what is known and unknown, simulate the situation, or search for a related example problem.
2. **Plan** your solution by searching for a formula or using the key words to translate to an equation.
3. **Execute** the plan by solving the equation/formula.
4. **Answer** the question. Look at the note about what you were to find and make sure you answer that question. Include appropriate units.
5. **Check** results.
   a. Try finding the solution in a different way, reversing the process, or estimating the answer and making sure the estimate and actual answer are reasonably close.
   b. Make sure the answer is reasonable.

As the course develops, we will explore the various strategies listed in the outline. In this section, we focus on using **formulas** to solve problems. Later in this chapter, we solve problems by using key words to translate to an equation.

Definition *Formula:* An equation that describes a mathematical relationship.

First, let's consider using formulas from geometry. The table on page 95 lists some common geometric formulas. Before we examine the formulas, let's review a few terms.

Definitions *Perimeter:* The distance around a figure.
Area: The total number of square units that fill a figure.
Volume: The total number of cubic units that fill a space.
Circumference: The distance around a circle.
Radius: The distance from the center of a circle to any point on the circle.
Diameter: The distance across a circle through its center.

Procedure **Using a Formula**

To use a formula,

1. Replace the variables with the corresponding given values.
2. Solve for the missing value.

Geometric Formulas

Plane Figures

Square

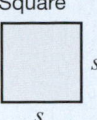

$P = 4s$
$A = s^2$

Rectangle

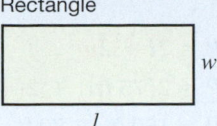

$P = 2l + 2w$
$A = lw$

Parallelogram

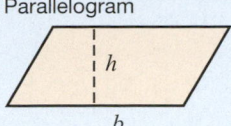

$A = bh$

Trapezoid

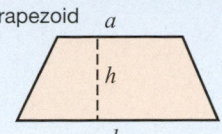

$A = \frac{1}{2}h(a + b)$

Note: A parallelogram is a four-sided figure with two pairs of parallel sides. A trapezoid has only one pair of parallel sides.

Triangle

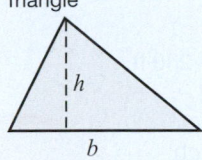

$A = \frac{1}{2}bh$

Circle

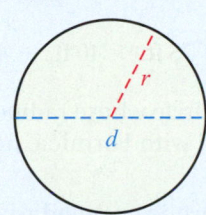

$C = \pi d$ or $C = 2\pi r$
$A = \pi r^2$

Note: A circle's radius is half its diameter. (Or its diameter is twice the radius.)

Note: Recall that π represents an irrational number. The π key on most calculators approximates the value as 3.141592654. Some people prefer to round the value to 3.14, which is simpler but less accurate.

Solids

Box

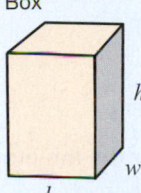

$V = lwh$
$SA = 2lw + 2lh + 2wh$

Pyramid

$V = \frac{1}{3}lwh$

Cylinder

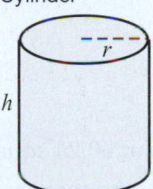

$V = \pi r^2 h$

Cone

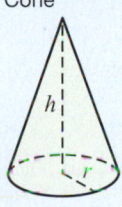

$V = \frac{1}{3}\pi r^2 h$

Sphere

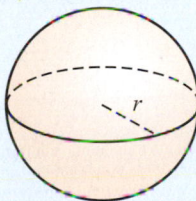

$V = \frac{4}{3}\pi r^3$

Connection Notice that the volume formulas for a box and a pyramid are similar. This is because a pyramid has $\frac{1}{3}$ the volume of a box with the same-size base and same height. The same relationship is true for cylinders and cones with the same-size base and same height.

Problems Involving a Single Formula

Example 2   **a.** A mason is building a rectangular foundation wall that is to be 75 feet by 40 feet. What is the total distance around the wall?

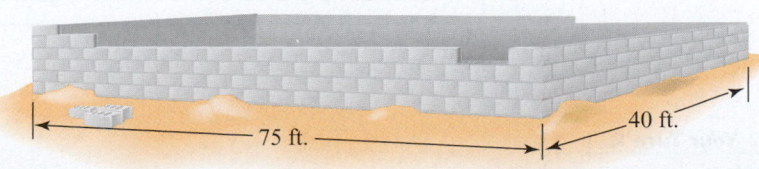

Understand   The "total distance around the wall" is the perimeter.

Plan   Because the shape is a rectangle, we can use the formula $P = 2l + 2w$.

Execute   Replace *l* with **75** feet and *w* with **40** feet; then calculate.

$$P = 2\boldsymbol{l} + 2\boldsymbol{w}$$
$$P = 2(\textbf{75 ft.}) + 2(\textbf{40 ft.})$$
$$P = 150 \text{ ft.} + 80 \text{ ft.}$$
$$P = 230 \text{ ft.}$$

Answer   The mason must have enough supplies to build a wall that is 230 feet long.

Check   In this case, the solution can be verified using an alternative method. One could add all four side lengths.

$$P = 75 \text{ ft.} + 75 \text{ ft.} + 40 \text{ ft.} + 40 \text{ ft.} = 230 \text{ ft.}$$

b. A tabletop is in the shape of a circle whose radius is 4 feet. If the tabletop is to be covered with Formica, how much Formica is needed?

4 ft.

Understand   The amount of Formica needed to cover the tabletop is the same as the area of the tabletop.

Plan   Because the tabletop is in the shape of a circle, we use the formula $A = \pi r^2$.

Execute   Replace π with **3.14** and *r* with **4 ft.** and simplify.

$$A = \boldsymbol{\pi r^2}$$
$$A = (\textbf{3.14})(\textbf{4 ft.})^2$$
$$A = (3.14)(16 \text{ ft.}^2)$$
$$A = 50.24 \text{ ft.}^2$$

Answer   About 50.24 square feet of Formica is needed to cover the tabletop.

Check   Verify the reasonableness of the answer using estimation. If we round π to 3, the answer is $(3)(4 \text{ ft.})(4 \text{ ft.}) = 48 \text{ ft.}^2$. Because π is a little more than 3 and our answer is a little more than 48 ft.^{2}, our answer is reasonable.

Your Turn 2   A wallpaper hanger is to paper a bedroom wall that is 18 feet long and 12 feet high. How much wallpaper is needed to cover the wall?

Problems Involving Combinations of Formulas

A *composite figure* contains more than one geometric figure and requires a combination of formulas.

Procedure   **Calculating the Area of Composite Figures**

For composite figures:

1. To calculate the area of a figure composed of two or more figures that are next to each other, add the areas of the individual figures.
2. To calculate the area of a region defined by a smaller figure within a larger figure, subtract the area of the smaller figure from the area of the larger figure.

Answer   **Your Turn 2**

216 square feet

Example 3   Following is a drawing of a room that is to be carpeted. Calculate the area of the room.

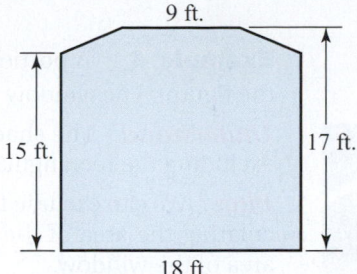

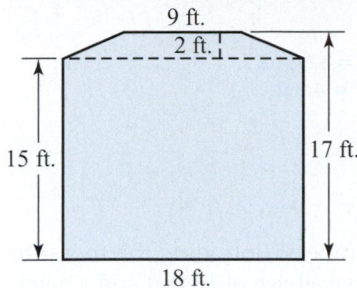

Understand   This figure can be viewed as a trapezoid and a rectangle. The height of the trapezoid is found by subtracting 15 feet from 17 feet.

Plan   Because the figure consists of a trapezoid and rectangle combined, we can add the areas of these shapes to find the total area of the figure. The area of a rectangle is found using the formula $A = lw$, and the area of a trapezoid is found using the formula $A = \frac{1}{2}h(a + b)$. The total area is

$$A = \text{Area of the rectangle} + \text{Area of the trapezoid}$$

$$A = \qquad lw \qquad + \qquad \frac{1}{2}h(a + b)$$

Execute   Replace the variables with the corresponding values and calculate.

$$A = lw + \frac{1}{2}h(a + b)$$

$$A = (15\,\text{ft.})(18\,\text{ft.}) + \frac{1}{2}(2\,\text{ft.})(18\,\text{ft.} + 9\,\text{ft.})$$

$l = 15$ ft., $w = 18$ ft., $h = 2$ ft., $a = 18$ ft., and $b = 9$ ft.

$$A = (15\,\text{ft.})(18\,\text{ft.}) + \frac{1}{2}(2\,\text{ft.})(27\,\text{ft.})$$

$$A = 270\,\text{ft.}^2 + 27\,\text{ft.}^2$$

$$A = 297\,\text{ft.}^2$$

Answer   The total area is 297 square feet.

Check   We can verify the reasonableness of the answer using estimation. Suppose the figure had been a rectangle measuring 18 feet by 17 feet. We would expect the area of this rectangle to be slightly greater than the area of the actual figure. The area of an 18-foot by 17-foot rectangle is $A = (18\,\text{ft.})(17\,\text{ft.}) = 306\,\text{ft.}^2$, which indicates that 297 square feet is reasonable.

Your Turn 3   Calculate the area of the following sign. (Use $\pi \approx 3.14$.)

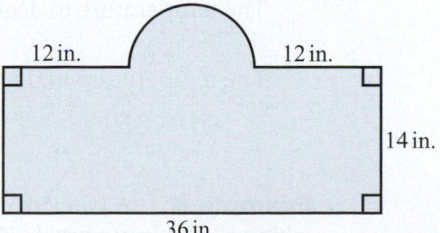

Answer   Your Turn 3

560.52 square inches

Sometimes, a figure may be placed within another figure and we must calculate the area of the region *between* the outside figure and inside figure.

Example 4   A portion of the front of a house is to be covered with siding (shaded in the figure). The window is 2.8 feet by 4.8 feet. Calculate the area to be covered.

Understand   The shaded area is the entire triangle excluding the rectangular window.

Plan   We can exclude the area of the window by calculating the area of the triangle and subtracting the area of the window.

$$A = \text{Area of the triangle} - \text{Area of the rectangle}$$

$$A = \frac{1}{2}bh \quad - \quad lw$$

Execute $A = \dfrac{1}{2}(12.5 \text{ ft.})(8 \text{ ft.}) - (2.8 \text{ ft.})(4.8 \text{ ft.})$ **$b = 12.5$ ft., $h = 8$ ft., $l = 2.8$ ft., $w = 4.8$ ft.**

$$A = 50 \text{ ft.}^2 - 13.44 \text{ ft.}^2$$

$$A = 36.56 \text{ ft.}^2$$

Answer   The area to be covered is 36.56 square feet.

Check   Estimate the area by rounding the decimal numbers. Rounding the measurements of the triangle to the nearest whole number, we have a base of 13 feet and a height of 8 feet, which means an area of 52 square feet. Rounding the window dimensions to the nearest whole number, we have a 3-foot by 5-foot window, which has an area of 15 square feet. Subtracting these areas, we have a final area of 37 square feet, which indicates that our answer of 36.56 square feet is reasonable. Because all of our estimates are greater than or equal to the actual amount, the estimated answer is larger than the actual answer.

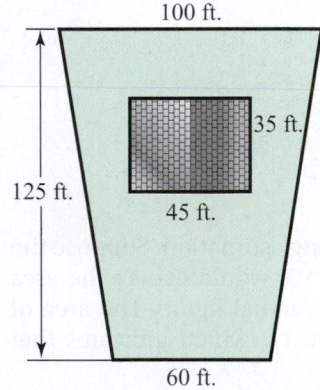

Your Turn 4   A house is to be situated on a lot as shown at the left. Once the house is built, the remaining area will be landscaped. Calculate the area to be landscaped.

Problems Involving Nongeometric Formulas

Following is a list of other formulas you may encounter.

The distance, d, an object travels given its rate, r, and the time of travel, t:	$d = rt$
The average rate of travel, r, given the total distance, d, and total time, t:	$r = \dfrac{d}{t}$
The voltage, V, in a circuit with a current, i, in amperes (A), and a resistance, R, in ohms (Ω):	$V = iR$
The temperature in degrees Celsius given degrees Fahrenheit:	$C = \dfrac{5}{9}(F - 32)$
The temperature in degrees Fahrenheit given degrees Celsius:	$F = \dfrac{9}{5}C + 32$

Example 5   A truck driver begins a delivery at 9 A.M. and travels 150 miles before taking a 30-minute break. He then travels another 128 miles, arriving at his destination at 2 P.M. What was his average driving rate?

Understand   We are given travel distances and times, and we are to find the average driving rate.

Answer   Your Turn 4

8425 ft.^{2}

Plan To find the average rate, we first need the total distance traveled and the total time spent driving. We can then use the formula $r = \dfrac{d}{t}$.

Execute Total distance $= 150 + 128 = 278$ miles $= d$

Total time driving $= 5 - 0.5 = 4.5$ hours $= t$

Now we can calculate the average rate.

$$r = \frac{278 \text{ miles}}{4.5 \text{ hours}}$$

$$r = 61.\overline{7} \text{ mph (miles per hour)}$$

> **Note:** From 9 A.M. to 2 P.M. is 5 hours, but we must deduct the 0.5-hour (30-minute) break because he was not traveling during that time.

Answer His average driving rate was $61.\overline{7}$ miles per hour.

Check We can use $d = rt$ to verify that if he traveled an average rate of $61.\overline{7}$ miles per hour for 4.5 hours, he would travel 278 miles.

$$d \approx 61.8(4.5)$$

$$d \approx 278.1 \text{ miles}$$

> **Note:** We rounded $61.\overline{7}$, so the distance is an approximation.

Your Turn 5 Hilda hikes 10 miles then rests for 20 minutes, after which she hikes another 8 miles. If she began her hike at 10:30 A.M. and stopped at 4:30 P.M., what was her average rate?

Answer **Your Turn 5**

$3\dfrac{3}{17}$ or ≈ 3.2 mph

2.1 Exercises For Extra Help MyMathLab

Note: Exercises marked with a ★ represent challenging exercises.

1. What symbol is present in an equation but not in an expression?

2. What is a solution to an equation?

3. Explain how to check a value to see if it is a solution to a given equation.

4. If the dimensions of a figure are given in inches, in what unit will the area be expressed?

For Exercises 5–20, check to see if the given number is a solution to the given equation. See Example 1.

5. $3x - 5 = 41; x = 2$

6. $4a + 7 = 51; a = 11$

7. $7y - 1 = y + 3; y = \dfrac{2}{3}$

8. $-8t - 3 = 2t - 15; t = -\dfrac{6}{5}$

9. $-4(x - 5) + 2 = 5(x - 1) + 3; x = -2$

10. $2(3m + 2) - 2 = 5m - 1; m = -3$

11. $-\dfrac{1}{2} + \dfrac{1}{3}y = \dfrac{3}{2}; y = 6$

12. $\dfrac{1}{2}p - \dfrac{1}{2} = \dfrac{2}{5}p + \dfrac{3}{2}; p = 20$

13. $4.3z - 5.71 = 2.1z + 0.07; z = 2.2$

14. $12.7a + 12.6 = a + 5.4a; a = -2$

★15. $b^2 - 4b = b^3 + 6; b = -1$

★16. $-x^3 + 9 = 2x^2 - 6x; x = -3$

★17. $|x^2 - 9| = -x + 3; x = 3$

★18. $-|2u - 3| = -3u + 8; u = 5$

★19. $\dfrac{4x - 3}{x - 4} = \sqrt{x + 8}; x = 17$

★20. $\dfrac{-y}{10 + y} = \dfrac{\sqrt{4 - y}}{3}; y = -5$

For Exercises 21–44, solve using geometric formulas. (Answers to exercises involving π were calculated using the π key on a calculator, which approximates the value as 3.141592654.) See Examples 2–4.

21. Janet wants to put a wallpaper border around her rectangular kitchen. The room is 13 feet by 20 feet.

   a. What is the total length of wallpaper border she needs?

   b. If the wallpaper border comes in packages of 12 feet, how many packages must she buy?

   c. If the packages are priced at $8.99 each, what will be the total cost of the wallpaper?

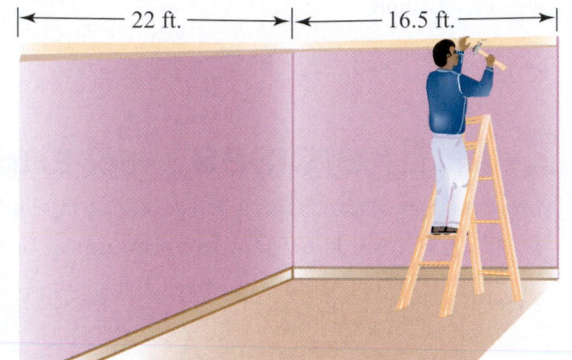

22. Jamal is planning to install crown molding where the walls and ceiling of his rectangular living room meet. The room is 16.5 feet by 22 feet.

   a. What is the total length of crown molding he needs?

   b. If the crown molding comes in strips of 8 feet, how many strips must he purchase?

   c. If the strips cost $9.99 each, what will be the cost of the crown molding?

23. Fermi National Accelerator Laboratory is a circular tunnel that is used to accelerate elementary particles. The radius of the tunnel is about 2 kilometers. If a particle travels one complete revolution around the circumference of the tunnel, what distance does it travel to the nearest hundredth of a kilometer? (*Source:* Cesare Emiliani, *The Scientific Companion*, 2nd ed., Wiley, 1995.)

24. The Chicxulub crater, which is buried partly beneath the Yucatan peninsula and partly beneath the Gulf of Mexico, is circular and is about 180 kilometers in diameter. What is the circumference of this crater to the nearest hundredth of a kilometer?

25. At one time, the Venetian Hotel and Casino in Las Vegas boasted having the largest hotel rooms in the world. If one of the rectangular rooms is 70 feet by 100 feet, what is the total square footage in the room?

26. The world's largest oil production barge was built by Kvaernel Oil and Gas, Inc. of Norway and Single Bouy Moorings, Inc. of Switzerland. The rectangular deck of the barge is 273 meters long and 50 meters wide. What is the area of the deck?

Of Interest

The Chicxulub crater was created when Earth was hit by an asteroid about 65 million years ago. The devastation of the impact is thought to have caused the extinction of about 50 percent of the species on Earth, including the dinosaurs.

27. A large wall in a home is to be painted with a custom accent color. The rectangular wall measures 32.5 feet by 12 feet. A painter charges $2.50 per square foot to paint a wall with a custom color. How much will it cost to have the painter paint the wall?

28. Hardwood floors are to be installed in a rectangular bedroom that measures 14 feet by 15 feet. If the contractor charges $34.50 per square foot to install the floors, what will be the cost of this improvement?

29. Tina needs to buy enough pine straw to cover a rectangular flower bed that measures 6 feet by 13 feet.

   a. What is the total area that she must cover?

   b. A landscape consultant tells Tina that she should use one bale of pine straw for every 10 square feet to be covered. If Tina follows this advice, how many bales must she purchase?

   c. If Tina's local garden center charges $4.50 per bale, what will be the cost of covering her flower bed?

30. Juan is considering installing carpet in his new office space. The office is rectangular and measures 42 feet by 36 feet. Juan has a budget of $3000.

   a. What is the total area that will be carpeted?

   b. How many square yards is this area? (There are 9 square feet in 1 square yard.)

   c. If the contractor quotes Juan a price of $22.50 per square yard to install the carpet, how much will it cost to carpet the space? Is this feasible with Juan's budget?

31. A city planning committee decides to install 20 historic markers around the city. The signs are triangular, measuring 26 inches wide at the base and 20 inches high.

   a. What is the area of each sign?

   b. If a contractor charges $0.75 per square inch to build and install the signs, how much will it cost to have them installed?

32. A parks department is going to install a flower bed in a triangular space created by three trails, as shown. The flower bed measures 32 feet at the base and is 24 feet high.

   a. What is the area of the flower bed?

   b. If a landscaper charges $6.50 per square foot to design and install the bed, what will the total cost be?

33. Conner and Ellyn purchase a new home. The area around an 8-foot by 9-foot storage building in the rectangular backyard is to be landscaped, as shown in the figure.

   a. What is the total area that is to be landscaped?

   b. If $\frac{2}{3}$ of this area is to be grass, what amount will be covered with grass?

   c. If each pallet of grass sod covers 504 square feet, how many pallets will be needed?

   d. Each pallet cost $106. What is the total cost of the sod?

34. A kitchen floor is to be recovered with granite tiles. There is a fixed island in the center of the floor, as shown in the figure.

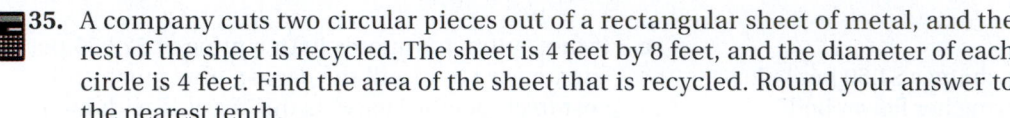

   a. What is the total area to be tiled?

   b. If each piece of tile covers $\frac{1}{4}$ of a square foot, how many pieces of tile will be needed?

   c. If each tile costs $3.95, what is the cost of the tile?

   d. If the contractor charges $8 per square foot to install the new floor, what will be the installation cost?

35. A company cuts two circular pieces out of a rectangular sheet of metal, and the rest of the sheet is recycled. The sheet is 4 feet by 8 feet, and the diameter of each circle is 4 feet. Find the area of the sheet that is recycled. Round your answer to the nearest tenth.

36. A compact disc has a diameter of $5\frac{3}{4}$ inches. On a CD, no information is placed within a circle in the center of the CD with a diameter of $1\frac{3}{4}$ inches. Find the area on a CD that can contain information. Round your answer to the nearest tenth.

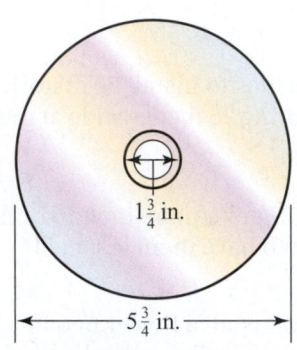

★37. The wall shown is to be painted. The door, which will not be painted, is 2.5 feet by 7.5 feet. What area will be painted?

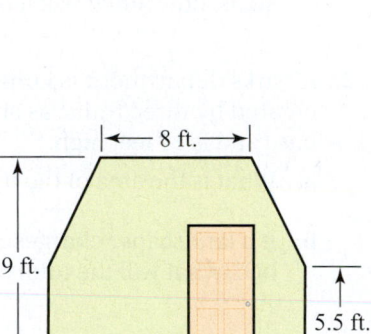

★38. The side of the house in the figure needs new siding. The window measures 3 feet by 4.5 feet. Find the area to be covered in siding.

39. A small storage building is shaped like a box that measures 2 meters by 5 meters by 4.75 meters. What is the storage building's volume?

40. A dorm refrigerator measures 2 feet by $1\frac{1}{2}$ feet by 4 feet. What is its volume?

41. A cylindrical bottle has a diameter of 8 centimeters and a height of 20 centimeters. Find the volume of the bottle. Round your answer to the nearest tenth.

42. Earth is roughly a sphere with a radius of approximately 6370 kilometers. What is the approximate volume of Earth?

43. The base of the Great Pyramid at Giza, Egypt, is 745 feet by 745 feet, and its height is 449 feet. Find the volume of the pyramid.

44. If a funnel is approximately a cone, approximate the volume of a funnel with a diameter of $8\frac{1}{2}$ inches and a height of 6 inches. Round your answer to the nearest tenth.

For Exercises 45–50, use the formulas relating distance, rate, and time
$(d = rt, r = \frac{d}{t})$. *See Example 5.*

45. A family began a trip of 516 miles at 8 A.M. They arrived at their final destination at 5:30 P.M. If they took two 15-minute breaks and an hour for lunch, what was their average driving rate?

46. A family began a trip at 7:30 A.M. At the beginning of the trip, the car's odometer read 45,362.6. When they arrived at their final destination at 6 P.M., the odometer read 45,785.2. If they took three 15-minute breaks and an hour for lunch, what was their average driving rate?

47. A truck driver is to make a delivery that requires him to drive 285 miles by 4 P.M. If he starts at 10 A.M. and makes three 15-minute stops at weigh stations along the route, what must his average driving rate be to make the delivery on time?

★48. In 2005, Lance Armstrong won the Tour de France, which covered a distance of 3606 kilometers, with a time of 82 hours, 34 minutes, and 5 seconds. What was his average rate in kilometers per hour for the race? (*Source: The World Almanac and Book of Facts 2005.*)

★49. A flight departs Atlanta at 8:30 A.M. EST to arrive in Las Vegas at 10:00 A.M. PST. If the plane flies at an average rate of $388\frac{1}{3}$ mph, what distance does it travel? (*Hint:* There is a three-hour time difference between EST and PST.)

★50. A flight departs at 7:30 A.M. EST from Philadelphia and travels to Dallas–Fort Worth, arriving at 10:10 A.M. CST. If the plane is averaging a speed of 368.2 miles per hour, what distance does it travel? (*Hint:* There is a one-hour time difference between EST and CST. Also express the time using fractions.)

For Exercises 51 and 52, use the formula relating voltage, current, and resistance
(V = iR). See Example 5.

51. A technician measures the current in a circuit to be −6.5 amperes, and the resistance is 8 ohms. Find the voltage.

52. The current in a circuit is measured to be 4.2 amperes. If the resistance is 16 ohms, find the voltage.

For Exercises 53–58, use the formulas for converting degrees Fahrenheit to degrees Celsius or degrees Celsius to degrees Fahrenheit $\left(C = \dfrac{5}{9}(F - 32), F = \dfrac{9}{5}C + 32\right).$
See Example 5.

53. On a drive to work, Tim notices that the temperature reading on a bank's digital thermometer is 34°C. What is this temperature in degrees Fahrenheit?

54. Iron melts at a temperature of 1535°C. What is this temperature in degrees Fahrenheit?

55. Liquid oxygen boils at a temperature of −183°C. What is this temperature in degrees Fahrenheit?

56. The average temperature on Neptune is approximately −360°F. What is this temperature in degrees Celsius?

57. At the South Pole, temperatures can get down to −76°F. What is this temperature in degrees Celsius?

58. The average temperature on Venus is approximately 890°F. What is this temperature in degrees Celsius?

Puzzle Problem

A contest has people guess how many marbles are inside a jar. The jar is a cylinder with a diameter of 9 inches and a height of 12 inches. Suppose each marble has a diameter of 0.5 inch. Calculate the number of marbles that might fit inside the jar. Discuss the accuracy of your calculation.

Review Exercises

Exercises 1–6 EXPRESSIONS

[1.5] *For Exercises 1 and 2, simplify.*

1. $6(-3 - 3) + 5(-3)$

2. $6(2 + 3^2) - 4(2 + 3)^2$

[1.7] *For Exercises 3 and 4, multiply using the distributive property.*

3. $\dfrac{1}{3}(x + 6)$

4. $-(4w - 6)$

[1.7] *For Exercises 5 and 6, combine like terms.*

5. $5x - 14 + 3x + 9$

6. $6.2y + 7 - 1.5 - 0.8y$

2.2 The Addition Principle of Equality

Objectives

1. Determine whether a given equation is linear.
2. Solve linear equations in one variable using the addition principle.
3. Solve equations with variables on both sides of the equal sign.
4. Solve identities and contradictions.
5. Solve application problems.

Objective 1 **Determine whether a given equation is linear.** There are many types of equations. In this section, we introduce and begin solving **linear equations**.

Definition *Linear equation:* An equation in which each variable term contains a single variable raised to an exponent of 1.

Equations that are not linear are *nonlinear equations*.

Connection Every equation has a corresponding graph. The names *linear* and *nonlinear* refer to the graphs of the equations that they describe. The graph of a linear equation is a line; the graph of a nonlinear equation is not.

Example 1 Determine whether the equation is linear or nonlinear.

a. $3x + 5 = 12$

Answer This equation is linear because the variable x has an exponent of 1.

b. $7x^3 + x = 1$

Answer This equation is nonlinear because the variable in the term $7x^3$ has an exponent of 3.

c. $3x + 2y = -10$

Answer This equation is linear because the variables x and y have exponents of 1.

Your Turn 1 Determine whether the given equation is linear.

a. $3x - 4 = 11$ **b.** $-6x + y = 30$ **c.** $n^2 = 5m + 4$ **d.** $y = 3x^2 + 1$

In this chapter, we focus on equations with the same variable throughout the equation. If a linear equation has the same variable throughout, we say it is a **linear equation in one variable**.

Definition *Linear equation in one variable:* An equation that can be written in the form $ax + b = c$, where a, b, and c are real numbers and $a \neq 0$.

$5x + 9 = -6$ is a linear equation in one variable.
$x^2 - 2x + 5 = 8$ is a nonlinear equation in one variable.

In Chapter 4, we consider linear equations in two variables, such as $x + y = 3$.

Objective 2 **Solve linear equations in one variable using the addition principle.** In Section 2.1, we discussed solutions to equations and how to check an equation's solution. We now consider a technique for solving an equation (which means to find its solution(s)). This technique is called the *balance technique*. Imagine an equation as scales with the equal sign as the fulcrum.

Answers **Your Turn 1**
a. yes **b.** yes **c.** no **d.** no

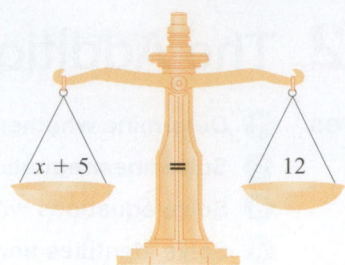

When weight is added or removed on one side of a balanced set of scales, the scales tip out of balance. In this case, we add 4 to the left side.

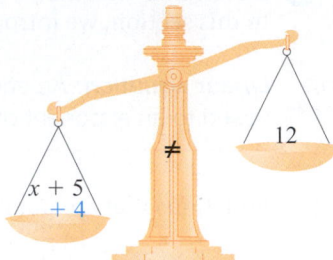

To maintain balance, the same weight must be added or removed on *both* sides of the scale. Adding 4 to the right side balances the scale.

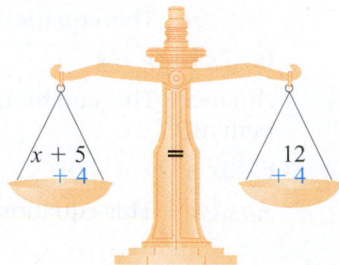

In the language of mathematics, this balance technique is stated in two principles: the *addition principle of equality* and the *multiplication principle of equality*. In this section, we consider the addition principle of equality. In Section 2.3, we consider the multiplication principle of equality. The addition principle of equality says that we can add the same amount to (or subtract the same amount from) both sides of an equation without affecting its solution(s).

Rule **The Addition Principle of Equality**

Note: We say that $a = b$ and $a + c = b + c$ are **equivalent equations** because they have the same solution.

If $a = b$, then $a + c = b + c$ is true for all real numbers a, b, and c.

In solving an equation, the goal is to write an equivalent equation in a simpler form, $x = c$, where the variable is alone, or isolated, on one side of the equal sign and the solution appears on the other side. For example, to isolate x in $x + 5 = 12$, we need to clear $+5$ from the left-hand side. To clear $+5$, we add -5 (or subtract 5) because $+5$ and -5 are additive inverses, which means that their sum is 0.

Note: To isolate x, we clear $+5$ by adding -5 to (or subtracting 5 from) the left side.

$$
\begin{aligned}
x + 5 &= 12 \\
-5 \quad &\;\; -5 \\
\hline
x + 0 &= 7 \\
x &= 7
\end{aligned}
$$

Note: To keep the equation balanced, we add -5 to (or subtract 5 from) the right side as well.

The statement $x = 7$ is the solution statement.

Procedure   **Using the Addition Principle of Equality**

To use the addition principle of equality to clear a term in an equation, add the additive inverse of that term to both sides of the equation. (That is, add or subtract appropriately so that the term you want to clear becomes 0.)

Example 2   Solve and check.

a. $x - 13 = -22$

SOLUTION   To isolate x, we need to clear -13.

Note: Add 13 to the left-hand side so that $-13 + 13 = 0$.

$$
\begin{aligned}
x - 13 &= -22 \\
+13 \quad\quad &+13 \\
x + 0 &= -9 \\
x &= -9
\end{aligned}
$$

Note: Because we added 13 to the left side, we *must* add 13 to the right side as well.

Check   Recall from Section 2.1 that to check, we replace x in the original equation with -9 and verify that the equation is true.

$$
\begin{aligned}
x - 13 &= -22 \\
-9 - 13 &\stackrel{?}{=} -22 \quad\quad \text{Replace } x \text{ with } -9. \\
-22 &= -22
\end{aligned}
$$

True; so -9 is the solution.

b. $\dfrac{3}{5} = y + \dfrac{1}{3}$

Note: Here, we illustrate another popular style of writing the addition principle, in which $-\dfrac{1}{3}$ is written beside the expressions on both sides of the equation instead of underneath the corresponding terms.

SOLUTION

$$\frac{3}{5} - \frac{1}{3} = y + \frac{1}{3} - \frac{1}{3} \quad\quad \text{Subtract } \tfrac{1}{3} \text{ from both sides to isolate } y.$$

$$\frac{9}{15} - \frac{5}{15} = y + 0 \quad\quad \text{Rewrite fractions with their LCD.}$$

$$\frac{4}{15} = y \quad\quad \text{Subtract } \tfrac{5}{15} \text{ from } \tfrac{9}{15}.$$

Check

$$\frac{3}{5} = y + \frac{1}{3}$$

$$\frac{3}{5} \stackrel{?}{=} \frac{4}{15} + \frac{1}{3} \quad\quad \text{Replace } y \text{ in the original equation with } \tfrac{4}{15} \text{ and verify that the equation is true.}$$

$$\frac{3}{5} \stackrel{?}{=} \frac{4}{15} + \frac{5}{15} \quad\quad \text{Rewrite fractions with their LCD.}$$

$$\frac{3}{5} \stackrel{?}{=} \frac{9}{15} \quad\quad \text{Add } \tfrac{4}{15} \text{ and } \tfrac{5}{15}.$$

$$\frac{3}{5} = \frac{3}{5} \quad\quad \text{Simplify to lowest terms.}$$

True; so $\dfrac{4}{15}$ is the solution.

LEARNING Strategy

VISUAL   If you are a visual learner, to become comfortable with the addition principle, consider writing the amount added or subtracted on both sides of an equation in a different color than the rest of the equation. Once you are comfortable seeing the amount added or subtracted, you may not need to use the color any longer.

Your Turn 2   Solve and check.

a. $4.7 = x - 9.8$

b. $n + \dfrac{3}{4} = \dfrac{1}{6}$

Simplifying before Isolating the Variable

Some equations have expressions that can be simplified. If like terms are on the same side of the equation, we combine the like terms before isolating the variable. If the equation to be solved contains parentheses, we use the distributive property to clear the parentheses before isolating the variable.

Example 3   Solve and check.

a. $8y - 2.1 - 7y = -3.7 + 9$

SOLUTION   Simplify the expressions; then isolate y.

$$8y - 2.1 - 7y = -3.7 + 9$$

Combine $8y$ and $-7y$ to equal y. $y - 2.1 = 5.3$ Combine -3.7 and 9 to equal 5.3.

$$\underline{+2.1 \quad +2.1}$$ Add 2.1 to both sides to isolate y.

$$y + 0 = 7.4$$

$$y = 7.4$$

Check   Replace y in the original equation with 7.4 and verify that the equation is true. We will leave this to the reader.

b. $6(n - 3) - 5n = -25 + 4$

SOLUTION   Use the distributive property, simplify, and then isolate n.

$$6(n - 3) - 5n = -25 + 4$$

Distribute 6 to clear parentheses. $6n - 18 - 5n = -21$ Combine -25 and 4 to equal -21.

Combine $6n$ and $-5n$ to equal n. $n - 18 = -21$

$$\underline{+18 \quad +18}$$ Add 18 to both sides to isolate n.

$$n + 0 = -3$$

$$n = -3$$

Check   Replace n with -3 in the original equation and verify that the equation is true. We will leave this to the reader.

Your Turn 3   Solve and check.

a. $9x + 4 - 8x = 1 - 6$

b. $6.2 - 0.4 = 1.5(m - 6) - 0.5m$

Objective ③ **Solve equations with variables on both sides of the equal sign.**   Some equations have variable terms on both sides of the equal sign. To isolate the variable, we must first get the variable terms together on the same side of the equal sign. To do this, we use the addition principle.

Answers   Your Turn 2

a. 14.5   **b.** $-\dfrac{7}{12}$

Answers   Your Turn 3

a. -9 **b.** 14.8

Example 4   Solve and check. $5x - 8 = 4x - 13$

SOLUTION   Use the addition principle to get the variable terms together on the same side of the equal sign. Then isolate the variable.

Note: Remember that x means $1 \cdot x$. It does *not* mean that $x = 1$.

$$
\begin{array}{rcl}
5x - 8 & = & 4x - 13 \\
\underline{-4x} & & \underline{-4x} \quad \text{Subtract } 4x \text{ from both sides.} \\
x - 8 & = & 0 - 13 \\
x - 8 & = & -13 \\
\underline{+8} & & \underline{+8} \quad \text{Add 8 to both sides to isolate } x. \\
x + 0 & = & -5 \\
x & = & -5
\end{array}
$$

Note: By choosing to subtract $4x$ on both sides, we clear the $4x$ term from the right side and combine it with the $5x$ on the left side. It doesn't matter which term you move first. (See the explanation following the example.)

Check $5x - 8 = 4x - 13$

$$
\begin{array}{rcl}
5(-5) - 8 & \overset{?}{=} & 4(-5) - 13 \\
-25 - 8 & \overset{?}{=} & -20 - 13 \\
-33 & = & -33
\end{array}
$$

Replace x in the original equation with -5 and verify that the equation is true.

True; so -5 is the solution.

When variable terms appear on both sides of the equal sign, it doesn't matter which term you choose to clear first, although some choices make the process easier. Suppose in Example 4 that we clear the $5x$ term first.

$$
\begin{array}{rcl}
5x - 8 & = & 4x - 13 \\
\underline{-5x} & & \underline{-5x} \\
0 - 8 & = & -x - 13 \\
-8 & = & -x - 13 \\
\underline{+13} & & \underline{+13} \\
5 & = & -x + 0 \\
5 & = & -x \\
-5 & = & x
\end{array}
$$

Note: Clearing the $5x$ term first gives $-x$, whereas when we cleared the $4x$ term, we got x. Most people prefer working with positive coefficients.

Note: $5 = -x$ means that we must find a number whose additive inverse is 5; so x must be equal to -5. In effect, we simply changed the signs of both sides. In Section 2.3, we will see another way to interpret $-x$.

Notice that the term we chose to clear in the first step did not affect the solution, but it did affect our approach to the rest of the problem.

Conclusion   In solving an equation that has variable terms on both sides of the equal sign, when we select a variable term to clear, we can avoid negative coefficients by clearing the term with the lesser coefficient.

Your Turn 4   Solve and check.

a. $7y - 3 = 6y + 10$

b. $2.6 + 3n = 9.5 + 4n$

Some equations that have variables on both sides require using the distributive property and combining like terms to solve.

Answers   Your Turn 4
a. 13   **b.** -6.9

Note: When we distribute a minus sign in an expression such as $-(3y + 11)$, we can think of the minus sign as -1; so we have

$$-(3y + 11)$$
$$= -1(3y + 11)$$
$$= -1(3y) + (-1)(11)$$
$$= -3y - 11$$

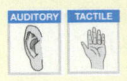
Example 5   Solve and check. $y - (3y + 11) = 5(y - 2) - 8y$

SOLUTION   Simplify both sides of the equation. Then isolate y.

$$y - (3y + 11) = 5(y - 2) - 8y$$

Distribute the minus sign. $y - 3y - 11 = 5y - 10 - 8y$ **Distribute 5.**

Note: The coefficient of $-3y$ is less than the coefficient of $-2y$, so we chose to clear $-3y$ by adding $3y$.

$$-2y - 11 = -3y - 10$$ **Combine like terms.**
$$\underline{+3y \qquad\qquad +3y}$$ **Add 3y to both sides.**
$$y - 11 = \quad 0 - 10$$
$$y - 11 = -10$$
$$\underline{+11 \qquad +11}$$ **Add 11 to both sides to isolate y.**
$$y + 0 = \quad 1$$
$$y = 1$$

Check   Replace y in the original equation with 1 and verify that the equation is true. We will leave this to the reader.

Following is an outline for solving equations based on everything you've learned so far.

Procedure   **Solving Linear Equations**

To solve linear equations requiring the addition principle only,

1. Simplify both sides of the equation as needed.
   a. Distribute to clear parentheses.
   b. Combine like terms.
2. Use the addition principle so that all variable terms are on one side of the equation and all constants are on the other side. Then combine like terms.

Tip: Clear the variable term that has the lesser coefficient to avoid negative coefficients.

Your Turn 5   Solve and check.

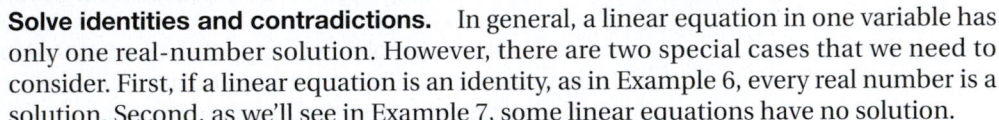

a. $7n - (n - 3) = 4(n + 4) + n$ b. $11 + 3(5m - 2) = 7(2m - 1) + 9$

Objective 4   **Solve identities and contradictions.**   In general, a linear equation in one variable has only one real-number solution. However, there are two special cases that we need to consider. First, if a linear equation is an identity, as in Example 6, every real number is a solution. Second, as we'll see in Example 7, some linear equations have no solution.

Equations with an Infinite Number of Solutions

Consider $3x + 9 = 3x + 9$. Because the expressions on each side of the equation are identical, each produces the same result no matter what number replaces the variable x. This means that every real number is a solution. Such an equation is called an **identity**.

Definition   *Identity:* An equation that has every real number as a solution (excluding any numbers that cause an expression in the equation to be undefined).

The equation $5y - 4 = 7y - 3$ is not an identity because not every real number is a solution. Consider the number 2, for example. If we replace y with 2 in the equation, we see that it does not check.

$$5(2) - 4 \stackrel{?}{=} 7(2) - 3$$
$$10 - 4 \stackrel{?}{=} 14 - 3 \qquad \text{The real number we picked, 2, is not a solution for}$$
$$6 \neq 11 \qquad\qquad 5y - 4 = 7y - 3 \text{; so the equation is not an identity.}$$

Sometimes it isn't obvious that an equation is an identity. Simplifying the expressions on each side of the equation can make the identity apparent.

Procedure

Recognizing an Identity

When solving a linear equation, if after simplifying each side of the equation the expressions are identical, the equation is an identity and every real number for which the equation is defined is a solution.

Example 6   Solve and check. $5(2t + 1) - 8 = 4t - 3(1 - 2t)$

SOLUTION   Simplify both sides of the equation. Then isolate the variable.

$$5(2t + 1) - 8 = 4t - 3(1 - 2t)$$

Distribute 5. $10t + 5 - 8 = 4t - 3 + 6t$ Distribute -3.

$$10t - 3 = 10t - 3 \qquad \text{Combine like terms.}$$

Note: We can stop here because the equation is obviously an identity. However, if we continue and apply the addition principle, the equation is still an identity.

$$10t - 3 = 10t - 3$$
$$\underline{-10t \qquad\quad -10t}$$
$$0 - 3 = \quad 0 - 3$$
$$-3 = -3$$

Because the linear equation is an identity, every real number is a solution.

Check   Every real number is a solution for an identity, so any number we choose should check in the original equation. We will choose to test the number 1.

$$5(2\mathbf{t} + 1) - 8 = 4\mathbf{t} - 3(1 - 2\mathbf{t})$$
$$5(2(\mathbf{1}) + 1) - 8 \stackrel{?}{=} 4(\mathbf{1}) - 3(1 - 2(\mathbf{1})) \qquad \text{Replace } t \text{ with 1.}$$
$$5(3) - 8 \stackrel{?}{=} 4 - 3(-1)$$
$$15 - 8 \stackrel{?}{=} 4 + 3$$
$$7 = 7$$

Because the equation is true, 1 is a solution, which supports but does not prove our conclusion that the equation is an identity. To be more certain, we could choose another number to check in the equation. (Remember, every number should work.) We will leave this to the reader.

Conclusion   If after simplifying, the linear equation is an identity, then every real number is a solution to the equation.

Equations with No Solution

Some equations, called **contradictions**, have no solution.

Definition   *Contradiction:*   An equation that has no real-number solution.

We recognize linear equations of this type by the fact that after the expressions on each side of the equation have been simplified, the variable terms will match but the constant terms will not. For example, $2x - 5 = 2x - 3$ is a contradiction. In Example 7, watch what happens when we try to solve the equation.

Example 7   Solve and check. $2x - 5 = 2x - 3$

SOLUTION
$$2x - 5 = 2x - 3$$
$$\underline{-2x \qquad -2x} \qquad \text{Subtract } 2x \text{ from both sides.}$$
$$0 - 5 = \ \ 0 - 3$$
$$-5 = -3$$

> **Note:** The variable terms are eliminated, and the resulting numeric equation is false, which indicates that the given equation is a contradiction.

> **Note:** The symbol $\varnothing$, which indicates the empty set, is often used to represent the solution set of an equation that has no solutions.

Because the equation is a contradiction, it has no solution.

Check   Because the variable terms, $2x$, are identical on both sides of the equal sign, replacing x with any number will yield an identical product. Because we subtract 5 from that product on the left side and we subtract 3 from that product on the right side, the equation cannot be true. Therefore, it has no solution.

Conclusion   When solving a linear equation in one variable, if applying the addition principle of equality causes the variable terms to be eliminated from the equation and the resulting equation is false, then the equation is a contradiction and has no solution.

Procedure   **Recognizing a Contradiction**

When solving a linear equation, if after simplifying the expressions on each side of the equation the expressions have the same variable term but different constant terms, the equation is a contradiction and has no solution. Equivalently, if the variable terms are eliminated and the resulting equation is false, the equation is a contradiction.

Your Turn 7   Solve and check.

   a. $4x - 3(x + 5) = 2(2x - 3) - 3(x + 5)$ **b.** $2(2x + 4) - 11 = 4(x - 2) + 5$

Objective ⑤ **Solve application problems.**   Now consider some application problems that can be solved using the addition principle.

Example 8   Laura wants to buy a car stereo that costs $275. She currently has $142. How much more is needed?

Understand   We are given the total required and the amount she currently has, and we must find how much she needs.

Plan   Let x represent the amount Laura needs. We will write an equation, then solve.

Execute   Current amount $+$ Needed amount $= 275$

$$142 \qquad + \qquad x \qquad = 275$$
$$142 + x = 275$$
$$\underline{-142 \qquad -142} \qquad \text{Subtract 142 from both sides}$$
$$0 + x = 133 \qquad \text{to isolate } x.$$
$$x = 133$$

Answers   **Your Turn 7**
a. contradiction—no solution
b. identity—all real numbers

Answer Laura needs $133 to buy the stereo.

Check Does $142 plus the additional $133 equal $275?

$$142 + 133 \overset{?}{=} 275$$
$$275 = 275 \qquad \text{It checks.}$$

Your Turn 8 Daryl has a balance of $-\$568$ on a credit card. How much must he pay to bring his balance to $-\$480$?

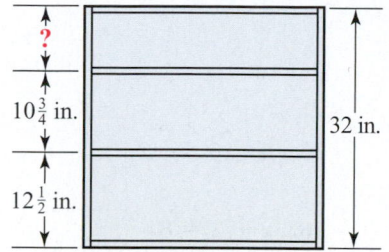

Example 9 The figure to the left shows the front-view design of a bookcase. Find the missing length.

Understand In the drawing, the sum of the dimensions on the left side must be equal to 32 inches.

Plan Let d represent the unknown distance. We will write an equation and solve for d.

Execute $12\dfrac{1}{2} + 10\dfrac{3}{4} + d = 32$

$$23\dfrac{1}{4} + d = 32 \qquad \text{Combine like terms.}$$

$$23\dfrac{1}{4} - 23\dfrac{1}{4} + d = 32 - 23\dfrac{1}{4} \qquad \text{Subtract } 23\dfrac{1}{4} \text{ from both sides to isolate } d.$$

$$d = 8\dfrac{3}{4}$$

Answer The unknown length is $8\dfrac{3}{4}$ inches.

Check Verify that the sum of the three lengths on the left side is equal to 32 inches. We will leave this to the reader.

Note: As in Example 2(b), because we are subtracting a fraction from both sides, we are writing the subtraction horizontally instead of vertically. Use whichever approach, horizontal or vertical, you prefer.

Answer   Your Turn 8

$88

Answer   Your Turn 9

157 students

Your Turn 9 In researching the grades given in a particular course during one semester, a department chair finds that 89 students received an A, 154 received a B, and 245 received a C. If the department does not give D's and 645 students enrolled in the course initially, how many students did not receive a passing grade in the course?

2.2 Exercises   For Extra Help *MyMathLab*

PRACTICE   WATCH   DOWNLOAD

Note: Exercises marked with a ★ represent challenging exercises.

1. In your own words, explain what the addition principle of equality says.

2. In the equation $5x - 9 = 4x + 1$, how do you use the addition principle to clear the $4x$ term?

3. In the equation $5x - 9 = 4x + 1$, how do you use the addition principle to clear the -9 term?

4. If an equation has like terms on the same side of the equation, what should you do first?

5. If an equation has variable terms on both sides of the equation, what should you do first?

6. How do you handle parentheses in an equation?

For Exercises 7–22, determine whether each equation is linear. See Example 1.

7. $4y + 7 = 2y - 8$

8. $6x - 5 = 3y + 64$

9. $m^2 + 4 = 20$

10. $-1 = w^2 - 5w$

11. $7u^4 + u^2 = 16$

12. $6n^3 - 9n^2 = 4n + 6$

13. $4x - 8 = 2(x + 3)$

14. $5(t - 2) + 7 = 3t - 1$

15. $6x - y = 11$

16. $3x + 2y = 12$

17. $x^2 + y^2 = 1$

18. $3x^2 + 2y^2 = 24$

19. $y = 8$

20. $x = -3$

21. $y = \dfrac{1}{4}x - 3$

22. $y = -0.5x + 2$

For Exercises 23–74, solve and check. See Examples 2–7.

23. $x - 7 = 2$

24. $a - 8 = 30$

25. $m - 8 = -3$

26. $n - 6 = -2$

27. $r - 2 = -9$

28. $a - 3 = -11$

29. $x + 7 = 12$

30. $x + 2 = 8$

31. $n + 12 = 2$

32. $y + 15 = 8$

33. $-16 = y + 9$

34. $-24 = n + 11$

35. $m + \dfrac{7}{8} = \dfrac{4}{5}$

36. $k + \dfrac{5}{9} = -\dfrac{1}{3}$

37. $-\dfrac{5}{8} = y - \dfrac{1}{6}$

38. $\dfrac{3}{4} = c - \dfrac{2}{3}$

39. $15.8 + y = 7.6$

40. $b + 8.8 = 5.4$

41. $-2.1 = n - 7.5 + 0.8$

42. $x + 0.4 - 1.6 = -12.5$

43. $6 - 18 = 7x - 3 - 6x$

44. $2z + 6 - z = 5 - 9$

45. $5m = 4m + 7$

46. $7y = 6y - 8$

47. $7x - 9 = 6x + 4$

48. $12y + 22 = 11y - 3$

49. $-2y - 11 = -3y - 5$

50. $-4t + 9 = -5t + 1$

51. $7x - 2 + x = 10x - 1 - x$

52. $3t + 6 + 4t = 9t - 2 - t$

53. $8n + 7 - 12n = 4n + 5 - 9n$

54. $-10x - 9 + 8x = -4x - 5 + 3x$

55. $2.6 + 7a + 5 = 8a - 5.6$

56. $9c + 4.8 = 7.5 + 4.8 + 8c$

57. $12 - 7(h + 6) + 8h = 14 - 8$

58. $19 - 3(m + 4) + 4m = 42 - 18$

59. $5 - \dfrac{1}{2}(4x + 6) = -3x - 8$

60. $6 - \dfrac{2}{3}(6b - 9) = -5b - 21$

61. $5y - (3y + 7) = y + 9$

62. $-15 - 2x = 16 - (3x - 9)$

63. $3(3x - 5) - 2(4x - 3) = 5 - 10$

64. $5(5x - 3) - 6(4x - 2) = 12 - 15$

65. $1.4(2.5x - 4.5) - (6.2 + 2.5x) = -7.3 + 4.5$

66. $0.5(3.8x - 6.2) - (0.9x - 4) = 2.9 - 4.7$

67. $4 + 3y + y - 12 = 5y + 9 - y - 17$

68. $-9 - 4v - 1 + v = -2v + 5 - v - 15$

69. $7.3 - 0.2x + 1.3 - 0.6x = 12 - 0.8x - 3.6$

70. $2.5y - 3.4 - 1.2y = 6.7 - 9.1 + 1.3y$

★**71.** $20z + \dfrac{2}{3}(6z - 48) - 9z = 14z + 0.2(50z - 250) - 9z$

★**72.** $6b - 1.5(8 + 2b) + 4 = 6b - \dfrac{1}{8}(24b + 48)$

★**73.** $8(2m - 4) + 20(m - 5) = 4m + 5(5m - 20) - 32 + 7m$

★**74.** $-3(2x + 5) + 8(x + 2) - 7 = 6(x - 5) - 4(x - 6)$

For Exercises 75–86, translate to an equation and then solve. See Examples 8 and 9.

75. Latonia is planning to buy a new car. The down payment is $2373. She has $1947 saved. How much more does she need?

76. Kent owes $12,412 on his Visa card. What payment should he make so he will owe $10,500?

77. Robert knows that the distance from his home to work is 42 miles. Unfortunately, he gets a flat tire 16 miles from work. How far did Robert drive before his tire went flat?

78. Susan is playing Yahtzee. The "chance" score is found by adding the value shown on five dice. If she has a total of 23 on four of the dice, what does the fifth die need to be so that her score will be 28?

79. On May 16, the balance in Nikki's checking account was $1741.62. Afterwards, she writes four checks and doesn't record the amount of the last check. The first check is for $16.82, the second is for $150.88, and the third is for $192.71. On May 21, she finds her balance to be $1286.65. If the checks were the only transactions that could have cleared the bank during that time, what was the amount of the fourth check?

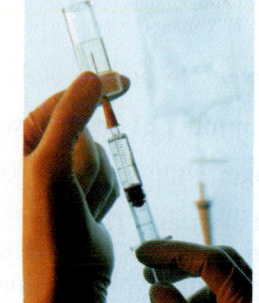

80. A patient must receive 350 cc of a medication in three injections. He has received two injections at 110 cc each. How much should the third injection be?

81. The perimeter of the trapezoid shown is 67.2 centimeters. Find the length of the missing side.

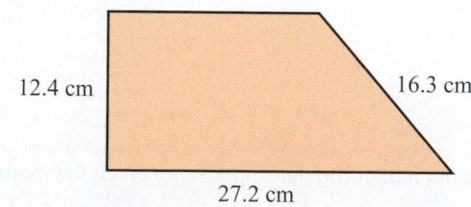

12.4 cm   16.3 cm

27.2 cm

82. The perimeter of the triangle shown is $84\dfrac{1}{2}$ inches. Find the length of the missing side.

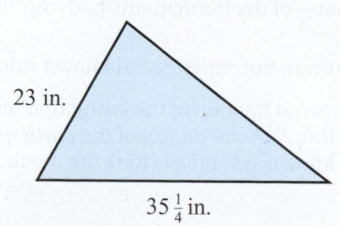

23 in.

$35\frac{1}{4}$ in.

83. In the blueprint shown, what is the distance x?

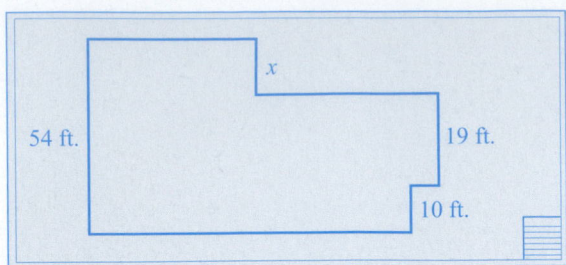

54 ft.

19 ft.

10 ft.

84. What is the distance x in the following figure?

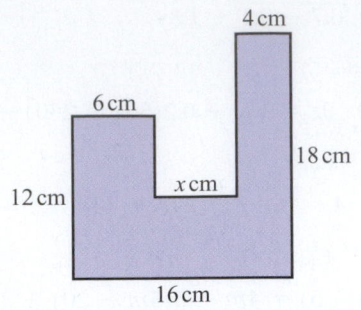

4 cm

6 cm

12 cm

x cm

18 cm

16 cm

85. John sells scanning devices. The company quota is set at $8500 each month. The spreadsheet shows John's sales as of the end of the second week of April. How much more does John need to sell to make the quota? Do you think John will make the quota? Why or why not?

Date	Item No.	Quantity	Price per Unit
4/2	32072	1	$900
4/9	17032	3	$1500
4/13	48013	2	$1245

86. Tamika works from 8 A.M. until 5 P.M. in the women's department of a major clothing store. She has set a goal for herself of $600 per day in sales. Thus far, she has sold 3 blouses for $25 each, 5 pairs of slacks for $30 each, and 2 pairs of shoes for $85 each. How much more in sales does she need to reach her goal? If it is now 4 P.M., is it likely that she will make her goal? Why or why not?

★**87.** In a survey, respondents were given a statement, and they could agree, disagree, or have no opinion regarding the statement. The results indicate that $\frac{1}{3}$ of the respondents agree with the statement while $\frac{2}{5}$ disagree. What fraction had no opinion?

★**88.** In a *Time*/CNN poll, $\frac{19}{50}$ of the respondents said that they believed that the increase in the number of divorces was due more to changes in women's attitude toward marriage, while $\frac{9}{50}$ said the increase was due more to changes in men's attitude toward marriage. The rest thought that the increase was a result of both men and women equally. What fraction of the respondents believed that the increase was due to men and women equally?

Puzzle Problem

In a study of an experimental medication, half of the participants are given the medication and half are given a placebo. The results indicate that $\frac{1}{3}$ of the participants had improvement in their condition, $\frac{1}{8}$ had no improvement in their condition but experienced known side effects of the medication, and the rest showed no effects at all. What fraction of the group that received the medication showed no discernible effects from it? (Hint: Assume that all of the participants that saw improvement in their condition or experienced known side effects took the medication.)

Review Exercises

Exercises 1–6 ▲ **EXPRESSIONS**

[1.4] **1.** What is the multiplicative inverse of $-\dfrac{4}{5}$?

For Exercises 2 and 3, simplify.

[1.4] **2.** $4(-8.3)$

[1.4] **3.** $27 \div (-3)$

[1.7] **4.** Simplify: $13 - 6(x + 2)$

[1.7] **5.** Combine like terms: $3x^2 - 6x + 5x^2 + x - 9$

[1.7] **6.** Use the distributive property to rewrite: $12\left(\dfrac{1}{3}x - \dfrac{3}{4}\right)$

2.3 The Multiplication Principle of Equality

Objectives

1 Solve linear equations using the multiplication principle.

2 Solve linear equations using both the addition and multiplication principles.

3 Use the multiplication principle to clear fractions and decimals from equations.

4 Solve application problems.

Objective 1 **Solve linear equations using the multiplication principle.** In Section 2.2, when we isolated the variable, the coefficient of that variable was always 1. What if we ended up with a coefficient other than 1? Look at the equation $6n - 11 = 4n + 3$. Using the methods of Section 2.2, we have

$$6n - 11 = 4n + 3$$
$$\underline{-4n \qquad\qquad -4n}$$
$$2n - 11 = 0 + 3$$
$$\underline{+11 \qquad +11}$$
$$2n + 0 = \quad 14$$
$$2n = 14$$

Note: The coefficient is not 1.

How can we solve $2n = 14$? To isolate n, we must clear the coefficient 2 so that we have $1n$, or simply n. Because coefficients multiply variables, we must undo that multiplication using its inverse operation, division. If we are going to divide, we must divide on both sides to keep the equation balanced. We write

$$2n \div 2 = 14 \div 2, \text{ or the more popular form } \frac{2n}{2} = \frac{14}{2}$$

Notice that the coefficient 2 divides out.

Note: The 2s divide out, leaving $1n$, which can be simplified as n.

$$\frac{\cancel{2}n}{\cancel{2}} = \frac{14}{2}$$
$$1n = 7$$
$$n = 7$$

You might remember that in Chapter 1, we discussed that multiplicative inverses are numbers whose product is 1. Multiplying the coefficient by its multiplicative inverse is another way to clear the coefficient. In the previous case, if we multiply the coefficient 2 by its multiplicative inverse $\dfrac{1}{2}$, we get the desired $1n$.

$$\frac{1}{2} \cdot 2n = 14 \cdot \frac{1}{2}$$

$$\frac{1}{\overset{1}{\cancel{2}}} \cdot \frac{\overset{1}{\cancel{2}}n}{1} = \frac{\overset{7}{\cancel{14}}}{1} \cdot \frac{1}{\underset{1}{\cancel{2}}}$$

$$1n = 7$$

$$n = 7$$

Note: Throughout the rest of the book, we omit this step of writing the resulting 1 coefficient.

Connection Remember that when we divide by a fraction, we write an equivalent multiplication using the multiplicative inverse. Multiplying by the multiplicative inverse is the same as dividing.

If we manipulate an equation using the addition principle and we still have a coefficient other than 1, we can clear that coefficient by dividing by the coefficient (or multiplying by its multiplicative inverse). This is the main purpose of the *multiplication principle*. The multiplication principle says that we can multiply (or divide) both sides of an equation by the same amount without affecting its solution(s).

Rule **The Multiplication Principle of Equality**

If $a = b$, then $ac = bc$ is true for all real numbers a, b, and c, where $c \neq 0$.

Procedure **Using the Multiplication Principle of Equality**

To use the multiplication principle of equality to clear a coefficient in an equation, multiply both sides of the equation by the multiplicative inverse of that coefficient or divide both sides by the coefficient.

Example 1 Solve and check.

a. $-\dfrac{4}{5} m = \dfrac{6}{7}$

SOLUTION $-\dfrac{4}{5} m = \dfrac{6}{7}$

$$-\frac{\overset{1}{\cancel{5}}}{\underset{1}{\cancel{4}}} \cdot -\frac{\overset{1}{\cancel{4}}}{\underset{1}{\cancel{5}}} m = \frac{\overset{3}{\cancel{6}}}{7} \cdot -\frac{5}{\underset{2}{\cancel{4}}}$$
Clear the coefficient $-\dfrac{4}{5}$ by multiplying both sides by its multiplicative inverse, $-\dfrac{5}{4}$.

$$m = -\frac{15}{14}$$

Check $-\dfrac{4}{5} m = \dfrac{6}{7}$

$$-\frac{\overset{2}{\cancel{4}}}{\underset{1}{\cancel{5}}} \left(-\frac{\overset{3}{\cancel{15}}}{\underset{7}{\cancel{14}}} \right) \overset{?}{=} \frac{6}{7}$$
Replace m in the original equation with $-\dfrac{15}{14}$ and verify that the equation is true.

$$\frac{6}{7} = \frac{6}{7}$$

True; therefore, $-\dfrac{15}{14}$ is correct.

b. $-1.9x = -4.56$

SOLUTION $-1.9x = -4.56$

$$\frac{\cancel{-1.9}x}{\cancel{-1.9}} = \frac{-4.56}{-1.9}$$
Clear the coefficient -1.9 by dividing both sides by -1.9.

$$x = 2.4$$

Note: If the equation contains fractions, the answer will be left as a fraction or an integer. If the equation contains decimals, the answer will be left as a decimal or an integer. If the equation contains integers only, the answer will be expressed as an integer, a fraction, or a decimal, as appropriate.

Check $-1.9x = -4.56$

$-1.9(2.4) = -4.56$ **Replace x with 2.4 and verify that the equation is true.**

$-4.56 = -4.56$

True; therefore, 2.4 is correct.

Your Turn 1   Solve and check.

a. $-56 = -7t$ **b.** $-\dfrac{3}{8}y = 9$ **c.** $0.4x = -0.92$

Objective ② Solve linear equations using both the addition and multiplication principles.   Now let's put the multiplication principle together with the addition principle. We follow the same outline as in Section 2.2, with one new step: using the multiplication principle to clear any remaining coefficient at the end.

Procedure   Solving Linear Equations

To solve linear equations in one variable,

1. Simplify both sides of the equation as needed.
   a. Distribute to clear parentheses.
   b. Combine like terms.
2. Use the addition principle so that all variable terms are on one side of the equation and all constants are on the other side. (Clear the variable term with the lesser coefficient to avoid negative coefficients.) Then combine like terms.
3. Use the multiplication principle to clear any remaining coefficient.

Example 2   Solve and check. $-3x - 11 = 7$

SOLUTION   Use the addition principle to separate the variable term and constant terms; then use the multiplication principle to clear any remaining coefficient.

Note: Multiplying both sides by $-\dfrac{1}{3}$ would also work:

$$\left(-\frac{1}{\underset{1}{\cancel{3}}}\right)\left(-\frac{\overset{1}{\cancel{3}}x}{1}\right) = \left(\frac{\overset{6}{\cancel{18}}}{1}\right)\left(-\frac{1}{\underset{1}{\cancel{3}}}\right)$$

$$x = -6$$

$$\begin{aligned} -3x - 11 &= 7 \qquad \text{\textbf{Add 11 to both sides to isolate the } } -3x \text{ \textbf{term.}}\\ \underline{+11 \quad +11}& \\ -3x + 0 &= 18 \\ \frac{-3x}{-3} &= \frac{18}{-3} \qquad \text{\textbf{Divide both sides by } } -3 \text{ \textbf{to clear the } } -3 \text{ \textbf{coefficient.}}\\ x &= -6 \end{aligned}$$

Check $-3x - 11 = 7$

$-3(-6) - 11 \overset{?}{=} 7$ **Replace x in the original equation with -6 and verify that the equation is true.**

$18 - 11 \overset{?}{=} 7$

$7 = 7$

True; therefore, -6 is correct.

Your Turn 2   Solve and check.

a. $6y - 19 = -22$ **b.** $-12 = 20 - 8t$

Answers   Your Turn 1

a. 8   **b.** -24 **c.** -2.3

Answers   Your Turn 2

a. $-\dfrac{1}{2}$ **b.** 4

Solving Equations with Variable Terms on Both Sides

Recall that when variable terms appear on both sides of the equal sign, we use the addition principle to get the variable terms on one side of the equal sign and the constant terms on the other side.

Example 3 Solve and check. $8y - 5 = 2y - 29$

SOLUTION Use the addition principle to get the variable terms on one side of the equation and the constant terms on the other side; then use the multiplication principle to clear the remaining coefficient.

$$8y - 5 = 2y - 29 \quad \text{Subtract } 2y \text{ from both sides. } (2y \text{ has the lesser coefficient.})$$

$$\underline{-2y \qquad -2y}$$

$$6y - 5 = 0 - 29$$

$$6y - 5 = -29$$

$$\underline{+5 \qquad +5} \quad \text{Add 5 to both sides to isolate the } 6y \text{ term.}$$

$$6y + 0 = -24$$

$$\frac{6y}{6} = \frac{-24}{6} \quad \text{Divide both sides by 6 to clear the 6 coefficient.}$$

$$y = -4$$

Check $\qquad 8y - 5 = 2y - 29$

$$8(-4) - 5 \overset{?}{=} 2(-4) - 29 \quad \begin{array}{l}\textbf{Replace } y \text{ in the original equation with } -4\\ \textbf{and verify that the equation is true.}\end{array}$$

$$-32 - 5 \overset{?}{=} -8 - 29$$

$$-37 = -37$$

True; therefore, -4 is correct.

Simplifying First

Recall that when an equation contains parentheses or like terms that appear on the same side of the equal sign, we first clear the parentheses and combine like terms. Then we use the addition principle to separate the variable terms and constant terms.

Example 4 Solve and check. $17 - (4n - 5) = 9n - 3(n + 7)$

SOLUTION $17 - (4n - 5) = 9n - 3(n + 7)$

$$17 - 4n + 5 = 9n - 3n - 21 \quad \text{Distribute to clear parentheses.}$$

$$22 - 4n = 6n - 21 \quad \text{Combine like terms.}$$

$$22 - 4n = 6n - 21$$

$$\underline{+4n \quad +4n} \quad \begin{array}{l}\textbf{Add } 4n \text{ to both sides.}\\ (-4n \text{ has the lesser coefficient.})\end{array}$$

$$22 + 0 = 10n - 21$$

$$22 = 10n - 21$$

$$\underline{+21 = \qquad +21} \quad \text{Add 21 to both sides to isolate } 10n.$$

$$43 = 10n + 0$$

$$\frac{43}{10} = \frac{10n}{10} \quad \begin{array}{l}\textbf{Divide both sides by 10 to clear}\\ \textbf{the 10 coefficient.}\end{array}$$

$$\frac{43}{10} \text{ or } 4.3 = n$$

Check
$$17 - (4n - 5) = 9n - 3(n + 7)$$
$$17 - (4(\mathbf{4.3}) - 5) \overset{?}{=} 9(\mathbf{4.3}) - 3(\mathbf{4.3} + 7)$$
$$17 - (17.2 - 5) \overset{?}{=} 38.7 - 3(11.3)$$
$$17 - 12.2 \overset{?}{=} 38.7 - 33.9$$
$$4.8 = 4.8$$

Replace n in the original equation with 4.3 and verify that the equation is true.

True; therefore, 4.3 is correct.

Your Turn 4 Solve and check.

a. $3t - 10 + 9t = 17 + 4t - 3$ b. $13 - 6(x + 2) = 3x - (11x + 5)$

Objective ③ **Use the multiplication principle to clear fractions and decimals from equations.**
To clear fractions and decimals from an equation, we can use the multiplication principle of equality. Although equations can be solved without clearing the fractions or decimals, most people find equations that contain only integers easier to solve.

Clearing Fractions in an Equation

If the equation contains fractions, we multiply both sides by a number that will clear all of the denominators. We could multiply both sides by any multiple of the denominators; however, using the LCD (least common denominator) results in the simplest equations.

Example 5 Solve and check. $\frac{1}{3}x - \frac{3}{4} = \frac{5}{6}x + 2$

SOLUTION Clear the fractions by multiplying through by the LCD. The LCD of 3, 4, and 6 is 12.

Note: We have chosen to clear the fractions in our first step. Remember, you can clear the fractions at any step in solving the equation or not clear them at all.

$$12\left(\frac{1}{3}x - \frac{3}{4}\right) = \left(\frac{5}{6}x + 2\right)12$$
Clear the fractions by multiplying both sides by the LCD, 12.

$$\frac{\overset{4}{\cancel{12}}}{1} \cdot \frac{1}{\cancel{3}}x - \frac{\overset{3}{\cancel{12}}}{1} \cdot \frac{3}{\cancel{4}} = \frac{\overset{2}{\cancel{12}}}{1} \cdot \frac{5}{\cancel{6}}x + 12 \cdot 2$$
Distribute 12; then divide out the denominators.

$$4x - 9 = 10x + 24$$
$$4x - 9 = 10x + 24$$
$$\underline{-4x \qquad\quad -4x}$$
Subtract $4x$ from both sides.
$$0 - 9 = 6x + 24$$
$$-9 = 6x + 24$$
$$\underline{-24 \qquad -24}$$
Subtract 24 from both sides.
$$-33 = 6x + 0$$
$$\frac{-33}{6} = \frac{6x}{6}$$
Divide both sides by 6 to clear the coefficient.
$$-\frac{11}{2} \text{ or } -5.5 = x$$
Simplify.

Answers **Your Turn 4**
a. 3 b. −3

Check

$$\frac{1}{3}x - \frac{3}{4} = \frac{5}{6}x + 2$$

$$\frac{1}{3}\left(-\frac{\mathbf{11}}{\mathbf{2}}\right) - \frac{3}{4} \overset{?}{=} \frac{5}{6}\left(-\frac{\mathbf{11}}{\mathbf{2}}\right) + 2$$ Replace x in the original equation with $-\frac{11}{2}$ and verify that the equation is true.

$$-\frac{11}{6} - \frac{3}{4} \overset{?}{=} -\frac{55}{12} + 2$$

$$-\frac{22}{12} - \frac{9}{12} \overset{?}{=} -\frac{55}{12} + \frac{24}{12}$$ Write equivalent fractions with their LCD.

$$-\frac{31}{12} = -\frac{31}{12}$$

True; therefore, $-\dfrac{11}{2}$ is correct.

Clearing Decimals in an Equation

The multiplication principle can also be used to clear decimals in an equation. Because multiplying by a power of 10 will cause the decimal point to move to the right, we can clear decimals by multiplying both sides of the equation by the appropriate power of 10. The power of 10 we use depends on the decimal number with the most decimal places.

Example 6   Solve and check. $0.2(y - 6) = 0.48y + 3$

SOLUTION   Decimals can be cleared at any step in the process of solving an equation. In this case, we will distribute to clear the parentheses first, then clear the decimals.

Connection Clearing decimals and fractions uses the same process. Decimal numbers represent fractions with denominators that are powers of 10. For example, $0.48 = \dfrac{48}{100}$ and $0.2 = \dfrac{2}{10}$. The LCD for these fractions is 100.

$$0.2(y - 6) = 0.48y + 3$$

$$0.2y - 1.2 = 0.48y + 3$$ Distribute to clear parentheses.

$$\mathbf{100}(0.2y - 1.2) = (0.48y + 3)\mathbf{100}$$

Note: The number 0.48 has two decimal places, which is more decimal places than any of the other decimal numbers; so we multiply both sides by 100 to clear the decimals.

$$20y - 120 = 48y + 300$$

$$20y - 120 = 48y + 300$$

$$\underline{-20y \qquad\qquad -20y}$$

$$0 - 120 = 28y + 300$$

$$-120 = 28y + 300$$

$$\underline{-300 \qquad\qquad -300}$$

$$-420 = 28y + 0$$

$$\frac{-420}{28} = \frac{28y}{28}$$

$$-15 = y$$

Check $$0.2(y - 6) = 0.48y + 3$$

$$0.2(\mathbf{-15} - 6) \overset{?}{=} 0.48(\mathbf{-15}) + 3$$ Replace y in the original equation with -15 and verify that the equation is true.

$$0.2(-21) \overset{?}{=} -7.2 + 3$$

$$-4.2 = -4.2$$

True; therefore, -15 is correct.

We can amend the outline for solving equations to include clearing fractions and decimals. Remember that this process is optional.

Procedure   **Solving Linear Equations**

To solve linear equations in one variable,

1. Simplify both sides of the equation as needed.
   a. Distribute to clear parentheses.
   b. Clear fractions or decimals by multiplying through by the LCD. In the case of decimals, the LCD is the power of 10 with the same number of zero digits as decimal places in the number with the most decimal places. (Clearing fractions and decimals is optional.)
   c. Combine like terms.
2. Use the addition principle so that all variable terms are on one side of the equation and all constants are on the other side. (Clear the variable term with the lesser coefficient to avoid negative coefficients.) Then combine like terms.
3. Use the multiplication principle to clear any remaining coefficient.

Your Turn 6   Solve and check.

a. $\dfrac{1}{3}(y - 2) = \dfrac{3}{5}y + \dfrac{1}{6}$

b. $2.5n - 1.04 = 0.15n - 0.1$

Objective ④ Solve application problems.   In Section 2.1, we began solving application problems using formulas. Recall that to use a formula, we replace the variables with the corresponding given numbers, then solve for the unknown amount.

Example 7   Solve.

a. The perimeter of the figure shown is 188 feet. Find the width and length.

Understand   The width is represented by w, and the length is represented by $w + 7$. We are given the perimeter, so we can use the formula $P = 2l + 2w$.

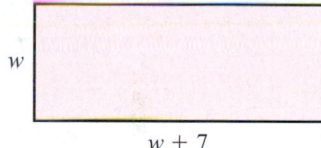

Plan   In the perimeter formula, replace P with **188** and l with $w + 7$ and solve for w.

Execute

$$P = 2l + 2w$$
$$\mathbf{188} = 2(\mathbf{w + 7}) + 2w$$
$$188 = 2w + 14 + 2w \qquad \text{Distribute.}$$
$$188 = 4w + 14 \qquad \text{Combine like terms.}$$
$$188 = 4w + 14$$
$$\underline{-14 \qquad\qquad -14} \qquad \text{Subtract 14 from both sides.}$$
$$174 = 4w + 0$$
$$\frac{174}{4} = \frac{4w}{4} \qquad \text{Divide both sides by 4.}$$
$$\frac{87}{2} \text{ or } 43.5 = w$$

Answer   The width is 43.5 feet. To find the length, we evaluate the expression that represents the length, $w + 7$, with $w = 43.5$.

$$\text{Length} = 43.5 + 7 = 50.5 \text{ ft.}$$

Answers Your Turn 6

a. $-\dfrac{25}{8}$ b. 0.4

Check First, verify that the length is 7 more than the width: 50.5 = 7 + 43.5. Second, verify that a rectangle with a length of 50.5 feet and a width of 43.5 feet has a perimeter of 188 feet.

$$P = 2l + 2w$$
$$P = 2(50.5) + 2(43.5)$$
$$P = 101 + 87$$
$$P = 188$$

It checks.

b. The total material allotted for the construction of a metal box is 4754 square inches. The length is to be 30 inches, and the width is to be 28 inches. Find the height.

Understand The material will be used to create the outer shell that is the box, which means that 4754 square inches is the surface area of the box. The formula for the surface area of a box is $SA = 2lw + 2lh + 2wh$.

Plan Replace SA with **4754**, l with **30**, and w with **28** and solve for h.

Execute

$$SA = 2lw + 2lh + 2wh$$
$$4754 = 2(30)(28) + 2(30)h + 2(28)h$$

$4754 = 1680 + 60h + 56h$	**Simplify.**
$4754 = 1680 + 116h$	**Combine like terms.**
$4754 = 1680 + 116h$	
$\underline{-1680 \quad -1680}$	**Subtract 1680 from both sides.**
$3074 = \quad 0 + 116h$	
$\dfrac{3074}{116} = \dfrac{116h}{116}$	**Divide both sides by 116.**
$26.5 = h$	

Answer The height is 26.5 inches.

Check Does a 30-inch by 28-inch by 26.5-inch box have a surface area of 4754 square inches?

$$4754 \overset{?}{=} 2(30)(28) + 2(30)(26.5) + 2(28)(26.5)$$
$$4754 \overset{?}{=} \quad 1680 \quad + \quad 1590 \quad + \quad 1484$$
$$4754 = 4754$$

It checks.

Your Turn 7 Solve.

a. The desired surface area for a box is 98 square feet. If the length of the box is to be 4 feet and the height is to be 6 feet, what will the width be?

b. The circumference of Venus at the equator is 24,170 miles. What is the diameter of Venus at the equator? (Use $C = \pi d$ with $\pi \approx 3.14$.)

Answers Your Turn 7
a. 2.5 ft. **b.** 7697 mi.

2.3 Exercises   For Extra Help *MyMathLab*

Note: Exercises marked with a ★ represent challenging exercises.

1. In your own words, explain what the multiplication principle of equality says.

2. How do you use the multiplication principle to clear a coefficient that is an integer?

3. How do you use the multiplication principle to clear a coefficient that is a fraction?

4. To clear a negative coefficient, do you multiply/divide by a positive or negative number? Why?

5. If an equation contains fractions, how do you transform the equation so that it contains only integers?

6. If an equation contains decimal numbers, how do you transform the equation so that it contains only integers?

For Exercises 7–18, solve and check. See Example 1.

7. $3x = 12$

8. $8x = -24$

9. $-6x = -18$

10. $-5y = 20$

11. $\dfrac{n}{2} = 8$

12. $\dfrac{t}{3} = -4$

13. $\dfrac{3}{4}y = -15$

14. $\dfrac{5}{6}x = 20$

15. $-\dfrac{4}{7}t = -\dfrac{2}{3}$

16. $-\dfrac{3}{8}a = \dfrac{5}{6}$

17. $-\dfrac{2}{5}t = \dfrac{8}{15}$

18. $-\dfrac{7}{9}t = -\dfrac{5}{12}$

For Exercises 19–40, solve and check. See Example 2.

19. $5n - 2n = 21$

20. $7t - 3t = 20$

21. $4x + 1 = 21$

22. $3x + 5 = 11$

23. $4a - 3 = 17$

24. $3x - 8 = 10$

25. $5x + 7 = -8$

26. $3x + 9 = -6$

27. $9 - 2n = 12$

28. $1 - 7y = -8$

29. $\dfrac{5}{8}x - 2 = 8$

30. $7 = \dfrac{3}{4}x + 13$

31. $2(x - 3) = -6$

32. $4(5x + 7) = 28$

33. $3(n + 7) = -6$

34. $4(n - 3) = -8$

35. $3a - 2a + 7 + 6a = -28$

36. $c + 2c + 3 + 4c = 24$

37. $12b - 8(2b - 3) = 16$

38. $2x - 6(x + 8) = -12$

39. $2(y - 33) + 3(y - 2) = 8$

40. $4(r - 8) + 2(r + 3) = -8$

For Exercises 41–62, solve and check. See Examples 3 and 4.

41. $8x + 5 = 2x + 17$

42. $10t + 1 = 6t + 13$

43. $5y - 7 = 2y + 13$

44. $9m + 1 = 3m - 14$

45. $9 - 4k = 15 - k$

46. $6 - 12m = -20m + 22$

47. $4k + 5 = 7k - 7 + 9k$

48. $-11b - 5 = -5b + 23 - 10$

49. $3a - 4a + 9 = -12a + 43 - 6a$

50. $17b - 11b - 17 = -4b + 13 - 5b$

51. $9x + 12 - 3x - 2 = x + 8 + 5x$

52. $12 - 6r - 14 = 9 - 4r - 7 - 2r$

53. $14(w - 2) + 13 = 4w + 5$

54. $2x + 2(3x - 4) = -23 + 5x$

55. $2n - (6n + 5) = 9 - (2n - 1)$

56. $2 - (17 - 5m) = 9m - (m + 7)$

57. $4 - (6x + 5) = 7 - 2(3x + 4)$

58. $-6 - (3z - 2) = 5z - 4(2z + 1)$

59. $5(a - 1) - 9(a - 2) = -3(2a + 1) - 2$

60. $-4(k + 4) + 13(k - 1) = -3(k - 2) + 13$

61. $3(2x - 5) - 4x = 2(x - 3) - 12$

62. $4(2x - 1) - 3(x + 5) = 5(x - 2) + 7$

For Exercises 63–82, use the multiplication principle of equality to eliminate the fractions or decimals; then solve and check. See Examples 5 and 6.

63. $\frac{2}{3}n - 1 = \frac{1}{4}$

64. $\frac{2}{5}n - 1 = \frac{3}{2}$

65. $\frac{3}{4}n - 2 = -\frac{7}{8}n + 4$

66. $-\frac{2}{5}t + 1 = \frac{3}{10}t - 3$

67. $\frac{3}{4}x + \frac{5}{6} = \frac{1}{4} + \frac{4}{3}x$

68. $\frac{7}{9}w - \frac{13}{6} = \frac{7}{2}w + \frac{5}{9}$

69. $\frac{1}{2}(y + 5) = \frac{3}{2} - y$

70. $\frac{2}{3}(x - 4) = \frac{4}{3} + 2x$

71. $\frac{1}{6}(m - 4) = \frac{2}{3}(m + 1) - \frac{1}{3}m$

72. $\frac{1}{5}(y - 3) = \frac{3}{10}(y + 5) - \frac{2}{5}y$

73. $-2.9u + 3.6u = 6.3$

74. $-4.6z + 2.2z = 4.8$

75. $0.2 - 1.3t - 0.8t = 1 - 2.1t - 0.8$

76. $4.2y - 8.2 + 2.3y = 0.9 + 6.5y - 9.1$

77. $0.3(a - 18) = 0.9a$

78. $0.6(w - 12) = 0.2w$

79. $0.08(15) + 0.4x = 0.05(36 + 7x)$

80. $0.06(25) + 0.27x = 0.3(4 + x)$

81. $9 - 0.2(v - 8) = 0.3(6v - 8)$

82. $0.5 - (t - 6) = 10.4 - 0.4(1 - t)$

Find
ⓧ *the mistake*

For Exercises 83–86, the check indicates that a mistake was made. Find and correct the mistake. See Examples 2–4.

83.
$$6x - 11 = 8x - 15$$
$$\underline{-6x \qquad\quad -6x}$$
$$11 = 2x - 15$$
$$\underline{+15 \qquad +15}$$
$$26 = 2x$$
$$\frac{26}{2} = \frac{2x}{2}$$
$$13 = x$$

Check $6x - 11 = 8x - 15$
$$6(13) - 11 \overset{?}{=} 8(13) - 15$$
$$78 - 11 \overset{?}{=} 104 - 15$$
$$67 \neq 89$$

84. $6n + 1 = -2n + 21$

$$\begin{array}{r} \underline{-1 \qquad\qquad -1} \\ 6n + 0 = -2n + 20 \\ 6n = -2n + 20 \\ \underline{+2n \qquad +2n} \\ 8n = 0 + 20 \\ \dfrac{8n}{8} = \dfrac{20}{8} \\ n = \dfrac{5}{2} \end{array}$$

Check $6n + 1 = -2n + 21$

$$\overset{3}{\underset{1}{\dfrac{\cancel{6}}{1}}}\left(\dfrac{5}{\underset{1}{\cancel{2}}}\right) + 1 \overset{?}{=} -\overset{1}{\underset{1}{\dfrac{\cancel{2}}{1}}}\left(\dfrac{5}{\underset{1}{\cancel{2}}}\right) + 21$$

$$3 + 1 \overset{?}{=} -5 + 21$$

$$4 \neq 16$$

85. $4 - (x + 2) = 2x + 3(x - 8)$

$$4 - x + 2 = 2x + 3x - 24$$

$$6 - x = 5x - 24$$

$$\underline{+x \qquad +x}$$

$$6 + 0 = 6x - 24$$

$$6 = 6x - 24$$

$$\underline{+24 \qquad\qquad +24}$$

$$30 = 6x + 0$$

$$\dfrac{30}{6} = \dfrac{6x}{6}$$

$$5 = x$$

Check $4 - (x + 2) = 2x + 3(x - 8)$

$$4 - (5 + 2) \overset{?}{=} 2(5) + 3(5 - 8)$$

$$4 - 7 \overset{?}{=} 10 + 3(-3)$$

$$-3 \overset{?}{=} 10 + (-9)$$

$$-3 \neq 1$$

86. $\dfrac{1}{2}n - 3 = \dfrac{2}{3}n + \dfrac{1}{4}$

$$\overset{6}{\underset{1}{\dfrac{\cancel{12}}{1}}} \cdot \dfrac{1}{\underset{1}{\cancel{2}}}n - 3 = \overset{4}{\underset{1}{\dfrac{\cancel{12}}{1}}} \cdot \dfrac{2}{\underset{1}{\cancel{3}}}n + \overset{3}{\underset{1}{\dfrac{\cancel{12}}{1}}} \cdot \dfrac{1}{\underset{1}{\cancel{4}}}$$

$$6n - 3 = 8n + 3$$

$$\underline{-6n \qquad\quad -6n}$$

$$0 - 3 = 2n + 3$$

$$\underline{-3 \qquad\qquad -3}$$

$$-6 = 2n + 0$$

$$\dfrac{-6}{2} = \dfrac{2n}{2}$$

$$-3 = n$$

Check $\dfrac{1}{2}n - 3 = \dfrac{2}{3}n + \dfrac{1}{4}$

$$\dfrac{1}{2}\left(-\dfrac{3}{1}\right) - 3 \overset{?}{=} \dfrac{2}{\underset{1}{\cancel{3}}}\left(-\dfrac{\overset{1}{\cancel{3}}}{1}\right) + \dfrac{1}{4}$$

$$-\dfrac{3}{2} - 3 \overset{?}{=} -2 + \dfrac{1}{4}$$

$$-\dfrac{3}{2} - \dfrac{6}{2} \overset{?}{=} -\dfrac{8}{4} + \dfrac{1}{4}$$

$$-\dfrac{9}{2} \neq -\dfrac{7}{4}$$

For Exercises 87–100, solve for the unknown amount. See Example 7.

87. The area of a rectangular deck is known to be 140 square feet. Find the length if the width must be 10 feet. (Use $A = lw$.)

88. A crate manufacturer receives an order for a crate with a volume of 128 cubic feet. The crate must be 5 feet 4 inches by 6 feet. What must the crate's height be? (Use $V = lwh$.)

89. The area of the metal plate shown in the figure is 45.6 square centimeters. What is the length of the side labeled h? (Use $A = \dfrac{1}{2}bh$.)

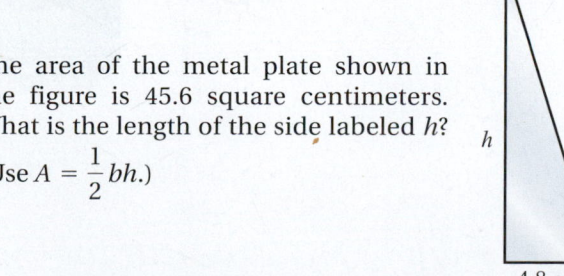

4.8 cm

90. The volume of a rectangular pyramid is 76.8 cubic inches, and the area of the base is 6.4 square inches. Find the height. (Use $V = \dfrac{1}{3}Ah$.)

91. A fence is to be installed around a rectangular field. The field's perimeter is 212 feet. Find the dimensions of the field if the length of the field is 2 feet more than the width, as shown in the figure. (Use $P = 2l + 2w$.)

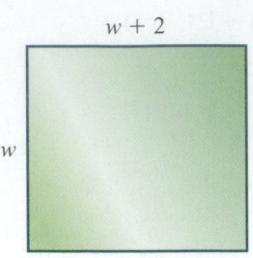

$w + 2$

w

92. The base of a building has a perimeter of 241 feet. The length is 16 feet more than the width, as shown in the figure. Find the dimensions of the base of the building. (Use $P = 2l + 2w$.)

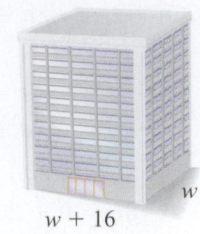

w

$w + 16$

93. The area of the trapezoidal plot of land shown in the figure is 7500 square feet. What is the length of the side labeled b?

(Use $A = \dfrac{1}{2}h(a + b)$.)

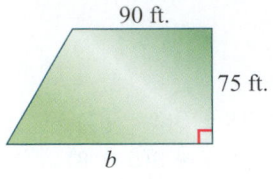

90 ft.

75 ft.

b

94. A tabletop is in the shape of a trapezoid and has an area of 1008 square inches. What is the length of the side labeled a?

(Use $A = \dfrac{1}{2}h(a + b)$.)

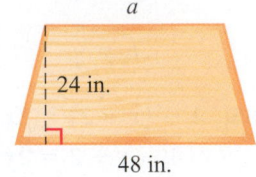

a

24 in.

48 in.

95. A small storage box is designed so that its surface area is 106 square inches. If the length is 7.5 inches and the width is 4 inches, find the height. (Use $SA = 2lw + 2lh + 2wh$.)

96. Jerry has 1992 square inches of wood to build a chest for storing musical equipment. He wants the length to be 22 inches and the width to be 18 inches. What will the height be? (Use $SA = 2lw + 2lh + 2wh$.)

97. The circumference of Earth along the equator is approximately 40,053.84 kilometers. What is the equatorial radius? (Use $C = 2\pi r$ and round your answer to the nearest tenth.)

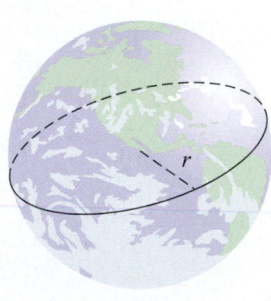

r

98. A giant sequoia is measured to have a circumference of 26.4 meters. What is the diameter? (Use $C = \pi d$ and round your answer to the nearest tenth.)

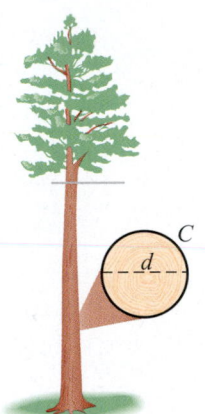

C

d

Of Interest

The giant sequoia tree is found along the western slopes of the Sierra Nevada range. These trees reach heights of nearly 100 meters, and some are over 4000 years old.

99. An engineer is designing a chemical storage tank in the shape of a cylinder. The volume of the tank is to be 60 cubic meters, and the radius is to be 2 meters. Find the height. (Use $V = \pi r^2 h$ and round your answer to the nearest tenth.)

2 m

100. The liquid in a standard 12-ounce soda can occupies a volume of 355 cubic centimeters. If the diameter is 6 centimeters, find the height of the liquid inside the can. (Use $V = \pi r^2 h$ and round your answer to the nearest tenth.)

6 cm

Lime-UP

12 ounces (355 mL)

★**101.** The area of the L-shaped room shown is 321 square feet. Find all of the missing dimensions.

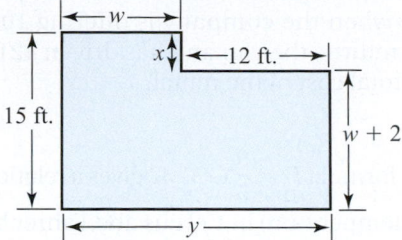

★**102.** The area of the figure below is 192 square inches. Find all of the missing dimensions.

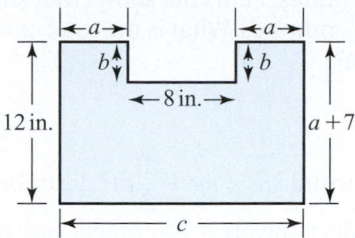

The formula V = iR describes the voltage, V, in a circuit when the current, i, passes through a resistance, R. Voltage is measured in volts (V), current is measured in amps (A), and resistance is measured in ohms (Ω). Use the formula for voltage to solve Exercises 103 and 104. See Example 7.

103. The voltage in a circuit is measured to be −76 V. If the resistance is 8 Ω, find the current.

104. A technician applies 15 V to a circuit. The current is measured to be 2.5 A. What is the resistance?

The formula F = ma describes the force that an object with mass m experiences when accelerated an amount a. If acceleration is measured in meters per second squared (m/sec.²) and the mass is measured in kilograms (kg), the force is in units of newtons (N). Use the force formula to solve Exercises 105–108. See Example 7.

105. A car with a mass of 1200 kilograms exerts a force of 4800 newtons. Find the acceleration.

106. An object weighs 44.1 newtons; that is, it exerts a downward force of −44.1 newtons. The object is dropped so that it accelerates at −9.8 meters per second squared. Find the mass of the object.

107. The Liberty Bell weighs 9265.9 newtons, so its downward force is −9265.9 newtons. If the acceleration due to gravity is about −9.8 meters per second squared, what is the mass of the bell?

108. Jacob weighs 160 pounds (a downward force of −160 pounds). If the acceleration due to gravity is about −32.2 feet per second squared, what is his mass? (The mass will be in terms of the unit *slugs*.)

> **Of Interest**
>
> Earth's gravity accelerates all objects at the same rate, −9.8 m/sec.² (or −32.2 ft./sec.²). The accelerations are negative because the direction is downward.

> **Of Interest**
>
> The Liberty Bell was cast in 1752 and arrived in Philadelphia in August of the same year. In September 1752, the bell cracked during a test while hanging in the yard of the Pennsylvania state house.

109. The formula $w = \dfrac{s}{f}$ describes the wavelength w of a musical note if f is the frequency and s is the speed of sound. A note is sounded with a wavelength of 0.94 meters/cycle and a frequency of 440 cycles/second. Find the speed of sound.

110. The formula $t = \dfrac{V}{g}$ gives the time that a free-falling object falls, where V is the velocity and g is the acceleration due to gravity. Find the velocity if the object has been falling for four seconds and the acceleration due to gravity is 9.8 meters per second squared.

111. The formula $C = 0.3m + 25$ describes the total cost in dollars of using a digital phone for 100 minutes, where m represents the number of minutes beyond 100 minutes. Pam's bill shows that she used the phone for 187 minutes. What is the total cost of her use of the phone?

112. A rental car agency rents a van for a weekend using the formula $C = 0.27m + 200$, where m represents the number of miles and C represents the cost in dollars. Laura rents a van from the company for a weekend when the company is offering 100 free miles. If she returns the van and has driven 421 miles, what is the total cost of the rental?

113. The formula $d = v_i t + \dfrac{1}{2}at^2$ describes the distance, d, an object travels if the object has an initial velocity (speed) v_i and an acceleration of a for a time t. An object travels 135 feet after being accelerated for three seconds with an acceleration of 20 feet per second squared. Find the object's initial velocity.

114. The formula $F = \dfrac{9}{5}C + 32$ gives a relationship between the temperature in Celsius and Fahrenheit. If the temperature is 82.4°F, find the temperature in Celsius.

Review Exercises

Exercises 1–3  **EXPRESSIONS**

[1.5] **1.** Simplify: $-6 + 3(5^2 + 2) - (6 - 1)$

[1.7] **2.** Combine like terms:
$$-3.2x^2 + 0.4x + 7x^2 - 8.03x + 1.5$$

[1.7] **3.** Use the distributive property to rewrite:
$$-\frac{3}{4}(6x + 8)$$

Exercises 4–6 **EQUATIONS AND INEQUALITIES**

For Exercises 4–6, solve and check.

[2.2] **4.** $6 + x = -14 + 2x$ **[2.3]** **5.** $-\dfrac{2}{5}x = 10$ **[2.3]** **6.** $2x + 3(x - 3) = 10x - (3x - 11)$

 2.4 # Applying the Principles to Formulas

Objective ① **Isolate a variable in a formula using the addition and multiplication principles.** So far when using formulas, we have substituted given numbers for all of the variables except one. We then used the addition and multiplication principles of equality to isolate that variable. However, we can actually isolate a particular variable in a formula without being given numbers for the other variables. We follow the same steps as if we had been given numbers.

Consider the area formula, $A = lw$. If we were given values for A and l and asked to isolate w, or solve for w, we would divide both sides by the value of l. We can perform this manipulation symbolically.

$$\frac{A}{l} = \frac{\cancel{l}w}{\cancel{l}} \qquad \text{To isolate } w, \text{ we clear } l \text{ by dividing both sides by } l.$$

$$\frac{A}{l} = w$$

This development suggests the following procedure:

Procedure **Isolating a Variable in a Formula**

To isolate a particular variable in a formula, treat all other variables like constants and isolate the desired variable using the outline for solving equations.

Isolate a Variable Using the Addition Principle

Recall that the addition principle is used to clear terms that are added or subtracted.

Example 1 Isolate R in the formula $P = R - C$.

SOLUTION

$$P = R - C$$
$$\underline{+C \qquad +C}$$
$$P + C = R + 0$$
$$P + C = R$$

To isolate R, we must clear C. Because C is subtracted from R, we add C to both sides.

Note: Because P and C are not like terms, they cannot be combined.

Your Turn 1 Isolate v_i in the formula that describes the velocity of an object after being accelerated from an initial velocity (v_i), $v = v_i + at$.

Isolate a Variable Using the Multiplication Principle

Now let's use the multiplication principle to isolate a variable in a formula.

Example 2

a. Isolate w in the formula for the volume of a box, $V = lwh$.

SOLUTION

$$\frac{V}{lh} = \frac{lwh}{lh}$$
$$\frac{V}{lh} = w$$

To isolate w, we must clear l and h. Because l and h are multiplying w, we divide both sides by l and h.

b. Isolate P in the formula $B = P(1 + rt)$.

SOLUTION

$$\frac{B}{1 + rt} = \frac{P(1 + rt)}{1 + rt}$$
$$\frac{B}{1 + rt} = P$$

To isolate P, we must clear $(1 + rt)$. Because $(1 + rt)$ is multiplying P, we divide both sides by $1 + rt$.

Answer **Your Turn 1**

$v - at = v_i$

Answers **Your Turn 2**

a. $\dfrac{A}{h} = b$ **b.** $\dfrac{V}{R + r} = i$

Your Turn 2

a. Isolate b in the formula for the area of a parallelogram, $A = bh$.

b. Isolate i in the voltage formula for a series circuit with two resistors, $V = i(R + r)$.

Isolate a Variable Using Both Principles

Now consider formulas that require the use of both principles to isolate the desired variable.

Example 3

a. Isolate l in the perimeter formula for rectangles, $P = 2l + 2w$.

SOLUTION First, use the addition principle to isolate the term that contains l. Then use the multiplication principle to clear the coefficient.

$$P = 2l + 2w$$

$$\underline{\quad -2w \qquad\quad -2w\quad} \qquad \text{Isolate } 2l \text{ by subtracting } 2w \text{ from both sides.}$$

$$P - 2w = 2l + \quad 0$$

$$\frac{P - 2w}{2} = \frac{2l}{2} \qquad \text{Isolate } l \text{ by dividing both sides by 2.}$$

$$\frac{P - 2w}{2} = l$$

b. Isolate d in the formula from physics, $v^2 = v_0^2 + \dfrac{2Fd}{m}$.

SOLUTION First, use the addition principle to isolate the term that contains d. Then use the multiplication principle to clear the coefficient.

$$v^2 = v_0^2 + \frac{2Fd}{m} \qquad \text{Isolate } \frac{2Fd}{m} \text{ by subtracting } v_0^2 \text{ on both sides.}$$

$$\underline{\quad -v_0^2 \qquad\quad -v_0^2\quad}$$

$$v^2 - v_0^2 = 0 + \frac{2Fd}{m}$$

$$v^2 - v_0^2 = \frac{2Fd}{m}$$

$$\frac{m}{2F} \cdot \frac{v^2 - v_0^2}{1} = \frac{\overset{1}{\cancel{m}}}{2F} \cdot \frac{2\overset{1}{\cancel{F}}d}{\underset{1}{\cancel{m}}} \qquad \begin{array}{l}\text{Clear the coefficient } \frac{2F}{m} \text{ by multiplying both sides}\\ \text{by its reciprocal.}\end{array}$$

$$\frac{m(v^2 - v_0^2)}{2F} = d$$

> ### Of Interest
>
> In this example, the 0 subscript is used to indicate that v_0 is an initial velocity. We have also seen v_i used to indicate initial velocity. Both forms are used in physics and engineering.

Your Turn 3

a. Isolate t in the formula that describes the velocity of an object after being accelerated from an initial velocity, $v = v_i + at$.

b. Isolate n in the formula $T = \dfrac{nt}{r} - R$.

Answers Your Turn 3

a. $\dfrac{v - v_i}{a} = t$

b. $\dfrac{r}{t}(T + R) = n$ or $\dfrac{r(T + R)}{t} = n$

2.4 Exercises   For Extra Help   *MyMathLab*

Note: Exercises marked with a ★ represent challenging exercises.

1. What would you do to isolate w in the formula $l = w - 2$? Why?

2. What would you do to isolate p in the formula $C = np$? Why?

3. To isolate y in $2x + 3y = 8$, what would you do first? Why?

4. How can you verify that your answer is correct after isolating a variable in a formula?

For Exercises 5–24, solve for the indicated variable. See Examples 1–3.

5. $t - 4u = v; t$

6. $x = a + 3y; a$

7. $-5y = x; y$

8. $2n = a; n$

9. $2x - 3 = b; x$

10. $3m + b = y; m$

11. $y = mx + b; m$

12. $ab + c = d; b$

13. $3x + 4y = 8; y$

14. $19 = 2l + 2w; w$

15. $\dfrac{mn}{4} - Y = f; Y$

16. $q = \dfrac{rs}{2} - p; p$

17. $6(c + 2d) = m - np; c$

18. $5(2n + a) = bn - c; a$

19. $\dfrac{x}{3} + \dfrac{y}{5} = 1; y$

20. $\dfrac{a}{4} + \dfrac{b}{6} = 3; a$

21. $\dfrac{3}{4} + 5a = \dfrac{n}{c}; n$

22. $\dfrac{x}{6} + 5y = \dfrac{m}{n}; m$

23. $t = \dfrac{A - P}{pr}; p$

24. $S = \dfrac{C}{1 - M}; M$

For Exercises 25–40, solve the geometric formula for the indicated variable. See Examples 1–3.

25. $P = a + b + c; a$

26. $a + b + c = 180; b$

27. $A = lw; l$

28. $C = 2\pi r; r$

29. $\dfrac{C}{d} = \pi; d$

30. $\dfrac{A}{b} = h; b$

31. $V = \dfrac{1}{3}\pi r^2 h; r^2$

32. $V = \dfrac{4}{3}\pi r^3; r^3$

33. $A = \dfrac{1}{2}bh; b$

34. $S = \dfrac{1}{2}gt^2; t^2$

35. $P = 2l + 2w; w$

36. $S = 2\pi r^2 - 2\pi rh; h$

37. $S = \dfrac{n}{2}(a + l); l$

38. $A = \dfrac{h}{2}(a + b); b$

★39. $V = h(\pi r^2 - lw); w$

★40. $V = h(\pi r^2 + lw); r^2$

For Exercises 41–48, solve the financial formula for the indicated variable. See Examples 1–3.

41. $P = R - C; C$

42. $P = R - C; R$

43. $I = Prt; r$

44. $I = Prt; t$

45. $P = \dfrac{C}{n}; C$

46. $P = \dfrac{C}{n}; n$

47. $A = P + Prt; r$

48. $A = P + Prt; t$

For Exercises 49–60, solve the physics formula for the indicated variable. See Examples 1–3.

49. $d = rt; t$

50. $F = ma; m$

51. $W = Fd; d$

52. $V = kT; k$

53. $P = \dfrac{W}{t}; t$

54. $D = \dfrac{M}{V}; V$

55. $v = -32t + v_0; t$

56. $x = x_0 + vt; v$

57. $F = \dfrac{9}{5}C + 32; C$

58. $C = \dfrac{5}{9}(F - 32); F$

59. $F = G\dfrac{Mm}{R^2}; m$

60. $F = \dfrac{kMn}{d^2}; k$

Find
✗ the mistake

For Exercises 61–64, find and explain the mistake in each solution and then solve correctly. See Examples 1–3.

61. $3n + 7t = 54$; isolate t

$$3n + 7t = 54$$
$$\underline{-3n \qquad\qquad -3n}$$
$$7t = 54 - 3n$$
$$7t = 54 - 3n$$
$$\underline{-7 \qquad -7}$$
$$t = 47 - 3n$$

62. $\dfrac{1}{4}kt = r$; isolate t

$$\frac{1}{4}kt = r$$
$$\frac{4}{1} \cdot \frac{1}{4}kt = 4r$$
$$\frac{1}{k} \cdot kt = 4r\frac{k}{1}$$
$$t = 4kr$$

63. $\dfrac{3m - 2}{5} = nk$; isolate m

$$\frac{3m - 2}{5} = nk$$
$$\frac{5}{1} \cdot \frac{3m - 2}{5} = nk \cdot 5$$
$$3m - 10 = 5nk$$
$$3m - 10 = 5nk$$
$$\underline{+10 \quad +10}$$
$$3m = 5nk + 10$$
$$\frac{3m}{3} = \frac{5nk + 10}{3}$$
$$m = \frac{5nk + 10}{3}$$

64. $7(y - 5) = xv$; isolate y

$$7(y - 5) = xv$$
$$7y - 5 = xv$$
$$7y - 5 = xv$$
$$\underline{+5 \qquad +5}$$
$$\frac{7y}{7} = \frac{xv + 5}{7}$$
$$y = \frac{xv + 5}{7}$$

Review Exercises

Exercises 1 and 2  **EXPRESSIONS**

[1.6] *For Exercises 1 and 2, translate each phrase to an expression.*

1. four more than seven times a number

2. nine less than three times the sum of a number and two

Exercises 3–6 **EQUATIONS AND INEQUALITIES**

[2.3] *For Exercises 3–6, solve.*

3. $4x - 7 = -5$

4. $\dfrac{3}{4}x = -\dfrac{5}{8}$

5. $3(n - 5) = 7n - 12$

6. $\dfrac{1}{4}y + 6 = \dfrac{1}{5}(y + 10)$

2.5 Translating Word Sentences to Equations

Objective ❶ Translate sentences to equations using key words, then solve.   In previous sections, we translated phrases to expressions (Section 1.6) and solved problems using formulas (Section 2.1). In this section, we further develop problem-solving strategies by translating sentences to equations. Because we have not learned how to solve equations containing exponents and roots, we will explore translations that involve only addition, subtraction, multiplication, and division. As in Section 1.6, we begin by translating algebraic equations to word sentences. Several key words or phrases indicate an equal sign.

Key words for an equal sign:	is equal to	is	yields
	is the same as	produces	results in

Translating Algebraic Equations to English Sentences

Equation	Translation
	Equations Involving Addition
$x + 3 = 6$	The sum of x and three is six.
	Three more than x is equal to six.
	x plus three results in six.
	Three added to x produces six.
	x increased by three is the same as six.
	Equations Involving Subtraction
$y - 2 = 7$	The difference of y and two is seven.
	y minus two equals seven.
	Two subtracted from y yields seven.
	Two less than a number results in seven.
	y minus two produces seven.
	A number decreased by two equals seven.
	Equations Involving Multiplication
$\frac{3}{4}m = 9$	The product of three-fourths and m is nine.
	Three-fourths times m is the same as nine.
	Three-fourths of a number equals nine.
	Equations Involving Division
$\frac{x}{4} = -2$	The quotient of x and four results in negative two.
	x divided by four produces negative two.
	Four divided into a number is equal to negative two.
	The ratio of x and four is the same as negative two.
	Equations That Involve More Than One Operation
$4n + 2 = 8$	Two more than the product of four and n is eight.
	Four times n increased by two is equal to eight.
	The sum of four times n and two is the same as eight.
	Two more than the product of four and a number equals eight.
	Two added to four times a number results in eight.
	Equations Involving Parentheses
$4(m - 3) = 6$	Four times the difference of m and three results in six.
	The product of four and the difference of m and three is six.

You also may want to review or reference the key words and their translations in Section 1.6 on pages 63–65.

Procedure **Translating Word Sentences**

To translate a word sentence to an equation, identify the unknown(s), constants, and key words; then write the corresponding symbolic form.

Problems Involving Addition or Subtraction

Example 1   Translate to an equation and then solve.

a. The sum of twenty-five and a number is equal to fourteen.

Understand   The key word *sum* indicates addition, *is equal to* indicates an equal sign, and *a number* indicates a variable.

Plan   Use the key words to translate to an equation, and then solve the equation. We'll use *n* as the variable.

Execute   Translate: The 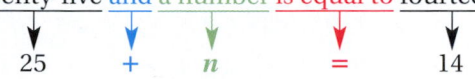 sum of twenty-five and a number is equal to fourteen.

$$25 \quad + \quad n \quad = \quad 14$$

Solve: $25 + n = 14$
$\quad\quad\quad\quad \underline{-25 \quad\quad\quad -25}$ **Subtract 25 from both sides to isolate *n*.**
$\quad\quad\quad\quad 0 + n = -11$

Answer $n = -11$

Check   In the original sentence, replace the unknown amount with -11 and verify that it makes the sentence true. Verify that the sum of twenty-five and a negative eleven is equal to fourteen.

$$25 + (-11) \stackrel{?}{=} 14$$
$$14 = 14$$

Yes, the sum of 25 and -11 is equal to 14.

b. Nineteen less than a number is seven.

Understand   The key words *less than* indicate subtraction. The key word *is* means an equal sign.

Plan   Use the key words to translate to an equation and then solve. We'll use *n* for the variable.

Execute   Translate: Nineteen less than a number is seven.

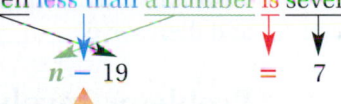

$$n - 19 \quad = \quad 7$$

> **Note:** The key words *less than* require careful translation. In the sentence, the word *nineteen* comes before the words *less than* and the words *a number* come after. In the translation, this order is reversed.

Solve: $n - 19 = 7$
$\quad\quad\quad\quad \underline{+19 \quad +19}$ **Add 19 to both sides.**
$\quad\quad\quad\quad n + \quad 0 = 26$

Answer $n = 26$

Check   Verify that 19 less than 26 is 7.

$$26 - 19 \stackrel{?}{=} 7$$
$$7 = 7$$

Yes, 19 less than 26 is 7.

Your Turn 1   Translate to an equation and then solve.

a. Fifteen more than a number is negative seven.

b. The difference of *x* and thirty-five is negative nine.

LEARNING Strategy

VISUAL If you are a visual learner, underline key words and phrases or write key words and phrases in different colors as is done in the text.

Answers   **Your Turn 1**

a. $n + 15 = -7$; $n = -22$

b. $x - 35 = -9$; $x = 26$

Problems Involving Multiplication

Example 2 Three-fourths of a number is negative five-eighths. Translate to an equation and then solve.

Understand When *of* is preceded by a fraction, it means multiply. The word *is* means an equal sign.

Plan Use the key words to translate to an equation and then solve.

Execute Translate: Three-fourths of a number is negative five-eighths.

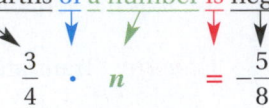

$$\frac{3}{4} \cdot n = -\frac{5}{8}$$

Solve: $\dfrac{\overset{1}{\cancel{4}}}{\underset{1}{\cancel{3}}} \cdot \dfrac{\overset{1}{\cancel{3}}}{\underset{1}{\cancel{4}}} n = -\dfrac{5}{\underset{2}{\cancel{8}}} \cdot \dfrac{\overset{1}{\cancel{4}}}{3}$ **Clear the coefficient $\dfrac{3}{4}$ by multiplying both sides by its reciprocal, $\dfrac{4}{3}$.**

Answer $n = -\dfrac{5}{6}$

Check Verify that $\dfrac{3}{4}$ of $-\dfrac{5}{6}$ is $-\dfrac{5}{8}$.

$$\frac{\overset{1}{\cancel{3}}}{4} \cdot -\frac{5}{\underset{2}{\cancel{6}}} \overset{?}{=} -\frac{5}{8}$$

$$-\frac{5}{8} = -\frac{5}{8}$$

Yes, $\dfrac{3}{4}$ of $-\dfrac{5}{6}$ is $-\dfrac{5}{8}$, so $-\dfrac{5}{6}$ is correct.

Your Turn 2 The product of 0.6 and a number is -25.08. Translate to an equation and then solve.

Problems Involving More Than One Operation

Example 3 Seven less than the product of four and a number is equal to negative five. Translate to an equation and then solve.

Understand *Less than* indicates subtraction in reverse order, *product* indicates multiplication, and *is equal to* indicates an equal sign.

Plan Translate to an equation using the key words and then solve the equation.

Execute Translate:

Seven less than the product of four and a number is equal to negative five.

$$4n - 7 \qquad\qquad = \qquad -5$$

Solve: $4n - 7 = -5$

$\underline{+7 \quad +7}$ **Add 7 to both sides to isolate $4n$.**

$4n + 0 = \quad 2$

$\dfrac{4n}{4} = \dfrac{2}{4}$ **Divide both sides by 4 to clear the coefficient.**

Answer $n = \dfrac{1}{2}$ or 0.5

Answer Your Turn 2

$0.6n = -25.08; n = -41.8$

Check Verify that 7 less than the product of 4 and $\frac{1}{2}$ is equal to -5.

$$4\left(\frac{1}{2}\right) - 7 \overset{?}{=} -5$$

$$2 - 7 \overset{?}{=} -5$$

$$-5 = -5$$

Yes, 7 less than the product of 4 and $\frac{1}{2}$ is equal to -5.

Your Turn 3 Thirteen more than the product of four and a number yields negative seven. Translate to an equation and then solve.

Translations Requiring Parentheses

Sometimes a translation requires parentheses. This occurs when a sum or difference is multiplied or divided. When we multiply or divide a sum or difference, the addition or subtraction is calculated first. Because addition and subtraction follow multiplication and division in the order of operations, parentheses must be inserted around the addition or subtraction so that these operations are performed first.

Example 4 Two-thirds of the difference of six and a number is the same as the number divided by four. Translate to an equation and then solve.

Understand *Of* means to multiply, and the word *difference* indicates subtraction. Because the difference is being multiplied, we write the difference expression in parentheses. *Divided by* indicates division.

Plan Use the key words to translate to an equation and then solve.

Execute Translate:

Two-thirds of the difference of six and a number is the same as the number divided by four.

$$\frac{2}{3} \cdot (6 - n) = \frac{n}{4}$$

Solve:

$$\frac{2}{3}(6 - n) = \frac{n}{4}$$

$$\frac{2}{\cancel{3}} \cdot \frac{\cancel{6}^{2}}{1} - \frac{2}{3}n = \frac{n}{4} \qquad \text{\color{blue}\textbf{Distribute to clear the parentheses.}}$$

$$4 - \frac{2}{3}n = \frac{n}{4} \qquad \text{\color{blue}\textbf{Simplify.}}$$

$$12\left(4 - \frac{2}{3}n\right) = \left(\frac{n}{4}\right)12 \qquad \text{\color{blue}\textbf{Multiply both sides by the LCD, 12, to clear the fractions.}}$$

$$12 \cdot 4 - \frac{\cancel{12}^{4}}{1} \cdot \frac{2}{\cancel{3}_{1}}n = \frac{n}{\cancel{4}_{1}} \cdot \frac{\cancel{12}^{3}}{1} \qquad \text{\color{blue}\textbf{Distribute 12 and simplify.}}$$

$$48 - 8n = 3n \qquad \text{\color{blue}\textbf{Add 8n to both sides to separate the constant term and variable terms.}}$$

$$\underline{+8n \qquad +8n}$$

$$48 + 0 = 11n$$

$$\frac{48}{11} = \frac{11n}{11} \qquad \text{\color{blue}\textbf{Clear the coefficient by dividing both sides by 11.}}$$

Answer $\dfrac{48}{11} = n$

Check Verify that $\dfrac{2}{3}$ of the difference of 6 and $\dfrac{48}{11}$ is the same as $\dfrac{48}{11}$ divided by 4.

$$\frac{2}{3}\left(6 - \frac{48}{11}\right) \overset{?}{=} \frac{1}{4}\left(\frac{48}{11}\right)$$

Write 6 as a fraction with a denominator of 11. $\dfrac{2}{3}\left(\dfrac{66}{11} - \dfrac{48}{11}\right) \overset{?}{=} \dfrac{1}{\overset{}{\underset{1}{4}}}\left(\dfrac{\overset{12}{\cancel{48}}}{11}\right)$ **Divide out 4.**

Subtract. $\dfrac{2}{3}\left(\dfrac{18}{11}\right) \overset{?}{=} \dfrac{12}{11}$ **Multiply.**

Divide out 3. $\dfrac{2}{\overset{}{\underset{1}{\cancel{3}}}}\left(\dfrac{\overset{6}{\cancel{18}}}{11}\right) \overset{?}{=} \dfrac{12}{11}$

Multiply. $\dfrac{12}{11} = \dfrac{12}{11}$

Yes, $\dfrac{2}{3}$ of the difference of 6 and $\dfrac{48}{11}$ is the same as $\dfrac{48}{11}$ divided by 4.

Answers Your Turn 4

a. $3(n - 5) = 7n - 12$;
$n = -\dfrac{3}{4}$

b. $\dfrac{1}{4}n + 6 = \dfrac{1}{5}(n + 10)$;
$n = -80$

Your Turn 4 Translate to an equation and then solve.

a. Three times the difference of a number and five is equal to seven times the number minus twelve.

b. The sum of one-fourth of a number and six is equal to one-fifth of the sum of the number and ten.

2.5 Exercises   For Extra Help   *MyMathLab*    PRACTICE   WATCH   DOWNLOAD

1. List three key words that indicate addition.

2. List three key words that indicate subtraction.

3. List three key words that indicate multiplication.

4. List three key words that indicate division.

5. Why are "four minus n" and "four less than n" different?

6. Write a sentence that requires parentheses when translated and explain why your sentence requires parentheses.

For Exercises 7–44, translate each sentence to an equation and then solve.
See Examples 1–4.

7. Three more than y is negative 8.

8. Six added to p is negative two.

9. Six decreased by the number x is equal to negative three.

10. Eighteen subtracted from a number is equal to negative three.

11. Eleven times m is negative ninety-nine.

12. The product of negative three and z is -18.

13. Four divided into m is 1.6.

14. The quotient of a number and -6.5 is 4.2.

15. Four-fifths of a number is five-eighths.

16. Three-sevenths of a number is negative nine-eighths.

17. Seven more than three times a number is equal to thirty-four.

18. Nineteen more than triple a number is one hundred.

19. Nine less than five times a number is seventy-six.

20. Twenty-five less than four times a number is eleven.

21. Eight times the sum of eight and a number is equal to one hundred sixty.

22. Four times the sum of a number and three is equal to sixteen.

23. Negative three times the difference of x and two is twelve.

24. Tripling the difference of x and five produces negative fifteen.

25. One-third of the sum of a number and two is one.

26. Half of the difference of a number and two results in four.

27. Eleven less than three times a number is equal to that number added to five.

28. Five more than four times a number is equal to seven subtracted from that number.

29. Ten is the result when one is subtracted from the ratio of a number to four.

30. Two less than the quotient of a number and five is six.

31. The product of three and the sum of a number and four subtracted from twice the number yields negative three.

32. Seven times the sum of a number and six subtracted from four times the number results in negative twenty-one.

33. The difference of a number and twelve is added to the difference of twice the number and eight so that the result is thirteen.

34. The sum of a number and six is added to the difference of the number tripled and four so that the result is fourteen.

35. Five times the sum of a number and two-thirds is equal to three times the number decreased by two-thirds.

36. The product of three and the sum of a number and two-fifths is four-fifths less than negative two times the number.

37. Two times a number decreased by four is equal to triple the number added to sixteen.

38. Six less than three times a number is equal to the difference of fourteen and twice the number.

39. Two-fifths of a number is equal to two less than one-half the number.

40. One-third of a number is the same as half of the number decreased by one.

41. The quotient of three less than a number and three is the same as the quotient of one less than the number and four.

42. The quotient of one more than a number and two is the same as the quotient of six more than twice the number and eight.

43. The product of negative four and the difference of two and a number is six less than the product of two and the difference of four and three times the number.

44. Negative two times the difference of one and three times a number is equal to four added to five times the difference of one and the number.

For Exercises 45–54, translate the equation to a word sentence. There is more than one correct translation.

45. $4x + 3 = 7$

46. $-3y + 8 = 10$

47. $6(y + 4) = -10y$

48. $8(w - 2) = 3w$

49. $\frac{1}{2}(x - 3) = \frac{2}{3}(x - 8)$

50. $\frac{1}{2}(t + 1) = \frac{1}{3}(t - 5)$

51. $0.05m + 0.06(m - 11) = 22$

52. $0.05v + 0.03(v - 4.5) = 0.465$

53. $\frac{2}{3}x + \frac{3}{4}x + \frac{1}{2}x = 10$

54. $\frac{1}{2}a + \frac{1}{3}a + \frac{1}{6}a = 5$

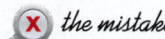 **Find the mistake** *For Exercises 55–60, explain the mistake in each translation; then translate correctly. See Examples 1–4.*

55. Ten less than a number is forty.
Translation: $10 - n = 40$

56. Twelve divided into a number is negative eight.
Translation: $12 \div n = -8$

57. Five times the difference of a number and six is equal to negative two.
Translation: $5x - 6 = -2$

58. Nine times the sum of a number and eight is the same as three subtracted from the number.
Translation: $9(y + 8) = 3 - y$

59. Twice a number minus the sum of the number and three is equal to negative six.
Translation: $2t - t + 3 = -6$

60. Four times the sum of a number and two is equal to the number minus the difference of the number and seven.
Translation: $4(y + 2) = y - (7 - y)$

For Exercises 61–68, translate to a formula; then use the formula to solve the problem.

61. The perimeter of a rectangle is equal to twice the sum of its length and width. Find the perimeter with the following length and width.

	Length	Width
a.	24 ft.	18 ft.
b.	12.5 cm	18 cm
c.	$8\frac{1}{4}$ in.	$10\frac{3}{8}$ in.

62. The surface area of a box is equal to twice the sum of its length times its width, its length times its height, and its width times its height. Find the surface area of a box with the following length, width, and height.

	Length	Width	Height
a.	15 in.	6 in.	4 in.
b.	9.2 cm	12 cm	6.5 cm
c.	8 ft.	$2\frac{1}{2}$ ft.	3 ft.

63. An isosceles triangle has two sides that are equal in length. The perimeter of an isosceles triangle is the sum of its base and twice the length of one of the other sides. Find the perimeter of each listed isosceles triangle.

	Base	Sides
a.	5 in.	11 in.
b.	1.5 m	0.6 m
c.	$15\frac{1}{4}$ in.	$9\frac{1}{2}$ in.

64. The perimeter of a semicircle is the sum of the diameter and π times the radius. Find the perimeter of each semicircle listed. Where appropriate, round your answer to the nearest tenth.

a. Diameter = 18 in. (Use $\pi \approx 3.14$.)
b. Radius = 4.5 m (Use $\pi \approx 3.14$.)
c. Diameter = $2\frac{3}{4}$ ft. (Use $\pi \approx \frac{22}{7}$.)

65. The simple interest earned after investing an amount of money, called principal, is equal to the product of the principal, the interest rate, and the time in years that the money remains invested. Use the formula to calculate the interest of each of the following investments.

	Principal	Rate	Time
a.	$4000	0.05	2 years
b.	$500	0.03	$\frac{1}{2}$ year
c.	$2000	0.06	$\frac{1}{4}$ year

66. Investors often consider a stock's price-to-earnings ratio to help them decide whether a stock is worth investing in. The price-to-earnings ratio is equal to the stock's price divided by its company's earnings per share over the last 12 months. Use your formula to calculate each company's price-to-earnings ratio. Round your answers to the nearest tenth. (*Source:* David and Tom Gardner, *The Motley Fool Investment Guide*, Simon & Schuster, 1996.)

	Company	Price	Earnings per Share
a.	Microsoft	$47.98	$1.41
b.	Intel	$18.79	$0.29
c.	General Motors	$46.55	$4.01

67. Skydivers often use a formula to estimate free-fall time. The amount of free-fall time in seconds is approximately equal to the difference of the exit altitude and the parachute deployment altitude divided by 153.8. Use this formula to calculate free-fall times if a skydiver deploys at each of the following altitudes after falling from an exit altitude of 12,500 feet. Round your answers to the nearest second.

   a. 5000 feet
   b. 4500 feet
   c. 3000 feet

68. A person's body mass index (BMI) is a measure of the amount of body fat based on height, in inches, and weight, in pounds. A person's BMI is equal to the ratio of the product of his or her weight and 704.5 to his or her height in inches multiplied by itself. Use your formula to find the BMI of each person listed below. Round your answers to the nearest tenth. (*Source:* National Institutes of Health.)

	Height	Weight
a.	5'4"	138 lb
b.	5'9"	155 lb
c.	6'2"	220 lb

Review Exercises

Exercises 1–6 **EQUATIONS AND INEQUALITIES**

[1.1]   **1.** Is $19 \geq 19$ true or false? Explain.

[1.1, 1.5] *For Exercises 2 and 3, use* $<, >,$ *or* $=$ *to make a true sentence.*

2. $-|-6|$ ▢ $-2(3)$

3. $5(4) - 23$ ▢ $2(3 - 8)$

[2.3] *For Exercises 4–6, solve and check.*

4. $7x + 9 = 4x - 6$

5. $\dfrac{5}{6} = -\dfrac{3}{4}x$

6. $\dfrac{1}{3}(y - 2) = \dfrac{3}{5}y + 6$

2.6 Solving Linear Inequalities

Objectives

1. Represent solutions to inequalities graphically and using set notation.
2. Solve linear inequalities.
3. Solve problems involving linear inequalities.

Objective 1 Represent solutions to inequalities graphically and using set notation. Not all problems translate to equations. Sometimes a problem can have a range of values as solutions. In mathematics, we can write inequalities to describe situations where a range of solutions is possible. Following are the inequality symbols and their meanings.

Symbol	Meaning
$<$	is less than
$>$	is greater than
$\leq$	is less than or equal to
$\geq$	is greater than or equal to

Throughout the chapter, we have explored techniques for solving linear equations. In this section, we explore how to solve **linear inequalities**.

Definition *Linear inequality:* An inequality containing expressions in which each variable term contains a single variable with an exponent of 1.

Following are some examples of linear inequalities:

$$x > 5 \qquad n + 2 < 6 \qquad 2(y - 3) \leq 5y - 9 \qquad 2x + 3y \geq 6$$

The solutions of an inequality are any numbers that can replace the variable(s) in the inequality and make it true. Because inequalities have a range of solutions, we often write those solutions in set-builder notation, which we introduced in Section 1.1. For example, the solution set for the inequality $x \geq 5$ contains 5 and every real number greater than 5, which we write as $\{x | x \geq 5\}$. We read this set-builder notation as shown here.

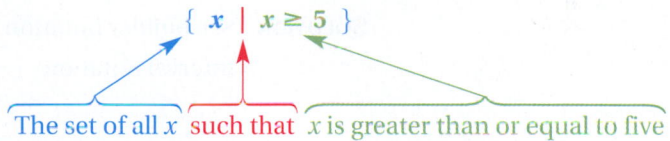

We can graph solution sets for inequalities on a number line. Because the solution set for $x \geq 5$ contains 5 and every real number to the right of 5, we draw a dot (or solid circle) at 5 and shade to the right of 5.

$x \geq 5$ is represented by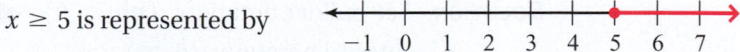

The solution set for $x < 2$ contains every real number to the left of 2, but not 2 itself; so it is written $\{x | x < 2\}$. To graph this solution set, we draw an open circle at 2, indicating that 2 is not included, and shade to the left of 2.

$x < 2$ is represented by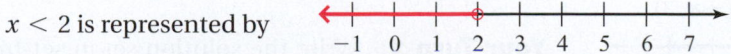

Another popular notation used to indicate ranges of values is *interval notation*, which uses parentheses and brackets to indicate whether end values are included. Parentheses are used for end values that are not included in the interval, and brackets are used for end values that are included. To write interval notation, we imagine traveling from left to right on the number line.

For example, as we travel from left to right on the graph for $x \geq 5$, we encounter 5 and then every real number to the right of 5; so the interval notation is $[5, \infty)$. The symbol ∞ means infinity and is used to indicate the positive extreme. Since ∞ can never be reached, it will always have a parenthesis indicating that it is not included as an end value. As we travel from left to right on the graph for $x < 2$, we go from the negative extreme up to 2, but not including 2; so the interval notation is $(-\infty, 2)$. Notice that $-\infty$ represents the leftmost extreme.

The parentheses and brackets from interval notation, instead of open and solid circles, can be used on the graphs.

$x \geq 5$ is represented by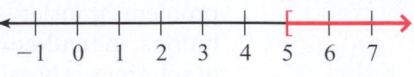

$x < 2$ is represented by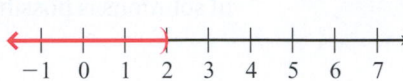

Your instructor may have a style preference. Because interval notation is a preferred notation in college algebra courses, we draw our graphs with the parentheses and brackets. Notice that the parentheses or brackets "open" in the direction that is shaded.

Procedure   **Graphing Inequalities**

To graph an inequality on a number line,

1. If the symbol is $\leq$ or $\geq$, draw a bracket (or solid circle) on the number line at the indicated number open to the left for $\leq$ and to the right for $\geq$. If the symbol is $<$ or $>$, draw a parenthesis (or open circle) on the number line at the indicated number open to the left for $<$ and to the right for $>$.
2. If the variable is greater than the indicated number, shade to the right of the indicated number. If the variable is less than the indicated number, shade to the left of the indicated number.

Example 1   Write the solution set in set-builder notation and interval notation; then graph the solution set.

a. $x \leq -6$

SOLUTION   Set-builder notation: $\{x | x \leq -6\}$
Interval notation: $(-\infty, -6]$

> **Note:** When writing interval notation, imagine traveling from left to right so this range of values is from $-\infty$ to -6, including -6.

Graph: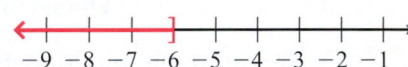

b. $n > 0$

SOLUTION   Set-builder notation: $\{n | n > 0\}$
Interval notation: $(0, \infty)$

> **Connection** $\{n | n > 0\}$ is a way of expressing the set of all positive real numbers.

Graph:

Your Turn 1   Write the solution set in set-builder notation and interval notation; then graph the solution set.

a. $x < 2$ **b.** $t \geq -1$

 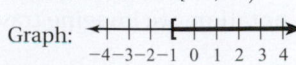

Compound Inequalities

Inequalities containing two inequality symbols are called *compound inequalities*. Compound inequalities are useful in writing a range of values between two numbers. For example, $1 < x < 6$ is a compound inequality meaning that x can be any number that is greater than 1 and less than 6. Its solution set contains every real number between 1 and 6, but not 1 and 6.

Set-builder notation: $\{x \mid 1 < x < 6\}$

Interval notation: $(1, 6)$

Graph:

Note: Because the end values 1 and 6 are not included in the solution set, we use parentheses in the interval notation and on the graph.

Example 2 Write the solution set for $-5 \leq x < 1$ in set-builder notation and interval notation; then graph the solution set.

SOLUTION Set-builder notation: $\{x \mid -5 \leq x < 1\}$

Interval notation: $[-5, 1)$

Graph:

Note: Because the end value -5 is included in the solution set, we use a bracket in the interval notation and on the graph. Because 1 is not included, we use a parenthesis.

Your Turn 2 Write the solution set for $-4 < y \leq -1$ in set-builder notation and interval notation; then graph the solution set.

Objective 2 Solve linear inequalities. To solve inequalities such as $n + 2 < 6$ and $2(x - 3) \leq 5x - 9$, we follow essentially the same process as for solving equations. The addition and multiplication principles of inequalities are similar to the principles of equality. First, let's examine the addition principle of inequality.

Consider the inequality $3 < 7$. According to the addition principle, if we add or subtract the same amount on both sides, the inequality should still be true. Let's test this by choosing some numbers to add and subtract on both sides.

Answer Your Turn 2

Set-builder notation:
$\{y \mid -4 < y \leq -1\}$
Interval notation: $(-4, -1]$

Graph:

Add **2** to both sides:	Subtract **9** from both sides:
$3 < 7$	$3 < 7$
$\underline{+2 \quad +2}$	$\underline{-9 \quad -9}$
$5 < 9$ Still true	$-6 < -2$ Still true

Rule   The Addition Principle of Inequality

If $a < b$, then $a + c < b + c$ is true for all real numbers a, b, and c. The principle also holds true when $<$ is replaced with $>$, $\leq$, or $\geq$.

Note: This principle indicates that we can add (or subtract) the same amount on both sides of an inequality without affecting its solution(s). Although the principle is written in terms of the $<$ symbol, the principle is true for any inequality symbol.

Example 3 Solve $x + 2 \leq -3$ and write the solution set in set-builder notation and interval notation; then graph the solution set.

SOLUTION

Note: Subtracting 2 from both sides does not affect the inequality.

$$x + 2 \leq -3$$
$$\underline{\quad -2 \quad\quad -2}$$
$$x \leq -5$$

Isolate x by subtracting 2 from both sides.

Set-builder notation: $\{x \mid x \leq -5\}$

Interval notation: $(-\infty, -5]$

Graph: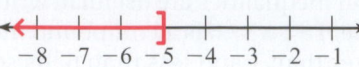

Your Turn 3 Solve $n - 6 > -5$ and write the solution set in set-builder notation and interval notation; then graph the solution set.

The multiplication principle, on the other hand, does not work as neatly as it did for equations. As we shall see, sometimes multiplying or dividing both sides of an inequality by the same number can turn a true inequality into a false inequality. Let's multiply both sides of $3 < 7$ by a positive number and then by a negative number.

Multiply both sides by 4:	**Multiply both sides by −2:**
$4(3) < 4(7)$	$-2(3) < -2(7)$
$12 < 28$ Still true	$-6 < -14$ Not true!

Conclusion An inequality remains true when we multiply both sides by a positive number, but not when we multiply both sides by a negative number.

Rule **The Multiplication Principle of Inequality**

If a and b are real numbers, where $a < b$, then $ac < bc$ is true if c is a positive real number.

If a and b are real numbers, where $a < b$, then $ac > bc$ is true if c is a negative real number.

The principle also holds true for $>$, $\leq$, and $\geq$.

Note: We can multiply (or divide) both sides of an inequality by the same positive amount without affecting its solution(s). However, if we multiply (or divide) both sides of an inequality by the same negative amount, we must reverse the direction of the inequality symbol to maintain the truth of the inequality.

Example 4 Solve and write the solution set in set-builder notation and interval notation; then graph the solution set.

a. $3x > -7$

SOLUTION

Note: Dividing both sides by a positive number does not affect the inequality.

$$\frac{3x}{3} > \frac{-7}{3}$$

To isolate x, we clear the 3 coefficient by dividing by 3 on both sides.

$$x > -\frac{7}{3}$$

Set-builder notation: $\left\{ x \mid x > -\frac{7}{3} \right\}$

Interval notation: $\left(-\frac{7}{3}, \infty \right)$

Graph:

b. $-8x \geq 24$

 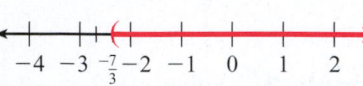
SOLUTION

Note: Because we divided both sides by a negative number, we reversed the direction of the inequality symbol.

$$\frac{-8x}{-8} \leq \frac{24}{-8}$$

To isolate x, we clear the -8 coefficient by dividing by -8 on both sides.

$$x \leq -3$$

Set-builder notation: $\{x | x \le -3\}$

Interval notation: $(-\infty, -3]$

Graph:

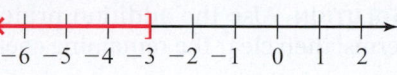

c. $\dfrac{5}{6} \le -\dfrac{3}{4}n$

SOLUTION

> **Note:** You may find it helpful to rewrite the inequality so that the variable is on the left side. The inequality $-\dfrac{10}{9} \ge n$ is the same as $n \le -\dfrac{10}{9}$.

$-\dfrac{\overset{2}{\cancel{4}}}{3}\left(\dfrac{5}{\underset{3}{\cancel{6}}}\right) \ge -\dfrac{\overset{1}{\cancel{4}}}{3}\left(-\dfrac{\overset{1}{\cancel{3}}}{\underset{1}{\cancel{4}}}n\right)$

$-\dfrac{10}{9} \ge n$

To isolate n, we clear the $-\dfrac{3}{4}$ coefficient by multiplying by its reciprocal $-\dfrac{4}{3}$ on both sides. Multiplying by a negative number means that we must reverse the direction of the inequality symbol.

Set-builder notation: $\left\{ n \middle| n \le -\dfrac{10}{9} \right\}$

Interval notation: $\left(-\infty, -\dfrac{10}{9} \right]$

Graph:

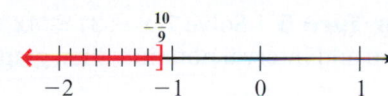

Your Turn 4   Solve $-\dfrac{2}{5}t < \dfrac{9}{10}$ and write the solution set in set-builder notation and in interval notation; then graph the solution set.

We can solve more complex linear inequalities using an outline similar to that for solving linear equations.

Procedure   **Solving Linear Inequalities**

To solve linear inequalities,

1. Simplify both sides of the inequality as needed.
   a. Distribute to clear parentheses.
   b. Clear fractions or decimals by multiplying through by the LCD just as we did for equations. (Clearing fractions and decimals is optional.)
   c. Combine like terms.

2. Use the addition principle so that all variable terms are on one side of the inequality and all constants are on the other side. Then combine like terms.

3. Use the multiplication principle to clear any remaining coefficient. If you multiply (or divide) both sides by a negative number, reverse the direction of the inequality symbol.

Notice that clearing the term with the smaller coefficient in step 2 results in a positive coefficient, which means that you won't have to reverse the direction of the inequality in step 3.

Answer   **Your Turn 4**

$t > -\dfrac{9}{4}$ or $-2\dfrac{1}{4}$

Set-builder notation: $\left\{ t \middle| t > -\dfrac{9}{4} \right\}$

Interval notation: $\left(-\dfrac{9}{4}, \infty \right)$

Graph:

Example 5   Solve $7x + 9 > 4x - 6$ and write the solution set in set-builder notation and interval notation; then graph the solution set.

SOLUTION   Use the addition principle to separate the variable terms and constant terms; then clear the remaining coefficient.

> **Note:** Clearing the $4x$ term results in a positive coefficient after combining like terms; so we won't have to reverse the inequality when we clear the 3 coefficient.

$$7x + 9 > 4x - 6$$
$$\underline{-4x \qquad -4x} \qquad \text{Subtract } 4x \text{ from both sides.}$$
$$3x + 9 > 0 - 6$$
$$3x + 9 > -6$$
$$\underline{-9 \qquad -9} \qquad \text{Subtract 9 from both sides.}$$
$$3x + 0 > -15$$
$$\frac{3x}{3} > \frac{-15}{3} \qquad \text{Divide both sides by 3 to isolate } x.$$
$$x > -5$$

Set-builder notation: $\{x \mid x > -5\}$

Interval notation: $(-5, \infty)$

Graph:

```
←――+――+――[――+――+――+――+――+――+→
  -7 -6 -5 -4 -3 -2 -1  0  1
```

Your Turn 5   Solve $7(x - 3) \leq 4x + 3$ and write the solution set in set-builder notation and interval notation; then graph the solution set.

Objective ③ Solve problems involving linear inequalities.   Problems requiring inequalities can be translated using key words much like those we used to translate sentences to equations. The following table lists common key words that indicate inequalities.

Less Than:		Greater Than:	
A number is less than seven.	$n < 7$	A number is greater than two.	$n > 2$
A number must be smaller than five.	$n < 5$	A number must be greater than three.	$n > 3$
		A number must be more than negative six.	$n > -6$

Less Than or Equal To:		Greater Than or Equal To:	
A number is at most nine.	$n \leq 9$	A number is at least two.	$n \geq 2$
The maximum is fourteen.	$n \leq 14$	The minimum is eighteen.	$n \geq 18$

Example 6   Three-fourths of a number is at least eighteen. Translate to an inequality; then solve.

Understand   Because the word *of* is preceded by a fraction, it means multiplication. The key words *at least* indicate a greater-than or equal-to symbol.

Plan   Translate the key words; then solve. We'll use n for the variable.

Execute   Translate.

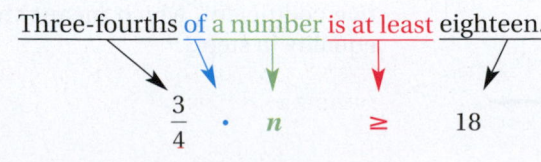

$$\frac{3}{4} \cdot n \geq 18$$

Solve: $\dfrac{3}{4} \cdot n \geq 18$

$$\dfrac{\overset{1}{\cancel{4}}}{\underset{1}{\cancel{3}}} \cdot \dfrac{\overset{1}{\cancel{3}}}{\underset{1}{\cancel{4}}} n \geq \dfrac{4}{\underset{1}{\cancel{3}}} \cdot \dfrac{\overset{6}{\cancel{18}}}{1}$$ **Multiply both sides by $\dfrac{4}{3}$ to isolate n.**

Answer $n \geq 24$

Check   Verify that 24 and any number greater than 24 satisfy the original sentence.

Test 24:	**Test a number greater than 24, such as 25:**
Is $\dfrac{3}{4}$ of 24 at least 18?	Is $\dfrac{3}{4}$ of 25 at least 18?
$\dfrac{3}{\underset{1}{\cancel{4}}} \cdot \dfrac{\overset{6}{\cancel{24}}}{1} \geq 18$	$\dfrac{3}{4} \cdot \dfrac{25}{1} \geq 18$
$18 \geq 18$ True	$\dfrac{75}{4} \geq 18$
	$18\dfrac{3}{4} \geq 18$ True

> **Your Turn 6**   0.2 times the sum of a number and eight is less than the difference of the number and five. Translate to an inequality; then solve.

Example 7   Darwin is planning a garden area, which he will enclose with a fence. Cost restricts him to a total of 240 feet of fencing materials. He wants the garden to span the entire width of his lot, which is 80 feet. What is the maximum length of the garden?

Understand   The fence surrounds the garden; so 240 feet is the maximum perimeter for the garden. The width of the garden is to be 80 feet, and we must find the length.

Plan   The formula for the perimeter of a rectangle is $P = 2l + 2w$. Because 240 feet is a *maximum* perimeter, we write the perimeter formula as an inequality so that the expression used to calculate perimeter ($2l + 2w$) is less than or equal to 240. Because the width is to be 80 feet, we replace w in the formula with 80.

Execute   Translate:

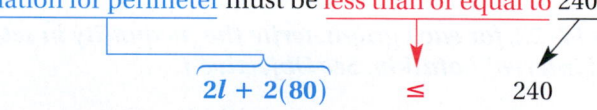

The calculation for perimeter must be less than or equal to 240.

$$2l + 2(80) \quad \leq \quad 240$$

Solve: $2l + 160 \leq 240$ **Subtract 160 from both sides.**
$$\underline{\; -160 \quad -160}$$
$$2l + 0 \leq 80$$
$$\dfrac{2l}{2} \leq \dfrac{80}{2}$$ **Divide both sides by 2 to isolate l.**
$$l \leq 40$$

Answer   The length must be less than or equal to 40 feet, which means that the maximum length is 40 feet.

Check   Verify that a garden with a length of 40 feet or less and a width of 80 feet has a perimeter that is 240 feet or less. This will be left to the reader.

Answer   Your Turn 6

$0.2(n + 8) < n - 5$;
$n > 8.25$

2.6 Exercises   For Extra Help   *MyMathLab*

Math XL PRACTICE   WATCH   DOWNLOAD

Note: Exercises marked with a ★ represent challenging exercises.

1. What is a solution for an inequality?

2. Explain the difference between $x < 8$ and $x \leq 8$.

3. In your own words, explain the meaning of $x \geq -5$.

4. What action causes the direction of an inequality symbol to change?

For Exercises 5–16: **a. Write the solution set in set-builder notation.**
b. Write the solution set in interval notation.   *See Examples 1 and 2.*
c. Graph the solution set.

5. $x \geq -3$

6. $n \leq -5$

7. $h < 6$

8. $x > 0$

9. $n < -\dfrac{2}{3}$

10. $a \geq \dfrac{4}{5}$

11. $t \geq 2.4$

12. $p \leq -0.6$

13. $-3 < x < 6$

14. $4 < c \leq 10$

15. $0 \leq n \leq 5$

16. $-1 < a < 5$

For Exercises 17–22, for each graph, write the inequality in set-builder notation and interval notation. See Objective 1.

17.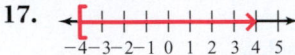

18.

19.

20.

21.

22.

For Exercises 23–56: **a.** *Solve.*
b. *Write the solution set in set-builder notation.*
c. *Write the solution set in interval notation.* } *See Examples 3–5.*
d. *Graph the solution set.*

23. $n - 3 > 2$ **24.** $m - 5 < 20$ **25.** $z + 2 \leq -4$ **26.** $x + 8 \geq -12$

27. $16y \geq 32$ **28.** $4z \leq -20$ **29.** $-3x \geq -12$ **30.** $-4x < -16$

31. $\frac{2}{3}x \geq 4$ **32.** $\frac{3}{5}a < -6$ **33.** $-\frac{3}{4}m < -6$ **34.** $-\frac{5}{3}p > -10$

35. $5y + 1 > 16$ **36.** $3y - 2 > 10$ **37.** $1 - 6x < 25$ **38.** $3 - 2c \geq 17$

39. $\frac{a}{2} + 1 < \frac{3}{2}$ **40.** $\frac{k}{3} - \frac{2}{3} > -2$ **41.** $10 + 2f \leq 2 - 2f$ **42.** $9 + 2n < 3 - 4n$

43. $3 - 6u \geq -5 - 2u$ **44.** $4 - 9k < -4k + 19$ **45.** $2(c - 3) < 3(c + 2)$ **46.** $4(d + 4) \geq 5(d + 2)$

47. $10(9w + 13) - 5(3w - 1) \leq 8(11w + 12)$ **48.** $6(n + 13) - 2(4n - 2) > 3(5n - 1)$

49. $\frac{1}{2}(3x + 1) \leq \frac{1}{3}(4x - 5)$ **50.** $\frac{1}{5}(6m - 7) > \frac{1}{2}(3m - 1)$

51. $\frac{1}{6}(2n + 4) - \frac{1}{3}(3n - 15) > 1$ **52.** $-\frac{1}{3}(2y - 5) + \frac{1}{6}(5y - 2) \geq 1$

53. $0.4l - 0.37 < 0.3(l + 2.1)$

54. $0.05 + 0.03(v - 4.5) \geq 0.465$

55. $0.6t + 0.2(t - 8) \leq 0.5t - 0.6$

56. $2.1s + 3(1.4s - 1.5) \geq 0.09s + 20.34$

For Exercises 57–68, translate to an inequality and then solve. See Example 6.

57. Four subtracted from a number is greater than twenty-four.

58. Six less than a number is less than or equal to negative four.

59. Four-ninths of a number is less than or equal to negative eight.

60. Three-fifths of a number is greater than negative twenty-one.

61. Eight times a number less thirty-six is at least sixty.

62. Three subtracted from a number times eleven is at least forty-one.

63. Five added to half of a number is at most two.

64. The sum of two-fifths of a number and two is at most eight.

65. The difference of four times a number and eight is less than two times the number.

66. Triple a number is less than the number decreased by eight.

67. Twenty-five is greater than or equal to seven more than six times a number.

68. Negative five is greater than fifteen less than five times a number.

For Exercises 69–82, solve. See Example 7. (Formulas are listed in the Summary on page 156.)

69. A school is planning a playground. The area of a rectangular playground may not exceed 170 square feet. If the width is 17 feet, what range of values can the length have?

70. Scotty needs to make a rectangular pool cover that is at least 128 square feet. If the pool length is 16 feet, what range of values must the width be?

71. The design of a storage box calls for a width of 27 inches and a length of 41 inches. If the surface area must be at least 4254 square inches, find the range of values for the height.

72. A company builds truck containers in the shape of rectangular solids. One such container has a length of 24 feet and width of 8 feet. What is the range of values for the height if the volume is to be at least 576 cubic feet?

73. Andre builds tables. One particular circular table requires him to glue a strip of veneer around the circumference of the top. He is designing a new circular table and has a strip of veneer that is 150 inches long that he wants to use. What range of values can the radius have? (Use $\pi \approx 3.14$.) Round your answer to the nearest hundredth.

74. The city parks department is putting in a circular flower garden that is to be enclosed using an edging material. If 125 feet of the edging material is available, what is the range of values for the diameter of the flower garden? Round your answer to the nearest hundredth.

75. Jon never exceeds 65 miles per hour when driving on the highway. If he is traveling 410 miles, what range of values can his time spent driving have?

76. A truck driver must make a delivery 298 miles away in five hours or less. What range of values can his average rate be so that he meets the schedule?

77. The fourth-quarter profit goal for a software company is at least $850,000. If it is known that the cost for the quarter will be $625,000, find the revenue that must be generated to achieve the goal. (Use $P = R - C$.)

78. Silver is a solid up to a temperature of 960.8°C. Write an inequality to describe this range of temperatures in degrees Fahrenheit. (Use $F = \frac{9}{5}C + 32$.)

79. To get an A in her math course, Tina needs the average of five tests to be at least 90. Her current test scores are 82, 91, 95, and 84. What range of scores on the fifth test would get her an A in the course?

80. In Aaron's English course, the final grade is determined by the average of five papers. The department requires any student whose average falls below 75 to repeat the course. Aaron's scores on the first four papers are 68, 78, 80, and 72. What range of scores on the fifth paper would result in him having to repeat the course?

81. The design of a circuit specifies that the voltage cannot exceed 12 volts. If the resistance of the circuit is 8 ohms, find the range of values that the current can have. (Use $V = iR$.)

82. A label in an elevator indicates that the maximum load is 9000 newtons; that is, the downward force should not exceed −9000 newtons. If the acceleration due to gravity is −9.8 meters per second squared, find the range of values of mass that can be loaded onto the elevator. Round your answer to the nearest hundredth. (Use $F = ma$.)

Puzzle Problem

Write the set of all values such that $|x| = -x$.

Review Exercises

Exercise 1 ▸ CONSTANTS AND VARIABLES

[1.1] **1.** Write a set representing the natural numbers up to 10.

Exercises 2–4 ▸ EXPRESSIONS

[1.4] **2.** Simplify: $\dfrac{2450}{140}$

[1.6] **3.** Translate to an expression: six more than twice the difference of a number and five.

[1.7] **4.** Evaluate $b^2 - 4ac$ when $a = 1$, $b = -2$, and $c = -3$.

Exercises 5 and 6 ▸ EQUATIONS AND INEQUALITIES

[2.3] *For Exercises 5 and 6, solve.*

   5. $3\frac{1}{5}n = -108$

   6. $5n = 7131.04$

Chapter 2 Summary

Defined Terms

Section 2.1
Equation (p. 92)
Solution (p. 92)
Formula (p. 94)
Perimeter (p. 94)
Area (p. 94)

Volume (p. 94)
Circumference (p. 94)
Radius (p. 94)
Diameter (p. 94)

Section 2.2
Linear equation (p. 105)
Linear equation in one
variable (p. 105)
Identity (p. 110)
Contradiction (p. 112)

Section 2.6
Linear inequality (p. 145)

Formulas

Perimeter of a rectangle: $P = 2l + 2w$ (p. 95)

Circumference of a circle: $C = \pi d$ or $C = 2\pi r$ (p. 95)

Area of a parallelogram: $A = bh$ (p. 95)

Area of a rectangle: $A = lw$ (p. 95)

Area of a triangle: $A = \dfrac{1}{2} bh$ (p. 95)

Area of a trapezoid: $A = \dfrac{1}{2} h(a + b)$ (p. 95)

Area of a circle: $A = \pi r^2$ (p. 95)

Surface area of a box: $SA = 2lw + 2lh + 2wh$ (p. 95)

Volume of a box: $V = lwh$ (p. 95)

Volume of a pyramid: $V = \dfrac{1}{3} lwh$ (p. 95)

Volume of a cylinder: $V = \pi r^2 h$ (p. 95)

Volume of a cone: $V = \dfrac{1}{3} \pi r^2 h$ (p. 95)

Volume of a sphere: $V = \dfrac{4}{3} \pi r^3$ (p. 95)

Distance, d, an object travels given its rate, r, and the time of travel, t: $d = rt$ (p. 98)

The average rate of travel, r, given the total time, t:

$r = \dfrac{d}{t}$ (p. 98)

Voltage, V, in a circuit with current, i, in amperes (A), and a resistance, R, in ohms (Ω): $V = iR$ (p. 98)

The temperature in degrees Celsius given degrees Fahrenheit: $C = \dfrac{5}{9} (F - 32)$ (p. 98)

The temperature in degrees Fahrenheit given degrees Celsius: $F = \dfrac{9}{5} C + 32$ (p. 98)

The profit, P, after cost, C, is deducted from revenue, R: $P = R - C$ (p. 131)

Procedure Problem-Solving Outline

1. **Understand** the problem.
   a. Read the question(s) (not the whole problem, just the question at the end) and write a note to yourself about what you are to find.
   b. Read the whole problem, underlining the key words.
   c. If possible or useful, draw a picture, make a list or table to organize what is known and unknown, simulate the situation, or search for a related example problem.

2. **Plan** your solution by searching for a formula or using the key words to translate to an equation.

3. **Execute** the plan by solving the equation/formula.

4. **Answer** the question. Look at the note about what you were to find and make sure you answer that question. Include appropriate units.

5. **Check** results.
   a. Try finding the solution in a different way, reversing the process, or estimating the answer and making sure the estimate and actual answer are reasonably close.
   b. Make sure the answer is reasonable.

LEARNING Strategy

  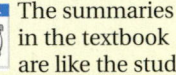 The summaries in the textbook are like the study sheet suggested in the To the Student section that precedes Chapter 1. Remember that your study sheet is a list of the rules, procedures, and formulas that you need to know. If you are a tactile or visual learner, spend a lot of time reviewing and writing the rules or procedures. Try to get to the point where you can write the essence of each rule and procedure from memory. If you are an audio learner, record yourself saying each rule and procedure; then listen to the recording over and over. Consider developing a clever rhyme or song for each rule or procedure to help you remember it.

Procedures, Rules, and Key Examples

Procedures/Rules	Key Examples

SECTION 2.1   Equations, Formulas, and the Problem-Solving Process

To determine whether a value is a solution to a given equation, replace the variable in the equation with the value. If the resulting equation is true, the value is a solution.

Example 1: Check to see if the given value is a solution to the given equation.

a. $7n - 13 = 3n + 11; n = 6$

$$7(6) - 13 \stackrel{?}{=} 3(6) + 11$$
$$42 - 13 \stackrel{?}{=} 18 + 11$$
$$29 = 29 \quad \text{Yes, 6 is a solution.}$$

b. $2(x - 3) = 5x + 4; x = 1$

$$2(1 - 3) \stackrel{?}{=} 5(1) + 4$$
$$2(-2) \stackrel{?}{=} 5 + 4$$
$$-4 \neq 9 \quad \text{No, 1 is not a solution.}$$

To use a formula:
1. Replace the variables with the corresponding given values.
2. Solve for the missing value.

Example 2: Find the volume of a box with a length of 4 centimeters, width of 5 centimeters, and height of 6 centimeters.

$$V = lwh$$
$$V = (4 \text{ cm})(5 \text{ cm})(6 \text{ cm})$$
$$V = 120 \text{ cm}^3$$

For composite figures:
1. To calculate the area of a figure composed of two or more figures that are next to each other, add the areas of the individual figures.
2. To calculate the area of a region defined by a smaller figure within a larger figure, subtract the area of the smaller figure from the area of the larger figure.

Example 3: Find the shaded area in the figure shown.

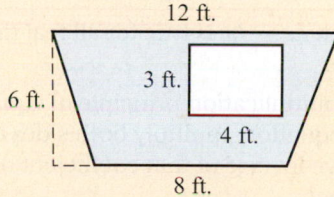

$$A = \text{Area of trapezoid} - \text{Area of rectangle}$$
$$A = \frac{1}{2}h(a + b) - lw$$
$$A = \frac{1}{2}(6)(12 + 8) - (3)(4)$$
$$A = \frac{1}{2}(6)(20) - 12$$
$$A = 60 - 12$$
$$A = 48 \text{ ft.}^2$$

SECTION 2.2   The Addition Principle of Equality

If $a = b$, then $a + c = b + c$ is true for all real numbers a, b, and c.

To use the addition principle of equality to clear a term in an equation, add the additive inverse of that term to both sides of the equation. (That is, add or subtract appropriately so that the term you want to clear becomes 0.)

Example 1: Solve.

a. $x + 9 = 15$
$$\underline{-9 \quad -9}$$
$$x + 0 = 6$$
$$x = 6$$

b. $y - 5 = 7$
$$\underline{+5 \quad +5}$$
$$y + 0 = 12$$
$$y = 12$$

(continued)

Procedures/Rules	Key Examples

SECTION 2.2 The Addition Principle of Equality (*continued*)

To solve linear equations requiring the addition principle only,
1. Simplify both sides of the equation as needed.
   a. Distribute to clear parentheses.
   b. Combine like terms.
2. Use the addition principle so that all variable terms are on one side of the equation and all constants are on the other side. Then combine like terms.

Tip: Clear the variable term that has the lesser coefficient to avoid negative coefficients.

c. $5(n - 2) + 3n = 7n - 4$
$$5n - 10 + 3n = 7n - 4$$
$$8n - 10 = 7n - 4$$
$$\underline{-7n \qquad\quad -7n}$$
$$n - 10 = 0 - 4$$
$$n - 10 = -4$$
$$\underline{+10 \quad +10}$$
$$n + 0 = \quad 6$$
$$n = 6$$

When solving a linear equation, if after simplifying each side of the equation the expressions are identical, the equation is an identity and every real number for which the equation is defined is a solution.

d. $5x - 9 - x = 4(x - 2) - 1$
$$4x - 9 = 4x - 8 - 1$$
$$4x - 9 = 4x - 9 \qquad \text{\color{blue}The equation is an}$$
$$\text{\color{blue}identity, so every real}$$
$$\text{\color{blue}number is a solution.}$$

When solving a linear equation, if after simplifying the expressions on each side of the equation the expressions have the same variable term but different constant terms, the equation is a contradiction and has no solution.

e. $4(2n - 3) - 5n + 2 = 9n - 2(3n - 6)$
$$\text{\color{blue}Distribute.}$$
$$8n - 12 - 5n + 2 = 9n - 6n + 12$$
$$\text{\color{blue}Combine like terms.}$$
$$3n - 10 = 3n + 12$$
$$\text{\color{blue}The equation is a contradiction,}$$
$$\text{\color{blue}so it has no solution.}$$

SECTION 2.3 The Multiplication Principle of Equality

If $a = b$, then $ac = bc$ is true for all real numbers a, b, and c, where $c \neq 0$.

To use the multiplication principle of equality to clear a coefficient in an equation, multiply both sides of the equation by the multiplicative inverse of that coefficient or divide both sides by the coefficient.

Example 1: Solve.

a. $-5x = 15$
$$\frac{-5x}{-5} = \frac{15}{-5}$$
$$x = -3$$

b. $\dfrac{3}{4}n = \dfrac{5}{6}$
$$\frac{\overset{1}{\cancel{4}}}{\underset{1}{\cancel{3}}} \cdot \frac{\overset{1}{\cancel{3}}}{\underset{1}{\cancel{4}}}n = \frac{\overset{5}{\cancel{5}}}{\underset{3}{\cancel{6}}} \cdot \frac{\overset{2}{\cancel{4}}}{3}$$
$$n = \frac{10}{9}$$

To solve linear equations in one variable,
1. Simplify both sides of the equation as needed.
   a. Distribute to clear parentheses.
   b. Clear fractions or decimals by multiplying by the LCD. In the case of decimals, the LCD is the power of 10 with the same number of zero digits as decimal places in the number with the most decimal places. (Clearing fractions and decimals is optional.)
   c. Combine like terms.
2. Use the addition principle so that all variable terms are on one side of the equation and all constants are on the other side. (Clear the variable term with the lesser coefficient to avoid negative coefficients.) Then combine like terms.
3. Use the multiplication principle to clear any remaining coefficient.

c. $7n - (n - 3) = 2n - 5$
$$7n - n + 3 = 2n - 5$$
$$6n + 3 = 2n - 5$$
$$\underline{-2n \qquad\quad -2n}$$
$$4n + 3 = 0 - 5$$
$$4n + 3 = -5$$
$$\underline{-3 \qquad -3}$$
$$4n + 0 = -8$$
$$\frac{4n}{4} = \frac{-8}{4}$$
$$n = -2$$

(*continued*)

Procedures/Rules	Key Examples

SECTION 2.4 Applying the Principles to Formulas

To isolate a particular variable in a formula, treat all other variables like constants and isolate the desired variable using the outline for solving equations.

Example 1: Isolate w in the formula $P = 2l + 2w$.

$$P = 2l + 2w$$
$$\underline{-2l \quad -2l}$$
$$P - 2l = 0 + 2w$$
$$\frac{P - 2l}{2} = \frac{2w}{2}$$
$$\frac{P - 2l}{2} = w$$

SECTION 2.5 Translating Word Sentences to Equations

To translate a word sentence to an equation, identify the unknown(s), constants, and key words; then write the corresponding symbolic form.

Example 1: Translate to an equation.
a. Five more than seven times a number is twelve.
Answer: $7n + 5 = 12$
b. Six times the difference of a number and nine is equal to four times the sum of a number and five.
Answer: $6(n - 9) = 4(n + 5)$

SECTION 2.6 Solving Linear Inequalities

To graph an inequality on a number line,
1. If the symbol is $\leq$ or $\geq$, draw a bracket (or solid circle) on the number line at the indicated number open to the left for $\leq$ and to the right for $\geq$. If the symbol is $<$ or $>$, draw a parenthesis (or open circle) on the number line at the indicated number open to the left for $<$ and to the right for $>$.
2. If the variable is greater than the indicated number, shade to the right of the indicated number. If the variable is less than the indicated number, shade to the left of the indicated number.

Example 1: Write the solution set in set-builder notation and interval notation; then graph the solution set.
a. $x \leq 6$
Set-builder notation: $\{x | x \leq 6\}$
Interval notation: $(-\infty, 6]$
Graph:
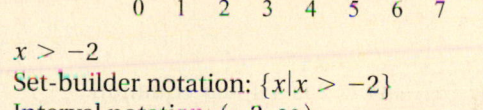

b. $x > -2$
Set-builder notation: $\{x | x > -2\}$
Interval notation: $(-2, \infty)$
Graph:
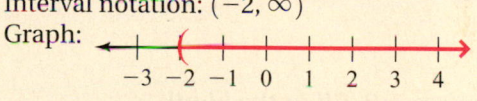

c. $-3 \leq x < 1$
Set-builder notation: $\{x | -3 \leq x < 1\}$
Interval notation: $[-3, 1)$
Graph:
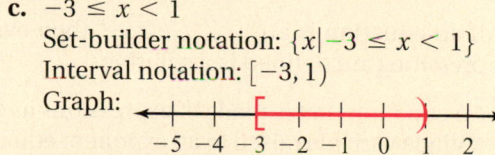

To solve linear inequalities:
1. Simplify both sides of the inequality as needed.
   a. Distribute to clear parentheses.
   b. Clear fractions or decimals by multiplying through by the LCD just as we did for equations. (Clearing fractions and decimals is optional.)
   c. Combine like terms.
2. Use the addition principle so that all variable terms are on one side of the inequality and all constants are on the other side. Then combine like terms. (Remember, clearing the term with the lesser coefficient results in a positive coefficient.)
3. Use the multiplication principle to clear any remaining coefficient. If you multiply (or divide) both sides by a negative number, reverse the direction of the inequality symbol.

Example 2: Solve and write the solution set in set-builder notation and interval notation. Then graph the solution set.
a. $-6x > 24$

$$\frac{-6x}{-6} < \frac{24}{-6} \qquad \text{Reverse the direction of the inequality symbol.}$$
$$x < -4$$

Set-builder notation: $\{x | x < -4\}$
Interval notation: $(-\infty, -4)$
Graph:
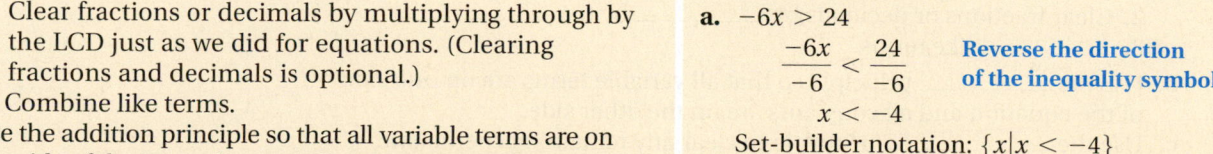

(continued)

SECTION 2.6 Solving Linear Inequalities (*continued*)

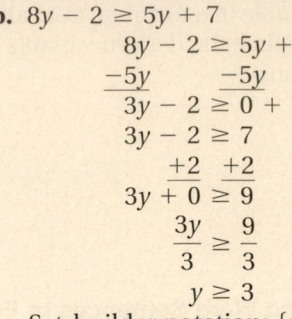

b. $8y - 2 \geq 5y + 7$

$$8y - 2 \geq 5y + 7$$
$$\underline{-5y \qquad\qquad -5y}$$
$$3y - 2 \geq 0 + 7$$
$$3y - 2 \geq 7$$
$$\underline{+2 \quad +2}$$
$$3y + 0 \geq 9$$
$$\frac{3y}{3} \geq \frac{9}{3}$$
$$y \geq 3$$

Set-builder notation: $\{y | y \geq 3\}$

Interval notation: $[3, \infty)$

Graph:

Chapter 2 Review Exercises

For Exercises 1–6, answer true or false.

[2.1]   **1.** $6x + 3y = 7x^2 - 4y$ is an equation.

[2.2]   **2.** $5(x + 2) = 10 - 3x + 8x$ is an identity.

[2.1]   **3.** 1 is a solution for $b^2 - 4b = 4$.

[2.5]   **4.** "Five less than a number is fourteen" translates to $5 - n = 14$.

[2.2]   **5.** $5x - 3y = 2$ is a linear equation.

[2.6]   **6.** $3 \leq x < 5$ is a compound inequality.

For Exercises 7–9, fill in the blank.

[2.2]   **7.** If an equation is a(n) _____, then every real number for which its expressions are defined is a solution.

[2.2]   **8.** An equation that is made of expressions in which each variable term contains a single variable raised to an exponent equal to 1 is a(n) _____ equation.

[2.3]   **9.** To solve equations,
   a. _____ both sides of the equation as needed.
      1. _____ to clear parentheses.
      2. Clear fractions or decimals by _____ by the _____.
      3. _____ like terms.
   b. Use the _____ principle so that all variable terms are on one side of the equation and all constants are on the other side.
   c. Use the _____ principle to clear any remaining coefficient.

[2.1]   **10.** In your own words, explain how to check a potential solution.

[2.1] *For Exercises 11–16, check to see if the given number is a solution to the equation.*

11. $y + 6 + 3y - 4 = 14$; $y = 3$

12. $\frac{1}{4}(n - 1) = \frac{1}{2}(n + 1)$; $n = 5$

13. $x^2 + 6.1 = 3x - 0.1$; $x = -0.1$

14. $\frac{5}{6}m - 1 = \frac{m}{3} + \frac{1}{2}$; $m = 3$

15. $\frac{3}{5}(a + 10) = -2$; $a = -20$

16. $6.1(7.2 + b) = 43.92$; $b = 0$

[2.2–2.3] *For Exercises 17–30, solve and check.*

17. $-6x = 30$

18. $-\frac{2}{3}y = -12$

19. $4n - 9 = 15$

20. $5(2a - 3) - 6(a + 2) = -7$

21. $12t + 9 = 4t - 7$

22. $2(c + 4) - 5c = 17 - 8c$

23. $1 - 2(3c - 5) = 10 - 6c$

24. $5(b - 1) - 3b = 7b - 5(1 + b)$

25. $5(u - 1) - (u - 2) = 10 - (2u + 1)$

26. $\frac{3}{4}m - \frac{1}{2} = \frac{2}{3}m + 1$

27. $\frac{1}{12}(4x + 9) = \frac{2}{3}x$

28. $\frac{5}{3} + 7x - 2x = \frac{1}{3}(9x + 2)$

29. $2 - \frac{2}{3}m = 6\left(m - \frac{1}{3}\right) - 4m + 2$

30. $1.4x - 0.5(9 - 6x) = 6 + 2.4x$

[2.4] *For Exercises 31–40, solve for the indicated variable.*

31. $x + y = 1$; for x

32. $a^2 + b^2 = c^2$; for a^2

33. $C = \pi d$; for d

34. $E = mc^2$; for m

35. $A = \frac{1}{2}bh$; for h

36. $F = \frac{mv^2}{r}$; for m

37. $V = \dfrac{1}{3}\pi r^2 h$; for h

38. $y = mx + b$; for m

39. $P = 2l + 2w$; for w

40. $A = \dfrac{1}{2}h(a + b)$; for h

[2.5] *For Exercises 41–44, translate to an equation and solve.*

41. Six times a number is the same as negative eighteen.

42. Three subtracted from half of a number is equal to nine less than one-fourth of the number.

43. One less than two times the sum of a number and four results in one.

44. Twice the difference of a number and one added to triple the number yields twenty less than six times the number.

[2.6] *For Exercises 45–50:* **a.** *Solve.*

b. *Write the solution set in set-builder notation.*

c. *Write the solution set in interval notation.*

d. *Graph the solution set.*

45. $-6x > 18$

46. $-3x + 2 \leq 5$

47. $1 - 2z + 4 \leq -3(z + 1) + 7$

48. $\dfrac{1}{4}v < 3 + v$

49. $\dfrac{1}{2}(m - 1) > 5 - \dfrac{7}{2}$

50. $2c - (5c - 7) + 4c \geq 8 - 10$

[2.6] *For Exercises 51–54, translate to an inequality and solve.*

51. Thirteen is greater than three minus ten times a number.

52. Three increased by twice a number is less than three times the difference of a number and five.

53. Negative one-half of a number is greater than or equal to four.

54. Negative two is less than or equal to one minus one-fourth of a number.

For Exercises 55–64, solve.

[2.1] **55.** Best Buy is selling an oak entertainment center. Elijah is concerned it will take up too much room in his den. Using the figure shown, determine the total area.

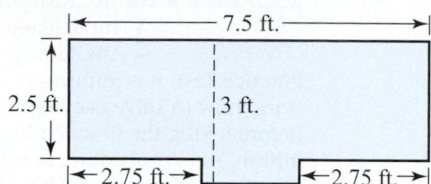

[2.1] **56.** Find the area of the figure shown.

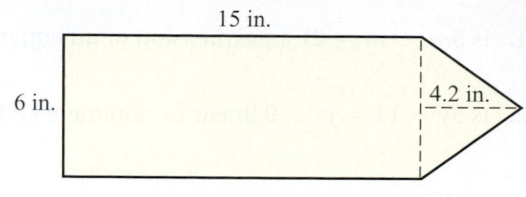

[2.1] **57.** A long-distance company uses the formula $C = 3 + 0.05m$ to describe a calling plan, where C is the total cost of a call lasting m minutes. If a customer uses 415 minutes, what is the total bill?

[2.3] **58.** Bonnie is making a shower curtain. The pattern says that the curtain will use 5760 square inches. If the length is to be 72 inches, how wide will the curtain be?

[2.3] **59.** The area of the trapezoid shown is 58 square inches. Find the missing length.

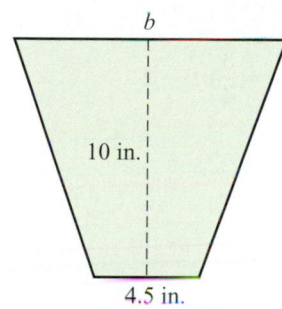

[2.3] **60.** Randy is snowskiing at Lake Tahoe. He knows that the trail is 1.8 miles long, and he is one-third of the way down. How far does he have left to ski?

[2.3] **61.** The surface area of a storage trunk is 2610 square inches. If the height of the trunk is 15 inches and the width is 36 inches, what is the length?

[2.6] **62.** The cost of border material limits the circumference of a circular flower garden to 10 feet. What is the range of values for the diameter of the garden? (Use $\pi \approx 3.14$.)

[2.6] **63.** Terrel has to travel at least 350 miles each day to stay on his schedule. If he averages 60 miles per hour, what is the range of values for his travel time?

[2.6] **64.** A company has a first-quarter profit goal of at least $350,000. If the projected cost is $475,000, what range of values can revenue have to meet the profit goal?

Chapter 2 Practice Test

For Extra Help

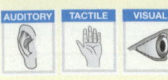

Step-by-step test solutions are found on the Chapter Test Prep Videos available via the Video Resources on DVD, in *MyMathLab*, and on YouTube (search "CarsonCombinedAlg" and click on "Channels").

1. Is $5m - 14 = 21$ an expression or an equation? Why?

2. Is $5y + 14 = y^2 - 9$ linear or nonlinear? Why?

3. Check to see if 3 is a solution for $3x^2 + x = 10x$.

4. Check to see if $\frac{2}{3}$ is a solution for $4 - 3(x + 2) = 0$.

For Exercises 5–10, solve and check.

5. $6x + 9 = -9$

6. $-5.6 + 8a = 7a + 7.6$

7. $8 + 5(x - 3) = 9x - (6x + 1)$

8. $1 - 3(c + 4) + 8c = 1 - 42 + 5(c + 7)$

9. $\frac{1}{2}x + \frac{5}{2} = \frac{3}{2} - x$

10. $-\frac{1}{8}(6m - 32) = \frac{3}{4}m - 2$

For Exercises 11–14, solve for the indicated variable.

11. $2x + y = 7; x$

12. $I = Prt; r$

13. $A = \frac{1}{2}bh; h$

14. $C = \frac{5}{9}(F - 32); F$

For Exercises 15 and 16, write the solution set in set-builder notation and interval notation; then graph the solution set.

15. $x \geq 3$

16. $-1 \leq x < 4$

For Exercises 17–20: a. *Solve.*
  b. *Write the solution set in set-builder notation.*
  c. *Write the solution set in interval notation.*
  d. *Graph the solution set.*

17. $-5m + 3 > -12$

18. $-2(2 - x) \geq -20$

19. $6(p - 2) - 5 > 3 - 4(p + 6)$

20. $\dfrac{1}{5}(l + 10) \geq \dfrac{1}{2}(l + 5)$

For Exercises 21–24, translate and then solve.

21. Two-thirds of a number minus one-sixth is the same as twice the number.

22. The product of negative seven and a number added to twelve yields five.

23. Three subtracted from five times the difference of a number and two is equal to ten minus four times the difference of the number and one.

24. One minus a number is greater than twice the number.

For Exercises 25–30, solve.

25. A carpenter needs to determine the amount of wood required to trim a window. If the window is 4 feet by 6 feet, what is the perimeter of the window?

26. Find the area of the figure shown.

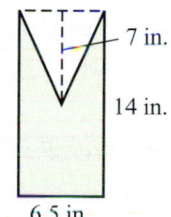

7 in.

14 in.

6.5 in.

27. The perimeter of the trapezoid shown is 63 centimeters. Find the length of the missing side.

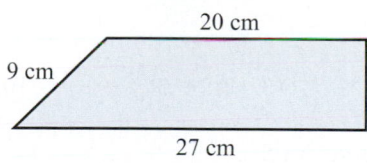

20 cm

9 cm

27 cm

28. A rent-to-own store advertises a television for $10 down with $10 payments each week. The formula $C = 10 + 10w$ represents the cost, C, after making payments for w weeks. If the customer signs a contract for one year, what is the total cost of the television?

29. The area of a sandbox cannot exceed 9 square feet. If the width of the sandbox is to be 3 feet, what range of values can the length have?

30. In Dan's psychology class, if his average on five tests is at least 92, he can be exempt from the final exam. His current scores are 90, 84, 89, and 96. What range of scores on the last test will allow him to be exempt from the final exam?

Chapters 1 and 2 Cumulative Review Exercises

Exercises 1–16 EXPRESSIONS

For Exercises 1–4, answer true or false.

[1.3] **1.** $(2 + 3) + 7 = 2 + (3 + 7)$

[1.1] **2.** $-16 < -19$

[1.2] **3.** 51 is a prime number.

[1.5] **4.** $-2^4 = -16$

For Exercises 5–8, answer.

[2.1] **5.** List the five steps for probem solving.

[1.5] **6.** List the order in which we perform operations.

[1.7] **7.** In your own words, explain how to evaluate an expression.

[1.1] **8.** Graph $-2\frac{3}{5}$ on a number line.

[1.5] *For Exercises 9–14, simplify*

9. $62 - (-40)$

10. $-7(9)$

11. $|6 - 3(-2)|$

12. $5 - 3^2$

13. $2(-4) - 15 \div (-5)$

14. $6^2 + [19 - 3(4)][6 + 5(2)]$

[1.7] **15.** Evaluate $3x - 4\sqrt{x + y}$ when $x = 9$ and $y = 16$.

[1.7] **16.** Combine like terms: $6x^2 - 4x - 9x + 7x^2 - 16$

Exercises 17–30 EQUATIONS AND INEQUALITIES

[2.3] *For Exercises 17–19, solve and check.*

17. $3b - 20 = 11$

18. $2(5 - n) = 7n - 17$

19. $2p + 3(p - 3) = 10p - (3p - 11)$

[2.4] *For Exercises 20 and 21, solve for the indicated variable.*

20. $B = P + Prt; r$

21. $V = \frac{1}{3} lwh; w$

[2.6] *For Exercises 22 and 23: a.* **Solve.**
   b. **Write the solution set in set-builder notation**
   c. **Write the solution set in interval notation.**
   d. **Graph the solution set.**

22. $-3h - 8 \leq 19$

23. $9m - (2m + 3) < 4(m - 5) + 2$

For Exercises 24 and 25, translate to an equation and solve.

[2.5]   24. Forty less than the product of three and y is equal to seven times y.

[2.5]   25. Twelve times a number minus the sum of the number and three is the same as negative eleven plus the product of two and the difference of the number and five.

For Exercises 26–30, solve.

[2.1]   26. Find the area of the shaded region in the figure.

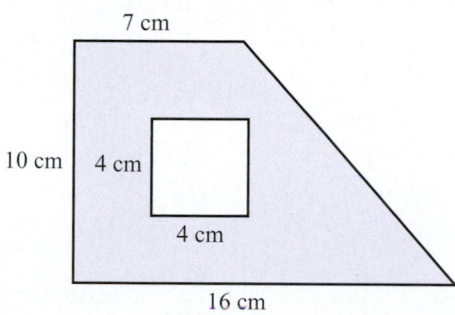

[2.1]   27. A company is producing cardboard boxes that are 2 feet by 3 feet by 1 foot. How many square feet of cardboard are needed to make the box?

[1.7]   28. The formula $h = -16t^2 + s_0$ describes the height in feet of a falling object, where s_0 represents the initial height from the ground and t represents the number of seconds after the object is dropped. If an object is dropped from 2400 feet, what will be its height after falling 8 seconds?

[2.6]   29. Fifteen less than twice a number is greater than or equal to five times the number.

[2.6]   30. Karen is building a rectangular flower bed, and her materials limit her to a maximum area of 24 square feet. If she makes the width 3 feet, what range of values can the length have?

Problem Solving

"Difficulty, my brethren, is the nurse of greatness—a harsh nurse, who roughly rocks her foster-children into strength and athletic proportion."

—William Cullen Bryant (1794–1878)
U.S. poet, editor

"It is one of man's curious idiosyncrasies to create difficulties for the pleasure of resolving them."

—Joseph de Maistre (1753–1821)
French diplomat, philosopher

In Chapter 2, we learned to solve linear equations and developed some techniques for solving problems, namely, using formulas and translating key words. In this chapter, we expand the problem-solving techniques introduced in Chapter 2 and solve problems that are a little more complex. These problems involve proportions, percents, and two unknown amounts.

Some of the exercises you encounter in this chapter will target a specific strategy such as translating key words or using a table to organize information. In the same way that an athlete uses exercise to target specific muscles, we use rather contrived problems to exercise specific problem-solving strategies and techniques. Most of the exercises an athlete performs in weight training are not sport-specific, meaning they are generic exercises designed to strengthen raw muscle tissue. Similarly, many of the problems in the chapter are not what you encounter in your job or everyday life. Rather, they are designed to exercise the strategies and techniques of problem solving.

3.1 Ratios and Proportions

Objectives

1. Solve problems involving ratios.
2. Solve for a missing number in a proportion.
3. Solve proportion problems.
4. Use proportions to solve for missing lengths in figures that are similar.

Objective 1 Solve problems involving ratios. In this section, we solve problems involving **ratios**.

Definition *Ratio:* A comparison of two quantities using a quotient.

Ratios are usually expressed in fraction form or with a colon. When a ratio is written in English, the word *to* separates the numerator and denominator quantities.

The ratio of **12** to **17** translates to $\frac{12}{17}$.

Numerator   Denominator

Note: The quantity preceding the word *to* is written in the numerator, and the quantity following *to* is written in the denominator.

Example 1 A small college has 2450 students and 140 faculty. Write the ratio of students to faculty in simplest form. Interpret the answer.

SOLUTION The ratio of students to faculty $= \frac{2450}{140} = \frac{35}{2}$.

This means that there are 35 students for every 2 faculty members.

Your Turn 1 In a local referendum, 6400 people voted for increased school taxes and 4800 people voted against the increase. In simplest form, write the ratio of those voting for higher taxes to those voting against. Interpret the answer.

As noted, the ratio in Example 1 indicates that there are 35 students for every 2 faculty members. Expressing a ratio as a **unit ratio** often makes the ratio easier to interpret.

Definition *Unit ratio:* A ratio with a denominator of 1.

Example 2 Express the ratio from Example 1 as a unit ratio. Interpret the answer.

SOLUTION $\frac{2450}{140} = \frac{35}{2} = \frac{17.5}{1}$ **Divide and write the denominator as 1.**

The unit ratio $\frac{17.5}{1}$ indicates that there are 17.5 students for every faculty member at the college.

Answer **Your Turn 1**

$\frac{6400}{4800} = \frac{4}{3}$; for every four people in favor of higher taxes, three are against them.

Your Turn 2 The price of a 12.5-ounce box of cereal is $2.85. Write the unit ratio of price to weight. Interpret the answer.

Objective 2 Solve for a missing number in a proportion. Some problems involve two equal ratios in a **proportion**.

Definition **Proportion:** An equation in the form $\dfrac{a}{b} = \dfrac{c}{d}$, where $b \neq 0$ and $d \neq 0$.

In Section 1.2, we discussed equivalent fractions and saw that fractions such as $\dfrac{3}{8}$ and $\dfrac{6}{16}$ are equivalent because $\dfrac{3}{8}$ can be rewritten as $\dfrac{6}{16}$ and $\dfrac{6}{16}$ can be reduced to $\dfrac{3}{8}$. Note that when fractions are equivalent, the cross products are equal.

$$16 \cdot 3 = 48 \qquad 8 \cdot 6 = 48$$

$$\dfrac{3}{8} \;\diagup\hspace{-1.1em}\diagdown\; \dfrac{6}{16}$$

Let's see why all proportions have equal cross products. We begin with the general form of a proportion from the definition $\dfrac{a}{b} = \dfrac{c}{d}$, where $b \neq 0$ and $d \neq 0$. We can use the multiplication principle of equality to clear the fractions by multiplying both sides by the LCD. In this case, the LCD is bd.

$$\dfrac{\overset{1}{\cancel{b}d}}{1} \cdot \dfrac{a}{\underset{1}{\cancel{b}}} = \dfrac{\overset{1}{b\cancel{d}}}{1} \cdot \dfrac{c}{\underset{1}{\cancel{d}}}$$

$$ad = bc$$

Note that ad and bc are the cross products.

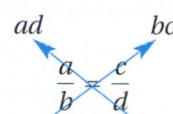

Conclusion If two ratios are equal, their cross products are equal.

Rule **Proportions and Their Cross Products**

If $\dfrac{a}{b} = \dfrac{c}{d}$, where $b \neq 0$ and $d \neq 0$, then $ad = bc$.

We can use cross products to solve for a missing number in a proportion.

Procedure **Solving a Proportion**

To solve a proportion using cross products,
1. Calculate the cross products.
2. Set the cross products equal to each other.
3. Use the multiplication principle of equality to isolate the variable.

Answer Your Turn 2

$\dfrac{0.228}{1}$; the cereal costs $0.228 per ounce.

Example 3   Solve.

a. $\dfrac{7}{16} = \dfrac{x}{10}$

SOLUTION

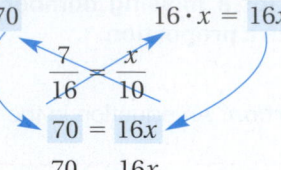

$10 \cdot 7 = 70 \qquad\qquad 16 \cdot x = 16x$ Calculate the cross products.

$\dfrac{7}{16} = \dfrac{x}{10}$

$70 = 16x$ Set the cross products equal to each other.

$\dfrac{70}{16} = \dfrac{16x}{16}$ Divide both sides by 16 to isolate x.

$\dfrac{35}{8} = x$

Note: We could have expressed the answer as 4375.

b. $\dfrac{3\frac{1}{5}}{12} = \dfrac{-9}{n}$

SOLUTION

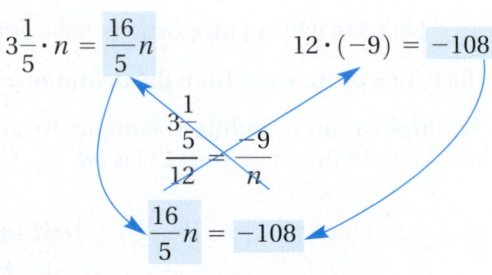

$3\frac{1}{5} \cdot n = \dfrac{16}{5}n \qquad\qquad 12 \cdot (-9) = -108$ Calculate the cross products.

$\dfrac{3\frac{1}{5}}{12} = \dfrac{-9}{n}$

$\dfrac{16}{5}n = -108$ Set the cross products equal to each other.

$\dfrac{\overset{1}{\cancel{5}}}{\cancel{16}} \cdot \dfrac{\cancel{16}}{\cancel{5}}n = \dfrac{\overset{27}{\cancel{-108}}}{1} \cdot \dfrac{5}{\underset{4}{\cancel{16}}}$ Multiply both sides by $\dfrac{5}{16}$ to isolate n.

$n = -\dfrac{135}{4} \text{ or } -33\dfrac{3}{4}$

Your Turn 3   Solve.

a. $\dfrac{9}{y} = -\dfrac{5}{6}$ **b.** $\dfrac{\frac{2}{3}}{n} = \dfrac{-5}{4\frac{1}{2}}$ **c.** $-\dfrac{2.4}{8.1} = -\dfrac{t}{12}$

Objective ③ Solve proportion problems.   In a typical proportion problem, you are given a ratio and then asked to find an equivalent ratio. The key to translating problems to a proportion is to write the proportion so that the numerators and denominators correspond to each other in a logical way.

Example 4   A company determines that it spends $254.68 every 5 business days on equipment failure. At this rate, how much would the company spend over 28 business days?

Understand   We are to find the amount spent on equipment failure over 28 days if the company spends $254.68 every 5 days. Translating $254.68 every 5 days to a ratio, we have $\dfrac{\$254.68}{5 \text{ days}}$.

LEARNING Strategy

VISUAL If you are a visual learner, draw the arrows as is done here to help visualize the pattern of cross products.

Answers Your Turn 3

a. $-\dfrac{54}{5}$ or -10.8 **b.** $-\dfrac{3}{5}$ **c.** $3.\overline{5}$

Plan *At this rate* indicates that $\dfrac{\$254.68}{5\text{ days}}$ stays the same for the 28 days; so we can use a proportion. If we let n represent the unknown amount spent over the 28 days, we have a ratio of $\dfrac{\$n}{28\text{ days}}$. We create the proportion by setting those ratios equal to each other.

Execute $\dfrac{\$254.68}{5\text{ days}} = \dfrac{\$n}{28\text{ days}}$

> **Note:** The units in the numerators and denominators correspond.
> **Tip:** One approach to setting up a proportion is to write the ratios so that the numerator units match and the denominator units match.

$28 \cdot 254.68 = 7131.04 \qquad\qquad 5 \cdot n = 5n$

$$\dfrac{254.68}{5} = \dfrac{n}{28}$$

$7131.04 = 5n$ **Set the cross products equal.**

$\dfrac{7131.04}{5} = \dfrac{5n}{5}$ **Divide both sides by 5 to isolate n.**

$1426.208 = n$

Answer Because the answer is in dollars, we round to the nearest hundredth (or cent). The company spends $1426.21 over 28 business days.

Check Do a quick estimate to verify the reasonableness of the answer. The company spends $254.68 every 5 business days. Since 28 days is a little less than 6 times the number of days, the company must spend a little less than 6 times the $254.68 spent in the 28 business days. To estimate the calculation, we round $254.68 to $250.00. We then have $6 \cdot \$250 = \$1500$, which is reasonably close to $1426.21.

Checking Proportion Problems

One of the nice things about proportion problems is that if you set up the proportion incorrectly, you usually get an unreasonable answer. A quick estimate check can be a fast way to recognize when you have set up a proportion incorrectly. For example, suppose we had incorrectly set up Example 4 this way:

$$\dfrac{\$254.68}{5\text{ days}} = \dfrac{28\text{ days}}{\$n}$$

> **Warning:** This setup is incorrect because the numerators and denominators do not correspond.

$254.68 \cdot n = 254.68n \qquad\qquad 5 \cdot 28 = 140$ **Cross multiply.**

$$\dfrac{254.68}{5} = \dfrac{28}{n}$$

$254.68n = 140$ **Set the cross products equal.**

$\dfrac{254.68n}{254.68} = \dfrac{140}{254.68}$ **Divide both sides by 254.68 to isolate n.**

$n \approx 0.55$

> **Warning:** This answer indicates that the company spends only $0.55 over 28 business days based on spending $254.68 every 5 business days, which is clearly too low.

There are many correct ways to set up a proportion for a particular problem. For example, following are some other correct ways we could have set up the proportion for Example 4. Note that the proportions involve rates that correspond with each other.

$$\frac{5 \text{ days}}{\$254.68} = \frac{28 \text{ days}}{\$n}$$

Note: Think:

In 5 days, $254.68 is spent. In 28 days, $n is spent.

$$\frac{5 \text{ days}}{28 \text{ days}} = \frac{\$254.68}{\$n}$$

Note: Think of this version as an analogy:

If 5 days means $254.68, then 28 days means $n.

$$\frac{28 \text{ days}}{5 \text{ days}} = \frac{\$n}{\$254.68}$$

Note: This version is a rearrangement of the previous version.

Procedure   **Solving Proportion Application Problems**

To solve proportion problems,

1. Set up a proportion in which the numerators and denominators of the ratios correspond in a logical manner.
2. Solve using cross products.

Your Turn 4   Pam drove 351.6 miles using 16.4 gallons of gasoline. At this rate, how much gasoline would she need to drive 1000 miles?

Objective 4   Use proportions to solve for missing lengths in figures that are similar.   Consider triangles *ABC* and *DEF*.

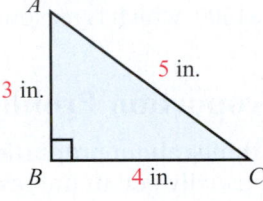
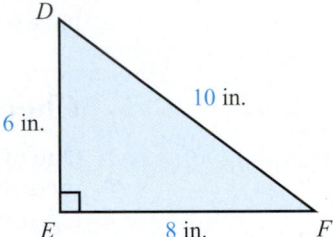

Although the triangles are different in size, their corresponding angle measurements are **congruent**.

Definition   *Congruent angles:* Angles that have the same measure. The symbol for congruent is ≅.

In symbols, we write $\angle B \cong \angle E$, $\angle A \cong \angle D$, and $\angle C \cong \angle F$.

Besides having all angles congruent, notice that the sides of triangle DEF are twice as long as the corresponding sides of triangle *ABC*. Or we could say that all of the sides of triangle *ABC* are half as long as the corresponding sides of triangle *DEF*. In other words, the corresponding side lengths are proportional.

Ratio of Triangle *DEF* to Triangle *ABC* $\longrightarrow \dfrac{6}{3} = \dfrac{8}{4} = \dfrac{10}{5} = \dfrac{2}{1}$

or

Ratio of Triangle *ABC* to Triangle *DEF* $\longrightarrow \dfrac{3}{6} = \dfrac{4}{8} = \dfrac{5}{10} = \dfrac{1}{2}$

Answer   **Your Turn 4**

≈ 46.6 gal.

Because the corresponding angles in the two triangles are congruent and the side lengths are proportional, we say that these triangles are **similar figures**.

Definition   *Similar figures:* Figures with congruent angles and proportional side lengths.

Example 5   The two trapezoids shown are similar. Find the missing lengths.

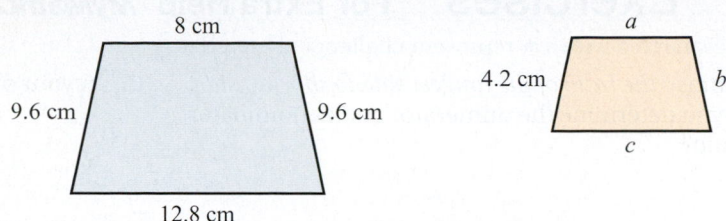

Understand   Because the figures are similar, the lengths of the corresponding sides are proportional. The corresponding sides can be shown as follows:

Ratio of larger trapezoid
to
smaller trapezoid
$\dfrac{9.6}{4.2} = \dfrac{8}{a} = \dfrac{9.6}{b} = \dfrac{12.8}{c}$

Plan   Write a proportion to find each missing side length.

Execute   It is not necessary to write a proportion to find b. Because two sides in the larger trapezoid measure 9.6 cm, the corresponding sides in the smaller trapezoid must be equal in length; so we can conclude that $b = 4.2$ cm.

To find a:

$$\frac{9.6}{4.2} = \frac{8}{a}$$

$a \cdot 9.6 = 9.6a \qquad 4.2 \cdot 8 = 33.6$

$$\frac{9.6}{4.2} \diagup \frac{8}{a}$$

$$9.6a = 33.6$$

$$\frac{9.6a}{9.6} = \frac{33.6}{9.6}$$

$$a = 3.5$$

To find c:

$$\frac{9.6}{4.2} = \frac{12.8}{c}$$

$c \cdot 9.6 = 9.6c \qquad 4.2 \cdot 12.8 = 53.76$

$$\frac{9.6}{4.2} \diagup \frac{12.8}{c}$$

$$9.6c = 53.76$$

$$\frac{9.6c}{9.6} = \frac{53.76}{9.6}$$

$$c = 5.6$$

Answer   The missing lengths are $a = 3.5$ cm, $b = 4.2$ cm, and $c = 5.6$ cm.

Check   We can verify the reasonableness of the answers by doing quick estimates of the missing side lengths. Comparing the smaller trapezoid to the larger trapezoid, we see that the 4.2-centimeter side is a little less than half of the corresponding 9.6-centimeter side in the larger trapezoid. Each of the corresponding sides should follow this same rule. The 8-centimeter side of the larger trapezoid should correspond to a length of a little less than 4 for side a of the smaller trapezoid, and the 12.8-centimeter side should correspond to a length of a little less than 6.4 for side c on the smaller trapezoid. These do check out.

Your Turn 5   Find the missing lengths in the similar figures.

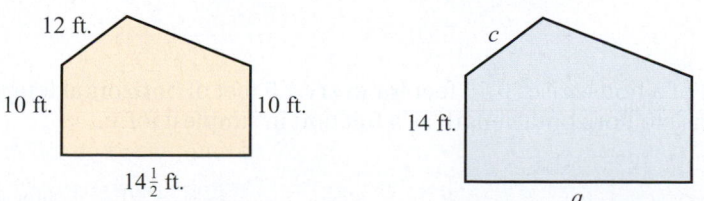

Answers   Your Turn 5

$a = 20\dfrac{3}{10}$ ft. or 20.3 ft.

$b = 14$ ft.

$c = 16\dfrac{4}{5}$ ft. or 16.8 ft.

Note: Exercises marked with a ★ represent challenging exercises.

1. In the phrase *the ratio of the longest side to shortest side,* how do you determine the numerator and denominator of the ratio?

2. In your own words, define a proportion.

3. If $\dfrac{a}{b} = \dfrac{c}{d}$, where $b \neq 0$ and $d \neq 0$, is true, what can we say about the cross products ad and bc?

4. Explain how to solve for a missing number in a proportion.

5. How do you set up and solve a word problem that translates to a proportion?

6. Why are $\dfrac{5}{8} \cdot \dfrac{x}{10}$ and $\dfrac{5}{8} = \dfrac{x}{10}$ different?

For Exercises 7–12, write the ratio in simplest form. See Example 1.

7. One molecule of lactose contains 12 carbon atoms, 22 hydrogen atoms, and 11 oxygen atoms.
   a. Write the ratio of carbon atoms to total atoms in the molecule.

   b. Write the ratio of carbon atoms to hydrogen atoms.

   c. Write the ratio of hydrogen atoms to oxygen atoms.

> **Of Interest**
>
> Lactose is a molecule found in milk. Its chemical formula is $C_{12}H_{22}O_{11}$. Notice that the subscripts indicate the number of atoms of each element in the molecule.

8. One molecule of hematcin contains 16 carbon atoms, 12 hydrogen atoms, and 6 oxygen atoms.
   a. What is the ratio of hydrogen atoms to the total number of atoms in the molecule?

   b. What is the ratio of carbon atoms to hydrogen atoms?

   c. What is the ratio of carbon atoms to oxygen atoms?

9. The roof of a house rises 0.75 feet for every 1.5 feet of horizontal length. Write the ratio of rise to horizontal length as a fraction in simplest form.

10. A steep mountain road has a grade of 8%, which means that the road rises 8 feet vertically for every 100 feet it travels horizontally. Write the grade as a ratio in simplest form.

11. In a certain gear on a bicycle, the back wheel rotates $1\frac{1}{3}$ times for every 2 rotations of the pedals. Write the ratio of rotations of the back wheel to pedal rotations as a fraction in simplest form.

12. A recipe calls for $2\frac{1}{2}$ cups of flour and $1\frac{1}{3}$ cups of milk. Write the ratio of milk to flour as a fraction in simplest form.

For Exercises 13–20, write each as a unit ratio and interpret the answer. See Example 2.

13. Jamial drove 254 miles in 4 hours. What is the unit ratio of miles to hours? Interpret the answer.

14. Janine drove 199.2 miles and used 6 gallons of gas. What is the unit ratio of miles to gallons? Interpret the answer.

15. Six bananas cost 75 cents. What is the unit ratio of cost to bananas? Interpret the answer.

16. An 8-pound bag of ice costs $1.96. What is the unit ratio of cost to pounds of ice. Interpret the answer.

17. In 2009, Omega Corporation's stock earned $0.92 per share. The current price of the stock is $15.15. Write a unit ratio that expresses the price to earnings. What does this ratio indicate?

18. The price of a 10.5-ounce can of soup is $1.68. Write the unit ratio that expresses the price to weight. What does this ratio indicate?

★ **19.** An 8-ounce container of sour cream costs $1.42. A 24-ounce container of the same brand of sour cream costs $3.59. Which is the better buy? Why?

★ **20.** The price of a 12-ounce box of corn flakes is $3.59, and an 18-ounce box of the same cereal is $4.59. Which is the better buy? Why?

For Exercises 21 and 22, use the graph to the right. See Example 2.

21. a. Write the unit ratio of convictions to total cases in 1999.

   b. Write the unit ratio of convictions to total cases in 2004.

   c. Compare the ratios for 1999 to 2004. What do these ratios indicate?

22. a. Write the unit ratio of incarcerated to convictions in 1999.

   b. Write the unit ratio of incarcerated to convictions in 2004.

   c. Compare the ratios for 1999 to 2004. What do these ratios indicate?

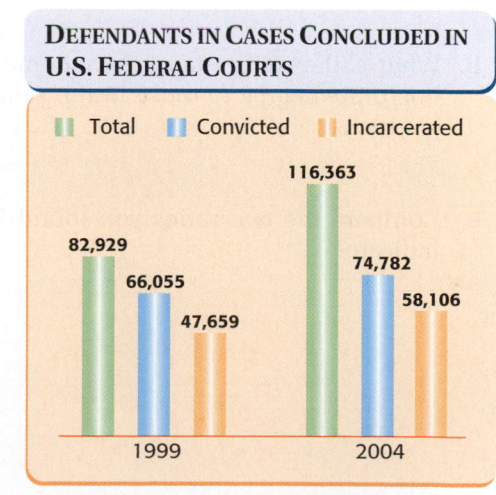

DEFENDANTS IN CASES CONCLUDED IN U.S. FEDERAL COURTS

Total   Convicted   Incarcerated

116,363
82,929
74,782
66,055
58,106
47,659

1999   2004

Source: Bureau of Federal Justice Statistics

For Exercises 23 and 24, use the following table. See Example 2.

Mean Earnings in the United States, 2005
Year-Round, Full-Time Workers by Educational Attainment and Age

	Not High School Graduate	High School Graduate or GED	Some College, No Degree	Associate Degree	Bachelor's Degree	Master's Degree	Professional Degree	Doctorate Degree
Male								
25 to 34 Years	$24,533	$33,509	$40,417	$42,200	$53,272	$69,601	$89,474	$72,164
35 to 44 Years	$29,059	$40,885	$49,499	$50,831	$75,812	$88,352	$151,687	$113,799
45 to 54 Years	$30,633	$43,638	$52,314	$53,700	$78,430	$87,248	$168,333	$108,212
55 to 64 Years	$32,225	$44,619	$54,629	$55,887	$79,764	$93,680	$171,879	$117,769
Female								
25 to 34 Years	$20,574	$26,132	$27,927	$32,144	$47,358	$49,694	$71,932	$67,582
35 to 44 Years	$21,980	$28,974	$33,814	$38,079	$51,028	$58,089	$105,850	$97,756
45 to 54 Years	$21,994	$29,729	$36,008	$40,785	$50,770	$62,874	$76,770	$78,396
55 to 64 Years	$21,803	$30,591	$36,032	$37,467	$47,571	$58,913	$78,242	$95,664

Source: Census 2005a

23. a. What is the unit ratio of the income for women in the 25–34 year age group with a bachelor's degree to men in the corresponding group? What does this ratio indicate?

b. Calculate the unit ratio for women in the 55–64 age group with a bachelor's degree with men in the corresponding group.

c. Compare the ratios you found for the two age groups in parts a and b. What do these ratios indicate?

24. a. What is the unit ratio of the income of men in the 45–54 age group with a bachelor's degree to those in the same age group that are not a high school graduate? What does this ratio indicate?

b. What is the unit ratio of the income of men in the 45–54 age group with a doctorate degree to those in the same age group that are not a high school graduate?

c. Compare the two ratios you found in parts a and b. What do these ratios indicate?

For Exercises 25–34, determine whether the ratios are equal. See Objective 2.

25. $\dfrac{15}{25} \overset{?}{=} \dfrac{3}{5}$

26. $\dfrac{6}{8} \overset{?}{=} \dfrac{15}{20}$

27. $\dfrac{20}{25} \overset{?}{=} \dfrac{25}{30}$

28. $\dfrac{12}{20} \overset{?}{=} \dfrac{18}{30}$

29. $\dfrac{7}{4} \overset{?}{=} \dfrac{15}{6.2}$

30. $\dfrac{15}{21} \overset{?}{=} \dfrac{130.2}{93}$

31. $\dfrac{10}{14} \overset{?}{=} \dfrac{2.5}{3.5}$

32. $\dfrac{6.5}{14} \overset{?}{=} \dfrac{10.5}{42}$

33. $\dfrac{\frac{2}{5}}{\frac{3}{10}} \overset{?}{=} \dfrac{15}{10\frac{1}{2}}$

34. $\dfrac{3\frac{1}{4}}{5\frac{1}{2}} \overset{?}{=} \dfrac{4\frac{2}{3}}{6\frac{1}{2}}$

For Exercises 35–50, solve for the missing number. See Example 3.

35. $\dfrac{2}{3} = \dfrac{7}{x}$

36. $\dfrac{7}{8} = \dfrac{y}{40}$

37. $\dfrac{-9}{2} = \dfrac{63}{n}$

38. $\dfrac{11}{m} = \dfrac{132}{-24}$

39. $\dfrac{21}{5} = \dfrac{h}{2.5}$

40. $\dfrac{29}{7} = \dfrac{k}{1.75}$

41. $\dfrac{-17}{b} = \dfrac{8.5}{4}$

42. $\dfrac{4.6}{3} = \dfrac{-23}{c}$

43. $\dfrac{-1.5}{60} = \dfrac{-2\frac{7}{8}}{t}$

44. $\dfrac{u}{-4\frac{7}{16}} = \dfrac{-60}{1.5}$

45. $\dfrac{2\frac{1}{4}}{6\frac{2}{3}} = \dfrac{d}{10\frac{2}{3}}$

46. $\dfrac{\frac{2}{5}}{4\frac{1}{2}} = \dfrac{7\frac{2}{3}}{j}$

★ 47. $\dfrac{2}{3} = \dfrac{8}{x+8}$

★ 48. $\dfrac{x+5}{6} = \dfrac{7}{3}$

★ 49. $\dfrac{4x-4}{8} = \dfrac{2x+1}{5}$

★ 50. $\dfrac{3}{4} = \dfrac{x+1}{x+3}$

For Exercises 51–74, solve. See Example 4.

51. LaTonia drives 110 miles in three hours. At this rate, how far will she go in five hours?

52. If 90 feet of wire weighs 18 pounds, what will 110 feet of the same type of wire weigh?

53. Jodi pays taxes of $1040 on her house, which is valued at $154,000. She is planning to buy a house valued at $200,000. At the same rate, how much will her taxes be on the new house?

54. Gary notices that his water bill was $24.80 for 600 cubic feet of water. At that rate, what would the charges be for 940 cubic feet of water?

55. A company determines that it spends $1575 every 20 business days (one month) for equipment maintenance. How much does the company spend in 120 business days (six months)?

56. A quality-control manager samples 1200 units of her company's product and finds seven defects. If the company produces about 50,000 units each year, how many of those units would she expect to have a defect?

57. Carson is planning a trip to London. He knows that it takes 1.8946 U.S. dollars to equal a British pound. If he has $800 to take with him, how much will he have in British pounds?

58. Sydney is planning a trip to Mexico. It takes 15.3827 Mexican pesos to equal 1 U.S. dollar. If she has $400 to take with her, how many pesos will she have?

59. Fletcher is building a model of an F-16 jet. If the scale is $1\frac{1}{2}$ inches to 1 foot, what length on the model represents 3.25 feet?

60. The legend on a map indicates that 1 inch = 500 feet. If a road measures $3\frac{3}{4}$ inches on the map, how long is the road in feet?

61. A recipe for pancakes calls for $1\frac{1}{2}$ teaspoons of sugar to make 12 pancakes. How many teaspoons of sugar should be used to make 20 pancakes?

62. A recipe for biscuits calls for $3\frac{1}{2}$ cups of flour to make 8 biscuits. How many cups of flour would be needed to make 12 biscuits?

63. Ford estimates that its 2008 Expedition will travel 504 highway miles on one tank of gas. If the gas tank of the Expedition holds 28 gallons, how far can a driver expect to travel on 20 gallons?

64. Chevrolet estimates that its 2008 Tahoe will travel 520 highway miles on one tank of gas. If the gas tank of the Tahoe holds 26 gallons, how far can a driver expect to travel on 20 gallons?

65. Gregory Arakelian of the United States holds the world record for fastest typing. In 1991, he typed 158 words per minute in the Key Tronic World Invitational Type-Off. If he could maintain that rate, how long would it take him to type a term paper with 2500 words? (*Source: Guinness Book of World Records*)

66. Sean Shannon of Canada holds the world record for the fastest recital of Hamlet's soliloquy "To be or not to be." On August 30, 1995, he recited the 260-word soliloquy in 23.8 seconds. If he could maintain that rate, how long would it take him to recite Shakespeare's entire script of Hamlet, which contains 32,241 words? (*Sources: Guinness Book of World Records*; http://william-shakespeare.info)

67. A consultant recommends that about 9 pounds of rye grass seed be used for every 1000 square feet. If the client's lawn is about 25,000 square feet, how many pounds of seed should he buy?

68. Mark used 3 gallons of paint to cover 1200 square feet of wall space in his home. Mark's neighbor wants to paint 2000 square feet. Based on Mark's rate, how many gallons should the neighbor buy?

★ 69. A carpet-cleaning company charges $49.95 to clean a rectangular living room that is 20 feet by 15 feet. The client asks the company also to clean the rectangular dining room, which measures 18 feet by 12 feet. If the company charges the same rate, how much should it cost to clean both the living room and dining room?

★ 70. A painter charged $320 to paint two walls that measured 12 feet by 9 feet and two walls that measured 10 feet by 9 feet. The client asks him to return to paint two walls that measure 15 feet by 12 feet and two walls that measure 18.5 feet by 12 feet. If the painter charges the same rate, what should the price be?

★ **71.** A state wildlife department wants to estimate the number of deer in a state forest. In one month, it captures 25 deer, tags them, and releases them. Later, it captures 45 deer and finds that 17 are tagged. Assuming that the ratio of tagged deer to total deer in the forest remains constant, estimate the number of deer in the preserve.

★ **72.** A marine biologist is studying the population of trout in a mountain lake. During one month, she catches 225 trout, tags them, and releases them. Later she captures 300 trout and finds that 48 are tagged. Assuming that the ratio of tagged trout to total trout in the lake remains constant, estimate the population of trout in the lake.

73. The Hoover Dam has 17 generators and can supply all of the electricity needed by a city of 750,000 people. If only 14 generators are running, how many people will be supplied with electricity?

74. Until the early 1950s, the erosion rate for Niagara Falls was an average of 3 feet per year. The suggested rate of erosion may now be as low as 1 foot per 10 years for Niagara Falls. Given this rate, how much erosion may be expected over the next 50 years?

For Exercises 75–78, find the missing lengths in the similar figures. See Example 5.

75.

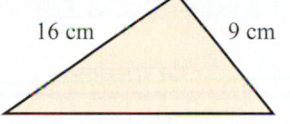

16 cm   9 cm

76.

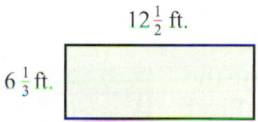

$12\frac{1}{2}$ ft.

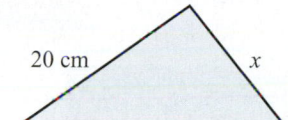

20 cm   x

$6\frac{1}{3}$ ft.

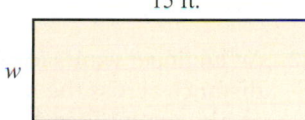

15 ft.

w

77.

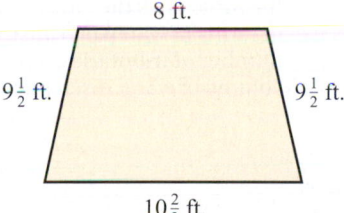

8 ft.

$9\frac{1}{2}$ ft. $9\frac{1}{2}$ ft.

$10\frac{2}{3}$ ft.

a

$5\frac{1}{4}$ ft.   b

c

78.

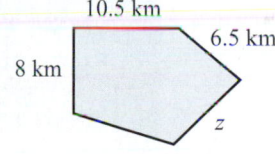

10.5 km

6.5 km

8 km

z

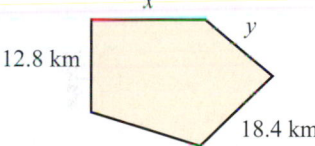

x

12.8 km   y

18.4 km

79. To estimate the height of the Eiffel Tower, a tourist uses the concept of similar triangles. The Eiffel Tower is found to have a shadow measuring 200 meters in length. The tourist has his own shadow measured at the same time. His shadow measures 1.2 meters. If he is 1.8 meters tall, how tall is the Eiffel Tower?

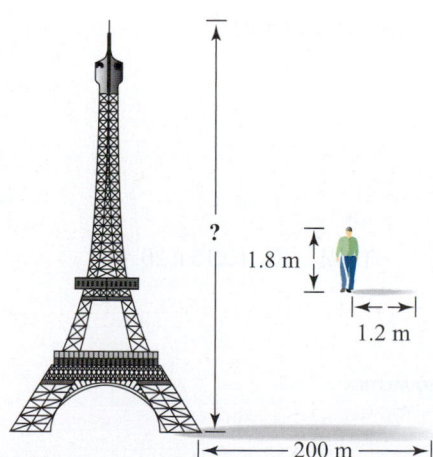

?

1.8 m

1.2 m

200 m

80. To estimate the height of the Great Pyramid in Egypt, the Greek mathematician Thales performed a procedure similar to that in Exercise 79, except that he supposedly used a staff instead of his body. Suppose the shadow from

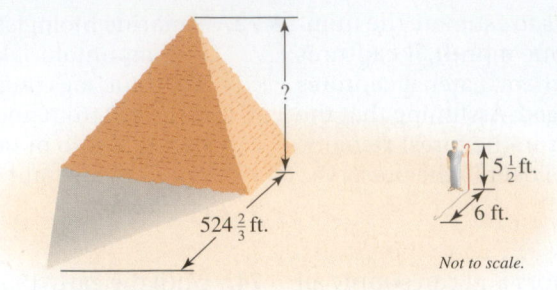

the staff was measured to be 6 feet and at the same time the shadow of the pyramid was measured to be $524\frac{2}{3}$ feet. If the staff was $5\frac{1}{2}$ feet tall, what was the height of the pyramid?

81. To estimate the distance across a river, an engineer creates similar triangles. The idea is to create two similar right triangles as shown. Calculate the width of the river.

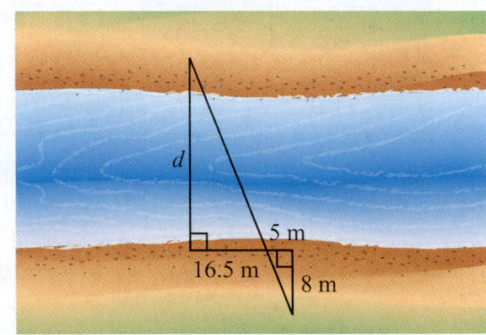

82. An engineer wants to estimate the distance across the Amazon River at his current location. The engineer creates two similar right triangles as shown. Calculate the distance across the river.

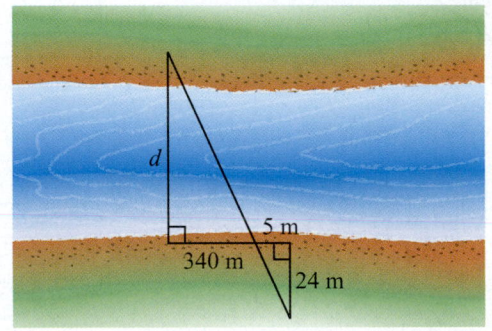

Review Exercises

Exercise 1 CONSTANTS AND VARIABLES

[1.1]   **1.** Is 0.58 a rational number? Why or why not?

Exercises 2–4 EXPRESSIONS

For Exercises 2–4, simplify.

[1.2]   **2.** $\dfrac{42}{350}$ [1.4]   **3.** $(0.15)(20.4)$ [1.4]   **4.** $45.2 \div 100$

Exercises 5 and 6 EQUATIONS AND INEQUALITIES

For Exercises 5 and 6, solve.

[2.3]   **5.** $0.75n = 240$ [2.3]   **6.** $60p = 24$

3.2 Percents

Objectives
1 Write a percent as a fraction or decimal number.
2 Write a fraction or decimal number as a percent.
3 Translate and solve percent sentences.
4 Solve problems involving percents.

Objective 1 Write a percent as a fraction or decimal number. The word *percent* is a compound word made from the prefix *per* and the suffix *cent*. The prefix *per* means "for each." The suffix *cent* comes from the Latin word *centum*, which means "100." Therefore, **percent** means "for each 100." In other words, percent is a ratio with the denominator 100.

Definition   *Percent:* A ratio representing some part out of 100.

The symbol for percent is %. For example, 20 percent, which means 20 out of 100, is written 20%. Note that the definition can be used to write percents as fractions or decimal numbers.

$$20\% = 20 \text{ out of } 100$$
$$= \frac{20}{100}$$
$$= \frac{1}{5} \text{ or } 0.2$$

Procedure   **Rewriting a Percent**

To write a percent as a fraction or decimal,

1. Write the percent as a ratio with 100 in the denominator.
2. Simplify to the desired form.

Note: When simplifying, remember that dividing a decimal number by 100 moves the decimal point two places to the left.

Example 1   Write each percent as a decimal and as a fraction in simplest form.

a. 20.5%

SOLUTION   Write 20.5 over 100; then simplify. In this case, it might be helpful to write the decimal number first and then write the equivalent fraction from that decimal number.

$$20.5\% = \frac{20.5}{100}$$ **Write as a ratio with 100 in the denominator.**

$$= 0.205$$ **Write the decimal form.**

$$= \frac{205}{1000}$$ **Write the fraction form.**

$$= \frac{41}{200}$$ **Simplify to lowest terms.**

b. $30\frac{1}{4}\%$

SOLUTION Write $30\frac{1}{4}$ over 100; then simplify. In contrast to part a, you could find the fraction form first and then find the decimal form.

$$30\frac{1}{4}\% = \frac{30\frac{1}{4}}{100} \qquad \text{Write as a ratio with 100 in the denominator.}$$

$$= 30\frac{1}{4} \div 100 \qquad \text{Rewrite the division.}$$

$$= \frac{121}{4} \cdot \frac{1}{100} \qquad \text{Write an equivalent multiplication.}$$

$$= \frac{121}{400} \text{ or } 0.3025 \qquad \text{Multiply.}$$

Your Turn 1 Write each percent as a fraction in simplest form and as a decimal.

a. 26% **b.** 9.25% **c.** $30\frac{2}{3}\%$

Objective 2 **Write a fraction or decimal number as a percent.** We have seen that $20\% = \frac{20}{100} = \frac{1}{5}$ or 0.2. To write a percent as a fraction or a decimal, we replace the % sign with division by 100. To write a fraction or decimal as a percent, we should multiply by 100% and simplify.

Procedure **Writing a Fraction or Decimal as a Percent**

To write a fraction or decimal number as a percent,

1. Multiply by 100%.
2. Simplify.

Note: Multiplying a decimal number by 100 moves the decimal point two places to the right.

Example 2 Write as a percent. If necessary, round your answer to the nearest tenth of a percent.

a. $\frac{5}{8}$

SOLUTION
$$\frac{5}{8} = \frac{5}{\overset{}{\underset{2}{8}}} \cdot \frac{\overset{25}{\cancel{100}}}{1}\% \qquad \text{Multiply by 100\%.}$$

$$= \frac{125}{2}\% \qquad \text{Simplify.}$$

$$= 62\frac{1}{2}\% \text{ or } 62.5\%$$

b. 0.267

SOLUTION $0.267 = 0.267 \cdot 100\%$ **Multiply by 100%.**
$$= 26.7\%$$

Connection Multiplying an amount by 100% is a way of rewriting the amount because multiplying by 100% is equivalent to multiplying by 1. The percent symbol means $\frac{1}{100}$; so 100% literally means

$$100 \cdot \% = 100 \cdot \frac{1}{100} = 1.$$

Your Turn 2 Write as a percent. If necessary, round your answer to the nearest tenth of a percent.

a. $\dfrac{5}{9}$ 　　　　　　　　　　　b. 2.4

Objective ③ Translate and solve percent sentences. To solve problems involving percents, it is often helpful to reduce the problem to the following simple percent sentence:

A **percent** of a **whole** is a **part** of the whole.

Note that there are three pieces in the simple sentence: the *percent*, the *whole*, and the *part*. Any one of those pieces could be unknown in a problem.

	A **percent** of	a **whole**	is a **part**.
Unknown part:	42% of	68	is what amount?
Unknown whole:	76% of	what number	is 63.84?
Unknown percent:	What percent of	72	is 63?

Note: The whole follows the word *of* because we are finding a percentage of the whole.

To solve for an unknown piece of a simple percent sentence, we can translate the sentence to an equation. We will look at two methods of translation: translating word for word and translating to a proportion.

Procedure **Translating Simple Percent Sentences**

Method 1. Translate the sentence word for word.

1. Select a variable for the unknown.
2. Translate the word *is* to an equal sign.
3. If *of* is preceded by the percent, translate it to multiplication. If *of* is preceded by a whole number, translate it to division.

Method 2. Translate to a proportion by writing the following form:

$$\text{Percent} = \frac{\text{Part}}{\text{Whole}},$$

where the percent is expressed as a fraction with a denominator of 100.

First, let's consider the case where the part is unknown.

Simple Percent Sentence with an Unknown Part

Example 3 42% of 68 is what number?

Note: This question can also be worded this way: "What is 42% of 68?"

SOLUTION Notice that the part is the unknown.

42% 　 of 　 68 　 is 　 what number?
Percent 　 of 　 the whole 　 is 　 the part.

Answers Your Turn 2

a. $55\dfrac{5}{9}\%$ or 55.6%

b. 240%

Method 1. Word-for-word translation: Because *of* is preceded by the percent, it means multiply, and *is* means equals.

$$42\% \text{ of } 68 \text{ is what number?}$$

Note: To calculate, we write the percent as a decimal (or fraction).

$$0.42 \cdot 68 = n$$
$$28.56 = n$$

LEARNING Strategy

AUDITORY If you are an auditory learner, when translating a percent sentence to a proportion, imagine or say to yourself that "*of* is underneath" to drive home the fact that the number following *of* is the denominator.

Method 2. Proportion: The percent, which is a ratio, is equivalent to the ratio of the part to the whole; so we can write a proportion in the following form:

$$\text{Percent} = \frac{\text{Part}}{\text{Whole}}$$

Note: $\frac{42}{100}$ is 42% written as a ratio.

$$\frac{42}{100} = \frac{n}{68} \quad \xleftarrow{\text{Part}} \quad \xleftarrow{\text{Whole}}$$

Note: The whole amount always follows *of*.

$$2856 = 100n \qquad \text{Equate the cross products.}$$
$$\frac{2856}{100} = \frac{100n}{100} \qquad \text{Divide both sides by 100 to isolate } n.$$
$$28.56 = n$$

Your Turn 3 What number is 15% of 36?

Simple Percent Sentence with an Unknown Whole

Now consider the situation in which the whole amount is unknown.

Example 4 76% of what number is 63.84?

Note: This question can be worded "63.84 is 76% of what number?"

SOLUTION Note the three pieces:

76%	of	what number	is	63.84?
Percent	of	the whole	is	the part.

Method 1. Word-for-word translation:

$$76\% \text{ of what number is } 63.84?$$

$$0.76 \cdot n = 63.84$$
$$0.76n = 63.84$$
$$\frac{0.76n}{0.76} = \frac{63.84}{0.76}$$
$$n = 84$$

Method 2. Proportion:

$$\text{Percent} = \frac{\text{Part}}{\text{Whole}}$$

Note: This is 76% expressed as a ratio.

$$\frac{76}{100} = \frac{63.84}{n} \quad \xleftarrow{\text{Part}} \quad \xleftarrow{\text{Whole}}$$

$$76n = 6384 \qquad \text{Equate the cross products.}$$
$$\frac{76n}{76} = \frac{6384}{76} \qquad \text{Divide both sides by 76 to isolate } n.$$
$$n = 84$$

Answer Your Turn 3

5.4

Your Turn 4 48 is 120% of what number?

Simple Percent Sentence with an Unknown Percent

Finally, consider the case in which the percent is the unknown.

Example 5 What percent of 72 is 63?

SOLUTION Note the pieces.

What percent of 72 is 63?

Percent of the whole is the part.

> **Note:** This question can be worded
> "63 is what percent of 72?"
> or
> "What percent is 63 out of 72?"
> or
> "What percent is 63 of 72?"
> Note how 72, the whole amount,
> always follows *of*.

Method 1. Word-for-word translation:

What percent of 72 is 63?

$$p \cdot 72 = 63$$
$$72p = 63$$
$$\frac{72p}{72} = \frac{63}{72}$$
$$p = 0.875$$

> **Note:** The division
> yields a decimal
> number that *must* be
> written as a percent.

To write 0.875 as a percent, we multiply by 100%.

Answer $0.875 \cdot 100\% = 87.5\%$

Method 2. Proportion:

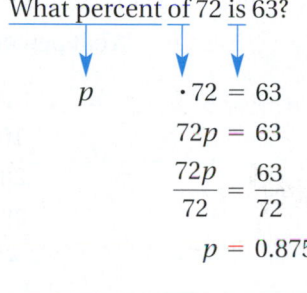

> **Note:** This is the unknown
> percent written as a ratio. We use
> an uppercase *P* here because the
> 100 in the proportion causes the
> solution to be slightly different
> from the solution to the equation
> in the word-for-word translation.

$$\frac{P}{100} = \frac{63}{72}$$ Part / Whole
$$72P = 6300$$ Equate the cross products.
$$\frac{72P}{72} = \frac{6300}{72}$$ Divide both sides by 72 to isolate *p*.
$$P = 87.5$$

> **Note:** In calculating the cross
> products, we multiplied by
> 100; so the solution is the
> percent value, and we need to
> add only the percent sign.

Answer 87.5%

As Example 5 shows, when the percent is unknown, the proportion method gives the answer as a percent; so you may want to consider using proportions in this case.

Your Turn 5 21.6 is what percent of 90?

Objective ④ **Solve problems involving percents.** If a problem involves a percent, it can be helpful to try to write the information in the form of a simple percent sentence.

Answer Your Turn 4

40

Answer Your Turn 5

24%

<blockquote>
Procedure **Solving Percent Application Problems**

To solve problems involving percent,

1. Determine whether the percent, the whole, or the part is unknown.
2. Write the problem as a simple percent sentence (if needed).
3. Translate to an equation (word for word or proportion).
4. Solve for the unknown.
</blockquote>

Example 6 In an English class, 12 of the 28 students received an A. What percent of the class received an A?

Understand The unknown is the percent of the class that received an A. The number of students in the class is 28, which is the whole. The number of students that received an A, which is the part, is 12.

Plan Write a simple percent sentence; then solve. We'll let P represent the unknown percent, and we'll translate to a proportion because the proportion's solution will be the percent value.

Execute What percent of 28 is 12?

$$\frac{P}{100} = \frac{12}{28} \quad \begin{array}{l} \leftarrow \text{Part} \\ \leftarrow \text{Whole} \end{array}$$

$$28P = 1200 \qquad \text{Equate the cross products.}$$

$$\frac{28P}{28} = \frac{1200}{28} \qquad \text{Divide both sides by 28 to isolate } P.$$

$$P \approx 42.9 \qquad \text{Note: We rounded to the nearest tenth. We need to add the \% sign.}$$

Answer About 42.9% of the students in the class received an A on their first paper.

Check Verify that 42.9% of the 28 students is about 12 students.

$$0.429 \cdot 28 = 12.012$$

Because 12.012 is reasonably close to 12, we can conclude that 42.9% is correct.

Your Turn 6 On Monday, January 18, 2010, the *Morning Sentinel* reported that it planned to cut 100 full-time jobs in a bid to return to profitability. This is 6% of the newspaper's workforce. How many employees does the *Sentinel* have prior to this layoff?

Percent Problems Involving Increase or Decrease

Sometimes, a percent problem involves an increase or decrease of some initial amount. Sales tax, interest, and discount are situations in which an initial amount of money is increased or decreased by an amount of money that is a percent of the initial amount.

Example 7 If the sales tax rate in a certain city is 6%, what is the total cost of the following items?

Raisin Bran	$3.79
Milk	$1.29
Orange Juice	$1.49
Yogurt	$2.59

Understand The total cost is the purchase price plus the sales tax, which is a percent of the purchase price.

Answer **Your Turn 6**

1667

<blockquote>
188 Chapter 3 Problem Solving
</blockquote>

Plan We must first calculate the purchase price. We can then write a simple percent sentence and solve for the sales tax. Finally, we add the tax to the purchase price.

Execute Purchase price $= 3.79 + 1.29 + 1.49 + 2.59 = \$9.16$

Simple percent sentence: The sales tax is 6% of $9.16. We will translate word for word, letting x represent the unknown sales tax.

$$x = 0.06 \cdot 9.16$$
$$x = 0.5496$$

Sales tax is always rounded up to the next cent, so the sales tax is $0.55.

Answer The total cost is $\$9.16 + \$0.55 = \$9.71$.

Check Determine whether the total cost of $9.71 represents a 6% increase in the total sale of $9.16. To determine the percent of increase, we need to calculate the amount of the increase by subtracting 9.16 from 9.71.

$$9.71 - 9.16 = 0.55 \qquad \text{This verifies that the final addition was correct.}$$

Now to determine the percent of the increase, we can set up a proportion.

Connection The steps in this check help us understand how to approach a type of problem in which the percent of increase or decrease is unknown. We will see this type of problem in Example 8.

$$\frac{P}{100} = \frac{0.55}{9.16} \longleftarrow \begin{array}{l} \text{Part} \\ \text{Whole} \end{array}$$

$$9.16P = 55 \qquad \textbf{Equate the cross products.}$$

$$\frac{9.16P}{9.16} = \frac{55}{9.16} \qquad \textbf{Divide both sides by 9.16 to isolate } P.$$

$$P \approx 6.004$$

Note: This does not match 6% exactly because we used the rounded sales tax amount, 0.55, instead of the exact value, 0.5496.

Your Turn 7 An advertisement indicates that a coat is on sale at 20% off the marked price. If the marked price is $79.95, what will be the price after the discount?

Connection A clever shortcut for Example 7 is to recognize that we pay 100% of the initial price *plus* 6% in sales tax, so the total is 106% of the initial price.

Total cost $= 1.06 \cdot 9.16 = 9.7096$, which rounds to $9.71.

A similar shortcut exists for the Your Turn problem involving a discount. If we take 30% off the marked price, we pay 70% of the marked price.

Price after discount $= 0.7 \cdot 79.95 = 55.965$, which rounds to $55.97.

The check in Example 7 is a common type of percent problem in which the percent of increase or decrease is unknown.

Example 8 On May 1, 2006, the median price of a house sold in Jacksonville, Florida, reached an all-time high of $322,990. On April 7, 2008, the median price was $253,990. What was the percent of decrease? (*Source:* www.housingtracker.net)

Understand We are to determine the percent of decrease given the initial price and the price after the decrease.

Plan To determine the percent of decrease, we need the amount of the decrease, which is found by subtracting the final amount from the initial amount.

Answer **Your Turn 7**

$63.96

Execute   Amount of decrease = 322,990 − 253,990 = 69,000

Simple percent sentence: What percent of 322,990 is 69,000?

> **Note:** The whole is the original amount (before the decrease).

Because the percent is unknown, we translate to a proportion, letting P represent the number of the unknown percent.

$$\frac{P}{100} = \frac{69,000}{322,990} \quad \longleftarrow \text{Part} \\ \quad\quad\quad\quad\quad\quad \longleftarrow \text{Whole}$$

$$322,990P = 6,900,000 \quad \text{Equate the cross products.}$$

$$\frac{322,990P}{322,990} = \frac{6,900,000}{322,990} \quad \begin{array}{l}\text{Divide both sides by 322,990} \\ \text{to isolate } P.\end{array}$$

$$P \approx 21.36$$

Answer   The percent of decrease in the median price of a house in Jacksonville was about 21.36%.

Check   Determine whether decreasing the initial median price of $322,990 by 21.36% results in a price of $253,990.

$$0.2136(322,990) = n$$

$$68,990.66 = n$$

$$69,000 \approx n$$

> **Note:** When rounded to the nearest thousand, this amount of decrease corresponds to our earlier calculation.

Now subtract the amount of the decrease: 322,990 − 69,000 = 253,990.

In general, when solving for a percent of increase or decrease, the *part* is the amount of increase or decrease and the *whole* is the initial amount.

$$\frac{P}{100} = \frac{\text{Amount of increase/decrease}}{\text{Initial amount}}$$

Your Turn 8   The Air Transportation Association reports that the per-gallon price for jet fuel jumped from $2.10 in 2007 to $2.85 in 2008. What is the percent of increase?

Answer   Your Turn 8

35.7%

3.2 Exercises   For Extra Help   *MyMathLab*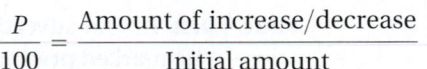

Note: Exercises marked with a ★ represent challenging exercises.

1. Explain how to write a percent as a decimal or fraction.

2. Explain how to write a decimal or fraction as a percent.

3. When preceded by a percent, the word *of* indicates what operation?

4. Explain how to translate a simple percent sentence to a proportion.

5. When using word-for-word translation to find a percent, the calculation will not be a percent, whereas if you use the proportion method, the result will be a percent. Why?

6. When given an initial amount and a final amount after an increase, how do you find the percent of the increase?

For Exercises 7–18, write each percent as a decimal and as a fraction in simplest form. See Example 1.

7. 20%

8. 30%

9. 15%

10. 85%

11. 14.8%

12. 18.6%

13. 3.75%

14. 6.25%

15. $45\frac{1}{2}\%$

16. $65\frac{1}{4}\%$

17. $33\frac{1}{3}\%$

18. $18\frac{1}{6}\%$

For Exercises 19–38, write as a percent rounded to the nearest tenth if necessary. See Example 2.

19. $\frac{3}{5}$

20. $\frac{1}{5}$

21. $\frac{3}{8}$

22. $\frac{5}{8}$

23. $\frac{5}{6}$

24. $\frac{4}{9}$

25. $\frac{2}{3}$

26. $\frac{5}{11}$

27. 0.96

28. 0.42

29. 0.8

30. 0.7

31. 0.09

32. 0.01

33. 1.2

34. 3.58

35. 0.028

36. 0.065

37. 4.051

38. 0.007

For Exercises 39–60, translate word for word or translate to a proportion; then solve. See Examples 3–5.

39. 80% of 35 is what number?

40. What number is 40% of 90?

41. 2.5% of 124 is what number?

42. 13.5% of 940 is what number?

43. What number is $9\frac{1}{4}\%$ of 64?

44. $3\frac{3}{4}\%$ of 24 is what number?

45. 120% of 86 is what number?

46. What number is 250% of 62.8?

47. 30% of what number is 15?

48. 45 is 5% of what number?

49. 16.4 is 20.5% of what number?

50. 30.2% of what number is 18.12?

51. 7.8 is $12\frac{1}{2}\%$ of what number?

52. $5\frac{1}{4}\%$ of what number is 1.26?

53. What percent of 45 is 9?

54. 15 is what percent of 40?

55. What percent of 38 is 39.9?

56. 73.44 is what percent of 68?

★ 57. What percent is 18 of 27?

★ 58. 50 of 60 is what percent?

★ 59. What percent is 12 of 15?

★ 60. 18 of 24 is what percent?

For Exercises 61–66, solve. See Example 6.

61. Angela answers 86% of the questions on a psychology multiple choice test correctly. If there were 50 questions on the test, how many did she answer correctly?

62. Terra is a server at a restaurant. She serves a table of seven people, for which the restaurant automatically adds a service charge of 15% of the cost of the meal. If the meal costs $127.40, how much is the service charge?

63. A county's annual property taxes are 4% of the market value of the property. How much is the tax on a property valued at $120,000? If the tax is paid in monthly installments, what amount must be paid each month?

64. How much is the tax on a house valued at $230,000 in the same county as that in Exercise 63? If the tax is paid in monthly installments, what amount must be paid each month?

65. In general, lenders do not want home buyers to spend more than 28% of their gross monthly income on a house payment. If a buyer has a monthly gross income of $3210, what is the most a lender will allow for a mortgage payment?

66. In general, lenders do not want home buyers to spend more than 36% of gross income a month on debts. Using the buyer in Exercise 65, what is the most a lender will accept in total debt payments per month? If she makes the highest mortgage payment allowed each month, what does she have left to spend on other debt and still be within the lender's guidelines?

Of Interest

In finance, the ratio of the house payment to the gross monthly income is known as the *front-end ratio*. The ratio of the monthly debts to the gross monthly income is known as the *back-end ratio*. The front-end and back-end ratios, along with a person's credit history, are primary factors in determining loan qualification.

For Exercises 67 and 68, solve using the circle graph. See Example 6.

67. How many teragrams of carbon dioxide were emitted by the United States?

68. How many teragrams of methane were emitted by the United States?

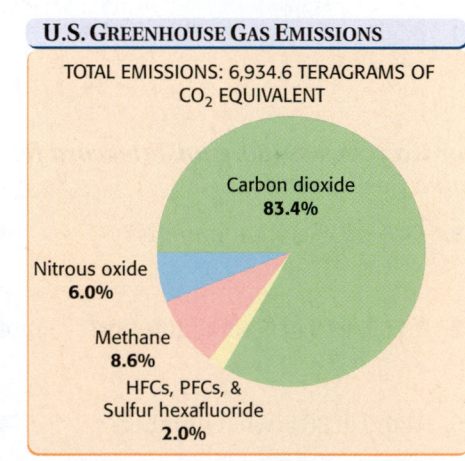

U.S. GREENHOUSE GAS EMISSIONS

TOTAL EMISSIONS: 6,934.6 TERAGRAMS OF CO_2 EQUIVALENT

Carbon dioxide 83.4%

Nitrous oxide 6.0%

Methane 8.6%

HFCs, PFCs, & Sulfur hexafluoride 2.0%

Source: U.S. Environmental Protection Agency

69. Terra, the server from Exercise 62, receives a tip of $8.50. If this was 20% of the cost of the meal, what was the cost of the meal?

70. When a camera is sold on eBay for $50.00 or less, the seller must pay eBay 6% of the selling price. If eBay charges $2.70 for the sale of a camera, what was the selling price of the camera?

71. A theater charges 2.5% of the ticket price as a handling fee to hold tickets at the will-call booth. If the handling fee for four tickets is $7.50, what is the total cost of the tickets?

72. Carol's checking account earns 0.025% of her average daily balance as a dividend every month. If she receives a dividend of 15 cents one month, what was her average daily balance?

73. During his career, Detroit Pistons player Isiah Thomas made 7194 field goals out of 15,904 attempts. What percent of his field goal attempts did he score?

74. In 2008, Chipper Jones of the Atlanta Braves had the number 1 batting average in the National League with 160 hits out of 439 at-bats. What percent of his at bats were hits?

75. According to the 2000 census, the total population of the United States was 281,421,906. If there were 196,899,193 people 21 years of age and older, what percent of the total population was 21 years of age and older? (*Source:* Bureau of the Census, Department of Commerce, 2000 Census)

76. In 2006, 55,521,868 households out of 111,617,402 households in the United States were headed by married couples. What percent of U.S. households was headed by married couples in 2006? (*Source:* Bureau of the Census, Department of Commerce)

> **Of Interest**
>
> Batting averages are expressed as decimal numbers rounded to the nearest thousandth. For example, a player's batting average might be 0.285, which means that he has had a hit 28.5% of the number of times he's been at bat.

For Exercises 77–80, use the following table, which shows the energy use in some of the countries that use the most energy. The unit used is the amount of energy produced by one metric ton of oil. (Source: Firefly's World of Facts, 2007)

Country	Oil	Gas	Coal	Nuclear	Hydroelectric	Total
United States	944.6	570.1	575.4	185.9	60.6	2336.6
China	327.3	42.3	1081.9	11.8	90.8	1554.0
Russia	130.0	364.6	111.6	33.9	39.6	679.6

77. What percent of the energy use in the United States comes from nuclear energy?

78. What percent of the energy use in China comes from coal?

79. For each country, find the percent of energy use that comes from gas. Which country has the greatest percentage?

80. For each country, find the percent of energy use that comes from the least polluting form of energy, hydroelectric. Which country has the smallest percentage?

81. A small company has the following costs in one quarter.

Category	Cost (in thousands)
Plant operations	124.2
Employee wages	248.6
Research and development	115.7

What percent of the company's total cost is spent on research and development?

82. A pharmaceutical company conducts a study on the side effects of a new drug. Following is a report of the data from the study.

Side Effect	Number of Subjects
No side effects	1462
Headache	35
Rash	24
Fever	65

What percent of the subjects in the study experienced fever?

★ **83.** Margaret mixes 50 milliliters of a solution that is 10% HCl with 100 milliliters of a solution that is 25% HCl. What percent of the resulting solution is HCl?

84. Leon invests $3400 in stock that returns 12% the first year and $2200 in stock that returns 15% during the same year. What is his total return as a percent of his total investment?
★

85. Carl purchases $124.45 in groceries. If the sales tax rate is 5%, what is the amount of the tax and the cost after tax is added?

86. Juanita purchases several books over the Internet that have a total price of $85.79. The online store charges a handling fee of 2.5% of the total cost. What is the handling fee and the price after the fee is added?

87. An appliance store has all refrigerators discounted 10% off the regular price. If the regular price of a refrigerator is $949, what is the amount of the discount and the price after the discount?

88. A furniture store has all sofas discounted 40% off. If the original price of a particular sofa is $785.99, what is the amount of the discount and the price after the discount?

★ **89.** Johanna paid a total of $2675 for a sofa, which included 7% sales tax. What was the price of the sofa before the sales tax was added?

90. Paul paid a total of $1908 for a refrigerator, which included 6% sales tax. What was the price of the refrigerator before the sales tax was added?
★

91. Video game sales in a department store increased from $2600 in November to $4550 in December. What was the amount of increase? What was the percent of increase?

92. Fifteen years ago James inherited a marble-top table. It was appraised at $420. The insurance company asked that he have it reappraised. It is now worth $1200. What was the amount of the increase? What was the percent of increase?

93. From 1989–1990, about 580,000 bachelor's degrees were awarded to women in the United States. From 2005–2006, about 845,000 bachelor's degrees were awarded to women. What was the percent of increase in degrees awarded to women? (*Source:* National Center of Education Statistics, U.S. Department of Education)

94. On May 14, 2009, the Dow Jones Industrial Average (DJIA) closed at 8331.72, and on May 5, it closed at 8268.64. What was the percent of decrease?

95. The price of a TI-84 calculator drops from $120 to $79. What is the percent of decrease?

96. After the Vallina family purchased a smaller car, their monthly gasoline bill dropped from $225 per month to $170 per month. What was the percent of decrease in the monthly gasoline bill?

97. In 1990, the carbon monoxide concentration in Baltimore was 7.2 parts per million, and in 2006, the concentration was 2.1 parts per million. What was the percent of decrease in the carbon monoxide concentration? (*Source:* U.S. Environmental Protection Agency, Office of Air Quality Planning and Standards)

98. In 1990, the population of Decatur, Illinois, was 83,900. In 2006, the population was 77,047. What was the percent of decrease in population? (*Source:* U.S. Bureau of the Census, Department of Commerce)

99. In 1938, the federal hourly minimum wage was $0.25. In 2009, the hourly minimum wage was $7.25. What was the percent of increase in the minimum wage from 1938 to 2009? (*Source:* U.S. Department of Labor)

100. A small business owner has an audit performed and finds that the net worth of the company is $58,000. After two years of poor business, an audit shows the net worth is −$20,000. What is the percent of the decrease in the net worth of the business?

101. The U.S. Bureau of the Census reported a 475% increase in the number of single fathers from 1970 to 2006. If the number of single fathers in 2006 was 2,300,000, how many single fathers were there in 1970? (*Source:* U.S. Census Report, June 18, 2006)

102. The projected population of Florida for the year 2030 is 30.1 million, which is a 986.25% increase from the population in 1950. What was the population in 1950? (*Source:* Floridians for a Sustainable Population, 2008)

103. On "Black Monday," October 19, 1987, the stock market closed at 1739, which was a 22.6% decrease from the previous day's closing price. What was the closing price on Friday, October 16?

104. On March 28, 2008, the closing price of Disney stock was at $31, which was a 24.3% decrease from March 28, 2000. What was the closing price on March 28, 2000?

Collaborative Exercises Occupation Growth

Complete the following table by calculating the amount of change and the percent of the increase for each occupation; then answer the questions. Round to the nearest tenth of a percent.

The 10 Fastest-Growing Occupations, 2002–2012

Occupation	Employment		Change	
	2002	2012	Amount	Percent
Medical assistants	364,600	579,400		
Network systems and data communications analysts	186,000	292,000		
Physician assistants	63,000	93,800		
Social and human service assistants	305,200	453,900		
Home health aides	579,700	858,700		
Medical records and health information technicians	146,900	215,600		
Physical therapist aides	37,000	54,100		
Computer software engineers, applications	394,100	573,400		
Computer software engineers, systems software	281,100	408,900		
Physical therapist assistants	50,200	72,600		

Source: Bureau of Labor Statistics, Office of Occupational Statistics and Employment Projections

1. In 2002, in which occupation were the greatest number of people employed?

2. By 2012, which occupation is projected to have the greatest number of people employed?

3. Explain why the two occupations you listed in Exercises 1 and 2 are not at the top of the list.

4. Which is a better indicator of the demand for people in a particular occupation: the number of people employed in a particular year, the amount of change projected from 2002 to 2012, or the percent of increase in employment from 2002 to 2012?

5. Based on your conclusions in Exercise 4, what occupation will have the greatest demand? What college majors might have the greatest potential for employment in that occupation? Write your conclusions and present them to the class.

Review Exercises

Exercises 1–3 **EXPRESSIONS**

[1.5] **1.** Simplify: $-3°$

[1.7] **2.** Multiply: $-3(x + 7)$

[1.7] **3.** Combine like terms: $7xy + 4x - 8xy + 3y$

Exercises 4–6 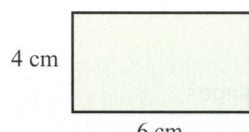 **EQUATIONS AND INEQUALITIES**

For Exercise 4, translate to an equation and then solve.

[2.5] **4.** Three times the sum of a number and nine is equal to eleven less than the number.

[3.1] **5.** Solve $\dfrac{1}{5} = \dfrac{n}{10.5}$.

[3.1] **6.** The figures shown are similar. Find the missing side length.

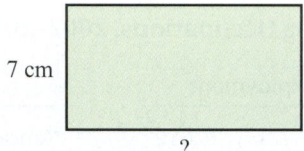

3.3 Problems with Two or More Unknowns

Objectives

1. Solve problems involving two unknowns.
2. Use a table in solving problems with two unknowns.

Objective ❶ **Solve problems involving two unknowns.** Sometimes problems have two or more unknown amounts. In general, if there are two unknowns, there will be two relationships. Our approach to solving these problems is to use the relationships to write an equation that we can solve. The following outline gives a more specific approach.

Procedure **Solving Problems with Two or More Unknowns**

To solve problems with two or more unknowns,

1. Determine which unknown will be represented by a variable.
   Tip: Let the unknown that is acted on be represented by the variable.
2. Use one of the relationships to describe the other unknown(s) in terms of the variable.
3. Use the other relationship to write an equation.
4. Solve the equation.

First, let's look at problems in which we simply translate key words.

Example 1 The greater of two numbers is twice the lesser number. If the sum of the two numbers is 57, find the numbers.

Understand There are two unknowns in the problem: We are to find the greater and lesser numbers. We are given two relationships.

Relationship 1: The greater number is twice the lesser number.

Relationship 2: The sum of the two numbers is 57.

Plan   Translate the relationships to an equation and then solve.

Execute   Use the first relationship to determine which unknown will be represented by a variable and represent the other unknown in terms of that variable.

Relationship 1: The greater number is twice the lesser number.

Greater number $= 2 \cdot$ lesser number

Greater number $= 2n$

> **Note:** We usually choose the unknown amount that is acted on to be the variable. Because the greater number is twice the lesser number, we let n represent the lesser number.

Now use the second relationship to write an equation.

Relationship 2: The sum of the two numbers is 57.

$$\text{Greater number } + \text{ lesser number } = 57$$
$$2n \qquad + \qquad n \qquad = 57$$
$$3n = 57 \qquad \text{\color{blue}{Combine like terms.}}$$
$$\frac{3n}{3} = \frac{57}{3} \qquad \text{\color{blue}{Divide both sides by 3.}}$$
$$n = 19$$

Answer   Because n represents the lesser number, the lesser number is 19. The greater number is twice the lesser number, so the greater number is $2 \cdot 19 = 38$.

Check   Verify that the sum of the two numbers is 57. $19 + 38 = 57$.

Your Turn 1   The greater of two numbers is 3 times the lesser number. If the sum of the two numbers is 36, find the numbers.

Example 2   Lindsey eats a snack consisting of 1 ounce of almonds and one oatmeal fruit bar. She is surprised to see that the ounce of almonds has 40 more calories than the fruit bar. If her snack has a total of 300 calories, how many calories are in the almonds and how many are in the fruit bar?

Understand   There are two unknowns in the problem: We must find the number of calories in 1 ounce of almonds and the number of calories in an oatmeal fruit bar. We are given two relationships.

Relationship 1: The number of calories in the almonds is 40 more than the number of calories in the fruit bar.

Relationship 2: The total calories in the snack are 300.

Plan   Translate the relationships to an equation and then solve.

Execute   Use the first relationship to determine which unknown will be represented by a variable.

Relationship 1: The number of calories in the almonds is 40 more than the number of calories in the fruit bar.

$$\text{Calories in almonds } = 40 \text{ calories } + \text{ Calories in the fruit bar}$$
$$\text{Calories in almonds } = \qquad 40 \qquad + \qquad n$$

> **Note:** We choose the unknown amount that is acted on to be the variable. Because 40 is added to the calories in the fruit bar, we let n represent the calories in a fruit bar.

Now use the second relationship to write an equation that we can solve for n.

Relationship 2: The total calories are 300.

Calories in the fruit bar + Calories in the almonds = 300

$$n \qquad + \qquad 40 + n \qquad = 300$$

$$2n + 40 = 300 \qquad \text{\color{blue}{Combine like terms.}}$$

$$2n + 40 = 300$$

$$\underline{-40 \qquad -40} \qquad \text{\color{blue}{Subtract 40 from both sides.}}$$

$$2n + 0 = 260$$

$$\frac{2n}{2} = \frac{260}{2} \qquad \text{\color{blue}{Divide both sides by 2.}}$$

$$n = 130$$

Answer   Because n represents the number of calories in the fruit bar, 1 fruit bar has 130 calories. The almonds contain 40 more calories than the fruit bar, so the almonds contain $40 + 130 = 170$ calories.

Check   Verify that the total calories is 300, which it is because $170 + 130 = 300$.

Your Turn 2   In a year, the average person eats 12 pounds more of chicken than pork. If the average person eats a total of 108 pounds of chicken and pork, how many pounds of each does the average person eat in a year?

Geometry Problems

Sometimes, problems do not give all of the needed relationships in an obvious manner. This often occurs in geometry problems in which the definition of a geometry term provides the needed relationship. For example, we must know that *perimeter* means "the sum of the lengths of the sides" in order to solve a problem that involves perimeter.

Example 3   A rectangular frame is to be built so that the length is 4 inches more than the width. The perimeter of the frame is to be 72 inches. Find the dimensions of the frame.

Understand   Draw a picture and list the relationships.

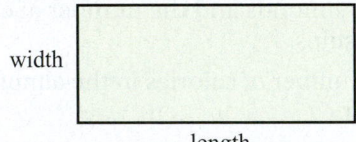

width

length

Relationship 1: The length is 4 inches more than the width.

Relationship 2: The perimeter must be 72 inches.

Plan   Translate to an equation using the key words and then solve the equation.

Execute   Relationship 1: The length is 4 inches more than the width.

$$\text{Length} = \text{Width} + 4$$

$$\text{Length} = w + 4$$

Note: Because 4 is added to the width and the length is isolated, we let w represent width.

Now use the other given relationship to write an equation to solve.

Answer   **Your Turn 2**

Pork: 48 pounds

Chicken: 60 pounds

Relationship 2: The perimeter is 72 inches. Recall that the formula for the perimeter of a rectangle is $P = 2l + 2w$, where l represents the length and w represents the width.

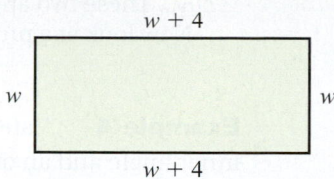

$$\text{Perimeter} = 72$$
$$2l + 2w = P$$

$$2(w + 4) + 2w = 72 \qquad \text{Substitute } w + 4 \text{ for } l \text{ and 72 for } P.$$
$$2w + 8 + 2w = 72 \qquad \text{Distribute.}$$
$$4w + 8 = 72 \qquad \text{Combine like terms.}$$
$$\underline{{-8} \quad -8} \qquad \text{Subtract 8 from both sides to isolate the}$$
$$4w + 0 = 64 \qquad \text{variable term.}$$
$$\frac{4w}{4} = \frac{64}{4} \qquad \text{Divide both sides by 4 to isolate } w.$$
$$w = 16$$

Answer   The width, w, is 16 inches. To determine the length, use Relationship 1. If the length $= w + 4$, then the length $= 16 + 4 = 20$ inches.

Check   Verify that both conditions in the problem are satisfied. First, the length must be 4 inches more than the width, and 20 inches is 4 inches more than 16 inches. Second, the perimeter must be 72 inches, which is true because $2 \cdot 16 + 2 \cdot 20 = 32 + 40 = 72$ inches.

Your Turn 3   Karen is constructing a rectangular flower garden with a wooden border. She wants the length to be twice the width and has 30 feet of border material. Find the dimensions of the garden.

In Example 3, the word *perimeter* was a key word that helped us write the equation that we solved. Other terms in geometry, by their definitions, give information to help us translate a problem into an equation. For example, a problem might involve angles that are **complementary** or **supplementary**.

Definition   ***Complementary angles:*** Two angles are complementary if the sum of their measures is 90°.

In the following figure, $\angle ABD$ and $\angle DBC$ are complementary because $32° + 58° = 90°$.

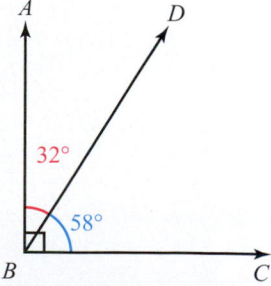

Note: The small square indicates a 90° angle.

Answer   **Your Turn 3**
5 ft. by 10 ft.

Definition   ***Supplementary angles:*** Two angles are supplementary if the sum of their measures is 180°.

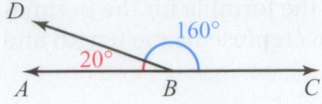

Inner angle   Outer angle

Note: The second relationship comes from the sketch. It shows that the inner and outer angles are supplementary, which means that the sum of their measures is 180°.

A straight line forms an angle that measures 180°. Any line that intersects a straight line will divide the 180° angle into two angles that are supplementary. In the figure in the margin, line segment DB joins line segment AC, forming two angles, $\angle ABD$ and $\angle DBC$. These two angles are supplementary angles because 20° + 160° = 180°.

Now look at a problem containing one of these terms.

Example 4   A steel beam is to be welded to a crossbeam, creating two angles, an inner angle and an outer angle. If the outer angle is to be 15° less than twice the inner angle, what are the angle measurements?

Understand   We must find the inner and outer angle measurements. The sketch for this situation is shown in the margin. (Notice that it is like the preceding figure.)

Next, list the relationships.

Relationship 1: The outer angle is 15° less than twice the inner angle.

Relationship 2: Inner angle + Outer angle = 180

Plan   Translate the relationships to an equation and then solve.

Execute   Relationship 1: The outer angle is 15° less than twice the inner angle.

$$\text{Outer angle} = 2 \cdot \text{Inner angle} - 15$$
$$\text{Outer angle} = 2a - 15$$

Note: Because the measure of the inner angle is multiplied by 2, we should let the variable represent the inner angle. Let's use the letter a.

Now use relationship 2 to write an equation that we can solve.

$$\text{Inner angle} + \text{Outer angle} = 180$$
$$a \quad + \quad 2a - 15 \quad = 180$$

$3a - 15 = 180$	**Combine like terms.**
$\underline{+15 \quad +15}$	**Add 15 to both sides.**
$3a + 0 = 195$	
$\dfrac{3a}{3} = \dfrac{195}{3}$	**Divide both sides by 3 to isolate a.**
$a = 65$	

Answer   The inner angle, represented by a, is 65°. According to relationship 1, the outer angle = $2a - 15$; so the outer angle = $2(65) - 15 = 115°$.

Check   Verify that all of the conditions in the problem are satisfied. The outer angle measuring 115° is 15° less than twice the inner angle measuring 65°. The sum of the inner and outer angles is 180°, which verifies that they are supplementary.

Your Turn 4   The wooden joists supporting a roof are connected as shown. The measure of $\angle DBC$ is 6° less than three times the measure of $\angle ABD$. Find the measures of $\angle DBC$ and $\angle ABD$.

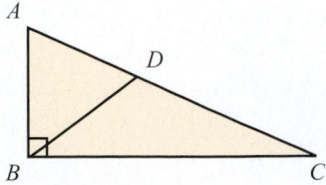

Problems Involving Consecutive Integers

Another group of problems in which the relationships are not obvious involve *consecutive integers*. The word *consecutive* means that the numbers are in sequence. For example, beginning with -3, consecutive integers are $-3, -2, -1, 0, 1, 2, \ldots$.

Answer   **Your Turn 4**

$\angle DBC = 66°$
$\angle ABD = 24°$

What relationship exists in the pattern of consecutive integers? Notice that to get from one integer in the sequence to the next integer, we add 1. For the sake of finding the general relationship, suppose the first integer in a sequence of consecutive integers is 1. By noting how we get from the number 1 to its consecutive integer, we can find a general pattern when the first integer is an unknown number, n, as the following table illustrates.

	With 1 as the First Integer	With n as the First Integer
First Integer	1	n
Next Integer	$1 + 1 = 2$	$n + 1$
Next Integer	$1 + 2 = 3$	$n + 2$

From the table, we see that the pattern for consecutive integers is $n, n + 1,$ $n + 2, \ldots$.

Now let's determine a general relationship for a sequence of consecutive odd integers, such as $1, 3, 5, 7, \ldots$ As we see in the following table, adding 2 to an odd integer gives the next consecutive odd integer.

	With 1 as the First Odd Integer	With n as the First Odd Integer
First Odd Integer	1	n
Next Odd Integer	$1 + 2 = 3$	$n + 2$
Next Odd Integer	$1 + 4 = 5$	$n + 4$

Warning: The pattern $n, n + 1, n + 3, \ldots$ is *not correct* for consecutive odd integers.

So the pattern for consecutive odd integers is $n, n + 2, n + 4, \ldots$.

As we see in the following table, the pattern for consecutive even integers is the same as for consecutive odd integers. Adding 2 to an even integer gives the next consecutive even integer.

	With 2 as the First Even Integer	With n as the First Even integer
First Even Integer	2	n
Next Even Integer	$2 + 2 = 4$	$n + 2$
Next Even Integer	$2 + 4 = 6$	$n + 4$

The pattern for consecutive even integers is $n, n + 2, n + 4, \ldots$ which is the same pattern as that for consecutive odd integers.

Example 5 The sum of three consecutive even integers is 78. Find the integers.

Understand The key word *sum* means to "add." If we let n represent the smallest of the even integers, the pattern for three unknown consecutive even integers is as follows:

Smallest unknown even integer: n

Next even integer: $n + 2$

Third even integer: $n + 4$

Plan Translate to an equation and then solve.

Execute Translate: The sum of three consecutive even integers is 78.

$$\boxed{\begin{array}{c}\text{Smallest even}\\\text{integer}\end{array}} + \boxed{\begin{array}{c}\text{Next even}\\\text{integer}\end{array}} + \boxed{\begin{array}{c}\text{Third even}\\\text{integer}\end{array}} = 78$$

$$
\begin{array}{llll}
n \quad + \quad n+2 \quad + \quad n+4 \;=\; 78 & \\
3n+6 = 78 & \textbf{Combine like terms.}\\
\underline{\;\;-6 \quad -6\;\;} & \textbf{Subtract 6 from both sides.}\\
3n+0 = 72 & \\
3n = 72 & \\
\dfrac{3n}{3} = \dfrac{72}{3} & \textbf{Divide both sides by 3 to}\\
& \textbf{isolate } n.\\
n = 24 &
\end{array}
$$

Answer The smallest of the three unknown integers, represented by n, is 24.

$$\text{Next even integer: } n+2 = 24+2 = 26$$
$$\text{Third even integer: } n+4 = 24+4 = 28$$

Check The numbers 24, 26, and 28 are three consecutive even integers, and $24 + 26 + 28 = 78$.

Your Turn 5 The sum of three consecutive integers is 96. Find the integers.

Objective ② **Use a table in solving problems with two unknowns.** We have seen how using key words, drawing pictures, making lists, and looking for patterns have helped us understand problems. Now we consider a group of problems in which a table is helpful in organizing information.

Example 6 A home-improvement store sells two sizes of cans of wood stain. A small can sells for $8.95, and a large can sells for $15.95. One day, the store sold twice as many of the large cans as the small cans. If the total revenue that day for these cans of wood stain was $694.45, how many cans of each size were sold?

Understand We know that the total revenue is $694.45. From this we can say

$$\boxed{\begin{array}{c}\text{Revenue from}\\\text{the small cans}\end{array}} + \boxed{\begin{array}{c}\text{Revenue from}\\\text{the large cans}\end{array}} = 694.45$$

We can also describe the revenue from the sale of each size can. We know the price of each size. If we knew the number of cans sold, we could multiply that number by the price per can to calculate the revenue from that particular size.

$$\boxed{\begin{array}{c}\text{Price}\\\text{per can}\end{array}} \cdot \boxed{\begin{array}{c}\text{Number of}\\\text{cans sold}\end{array}} = \boxed{\begin{array}{c}\text{Revenue for}\\\text{that size can}\end{array}}$$

With so much information, it is helpful to use a table to list it all. We use a four-column table with the labels *Categories*, *Price per Can*, *Number of Cans*, and *Revenue*. The categories are the two sizes of wood stain.

Answer **Your Turn 5**
31, 32, 33

202 Chapter 3 Problem Solving

Categories	Price per Can	Number of Cans	Revenue
Small	8.95	n	$8.95n$
Large	15.95	$2n$	$15.95(2n)$

Note: We were given the price.

Note: We selected n to represent the number of small cans, then translated "twice as many of the large cans as the small cans."

Note: We multiplied straight across because

$$\boxed{\text{Price per can}} \cdot \boxed{\text{Number of cans}} = \boxed{\text{Revenue from each size}}$$

The expressions in the last column in the table give the revenue from the sale of each size can.

Plan Translate the information to an equation and then solve.

Execute Now we can use our initial relationship.

$$\boxed{\text{Revenue from the small cans}} + \boxed{\text{Revenue from the large cans}} = 694.45$$

$$
\begin{aligned}
8.95n \quad + \quad 15.95(2n) &= 694.45 \\
8.95n + 31.9n &= 694.45 \qquad \text{\color{blue}{Multiply } } 2n \text{ \color{blue}{by 15.95.}} \\
40.85n &= 694.45 \qquad \text{\color{blue}{Combine like terms.}} \\
\frac{40.85n}{40.85} &= \frac{694.45}{40.85} \qquad \text{\color{blue}{Divide both sides by 40.85 to isolate } } n. \\
n &= 17
\end{aligned}
$$

Answer The number of small cans sold, n, is 17. The number of large cans sold, $2n$, is 34 (found by multiplying $2 \cdot 17$).

Check First, verify that twice the number of small cans sold, 17, is 34, which is true. Second, verify that the total revenue from the sale of wood stain is $694.45.

$$8.95(17) + 15.95(34) = 152.15 + 542.30 = 694.45 \qquad \text{\color{blue}{It checks.}}$$

The table used in Example 6 is a common type of table. In general, when a problem involves two or more categories, we can construct a table with a column labeled *Categories* and a row for each category in the problem. We then create a column for each parameter described in the problem. For example, parameters in a problem might be the *value* of each item, the *number* of items, and the total *amount* of money in each item. The table would look like this:

Categories	Value	Number	Amount

Because *value* · *number* = *amount*, the column labeled *Amount* contains the products of the expressions in the Value and Number columns.

Your Turn 6 Complete a table, write an equation, and then solve.

A computer store sells two different ink cartridges for a particular printer. The black ink cartridge sells for $12.75, and the color cartridge sells for $24.95. A company orders five more black ink cartridges than color cartridges. If the total cost is $629.25, how many of each cartridge were purchased?

Answer   Your Turn 6

20 black ink and 15 color

Note: Exercises marked with a ★ represent challenging exercises.

1. Suppose an investor has $10,000 to invest in two accounts. The table lists some ways to split the principal that he is considering.
   a. Complete the table.
   b. How did you determine the unknown principals in the table?

Account 1	Account 2
$9000	$1000
$8500	
	$6400
$3200	

   c. In general, if he invests P in one of the accounts, how do you describe the amount invested in the other account?

2. An investor always invests in two accounts according to the pattern shown in the table.
   a. Complete the table.
   b. In general, if he invests P in account 1, how would you describe the amount invested in account 2?

Account 1	Account 2
$500	$1000
$700	$1400
$900	
	$2400

   c. If he invests P in account 2, how would you describe the amount invested in account 1?

3. Complete the following table.

Principal	Interest Rate	Annual Interest
$5500	4%	
$10,450	3.75%	
P	5%	
$(15,000 - P)$	6.5%	

4. Complete the following table.

Principal	Interest Rate	Annual Interest
$6800	6%	
$12,650	$4\frac{1}{2}\%$	
P	6.5%	
$(50,000 - P)$	4.3%	

5. Complete the table, which contains a description of two unknowns and a variable for one of the unknowns. Use the description to describe the second unknown in terms of the variable.

Description	Account 1	Account 2
Carol invests twice as much in account 2 as in account 1.	P	
Jennifer invests $1200 more in account 1 as in account 2.		P
Tim invests $6000 in the two accounts.	P	
Johnny invests $4000 in the two accounts.		P

6. Complete the table, which contains a description of two unknowns and a variable for one of the unknowns. Use the description to describe the second unknown in terms of the variable.

Description	Account 1	Account 2
Ahmad invests two-thirds as much in account 2 as in account 1.	P	
Johanna invests $3500 less in account 1 than in account 2.		P
Franz invests $5000 more in account 2 than in account 1.	P	
Murlene invests a total of $10,000 in the two accounts.		P

7. Given the total interest from an investment in two plans, once you have the expressions for the two interest amounts, how do you write the equation?

8. When completing a table for mixture problems, what are the three categories?

9. The following table shows some possible combinations of two solutions. Look for a pattern; then complete the table.

Volume of Solution 1	Volume of Solution 2	Combined Solution
25 ml	50 ml	75 ml
40 ml	80 ml	
100 ml		
n ml		

10. The following table shows possible combinations of two solutions. Look for a pattern and complete the table.

Volume of Solution 1	Volume of Solution 2	Combined Solution
10 oz.	30 oz.	40 oz.
40 oz.	120 oz.	
75 oz.		
x oz.		

11. Complete the following table.

Volume of Solution	Percent Pure Alcohol	Volume of Alcohol in Solution
130 oz.	5%	
65 oz.	16.5%	
x oz.	$12\frac{1}{2}$%	
$(50 - x)$ oz.	8.2%	

12. Complete the following table.

Volume of Solution	Percent Pure Sulfuric Acid	Volume of Sulfuric Acid in Solution
30 ml	45%	
150 ml	$5\frac{3}{4}$%	
y ml	3.5%	
$(125 - y)$ ml	27.6%	

Exercises 13–28 involve investment. Complete a table, write an equation, and then solve. See Example 1.

13. Janice invests in a plan that has an APR of 2%. She invests three times as much in a plan that has an APR of 4%. If the total interest after one year from the investments is $644, how much was invested in each plan?

14. Gayle invests in a plan that has an APR of 3%. She invests twice as much in a plan that has an APR of 5%. If the total interest after one year from the investments is $637, how much was invested in each plan?

15. Tory invests money in two plans. She invests two-fifths of the money at an annual return rate of 2% in a money market account. She invests the remainder of the money in a short-term mutual fund account with a return rate of 3%. If the total annual interest from the investments is $104, how much was invested in each plan?

16. Dwight invests money in two plans. He invests two-fifths of the money at an APR of 8% and the rest at an APR of 7.6%. If the total interest after one year from the investments is $620.80, how much was invested in each plan?

17. Boyd invests in a plan that has an APR of 8%. He invests in a 12% APR account $650 more than what he invested in the 8% account. If the total interest after one year from the investments is $328, how much was invested in each plan?

18. Deon invests in an account that has an APR of 2.5%. He invests in a 3.5% APR account $400 more than what he invests in the 2.5% account. If the total annual interest from the investments is $242, how much was invested in each account?

19. Agnes invests $4500 in two plans. Plan 1 is at an APR of 4%, and plan 2 is at an APR of 6%. If the total interest after one year is $234, what principal was invested in each plan?

20. Juan invests $12,800 in two plans. Plan 1 is at an APR of 7%, and plan 2 is at an APR of 5%. If the total interest after one year is $706, what principal was invested in each plan?

21. Kennon invested $1600. She was able to put some of this money into a CD at a 5% APR and the rest into a CD at only 4%. If she achieved an annual interest income of $73.40 on these investments, how much was invested in each plan?

22. Shanequa has $4000. She put some of the money into savings that pays 6% annually and the rest in an account that pays 7%. If her total interest for the year is $264, how much did she invest at each rate?

23. Annette invested $5000. She put some into an investment earning 5.5% annually and the rest earning 7%. If her interest income during this year was $335, how much did she invest at each rate?

24. Buddy inherited $12,000. He invested part of the money at a 4% APR and the rest at 5.5%. If his annual interest income from these investments is $532.50, how much did he invest at each rate?

★ **25.** Deloris invests in a plan that has an APR of 6%. She invests twice as much in a plan that ends up as a loss of 2% (that is, a −2% APR). If the net gain from the investments is $84, how much was invested in each plan?

★ **26.** Dominique invests a total of $10,000 in two plans. Plan 1 is at an APR of 5%, while plan 2 ends up returning a loss of 9% (that is, a −9% APR). If the return on the investments after one year is a net loss of $312, what principal was invested in each plan?

★ **27.** Cody has two investments totaling $10,000. Plan A is at an APR of 4%, and plan B is at an APR of 8%. After one year, the interest earned from plan B is $320 more than the interest from plan A. How much did Cody invest in each plan?

★ **28.** Curtis has two investments totaling $12,000. Plan A is at an APR of 5%, and plan B is at an APR of 7%. After one year, the interest earned from plan B is $360 more than the interest earned from plan A. How much did Curtis invest in each plan?

Exercises 29–46 involve mixtures. Complete a table, write an equation, and then solve. See Example 2.

29. How many liters of a 40% solution of HCl are added to 2000 liters of a 20% solution to obtain a 35% solution?

30. Charles has 80 milliliters of a 15% acid solution. How much of a 20% acid solution must be added to create a solution that is 18% acid?

31. Wilson has a bottle containing 35 milliliters of 15% saline solution and a bottle of 40% saline solution. He wants a 30% solution. How much of the 40% solution must be added to the 15% solution so that a 30% concentration is created?

32. Sharon has a bottle containing 45 milliliters of 15% HCl solution and a bottle of 35% HCl solution. She wants a 25% solution. How much of the 35% solution must be added to the 15% solution so that a 25% concentration is created?

33. A dairy is making a 30% buttermilk cream. If it mixes a 26% buttermilk cream with a 35% buttermilk cream, how much of each does it need to use to produce 300 pounds of the 30% buttermilk cream?

34. A pharmacist has a 45% acid solution and a 35% acid solution. How many liters of each must be mixed to form 80 liters of a 40% acid solution?

35. To form a 10% copper alloy weighing 75 grams, a 6% copper alloy is combined with an 18% copper alloy. How much of each type should be used?

36. A jeweler is mixing two silver alloys. How many ounces of an alloy containing 25% silver must be mixed with an alloy containing 20% silver to obtain 50 ounces of a 22% silver alloy?

37. A cough medicine contains 25% alcohol. How much nonalcoholic liquid should a pharmacist add to 120 milliliters of the cough medicine so that it contains only 20% alcohol?

38. A solution of gasoline and oil used for a gas-powered weed trimmer is 8% oil. How much gasoline must be added to 3 gallons of the solution to obtain a new solution that is 5% oil?

39. How many quarts of pure antifreeze must be added to 6 quarts of a 40% antifreeze solution to obtain a 50% antifreeze solution?

40. A 30-milliliter solution of alcohol and water is 10% alcohol. How much pure alcohol must be added to yield a 40% solution?

41. A candy store is selling two types of candy, one at $1.20 per pound and the other at $0.90 per pound. To sell a mixture at $1.11 a pound, how many pounds of each must the store mix to make 80 pounds?

42. Daly's Farm Feed sells two kinds of horse feed. One is 20% oats, and the other is 28% oats. How many pounds of each should be mixed to form 50 pounds of a feed that is 25% oats?

43. The Market Street Sweets sells two types of pralines: a caramel version for $1.65 a pound and a chocolate version for $1.25 a pound. A mixture of these pralines can be sold for $1.49 a pound. How many pounds of each kind should be used to make 5 pounds of the $1.49 mixture?

44. Cromer's sells salted peanuts for $2.50 per pound and cashews for $6.75 per pound. How many pounds of each are needed to obtain a 10-pound mixture costing $4.20 per pound?

45. The Candy Shoppe wants to mix 115 pounds of candy to sell for $0.80 per pound. How many pounds of $0.60 candy must be mixed with a candy costing $1.20 per pound to make the desired mix?

46. A coffee shop wants to mix 20 pounds of coffee that will sell for $7 per pound. How many pounds of coffee at $5 per pound must be mixed with coffee costing $8 per pound to make the mix?

Puzzle Problem

Herschal has $18,000 to invest. He invests $\frac{2}{3}$ in one account and the rest in a second account.

The first account returned 2% more than the second account. The total return on his investment was $1050. What were the percentage rates of the two accounts?

Review Exercises

Exercises 1–3 EXPRESSIONS

[1.5] *For Exercises 1 and 2, simplify.*

1. $\dfrac{3}{4} - \dfrac{5}{6} \cdot \left(\dfrac{2}{3}\right)^2$

2. $(-4)^3$

$\begin{bmatrix}1.6\\1.7\end{bmatrix}$ **3.** Write an expression in simplest form that describes the perimeter of the rectangle shown.

x

$5x$

[1.1] **4.** Use $<$, $>$, or $=$ to make a true statement:

$-|-14|$ ⬛ 14

Exercises 4–6 EQUATIONS AND INEQUALITIES

[2.3] *For Exercises 5 and 6, solve.*

5. $4(y - 3) + y = 7y - 8$

6. $\dfrac{3}{4}x - 5 = \dfrac{2}{3}x + 1$

Chapter 3 Summary

Defined Terms

Section 3.1
Ratio (p. 170)
Unit ratio (p. 170)
Proportion (p. 171)
Congruent angles (p. 174)
Similar figures (p. 174)

Section 3.2
Percent (p. 183)

Section 3.3
Complementary angles (p. 199)
Supplementary angles (p. 199)

Formulas

Simple percent sentence: A percent of a whole is a part of the whole. (p. 185)

$\text{Percent} = \dfrac{\text{Part}}{\text{Whole}}$ (p. 185)

Perimeter of a rectangle: $P = 2l + 2w$ (p. 199)

Distance, rate, and time: $d = rt$ (p. 211)

Simple interest: $I = Pr$ (p. 220)

Procedures, Rules, and Key Examples

Procedures/Rules	Key Examples

SECTION 3.1 Ratios and Proportions

If $\dfrac{a}{b} = \dfrac{c}{d}$, where $b \neq 0$ and $d \neq 0$, then $ad = bc$.

Example 1: Determine whether the ratios are equal.

$$\frac{5}{6} \overset{?}{} \frac{15}{18}$$

$18 \cdot 5 = 90 \qquad\qquad 6 \cdot 15 = 90$

$$\frac{5}{6} \bowtie \frac{15}{18}$$

Because the cross products are equal, the ratios are equal.

To solve a proportion using cross products,
1. Calculate the cross products.
2. Set the cross products equal to each other.
3. Use the multiplication principle of equality to isolate the variable.

Example 2: Solve for the missing number in the proportion.

$$\frac{5}{12} = \frac{x}{18}$$

$90 = 12x$ **Cross products**

$\dfrac{90}{12} = \dfrac{12x}{12}$ **Divide by 12.**

$7.5 = x$

(continued)

Procedures/Rules	Key Examples

SECTION 3.1   Ratios and Proportions (*continued*)

To solve proportion problems,

1. Set up a proportion in which the numerators and denominators of the ratios correspond in a logical manner.
2. Solve using cross products.

Example 3: A car can travel about 350 miles on 16 gallons of gasoline. How many gallons would be needed to travel 600 miles?

$$\frac{350}{16} = \frac{600}{x}$$

$$350x = 9600 \qquad \text{Cross products}$$

$$\frac{350x}{350} = \frac{9600}{350} \qquad \text{Divide by 350.}$$

$$x \approx 27.4 \text{ gal.}$$

SECTION 3.2   Percents

To write a percent as a fraction or decimal,

1. Write the percent as a ratio with 100 in the denominator.
2. Simplify to the desired form.

Note: When simplifying, remember that dividing a decimal number by 100 moves the decimal point two places to the left.

Example 1: Write each percent as a fraction in lowest terms and as a decimal.

a. $42\% = \dfrac{42}{100} = \dfrac{21}{50} = 0.42$

b. $8.5\% = \dfrac{8.5}{100} = \dfrac{85}{1000} = \dfrac{17}{200} = 0.085$

c. $20\frac{1}{2}\% = \dfrac{20\frac{1}{2}}{100} = \dfrac{41}{200} = 0.205$

To write a fraction or decimal number as a percent,

1. Multiply by 100%.
2. Simplify.

Note: Multiplying a decimal number by 100 moves the decimal point two places to the right.

Example 2: Write as a percent.

a. $0.453 = 0.453 \cdot 100\% = 45.3\%$

b. $\dfrac{5}{8} = \dfrac{5}{\underset{2}{\cancel{8}}} \cdot \dfrac{\overset{25}{\cancel{100}}}{1}\% = \dfrac{125}{2}\% = 62\frac{1}{2}\%$

To translate simple percent sentences:

Method 1. Translate the sentence word for word.

1. Select a variable for the unknown.
2. Translate the word *is* to an equal sign.
3. If *of* is preceded by the percent, translate it to multiplication. If *of* is preceded by a whole number, translate it to division.

Method 2. Translate to a proportion by writing the following form:

$$\text{Percent} = \frac{\text{Part}}{\text{Whole}},$$

where the percent is expressed as a fraction with a denominator of 100.

Example 3: 15% of what number is 33?

Method 1: Direct translation

$$0.15n = 33$$

$$\frac{0.15n}{0.15} = \frac{33}{0.15}$$

$$n = 220$$

Method 2: Proportion

$$\frac{15}{100} = \frac{33}{n}$$

$$15n = 3300$$

$$\frac{15n}{15} = \frac{3300}{15}$$

$$n = 220$$

(*continued*)

Procedures/Rules	Key Examples

SECTION 3.2 Percents (*continued*)

To solve problems involving percent,
1. Determine whether the percent, the whole, or the part is unknown.
2. Write the problem as a simple percent sentence (if needed).
3. Translate to an equation (word for word or proportion).
4. Solve for the unknown.

Example 4: A student answered 24 questions out of 30 correctly. What percent of the questions did the student answer correctly?

Simple percent sentence: What percent of 30 is 24?

Direct Translation:

$$p \cdot 30 = 24$$
$$\frac{30p}{30} = \frac{24}{30}$$
$$p = 0.8$$

Answer:
$$0.8 \cdot 100\% = 80\%$$

Proportion:

$$\frac{P}{100} = \frac{24}{30}$$
$$30P = 2400$$
$$\frac{30p}{30} = \frac{2400}{30}$$
$$P = 80$$

Answer: 80%

SECTION 3.3 Problems with Two or More Unknowns

To solve problems with two or more unknowns,
1. Determine which unknown will be represented by a variable. *Tip:* Let the unknown that is acted on be represented by the variable.
2. Use one of the relationships to describe the other unknown(s) in terms of the variable.
3. Use the other relationship to write an equation.
4. Solve the equation.

Example 1: One number is 15 more than twice another. The sum of the numbers is 39. Find the numbers.

Relationship 1: "One number is fifteen more than twice another."

Translation: $2x + 15$, where x represents the smaller number.

Relationship 2: "The sum of the numbers is 39."

Translation:
$$x + 2x + 15 = 39$$
$$3x + 15 = 39$$
$$\underline{\quad -15 \quad -15}$$
$$3x + 0 = 24$$
$$\frac{3x}{3} = \frac{24}{3}$$
$$x = 8$$

The smaller number is 8, and the other number is $2(8) + 15 = 31$.

SECTION 3.4 Rates

To solve for time when two objects are moving in opposite directions,
1. Use a table with columns for categories, rate, time, and distance. Use the fact that rate · time = distance.
2. Write an equation that is the sum of the individual distances equal to the total distance of separation.
3. Solve the equation.

Example 1: Mark and Latonia are 17 miles apart traveling toward each other. Mark is traveling at a rate of 45 miles per hour, and Latonia is traveling at a rate of 40 miles per hour. How long will it be until they meet?

Categories	Rate	Time	Distance
Mark	45	t	$45t$
Latonia	40	t	$40t$

Equation:
$$45t + 40t = 17$$
$$85t = 17$$
$$\frac{85t}{85} = \frac{17}{85}$$
$$t = 0.2 \text{ hr. or } 12 \text{ min.}$$

(*continued*)

Procedures/Rules	Key Examples

SECTION 3.4 Rates (continued)

To solve problems involving two objects traveling in the same direction in which the objective is to determine the time for one object to catch up to the other,

1. Use a table to organize the rates and times. Let t represent the time for the object to catch up and add the time difference to t to represent the other object's time.
2. Set the expressions for the individual distances equal.
3. Solve the equation.

Example 2: Frank is traveling at a rate of 64 miles per hour and passes Exit 43 at 4:35 P.M. Juanita is traveling at a rate of 72 miles per hour in the same direction and passes the same exit at 4:50 P.M. At what time does she catch up with Frank?

Categories	Rate	Time	Distance
Frank	64	$t + 0.25$	$64(t + 0.25)$
Juanita	72	t	$72t$

Equation:
$$64(t + 0.25) = 72t$$
$$64t + 16 = 72t$$
$$\underline{-64t \qquad\quad -64t}$$
$$0 + 16 = 8t$$
$$16 = 8t$$
$$\frac{16}{8} = \frac{8t}{8}$$
$$2 = t$$

Answer: Juanita catches up at 6:50 P.M.

SECTION 3.5 Investment and Mixture

To solve problems involving two interest rates in which the total interest is given,

1. Use a table to organize the interest rates and principals. Multiply the individual rates and principals to get expressions for the individual interests.
2. Write an equation that is the sum of the expressions for interest set equal to the given total interest.
3. Solve the equation.

Example 1: Dan invests a total of $4000 in two different plans. The first plan returns 9% annually, while the second plan returns 6%. If the total interest earned after one year is $285, what principal was invested in each plan?

Categories	Rate	Principal	Interest
Plan 1	0.09	p	$0.09p$
Plan 2	0.06	$4000 - p$	$0.06(4000 - p)$

Equation:
$$0.09p + 0.06(4000 - p) = 285$$
$$0.09p + 240 - 0.06p = 285$$
$$0.03p + 240 = 285$$
$$\underline{-240 \quad -240}$$
$$0.03p + 0 = 45$$
$$\frac{0.03p}{0.03} = \frac{45}{0.03}$$
$$p = 1500$$

Answer: $1500 in plan 1 and
$4000 - $1500 = $2500 in plan 2

(continued)

Chapters 1–3 Cumulative Review Exercises

For Exercises 1–4, answer true or false.

[1.1] **1.** $\sqrt{2}$ is an irrational number.

[1.5] **2.** When evaluating an expression using the order of operations, you must always do multiplication before division.

[1.6] **3.** "Four less than a number" can be translated to $4 - n$.

[2.2] **4.** The equation $2(x - 1) - 3x = 5 - x - 7$ is an identity.

For Exercises 5 and 6, fill in the blank.

[1.2] **5.** A prime number is a natural number that has exactly two different factors, 1 and _____.

[2.3] **6.** To clear fractions from an equation, we can use the multiplication principle of equality and multiply both sides by the _____.

[1.2] **7.** Find the prime factorization of 120.

[1.7] **8.** Determine the values that make $\dfrac{4x}{x - 3}$ undefined.

[1.7] **9.** Which property of arithmetic is illustrated by $4(x + 3) = 4x + 12$?

[3.1] **10.** Explain in your own words how to solve a proportion.

Exercises 11–17 **EXPRESSIONS**

[1.5] For Exercises 11–16, simplify.

11. $\left(-\dfrac{2}{3}\right)^2$

12. $\sqrt{16} + \sqrt{9}$

13. $\dfrac{5 - 2(-1)^2}{3\sqrt{64} + 36}$

14. $-|3 \cdot (-5)| + |(-1)^3|$

15. $2\left(-\dfrac{1}{2}\right)^2 - 4\left(\dfrac{3}{5}\right) + 3$

16. $-6 - 2(3 + 0.2)^2$

[1.7] **17.** Use the distributive property to simplify $-3\left(\dfrac{1}{6}x - 9\right)$.

Exercises 18–30 **EQUATIONS AND INEQUALITIES**

[2.2 and 2.3] For Exercises 18–21, solve and check.

18. $3.23c - 8.75 = 1.41c + 7.63$

19. $2m + \dfrac{1}{2} = 4m - 3\dfrac{1}{2} + 8m$

20. $6x - 2 = 3\left(2x - \dfrac{2}{3}\right)$

21. $-7 = -6(p + 2) + 5(2p - 3)$

22. $A = \dfrac{1}{2}bh$ for b

23. $P = a + b + c$ for c

For Exercises 24–30, solve.

[3.2] **24.** What is 15% of 42.3?

[3.3] **25.** The sum of three consecutive odd integers is -3. What are the integers?

[2.6] **26.** Jon scored an 82 on his first test and a 70 on his second test. What range of scores does he need on his third test to have an average of at least 80?

[3.5] **27.** How many liters of a 60% solution of boric acid should be added to 10 liters of a 30% solution to obtain a 50% solution?

[3.4] **28.** Archie and Veronica are both driving east on Riverdale Drive. At 3 P.M., Archie, driving at 50 miles per hour, was 5 miles east of Veronica. A little later Veronica, driving at 55 miles per hour, passed Archie. At what time did Veronica pass Archie?

[3.5] **29.** Cromers sells cashews for $7.00 per pound and peanuts for $3.00 per pound. If a 10-pound bag is mixed selling at $4.50 per pound, how many pounds of each type nut should be used?

[3.1] **30.** If a car travels 120 miles on 4 gallons of gasoline, how far can it travel on 10 gallons?

Graphing Linear Equations and Inequalities

4

All equations and inequalities have corresponding pictures, or graphs. The graph of an equation or inequality offers information about the solutions to the equation or inequality. Graphs can be helpful in solving problems. For example, we may be given an equation or inequality that describes a particular problem. The graph of that equation may offer additional insight about possible solutions to the problem. Or a problem may have some data that we can graph. We can often use such a graph to help us write an equation or inequality that describes the situation. Both approaches can be quite useful in solving problems.

4.1 The Rectangular Coordinate System

Objectives

1. Determine the coordinates of a given point.
2. Plot points in the coordinate plane.
3. Determine the quadrant for a given ordered pair.
4. Determine whether the graph of a set of data points is linear.

Objective 1 Determine the coordinates of a given point. In 1619, René Descartes, the French philosopher and mathematician, recognized that positions of points in a plane could be described using two number lines that intersect at a right angle. Each number line is called an *axis*.

Two perpendicular axes form the *rectangular*, or Cartesian, *coordinate system*, named in honor of René Descartes. Usually, we call the horizontal axis the *x-axis* and the vertical axis the *y-axis*.

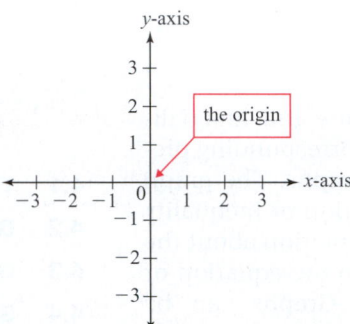

The point where the axes intersect is 0 for both the *x*-axis and *y*-axis. This position is called the *origin*. The positive numbers are to the right and up from the origin, whereas negative numbers are to the left and down from the origin.

Any point in the plane can be described using two numbers, one number from each axis. To avoid confusion, these two numbers are written in a specific order. The number representing a point's *horizontal distance* from the origin is given *first*, and the number representing a point's *vertical distance* from the origin is given *second*. Because the order in which we say or write these two numbers matters, we say that they form an *ordered pair*. Each number in an ordered pair is called a *coordinate* of the ordered pair. The notation for writing ordered pairs is (horizontal coordinate, vertical coordinate).

Consider the point labeled *A* in the coordinate plane shown. The point is drawn at the intersection of the 3rd line to the right of the origin and the 4th line up from the origin. The ordered pair that describes point *A* is (3, 4).

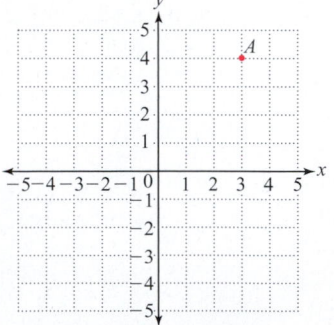

Connection Think of the intersecting lines in the grid as avenues and streets in a city such as New York, where avenues are north–south and streets are east–west. To describe an intersection, a person would say the avenue first, then the street. Point *A* is at the intersection of 3rd Avenue and 4th Street, or, as a New Yorker would say, "3rd and 4th."

Procedure Identifying the Coordinates of a Point

To determine the coordinates of a given point in the rectangular system,

1. Follow a vertical line from the point to the *x*-axis (horizontal axis). The number at this position on the *x*-axis is the first coordinate.
2. Follow a horizontal line from the point to the *y*-axis (vertical axis). The number at this position on the *y*-axis is the second coordinate.

Example 1   Write the coordinates for each point shown.

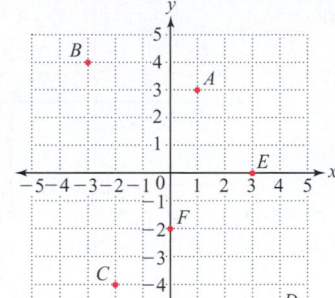

Answers

A: $(1, 3)$ From point A, a vertical line intersects the x-axis at 1 and a horizontal line intersects the y-axis at 3.

B: $(-3, 4)$ From point B, a vertical line intersects the x-axis at -3 and a horizontal line intersects the y-axis at 4.

C: $(-2, -4)$ From point C, a vertical line intersects the x-axis at -2 and a horizontal line intersects the y-axis at -4.

D: $(4, -5)$ From D, a vertical line intersects the x-axis at 4 and a horizontal line intersects the y-axis at -5.

E: $(3, 0)$ From E, a vertical line intersects the x-axis at 3 and a horizontal line intersects the y-axis at 0.

F: $(0, -2)$ From F, a vertical line intersects the x-axis at 0 and a horizontal line intersects the y-axis at -2.

Your Turn 1   Write the coordinates for each point shown.

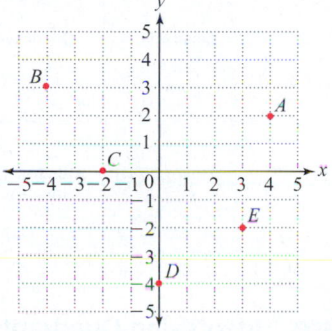

Objective 2   Plot points in the coordinate plane.   Any point can be plotted in the coordinate plane using its coordinates. Remember that the first coordinate in an ordered pair indicates the distance to move right or left and that the second coordinate indicates the distance to move up or down. To plot the ordered pair $(3, 4)$, we begin at the origin, move to the right 3, move up 4, and draw a dot to indicate the point.

Procedure   **Plotting a Point**

To graph or plot a point given its coordinates,

1. Beginning at the origin, $(0, 0)$, move to the right or left along the x-axis the amount indicated by the first coordinate.
2. From that position on the x-axis, move up or down the amount indicated by the second coordinate.
3. Draw a dot to represent the point described by the coordinates.

Example 2   Plot the point described by the coordinates.

a. $(3, -4)$ **b.** $(-4, 2)$ **c.** $(0, 5)$

Answers   Your Turn 1

A: $(4, 2)$ D: $(0, -4)$
B: $(-4, 3)$ E: $(3, -2)$
C: $(-2, 0)$

SOLUTION

Note: For $(-4, 2)$, we begin at the origin and move to the left 4, then up 2.

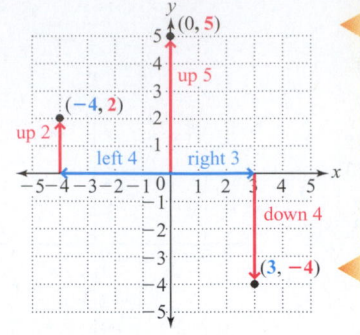

Note: For $(0, 5)$, because the first coordinate is 0, we do not move right or left. Because the second coordinate is 5, we move straight up 5, drawing the point on the *y*-axis.

Note: For $(3, -4)$, we begin at the origin and move to the right 3, then down 4.

Your Turn 2 Plot the point described by the coordinates.

$A: (-3, 2)$ $B: (1, 4)$ $C: (-2, -4)$ $D: (3, 0)$

Objective ❸ Determine the quadrant for a given ordered pair. The two perpendicular axes divide the coordinate plane into four regions called *quadrants*.

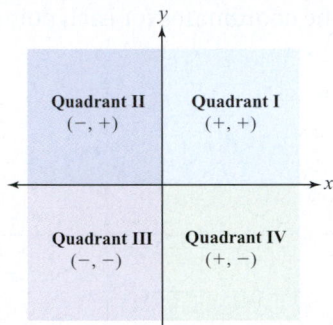

The quadrants are numbered using Roman numerals, as shown to the left. Note that the signs of the coordinates determine the quadrant in which a point lies. For points in quadrant I, both coordinates are positive. Every point in quadrant II has a negative first coordinate (horizontal) and a positive second coordinate (vertical). Points in quadrant III have both coordinates negative. Every point in quadrant IV has a positive first coordinate and a negative second coordinate. Points on axes are not in any quadrant.

Procedure | **Identifying Quadrants**

To determine the quadrant for a given ordered pair, consider the signs of the coordinates.

$(+, +)$ means the point is in quadrant I.

$(-, +)$ means the point is in quadrant II.

$(-, -)$ means the point is in quadrant III.

$(+, -)$ means the point is in quadrant IV.

Note: If either coordinate is 0, the point is on an axis and not in a quadrant.

Answer   Your Turn 2

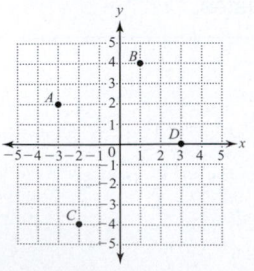

Example 3 State the quadrant in which each point is located. If the point is on an axis, state which axis.

a. $(-61, 23)$

Answer Quadrant II because the first coordinate is negative and the second coordinate is positive.

b. $\left(14, -37\frac{2}{3} \right)$

Answer Quadrant IV because the first coordinate is positive and the second coordinate is negative.

c. $(0, -12)$

Answer   Because the x-coordinate is 0, this point is on the y-axis and is not in a quadrant.

Your Turn 3   State the quadrant in which each point is located. If the point is on an axis, state which axis.

a. $(-42, -109)$ **b.** $(37, -15.9)$ **c.** $(4.7, 0)$ **d.** $\left(-16\frac{3}{4}, 124\right)$ **e.** $\left(4.3, 1\frac{7}{8}\right)$

Objective ④ Determine whether the graph of a set of data points is linear.   In many problems, data are listed as ordered pairs. For example, in a report on stock prices, we might have the listing in the margin of the closing price of a particular stock for each day over four days.

Day	Closing Price
1	$24.50
2	$25.75
3	$27.00
4	$28.25

Note: Because the vertical axis has dollar values that involve fractions, there are two marks for every whole dollar. This allows halves to be plotted accurately and quarters to be estimated as well.

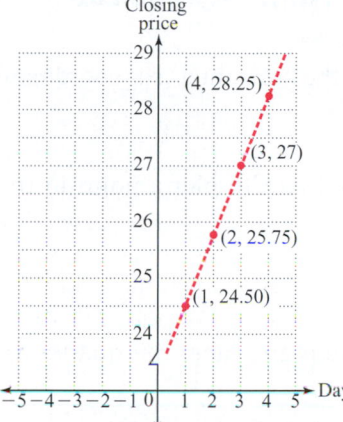

If we plot each pair of data as an ordered pair in the form (day, closing price), we see that the points lie on a straight line.

Because the points can be connected to form a straight line, they are said to be *linear*. Points that do not form a straight line are *nonlinear*.

Note: The broken line in the vertical axis indicates that values between 0 and 24 were skipped to conserve space.

Time (in seconds)	Height (in feet)
0	6.00
0.5	7.50
1.0	8.25
1.5	7.50
2.0	6.00
2.5	2.00

Example 4   The data points in the margin track the trajectory, or path, of an object over a period of time. Plot the points with the time along the horizontal axis and the height along the vertical axis. Then state whether the trajectory is linear or nonlinear.

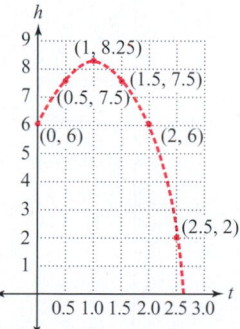

Solution   Because the data points do not form a straight line when connected, the trajectory is nonlinear.

Note: The graph for Example 4 does not have the origin at the center of the graph. It is often necessary to adjust the location of the origin, and consequently the axes, to fit the data. This frequently happens when the values plotted on the vertical and/or horizontal axes are nonnegative.

Note: In general, when a graph involves time, the time values are plotted along the horizontal axis.

Answers   Your Turn 3

a. III   **b.** IV   **c.** on the x-axis

d. II   **e.** I

Your Turn 4 The following data points show the velocity of an object as time passes. Plot the points with the time along the horizontal axis and the velocity along the vertical axis. Then state whether the points are linear or nonlinear.

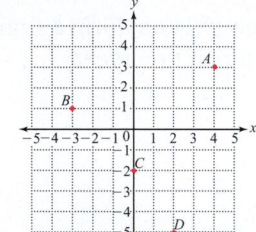
Time (in seconds)	Height (in feet)
0	2.0
1	2.5
2	3.0
3	3.5

4.1 Exercises   For Extra Help   *MyMathLab*

Note: Exercises marked with a ★ represent challenging exercises.

1. When writing an ordered pair, which is written first, the horizontal-axis coordinate or the vertical-axis coordinate?

2. Describe how to use the coordinates of an ordered pair to plot a point in the rectangular coordinate system.

3. Draw the axes in the rectangular coordinate system and label the quadrants. Include the signs for coordinates in each quadrant.

4. Describe how to determine whether a set of data points is linear.

For Exercises 5–8, write the coordinates for each point. See Example 1.

5.
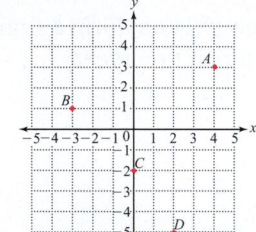

6.
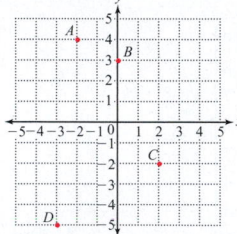

7.
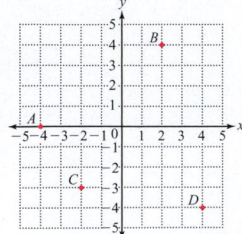

8.
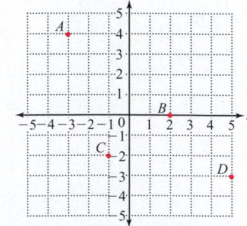

For Exercises 9–12, on a piece of graph paper, draw and label x- and y-axes. Then plot and label the points indicated by the coordinate pairs. See Example 2.

9. $(5, 4), (2, -3), (-1, -3), (2, 0)$

10. $(-3, 0), (0, 1), (-2, -2), (4, -1)$

11. $(-2, 0), (3, -3), (-1, -5), (2, 2)$

12. $(-1, -4), (0, -2), (4, -2), (-4, 2)$

For Exercises 13–24, state the quadrant in which the point is located. If the point lies on an axis, state which axis. See Example 3.

13. $(-6.5, 1050)$

14. $\left(-42\frac{1}{3}, -500\right)$

15. $(620, 50)$

16. $(-42, 50)$

17. $(57, -82.71)$

18. $(16, 27)$

19. $\left(-41\frac{1}{2}, -82\right)$

20. $\left(47, -51\frac{5}{8}\right)$

21. $(0, -9)$

22. $(0, 5.3)$

23. $(-0.6, 0)$

24. $(8.2, 0)$

For Exercises 25–34, determine whether the set of points is linear or nonlinear. See Example 4.

25. $(-2, -9), (0, -5), (2, -1), (4, 3), (5, 5)$

26. $(-5, 7), (-1, 3), (2, 0), (4, -2), (6, -4)$

27. $(-4, 8), (-2, 2), (0, 0), (1, 0.5), (3, 4.5)$

28. $(-1, -3), (0, 0), (2, 0), (3, -3)$

29. (pounds of chicken, cost): $(1, 2.5), (2, 5), (3, 7.5), (4, 10)$

30. (time, distance): $(0, 2), (1, 5), (2, 8), (3, 11), (4, 14)$

31. (time, height): $(0, 8.0), (0.5, 9.5), (1.0, 11.0), (1.5, 9.5), (2.0, 8.0), (2.5, 6.5)$

32. (days, stock closing price): $(1, 23), (2, 21), (3, 19), (4, 17)$

★33. (copy costs, number of copies): $(15, 300), (18, 400), (21, 500), (24, 600)$

★34. (number of hours, plumbing cost): $(1, 120), (2, 160), (3, 200), (4, 240)$

35. List the coordinates of three points on the line shown.

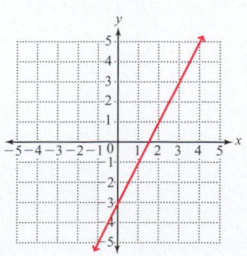

36. List the coordinates of three points on the line shown.

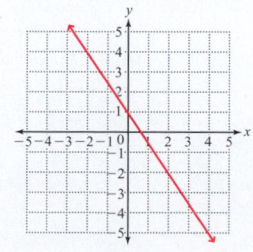

A vertex on a figure is a point where two line segments form a corner or joint on the figure. For Exercises 37 and 38, find the coordinates of each labeled vertex. See Example 1.

37.

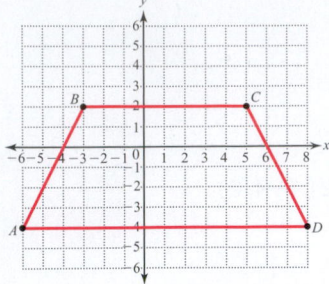

38.

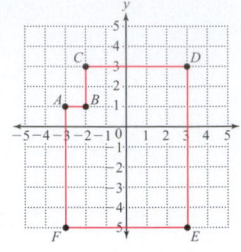

★ **39.** A parallelogram has been moved from its original position, shown with the dashed lines, to a new position, shown with solid lines.

   a. List the coordinates of each vertex of the parallelogram in its original position.

   b. List the coordinates of each vertex of the parallelogram in its new position.

   c. Write a rule that describes in mathematical terms how to move each point on the parallelogram in its original position to each point on the parallelogram in its new position.

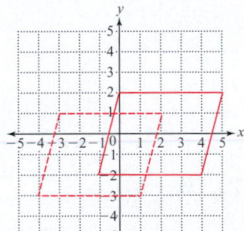

40. A trapezoid has been moved from its original position, shown with the dashed lines, to a new position, shown with solid lines.

   a. List the coordinates of each vertex of the trapezoid in its original position.

   b. List the coordinates of each vertex of the trapezoid in its new position.

   c. Write a rule that describes in mathematical terms how to move each point on the trapezoid in its original position to each point on the trapezoid in its new position.

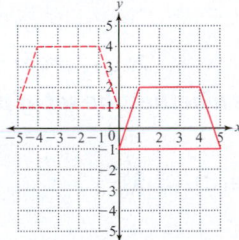

41. The ordered pairs $(-3, 1)$, $(-1, 1)$, $(-1, 3)$, $(2, 3)$, $(2, -2)$, and $(-3, -2)$ form the vertices of a figure.

   a. Plot the points and connect them to form the figure.

   b. Find the perimeter of the figure.

   c. Find the area of the figure.

42. The ordered pairs $(-4, 2)$, $(-2, 2)$, $(-2, 4)$, $(1, 4)$, $(1, 3)$, $(3, 3)$, $(3, -4)$, and $(-4, -4)$ form the vertices of a figure.

   a. Plot the points and connect them to form the figure.

   b. Find the perimeter of the figure.

   c. Find the area of the figure.

Review Exercises

Exercise 1 CONSTANTS AND VARIABLES

[1.1] 1. Graph -3 on a number line.

Exercises 2–4 EXPRESSIONS

[1.7] 2. Evaluate $2x - 3y$ when $x = 4$ and $y = 2$. [1.7] 3. Evaluate $-4x - 5y$ when $x = -3$ and $y = 0$.

[1.7] 4. Evaluate $-\frac{1}{3}x + 2$ when $x = 6$.

Exercises 5 and 6 EQUATIONS AND INEQUALITIES

[2.3] *For Exercises 5 and 6, solve.*

5. $2x - 5 = 10$

6. $-\frac{5}{6}x = \frac{3}{4}$

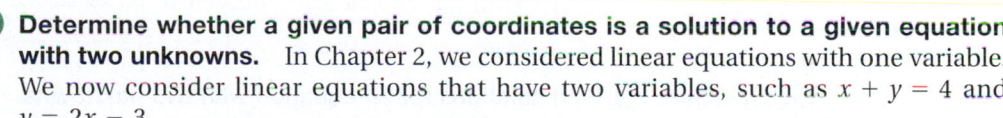

4.2 Graphing Linear Equations

Objectives

1. Determine whether a given pair of coordinates is a solution to a given equation with two unknowns.
2. Find solutions for an equation with two unknowns.
3. Graph linear equations.

Objective 1 Determine whether a given pair of coordinates is a solution to a given equation with two unknowns. In Chapter 2, we considered linear equations with one variable. We now consider linear equations that have two variables, such as $x + y = 4$ and $y = 2x - 3$.

A solution for an equation with two variables is a pair of numbers, one number for each variable, that can replace the corresponding variables and make the equation true. These solutions are ordered pairs, so we can write them using coordinates. The ordered pair $(1, 3)$ is a solution for $x + y = 4$ because replacing x with 1 and y with 3 makes the equation true.

Note: Unless otherwise specified, the coordinates are written in alphabetical order; so $(1, 3)$ means that $x = 1$ and $y = 3$.

$$x + y = 4$$
$$1 + 3 = 4$$

The equation is true, so the ordered pair is a solution.

Procedure **Checking a Potential Solution for an Equation with Two Variables**

To determine whether a given ordered pair is a solution for an equation with two variables,

1. Replace the variables in the equation with the corresponding coordinates.
2. Verify that the equation is true.

Example 1   Determine whether the ordered pair is a solution for the equation.

a. $(-4, -10); y = 2x - 3$

SOLUTION $-10 \stackrel{?}{=} 2(-4) - 3$ Replace x with -4 and y with -10 and see if the equation is true.

$-10 \stackrel{?}{=} -8 - 3$

$-10 \neq -11$ Because the equation is not true, $(-4, -10)$ is not a solution for $y = 2x - 3$.

b. $(4, -1); 2m - n = 9$

SOLUTION $2(4) - (-1) \stackrel{?}{=} 9$ Unless otherwise specified, ordered pairs are expressed in alphabetical order; therefore, replace m with 4 and n with -1 and see if the equation is true.

$8 + 1 \stackrel{?}{=} 9$

$9 = 9$ The equation is true, so $(4, -1)$ is a solution for $2m - n = 9$.

c. $\left(-2, -\dfrac{1}{3}\right); y = \dfrac{2}{3}x + 1$

SOLUTION $-\dfrac{1}{3} \stackrel{?}{=} \dfrac{2}{3}(-2) + 1$ Replace x with -2 and y with $-\dfrac{1}{3}$ and see if the equation is true.

$-\dfrac{1}{3} \stackrel{?}{=} -\dfrac{4}{3} + \dfrac{3}{3}$

$-\dfrac{1}{3} = -\dfrac{1}{3}$ The equation is true, so $\left(-2, -\dfrac{1}{3}\right)$ is a solution for $y = \dfrac{2}{3}x + 1$.

Your Turn 1   Determine whether the ordered pair is a solution for the equation.

a. $(-8, -2); x - 3y = -2$ **b.** $(1, -6); q = -5p + 1$ **c.** $\left(\dfrac{3}{4}, \dfrac{9}{4}\right); y = \dfrac{x}{3} + 2$

Objective ② Find solutions for an equation with two unknowns.   Consider the equation $x + y = 4$. We have already seen that $(1, 3)$ is a solution. But there are other solutions. We list some of those solutions in the following table.

x	y	Ordered Pair
0	4	$(0, 4)$
1	3	$(1, 3)$
2	2	$(2, 2)$
3	1	$(3, 1)$
4	0	$(4, 0)$
5	-1	$(5, -1)$

Note: Every solution for $x + y = 4$ is an ordered pair of numbers whose sum is 4.

In fact, $x + y = 4$, and every other linear equation, has an infinite number of solutions. For every x-value, there is a corresponding y-value that will add to the x-value to equal 4 and vice versa, which gives a clue about how to find solutions. We can simply choose a value for x or y and solve for the corresponding value of the other variable. (In this case, our equation was easy enough that we could solve for y mentally.)

Answers   Your Turn 1
a. yes   **b.** no   **c.** yes

Procedure **Finding Solutions to Linear Equations with Two Variables**

To find a solution to a linear equation with two variables,

1. Choose a value for one of the variables (any value).
2. Replace the corresponding variable with your chosen value.
3. Solve the equation for the value of the other variable.

Example 2 Find three solutions for the equation.

a. $3x + y = 2$

SOLUTION To find a solution, we replace one of the variables with a chosen value and then solve for the value of the other variable.

For the first solution, we will choose x to be 0.	For the second solution, we will choose x to be 1.	For the third solution, we will choose y to be -4.
$3x + y = 2$	$3x + y = 2$	$3x + y = 2$
$3(0) + y = 2$	$3(1) + y = 2$	$3x + (-4) = 2$
$y = 2$	$3 + y = 2$	$3x = 6$
Solution $(0, 2)$	$y = -1$	$x = 2$

Note: Choosing x (or y) to equal 0 usually makes the equation very easy to solve.

For the second solution: Subtract 3 from both sides.

Solution $(1, -1)$

For the third solution: Divide both sides by 3.

Solution $(2, -4)$

We can summarize the solutions in a table.

If we choose x to be 0, then y is equal to 2.
If we choose x to be 1, then y is equal to -1.
If we choose y to be -4, then x is equal to 2.

x	y	Ordered Pair
0	2	$(0, 2)$
1	-1	$(1, -1)$
2	-4	$(2, -4)$

Keep in mind that there are an infinite number of correct solutions for a given equation in two variables; so your solutions may be different from the solutions someone else finds.

b. $y = -\dfrac{1}{3}x + 2$

SOLUTION Notice that this equation has y isolated. If we select values for x, we won't have to isolate y as we did in part a. We simply calculate the y value. Also notice that the coefficient for x is a fraction. Because we can choose any value for x, let's choose values such as 3 and 6 that will divide out nicely with the denominator of 3.

For the first solution, we will choose x to be 0.	For the second solution, we will choose x to be 3.	For the third solution, we will choose x to be 6.
$y = -\dfrac{1}{3}x + 2$	$y = -\dfrac{1}{3}x + 2$	$y = -\dfrac{1}{3}x + 2$
$y = -\dfrac{1}{3}(0) + 2$	$y = -\dfrac{1}{3}(3) + 2$	$y = -\dfrac{1}{3}(6) + 2$
$y = 2$	$y = -\dfrac{1}{\overset{1}{\cancel{3}}}\left(\dfrac{\overset{1}{\cancel{3}}}{1}\right) + 2$	$y = -\dfrac{1}{\overset{1}{\cancel{3}}}\left(\dfrac{\overset{2}{\cancel{6}}}{1}\right) + 2$

Solution $(0, 2)$

$y = -1 + 2$
$y = 1$

Solution $(3, 1)$

$y = -2 + 2$
$y = 0$

Solution $(6, 0)$

SOLUTION In summary,

If we choose x to be 0, then y is equal to 2.

If we choose x to be 3, then y is equal to 1.

If we choose x to be 6, then y is equal to 0.

x	y	Ordered Pair
0	2	$(0, 2)$
3	1	$(3, 1)$
6	0	$(6, 0)$

Your Turn 2 Find three solutions for each equation. (Answers may vary.)

a. $x + 4y = 8$

b. $3x - 2y = 6$

c. $y = \frac{1}{2}x + 5$

Objective ③ Graph linear equations. We have learned that equations in two variables have an infinite number of solutions. Because of this fact, there is no way that all solutions to an equation can be found, much less listed. However, all of the solutions can be represented using a graph. Consider some of the solutions for $x + y = 4$, which we listed in the table on page 250: $(0, 4)$, $(1, 3)$, $(2, 2)$, $(3, 1)$, $(4, 0)$, and $(5, -1)$.

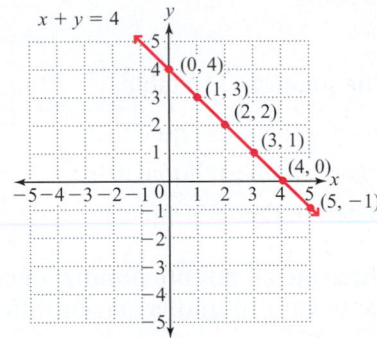

Notice that when we plot our ordered pair solutions for $x + y = 4$, the corresponding points lie in a straight line. In fact, all solutions for $x + y = 4$ are on that same line, which is why we say it is a *linear* equation. By connecting the points, we are graphically representing every possible solution for $x + y = 4$.

The graph of the solutions of every linear equation will be a straight line. Because two points determine a line in a plane, we need a minimum of two ordered pairs. This means that we must find at least two solutions in order to graph the line. However, it is wise to find three solutions, using the third solution as a check. If we plot all three points and they cannot be connected with a straight line, then we know something is wrong.

Procedure **Graphing Linear Equations**

To graph a linear equation,

1. Find at least two solutions to the equation.
2. Plot the solutions as points in the rectangular coordinate system.
3. Connect the points to form a straight line.

Answers **Your Turn 2**

Remember, your solutions may be different.

a. $(0, 2), (4, 1), (8, 0)$

b. $(0, -3), (2, 0), \left(1, -\frac{3}{2}\right)$

c. $(0, 5), (2, 6), (4, 7)$

Example 3 Graph each equation.

a. $3x + y = 2$

SOLUTION We found three solutions to this equation in Example 2(a). Those solutions are listed in the following table. We plot each solution as a point in the rectangular coordinate system and then connect the points to form a straight line.

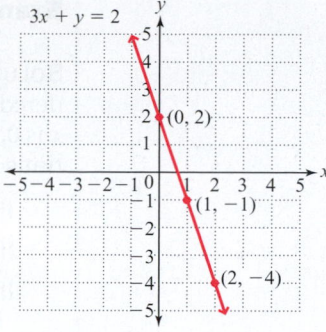

x	y	Ordered Pair
0	2	$(0, 2)$
1	-1	$(1, -1)$
2	-4	$(2, -4)$

b. $y = -\dfrac{1}{3}x + 2$

SOLUTION We found three solutions to this equation in Example 2(b). Those solutions are listed in the following table. We plot each solution as a point in the rectangular coordinate system and then connect the points to form a straight line.

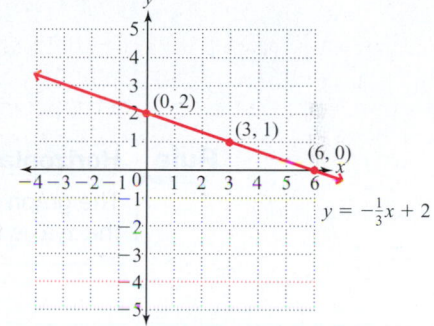

x	y	Ordered Pair
0	2	$(0, 2)$
3	1	$(3, 1)$
6	0	$(6, 0)$

Your Turn 3 Graph each equation.

a. $x + 2y = 4$

b. $y = \dfrac{1}{2}x - 1$

The graph of an equation is a visual representation of the entire solution set of the equation. The graph of every ordered pair that solves the equation lies on the line, and the coordinates of every point on the line is a solution of the equation. Looking at the graph of Example 3(b), we see that $(-3, 3)$ are the coordinates of a point on the line; so $(-3, 3)$ is a solution of the equation $y = -\dfrac{1}{3}x + 2$. We verify by substituting and simplifying.

$$y = -\frac{1}{3}x + 2$$

$$3 = -\frac{1}{3}(-3) + 2$$

$$3 = 1 + 2$$

$$3 = 3$$

We have graphed linear equations in two variables. Now let's consider the graphs of equations in one variable, such as $y = 3$ or $x = 4$, in which the variable is equal to a constant.

Answers Your Turn 3

a.

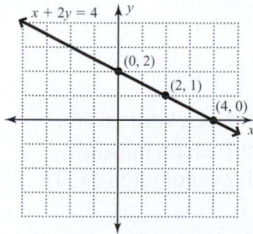

b.

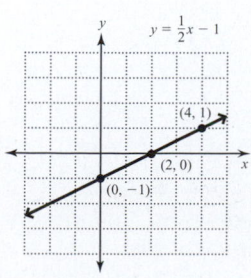

Example 4   Graph $y = 3$.

SOLUTION   The equation $y = 3$ indicates that y is equal to a constant, 3. To establish ordered pairs, we can rewrite the equation $y = 3$ as $0x + y = 3$. Because the coefficient of x is 0, y is always 3 no matter what we choose for x. We could complete a table of solutions like this:

If we choose x to be 0, then y equals 3.

If we choose x to be 2, then y is 3.

If we choose x to be 4, then y is still 3.

x	y	Ordered Pair
0	3	$(0, 3)$
2	3	$(2, 3)$
4	3	$(4, 3)$

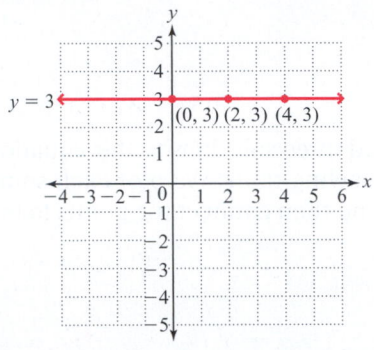

Note: The graph of $y = 3$ is a horizontal line parallel to the x-axis that passes through the y-axis at a point with coordinates $(0, 3)$.

Rule   **Horizontal Lines**

The graph of $y = c$, where c is a real-number constant, is a horizontal line parallel to the x-axis that passes through the y-axis at a point with coordinates $(0, c)$.

Example 5   Graph $x = 4$.

SOLUTION   The equation $x = 4$ indicates that x is equal to a constant, 4. If the y-variable is missing from an equation, it means that its coefficient is 0; so we can write $x = 4$ as $x + 0y = 4$. Notice that x is always 4 no matter what we choose for y. We could complete a table of solutions like this:

If we choose y to be 0, then x equals 4.

If we choose y to be 1, then x is 4.

If we choose y to be 2, then x is still 4.

x	y	Ordered Pair
4	0	$(4, 0)$
4	1	$(4, 1)$
4	2	$(4, 2)$

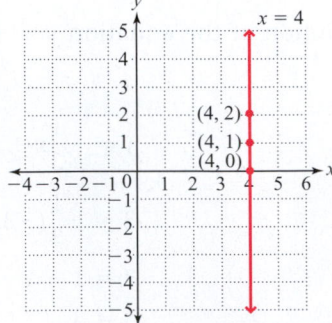

Note: The graph of $x = 4$ is a vertical line parallel to the y-axis that passes through the x-axis at a point with coordinates $(4, 0)$.

Rule **Vertical Lines**

The graph of $x = c$, where c is a real-number constant, is a vertical line parallel to the y-axis that passes through the x-axis at a point with coordinates $(c, 0)$.

Your Turn 5 Graph.

a. $y = -4$

b. $x = -3$

Answers **Your Turn 5**

a.
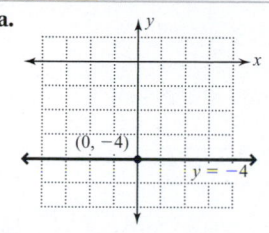

b. $x = -3$
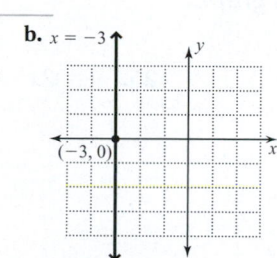

4.2 Exercises For Extra Help *MyMathLab*

Math XL
PRACTICE   WATCH   DOWNLOAD

Note: Exercises marked with a ★ represent challenging exercises.

1. How do you determine whether an ordered pair is a solution for a given equation?

2. If a given ordered pair is not a solution for a linear equation with two variables, does that mean the equation has no solutions? Explain.

3. How do you find a solution for a given equation with two unknowns?

4. What does the graph of an equation represent?

5. What is the minimum number of points needed to draw a straight line? Explain.

6. In your own words, explain the process of graphing a linear equation.

For Exercises 7–22, determine whether the given ordered pair is a solution for the equation. See Example 1.

7. $(2, 3); x + 2y = 8$

8. $(3, 1); 2x - y = 5$

9. $(4, 9); 3x - 2y = -6$

10. $(-5, 10); 5x + 3y = 5$

11. $(-5, 2); y - 4x = 3$

12. $(4, -6); y - 3x = -1$

13. $(9, 0); y = -2x + 18$

14. $(0, -1); y = 2x + 1$

15. $(6, -2); y = -\dfrac{2}{3}x$

16. $(-10, -4); y = \dfrac{2}{5}x$

17. $(-8, -8); y = \dfrac{3}{4}x - 2$

18. $(-5, 0); y = -\dfrac{3}{5}x + 3$

19. $\left(-1\dfrac{2}{5}, 0\right); y - 3x = 5$

20. $\left(0, -2\dfrac{1}{3}\right); 4x - 3y = 7$

21. $(2.2, -11.2); y + 6x = -2$

22. $(-1.5, -1.3); y = 0.2x - 1$

For Exercises 23–58, find three solutions for the given equation. Then graph. (Answers may vary for the three solutions.) See Examples 2–5.

23. $y = x$

24. $y = -x$

25. $y = 2x$

26. $y = 3x$

27. $y = -5x$

28. $y = -2x$

29. $y = x - 3$

30. $y = x + 4$

31. $y = -x - 2$

32. $y = -x + 5$

33. $y = 2x - 5$

34. $y = 3x + 2$

35. $y = -2x + 4$

36. $y = -5x - 1$

37. $y = \dfrac{1}{2}x$

38. $y = \dfrac{1}{3}x$

39. $y = -\dfrac{2}{3}x$

40. $y = -\dfrac{3}{4}x$

41. $y = -\dfrac{2}{3}x + 4$

42. $y = \dfrac{4}{5}x - 1$

43. $x - y = 8$

44. $x + y = -5$

45. $2x + y = 6$

46. $3x - y = 9$

47. $2x + 3y = 12$

48. $3x + 5y = 15$

49. $4x - 3y = -12$

50. $3x - 2y = -6$

51. $x - \dfrac{1}{4}y = 2$

52. $\dfrac{1}{3}x + y = -1$

53. $y = -5$

54. $y = 4$

55. $x = 7$

56. $x = -6$

★**57.** $y = 0.4x - 2.5$

★**58.** $1.2x + 0.5y = 6$

59. Compare the graphs of $y = x$, $y = 2x$, and $y = 3x$ from Exercises 23, 25, and 26. For an equation in the form $y = ax$, where $a > 0$, what effect does increasing a seem to have on the graph?

60. Compare the graphs of $y = x$ and $y = -x$ from Exercises 23 and 24; then compare the graphs of $y = 2x$ and $y = -2x$ from Exercises 25 and 28. For an equation in the form $y = ax$, what can you conclude about the graph when a is positive versus when a is negative?

61. Compare the graphs of $y = 3x$ and $y = 3x + 2$ from Exercises 26 and 34; then compare the graphs of $y = -2x$ and $y = -2x + 4$ from Exercises 28 and 35. For an equation in the form $y = ax + b$, where $b > 0$, what effect does adding b to $y = ax$ seem to have on the graph?

62. Compare the graphs of $y = x$ and $y = x - 3$ from Exercises 23 and 29; then compare the graphs of $y = -5x$ and $y = -5x - 1$ from Exercises 27 and 36. For an equation in the form $y = ax - b$, where $b > 0$, what effect does subtracting b from $y = ax$ seem to have on the graph?

For Exercises 63–70, solve.

63. A plumber charges $80 plus $40 per hour of labor. The equation $c = 40n + 80$ describes the total cost that she would charge for a service visit, where n represents the number of hours of labor.
   a. Find the total cost if labor is 2 hours.

   b. If a client's total charges are $240, for how many hours of labor was the client charged?

   c. Graph the equation with n along the horizontal axis and c along the vertical axis. (*Hint:* Let each grid mark on the c-axis represent $20.) Because n and c are nonnegative, the graph is restricted to the first quadrant only.
   d. What does the point where the graph intersects the c-axis represent?

64. An academic tutor charges $20 for supplies and $25 per hour of tutoring. The equation $c = 25n + 20$ describes the total cost that she would charge for tutoring, where n represents the number of hours of tutoring.
   a. Find the total cost if the tutor works 3 hours.

   b. If a client's total charges are $145, for how many hours of tutoring was the client charged?

   c. Graph the equation with n along the horizontal axis and c along the vertical axis. Because n and c are nonnegative, the graph is restricted to the first quadrant only.
   d. What does the point where the graph intersects the c-axis represent?

65. A copy center charges $15.00 for the first 300 copies plus $0.03 per copy after that. The equation $c = 0.03n + 15$ describes the total cost that would be charged for a copy service, where n represents the number of copies above 300.
   a. Find the total cost for 400 copies.

   b. If a client's total charges are $21, for how many copies was the client charged?

   c. Graph the equation with n along the horizontal axis and c along the vertical axis. Because n and c are nonnegative, the graph is restricted to the first quadrant only. (*Hint:* Let each grid mark on the n-axis represent 50 copies and each grid mark on the c-axis represent $3.)

66. A wireless company charges $20.00 per month for 300 anytime minutes plus $0.10 per minute above 300. The equation $c = 0.10n + 20$ describes the total cost that the company would charge for a monthly bill, where n represents the number of minutes used above 300.

  a. Find the total cost if a customer uses 400 anytime minutes.

  b. If a client's total charges are $22, for how many additional minutes was the client charged?

  c. Graph the equation with n along the horizontal axis and c along the vertical axis. Because n and c are nonnegative, the graph is restricted to the first quadrant only.

67. It is recommended that a hot tub be emptied and cleaned regularly. The hot tub Benjamin owns holds approximately 500 gallons of water. He can drain 3.5 gallons of water per minute using a garden hose. The equation $g = -3.5m + 500$ describes the amount of water in the hot tub in gallons, where g represents the number of gallons remaining after pumping out water for m minutes.

  a. After 25 minutes, what is the amount of water in the hot tub?

  b. How long does it take to pump half of the water out of the hot tub?

  c. How long does it take to pump all of the water out of the hot tub?

  d. Graph the equation. Because m and g are nonnegative, the graph is restricted to the first quadrant only.

  e. What does the point where the graph intersects the g-axis represent?

  f. What does the point where the graph intersects the m-axis represent?

68. The battery in Jenny's laptop is fully charged (100%). While using the laptop continuously on a flight from Kansas to California, she depletes the battery at a rate of 0.75% per minute. The equation $P = -0.75m + 100$ describes the battery's power, where P represents the percent of the battery's power that remains after using the laptop for m minutes.

  a. After 1 hour, what percent of the battery's power remains?

  b. How long does it take for the battery to be at one-fourth of its full power?

  c. How long does it take for the battery to reach 0%?

  d. Graph the equation. Because m and P are nonnegative, the graph is restricted to the first quadrant only.

  e. What does the point where the graph intersects the P-axis represent?

  f. What does the point where the graph intersects the m-axis represent?

69. A small business owner buys a new computer for $2400. Each year the computer is in use, she can deduct its depreciated value when calculating her taxable income. The equation $c = -300n + 2400$ describes the value of the computer, where c is the value after n years of use.

  a. Find the value of the computer after 4 years.
  b. In how many years will the computer be worth half of its initial value?
  c. After how many years will the computer have a value of $0?
  d. Graph the equation with n along the horizontal axis and c along the vertical axis. Because n and c are nonnegative, the graph is restricted to the first quadrant only.

70. As of 2005, it is estimated that Earth is being deforested at a rate of about 40,000 square miles per year. The equation $A = -40,000n + 2,000,000$ describes the area of forest remaining, where A represents the area remaining n years after 2005. (*Source:* www.wikipedia.org)

    a. Find the area of forest remaining in 2015.

    b. If the rate does not change, in what year will Earth have half of the area of the forest that was present in 2005?

    c. If the rate does not change, in what year will forest be gone from Earth?

    d. Graph the equation with n along the horizontal axis and A along the vertical axis. Because n and A are nonnegative, the graph is restricted to the first quadrant only.

Puzzle Problem

Without lifting your pencil from the paper, connect the dots using only four straight lines:

:::
• • •

• • •

• • •
:::

Review Exercises

Exercises 1–3 **EXPRESSIONS**

[1.7]   **1.** Evaluate $\frac{2}{3}x + 2$ when $x = 3$.

[2.1]   **2.** Calculate the perimeter of a rectangle with a length of 8.5 inches and a width of 7.25 inches.

[2.1]   **3.** Calculate the area of a triangle with a base of 4.2 inches and a height of 2.7 inches.

Exercises 4–6 **EQUATIONS AND INEQUALITIES**

[2.3]   **4.** Solve $0 = 3x - 5$.

[2.4]   **5.** Isolate x in the formula $v = mx + k$.

[4.2]   **6.** Graph the line that passes through the points $(3, 0)$ and $(0, -2)$.

4.3 Graphing Using Intercepts

Objectives

1 Given an equation, find the coordinates of the *x*- and *y*-intercepts.

2 Graph linear equations using intercepts.

Objective 1 Given an equation, find the coordinates of the *x*- and *y*-intercepts. Look at the following graph.

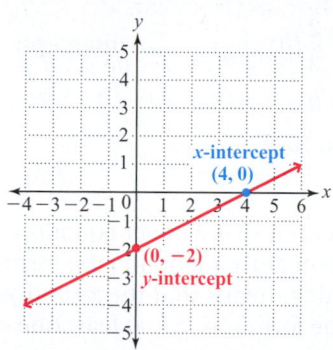

Notice that the line intersects the *x*-axis at $(4, 0)$ and the *y*-axis at $(0, -2)$. These points are called *intercepts*. A point where a graph intersects the *x*-axis is called an **x-intercept**. A point where a graph intersects the *y*-axis is called a **y-intercept**.

Definitions

x-intercept: A point where a graph intersects the *x*-axis.

y-intercept: A point where a graph intersects the *y*-axis.

Note that the *y*-coordinate of any *x*-intercept is always 0. Similarly, the *x*-coordinate of any *y*-intercept is always 0. These facts about intercepts suggest the following procedures.

Procedure Finding the *x*- and *y*-intercepts

To find an *x*-intercept,	To find a *y*-intercept,
1. Replace *y* with 0 in the given equation.	**1.** Replace *x* with 0 in the given equation.
2. Solve for *x*.	**2.** Solve for *y*.

Example 1 For the equation $5x - 2y = 10$, find the *x*- and *y*-intercepts.

SOLUTION

Note: We no longer show the steps containing the addition or multiplication principles as we did in Chapters 2 and 3.

For the *x*-intercept, replace *y* with 0 and solve for *x*.

$$5x - 2(0) = 10$$
$$5x = 10$$
$$x = 2 \quad \text{Divide both sides by 5.}$$

x-intercept: $(2, 0)$

For the *y*-intercept, replace *x* with 0 and solve for *y*.

$$5(0) - 2y = 10$$
$$-2y = 10$$
$$y = -5 \quad \text{Divide both sides by } -2.$$

y-intercept: $(0, -5)$

Your Turn 1 For the equation $3x - 4y = 8$, find the *x*- and *y*-intercepts.

Answers Your Turn 1

x-intercept: $\left(\frac{8}{3}, 0\right)$

y-intercept: $(0, -2)$

The equation in Example 1 is in the form $Ax + By = C$, where *A*, *B*, and *C* are real-number constants with both *A* and *B* not equal to 0. A linear equation written in this form is said to be in *standard form*. One of the advantages of having a linear equation in standard form is that it is easy to find the intercepts. Throughout this section, we introduce other forms for linear equations and explore some of the information that can be discerned from equations written in various forms.

Example 2 For the equation $y = \dfrac{2}{3}x$, find the x- and y-intercepts.

SOLUTION Replace x with 0 and solve for y. Then replace y with 0 and solve for x.

y-intercept: $y = \dfrac{2}{3}(0)$ x-intercept: $0 = \dfrac{2}{3}x$

$\phantom{y\text{-intercept:}\ \ }y = 0$ $\phantom{x\text{-intercept:}\ \ }0 = x$ **Multiply both sides by $\dfrac{3}{2}$.**

x- and y-intercept: $(0, 0)$

The equation in Example 2 is a linear equation in the form $y = mx$, where m is any real number. The graph of any equation in the form $y = mx$ is always a line that passes through the origin, which means that the x- and y-intercepts are both at $(0, 0)$. So to graph an equation of the form $y = mx$, at least one other point must be found.

Rule **Intercepts for $y = mx$**

If an equation can be written in the form $y = mx$, where m is a real number other than 0, then the x- and y-intercepts are at the origin, $(0, 0)$.

Example 3 For the equation $y = 2x + 1$, find the x- and y-intercepts.

SOLUTION Replace y with 0 and solve for x. Then replace x with 0 and solve for y.

x-intercept: $0 = 2x + 1$ y-intercept: $y = 2(0) + 1$

$\phantom{x\text{-intercept:}\ \ }-1 = 2x$ **Subtract 1 from both sides.** $y = 1$

$\phantom{x\text{-intercept:}\ \ }-\dfrac{1}{2} = x$ **Divide both sides by 2.** y-intercept: $(0, 1)$

x-intercept: $\left(-\dfrac{1}{2}, 0\right)$

Note: When we replace x with 0, we are left with the constant 1 for the y-intercept.

The equation in Example 3 is in the form $y = mx + b$, where m and b can be any real number. In this form, when we replace x with 0 to find the y-intercept, we are left with the constant b.

Rule **The y-intercept for $y = mx + b$**

If an equation is in the form $y = mx + b$, where m and b are nonzero real numbers, then the y-intercept will be $(0, b)$.

Your Turn 3 Find the x- and y-intercepts for the following equations.

a. $y = -\dfrac{3}{4}x$ b. $y = 3x - 5$

Answers Your Turn 3
a. x- and y-intercept: $(0, 0)$
b. x-intercept: $\left(\dfrac{5}{3}, 0\right)$
y-intercept: $(0, -5)$

Example 4 For the equation $y = 3$, find the x- and y-intercepts.

SOLUTION Remember from Section 4.2, Example 4, that the graph of $y = 3$ is a horizontal line parallel to the x-axis that passes through the y-axis at the point $(0, 3)$. Notice that this point is the y-intercept. Because the line is parallel to the x-axis, it will never intersect the x-axis. Therefore, there is no x-intercept.

Example 4 suggests the following rule.

Rule **Intercepts for $y = c$**

The graph of an equation in the form $y = c$, where c is a nonzero real number, has no x-intercept and the y-intercept is $(0, c)$. The graph of $y = 0$ is the x-axis.

We can draw a similar conclusion about the graph of an equation in the form $x = c$.

Rule **Intercepts for $x = c$**

The graph of an equation in the form $x = c$, where c is a nonzero real number, has no y-intercept and the x-intercept is $(c, 0)$. The graph of $x = 0$ is the y-axis.

Your Turn 4 Find the x- and y-intercepts for the following equations.

a. $y = -5$
b. $x = -2$

Objective ② **Graph linear equations using intercepts.** Because two points determine a line in a plane, we can use the x- and y-intercepts to draw the graph of a linear equation.

Example 5 Graph using the x- and y-intercepts: $5x - 2y = 10$

SOLUTION In Example 1, the intercepts were found to be $(2, 0)$ and $(0, -5)$. We plot these intercepts and connect them to graph the line. It is helpful to find a third solution and verify that all three points can be connected to form a straight line. For this third point, we will choose $x = 3$ and solve for y.

Third solution: (check) If $x = 3$, then

$$5(3) - 2y = 10$$
$$15 - 2y = 10$$
$$-2y = -5 \quad \text{Subtract 15 from both sides.}$$
$$y = \frac{5}{2} \quad \text{Divide both sides by } -2. \quad \textbf{Third solution:} \left(3, \frac{5}{2}\right)$$

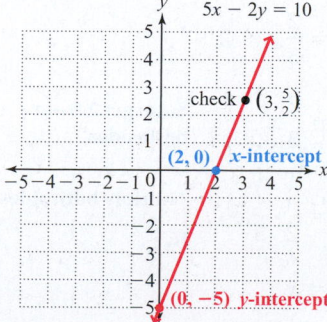

Table of Solutions

	x	y
x-intercept:	2	0
y-intercept:	0	-5
Third solution: (check)	3	$\frac{5}{2}$

As previously mentioned, if the *x*- and *y*-intercepts are both at the origin, we must find an additional point in order to graph the linear equation.

Example 6 Graph using the *x*- and *y*-intercepts.

a. $y = \dfrac{2}{3}x$

SOLUTION In Example 2 we found $(0, 0)$ to be both the *x*- and the *y*-intercept. Because this single point is not enough to determine the line in the plane, we must find at least one more solution. We will also find a third solution as a check.

Second solution:
Choose *x* to be 3.

$$y = \frac{2}{3}(3)$$

$$y = \frac{2}{\overset{1}{\cancel{3}}}\left(\frac{\overset{1}{\cancel{3}}}{1}\right)$$

$$y = 2$$

Second solution: $(3, 2)$

Third solution (check):
Choose *x* to be -3.

$$y = \frac{2}{3}(-3)$$

$$y = \frac{2}{\overset{1}{\cancel{3}}}\left(-\frac{\overset{1}{\cancel{3}}}{1}\right)$$

$$y = -2$$

Third solution: $(-3, -2)$

Now plot the solutions and connect the points to form the line.

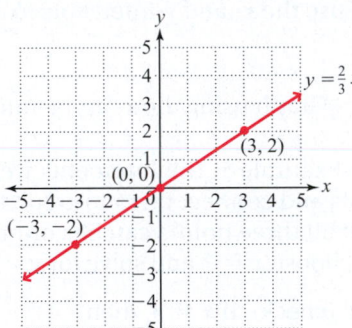

Table of Solutions

	x	*y*
x- and *y*-intercept:	0	0
Second solution:	3	2
Third solution: (check)	-3	-2

b. $y = \dfrac{2}{3}x + 2$

SOLUTION Notice that this equation is in the form $y = mx + b$. Because $b = 2$ in this case, we can conclude that the *y*-intercept is $(0, 2)$. For the *x*-intercept, let $y = 0$ and solve for *x*. Also, as a check, we will choose $x = 3$ to generate a third point.

x-intercept: $0 = \dfrac{2}{3}x + 2$

$$-2 = \frac{2}{3}x \qquad \begin{array}{l}\text{\textcolor{blue}{\textbf{Subtract 2 from}}}\\ \text{\textcolor{blue}{\textbf{both sides.}}}\end{array}$$

$$\frac{3}{\underset{1}{\cancel{2}}} \cdot \frac{-\overset{1}{\cancel{2}}}{1} = \frac{\overset{1}{\cancel{3}}}{\underset{1}{\cancel{2}}} \cdot \frac{\overset{1}{\cancel{2}}}{\underset{1}{\cancel{3}}}x \qquad \begin{array}{l}\text{\textcolor{blue}{\textbf{Multiply both sides}}}\\ \text{\textcolor{blue}{\textbf{by }}}\textcolor{blue}{\dfrac{3}{2}.}\end{array}$$

$$-3 = x$$

x-intercept: $(-3, 0)$

Third solution: (check) $y = \dfrac{2}{3}(3) + 2$

$$y = \frac{2}{\underset{1}{\cancel{3}}} \cdot \frac{\overset{1}{\cancel{3}}}{1} + 2$$

$$y = 2 + 2$$

$$y = 4$$

Third solution: $(3, 4)$

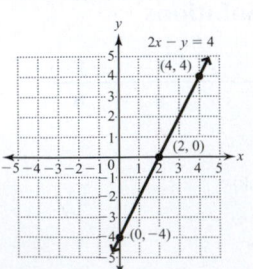

Now plot the solutions and connect the points to form the line.

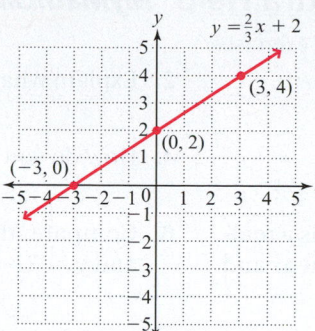

Table of Solutions

	x	y
x-intercept:	-3	0
y-intercept:	0	2
Third solution: (check)	3	4

Your Turn 6   Graph using the x- and y-intercepts.

a. $y = \dfrac{1}{2}x$

b. $y = \dfrac{2}{3}x - 2$

Compare the graphs in Examples 6(a) and 6(b). If we place both lines on the same grid (shown here), we see that they are parallel, which means that they do not intersect at any point. Further, each point on the graph of $y = \dfrac{2}{3}x + 2$ is 2 units higher than each point with the same x-coordinate on the graph of $y = \dfrac{2}{3}x$.

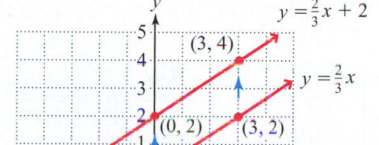

Table of Solutions

x	$y = \dfrac{2}{3}x$	$y = \dfrac{2}{3}x + 2$
-3	-2	0
0	0	2
3	2	4

From the table, we see that the difference in an equation in the form $y = mx$ and an equation in the form $y = mx + b$ is the value of the y-intercept. In other words, the line $y = mx$ shifts upward or downward by b units. The graph of $y = \dfrac{2}{3}x - 2$ is a line parallel to $y = \dfrac{2}{3}x$, only shifted down 2 units so that its y-intercept is $(0, -2)$. We will explore this further in Section 4.5.

Answers Your Turn 6

a.

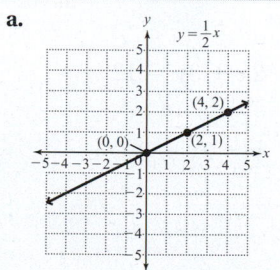

b.

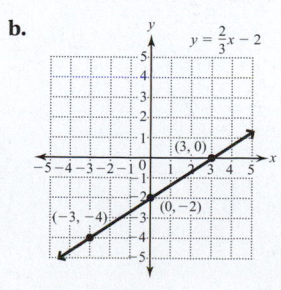

Note: Exercises marked with a ★ represent challenging exercises.

1. Explain what an *x*-intercept is.

2. Explain what a *y*-intercept is.

3. How do you find an *x*-intercept?

4. How do you find a *y*-intercept?

5. If an equation is in the form $y = mx$, where m is a real-number constant, what are the coordinates of the *x*- and *y*-intercepts? Explain.

6. Compare the graphs of $y = 2x$ and $y = 2x + 5$.

For Exercises 7–14, find the x- and y-intercepts of each graph. See Objective 1.

7.
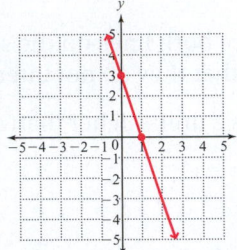

8.
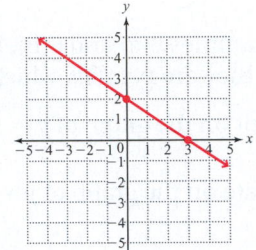

9.
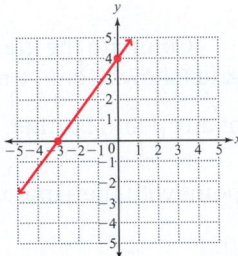

10.
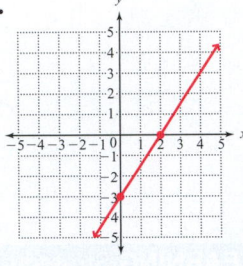

11.
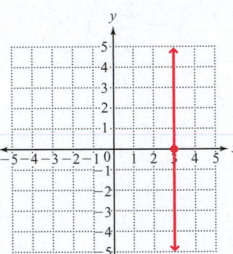

12.
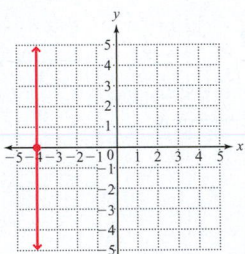

13.
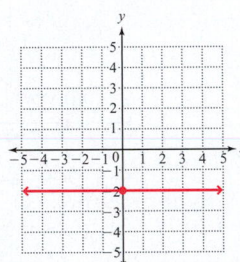

14.
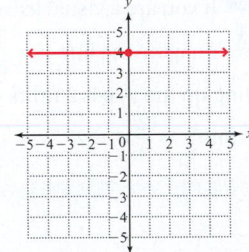

For Exercises 15–32, find the x- and y-intercepts. See Examples 1–4.

15. $x - y = 4$

16. $x + y = 7$

17. $2x + 3y = 6$

18. $2x + 3y = 12$

19. $3x - 4y = -10$

20. $3x - 4y = 2$

21. $y = 3x - 6$

22. $y = 2x + 8$

23. $y = 2x + 5$

24. $y = 3x - 1$

25. $y = 2x$

26. $y = \dfrac{2}{5}x$

27. $2x + 3y = 0$

28. $x + 3y = 0$

29. $x = 5$

30. $x = -2$

31. $y = -2$

32. $y = 7$

For Exercises 33–64, graph using the x- and y-intercepts. See Examples 5 and 6.

33. $x + y = 7$ **34.** $x + y = 10$ **35.** $3x - y = 6$ **36.** $x - 5y = 15$

37. $2x + 3y = 6$ **38.** $3x + 4y = 12$ **39.** $5x + 2y = 10$ **40.** $3x - 2y = 18$

41. $4x + 3y = -12$ **42.** $3x + 2y = -6$ **43.** $5x - 3y = -15$ **44.** $2x - 5y = -10$

45. $y = 2x$ **46.** $y = -3x$ **47.** $y = \dfrac{2}{5}x$ **48.** $-\dfrac{1}{4}x = y$

49. $y = x - 4$ **50.** $y = x + 2$ **51.** $y = 2x - 6$ **52.** $y = 3x + 9$

53. $6x + 3y = 9$ **54.** $2x - y = -1$ **55.** $4x + 2y = 5$ **56.** $3x + 2y = 7$

57. $x = -4$ **58.** $x = 5$ **59.** $y = -2$ **60.** $y = 5$

61. $y - 2 = 0$ **62.** $y + 4 = 0$ **63.** $x - 2 = 0$ **64.** $x + 3 = 0$

★**65.** Which of the following could be the graph of $1.4x - 0.9y = 2.7$? Explain.

a.
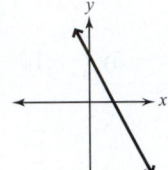

b.
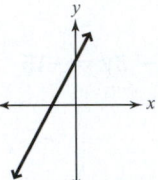

c.
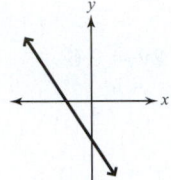

d.
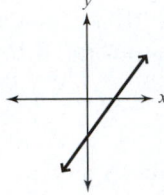

★**66.** Which of the following could be the graph of $0.65x - 2.9y = -3.5$? Explain.

a.
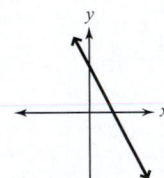

b.
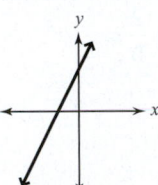

c.
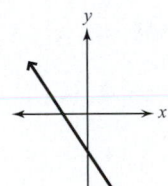

d.
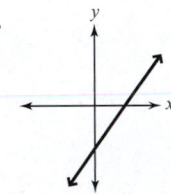

★**67.** Which of the following could be the graph of $\frac{2}{3}x - \frac{3}{4}y = \frac{10}{3}$? Explain.

a.
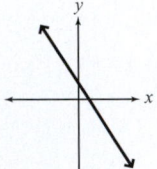

b.
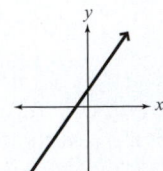

c.
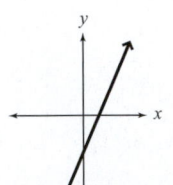

d.
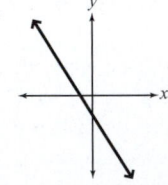

★**68.** Which of the following could be the graph of $\frac{3}{2}x + \frac{6}{5}y = -\frac{8}{5}$? Explain.

a.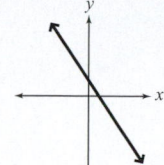

b.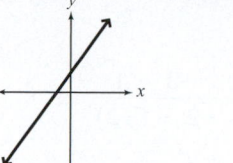

c.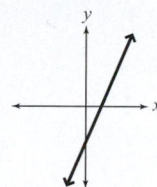

d.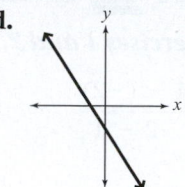

69. If a circle with a radius of 2.7 units was to be drawn in the coordinate plane with its center at the origin, what would be the coordinates of its x- and y-intercepts?

70. If a circle with a diameter of 6.8 units was to be drawn in the coordinate plane with its center at the origin, what would be the coordinates of its x- and y-intercepts?

★**71.** A salesperson sells two different sizes of facial cleansing lotion. She sells the small size for $15 and the large size for $20.

  a. If x represents the number of units of the small size sold and y represents the number of units of the large size sold, write an expression that describes her total sales.

  b. If her supervisor sets a goal of selling $2000 worth of the cleansing lotion, use the expression from part a to write an equation describing the sales goal.

  c. Find the x- and y-intercepts.

  d. What do the x- and y-intercepts mean in terms of sales?

  e. Find a third combination of sales numbers that would equal the sales goal.

  f. Graph the equation. Because x and y are nonnegative, the graph is restricted to the first quadrant only.

★**72.** A comic book collector will pay $50 for any Superman comic published prior to 1970 if it is in mint condition and $30 if it is not in mint condition.

  a. If x represents the number of comic books in mint condition and y represents the number of comic books not in mint condition, write an expression that describes the amount the collector would pay for a combination of comic books in mint and nonmint condition.

  b. Suppose the collector visits a convention for collectors and has a budget of $1800 to spend. Use the expression from part a to write an equation describing his expenditure on a combination of the comics if he spends all of his budget.

  c. Find the x- and y-intercepts.

  d. What do the x- and y-intercepts mean in terms of purchases?

  e. Find a third combination of comic books he could purchase to equal his budget amount.

  f. Graph the equation. Because x and y are nonnegative, the graph is restricted to the first quadrant only.

Review Exercises

Exercises 1 and 2 EXPRESSIONS

[1.5] *For Exercises 1 and 2, evaluate.*

1. $\dfrac{4 - (-2)}{-2 - 1}$

2. $\dfrac{-3 - 4}{-2 - (-2)}$

Exercises 3–6 EQUATIONS AND INEQUALITIES

[2.4] 3. Solve for w in the equation $P = 2l + 2w$.

[2.5] 4. Translate to an equation and then solve. Four more than three times a number is equal to five times the sum of the number and six.

[4.1] 5. Where is the point located whose coordinates are $(0, 4)$?

[4.2] 6. Determine whether $(-8, -2)$ is a solution for $x + 3y = -5$.

4.4 Slope–Intercept Form

Objectives

1 Compare lines with different slopes.

2 Graph equations in slope–intercept form.

3 Use slope–intercept form to write the equation of a line.

4 Find the slope of a line given two points on the line.

In Section 4.3, we discussed various forms of linear equations, one of which was $y = mx + b$, where m and b are real-number constants. Further, we discovered that the graph of an equation in this form is a line with its y-intercept at $(0, b)$. In this section, we explore how the coefficient m affects the graph.

 Objective 1 Compare lines with different slopes. First, we consider graphs of equations in which $b = 0$ so that the equation is $y = mx$ and the graphs all have the origin $(0, 0)$ as the x- and y-intercepts.

Example 1 Graph each of the following equations on the same grid.

$$y = x \qquad y = 2x \qquad y = 3x$$

SOLUTION In the following table, the same choice for x has been substituted into each equation and the corresponding y-coordinate has been found. We plot the ordered pairs and connect the points to graph the lines.

If x is	and $y = x$, then y is	and $y = 2x$, then y is	and $y = 3x$, then y is
0	0	0	0
1	1	2	3
2	2	4	6

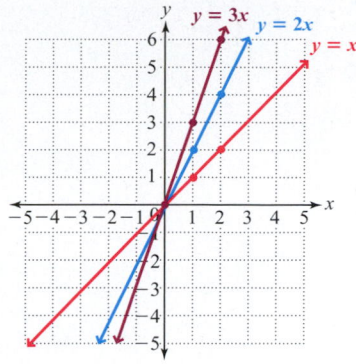

From Example 1, we see that for equations of the form $y = mx$, as the coefficient m increases, the graphs get steeper. Because the coefficient m affects how steep a line is, m is called the *slope* of the line.

For $y = x$, the slope is 1 ($m = 1$).

For $y = 2x$, the slope is 2 ($m = 2$).

For $y = 3x$, the slope is 3 ($m = 3$).

We have seen that if the slope is greater than 1, the corresponding line has a steeper incline than the graph of $y = x$. What values for m would cause a line to be less inclined than $y = x$? The pattern in the graphs suggests that we should explore values less than 1.

First, consider values less than 1 but greater than 0, such as $\frac{1}{2}$ and $\frac{1}{3}$.

Example 2   Graph each of the following equations on the same grid.

$$y = x \qquad y = \frac{1}{2}x \qquad y = \frac{1}{3}x$$

SOLUTION   For consistency, we use the same three choices for x that we used in Example 1. After finding the corresponding y-coordinates for each equation, we plot the ordered pairs and connect the points to graph the lines.

If x is	and $y = x$, then y is	and $y = \frac{1}{2}x$, then y is	and $y = \frac{1}{3}x$, then y is
0	0	0	0
1	1	$\frac{1}{2}$	$\frac{1}{3}$
2	2	1	$\frac{2}{3}$

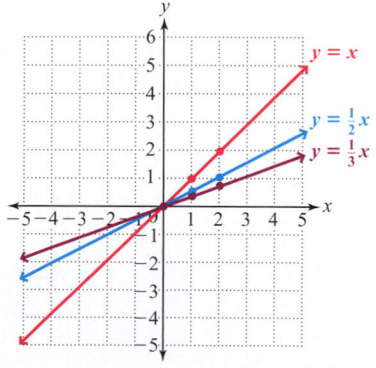

Notice that if the slope is a fraction between 0 and 1, the smaller the fraction, the less inclined or flatter the lines. So far, we have considered only positive slope values. If we "travel" from left to right along lines with positive slopes, we travel uphill. What would a line with a negative slope look like?

Example 3 Graph each of the following equations on the same grid.

$$y = -x \qquad y = -2x \qquad y = -\frac{1}{2}x$$

SOLUTION Again, we make a table of ordered pairs, then plot those ordered pairs and connect the points to graph the lines.

If x is	and $y = -x$, then y is	and $y = -2x$, then y is	and $y = -\frac{1}{2}x$, then y is
0	0	0	0
1	−1	−2	$\frac{1}{2}$
2	−2	−4	−1

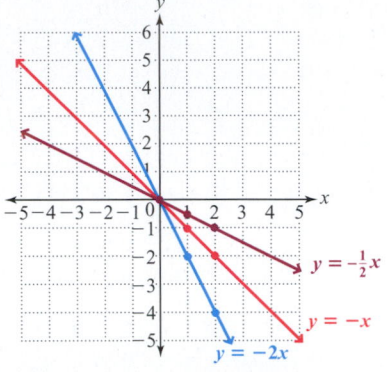

Note: When we speak of traveling along a graph, the convention is to travel from left to right, just as we do when we read English.

Notice that if we "travel" from left to right along lines with negative slopes, we travel downhill. As the slope becomes more negative, the downhill incline becomes steeper.

Rule **Graphs of $y = mx$**

Given an equation of the form $y = mx$, the graph of the equation is a line passing through the origin and having the following characteristics:

If $m > 0$ (slope is positive), the graph is a line that slants uphill from left to right.

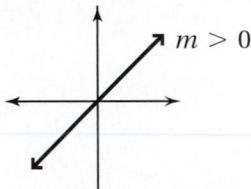

If $m < 0$ (slope is negative), the graph is a line that slants downhill from left to right.

The greater the absolute value of m, the steeper the line.

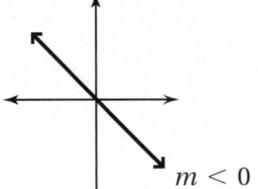

Your Turn 3

a. Is $y = 5x$ steeper or less inclined than $y = x$?

b. Is the graph of $y = -\frac{2}{3}x$ uphill or downhill from left to right?

Answers   Your Turn 3

a. $y = 5x$ is steeper than $y = x$.

b. downhill

incline of a line. What more can we discover about slope? Notice that if we isolate m

Note: A precise formula for the slope of a line will be given in Objective 4.

in $y = mx$, we get $m = \dfrac{y}{x}$, which suggests that **slope** is a ratio of the amount of vertical

change to the amount of horizontal change. A change up or to the right is a positive change, and a change down or to the left is a negative change.

Definition   *Slope:* The ratio of the vertical change between any two points on a line to the horizontal change between these points.

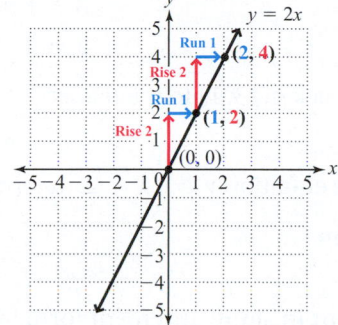

Note: If m is an integer, think of it as a fraction whose denominator is 1, as in $2 = \dfrac{2}{1}$.

For example, the slope of $y = 2x$ is 2, which can be written as $\dfrac{2}{1}$. This slope value means that rising vertically 2 units and then running horizontally 1 unit from any point on the line locates a second point on the line. For example, if we rise 2 units and run 1 unit from $(0, 0)$, we end up at $(1, 2)$. Similarly, rising 2 units and running 1 unit from $(1, 2)$ puts us at $(2, 4)$.

We have learned that in an equation of the form $y = mx + b$, the slope is m and the y-intercept is $(0, b)$. For this reason, we say that equations of the form $y = mx + b$ are in *slope–intercept* form. We can graph equations in slope–intercept form using the y-intercept as a starting point and then using the slope to locate other points on the line.

Procedure   **Graphing Equations in Slope–Intercept Form**

To graph an equation in slope–intercept form, $y = mx + b$,

Note: The "rise" will be a "fall" if the slope is negative and the run is positive.

1. Plot the y-intercept, $(0, b)$.
2. Plot a second point by rising the number of units indicated by the numerator of the slope, m, then running the number of units indicated by the denominator of the slope, m.
3. Draw a straight line through the two points.

Note: You can check by locating additional points using the slope. Every point you locate using the slope should be on the line.

Example 4   For the equation $y = -\dfrac{1}{3}x + 2$, determine the slope and the y-intercept. Then graph the equation.

SOLUTION   Because the equation is in the form $y = mx + b$, where $m = -\dfrac{1}{3}$ and $b = 2$,

the slope of the line is $-\dfrac{1}{3}$ and the y-intercept is $(0, 2)$. To graph the line, we can plot the

y-intercept and then use the slope to find other points. To interpret $-\dfrac{1}{3}$ in terms of "rise"

and "run," we place the negative sign in either the numerator or denominator. We can "rise" -1 and "run" 3, which gives the point $(3, 1)$. Or we can "rise" 1 and "run" -3, which gives the point $(-3, 3)$. Either interpretation gives another point on the same line.

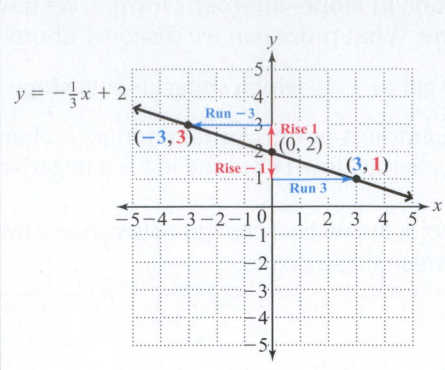

The points we found using the slope can be verified in the equation. We can check the ordered pair $(-3, 3)$ in the equation.

$$y = -\frac{1}{3}x + 2$$

$$3 \overset{?}{=} -\frac{1}{3}(-3) + 2$$

$$3 \overset{?}{=} -\frac{1}{\cancel{3}} \cdot -\frac{\cancel{3}^{1}}{1} + 2$$

$$3 \overset{?}{=} 1 + 2$$

$$3 = 3 \qquad \text{This checks.}$$

We will leave the check of $(3, 1)$ to the reader.

Your Turn 4 For the equation $y = \frac{2}{5}x - 2$, determine the slope and the y-intercept. Then graph the equation.

If an equation is not in slope–intercept form, as in $3x + 4y = 8$, and we need to determine the slope and y-intercept, we can write the equation in slope–intercept form by isolating y.

Example 5 For the equation $3x + 4y = 8$, determine the slope and the y-intercept. Then graph the equation.

SOLUTION Write the equation in slope–intercept form by isolating y.

$$3x + 4y = 8$$

$$4y = -3x + 8 \qquad \textbf{Subtract } 3x \textbf{ from both sides to isolate } 4y.$$

$$\frac{4y}{4} = \frac{-3x + 8}{4} \qquad \textbf{Divide both sides by 4 to isolate } y.$$

$$y = -\frac{3}{4}x + \frac{8}{4} \qquad \textbf{Simplify.}$$

$$y = -\frac{3}{4}x + 2$$

The slope is $-\frac{3}{4}$, and the y-intercept is $(0, 2)$. To graph the line, we begin at $(0, 2)$ and rise -3 and run 4, which gives the point $(4, -1)$. Or we can begin at $(0, 2)$ and rise 3 and run -4, which gives the point $(-4, 5)$.

Answer   **Your Turn 4**

$m = \frac{2}{5}$; y-intercept: $(0, -2)$

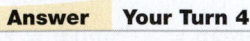

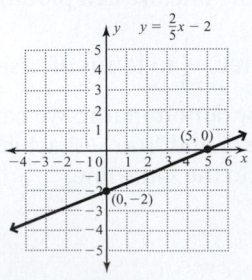

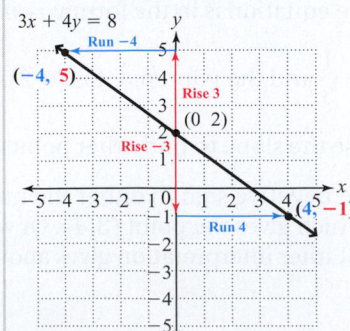

We can check the ordered pairs in the equation.

For $(-4, 5)$:

$$3x + 4y = 8$$
$$3(-4) + 4(5) \overset{?}{=} 8$$
$$-12 + 20 \overset{?}{=} 8$$
$$8 = 8$$

For $(4, -1)$:

$$3x + 4y = 8$$
$$3(4) + 4(-1) \overset{?}{=} 8$$
$$12 + (-4) \overset{?}{=} 8$$
$$8 = 8$$

Both ordered pairs check.

Your Turn 5 For the equation $3x - 2y = 6$, determine the slope and the y-intercept. Then graph the equation.

Objective ③ Use slope–intercept form to write the equation of a line. If you are given the slope, m, and the y-intercept, $(0, b)$, of a line, use the slope–intercept form of the equation to write the equation of the line.

Procedure **Equation of a Line Given Its Slope and y-Intercept**

To write the equation of a line given its slope, m, and its y-intercept, $(0, b)$, use the slope–intercept form of the equation, $y = mx + b$.

Example 6 A line with a slope of 5 crosses the y-axis at the point $(0, 3)$. Write the equation of the line.

SOLUTION We use $y = mx + b$, the slope–intercept form of the equation, replacing m with the slope, 5, and b with the y-coordinate of the y-intercept, 3.

$$y = 5x + 3$$

Your Turn 6 A line has a slope of -3. If the y-intercept is $(0, -4)$, write an equation of the line in slope–intercept form.

Objective ④ Find the slope of a line given two points on the line. Given two points on a line, we can determine the slope of the line. Consider the points $(-3, 0)$ and $(3, 4)$ on the graph of $y = \frac{2}{3}x + 2$, which we graphed in Example 6(b) of Section 4.3. In the equation, we see that the slope is $\frac{2}{3}$. Recall that slope is the ratio of the vertical change between any two points on a line to the horizontal change between those points. Therefore, the ratio of the vertical distance between $(-3, 0)$ and $(3, 4)$ to the horizontal distance between those points should be $\frac{2}{3}$. Let's verify.

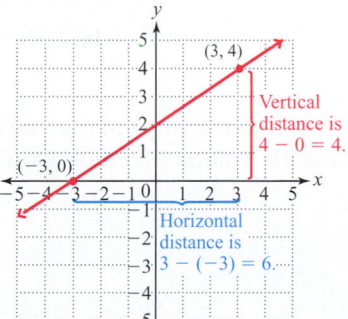

To find the vertical distance between the two points, we can subtract their y-coordinates: $4 - 0 = 4$. Similarly, the horizontal distance between the points is the difference of the x-coordinates: $3 - (-3) = 6$. Now we can write the ratio.

$$\frac{\text{Vertical distance}}{\text{Horizontal distance}} = \frac{4 - 0}{3 - (-3)} = \frac{4}{6} = \frac{2}{3}$$

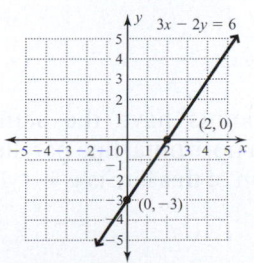

Our example suggests the following formula:

Procedure   The Slope Formula

Warning: When using the slope formula, make sure your x_1 is from the same ordered pair as your y_1.

Given two points (x_1, y_1) and (x_2, y_2), where $x_2 \neq x_1$, the slope of the line connecting the two points is given by the formula $m = \dfrac{y_2 - y_1}{x_2 - x_1}$.

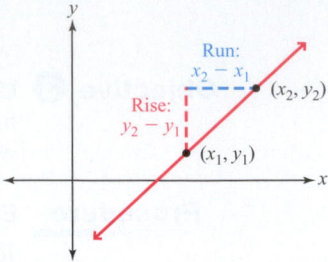

Example 7   Find the slope of the line connecting the given points.

a. $(3, 5)$ and $(-1, 7)$

SOLUTION   Using $m = \dfrac{y_2 - y_1}{x_2 - x_1}$, replace the variables with their corresponding values and then simplify. We'll let $(3, 5)$ be (x_1, y_1) and $(-1, 7)$ be (x_2, y_2).

$$m = \frac{7 - 5}{-1 - 3} = \frac{2}{-4} = -\frac{1}{2}$$

It doesn't matter which ordered pair is (x_1, y_1) and which is (x_2, y_2). If we let $(-1, 7)$ be (x_1, y_1) and $(3, 5)$ be (x_2, y_2), we get the same slope.

$$m = \frac{5 - 7}{3 - (-1)} = \frac{-2}{4} = -\frac{1}{2}$$

b. $(-3, -4)$ and $(1, -4)$

SOLUTION   We'll let $(-3, -4)$ be (x_1, y_1) and $(1, -4)$ be (x_2, y_2).

$$m = \frac{-4 - (-4)}{1 - (-3)} = \frac{0}{4} = 0$$

LEARNING Strategy

VISUAL Many people, especially visual learners, find it helpful to write the coordinate labels beneath their corresponding value.

$\begin{array}{cc} (3, 5) & (-1, 7) \\ (x_1, y_1) & (x_2, y_2) \end{array}$

Also, some people prefer to stack the ordered pairs.

$\begin{array}{c} (-1, 7) \\ (3, 5) \end{array}$

Your Turn 7   Find the slope of the line connecting the given points.

a. $(2, 5)$ and $(6, 1)$ **b.** $(-5, -2)$ and $(7, -5)$

In Example 7(b), we found the slope of the line connecting $(-3, -4)$ and $(1, -4)$ to be 0. What does a line with a slope of 0 look like? Let's plot the points $(-3, -4)$ and $(1, -4)$ and see.

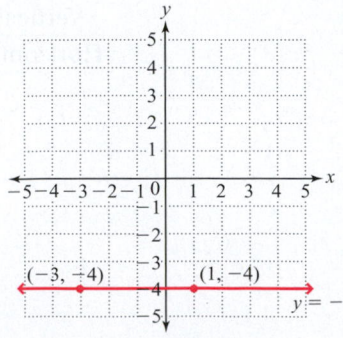

Because the two points have the same y-coordinate, they form a horizontal line parallel to the x-axis. The equation of this line is $y = -4$.

Answers   Your Turn 7

a. -1 **b.** $-\dfrac{1}{4}$

Rule   Zero Slope

Two points with different x-coordinates and the same y-coordinates, (x_1, c) and (x_2, c), will form a line with a slope of 0 (a horizontal line) and the equation $y = c$.

What is the slope of a vertical line? Two points on a vertical line have the same x-coordinates but different y-coordinates, as in $(2, 4)$ and $(2, -3)$.

Example 8   Find the slope of the line connecting the points $(2, 4)$ and $(2, -3)$.

Solution   Note that $(2, 4)$ and $(2, -3)$ connect to form a vertical line whose equation is $x = 2$. The slope of this line is found using the slope formula.

$$m = \frac{-3 - 4}{2 - 2} \qquad \text{Replace the variables in } m = \frac{y_2 - y_1}{x_2 - x_1} \text{ with their corresponding values.}$$

$$m = \frac{-7}{0} \qquad \text{Simplify.}$$

Because we never divide by 0, the slope is *undefined*.

Rule   Undefined Slope

Two points with the same x-coordinates and different y-coordinates, (c, y_1) and (c, y_2), will form a line with a slope that is undefined (a vertical line) and the equation $x = c$.

Your Turn 8   Find the slope of the line connecting the given points.

a. $(4, -1)$ and $(-3, -1)$ **b.** $(3, 2)$ and $(3, -4)$

Answers   Your Turn 8
a. 0   **b.** undefined

🖩 Calculator Tips

To graph an equation on a graphing calculator, you first enter the equation using the $\boxed{Y=}$ *key. When you press the* $\boxed{Y=}$ *key, you will see a list of Y's: Y1, Y2, Y3, and so on. Each of the Y's can be used to enter an equation, which means you can graph multiple equations on the same grid if you like.*

To enter the equation $y = 3x + 2$*, press* $\boxed{Y=}$*; at Y1, type* $\boxed{3}$ $\boxed{X, T, \theta, n}$ $\boxed{+}$ $\boxed{2}$*.*

Notice that the $\boxed{X, T, \theta, n}$ *key is used to type the variable x. To have the calculator draw the graph, press* $\boxed{\text{GRAPH}}$*.*

Sometimes you may need to adjust or change your viewing window. To change the range of tick marks on your axes, press $\boxed{\text{WINDOW}}$*. For example, if you want the x-axis to range from* -10 *to 10, press* $\boxed{\text{WINDOW}}$*; then at Xmin, enter* $\boxed{(-)}$ $\boxed{1}$ $\boxed{0}$ *and at Xmax, enter* $\boxed{1}$ $\boxed{0}$*. Xscl sets the frequency of the marks in the range. For example, setting the Xscl to 1 places a mark at each integer in the range; so if the range is* -10 *to 10, a mark will appear at* $-10, -9, -8, \ldots, 9, 10$*. If the Xscl is 2 and the range is* -10 *to 10, a mark will appear at* $-10, -8, -6, \ldots, 8, 10$*. Similarly, Ymax and Ymin change the range on the y-axis and Yscl affects how many marks are shown in that range.*

1. In your own words, what is the slope of a line?

2. What is the formula for finding the slope of a line connecting two points (x_1, y_1) and (x_2, y_2)?

3. Does the graph of $y = -2x + 1$ go uphill from left to right or downhill? Why?

4. Does the graph of $3x - 5y = 10$ go uphill from left to right or downhill? Why?

5. Describe a line with a slope of 0. Explain why a slope of 0 produces this type of line.

6. Describe a line with a slope that is undefined. Explain why an undefined slope produces this type of line.

For Exercises 7–14, graph each set of equations on the same grid. For each set of equations, compare the slopes, y-intercepts, and their effects on the graphs. See Examples 1–3.

7. $y = \dfrac{1}{2}x$

   $y = x$

   $y = 2x$

8. $y = \dfrac{1}{3}x$

   $y = x$

   $y = 3x$

9. $y = x$

   $y = \dfrac{1}{2}x$

   $y = \dfrac{1}{5}x$

10. $y = x$

    $y = \dfrac{1}{3}x$

    $y = \dfrac{1}{6}x$

11. $y = -\dfrac{1}{2}x$

    $y = -x$

    $y = -2x$

12. $y = -\dfrac{1}{3}x$

    $y = -x$

    $y = -3x$

13. $y = x$

    $y = x + 2$

    $y = x - 2$

14. $y = 2x$

    $y = 2x - 1$

    $y = 2x - 3$

For Exercises 15–38, determine the slope and the y-intercept. Then graph the equation. See Examples 4 and 5.

15. $y = 2x + 3$

16. $y = 4x - 3$

17. $y = -2x - 1$

18. $y = -3x + 5$

19. $y = \dfrac{1}{3}x + 2$

20. $y = \dfrac{1}{2}x - 4$

21. $y = \dfrac{3}{4}x - 2$

22. $y = \dfrac{2}{5}x - 5$

23. $y = -\dfrac{2}{3}x + 8$

24. $y = -\dfrac{4}{3}x + 7$

25. $2x + y = 4$

26. $3x + y = 6$

27. $3x - y = -1$

28. $2x - y = 5$

29. $2x + 3y = 6$

30. $4x - 3y = 12$

31. $x + 2y = -4$

32. $2x + 3y = -6$

33. $3x - 2y + 7 = 0$

34. $2x - 3y + 8 = 0$

35. $2y = -3x + 5$

36. $3y = 2x - 4$

37. $0.6x - 0.2y = 1$

38. $0.3x + 1.5y = -6$

For Exercises 39–50, write the equation of the line in slope–intercept form given the slope and the coordinates of the y-intercept. See Example 6.

39. $m = 2; (0, 3)$

40. $m = 3; (0, -2)$

41. $m = -3; (0, -9)$

42. $m = -4; (0, -5)$

43. $m = \dfrac{3}{4}; (0, 5)$

44. $m = \dfrac{3}{7}; (0, -1)$

45. $m = -\dfrac{2}{5}; \left(0, \dfrac{7}{8}\right)$

46. $m = \dfrac{5}{2}; \left(0, \dfrac{3}{8}\right)$

47. $m = 0.8; (0, -5.1)$

48. $m = -0.75; (0, 2.5)$

49. $m = -3; (0, 0)$

50. $m = 1; (0, 0)$

For Exercises 51–66, find the slope of the line through the given points. See Examples 7 and 8.

51. $(4, 1), (8, 11)$

52. $(1, 5), (3, 8)$

53. $(2, 4), (5, 2)$

54. $(6, 2), (3, 7)$

55. $(-3, 5), (4, 7)$

56. $(3, -2), (5, 5)$

57. $(10, -12), (4, -4)$

58. $(-8, 14), (4, 6)$

59. $(3, -3), (-15, -15)$

60. $(8, -12), (-4, -18)$

61. $(0, 5), (4, 0)$

62. $(-3, 0), (0, 5)$

63. $(-3, 5), (4, 5)$

64. $(-2, -8), (4, -8)$

65. $(6, 1), (6, -8)$

66. $(-3, 2), (-3, 5)$

For Exercises 67–76, match the equation with the appropriate graph. See Examples 4 and 5.

67. $y = 2x + 4$

68. $3x + 2y = 6$

69. $x = y$

70. $y = 3$

71. $x - 2 = 0$

72. $x - 2y = 8$

73. $y = \dfrac{2}{3}x - 2$

74. $y = -\dfrac{3}{4}x$

75. $x = -3$

76. $y + 4 = 0$

a.
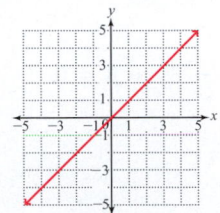

b.
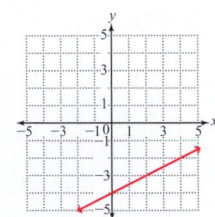

c.
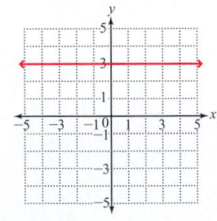

d.
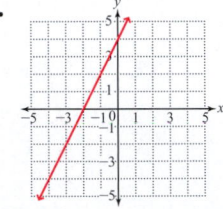

e.
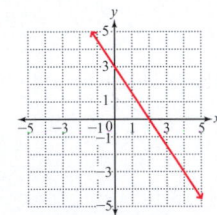

f.
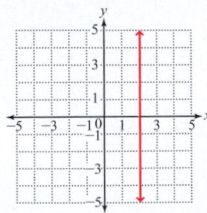

g.
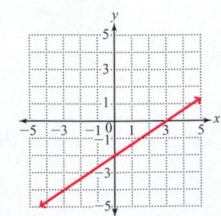

h.
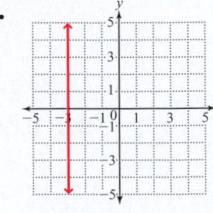

i.
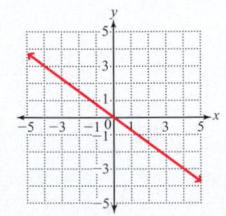

j.
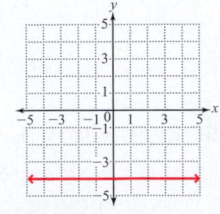

For Exercises 77–88, find the slope of the line and the y-intercept and write the equation of the line in slope–intercept form. See Examples 4 and 8.

77.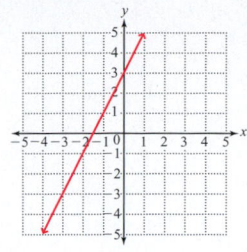

78.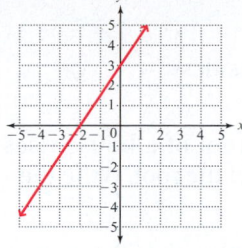

79.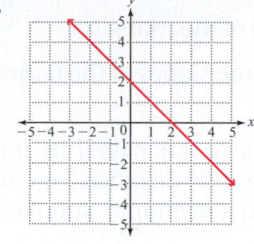

80.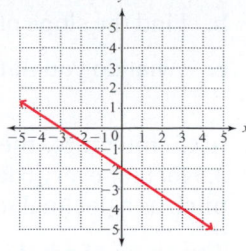

81.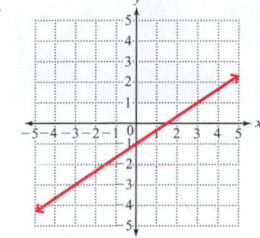

82.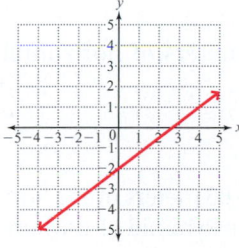

83.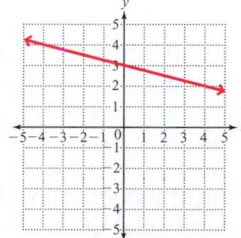

84.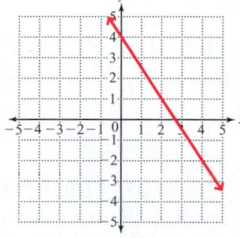

85.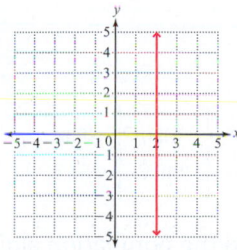

86.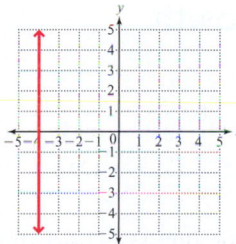

87.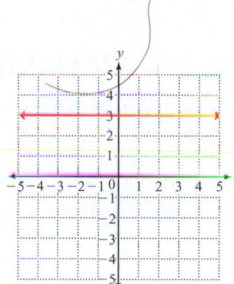

88.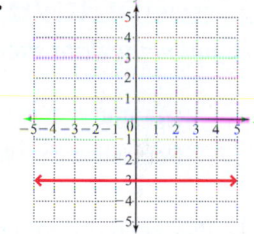

89. What is the equation of a horizontal line through $(0, -2)$? A vertical line through $(3, 0)$?

90. What is the equation of a vertical line through $(-3.5, 0)$? A horizontal line through $(0, 2.6)$?

91. What equation describes the x-axis?

92. What equation describes the y-axis?

93. A parallelogram has vertices at $(1, -2)$, $(3, 4)$, $(-3, -2)$, and $(-1, 4)$.
   a. Plot the vertices in a coordinate plane; then connect them to form the parallelogram.
   b. Find the slope of each side of the parallelogram.

   c. What do you notice about the slopes of the parallel sides?

94. A right triangle has vertices at $(-2, 3)$, $(2, 1)$, and $(3, 3)$.

    a. Plot the vertices in a coordinate plane; then connect them to form the right triangle with the right angle at $(2, 1)$.

    b. Find the slope of each side of the triangle.

    c. What do you notice about the slopes of the perpendicular sides?

95. In architecture, the slope of a roof is called its *pitch*. The roof of a house rises 3.5 feet for every 6 feet of horizontal distance. Find the pitch of the roof. Write the pitch as a decimal number and a fraction in simplest form.

96. The slope of a hill is referred to as the *grade* of the hill. Heartbreak Hill, which is part of the Boston Marathon, rises from an elevation of 150 feet to an elevation of 225 feet over a horizontal distance of about 5000 feet. Find the grade of Heartbreak Hill. (*Source:* Boston Athletic Association)

97. The Great Pyramid at Giza rises 29.3 meters for every 23 meters of horizontal distance. Find the slope of the pyramid.

98. The steps of the Pyramid of the Sun at Teotihuacan, Mexico, rise 65 meters over a horizontal distance of 112.5 meters. Find the slope of the steps of the Pyramid of the Sun.

Collaborative Exercises Population Growth

The following data show the total U.S. population by decade.

Decade	1900	1910	1920	1930	1940	1950	1960	1970	1980	1990	2000
Population (in millions)	76.2	92.2	106.0	123.2	132.2	151.3	179.3	203.3	226.5	248.7	281.4

1. Plot each ordered pair as a point on a graph with the decades along the horizontal axis and the population along the vertical axis.
2. Do the points have a linear relationship?
3. Draw a straight line that connects the point at 1900 to the point at 1950. Find the slope of this line.
4. Draw a straight line that connects the point at 1950 to the point at 2000. Find the slope of this line.
5. Compare the slopes of the two lines you have drawn. Which has the greater slope? What does this indicate about population growth in the United States during the 20th century?

6. Use the line you drew from 1950 to 2000 to predict the population at 2110.

Review Exercises

4.5 Point–Slope Form

Objectives

1. Use point–slope form to write the equation of a line.

2. Write the equation of a line parallel to a given line.

3. Write the equation of a line perpendicular to a given line.

In this section, we explore how to write an equation of a line given information about the line, such as its slope and the coordinates of a point on the line.

Objective 1   **Use point–slope form to write the equation of a line.**   In the previous section, we used the slope–intercept form to write the equation of a line because we were given the slope and y-intercept. However, we need a more general approach so that we can write the equation of a line given *any* two points on the line. For this more general approach, we can use the slope formula to derive a new form of the equation of a line called the *point–slope* form. Recall the slope formula.

Connection We also can view the formula for slope as a proportion and cross multiply to get the point–slope form.

$$\frac{m}{1} = \frac{y_2 - y_1}{x_2 - x_1}$$

$$(x_2 - x_1)m = y_2 - y_1$$

$$m = \frac{y_2 - y_1}{x_2 - x_1}$$

$$(x_2 - x_1) \cdot m = \left(\frac{y_2 - y_1}{x_2 - x_1}\right) \cdot \left(\frac{x_2 - x_1}{1}\right) \qquad \text{Multiply both sides by } x_2 - x_1 \text{ to isolate the } y\text{'s.}$$

$$(x_2 - x_1)m = y_2 - y_1 \qquad \text{Simplify.}$$

$$y_2 - y_1 = m(x_2 - x_1) \qquad \text{Rewrite with } y\text{'s on the left side to resemble slope–intercept form.}$$

To write the equation of a line with this formula, we replace m with the slope and substitute one of the given ordered pairs for x_1 and y_1, leaving x_2 and y_2 as variables. To indicate that x_2 and y_2 remain variables, we remove their subscripts so that we have $y - y_1 = m(x - x_1)$, which is called the point–slope form of the equation of a line.

Procedure   **Using the Point–Slope Form of the Equation of a Line**

To write the equation of a line given its slope and any point, (x_1, y_1), on the line, use the point–slope form of the equation of a line, $y - y_1 = m(x - x_1)$. If given a second point, (x_2, y_2), and not the slope, we first calculate the slope using $m = \dfrac{y_2 - y_1}{x_2 - x_1}$; then we use $y - y_1 = m(x - x_1)$.

Example 1   Write the equation of a line with a slope of 2 passing through the point $(3, 5)$. Write the equation in slope–intercept form.

SOLUTION   Because we are given the coordinates of a point and the slope of a line passing through the point, we begin with the point–slope form. After replacing m, x_1, and y_1 with their corresponding values, we isolate y to get slope–intercept form.

$$y - y_1 = m(x - x_1)$$
$$y - 5 = 2(x - 3) \qquad \text{Replace } m \text{ with 2, } x_1 \text{ with 3, and } y_1 \text{ with 5.}$$
$$y - 5 = 2x - 6 \qquad \text{Simplify.}$$
$$y = 2x - 1 \qquad \text{Add 5 to both sides to isolate } y.$$
$$\text{This is now slope–intercept form.}$$

Your Turn 1   Write the equation of the line in slope–intercept form with the given slope passing through the given point.

$$m = 0.6; \, (-4, 2)$$

Example 2   Write the equation of a line passing through the points $(3, 2)$ and $(-3, 6)$. Write the equation in slope–intercept form.

SOLUTION   Because we do not have the slope, we calculate it.

$$m = \frac{6 - 2}{-3 - 3} = \frac{4}{-6} = -\frac{2}{3} \qquad \text{Use } m = \frac{y_2 - y_1}{x_2 - x_1}.$$

Now we can use the point–slope form, then isolate y to write the slope–intercept form. Because we are given two points, we can use either point for (x_1, y_1) in the point–slope equation. We select $(3, 2)$ to be (x_1, y_1).

Note: If one of the given points is the y-intercept, after finding the slope, you can use the slope–intercept form to write the equation of the line.

$$y - y_1 = m(x - x_1)$$
$$y - 2 = -\frac{2}{3}(x - 3) \qquad \text{Replace } m \text{ with } -\frac{2}{3}, x_1 \text{ with 3, and } y_1 \text{ with 2.}$$
$$y - 2 = -\frac{2}{3}x + 2 \qquad \text{Simplify.}$$
$$y = -\frac{2}{3}x + 4 \qquad \text{Add 2 to both sides to isolate } y.$$

Answer   Your Turn 1

$y = 0.6x + 4.4$

Your Turn 2 Write the equation of the line passing through the given points. Write the equation in slope–intercept form.

$$(-2, -4), (4, -1)$$

Writing Linear Equations in Standard Form

Equations can also be written in standard form. Recall from Section 4.3 that standard form is $Ax + By = C$, where A, B, and C are real numbers. We will manipulate the point–slope form of the equation so that the x and y terms are on the same side of the equation and a constant appears on the other side of the equation.

Example 3 The following data points relate the velocity of an object as time passes. (You may recall this data from Your Turn 4 in Section 4.1.) The graph shows that the points are in a line. Write the equation of the line in standard form.

Time (x) (in seconds)	Velocity (y) (in ft./sec.)
0	2.0
1	2.5
2	3.0
3	3.5

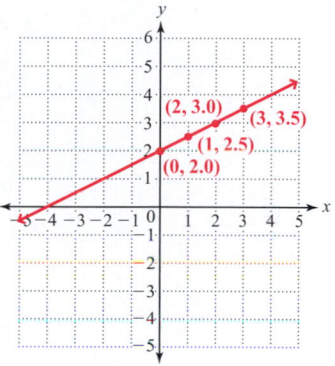

SOLUTION First, we find the slope of the line using any two ordered pairs in the slope formula. We will use $(0, 2)$ and $(1, 2.5)$.

$$m = \frac{2.5 - 2}{1 - 0} = 0.5 \text{ or } \frac{1}{2} \qquad \text{Use } m = \frac{y_2 - y_1}{x_2 - x_1}.$$

Because $(0, 2)$ is the y-intercept, we can write the equation in slope–intercept form, then manipulate the equation so that x and y appear on the same side of the equal sign to get standard form.

$$y = \frac{1}{2}x + 2 \qquad \textbf{Slope–intercept form}$$

$$y - \frac{1}{2}x = 2 \qquad \textbf{Subtract } \frac{1}{2}x \textbf{ from both sides to get } x \textbf{ and } y \textbf{ together.}$$

When we write an equation in standard form ($Ax + By = C$), it is customary to write the equation so that the x term is first with a positive coefficient and, if possible, so that A, B, and C are all integers. Let's manipulate $y - \frac{1}{2}x = 2$ to this more polished form.

$$-\frac{1}{2}x + y = 2 \qquad \textbf{Use the commutative property of addition to write the } x \textbf{ term first.}$$

$$(-2)\left(-\frac{1}{2}x + y\right) = (-2)(2) \qquad \textbf{Multiply both sides by } -2 \textbf{ so that the coefficient of } x \textbf{ is a positive integer.}$$

$$x - 2y = -4 \qquad \textbf{Standard form with } A, B, \textbf{ and } C \textbf{ integers and } A > 0$$

Answer **Your Turn 2**

$$y = \frac{1}{2}x - 3$$

Example 4   A line connects the points $(1, 5)$ and $(-3, 2)$. Write the equation of the line in the form $Ax + By = C$, where A, B, and C are integers and $A > 0$.

SOLUTION   Find the slope; then write the equation of the line using point–slope form. To finish, rewrite the equation in standard form.

$$\text{Find the slope:} \quad m = \frac{2 - 5}{-3 - 1} = \frac{-3}{-4} = \frac{3}{4} \qquad \text{Use } m = \frac{y_2 - y_1}{x_2 - x_1}.$$

Because we were not given the y-intercept, we use the point–slope form of the linear equation.

$$y - 5 = \frac{3}{4}(x - 1) \qquad \text{Use } y - y_1 = m(x - x_1) \text{ and } (1, 5) \text{ for } (x_1, y_1).$$

Now manipulate the equation to get the form $Ax + By = C$, where A, B, and C are integers and $A > 0$.

$y - 5 = \dfrac{3}{4}x - \dfrac{3}{4}$	**Distribute $\dfrac{3}{4}$ to clear parentheses.**
$4(y - 5) = 4\left(\dfrac{3}{4}x - \dfrac{3}{4}\right)$	**Multiply both sides by the LCD, 4.**
$4y - 20 = 3x - 3$	
$4y - 3x - 20 = -3$	**Subtract $3x$ from both sides to get x and y together.**
$4y - 3x = 17$	**Add 20 to both sides to get the constant terms on the right-hand side of the equation.**
$-3x + 4y = 17$	**Use the commutative property of addition to write the x term first.**
$(-1)(-3x + 4y) = (-1)(17)$	**Multiply both sides by -1 so that the coefficient of x is positive.**
$3x - 4y = -17$	**Standard form with A, B, and C integers and $A > 0$**

Your Turn 4   Write the equation of the line through the given points in the form $Ax + By = C$, where A, B, and C are integers and $A > 0$.

$$(-3, 6), (1, -4)$$

Objective ② Write the equation of a line parallel to a given line.   Consider the graphs of $y = 2x - 3$ and $y = 2x + 1$.

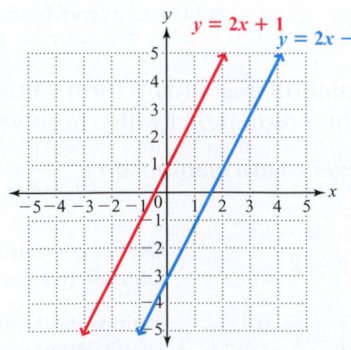

These two lines are parallel, which means that they will never intersect. Notice in the equations that their slopes are equal, which is why they are parallel.

Answer   Your Turn 4

$5x + 2y = -3$

Parallel Lines

Nonvertical parallel lines have equal slopes and different y-intercepts. Vertical lines are parallel.

Example 5 Write the equation of a line that passes through $(1, -4)$ and is parallel to $y = -3x + 5$. Write the equation in slope–intercept form.

SOLUTION In the given equation, we see that the slope of the line is -3; so the slope of the parallel line will also be -3. We can now use the point–slope form to write the equation of the parallel line.

$$y - y_1 = m(x - x_1)$$
$$y - (-4) = -3(x - 1) \qquad \text{Replace } m \text{ with } -3, x_1 \text{ with 1, and } y_1 \text{ with } -4.$$
$$y + 4 = -3x + 3 \qquad \text{Simplify.}$$
$$y = -3x - 1 \qquad \text{Subtract 4 from both sides to isolate } y.$$

Your Turn 5 Write the equation of the line that passes through the given point and is parallel to the given line. Write the equation in slope–intercept form.

$$(-4, -2); y = \frac{3}{4}x - 7$$

Objective ③ Write the equation of a line perpendicular to a given line. Consider the graphs of $y = \frac{2}{3}x - 4$ and $y = -\frac{3}{2}x + 1$.

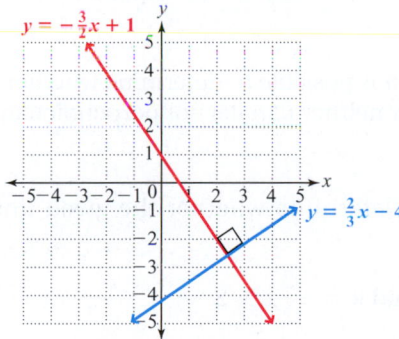

These two lines are perpendicular, which means that they intersect at a 90° angle. Notice in the equations that their slopes are negative reciprocals. In other words, $-\frac{3}{2}$ is the negative reciprocal of $\frac{2}{3}$.

Rule Perpendicular Lines

The slope of a line perpendicular to a line with a slope of $\frac{a}{b}$ will be $-\frac{b}{a}$. Horizontal and vertical lines are perpendicular.

Another way of stating the rule is that the slopes of perpendicular lines are opposites *and* reciprocals. Also note that $\left(\frac{a}{b}\right)\left(-\frac{b}{a}\right) = -1$; so two lines are perpendicular if and only if the product of their slopes is -1.

Answer Your Turn 5

$$y = \frac{3}{4}x + 1$$

Example 6 Write the equation of a line that passes through $(3, -1)$ and is perpendicular to $3x + 4y = 8$. Write the equation in slope–intercept form.

Solution To write the equation of the perpendicular line, we need its slope. To determine its slope, we need the slope of the given line. Because the equation of the given line is in standard form, to determine its slope, we rewrite it in slope–intercept form.

$$3x + 4y = 8$$
$$4y = -3x + 8 \qquad \text{Subtract } 3x \text{ from both sides.}$$
$$y = -\frac{3}{4}x + 2 \qquad \text{Divide both sides by 4.}$$

We see that the slope of the given line is $-\frac{3}{4}$; so the slope of any perpendicular line will be $\frac{4}{3}$. We can now use point–slope form to write the equation of the perpendicular line through $(3, -1)$.

$$y - y_1 = m(x - x_1)$$
$$y - (-1) = \frac{4}{3}(x - 3) \qquad \text{Replace } m \text{ with } \frac{4}{3}, x_1 \text{ with 3, and } y_1 \text{ with } -1.$$
$$y + 1 = \frac{4}{3}x - 4 \qquad \text{Simplify.}$$
$$y = \frac{4}{3}x - 5 \qquad \text{Subtract 1 from both sides to isolate } y.$$

Your Turn 6 Write the equation of a line that passes through $(-10, 3)$ and is perpendicular to $5x - 2y = 6$. Write the equation in slope–intercept form.

By using slopes, it is possible to determine whether the graphs of two lines are parallel, perpendicular, or neither parallel nor perpendicular.

Example 7 Determine whether the given lines are parallel, perpendicular, or neither.

a. $y = \frac{2}{3}x - 4$ and $y = -\frac{3}{2}x + 2$

Solution The slope of $y = \frac{2}{3}x - 4$ is $\frac{2}{3}$, and the slope of $y = -\frac{3}{2}x + 2$ is $-\frac{3}{2}$. Because the slopes are opposites (negatives) and reciprocals, the lines are perpendicular.

b. $2x + 4y = 7$ and $3x + 6y = 8$

Solution To determine the slopes of the lines, write the equations in slope–intercept form.

$$2x + 4y = 7 \qquad\qquad 3x + 6y = 8$$
$$4y = -2x + 7 \qquad\qquad 6y = -3x + 8$$
$$y = -\frac{1}{2}x + \frac{7}{4} \qquad\qquad y = -\frac{1}{2}x + \frac{4}{3}$$

Because the slope of both lines is $-\frac{1}{2}$ and the y-intercepts differ, the lines are parallel.

Answer   Your Turn 6

$y = -\frac{2}{5}x - 1$

Note: If the y-intercepts also were the same, these would be the same line, not two parallel lines.

Your Turn 7 Determine whether the given lines are parallel, perpendicular, or neither.

Answer   Your Turn 7

a. neither   **b.** parallel

Your Turn 7 Determine whether the given lines are parallel, perpendicular, or neither.

a. $y = 2x + 4$ and $y = -2x - 1$ **b.** $x - 2y = 6$ and $3x - 6y = 12$

4.5 Exercises   For Extra Help   *MathXL*
PRACTICE   WATCH   DOWNLOAD

Note: Exercises marked with a ★ represent challenging exercises.

1. If given the slope and the *y*-intercept and asked to write the equation of the line, which form of the equation would be easiest to use? Why?

2. If given two ordered pairs, neither of which is the *y*-intercept, and asked to write the equation of the line, which form of the equation would be easiest to use? Why?

3. What is the first step when you are given two ordered pairs and asked to write the equation of the line connecting the two points?

4. When given two ordered pairs and asked to write an equation of a line, does it matter which ordered pair is used in the point–slope equation? Explain.

5. How do you rewrite an equation that is in standard form in slope–intercept form?

6. How do you rewrite an equation that is in point–slope form in standard form?

For Exercises 7–18, write the equation of the line in slope–intercept form with the given slope passing through the given point. See Example 1.

7. $m = 3; (5, 2)$

8. $m = 2; (3, -5)$

9. $m = -1; (-1, -1)$

10. $m = -5; (2, 0)$

11. $m = \dfrac{2}{3}; (6, 3)$

12. $m = \dfrac{3}{2}; (-2, 3)$

13. $m = -\dfrac{3}{4}; (-1, -5)$

14. $m = -\dfrac{2}{5}; (-2, -1)$

15. $m = -\dfrac{4}{3}; (-2, 4)$

16. $m = \dfrac{2}{9}; (-1, -7)$

17. $m = 2; (0, 0)$

18. $m = -3; (0, 0)$

For Exercises 19–22, write the equation of a line. Leave your answer in standard form. (Hint: Find the slope and a point or two points on the line.) See Examples 1 and 2.

19.
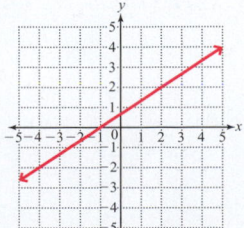

20.
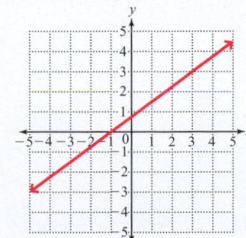

21.
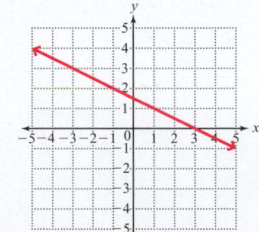

22.
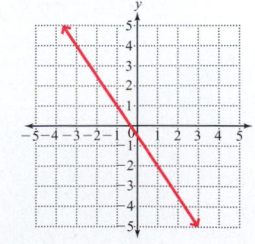

For Exercises 23–34, write the equation of a line passing through the given points. Write the equation in slope–intercept form. See Example 2.

23. $(4, -3), (-1, 7)$

24. $(-1, -5), (-4, 1)$

25. $(3, 1), (5, 6)$

26. $(-5, -2), (7, 5)$

27. $(0, 0), (-1, -7)$

28. $(0, 0), (-5, 1)$

29. $(8, -1), (8, -10)$

30. $(-3, 2), (-3, 5)$

31. $(-9, -1), (2, -1)$

32. $(-2, 5), (6, 5)$

★ **33.** $(1.4, 5), (3, 9.8)$

★ **34.** $(3.9, -2.1), (4.1, -1)$

For Exercises 35–38, write the equation of the line in slope–intercept form given the y-intercept and one other point. See Objective 1.

35. $(3, 4), (0, -2)$

36. $(0, 2), (4, 10)$

37. $(0, 3), (5, -3)$

38. $(-3, -1), (0, -5)$

For Exercises 39–50, write the equation of a line through the given points in the form $Ax + By = C$, where A, B, and C are integers and $A > 0$. See Example 4.

39. $(2, 3), (4, 8)$

40. $(4, 1), (8, 4)$

41. $(2, 0), (0, -3)$

42. $(-6, 0), (0, 6)$

43. $(-4, -1), (7, -5)$

44. $(5, -2), (-3, -3)$

45. $(-3, 2), (-6, 4)$

46. $(4, 8), (-4, -8)$

47. $(-1, 0), (-1, 8)$

48. $(-2, 5), (-2, -6)$

49. $(3, -1), (4, -1)$

50. $(2, -4), (3, -4)$

For Exercises 51–62, write the equation of a line that passes through the given point and is parallel to the given line. See Example 5.
a. Write the equation in slope–intercept form.
b. Write the equation in the form $Ax + By = C$, where A, B, and C are integers and $A > 0$.

51. $(4, 2); y = 4x + 1$

52. $(0, -1); y = 3x + 2$

53. $(-2, 1); y = -3x + 1$

54. $(0, 3); y = -2x + 4$

55. $(5, 2); y = \dfrac{1}{3}x + 4$

56. $(-2, 4); y = \dfrac{3}{4}x - 3$

57. $(-1, -3); y = -\dfrac{3}{4}x + 1$

58. $(-1, -1); y = -\dfrac{2}{3}x + 6$

59. $(2, -3); 2x + y = 5$

60. $(5, 1); 6x + 3y = 8$

61. $(-4, 2); 3x - 4y = 7$

62. $(2, -6); 3x - 2y = 5$

For Exercises 63–74, write the equation of a line that passes through the given point and is perpendicular to the given line. See Example 6.
a. Write the equation in slope–intercept form.
b. Write the equation in the form $Ax + By = C$, where A, B, and C are integers and $A > 0$.

63. $(-4, -1); y = 2x + 3$ **64.** $(3, -3); y = 3x - 1$ **65.** $(3, -4); y = \frac{1}{4}x + 2$ **66.** $(2, 3); y = \frac{1}{5}x - 4$

67. $(2, -4); y = -2x + 7$ **68.** $(-1, -1); y = -3x + 4$ **69.** $(-2, 3); y = \frac{3}{2}x - \frac{5}{2}$ **70.** $(4, 2); y = \frac{3}{4}x - 11$

71. $(-3, -1); 2x - 5y = 10$ **72.** $(-2, -4); 3x - 4y = 8$ **73.** $(3, -4); 2x + 3y = 12$ **74.** $(3, 7); 5x + 2y = 3$

For Exercises 75–90, determine whether the given lines are parallel, perpendicular, or neither. See Example 7.

75. $y = 4x - 7$
$y = 4x + 2$

76. $y = 3x + 2$
$y = 3x - 2$

77. $y = \frac{1}{2}x$
$y = -2x + 4$

78. $y = -x + 3$
$y = x - 2$

79. $y = \frac{1}{2}x + 3$
$y = 2x - 1$

80. $y = 3x + 4$
$y = \frac{1}{3}x - 3$

81. $4x + 6y = 5$
$8x + 12y = 9$

82. $3x + 6y = 10$
$2x + 4y = 9$

83. $5x + 10y = 12$
$4x - 2y = 5$

84. $5x - 3y = 11$
$3x + 5y = 8$

85. $2x + 3y = -6$
$3x + 2y = 4$

86. $4x - 2y = 5$
$2x - 4y = 3$

87. $y = 1$
$x = -3$

88. $x = 2$
$y = -4$

89. $x = 4$
$x = -2$

90. $y = 4$
$y = -3$

91. From Logan Airport in Boston, a taxi charges a $7.75 initial fee plus $0.30 for each one-eighth of a mile.
a. Let n represent the number of eighths of a mile and c represent the total cost of the taxi. Write an equation in slope–intercept form that describes how much it costs to hire a taxi from Logan Airport.

b. What will be the total cost of traveling 5 miles from the airport?

c. Graph the equation with n plotted along the horizontal axis and c along the vertical axis. Because n and c are nonnegative, restrict the graph to the first quadrant only.

92. A company manager is examining profits. He sees that the profit in May was $52,000. He then discovers that his company's profits have been declining $1000 each month thereafter.
a. Let n represent the number of months and p represent the profit. Write an equation in slope–intercept form that describes the profit n months after May.

b. If the decline continues at the same rate, what will be the profit in nine months?

c. Graph the equation with n plotted along the horizontal axis and p along the vertical axis. Because n and p are nonnegative, restrict the graph to the first quadrant only.

93. In the 1996–1997 school year, about 2.47 million females were participating in high school athletics. Over the next ten years, participation increased in an approximately linear pattern so that in 2006–2007, 3.02 million females were participating in high school athletics. (*Source:* 2007 High School Participation Survey)

a. Consider 1996–1997 to be year 0 so that 2006–2007 is year 10. Plot the two ordered pairs with the number of years along the horizontal axis and the number of females participating in athletics along the vertical axis. Then draw a line through the two points. Because the number of athletes and years are both nonnegative, restrict the graph to the first quadrant.

b. Find the slope of the line.

c. Let n represent the number of years after the 1996–1997 school year and p represent the number of female participants in high school athletics. Write an equation of the line in slope–intercept form.

d. Using your equation, how many females participated in high school athletics in the 2001–2002 school year?

e. If the linear trend continues, how many females will participate in high school athletics in the 2012–2013 school year?

94. The population of the United States was 274 million in 1997 and continued to increase in a linear pattern until 2007, when the population was 303 million.

a. Consider 1997 to be year 0 so that 2007 is year 10. Plot the two ordered pairs with the year along the horizontal axis and the population (in millions) along the vertical axis. Then draw a line through the two points. Because the years and the population are both nonnegative, restrict the graph to the first quadrant.

b. Find the slope of the line.

c. Let n represent the number of years after 1997 and P represent the population in millions. Write an equation of the line in slope–intercept form.

d. Using your equation, what was the population of the United States in 2003?

e. If the linear trend continues, what will be the population in 2015?

95. A stock begins a decline in price following a linear pattern of depreciation. The stock's initial price before the decline began was $36. On the 5th day, the closing price was $28.50.

a. Graph the line with the number of days after the decline begins plotted along the horizontal axis and the price along the vertical axis. Because the number of days and price are nonnegative, the graph is restricted to the first quadrant only.

b. Find the slope of the line.

c. Write an equation of the line in slope–intercept form with n representing the number of days of decline in price and p representing the price.

d. Find the p-intercept. Explain what the p-intercept represents in the problem.

e. Find the n-intercept. Explain what the n-intercept represents in the problem.

f. If the decline in price were to continue at the same rate, what would be the price of the stock after eight days of decline?

96. Keith deducts the value of his computer on his taxes. The amount of the deduction is based on the value of the computer, which depreciates each year. The initial value of the computer was $2100. After two years in service, the computer's value is $1500.

a. Graph the line with the number of years plotted along the horizontal axis and the value along the vertical axis. Because the number of years and the value are nonnegative, the graph is restricted to the first quadrant only.

b. Find the slope of the line.

c. Write an equation of the line in slope–intercept form with n representing the number of years the computer is in service and v representing the value.

d. Find the v-intercept. Explain what the v-intercept represents in the problem.

e. Find the n-intercept. Explain what the n-intercept represents in the problem.

f. If the decline in value were to continue at the same rate, what would be the computer's value after five years in service?

Review Exercises

Exercises 1–6 EQUATIONS AND INEQUALITIES

[1.1] **1.** Use $<$, $>$, or $=$ to write a true statement: **[2.3]** **2.** Solve: $\dfrac{3}{4}x - \dfrac{4}{5} = \dfrac{1}{2}x + 1$

$0.6 \ ? \ \dfrac{3}{4}$

[2.6] *For Exercises 3 and 4, solve and graph the solution set on the number line.*

3. $4x - 9 \geq 6x + 5$ **4.** $2(x + 1) - 5x \geq 3x - 6$

[4.4] **5.** What are the slope and y-intercept for **[4.3]** **6.** Find the x- and y-intercepts for $2x - 7y = 14$.
$y = -\dfrac{3}{4}x + 2$?

4.6 Graphing Linear Inequalities

Objectives **1** Determine whether an ordered pair is a solution for a linear inequality with two variables.

2 Graph linear inequalities.

Objective 1 **Determine whether an ordered pair is a solution for a linear inequality with two variables.** Now that we have learned how to graph linear equations, let's turn our attention to linear inequalities. Linear inequalities have the same form as linear equations except that they contain an inequality symbol instead of an equal sign. Examples of linear inequalities are as follows:

$$2x + 3y > 6$$
$$y \le x - 3$$

Remember that a solution for a linear equation in two variables is an ordered pair that makes the equation true. Similarly, a solution to a linear inequality in two variables is an ordered pair that makes the inequality true. For example, the ordered pair $(2, 1)$ is a solution for $2x + 3y > 6$.

$$2x + 3y > 6$$
$$2(2) + 3(1) \overset{?}{>} 6 \qquad \text{Replace } x \text{ with 2 and } y \text{ with 1.}$$
$$4 + 3 \overset{?}{>} 6$$
$$7 > 6 \qquad \text{This is true, so } (2, 1) \text{ is a solution.}$$

Procedure   **Checking an Ordered Pair**

To determine whether an ordered pair is a solution for an inequality, replace the variables with the corresponding coordinates and see if the resulting inequality is true. If it is, the ordered pair is a solution.

Example 1   Determine whether $(-4, 7)$ is a solution for $y \ge -2x + 3$.

SOLUTION $y \ge -2x + 3$

$$7 \overset{?}{\ge} -2(-4) + 3 \qquad \begin{array}{l}\text{Replace } x \text{ with } -4 \text{ and } y \text{ with 7; then}\\ \text{determine whether the inequality is true.}\end{array}$$
$$7 \overset{?}{\ge} 8 + 3$$
$$7 \ge 11$$

This statement is false, so $(-4, 7)$ is not a solution.

Your Turn 1   Determine whether the given ordered pair is a solution for the given inequality.

   a. $3x - 5y < 6; (-1, -2)$ **b.** $y \le -4x + 1; (2, -7)$

Objective ❷ Graph linear inequalities.   Graphing linear inequalities is very much like graphing linear equations. Consider $2x + 3y \ge 6$. The greater than or equal to sign indicates that solutions to the inequality $2x + 3y \ge 6$ will be ordered pairs that satisfy $2x + 3y > 6$ and $2x + 3y = 6$. The graph of $2x + 3y = 6$ is a line. Let's graph this line using intercepts.

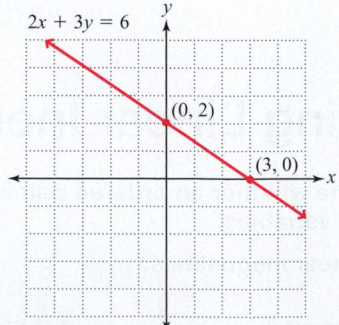

Find the x- and y-intercepts.

x-intercept:

$$2x + 3y = 6$$
$$2x + 3(0) = 6$$
$$2x = 6$$
$$x = 3$$

y-intercept:

$$2x + 3y = 6$$
$$2(0) + 3y = 6$$
$$3y = 6$$
$$y = 2$$

x-intercept: $(3, 0)$
y-intercept: $(0, 2)$

Answers   **Your Turn 1**

a. no   **b.** yes

The "greater than" portion of the inequality consists of ordered pairs that satisfy $2x + 3y > 6$. To find one of these pairs, we select a point on one side of the line and see if its coordinates satisfy $2x + 3y > 6$. Let's select the point with coordinates $(1, 4)$.

$$2x + 3y > 6$$

$$2(1) + 3(4) \overset{?}{>} 6 \qquad \text{Replace } x \text{ with 1 and } y \text{ with 4.}$$

$$2 + 12 \overset{?}{>} 6$$

$$14 > 6 \qquad \text{This statement is true; therefore, } (1, 4) \text{ is a solution.}$$

It can be shown that every ordered pair in the region containing the point $(1, 4)$, which is above the line, will satisfy $2x + 3y > 6$. It also can be shown that no ordered pair on the other side of the line will satisfy $2x + 3y > 6$. For example, consider the origin $(0, 0)$.

$$2x + 3y > 6 \qquad \text{Replace } x \text{ with 0 and } y \text{ with 0.}$$

$$2(0) + 3(0) \overset{?}{>} 6$$

$$0 > 6 \qquad \text{This statement is false, so } (0, 0) \text{ is not a solution.}$$

The line that is the graph of $2x + 3y = 6$ is called a *boundary*. The region on one side of this boundary line contains all of the ordered pair solutions satisfying $2x + 3y > 6$. The region on the other side contains all ordered pairs that satisfy $2x + 3y < 6$. To indicate that every ordered pair above the boundary line is a solution to the inequality, we shade the region above the line.

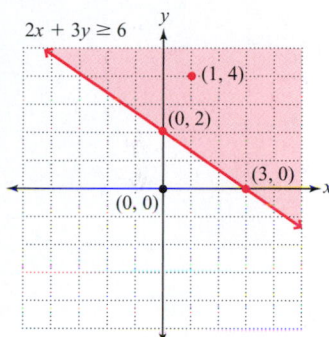

If the inequality had been $2x + 3y > 6$, the ordered pairs that satisfy $2x + 3y = 6$ would not be solutions. Because these ordered pairs form the boundary line, we draw a dashed line instead of a solid line to indicate that those ordered pairs are not solutions. The graph of $2x + 3y > 6$ is shown here.

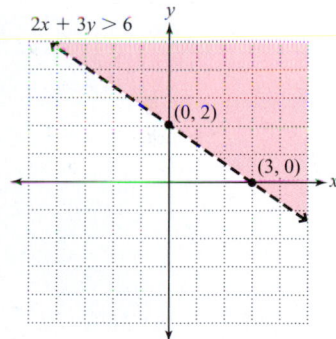

Connection When graphing inequalities with the symbols $\leq$ or $\geq$ on a number line, we used a bracket (or closed circle) to indicate that equality is included. When the inequalities have $<$ or $>$, we used a parenthesis (or open circle) to indicate that equality is excluded. When linear inequalities are graphed in the rectangular coordinate system, a solid line is like a bracket and a dashed line is like a parenthesis.

For the inequalities $2x + 3y \leq 6$ and $2x + 3y < 6$, we shade the region on the other side of the boundary line.

The graph of $2x + 3y \leq 6$:

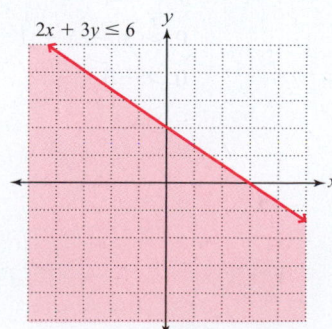

The graph of $2x + 3y < 6$:

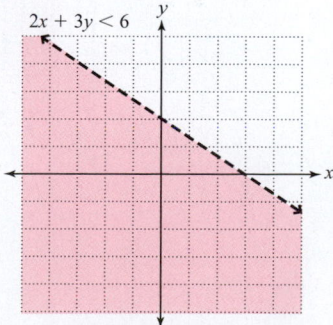

To graph a linear inequality in two variables,

1. Graph the related equation (the boundary line). The related equation has an equal sign in place of the inequality symbol. If the inequality symbol is $\leq$ or $\geq$, draw a solid line. If the inequality symbol is $<$ or $>$, draw a dashed line.
2. Choose an ordered pair on one side of the boundary line and test this ordered pair in the inequality. If the ordered pair satisfies the inequality, shade the region that contains it. If the ordered pair does not satisfy the inequality, shade the region on the other side of the boundary line.

Example 2 Graph the linear inequalities.

a. $y > 3x + 1$

SOLUTION First, graph the related equation $y = 3x + 1$. Two ordered pairs that satisfy $y = 3x + 1$ are $(0, 1)$ and $(1, 4)$.

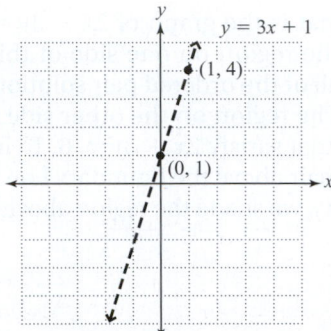

Note: The equation $y = 3x + 1$ could also be graphed using the slope and y-intercept.

Because the inequality is strictly greater than, we draw a dashed line to indicate that ordered pairs on this boundary line are not solutions.

Now we choose an ordered pair on one side of the line and test this ordered pair in the inequality. We can choose any point, but for easier calculations, let's choose $(0, 0)$.

Note: You cannot choose the origin if it lies on the boundary line.

$$y > 3x + 1$$
$$0 \overset{?}{>} 3(0) + 1 \qquad \text{Replace } x \text{ with 0 and } y \text{ with 0.}$$
$$0 > 1 \qquad \text{This statement is false, so } (0, 0) \text{ is}$$
not a solution for the inequality.

Because $(0, 0)$ did not satisfy the inequality, we shade the region on the other side of the boundary line.

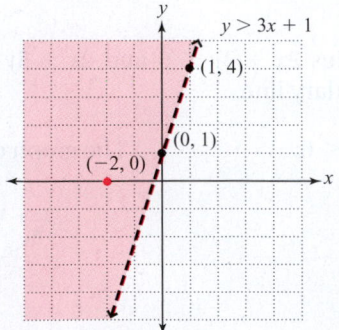

We can confirm that our shading is correct by choosing an ordered pair in this region. We will choose the ordered pair $(-2, 0)$.

$$y > 3x + 1$$
$$0 \overset{?}{>} 3(-2) + 1 \qquad \text{Replace } x \text{ with } -2 \text{ and } y \text{ with 0.}$$
$$0 > -5 \qquad \text{This statement is true, which con-}$$
firms that we have shaded the correct side of the boundary line.

b. $x - 2y \leq 4$

SOLUTION First, graph the related equation $x - 2y = 4$. Two ordered pairs that satisfy $x - 2y = 4$ are $(0, -2)$ and $(4, 0)$.

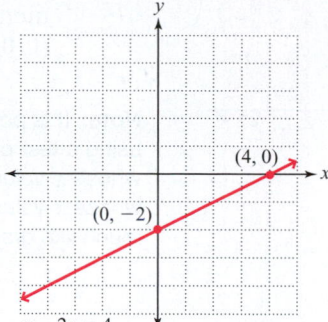

Because the inequality symbol is less than or equal to, we draw a solid line to indicate that ordered pairs on the boundary line are solutions.

Now we choose an ordered pair on one side of the line and test this ordered pair in the inequality. We will choose the origin $(0, 0)$.

$$x - 2y \leq 4$$
$$0 - 2(0) \overset{?}{\leq} 4 \qquad \text{Replace } x \text{ with 0 and } y \text{ with 0.}$$
$$0 \leq 4 \qquad \text{This statement is true; therefore, } (0, 0) \text{ is a solution for the inequality.}$$

Because $(0, 0)$ satisfies the inequality, we shade the region that contains it.

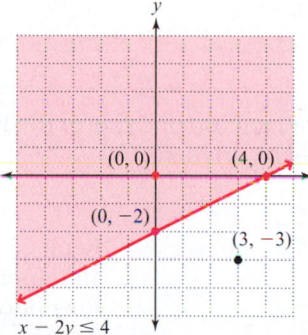

We can confirm that the region on the other side of the line should not be shaded by choosing a point in that region, such as $(3, -3)$.

$$x - 2y \leq 4$$
$$3 - 2(-3) \overset{?}{\leq} 4 \qquad \text{Replace } x \text{ with 3 and } y \text{ with } -3.$$
$$3 + 6 \overset{?}{\leq} 4$$
$$9 \leq 4$$

This statement is false, which confirms that the region containing the point with coordinates $(3, -3)$ should not be shaded.

c. $x > 2$

SOLUTION First, graph the related line $x = 2$, which is a vertical line intersecting the x-axis at $(2, 0)$.

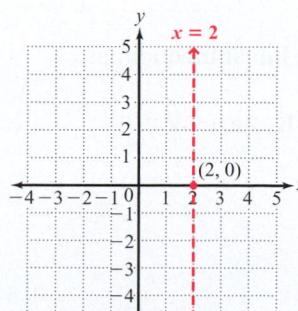

Because the inequality symbol is greater than, we draw a dashed line to indicate that the ordered pairs on the boundary line are not solutions.

Now we choose an ordered pair on one side of the line and test this ordered pair in the inequality. We will choose the origin $(0, 0)$.

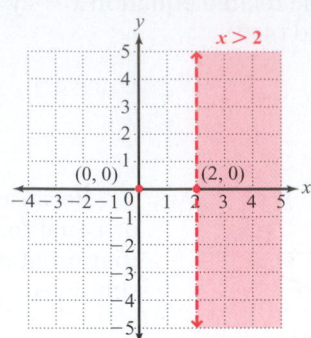

$x > 0$

$0 > 0$ Replace x with 0, which results in a false statement. So shade the side of the line opposite $(0, 0)$.

Note: It is possible to determine the solutions without using a test point. We are looking for all ordered pairs whose x-values are greater than 2. The ordered pairs whose x-values are greater than 2 are to the right of 2, much like graphing on a number line.

d. $y \leq -3$

SOLUTION   First, graph the related line $y = -3$, which is a horizontal line intersecting the y-axis at $(0, -3)$.

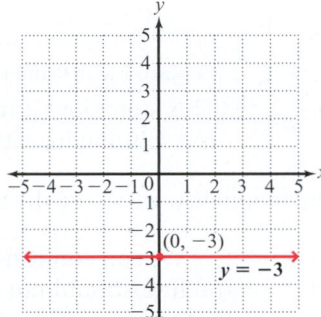

Because the inequality symbol is less than or equal to, we draw a solid line to indicate that the ordered pairs on the boundary line are solutions.

Answers   **Your Turn 2**

a. $y < -\frac{1}{3}x + 2$

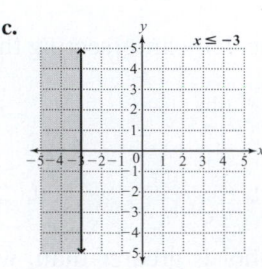

b. $2x + 3y \geq 6$

c. $x \leq -3$

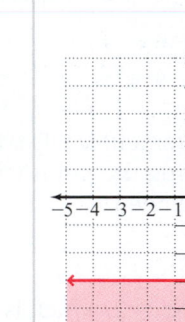

d. $y > 2$

Now we choose an ordered pair on one side of the line and test this ordered pair in the inequality. We will choose the origin $(0, 0)$.

$y \leq -3$

$0 \leq -3$ Replace y with 0, which results in a false statement. So shade the side of the line on the opposite side from $(0, 0)$.

Note: Again, it is possible to determine the solutions without using a test point. We are looking for all ordered pairs whose y-values are less than or equal to -3. The ordered pairs whose y-values are less than or equal to -3 are either on or below the line.

Your Turn 2   Graph the linear inequalities.

a. $y < -\frac{1}{3}x + 2$ **b.** $2x + 3y \geq 6$ **c.** $x \leq -3$ **d.** $y > 2$

4.6 Exercises For Extra Help *MyMathLab*

1. How do you determine whether an ordered pair is a solution for an inequality?

2. How do you determine the related equation for a given inequality?

3. In graphing a linear inequality, when do you draw a dashed line and when do you draw a solid line?

4. When graphing a linear inequality, how do you determine which side of the boundary line to shade?

For Exercises 5–12, determine whether the ordered pair is a solution for the linear inequality. See Example 1.

5. $(5, 5); y > -x + 2$

6. $(0, 1); y > -x + 2$

7. $(2, 0); y \geq \frac{1}{2}x + 1$

8. $(0, 0); y < -\frac{2}{3}x + 2$

9. $(-1, 2); 3x - y \leq -8$

10. $(3, -1); x - y < -2$

11. $(-1, -2); x - 2y > -7$

12. $(0, -2); x + 4y \geq 8$

For Exercises 13–40, graph the linear inequality. See Example 2.

13. $y \leq -x + 1$

14. $y \leq x - 8$

15. $y > -2x + 5$

16. $y < -3x + 2$

17. $y > 2x$

18. $y < x$

19. $y > \frac{1}{3}x$

20. $y < \frac{2}{5}x$

21. $y \geq \frac{3}{4}x + 2$

22. $y \geq -\frac{2}{3}x - 3$

23. $3x + y \leq 9$

24. $2x + y > 6$

25. $5x - 2y < 10$ **26.** $3x - 2y \geq 6$ **27.** $3x + y \geq 8$ **28.** $5x + y \geq -10$

29. $2x + 3y < 9$ **30.** $4x + 2y \leq 3$ **31.** $3x + y \geq 0$ **32.** $-3x - 2y < 0$

33. $x > -4$ **34.** $x \leq 6$ **35.** $y \leq 2$ **36.** $y > -1$

37. $x + 2 \geq 0$ **38.** $x - 4 < 0$ **39.** $y - 3 < 0$ **40.** $y + 2 \geq 0$

41. A company sells two sizes of bottles of hand lotion. The revenue for each unit of the large bottle is $2.00. The revenue for each unit of the small bottle is $1.50. For the company to break even during the first quarter, the company must generate $280,000 in revenue. The inequality $2x + 1.5y \geq 280,000$ describes the amount of revenue that must be generated from the sale of both size bottles for the company to break even or turn a profit.
   a. What do x and y represent?

   b. Graph the inequality. Because x and y are both nonnegative, restrict the graph to the first quadrant.
   c. What does the boundary line represent? What does the shaded region represent?

   d. List three combinations of sales numbers of each size that allow the company to break even.

   e. List three combinations of sales numbers of each size that allow the company to turn a profit.

   f. In reality, is every combination of units sold represented by the line and shaded region a possibility? Explain.

42. A company produces two versions of tax preparation software. The regular version costs $12.50 per unit to produce, and the deluxe version costs $15.00 per unit to produce. Management plans for the cost of production to be a maximum of $300,000 for the first quarter. The inequality $12.50x + 15y \leq 300,000$ describes the total cost as prescribed by management.

    a. What do x and y represent?

    b. Graph the inequality. Because x and y are both nonnegative, restrict the graph to the first quadrant.

    c. What does the line represent?

    d. What does the shaded region represent?

    e. List three combinations of numbers of units produced of each version of software that yield a total cost of $300,000.

    f. List three combinations of numbers of units produced of each version of software that yield a total cost less than $300,000.

    g. In reality, is every combination of units produced represented by the line and shaded region a possibility? Explain.

43. Andrea visits a garden center to purchase new plants for her garden. She has a $100 gift certificate. The plants that she is considering come in two sizes and are priced according to size. Plants in 1-gallon containers are priced at $5.50 each, and plants in 5-gallon containers are priced at $12.50 each.

    a. Select a variable for the number of each plant size and write an inequality that has Andrea's total cost within the amount of the gift certificate.

    b. Graph the inequality. Because both quantities are nonnegative, restrict the graph to the first quadrant.

    c. What does the line represent?

    d. What does the shaded region represent?

    e. What are some possible combinations of containers she could purchase?

    f. In reality, could she purchase any combinations that would yield a total in the exact amount of the gift certificate? Explain.

44. Michelle is coordinating a fund-raiser that will sell two different cookbooks. One cookbook sells for $10; the other, for $12. The goal is to raise at least $30,000.

    a. Select variables for each type of cookbook and write an inequality in which total sales is at least $30,000.

    b. Graph the inequality. Because both quantities are nonnegative, restrict the graph to the first quadrant.

    c. What does the line represent?

    d. What does the shaded region represent?

e. Find two combinations of book sales that yield exactly $30,000 in total sales.

f. Find two combinations of units sold that yield more than $30,000.

45. A building with a rectangular base is to be designed so that the maximum perimeter is 180 feet.

    a. Select variables for the length and width and write an inequality in which the maximum perimeter is 180 feet.

    b. Graph the inequality. Because the length and width are both positive, restrict the graph to the first quadrant.

    c. What does the line represent?

    d. What does the shaded region represent?

    e. Find three combinations of length and width that yield a perimeter of exactly 180 feet.

    f. Find three combinations of length and width that yield a perimeter of less than 180 feet.

46. Steel beams are to be welded together to form a frame that is an isosceles triangle (two sides of equal length). The perimeter of the frame cannot exceed 60 feet.

    a. Select a variable for the length of the base and a second variable for the sides of equal length. Then write an inequality in which the maximum perimeter is 60 feet.

    b. Graph the inequality. Because the lengths are positive, restrict the graph to the first quadrant.

    c. What does the line represent?

    d. What does the shaded region represent?

    e. Find sets of dimensions that yield a perimeter of exactly 60 feet.

    f. Find sets of dimensions that yield a perimeter of less than 60 feet.

> **Puzzle Problem**
>
> The sum of two different whole numbers is equal to 8, while the difference of the two numbers is greater than or equal to 1. Find all possible ordered pairs that satisfy both constraints.

Review Exercises

Exercise 1 **CONSTANTS AND VARIABLES**

[1.1]   **1.** Write a set containing all vowels in the English alphabet.

Exercises 2–6 **EQUATIONS AND INEQUALITIES**

[1.7]   **2.** Evaluate $\frac{3}{4}x - 5$ when $x = 8$.   **[2.3]**   **3.** Solve and check: $3x - 4(x + 7) = -15$

[3.3]   **4.** The length of a rectangle is four more than the width. If the perimeter is 88 feet, find the dimensions.

[4.3]   **5.** Find the x- and y-intercepts for $3x - 2y = 8$.   **[4.3]**   **6.** Graph $-x + y = 7$.

4.7 Introduction to Functions and Function Notation

Objectives

Objectives

1 Identify the domain and range of a relation.

2 Identify functions and their domains and ranges.

3 Find the value of a function.

4 Graph linear functions.

In this section, we study relations and functions, which develop naturally from our work with graphing because they involve ordered pairs.

Objective 1 **Identify the domain and range of a relation.** To formally define a function, we first discuss **relations**.

Definition *Relation:* A set of ordered pairs.

A linear equation such as $y = 2x$ is a relation because it pairs a given x-value with a corresponding y-value. You might recall that with an equation in this form, we thought of the x-values as input values and the y-values as output values. In a relation, the set of all of the input values (x-values) is called the **domain** and the set of all of the output values (y-values) is called the **range**.

Definitions *Domain:* The set of all input values (x-values) for a relation.

Range: The set of all output values (y-values) for a relation.

The following table lists some values in the domain and range for $y = 2x$.

Domain (x-values)	Range (y-values)
0	0
1	2
2	4
3	6

Example 1 Determine the domain and range of the relation $\{(2, -1), (3, 0), (4, 5), (5, 8)\}$.

SOLUTION The domain is the set containing all of the x-values $\{2, 3, 4, 5\}$, and the range is the set containing all of the y-values $\{-1, 0, 5, 8\}$.

Domain and Range with a Graph

We can determine the domain and range of a relation from its graph. Look at the graph of $y = 2x$.

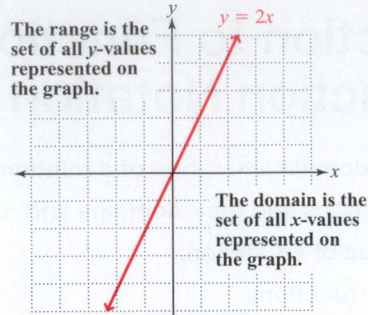

The range is the set of all *y*-values represented on the graph.

$y = 2x$

The domain is the set of all *x*-values represented on the graph.

It is convenient to imagine traveling along a graph from left to right. As we travel along $y = 2x$ from left to right, every *x*-value along the *x*-axis has a corresponding *y*-value. Thus, the domain is a set containing all real numbers. Similarly, every *y*-value is paired with an *x*-value; so the range also contains all real numbers.

Procedure **Determining Domain and Range**

To determine the domain of a relation given its graph, answer this question: What are all of the *x*-values that have a corresponding *y*-value? To determine the range, answer this question: What are all of the *y*-values that have a corresponding *x*-value?

Example 2 Determine the domain and range of the relation.

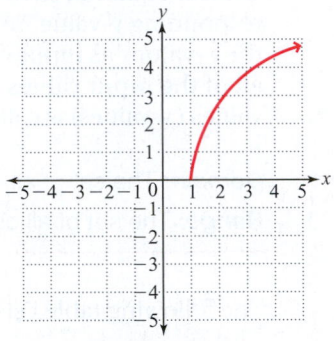

SOLUTION From left to right, this graph "begins" at the point $(1, 0)$. Values along the *x*-axis begin at 1 and continue infinitely; so the domain is $\{x \,|\, x \geq 1\}$ in set-builder notation and $[1, \infty)$ in interval notation. For the range, the *y*-values begin at 0 and continue infinitely; so the range is $\{y \,|\, y \geq 0\}$ in set-builder notation and $[0, \infty)$ in interval notation.

Your Turn 2 Determine the domain and range of the relation.

a. $\{(-2, -4), (0, -1), (3, 6), (7, 5)\}$

b.

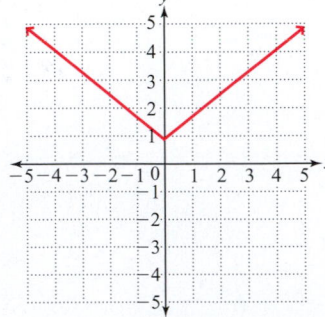

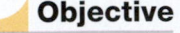

Answers **Your Turn 2**

a. Domain: $\{-2, 0, 3, 7\}$
Range: $\{-4, -1, 6, 5\}$

b. Domain: All real numbers or $(-\infty, \infty)$
Range: $\{y \,|\, y \geq 1\}$ or $[1, \infty)$

Objective ❷ Identify functions and their domains and ranges. Now we turn to a special type of relation called a *function*. Let's reconsider the trajectory data from Example 4 of Section 4.1 (p. 245), which we have listed here as a relation in the following table.

Time in Seconds (domain)	Height in Feet (range)
0	6.00
0.5	7.50
1.0	8.25
1.5	7.50
2.0	6.00
2.5	2.00

Note that each value for time is paired with one and only one height value. This is the key feature that makes this relation a **function**.

Definition   *Function:* A relation in which every value in the domain is paired with (or assigned to) exactly one value in the range.

From the data in the preceeding table, we see that a function can have different values in the domain assigned to the same value in the range. For example, domain values 0.5 seconds and 1.5 seconds each correspond to the same range value 7.50 feet. Also, the domain values 0 seconds and 2.0 seconds each correspond to 6.00 feet in the range. Using arrows to match the values in the domain with their corresponding values in the range can be a helpful technique in determining whether a relation is a function. Such a map might look like this:

Domain: {0,   0.5,   1.0,   1.5,   2.0,   2.5}

Note: Each element in the domain has a single arrow pointing to an element in the range.

Range: {2.00,   6.00,   7.50,   8.25}

Note that every function is a relation but not every relation is a function. For example, consider the following relation that assigns baseball outfield positions (the domain) to the 2008 Atlanta Braves outfielders (the range).

Note: Elements in the domain have more than one arrow pointing to an element in the range.

Domain: {left field, center field, right field}

Range: {Diaz, Anderson, Kotsay, Anderson, Francoeur, Anderson}

Notice that all three positions are assigned to more than one player. Since an element in the domain is assigned to more than one element in the range, this relation is not a function.

Conclusion   If any value in the domain is assigned to more than one value in the range, the relation is not a function.

Example 3   Determine whether the relation is a function.

a. The table to the right indicates the actual and projected national health expenditures, in billions of dollars, for selected years. (*Source:* www.healthcare.org)

Year (domain)	Expenditure (range)
2000	1310.0
2001	1424.5
2002	1547.6
2003	1660.5
2008	2354.6
2010	3079.8

SOLUTION This relation is a function because each year value in the domain has only one expenditure value in the range.

b. The following relation assigns a birth date to the corresponding person.

Domain:　{May 1, June 4, July 8, August 2}

Range:　{Danielle, Gerard, Juan, René, Candice}

SOLUTION This relation is not a function because an element in the domain, July 8, is assigned to two people in the range.

c. $\{(1, 4), (3, -2), (-5, 0) \ (3, 4)\}$

SOLUTION This relation is not a function because an element in the domain, 3, is paired with two elements, -2 and 4, in the range.

> **Note:** The domain of Example 3(c) is $\{1, 3, -5\}$. Even though 3 occurs twice, it is listed only once.

Your Turn 3 Determine whether the relation is a function.

a.

Index Topic (domain)	Page Number(s) (range)
Celestial sphere	56
Celsius	36
Centigrade	36
Centimeter	6
Centripetal force	239, 305

b. The table shows the top five finishers in the 2009 Daytona 500 race.

Final Position (domain)	Car Number (range)
1st	17 (Matt Kenseth)
2nd	29 (Kevin Harvick)
3rd	44 (A. J. Allmendinger)
4th	33 (Clint Bowyer)
5th	19 (Elliott Sadler)

c. $\{(-2, -1), (5, -2), (0, 0), (3, -2)\}$

We can determine whether a relation is a function by its graph. Remember that in Example 4 of Section 4.1 (p. 245), we created a graph of the trajectory data (that data also appears in this section on p. 305) by plotting the ordered pairs as points in the coordinate plane.

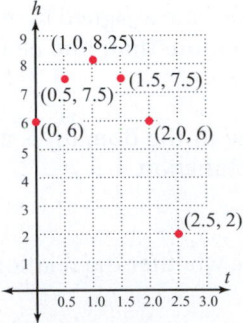

Notice that the domain (time in this case) is plotted along the horizontal axis and the range (height) is plotted along the vertical axis.

To determine whether a relation is a function from its graph, we can perform a test called the *vertical line test*. The nature of the test is to draw or imagine a vertical line through every point in the domain.

Answers　Your Turn 3

a. not a function

b. function

c. function

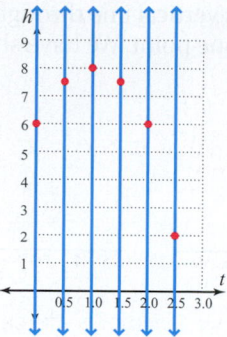

Here, we've drawn lines in blue through each value in the domain. Note that each vertical line intersects the graph at one and only one point, which means that each value in the domain corresponds to exactly one value in the range. This relation is a function.

Procedure **The Vertical Line Test**

To determine whether a relation is a function from its graph, perform a vertical line test.

1. Draw or imagine vertical lines through each point in the domain.
2. If each vertical line intersects the graph in at most one point, the graph is the graph of a function.
3. If any vertical line intersects the graph at two or more different points, the graph is not the graph of a function.

LEARNING Strategy

TACTILE If you are a tactile learner, perform the vertical line test by using a pencil as your vertical line and sweep it from left to right across the graph. If your pencil (held vertically) ever touches the graph in two or more places, the graph is *not* a function.

Example 4 For each graph, determine the domain and range. Then state whether the relation is a function.

a.

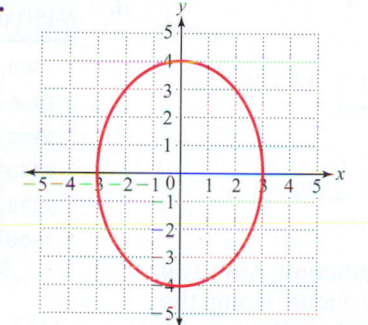
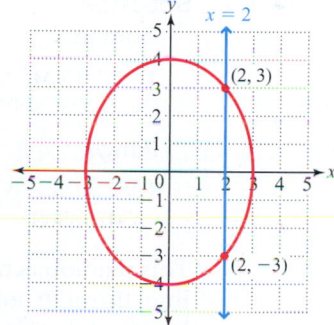

SOLUTION Domain: $\{x \mid -3 \leq x \leq 3\}$ or $[-3, 3]$

Range: $\{y \mid -4 \leq y \leq 4\}$ or $[-4, 4]$

This relation is not a function because a vertical line can be drawn that intersects the graph at two different points.

For example, the vertical line $x = 2$ (shown in blue) passes through two different points on the graph, $(2, 3)$ and $(2, -3)$.

b.

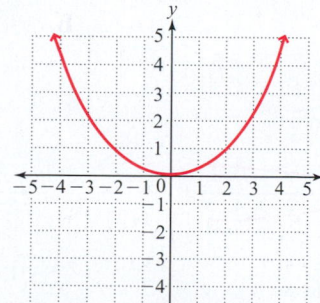
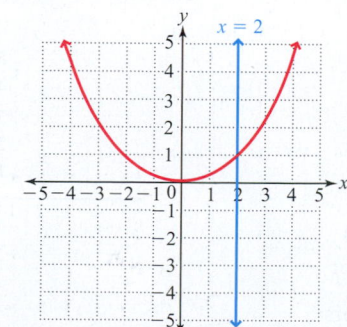

SOLUTION Domain: all real numbers

Range: $\{y \mid y \geq 0\}$ or $[0, \infty)$

4.7 Introduction to Functions and Function Notation **307**

This relation is a function. A vertical line through any value along the horizontal axis will intersect the graph at only one point. We have shown $x = 2$ to illustrate one such line.

c.

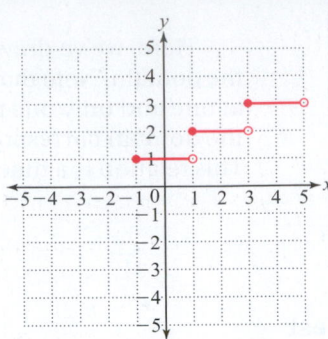

SOLUTION Domain: $\{x \mid -1 \le x < 5\}$ or $[-1, 5)$

Range: $\{y \mid y = 1, 2, 3\}$

This relation is a function. It may seem that the vertical lines $x = 1$ and $x = 3$ (shown in blue) intersect the graph at two different points. However, the open circles at $(1, 1)$, $(3, 2)$, and $(5, 3)$ indicate that those points are not part of the graph. The closed circles at $(-1, 1)$, $(1, 2)$, and $(3, 3)$ are part of the graph. The vertical line $x = 1$ "passes through" the open circle at $(1, 1)$ and intersects the graph at only one point, the closed circle at $(1, 2)$. Similarly, $x = 3$ passes through the open circle at $(3, 2)$ and intersects the graph at only one point, $(3, 3)$.

SOLUTION

Domain:

$\{t \mid 9{:}30 \text{ A.M.} \le t \le 4{:}00 \text{ P.M.}\}$ or $[9{:}30 \text{ A.M.}, 4{:}00 \text{ P.M.}]$

Range:

$\{N \mid 2507 \le N \le 2569\}$ or $[2507, 2569]$

This relation is a function. A vertical line through any value along the horizontal axis will intersect the graph at only one point. In the terminology from the graph, at each time during this business day, the Nasdaq had only one value.

d.

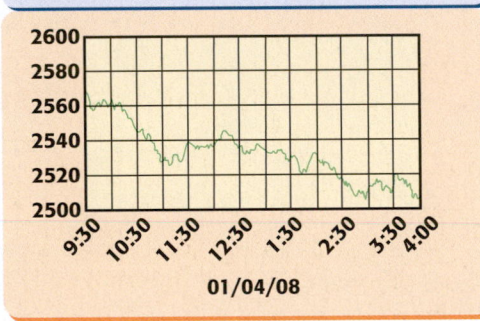

NASDAQ Composite Index Intraday stock chart

01/04/08

Your Turn 4 For each graph, determine the domain and range. Then state whether the relation is a function.

a.

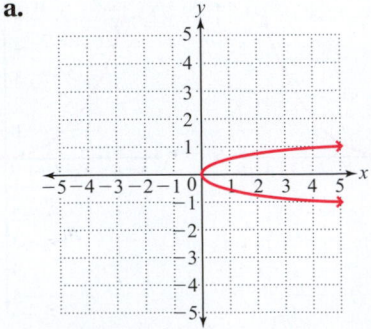

b.

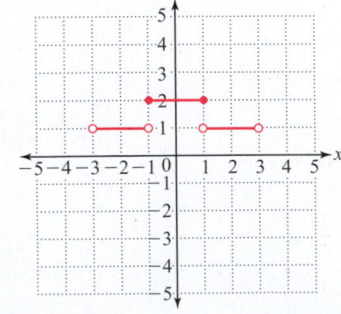

Function Notation

When written as an equation, the notation for a function is a modification of an equation in two variables. Instead of $y = 2x + 5$, in function notation, we would write $f(x) = 2x + 5$ to mean the same equation. Notice that the notation $f(x)$, which is read "a function in terms of x," or "f of x," replaces the variable y. The following information shows how equations in two variables can be written using function notation.

Equation in two variables: Function notation:

$$y = 3x \qquad\qquad f(x) = 3x$$
$$y = -4x + 1 \qquad\qquad f(x) = -4x + 1$$
$$y = \frac{2}{3}x - 5 \qquad\qquad f(x) = \frac{2}{3}x - 5$$

Warning: The notation $f(x)$ does *not* mean multiply f and x.

Finding the Value of a Function

Remember that when we found solutions to equations with y isolated, we often spoke of the x-value as an input value and the corresponding y-value as the output value. Finding the value of a function is the same as finding a solution to an equation in two variables: We input a value for x and calculate the output value, $f(x)$. The following tables list the same ordered pairs. The table on the left is for $y = 2x + 5$, and the table on the right is for $f(x) = 2x + 5$. Notice that the only difference is in labeling the output values.

(input) x	(output) y	(input) x	(output) $f(x)$
0	$2(0) + 5 = 5$	0	$2(0) + 5 = 5$
1	$2(1) + 5 = 7$	1	$2(1) + 5 = 7$
2	$2(2) + 5 = 9$	2	$2(2) + 5 = 9$

Function notation offers a clever way to indicate that a specific x-value is to be used. For example, given a function $f(x) = 3x$, the notation $f(2)$ means to find the value of the function where $x = 2$ so that $f(2) = 3(2) = 6$, which is another way of representing the ordered pair $(2, 6)$.

Procedure **Finding the Value of a Function**

Given a function $f(x)$, to find $f(a)$, where a is a real number in the domain of f, replace x in the function with a and calculate the value.

Example 5 For the function $f(x) = \dfrac{4}{x - 3}$, find the following.

a. $f(2)$

SOLUTION $f(2) = \dfrac{4}{2 - 3}$ Replace x in $\dfrac{4}{x-3}$ with 2; then calculate.

$$= \frac{4}{-1}$$

$$= -4$$

b. $f(-5)$

SOLUTION $f(-5) = \dfrac{4}{-5-3}$ Replace x in $\dfrac{4}{x-3}$ with -5; then calculate.

$$= \dfrac{4}{-8}$$

$$= -\dfrac{1}{2}$$

c. $f(3)$

SOLUTION $f(3) = \dfrac{4}{3-3}$ Replace x in $\dfrac{4}{x-3}$ with 3; then calculate.

$$= \dfrac{4}{0}$$ Therefore, $f(3)$ is undefined.

Your Turn 5   For the function $\sqrt{x+4}$, find the following.

   a. $f(5)$ **b.** $f(3)$ **c.** $f(-6)$

Finding the Value of a Function Given Its Graph

We can also find the value of a function given its graph. For a given value in the domain (x-value), we look on the graph for the corresponding value in the range (y-value).

Example 6   Using the graph, find the value of the function.

a. $f(0)$

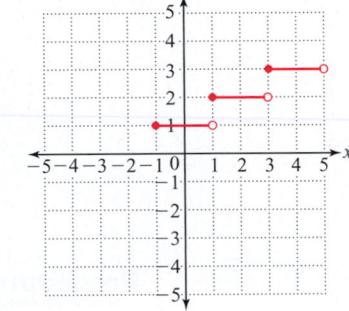

SOLUTION The notation $f(0)$ means to find the value of the function (y-value), where $x = 0$. On the graph, we see that when $x = 0$, the corresponding y-value is 1; so we say that $f(0) = 1$.

b. $f(1)$

SOLUTION When $x = 1$, we see that $y = 2$; so $f(1) = 2$.

c. $f(-4)$

SOLUTION Remember from Example 4(c) that the domain is $\{x \mid -1 \le x < 5\}$ or $[-1, 5)$. Because the value -4 is not in the domain, we say that $f(-4)$ is undefined.

Your Turn 6   Use the graph in the margin to find the value of the function.

   a. $f(-1)$ **b.** $f(0)$ **c.** $f(5)$ **d.** $f(-3)$

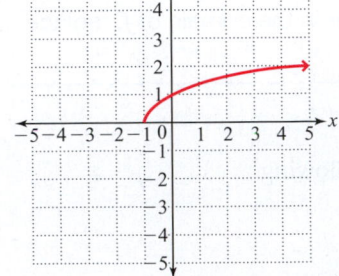

Objective 4   **Graph linear functions.** We create the graph of a function the same way we create the graph of an equation in two variables. In this chapter, we graph only linear functions. Linear functions have a form similar to linear equations in slope–intercept form.

     Slope–intercept form: $y = mx + b$

     Linear function: $f(x) = mx + b$

Example 7   Graph: $f(x) = -2x + 3$

SOLUTION   Think of the function as the equation $y = -2x + 3$. We can make a table of ordered pairs or use the fact that the slope is -2 and the y-intercept is 3.

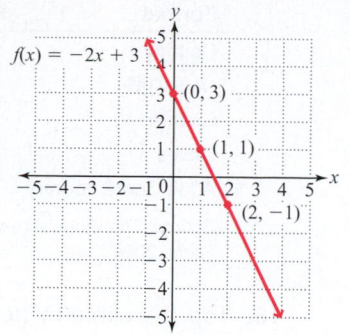

Table of ordered pairs

x	$f(x)$
0	3
1	1
2	-1

Answer   **Your Turn 7**

Your Turn 7   Graph $f(x) = 3x + 1$.

4.7 Exercises   For Extra Help   *MyMathLab* |

Note: Exercises marked with a ★ represent challenging exercises.

1. Explain domain of a relation in your own words.

2. Explain range of a relation in your own words.

3. Given a graph of a relation, how do you determine the domain?

4. Given a graph of a relation, how do you determine the range?

5. When using the vertical line test on a graph, how do you determine whether the graph is the graph of a function?

6. True or false: All relations are functions. Explain.

For Exercises 7–12, determine the domain and range of the relation. See Example 1.

7. $\{(2, 1), (3, -2), (1, 4), (-1, -1)\}$

8. $\{(-2, 1), (4, 3), (2, 5), (-3, -2)\}$

9. $\{(4, 1), (2, 1), (3, -5), (-4, 0)\}$

10. $\{(-2, 1), (8, 11), (-3, 5), (4, -4)\}$

11. $\{(0, 0), (-3, -3), (2, 9), (8, -16)\}$

12. $\{(5, 4), (-3, 8), (10, -2), (0, -5)\}$

For Exercises 13–20, determine whether the relation is a function. In each table, the domain is in the left column and the range is in the right column. See Example 3.

13. Average college tuition in the United States for 2005–2006:

Sector	Cost ($)
Two-year public	1935
Two-year private	11,180
Four-year public	5351
Four-year private	19,292

Source: National Center for Education Statistics, U.S. Department of Education

14. The number of U.S. radio stations by format in 2007:

Format	Radio Stations
Country	2034
News/Talk	1370
Spanish	777
Oldies	711
Adult Contemporary	661

Source: The World Almanac and Book of Facts, 2008

15. Results of the 2007 Masters golf tournament:

Position	Name
1st	Zach Johnson
2nd	Rory Sabbatini; Retief Goosen, Tiger Woods
5th	Jerry Kelly, Justin Rose
7th	Padrig Harrington, Stuart Appleby

(*Source:* www.masters.org)

16. Top-five cable TV networks in 2006:

Rank	Channel
1	Discovery Channel
2	ESPN and CNN
3	TNT, LIFE, and USA
4	Weather
5	Nickelodeon and History

Source: The World Almanac and Book of Facts, 2007

17. Highest average second-quarter home prices by metropolitan area in 2007:

Metropolitan Area	Average Price ($)
San Jose-Sunnyvale-Santa Clara	865,000
San Francisco-Oakland-Fremont	846,800
Anaheim-Santa Ana	727,000
Honolulu	665,000
San Diego-Carlsbad-San Marcos	614,000
Los Angeles-Long Beach-Santa Ana	593,000
New York-Wayne-White Plains	557,500

Source: National Association of Realtors

18. Top brands of ice cream by total sales:

Brand	Sales ($)
Private Label	846,152,600
Breyers	618,797,200
Dreyer's Edy's Grand	426,366,100
Dreyer's Edy's Slow-Churned	342,514,200
Haagen-Dazs	290,367,400

Source: Information Resources Inc.

19. Average high temperature for Orlando, Florida:

Average High (°F)	Month
92	July, August
91	June
89	September
88	May
84	October
83	April
77	March, November
73	February, December
72	January

20. Average rainfall for Seattle:

Avg. Rainfall (inches)	Month
6	January, December
4	February, March
3	October, November
2	April, May, September
1	June, July, August

For Exercises 21–36, determine whether the relation is a function and explain your answer. See Examples 3 and 4.

21. $\{(2,7),(3,4),(-1,4),(0,5)\}$

22. $\{(1,2),(4,-7),(9,5),(6,14)\}$

23. $\{(8,2),(4,-5),(3,0),(8,-1)\}$

24. $\{(-1,5),(2,6),(2,9),(-3,-2)\}$

25. The grades on a recent calculus test:

Student	Score
Jonathon	54
Nathan	79
Rebecca	84
Terri	97
Allen	97

26. The sources of distraction among inattentive drivers:

Distraction	Percentage of Drivers
Outside person, object, or event	29.4
Adjusting radio/CD	11.4
Other occupant	10.9
Using/dialing cell phone	1.5
Smoking related	0.9

Source: AAA Foundation for Traffic Safety/UNC Safety Research Center

27. Ten most home runs hit in a single season:

Player	Home Runs
Barry Bonds	73
Mark McGwire	70
Sammy Sosa	66
Mark McGwire	65
Sammy Sosa	64
Sammy Sosa	63
Roger Maris	61
Babe Ruth	60
Babe Ruth	59
Mark McGwire	58

28. Number of NASCAR championships (3 or more):

Number	Driver
7	Richard Perry
	Dale Earnhardt
4	Jeff Gordon
3	David Pearson
	Darrell Waltrip
	Cale Yarborough
	Lee Petty
	Jimmie Johnson

29.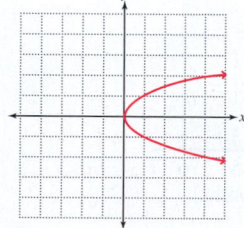

30.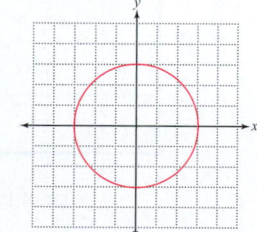

31.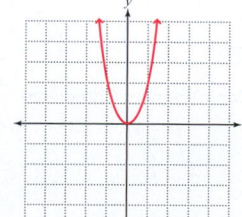

32.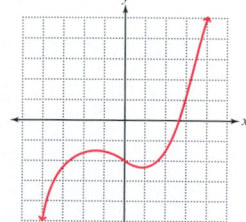

33.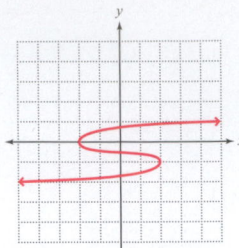

34.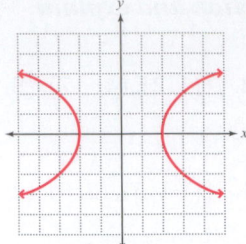

35.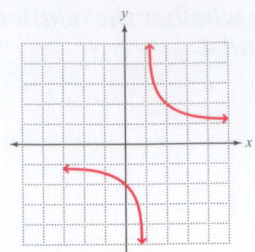

36.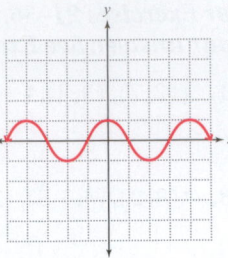

For Exercises 37–46, determine the domain and range. Then state whether the relation is a function. See Examples 2 and 4.

37. This graph shows the number of Catholics in the United States from 1900 to 2005.

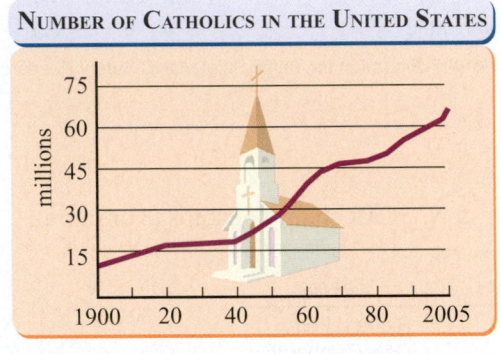

Source: BBC, 2005

38. The line graph represents postage costs for first-class letters from 1991 to 2009.

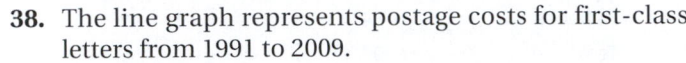
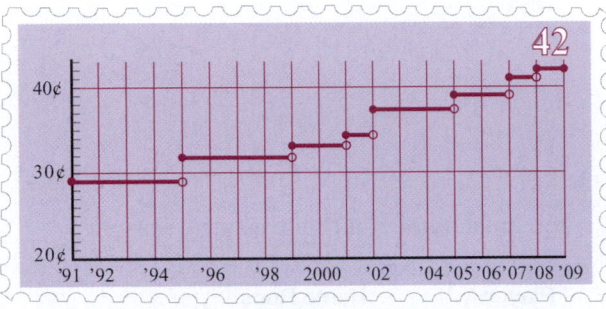

Source: Postal Rate Commission

39. This graph shows the percentage of the population that was foreign-born at each census.

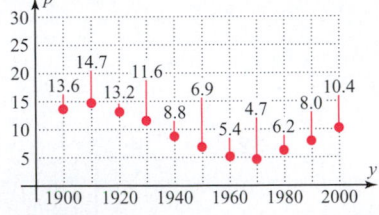

Source: Bureau of the Census, U.S. Department of Commerce

40. This graph shows the total number of people enrolled in a program each month after the program began.

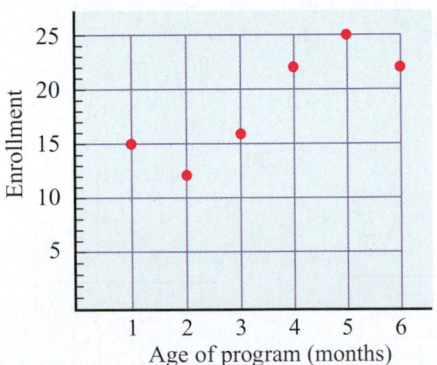

41.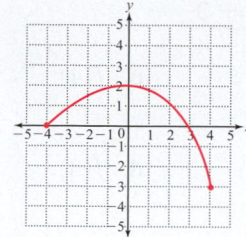

42.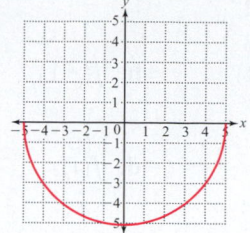

43.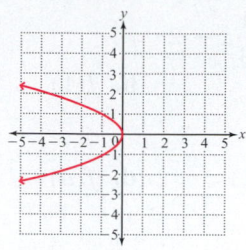

44.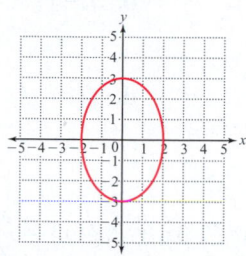

45.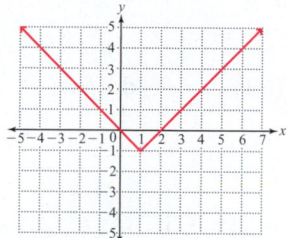

46.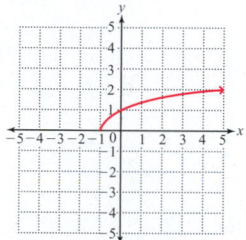

For Exercises 47–70, find the value of the function. See Example 5.

47. $f(x) = 2x - 5$
   a. $f(0)$ **b.** $f(1)$

   c. $f(-2)$ **d.** $f\left(\dfrac{1}{2}\right)$

48. $f(x) = 5x + 3$
   a. $f(1)$ **b.** $f(0)$

   c. $f(-1)$ **d.** $f(t)$

49. $f(x) = x^2 - 3x + 7$
   a. $f(0)$ **b.** $f(1)$

   c. $f(-1)$ **d.** $f(-2)$

50. $f(x) = 3x^2 + 2x - 1$
   a. $f(-2)$ **b.** $f(0.2)$

   c. $f(a)$ **d.** $f(3)$

51. $f(x) = \sqrt{x + 3}$
   a. $f(0)$ **b.** $f(-4)$

   c. $f(-1)$ **d.** $f(a)$

52. $f(x) = \sqrt{x + 5}$
   a. $f(-2)$ **b.** $f(6)$

   c. $f(-5)$ **d.** $f(a)$

53. $f(x) = \sqrt{x^2 - 4}$
   a. $f(2)$ **b.** $f(1)$

   c. $f(-2)$ **d.** $f(3)$

54. $f(x) = \sqrt{x^2 + 2x}$
   a. $f(-2)$ **b.** $f(-1)$

   c. $f(0)$ **d.** $f(1)$

55. $f(x) = \dfrac{1}{x - 1}$
   a. $f(0)$ **b.** $f(1)$

   c. $f(-1)$ **d.** $f(a)$

56. $f(x) = \dfrac{2}{x + 2}$
   a. $f(-2)$ **b.** $f(-1)$

   c. $f(3)$ **d.** $f(a)$

57. $f(x) = x^2 - 2x - 3$
   a. $f(0)$ **b.** $f(0.1)$

   c. $f(a)$ **d.** $f(-a)$

58. $f(x) = 2x^2 + x - 3$
   a. $f\left(\dfrac{3}{4}\right)$ **b.** $f(-1)$

   c. $f(a)$ **d.** $f(-a)$

59. $f(x) = \dfrac{1}{3}x - 2$

   a. $f(0)$ **b.** $f(1)$

   c. $f(-1)$ **d.** $f(a)$

60. $f(x) = \dfrac{2}{3}x + 1$

   a. $f(-3)$ **b.** $f(-1)$

   c. $f(3)$ **d.** $f(t)$

61. $f(x) = \dfrac{x + 3}{x - 4}$

   a. $f(0)$ **b.** $f(4)$

   c. $f(-2)$ **d.** $f(3)$

62. $f(x) = \dfrac{9 - x}{x + 2}$

   a. $f(0)$ **b.** $f(-2)$

   c. $f(3)$ **d.** $f(-1)$

63. $f(x) = \dfrac{1}{\sqrt{x + 2}}$

   a. $f(-2)$ **b.** $f(1)$

   c. $f(-3)$ **d.** $f(a)$

64. $f(x) = \dfrac{2}{\sqrt{2 - x}}$

   a. $f(-2)$ **b.** $f(-1)$

   c. $f(3)$ **d.** $f(a)$

65. $f(x) = |4x - 2|$

   a. $f(0)$ **b.** $f(1.5)$

   c. $f(-1)$ **d.** $f(-2)$

66. $f(x) = |3x - 5|$

   a. $f(-2)$ **b.** $f(-1)$

   c. $f(2)$ **d.** $f(3)$

67. $f(x) = x^2 - 3x + 2, g(x) = \dfrac{x - 4}{x + 2}$

   a. $f(2)$ **b.** $g(2)$

   c. $f(-3)$ **d.** $g(-3)$

68. $f(x) = \sqrt{x^2 - 3x}, g(x) = \dfrac{2}{x + 4}$

   a. $g(-3)$ **b.** $f(-1)$

   c. $g(6)$ **d.** $f(4)$

★**69.** $f(x) = 2x^2 + 3x + 1$

   a. $f(2a)$ **b.** $f(-3a)$

   c. $f\left(\dfrac{1}{2}a\right)$ **d.** $f(0.1a)$

★**70.** $f(x) = 3x^2 - 2x - 4$

   a. $f(3a)$ **b.** $f(-2a)$

   c. $f\left(\dfrac{2}{3}a\right)$ **d.** $f(0.2a)$

For Exercises 71–74, use the given graph to find the value of the function. See Example 6.

71.
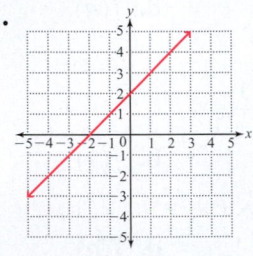

   a. $f(1)$

   b. $f(2)$

   c. $f(0)$

72.
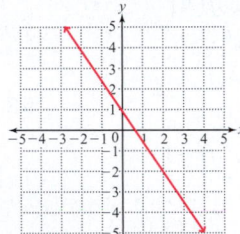

   a. $f(4)$

   b. $f(2)$

   c. $f(0)$

73.
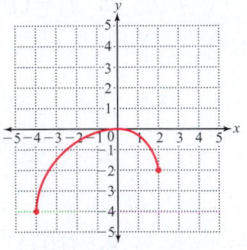

   a. $f(0)$

   b. $f(2)$

   c. $f(3)$

74.
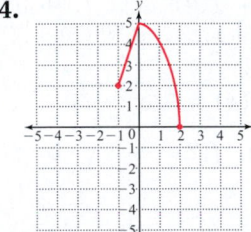

   a. $f(0)$

   b. $f(2)$

   c. $f(-2)$

For Exercises 75–84, graph. See Example 7.

75. $f(x) = 2x + 1$

76. $f(x) = 3x - 5$

77. $f(x) = \frac{1}{3}x + 3$

78. $f(x) = \frac{1}{2}x - 3$

79. $f(x) = -4x + 1$

80. $f(x) = -x - 1$

81. $f(x) = -\frac{2}{3}x$

82. $f(x) = \frac{2}{5}x$

83. $f(x) = -\frac{1}{4}x - 2$

84. $f(x) = -\frac{1}{3}x + 4$

85. The cost, $C(x)$, as a function of the number of items produced, x, is given by $C(x) = 20x + 150$.
  a. Find the cost of producing ten items.

  b. Find the cost of producing fifteen items.

  c. What does $C(30) = 750$ mean?

86. For a rental car, the cost, $C(x)$, as a function of the number of miles driven, x, is given by $C(x) = 0.12x + 22$.
  a. Find the cost of driving 100 miles.

  b. Find the cost of driving 350 miles.

  c. What does $C(250) = 52$ mean?

87. The value, $V(x)$, of a car as a function of the number of years after it was purchased, x, is given by $V(x) = -1500x + 18,000$.
  a. Find the value of the car after three years.

  b. Find the value of the car after six years.

  c. What does $V(10) = 3000$ mean?

88. The monthly salary, $S(x)$, of a salesperson as a function of her sales, x, is given by $S(x) = 0.10x + 175$.
  a. Find her salary for sales of $15,000.

  b. Find her salary for sales of $10,000.

  c. What does $S(20,000) = 2175$ mean?

Review Exercises

For Exercises 1 and 2, simplify.

[1.5]   1. $\dfrac{2}{5} - \dfrac{5}{8} \div \left(\dfrac{3}{4}\right)$

[1.7]   2. $x + 3y - 5x + 7 - 3y - 9$

[1.3]   3. Which property is illustrated in the equation $4(x + 9) = 4(9 + x)$?

[4.3]   4. Graph $2x - y = 4$ using the intercepts.

[4.4]   5. Solve $3x - 2y = 6$ for y.

[4.5]   6. Are the graphs of $2x + 3y = 9$ and $4x + 6y = 12$ parallel, perpendicular, or neither?

Chapter 4 Summary

Defined Terms

Section 4.3
x-intercept (p. 261)
y-intercept (p. 261)

Section 4.4
Slope (p. 273)

Section 4.7
Relation (p. 303)
Domain (p. 303)
Range (p. 303)
Function (p. 305)

Forms of a Linear Equation

Slope–intercept form: $y = mx + b$ (p. 273)

Point–slope form: $y - y_1 = m(x - x_1)$ (p. 284)

Standard form: $Ax + By = C$, where A, B, and C are integers and $A > 0$ (p. 286)

Formula

Slope: $m = \dfrac{y_2 - y_1}{x_2 - x_1}, x_2 \neq x_1$ (p. 276)

Procedures, Rules, and Key Examples

Procedures/Rules	Key Examples

SECTION 4.1 The Rectangular Coordinate System

To determine the coordinates of a given point in the rectangular system,
1. Follow a vertical line from the point to the x-axis (horizontal axis). The number at this position on the x-axis is the first coordinate.
2. Follow a horizontal line from the point to the y-axis (vertical axis). The number at this position on the y-axis is the second coordinate.

To graph or plot a point given its coordinates,
1. Beginning at the origin, $(0, 0)$, move to the right or left along the x-axis the amount indicated by the first coordinate.
2. From that position on the x-axis, move up or down the amount indicated by the second coordinate.
3. Draw a dot to represent the point described by the coordinates.

To determine the quadrant for a given ordered pair, consider the signs of the coordinates.

$(+, +)$ means the point is in quadrant I.
$(-, +)$ means the point is in quadrant II.
$(-, -)$ means the point is in quadrant III.
$(+, -)$ means the point is in quadrant IV.

Note: If either coordinate is 0, the point is on an axis and not in a quadrant.

Example 1: Plot each of the following:
$(-2, 4), (5, 0), (0, -2), (-4, -3), (3, -4), (2, 3)$

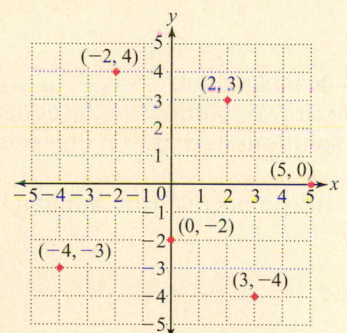

Example 2: State the quadrant in which each point is located. If the point lies on an axis, state which axis.

$(19, 78)$:	quadrant I
$(-67, 45)$:	quadrant II
$(-107, -36)$:	quadrant III
$(58, -92)$:	quadrant IV
$(0, -16)$:	This point is on the y-axis.

(continued)

Procedures/Rules	Key Examples

SECTION 4.2 Graphing Linear Equations

To determine whether a given ordered pair is a solution for an equation with two variables,
1. Replace the variables in the equation with the corresponding coordinates.
2. Verify that the equation is true.

Example 1: Determine whether $(3, 5)$ is a solution for $4x - y = 7$.

Solution: $4(3) - 5 \overset{?}{=} 7$ **Replace x with 3 and y with 5.**

$$12 - 5 \overset{?}{=} 7$$
$$7 = 7$$

$(3, 5)$ is a solution.

Example 2: Determine whether $(4, -9)$ is a solution for $4x - y = 7$.

Solution: $4(4) - (-9) \overset{?}{=} 7$ **Replace x with 4 and y with -9.**

$$16 + 9 \overset{?}{=} 7$$
$$25 \neq 7$$

$(4, -9)$ is not a solution.

To find a solution to a linear equation with two variables,
1. Choose a value for one of the variables (any value).
2. Replace the corresponding variable with your chosen value.
3. Solve the equation for the value of the other variable.

Example 3: Find two solutions for the equation $y = 2x - 3$.

First solution:
Let $x = 0$

Second solution:
Let $x = 1$

$y = 2(0) - 3$
$y = -3$

$y = 2(1) - 3$
$y = 2 - 3$
$y = -1$

Solution: $(0, -3)$ Solution: $(1, -1)$

To graph a linear equation,
1. Find at least two solutions to the equation.
2. Plot the solutions as points in the rectangular coordinate system.
3. Connect the points to form a straight line.

Example 4: Graph $y = 2x - 3$.
We found two solutions above: $(0, -3)$ and $(1, -1)$.

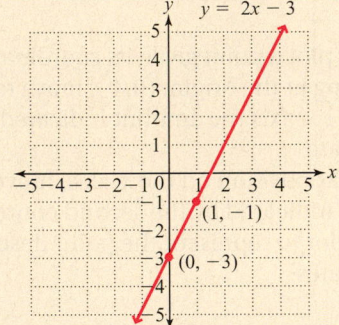

The graph of $y = c$, where c is a real-number constant, is a horizontal line parallel to the x-axis that passes through the y-axis at a point with coordinates $(0, c)$.

The graph of $x = c$, where c is a real-number constant, is a vertical line parallel to the y-axis that passes through the x-axis at a point with coordinates $(c, 0)$.

Example 5: Graph **a.** $y = 4$
                  b. $x = -3$

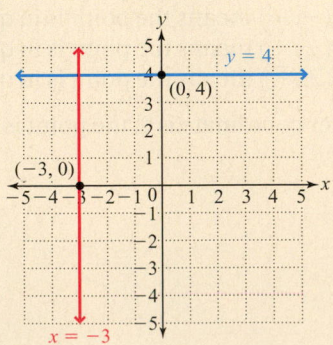

(continued)

SECTION 4.3 Graphing Using Intercepts

To find an x-intercept,
1. Replace y with 0 in the given equation.
2. Solve for x.

To find a y-intercept,
1. Replace x with 0 in the given equation.
2. Solve for y.

Example 1: Find the x- and y-intercepts for $2x - 3y = 18$.

Solution:

x-intercept:	y-intercept:
$2x - 3(0) = 18$	$2(0) - 3y = 18$
$2x = 18$	$-3y = 18$
$x = 9$	$y = -6$

x-intercept: $(9, 0)$ y-intercept: $(0, -6)$

If an equation can be written in the form $y = mx$, where m is a real number other than 0, then the x- and y-intercepts are at the origin, $(0, 0)$.

Example 2: The graph of the equation $y = 3x$ has $(0, 0)$ as both the x- and y-intercept.

If an equation is in the form $y = mx + b$, where m and b are nonzero real numbers, then the y-intercept will be $(0, b)$.

Example 3: The y-intercept of the graph of $y = 3x - 2$ is $(0, -2)$.

The graph of an equation in the form $y = c$, where c is a nonzero real number, has no x-intercept and the y-intercept is $(0, c)$. The graph of $y = 0$ is the x-axis.

Example 4: The graph of the equation $y = 4$ has no x-intercept, and its y-intercept is $(0, 4)$.

The graph of an equation in the form $x = c$, where c is a nonzero real number, has no y-intercept and the x-intercept is $(c, 0)$. The graph of $x = 0$ is the y-axis.

Example 5: The graph of the equation $x = -3$ has no y-intercept, and its x-intercept is $(-3, 0)$.

SECTION 4.4 Slope–Intercept Form

Given an equation of the form $y = mx + b$,

If $m > 0$ (slope is positive) the graph is a line that slants uphill from left to right.

If $m < 0$ (slope is negative) the graph is a line that slants downhill from left to right.

To graph an equation in slope–intercept form, $y = mx + b$,
1. Plot the y-intercept, $(0, b)$.
2. Plot a second point by rising the number of units indicated by the numerator of the slope, m, then running the number of units indicated by the denominator of the slope, m.
3. Draw a straight line through the two points.

Note: You can check by locating additional points using the slope. Every point you locate using the slope should be on the line.

Example 1: Graph.
a. $y = 2x + 1$ **b.** $y = -\dfrac{3}{4}x - 2$

Solution: For $y = 2x + 1$, the slope is 2 and the y-intercept is $(0, 1)$. The positive slope indicates an uphill line from left to right. To graph, we can rise 2, then run 1 from the y-intercept, $(0, 1)$, to get a second point $(1, 3)$.

For $y = -\dfrac{3}{4}x - 2$, the slope is $-\dfrac{3}{4}$ and the y-intercept is $(0, -2)$. The negative slope indicates a downhill line from left to right. To graph, we rise -3, then run 4 from the y-intercept, $(0, -2)$, to get a second point $(4, -5)$.

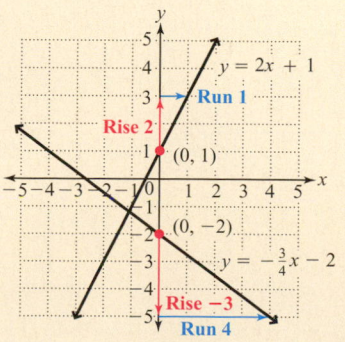

(*continued*)

Procedures/Rules	Key Examples

SECTION 4.4 Slope–Intercept Form (*continued*)

To write the equation of a line given its slope, m, and its y-intercept, $(0, b)$, use the slope–intercept form of the equation, $y = mx + b$.

Example 2: Write the slope–intercept form of the equation of a line with slope $\frac{3}{4}$ and y-intercept $(0, -1)$.

Solution: $y = \frac{3}{4}x - 1$

Given two points (x_1, y_1) and (x_2, y_2) where $x_2 \neq x_1$, the slope of the line connecting the two points is given by the formula

$$m = \frac{y_2 - y_1}{x_2 - x_1}$$

Example 3: Find the slope of a line connecting the points $(2, 7)$ and $(-8, 3)$.

Solution: $m = \dfrac{3 - 7}{-8 - 2} = \dfrac{-4}{-10} = \dfrac{2}{5}$

Two points with different x-coordinates and the same y-coordinates, (x_1, c) and (x_2, c), will form a line with a slope of 0 (a horizontal line) and the equation $y = c$.

Example 4: The slope of a line connecting the points $(2, 4)$ and $(1, 4)$ is 0, and the equation of the line is $y = 4$.

Two points with the same x-coordinates and different y-coordinates, (c, y_1) and (c, y_2), will form a line with a slope that is undefined (a vertical line) and the equation $x = c$.

Example 5: The slope of a line connecting the points $(-3, 7)$ and $(-3, -2)$ is undefined and the equation of the line is $x = -3$.

SECTION 4.5 Point–Slope Form

To write the equation of a line given its slope and any point, (x_1, y_1), on the line, use the point–slope form of the equation of a line, $y - y_1 = m(x - x_1)$. If given a second point, (x_2, y_2), and not the slope, we first calculate the slope using $m = \dfrac{y_2 - y_1}{x_2 - x_1}$; then we use $y - y_1 = m(x - x_1)$.

Example 1: Write the slope–intercept form of the equation of a line with slope -3 that passes through $(2, 5)$.

Solution:
$$\begin{aligned}
y - y_1 &= m(x - x_1) \\
y - 5 &= -3(x - 2) \qquad x_1 = 2, y_1 = 5 \\
y - 5 &= -3x + 6 \\
y &= -3x + 11
\end{aligned}$$

Example 2: Write the equation of a line connecting $(2, 3)$ and $(-3, -1)$ in the form $Ax + By = C$, where A, B, and C are integers and $A > 0$.

Solution: Find the slope:

$$m = \frac{-1 - 3}{-3 - 2} = \frac{-4}{-5} = \frac{4}{5}$$

Write the equation: $y - 3 = \dfrac{4}{5}(x - 2)$

$5(y - 3) = 5 \cdot \dfrac{4}{5}(x - 2)$ { Multiply by 5 to clear the fraction.

$5y - 15 = 4(x - 2)$

$5y - 15 = 4x - 8$ Distribute 4.

$5y - 4x - 15 = -8$ Subtract $4x$ from both sides.

$5y - 4x = 7$ Add 15 to both sides.

$-4x + 5y = 7$ Rearrange so that the $-4x$ term is first.

$-1(-4x + 5y) = -1 \cdot 7$ { Multiply by -1 so that the x term is positive.

$4x - 5y = -7$ Simplify.

(*continued*)

Procedures/Rules	Key Examples

SECTION 4.5 Point–Slope Form (*continued*)

Nonvertical parallel lines have equal slopes and different y-intercepts. Vertical lines are parallel.

The slope of a line perpendicular to a line with a slope of $\dfrac{a}{b}$ will be $-\dfrac{b}{a}$. Horizontal and vertical lines are perpendicular.

Example 3: Find the slope of a line parallel and perpendicular to $3x + 2y = 4$.

Solution: Write $3x + 2y = 4$ in slope–intercept form.

$$2y = -3x + 4 \qquad \text{Subtract } 3x \text{ from both sides.}$$
$$y = -\frac{3}{2}x + 2 \qquad \text{Divide both sides by 2.}$$

The slope of $3x + 2y = 4$ is $-\dfrac{3}{2}$.

The slope of a line parallel to $3x + 2y = 4$ is $-\dfrac{3}{2}$. The slope of a line perpendicular to $3x + 2y = 4$ is $\dfrac{2}{3}$.

SECTION 4.6 Graphing Linear Inequalities

To determine whether an ordered pair is a solution for an inequality, replace the variables with the corresponding coordinates and see if the resulting inequality is true. If it is, the ordered pair is a solution.

Example 1: Determine whether $(4, 3)$ is a solution for $2x - 4y \leq 8$.

$$2x - 4y \leq 8$$
$$2(4) - 4(3) \overset{?}{\leq} 8 \qquad x = 4, y = 3$$
$$8 - 12 \overset{?}{\leq} 8$$
$$-4 \leq 8 \qquad \text{This is true, so } (4, 3) \text{ is a solution.}$$

Example 2: Determine whether $(3, -1)$ is a solution for $2x - 4y \leq 8$.

$$2x - 4y \leq 8$$
$$2(3) - 4(-1) \overset{?}{\leq} 8 \qquad x = 3, y = -1$$
$$6 + 4 \overset{?}{\leq} 8$$
$$10 \leq 8 \qquad \text{This is false, so } (3, -1) \text{ is not a solution.}$$

To graph a linear inequality in two variables,
1. Graph the related equation (the boundary line). The related equation has an equal sign in place of the inequality symbol. If the inequality symbol is $\leq$ or $\geq$, draw a solid line. If the inequality symbol is $<$ or $>$, draw a dashed line.
2. Choose an ordered pair on one side of the boundary line and test this ordered pair in the inequality. If the ordered pair satisfies the inequality, shade the region that contains it. If the ordered pair does not satisfy the inequality, shade the region on the other side of the boundary line.

Example 3: Graph $2x - 4y \leq 8$.

Solution: Graph $2x - 4y = 8$ as a solid line. We found $(0, 0)$ to be a solution; therefore, shade the region containing $(0, 0)$.

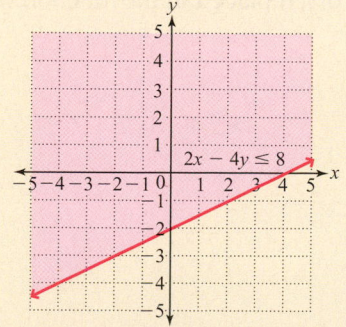

(*continued*)

SECTION 4.7 Introduction to Functions and Function Notation

To determine the domain of a relation given its graph, answer this question: What are all of the *x*-values that have a corresponding *y*-value?

To determine the range, answer this question: What are all of the *y*-values that have a corresponding *x*-value?

Example 1: Give the domain and range of the relation shown.

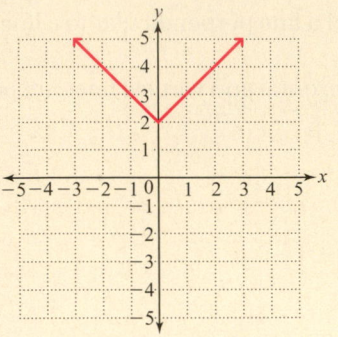

Domain: all real numbers

Range: $\{ y \mid y \geq 2 \}$ or $[2, \infty)$

To determine whether a graph of a relation is a graph of a function, perform a vertical line test.
1. Draw or imagine vertical lines through each point in the domain.
2. If each vertical line intersects the graph at only one point, the graph is the graph of a function.
3. If any vertical line intersects the graph at two or more different points, the graph is not the graph of a function.

Example 2: Determine whether the graph of the relation is the graph of a function.

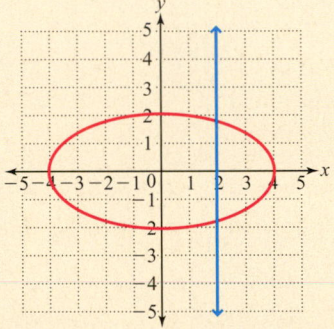

Solution: The relation in the graph is not a function because a vertical line can be drawn that intersects the graph at two points. This means that there is a domain value that corresponds to two values in the range.

Given a function $f(x)$, to find $f(a)$, where a is a real number in the domain of f, replace x in the function with a and calculate the value.

Example 3: For the function $f(x) = x^2 - 3x + 1$, find $f(-2)$.

$$f(-2) = (-2)^2 - 3(-2) + 1$$
$$= 4 + 6 + 1$$
$$= 11$$

Chapter 4 Review Exercises

For Exercises 1–6, answer true or false.

[4.1] **1.** When writing an ordered pair, the horizontal coordinate is written first.

[4.3] **2.** An x-intercept is a point where a graph intersects the horizontal axis.

[4.4] **3.** The slope formula is $m = \dfrac{x_2 - x_1}{y_2 - y_1}$.

[4.6] **4.** A linear inequality has just one solution.

[4.5] **5.** Given any two points on a line, the equation of the line may be found.

[4.5] **6.** Given the slope and any point on a line, the equation of the line may be found.

For Exercises 7–10, complete the rule.

[4.3] **7.** To find an x-intercept,
  a. Replace _____ with 0 in the given equation.
  b. Solve for _____ .

[4.3] **8.** To find a y-intercept,
  a. Replace _____ with 0 in the given equation.
  b. Solve for _____ .

[4.4] **9.** If an equation is in the form $y = mx + b$, where m and b are nonzero real numbers, the _____ will be $(0, b)$.

[4.4] **10.** Given an equation of the form $y = mx + b$,
  if $m > 0$, the graph will be a line that slants _____ from left to right.
  if $m < 0$, the graph will be a line that slants _____ from left to right.

[4.1] **For Exercises 11 and 12, write the coordinates for each point.**

11.

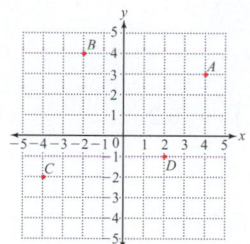

12.

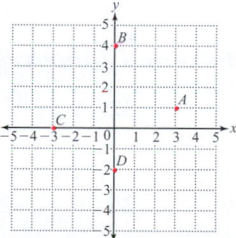

[4.1] **For Exercises 13 and 14, plot and label the points indicated by the coordinate pairs.**

13. $(3, 4), (-2, 0), (0, 1), (-1, -5)$

14. $(-2, 2), (0, -3), (4, 1), (5, 0)$

[4.1] *For Exercises 15–18, state the quadrant in which the point is located. If the point lies on an axis, state which axis.*

15. $(-2.8, 1203)$ **16.** $\left(-\dfrac{1}{8}, -16\right)$ **17.** $(421, 5300)$ **18.** $\left(52\dfrac{2}{3}, -0.36\right)$

Exercises 19–85   EQUATIONS AND INEQUALITIES

[4.2] *For Exercises 19–24, determine whether the given ordered pair is a solution for the equation.*

19. $(-1, 3); x + 3y = 8$ **20.** $(2, 1); 2x - y = 3$ **21.** $(0, 0); 6x = 2y + 1$

22. $(-2, -2); y = \dfrac{2}{3}x$ **23.** $\left(\dfrac{3}{2}, -\dfrac{1}{2}\right); x + y = 2$ **24.** $(2.1, 3.2); y - 3x = 5$

[4.2] *For Exercises 25–30, find three solutions for the given equation. Then graph.*

25. $y = -2x$ **26.** $x = y$ **27.** $y = -x + 7$

28. $x + 4y = 5$ **29.** $2x + 3y = 6$ **30.** $y + \dfrac{1}{3}x = 3$

[4.3] *For Exercises 31–36, find the x- and y-intercepts.*

31. $x + 2y = 12$ **32.** $y = -3x$ **33.** $2x - 3.1y = 6.2$

34. $y = \dfrac{2}{3}x + 5$ **35.** $x = -2$ **36.** $y = 7$

[4.3] *For Exercises 37–42, graph using the x- and y-intercepts.*

37. $3x + 2y = 6$ **38.** $y = -2x + 1$ **39.** $y = -\dfrac{1}{5}x + 3$

40. $2x = 5y + 10$ **41.** $7 = y$ **42.** $x = -1$

[4.4] *For Exercises 43 and 44, graph each equation on the same grid.*

43. $y = x$
$y = 2x$
$y = 4x$

44. $y = -x$
$y = -2x$
$y = -3x$

[4.4] *For Exercises 45–50, determine the slope and the y-intercept and then graph.*

45. $y = -2x + 3$ **46.** $y = -\dfrac{1}{4}x + 2$ **47.** $y = 3x$

48. $x + y = 5$ **49.** $x - 3y = 7$ **50.** $2x - 3y = 8$

[4.4] *For Exercises 51–56, find the slope of the line connecting the given points.*

51. $(2, 7), (-1, -2)$ **52.** $(0, -1), (3, 2)$ **53.** $(-3, 1), (3, -1)$

54. $(-6, 2), (-1, -1)$ **55.** $(2, 8), (-1, 8)$ **56.** $(-7, 3), (-7, -5)$

[4.4] *For Exercises 57–60, write the equation of the line in slope–intercept form given the slope and the y-intercept.*

57. $m = -1; (0, 7)$ **58.** $m = -\dfrac{1}{5}; (0, -8)$ **59.** $m = 0.2; (0, 6)$ **60.** $m = 1; (0, 0)$

[4.5] *For Exercises 61–66, write the equation of a line in slope–intercept form with the given slope that passes through the given point.*

61. $m = -2; (1, 7)$

62. $m = 1; (3, -6)$

63. $m = \dfrac{1}{3}; (-5, 0)$

64. $m = -\dfrac{2}{5}; (3, 3)$

65. $m = 6.2; (-2, -3)$

66. $m = -0.4; (-1, 2)$

[4.5] *For Exercises 67–70, write the equation of a line connecting the given points in slope–intercept form and in standard form.*

67. $(7, -3), (2, 2)$

68. $(5, -3), (9, 2)$

69. $(3, 2), (5, 1)$

70. $(-3, -2), (-8, -1)$

[4.5] *For Exercises 71–74, write the equations in slope–intercept form.*

71. Find the equation of a line that passes through $(0, -2)$ and is parallel to the line $y = 2x + 7$.

72. Find the equation of a line that passes through $(1, 5)$ and is parallel to the line $2x + 3y = 6$.

73. Find the equation of a line that passes through $(-1, -2)$ and is perpendicular to the line $y = \dfrac{3}{5}x - 1$.

74. Find the equation of a line that passes through $(-2, 4)$ and is perpendicular to the line $y = -\dfrac{2}{5}x - 6$.

[4.2] 75. A salesperson receives $1000 per month plus a commission of 5% of his total sales. The equation $p = 0.05s + 1000$ describes the salesperson's gross pay each month, where p represents the gross pay and s represents the total sales in dollars.
   a. Find the gross pay if the person sells $24,000 worth of merchandise.

   b. Find the point where the graph intersects the p-axis. Explain what it indicates.

   c. Graph the equation.

[4.5] 76. Andrew purchased a photocopier for his business in 2004 for $18,000. In 2008, he considers selling the copier on eBay and notes that others like it are selling for $12,000.
   a. Assuming that the depreciation is linear, plot the two given data points with 2004 being year 0 so that 2008 is year 4. Draw a line connecting the two points.
   b. What is the slope of the line?
   c. Let n represent the number of years the copier is in service and v represent the value of the copier. Write the equation of the line in slope–intercept form.

   d. If the depreciation continues at the same rate, in what year was the copier worth half of its original value?

   e. In what year will the copier be worth $0?

[4.6] *For Exercises 77–80, determine whether the ordered pair is a solution for the linear inequality.*

77. $(3, 2); x + 2y > 5$ **78.** $(-1, 0); y < x + 2$ **79.** $(-5, 8); 2x - 6y \leq 17$ **80.** $(0, 0); x + y \geq 5$

[4.6] *For Exercises 81–85, graph the linear inequality.*

81. $y < 3x - 5$ **82.** $y \geq -\dfrac{2}{3}x$ **83.** $x - y \geq 3$

84. $-3x - 5y < -15$ **85.** $x \geq -1$

[4.6] **86.** A company produces two versions of its product. The lower-priced package costs $6 to make, whereas the higher-priced package costs $8 to make.
   a. Let a represent the number of the lower-priced packages produced and b represent the number of the higher-priced packages produced. Write an inequality in which the total cost is at most $12,000.

   b. Graph the inequality.
   c. Give a combination of the number of each package that the company could produce that has a cost equal to $12,000.

   d. Give a combination of the number of each package that the company could produce that has a cost less than $12,000.

[4.7] *For Exercises 87 and 88, find the domain and range of the relation.*

87. $\{(2, 3), (-1, 5), (3, 6), (-3, -4)\}$ **88.** Cars holding their value for resale:

Automobile	Percentage of Original Value
Volkswagen	52.2
Honda	49.7
Toyota	49.0
Subaru	47.8
Nissan	45.8
Source: Automotive Lease Guide	

[4.7] *For Exercises 89 and 90, determine whether the relation is a function.*

89. The following relation shows the courses taught by each instructor during one semester.

Domain (instructor name)	Range (courses taught)
Hames	Math 100
Carson	Math 100, Math 102
Pritchard	Math 035
Webb	Math 100, Math 110

90. The following relation shows the price of a particular brand of dog food based on the size of the bag.

Domain (size of bag)	Range (price)
5 lb.	$3.95
10 lb.	$7.90
20 lb.	$15.00
40 lb.	$29.95

[4.7] *For Exercises 91–93, give the domain and range. Then determine whether the graph is the graph of a function.*

91.

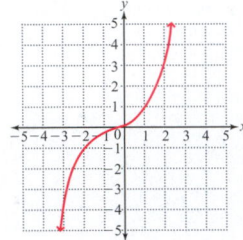

92.

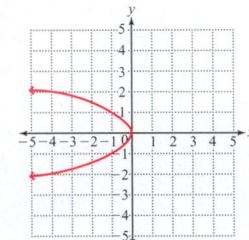

93.

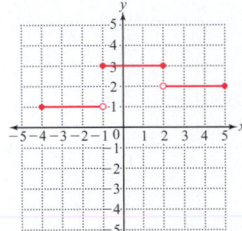

[4.7] 94. Use the graph in Exercise 93 to find the value of the function at the indicated values.

    a. $f(-3)$ **b.** $f(-1)$

    c. $f(0)$ **d.** $f(3)$

[4.7] *For Exercises 95 and 96, find the value of the function.*

95. $f(x) = x^3 - 5x + 2$

    a. $f(2)$ **b.** $f(0)$ **c.** $f(-3)$

96. $f(x) = \dfrac{x}{x - 3}$

    a. $f(6)$ **b.** $f(3)$ **c.** $f(-5)$

Chapter 4 Practice Test

For Extra Help — Step-by-step test solutions are found on the Chapter Test Prep Videos available via the Video Resources on DVD, in **MyMathLab**, and on YouTube (search "CarsonCombinedAlg" and click on "Channels").

1. Determine the coordinates for each point.

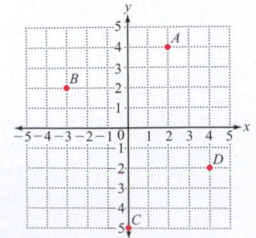

2. Plot and label the points indicated by the coordinate pairs.

$(0, -3), (4, -1), (-3, 2), (2, 0), (-1, -5)$

3. State the quadrant in which $\left(4\frac{2}{3}, 5115\right)$ is located. If the point lies on an axis, state which axis.

4. Determine whether $(1, 7)$ is a solution for $y = -\frac{2}{5}x - 8$.

For Exercises 5–8, find the x- and y-intercepts and then graph.

5. $x - 2y = 4$

6. $2x + 5y = 10$

7. $y = -2x$

8. $y = \frac{3}{4}x - 1$

For Exercises 9 and 10, determine the slope and the y-intercept.

9. $y = \frac{3}{4}x + 11$

10. $5x - 3y = 8$

For Exercises 11 and 12, determine the slope of the line connecting the given points.

11. $(-5, 6), (-2, -4)$

12. $(6, 2), (3, 2)$

For Exercises 13 and 14, write the equation of the line in slope–intercept form.

13. $m = \frac{3}{5}$; y-intercept $(0, 4)$

14. Passing through the points $(8, -1), (-7, 5)$

15. Write the equation of a line in the form $Ax + By = C$ through the points $(2, 1)$ and $(5, 3)$.

16. Write the equation of a line in the form $Ax + By = C$ through the point $(4, -2)$ and perpendicular to the line $y = -3x + 4$.

17. From 1999 to 2006, the number of people traveling more than 50 miles for leisure increased at a linear rate. In 1999, the number of people traveling 50 miles or more for leisure was 848.6 million. In 2006, that number was 990.2 million people. (*Source:* Travel Industry Association of America, TravelScope)
   a. Let 1999 be year 0 so that 2006 is year 7. Let x represent the number of years after 1999 and y represent the number of people traveling 50 miles or more. Plot the two data points in the coordinate plane; then draw a line connecting them.
   b. Find the slope of the line.

c. Write the equation of the line in slope–intercept form.

d. If the trend continues, predict the number of people traveling more than 50 miles for leisure in 2010. Round to the nearest tenth.

18. Determine whether $(2, 7)$ is a solution for the linear inequality $y \geq 2x - 9$.

For Exercises 19 and 20, graph the linear inequality.

19. $y \geq \dfrac{2}{5}x - 1$

20. $x - 3y < 8$

21. Alex works two part-time jobs. He receives $10 per hour when working as a cook in a restaurant and $8 per hour when working at a music store. To pay all of his monthly expenses, he needs to make at least $360 per week.

    a. Let x represent the number of hours he works at the restaurant and y represent the number of hours he works at the music store. Write an inequality in which his total weekly income is at least $360.

    b. Graph the inequality. Because he can work only a positive number of hours, restrict the graph to the first quadrant.

    c. Give a combination of hours that provides him with an income of exactly $360.

    d. Give a combination of hours that provides him with an income of more than $360.

22. The following relation shows U.S. per capita income for each year. Is the relation a function?

Year	Income ($)
2000	29,469
2001	30,413
2002	30,906
2003	31,632
2004	33,050
2005	34,586
2006	36,276

Source: U.S. Department of Commerce

23. Determine whether the graph is the graph of a function.

24. Find the indicated value the function $f(x) = \dfrac{x^2}{x - 4}$.

    a. $f(2)$ **b.** $f(4)$ **c.** $f(-3)$

25. a. Give the domain and range of the function graphed below.

    b. Find $f(1)$.

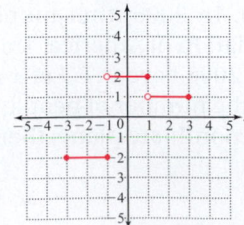

Chapters 1–4 Cumulative Review Exercises

For Exercises 1–4, answer true or false.

[1.1] **1.** π is a rational number.

[4.4] **2.** $y = mx + b$ is the equation of a line in slope–intercept form.

[1.5] **3.** $6° = 6$

[1.5] **4.** $\sqrt{-36} = -6$

For Exercises 5 and 6, fill in the blank.

[2.2] **5.** An equation that has every real number for a solution is a(n) _____.

[3.2] **6.** To write a percent as a fraction or decimal,
    a. Write the percent as a ratio with _____ in the denominator.
    b. Simplify to the desired form.

[3.2] **7.** Write 16.5% as a fraction.

Exercises 8–14   EXPRESSIONS

[2.3] **8.** Explain in your own words how to clear decimals from an equation.

[1.5] For Exercises 9–14, simplify.

9. $(-2)^3 - 3\sqrt{9} + 16$

10. $\{4 - 2[6 + (-10)]\} + 3$

11. $-|-12 - (-2)(-3)|$

12. -3^4

13. $-5\sqrt{25} + 20 \div (-4) \cdot 2 - 6$

14. $4^2 - 7 \cdot 3 + \sqrt{144} - 25 \div (-5)$

Exercises 15–30   EQUATIONS AND INEQUALITIES

[4.3] **15.** Find the x- and y-intercepts for $2x - 3y = 6$.

[4.2] **16.** Find three solutions for $y = -3x + 4$.

[4.3] **17.** Graph $2y - 3x = 9$.

[4.4] **18.** Determine the slope of the line $x + 3y = 8$.

[4.5] **19.** Write the equation of a line through the points $(5, 3)$ and $(-3, -1)$.

[4.6] **20.** Graph $2x + 5y > 10$.

For Exercises 21–23, solve and check:

[2.3] **21.** $\dfrac{3}{5}a - 8 = \dfrac{1}{2}$

[3.1] **22.** $\dfrac{u}{14} = -\dfrac{6}{7}$

[2.3] **23.** $1.6x - 14 = 8 + 2.4(x - 5)$

[2.4] **24.** Solve $C = 2\pi r$ for r.

For Exercises 25–30, solve.

[3.2] **25.** Using the pie chart, if 1000 Internet users were polled, how many would check their e-mail weekly?

[3.2] **26.** If a married couple makes a combined income of $48,241 per year, what percent does the couple spend on holiday gifts?

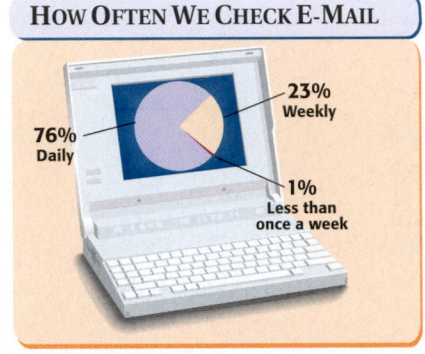

Source: UCLA Center for Communication Policy

Source: The Gallup Organization

[3.3] **27.** Two angles are complementary. The smaller angle is $\dfrac{2}{3}$ the measure of the larger angle. What are the measures of the two angles?

[3.2] **28.** The ticket prices for the 2008 Super Bowl in Glendale, Arizona, ranged from $3270 for end zone tickets to $7725 for lower-level 30–50 yard line tickets. What is the percent of increase from an end zone ticket to a lower-level 30–50 yard line ticket?

[3.3] **29.** Angela has been saving her quarters and dimes for several months. If she has 190 coins totaling $30.70, how many of each coin does she have?

[3.5] **30.** Juanita invests in a plan that has an APR of 8%. She invests twice as much in a plan that has an APR of 10%. If the total interest from the investments is $5600, how much did she invest in each plan?

Polynomials

The ability to simplify means to eliminate the unnecessary so that the necessary may speak.

—Hans Hofmann (1880–1966)
teacher and painter

Everything should be made as simple as possible, but not simpler.

—Albert Einstein (1876–1955)
physicist

In Chapter 1, we reviewed real numbers and introduced algebraic expressions. We learned that there are two actions we can perform on an expression: (1) We can evaluate an expression, or (2) we can rewrite an expression. In this chapter, we focus on polynomial expressions, exploring how to evaluate and rewrite these expressions. We also see that we can add, subtract, multiply, and divide polynomial expressions in much the same way that we add, subtract, multiply, and divide numbers.

5.1 Exponents and Scientific Notation

Objectives

1. Evaluate exponential forms with integer exponents.
2. Write scientific notation in standard form.
3. Write standard form numbers in scientific notation.

Objective 1 Evaluate exponential forms with integer exponents.

Positive Integer Exponents

We learned in Section 1.5 that exponents indicate repeated multiplication of a base number.

$$2^4 = 2 \cdot 2 \cdot 2 \cdot 2 = 16 \qquad (-5)^3 = (-5)(-5)(-5) = -125$$

four factors of two

three factors of negative five

We also discovered some rules for evaluating exponential forms with negative bases.

Rule Evaluating Exponential Forms with Negative Bases

If the base of an exponential form is a negative number and the exponent is even, then the product is positive.

If the base is a negative number and the exponent is odd, then the product is negative.

What if the base is a fraction, as in $\left(\dfrac{2}{3}\right)^4$?

$$\left(\frac{2}{3}\right)^4 = \frac{2}{3} \cdot \frac{2}{3} \cdot \frac{2}{3} \cdot \frac{2}{3} = \frac{2^4}{3^4} = \frac{16}{81}$$

This example suggests the following conclusion and rule.

Conclusion If a fraction is raised to a power, we can write its numerator and denominator to that power.

Rule Raising a Quotient to a Power

If a and b are real numbers, where $b \neq 0$ and n is a natural number, then $\left(\dfrac{a}{b}\right)^n = \dfrac{a^n}{b^n}$.

Example 1 Evaluate each exponential form.

a. $(-3)^5$

SOLUTION $(-3)^5 = (-3)(-3)(-3)(-3)(-3) = -243$

b. -2^4

SOLUTION $-2^4 = -[2 \cdot 2 \cdot 2 \cdot 2] = -16$

Warning: Notice that -2^4 does not mean that four -2's are to be multiplied. Rather, it means find the additive inverse after multiplying four 2's.

c. $\left(\dfrac{4}{5}\right)^3$

SOLUTION $\left(\dfrac{4}{5}\right)^3 = \dfrac{4^3}{5^3} = \dfrac{64}{125}$

Your Turn 1 Evaluate each exponential form.

a. $(-4)^4$ **b.** -4^4 **c.** $\left(\dfrac{5}{8}\right)^2$

Nonpositive Integer Exponents

So far, we have considered evaluating exponential forms only with positive exponents. However, exponents can be positive, zero, or negative. What does an exponent of 0 indicate? Consider the following pattern.

$$2^4 = 2 \cdot 2 \cdot 2 \cdot 2 = 16$$
$$2^3 = 2 \cdot 2 \cdot 2 \quad = 8 \quad \longleftarrow$$
$$2^2 = 2 \cdot 2 \quad\quad = 4$$
$$2^1 = 2 \quad\quad\quad = 2$$
$$2^0 = \quad\quad\quad\quad = 1 \quad \longleftarrow$$

Notice that as the exponent decreases, we can divide by 2 to determine the next result; that is, because 2^4 is 16, 2^3 is $16 \div 2$, which is 8.

To continue the pattern, when we go from 2^1 to 2^0, we divide the result for 2^1 by 2 to determine the result for 2^0, and $2 \div 2 = 1$.

The pattern suggests the following conclusion and rule.

Conclusion A nonzero base number raised to the 0 power is equal to 1. It can be shown that any nonzero number raised to the 0 power equals 1.

Rule **Zero as an Exponent**

If a is a real number and $a \neq 0$, then $a^0 = 1$.

If we continue the pattern, we can see what a negative exponent means.

$$2^0 = 1$$
$$2^{-1} = \dfrac{1}{2} \quad \longleftarrow$$
$$2^{-2} = \dfrac{1}{2^2} = \dfrac{1}{4} \quad \longleftarrow$$
$$2^{-3} = \dfrac{1}{2^3} = \dfrac{1}{8}$$

To continue the pattern, when we go from 2^0 to 2^{-1}, we divide the result for 2^0 by 2 to determine the result for 2^{-1}, and $1 \div 2 = \dfrac{1}{2}$.

To continue the pattern, when we go from 2^{-1} to 2^{-2}, we divide the result for 2^{-1} by 2 to determine the result for 2^{-2}, and

$$\dfrac{1}{2} \div 2 = \dfrac{1}{2} \cdot \dfrac{1}{2} = \dfrac{1}{2^2} = \dfrac{1}{4}.$$

The pattern suggests the following conclusion.

Conclusion An exponential form with a negative exponent is equal to its reciprocal with the exponent made positive.

It can be shown that our conclusion is true for all nonzero real-number bases.

Rule If a is a real number, where $a \neq 0$ and n is a natural number, then $a^{-n} = \dfrac{1}{a^n}$.

Answers **Your Turn 1**

a. 256 **b.** -256 **c.** $\dfrac{25}{64}$

Example 2   Rewrite with a positive exponent; then if the expression is numeric, evaluate it.

a. 10^0

Warning: A common error is to confuse negative exponents with negative numbers. The value of $4^{-2} \neq -16$. Also, $4^{-2} \neq (-2)(4)$. The value of $4^{-2} = \dfrac{1}{4^2} = \dfrac{1}{16}$.

SOLUTION $10^0 = 1$ $a^0 = 1$ for $a \neq 0$.

b. $(-2)^{-4}$

SOLUTION $(-2)^{-4} = \dfrac{1}{(-2)^4} = \dfrac{1}{(-2)(-2)(-2)(-2)} = \dfrac{1}{16}$

c. $(-5)^{-3}$

SOLUTION $(-5)^{-3} = \dfrac{1}{(-5)^3} = \dfrac{1}{(-5)(-5)(-5)} = \dfrac{1}{-125} = -\dfrac{1}{125}$

d. x^{-6}

SOLUTION $x^{-6} = \dfrac{1}{x^6}$ **Rewrite using $a^{-n} = \dfrac{1}{a^n}$.**

e. $2a^{-3}$

SOLUTION $2a^{-3} = 2 \cdot \dfrac{1}{a^3} = \dfrac{2}{a^3}$ **Note:** Only a is raised to the -3 power.

f. -6^{-2}

SOLUTION $-6^{-2} = -\dfrac{1}{6^2} = -\dfrac{1}{36}$ **Note:** Only 6 is raised to the -2 power, not -6.

Your Turn 2   If necessary, rewrite with a positive exponent; then if the expression is numeric, evaluate it.

a. $(-8)^0$ **b.** $(-3)^{-4}$ **c.** $(-2)^{-3}$ **d.** y^{-5} **e.** $4b^{-5}$ **f.** -2^{-4}

🖩 Calculator Tips

To evaluate exponential forms with negative exponents, such as $(-2)^{-4}$ in part b of Example 2, using a calculator that has ⌃ *for exponents and* (−) *for negatives, type*

((−) 2) ⌃ (−) 4 ENTER .

The result will be a decimal number, 0.0625. To see the fraction form, press MATH *and select 1: Frac from the menu; then press* ENTER *. On some scientific calculators, use the* F ◄► D *function.*

If the calculator has y^x *and* +/− *keys, type*

(2 +/−) y^x 4 +/− = .

Answers   Your Turn 2

a. 1   **b.** $\dfrac{1}{81}$ **c.** $-\dfrac{1}{8}$ **d.** $\dfrac{1}{y^5}$

e. $\dfrac{4}{b^5}$ **f.** $-\dfrac{1}{16}$

Connection Remember that $\dfrac{1}{\frac{1}{8}}$ means $1 \div \dfrac{1}{8}$ and that

$1 \div \dfrac{1}{8} = 1 \cdot \dfrac{8}{1} = \dfrac{8}{1} = 8.$

What if a negative exponent is in the denominator of a fraction, as in $\dfrac{1}{2^{-3}}$? If we rewrite 2^{-3} in the denominator using $a^{-n} = \dfrac{1}{a^n}$, we have

$$\dfrac{1}{2^{-3}} = \dfrac{1}{\frac{1}{2^3}} = \dfrac{1}{\frac{1}{8}} = \dfrac{8}{1} = 8$$ **Rewrite the denominator using $a^{-n} = \dfrac{1}{a^n}$.**

Our example suggests the following rule.

Rule If a is a real number, where $a \neq 0$ and n is a natural number, then $\dfrac{1}{a^{-n}} = a^n$.

Example 3 Rewrite with a positive exponent; then if the expression is numeric, evaluate it.

a. $\dfrac{1}{3^{-4}}$

SOLUTION $\dfrac{1}{3^{-4}} = 3^4 = 3 \cdot 3 \cdot 3 \cdot 3 = 81$ Rewrite using $\dfrac{1}{a^{-n}} = a^n$; then simplify.

b. $\dfrac{1}{x^{-5}}$

SOLUTION $\dfrac{1}{x^{-5}} = x^5$ Rewrite using $\dfrac{1}{a^{-n}} = a^n$.

Your Turn 3 Rewrite with a positive exponent; then if the expression is numeric, evaluate it.

a. $\dfrac{1}{4^{-3}}$ **b.** $\dfrac{1}{x^{-8}}$

What if a fraction is raised to a negative exponent, as in $\left(\dfrac{2}{5}\right)^{-2}$? If we rewrite the expression using $a^{-n} = \dfrac{1}{a^n}$, we have

$$\left(\frac{2}{5}\right)^{-2} = \frac{1}{\left(\frac{2}{5}\right)^2} = \frac{1}{\frac{4}{25}} = \frac{25}{4} = \left(\frac{5}{2}\right)^2.$$

Our example suggests that when we evaluate a fraction raised to a negative exponent, we can write the reciprocal of the fraction and change the sign of the exponent.

Rule If a and b are real numbers, where $a \neq 0$ and $b \neq 0$ and n is a natural number, then
$$\left(\frac{a}{b}\right)^{-n} = \left(\frac{b}{a}\right)^n.$$

Example 4 Rewrite with a positive exponent; then if the expression is numeric, evaluate it.

a. $\left(\dfrac{3}{4}\right)^{-3}$

SOLUTION $\left(\dfrac{3}{4}\right)^{-3} = \left(\dfrac{4}{3}\right)^3 = \dfrac{4^3}{3^3} = \dfrac{64}{27}$ Rewrite using $\left(\dfrac{a}{b}\right)^{-n} = \left(\dfrac{b}{a}\right)^n$; then simplify.

b. $\left(\dfrac{x}{y}\right)^{-4}$

SOLUTION $\left(\dfrac{x}{y}\right)^{-4} = \left(\dfrac{y}{x}\right)^4 = \dfrac{y^4}{x^4}$ Rewrite using $\left(\dfrac{a}{b}\right)^{-n} = \left(\dfrac{b}{a}\right)^n$; then simplify.

Answers Your Turn 3
a. 64   **b.** x^8

Your Turn 4 Rewrite with a positive exponent; then if the expression is numeric, evaluate it.

a. $\left(-\dfrac{4}{5}\right)^{-3}$

b. $\left(\dfrac{a}{b}\right)^{-5}$

Objective 2 Write scientific notation in standard form. Sometimes, we use very large or very small numbers, such as when we describe the vast distances between stars and galaxies or the tiny size of bacteria and atomic structures. For example, the distance from the Sun to the next nearest star, Proxima Centauri, is about 24,700,000,000,000 miles; a single streptococcus bacterium (photo at left) is about 0.00000075 meter in diameter. The large number of zero digits in these numbers makes them tedious to write. **Scientific notation** gives us a shorthand way to write such numbers.

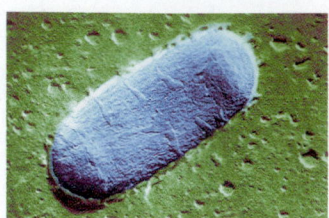

Definition **Scientific notation:** A number expressed in the form $a \times 10^n$, where a is a decimal number with $1 \le |a| < 10$ and n is an integer.

The number 3.58×10^4 is in scientific notation because 3.58 is greater than or equal to 1 but less than 10 and the exponent 4 is an integer. The number 35.8×10^3 is not in scientific notation because 35.8 is greater than 10. The number 0.358×10^5 also is not in scientific notation because 0.358 is less than 1.

What does 3.58×10^4 mean?

$$
\begin{aligned}
3.58 \times 10^4 &= 3.58 \times 10 \times 10 \times 10 \times 10 \\
&= 35.8 \times 10 \times 10 \times 10 \\
&= 358 \times 10 \times 10 \\
&= 3580 \times 10 \\
&= 35,800
\end{aligned}
$$

Note: 3.58×10^4, 35.8×10^3, and 0.358×10^5 all represent the same number, 35,800. Only 3.58×10^4 is in scientific notation.

Notice that each multiplication by 10 moves the decimal point one place to the right. Because we multiplied by four factors of 10, the decimal point moved a total of four places to the right from where it started in 3.58.

$$3.58 \times 10^4 = 3\,5\,8\,0\,0 = 35,800 \qquad \textbf{\textcolor{blue}{The decimal point moves four places to the right.}}$$

Our example suggests that the exponent of the power of 10 determines the number of places that the decimal point moves.

Procedure **Changing Scientific Notation (Positive Exponent) to Standard Form**

To change from scientific notation with a positive integer exponent to standard form, move the decimal point to the right the number of places indicated by the exponent.

Example 5 Write each number in standard form.

a. 4.5×10^6

SOLUTION Multiplying 4.5 by 10^6 means that the decimal point will move six places to the right.

$$4.5 \times 10^6 = 4,500,000$$

Connection Because 10^6 is another way to represent 1,000,000, it is popular to read 4.5×10^6 as 4.5 million.

Answers Your Turn 4

a. $-\dfrac{125}{64}$ b. $\dfrac{b^5}{a^5}$

b. 9×10^5

SOLUTION Multiplying 9 by 10^5 means that the decimal point will move five places to the right.

$$9 \times 10^5 = 900{,}000$$

Your Turn 5 Write each number in standard form.

a. 7.305×10^6 **b.** 8×10^7

What if the scientific notation contains a negative exponent, as in 8.45×10^{-2}? To evaluate 10^{-2}, we write it with a positive exponent, $\dfrac{1}{10^2}$.

$$8.45 \times 10^{-2} = 8.45 \times \frac{1}{10^2}$$

$$= \frac{8.45}{10^2}$$

$$= \frac{8.45}{100}$$

$$= 0.0845$$

Note: Multiplying 8.45 by 10^{-2} is equivalent to dividing 8.45 by 10^{-2}. Dividing by 10^2, which is 100, causes the decimal point to move two places to the left.

0.0 8 4 5

Our example suggests the following procedure.

Procedure   **Changing Scientific Notation (Negative Exponent) to Standard Form**

To change from scientific notation with a negative exponent to standard form, move the decimal point to the left the same number of places as the absolute value of the exponent.

Example 6   Write each number in standard form.

a. 4.2×10^{-5}

SOLUTION Multiplying by 10^{-5} is equivalent to dividing by 10^5, which causes the decimal point to move five places to the *left*.

$$4.2 \times 10^{-5} = \frac{4.2}{10^5} = 0.000042$$

b. 7×10^{-9}

SOLUTION Multiplying by 10^{-9} is equivalent to dividing by 10^9, which causes the decimal point to move nine places to the left.

$$7 \times 10^{-9} = \frac{7}{10^9} = 0.000000007$$

Answers   **Your Turn 5**
a. 7,305,000   **b.** 80,000,000

Your Turn 6 Write each number in standard form.

a. 7.24×10^{-6} **b.** 5×10^{-7}

Answers   **Your Turn 6**
a. 0.00000724   **b.** 0.0000005

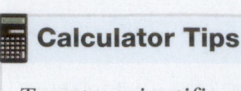 **Calculator Tips**

To enter scientific notation on a graphing or scientific calculator, we can use the [EE] *function. For example, to enter* 4.2×10^{-5}, *type*

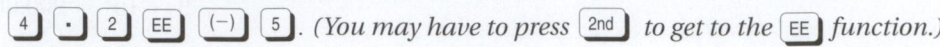

[4] [.] [2] [EE] [(−)] [5] . *(You may have to press* [2nd] *to get to the* [EE] *function.)*

Notice that the base 10 is understood and is not entered when the [EE] *function is used.*

We can also enter scientific notation without using [EE] , *like this:*

[4] [.] [2] [×] [1] [0] [^] [(−)] [5]

After entering scientific notation, if you press [ENTER] , *the calculator will change the number to standard form. However, if the number has too many digits for the calculator to display, the calculator will leave the number in scientific notation. Similarly, if the answer to a calculation is too large for the calculator to display, it will show the result in scientific notation.*

Objective 3 **Write standard form numbers in scientific notation.** We mentioned that the distance from the Sun to the next nearest star, Proxima Centauri, is about 24,700,000,000,000 miles. Suppose we want to express 24,700,000,000,000 in scientific notation. Because scientific notation begins with a decimal number greater than or equal to 1 but less than 10, our first step is to determine the position of the decimal point. The only place the decimal point can go to satisfy the criteria for scientific notation is between the 2 and 4 digits.

$$24{,}700{,}000{,}000{,}000$$

The decimal goes here to express a decimal number greater than or equal to 1 but less than 10.

Notice that there are 13 place values to the right of the decimal point in its new position between the 2 and the 4. To make the scientific notation equal to the original number, we must account for the 13 place values by expressing them as a power of 10.

$$24{,}700{,}000{,}000{,}000 = 2.4700000000000 \times 10^{13}$$

The 13 places to the right of the new decimal position are expressed as a power of 10.

$$24{,}700{,}000{,}000{,}000 = 2.4700000000000 \times 10^{13} = 2.47 \times 10^{13}$$

These zeros to the right of the 7 are no longer needed.

Procedure **Changing Standard Form to Scientific Notation**

To write a number greater than 1 in scientific notation,

1. Move the decimal point so that the number is greater than or equal to 1 but less than 10. (*Tip: Place the decimal point to the right of the first nonzero digit*).
2. Write the decimal number multiplied by 10^n, where n is the number of places between the new decimal position and the original decimal position.
3. Delete zeros to the right of the last nonzero digit.

Example 7   Write 986,000 in scientific notation.

SOLUTION   Place the decimal to the right of the first nonzero digit, 9. Next, count the places to the right of this decimal position, which is five places. Write the 5 as the exponent of 10. Finally, delete all of the 0's to the right of the last nonzero digit, which is the 6 in this case.

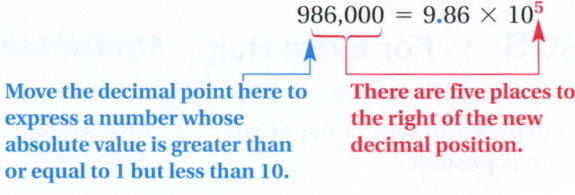

$$986,000 = 9.86 \times 10^5$$

Move the decimal point here to express a number whose absolute value is greater than or equal to 1 but less than 10.

There are five places to the right of the new decimal position.

Your Turn 7   Write the number in scientific notation.

a. 603,000,000

b. 92,500,000,000,000

We learned that a number written in scientific notation that has a negative exponent, such as 5×10^{-2}, has an absolute value less than 1. Therefore, a positive number less than 1, such as 0.00000075, has a negative exponent when written in scientific notation. To write 0.00000075 in scientific notation, we place the decimal between the 7 and 5 digits so that the number is greater than or equal to 1 but less than 10. Moving the decimal point to this position means that we have seven places to account for, which we indicate with a −7 exponent.

$$0.00000075 = 7.5 \times 10^{-7}$$

This suggests the following procedure.

Procedure   **Changing Standard Form to Scientific Notation**

To write a positive decimal number that is less than 1 in scientific notation,

1. Move the decimal point so that the number is greater than or equal to 1 but less than 10. (*Tip: Place the decimal point to the right of the first nonzero digit.*)
2. Write the decimal number multiplied by 10^n, where n is a negative integer whose absolute value is the number of places between the new decimal position and the original decimal position.
3. Delete zeros to the left of the first nonzero digit.

Example 8   Write 0.0000608 in scientific notation.

SOLUTION   Place the decimal point between the 6 and 0 digits so that you have 6.08, which is a decimal number greater than 1 but less than 10. Because there are five decimal places between the original decimal position and the new position, the exponent is −5.

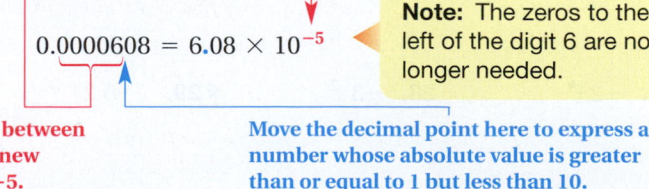

Note: The zeros to the left of the digit 6 are no longer needed.

$$0.0000608 = 6.08 \times 10^{-5}$$

There are five decimal places between the original position and the new position, so the exponent is −5.

Move the decimal point here to express a number whose absolute value is greater than or equal to 1 but less than 10.

Your Turn 8   Write in scientific notation.

a. 0.000172   b. 0.0000009034

5.1 Exercises   For Extra Help   *MyMathLab*

Note: Exercises marked with a ★ represent challenging exercises.

1. Explain how to rewrite an expression with a negative exponent so that its exponent is positive.

2. After simplifying -5^2, is the result positive or negative? Explain.

3. After simplifying 2^{-3}, is the result positive or negative? Explain.

4. For what kinds of numbers is scientific notation a useful notation?

5. How do we write a number that is greater than 1 in scientific notation?

6. How do we write a positive decimal number that is less than 1 in scientific notation?

For Exercises 7–20, evaluate the exponential form. See Example 1.

7. 4^0 8. 3^0 9. -2^3 10. -3^2 11. $(-5)^3$ 12. $(-2)^5$ 13. -4^2

14. -3^4 15. $\left(\dfrac{3}{4}\right)^4$ 16. $\left(\dfrac{5}{6}\right)^2$ 17. $(-0.2)^3$ 18. $(-0.04)^3$ 19. $(-2.3)^2$ 20. $(-1.2)^2$

For Exercises 21–42, rewrite each expression with positive exponents; then if the expression is numeric, evaluate it. See Examples 2–4.

21. a^{-5} 22. x^{-4} 23. 2^{-3} 24. 5^{-1} 25. $(-4)^{-3}$ 26. $(-1)^{-5}$

27. -2^{-4} 28. -3^{-2} ★29. $(-0.2)^{-3}$ ★30. $(-0.1)^{-4}$ 31. $5m^{-2}$ 32. $6r^{-4}$

33. $-6x^{-7}$ **34.** $-8y^{-6}$ **35.** $\left(\dfrac{m}{n}\right)^{-5}$ **36.** $\left(\dfrac{p}{q}\right)^{-6}$ **37.** $\left(\dfrac{5}{7}\right)^{-2}$ **38.** $\left(\dfrac{1}{6}\right)^{-3}$

39. $\dfrac{1}{b^{-6}}$ **40.** $\dfrac{1}{c^{-2}}$ **41.** $\dfrac{1}{4^{-3}}$ **42.** $\dfrac{1}{3^{-4}}$

For Exercises 43–50, use a scientific or graphing calculator to evaluate each exponential form.

43. $(-24)^{-5}$ **44.** $(-45)^{-4}$ **45.** $\left(-\dfrac{5}{8}\right)^{-3}$ **46.** $\left(-\dfrac{4}{9}\right)^{-4}$

47. -0.05^{-4} **48.** -0.02^{-2} **49.** $(3.7)^{-6}$ **50.** $(2.3)^{-4}$

Find ⊗ **the mistake**   For Exercises 51–54, find and explain the mistake; then work the problem correctly. See Examples 1 and 2.

51. $-3^4 = (-3)(-3)(-3)(-3) = 81$ **52.** $(-5)^3 = -75$

53. $5^{-2} = -25$ **54.** $\left(\dfrac{3}{4}\right)^{-2} = \dfrac{9}{16}$

For Exercises 55–70, write the number in standard form. See Examples 5 and 6.

55. Scientists speculate that the universe is about 1.65×10^{10} years old.

56. The solar system formed approximately 4.6×10^9 years ago.

57. In empty space, light travels at a constant speed of approximately 2.998×10^8 meters per second.

58. The equatorial radius of the Sun is approximately 6.96×10^8 meters.

59. The equatorial radius of Earth is approximately 6.378×10^6 meters.

60. A light-year is the distance that light travels in a year's time and is equal to about 5.76×10^{12} miles.

61. Earth is about 9.292×10^7 miles from the Sun.

62. The next nearest star to ours (the Sun) is Proxima Centauri, which is about 2.47×10^{13} miles from the Sun.

63. Mitochondria are organelles that are about 2.5×10^{-6} meter across.

64. The diameter of human DNA is about 2×10^{-9} meter.

65. The rest mass of a proton is about 1.67×10^{-27} kilogram.

66. Light with a wavelength of 7.6×10^{-11} meter is in the X-ray portion of the spectrum.

67. Light with a wavelength of 2.95×10^{-9} meter is in the ultraviolet portion of the spectrum.

68. Light with a wavelength of 4.5×10^{-7} meter is in the visible spectrum and is blue in color.

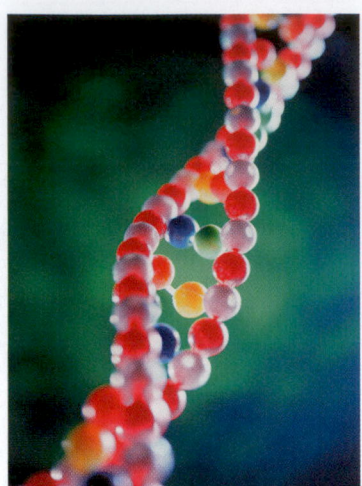

69. The mass of an alpha particle, which is emitted in the radioactive decay of plutonium 239, is 6.645×10^{-27} kilogram.

70. Protons have a positive electrical charge of 1.602×10^{-19} coulomb.

For Exercises 71–86, write the number in scientific notation. See Examples 7 and 8.

71. There are approximately 25,000,000,000,000 red blood cells in the average person's bloodstream.

72. A single red blood cell contains about 250,000,000 molecules of hemoglobin.

73. Human DNA consists of about 5,300,000,000 nucleotide pairs.

74. The Andromeda galaxy is about 2,140,000 light-years from the Sun.

75. Uranium 236 is a radioactive isotope of uranium and has a half-life of 23,420,000 years.

76. The population of the United States is about 304,000,000.

77. The most populated country in the world is China, with a population of about 1,300,000,000.

78. In 2008, the population of Earth was about 6,678,000,000.

79. Light with a wavelength of 0.000000586 meter is yellow in color.

80. Light with a wavelength of 0.000000712 meter is red in color.

81. Electrons have a mass of approximately 0.00000000000000000000000000000091094 kilogram.

82. The size of the HIV virus that causes AIDS is about 0.0000001 meter.

83. The most common diameter for the core of a fiber-optic cable is 0.0000625 meter.

84. A human hair is about 0.00005 meter wide.

85. Intel introduced an improved process for manufacturing computer chips, which allows the company to make transistors that are 0.00000005 meter in length.

86. On April 2, 2008, 0.00975 U.S. dollar equaled 1 japanese yen.

For Exercises 87 and 88, write the numbers in order from smallest to largest.

87. 7.5×10^6, 8.95×10^5, 9×10^7, 1.3×10^8, 7.2×10^6

88. 2.8×10^{-5}, 3.7×10^{-7}, 6.2×10^{-2}, 9.8×10^{-5}, 6.1×10^{-2}

★ **89.** In Exercises 66 and 67, we saw that light with a wavelength of 7.6×10^{-11} meter is in the X-ray portion of the spectrum and light with a wavelength of 2.95×10^{-9} meter is in the ultraviolet portion of the spectrum. About how many times greater is ultraviolet light's wavelength compared to X-rays?

★ **90.** In Exercises 65 and 69, we saw that a proton has a mass of 0.00000000000000000000000000167 kilogram and an alpha particle has a rest mass of about 6.645×10^{-27} kilogram, respectively. Which has the greater mass? About how many times greater is the mass of the larger of the two particles?

Review Exercises

Exercises 1 and 2 **EXPRESSIONS**

[1.7]   **1.** Use the distributive property to rewrite the expression $-4(y - 3)$.

[1.3]   **2.** Simplify $-13 + 2$.

Exercises 3–6 **EQUATIONS AND INEQUALITIES**

[3.2]   **3.** 40 is 15% of what?

[4.4]   **4.** Find the slope of the line that passes through $(2, -1)$ and $(-3, -2)$.

[4.4]   **5.** Find the slope and y-intercept for $2x - 3y = 6$.

[4.7]   **6.** Given the function $f(x) = x^3 - 5x^2 + 7$, find $f(-3)$.

5.2 Introduction to Polynomials

Objectives

1. Identify monomials.
2. Identify the coefficient and degree of a monomial.
3. Classify polynomials.
4. Identify the degree of a polynomial.
5. Evaluate polynomials.
6. Write polynomials in descending order of degree.
7. Combine like terms.

Objective 1 Identify monomials. We have explored some simple algebraic expressions (Chapter 1) and linear equations and inequalities (Chapters 2–4). We now move back to the expression level of our Algebra Pyramid and explore a particular class of expressions known as *polynomials*.

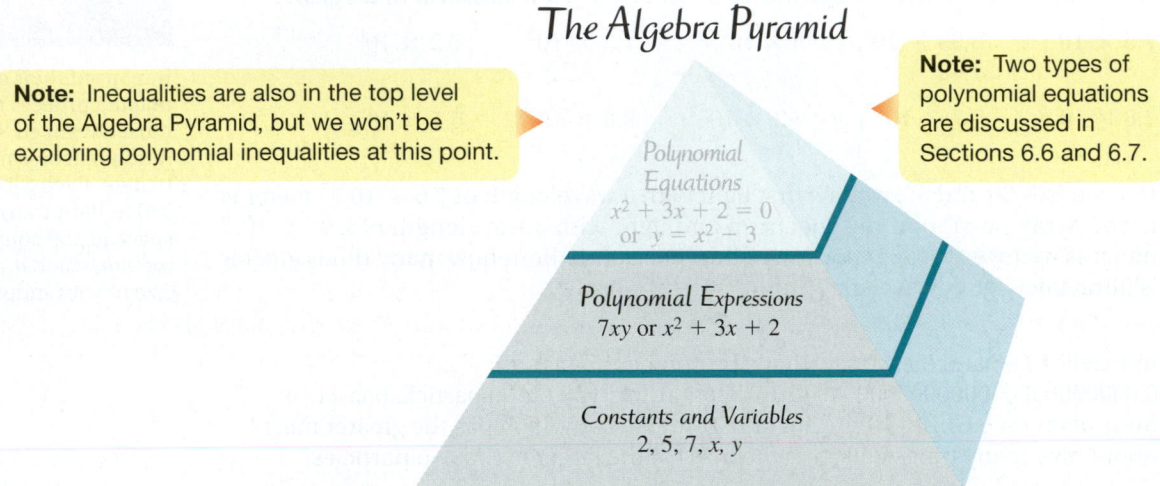

The Algebra Pyramid

Note: Inequalities are also in the top level of the Algebra Pyramid, but we won't be exploring polynomial inequalities at this point.

Note: Two types of polynomial equations are discussed in Sections 6.6 and 6.7.

Polynomial Equations
$x^2 + 3x + 2 = 0$
or $y = x^2 - 3$

Polynomial Expressions
$7xy$ or $x^2 + 3x + 2$

Constants and Variables
$2, 5, 7, x, y$

First, we consider a special type of polynomial expression called a **monomial**.

Definition   *Monomial:* An expression that is a constant, a variable, or a product of a constant and variable(s) that are raised to whole number powers.

Implied in the definition of a monomial is that there is no addition or subtraction; so there is only one term.

Connection Although no variable factor appears with the constant 7, it can be expressed with a variable factor. Recall from Section 5.1 that any nonzero number raised to the 0 power simplifies to 1. Therefore, 7 can be written as $7x^0$ because $x^0 = 1$ (assuming that $x \neq 0$). We use this fact to help us understand the next objective concerning degree.

Example 1   Is the given expression a monomial? Explain.

a. 7

Answer   7 is a monomial because it is a constant.

b. $3x^2$

Answer $3x^2$ is a monomial because it is a product of a constant, 3, and a variable, x, which has a whole-number exponent of 2.

c. $9y$

Answer   Remember that a number or variable with no apparent exponent has an understood 1 exponent; so $9y = 9y^1$. Therefore, $9y$ is a monomial because it is a product of a constant, 9, and a variable, y, which has a whole-number exponent of 1.

d. $-0.6xy^2$

Answer $-0.6xy^2$ is a monomial because it is a product of a constant, -0.6, and variables, x and y^2, which have whole-number exponents. As we saw in Example 1(c), we could write $-0.6xy^2$ as $-0.6x^1y^2$. Both of the variables' exponents, 1 and 2, are whole numbers.

e. y^4

Answer It might seem that y^4 is not a monomial because no numerical factor is visible. However, y^4 is the same as $1y^4$. We don't need to write the 1 because 1 multiplied by an amount does not affect the amount.

f. $\dfrac{5}{x^2}$

> **Note:** This means that a monomial can't have a variable as a divisior.

Answer The expression $\dfrac{5}{x^2}$ can be written as a product by writing the divisor with a negative exponent, $5x^{-2}$. However, when written as a product, the variable's exponent is not a whole number, which means it is not a monomial.

g. $4y^2 + 9y - 3$

Answer The expression is not a monomial because it is not a product of a constant and variables. Instead, addition and subtraction are involved.

Your Turn 1 Is the given expression a monomial? Explain.

a. x^3 **b.** $-y^2$ **c.** $-\dfrac{2}{3}tu$ **d.** $\dfrac{4m^2}{n}$ **e.** 0.56 **f.** $3y + 7$

Objective ❷ Identify the coefficient and degree of a monomial. In our introduction to expressions in Section 1.7, we defined a **coefficient** as the numerical or constant factor in a term. The same is true for monomials. Another piece of information about a monomial is its **degree**, which relates to the exponents of the variables.

Definitions

Coefficient of a monomial: The numerical factor in a monomial.

Degree of a monomial: The sum of the exponents of all variables in a monomial.

Answers Your Turn 1

a. x^3 is a monomial because it is a product of a constant, 1, and a variable, x, which has a whole-number exponent of 3.

b. $-y^2$ is a monomial because it is a product of a constant, -1, and a variable y, with a whole-number exponent of 2.

c. $-\dfrac{2}{3}tu$ is a monomial because it is a product of a constant, $-\dfrac{2}{3}$, and variables, t and u, both of which have the whole number 1 as an exponent.

d. $\dfrac{4m^2}{n}$ is not a monomial because it has a variable, n, as a divisor.

e. 0.56 is a monomial because 0.56 is a constant.

f. $3y + 7$ is not a monomial because it contains addition.

Example 2 Identify the coefficient and degree of each monomial.

a. $-5t^2$

Answer The coefficient is -5 because it is the numerical factor. The degree is 2 because it is the exponent for the single variable t.

b. $-mn^4$

Answer We can express $-mn^4$ as $-1m^1n^4$. In this form, we can see that the coefficient is -1 and that the exponents for the variables are 1 and 4. Because the degree is the sum of the variables' exponents, the degree is 5.

c. 7

Answer Remember that $7 = 7x^0$, where x is any real number except 0. In this alternative form, we can see that 7 is the coefficient and 0 is the degree because it is the variable's exponent.

Your Turn 2 Identify the coefficient and degree of each monomial.

a. $-4x^4$ **b.** $-0.9n^2p^5$ **c.** $\dfrac{5}{8}ab$ **d.** 13

Answers Your Turn 2

a. c: -4; d: 4 **b.** c: -0.9; d: 7

c. c: $\dfrac{5}{8}$; d: 2 **d.** c: 13; d: 0

Objective ③ Classify polynomials. Now we are ready to formally define a **polynomial**.

Definition   *Polynomial:* A monomial or an expression that can be written as a sum of monomials.

Examples of polynomials: $4x$ $4x + 8$ $2x^2 + 5xy + 8y$ $2x^3 - 5x^2 + 3x - 9$

> **Note:** Although it seems that the subtractions violate the word *sum* in the definition, the expression $2x^3 - 5x^2 + 3x - 9$ is a polynomial because it can be written as a sum of monomials by rewriting the subtractions as equivalent additions.
>
> $$2x^3 - 5x^2 + 3x - 9 = 2x^3 + (-5x^2) + 3x + (-9)$$
>
> Although those two forms are equivalent, $2x^3 - 5x^2 + 3x - 9$ is preferable because it is the simplest form. That is, it has the fewest symbols.

Notice that the polynomial $2x^3 - 5x^2 + 3x - 9$ has the same variable x in each variable term, whereas $2x^2 + 5xy + 8y$ has a mixture of two variables, x and y. We call a polynomial such as $2x^3 - 5x^2 + 3x - 9$ a **polynomial in one variable**.

Definition   *Polynomial in one variable:* A polynomial in which every variable term has the same variable.

$x^2 - 5x + 2$ is a polynomial in one variable, x.

$x^2 + 8y$ is a polynomial in two variables, x and y.

Special names have been given to some polynomials. We have already discussed monomials such as $4x$, which is a single-term polynomial. Notice that the prefix *mono* in the word *monomial* indicates "one" term and the prefix *poly* in *polynomial* indicates "many" terms. Continuing with the prefixes, a two-term polynomial such as $4x + 8$ is called a **binomial**, and a three-term polynomial such as $2y^2 + 5y + 8$ is called a **trinomial**. No special names are given to polynomials with more than three terms.

Definitions   *Binomial:* A polynomial containing two terms.

Trinomial: A polynomial containing three terms.

Example 3   Indicate whether the expression is a monomial, binomial, or trinomial, has no special polynomial name or is not a polynomial.

a. $2.9xy^3z$

Answer $2.9xy^3z$ is a monomial because it has a single term.

b. $-7a^2 + b$

Answer $-7a^2 + b$ is a biomial because it contains two terms.

c. $9t^2 + 3.4t - 8.2$

Answer $9t^2 + 3.4t - 8.2$ is a trinomial because it contains three terms.

d. $y^3 + 6y^2 - y + 15$

Answer   Although $y^3 + 6y^2 - y + 15$ is a polynomial, it has no special name because it has more then three terms.

e. $6x + \dfrac{3}{y}$

Answer $6x + \dfrac{3}{y}$ is not a polynomial because $\dfrac{3}{y}$ is not a monomial.

Your Turn 3 Indicate whether the expression is a monomial, binomial, or trinomial, has no special polynomial name or is not a polynomial.

a. $b + 2$ **b.** $7r^3 - 9r^2 + r - 6$ **c.** $5.7x^4y$ **d.** $x^2 - 7xy - y^2$ **e.** $4m - \dfrac{3}{n^2}$

Objective ④ **Identify the degree of a polynomial.** Previously, we defined the degree of a monomial to be the sum of the exponents of the variables. We can also talk about the **degree of a polynomial**.

Definition *Degree of a polynomial:* The greatest degree of any of the terms in the polynomial.

Example 4 Identify the degree of each polynomial.

a. $x^3 + 12x^7 - 10x^2 + 9x - 14$

Answer The degree is 7 because it is the greatest degree of all of the terms.

b. $12x^6 - 3x^5y^3 - xy^3 + 4y^2 - 16$

Answer To determine the degree of the monomial $-3x^5y^3$, we add the exponents of its variables; so its degree is 8. Likewise, we add the exponents of the variables in $-xy^3$ and find that its degree is 4. Comparing these degrees with the degrees of the other terms, we see that 8 is the greatest degree. Therefore, 8 is the degree of the polynomial.

Your Turn 4 Identify the degree of each polynomial.

a. $y^7 - 3y^2 - 8y^9 + 4y + 5$ **b.** $2x^5 - 9x^6y + x^3y^3 - 15y - 7$

Objective ⑤ **Evaluate polynomials.** In Chapter 1, we learned that we can evaluate an expression. Let's now evaluate a polynomial expression. To evaluate an expression, we replace the variable(s) with given values and calculate a numerical result.

Example 5 Evaluate $x^3 - 5x^2 + 7$ when $x = -3$.

SOLUTION $x^3 - 5x^2 + 7$

$(-3)^3 - 5(-3)^2 + 7$ **Replace each x with -3.**

$= -27 - 5(9) + 7$ **Simplify.**

$= -27 - 45 + 7$

$= -65$

Example 6 If we neglect air resistance, the polynomial $-16t^2 + h_0$ describes the height of a falling object after falling from an initial height h_0 for t seconds. It is said that Galileo Galilei dropped cannonballs from the Leaning Tower of Pisa to study gravitational effects. If the tower is 180 feet tall, what is the height of a cannonball after it falls 3 seconds?

Answers   Your Turn 3

a. binomial
b. no special polynomial name
c. monomial
d. trinomial
e. not a polynomial

Answers   Your Turn 4

a. 9   **b.** 7

SOLUTION Evaluate $-16t^2 + h_0$ when $t = 3$ and $h_0 = 180$.

$$-16(3)^2 + 180 \qquad \text{Replace } t \text{ with 3 and } h_0 \text{ with 180.}$$
$$= -16(9) + 180$$
$$= -144 + 180$$
$$= 36$$

After 3 seconds, the height of the cannonball is 36 feet.

Your Turn 6 Use the polynomial in Example 6 to find the height of an object after it falls 4 seconds from an initial height of 300 feet.

Objective 6 **Write polynomials in descending order of degree.** Now let's consider some ways to rewrite a polynomial. Recall from Chapter 1 that rewriting an expression means that we write an equivalent expression by applying properties of arithmetic. For example, we can apply the commutative property of addition and rearrange the terms in $-8 + 4x^3 - 7x^2 + 2x$ to get an equivalent expression, $4x^3 - 7x^2 + 2x - 8$. Notice that the signs move with the terms. Also notice that the terms in $4x^3 - 7x^2 + 2x - 8$ are written in order from the greatest degree to the least degree.

Connection Because $4x^3 - 7x^2 + 2x - 8$ is a rearrangement of $-8 + 4x^3 - 7x^2 + 2x$, equating them forms the identity $4x^3 - 7x^2 + 2x - 8 = -8 + 4x^3 - 7x^2 + 2x$. This means that we can evaluate both expressions using any real number and both will give the same result.

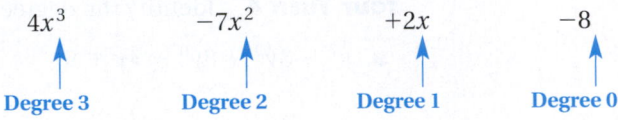

A polynomial written this way is said to be in *descending order of degree*.

Procedure **Writing a Polynomial in Descending Order of Degree**

To write a polynomial in descending order of degree, place the highest degree term first, then the next highest degree term, and so on.

Writing a polynomial in more than one variable in descending order of degree can be complicated, so we will write only polynomials in one variable in descending order of degree.

Example 7 Write each polynomial in descending order of degree.

a. $9x^2 + 7x^4 - 2x^3 - 5 + 3x$

SOLUTION Rearrange the terms so that the highest degree term is first, then the next highest degree, and so on.

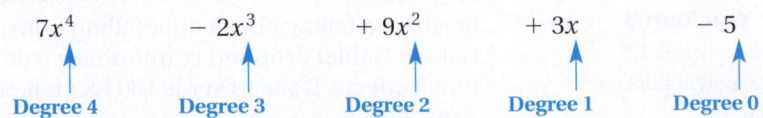

Answer $7x^4 - 2x^3 + 9x^2 + 3x - 5$

b. $-9y^3 + 2y^6 + y - 3y^4 + 12 - 6y^2$

Answer $2y^6 - 3y^4 - 9y^3 - 6y^2 + y + 12$

Answer   Your Turn 6

44 feet

Your Turn 7 Write each polynomial in descending order of degree.

a. $2n^3 + n^5 - 9n^2 + 8 - 3n^4 + 4n$ **b.** $-4t^2 + 2t + 16 + 5t^3 - 13t^6$

Objective ⑦ **Combine like terms.** Another way we can rewrite a polynomial expression is by combining like terms. We learned in Section 1.7 that combining like terms is one way to rewrite an expression in simplest form. We also learned that like terms have the same variables raised to the same powers.

Example 8 Combine like terms and write the resulting polynomial in descending order of degree.

$$12x^3 + 5x^2 - 2x^3 - 13 + 6x + x^2 + 2 - 6x$$

SOLUTION $12x^3 + 5x^2 - 2x^3 - 13 + 6x + x^2 + 2 - 6x$

$= 12x^3 - 2x^3 + 5x^2 + x^2 + 6x - 6x - 13 + 2$ Collect like terms as needed.

$= 10x^3 + 6x^2 + 0 - 11$ Combine like terms.

$= 10x^3 + 6x^2 - 11$ Drop 0 and bring the terms together.

> **Warning:** When combining like terms, we *add only the coefficients*, keeping the exponents the same.

ALTERNATIVE SOLUTION Some people prefer to skip the collecting step and strike through like terms as they are combined to keep track of the terms that have been combined. It is helpful to combine the terms in descending order of degree so that the answer will be in descending order of degree.

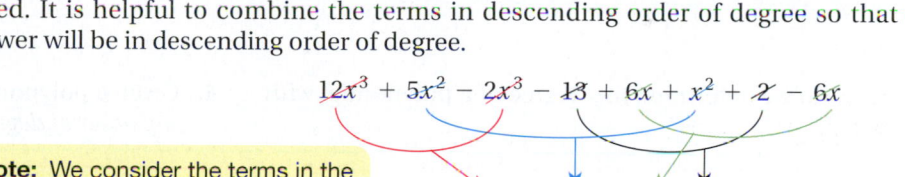

> **Note:** We consider the terms in the following order:
> Degree 3 terms: $12x^3 - 2x^3 = 10x^3$
> Degree 2 terms: $5x^2 + x^2 = 6x^2$
> Degree 1 terms: $6x - 6x = 0$
> Degree 0 terms: $-13 + 2 = -11$

$= 10x^3 + 6x^2 + 0 - 11$
$= 10x^3 + 6x^2 - 11$

Your Turn 8 Combine like terms and write the resulting polynomial in descending order of degree.

a. $10x^3 - 4x^2 - 3x^3 + 8 - 5x^2 - 9 + 6x$

b. $3y^4 - 12y - y^2 + y - 6 + y^2 + 5y^4$

Remember that polynomials may have multiple variables. Multivariable monomials are like terms if they have the same variables raised to the same exponents.

Example 9 Combine like terms.

$$x^4 + 8x^2y - 7 - 13x^2y + 4y - 2xy^2 + 6y - x^4 + 2$$

SOLUTION Collecting the like terms and then combining them gives us the following:

$$x^4 + 8x^2y - 7 - 13x^2y + 4y - 2xy^2 + 6y - x^4 + 2$$

$= x^4 - x^4 + 8x^2y - 13x^2y - 2xy^2 + 4y + 6y + 2 - 7$

$= \quad 0 \quad\quad - 5x^2y \quad - 2xy^2 \quad + 10y \quad - 5$

$= -5x^2y - 2xy^2 + 10y - 5$

Answers **Your Turn 7**
a. $n^5 - 3n^4 + 2n^3 - 9n^2 + 4n + 8$
b. $-13t^6 + 5t^3 - 4t^2 + 2t + 16$

Answers **Your Turn 8**
a. $7x^3 - 9x^2 + 6x - 1$
b. $8y^4 - 11y - 6$

ALTERNATIVE SOLUTION Instead of collecting the like terms, we strike through like terms in the given polynomial as they are combined:

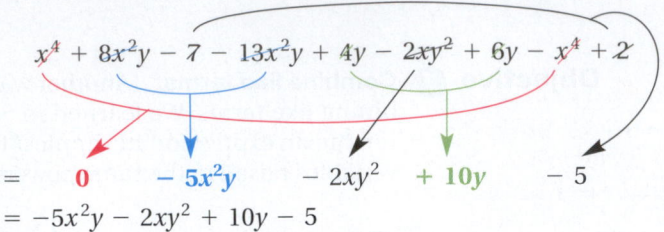

$$= 0 \quad - 5x^2y \quad - 2xy^2 \quad + 10y \quad - 5$$
$$= -5x^2y - 2xy^2 + 10y - 5$$

Your Turn 9 Combine like terms.

a. $-x^5 + 6x^3y^2 + 2y - 9x^3y^2 + 3x^5 - 4 - 2y$

b. $-6t^3u + 12tu - 4t^2u + 9t^3u - u + 3t^2u - 12tu - u$

Answers Your Turn 9

a. $2x^5 - 3x^3y^2 - 4$

b. $3t^3u - t^2u - 2u$

5.2 Exercises   For Extra Help   *MyMathLab* |

Note: Exercises marked with a ★ represent challenging exercises.

1. How do we determine the degree of a monomial?

2. Explain the differences between a monomial, a binomial, and a trinomial.

3. How do we determine the degree of a polynomial with two or more terms?

4. Given a polynomial in one variable, what does *descending order of degree* mean?

5. Explain how to combine like terms.

6. Why do we say that combining the like terms in a polynomial simplifies it?

For Exercises 7–18, determine whether the expression is a monomial. See Example 1.

7. $-3x^2$

8. $5y^3$

9. $\dfrac{1}{5}$

10. $\dfrac{4}{7}$

11. $\dfrac{6m}{4y^3}$

12. $\dfrac{-8t}{5x}$

13. $4x^2 - 3x + 7$

14. $p^2 - q^2$

15. y

16. 0.76

17. $3m^4n^2$

18. $6u^3v^4w$

For Exercises 19–30, identify the coefficient and degree of each monomial. See Example 2.

19. $-5m^3$

20. $8p^2$

21. $-xy^4$

22. $-m^2n^6$

23. -9 **24.** 18 **25.** $4.2n^3p$ **26.** $-6.7uv^7$

27. $16abc$ **28.** $-8.1lkm$ **29.** y **30.** w

For Exercises 31–42, indicate whether the expression is a monomial, binomial, or trinomial or has no special polynomial name. If the expression is a polynomial, give the degree. See Examples 3 and 4.

31. $6x + 8y + 3z$ **32.** $17.3x^2 - 3x + 2.1$ **33.** $-7m^2n$ **34.** $18xy^3$

35. $5.2x^3 - 3x^2 + 4.1x - 11$ **36.** $x^3 - 4x^2 + 4x + 5$ **37.** $5u^3 - 16u$ **38.** $7k + 5k^3$

39. -21 **40.** 36 **41.** $\dfrac{25}{x} - x^2$ **42.** $y^2 - \dfrac{16}{y}$

For Exercises 43–52, identify the degree of each polynomial. See Example 4.

43. $19 - 7y^4 + 3y - 2y^3 - 7$ **44.** $6a^5 - 19a^3 + a^9 + 5a - 14$ **45.** $16z^4 + 7z^2 - z^5 - 4z^3 + z$

46. $11 + 4t^5 - 5t + 7t^3 - 18t^2 + t^8$ **47.** $2u^2 + 5u^3 - u^7 + 13u^4 - 3u$ **48.** $22j^3 + 5j^2 - 16j^4 + 21 - 14j$

49. $3ab^2 - 6a^2b^2 + 4a - 7b^3$ **50.** $6mn - 7m^3n^2 + 4m^2n^5 - 3m^4$

51. $-8x^4y + 5x^3y^3 - 5x^2y^6 + 3xy^7 + 2x^6$ **52.** $-2p^3q^3 + 5p^6q - 3p^2q^4 + 7p^4q^4 - 8q^5$

For Exercises 53–60, evaluate the polynomial using the given values. See Example 5.

53. $-3xy^2;\ x = -5, y = 2$ **54.** $-2x^2y;\ x = -1, y = 4$ **55.** $x^2 - 6x + 1;\ x = 3$

56. $n^2 - 8n - 3;\ n = -4$ **57.** $a^3 + 0.5ab + 2.4b;$ **58.** $m^3 + 3.6mn - 0.2n^2;$
                                                       $a = -1, b = -2$ $m = -3, n = 4$

59. $\dfrac{1}{2}x^2 + \dfrac{2}{3}x + 3;\ x = 6$ **60.** $\dfrac{1}{4}a^2 + \dfrac{3}{2}a + 2;\ a = 4$

For Exercises 61–68, evaluate. See Example 6.

61. If we neglect air resistance, the polynomial $-16t^2 + h_0$ describes the height of a falling object after falling from an initial height h_0 for t seconds. A marble is dropped from a tower at a height of 50 feet.
   a. What is its height after 0.5 seconds?

   b. What is its height after 1.2 seconds?

62. The polynomial $2lw + 2lh + 2wh$ describes the surface area of a box.
   a. Find the surface area of a box with a length of 8 inches, a width of 6.5 inches, and a height of 4 inches.

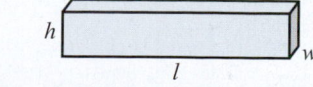

   b. Find the surface area of a box with a length of 4 inches, a width of 3 inches, and a height of 2.5 inches.

63. The polynomial $5p^2 - 6p + 3$ describes the number of units sold for every 1000 people in a particular region based on the price of the product, which is represented by p.
  a. Find the number of units sold for every 1000 people if the price of each unit is $3.

  b. Find the number of units sold for every 1000 people if the price of each unit is $3.50.

64. The polynomial $7r^2 - 2r + 6$ describes the voltage in a circuit, where r represents the resistance in the circuit.
  a. Find the voltage if the resistance is 6 ohms.

  b. Find the voltage if the resistance is 8 ohms.

65. An engineer is designing a chemical storage tank that is capsule-shaped. The polynomial $\frac{4}{3}\pi r^3 + \pi r^2 h$ describes the volume of the tank.

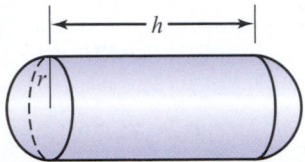

  a. Find the volume of the tank if the radius is 4 feet and the height is 20 feet. Round the result to the nearest tenth.

  b. Find the volume of the tank if the radius is 3 feet and the height is 15 feet. Round the result to the nearest tenth.

66. The polynomial $lwh - \pi r^2 h$ describes the volume of metal remaining in a block after a cylinder has been bored into the block of metal.

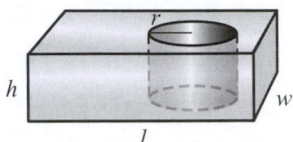

  a. Find the volume of metal remaining if the length is 15 inches, the width is 7 inches, the height is 4 inches, and the radius of the cylinder is 3 inches. Round to the nearest tenth.

  b. Find the volume of metal remaining if the length is 15 inches, the width is 7 inches, the height is 4 inches, and the radius of the cylinder is 2 inches. Round to the nearest tenth.

67. The polynomial $-0.2x^2 + 4.25x + 26$ describes the number of international visitors (in millions) to the United States each year from 1986 on, where x represents the number of years after 1986. ($x = 0$ means 1986, $x = 1$ means 1987, and so on.) (*Source:* Tourism Industries, International Trade Administration, Department of Commerce)
  a. How many international visitors came to the United States in 1986?

  b. How many international visitors came to the United States in 1993?

  c. How many international visitors came to the United States in 2003?

  d. If the model holds, predict the number of international visitors in 2012. Do you think the prediction is reasonable? Explain.

68. The polynomial $28x^3 + 25x^2 + 100x + 340$ describes the number of cell phone subscribers in the United States (in thousands of subscribers) each year since 1985, where x represents the number of years since 1985. ($x = 0$ means 1985, $x = 1$ means 1986, and so on.) (*Source:* The CTIA Semiannual Wireless Industry Survey)
  a. How many subscribers were in the United States in 1985?

  b. How many subscribers were in the United States in 1995?

  c. How many subscribers were in the United States in 2001?

  d. If the model holds, predict the number of subscribers in 2015. Do you think the prediction is reasonable? Explain.

For Exercises 69–74, write each polynomial in descending order of degree. See Example 7.

69. $7x^4 + 5x^7 - 8x + 14 - 3x^5$

70. $-4y^3 + 7y - 8y^5 - 6y^2 + 9 + 2y^4$

71. $4r - 3r^2 + 18r^5 + 7r^3 - 8r^6$

72. $u^4 + 7u^2 - u^9 + 15u + 27 - 3u^5 - 8u^3$

73. $20 - w^3 + 5w + 11w^2 - 12w^4$

74. $a - 6 + 7a^5 + 3a^2 - 4a^3$

For Exercises 75–94, combine like terms and write the resulting polynomial in descending order of degree. See Example 8.

75. $3x + 4 + 2x - 7$

76. $2y - 9 + 5y + 3$

77. $7 - 6a + 2a - 2$

78. $9 + 4b - 8b - 3$

79. $7x^2 + 2x + 4x - 9x^2$

80. $4a^3 + 3a^2 - 2a^2 - 7a^3$

81. $\frac{1}{2}x^2 + \frac{1}{3}x + \frac{4}{3}x + \frac{3}{2}x^2$

82. $\frac{2}{3}y^2 + \frac{1}{4}y + \frac{7}{4}y + \frac{5}{3}y^2$

83. $\frac{5}{3}a^2 + \frac{2}{5}a - \frac{1}{2}a^2 - \frac{5}{3}a$

84. $\frac{9}{4}n^2 - \frac{7}{6}n - \frac{2}{3}n^2 + \frac{3}{4}n$

85. $2x^2 + 5x - 7x^2 + 8 - 5x + 6$

86. $3m^5 + 7m^2 - 8m + 9m^5 - 7m - 7m^2$

87. $15 - 4y + 8y^2 - 4y - 3y^2 + 7 - y$

88. $6l - 5l^3 - 2l^4 + 7l^4 + 5l - l^3 + 4l^3$

89. $9k - 5k^2 + 6k^3 + 2k^2 + k^4 - 3k - 3k^4$

90. $11p - p^2 + 12p^3 + 5p^2 - 7p^3 + 20 - 4p$

91. $7a^2 - 5a^4 + 6a^3 - 5a^4 + 12 - 6a^3 + a + 4$

92. $-6c^9 + 7c^3 - 4c^5 + 8 - 14c^2 + 4c^5 - 7 - c^9$

93. $12v^3 + 19v^2 - 20 - v^5 - 5v^3 + 16v^2 - 20v^5 + v^3$

94. $b^3 - 13b^2 + b^4 + 6 - 2b^3 - 15b^4 - 17b^3 + 8$

For Exercises 95–106, combine like terms. See Example 9.

95. $6a - 3b - 2a + b$

96. $5x - y - 3x - 2y$

97. $3y^2 + 5y - 2y - 7y^2$

98. $2z^3 - 4z - 6z - 10z^3$

99. $2x^2 + 3y - 4x^2 + 6y - 5x^2 - 2x^2$

100. $3a^2 - 5b - 4a^2 + 7b - 5b + 6b$

★101. $\frac{1}{3}a^2 + \frac{3}{5}a - \frac{3}{2}a + \frac{5}{4}a^2 + \frac{1}{3}a$

★102. $\frac{7}{6}x^2 - \frac{3}{4}x - \frac{2}{5}x^2 - \frac{4}{5}x + \frac{4}{3}x$

103. $y^6 + 2yz^4 - 10yz + 3yz^4 - 3z^5 - 7z^3 + 4yz - 10$ **104.** $x^4 + xy^3 - 6xy + xy^3 + 3y^4 - 4y^2 + 3xy - 8$

105. $-3w^2z - 9 + 6wz^2 + 3w^2 - w^2z + 8 + 9w^2 - 7wz^2$ **106.** $-m^3n - 8mn + 4 - 5m^2n^2 + 3m^3n - 7 + mn - 16m^2n^2$

Puzzle Problem

Rearrange the numbers in the figure shown so that no two consecutive numbers are next to each other horizontally, vertically, or diagonally.

Review Exercises

Exercises 1–4   **EXPRESSIONS**

[1.3] *For Exercises 1 and 2, simplify.*

1. $-\dfrac{3}{4} + \dfrac{1}{6}$

2. $-14.6 - (-10.3)$

[5.1] **3.** Evaluate: -7^0

[5.1] **4.** The Andromeda galaxy (the closest one to our Milky Way galaxy) contains at least 200,000,000,000 stars. Write this number in scientific notation.

Exercises 5 and 6    **EQUATIONS AND INEQUALITIES**

[2.3] **5.** Solve and check: $\dfrac{1}{4}x = 3$

[3.5] **6.** How many liters of a 40% solution must be added to 200 liters of a 20% solution to obtain a 35% solution?

5.3 Adding and Subtracting Polynomials

Objectives

1 Add polynomials.

2 Subtract polynomials.

Objective 1 Add polynomials.

Understanding Polynomial Addition

We can add and subtract polynomials in the same way that we add and subtract numbers. In fact, polynomials are like whole numbers that are in an expanded form. Consider the polynomial $4x^2 + 3x + 6$. If we replace the x's with the number 10, we have the expanded form for the number 436.

$$4x^2 + 3x + 6$$

$$4 \cdot 10^2 + 3 \cdot 10 + 6 \qquad \text{\textcolor{blue}{Replacing } } x \text{ with 10}$$
$$= 400 + 30 + 6$$
$$= 436$$

In our base-ten number system, each place value is a power of 10. We can think of polynomials as a variable-base number system. In other words, the place values are variables, where x^2 is like the hundreds place (10^2) and x is like the tens place (10^1). To add whole numbers, we add the digits in like place values. Polynomials are added in a similar way. However, instead of adding digits in like place values, we add like terms. Consider the following comparison.

Numeric addition:

$436 + 251 = 687$

$$\begin{array}{r} 436 \\ + 251 \\ \hline 687 \end{array}$$

Note: In numeric addition, like place values are added.

Polynomial addition:

$(4x^2 + 3x + 6) + (2x^2 + 5x + 1) = 6x^2 + 8x + 7$

$$\begin{array}{r} 4x^2 + 3x + 6 \\ + 2x^2 + 5x + 1 \\ \hline 6x^2 + 8x + 7 \end{array}$$

Note: In polynomial addition, like terms are added.

From this simple case, we see that polynomials are added by combining the like terms.

Procedure **Adding Polynomials**

To add polynomials, combine like terms.

Adding Polynomials

Although we stacked the preceding polynomials, we do not actually have to stack polynomials in order to add them.

Example 1 Add and write the resulting polynomial in descending order of degree.

a. $(2x^3 + 5x^2 + 4x + 1) + (6x^3 + 3x + 8)$

SOLUTION Combine like terms. Notice that combining in order of degree places the resulting polynomial in descending order of degree.

$$(2x^3 + 5x^2 + 4x + 1) + (6x^3 + 3x + 8)$$

$$= 8x^3 + 5x^2 + 7x + 9$$

b. $(y^5 + 9y^3 - 3y - 7) + (2y^5 - 4y^3 + 3y + 5)$

SOLUTION $(y^5 + 9y^3 - 3y - 7) + (2y^5 - 4y^3 + 3y + 5)$ **Combine like terms.**

$$= 3y^5 + 5y^3 \qquad + 0 \qquad -2$$
$$= 3y^5 + 5y^3 - 2$$

If we had stacked the polynomials in Example 1(a), we would have had to leave a blank space under the $5x^2$ because there was no x^2 term in the second polynomial. It is the same as having a 0 in that place.

$$\begin{array}{r} 2x^3 + 5x^2 + 4x + 1 \\ +6x^3 + 0 + 3x + 8 \\ \hline 8x^3 + 5x^2 + 7x + 9 \end{array}$$

As the polynomials get more complex, these blanks become more prevalent. This is why the stacking method is not a preferred method for adding polynomials.

Your Turn 1   Add and write the resulting polynomial in descending order of degree.

a. $(x^4 + 3x^3 + 4x + 2) + (5x^3 + 2x + 7)$

b. $(14n^5 - 6n^3 - 12n + 15) + (2n^5 - 4n^3 + 8n - 5)$

Sometimes the polynomials contain terms that have several variables or coefficients that are fractions or decimals.

Example 2   Add and write the resulting polynomial in descending order of degree:

$$\left(7t^4 + 2.9tu^2 - \frac{3}{4}tu + 11\right) + \left(-8t^4 - 7tu^2 + \frac{1}{6}tu - 5\right)$$

SOLUTION $\left(7t^4 + 2.9tu^2 - \dfrac{3}{4}tu + 11\right) + \left(-8t^4 - 7tu^2 + \dfrac{1}{6}tu - 5\right)$ **Combine like terms.**

$$= -t^4 - 4.1tu^2 - \frac{7}{12}tu + 6$$

Your Turn 2   Add and write the resulting polynomial in descending order of degree.

a. $(12x^3 - 6.2xy^2 - 0.3xy - y^2) + (x^3 + 7xy^2 - 0.9xy + 0.4y^2)$

b. $\left(a^4 - \dfrac{4}{5}ab^3 - \dfrac{2}{3}ab + 9\right) + \left(a^4 + \dfrac{1}{4}ab^3 + ab - 5\right)$

Example 3 Write an expression in simplest form for the perimeter of the rectangle shown.

$3x + 7$ | $5x - 1$

Understand *Perimeter* means the total distance around the shape. Therefore, we need to add the lengths of all of the sides of the shape.

Plan In this case, the lengths of the sides are represented by polynomials. Therefore, we add the polynomials to represent the perimeter.

Execute Perimeter = Length + Width + Length + Width

$$= (5x - 1) + (3x + 7) + (5x - 1) + (3x + 7)$$
$$= 5x + 3x + 5x + 3x - 1 + 7 - 1 + 7$$
$$= 16x + 12$$

Answer The expression for the perimeter is $16x + 12$.

Check To check, we (1) choose a value for x and evaluate the original expressions for length and width, (2) determine the corresponding numeric perimeter, and (3) evaluate the perimeter expression using the same value for x and verify that we get the same numeric perimeter. Let's choose $x = 2$.

Length: $5x - 1$ | Width: $3x + 7$ | Perimeter:
$5(2) - 1$ | $3(2) + 7$ | The perimeter of the rectangle with a
$= 10 - 1$ | $= 6 + 7$ | length of 9 and width of 13 is
$= 9$ | $= 13$ | $9 + 13 + 9 + 13 = 44$.

Now evaluate the perimeter expression where $x = 2$; you should find that the result is 44.

Perimeter expression: $16x + 12$
$$16(2) + 12$$
$$= 32 + 12$$
$$= 44$$ **This agrees with our calculation above.**

Your Turn 3 Write an expression in simplest form for the perimeter of the following shape.

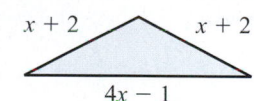

Objective ② Subtract polynomials.

Understanding Polynomial Subtraction

Subtracting polynomials is similar to subtracting signed numbers. When we subtract signed numbers, it is often simpler to write the subtraction statement as an equivalent addition statement. Consider the following comparison of numeric subtraction and polynomial subtraction.

Numeric subtraction:

$975 - 621$

We can stack by place:

$$
\begin{array}{r}
975 \\
-621 \\
\hline
354
\end{array}
$$

Polynomial subtraction:

$(9x^2 + 7x + 5) - (6x^2 + 2x + 1)$

We can stack like terms:

$$
\begin{array}{r}
9x^2 + 7x + 5 \\
-(6x^2 + 2x + 1) \\
\hline
3x^2 + 5x + 4
\end{array}
$$

Answer   Your Turn 3
$6x + 3$

5.3 Adding and Subtracting Polynomials **361**

In numeric subtraction, we subtract digits in the same place value, whereas in polynomial subtraction, we subtract like terms. Note that subtracting 1 from 5 is equivalent to combining 5 and -1, subtracting $2x$ from $7x$ is equivalent to combining $7x$ and $-2x$, and subtracting $6x^2$ from $9x^2$ is equivalent to combining $9x^2$ and $-6x^2$. This suggests that we can write polynomial subtraction as equivalent polynomial addition by changing the sign of each term in the second polynomial (the subtrahend), like this:

$$(9x^2 + 7x + 5) - (6x^2 + 2x + 1)$$
$$= (9x^2 + 7x + 5) + (-6x^2 - 2x - 1)$$
$$= 3x^2 + 5x + 4$$

Note: We are allowed to change the signs because of the distributive property. If we disregard the initial polynomial, we have

$$-(6x^2 + 2x + 1) = -1(6x^2 + 2x + 1).$$

The distributive property tells us that we can distribute the -1 (or minus sign) to each term inside the parentheses.

$$= -1 \cdot 6x^2 - 1 \cdot 2x - 1 \cdot 1$$
$$= -6x^2 - 2x - 1$$

Subtracting Polynomials

Our exploration suggests the following procedure for subtracting polynomials.

Procedure **Subtracting Polynomials**

To subtract polynomials,

1. Write the subtraction statement as an equivalent addition statement.
   a. Change the operation symbol from a minus sign to a plus sign.
   b. Change the subtrahend (second polynomial) to its additive inverse. To get the additive inverse, change the sign of each term in the polynomial.
2. Combine like terms.

Connection The procedure for subtracting polynomials is the same as the procedure for subtracting signed numbers (Section 1.3). In both cases, we write subtraction as equivalent addition by changing the subtrahend to its additive inverse.

Example 4 Subtract and write the resulting polynomial in descending order of degree: $(7x^3 + 8x^2 + 6x + 9) - (3x^3 + 6x^2 + 5x + 1)$

SOLUTION Write an equivalent addition statement; then combine like terms.

$$(7x^3 + 8x^2 + 6x + 9) - (\ 3x^3 + 6x^2 + 5x + 1)$$

Change the minus sign to a plus sign.   Change all signs in the subtrahend.

$$= (7x^3 + 8x^2 + 6x + 9) + (-3x^3 - 6x^2 - 5x - 1)$$
$$= 4x^3 + 2x^2 + x + 8 \qquad \text{Combine like terms.}$$

Connection The polynomial subtraction in Example 4 is equivalent to the following numeric subtraction:

$$
\begin{array}{r}
7869 \\
-3651 \\
\hline
4218
\end{array}
$$

Notice that the numeric result 4218 corresponds to the polynomial result $4x^3 + 2x^2 + x + 8$. We could use the stacking method for subtracting polynomials; however, as we saw with adding polynomials, it is not the best method for all cases because we often have missing terms or terms that are not like terms.

You may have noticed that the polynomials in Example 4 contained only plus signs. It is important to recognize that the polynomials may contain a mixture of signs and multiple variables. This is shown in Example 5.

Example 5 Subtract and write the resulting polynomial in descending order of degree.

a. $(15y^3 + 3y^2 + y - 2) - (7y^3 - 6y^2 + 5y - 8)$

SOLUTION Write an equivalent addition statement; then combine like terms.

$$(15y^3 + 3y^2 + y - 2) - (\ 7y^3 - 6y^2 + 5y - 8)$$

Change the minus sign to a plus sign. Change all signs in the subtrahend.

$$= (15y^3 + 3y^2 + y - 2) + (-7y^3 + 6y^2 - 5y + 8)$$
$$= 8y^3 + 9y^2 - 4y + 6 \qquad \text{Combine like terms.}$$

b. $(9.7x^5 - 2x^2y + xy^2 - 14.6xy - 7y^2) - (x^5 - 5.8xy^2 + 10.3xy - 15y^2)$

SOLUTION We write an equivalent addition statement, then combine like terms.

$$(9.7x^5 - 2x^2y + xy^2 - 14.6xy - 7y^2) - (\ x^5 - 5.8xy^2 + 10.3xy - 15y^2)$$

Change the minus sign to a plus sign. Change all signs in the subtrahend.

$$= (9.7x^5 - 2x^2y + xy^2 - 14.6xy - 7y^2) + (-x^5 + 5.8xy^2 - 10.3xy + 15y^2)$$
$$= 8.7x^5 - 2x^2y + 6.8xy^2 - 24.9xy + 8y^2$$

Note: The $-2x^2y$ term had no like term, so we rewrote it in the final expression.

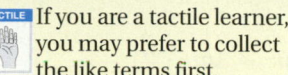
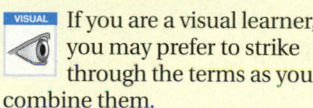

LEARNING Strategy

TACTILE If you are a tactile learner, you may prefer to collect the like terms first.

VISUAL If you are a visual learner, you may prefer to strike through the terms as you combine them.

Your Turn 5 Subtract and write the resulting polynomial in descending order of degree.

a. $(8t^4 + 5t^2 + 9t + 7) - (2t^4 + 5t^2 + t + 4)$

b. $(12.5x^4 - 15x^3 + 2x^2 - 9) - (13.1x^4 + 9x^3 - 6x^2 - 14)$

c. $(7a^5 - a^3b^2 + 9a^2b^2 - 2ab^2 + b - 18) - (2a^5 + a^3b^2 - ab^2 + 6b - 15)$

Answers Your Turn 5

a. $6t^4 + 8t + 3$

b. $-0.6x^4 - 24x^3 + 8x^2 + 5$

c. $5a^5 - 2a^3b^2 + 9a^2b^2 - ab^2 - 5b - 3$

5.3 Exercises   For Extra Help   *MyMathLab*

Note: Exercises marked with a ★ represent challenging exercises.

1. How do we add two polynomials?

2. What is the function of parentheses in the addition problem $(3x + 9) + (4x - 2)$?

3. How do we find the additive inverse of a polynomial?

4. How do we subtract two polynomials?

For Exercises 5–28, add and write the resulting polynomial in descending order of degree. See Examples 1 and 2.

5. $(3x + 2) + (5x - 1)$

6. $(5y + 4) + (3y + 1)$

7. $(8y + 7) + (2y + 5)$

8. $(3m - 4) + (5m + 1)$

9. $(2x + 3y) + (5x - 3y)$

10. $(4a - 3b) + (6a + 3b)$

11. $(2x + 5) + (3x^2 - 4x + 7)$

12. $(5p^2 + 3p - 1) + (4p + 8)$

13. $(z^2 - 3z + 7) + (5z^2 - 8z - 9)$

14. $(2w^2 - 5w - 1) + (4w^2 - 5w + 1)$

15. $(3r^2 - 2r + 10) + (2r^2 - 5r - 11)$

16. $(7m^2 + 8m - 1) + (-4m^2 - 3m - 2)$

17. $(4y^2 - 8y + 1) + (5y^2 + 8y + 2)$

18. $(5k^2 - k - 1) + (4k^2 + k + 7)$

19. $\left(x^2 - \dfrac{2}{3}x + 3\right) + \left(2x^2 + \dfrac{3}{4}x - 2\right)$

20. $\left(3a^2 + 3a - \dfrac{5}{6}\right) + \left(2a^2 - 5a + \dfrac{3}{4}\right)$

21. $(4r^2 + 2.7r - 3.6) + (-2r^2 - 1.3r - 4.2)$

22. $(5b^2 - 4.6b - 1.7) + (-4b^2 + 3.3b + 5.6)$

23. $(9a^3 + 5a^2 - 3a - 1) + (-4a^3 - 2a^2 + 6a - 2)$

24. $(4u^3 - 6u^2 + u + 11) + (-5u^3 - 3u^2 + u - 5)$

25. $(7p^3 - 9p^2 + 5p - 1) + (-4p^3 + 8p^2 + 2p + 10)$

26. $(12r^4 - 5r^2 + 8r - 15) + (-7r^4 + 3r^3 + 2r - 9)$

27. $(-5w^4 - 3w^3 - 8w^2 + w - 14) + (-3w^4 + 6w^3 + w^2 + 12w + 5)$

28. $(-r^4 + 3r^2 - 12r - 14) + (5r^4 - 2r^3 - 3r^2 + 10r - 5)$

★ *For Exercises 29–38, add. See Example 2.*

29. $(a^3b^2 - 6ab^2 - 6a^2b + 5ab + 5b^2 - 6) + (-4a^3b^2 + 4ab^2 - 2a^2b - ab - 2b^2 - 2)$

30. $(x^2y^2 + 8xy^2 - 12x^2y - 4xy + 7y^2 - 9) + (-3x^2y^2 + 2xy^2 + 4x^2y - xy + 6y^2 + 4)$

31. $(-2u^4 + 6uv^3 - 8u^2v^3 + 7u^3v - u + v^2 + 9) + (-5u^4 - 6uv^3 + 9u^2v^3 + 8u - 14v^2 - 12)$

32. $(-13a^6 + a^3b^2 + 3ab^3 - 8b^3) + (11a^6 - 3a^2b^3 - 4ab^3 + 10ab^2 + 4b^3)$

33. $(-8mnp - 6m^2n^2 - 13mn^2p + 14m^2n - 12n + 6) + (-4nmp - 10m^2n^2 + mn^2p - 19n + 3)$

34. $(14abc - 2a^2b^2 - ab^2c + 4b - 9c - 3) + (-15abc + 5a^2b^2 - 13ab^2c + 7a - 12b + 6)$

35. $\left(4a^4 - \dfrac{2}{3}a^3b + \dfrac{3}{5}ab^3 - 3b^4\right) + \left(-2a^4 + \dfrac{1}{4}a^3b + \dfrac{1}{3}ab^3 + 2b^4\right)$

36. $\left(-5y^4 + \dfrac{3}{4}y^2z^2 - \dfrac{5}{6}yz^3 + 2z^4\right) + \left(3y^4 - \dfrac{1}{6}y^2z^2 - \dfrac{3}{8}yz^3 - 5z^4\right)$

37. $(3.6a^2bc + 2.3ab^2c^2 - 5.7a^2b^2c^2) + (-1.8a^2bc - 4.1ab^2c^2 + 3.3a^2b^2c^2)$

38. $(-6.8x^3yz^2 + 4.1x^2y^2z - 8.2xy^2z^2) + (5.2x^3yz^2 - 3.6x^2y^2z - 3.2xy^2z^2)$

For Exercises 39–42, write an expression for the perimeter in simplest form.
See Example 3.

39.
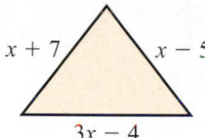

40.
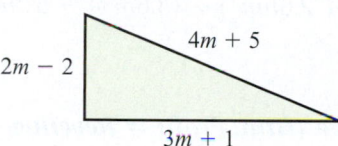

41.
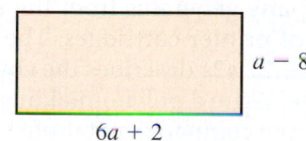

42.
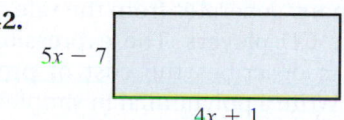

For Exercises 43–60, subtract and write the resulting polynomial in descending
order of degree. See Examples 4 and 5.

43. $(6x + 3) - (2x + 1)$ **44.** $(5m + 8) - (2m + 3)$ **45.** $(18a^2 + 3) - (5a^2 + 4)$

46. $(12b^2 + 4b) - (5b^2 + b)$ **47.** $(2x^2 - 3x + 4) - (6x + 2)$ **48.** $(7y^2 - 3y + 2) - (8y + 5)$

49. $(6z^2 - 3z + 1) - (4z^2 + 4z - 8)$ **50.** $(5y^2 - 3y - 10) - (5y^2 + 2y - 10)$

51. $(8a^2 + 11a - 12) - (-4a^2 + a + 3)$ **52.** $(y^2 + 3y + 6) - (-5y^2 + 3y - 8)$

53. $(6u^3 + 3u^2 - 8u + 5) - (7u^3 + 5u^2 - 2u - 3)$ **54.** $(-8t^3 + 3t^2 - 7t - 9) - (-10t^3 - 5t^2 + 3t - 11)$

55. $(-w^5 + 3w^4 + 8w^2 - w - 1) - (10w^5 + 3w^2 - w - 3)$

56. $(-s^4 + 7s^3 + 6s^2 - 13s - 11) - (3s^4 - 2s^2 - s + 5)$

57. $(-5p^4 + 8p^3 - 9p^2 + 3p + 1) - (6p^4 + 5p^3 + 4p^2 - 2p + 7)$

58. $(-6r^5 - 9r^4 + 5r^3 - 2r^2 + 8) - (5r^5 + 2r^4 + 7r^3 - 8r^2 - 9)$

59. $(16v^4 + 21v^2 + 4v - 8) - (5v^4 - 3v^3 - 2v^2 + 9v + 1)$

60. $(12q^4 + 5q^3 - 2q + 15) - (8q^4 - 14q^3 - 3q^2 + 15q + 8)$

For Exercises 61–68, subtract. See Example 5.

61. $(x^2 + xy + y^2) - (x^2 - xy + y^2)$

62. $(a^2 - 4ab + b^2) - (a^2 + 4ab + b^2)$

63. $(4xy + 5xz - 6yz) - (10xy - xz - 8yz)$

64. $(4ab - 5bc + 6ac) - (10ab - bc - ac)$

65. $(18p^3q^2 - p^3q^3 + 6pq^2 - 4pq + 16q^2 - 4) - (-p^3q^2 - p^3q^3 + 4pq^2 - 3pq - 6)$

66. $(15a^3y^4 + a^2y^2 + 5ay^3 - 7ay + 12y^2 - 11) - (-a^3y^4 + a^2y^2 + 2ay^3 - ay + 9)$

67. $(5.6a^2b - 2.3ab^2 - 4.3a^2b^2) - (3.7a^2b - 3.4ab^2 + 2.2a^2b^2)$

68. $(7.4m^3n - 6.7m^2n^2 + 2.6mn^3) - (3.6m^3n - 8.3m^2n^2 - 4.7mn^3)$

For Exercises 69–72, solve. (Hint: Profit = Revenue − Cost)

69. The polynomial $54.95x + 92.20y + 25.35z$ describes the revenue that a company generates from the sale of three different portable CD players. The expression $22.85x + 56.75y + 19.34z$ describes the cost of producing each CD player. Write a polynomial in simplest form that describes the company's net profit.

70. The polynomial $27.50x + 14.70y + 42.38z$ describes the revenue that a company generates from the sale of three different types of printer cartridges. The expression $12.75x + 5.25y + 27.42z$ describes the cost of producing each cartridge. Write a polynomial in simplest form that describes the company's net profit.

71. The polynomial $6.50a + 3.38b + 25.00c$ describes the revenue that a department store generates from the sale of three different candles. The expression $2.28a + 1.75b + 12.87c$ describes the cost the store pays to sell each candle. Write a polynomial in simplest form that describes the store's net profit.

72. The polynomial $10.55m + 14.75n + 27.50p$ describes the revenue that a pet store generates from the sale of three different litter boxes. The expression $5.73m + 8.26n + 15.22p$ describes the cost the store pays to sell each product. Write a polynomial in simplest form that describes the store's net profit.

Find (X) the mistake *For Exercises 73 and 74, find and explain the mistake; then work the problem correctly.*

73. $(3x^2 - 5x + 10) - (7x^2 - 3x + 5) = (3x^2 - 5x + 10) + (-7x^2 - 3x + 5)$

$$= 3x^2 - 5x + 10 - 7x^2 - 3x + 5$$

$$= -4x^2 - 8x + 15$$

74. $(7y^2 - 3y + 1) - (y^2 + 4y - 1) = (-7y^2 + 3y - 1) + (-y^2 - 4y + 1)$

$$= -7y^2 + 3y - 1 - y^2 - 4y + 1$$

$$= -8y^2 - y$$

Collaborative Exercises   Building Furniture and Profits

Suppose your group operates a furniture-manufacturing company that produces two different sizes of coffee table. Each larger coffee table sells for $350, and each smaller table sells for $280.

1. Write a monomial expression for the revenue if x is the number of large tables sold per month.

2. Write a monomial expression for the revenue if y is the number of small tables sold per month.

3. Write a polynomial expression that describes the total revenue generated from the sale of the two tables per month.

Suppose it costs $225 to produce each large table and $135 to produce each small table.

4. Write a monomial expression for the cost of producing x number of large tables per month.

5. Write a monomial expression for the cost of producing y number of small tables per month.

6. Your company also spends $18,500 per month to pay for the lease, utilities, and salaries. Write a polynomial expression that describes the total cost involved in production.

7. Using your polynomials for revenue and cost, write a polynomial in simplest form for the profit if x number of large tables are produced and sold and y number of small tables are produced and sold.

8. What does each coefficient in the polynomial for profit indicate?

9. Suppose in one month, 200 large tables are produced and sold and 400 small tables are produced and sold. Find the revenue, cost, and profit for the month.

10. Suppose in one month, 56 large tables are produced and sold and 65 small tables are produced and sold. Find the revenue, cost, and profit for the month. Explain the meaning of your answer.

11. Suppose in one month, only large tables are produced and sold. How many would have to be produced and sold to break even (profit is 0)?

12. Suppose only small tables are produced and sold in one month. How many would have to be produced and sold to break even?

Review Exercises

Exercises 1–5 EXPRESSIONS

For Exercises 1–4, evaluate.

[1.4]   1. $(2.3)(-7.5)$ [1.4]   2. $(-5)(-2)(1.5)$ [5.1]   3. 10^{-4} [1.5]   4. $10^3 \cdot 10^4$

[5.1]   5. Write 5.89×10^7 in standard form.

Exercise 6   EQUATIONS AND INEQUALITIES

[2.1]   6. The surface of the water in a rectangular swimming pool measures 20 meters by 15 meters. What is the surface area of the water?

5.4 Exponent Rules and Multiplying Monomials

Objectives

1. Multiply monomials.
2. Multiply numbers in scientific notation.
3. Simplify a monomial raised to a power.

In this section, we learn how to multiply monomials, multiply numbers in scientific notation, and simplify a monomial raised to a power. These simplifications require some new rules for exponents.

Objective 1 **Multiply monomials.** Because monomials contain variables in exponential form, to multiply monomials, we must develop a rule for multiplying exponential forms.

Multiplying Exponential Forms

Consider $2^3 \cdot 2^4$, which is a product of exponential forms. To simplify $2^3 \cdot 2^4$, we could follow the order of operations and evaluate the exponential forms first, then multiply.

$$2^3 \cdot 2^4 = 8 \cdot 16 = 128$$

However, there is an alternative. We can write the result in exponential form by first writing 2^3 and 2^4 in their factored forms.

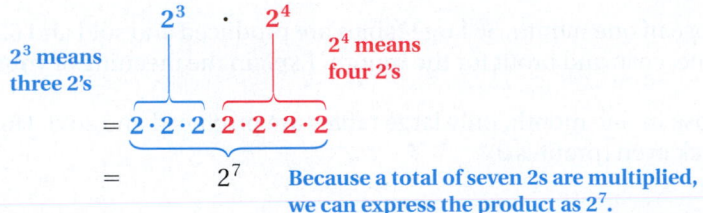

Notice that $2^7 = 128$, which agrees with our earlier calculation. Also notice that in the alternative method, the resulting exponent is the sum of the original exponents.

$$2^3 \cdot 2^4 = 2^{3+4} = 2^7$$

Conclusion To multiply exponential forms that have the same base, we can add the exponents and keep the same base.

This suggests the following rule.

Rule **Product Rule for Exponents**

If a is a real number and m and n are integers, then $a^m \cdot a^n = a^{m+n}$.

We apply this rule when multiplying monomials.

Example 1 Multiply $(x^5)(x^3)$.

SOLUTION Because the bases are the same, we can add the exponents and keep the same base.

$$(x^5)(x^3) = x^{5+3} = x^8$$

Your Turn 1 Multiply.

a. $3^2 \cdot 3^5$ **b.** $(c^7)(c^3)$ **c.** $t^5(t^3)$

Multiplying Monomials

Now let's consider multiplying monomials that have coefficients, such as $(4x^3)(2x^6)$. Because monomials are products by definition, we can use the commutative property of multiplication and write the following:

$$
\begin{aligned}
(4x^3)(2x^6) &= 4 \cdot x^3 \cdot 2 \cdot x^6 \\
&= 4 \cdot 2 \cdot x^3 \cdot x^6 \\
&= 8 \cdot x^{3+6} \\
&= 8x^9
\end{aligned}
$$

We multiply the coefficients and use the product rule for exponents to simplify $x^3 \cdot x^6$.

What if the monomial factors have some variables that are different, as in $(5x^2)(7xy)$? If we expand the monomials in factored form and use the commutative property of multiplication, we have the following:

$$
\begin{aligned}
(5x^2)(7xy) &= 5 \cdot x^2 \cdot 7 \cdot x \cdot y \\
&= 5 \cdot 7 \cdot x^2 \cdot x \cdot y \\
&= 35 \cdot x^{2+1} \cdot y \\
&= 35x^3 y
\end{aligned}
$$

Notice that the unlike variable base, y, is simply written unchanged in the result. Our examples suggest the following procedure.

Procedure   **Multiplying Monomials**

To multiply monomials,

1. Multiply coefficients.
2. Add the exponents of the like bases.
3. Write any unlike variable bases unchanged in the product.

Example 2   Multiply.

a. $(9y^5)(3y^8)$

SOLUTION $(9y^5)(3y^8) = 9 \cdot 3y^{5+8}$ Multiply the coefficients and add the exponents of the like bases.

$$= 27y^{13}$$ Simplify the exponents.

b. $\left(-\dfrac{5}{6}a^2bc^3\right)\left(\dfrac{2}{3}a^5b^4\right)$

Note: Remember that a variable or number with no apparent exponent has an understood exponent of 1.

$$-\frac{5}{6}a^2bc^3 = -\frac{5}{6}a^2b^1c^3$$

SOLUTION $\left(-\dfrac{5}{6}a^2bc^3\right)\left(\dfrac{2}{3}a^5b^4\right) = -\dfrac{5}{\overset{}{\underset{3}{6}}} \cdot \dfrac{\overset{1}{2}}{3}a^{2+5}b^{1+4}c^3$

Multiply the coefficients, add the exponents of the like bases, and write the unlike variable base c unchanged in the product.

$$= -\frac{5}{9}a^7b^5c^3$$ Simplify.

Answers Your Turn 1

a. 3^7 **b.** c^{10} **c.** t^8

Your Turn 2   Multiply.

a. $(9n^4)(5n^8)$ **b.** $-2.5x^2y^3(-6xy^2z^3)$

Answers Your Turn 2

a. $45n^{12}$ **b.** $15x^3y^5z^3$

Connection Multiplying distance measurements in area and volume is like multiplying monomials.

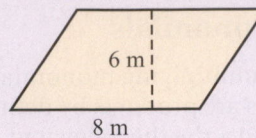

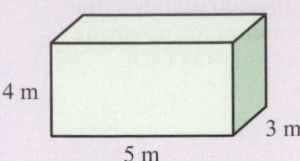

The area of the parallelogram:

$A = bh$

$A = (8\,\text{m})\,(6\,\text{m})$

$A = 48\,\text{m}^2$

The area calculation is similar to multiplying two monomials with m as the variable.

Multiply: $(8m)(6m) = 48m^2$

The volume of the box:

$V = lwh$

$V = (5\,\text{m})(3\,\text{m})(4\,\text{m})$

$V = 60\,\text{m}^3$

The volume calculation is similar to multiplying three monomials with m as the variable.

Multiply: $(5m)(3m)(4m) = 60m^3$

Example 3   Write an expression in simplest form for the volume of the box shown.

Understand   We are given a box with side lengths that are monomial expressions.

Plan   The volume of a box is found by multiplying the length, width, and height.

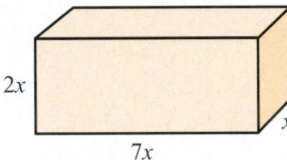

Execute $V = lwh$

$V = (7x)\,(x)\,(2x)$

$V = 14x^3$

Answer   The expression for volume is $14x^3$.

Check   Because $(7x)(x)(2x) = 14x^3$ is an identity, we can evaluate $(7x)$, (x), $(2x)$, and $14x^3$ using any chosen value for x and get the same result. Let's choose a value for x, such as 3, and try it.

$$(7x)(x)(2x) = 14x^3$$
$$[7(3)][(3)][2(3)] \overset{?}{=} 14(3)^3$$
$$(21)(3)(6) \overset{?}{=} 14(27)$$
$$378 = 378 \qquad \text{This is true.}$$

> **Note:** Choosing only one value does not guarantee that our simplification is correct, but it is a good indicator. Testing several values provides more conclusive assurance.

Objective ❷ Multiply numbers in scientific notation.   We multiply numbers in scientific notation using the same procedure we used to multiply monomials. If we replace a with 10 in $(4a^3)(2a^6)$, we have a product of two numbers in scientific notation: $(4 \times 10^3)(2 \times 10^6)$. Following the same procedure for multiplying monomials, we multiply $4 \cdot 2$ to get 8 and then add the exponents of the like bases.

Monomials:

$$(4a^3)(2a^6) = 4 \cdot 2a^{3+6}$$
$$= 8a^9$$

Scientific notation:

$$(4 \times 10^3)(2 \times 10^6) = 4 \times 2 \times 10^{3+6}$$
$$= 8 \times 10^9$$

Example 4 Multiply and write the answers in scientific notation.

a. $(2.3 \times 10^5)(7.5 \times 10^6)$

SOLUTION Multiply 2.3 and 7.5; then add the exponents for base 10s.

$$\begin{aligned}(2.3 \times 10^5)(7.5 \times 10^6) &= 2.3 \times 7.5 \times 10^{5+6}\\ &= 17.25 \times 10^{11}\\ &= 1.725 \times 10^{12}\end{aligned}$$

b. $(4.2 \times 10^4)(6.5 \times 10^{-9})$

SOLUTION Multiply 4.2 and 6.5 and add the exponents for the base 10s.

$$\begin{aligned}(4.2 \times 10^4)(6.5 \times 10^{-9}) &= 4.2 \times 6.5 \times 10^{4+(-9)}\\ &= 27.3 \times 10^{-5}\\ &= 2.73 \times 10^{-4} \quad \text{\color{blue}Rewrite in scientific notation.}\end{aligned}$$

Your Turn 4 Multiply and write the answers in scientific notation.

a. $(4.2 \times 10^5)(2.8 \times 10^8)$ **b.** $(2.7 \times 10^{-3})(7.3 \times 10^{10})$

Objective ③ Simplify a monomial raised to a power. We now discuss how to simplify an expression such as $(3x^2)^4$, which is a monomial raised to a power. To simplify such an expression, we need to discuss two new rules of exponents: (1) raising a power to a power and (2) raising a product to a power.

Raising a Power to a Power

Consider a power raised to a power, such as $(2^3)^2$. To calculate $(2^3)^2$, we could follow the order of operations and evaluate the exponential form 2^3 within the parentheses first, then square the result.

$$(2^3)^2 = (2 \cdot 2 \cdot 2)^2 = 8^2 = 64$$

However, there is an alternative. The outside exponent, 2, indicates to multiply the inside exponential form 2^3 by itself.

$$\begin{aligned}(2^3)^2 &= 2^3 \cdot 2^3\\ &= 2^{3+3}\\ &= 2^6\end{aligned}$$ **\color{blue}Because this is a multiplication of exponential forms that have the same base, we can add the exponents.**

Notice that $2^6 = 64$. Also, in comparing $(2^3)^2$ with 2^6, notice that the exponent in 2^6 is the product of the original exponents.

$$(2^3)^2 = 2^{3 \cdot 2} = 2^6$$

Conclusion If an exponential form is raised to a power, we can multiply the exponents and keep the same base.

Rule   **A Power Raised to a Power**

If a is a real number and m and n are integers, then $(a^m)^n = a^{mn}$.

Answers Your Turn 4
a. 1.176×10^{14} **b.** 1.971×10^8

Now let's consider a product raised to a power, such as $(3 \cdot 4)^2$. We could write the factored form twice and use the commutative property to rearrange the like bases.

$$(3 \cdot 4)^2 = (3 \cdot 4)(3 \cdot 4)$$

Because there are two factors of 3 and two factors of 4, we can write each of these in exponential form.

$$= 3 \cdot 3 \cdot 4 \cdot 4$$
$$= 3^2 \cdot 4^2$$

Conclusion If a product is raised to a power, we can evaluate the factors raised to that power.

Rule **Raising a Product to a Power**

If a and b are real numbers and n is an integer, then $(ab)^n = a^n b^n$.

Raising a Monomial to a Power

We apply these exponent rules to simplify an expression such as $(3x^2)^4$, which is a monomial raised to a power. Because the monomial $3x^2$ expresses a product, we can use the rule for raising a product to a power and raise the factors 3 and x^2 to the 4th power.

$$(3x^2)^4 = (3)^4(x^2)^4$$

To simplify $(x^2)^4$, we use the rule for raising a power to a power.

$$= 81x^{2 \cdot 4}$$
$$= 81x^8$$

Notice that the outside exponent is distributed to the coefficient and the variable(s) within the monomial. This suggests the following procedure.

Procedure **Simplifying a Monomial Raised to a Power**

To simplify a monomial raised to a power,

1. Evaluate the coefficient raised to that power.
2. Multiply each variable's exponent by the power.

Example 5 Simplify.

a. $(5a^6)^3$

SOLUTION $(5a^6)^3 = (5)^3 a^{6 \cdot 3}$

Write the coefficient, 5, raised to the 3rd power and multiply the variable's exponent by 3.

$$= 125a^{18}$$

Simplify.

b. $\left(-\frac{2}{3}xy^3z^5\right)^4$

Note: Recall that $x = x^1$.

SOLUTION $\left(-\frac{2}{3}xy^3z^5\right)^4 = \left(-\frac{2}{3}\right)^4 x^{1 \cdot 4}y^{3 \cdot 4}z^{5 \cdot 4}$

Write the coefficient, $-\frac{2}{3}$, raised to the 4th power and multiply each variable's exponent by 4.

$$= \frac{16}{81}x^4 y^{12} z^{20}$$

Simplify.

c. $(-0.4mn^6)^3$

SOLUTION $(-0.4mn^6)^3 = (-0.4)^3 m^{1 \cdot 3} n^{6 \cdot 3}$
$$= -0.064m^3 n^{18}$$

Your Turn 5 Simplify.

a. $(2y^4)^5$　　　　**b.** $\left(-\dfrac{5}{3}t^2u^6\right)^3$　　　　**c.** $(-0.2a^4b^5c)^6$

Now let's consider expressions that require us to use all of these exponent rules.

Example 6 Simplify.

a. $(3x^2)(4x^5)^3$

SOLUTION Because the order of operations is to simplify exponents before multiplying, we simplify $(4x^5)^3$ first, then multiply the result by $3x^2$.

$$
\begin{aligned}
(3x^2)(4x^5)^3 &= (3x^2)(4^3x^{5\cdot3}) \\
&= (3x^2)(64x^{15}) \quad &&\text{Simplify.} \\
&= 3\cdot64x^{2+15} \quad &&\text{Multiply coefficients and add} \\
&= 192x^{17} \quad &&\text{exponents of like variables.}
\end{aligned}
$$

b. $(-0.2a)^3(4ab)(1.5a^3c)^2$

SOLUTION We follow the order of operations and simplify the monomials raised to a power first, then multiply the monomials.

$$
\begin{aligned}
(-0.2a)^3(4ab)(1.5a^3c)^2 &= ((-0.2)^3a^{1\cdot3})(4ab)((1.5)^2a^{3\cdot2}c^{1\cdot2}) \\
&= (-0.008a^3)(4ab)(2.25a^6c^2) \quad &&\text{Simplify.} \\
&= -0.008\cdot4\cdot2.25a^{3+1+6}bc^2 \quad &&\text{Multiply coefficients} \\
&= -0.072a^{10}bc^2 \quad &&\text{and add exponents of} \\
& &&\text{like variables.}
\end{aligned}
$$

Answers Your Turn 5

a. $32y^{20}$ **b.** $-\dfrac{125}{27}t^6u^{18}$

c. $0.000064a^{24}b^{30}c^6$

Your Turn 6 Simplify.

a. $\left(\dfrac{1}{4}y^3\right)^2\left(\dfrac{8}{9}y^5\right)$　　　　　　**b.** $(-5mn^4)(-2mp^2)^3(1.5m^2n)$

Answers Your Turn 6

a. $\dfrac{1}{18}y^{11}$ **b.** $60m^6n^5p^6$

5.4 Exercises For Extra Help MyMathLab

 PRACTICE　 WATCH　 DOWNLOAD

Note: Exercises marked with a ★ represent challenging exercises.

1. What rule can be applied when two exponential forms that have the same base are multiplied?

2. Explain why $2^3\cdot5^2\neq10^5$.

3. Explain how to multiply monomials.

4. In multiplying the powers of 10 in $(8\times10^4)(6\times10^5)$, we add the exponents to get 10^9. However, the correct answer in scientific notation is 4.8×10^{10}. Why is the exponent 10 instead of 9?

5. How do we simplify $(x^3)^8$?

6. Explain how to simplify $(3x^4)^2$.

For Exercises 7–32, multiply. See Examples 1 and 2.

7. $a^3 \cdot a^4$

8. $b^2 \cdot b^3$

9. $2^3 \cdot 2^{10}$

10. $5^4 \cdot 5^3$

11. $a^3 \cdot a^2 b$

12. $y^2 \cdot y^3 z$

13. $2x \cdot 3x^2$

14. $5b \cdot 2b^3$

15. $(-2mn)(4m^2n)$

16. $(-3uv^2)(2u^3v)$

17. $\left(\dfrac{5}{8}st\right)\left(-\dfrac{2}{7}s^2t^7\right)$

18. $\left(\dfrac{3}{8}x^2y\right)\left(\dfrac{4}{9}xy^2\right)$

19. $(2.3ab^2c)(1.2a^2b^3c^5)$

20. $(-3.1mn)(2.4m^2n)$

21. $(r^2)(2r^2)(-3r)$

22. $(-2j)(-5j^3)(7j)$

23. $3xz^2(-4x^3z^2)(z^5)$

24. $-5mn^4(-m^3n^9)(4m^3n)$

25. $(5xyz)(9x^2y^4)(2x^3z^2)$

26. $(6hj^3k)(7h^5k)(hkj)$

27. $(5q^2r^4s^9)(-q^3rs^2)(-r^5s^{10})$

28. $(a^3b^3c^3)(-6a^3b^3c^3)(-a^4c^2)$

★ **29.** $-5a^2\left(-\dfrac{1}{15}abc\right)\left(-\dfrac{3}{4}a^3b^2c\right)$

★ **30.** $\left(-\dfrac{1}{4}l^2m\right)\left(\dfrac{5}{6}lm^3n\right)\left(-\dfrac{2}{3}l^5\right)$

★ **31.** $(0.4wxy)(w^2y^2)(2.5x^2y^2)$

★ **32.** $(0.4u^2v)(-3.2u^2v^3w^3)(0.2vw^9)$

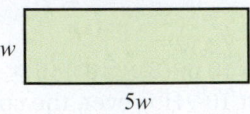

Find the mistake *For Exercises 33–36, find and explain the mistake; then work the problem correctly. See Examples 1 and 2.*

33. $7x^3 \cdot 5x^4y = 35x^{12}y$

34. $-3m^2n \cdot 10m^5n = -30m^7n$

35. $(9x^3y^2z)(6xy^4) = 54x^3y^6z$

36. $(a^3b^2c)(ab^5) = a^4b^7$

For Exercises 37 and 38, write an expression in simplest form for the area of the figure. See Example 3.

37.

w

$5w$

38.

h

$7h$

For Exercises 39 and 40, write an expression in simplest form for the volume of the figure. See Example 3.

39.

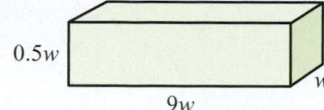

0.5w

9w

w

40.

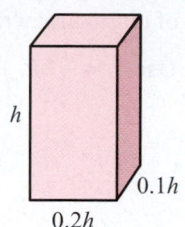

h

0.1h

0.2h

★ **41.** The height of a trapezoid is half the base. The top side of the trapezoid is one-fourth the base.

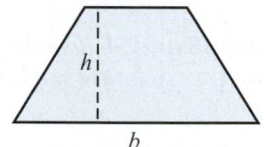

h

b

   a. If b represents the length of the base, write expressions for the height and length of the top side of the trapezoid in terms of b.

   b. Using your expressions from part a, write an expression in simplest form for the area of the trapezoid. $\left[\text{Use } A = \dfrac{1}{2}h(a + b).\right]$

★ **42.** The height of a box is three times the width. The length is five times the width.

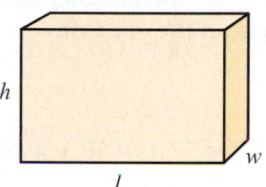

h

l

w

   a. If w represents the width, write expressions for the height and length of the box in terms of w.

   b. Using your expressions from part a, write an expression in simplest form for the volume of the box.

   c. Using your expressions from part a, write an expression in simplest form for the surface area of the box. (Use $SA = 2lw + 2lh + 2wh$.)

For Exercises 43–52, multiply and write your answer in scientific notation. See Example 4.

43. $(2 \times 10^5)(6 \times 10^3)$

44. $(9 \times 10^6)(5 \times 10^5)$

45. $(3.2 \times 10^5)(7.5 \times 10^8)$

46. $(4.5 \times 10^{10})(8.4 \times 10^7)$

47. $(2.9 \times 10^8)(6.3 \times 10^{-4})$

48. $(7.2 \times 10^{-2})(4.6 \times 10^7)$

49. $(4.23 \times 10^{-8})(2.7 \times 10^3)$

50. $(3.75 \times 10^6)(4.3 \times 10^{-12})$

51. $(9.2 \times 10^{-3})(6.1 \times 10^{-8})$

52. $(8.4 \times 10^{-6})(7.3 \times 10^{-8})$

53. Tennessee is roughly shaped like a parallelogram with a base of approximately 350 miles and a height of approximately 120 miles. Calculate the area of Tennessee and write your answer in scientific notation. (Use $A = bh$.)

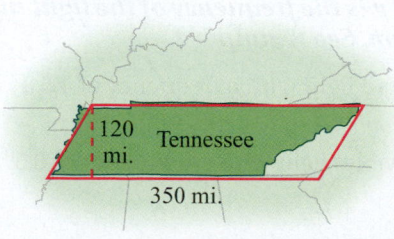

120 mi.

Tennessee

350 mi.

54. Idaho is roughly shaped like a triangle with a base of approximately 320 miles and a height of approximately 520 miles. Calculate the area of Idaho and write your answer in scientific notation. $\left(\text{Use } A = \frac{1}{2}bh.\right)$

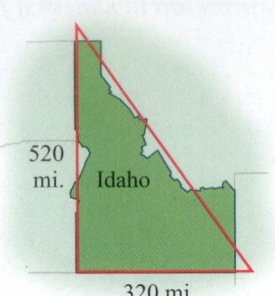

520 mi.   Idaho

320 mi.

For Exercises 55 and 56, use the formula for the circumference of a circle: $C = 2\pi r$. (Use 3.14 to approximate π.) Write your answers in scientific notation. See Example 4.)

★ **55.** The path of Earth's orbit around the Sun is approximately a circle with a radius of about 9.3×10^7 miles. What is the circumference of Earth's orbit?

★ **56.** The radius of Earth is approximately 6.37×10^6 meters. Calculate the circumference of the equator.

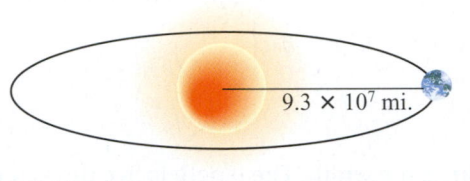

9.3×10^7 mi.

Not to scale.

For Exercises 57 and 58, use the formula for the volume of a sphere: $V = \frac{4}{3}\pi r^3$.

(Use 3.14 to approximate π.) Write your answers in scientific notation. See Example 4.

★ **57.** The radius of the Sun is approximately 6.96×10^8 meters. Calculate the volume of the Sun.

★ **58.** The radius of Earth is approximately 6.37×10^6 meters. Calculate the volume of Earth.

Earth   Sun

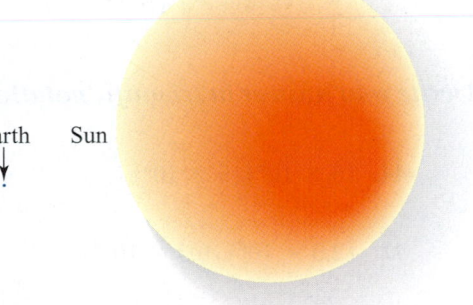

For Exercises 59 and 60, use the following information. The energy, in joules, of a single photon of light can be determined by $E = hv$, where h is the constant 6.626×10^{-34} joule-seconds and v is the frequency of the light in hertz. Write your answers in scientific notation. See Example 4.

59. Find the energy of a photon of red light with a frequency of 4.5×10^{14} Hz.

60. Find the energy of a photon of violet light with a frequency of 8.5×10^{14} Hz.

For Exercises 61 and 62, use the following information. *Albert Einstein discovered that if a mass of m kilograms is converted to pure energy (as in a nuclear reaction), the amount of energy, in joules, that is released is described by* $E = mc^2$, *where c represents a constant speed of light, which is approximately* 3×10^8 *meters per second. Write your answers in scientific notation. See Example 4.*

61. Suppose 2.4×10^{-10} kilogram of uranium is converted to energy in a nuclear reaction. How much energy is released?

62. Suppose 4.5×10^{-8} kilogram of uranium is converted to energy in a nuclear reaction. How much energy is released?

For Exercises 63–98, simplify. See Examples 5 and 6.

63. $(x^2)^3$

64. $(y^3)^5$

65. $(3^2)^4$

66. $(2^3)^2$

67. $(-h^2)^4$

68. $(-j^3)^2$

69. $(-y^2)^3$

70. $(-x^4)^5$

71. $(xy)^3$

72. $(ab)^3$

73. $(3x^2)^3$

74. $(4x^4)^2$

75. $(-2x^3y^2)^3$

76. $(-5a^4b^2)^2$

77. $\left(\dfrac{1}{4}m^2n\right)^3$

78. $\left(\dfrac{5}{6}u^3v\right)^2$

79. $(-2p^4q^4r^2)^6$

80. $(-6u^2v^4w^2)^3$

81. $(-0.2x^2y^5z)^3$

82. $(-0.3r^4st^3)^4$

83. $(2x^3)(3x^2)^2$

84. $(-3y^2)(2y^2)^3$

85. $(rs^2)(r^2s)^2$

86. $(m^2n)^3(mn)$

87. $(2ab)^2(5a^2b)^2$

88. $(4xy^3)^2(2x^2y)^2$

89. $\left(\dfrac{1}{4}a^2b^3c\right)^2\left(\dfrac{1}{4}a^2b^3c\right)^3$

90. $\left(-\dfrac{1}{2}s^2tu^4\right)^2\left(-\dfrac{1}{2}s^2tu^4\right)^5$

91. $5(3a)^2(2b)^3$

92. $2(2x)^4(4y)^3$

93. $-3(-2a^4b^2)^3(3a^3b)^2$

94. $-6(-2m^2n^4)^3(-4mn^3)^2$

95. $(6u)^2(u^2v)(-3uv)^2$

96. $(3x)^2(xy^3)(2x^3yz)^3$

97. $(0.6u)^2(u^2v)(-3uv)^2$

98. $(-0.2w)^4(-w^2z^3)^3(-4w^4z)^2$

Find
✗ the mistake

For Exercises 99–102, find and explain the mistake; then work the problem correctly. See Example 5.

99. $(5x^3)^2 = 25x^5$

100. $(-4mn^5)^3 = -12m^3n^{15}$

101. $(2x^4y)^4 = 8x^8y$

102. $(-2ab^5)^3 = -8ab^{15}$

Review Exercises

Exercises 1–6 **EXPRESSIONS**

[1.7] **1.** Simplify $2(3x + 4)$ using the distributive property.

[5.4] **2.** Simplify $-0.3t^2u \cdot 6t^3u$.

[1.7] **3.** Simplify $-10x - 12x$.

[5.3] **4.** Add $(3x + 4y - z) + (-7x - 4y - 3z)$.

[1.3] **5.** Add $(-4) + (-57)$.

[5.2] **6.** What is the degree of the monomial $-2r^5s$?

 5.5 Multiplying Polynomials; Special Products

Objectives

❶ Multiply a polynomial by a monomial.

❷ Multiply binomials.

❸ Multiply polynomials.

❹ Determine the product when given special polynomial factors.

In this section, we use the rules for multiplying monomials to help us multiply multiple-term polynomials.

 Objective ❶ **Multiply a polynomial by a monomial.** To multiply polynomials, we use the distributive property, which we introduced in Section 1.7 and used in solving linear equations and inequalities in Chapter 2. For example, we have applied the distributive property in situations such as $2(3x + 4)$.

$$2(3x + 4) = 2 \cdot 3x + 2 \cdot 4$$
$$= 6x + 8$$

Notice that $2(3x + 4)$ is a product of a monomial 2 and a binomial $3x + 4$. This suggests the following procedure.

Procedure **Multiplying a Polynomial by a Monomial**

To multiply a polynomial by a monomial, use the distributive property to multiply each term in the polynomial by the monomial.

Connection Multiplying a polynomial by a monomial is like multiplying a multidigit number by a single-digit number. Consider $2 \cdot 34$.

$$\begin{array}{r} 34 \\ \times\ 2 \\ \hline 68 \end{array}$$

Notice that in the numerical multiplication, we distribute the 2 to each digit in the 34. In fact, if we expand 34 by expressing it as $30 + 4$, the problem takes the same form as $2(3x + 4)$, where $x = 10$.

$$2(30 + 4) = 2 \cdot 30 + 2 \cdot 4 \qquad\qquad 2(3x + 4) = 2 \cdot 3x + 2 \cdot 4$$
$$= 60 + 8 \qquad\qquad\qquad\qquad\qquad = 6x + 8$$
$$= 68$$

Now let's multiply a polynomial by a monomial that contains variables.

Example 1 Multiply.

a. $2y(3y^2 + 4y + 1)$

SOLUTION $2y \qquad (3y^2 + 4y + 1)$ **Multiply each term in the polynomial by $2y$.**

$2y \cdot 3y^2$

$2y \cdot 4y$

$2y \cdot 1$

$= 2y \cdot 3y^2 + 2y \cdot 4y + 2y \cdot 1$

$= 6y^3 + 8y^2 + 2y$

> **Note:** When multiplying the individual terms by the monomial, we multiply the coefficients and add the exponents of the like bases.

b. $-7x^2(5x^2 + 3x - 4)$

> **Note:** When we multiply a polynomial by a negative monomial, the signs of the resulting polynomial are opposite the signs in the original polynomial.
>
> $-7x^2(\ 5x^2\ +\ 3x\ -\ 4)$
>
> $= -35x^4 - 21x^3 + 28x^2$

SOLUTION $-7x^2 \qquad (5x^2 + 3x - 4)$ **Multiply each term in the polynomial by $-7x^2$.**

$-7x^2 \cdot 5x^2$

$-7x^2 \cdot 3x$

$-7x^2 \cdot (-4)$

$= -7x^2 \cdot 5x^2 - 7x^2 \cdot 3x - 7x^2 \cdot (-4)$

$= -35x^4 - 21x^3 + 28x^2$

c. $-0.3t^2u(6t^3u + tu^2 - 3t^3 + 0.2uv)$

> **LEARNING Strategy**
>
> **VISUAL** If you are a visual learner, try drawing lines connecting the terms that are to be multiplied.

SOLUTION $-0.3t^2u \quad (\quad 6t^3u\quad +\quad tu^2\quad -\quad 3t^3\quad +\quad 0.2uv)$ **Multiply each term in the polynomial by $-0.3t^2u$.**

$-0.3t^2u \cdot 6t^3u$

$-0.3t^2u \cdot tu^2$

$-0.3t^2u \cdot (-3t^3)$

$-0.3t^2u \cdot 0.2uv$

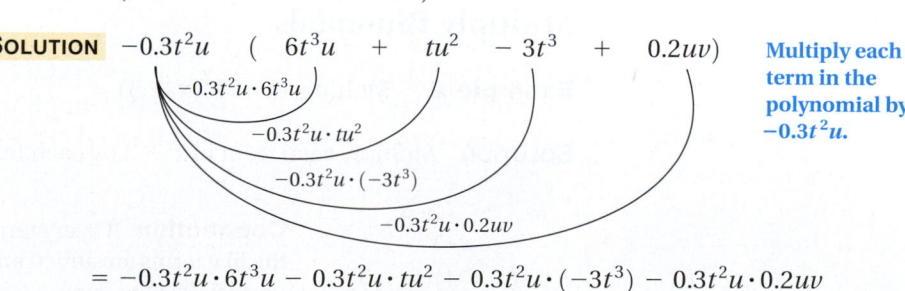

$= -0.3t^2u \cdot 6t^3u - 0.3t^2u \cdot tu^2 - 0.3t^2u \cdot (-3t^3) - 0.3t^2u \cdot 0.2uv$

$= \quad -1.8t^5u^2 \quad - \quad 0.3t^3u^3 \quad + \quad 0.9t^5u \quad - \quad 0.06t^2u^2v$

> **Note:** When multiplying multivariable terms, it is helpful to multiply the coefficients first, then the variables in alphabetical order.
>
> For $-0.3t^2u \cdot 6t^3u$, think $-0.3 \cdot 6 = -1.8$; $t^2 \cdot t^3 = t^5$; $u \cdot u = u^2$.
>
> Result: $-1.8t^5u^2$

Your Turn 1 Multiply.

   a. $4x^3(9x^2 - 7x + 3)$ **b.** $-3x^3yz(5x^2z - y^3z^2 + 8xz)$

Objective 2 Multiply binomials.

Understanding Binomial Multiplication

Now that we have seen how to multiply a polynomial by a monomial, we are ready to multiply two binomials, which is like multiplying a pair of two-digit numbers. For example, consider $(12)(13)$ and compare this with $(x + 2)(x + 3)$.

$$
\begin{array}{r}
12 \\
\times\, 13 \\
\hline
36 \\
+\, 12 \\
\hline
156
\end{array}
$$

Note: We think to ourselves "3 times 2 is 6, then 3 times 1 is 3," which creates the 36. We then move to the 1 in the tens place of the 13 and do the same thing. Because this 1 digit is in the tens place, it means 10; so when we multiply this 10 times 12, it makes 120. We usually omit writing the 0 in the ones place and write 12 in the next two places.

Again, the distributive property is the governing principle. We multiply each digit in one number by each digit in the other number and shift underneath as we move to each new place. The same process applies to the binomials. We can stack them as we just did. Notice, however, that we stack this only once to make the connection to numeric multiplication. As the polynomials get more complex, the stacking method becomes too tedious.

$$
\begin{array}{r}
x + 2 \\
x + 3 \\
\hline
3x + 6 \\
x^2 + 2x \\
\hline
x^2 + 5x + 6
\end{array}
$$

Note: We think "3 times 2 is 6, then 3 times x is $3x$." Now move to the x and think "x times 2 is $2x$, then x times x is x^2." Notice how we shifted so that the $2x$ and $3x$ line up. This is because they are like terms. It is the same as lining up the tens column when multiplying numbers. Note that the numeric result, 156, is basically the same as the algebraic result, $x^2 + 5x + 6$.

Notice that each term in $x + 2$ is multiplied by each term in $x + 3$. In general, this is how we multiply two polynomials.

Procedure   **Multiplying Polynomials**

To multiply two polynomials,

1. Multiply each term in the second polynomial by each term in the first polynomial.
2. Combine like terms.

Multiply Binomials

Example 2   Multiply $(x + 4)(x + 3)$.

SOLUTION   Multiply each term in $x + 3$ by each term in $x + 4$.

$$(x + 4)(x + 3)$$

Connection If every term in both binomials is positive, the like terms are added and both signs in the resulting trinomial are plus signs. Noting these sign patterns will help when we factor in Chapter 6.

$$= x \cdot x + x \cdot 3 + 4 \cdot x + 4 \cdot 3$$
$$= x^2 + 3x + 4x + 12$$
$$= x^2 + 7x + 12 \qquad \textcolor{red}{\textbf{Combine like terms: } 3x + 4x = 7x.}$$
$$= x^2 + 7x + 12$$

Answers   **Your Turn 1**

a. $36x^5 - 28x^4 + 12x^3$
b. $-15x^5yz^2 + 3x^3y^4z^3 - 24x^4yz^2$

The word *FOIL* is a popular way to remember the process of multiplying two binomials. *FOIL* stands for **First Outer Inner Last.** We will use Example 2 to demonstrate.

$$(x + 4)(x + 3) \qquad \text{First terms: } x \cdot x = x^2$$

$$(x + 4)(x + 3) \qquad \text{Outer terms: } x \cdot 3 = 3x$$

$$(x + 4)(x + 3)$$
$$\qquad \text{Inner terms: } 4 \cdot x = 4x$$

$$(x + 4)(x + 3)$$
$$\qquad \text{Last terms: } 4 \cdot 3 = 12$$

Example 3   Multiply $(3x + 5)(x - 4)$.

SOLUTION   Multiply each term in $x - 4$ by each term in $3x + 5$. (Think FOIL.)

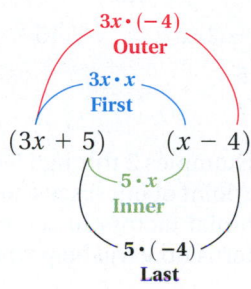

Connection   Notice that $3x + 5$ has a plus sign, whereas $x - 4$ has a minus sign. When multiplying binomials such as these, the last term of the resulting trinomial is always negative. The middle term can be positive or negative depending on the coefficients of the like terms.

$$\begin{array}{cccc} \textbf{First} & \textbf{Outer} & \textbf{Inner} & \textbf{Last} \end{array}$$
$$= 3x \cdot x + 3x \cdot (-4) + 5 \cdot x + 5 \cdot (-4)$$
$$= 3x^2 \qquad - 12x \quad + 5x \qquad - 20$$
$$= 3x^2 \qquad\qquad - 7x \qquad\qquad - 20 \qquad \text{Combine like terms: } -12x + 5x = -7x.$$
$$= 3x^2 - 7x - 20$$

In Example 4, we switch the signs in the binomials from Example 3. Look at how this changes the sign of the middle term in the resulting trinomial.

Example 4   Multiply $(3x - 5)(x + 4)$.

SOLUTION   Multiply each term in $x + 4$ by each term in $3x - 5$.

$$\begin{array}{c} 3x \cdot 4 \\ 3x \cdot x \\ (3x - 5) \quad (x + 4) \\ (-5) \cdot x \\ (-5) \cdot 4 \end{array}$$

$$\begin{array}{cccc} \text{First} & \text{Outer} & \text{Inner} & \text{Last} \end{array}$$
$$= 3x \cdot x + 3x \cdot 4 + (-5) \cdot x + (-5) \cdot 4$$
$$= 3x^2 \qquad + 12x \qquad - 5x \qquad - 20$$
$$= 3x^2 \qquad\qquad + 7x \qquad\qquad - 20 \qquad \text{Combine like terms: } 12x - 5x = 7x.$$
$$= 3x^2 + 7x - 20$$

Now consider a case when the signs between terms in the binomials are both negative.

Example 5 Multiply $(2x - 3)(4x - 5)$.

SOLUTION Multiply each term in $4x - 5$ by each term in $2x - 3$. (Think *FOIL*.)

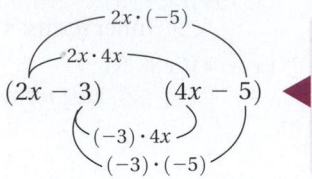

Connection When multiplying binomials such as $2x - 3$ and $4x - 5$, notice that the last term in the resulting trinomial is positive because the product of two negatives is positive and the middle term is negative because the sum of two negatives is negative.

First	Outer	Inner	Last

$$= 2x \cdot 4x + 2x \cdot (-5) + (-3) \cdot 4x + (-3) \cdot (-5)$$
$$= 8x^2 \qquad\quad - 10x \qquad - 12x \qquad\quad + 15$$
$$= 8x^2 \qquad\qquad\quad - 22x \qquad\qquad + 15 \qquad \text{\color{blue}{Combine like terms: } } -10x - 12x = -22x.$$
$$= 8x^2 - 22x + 15$$

Keep in mind that Examples 2 through 5 do not illustrate all sign combinations that you could be given. The point of our discussion of signs is to note some patterns with the signs in the given binomial factors to see how they affect the resulting polynomial. Noticing these sign patterns now will help when we factor in Chapter 6.

Your Turn 5 Multiply.

a. $(n + 6)(n - 2)$

b. $(2y - 5)(y - 3)$

Connection The product of two binomials can be shown in terms of geometry. Consider the rectangle shown with a length of $x + 7$ and a width of $x + 5$. We can describe its area in two ways: (1) the product of the length and width, $(x + 7)(x + 5)$, and (2) the sum of the areas of the four internal rectangles, $x^2 + 5x + 7x + 35$. Because these two approaches describe the same area, the two expressions must be equal.

	x	7
x	x^2	$7x$
5	$5x$	35

$$\text{Length} \cdot \text{width} = \text{Sum of the areas of the}$$
$$\text{four internal rectangles}$$
$$(x + 7)(x + 5) = x^2 + 5x + 7x + 35$$
$$= x^2 + 12x + 35 \qquad \text{\color{blue}{Combine like terms.}}$$

Objective ③ Multiply polynomials. Now consider an example of multiplication involving a larger polynomial, such as a trinomial. Remember that no matter how many terms are in the polynomials, the process is to multiply every term in the second polynomial by every term in the first polynomial.

Answers **Your Turn 5**
a. $n^2 + 4n - 12$
b. $2y^2 - 11y + 15$

Example 6 Multiply.

a. $(x + 2)(3x^2 + 5x - 4)$

SOLUTION Multiply each term in $3x^2 + 5x - 4$ by each term in $x + 2$.

$$
\begin{array}{c}
\overset{\displaystyle x \cdot (-4)}{\overset{\displaystyle x \cdot 5x}{\overset{\displaystyle x \cdot 3x^2}{}}}\\
(x + 2)\quad (3x^2 + 5x - 4)\\
\underset{\displaystyle 2 \cdot 3x^2}{\underset{\displaystyle 2 \cdot 5x}{\underset{\displaystyle 2 \cdot (-4)}{}}}
\end{array}
$$

Warning: FOIL does not make sense here because there are too many terms. FOIL handles only the four terms from two binomials.

$$
\begin{aligned}
&= x \cdot 3x^2 + x \cdot 5x + x \cdot (-4) + 2 \cdot 3x^2 + 2 \cdot 5x + 2 \cdot (-4)\\
&= 3x^3 \qquad\quad + 5x^2 \quad - 4x \qquad + 6x^2 \quad + 10x \qquad - 8\\
&= 3x^3 + 11x^2 + 6x - 8 \qquad \text{\textcolor{blue}{Combine like terms.}}
\end{aligned}
$$

b. $(2x - 3)(4x^2 + 6x + 9)$

SOLUTION Multiply each term in $4x^2 + 6x + 9$ by each term in $2x - 3$.

$$
\begin{aligned}
&(2x - 3)(4x^2 + 6x + 9)\\
&= 2x(4x^2 + 6x + 9) - 3(4x^2 + 6x + 9)\\
&= 8x^3 + 12x^2 + 18x - 12x^2 - 18x - 27 \qquad \text{\textcolor{blue}{Distribute.}}\\
&= 8x^3 - 27 \qquad\qquad\qquad\qquad\qquad\qquad \text{\textcolor{blue}{Combine like terms.}}
\end{aligned}
$$

c. $(x^2 - 3x + 4)(x^2 + 2x - 5)$

SOLUTION Multiply each term in $x^2 + 2x - 5$ by each term in $x^2 - 3x + 4$.

$$
\begin{aligned}
&(x^2 - 3x + 4)(x^2 + 2x - 5)\\
&= x^2(x^2 + 2x - 5) - 3x(x^2 + 2x - 5) + 4(x^2 + 2x - 5)\\
&= x^4 + 2x^3 - 5x^2 - 3x^3 - 6x^2 + 15x + 4x^2 + 8x - 20 \qquad \text{\textcolor{blue}{Distribute.}}\\
&= x^4 - x^3 - 7x^2 + 23x - 20 \qquad\qquad\qquad\qquad\qquad \text{\textcolor{blue}{Combine like terms.}}
\end{aligned}
$$

Your Turn 6 Multiply.

a. $(x - 3)(x^2 - 6x + 2)$
b. $(x + 5)(x^2 - 5x + 25)$
c. $(y^2 + 4y - 2)(y^2 - 3y - 5)$

Answers Your Turn 6

a. $x^3 - 9x^2 + 20x - 6$
b. $x^3 + 125$
c. $y^4 + y^3 - 19y^2 - 14y + 10$

Objective ④ Determine the product when given special polynomial factors.

Multiplying Conjugates

In Objective 2, we discussed how the signs in the binomial factors affect the signs in the product in preparing for factoring in Chapter 6. To further prepare for factoring, it is helpful to note some patterns that occur when certain special polynomial factors are multiplied. First, we will multiply **conjugates**.

Definition *Conjugates:* Binomials that differ only in the sign separating the terms.

The following binomial pairs are conjugates.

$$x + 9 \text{ and } x - 9 \qquad 2x + 3 \text{ and } 2x - 3 \qquad -6x + 5 \text{ and } -6x - 5$$

Let's multiply the conjugates $2x + 3$ and $2x - 3$ to see what pattern emerges.

$$
\begin{array}{ll}
 & \qquad\quad \text{First} \quad\;\; \text{Outer} \quad\;\; \text{Inner} \quad\;\; \text{Last} \\
(2x + 3)(2x - 3) = 2x \cdot 2x + 2x \cdot (-3) + 3 \cdot 2x + 3 \cdot (-3) \\
\qquad\qquad\qquad\;\; = \quad 4x^2 \qquad -6x \qquad\; + 6x \qquad\;\; -9 \\
\qquad\qquad\qquad\;\; = \quad 4x^2 + 0 - 9 \qquad \textcolor{blue}{\textbf{Combine like terms: } -6x + 6x = 0.} \\
\qquad\qquad\qquad\;\; = \quad 4x^2 - 9
\end{array}
$$

Note: The sum of the like terms is 0, so the resulting binomial is the square of the first term minus the square of the last term.

When conjugates are multiplied, the like terms are always additive inverses; so their sum is always 0. This suggests the following rule.

Rule **Multiplying Conjugates**

If a and b are real numbers, variables, or expressions, then $(a + b)(a - b) = a^2 - b^2$.

Using this rule allows us to multiply conjugates quickly.

Example 7 Multiply.

a. $(x + 3)(x - 3)$

SOLUTION $(x + 3)(x - 3) = x^2 - 3^2$ $\qquad$ **Use $(a + b)(a - b) = a^2 - b^2$.**

$\qquad\qquad\qquad\qquad\qquad = x^2 - 9$ $\qquad$ **Simplify.**

b. $(3t - 8)(3t + 8)$

SOLUTION $(3t - 8)(3t + 8) = (3t)^2 - 8^2$ $\qquad$ **Use $(a + b)(a - b) = a^2 - b^2$.**

$\qquad\qquad\qquad\qquad\qquad\;\; = 9t^2 - 64$ $\qquad$ **Simplify.**

Your Turn 7 Multiply.

$\quad$ **a.** $(m + 7)(m - 7)$ $\qquad\qquad\qquad$ **b.** $(5y - 3)(5y + 3)$

Squaring a Binomial

Now let's look for a pattern when we square a binomial, as in $(3x + 4)^2$. Recall that squaring a number or an expression means to multiply that number or expression by itself; so $(3x + 4)^2 = (3x + 4)(3x + 4)$.

$$
\begin{array}{ll}
 & \qquad\qquad\qquad\quad \text{First} \quad\;\; \text{Outer} \quad\; \text{Inner} \quad\;\; \text{Last} \\
(3x + 4)^2 = (3x + 4)(3x + 4) = 3x \cdot 3x + 3x \cdot 4 + 4 \cdot 3x + 4 \cdot 4 \\
\qquad\qquad\qquad\qquad\qquad\;\;\; = 9x^2 + 12x + 12x + 16 \\
\qquad\qquad\qquad\qquad\qquad\;\;\; = 9x^2 + 24x + 16 \qquad \textcolor{blue}{\textbf{Combine like terms. } 12x + 12x = 24x}
\end{array}
$$

Warning: Notice that $(3x + 4)^2$ does not equal $9x^2 + 16$.

Notice the pattern.

$$(3x + 4)^2 = 9x^2 + 24x + 16$$

This first term is the square of the first term in the binomial: $(3x)^2 = 9x^2$.

This last term is the square of the second term in the binomial: $(4)^2 = 16$.

This middle term is twice the product of the two terms in the binomial: $2(3x)(4) = 24x$.

Connection The square of a binomial can also be shown in terms of geometry. Consider the square shown, each side being $x + 3$. We can describe the area in two ways: (1) the length of a side squared, $(x + 3)^2$, or (2) the sum of the areas of the four internal rectangles, $x^2 + 3x + 3x + 3^2 = x^2 + 2 \cdot 3x + 9 = x^2 + 6x + 9$.

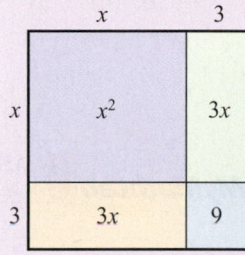

Side squared $=$ Sum of the areas of the four internal rectangles.

$$(x + 3)^2 = x^2 + 2 \cdot 3x + 9$$
$$= x^2 + 6x + 9$$

Squaring a binomial with a minus sign, as in $(3x - 4)^2$, yields a similar pattern.

First   Outer   Inner   Last

$$(3x - 4)^2 = (3x - 4)\ (3x - 4) = 3x \cdot 3x + 3x \cdot (-4) + (-4) \cdot 3x + (-4) \cdot (-4)$$
$$= 9x^2 - 12x - 12x + 16$$
$$= 9x^2 - 24x + 16 \qquad \text{Combine like terms: } -12x - 12x = -24x.$$

$3x \cdot (-4)$
$3x \cdot 3x$
$(-4) \cdot 3x$
$(-4) \cdot (-4)$

Warning: Notice that $(3x - 4)^2$ does *not* equal $9x^2 - 16$ or $9x^2 + 16$.

Notice the pattern.

$$(3x - 4)^2 = 9x^2 - 24x + 16$$

This first term is the square of the first term in the binomial: $(3x)^2 = 9x^2$.

This last term is the square of the second term in the binomial: $(-4)^2 = 16$.

This middle term is twice the product of the two terms in the binomial: $2(3x)(-4) = -24x$.

The patterns from our two examples suggest the following rules.

Rules   **Squaring a Binomial**

If a and b are real numbers, variables, or expressions, then

$$(a + b)^2 = a^2 + 2ab + b^2$$
$$(a - b)^2 = a^2 - 2ab + b^2$$

Example 8 Multiply.

a. $(5x + 3)^2$

> **SOLUTION** $(5x + 3)^2 = (5x)^2 + 2(5x)(3) + (3)^2$ Use $(a + b)^2 = a^2 + 2ab + b^2$.
> $\qquad\qquad\qquad = 25x^2 + 30x + 9$ Simplify.

b. $(6y - 7)^2$

> **SOLUTION** $(6y - 7)^2 = (6y)^2 - 2(6y)(7) + (7)^2$ Use $(a - b)^2 = a^2 - 2ab + b^2$.
> $\qquad\qquad\qquad = 36y^2 - 84y + 49$ Simplify.

Answers Your Turn 8
a. $9m^2 + 42m + 49$
b. $16t^2 - 40t + 25$

Your Turn 8 Multiply.

a. $(3m + 7)^2$ **b.** $(4t - 5)^2$

5.5 Exercises   For Extra Help   *MyMathLab*

Note: Exercises marked with a ★ represent challenging exercises.

1. What is the mathematical property applied in the first step when multiplying polynomials?

2. Explain, in general, how to multiply two polynomials.

3. Explain how to recognize that two binomials are conjugates.

4. What happens when conjugates are multiplied?

5. Explain why $(a + b)^2 \neq a^2 + b^2$.

6. Explain why $(a - b)^2 \neq a^2 - b^2$.

For Exercises 7–34, multiply the polynomial by the monomial. See Example 1.

7. $5(x + 3)$

8. $2(y + 7)$

9. $-6(2x - 4)$

10. $-4(3y - 5)$

11. $4n(7n + 2)$

12. $6n(2n + 5)$

13. $9a(a - b)$

14. $3y(y - z)$

15. $\frac{1}{8}m(2m - 5n)$

16. $\frac{1}{4}k(2k - 3l)$

17. $-0.3p^2(p^3 - 2q^2)$

18. $-0.4b^2(b^2 - 3c^2)$

19. $5x(4x^2 + 2x - 5)$

20. $5a(2a^2 - a + 6)$

21. $-3x^2(7x^3 - 5x + 1)$

22. $-4w^3(3w^3 + 7w^2 - 2)$

23. $-r^2s(r^2s^3 + 3rs^2 - s)$

24. $-x^2y^3(-2xy^2 + 3x^2y^5 - y)$

25. $-2a^2b^2(2a^3 - 6b + 3ab - a^2b^2)$

26. $-5hk^3(3k^3 - 2h^2k - 3hk^4 + 4h)$

27. $3abc(4a^2b^2c^2 - 2abc + 5)$

28. $5xy^2z(2xy^3 - 4x^2z^2 + 3)$

29. $4b\left(b^5 - \dfrac{1}{4}b^3 + \dfrac{1}{20}b^2 - \dfrac{1}{8}b + 4\right)$

30. $5r^2\left(r^4 - \dfrac{1}{20}r^3 - \dfrac{1}{5}r^2 + \dfrac{1}{10}r - 1\right)$

31. $-0.4rt^2(2.5r^3t - 8r^2t^2 + 4.1rt^3 - 2)$

32. $-0.25suv^3(10s^2u - 2u^2v - 18sv^4 + 4)$

33. $\dfrac{5}{6}x^2(30x^4y - 18x^3y^2 + 60xy^9)$

34. $\dfrac{2}{3}b^4(-30a^2 + 18ab^3 - 12a^2b^2)$

For Exercises 35–38, a larger rectangle is formed out of smaller rectangles.

    a. Write an expression in simplest form for the length (along the top).
    b. Write an expression in simplest form for the width (along the side).
    c. Write an expression that is the product of the length and width that you found in parts a and b.
    d. Write an expression in simplest form that is the sum of the areas of each of the smaller rectangles.
    e. Explain why the expressions in parts c and d are equivalent.

35.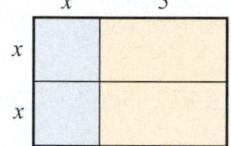

36.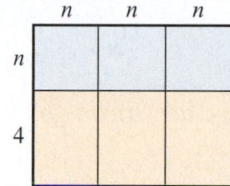

37.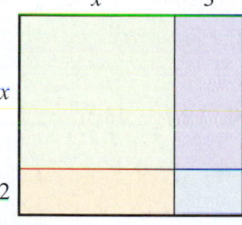

38.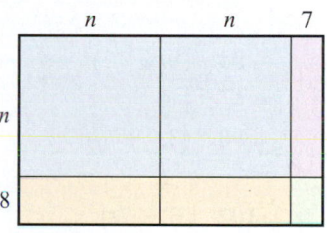

For Exercises 39–56, multiply the binomials. (Use FOIL.) See Examples 2–5.

39. $(x + 3)(x + 4)$ **40.** $(a + 4)(a + 9)$ **41.** $(x - 7)(x + 2)$ **42.** $(n + 7)(n - 5)$

43. $(y - 3)(y - 6)$ **44.** $(r - 4)(r - 5)$ **45.** $(2y + 5)(3y + 2)$ **46.** $(2x + 4)(4x + 3)$

47. $(5m - 3)(3m + 4)$ **48.** $(2x + 3)(3x - 2)$ **49.** $(3t - 5)(4t - 2)$ **50.** $(2t - 7)(3t - 1)$

51. $(5q - 3t)(3q + 4t)$ **52.** $(7k - 2j)(3k + 4j)$ **53.** $(7y + 3x)(2y + 4x)$ **54.** $(2a + 5b)(3a + 4b)$

55. $(a^2 + b)(a^2 - 2b)$ **56.** $(x^2 + 4y)(x^2 - 5y)$

For Exercises 57–72, multiply the polynomials. See Example 6.

57. $(x + 2)(x^2 - 2x + 1)$ **58.** $(x + 3)(x^2 - 3x + 2)$ **59.** $(a - 1)(3a^2 - a - 2)$

60. $(x - 1)(3x^2 - 2x + 1)$ **61.** $(3c + 2)(2c^2 - c - 3)$ **62.** $(2x + 3)(4x^2 + 3x - 5)$

63. $(2f + 3g)(2f^2 - 3fg - 9g^2)$ **64.** $(2m - 3n)(4m^2 - 6mn + 8n^2)$ **65.** $(3m - 2n)(9m^2 + 6mn + 4n^2)$

66. $(4a - b)(16a^2 + 4ab + b^2)$ **67.** $(2a + 5b)(4a^2 - 10ab + 25b^2)$ **68.** $(5m + 3n)(25m^2 - 15mn + 9n^2)$

69. $(x^2 + xy + y^2)(x^2 + 2xy - y^2)$ **70.** $(u^2 - u - 1)(u^2 + 2u + 1)$ **71.** $(x^2 + 10x + 25)(x^2 + 4x + 4)$

72. $(a^2 + 6a + 9)(a^2 - 8a + 16)$

For Exercises 73–80, state the conjugate of the given binomial. See Objective 4.

73. $x + 3$ **74.** $y + 8$ **75.** $4x - 2y$ **76.** $2g - 3h$

77. $4d + 3c$ **78.** $5m + 2n$ **79.** $-3j - k$ **80.** $-3x - 5y$

For Exercises 81–102, multiply using the rules for special products. See Examples 7 and 8.

81. $(x - 5)(x + 5)$ **82.** $(y + 4)(y - 4)$ **83.** $(2m - 5)(2m + 5)$ **84.** $(3p + 2)(3p - 2)$

85. $(x + y)(x - y)$ **86.** $(a + b)(a - b)$ **87.** $(8r - 10s)(8r + 10s)$ **88.** $(4b - 5c)(4b + 5c)$

89. $(-2x - 3)(-2x + 3)$ **90.** $(-h + 2k)(-h - 2k)$ **91.** $(x + 3)^2$ **92.** $(y + 5)^2$

93. $(4t - 1)^2$ **94.** $(9k - 2)^2$ **95.** $(m + n)^2$ **96.** $(a + b)^2$

97. $(2u + 3v)^2$ **98.** $(3r + 7s)^2$ **99.** $(9w - 4z)^2$ **100.** $(2q - 11c)^2$

101. $(9 - 5y)^2$ **102.** $(6 - 7t)^2$

For Exercises 103–106, write an expression in simplest form for the area.

103.
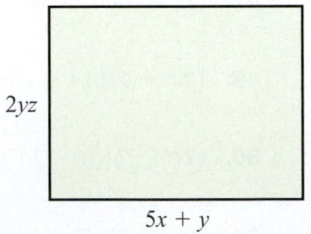
$2yz$

$5x + y$

104.
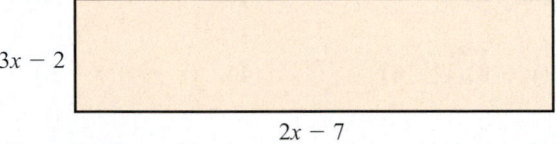
$3x - 2$

$2x - 7$

105.
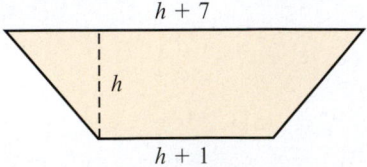
$h + 7$

h

$h + 1$

106.

h

$3h - 5$

107. A rectangular room has a length that is 3 feet more than the width. Using w to represent the width, write an expression in simplest form for the area of the room in terms of w.

★**108.** A circular metal plate for a machine has radius r. A smaller plate with radius $r - 2$ is to be used in another part of the machine. Write an expression in simplest form for the sum of the areas of the two circles in terms of r.

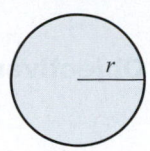 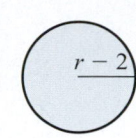

For Exercises 109 and 110, write an expression in simplest form for the volume.

109.

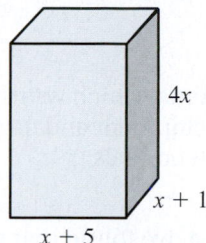

$4x$

$x + 1$

$x + 5$

110.

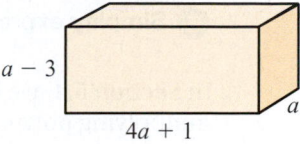

$a - 3$

a

$4a + 1$

111. A crate is designed so that the length is twice the width and the height is 3 feet more than the length. Let w represent the width. Write an expression for the volume in terms of w.

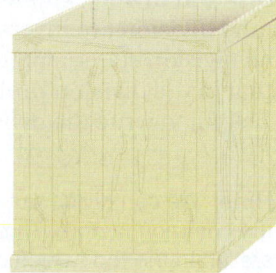

112. A fish tank's length is 8 inches more than its width. The height of the tank is 5 inches greater than its width. Let w represent the width. Write an expression for the volume of the tank in terms of w.

★**113.** A cylinder has a height that is 2 inches more than the radius. Write an expression for the volume of the cylinder in terms of the radius.

114. A cone has a radius that is 5 centimeters less than the height. Write an expression for the volume of the cone in terms of the height.
★

Review Exercises

EXERCISES 1–4   EXPRESSIONS

For Exercises 1 and 2, simplify.

[1.4]   **1.** $1.08 \div 2.4$

[1.3]   **2.** $-6 - (-1)$

[5.1]   **3.** Write 6.304×10^6 in standard form.

[5.4]   **4.** Multiply $x^2 \cdot x^5$.

EXERCISES 5 AND 6   EQUATIONS AND INEQUALITIES

[2.4]   **5.** Isolate r in the formula $\dfrac{5 - d}{r} = m$.

[3.3]   **6.** The greater of two positive numbers is four times the smaller number. The difference between the numbers is 27. What are the numbers?

5.6 Exponent Rules and Dividing Polynomials

Objectives

1. Divide exponential forms with the same base.
2. Divide numbers in scientific notation.
3. Divide monomials.
4. Divide a polynomial by a monomial.
5. Use long division to divide polynomials.
6. Simplify expressions using rules of exponents.

In Section 5.4, we began by learning the product rule of exponents, which we used in multiplying polynomials. In this section, we follow a similar development and begin by learning the quotient rule of exponents so that we can divide polynomials.

Objective 1 **Divide exponential forms with the same base.** In Section 5.4, we found that when we multiply exponential forms that have the same base, we can add the exponents and keep the same base. For example,

$$2^3 \cdot 2^4 = 2^{3+4} = 2^7.$$

Now consider how to reverse this process and divide exponential forms with the same base. Remember that multiplication and division are inverse operations, which means that they "undo" each other. A related division statement for the multiplication just shown is $2^7 \div 2^4 = 2^3$, which implies that in the division problem, the exponents can be subtracted. To verify this result, it is helpful to write the division in fraction form and use the definition of exponents.

$$\frac{2^7}{2^4} = \frac{2 \cdot 2 \cdot 2 \cdot 2 \cdot 2 \cdot 2 \cdot 2}{2 \cdot 2 \cdot 2 \cdot 2}$$

$$= \frac{2 \cdot 2 \cdot 2}{1}$$

$$= 2^3$$

Note: We can divide out four of the common factors. Three 2's are left in the top, which can be expressed in exponential form.

This result supports our thinking that we can subtract the exponents.

$$2^7 \div 2^4 = 2^{7-4} = 2^3$$

Notice that the divisor's exponent is subtracted from the dividend's exponent. This order is important because subtraction is not a commutative operation.

Conclusion To divide exponential forms that have the same base, we can subtract the divisor's exponent from the dividend's exponent and keep the same base.

Rule **Quotient Rule for Exponents**

If m and n are integers and a is a real number, where $a \neq 0$, then $\dfrac{a^m}{a^n} = a^{m-n}$.

Note: We cannot let a equal 0 because if a were replaced with 0, we would have $0^m \div 0^n$, and in Section 1.4, we concluded that $0 \div 0$ is indeterminate. Recall also that $\dfrac{a^m}{a^n}$ can be written as $a^m \div a^n$.

Example 1 Divide. $\dfrac{x^6}{x^2}$

Note: Throughout the remainder of this section, assume that variables are not replaced by values that cause denominators to be 0.

SOLUTION Because the exponential forms have the same base, we can subtract the exponents and keep the same base.

$$\frac{x^6}{x^2} = x^6 \div x^2 = x^{6-2} = x^4$$

Your Turn 1 Divide.

a. $\dfrac{t^8}{t^2}$
b. $\dfrac{n^5}{n}$
c. $5^9 \div 5^2$

We can apply the quotient rule for exponential forms when the exponents are negative. After simplifying, if the exponent in the result is negative, we will rewrite the exponential form so that the exponent is positive. Recall the rule from Section 5.1:

If a is any real number not equal to 0 and n is a natural number, then $a^{-n} = \dfrac{1}{a^n}$.

Example 2 Divide and write the result with a positive exponent.

a. $\dfrac{x^6}{x^{-2}}$

SOLUTION $\dfrac{x^6}{x^{-2}} = x^{6-(-2)}$ **Subtract exponents and keep the same base.**

Note: To evaluate the subtraction, we write an equivalent addition.

$= x^{6+2}$

$= x^8$

b. $n^{-5} \div n^{-3}$

SOLUTION $n^{-5} \div n^{-3} = n^{-5-(-3)}$ **Subtract the exponents and keep the same base.**

$= n^{-5+3}$ **Rewrite the subtraction as addition.**

$= n^{-2}$ **Simplify.**

$= \dfrac{1}{n^2}$ **Write with a positive exponent.**

Your Turn 2 Divide and write the result with a positive exponent.

a. $\dfrac{y^3}{y^7}$
b. $m^{-2} \div m^{-6}$
c. $\dfrac{7^{-4}}{7^5}$

Objective ② Divide numbers in scientific notation. We can use the quotient rule of exponents to divide numbers in scientific notation.

Example 3 Divide and write the result in scientific notation: $\dfrac{1.08 \times 10^9}{2.4 \times 10^4}$

Answers Your Turn 1
a. t^6 b. n^4 c. 5^7

Answers Your Turn 2
a. $\dfrac{1}{y^4}$ b. m^4 c. $\dfrac{1}{7^9}$

SOLUTION The decimal factors and powers of 10 can be separated into a product of two fractions. This allows us to calculate the decimal division and divide the powers of 10 separately.

$$\frac{1.08 \times 10^9}{2.4 \times 10^4} = \frac{1.08}{2.4} \times \frac{10^9}{10^4}$$ Separate the decimal numbers and powers of 10.

$$= 0.45 \times 10^{9-4}$$ Divide decimal numbers and subtract exponents.

$$= 0.45 \times 10^5$$

$$= 4.5 \times 10^4$$

Your Turn 3 Divide $\dfrac{1.82 \times 10^{-9}}{6.5 \times 10^4}$ and write the result in scientific notation.

Objective ③ Divide monomials. Dividing monomials is much like dividing numbers in scientific notation. Let's consider $\dfrac{24x^5}{6x^2}$. We can separate the coefficients and variables in the same way that we separated the decimal factors and powers of 10 when dividing numbers in scientific notation.

$$\frac{24x^5}{6x^2} = \frac{24}{6} \cdot \frac{x^5}{x^2}$$

We can now divide the coefficients and subtract the exponents of the like bases.

$$\frac{24x^5}{6x^2} = \frac{24}{6} \cdot \frac{x^5}{x^2} = 4x^{5-2} = 4x^3$$

This suggests the following procedure.

Procedure **Dividing Monomials**

To divide monomials,

1. Divide the coefficients or simplify them to fractions in lowest terms.
2. Use the quotient rule for the exponents with like bases.
3. Do not change unlike variable bases in the quotient.
4. Write the final expression so that all exponents are positive.

Example 4 Divide.

a. $\dfrac{16a^4b^7c}{20a^5b^2}$

SOLUTION $\dfrac{16a^4b^7c}{20a^5b^2} = \dfrac{16}{20} \cdot \dfrac{a^4}{a^5} \cdot \dfrac{b^7}{b^2} \cdot \dfrac{c}{1}$

$$= \frac{16}{20} \cdot a^{4-5} \cdot b^{7-2} \cdot c$$ Divide coefficients and subtract exponents of the like bases. The unlike base, c, will remain unchanged.

$$= \frac{4}{5} a^{-1} b^5 c$$

$$= \frac{4}{5} \cdot \frac{1}{a} \cdot \frac{b^5}{1} \cdot \frac{c}{1}$$

$$= \frac{4b^5c}{5a}$$

b. $\dfrac{10x^3y^2}{24x^3y}$

SOLUTION $\dfrac{10x^3y^2}{24x^3y} = \dfrac{10}{24} \cdot \dfrac{x^3}{x^3} \cdot \dfrac{y^2}{y}$

$= \dfrac{10}{24} \cdot x^{3-3} \cdot y^{2-1}$ **Divide coefficients and subtract exponents of like bases.**

$= \dfrac{5}{12}x^0y^1$

$= \dfrac{5}{12}y$

Your Turn 4   Divide.

a. $\dfrac{-32m^2n}{8mn}$

b. $\dfrac{56t^2u^6v}{16t^2u^4}$

Objective 4   **Divide a polynomial by a monomial.**   Let's now consider how to divide a polynomial by a monomial, as in $\dfrac{24x^5 + 18x^3}{6x^2}$. Recall that when fractions with a common denominator are added (or subtracted), the numerators are added and the denominator stays the same.

$$\dfrac{1}{7} + \dfrac{2}{7} = \dfrac{1+2}{7}$$

This process can be reversed so that a sum in the numerator of a fraction can be broken into fractions, with each addend over the same denominator.

$$\dfrac{1+2}{7} = \dfrac{1}{7} + \dfrac{2}{7}$$

Rule   **Dividing a Polynomial by a Monomial**

If a, b, and c are real numbers, variables, or expressions with $c \neq 0$, then

$$\dfrac{a+b}{c} = \dfrac{a}{c} + \dfrac{b}{c}.$$

We can apply this rule to the division problem $\dfrac{24x^5 + 18x^3}{6x^2}$.

$\dfrac{24x^5 + 18x^3}{6x^2} = \dfrac{24x^5}{6x^2} + \dfrac{18x^3}{6x^2}$ **We now have a sum of monomial divisions, which we can simplify separately.**

$= 4x^{5-2} + 3x^{3-2}$

$= 4x^3 + 3x$

Our illustration suggests the following procedure.

Procedure   **Dividing a Polynomial by a Monomial**

To divide a polynomial by a monomial, divide each term of the polynomial by the monomial.

Answers   **Your Turn 4**

a. $-4m$ **b.** $\dfrac{7}{2}u^2v$

Example 5   Divide.

a. $\dfrac{32y^6 + 24y^4 - 48y^2}{8y^2}$

SOLUTION $\quad \dfrac{32y^6 + 24y^4 - 48y^2}{8y^2} = \dfrac{32y^6}{8y^2} + \dfrac{24y^4}{8y^2} - \dfrac{48y^2}{8y^2}$ **Divide each term in the polynomial by the monomial.**

$$= 4y^4 + 3y^2 - 6y^0$$

$$= 4y^4 + 3y^2 - 6$$

b. $\dfrac{42t^7u^2 - 18t^4u - 10t}{2t^3u}$

SOLUTION $\quad \dfrac{42t^7u^2 - 18t^4u - 10t}{2t^3u} = \dfrac{42t^7u^2}{2t^3u} - \dfrac{18t^4u}{2t^3u} - \dfrac{10t}{2t^3u}$

> **Warning:** A common error is to write the answer to Example 5(b) as $\dfrac{21t^4u - 9t - 5}{ut^2}$.

> **Note:** When t is divided by t^3, because the denominator's exponent is greater than the numerator's exponent, the result is a negative exponent. Because u has no like base in the third term, it is written unchanged.

$$= 21t^4u - 9t^1u^0 - \dfrac{5t^{-2}}{u}$$

$$= 21t^4u - 9t - \dfrac{5}{ut^2}$$ **Rewrite with positive exponents.**

Your Turn 5   Divide.

a. $\dfrac{45x^6 - 27x^5 + 9x^3}{9x^3}$ b. $\dfrac{28m^2n^6 + 36m^4n - 40m^2}{4m^2n}$

Objective ⑤ Use long division to divide polynomials. To divide a polynomial by a polynomial, we can use long division. Consider the following numeric long division and think about the process.

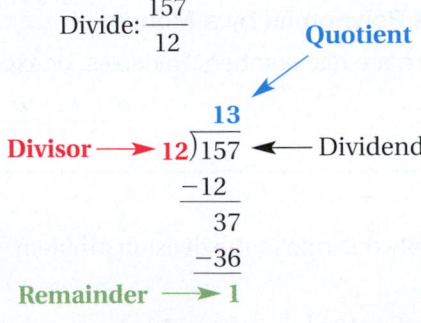

Divide: $\dfrac{157}{12}$

Quotient

Divisor ⟶ $12\overline{)157}$ ⟵ Dividend

$\quad\quad \underline{-12}$

$\quad\quad\quad 37$

$\quad\quad\quad \underline{-36}$

Remainder ⟶ 1

Because the answer has a remainder, we write the result as a mixed number, $13\dfrac{1}{12}$. Notice that we could check the answer by multiplying the divisor by the quotient, then adding the remainder. The result should be the dividend.

Quotient · Divisor + Remainder = Dividend

$\quad 13 \quad \cdot \quad 12 \quad + \quad 1 \quad = 157$

Now let's consider polynomial division, which follows the same long division process.

Example 6   Divide $\dfrac{x^2 + 5x + 7}{x + 2}$.

SOLUTION   Use long division. First, we determine what term will multiply by the first term in the divisor to equal the first term in the dividend. A clever way to determine this first term in the quotient is to divide the first term in the dividend by the first term in the divisor.

$$
x + 2 \overline{\smash{)}x^2 + 5x + 7} \quad \overset{x}{}
$$

Divide these first terms to determine the first term in the quotient: $x^2 \div x = x.$

Next, we multiply the divisor $x + 2$ by the x in the quotient.

multiply

$$
\begin{array}{r}
x \\
x + 2 \overline{\smash{)}x^2 + 5x + 7} \\
x^2 + 2x
\end{array}
$$

Next, we subtract. Note that to subtract the binomial, we change the signs of the terms and then combine like terms. After combining terms, we bring down the next term in the dividend, which is 7.

$$
\begin{array}{r}
x \\
x + 2 \overline{\smash{)}x^2 + 5x + 7} \\
-(x^2 + 2x)
\end{array}
$$

Change signs. →

$$
\begin{array}{r}
x \\
x + 2 \overline{\smash{)}x^2 + 5x + 7} \\
-x^2 - 2x \\
\hline
3x + 7
\end{array}
$$

Combine like terms and bring down the next term.

Now we repeat the process with $3x + 7$ as the dividend. The next term in the quotient is found by dividing the first term of $3x + 7$ by the first term in the divisor, $x + 2$. This gives $3x \div x = 3$. We then multiply the divisor by this 3.

multiply

$$
\begin{array}{r}
x + 3 \\
x + 2 \overline{\smash{)}x^2 + 5x + 7} \\
-x^2 - 2x \\
\hline
3x + 7 \\
3x + 6
\end{array}
$$

Then we subtract. As before, we change the signs of the binomial to be subtracted.

$$
\begin{array}{r}
x + 3 \\
x + 2 \overline{\smash{)}x^2 + 5x + 7} \\
-x^2 - 2x \\
\hline
3x + 7 \\
-(3x + 6)
\end{array}
$$

Change signs. →

$$
\begin{array}{r}
x + 3 \\
x + 2 \overline{\smash{)}x^2 + 5x + 7} \\
-x^2 - 2x \\
\hline
3x + 7 \\
-3x - 6 \\
\hline
1
\end{array}
$$

Combine like terms.

Notice that we have a remainder, 1. Recall that in the numeric version, we wrote the answer as a mixed number, $13\frac{1}{12}$, which means $13 + \frac{1}{12}$ and is in the form quotient $+ \dfrac{\text{remainder}}{\text{divisor}}$. With polynomials, we write a similar expression:

$$\underbrace{x + 3}_{\text{quotient}} + \underbrace{\frac{1}{x + 2}}_{}$$

remainder ← 1
divisor ← x + 2

To check, we can multiply the quotient and the divisor, then add the remainder. The result should be the dividend.

$$(x + 3)(x + 2) + 1 = x^2 + 2x + 3x + 6 + 1$$
$$= x^2 + 5x + 7 \qquad \text{It checks.}$$

Procedure | **Dividing a Polynomial by a Polynomial**

To divide a polynomial by a polynomial, use long division. If there is a remainder, write the result in the following form:

$$\text{quotient} + \frac{\text{remainder}}{\text{divisor}}$$

In Example 6, the coefficients of the first term in the divisor and dividend were both 1. What if those coefficients are other than 1?

Example 7 Divide $\dfrac{15x^2 - 22x + 14}{3x - 2}$.

SOLUTION Begin by dividing the first term in the dividend by the first term in the divisor: $15x^2 \div 3x = 5x$.

$$\begin{array}{r} 5x \\ 3x - 2 \overline{)\, 15x^2 - 22x + 14} \\ -(15x^2 - 10x) \end{array}$$

Change signs. →

$$\begin{array}{r} 5x \\ 3x - 2 \overline{)\, 15x^2 - 22x + 14} \\ -15x^2 + 10x \\ \hline -12x + 14 \end{array}$$

Combine like terms; then bring down the next term.

Determine the next part of the quotient by dividing $-12x$ by $3x$, which is -4, and repeat the multiplication and subtraction steps.

$$\begin{array}{r} 5x - 4 \\ 3x - 2 \overline{)\, 15x^2 - 22x + 14} \\ -15x^2 + 10x \\ \hline -12x + 14 \\ -(-12x + 8) \end{array}$$

Change signs. →

$$\begin{array}{r} 5x - 4 \\ 3x - 2 \overline{)\, 15x^2 - 22x + 14} \\ -15x^2 + 10x \\ \hline -12x + 14 \\ +12x - 8 \\ \hline 6 \end{array}$$

Combine like terms.

Answer $\quad 5x - 4 + \dfrac{6}{3x - 2}$

Your Turn 7 Divide.

a. $\dfrac{x^2 + 6x + 8}{x + 2}$

b. $\dfrac{24x^2 + 14x + 4}{6x - 1}$

Using a Place Marker in Long Division

In polynomial division, it is important that we write the terms in descending order of degree. If a term is missing, we write that term with a 0 coefficient as a place marker.

Example 8 Divide $\dfrac{-14x^2 - 9 + 8x^4}{2x - 1}$.

SOLUTION First, write the dividend in descending order of degree. Also, because degree 3 and degree 1 terms are missing in the dividend, we write those terms with 0 coefficients as placeholders. Divide $8x^4$ by $2x$ to get the first term in the quotient.

$$
\begin{array}{r}
4x^3 \\
2x - 1 \overline{)8x^4 + 0x^3 - 14x^2 + 0x - 9} \\
\underline{-(8x^4 - 4x^3)}
\end{array}
$$

Change signs. →

$$
\begin{array}{r}
4x^3 \\
2x - 1 \overline{)8x^4 + 0x^3 - 14x^2 + 0x - 9} \\
\underline{-8x^4 + 4x^3} \\
4x^3 - 14x^2
\end{array}
$$

Combine like terms; then bring down the next term.

Determine the next part of the quotient by dividing $4x^3$ by $2x$ and repeat the multiplication and subtraction steps.

$$
\begin{array}{r}
4x^3 + 2x^2 \\
2x - 1 \overline{)8x^4 + 0x^3 - 14x^2 + 0x - 9} \\
\underline{-8x^4 + 4x^3} \\
4x^3 - 14x^2 \\
\underline{-(4x^3 - 2x^2)}
\end{array}
$$

Change signs. →

$$
\begin{array}{r}
4x^3 + 2x^2 \\
2x - 1 \overline{)8x^4 + 0x^3 - 14x^2 + 0x - 9} \\
\underline{-8x^4 + 4x^3} \\
4x^3 - 14x^2 \\
\underline{-4x^3 + 2x^2} \\
-12x^2 + 0x
\end{array}
$$

Combine like terms; then bring down the next term.

Determine the next part of the quotient by dividing $-12x^2$ by $2x$ and repeat the multiplication and subtraction steps.

$$
\begin{array}{r}
4x^3 + 2x^2 - 6x \\
2x - 1 \overline{)8x^4 + 0x^3 - 14x^2 + 0x - 9} \\
\underline{-8x^4 + 4x^3} \\
4x^3 - 14x^2 \\
\underline{-4x^3 + 2x^2} \\
-12x^2 + 0x \\
\underline{-(-12x^2 + 6x)}
\end{array}
$$

Change signs. →

$$
\begin{array}{r}
4x^3 + 2x^2 - 6x \\
2x - 1 \overline{)8x^4 + 0x^3 - 14x^2 + 0x - 9} \\
\underline{-8x^4 + 4x^3} \\
4x^3 - 14x^2 \\
\underline{-4x^3 + 2x^2} \\
-12x^2 + 0x \\
\underline{+12x^2 - 6x} \\
-6x - 9
\end{array}
$$

Combine like terms; then bring down the next term.

Determine the final term of the quotient by dividing $-6x$ by $2x$, which is -3.

$$
\begin{array}{r}
4x^3 + 2x^2 - 6x - 3 \\
2x - 1 \overline{)8x^4 + 0x^3 - 14x^2 + 0x - 9} \\
\underline{-8x^4 + 4x^3} \\
4x^3 - 14x^2 \\
\underline{-4x^3 + 2x^2} \\
-12x^2 + 0x \\
\underline{12x^2 - 6x} \\
-6x - 9 \\
\underline{-(-6x + 3)}
\end{array}
$$

Change signs. →

$$
\begin{array}{r}
4x^3 + 2x^2 - 6x - 3 \\
2x - 1 \overline{)8x^4 + 0x^3 - 14x^2 + 0x - 9} \\
\underline{-8x^4 + 4x^3} \\
4x^3 - 14x^2 \\
\underline{-4x^3 + 2x^2} \\
-12x^2 + 0x \\
\underline{12x^2 - 6x} \\
-6x - 9 \\
\underline{+6x - 3} \\
-12
\end{array}
$$

Combine like terms.

Answer $4x^3 + 2x^2 - 6x - 3 + \dfrac{-12}{2x - 1}$, or $4x^3 + 2x^2 - 6x - 3 - \dfrac{12}{2x - 1}$

Your Turn 8 Divide $\dfrac{7 + 8x^3 - 18x}{4x - 2}$.

Answer   **Your Turn 8**

$2x^2 + x - 4 + \dfrac{-1}{4x - 2}$ or

$2x^2 + x - 4 - \dfrac{1}{4x - 2}$

Objective 6 Simplify expressions using rules of exponents. The quotient rule for exponents completes our rules of exponents in this chapter. So far, the problems involving exponent rules have required only that we use one or two rules. Let's now consider problems that require us to use several of the exponent rules. Following is a summary of all of the rules of exponents.

Rule Exponents Summary

Assume that no denominators are 0, that a and b are real numbers, and that m and n are integers.

Zero as an exponent: $a^0 = 1$, where $a \neq 0$

0^0 is indeterminate.

Negative exponents: $a^{-n} = \dfrac{1}{a^n}$

$\dfrac{1}{a^{-n}} = a^n$

$\left(\dfrac{a}{b}\right)^{-n} = \left(\dfrac{b}{a}\right)^n$

Product rule for exponents: $a^m \cdot a^n = a^{m+n}$

Quotient rule for exponents: $\dfrac{a^m}{a^n} = a^{m-n}$

Raising a power to a power: $(a^m)^n = a^{mn}$

Raising a product to a power: $(ab)^n = a^n b^n$

Raising a quotient to a power: $\left(\dfrac{a}{b}\right)^n = \dfrac{a^n}{b^n}$

Example 9 Simplify. Write all answers with positive exponents.

a. $\left(\dfrac{n^{-5}}{n^2}\right)^3$

SOLUTION Following the order of operations, we simplify within the parentheses first, then consider the exponent outside the parentheses.

$\left(\dfrac{n^{-5}}{n^2}\right)^3 = (n^{-5-2})^3$ **Use the quotient rule for exponents.**

$= (n^{-7})^3$

$= n^{-7 \cdot 3}$ **Use the rule for raising a power to a power.**

$= n^{-21}$

$= \dfrac{1}{n^{21}}$ **Write with a positive exponent.**

b. $\dfrac{(m^2)^{-3}}{m^4 \cdot m^{-5}}$

SOLUTION $\dfrac{(m^2)^{-3}}{m^4 \cdot m^{-5}} = \dfrac{m^{2 \cdot (-3)}}{m^{4+(-5)}}$ **In the numerator, use the rule for raising a power to a power. In the denominator, use the product rule for exponents.**

$= \dfrac{m^{-6}}{m^{-1}}$

$= m^{-6-(-1)}$ **Use the quotient rule for exponents.**

$= m^{-6+1}$ **Write the subtraction as an equivalent addition.**

$= m^{-5}$

$= \dfrac{1}{m^5}$ **Write with a positive exponent.**

c. $\dfrac{(4y^3)^2}{(2y^4)^5}$

SOLUTION $\dfrac{(4y^3)^2}{(2y^4)^5} = \dfrac{4^2 y^{3\cdot 2}}{2^5 y^{4\cdot 5}}$ **Use the rule for raising a product to a power.**

$$= \dfrac{16y^6}{32y^{20}}$$

$$= \dfrac{16}{32} \cdot \dfrac{y^6}{y^{20}}$$ **Separate the coefficients and variables.**

$$= \dfrac{16}{32} \cdot y^{6-20}$$ **Use the quotient rule for exponents.**

$$= \dfrac{1}{2} \cdot y^{-14}$$ **Simplify the coefficient.**

$$= \dfrac{1}{2} \cdot \dfrac{1}{y^{14}}$$ **Write with a positive exponent.**

$$= \dfrac{1}{2y^{14}}$$ **Simplify.**

Your Turn 9 Simplify. Write all answers with positive exponents.

Answers Your Turn 9

a. $\dfrac{1}{x^{28}}$ **b.** y^{35} **c.** $\dfrac{y^6}{8x^8}$

a. $\left(\dfrac{x^2}{x^{-5}}\right)^{-4}$

b. $(y^{-2} \cdot y^{-3})^{-4} \div y^{-15}$

c. $\dfrac{(2x^4)^{-3}}{(x^2 y^3)^{-2}}$

5.6 Exercises

For Extra Help MyMathLab

Note: Exercises marked with a ★ represent challenging exercises.

1. Explain how to divide exponential forms that have the same base.

2. What is the rule for rewriting a^{-n}, where a is a real number, $a \neq 0$, and n is a natural number, so that it has a positive exponent?

3. How do we divide two numbers in scientific notation?

4. How do we divide two monomials?

5. Explain how to divide a polynomial by a monomial.

6. When do we use a 0 placeholder in the process of dividing two polynomials by long division?

For Exercises 7–32, simplify using the rules of exponents. Write all answers with positive exponents. See Examples 1 and 2.

7. a^{-3}

8. n^{-7}

9. 2^{-5}

10. 3^{-4}

11. $\dfrac{y^8}{y^3}$

12. $\dfrac{n^4}{n}$

13. $\dfrac{3^4}{3^3}$

14. $\dfrac{6^{11}}{6^5}$

15. $\dfrac{4^3}{4^9}$

16. $\dfrac{5}{5^9}$

17. $\dfrac{x^3}{x^5}$

18. $\dfrac{m^4}{m^{10}}$

19. $a^8 \div a^6$

20. $u^5 \div u^3$

21. $w^7 \div w^{10}$

22. $j^6 \div j^{14}$

23. $\dfrac{a^{-2}}{a^5}$

24. $\dfrac{m^{-6}}{m^3}$

25. $\dfrac{r^6}{r^{-4}}$

26. $\dfrac{u^5}{u^{-12}}$

27. $\dfrac{y^{-7}}{y^{-15}}$

28. $\dfrac{x^{-3}}{x^{-8}}$

29. $\dfrac{p^{-5}}{p^{-2}}$

30. $\dfrac{n^{-11}}{n^{-5}}$

31. $\dfrac{t^{-4}}{t^{-4}}$

32. $\dfrac{x^{-6}}{x^{-6}}$

For Exercises 33–40, divide and write your answers in scientific notation. See Example 3.

33. $\dfrac{8.32 \times 10^5}{3.2 \times 10^6}$

34. $\dfrac{3.6 \times 10^6}{2.4 \times 10^9}$

35. $\dfrac{1.26 \times 10^{-4}}{2.1 \times 10^3}$

36. $\dfrac{1.32 \times 10^{-5}}{6.6 \times 10^{15}}$

37. $\dfrac{9.088 \times 10^1}{1.28 \times 10^{-9}}$

38. $\dfrac{9.964 \times 10^4}{1.88 \times 10^{-2}}$

39. $\dfrac{8.64 \times 10^{-3}}{3.2 \times 10^{-7}}$

40. $\dfrac{7.92 \times 10^{-3}}{2.2 \times 10^{-7}}$

For Exercises 41–44, solve. See Example 3.

41. If light travels at 3×10^8 meters per second, how long does it take light to travel the 1.5×10^{11} meters from the Sun to Earth?

42. In 2008, the population of the United States was approximately 3.04×10^8. Calculate the approximate number of people per square mile in the United States in 2008 if the estimated amount of land in the United States is 3.8×10^6 square miles. (*Source:* U.S. Bureau of the Census)

43. As of March 11, 2009, the national debt was about $\$1.128 \times 10^{13}$. If each of the 3.06×10^8 people in the United States contributed an equal share toward the debt, what would each person have to contribute to pay off the debt? (*Source:* U.S. National Debt Clock)

44. There are about 2.3×10^6 stone blocks in the Great Pyramid. If the total weight of the pyramid is about 1.3×10^{10} pounds and we assume that each block is of equal size, how much does each block weigh? (*Source:* Nova Online)

⭐ **For Exercises 45 and 46, use Einstein's formula $E = mc^2$, which describes the rest energy contained within a mass m, where c represents the speed of light, which is a constant with a value of 3×10^8 meters per second. The units for energy are joules, which are equivalent to $\dfrac{kg \cdot m^2}{s^2}$. See Example 3.**

45. The first atom bomb released an energy equivalent of about 8.4×10^{13} joules. What amount of mass was converted to energy? Write your answer in scientific notation.

46. The largest atom bomb detonated was a hydrogen bomb tested by the Soviet Union in 1961. The bomb released 2.4×10^{17} joules of energy. What amount of mass was converted to energy in that explosion? (*Source*: Cesare Emiliani, *The Scientific Companion*; Wiley Popular Science, 1995)

For Exercises 47–58, divide the monomials. See Example 4.

47. $\dfrac{15x^5}{3x^2}$

48. $\dfrac{14a^6}{2a^4}$

49. $\dfrac{24x^2}{-6x^5}$

50. $\dfrac{-18a^6}{9a^{12}}$

51. $\dfrac{9m^3n^5}{-3mn}$

52. $\dfrac{-42x^5y^3}{6xy^2}$

53. $\dfrac{-24p^3q^5}{15p^3q^2}$

54. $\dfrac{48t^4u^7}{-18t^4u^2}$

55. $\dfrac{56x^2y^6}{42x^7y^2}$

56. $\dfrac{60a^8b^4}{66a^2b^9}$

57. $\dfrac{12a^4bc^3}{9a^6bc^2}$

58. $\dfrac{28p^5q^4r}{42p^2q^7r}$

For Exercises 59–76, divide the polynomial by the monomial. See Example 5.

59. $\dfrac{7a + 14b}{7}$

60. $\dfrac{4m + 16y}{4}$

61. $\dfrac{12x^3 - 6x^2}{3x}$

62. $\dfrac{18m^4 - 27m^2}{9m}$

63. $\dfrac{12x^3 + 8x}{4x^2}$

64. $\dfrac{24y^3 + 16y}{8y^2}$

65. $\dfrac{5x^2y^2 - 15xy^3}{5xy}$

66. $\dfrac{12k^4l^2 + 15k^2l^3}{3k^2l}$

67. $\dfrac{6abc^2 - 24a^2b^2c}{-3abc}$

68. $\dfrac{30x^3yz^5 - 15xyz^2}{-5xyz^2}$

69. $\dfrac{24x^3 + 16x^2 - 8x}{8x}$

70. $\dfrac{3x^3 + 6x^2 - 3x}{3x}$

71. $\dfrac{6x^3 - 12x^2 + 9x}{3x^2}$

72. $\dfrac{12y^4 - 16y^3 + 8y}{4y^2}$

73. $\dfrac{36u^3v^4 + 12uv^5 - 15u^2v^2}{3u^2v}$

74. $\dfrac{16hk^4 - 28hk - 4h^2k^2}{4hk^2}$

75. $\dfrac{y^5 + y^7 - 3y^8 + y}{y^3}$

76. $\dfrac{x^6 + x^8 - 4x^{10} + x^2}{x^4}$

77. The area of the parallelogram shown is described by the monomial $35mn^2$. Find the height.

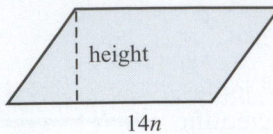

height

$14n$

78. The area of the triangle shown is $18x^2y$. Find the base.

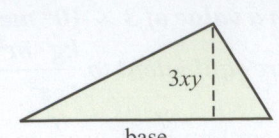

$3xy$

base

79. An engineer is designing a steel cover plate in the shape of a rectangle. The area of the plate is described by the polynomial $9x^2 - 15x + 12$, and the length must be $3x$.
  a. Find an expression for the width.

  b. Find the area, length, and width if $x = 2$.

80. The voltage in a circuit is described by the polynomial $36s^3 - 24s^2 + 42s$. The resistance is described by $6s$.
  a. Find an expression for the current.
  (*Hint:* Voltage = Current × Resistance)

  b. Find the voltage, current, and resistance if $s = 3$.

For Exercises 81–104, use long division to divide the polynomials. See Examples 6–8.

81. $\dfrac{x^2 + 7x + 12}{x + 3}$

82. $\dfrac{y^2 + 4y + 4}{y + 2}$

83. $\dfrac{m^2 - 8m + 15}{m - 3}$

84. $\dfrac{u^2 - 6u + 9}{u - 3}$

85. $\dfrac{x^2 - 17x + 64}{x - 5}$

86. $\dfrac{x^2 + 6x + 10}{x + 3}$

87. $\dfrac{3x^3 - 73x - 10}{x - 5}$

88. $\dfrac{2x^3 - 35x - 12}{x + 4}$

89. $\dfrac{2x^3 - x^2 + 5}{x - 2}$

90. $\dfrac{2x^3 - 2x^2 + 5}{x + 3}$

91. $\dfrac{p^3 + 125}{p + 5}$

92. $\dfrac{x^3 + 8}{x + 2}$

93. $\dfrac{12x^2 - 7x - 10}{4x - 5}$

94. $\dfrac{6a^2 + 2a - 28}{3a + 7}$

95. $\dfrac{14y^2 - 8y + 17}{7y + 3}$

96. $\dfrac{6x^2 + 7x + 5}{3x - 1}$

97. $\dfrac{4k^3 + 8k^2 + 6 - k}{k + 2}$

98. $\dfrac{y^3 + 2y^2 + 3 - 11y}{y + 2}$

99. $\dfrac{21u^3 - 19u^2 + 14u - 6}{3u + 2}$

100. $\dfrac{2v^3 + 5v^2 - 10v + 2}{2v - 3}$

101. $\dfrac{14b + b^3 - 6b^2 - 12}{b - 3}$

102. $\dfrac{-14x + x^2 + x^3 - 5}{x + 4}$

103. $\dfrac{y^3 - y + 6}{y + 2}$

104. $\dfrac{2b^4 + 3b^3 - 4b - 6}{2b + 3}$

105. The volume of a restaurant's walk-in refrigerator is described by $(6x^2 + 42x + 72)$ cubic feet. The width has to be 6 feet, and the length is described by $(x + 3)$ feet.
  a. Find an expression for the height of the refrigerator.

  b. Find the length, height, and volume of the refrigerator if $x = 4$.

106. In an architectural project, the specifications for a rectangular room call for the area to be $(10x^2 + 3x - 18)$ square feet. It is decided that the length should be described by the binomial $(2x + 3)$ feet.
  a. Find an expression for the width.

  b. Find the length, width, and area of the room if $x = 5$.

For Exercises 107–130, simplify. Write all answers with positive exponents.
See Example 9.

107. $(6m^2n^{-2})^2$

108. $(-4a^3b^{-2}c)^2$

109. $(4x^{-3}y^2)^{-3}$

110. $(3m^4n^{-3})^{-4}$

111. $(4rs^3)^3(2r^3s^{-1})^2$

112. $(-3abc^{-2})^2(2a^2b^{-3})^3$

113. $\left(\dfrac{m^6}{m^2}\right)^4$

114. $\left(\dfrac{p^7}{p^4}\right)^5$

115. $\left(\dfrac{x^{-3}}{x^4}\right)^2$

116. $\left(\dfrac{r^{-4}}{r^2}\right)^5$

117. $\left(\dfrac{x^2y^7}{x^5y^{-2}}\right)^3$

118. $\left(\dfrac{m^6n^5}{m^{-3}n^8}\right)^2$

119. $\left(\dfrac{x^{-3}}{x^4}\right)^{-2}$

120. $\left(\dfrac{p^{-5}}{p^3}\right)^{-3}$

121. $\left(\dfrac{3x^{-1}y^2}{z^2}\right)^3$

122. $\left(\dfrac{2x^{-2}y}{z^{-1}}\right)^4$

123. $\dfrac{(x^4)^{-3}}{x^{-5}\cdot x^3}$

124. $\dfrac{(y^3)^{-5}}{y^6\cdot y^{-8}}$

125. $\dfrac{4xy^{-2}z^2}{x^{-3}y^3z^{-1}}$

126. $\dfrac{3m^6n^{-1}p^{-2}}{m^{-1}n^{-3}p}$

127. $\dfrac{(2ab)^3}{12a^4b^5}$

128. $\dfrac{(2xy^2)^2}{16x^2y^3}$

129. $\dfrac{(8x^3)^3}{(4x^4)^5}$

130. $\dfrac{(3y^4)^2}{(6y^7)^3}$

Puzzle Problem

Fill the following blanks with the numbers 1, 2, 3, 4, 5, 6, 7, 8, and 9, using each number once so that the equation is true.

$$\square\square \div \square = \square\square \div \square = \square\square \div \square$$

Review Exercises

Exercises 1–4 **EXPRESSIONS**

[1.2] **1.** Find the prime factorization of 3024.

[5.4] **2.** Simplify: $-3x^2y\cdot 6x^2y^2$

[5.5] **3.** Simplify: $y(3x-4)$

[1.6] **4.** Translate to an algebraic expression: one number plus half of a second
number.

Exercises 5 and 6 **EQUATIONS AND INEQUALITIES**

[2.1] **5.** Find the area of a playground that is 17 feet by 20 feet.

[2.3] **6.** Solve and check: $3-y=27$

Chapter 5 Summary

Defined Terms

Section 5.1
Scientific notation (p. 340)

Section 5.2
Monomial (p. 348)
Coefficient of a monomial (p. 349)
Degree of a monomial (p. 349)
Polynomial (p. 350)

Polynomial in one variable (p. 350)
Binomial (p. 350)
Trinomial (p. 350)
Degree of a polynomial (p. 351)

Section 5.5
Conjugates (p. 383)

Procedures, Rules, and Key Examples

Procedures/Rules	Key Examples

SECTION 5.1 Exponents and Scientific Notation

If the base of an exponential form is a negative number and the exponent is even, then the product is positive.

If the base is a negative number and the exponent is odd, then the product is negative.

Example 1: Evaluate.
a. $(-2)^4 = (-2)(-2)(-2)(-2) = 16$
b. $(-2)^5 = (-2)(-2)(-2)(-2)(-2) = -32$

If a and b are real numbers, where $b \neq 0$ and n is a natural number, then $\left(\dfrac{a}{b}\right)^n = \dfrac{a^n}{b^n}$.

If a is a real number and $a \neq 0$, then $a^0 = 1$.

If a is a real number, where $a \neq 0$ and n is a natural number, then $a^{-n} = \dfrac{1}{a^n}$ and $\dfrac{1}{a^{-n}} = a^n$.

If a and b are real numbers, where $a \neq 0$ and $b \neq 0$ and n is a natural number, then $\left(\dfrac{a}{b}\right)^{-n} = \left(\dfrac{b}{a}\right)^n$.

Example 2: Simplify.
a. $\left(\dfrac{3}{4}\right)^2 = \dfrac{3^2}{4^2} = \dfrac{9}{16}$
b. $7^0 = 1$
c. $2^{-4} = \dfrac{1}{2^4} = \dfrac{1}{16}$
d. $\dfrac{1}{3^{-2}} = 3^2 = 9$
e. $\left(\dfrac{2}{3}\right)^{-4} = \left(\dfrac{3}{2}\right)^4 = \dfrac{81}{16}$

To change from scientific notation with a positive integer exponent to standard form, move the decimal point to the right the number of places indicated by the exponent.

Example 3: Write in standard form.
a. $3.4 \times 10^5 = 340{,}000$

To change from scientific notation with a negative exponent to standard form, move the decimal point to the left the same number of places as the absolute value of the exponent.

b. $4.2 \times 10^{-5} = 0.000042$

To write a number greater than 1 in scientific notation,
1. Move the decimal point so that the number is greater than or equal to 1 but less than 10. (Place the decimal point to the right of the first nonzero digit.)
2. Write the decimal number multiplied by 10^n, where n is the number of places between the new decimal position and the original decimal position.
3. Delete zeroes to the right of the last nonzero digit.

Example 4: Write in scientific notation.
$$7{,}230{,}000 = 7.23 \times 10^6$$

(continued)

Procedures/Rules	**Key Examples**

SECTION 5.1 **Exponents and Scientific Notation** (*continued*)

To write a positive decimal number that is less than 1 in scientific notation,	**Example 5:** Write in scientific notation.
1. Move the decimal point so that the number is greater than or equal to 1 but less than 10. (Place the decimal point to the right of the first nonzero digit.)	$$0.00000056 = 5.6 \times 10^{-7}$$
2. Write the decimal number multiplied by 10^n, where n is a negative integer whose absolute value is the number of places between the new decimal position and the original decimal position.	
3. Delete zeroes to the left of the first nonzero digit.	

SECTION 5.2 **Introduction to Polynomials**

To write a polynomial in descending order of degree, place the highest degree term first, then the next highest degree term, and so on.	**Example 1:** Write in descending order.
	$$2x^2 + 4x^5 + 9 - 6x^3 = 4x^5 - 6x^3 + 2x^2 + 9$$

SECTION 5.3 **Adding and Subtracting Polynomials**

To add polynomials, combine like terms.	**Example 1:** Add.
	$$(5x^3 + 12x^2 - 9x + 1) + (7x^2 - x - 13)$$ $$= 5x^3 + 12x^2 + 7x^2 - 9x - x + 1 - 13$$ $$= 5x^3 + 19x^2 - 10x - 12$$
To subtract polynomials,	**Example 2:** Subtract.
1. Write the subtraction statement as an equivalent addition statement.	$$(6y^3 - 5y + 18) - (y^3 - 5y + 7)$$
a. Change the operation symbol from a minus sign to a plus sign.	$$= (6y^3 - 5y + 18) + (-y^3 + 5y - 7)$$
b. Change the subtrahend (second polynomial) to its additive inverse. To get the additive inverse, change the sign of each term in the polynomial.	$$= 6y^3 - y^3 - 5y + 5y + 18 - 7$$ $$= 5y^3 + 11$$
2. Combine like terms.	

SECTION 5.4 **Exponent Rules and Multiplying Monomials**

Product rule for exponents: If a is a real number and m and n are integers, then $a^m \cdot a^n = a^{m+n}$.	**Example 1:** Multiply. a. $x^3 \cdot x^4 = x^{3+4} = x^7$
To multiply monomials, 1. Multiply coefficients. 2. Add the exponents of the like bases. 3. Write any unlike variable bases unchanged in the product.	b. $-6x^4yz^2 \cdot 5x^5y^3 = -30x^{4+5}y^{1+3}z^2$ $ = -30x^9y^4z^2$
If a is a real number and m and n are integers, then $(a^m)^n = a^{mn}$.	**Example 2:** Simplify. a. $(x^3)^4 = x^{3 \cdot 4} = x^{12}$
If a and b are real numbers and n is an integer, then $$(ab)^n = a^n b^n.$$	b. $(m^2n^3)^5 = m^{2 \cdot 5}n^{3 \cdot 5} = m^{10}n^{15}$
To simplify a monomial raised to a power, 1. Evaluate the coefficient raised to that power. 2. Multiply each variable's exponent by the power.	c. $(3x^2y^5)^4 = 3^4x^{2 \cdot 4}y^{5 \cdot 4} = 81x^8y^{20}$

(*continued*)

Procedures/Rules	Key Examples

SECTION 5.5 Multiplying Polynomials; Special Products

To multiply a polynomial by a monomial, use the distributive property to multiply each term in the polynomial by the monomial.

Example 1: Multiply.

a. $3x(x^4 - 7x^2y + 9y^2 - 2)$
$= 3x \cdot x^4 - 3x \cdot 7x^2y + 3x \cdot 9y^2 - 3x \cdot 2$
$= 3x^5 - 21x^3y + 27xy^2 - 6x$

To multiply two polynomials,
1. Multiply each term in the second polynomial by each term in the first polynomial.
2. Combine like terms.

b. $(3x + 5)(2x - 1)$
$= 3x \cdot 2x + 3x \cdot (-1) + 5 \cdot 2x + 5(-1)$
$= 6x^2 - 3x + 10x - 5$
$= 6x^2 + 7x - 5$

Special Products (a and b are real numbers, variables, or expressions):

Conjugates: $(a + b)(a - b) = a^2 - b^2$

c. $(4x + 3)(4x - 3) = 16x^2 - 9$

Squaring a sum: $(a + b)^2 = a^2 + 2ab + b^2$

d. $(3y + 2)^2 = 9y^2 + 12y + 4$

Squaring a difference: $(a - b)^2 = a^2 - 2ab + b^2$

e. $(3y - 2)^2 = 9y^2 - 12y + 4$

SECTION 5.6 Exponent Rules and Dividing Polynomials

Quotient rule for exponents: If m and n are integers and a is a real number, where $a \neq 0$, then $\dfrac{a^m}{a^n} = a^{m-n}$.

Example 1: Divide.

a. $\dfrac{x^7}{x^2} = x^{7-2} = x^5$

To divide monomials,
1. Divide the coefficients or simplify them to fractions in lowest terms.
2. Use the quotient rule for the exponents with like bases.
3. Do not change unlike variable bases in the quotient.
4. Write the final expression so that all exponents are positive.

b. $\dfrac{-39a^6b^5c^2}{13a^4b^5c}$
$= -3a^{6-4}b^{5-5}c^{2-1}$
$= -3a^2b^0c^1$
$= -3a^2(1)c$
$= -3a^2c$

If a, b, and c are real numbers, variables, or expressions with $c \neq 0$, then

$$\frac{a + b}{c} = \frac{a}{c} + \frac{b}{c}.$$

c. $\dfrac{30y^6 + 45y^3}{5y^2} = \dfrac{30y^6}{5y^2} + \dfrac{45y^3}{5y^2}$
$= 6y^{6-2} + 9y^{3-2}$
$= 6y^4 + 9y$

To divide a polynomial by a monomial, divide each term in the polynomial by the monomial.

To divide a polynomial by a polynomial, use long division. If there is a remainder, write the result in the following form:

$$\text{quotient} + \frac{\text{remainder}}{\text{divisor}}$$

d. $\dfrac{6x^2 + 7x - 17}{3x - 4}$

$$
\begin{array}{r}
2x + 5 \\
3x - 4 \overline{)6x^2 + 7x - 17} \\
\underline{-6x^2 + 8x} \\
15x - 17 \\
\underline{-15x + 20} \\
3
\end{array}
$$

Answer: $2x + 5 + \dfrac{3}{3x - 4}$

(continued)

Procedures/Rules	Key Examples

Exponents Summary

Assume that no denominators are 0, that a and b are real numbers, and that m and n are integers.

Zero as an exponent:

$a^0 = 1$, where $a \neq 0$

0^0 is indeterminate.

Negative exponents:

$$a^{-n} = \frac{1}{a^n}$$

$$\frac{1}{a^{-n}} = a^n$$

$$\left(\frac{a}{b}\right)^{-n} = \left(\frac{b}{a}\right)^n$$

Product rule for exponents:

$a^m \cdot a^n = a^{m+n}$

Quotient rule for exponents:

$$\frac{a^m}{a^n} = a^{m-n}$$

Raising a power to a power:

$(a^m)^n = a^{mn}$

Raising a product to a power:

$(ab)^n = a^n b^n$

Raising a quotient to a power:

$$\left(\frac{a}{b}\right)^n = \frac{a^n}{b^n}$$

Example 1: Evaluate.

a. $3^0 = 1$ **b.** $x^0 = 1, x \neq 0$

c. $3^{-4} = \dfrac{1}{3^4} = \dfrac{1}{81}$

d. $\dfrac{1}{x^{-5}} = x^5$

e. $\left(\dfrac{3}{x}\right)^{-2} = \left(\dfrac{x}{3}\right)^2$

f. $x^3 \cdot x^5 = x^{3+5} = x^8$

g. $\dfrac{r^4}{r^{-2}} = r^{4-(-2)} = r^{4+2} = r^6$

h. $(y^3)^5 = y^{3 \cdot 5} = y^{15}$

i. $(3x)^4 = 3^4 x^4 = 81x^4$

j. $\left(\dfrac{2}{a}\right)^5 = \dfrac{2^5}{a^5} = \dfrac{32}{a^5}$

Chapter 5 Review Exercises

For Exercises 1–6, answer true or false.

[5.1] **1.** A negative exponent will always make the coefficient negative.

[5.2] **2.** The degree of a binomial is two.

[5.4] **3.** To raise an exponential form to a power, multiply the exponents.

[5.5] **4.** Conjugates occur when both binomials contain subtraction.

[5.5] **5.** FOIL can be used for all types of polynomial multiplication.

[5.5] **6.** $(x - 4)^2 = x^2 + 16$

For Exercises 7–10, complete the rule.

[5.3] **7.** To add polynomials, _____ like terms.

[5.4] **8.** To multiply monomials:
   a. Multiply the _____.
   b. Add the _____ of the like bases.
   c. Rewrite any unlike variable bases in the product.

[5.4] **9.** $(a^m)^n = a^?$, where a is a real number and m and n are integers.

[5.5] **10.** $(a + b)^2 = a^2 + $ _____ $ + b^2$

[5.1] *For Exercises 11–14, evaluate the exponential form.*

11. $\left(\dfrac{2}{5}\right)^3$

12. -4^2

13. 5^{-2}

14. 13^{-1}

[5.1] *For Exercises 15–18, write the number in standard form.*

15. The radioactive half-life of uranium 238 is 4.5×10^9 years.

16. The temperature at the Sun's core is 1.38×10^7 degrees Celsius.

17. The mass of a hydrogen atom is 1.663×10^{-24} gram.

18. The mass of an electron is 2.006×10^{-30} pound.

[5.1] *For Exercises 19–22, write the number in scientific notation.*

19. Because atoms are so small, their weight is measured in atomic mass units (AMU). One atomic mass unit is 0.000000000000000000000001661 gram.

20. One molecule of NaCl, sodium chloride (table salt) is 0.00000000000000000000000963 gram.

21. The speed of light is about 300,000,000 meters per second.

22. The approximate radius of Earth is 6,370,000 meters.

[5.2] *For Exercises 23–26, determine whether the expression is a monomial.*

23. $3xy^5$

24. $-\dfrac{1}{2}x$

25. $\dfrac{4}{ab^3}$

26. $2x^2 - 9$

[5.2] *For Exercises 27–30, identify the coefficient and degree of each monomial.*

27. $6x^4$

28. 27

29. $-2.6xy^3$

30. $-m$

[5.2] *For Exercises 31–34, indicate whether the expression is a monomial, binomial, or trinomial, has no special polynomial name or is not a polynomial.*

31. $4x^2 - 25$

32. $-st^5$

33. $3x - \dfrac{4}{y^2}$

34. $5x^3 - 6x^2 - 3x + 11$

[5.2] *For Exercises 35–38, identify the degree of each polynomial and write each polynomial in descending order of degree.*

35. $15 - 2x^9 - 3x + 21x^5 - 19x^3$

36. $22j^2 - 19 - j^4 + 5j$

37. $4u^5 - 18u^3 + 13u^4 - u - 21$

38. $16v^3 + 21v^4 - 2v + 6v^8 - 19 - v^2$

[5.2] *For Exercises 39–49, combine like terms. If possible, write the resulting polynomial in descending order of degree.*

39. $8y^5 - 3y^4 + 2y - 4y^5 - 2y + 8 + 7y^4$

40. $8l + 5l^4 - 2l^3 - l + 18l^5 + 6l^2 + 20$

41. $3m - 4 - m^2 - 2m^3 + 7 - 2m + 8m^2$

42. $-9 + 7y - 2y - 6 + 2y - 3y^2$

43. $a^2bc - 3ab^2c + 2ab^2c - 6$

44. $5xyz^2 - 4x^2yz - 7xyz^2 - 5x^2yz - 2x^2zy$

45. $6cd - 4cd^2 + 3cd^2 + 8c^3 + 5cd - 12c^3 + 2d + 8$

46. $4a^2bc + 5abc^3 - 8abc - 9a^5 - 2abc^3 - 4a^2bc + 8 + a^5$

47. $18jk - 2j^4 + 12jk^3 - j^4 - 8jk - 9jk^3 + 12 + k^2$

48. $4m^2 - 3mn + 2mn^2 - 8n^2 - 6mn - 4m^2 + 7 + mn^2$

49. $2a + 3b - 4ab - c - 18c - 20a + b - 4b + 8$

[5.3] *For Exercises 50–59, add or subtract and write the resulting polynomial in descending order of degree.*

50. $(5y - 2) + (6y - 8)$

51. $(8x^2 - 3) + (2x^2 + 3x - 1)$

52. $(5m - 1) - (2m + 7)$

53. $(5n + 8) - (2n^3 - 3n^2 + n + 5)$

54. $(2y - 4) + (4y + 8) - (6y - 5)$

55. $(3x^3 - 2x + 8) + (4x^2 - 3x - 1) - (4x^3 - 5x - 10)$

56. $(4p^3 + 2p^2 - 8p - 1) + (-5p^3 + 8p^2 - 2p + 10)$

57. $(2x^3 - 3x^2 - 2x - 1) - (18x^3 - 3x^2 + 2x + 1)$

58. $(5x^2 - 2xy + y^2) + (3x^2 + xy + 5y^2)$

59. $(m^2 + 5mn + 6) - (4m^2 - 3mn + 8)$

[5.4, 5.6] *For Exercises 60–74, simplify. Write all answers with positive exponents.*

60. $m \cdot m^4$

61. $2a \cdot 5a^8$

62. $(-3x^2y)(2x^4y^5)$

63. $(5x^2y)^2$

64. $(-6u^3)^2(2u^4)$

65. $(2x^2)(2x^2)^4$

66. $\dfrac{x^3}{x^{-5}}$

67. $\dfrac{u^{-1}}{u^{-8}}$

68. $\dfrac{s^{-5}}{s}$

69. $\left(\dfrac{1}{x^3}\right)^{-2}$

70. $(2a^3b^{-2})^{-3}$

71. $\dfrac{4hj^4k^{-2}}{(3hj^{-2})^2}$

72. $-5x^{-2}$

73. $8x^0 - (2x)^0$

74. $(-18x^4y^{-5})^0$

[5.5] *For Exercises 75–92, multiply.*

75. $2(x - 21)$

76. $-8(2x - 1)$

77. $4a(a - 3)$

78. $-6b(2b^2 - 4b - 1)$

79. $4m^3(-2m^4 - 3m^2 + m + 5)$

80. $-4abc^2(3a - 4ab^3 + 2abc^2 + 8)$

81. $(x + 5)(x - 1)$

82. $(y - 3)(y - 8)$

83. $(2m + 5)(6m - 1)$

84. $(5a + 2b)(3a - 2b)$

85. $(2x + 1)(2x - 1)$

86. $(x - 3)(x^2 + 2x + 5)$

87. $(3y - 1)(y^2 - 2y + 4)$

88. $(a^2 + ab + b^2)(a^2 - 2ab - b^2)$

89. $(4 - x)(4 + x)$

90. $(x - 6)^2$

91. $(3r - 5)^2$

92. $(6s + 2r)(6s - 2r)$

[5.5] **93. a.** Write an expression in simplest form for the area of the figure shown.

b. Calculate the area if $x = 3$ feet.

$5x - 7$

$8x$

[5.4] **94.** A park ranger is studying the number of trees per square meter in a national forest. The area in the study is rectangular and measures 9×10^3 meters by 2.5×10^4 meters. Find the area of this region. Write your answer in scientific notation.

[5.4] **95.** In a circuit, the current is measured to be 3.5×10^{-3} ampere. If the circuit has a resistance of 8.4×10^{-2} ohm, find the voltage. (*Hint*: Voltage = Current × Resistance) Write your answer in scientific notation.

[5.6] *For Exercises 96–101, divide.*

96. $\dfrac{4x - 20}{4}$

97. $\dfrac{5y^2 - 10y + 15}{-5}$

98. $\dfrac{2st + 20s^2t^4 - 4st^5}{2st}$

99. $\dfrac{a^3bc^3 - abc^2 + 2a^2bc^4}{abc^3}$

100. $\dfrac{6a^2b^2 + 3a^2b^3}{3a^2b^2}$

101. $\dfrac{2xyz^2 - 5x^2y^2z^2 + 10xy^2z}{5xy^3z}$

[5.6] **102. a.** The area of the parallelogram shown is described by the polynomial $54x^2 - 60x$. Find the height.

b. Find the base, height, and area of this parallelogram if $x = 4$ inches.

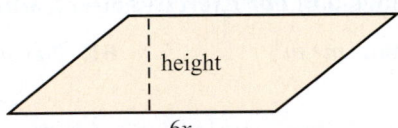

height

$6x$

[5.6] **103.** Proxima Centauri is approximately 2.47×10^{13} miles from the Sun. If light travels 5.881×10^{12} miles per year, how long does the light from Proxima Centauri take to reach Earth? (*Hint*: Distance = Rate × Time)

[5.6] *For Exercises 104–111, use long division to divide the polynomials.*

104. $\dfrac{z^2 + 8z + 15}{z + 3}$

105. $\dfrac{3m^2 - 17m + 11}{m - 5}$

106. $\dfrac{2y^2 - 7y - 8}{2y + 1}$

107. $\dfrac{6m^2 - 10m + 5}{3m - 2}$

108. $\dfrac{x^3 - 1}{x + 1}$

109. $\dfrac{8x^3 - 27}{2x - 3}$

110. $\dfrac{2s^3 - 2s - 3}{s - 1}$

111. $\dfrac{10y^4 - 5y^3 - 6y^2 + 7y - 2}{2y - 1}$

[5.6] 112. A storage room is to have an area described by $4x^2 + 23x - 35$. If the length is described by $x + 7$, find an expression for the width.

Chapter 5 Practice Test

For Extra Help

Step-by-step test solutions are found on the Chapter Test Prep Videos available via the Video Resources on DVD, in **MyMathLab**, and on YouTube (search "CarsonCombinedAlg" and click on "Channels").

For Exercises 1 and 2, evaluate the exponential form.

1. 2^{-3}

2. $\left(\dfrac{2}{3}\right)^{-2}$

3. Write 6.201×10^{-3} in standard form.

4. Write 275,000,000 in scientific notation.

For Exercises 5 and 6, identify the degree.

5. $-7x^2y$

6. $4x^2 - 9x^4 + 8x - 7$

7. Evaluate $-6mn - n^3$, where $m = 4$ and $n = -2$.

8. Combine like terms and write the resulting polynomial in descending order of degree.
$-5x^4 + 7x^2 + 6x^2 - 5x^4 + 12 - 6x^3 + x^2$

For Exercises 9 and 10, add or subtract and write the resulting polynomial in descending order of degree.

9. $(3x^2 + 4x - 2) + (5x^2 - 3x - 2)$

10. $(7x^4 - 3x^2 + 4x + 1) - (2x^4 + 5x - 7)$

For Exercises 11–18, multiply.

11. $(4x^2)(3x^5)$

12. $(2ab^3c^7)(-a^5b)$

13. $(4xy^3)^2$

14. $3x(x^2 - 4x + 5)$

15. $-6t^2u(4t^3 - 8tu^2)$

16. $(n - 1)(n + 4)$

17. $(2x - 3)^2$

18. $(x + 2)(x^2 - 4x + 3)$

19. Write an expression in simplest form for the area of the shape shown.

$2n - 5$ [rectangle] $3n + 4$

For Exercises 20–25, simplify.

20. $\dfrac{x^9}{x^4}$

21. $\dfrac{(x^3)^{-2}}{x^4 \cdot x^{-5}}$

22. $\dfrac{(3y)^{-2}}{(x^3 y^2)^{-3}}$

23. $\dfrac{24x^5 + 18x^3}{6x^2}$

24. $\dfrac{x^2 - x - 12}{x + 3}$

25. $\dfrac{15x^2 - 22x + 14}{3x - 2}$

Chapters 1–5 Cumulative Review Exercises

For Exercises 1–4, answer true or false.

[4.3] **1.** The graph of $3x - 4y = -24$ has an x-intercept of $(0, 8)$ and y-intercept of $(-6, 0)$.

[1.6] **2.** 6 less than a number can be translated to $6 - x$.

[4.5] **3.** $y - 2 = -3(x - 5)$ is an example of the point–slope form of a line.

[4.2] **4.** There is only one solution to any linear equation in two variables.

For Exercises 5 and 6, fill in the blank.

[2.3] **5.** To use the multiplication principle to clear a coefficient,
  a. Determine the coefficient you want to clear.
  b. Multiply both sides by the multiplicative inverse of that _____ or divide both sides by the _____.

[4.2] **6.** If an equation can be written in the form $y = mx$, where m is a real number other than 0, the x- and y-intercept will be at _____.

Exercises 7–18   **EXPRESSIONS**

[1.5] *For Exercises 7 and 8, simplify.*

7. $5 + 2(6 - 3^2)^2$

8. $|5 - 10| + 3(2) \div (-1)$

[3.2] **9.** 30.2% of what number is 18.12?

For Exercises 10–15, simplify.

[5.3] **10.** $(3y^2 - 2y + 10) + (2y^2 - 2y - 11)$

[5.3] **11.** $(2x + 1) - (-4x + 7)$

[5.5] **12.** $(2x + 3)(x - 7)$

[5.5] **13.** $(2x - 3)^2$

[5.6] **14.** $(3m^2 - 6m + 1) \div (3m)$

[5.6] **15.** $(x^2 + 5x + 7) \div (x + 2)$

[5.6] For Exercises 16–18, simplify. Leave your answers with positive exponents only.

16. $(3x^{-2}y^3)^{-3}$

17. $\left(\dfrac{2x^3}{y}\right)^2$

18. $(2x^2y^{-3})^2(-3x^4y^2)^3$

Exercises 19–30 **EQUATIONS AND INEQUALITIES**

[4.3] 19. Graph: $3x + 2y = -6$

[4.6] 20. Graph: $y > \dfrac{2}{3}x - 2$

For Exercises 21–24, solve.

[2.3] 21. $10x - 3 + 5 = 8x + 26 + 4x$

[2.6] 22. $3(x - 2) > 4(x + 1)$

[3.1] 23. $\dfrac{16}{9} = \dfrac{x}{20.25}$

[2.3] 24. $-\dfrac{3}{7}x = \dfrac{12}{35}$

[2.4] 25. Solve for l in the equation $P = 2l + 2w$.

For Exercises 26–30, solve.

[3.5] 26. The butcher at Kroger is combining 60 pounds of hamburger costing $2.50 per pound with some hamburger costing $3.10 per pound. How many pounds of the hamburger costing $3.10 does he need to add to make a mixture costing $2.65 per pound?

[3.5] 27. A broker invests $\dfrac{1}{4}$ of a client's money in a government security paying 9% interest annually and the rest in a money market fund paying 13% interest annually. Find the amount invested in each if the total interest earned is $5400.

[3.2] 28. Use the accompanying bar graph. If 1600 parents at a local high school were polled, how many could be expected to think that teens should not work?

[3.2] 29. In Westchester County, New York, the average temperature is 68° in July and only 31.5° in January. What is the percent of decrease in temperature from July to January?

[3.3] 30. The cashier's office at the college has collected $7020 from students for a total of 42 part-time classes. If each credit hour costs $45, how many three-hour classes and how many five-hour classes were paid for?

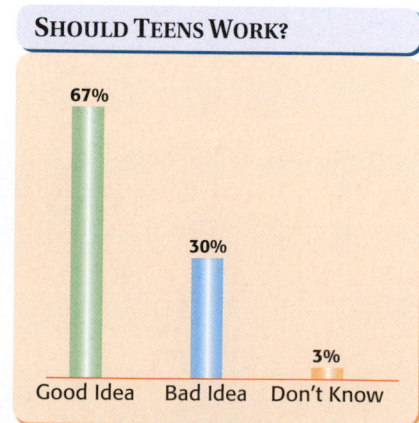

SHOULD TEENS WORK?

67% Good Idea
30% Bad Idea
3% Don't Know

Source: Yankelovich Partners for Lutheran Brotherhood, U.S. Department of Labor

Factoring

6.1 Greatest Common Factor and Factoring by Grouping

> **Discipline is the keynote to learning. Discipline has been the great factor in my life.**
>
> —Rose Hoffman
> Elementary schoolteacher

> **The higher processes are all simplification.**
>
> —Willa Cather (1873–1947)
> U.S. novelist

In Chapter 5, we learned about a specific class of expressions called *polynomials*. We learned that polynomials can be added, subtracted, multiplied, and divided in much the same way as numbers. In Chapter 6, we explore polynomials further, and in Sections 6.1–6.5, we learn how to factor polynomial expressions. Factoring is another way to rewrite an expression and is one of the most important skills you will acquire in algebra. It is used extensively in simplifying the more complex expressions that we will study in future chapters. In Section 6.6, we use factoring to solve certain equations that contain polynomial expressions. Finally, in Section 6.7, we graph equations and functions containing polynomial expressions.

6.1 Greatest Common Factor and Factoring by Grouping

Objectives

1. List all possible factors for a given number.
2. Find the greatest common factor of a set of numbers or monomials.
3. Write a polynomial as a product of a monomial GCF and a polynomial.
4. Factor by grouping.

Often in mathematics, we need to consider the factors of a number or an expression. A number or an expression written as a product of factors is said to be in **factored form**.

Definition   *Factored form:* A number or an expression written as a product of factors.

Following are some examples of factored form.

An integer written in factored form with integer factors: $28 = 2 \cdot 14$
A monomial written in factored form with monomial factors: $8x^5 = 4x^2 \cdot 2x^3$
A polynomial written in factored form with a monomial factor and a polynomial factor: $2x + 8 = 2(x + 4)$
A polynomial written in factored form with two polynomial factors:
$x^2 + 5x + 6 = (x + 2)(x + 3)$

Notice that we can check the factored form by multiplying the factors to equal the product. The process of writing an expression in factored form is called *factoring*.

Objective 1 List all possible factors for a given number.   We begin our exploration of factoring by listing natural number factors when given an integer.

Example 1   List all natural number factors of 24.

SOLUTION   To list all natural number factors, we can divide 24 by 1, 2, 3, and so on, writing each divisor and quotient pair as a product until we have all possible combinations.

> **Note:** The number 5 does not divide 24 evenly, and 6 is already in the $4 \cdot 6$ pair. When you reach a natural number that you already have listed, you can stop.

$1 \cdot 24$
$2 \cdot 12$
$3 \cdot 8$
$4 \cdot 6$

The natural number factors of 24 are 1, 2, 3, 4, 6, 8, 12, 24.

Your Turn 1   List all natural number factors of each given number.

a. 30

b. 42

Objective 2 Find the greatest common factor of a set of numbers or monomials.   When factoring polynomials, the first step is to determine whether there is a monomial factor that is common to all of the terms in the polynomial. Further, we want that monomial factor to be the **greatest common factor** of the terms. Let's first consider how to find the greatest common factor of numbers.

Answers   Your Turn 1

a. 1, 2, 3, 5, 6, 10, 15, 30
b. 1, 2, 3, 6, 7, 14, 21, 42

Greatest common factor (GCF): The largest natural number that divides all given numbers with no remainder.

For example, the greatest common factor of 12 and 18 is 6 because 6 is the largest number that divides into both 12 and 18 evenly.

Using Listing to Find the GCF of Numbers

One way to find the GCF of a given set of numbers is by listing factors.

Procedure **Listing Method for Finding GCF**

To find the GCF of a set of numbers by listing,

1. List all possible factors for each given number.
2. Search the lists for the largest factor common to all lists.

Example 2 Find the GCF of 24 and 60.

SOLUTION Factors of 24: 1, 2, 3, 4, 6, 8, 12, 24
Factors of 60: 1, 2, 3, 4, 5, 6, 10, 12, 15, 20, 30, 60
The GCF of 24 and 60 is 12.

Using Prime Factorization to Find the GCF

The listing method is a good method for smaller numbers, but for larger numbers, it is not the most efficient method to use. It turns out that we can use prime factorization, which we explained in Section 1.2, to find the GCF of a given set of numbers. Consider the prime factorizations for the numbers from Example 2.

$$24 = 2 \cdot 2 \cdot 2 \cdot 3 = 2^3 \cdot 3$$
$$60 = 2 \cdot 2 \cdot 3 \cdot 5 = 2^2 \cdot 3 \cdot 5$$

In Example 2, we found the GCF to be 12. Let's look at the prime factorization of 12 to see if we can discover a rule for finding the GCF.

$$12 = 2 \cdot 2 \cdot 3 = 2^2 \cdot 3$$

Notice that the 12 contains two factors of 2 and one factor of 3, which are the factors that are common to 24's and 60's prime factorizations. This suggests that we use only primes that are common to all factorizations involved. The factor 5 is not common to both 24's and 60's prime factorizations, so it is not included in the GCF.

$$24 = 2 \cdot 2 \cdot 2 \cdot 3 = 2^3 \cdot 3$$
$$60 = 2 \cdot 2 \cdot 3 \cdot 5 = 2^2 \cdot 3 \cdot 5$$

$$GCF = 2 \cdot 2 \cdot 3 = 2^2 \cdot 3 = 12$$

LEARNING Strategy

VISUAL If you are a visual learner, you may prefer to continue circling the common prime factors.

Notice that we could also compare exponents for a given common prime factor to determine how many of that factor should be included in the GCF. For example, 2^2 is included instead of 2^3 because 2^2 has the smaller exponent. This suggests the following procedure on the following page.

Prime Factorization Method for Finding GCF

To find the GCF of a given set of numbers by prime factorization,

1. Write the prime factorization of each given number in exponential form.
2. Create a factorization for the GCF that includes only those prime factors common to all factorizations, each raised to its smallest exponent in the factorizations.
3. Multiply the factors in the factorization created in step 2.

Note: If there are no common prime factors, the GCF is 1.

Example 3   Find the GCF.

a. 54 and 90

SOLUTION   Write the prime factorization of 54 and 90 in factored form.

$$54 = 2 \cdot 3^3$$
$$90 = 2 \cdot 3^2 \cdot 5$$

The common prime factors are 2 and 3. Compare the exponents on each of these common factors. The smallest exponent of 2 is 1, and the smallest exponent of 3 is 2. The GCF is the product of 2^1 and 3^2.

$$GCF = 2^1 \cdot 3^2 = 2 \cdot 9 = 18$$

b. 3024 and 2520

SOLUTION   Write the prime factorization of 3024 and 2520 in factored form.

$$3024 = 2^4 \cdot 3^3 \cdot 7$$
$$2520 = 2^3 \cdot 3^2 \cdot 5 \cdot 7$$

The common prime factors are 2, 3, and 7. Compare the exponents of each of these common factors. The smallest exponent of 2 is 3. The smallest exponent of 3 is 2. The smallest exponent of 7 is 1. The GCF is the product of $2^3, 3^2$, and 7.

$$GCF = 2^3 \cdot 3^2 \cdot 7 = 8 \cdot 9 \cdot 7 = 504$$

Your Turn 3   Find the GCF.

a. 180 and 600   **b.** 63, 84, and 105

GCF of Monomials

We use a similar approach to find the GCF of a set of monomials that have variables. The variables in the monomials are treated like prime factors.

Example 4   Find the GCF.

a. $x^3y^3z^4$ and x^2y^6

SOLUTION   By treating the variables like prime factors, each monomial is already written as a prime factorization. The common prime factors are x and y. The smallest exponent for the factor x is 2, and the smallest exponent for the factor y is 3. The GCF is the product of x^2 and y^3.

$$GCF = x^2y^3$$

Answers **Your Turn 3**

a. 60   **b.** 21

b. $24x^2y$ and $60x^3$

SOLUTION Write the prime factorization of each monomial, treating the variables like prime factors.

$$24x^2y = 2^3 \cdot 3 \cdot x^2 \cdot y$$
$$60x^3 = 2^2 \cdot 3 \cdot 5 \cdot x^3$$

The common prime factors are 2, 3, and x. The smallest exponent of 2 is 2. The smallest exponent of 3 is 1. The smallest exponent of x is 2. The GCF is the product of 2^2, 3, and x^2.

$$\text{GCF} = 2^2 \cdot 3 \cdot x^2 = 12x^2$$

Your Turn 4   Find the GCF.

a. $a^3b^4c^3$ and $a^2c^5d^3$ **b.** $32a^2b$ and $40abc$ **c.** $35x^2$ and $18y$

Objective ③ Write a polynomial as a product of a monomial GCF and a polynomial.   Recall that when we multiply a monomial by a polynomial, we apply the distributive property.

$$2(x + 4) = 2 \cdot x + 2 \cdot 4$$
$$= 2x + 8$$

Now let's reverse this process. Suppose we are given the product, $2x + 8$, and we want to write the factored form, $2(x + 4)$. Notice that terms in $2x + 8$ have 2 as a common factor. When writing factored form, we first determine the GCF of the terms. We then create a missing factor statement like this:

$$2x + 8 = 2(?)$$

Notice that the missing factor must be $x + 4$. To determine this missing factor, we divide the product $2x + 8$ by the known factor, 2.

$$2x + 8 = 2\,(?)$$
$$\frac{2x + 8}{2} = (?)$$
$$\frac{2x}{2} + \frac{8}{2} = (?)$$
$$x + 4 = (?)$$

> **Connection** In Section 5.6, we learned that we divide a polynomial by a monomial by dividing each term in the polynomial by the monomial.

The factored form of $2x + 8$ is $2(x + 4)$, which suggests the following procedure.

Procedure   **Factoring a Monomial GCF Out of a Polynomial**

To factor a monomial GCF out of a given polynomial,

1. Find the GCF of the terms in the polynomial.
2. Rewrite the given polynomial as a product of the GCF and the quotient of the polynomial and the GCF.

$$\text{Polynomial} = \text{GCF}\left(\frac{\text{Polynomial}}{\text{GCF}}\right)$$

Example 5   Factor $12x^2 + 18x$.

SOLUTION

1. Find the GCF of $12x^2$ and $18x$.

   The GCF of $12x^2$ and $18x$ is $6x$. ◀ **Note:** 6 is the largest number that divides both 12 and 18 evenly, and x has the smaller exponent of the x^2 and x.

Note: An alternative way of doing Example 5 is by inspection using the distributive property. Think of $12x^2 + 18x = 6x(\underline{} + \underline{})$, where $6x$ times the first blank gives $12x^2$ and $6x$ times the second blank gives $18x$. Because $6x \cdot 2x = 12x^2$ and $6x \cdot 3 = 18x$, the blanks are filled with $2x$ and 3. So $12x^2 + 18x = 6x(2x + 3)$.

2. Write the given polynomial as the product of the GCF and the quotient of the polynomial and the GCF.

$$12x^2 + 18x = 6x\left(\frac{12x^2 + 18x}{6x}\right)$$

$$= 6x\left(\frac{12x^2}{6x} + \frac{18x}{6x}\right) \qquad \text{Separate the terms.}$$

$$= 6x(2x + 3) \qquad \text{Divide the terms by the GCF.}$$

Check   We can check by multiplying the factored form using the distributive property.

$$6x(2x + 3) = 6x \cdot 2x + 6x \cdot 3 \qquad \text{Distribute } 6x.$$

$$= 12x^2 + 18x \qquad \text{The product is the original polynomial.}$$

Connection Keep in mind that when we factor, we are simply writing the original expression in a different form called *factored form*. When written equal to each other, the factored form and product make an identity, which means that every real number is a solution to the equation. Consider $12x^2 + 18x = 6x(2x + 3)$. Let's choose a value such as $x = 2$ and verify that it satisfies the equation. (Remember, we can select any real number.)

$$12x^2 + 18x = 6x(2x + 3)$$

$$12(2)^2 + 18(2) \overset{?}{=} 6(2)(2(2) + 3)$$

$$12(4) + 36 \overset{?}{=} 12(4 + 3)$$

$$48 + 36 \overset{?}{=} 12(7)$$

$$84 = 84 \qquad \text{It checks.}$$

Note: Again, we could factor using the distributive property by thinking of $24x^2y - 60x^3 = 12x^3(\underline{} + \underline{})$. Because $12x^2 \cdot 2y = 24x^2y$ and $12x^2(-5x) = -60x^3$, the blanks represent $2y$ and $-5x$. So $24x^2y - 60x^3 = 12x^2(2y - 5x)$.

Example 6   Factor $24x^2y - 60x^3$.

SOLUTION

1. Find the GCF of $24x^2y$ and $60x^3$.

   We found this GCF in Example 4(b) to be $12x^2$.

2. Write the given polynomial as the product of the GCF and the quotient of the polynomial and the GCF.

$$24x^2y - 60x^3 = 12x^2\left(\frac{24x^2y - 60x^3}{12x^2}\right)$$

$$= 12x^2\left(\frac{24x^2y}{12x^2} - \frac{60x^3}{12x^2}\right)$$

$$= 12x^2(2y - 5x)$$

◀ **Connection** Remember that when we divide exponential forms that have the same base, we subtract exponents.

Check Multiply the factored form using the distributive property.

$$12x^2(2y - 5x) = 12x^2 \cdot 2y - 12x^2 \cdot 5x$$
$$= 24x^2y - 60x^3$$

Your Turn 6 Factor.

  a. $30xy - 45x$

  b. $16r^2s^4 - 8r^3s^4 + 12rs^3$

Factoring When the First Term Is Negative

Consider factoring the expression $-6x + 10y$. Because the first term of the polynomial is negative, when we factor out the GCF, 2, the first term inside the parentheses is also negative, so that we have $2(-3x + 5y)$. However, it is considered undesirable to have a negative first term inside parentheses. We can avoid a negative first term in the parentheses by factoring the negative of the GCF; so in our example, we factor out -2. This changes the sign of each term inside the parentheses so that we have $-2(3x - 5y)$.

Example 7 Factor by factoring out the negative of the GCF:

$$-18x^4y^3 + 9x^2y^2z - 12x^3y$$

SOLUTION

  1. Find the GCF of $-18x^4y^3$, $9x^2y^2z$, and $-12x^3y$.

   Because the first term in the polynomial is negative, we factor out the negative of the GCF to avoid a negative first term inside the parentheses. We factor out $-3x^2y$.

  2. Write the given polynomial as the product of the GCF and the quotient of the polynomial and the GCF.

$$-18x^4y^3 + 9x^2y^2z - 12x^3y = -3x^2y\left(\frac{-18x^4y^3 + 9x^2y^2z - 12x^3y}{-3x^2y}\right)$$

$$= -3x^2y\left(-\frac{18x^4y^3}{-3x^2y} + \frac{9x^2y^2z}{-3x^2y} - \frac{12x^3y}{-3x^2y}\right)$$

$$= -3x^2y(6x^2y^2 - 3yz + 4x)$$

Check Multiply the factored form using the distributive property.

$$-3x^2y(6x^2y^2 - 3yz + 4x) = -3x^2y \cdot 6x^2y^2 - 3x^2y \cdot (-3yz) - 3x^2y \cdot 4x$$
$$= -18x^4y^3 + 9x^2y^2z - 12x^3y$$

Your Turn 7 Factor by factoring out the negative of the GCF.

  a. $-10x^3 - 15x^2 + 25x$

  b. $-48a^4b^5 - 24a^3b^4 + 16ab^2c$

Before we discuss additional techniques of factoring, it is important to state that no matter what type of polynomial we are asked to factor, we always consider whether a monomial GCF (other than 1) can be factored out of the polynomial.

Factoring When the GCF Is a Polynomial

Sometimes in factoring, the GCF is a polynomial with more than one term.

Answers **Your Turn 6**

a. $15x(2y - 3)$

b. $4rs^3(4rs - 2r^2s + 3)$

Answers **Your Turn 7**

a. $-5x(2x^2 + 3x - 5)$

b. $-8ab^2(6a^3b^3 + 3a^2b^2 - 2c)$

Example 8   Factor $y(x + 2) + 7(x + 2)$.

SOLUTION   Notice that this expression is a sum of two products, y and $(x + 2)$, and 7 and $(x + 2)$. Further, note that $(x + 2)$ is the GCF of the two products.

$$y(x + 2) + 7(x + 2) = (x + 2)\left(\frac{y(x + 2) + 7(x + 2)}{x + 2}\right)$$

$$= (x + 2)\left(\frac{y(x + 2)}{x + 2} + \frac{7(x + 2)}{x + 2}\right)$$

$$= (x + 2)(y + 7)$$

Note: The procedure is the same. We rewrite the given expression as the product of the GCF and the quotient of the expression and the GCF.

Your Turn 8   Factor $4a(a - 3) - b(a - 3)$.

Objective ④ **Factor by grouping.**   The process of factoring out a polynomial GCF, as we did in Example 8, is an intermediate step in a process called *factoring by grouping*, which is a technique that we try when factoring a four-term polynomial such as $xy + 2y + 7x + 14$. The method is called *grouping* because we group pairs of terms and look for a common factor within each group or pair. We begin by pairing the first two terms as one group and the last two terms as a second group.

$$xy + 2y + 7x + 14 = (xy + 2y) + (7x + 14)$$

Note that the first two terms have a common factor of y and the last two terms have 7 as a common factor. If we factor y out of the first two terms and 7 out of the last two terms, we have the same expression that we factored in Example 8.

$$xy + 2y + 7x + 14 = (xy + 2y) + (7x + 14)$$

$$= y(x + 2) + 7(x + 2)$$

$$= (x + 2)(y + 7) \qquad \text{Factor out } (x + 2).$$

Procedure   **Factoring by Grouping**

To factor a four-term polynomial by grouping,

1. Factor out any monomial GCF (other than 1) that is common to all four terms.
2. Group pairs of terms and factor the GCF out of each pair or group.
3. If there is a common binomial factor, factor it out.
4. If there is no common binomial factor, interchange the middle two terms and repeat the process. If there is still no common binomial factor, the polynomial cannot be factored by grouping.

Note: Sometimes the GCF is 1 or -1.

Example 9   Factor.

a. $6x^3 - 8x^2 + 3xy - 4y$

SOLUTION   First, we look for a monomial GCF (other than 1). This polynomial does not have one. Because the polynomial has four terms, we now try to factor by grouping.

$$6x^3 - 8x^2 + 3xy - 4y = (6x^3 - 8x^2) + (3xy - 4y)$$

$$= 2x^2(3x - 4) + y(3x - 4)$$

$$= (3x - 4)(2x^2 + y)$$

Factor $2x^2$ out of $6x^3$ and $8x^2$; then factor y out of $3xy$ and $4y$.

Factor out $3x - 4$.

Answer   **Your Turn 8**
$(a - 3)(4a - b)$

b. $12mn^2 - 20mn - 24n^2 + 40n$

Note: When factoring by grouping and the third term is negative, the GCF of the third and fourth terms is usually negative.

SOLUTION Note that in this case, there is a monomial GCF, $4n$; so we first factor this GCF out of all four terms.

$$12mn^2 - 20mn - 24n^2 + 40n = 4n(3mn - 5m - 6n + 10)$$

Because the polynomial in the parentheses has four terms, we try to factor by grouping.

$$= 4n[(3mn - 5m) + (-6n + 10)]$$ Factor m out of $3mn$ and $5m$; then
$$= 4n[m(3n - 5) - 2(3n - 5)]$$ factor -2 out of $-6n$ and 10.
$$= 4n(3n - 5)(m - 2)$$ Factor out $3n - 5$.

Note: Now that the expression is in factored form, we no longer need the brackets.

Your Turn 9   Factor.

    a. $6x^2 + 15xz + 2xy + 5yz$ **b.** $42a^2b - 56a^2 - 12ab + 16a$

Answers   **Your Turn 9**

a. $(2x + 5z)(3x + y)$
b. $2a(3b - 4)(7a - 2)$

6.1 Exercises   For Extra Help   *MyMathLab*

Note: Exercises marked with a ★ represent challenging exercises.

1. Explain in your own words how to list all possible natural number factors of a number.

2. Define the GCF of a given set of numbers in your own words.

3. Given a set of terms, after finding the prime factorization of each term, how do you use those prime factors to create the GCF's factorization?

4. When factoring a monomial GCF out of the terms of a polynomial, after finding the GCF, how do you rewrite the polynomial?

5. What kinds of polynomials are factored by grouping?

6. When factoring by grouping, after factoring out a monomial GCF, what is the next step?

For Exercises 7–22, list all natural number factors of the given number. See Example 1.

7. 9   8. 49   9. 33   10. 21

11. 18   12. 12   13. 16   14. 81

15. 44   **16.** 45   **17.** 60   **18.** 36

19. 56   **20.** 64   **21.** 90   **22.** 84

For Exercises 23–34, find the GCF. See Examples 2–4.

23. 21, 30   **24.** 21, 35   **25.** 72, 80   **26.** 35, 60   **27.** 12, 42, 60   **28.** 10, 18, 36

29. $4xy, 6xy$ **30.** $6ab, 9ab$ **31.** $25h^2, 60h^4$ **32.** $25x^4, 10x^3$ **33.** $6a^5b, 15a^4b^2$ **34.** $7m^6n, 21m^5n^4$

For Exercises 35–62, factor by factoring out the GCF. See Examples 5 and 6.

35. $5c - 20$ **36.** $8y - 24$ **37.** $8x - 12$ **38.** $15x - 10$

39. $x^2 - x$ **40.** $y^2 + y$ **41.** $18z^6 - 12z^4$ **42.** $15k^3 - 24k^2$

43. $18p^3 - 15p^5$ **44.** $18r^4 - 27r^6$ **45.** $6a^2b - 3ab^2$ **46.** $9m^2n - 3mn^2$

47. $14uv^2 - 7uv$ **48.** $12x^2y^3 - 6xy^2$ **49.** $25xy - 50xz + 100x$ **50.** $8ab - 32ac + 40a$

51. $x^2y + xy^2 + x^3y^3$ **52.** $w^3v^2 + w^2v + wv^2$ **53.** $28ab^3c - 36a^2b^2c$ **54.** $30x^3yz^2 - 24x^2y^3z$

55. $20p^2q + 24pq - 16pq^2$ **56.** $6a^2c + 18abc - 12ac^2$ **57.** $3mn^5p^2 + 18mn^3p - 6mnp$

58. $4ax^3y^3 - 6ax^2y^2 + 4axy^2$ **59.** $105a^3b^2 - 63a^2b^3 + 84a^6b^4$ **60.** $21t^4u^3 - 105t^5u^3v + 63t^6u^2v^4$

61. $18x^4 - 9x^3 + 30x^2 - 12x$ **62.** $10g + 40g^3 - 100g^4 + 120g^5$

For Exercises 63–72, factor by factoring out the negative of the GCF.
See Example 7.

63. $-3x + 6y$ **64.** $-8x + 4y$ **65.** $-20a^2 - 15a$ **66.** $-30b^2 - 24b$

67. $-12a^4b + 20a^3b^3$ **68.** $-15x^4y + 21x^2y^2$ **69.** $-4x^2 - 8x + 16$ **70.** $-6x^2 - 18x + 24$

71. $-24x^3y^2z - 30x^2y^3z^4 + 12x^4y^5z^2$ **72.** $-21a^4bc^3 - 15a^3b^3c^3 + 27a^2b^2c^4$

For Exercises 73–80, factor out the polynomial GCF. See Example 8.

73. $y(a - 3) + 2(a - 3)$ **74.** $a(x - 4) + 5(x - 4)$ **75.** $4a(2m + 3n) + b(2m + 3n)$

76. $3m(3a + 5b) + n(3a + 5b)$ **77.** $5x(4m - 5) - 2(4m - 5)$ **78.** $3m(3a - 2) - 4(3a - 2)$

★**79.** $2r(6p + 5q) - 4s(6p + 5q)$ ★**80.** $6x(3y + 5z) - 3w(3y + 5z)$

For Exercises 81–104, factor by grouping. See Example 9(a).

81. $bx + 2b + cx + 2c$

82. $cx + cy + bx + by$

83. $am - an - bm + bn$

84. $xy - xw - yz + wz$

85. $x^3 + 2x^2 - 3x - 6$

86. $y^4 + 4y^3 - by - 4b$

87. $1 - m + m^2 - m^3$

88. $1 + x + x^2 + x^3$

89. $3xy + 5y + 6x + 10$

90. $2ab + 8a + 3b + 12$

91. $4b^2 - b + 4b - 1$

92. $x^3 - 4x^2 + x - 4$

93. $6 + 2b - 3a - ab$

94. $12 + 3y - 4x - xy$

95. $3ax + 6ay + 8by + 4bx$

96. $ac + 2ad + 2bc + 4bd$

97. $x^2y - x^2s - ry + rs$

98. $a^3m - a^3n - bm + bn$

99. $w^2 + 3wz + 5w + 15z$

100. $a^2 + 4ab + 3a + 12b$

101. $3st + 3ty - 2s - 2y$

102. $5ab + 5ac - 3b - 3c$

103. $ax^2 - 5y^2 + ay^2 - 5x^2$

104. $m^2p - 3n^2 + n^2p - 3m^2$

For Exercises 105–114, factor completely. See Example 9(b).

105. $2a^2 + 2ab + 6a + 6b$

106. $3m^2 + 3mn + 12m + 12n$

107. $12ab + 18a + 16b + 24$

108. $12pq + 8q + 30p + 20$

109. $24xz - 12xw - 6yz + 3yw$

110. $12ac + 8ad - 18bc - 12bd$

111. $3a^2y - 12a^2 + 9ay - 36a$

112. $4x^2y - 8x^2 + 20xy - 40x$

113. $2x^3 + 6x^2y + 10x^2 + 30xy$

114. $10uv^2 - 5v^2 + 30uv - 15v$

For Exercises 115 and 116, write an expression for the area of the shaded region; then factor completely.

115.

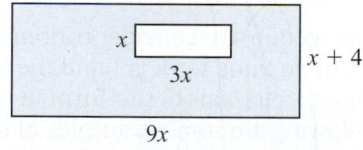

116.

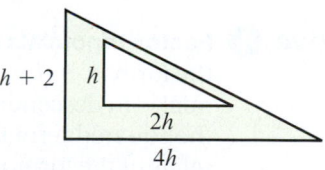

★**117.** The diagram shows the floor plan of a room with the hearth of a fireplace. Write an expression in factored form of the area of the room excluding the hearth, which is 4 feet by 1 foot.

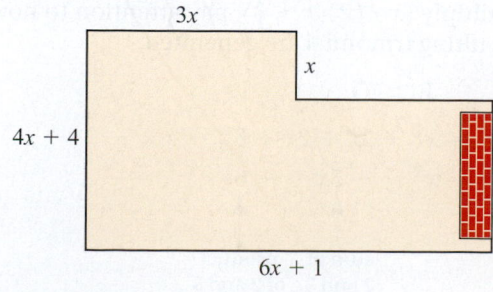

★**118.** Write an expression in factored form for the area of the side of the house shown excluding the window.

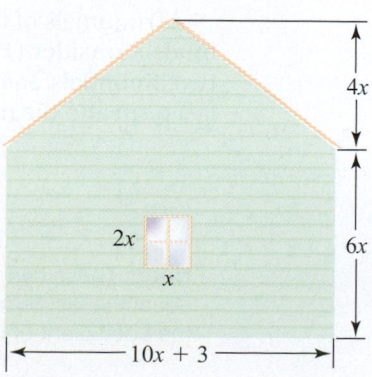

★ For Exercises 119 and 120, write an expression for the volume of the object shown; then factor completely.

119.

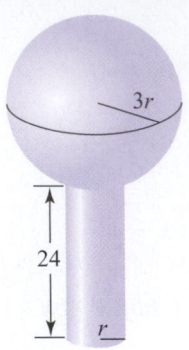

120.

Review Exercises

Exercises 1–6 EXPRESSIONS

[5.1]   1. Write 3.74×10^9 in standard form.

[5.1]   2. Write 45,600,000 in scientific notation.

[5.5] *For Exercises 3–6, multiply.*

3. $(x + 2)(x + 5)$

4. $(x - 4)(x + 3)$

5. $(x - 5)(x - 3)$

6. $x(x - 7)$

6.2 Factoring Trinomials of the Form $x^2 + bx + c$

Objectives

① Factor trinomials of the form $x^2 + bx + c$.

② Factor out a monomial GCF; then factor the trinomial of the form $x^2 + bx + c$.

Objective ① **Factor trinomials of the form $x^2 + bx + c$.** In this section, we consider trinomials of the form $x^2 + bx + c$. Notice that the coefficient of the squared term is 1 and the trinomial is in descending order. In Section 6.3, we consider trinomials of this form in which the squared term has a coefficient other than 1. Following are some examples of trinomials of the form $x^2 + bx + c$.

$$x^2 + 5x + 6 \quad \text{or} \quad x^2 - 7x + 12 \quad \text{or} \quad x^2 + 2x - 15 \quad \text{or} \quad x^2 - 5x - 24$$

Trinomials of the form $x^2 + bx + c$ are often the result of the product of two binomials. Consider $(x + 2)(x + 3)$. Recall from Section 5.5 that we use FOIL to multiply two binomials. In the use of FOIL to multiply $(x + 2)(x + 3)$, pay attention to how the last term and the middle term of the resulting trinomial are generated.

$$\begin{array}{cccc} \text{F} & \text{O} & \text{I} & \text{L} \end{array}$$
$$(x + 2)(x + 3) = x^2 + 3x + 2x + 6$$
$$= x^2 \quad\quad + 5x \quad\quad + 6$$

Note: These two numbers multiply to equal the last term, 6, and add to equal the coefficient of the middle term, 5.

sum of   product
2 and 3   of 2 and 3

When we factor a trinomial of the form $x^2 + bx + c$, we reverse the FOIL process, using the fact that b is the sum of the last terms in the binomials and c is the product of the last terms in the binomials. For example, to factor $x^2 + 5x + 6$, we must find two binomials with last terms whose product is 6 and whose sum is 5.

$$x^2 + 5x + 6 = (x + \boxed{})(x + \boxed{})$$

The product of these numbers must be 6, and their sum must be 5.

If unsure of what those two numbers are, we can list all factor pairs of 6 and then find the pair whose sum is 5.

Factors Pairs of 6	Corresponding Sum	
$1 \cdot 6$	$1 + 6 = 7$	
$2 \cdot 3$	$2 + 3 = 5$	⟵ This is the correct pair.

Notice that the 2 and 3 become the last terms in the binomials so that the factored form for $x^2 + 5x + 6$ looks like this:

$$x^2 + 5x + 6 = (x + 2)(x + 3)$$

This suggests the following procedure.

Procedure   **Factoring $x^2 + bx + c$**

To factor a trinomial of the form $x^2 + bx + c$,

1. Find two numbers with a product equal to c and a sum equal to b.
2. The factored trinomial will have the form $(x + \text{first number})(x + \text{second number})$.

Note: The signs in the binomial factors can be minus signs depending on the signs of b and c.

In the following examples, we explore cases where minus signs occur. Pay attention to how we determine the signs in the binomial factors.

Example 1   Factor.

a. $x^2 - 7x + 12$

SOLUTION   We must find a pair of numbers whose product is 12 and whose sum is -7. If two numbers have a positive product and negative sum, both must be negative. Following is a table listing the products and sums.

Product	Sum	
$(-1)(-12) = 12$	$-1 + (-12) = -13$	This is the correct combination,
$(-2)(-6) = 12$	$-2 + (-6) = -8$	so -3 and -4 are the second terms in each binomial of the
$(-3)(-4) = 12$	$-3 + (-4) = -7$	⟵ factored form.

Answer $x^2 - 7x + 12 = (x - 3)(x - 4)$ ◀ **Note:** We could have first written $[x + (-3)][x + (-4)]$ to follow the procedure and then simplify.

Check We can check by multiplying the binomial factors to see if their product is the original polynomial.

$$(x - 3)(x - 4) = x^2 - 4x - 3x + 12 \qquad \text{Multiply the factors using FOIL.}$$
$$= x^2 - 7x + 12 \qquad \text{The product is the original polynomial.}$$

b. $x^2 + 2x - 15$

SOLUTION We must find a pair of numbers whose product is -15 and whose sum is 2. Because the product is negative, the two numbers must have different signs. Because the sum is positive, the number with the greater absolute value will be positive.

Product	Sum
$(-1)(15) = -15$	$-1 + 15 = 14$
$(-3)(5) = -15$	$-3 + 5 = 2$

This is the correct combination, so -3 and 5 are the second terms in each binomial of the factored form.

Answer $x^2 + 2x - 15 = (x - 3)(x + 5)$

Check $(x - 3)(x + 5) = x^2 + 5x - 3x - 15 \qquad \text{Multiply the factors using FOIL.}$
$$= x^2 + 2x - 15 \qquad \text{The product is the original polynomial.}$$

c. $x^2 - 5x - 24$

SOLUTION We must find a pair of numbers whose product is -24 and whose sum is -5. Because the product is negative, the two numbers must have different signs. Because the sum also is negative, the number with the greater absolute value will be negative.

Product	Sum
$(1)(-24) = -24$	$1 + (-24) = -23$
$(2)(-12) = -24$	$2 + (-12) = -10$
$(3)(-8) = -24$	$3 + (-8) = -5$

This is the correct combination, so 3 and -8 are the second terms in each binomial of the factored form.

Note: In Examples 1(c)–3, we leave the checks to the reader.

Answer $x^2 - 5x - 24 = (x + 3)(x - 8)$

d. $x^2 + 3x + 4$

SOLUTION We must find a pair of numbers whose product is 4 and whose sum is 3. If two numbers have a positive product and a positive sum, both must be positive.

Product	Sum
$(1)(4) = 4$	$1 + 4 = 5$
$(2)(2) = 4$	$2 + 2 = 4$

Note: We listed all positive factors of 4 and found no combination whose sum is 3. This means that $x^2 + 3x + 4$ has no binomial factors with integer coefficients.

A polynomial such as $x^2 + 3x + 4$ that cannot be factored is like a prime number in that its factors are only 1 and the polynomial itself. We say that such a polynomial is *prime*. Notice that the factorization $(x + 4)(x - 1)$ gives the correct middle term, $3x$, but the last term is -4 instead of $+4$.

Answers Your Turn 1

a. $(y - 2)(y - 9)$
b. $(t - 4)(t + 7)$
c. $(n + 5)(n - 6)$
d. prime

Your Turn 1 Factor.

a. $y^2 - 11y + 18$ **b.** $t^2 + 3t - 28$ **c.** $n^2 - n - 30$ **d.** $x^2 - 2x - 6$

Now let's see how to factor trinomials containing two variables.

Example 2 Factor.

a. $x^2 - 10xy + 25y^2$

SOLUTION Because y is in the last term, it is helpful to rearrange the order of the variables in the middle term, writing the polynomial as $x^2 - 10yx + 25y^2$, so that we view the "coefficient" of $-10yx$ as $-10y$. We must find a pair of terms whose product is $25y^2$ and whose sum is $-10y$. These terms would have to be $-5y$ and $-5y$.

Answer $x^2 - 10xy + 25y^2 = (x - 5y)(x - 5y)$ or $(x - 5y)^2$

b. $a^2 + 2ab - 15b^2$

SOLUTION As in Example 2(a), we can view the coefficient of a in the middle term as $2b$. We must find a pair of terms whose product is $-15b^2$ and whose sum is $2b$. These terms would have to be $-3b$ and $5b$.

Answer $a^2 + 2ab - 15b^2 = (a - 3b)(a + 5b)$

Your Turn 2 Factor.

    a. $m^2 + 7mn - 18n^2$ **b.** $x^2 - xy - 12y^2$

Objective ② Factor out a monomial GCF; then factor the trinomial of the form $x^2 + bx + c$.
Sometimes there is a monomial GCF (other than 1) in the three terms of a trinomial. Consider $2x^3 + 10x^2 + 12x$. Notice that the monomial $2x$ is the GCF of the terms. Factoring out this monomial, we have

$$2x^3 + 10x^2 + 12x = 2x(x^2 + 5x + 6).$$

Now we try to factor the trinomial within the parentheses. We look for two numbers whose product is 6 and whose sum is 5. Note that 2 and 3 work, so we can write the following:

$$2x^3 + 10x^2 + 12x = 2x(x^2 + 5x + 6)$$
$$= 2x(x + 2)(x + 3) \longleftarrow \boxed{\text{Factored form}}$$

Whenever factoring polynomials, the first step always should be to look for a monomial GCF among the terms.

Example 3 Factor.

a. $3mn^3 + 12mn^2 - 96mn$

SOLUTION First, we look for a monomial GCF (other than 1). Notice that the GCF of the terms is $3mn$. Factoring out this monomial, we have

$$3mn^3 + 12mn^2 - 96mn = 3mn(n^2 + 4n - 32).$$

Now try to factor the trinomial to two binomials. We must find a pair of numbers whose product is -32 and whose sum is 4.

Product	Sum	
$(-1)(32) = -32$	$-1 + 32 = 31$	
$(-2)(16) = -32$	$-2 + 16 = 14$	This is the correct combination, so -4 and 8 are the second terms in
$(-4)(8) = -32$	$-4 + 8 = 4$	$\longleftarrow$ each binomial.

Answer $3mn^3 + 12mn^2 - 96mn = 3mn(n - 4)(n + 8)$

b. $x^4 + 6x^3 + 14x^2$

SOLUTION First, we factor out the monomial GCF, x^2.

$$x^4 + 6x^3 + 14x^2 = x^2(x^2 + 6x + 14)$$

Now we try to factor the trinomial to two binomials. We must find a pair of numbers whose product is 14 and whose sum is 6.

Product	Sum
$(1)(14) = 14$	$1 + 14 = 15$
$(2)(7) = 14$	$2 + 7 = 9$
$(-1)(-14) = 14$	$-1 + (-14) = -15$
$(-2)(-7) = 14$	$-2 + (-7) = -9$

Answers Your Turn 3

a. $y^2(y - 9)(y - 1)$

b. $4ab(a - 7)(a + 2)$

c. $4x(y^2 - 2y + 3)$

Your Turn 3 Factor.

a. $y^4 - 10y^3 + 9y^2$

b. $4a^3b - 20a^2b - 56ab$

c. $4xy^2 - 8xy + 12x$

6.2 Exercises   For Extra Help   *MyMathLab*

 PRACTICE   WATCH   DOWNLOAD

Note: Exercises marked with a ★ represent challenging exercises.

1. If given a trinomial in the form $x^2 + bx + c$, where $c > 0$ and $b > 0$, what can you conclude about the pair of numbers in the binomial factors?

2. If given a trinomial in the form $x^2 - bx + c$, where $c > 0$ and $b > 0$, what can you conclude about the pair of numbers in the binomial factors?

3. If given a trinomial in the form $x^2 + bx - c$, where $c > 0$ and $b > 0$, what can you conclude about the pair of numbers in the binomial factors?

4. If given a trinomial in the form $x^2 - bx - c$, where $c > 0$ and $b > 0$, what can you conclude about the pair of numbers in the binomial factors?

For Exercises 5–12, fill in the missing values in the factors. See Objective 1.

5. $x^2 + 7x + 10 = (x + 2)(x +)$

6. $y^2 + 8y + 15 = (y + 5)(y +)$

7. $n^2 - 12n + 20 = (n - 10)(n -)$

8. $t^2 - 9t + 18 = (t -)(t - 3)$

9. $m^2 - 2m - 24 = (m -)(m + 4)$

10. $a^2 + a - 12 = (a + 4)(a -)$

11. $u^2 + 13u - 30 = (u -)(u + 15)$

12. $n^2 - 9n - 36 = (n -)(n + 3)$

For Exercises 13–44, factor. If the polynomial is prime, so state. See Example 1.

13. $r^2 + 4r + 3$

14. $t^2 + 8t + 7$

15. $x^2 - 8x + 7$

16. $x^2 - 4x + 3$

17. $z^2 - 2z - 3$

18. $y^2 - 4y - 5$

19. $y^2 + 5y + 6$

20. $n^2 + 8n + 15$

21. $u^2 - 6u + 8$ **22.** $b^2 - 10b + 21$ **23.** $u^2 + u - 6$ **24.** $k^2 - 3k - 10$

25. $a^2 + 7a + 12$ **26.** $x^2 + 9x + 18$ **27.** $y^2 - 6y + 9$ **28.** $x^2 - 10x + 25$

29. $w^2 - w - 12$ **30.** $b^2 - 3b - 40$ **31.** $x^2 - x - 30$ **32.** $b^2 - 6b - 16$

33. $n^2 - 11n + 30$ **34.** $y^2 - 7y + 12$ **35.** $r^2 - 9r + 18$ **36.** $x^2 - 10x + 24$

37. $x^2 - 5x - 24$ **38.** $a^2 - 10a - 24$ **39.** $p^2 - 5p - 36$ **40.** $u^2 - 8u - 20$

41. $m^2 - 4m + 6$ **42.** $z^2 + 7z + 8$ **43.** $x^2 - 6x - 8$ **44.** $a^2 + 3a - 20$

For Exercises 45–52, factor the trinomials containing two variables. If the polynomial is prime, so state. See Example 2.

45. $p^2 - 10pq + 21q^2$ **46.** $x^2 - 9xy + 14y^2$ **47.** $a^2 - 6ab - 27b^2$ **48.** $m^2 + 2mn - 8n^2$

49. $x^2 - 14xy + 24y^2$ **50.** $t^2 - 11tu + 24u^2$ **51.** $r^2 - rs - 30s^2$ **52.** $h^2 - 8hk - 20k^2$

For Exercises 53–72, factor completely. See Example 3.

53. $4x^2 - 40x + 84$ **54.** $3m^2 - 33m + 54$ **55.** $2m^3 - 14m^2 + 12m$ **56.** $2k^2y - 18ky + 28y$

57. $3a^2b - 15ab - 72b$ **58.** $3a^2y + 6ay - 72y$ **59.** $4x^2 - 24x + 36$ **60.** $6a^2r + 12ar + 6r$

61. $n^4 + 5n^3 + 6n^2$ **62.** $r^4 + 6r^3 + 8r^2$ **63.** $7u^4 + 42u^3 + 35u^2$ **64.** $5x^5 + 20x^4 + 15x^3$

65. $6a^2b^2c - 36ab^2c + 48b^2c$ **66.** $3a^2b^2c - 24ab^2c + 45b^2c$ **67.** $3x^2 - 21xy + 30y^2$

68. $5m^2 - 30mn + 40n^2$ **69.** $2a^2b - 6ab^2 - 36b^3$ **70.** $7h^2k + 14hk^2 - 168k^3$

71. $4x^4y - 12x^3y^2 - 60x^2y^3$ **72.** $3a^3b^2 + 15a^2b^3 - 12ab^4$

Find

(X) the mistake **For Exercises 73–76, find and correct the mistake. See Example 1.**

73. $x^2 - 5x - 6 = (x - 3)(x - 2)$ **74.** $x^2 - 5x + 6 = (x - 6)(x + 1)$

75. $x^2 - 3x - 4 = (x - 1)(x + 4)$ **76.** $x^2 + x + 2 = (x + 2)(x + 1)$

★ **For Exercises 77–80, find all natural number values of b that make the trinomial factorable.**

77. $x^2 + bx - 21$ **78.** $x^2 + bx - 10$ **79.** $x^2 + bx + 12$ **80.** $x^2 + bx + 18$

★ *For Exercises 81–84, find all natural numbers c that make the trinomial factorable.*

81. $x^2 - 7x + c$ **82.** $x^2 - 5x + c$ **83.** $x^2 + 10x + c$ **84.** $x^2 + 11x + c$

85. The expression $h^2 + 6h + 8$ describes the area of the top (or bottom) of the crate shown, where h represents its height. The unknown expression for the length is the sum of h and an integer, and the expression for the width is the sum of h and a different integer. Find expressions for the length and width.

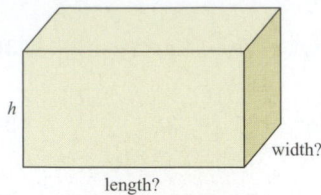

86. The expression $w^2 + w - 2$ describes the area of the top of the bar of gold shown, where w represents the width of the base of the bar. The unknown expression for the length of the top of the bar is the sum of w and an integer. The expression for the width of the top of the bar is the difference of w and an integer. Find the expressions for the length and width of the top of the bar.

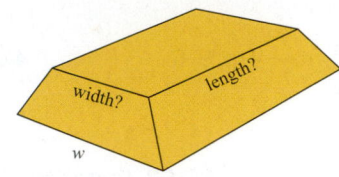

87. The expression $w^2 - 16w + 60$ describes the viewing area in the picture frame shown, where w represents the width of the frame. The unknown expression for the length of the viewing area is the difference of w and an integer. The expression for the width of the viewing area is the difference of w and a different integer. Find expressions for the length and width of the viewing area in the frame.

★ **88.** The expression $\pi r^2 - 2\pi r + \pi$ describes the area occupied by the circular base of the papasan chair shown, where r represents the radius of the circle on which the chair rests. The expression that describes the radius of the circle that touches the floor is the difference of r and an integer. Find the expression that describes the radius of the circle that touches the floor.

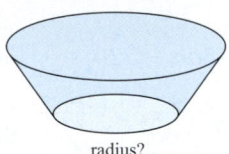

Review Exercises

Exercises 1–10 **EXPRESSIONS**

[6.1] **1.** What is the GCF of $6x$ and $35y$?

[5.5] *For Exercises 2–4, multiply.*

2. $(2x + 3)(5x + 1)$ **3.** $(3y - 4)(5y + 6)$ **4.** $4n(6n - 1)(3n - 2)$

[6.1] *For Exercises 5–10, factor completely.*

5. $10x^2y^3 - 5x^3y - 20x^2y^2$ **6.** $8a^2b^4 - 16ab^2 - 12a^2b^3$

7. $x(x + 2) - 3(x + 2)$ **8.** $z(z + 5) - 6(z + 5)$

9. $15ac - 5ad - 3bc + bd$ **10.** $6ac + 4ad - 9bc - 6bd$

6.3 Factoring Trinomials of the Form $ax^2 + bx + c$, where $a \neq 1$

Objectives

1. Factor trinomials of the form $ax^2 + bx + c$, where $a \neq 1$, by trial.
2. Factor trinomials of the form $ax^2 + bx + c$, where $a \neq 1$, by grouping.

Objective ① **Factor trinomials of the form $ax^2 + bx + c$, where $a \neq 1$, by trial.** In Section 6.2, we factored trinomials in which the coefficient of the squared term was 1. Now we focus on factoring trinomials in which the coefficient of the squared term is other than 1, such as the following:

$$3x^2 + 17x + 10 \qquad 8x^2 + 29x - 12$$

In general, like trinomials of the form $x^2 + bx + c$, trinomials of the form $ax^2 + bx + c$, where $a \neq 1$, may also have two binomial factors. When a is not 1, we must consider the factors of the ax^2 term and the factors of the c term. First, we develop a trial-and-error method for finding the factored form. Consider $3x^2 + 17x + 10$. Because all of the terms are positive, we know that all of the terms in the binomial factors will be positive.

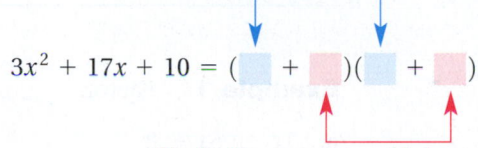

The *first* terms must multiply to equal $3x^2$. The logical factors are $3x$ and x.

$$3x^2 + 17x + 10 = (\ \ + \ \ )(\ \ + \ \ )$$

The *last* terms must multiply to equal 10. These factors could be 1 and 10, or 2 and 5.

We now try various combinations of these *first* and *last* terms in binomial factors and consider the resulting trinomial products. If the sum of the inner and outer terms matches the original trinomial, we know we have the correct combination in the binomial factors.

$$\left.\begin{array}{l}(3x + 10)(x + 1) = 3x^2 + 3x + 10x + 10 = 3x^2 + 13x + 10 \\ (3x + 1)(x + 10) = 3x^2 + 30x + x + 10 = 3x^2 + 31x + 10 \\ (3x + 5)(x + 2) = 3x^2 + 6x + 5x + 10 = 3x^2 + 11x + 10\end{array}\right\} \begin{array}{l}\text{Incorrect} \\ \text{combinations}\end{array}$$

$$(3x + 2)(x + 5) = 3x^2 + 15x + 2x + 10 = 3x^2 + 17x + 10 \qquad \text{Correct combination}$$

You may find that drawing lines to connect each product is helpful.

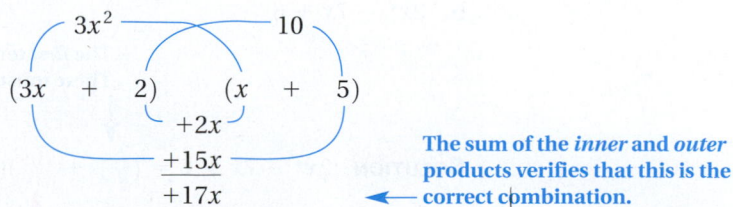

The sum of the *inner* and *outer* products verifies that this is the ⟵ correct combination.

Our example suggests the following procedure.

Procedure | **Factoring by Trial and Error**

To factor a trinomial of the form $ax^2 + bx + c$, where $a \neq 1$, by trial and error,

1. Look for a monomial GCF in all of the terms. If there is one, factor it out.

2. Write a pair of *first* terms whose product is ax^2.

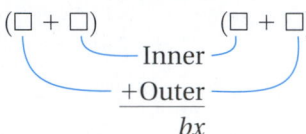

3. Write a pair of *last* terms whose product is c.

$$(\square + \square)(\square + \square)$$

4. Verify that the sum of the *inner* and *outer* products is bx (the middle term of the trinomial).

$$(\square + \square) \qquad (\square + \square)$$
$$\underline{\text{Inner} \atop +\text{Outer}}$$
$$bx$$

If the sum of the inner and outer products is *not* bx, try the following:

a. Exchange the last terms of the binomials from step 3; then repeat step 4.
b. For each additional pair of last terms, repeat steps 3 and 4.
c. For each additional pair of first terms, repeat steps 2–4.

Example 1 Factor.

a. $3x^2 + 5x + 2$

The *first* terms must multiply to equal $3x^2$.
These must be $3x$ and x.

SOLUTION $3x^2 + 5x + 2 = (\ \ \ + \ \ \ )(\ \ \ + \ \ \ )$

The *last* terms must multiply to equal 2, which can only be 1 and 2. Because the product of the last terms is positive and the sum of the outside and inside products must be $+$, both signs of the binomial must be $+$.

$(3x + 1)(x + 2) = 3x^2 + 6x + x + 2 = 3x^2 + 7x + 2$ **Incorrect combination**
$(3x + 2)(x + 1) = 3x^2 + 3x + 2x + 2 = 3x^2 + 5x + 2$ **Correct combination**

Answer $3x^2 + 5x + 2 = (3x + 2)(x + 1)$

b. $2x^2 - 7x + 6$

The *first* terms must multiply to $2x^2$.
These must be $2x$ and x.

SOLUTION $2x^2 - 7x + 6 = (\ \ \ + \ \ \ )(\ \ \ - \ \ \ )$

Note: Because the original polynomial does not have a common factor, no factors can have a common factor. This means that $2x - 6$ and $2x - 2$ cannot be factors because both have a common factor of 2.

The *last* terms must multiply to equal $+6$, which could be 1 and 6, or 2 and 3. Because the product of the last terms is positive and the the sum of the outside and inside products is $-$, both signs of the binomial must be $-$.

$(2x - 1)(x - 6) = 2x^2 - 12x - x + 6 = 2x^2 - 13x + 6$ **Incorrect combination**
$(2x - 6)(x - 1) = 2x^2 - 2x - 6x + 6 = 2x^2 - 8x + 6$ **Incorrect combination**
$(2x - 2)(x - 3) = 2x^2 - 6x - 2x + 6 = 2x^2 - 8x + 6$ **Incorrect combination**
$(2x - 3)(x - 2) = 2x^2 - 4x - 3x + 6 = 2x^2 - 7x + 6$ **Correct combination**

Answer $2x^2 - 7x + 6 = (2x - 3)(x - 2)$

c. $8x^2 + 29x - 12$

SOLUTION

The *first* terms must multiply to equal $8x^2$. These could be x and $8x$, or $2x$ and $4x$.

$$8x^2 + 29x - 12 = (\,\square + \square\,)(\,\square - \square\,)$$

The *last* terms must multiply to equal -12. Because -12 is negative, the last terms in the binomials must have different signs. We have already written the appropriate signs, so these factor pairs could be 1 and 12, 2 and 6, or 3 and 4.

Now we multiply binomials with various combinations of these first and last terms until we find a combination whose inner and outer products combine to equal $29x$.

$$(8x + 1)(x - 12) = 8x^2 - 96x + x - 12 = 8x^2 - 95x - 12$$
$$(8x - 1)(x + 12) = 8x^2 + 96x - x - 12 = 8x^2 + 95x - 12$$
$$(8x + 12)(x - 1) = 8x^2 - 8x + 12x - 12 = 8x^2 + 4x - 12$$
$$(8x + 2)(x - 6) = 8x^2 - 48x + 2x - 12 = 8x^2 - 46x - 12$$
$$(8x + 6)(x - 2) = 8x^2 - 16x + 6x - 12 = 8x^2 - 10x - 12$$
$$(8x + 3)(x - 4) = 8x^2 - 32x + 3x - 12 = 8x^2 - 29x - 12$$

Incorrect combinations

Note: The products $(8x + 1)(x - 12)$ and $(8x - 1)(x + 12)$ differ only in the sign of the middle term. So unless a factorization is correct except for the sign of the middle term of the product, switching the signs of the binomial factors will not result in the correct factorization.

Note: In the above factorizations, the factors $8x + 12, 8x + 2,$ and $8x + 6$ are not possible because each has a common factor.

Note: This middle term has the correct absolute value but the incorrect sign. (We are looking for $+29x$.) Switching the $+$ and $-$ signs in the binomials will give us $+29x$.

$$(8x - 3)(x + 4) = 8x^2 + 32x - 3x - 12 = 8x^2 + 29x - 12 \qquad \text{Correct combination}$$

Answer $\;8x^2 + 29x - 12 = (8x - 3)(x + 4)$

Your Turn 1 Factor.

a. $5x^2 + 9x + 4$ **b.** $6c^2 - 19cd + 15d^2$ **c.** $6y^2 + 13y - 5$

Factoring Out a Monomial GCF First

Remember that we should always look to see if a monomial GCF can be factored out of the terms in the trinomial. This is step 1 for every factoring problem.

Example 2   Factor $40x^3 - 68x^2 + 12x$.

SOLUTION   First, we factor out the monomial GCF, $4x$.

$$40x^3 - 68x^2 + 12x = 4x(10x^2 - 17x + 3)$$

Answers   **Your Turn 1**

a. $(5x + 4)(x + 1)$

b. $(3c - 5d)(2c - 3d)$

c. $(2y + 5)(3y - 1)$

Now we factor the trinomial within the parentheses. The fact that 3 is positive and $-17x$ is negative indicates that both signs of the last terms in the binomial factors will be minus signs.

The *first* terms must multiply to equal $10x^2$. These could be x and $10x$, or $2x$ and $5x$.

$$4x(10x^2 - 17x + 3) = 4x(\boxed{} - \boxed{})(\boxed{} - \boxed{})$$

The *last* terms must multiply to equal 3. Because 3 is a prime number, its factors are 1 and 3.

Now we multiply binomials with various combinations of these first and last terms until we find a combination whose inner and outer products combine to equal $-17x$.

$$\left.\begin{aligned}
4x(10x - 3)(x - 1) &= 4x(10x^2 - 10x - 3x + 3) = 4x(10x^2 - 13x + 3) \\
4x(10x - 1)(x - 3) &= 4x(10x^2 - 30x - x + 3) = 4x(10x^2 - 31x + 3) \\
4x(5x - 3)(2x - 1) &= 4x(10x^2 - 5x - 6x + 3) = 4x(10x^2 - 11x + 3)
\end{aligned}\right\} \begin{array}{l}\text{Incorrect} \\ \text{combinations}\end{array}$$

$$4x(5x - 1)(2x - 3) = 4x(10x^2 - 15x - 2x + 3) = 4x(10x^2 - 17x + 3) \quad \begin{array}{l}\text{Correct} \\ \text{combination}\end{array}$$

Answer $\quad 40x^3 - 68x^2 + 12x = 4x(5x - 1)(2x - 3)$

Your Turn 2   Factor.

a. $16y^3 - 102y^2 + 36y$

b. $36m^2n - 120mn^2 - 21n^3$

Objective ❷ Factor trinomials of the form $ax^2 + bx + c$, where $a \neq 1$, by grouping. The major drawback about using trial and error to factor is that it can be tedious to find the correct combination. An alternative method is to factor by grouping, which we introduced in Section 6.1. Recall that in the grouping method, we group pairs of terms in a four-term polynomial, then factor out the GCF from each pair of terms. Because a trinomial of the form $ax^2 + bx + c$ has only three terms, we split the bx term into two like terms to create a four-term polynomial that we can factor by grouping. To determine how to split the bx term, we use the fact that if $ax^2 + bx + c$ is factorable, then b will always equal the sum of a pair of factors of the product of a and c.

Consider again the trinomial $3x^2 + 17x + 10$, which we factored by trial and error at the beginning of this section. Notice that $a = 3$, $b = 17$, and $c = 10$; so the product of a and c is $(3)(10) = 30$. To split the bx term, which is $17x$ in this case, we look for a pair of factors of 30 whose sum is 17. It is helpful to list the factor pairs and their corresponding sums in a table.

Factors of ac	Sum of Factors of ac
$(1)(30) = 30$	$1 + 30 = 31$
$(2)(15) = 30$	$2 + 15 = 17$
$(3)(10) = 30$	$3 + 10 = 13$
$(5)(6) = 30$	$5 + 6 = 11$

Notice that 2 and 15 is the only factor pair of 30 whose sum is 17.

Now we can write $17x$ as $15x + 2x$ and then factor by grouping.

$$\begin{aligned}
3x^2 + 17x + 10 &= 3x^2 + 15x + 2x + 10 \\
&= 3x(x + 5) + 2(x + 5) \\
&= (x + 5)(3x + 2)
\end{aligned}$$

Note: We also could have written $17x$ as $2x + 15x$ to get the same result.

Every trinomial factorable by grouping has only one factor pair of ac whose sum is b, which suggests the following procedure.

Answers   **Your Turn 2**

a. $2y(y - 6)(8y - 3)$

b. $3n(6m + n)(2m - 7n)$

Procedure **Factoring $ax^2 + bx + c$, where $a \neq 1$, by Grouping**

To factor a trinomial of the form $ax^2 + bx + c$, where $a \neq 1$, by grouping,

1. Look for a monomial GCF in all of the terms. If there is one, factor it out.
2. Multiply a and c.
3. Find two factors of this product whose sum is b.
4. Write a four-term polynomial in which bx is written as the sum of two like terms whose coefficients are the two factors you found in step 3.
5. Factor by grouping.

Example 3 Factor by grouping.

a. $8x^2 - 22x + 5$

SOLUTION Notice that for this trinomial, $a = 8$, $b = -22$, and $c = 5$. We begin by multiplying a and c: $(8)(5) = 40$. Now we find two factors of 40 whose sum is -22. Notice that both factors must be negative. It is helpful to list the combinations in a table.

Factors of ac	Sum of Factors of ac	
$(-1)(-40) = 40$	$-1 + (-40) = -41$	
$(-2)(-20) = 40$	$-2 + (-20) = -22$	◄── Correct
$(-4)(-10) = 40$	$-4 + (-10) = -14$	
$(-5)(-8) = 40$	$-5 + (-8) = -13$	

You do not need to list all possible combinations as we did here. We listed them to illustrate that only one combination is correct. We now write $-22x$ as $-2x - 20x$ and then factor by grouping.

$$8x^2 - 22x + 5 = 8x^2 - 2x - 20x + 5 \qquad \text{Write } -22x \text{ as } -2x - 20x.$$
$$= 2x(4x - 1) - 5(4x - 1) \qquad \text{Factor } 2x \text{ out of } 8x^2 - 2x; \\ \text{factor } -5 \text{ out of } -20x + 5.$$
$$= (4x - 1)(2x - 5) \qquad \text{Factor out } (4x - 1).$$

Check $(4x - 1)(2x - 5) = 8x^2 - 20x - 2x + 5 = 8x^2 - 22x + 5$

b. $6a^2 + 13a + 6$

SOLUTION For this trinomial, $a = 6$, $b = 13$, and $c = 6$. The product of $ac = 36$, so we need two factors of 36 whose sum is 13. Because the product is positive, the two factors have the same sign. Because the sum is positive, both factors are positive.

Factors of ac	Sum of Factors of ac	
$(1)(36) = 36$	$1 + 36 = 37$	
$(2)(18) = 36$	$2 + 18 = 20$	
$(3)(12) = 36$	$3 + 12 = 15$	
$(4)(9) = 36$	$4 + 9 = 13$	◄── Correct

Now write $13a$ as $4a + 9a$ and then factor by grouping.

$$6a^2 + 13a + 6 = 6a^2 + 4a + 9a + 6 \qquad \text{Write } 13a \text{ as } 4a + 9a.$$
$$= 2a(3a + 2) + 3(3a + 2) \qquad \text{Factor } 2a \text{ out of } 6a^2 + 4a; \\ \text{factor } 3 \text{ out of } 9a + 6.$$
$$= (3a + 2)(2a + 3) \qquad \text{Factor out } (3a + 2).$$

Check $(3a + 2)(2a + 3) = 6a^2 + 9a + 4a + 6 = 6a^2 + 13a + 6$

c. $18y^4 + 15y^3 - 12y^2$

SOLUTION Notice that there is a monomial GCF, $3y^2$, that we can factor out.

$$18y^4 + 15y^3 - 12y^2 = 3y^2(6y^2 + 5y - 4)$$

Now we factor the trinomial within the parentheses. For this trinomial, $a = 6$, $b = 5$, and $c = -4$. The product of ac is $(6)(-4) = -24$. Now find two factors of -24 whose sum is 5. Because the product is negative, the two factors will have different signs. Because the sum is positive, the factor with the greater absolute value must be positive.

Factors of ac	Sum of Factors of ac	
$(-1)(24) = -24$	$-1 + 24 = 23$	
$(-2)(12) = -24$	$-2 + 12 = 10$	
$(-3)(8) = -24$	$-3 + 8 = 5$	← Correct

Now write $5y$ as $-3y + 8y$ and then factor by grouping.

$$3y^2(6y^2 + 5y - 4) = 3y^2(6y^2 - 3y + 8y - 4) \qquad \text{Write } 5y \text{ as } -3y + 8y.$$
$$= 3y^2\,[3y(2y - 1) + 4(2y - 1)] \qquad \text{Factor } 3y \text{ out of } 6y^2 - 3y; \\ \text{factor 4 out of } 8y - 4.$$
$$= 3y^2(2y - 1)(3y + 4) \qquad \text{Factor out } (2y - 1).$$

Check The check is left to the reader.

Answers   Your Turn 3

a. $(3x - 2)(2x - 5)$
b. $(3x - 4)(2x + 3)$
c. $4y(2x - 9)(x + 2)$

Your Turn 3 Factor by grouping.

a. $6x^2 - 19x + 10$ **b.** $6x^2 + x - 12$ **c.** $8x^2y - 20xy - 72y$

6.3 Exercises   For Extra Help   *MyMathLab*    PRACTICE   WATCH   DOWNLOAD

Note: Exercises marked with a ★ represent challenging exercises.

1. In factoring a polynomial, your first step should be to look for a(n) _____.

2. Given a trinomial of the form $ax^2 + bx + c$, where $a \neq 1$, how do you determine the first terms in its binomial factors?

3. Given a trinomial of the form $ax^2 + bx + c$, where $a \neq 1$, how do you determine the last terms in its binomial factors?

4. When factoring a trinomial of the form $ax^2 + bx + c$, assuming there is no monomial GCF, what is the first step?

For Exercises 5–10, fill in the missing values in the factors. See Objective 1.

5. $3y^2 + 14y + 8 = (3y + \quad)(y + \quad)$

6. $6x^2 + 11x + 4 = (2x + \quad)(3x + \quad)$

7. $4t^2 + 19t - 30 = (\quad + 6)(\quad - 5)$

8. $8m^2 - 2m - 15 = (\quad - 3)(\quad + 5)$

9. $6x^2 - 25x + 14 = (2x - \quad)(\quad - 2)$

10. $9n^2 - 56n + 12 = (9n - \quad)(\quad - 6)$

For Exercises 11–42, factor completely. If prime, so indicate. See Examples 1 and 2.

11. $2j^2 + 5j + 2$

12. $2x^2 + 7x + 3$

13. $2y^2 - 3y - 5$

14. $3m^2 - 10m + 3$

15. $3m^2 - 10m + 8$

16. $3y^2 - 17y + 10$

17. $6a^2 + 13a + 7$

18. $4y^2 + 8y + 3$

19. $6p^2 + 2p + 1$

20. $6u^2 + 3u - 7$

21. $4a^2 - 19a + 12$

22. $4u^2 - 23u + 15$

23. $6x^2 + 19x + 15$

24. $8x^2 + 18x + 9$

25. $16d^2 - 14d - 15$

26. $18a^2 + 9a - 35$

27. $3p^2 + 13pq + 4q^2$

28. $2u^2 + 7uv + 6v^2$

29. $5k^2 - 7kh - 12h^2$

30. $3a^2 - 13ab - 30b^2$

31. $12a^2 - 40ab + 25b^2$

32. $6w^2 - 19wv + 10v^2$

33. $8m^2 - 27mn + 9n^2$

34. $9x^2 - 21xy + 10y^2$

35. $8x^2y - 4xy - y$

36. $6a^2b + 3ab + b$

37. $12y^2 + 24y + 9$

38. $6a^2 - 20a + 16$

39. $6ab^2 - 20ab + 14a$

40. $20xy^2 - 50xy + 30x$

41. $6w^3 + 16w^2v + 8wv^2$

42. $6x^3 + 51x^2y + 90xy^2$

For Exercises 43–74, factor by grouping. If prime, so state. See Example 3.

43. $3a^2 + 4a + 1$

44. $5m^2 + 11m + 2$

45. $2t^2 - 3t + 1$

46. $3h^2 - 5h + 2$

47. $3x^2 - 4x - 7$

48. $3l^2 + 2l - 5$

49. $2y^2 - y - 6$

50. $3a^2 + 2a - 8$

51. $10a^2 - 19a + 7$

52. $8m^2 - 10m + 3$

53. $8r^2 - 6r - 9$

54. $6x^2 + 5x - 6$

55. $20x^2 - 23x + 6$

56. $12c^2 - 28c + 15$

57. $6k^2 + 7jk + 2j^2$

58. $9x^2 + 9xy + 2y^2$

59. $5x^2 - 26xy + 5y^2$

60. $2t^2 - 3tu + u^2$

61. $10s^2 + st - 2t^2$

62. $10x^2 - 33xy - 7y^2$

63. $6u^2 + 5uv - 6v^2$

64. $14u^2 + uv - 4v^2$

65. $15y^2 - 13y - 8$

66. $8n^2 - 5n - 9$

67. $4k^2 - 14k + 12$

68. $12a^2 + 34a + 20$

69. $12m^3 + 10m^2 - 12m$

70. $36x^2y - 78xy + 36y$

71. $24x^3 - 64x^2y - 24xy^2$

72. $36x^3 - 93x^2y + 60xy^2$

73. $24x^3 - 20x^2y - 24xy^2$

74. $28m^2n + 2mn - 8n$

Find
X *the mistake*

For Exercises 75–78, explain the mistake; then factor the trinomial correctly. See Examples 1 and 2.

75. $6x^2 + 13x + 5 = (6x + 1)(x + 5)$

76. $2x^2 + x - 3 = (2x - 3)(x + 1)$

77. $4n^2 + 12n + 8 = (4n + 8)(n + 1)$

78. $4h^2 - 3hk - 10k^2 = (h - 2)(4h + 5)$

For Exercises 79 and 80, given the area of the figure, factor to find possible expressions for the length and width. See Example 1.

79.

Area = $15x^2 - 11x + 2$

80.

Area = $2x^2 + 13x + 20$

81. Two adjacent plots of land are for sale. The area of the larger plot is described by $6w^2 - w - 2$ square feet, where w is the width of the smaller plot. Factor to find possible expressions for the dimensions of the larger plot.

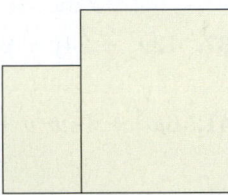

82. A cookie sheet has an area of $5h^2 + 8h - 85$ square inches, where h represents the height of the lip of the sheet. Factor to find possible expressions for the width and length of the cookie sheet.

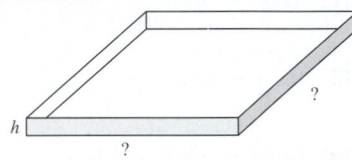

★ *For Exercises 83–88, find all natural numbers that can replace b and make the expression factorable. See Example 1.*

83. $2x^2 + bx + 6$

84. $3x^2 + bx + 10$

85. $5x^2 + bx - 9$

86. $2x^2 + bx - 15$

87. $6x^2 + bx - 7$

88. $4x^2 + bx - 5$

Review Exercises

Exercises 1–7   EXPRESSIONS

[5.5] *For Exercises 1–3, multiply.*

1. $(3x + 5)(3x - 5)$

2. $(x - 2)(x^2 + 2x + 4)$

3. $(2x - 5)^2$

For Exercises 4–7, factor completely.

[6.1]   **4.** $16c^3d^2 - 24c^2d^4 + 36cd^2$

[6.1]   **5.** $ab + 5a + 3b + 15$

[6.2]   **6.** $m^2 + 2m - 24$

[6.2]   **7.** $3x^3 + 9x^2 - 30x$

6.4 Factoring Special Products

Objectives

1. Factor perfect square trinomials.
2. Factor a difference of squares.
3. Factor a difference of cubes.
4. Factor a sum of cubes.

In Section 5.5, we explored some special products found by multiplying conjugates and squaring binomials. In this section, we see how to factor those special products.

Objective 1 Factor perfect square trinomials. First, we consider the product that is a result of squaring a binomial. The trinomial product is a perfect square. Recall the following rules for squaring a binomial that we developed in Section 5.5.

$$(a + b)^2 = a^2 + 2ab + b^2$$
$$(a - b)^2 = a^2 - 2ab + b^2$$

Consider $(2x + 3)^2$ and $(2x - 3)^2$. Using the rules, we have

$$(2x + 3)^2 = (2x)^2 + 2(2x)(3) + (3)^2 \qquad (2x - 3)^2 = (2x)^2 - 2(2x)(3) + (3)^2$$
$$= 4x^2 + 12x + 9 \qquad\qquad\qquad = 4x^2 - 12x + 9$$

To use these rules when factoring, we look at the terms in the given trinomial to determine whether the trinomial fits the form of a perfect square. If we were asked to factor $4x^2 + 12x + 9$, we would note that the first and last terms are perfect squares and that twice the product of their square roots equals the middle term.

The square root of $4x^2$ is $2x$.   The square root of 9 is 3.

$$4x^2 + 12x + 9$$

Twice the product of $2x$ and 3 is the middle term, $12x$.

Because $4x^2 + 12x + 9$ is a perfect square trinomial fitting the form $a^2 + 2ab + b^2$, where $a = 2x$ and $b = 3$, we can write the factored form as $(a + b)^2$, which is $(2x + 3)^2$.

Rule   **Factoring Perfect Square Trinomials**

$$a^2 + 2ab + b^2 = (a + b)^2$$
$$a^2 - 2ab + b^2 = (a - b)^2$$

Example 1 Factor.

a. $9y^2 + 30y + 25$

SOLUTION This trinomial is a perfect square because the first and last terms are perfect squares and twice the product of their roots is the middle term.

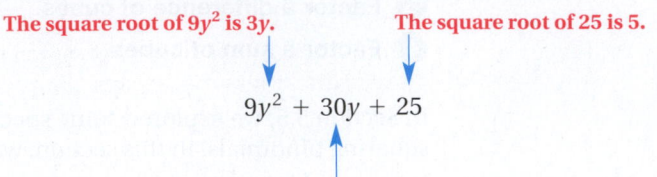

The square root of $9y^2$ is $3y$. The square root of 25 is 5.

$$9y^2 + 30y + 25$$

Twice the product of $3y$ and 5 is
$2(3y)(5) = 30y$, which is the middle term.

Answer $9y^2 + 30y + 25 = (3y + 5)^2$ Use $a^2 + 2ab + b^2 + (a + b)^2$, where $a = 3y$ and $b = 5$.

b. $36x^2 - 84x + 49$

SOLUTION This trinomial is a perfect square.

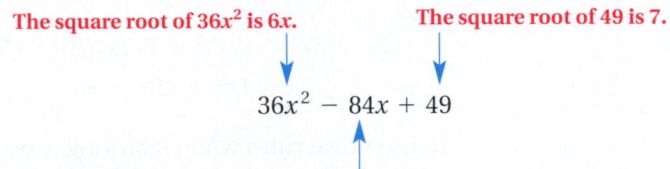

The square root of $36x^2$ is $6x$. The square root of 49 is 7.

$$36x^2 - 84x + 49$$

Twice the product of $6x$ and 7 is
$2(6x)(7) = 84x$, which is the middle term.

Answer $36x^2 - 84x + 49 = (6x - 7)^2$ Use $a^2 - 2ab + b^2 = (a - b)^2$, where $a = 6x$ and $b = 7$.

Your Turn 1 Factor.

a. $25n^2 + 20n + 4$ **b.** $49x^2 - 14x + 1$

Note that in the perfect square trinomial forms, both a and b can have variables. Also don't forget that as a first step, you should always look for a monomial GCF in the terms.

Example 2 Factor.

a. $16x^2 - 8xy + y^2$

SOLUTION $16x^2 - 8xy + y^2 = (4x - y)^2$ Use $a^2 - 2ab + b^2 = (a - b)^2$, where $a = 4x$ and $b = y$.

b. $4x^3y - 36x^2y + 81xy$

SOLUTION

$$4x^3y - 36x^2y + 81xy = xy(4x^2 - 36x + 81)$$ Factor out the monomial GCF, xy.
$$= xy(2x - 9)^2$$ Factor $4x^2 - 36x + 81$ using $a^2 - 2ab + b^2 = (a - b)^2$, where $a = 2x$ and $b = 9$.

Your Turn 2 Factor.

   a. $9m^2 + 48mn + 64n^2$
           b. $2xy^4 + 24xy^3 + 72xy^2$

Objective 2 **Factor a difference of squares.** Another special product that we considered in Section 5.5 was found by multiplying conjugates. Recall that conjugates are binomials that differ only in the sign separating the terms. For example, $3x + 2$ and $3x - 2$ are conjugates. Note that multiplying conjugates produces a product that is a difference of squares.

$$(3x + 2)(3x - 2) = 9x^2 - 4$$

This term is the square of $3x$.   This term is the square of 2.

The rule for multiplying conjugates that we developed in Section 5.5 is:

$$(a + b)(a - b) = a^2 - b^2.$$

To turn that rule into a rule for factoring a difference of squares, we reverse it.

Rule   **Factoring a Difference of Squares**

$$a^2 - b^2 = (a + b)(a - b)$$

Warning: A *sum* of squares $a^2 + b^2$ is prime and *cannot* be factored.

Example 3 Factor.

a. $25x^2 - 36y^2$

SOLUTION This binomial is a difference of squares because $25x^2 - 36y^2 = (5x)^2 - (6y)^2$. To factor it, we use the rule $a^2 - b^2 = (a + b)(a - b)$.

$$a^2 - b^2 = (a + b)(a - b)$$
$$25x^2 - 36y^2 = (5x)^2 - (6y)^2 = (5x + 6y)(5x - 6y)$$

Replace a with $5x$ and b with $6y$.

b. $9x^4 - 16$

SOLUTION This binomial is a difference of squares because $9x^4 - 16 = (3x^2)^2 - (4)^2$.

$$9x^4 - 16 = (3x^2 + 4)(3x^2 - 4)$$

Use $a^2 - b^2 = (a + b)(a - b)$ with $a = 3x^2$ and $b = 4$.

c. $25y^7 - 64y^9$

SOLUTION The terms in this binomial have a monomial GCF, y^7.

$$25y^7 - 64y^9 = y^7(25 - 64y^2)$$
$$= y^7(5 + 8y)(5 - 8y)$$

Factor out the monomial GCF, y^7.
Factor $25 - 64y^2$ using $a^2 - b^2 = (a + b)(a - b)$ with $a = 5$ and $b = 8y$.

Answers **Your Turn 2**
a. $(3m + 8n)^2$
b. $2xy^2(y + 6)^2$

Your Turn 3 Factor.

   a. $n^2 - 49$
     b. $9t^6 - 49u^4$
     c. $50x^3 - 18x$

Answers **Your Turn 3**
a. $(n + 7)(n - 7)$
b. $(3t^3 + 7u^2)(3t^3 - 7u^2)$
c. $2x(5x + 3)(5x - 3)$

Sometimes the factors themselves can be factored.

Example 4   Factor $x^4 - 81$.

SOLUTION   This binomial is a difference of squares, where $a = x^2$ and $b = 9$.

$$x^4 - 81 = (x^2 + 9)(x^2 - 9) \qquad \text{Use } a^2 - b^2 = (a + b)(a - b).$$
$$= (x^2 + 9)(x + 3)(x - 3) \qquad \text{Factor } x^2 - 9 \text{ using } a^2 - b^2 = (a + b)(a - b)$$
$$\text{with } a = x \text{ and } b = 3.$$

Note: The factor $x^2 + 9$ is a sum of squares, which is prime.

Your Turn 4   Factor $y^4 - 16$.

Objective ③ Factor a difference of cubes.   Another common binomial form that we can factor is a difference of cubes. A difference of cubes is a result of a multiplication such as the following:

$$(x - 2)(x^2 + 2x + 4) = x^3 + 2x^2 + 4x - 2x^2 - 4x - 8$$
$$= x^3 - 8 \longleftarrow \text{Difference of cubes}$$

Writing factored form reverses this process so that we have

$$x^3 - 8 = (x - 2)(x^2 + 2x + 4).$$

This suggests the following rule for factoring a difference of cubes.

Rule   **Factoring a Difference of Cubes**

$$a^3 - b^3 = (a - b)(a^2 + ab + b^2)$$

Example 5   Factor.

a. $64y^3 - 27$

SOLUTION   This binomial is a difference of cubes because $64y^3 - 27 = (4y)^3 - (3)^3$. To factor, we use the rule $a^3 - b^3 = (a - b)(a^2 + ab + b^2)$ with $a = 4y$ and $b = 3$.

$$a^3 \ - \ b^3 \ = \ (a - b) \ (a^2 \ + \ a \ b \ + \ b^2)$$
$$64y^3 - 27 = (4y)^3 - (3)^3 = (4y - 3)((4y)^2 + (4y)(3) + (3)^2) \qquad \begin{array}{l}\text{Replace } a \text{ with} \\ 4y \text{ and } b \text{ with 3.}\end{array}$$

$$= (4y - 3)(16y^2 + 12y + 9) \qquad \text{Simplify.}$$

Note: This trinomial may seem like a perfect square. However, to be a perfect square, the middle term should be $2ab$. In this trinomial, we have only ab; so it cannot be factored.

b. $27m^4 - 8mn^3$

SOLUTION   The terms in this binomial have a monomial GCF, m.

$$27m^4 - 8mn^3 = m(27m^3 - 8n^3) \qquad \begin{array}{l}\text{Factor out the} \\ \text{monomial GCF, } m.\end{array}$$
$$= m(3m - 2n)((3m)^2 + (3m)(2n) + (2n)^2) \qquad \begin{array}{l}\text{Factor } 27m^3 - 8n^3, \text{ which} \\ \text{is a difference of cubes with}\end{array}$$
$$= m(3m - 2n)(9m^2 + 6mn + 4n^2) \qquad \begin{array}{l}a = 3m \text{ and } b = 2n. \\ \text{Simplify.}\end{array}$$

Answer   Your Turn 4

$(y^2 + 4)(y + 2)(y - 2)$

Your Turn 5 Factor.

a. $27 - u^3$ 　　　　　　　　　　　　　　b. $xy^3 - 64x$

Objective 4 **Factor a sum of cubes.** A sum of cubes can be factored using a pattern similar to the difference of cubes. A sum of cubes is a result of a multiplication such as the following:

$$(x + 2)(x^2 - 2x + 4) = x^3 - 2x^2 + 4x + 2x^2 - 4x + 8$$
$$= x^3 + 8$$

Writing factored form reverses this process so that we have

$$x^3 + 8 = (x + 2)(x^2 - 2x + 4).$$

In general, we can write the following rule for factoring a sum of cubes.

Rule **Factoring a Sum of Cubes**

$$a^3 + b^3 = (a + b)(a^2 - ab + b^2)$$

The factored form of the sum or difference of cubes always has a binomial and a trinomial factor. The binomial factor consists of the cube roots of the terms of the sum or difference of cubes with the same sign. The first term of the trinomial factor is the square of the first term of the binomial factor, the second term of the trinomial factor is the opposite of the product of the terms of the binomial factor, and the third term of the trinomial factor is the square of the second term of the binomial factor.

Example 6 Factor.

a. $27x^3 + 8$

SOLUTION This binomial is a sum of cubes because $27x^3 + 8 = (3x)^3 + (2)^3$. To factor, we use the rule $a^3 + b^3 = (a + b)(a^2 - ab + b^2)$ with $a = 3x$ and $b = 2$.

$$a^3 + b^3 = (a + b)(a^2 - ab + b^2)$$
$$27x^3 + 8 = (3x)^3 + (2)^3 = (3x + 2)((3x)^2 - (3x)(2) + (2)^2)$$ 　　Replace a with $3x$ and b with 2.
$$= (3x + 2)(9x^2 - 6x + 4)$$ 　　Simplify.

Note: This trinomial cannot be factored.

Note: Because $1^2 = 1$, $1^3 = 1$, and so on, 1 is a perfect square, a perfect cube, and so on.

b. $5h + 40hk^3$

SOLUTION The terms in this binomial have a monomial GCF, $5h$.

$$5h + 40hk^3 = 5h(1 + 8k^3)$$ 　　Factor out the monomial GCF, $5h$.
$$= 5h(1 + 2k)((1)^2 - (1)(2k) + (2k)^2)$$ 　　Factor $1 + 8k^3$, which is a sum of cubes with $a = 1$ and $b = 2k$.
$$= 5h(1 + 2k)(1 - 2k + 4k^2)$$

Answers Your Turn 5

a. $(3 - u)(9 + 3u + u^2)$
b. $x(y - 4)(y^2 + 4y + 16)$

Answers Your Turn 6

a. $(4 + m)(16 - 4m + m^2)$
b. $x^2(5x + 3)(25x^2 - 15x + 9)$

Your Turn 6 Factor.

a. $64 + m^3$ 　　　　　　　　　　　　　　b. $125x^5 + 27x^2$

Note: Exercises marked with a ★ represent challenging exercises.

1. Explain in your own words how to recognize a perfect square trinomial.

2. How is the sign in the factored form of a perfect square trinomial determined?

3. The factors of a difference of squares are binomials known as _____.

4. There is only one minus sign in the factors of a difference of cubes. Where is that minus sign placed in the factors?

5. There is only one minus sign in the factors of a sum of cubes. Where is that minus sign placed in the factors?

6. How are the a and b terms determined in a sum or difference of cubes?

For Exercises 7–24, factor the trinomials that are perfect squares. If the trinomial is not a perfect square, write **not a perfect square.** *See Examples 1 and 2.*

7. $x^2 + 14x + 49$

8. $y^2 + 6y + 9$

9. $b^2 - 8b + 16$

10. $m^2 - 6m + 9$

11. $n^2 + 12n + 144$

12. $y^2 + 6y + 36$

13. $25u^2 - 30u + 9$

14. $9m^2 - 24m + 16$

15. $100w^2 + 20w + 1$

16. $25a^2 + 10a + 1$

17. $y^2 + 2yz + z^2$

18. $w^2 + 2wv + v^2$

19. $4p^2 - 28pq + 49q^2$

20. $9r^2 - 12rs + 4s^2$

21. $16g^2 + 24gh + 9h^2$

22. $4y^2 - 12by + 9b^2$

23. $16t^2 + 80t + 100$

24. $54q^2 - 72q + 24$

For Exercises 25–44, factor the binomials that are the difference of squares. If prime, so state. See Examples 3 and 4.

25. $x^2 - 4$

26. $a^2 - 121$

27. $16 - y^2$

28. $49 - a^2$

29. $p^2 - q^2$

30. $m^2 - n^2$

31. $25u^2 - 16$

32. $16x^2 - 49$

33. $9x^2 + b^2$

34. $16x^2 + y^2$

35. $64m^2 - 25n^2$

36. $121p^2 - 9q^2$

37. $50x^2 - 32y^2$

38. $27a^2 - 48b^2$

39. $4x^2 + 100y^2$

40. $16x^2 + 36y^2$

41. $x^4 - y^4$

42. $a^4 - b^4$

43. $x^4 - 16$

44. $b^4 - 1$

For Exercises 45–64, factor the sum or difference of cubes. See Examples 5 and 6.

45. $n^3 - 27$

46. $y^3 - 64$

47. $x^3 + 27$

48. $y^3 + 8$

49. $x^3 - 1$

50. $r^3 - 1$

51. $m^3 + n^3$

52. $p^3 + q^3$

53. $27k^3 - 8$

54. $125a^3 - 64$

55. $27k^3 + 8$

56. $125b^3 + 27$

57. $c^3 - 64d^3$

58. $a^3 - 8b^3$

59. $125x^3 + 64y^3$

60. $27a^3 + 1000b^3$

61. $27x^3 - 64y^3$

62. $1000a^3 - 27b^3$

63. $8p^3 + q^3z^3$

64. $x^3y^3 + 27z^3$

For Exercises 65–80, factor. If prime, so state. See Examples 1–6.

65. $2x^2 - 50$

66. $5y^2 - 20$

67. $16x^2 - \dfrac{25}{49}$

68. $25y^2 - \dfrac{1}{36}$

69. $2u^3 - 2u$

70. $4x - 16x^3$

71. $y^5 - 16y^3b^2$

72. $a^4b^2 - a^6$

73. $50x^3 + 2x$

74. $27m^3 + 363mn^2$

75. $3y^3 - 24z^3$

76. $2x^3 - 54$

77. $c^3 - \dfrac{8}{27}$

78. $y^3 - \dfrac{1}{125}$

79. $16c^4 + 2cd^3$

80. $8d^4 - 64c^3d$

★*For Exercises 81–88, use the rules for a difference of squares, sum of cubes, or difference of cubes to factor completely. See Examples 3–6.*

81. $(2a - b)^2 - c^2$

82. $(3x - y)^2 - z^2$

83. $16 - 9(x - y)^2$

84. $25a^2 - 16(b + c)^2$

85. $x^3 + 27(y + z)^3$

86. $m^3 + 8(y + b)^3$

87. $(x + y)^3 - 64d^3$

88. $(t + u)^3 - 64$

For Exercises 89–92, find a natural number, b, that makes the expression a perfect square trinomial. See Objective 1.

89. $16x^2 + bx + 9$

90. $25x^2 + bx + 4$

91. $4x^2 - bx + 81$

92. $36x^2 - bx + 49$

For Exercises 93–96, find the natural number, c, that completes the perfect square trinomial. See Objective 1.

93. $x^2 + 10x + c$

94. $x^2 + 8x + c$

95. $9x^2 - 24x + c$

96. $4x^2 - 28x + c$

For Exercises 97–98, write a polynomial for the area of the shaded region; then factor completely.

★ **97.**

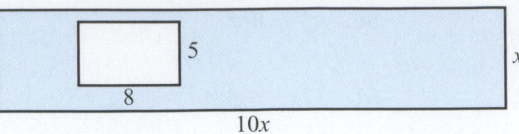

★ **98.**

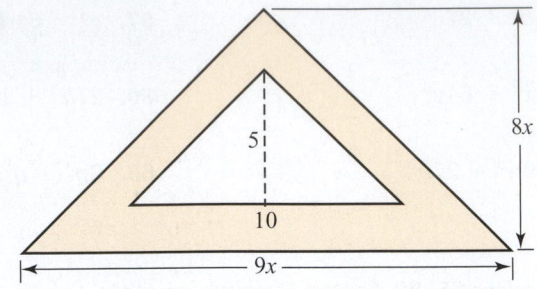

For Exercises 99–100, write a polynomial for the volume of the shaded region; then factor completely.

★ **99.**

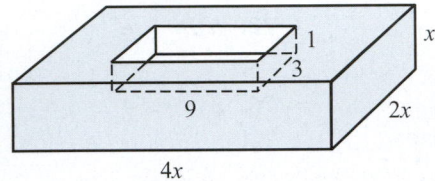

★ **100.**

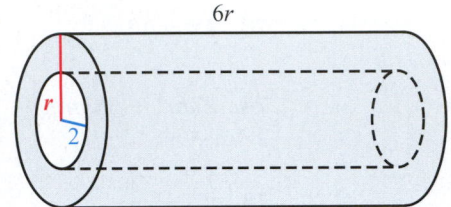

Review Exercises

Exercises 1 and 2 CONSTANTS AND VARIABLES

[6.1]   **1.** List all natural number factors of 36.

[1.2]   **2.** Find the prime factorization of 100.

Exercises 3–8   EXPRESSIONS

For Exercises 3–8, factor completely.

[6.1]   **3.** $14x^3y^3 - 21x^2y^4 - 7xy^2$

[6.1]   **4.** $x^2 + 4xy + 3x + 12y$

[6.2]   **5.** $a^2 - 3ab - 18b^2$

[6.2]   **6.** $3a^2b + 6ab - 45b$

[6.3]   **7.** $6n^2 + 5n - 6$

[6.3]   **8.** $12x^3y - 38x^2y^2 + 20xy^3$

6.5 Strategies for Factoring

Objective ① **Factor polynomials.** One of the difficulties in factoring expressions is determining which technique to use. In this section, we use the following general outline for factoring.

Procedure | **Factoring a Polynomial**

To factor a polynomial, first factor out any monomial GCF; then consider the number of terms in the polynomial. If the polynomial has

I. Four terms, try to factor by grouping.

II. Three terms, determine whether the trinomial is a perfect square.
   A. If the trinomial is a perfect square, consider its form.
   1. If in form $a^2 + 2ab + b^2$, the factored form is $(a + b)^2$.
   2. If in form $a^2 - 2ab + b^2$, the factored form is $(a - b)^2$.
   B. If the trinomial is not a perfect square, consider its form.
   1. If in form $x^2 + bx + c$, find two factors of c whose sum is b and write the factored form as $(x + \text{first number})(x + \text{second number})$.
   2. If in form $ax^2 + bx + c$, where $a \neq 1$, use trial and error. Or find two factors of ac whose sum is b; write these factors as coefficients of two like terms that, when combined, equal bx; and factor by grouping.

III. Two terms, determine whether the binomial is a difference of squares, sum of cubes, or difference of cubes.
   A. If given a binomial that is a difference of squares, $a^2 - b^2$, the factors are conjugates and the factored form is $(a + b)(a - b)$. Note that a sum of squares cannot be factored.
   B. If given a binomial that is a sum of cubes, $a^3 + b^3$, the factored form is $(a + b)(a^2 - ab + b^2)$.
   C. If given a binomial that is a difference of cubes, $a^3 - b^3$, the factored form is $(a - b)(a^2 + ab + b^2)$.

Note: Always check to see if any of the factors can be factored further.

Example 1   Factor.

a. $12x^2 + 28x - 5$

SOLUTION   There is no monomial GCF. Because this expression is a trinomial, we check to see if it is a perfect square. It is not a perfect square because the first and last terms are not perfect squares. It is a trinomial of the form $ax^2 + bx + c$, where $a \neq 1$. Therefore, we use trial and error or the alternative method that involves factoring by grouping. We will use trial and error here.

The *first* terms multiply to equal $12x^2$. The factors could be x and $12x$, $2x$ and $6x$, or $3x$ and $4x$.

$$12x^2 + 28x - 5 = (+)(-)$$

The *last* terms multiply to equal -5. We have already written in the appropriate signs, so these factors could be 1 and 5.

Now list possible combinations of these first and last factor pairs and check by multiplying to see if the sum of the products of the inner and outer terms equals $28x$.

$$
\left.
\begin{aligned}
(x + 5)(12x - 1) &= 12x^2 - x + 60x - 5 = 12x^2 + 59x - 5 \\
(12x + 5)(x - 1) &= 12x^2 - 12x + 5x - 5 = 12x^2 - 7x - 5
\end{aligned}
\right\}
$$
Incorrect combinations

$$(2x + 5)(6x - 1) = 12x^2 - 2x + 30x - 5 = 12x^2 + 28x - 5$$
Correct combination

Answer $12x^2 + 28x - 5 = (2x + 5)(6x - 1)$

b. $7x^3 + 14x^2 - 105x$

SOLUTION

$$7x^3 + 14x^2 - 105x = 7x(x^2 + 2x - 15)$$ **Factor out the monomial GCF, $7x$.**

The trinomial in the parentheses is not a perfect square and is in the form $x^2 + bx + c$; so we look for two numbers whose product is -15 and whose sum is 2. Because the product is negative, the two factors must have different signs. Because the sum is positive, the factor with the greater absolute value will be the positive factor.

Product	Sum	
$(-1)(15) = -15$	$-1 + 15 = 14$	
$(-3)(5) = -15$	$-3 + 5 = 2$	**Correct combination**

Answer $7x^3 + 14x^2 - 105x = 7x(x - 3)(x + 5)$

c. $12m^4 - 48n^2$

SOLUTION

$$12m^4 - 48n^2 = 12(m^4 - 4n^2)$$ **Factor out the monomial GCF, 12.**

The binomial in the parentheses is a difference of squares in the form $a^2 - b^2$, where $a = m^2$ and $b = 2n$. Using $a^2 - b^2 = (a + b)(a - b)$ gives us

$$= 12(m^2 + 2n)(m^2 - 2n).$$ ◀ **Note:** Neither of these binomial factors can be factored further.

d. $10y^5 + 40y^3$

SOLUTION

$$10y^5 + 40y^3 = 10y^3(y^2 + 4)$$ **Factor out the monomial GCF, $10y^3$.**

The binomial in the parentheses is a sum of squares, which cannot be factored.

e. $45x^3y^3 - 120x^2y^4 + 80xy^5$

SOLUTION

$$45x^3y^3 - 120x^2y^4 + 80xy^5 = 5xy^3(9x^2 - 24xy + 16y^2)$$ **Factor out the monomial GCF, $5xy^3$.**

The trinomial in the parentheses is a perfect square in the form $a^2 - 2ab + b^2$, where $a = 3x$ and $b = 4y$. Using $a^2 - 2ab + b^2 = (a - b)^2$ gives us

$$= 5xy^3(3x - 4y)^2.$$

f. $x^5 - x^3 - 8x^2 + 8$

SOLUTION There is no monomial GCF common to all terms. Because this polynomial has four terms, we try to factor by grouping.

$$x^5 - x^3 - 8x^2 + 8 = x^3(x^2 - 1) - 8(x^2 - 1)$$
$$= (x^2 - 1)(x^3 - 8)$$

The two binomial factors can be factored further. The first binomial, $x^2 - 1$, is a difference of squares with $a = x$ and $b = 1$. The second binomial, $x^3 - 8$, is a difference of cubes with $a = x$ and $b = 2$.

$$= (x + 1)(x - 1)(x - 2)(x^2 + 2x + 4)$$

Answers Your Turn 1
a. $(3y - 7)(8y + 1)$
b. $2t(t - 6)(t - 3)$
c. $5(3a + b)(3a - b)$
d. $3m^2n(n^2 + 5n + 2)$
e. $4a^2b(2a - 5b)^2$
f. $(2x + 3)(2x - 3)(3x - 2)$

Your Turn 1 Factor.

a. $24y^2 - 53y - 7$

b. $2t^3 - 18t^2 + 36t$

c. $45a^2 - 5b^2$

d. $3m^2n^3 + 15m^2n^2 + 6m^2n$

e. $16a^4b - 80a^3b^2 + 100a^2b^3$

f. $12x^3 - 8x^2 - 27x + 18$

6.5 Exercises   For Extra Help   MyMathLab

1. What should you look for first when factoring?

2. What are three different types of factorable binomials?

3. Describe one type of binomial that cannot be factored.

4. What are the two forms that a perfect square trinomial can have?

5. When do you factor by grouping?

6. How might you check a factored form to see if it is correct?

For Exercises 7–82, factor completely. If prime, so state. See Example 1.

7. $3xy^2 + 6x^2y$

8. $2a + ab$

9. $7a(x + y) - b(x + y)$

10. $(u + v)r + (u + v)s$

11. $2x^2 - 32$

12. $2x^2 - 2y^2$

13. $ax + ay + bx + by$

14. $ax + xy + ay + y^2$

15. $12a^3b^2c + 3a^2b^2c^2 + 5abc^3$

16. $15m^2n^3p + 2m^2n^2p^2 - 2mn^3p$

17. $x^2 + 8x + 15$

18. $b^2 + 9b + 14$

19. $x^4 - 16$

20. $t^4 - 81$

21. $x^2 + 25$

22. $a^2 + 49$

23. $15x^2 + 7x - 2$

24. $2a^2 - 5a - 12$

25. $ax^2 + 4ax + 4a$

26. $9a^2x + 12ax + 4x$

27. $6ab - 36ab^2$

28. $5ax - 25a^2x$

29. $x^2 + x + 2$

30. $x^2 + 3x + 4$

31. $x^2 - 49$

32. $x^2 - 25$

33. $p^2 + p - 30$

34. $y^2 - 6y - 40$ **35.** $u^3 - u$ **36.** $5m^2 - 45$ **37.** $2b^2 + 14b + 24$

38. $4y^2 + 36y + 80$ **39.** $6r^2 - 15r^3$ **40.** $18y^3 - 24y^4$ **41.** $14u^2 + 7u - 105$

42. $6r^2 - 26r - 20$ **43.** $4h^2 + 12h + 4$ **44.** $3m^2 + 9m + 3$ **45.** $5p^2 - 80$

46. $3r^2 - 108$ **47.** $3w^2 + 5w + 2$ **48.** $3m^2 - 8m + 5$ **49.** $8q^2 - 10q - 3$

50. $10r^2 + 9r - 7$ **51.** $12v^2 + 23v + 10$ **52.** $6x^2 + 13x + 6$ **53.** $2 - 50x^2$

54. $6 - 24x^2$ **55.** $80k^2 - 20k^2l^2$ **56.** $27x^3 - 3x^3y^2$ **57.** $2j^2 + j + 3$

58. $3k^2 - 2k + 1$ **59.** $50 - 20t + 2t^2$ **60.** $243 - 54x + 3x^2$ **61.** $3ax - 6ay - 8by + 4bx$

62. $2ax - 6bx - 3by + ay$ **63.** $x^3 - x^2 - x + 1$ **64.** $x^3 - 2x^2 - x + 2$ **65.** $4x^2 - 28x + 49$

66. $4m^2 + 20m + 25$ **67.** $2x^4 - 162$ **68.** $5a^4 - 3125$ **69.** $a^2 + 2ab + ab + 2b^2$

70. $x^2 - 3xy + xy - 3y^2$ **71.** $9 - 4m^2$ **72.** $64 - 9r^2$ **73.** $x^5 - 4xy^2$

74. $m^5 - 25mn^2$ **75.** $20x^2 + 3xy + 2y^2$ **76.** $20a^2 - 9a + 20$ **77.** $b^3 + 125$

78. $x^3 + 216$ **79.** $3y^3 - 24$ **80.** $4x^3 - 4$ **81.** $54x - 2xy^3$

82. $81n - 3m^3n$

For Exercises 83 and 84, given an expression for the area of each rectangle,
find the length and width.

83.

Area $= 6x^2 - 11x - 7$

84.

Area $= 25n^2 - 49$

For Exercises 85 and 86, given an expression for the volume of each box, find the
length, width, and height.

85.

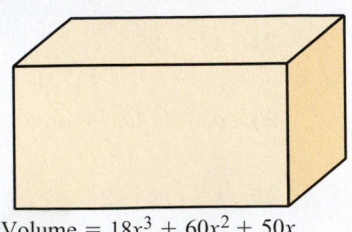

Volume $= 18x^3 + 60x^2 + 50x$

86.

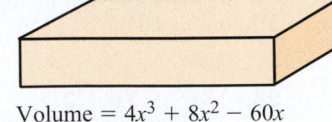

Volume $= 4x^3 + 8x^2 - 60x$

87. A swimmer's goggles fall off a diving board that is 100 meters high. The height of the goggles after t seconds is given by $100 - 16t^2$. Factor $100 - 16t^2$.

88. An object is dropped from a cliff that is 400 feet high. The expression $400 - 16t^2$ gives the height of the falling object after t seconds. Factor $400 - 16t^2$.

89. The voltage in a circuit is the product of two factors, the resistance in the circuit and the current. If the voltage in a circuit is described by the expression $6ir + 15i + 8r + 20$, find the expressions for the current and resistance. (The expression for current will contain i, and the expression for resistance will contain r.)

90. If the voltage in the circuit described in Exercise 89 is changed so that it is described by the expression $35ir + 15i - 7r - 3$, what are the expressions for the current and resistance?

Review Exercises

Exercises 1 and 2 **EXPRESSIONS**

[5.6] **1.** Divide: $\dfrac{2x^2 - 7x - 8}{2x + 1}$

[6.3] **2.** Factor $6y^2 + 23y - 4$.

Exercises 3–5 **EQUATIONS AND INEQUALITIES**

[2.1] **3.** Find the area of a rectangular garden that is 14 feet by 16 feet.

[2.5] **4.** Translate to an equation; then solve. The product of four and the sum of a number and three is equal to twenty. Find the unknown number.

[3.3] **5.** Find three consecutive integers whose sum is 54.

6.6 Solving Quadratic Equations by Factoring

Objectives

1 Use the zero-factor theorem to solve equations containing expressions in factored form.

2 Solve quadratic equations by factoring.

3 Solve problems involving quadratic equations.

4 Use the Pythagorean theorem to solve problems.

Objective 1 **Use the zero-factor theorem to solve equations containing expressions in factored form.** So far in Chapters 5 and 6, we have rewritten polynomial expressions. We are now ready to work with polynomial equations, which we will solve using factoring. Notice that we are moving to the equation level of our Algebra Pyramid, which is built upon the foundation of expressions, variables, and constants.

The Algebra Pyramid

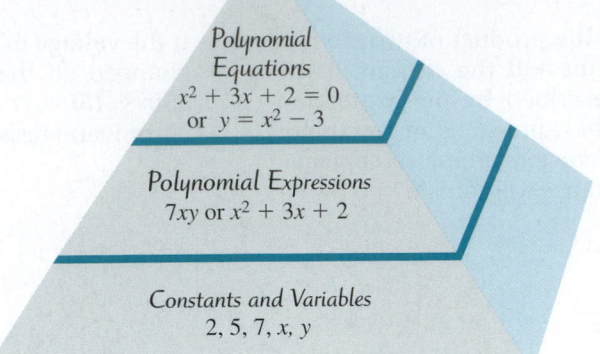

Note: Inequalities are also in the top level of the Algebra Pyramid, but we will not be exploring polynomial inequalities at this point.

Connection

Throughout this book, after developing new types of expressions and ways of rewriting those expressions, we put those skills into the context of solving equations.

When solving polynomial equations by factoring, we make use of the **zero-factor theorem**, which states that if the product of two factors is 0, then one or the other of the two factors or both factors are equal to 0.

Rule | **Zero-Factor Theorem**

If a and b are real numbers and $ab = 0$, then $a = 0$ or $b = 0$.

Example 1   Solve $(x + 2)(x - 3) = 0$.

SOLUTION   In this equation, the product of two factors, $x + 2$ and $x - 3$, is equal to 0. According to the zero-factor theorem, one of the two factors or both factors must equal 0. In solving, we determine what values for x cause each factor to equal 0.

$$x + 2 = 0 \quad \text{or} \quad x - 3 = 0$$

Now solve each of the linear equations to get two solutions.

$$x = -2 \qquad x = 3$$

Check   Verify that -2 and 3 satisfy the original equation, $(x + 2)(x - 3) = 0$.

For $x = -2$: $(-2 + 2)(-2 - 3) \stackrel{?}{=} 0$ For $x = 3$: $(3 + 2)(3 - 3) \stackrel{?}{=} 0$

$(0)(-5) = 0$ $(5)(0) = 0$

Both -2 and 3 check; therefore, both are solutions.

Your Turn 1   Solve $(n - 6)(n + 1) = 0$.

Example 1 suggests the following procedure.

Procedure | **Solving Equations with Two or More Factors Equal to 0**

To solve an equation in which two or more factors are equal to 0, use the zero-factor theorem.

1. Set each factor equal to zero.
2. Solve each of those equations.

Answer   **Your Turn 1**

$n = 6$ or -1

Example 2   Solve.

a. $y(3y + 4) = 0$

SOLUTION   In this equation, the product of two factors, y and $3y + 4$, is equal to 0. We set each factor equal to 0 and then solve each of those equations.

$$y = 0 \quad \text{or} \quad 3y + 4 = 0$$

Note: This equation is already solved.

$$3y = -4 \qquad \text{Subtract 4 from both sides to isolate } 3y.$$

$$y = -\frac{4}{3} \qquad \text{Divide both sides by 3 to isolate } y.$$

Check   Verify that 0 and $-\frac{4}{3}$ satisfy the original equation, $y(3y + 4) = 0$.

$$\text{For } y = 0: (0)(3(0) + 4) \overset{?}{=} 0 \qquad\qquad \text{For } y = -\frac{4}{3}: \left(-\frac{4}{3}\right)\left[\overset{1}{\frac{3}{1}}\left(-\frac{4}{\underset{1}{3}}\right) + 4\right] \overset{?}{=} 0$$

$$(0)(4) = 0 \qquad\qquad\qquad\qquad\qquad \left(-\frac{4}{3}\right)[-4 + 4] \overset{?}{=} 0$$

$$\left(-\frac{4}{3}\right)(0) = 0$$

Both 0 and $-\frac{4}{3}$ check; therefore, both are solutions.

b. $(x - 4)^2 = 0$

SOLUTION   Note that $(x - 4)^2 = 0$ can be written as $(x - 4)(x - 4) = 0$. Because the two factors are identical, there is no need to write two separate equations. We set $x - 4$ equal to 0 and then solve.

$$x - 4 = 0$$

$$x = 4 \qquad \text{Add 4 to both sides to isolate } x.$$

Check   Verify that 4 satisfies the original equation, $(x - 4)^2 = 0$.

$$(4 - 4)^2 \overset{?}{=} 0$$

$$(0)^2 = 0 \qquad \text{4 checks; therefore, it is a solution.}$$

c. $n(n + 6)(5n - 3) = 0$

SOLUTION   In this equation, the product of three factors, n, $n + 6$, and $5n - 3$, is 0. We set each factor equal to 0 and then solve each of those equations.

$$n = 0 \quad \text{or} \quad n + 6 = 0 \quad \text{or} \quad 5n - 3 = 0$$

$$n = -6 \qquad\qquad 5n = 3$$

$$n = \frac{3}{5}$$

Check   To check, we verify that 0, -6, and $\frac{3}{5}$ satisfy the original equation. We leave this check to the reader.

Your Turn 2   Solve.

a. $5x(3x - 2) = 0$ **b.** $(y + 3)^2 = 0$ **c.** $t(4t - 3)(t + 1) = 0$

Answers   Your Turn 2

a. $x = 0$ or $\frac{2}{3}$

b. $y = -3$

c. $t = 0, \frac{3}{4},$ or -1

Objective ② **Solve quadratic equations by factoring.** Now that we have seen how to use the zero-factor theorem to solve equations containing expressions that are in factored form, we are ready to consider equations that have an expression that is not in factored form. More specifically, we will use factoring to solve **quadratic equations in one variable**.

Definition   *Quadratic equation in one variable:* An equation that can be written in the form $ax^2 + bx + c = 0$, where a, b, and c are real numbers and $a \neq 0$.

The fact that a cannot equal zero means that the ax^2 term must be present for the equation to be a quadratic equation. A quadratic equation written in the form $ax^2 + bx + c = 0$ is said to be in *standard form*. Following are some examples of quadratic equations.

$$x^2 + 5x + 6 = 0 \qquad 3x^2 - 48 = 0 \qquad x^2 = x + 6$$

Note: This quadratic equation is in standard form.

Note: This quadratic equation is in standard form. It has no bx term because $b = 0$.

Note: Although not in standard form, this equation is still quadratic. Written in standard form, the equation would be $x^2 - x - 6 = 0$.

In Objective 1, we learned that to use the zero-factor theorem, we need an expression in factored form set equal to 0. So to solve a quadratic equation using factoring, it needs to be in standard form. If given an equation that is not in standard form, such as $x^2 = x + 6$, we first write it in standard form.

$$x^2 - x - 6 = 0 \qquad \text{Subtract } x \text{ and 6 from both sides to get standard form.}$$

After factoring $x^2 - x - 6$, we have $(x + 2)(x - 3) = 0$, which is the same equation that we solved in Example 1. This suggests the following procedure.

Procedure   **Solving Quadratic Equations Using Factoring**

To solve a quadratic equation using factoring,
1. Write the equation in standard form ($ax^2 + bx + c = 0$).
2. Write the variable expression in factored form.
3. Use the zero-factor theorem to solve.

Example 3   Solve $2x^2 + 9x - 5 = 0$.

Solution   This equation is in standard form, so we can simply factor the variable expression.

$$(2x - 1)(x + 5) = 0 \qquad \text{Factor } 2x^2 + 9x - 5.$$

Now use the zero-factor theorem to solve.

$$2x - 1 = 0 \quad \text{or} \quad x + 5 = 0$$
$$2x = 1 \qquad\qquad\quad x = -5$$
$$x = \frac{1}{2}$$

Check   To check, we verify that $\frac{1}{2}$ and -5 satisfy the original equation. We leave this check to the reader.

Your Turn 3 Solve $12y^2 + 11y + 2 = 0$.

Sometimes we may have to factor out a monomial GCF.

Example 4 Solve $x^3 - 3x^2 - 10x = 0$.

SOLUTION

Note: Although $x^3 - 3x^2 - 10x = 0$ is not a quadratic equation, after we factor out the x, the parentheses contain a quadratic form: $x^2 - 3x - 10$.

$$x(x^2 - 3x - 10) = 0 \qquad \text{Factor out the monomial GCF, } x.$$
$$x(x - 5)(x + 2) = 0 \qquad \text{Factor } x^2 - 3x - 10.$$

Now use the zero-factor theorem to solve.

$$x = 0 \quad \text{or} \quad x - 5 = 0 \quad \text{or} \quad x + 2 = 0$$
$$x = 5 \qquad\qquad x = -2$$

Check Verify that 0, 5, and −2 each satisfy the original equation. We leave the check to the reader.

Your Turn 4 Solve $12x^3 + 20x^2 - 8x = 0$.

Rewrite Quadratic Equations in Standard Form

If a quadratic equation is not in standard form ($ax^2 + bx + c = 0$), we put it in standard form, then factor and use the zero-factor theorem.

Example 5 Solve.

a. $6y^2 - 5y = 4 - 28y$

SOLUTION We first need to manipulate the equation so that it is in standard form.

Answer Your Turn 3

$y = -\dfrac{2}{3}$ or $-\dfrac{1}{4}$

$$6y^2 + 23y = 4 \qquad \text{Add } 28y \text{ to both sides.}$$
$$6y^2 + 23y - 4 = 0 \qquad \text{Subtract 4 from both sides to get standard form.}$$
$$(6y - 1)(y + 4) = 0 \qquad \text{Factor using trial and error.}$$
$$6y - 1 = 0 \quad \text{or} \quad y + 4 = 0 \qquad \text{Use the zero-factor theorem to solve.}$$
$$6y = 1 \qquad\qquad y = -4$$
$$y = \frac{1}{6}$$

Answer Your Turn 4

$x = 0, \dfrac{1}{3}$, or -2

Check We leave the check to the reader.

b. $x(x - 5) = -6$

Warning: We cannot solve $x(x - 5) = -6$ by setting $x = -6$ or $x - 5 = -6$. For the *zero-factor* theorem to be used, one side of the equation must be 0.

SOLUTION

Note: We could have multiplied $x(x - 5)$ first and then moved the 6.

$$x(x - 5) + 6 = 0 \qquad \text{Add 6 to both sides so that the right-hand side is 0.}$$
$$x^2 - 5x + 6 = 0 \qquad \text{Multiply } x(x - 5) \text{ to get standard form.}$$
$$(x - 2)(x - 3) = 0 \qquad \text{Factor by looking for two numbers whose product is 6 and sum is } -5.$$
$$x - 2 = 0 \quad \text{or} \quad x - 3 = 0 \qquad \text{Use the zero-factor theorem to solve.}$$
$$x = 2 \qquad\qquad x = 3$$

Check   We leave the check to the reader.

c. $(x + 3)(x - 5) = 20$

SOLUTION

$$(x + 3)(x - 5) = 20$$
$$x^2 - 2x - 15 = 20 \qquad \text{Multiply } x + 3 \text{ and } x - 5.$$
$$x^2 - 2x - 35 = 0 \qquad \text{Subtract 20 from both sides.}$$
$$(x + 5)(x - 7) = 0 \qquad \text{Factor.}$$
$$x + 5 = 0 \quad \text{or} \quad x - 7 = 0 \qquad \text{Use the zero-factor theorem to solve.}$$
$$x = -5 \qquad\qquad x = 7$$

Check   We leave the check to the reader.

Your Turn 5   Solve.

a. $15n^2 + 20n = -8 - 6n$ **b.** $x(x + 2) = 48$ **c.** $(x + 4)(x - 7) = -18$

Objective ③ Solve problems involving quadratic equations.   Many problems and applications are solved using quadratic equations. See Section 2.5 for a review of key words and their translations.

Example 6   The product of two consecutive odd natural numbers is 195. Find the numbers.

Understand   Odd natural numbers are $1, 3, 5, \ldots$. Note that adding 2 to a given odd natural number gives a consecutive odd natural number, which suggests this pattern:

First odd natural number: x

Consecutive odd natural number: $x + 2$

The word *product* means that two numbers are multiplied to equal 195.

Plan   Translate to an equation and then solve.

Execute $x(x + 2) = 195$ The equation has the product of x and $x + 2$ set equal to 195.

$$x(x + 2) - 195 = 0 \qquad \text{Subtract 195 from both sides.}$$
$$x^2 + 2x - 195 = 0 \qquad \text{Multiply } x(x + 2) \text{ to get the form } ax^2 + bx + c = 0.$$
$$(x - 13)(x + 15) = 0 \qquad \text{Factor by finding a pair of numbers whose product is } -195 \text{ and sum is 2.}$$
$$x - 13 = 0 \quad \text{or} \quad x + 15 = 0 \qquad \text{Use the zero-factor theorem to solve.}$$
$$x = 13 \qquad\qquad x = -15$$

Answer   Because -15 is not a natural number and 13 is a natural number, the first number is 13. This means that the consecutive odd natural number is 15.

Check   13 and 15 are consecutive odd natural numbers, and their product is 195; so the answer is correct.

Connection   In Section 3.3, we studied problems that dealt with the sum or difference of consecutive integers. The pattern for consecutive odd natural numbers is the same as for consecutive odd integers because every natural number is an integer.

Your Turn 6   The product of two consecutive natural numbers is 462. Find the numbers.

Answers   Your Turn 5

a. $n = -\dfrac{2}{5}$ or $-\dfrac{4}{3}$

b. $x = 6$ or -8

c. $x = -2$ or $x = 5$

Answer   Your Turn 6

21 and 22

Example 7   An architect is designing an addition to a house. The addition will be in the shape of a rectangle, and the room's floor will have an area of 270 square feet. The width of the room is to be 3 feet less than the length. Find the dimensions of the floor.

Understand   The area of a rectangle is found by multiplying the length and width. We are given a relationship about the width and length, which we can translate as follows:

"The width is to be 3 feet less than the length."
Width = $l - 3$, where l represents length.

Plan   Translate to an equation and then solve.

Execute $l(l - 3) = 270$ The equation has the product of the length, l, and width, $l - 3$, set equal to 270.

$l(l - 3) - 270 = 0$ Subtract 270 from both sides.

$l^2 - 3l - 270 = 0$ Multiply $l(l - 3)$ to get standard form.

$(l - 18)(l + 15) = 0$ Factor.

$l - 18 = 0$ or $l + 15 = 0$ Use the zero-factor theorem to solve.

$l = 18$ $l = -15$

Answer   Because the problem involves room dimensions, only the positive number makes sense as an answer. If the length is 18 feet and the width is 3 feet less than that, the width is 15 feet.

Check   The width, 15 feet, is 3 feet less than the length, 18 feet. Also, the area of a 15-foot by 18-foot rectangle is 270 square feet.

Area = 113.04 cm^2

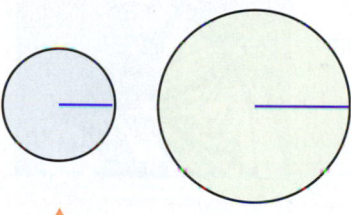

Note: This radius is 8 less than the other disc's radius.

Your Turn 7   The two discs in the margin are designed so that the smaller of the two has an area of 113.04 square centimeters and a radius that is 8 centimeters less than the larger disc's radius. Find the radius of the larger disc. (Use 3.14 to approximate π.)

Example 8   The equation $h = -16t^2 + h_0$ describes the height, h, in feet of an object after falling from an initial height h_0 for a time, t seconds. An object is dropped from a height of 600 feet. When the object reaches 344 feet, how much time has passed?

Understand   We are given a formula, and we are to find the time it takes an object to drop from an initial height of 600 feet to a height of 344 feet.

Plan   Replace the variables in the formula with the given values to get an equation in terms of t; then solve for t.

Execute $344 = -16t^2 + 600$ Replace h with 344 and h_0 with 600.

$-256 = -16t^2 + 0$ Subtract 600 from both sides.

$16t^2 - 256 = 0$ Add $16t^2$ to both sides.

$16(t^2 - 16) = 0$ Factor out the monomial GCF, 16.

$t^2 - 16 = 0$ Divide out the constant factor, 16.

$(t - 4)(t + 4) = 0$ Factor.

$t - 4 = 0$ or $t + 4 = 0$ Use the zero-factor theorem to solve.

$t = 4$ $t = -4$

Answer   Because the problem involves the amount of time elapsed after an object was dropped, only the positive value, 4, makes sense. This means that it took the object 4 seconds to fall from 600 feet to 344 feet.

Check   Verify that after 4 seconds, an object dropped from an initial height of 600 feet will descend to a height of 344 feet.

$$h = -16(4)^2 + 600$$
$$h = -256 + 600$$
$$h = 344$$

Answer   Your Turn 7

14 cm

Your Turn 8 An object is dropped from a height of 225 feet above the ground. When the object is 81 feet from the ground, how much time has passed?

Objective 4 Use the Pythagorean theorem to solve problems. One of the most popular theorems in mathematics is the Pythagorean theorem, named after the Greek mathematician Pythagoras. The theorem relates the side lengths of all right triangles. Recall that in a right triangle, the two sides that form the 90° angle are *legs* and the side directly across from the 90° angle is the *hypotenuse*.

Rule **Pythagorean Theorem**

Given a right triangle, where a and b represent the lengths of the legs and c represents the length of the hypotenuse, $a^2 + b^2 = c^2$. Further, if a, b, and c represent the lengths of the sides of a triangle and $a^2 + b^2 = c^2$, then the triangle is a right triangle with hypotenuse c.

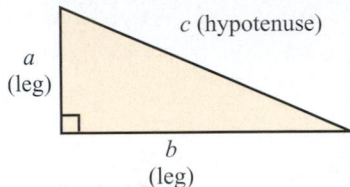

Of Interest

Pythagoras was a Greek mathematician who is believed to have been born around 569 B.C. and died in 500 B.C. The Pythagorean theorem is named after Pythagoras not because he discovered the relationship, as commonly thought, but because he and his followers are the first to have proved that the relationship is true for all right triangles. The relationship was known and used by other cultures prior to the Greeks; however, Pythagoras and others in the Greek culture believed that it was important to prove mathematical relationships.

We can use the Pythagorean theorem to find a missing length in a right triangle if we know the other two lengths.

Example 9 Find the length of the missing side.

SOLUTION Use the Pythagorean theorem, $a^2 + b^2 = c^2$. The hypotenuse, c, is the missing side length.

$$3^2 + 4^2 = c^2 \qquad \text{Substitute } a = 3 \text{ and } b = 4.$$
$$9 + 16 = c^2 \qquad \text{Simplify exponential forms.}$$
$$25 = c^2 \qquad \text{Add.}$$
$$0 = c^2 - 25 \qquad \text{Subtract 25 from both sides to get standard form.}$$
$$0 = (c - 5)(c + 5) \qquad \text{Factor.}$$
$$c - 5 = 0 \quad \text{or} \quad c + 5 = 0 \qquad \text{Use the zero-factor theorem.}$$
$$c = 5 \qquad\qquad c = -5$$

Because we are dealing with lengths, only the positive solution is sensible; so the missing length is 5.

Your Turn 9 In the construction of a roof, three beams are used to form a right triangle frame. A 12-foot beam and a 5-foot beam are brought together to form a 90° angle. How long must the third beam be?

5 ft.

?

12 ft.

6.6 Exercises   For Extra Help   *MyMathLab*

PRACTICE   WATCH   DOWNLOAD

Note: Exercises marked with a ★ represent challenging exercises.

1. Explain the zero-factor theorem in your own words.

2. How do you use the zero-factor theorem to solve an equation that has an expression in factored form set equal to 0?

3. What is the first step in solving a quadratic equation?

4. Explain why we cannot write $(x + 2)(x - 2) = 32$ as $x + 2 = 32$ and $x - 2 = 32$.

For Exercises 5–16, solve using the zero-factor theorem. See Examples 1 and 2.

5. $(x + 5)(x + 2) = 0$

6. $(x - 7)(x + 3) = 0$

7. $(a + 3)(a - 4) = 0$

8. $(b - 7)(b + 2) = 0$

9. $(3x - 4)(2x + 3) = 0$

10. $(3a + 2)(2a - 5) = 0$

11. $x(x - 7) = 0$

12. $y(y + 3) = 0$

13. $m(m - 1)(m + 2) = 0$

14. $r(r - 2)(r + 3) = 0$

15. $b(b - 2)^2 = 0$

16. $m(m - 4)^2 = 0$

For 17–56, solve the quadratic equations. See Examples 3–5.

17. $y^2 - 4y = 0$

18. $p^2 - 3p = 0$

19. $6r^2 + 10r = 0$

20. $12c^2 + 9c = 0$

21. $x^2 - 4 = 0$

22. $y^2 - 25 = 0$

23. $x^2 + x - 6 = 0$

24. $b^2 - 2b - 8 = 0$

25. $n^2 - 6n + 9 = 0$

26. $p^2 - 16p + 64 = 0$

27. $3a^2 - 11a - 4 = 0$

28. $2d^2 + 7d - 4 = 0$

29. $6r^2 + 11r - 10 = 0$

30. $8c^2 - 6c - 9 = 0$

31. $x^2 - 4x = 21$

32. $u^2 + 6u = 27$

33. $p^2 = 3p - 2$

34. $x^2 = 6x - 8$

35. $v^2 = 9$

36. $m^2 = 16$

37. $2a^2 + 18a + 28 = 0$

38. $3x^2 - 6x - 45 = 0$

39. $x^3 - x^2 - 12x = 0$

40. $x^3 + 7x^2 + 10x = 0$

41. $12r^3 + 22r^2 - 20r = 0$

42. $16c^3 - 12c^2 - 18c = 0$

43. $3y^2 + 8 = 14y$

44. $6k^2 + 1 = 7k$

45. $4x^2 = 60 - x$ **46.** $4n^2 = 5 - 19n$ **47.** $7t = 12 + t^2$ **48.** $-9x = -20 - x^2$

49. $b(b - 5) = 14$ **50.** $m(m + 6) = -9$ **51.** $4x(x + 7) = -49$ **52.** $c(2c - 11) = -5$

53. $(x + 1)(x + 5) = -3$ **54.** $(a + 3)(a + 6) = -2$ **55.** $(a - 1)(a + 4) = 14$ **56.** $(b + 4)(b - 2) = 16$

For Exercises 57–72, translate to an equation and then solve. See Examples 6 and 7.

57. Find every number such that the square of the number added to 55 is the same as sixteen times the number.

58. Find every number such that triple the square of the number is equal to four times that number.

59. The product of two consecutive odd natural numbers is 143. Find the numbers.

60. The product of two consecutive natural numbers is 306. Find the numbers.

★**61.** The sum of the squares of two consecutive natural numbers is 365. Find the numbers.

★**62.** The difference of the squares of two consecutive even natural numbers is 60. Find the numbers.

63. The length of a rectangular garden is 9 meters more than the width. If the area is 252 square meters, find the dimensions of the garden.

64. The floor of the central chamber of the Lincoln Memorial has an area of 4440 square feet. If the length of the chamber is 14 feet more than the width, find the dimensions of the chamber floor. (*Source:* National Parks Service)

65. The largest billboard in Times Square's history has a length that is 10 feet more than three times its width. The area of the billboard is 9288 square feet. Find the dimensions of the billboard.

66. The length of a football field is 40 yards less than three times the width. The total area of the football field is 6400 square yards. Find the dimensions of the football field.

67. The design of the base of a small building calls for a rectangular shape with dimensions of 22 feet by 28 feet. The architect decides to change the shape of the building to a circle, but wants the base to have the same area. If we use $\frac{22}{7}$ to approximate π, what is the radius of the circular building?

68. A rectangle has a length of 14 centimeters and a width of 11 centimeters. Using $\frac{22}{7}$ as an approximation for π, find the radius of a circle with the same area as the rectangle.

69. A design on the front of a marketing brochure calls for a triangle with a base that is 6 centimeters less than the height. If the area of the triangle is to be 216 square centimeters, what are the lengths of the base and height?

70. The base of a triangular sign is 4 inches more than the height. If the area of the sign is 160 square inches, find the base and height.

71. A steel plate in the shape of a trapezoid has an area of 85.5 square inches. The dimensions are shown. Note that the length of the base is equal to the height. Calculate the height.

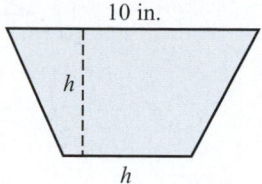

10 in.

h

h

72. The front elevation of one wing of a house is shown. Because of budget constraints, the total area of the front of this wing must be 352 square feet. The height of the triangular portion is 14 feet less than the base. Find the base length.

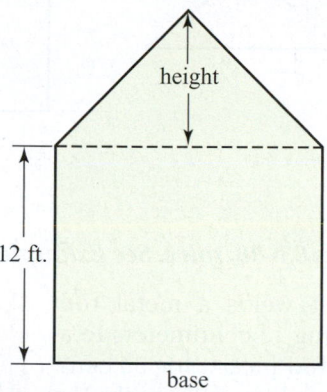

height

12 ft.

base

73. Find three consecutive even integers such that the product of the largest and second largest integers is 168.

74. Find three consecutive odd integers such that the product of the largest and second largest integers is 255.

For Exercises 75 and 76, use the formula $h = -16t^2 + v_0t + h_0$, where h is the final height in feet, t is the time of travel in seconds, v_0 is the initial velocity in feet per second, and h_0 is the initial height in feet of an object traveling upward. See Example 8.

★ **75.** A ball is thrown upward at 4 feet per second from a building 29 feet high. When will the ball be 9 feet above the ground?

★ **76.** A toy rocket is fired from the ground with an upward velocity of 200 feet per second. How many seconds will the rocket take to return to the ground?

For Exercises 77 and 78, use the formula $B = P\left(1 + \frac{r}{n}\right)^{nt}$, which is used to calculate the final balance of an investment or a loan after being compounded. Following is a list of what each variable represents: See Example 8.

    B represents the final balance.

    P represents the principal, which is the amount invested.

    r represents the annual interest rate as a decimal.

    t represents the number of years the principal is compounded.

    n represents the number of times per year the principal is compounded.

77. Carlita invests $4000 in an account that is compounded annually. If after two years her balance is $4840, what was the interest rate of the account?

78. Donovan invests $1000 in an account that is compounded semiannually (every six months). If after one year his balance is $1210, what is the interest rate of the account?

For Exercises 79–82, find the length of the hypotenuse. See Example 9.

79.

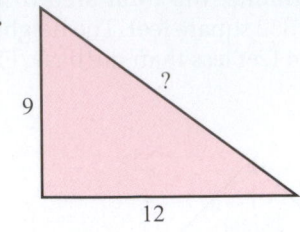

9   ?   12

80.

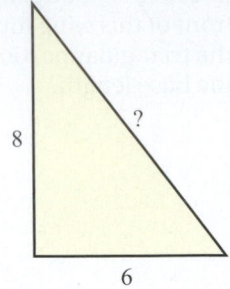

8   ?   6

81.

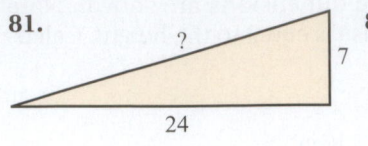

?   7   24

82.

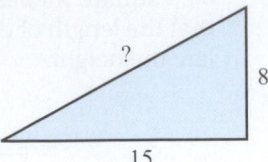

?   8   15

For Exercises 83–88, solve. See Example 9.

83. An artist welds a metal rod measuring 12 centimeters to a second rod measuring 35 centimeters to form a 90° angle. He wants a third rod to be welded to the ends of the first two rods to form a right triangle. What length must the third rod be?

84. A rectangular screen measures 27 inches by 36 inches. Find the diagonal distance between two opposing corners.

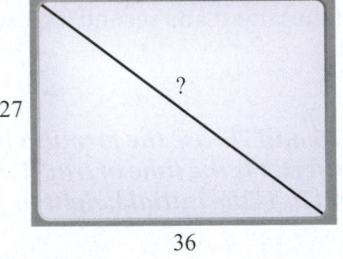

27   ?   36

85. A person drives south 18 blocks then turns east and drives 24 blocks. Assuming that all city blocks are equal size and form a grid of streets that intersect at 90°, find the shortest distance in blocks that the person is from her original location.

86. To support a power pole, a wire is to be attached 24 feet from the ground, then staked 10 feet from the base of the pole. How long must the wire be?

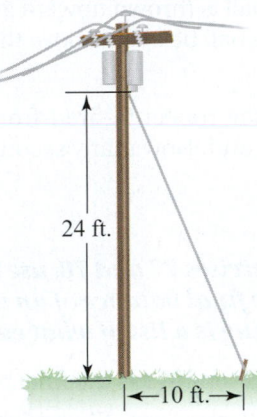

24 ft.   ←10 ft.→

★ **87.** When hurricanes threaten, a common practice is to tape windows to keep the glass from shattering. If a rectangular window is 24 inches wide and 51 inches diagonally, how long is the window?

★ **88.** A sign is in the shape of a rectangle with a stripe painted diagonally across the sign. If the length of the stripe is 26 inches and the sign is 24 inches long, how wide is the sign?

Review Exercises

Exercises 1–6   **EQUATIONS AND INEQUALITIES**

[4.2]   **1.** Is the equation $y = x^2 - 1$ a linear equation? Explain.

[4.2]   **2.** Is $(1, -4)$ a solution for $4x + 3y = -8$?

[4.3]   **3.** Find the x- and y-intercepts for $2x - 5y = -20$.

[4.5]   **4.** Write the equation of the line passing through $(-1, 4)$ and $(-3, -2)$ in the form $Ax + By = C$, where A, B, and C are integers and $A > 0$.

[4.7]   **5.** For the function $f(x) = x^2 + 2$, find each of the following.
      a. $f(0)$ **b.** $f(-2)$

[4.7]   **6.** Graph: $f(x) = 2x - 3$

6.7 Graphs of Quadratic Equations and Functions

Objectives

① Graph quadratic equations in the form $y = ax^2 + bx + c$.

② Graph quadratic functions.

In Chapter 4, we graphed linear equations in two variables. Now we will graph **quadratic equations in two variables**.

Definition   *Quadratic equation in two variables:* An equation that can be written in the form $y = ax^2 + bx + c$, where a, b, and c are real numbers and $a \neq 0$.

Following are some examples of quadratic equations in two variables.

$$y = x^2 \qquad y = 3x^2 + 2 \qquad y = x^2 - 4x + 1$$

We learned in Section 2.2 that every term in a linear equation has a degree of 1 or 0. Because quadratic equations always have a term with a degree of 2, they are non-linear, which means that their graphs will not be straight lines. What might their graphs look like?

Objective ①   **Graph quadratic equations in the form $y = ax^2 + bx + c$.** Recall that one way to graph an equation is to plot solutions to the equation in the coordinate plane. In Section 4.1, we said that a solution to an equation in two variables is an ordered pair (x, y) that satisfies the equation. For example, $(3, 9)$ is a solution to $y = x^2$ because $9 = (3)^2$ is true. Let's make a table of solutions for $y = x^2$, plot those solutions, and connect the points to see what the graph looks like.

x	y
-3	9
-2	4
-1	1
0	0
1	1
2	4
3	9

Note: Because we know that quadratic equations are nonlinear, we find many solutions to get a good sense of the shape of the graph.

The graph of $y = x^2$ is a *parabola*. In fact, the graph of every quadratic equation in two variables is a parabola.

In the table of solutions for $y = x^2$, notice that when $x = 1$ or -1, $y = 1$; when $x = 2$ or -2, $y = 4$; and when $x = 3$ or -3, $y = 9$. In general, each x-coordinate and its additive inverse have the same y-coordinate. Looking at the graph, we see that the left side of the y-axis is the mirror image of the right side of the y-axis. Consequently, we say that this graph is symmetrical about the line $x = 0$ (the y-axis). Every parabola will have symmetry about a line called its **axis of symmetry**. Further, a parabola's axis of symmetry always passes through a point called the **vertex**.

Definitions   *Axis of symmetry:* A line that divides a graph into two symmetrical halves.

Vertex of a parabola: The lowest point on a parabola that opens up or the highest point on a parabola that opens down.

For $y = x^2$, the axis of symmetry is the line $x = 0$ (y-axis) and the vertex is $(0, 0)$. In Section 11.4, we will learn more about the vertex and axis of symmetry. For now, we graph by finding enough solutions to see where the parabola turns.

The axis of symmetry divides the graph into two halves that are mirror images of each other.

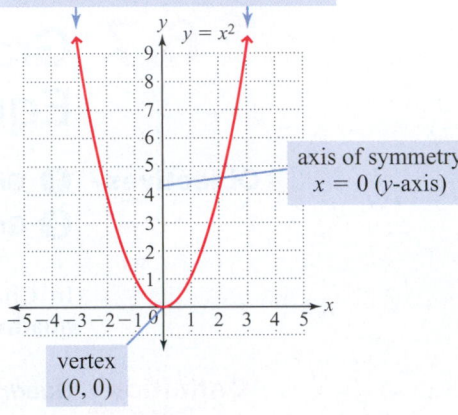

axis of symmetry $x = 0$ (y-axis)

vertex $(0, 0)$

Procedure   **Graphing Quadratic Equations**

To graph a quadratic equation,

1. Find ordered pair solutions and plot them in the coordinate plane. Continue finding and plotting solutions until the shape of the parabola can be clearly seen.
2. Connect the points to form a parabola.

Example 1   Graph.

a. $y = 3x^2 + 2$

SOLUTION   We complete a table of solutions, plot the solutions in the coordinate plane, and connect the points to form the graph.

Connection In a quadratic equation in the form $y = ax^2 + bx + c$, the constant c indicates the y-intercept just as the constant b does in a linear equation in the form $y = mx + b$.

Note: The constant 2 in $y = 3x^2 + 2$ indicates the y-intercept because replacing x with 0 leaves us with $y = 3(0)^2 + 2 = 2$.

x	y
-2	14
-1	5
0	2
1	5
2	14

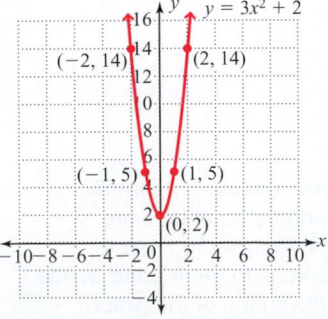

b. $y = x^2 + 2x - 8$

SOLUTION

Connection To find the x-intercepts of a graph, let $y = 0$. Thus, the solutions of $0 = x^2 + 2x - 8$ are the x-intercepts of the graph of $y = x^2 + 2x - 8$, which are $(-4, 0)$ and $(2, 0)$.

x	y
-4	0
-3	-5
-2	-8
-1	-9
0	-8
1	-5
2	0

Notice that in each equation we have graphed so far ($y = x^2$, $y = 3x^2 + 2$, and $y = x^2 + 2x - 8$), the coefficient of x^2 is a positive number. Also notice that the graph of each of those equations was a parabola that opens upward. Now let's consider quadratic equations in which the coefficient of x^2 is a negative number, as in $y = -2x^2 + 6x + 3$. We will see that the graphs of these equations open downward.

Example 2 Graph $y = -2x^2 + 6x + 3$.

SOLUTION Complete a table of solutions, plot the solutions, and connect the points to form a parabola.

x	y
-1	-5
0	3
1	7
2	7
3	3
4	-5

Note: The coefficient of x^2 is a negative number, -2, and the graph opens downward. In an equation of the form $y = ax^2 + bx + c$, if a is negative, the parabola opens downward.

Note: Because $x = 1$ and $x = 2$ have the same y-values, the x-coordinate of the vertex will be halfway between 1 and 2 at 1.5.

Rule   **Opening of a Parabola**

Given an equation in the form $y = ax^2 + bx + c$, if $a > 0$, then the parabola opens upward; if $a < 0$, then the parabola opens downward.

Your Turn 2 Graph.

a. $y = x^2 - 5$ **b.** $y = 2x^2 - x + 1$ **c.** $y = -3x^2 + 6x + 2$

Objective ② Graph quadratic functions. In Section 4.7, we learned how to graph linear functions. Now let's graph quadratic functions. A quadratic *equation* in two variables, $y = ax^2 + bx + c$, is a quadratic *function* that can be written in the form $f(x) = ax^2 + bx + c$. Notice that to change the equation in two variables to function notation, we replace y with $f(x)$.

Quadratic equation in two variables: $y = -2x^2 + 6x + 3$

Using function notation: $f(x) = -2x^2 + 6x + 3$

The graph of $f(x) = -2x^2 + 6x + 3$ is the same as $y = -2x^2 + 6x + 3$, which we graphed in Example 2. Recall from Section 4.7 that with function notation, we find ordered pairs by evaluating the function using various values of x. For example, the notation $f(1)$ means to find the value of the function when $x = 1$.

$$f(1) = -2(1)^2 + 6(1) + 3$$
$$= -2 + 6 + 3$$
$$= 7 \qquad \text{The ordered pair is } (1, 7).$$

Connection Finding $f(1)$ for $f(x) = -2x^2 + 6x + 3$ is the same as finding the y-value for $y = -2x^2 + 6x + 3$ when $x = 1$.

We would continue finding ordered pairs in this manner to produce the same table and graph that we produced in Example 2, which suggests the following procedure.

Procedure   Graphing Quadratic Functions

To graph a quadratic function,

1. Find enough ordered pairs by evaluating the function for various values of x so that when those ordered pairs are plotted, the shape of the parabola can be clearly seen.
2. Connect the points to form the parabola.

Answers   Your Turn 2

a.
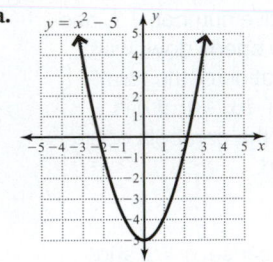

b.
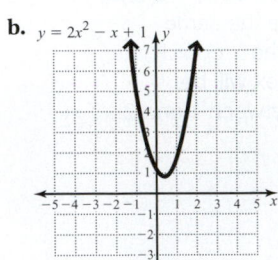

c.
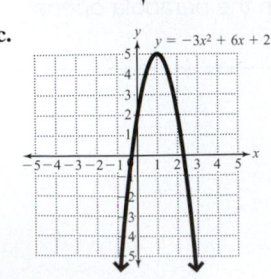

Example 3   Graph.

a. $f(x) = 3x^2 - 5$

SOLUTION   Find enough ordered pairs to clearly see the graph; then connect the points to form the parabola.

x	y
-2	7
-1	-2
0	-5
1	-2
2	7

Connection This parabola opens upward. Given a quadratic function in the form $f(x) = ax^2 + bx + c$, if $a > 0$, the parabola opens upward.

b. $f(x) = -x^2 + 2x + 4$

SOLUTION

x	y
-2	-4
-1	1
0	4
1	5
2	4
3	1
4	-4

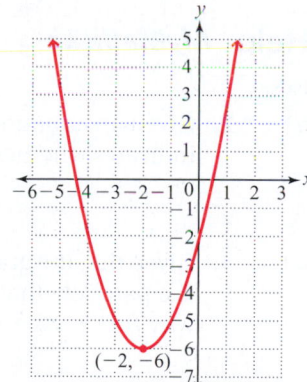

Connection This parabola opens downward. Given a quadratic function in the form $f(x) = ax^2 + bx + c$, if $a < 0$, the parabola opens downward.

Your Turn 3 Graph.

a. $f(x) = x^2 - 4$

b. $f(x) = -2x^2 - 8x - 3$

In Section 4.7, we learned that a graph represents a function if it passes the vertical line test. (A vertical line through any point in the domain touches the graph at only one point.) Because graphs of quadratic functions are parabolas that open up or down, they pass the vertical line test. There are also parabolas that open right or left, which do not pass the vertical line test, and therefore are not functions.

Example 4 Determine whether the graph is the graph of a function. Give the domain and range.

a.

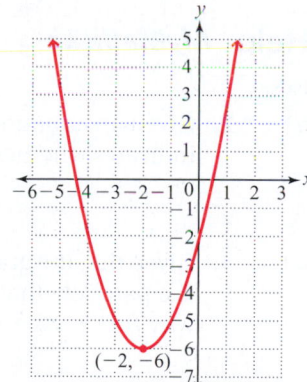

b.

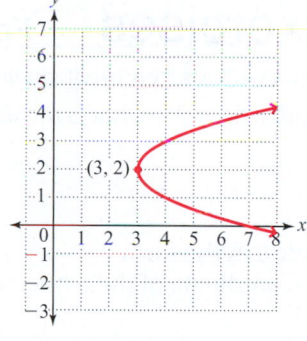

Answers Your Turn 3

a. $f(x) = x^2 - 4$

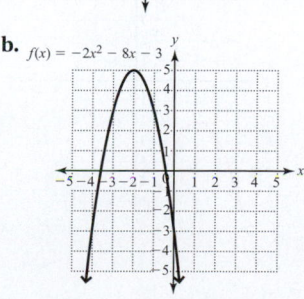

b. $f(x) = -2x^2 - 8x - 3$

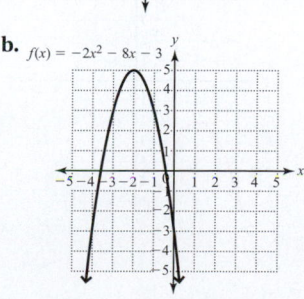

SOLUTION a. This is a function because any vertical line intersects the graph in at most one point as illustrated with $x = -4$.

Domain: all real numbers or $(-\infty, \infty)$

Range: $\{y | y \geq -6\}$ or $[-6, \infty)$

SOLUTION b. This is not a function because a vertical line can be drawn that intersects the graph in more than one point as illustrated by $x = 5$.

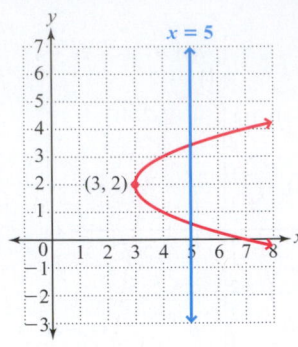

Domain: $\{x \mid x \geq 3\}$ or $[3, \infty)$

Range: all real numbers or $(-\infty, \infty)$

Your Turn 4 Determine whether the graph is the graph of a function. Give the domain and range.

a.

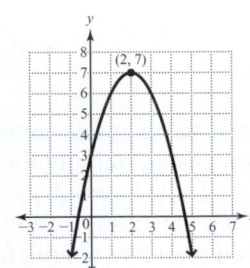

b.

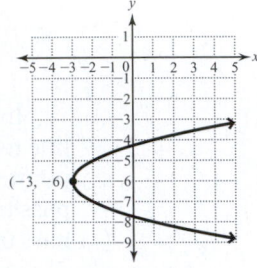

Answers **Your Turn 4**

a. is a function; domain: all real numbers or $(-\infty, \infty)$; range: $\{y \mid y \leq 7\}$ or $(-\infty, 7]$

b. not a function; domain: $\{x \mid x \geq -3\}$ or $[-3, \infty)$; range: all real numbers or $(-\infty, \infty)$

6.7 Exercises   For Extra Help   *MyMathLab*

Note: Exercises marked with a ★ represent challenging exercises.

1. In your own words, describe what is meant by *symmetry* in a parabola.

2. Given an equation in the form $y = ax^2 + bx + c$, what indicates whether the graph opens up or down?

3. Given an equation in the form $y = ax^2 + bx + c$, what indicates the y-coordinate of the y-intercept?

4. Given that the graph of $y = ax^2 + bx + c$, where $a \neq 0$, is a parabola that opens upward or downward, explain why $f(x) = ax^2 + bx + c$ is a function.

For Exercises 5–8, use the graph of the parabola to determine the coordinates of the vertex. See Objective 1.

5.

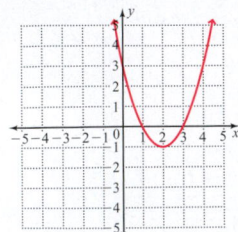

6.

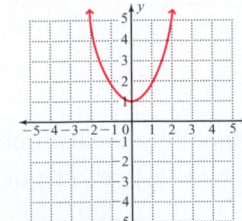

7.

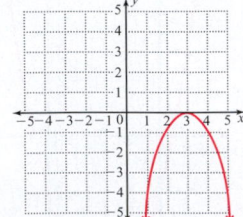

8.

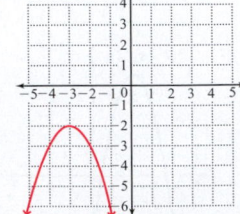

For Exercises 9–20, complete the table of solutions and then graph.
See Examples 1 and 2.

9. $y = 2x^2$

x	y
-2	
-1	
0	
1	
2	

10. $y = 3x^2$

x	y
-2	
-1	
0	
1	
2	

11. $y = -x^2$

x	y
-2	
-1	
0	
1	
2	

12. $y = -x^2 + 2$

x	y
-2	
-1	
0	
1	
2	

13. $y = x^2 - 3$

x	y
-2	
-1	
0	
1	
2	

14. $y = x^2 + 3$

x	y
-2	
-1	
0	
1	
2	

15. $y = -x^2 + 2x$

x	y
-1	
0	
1	
2	
3	

16. $y = -x^2 + 4x$

x	y
0	
1	
2	
3	
4	

17. $y = 2x^2 - 4x + 1$

x	y
-1	
0	
1	
2	
3	

18. $y = 2x^2 + 4x - 2$

x	y
-3	
-2	
-1	
0	
1	

19. $y = -3x^2 + 6x + 4$

x	y
-1	
0	
1	
2	
3	

20. $y = -2x^2 + 4x + 2$

x	y
-1	
0	
1	
2	
3	

For Exercises 21–36, graph. See Examples 1 and 2.

21. $y = x^2 + 1$

22. $y = x^2 - 4$

23. $y = -x^2 + 3$

24. $y = -x^2 + 2$

25. $y = 2x^2 - 5$ **26.** $y = 3x^2 - 4$ **27.** $y = -3x^2 + 2$ **28.** $y = -4x^2 + 5$

29. $y = x^2 + 4x - 1$ **30.** $y = x^2 - 6x + 3$ **31.** $y = -x^2 + 6x - 5$ **32.** $y = -x^2 + 4x - 5$

33. $y = 2x^2 - 4x - 3$ **34.** $y = 3x^2 + 6x - 1$ **35.** $y = -2x^2 + 8x - 5$ **36.** $y = -4x^2 - 12x - 3$

For Exercises 37–48, graph the function. See Example 3.

37. $f(x) = 3x^2$ **38.** $f(x) = 2x^2$ **39.** $f(x) = -x^2 + 1$ **40.** $f(x) = -x^2 - 1$

41. $f(x) = x^2 + 4x - 1$ **42.** $f(x) = x^2 - 6x + 5$ **43.** $f(x) = -x^2 - 4x - 3$ **44.** $f(x) = -x^2 + 6x - 2$

45. $f(x) = 3x^2 - 12x + 5$ **46.** $f(x) = 2x^2 + 6x - 1$ **47.** $f(x) = -2x^2 - 4x - 1$ **48.** $f(x) = -4x^2 + 8x - 3$

For Exercises 49–56, state whether the graph is the graph of a function. Give the
domain and range. See Example 4.

49.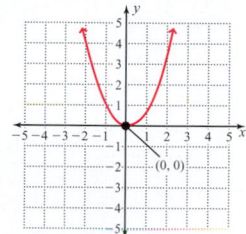

50.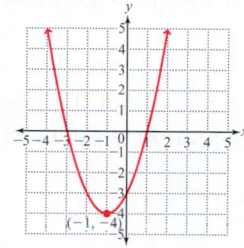

51.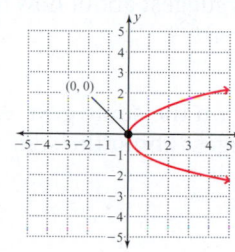

52.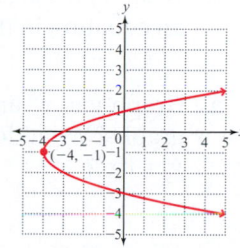

53.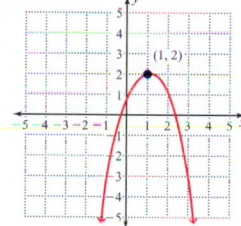

54.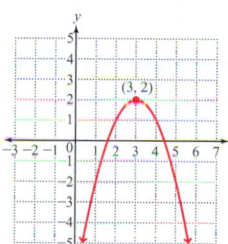

55.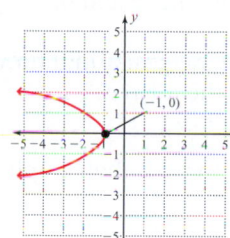

56.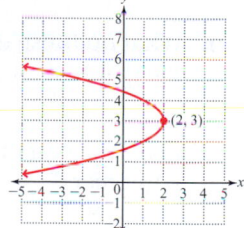

For Exercises 57–60, use a graphing calculator.

57. a. Enter the equation $y = x^2$; then graph. What are the coordinates of the
y-intercept?

b. Enter a second equation, $y = x^2 + 2$; then graph on the same grid. What are
the coordinates of the y-intercept?

c. On the same grid, graph $y = x^2 + 4$ and compare all three graphs. What do
your explorations suggest about how c affects the graph of an equation in the
form $y = x^2 + c$?

d. On the same grid, graph $y = x^2 - 3$. Comparing this graph to the first three
graphs, how did changing c to a negative value affect the graph?

58. a. Enter the equation $y = x^2$; then graph. What are the coordinates of the vertex?

b. Enter a second equation, $y = 2x^2$. Compare the graph of $y = 2x^2$ with the graph of $y = x^2$; then discuss their similarities and differences.

c. On the same grid, graph $y = 3x^2$ and compare all three graphs. What does this exploration suggest about how the size of a in $y = ax^2$ affects the graph?

d. In $y = ax^2$, if $a = \dfrac{1}{2}$, how will the graph be different from the graphs in parts a–c? Graph $y = \dfrac{1}{2}x^2$.

e. What do these explorations suggest about how the size of a affects the graph of an equation in the form $y = ax^2$?

59. a. Graph $y = x^2$ and $y = -x^2$ on the same grid. What is different about the two graphs? What does this suggest about the sign of a in an equation of the form $y = ax^2$?

b. Clearing the equations from part a, graph $y = 2x^2$ and $y = -2x^2$ together. Do these graphs confirm your conclusion from part a?

c. On a new grid, graph $y = \dfrac{1}{2}x^2$ and $y = -\dfrac{1}{2}x^2$. What is different about these graphs compared to those in parts a and b?

d. What do these explorations suggest about how the absolute value of a affects the graph of an equation in the form $y = ax^2$?

60. a. Enter the equation $y = x^2 + x + 1$; then graph. What are the coordinates of the y-intercept?

b. On the same grid, graph $y = x^2 + 2x + 1$. Compare the graph of $y = x^2 + x + 1$ with the graph of $y = x^2 + 2x + 1$; then discuss their similarities and differences.

c. Using the same grid, graph $y = x^2 + 3x + 1$. Compare this graph with the graphs of $y = x^2 + x + 1$ and $y = x^2 + 2x + 1$; then discuss similarities and differences.

d. On the same grid, graph $y = x^2 - x + 1$, $y = x^2 - 2x + 1$, and $y = x^2 - 3x + 1$. Compare these graphs with your graphs from parts a–c. What do these explorations suggest about how b affects the graph of an equation in the form $y = ax^2 + bx + c$?

Collaborative Exercises What Goes Up Must Come Down

Many sets of variables have a relationship that is roughly modeled by a parabolic shape. The following data give the high temperature (in degrees Fahrenheit, °F) and peak power load (in megawatts, MW) for 31 days throughout the calendar year.

1. On a sheet of graph paper, draw and label axes with the temperature along the horizontal axis and the peak load on the vertical axis. Then plot the points.

2. Draw a smooth parabola through the points that might best fit the data.

3. At what temperature does the minimum peak load seem to occur?

4. In your own words, describe what happens to the peak load as the temperature increases and decreases and explain why.

5. Use the model that you drew in part 2 to estimate the peak load on a day when the maximum temperature is 85.

6. Give another example of a situation that might be best modeled by a parabola.

Temp (°F)	Peak Load (MW)	Temp (°F)	Peak Load (MW)
69	94	95	141
71	95	46	142
59	100	43	147
65	100	98	150
67	102	96	151
60	104	97	155
74	105	100	158
79	105	43	165
56	112	38	173
84	113	102	176
52	114	36	193
88	115	103	199
50	120	105	204
90	123	106	226
92	132	34	228
92	132		

Review Exercises

Exercises 1–4 **EXPRESSIONS**

For Exercises 1 and 2, multiply.

[5.4] **1.** $3mn(m^2 - 4mn + 5)$

[5.5] **2.** $(x - 3)(2x + 1)$

For Exercises 3 and 4, factor.

[6.3] **3.** $3n^2 - 17n + 20$

[6.4] **4.** $x^2 - 25$

Exercises 5 and 6 **EQUATIONS AND INEQUALITIES**

[6.6] *For Exercises 5 and 6, solve.*

5. $h(h - 3) = 0$

6. $y^2 + 3y - 18 = 0$

Chapter 6 Summary

Defined Terms

Section 6.1
Factored form (p. 416)
Greatest common factor (GCF) (p. 417)

Section 6.6
Quadratic equation in one variable (p. 456)

Section 6.7
Quadratic equation in two variables (p. 465)

Axis of symmetry (p. 466)
Vertex of a parabola (p. 466)

Procedures, Rules, and Key Examples

Procedures/Rules	Key Examples

SECTION 6.1 Greatest Common Factor and Factoring by Grouping

Procedures/Rules	Key Examples
To find the GCF of a set of numbers by listing, 1. List all possible factors for each given number. 2. Search the lists for the largest factor common to all lists.	**Example 1:** Find the GCF of 54 and 180 by listing. Factors of 54: 1, 2, 3, 6, 9, 18, 27, 54 Factors of 180: 1, 2, 3, 4, 5, 6, 9, 10, 12, 15, 18, 20, 30, 36, 45, 60, 90, 180 GCF = 18
To find the GCF of a given set of numbers by prime factorization, 1. Write the prime factorization of each given number in exponential form. 2. Create a factorization for the GCF that includes only those prime factors common to all factorizations, each raised to its smallest exponent in the factorizations. 3. Multiply the factors in the factorization created in step 2. *Note:* If there are no common prime factors, the GCF is 1.	**Example 2:** Find the GCF of 54 and 180 by prime factorization. $$54 = 2 \cdot 3^3$$ $$180 = 2^2 \cdot 3^2 \cdot 5$$ $$GCF = 2 \cdot 3^2 = 18$$
To factor a monomial GCF out of a given polynomial, 1. Find the GCF of the terms in the polynomial. 2. Rewrite the given polynomial as a product of the GCF and the quotient of the polynomial and the GCF. $$\text{Polynomial} = \text{GCF}\left(\frac{\text{Polynomial}}{\text{GCF}}\right)$$	**Example 3:** Factor $54x^3y - 180x^2yz$. $54x^3y - 180x^2yz$ $$= 18x^2y\left(\frac{54x^3y - 180x^2yz}{18x^2y}\right)$$ $$= 18x^2y\left(\frac{54x^3y}{18x^2y} - \frac{180x^2yz}{18x^2y}\right)$$ $$= 18x^2y(3x - 10z)$$
To factor a four-term polynomial by grouping, 1. Factor out any monomial GCF (other than 1) that is common to all four terms. 2. Group pairs of terms and factor the GCF out of each pair or group. 3. If there is a common binomial factor, factor it out. 4. If there is no common binomial factor, interchange the middle two terms and repeat the process. If there is still no common binomial factor, the polynomial cannot be factored by grouping.	**Example 4:** Factor by grouping: $$8x^2 - 20x - 6xy + 15y$$ $8x^2 - 20x - 6xy + 15y$ $$= (8x^2 - 20x) + (-6xy + 15y)$$ $$= 4x(2x - 5) - 3y(2x - 5)$$ $$= (2x - 5)(4x - 3y)$$

(continued)

Procedures/Rules	Key Examples

SECTION 6.2 Factoring Trinomials of the Form $x^2 + bx + c$

To factor a trinomial of the form $x^2 + bx + c$,
1. Find two numbers with a product equal to c and a sum equal to b.
2. The factored trinomial will have the form $(x + \text{first number})(x + \text{second number})$.

Note: The signs in the binomial factors can be minus signs depending on the signs of b and c.

Example 1: Factor $x^2 + 9x + 20$.
Find two numbers whose product is 20 and whose sum is 9.

Product:	Sum:	
$(1)(20) = 20$	$1 + 20 = 21$	
$(2)(10) = 20$	$2 + 10 = 12$	
$(4)(5) = 20$	$4 + 5 = 9$	**Correct combination**

$$x^2 + 9x + 20 = (x + 4)(x + 5)$$

Example 2: Factor $x^2 + 2x - 15$.

Product:	Sum:	
$(-1)(15) = -15$	$-1 + 15 = 14$	
$(-3)(5) = -15$	$-3 + 5 = 2$	**Correct combination**

$$x^2 + 2x - 15 = (x - 3)(x + 5)$$

SECTION 6.3 Factoring Trinomials of the Form $ax^2 + bx + c$, where $a \neq 1$

To factor a trinomial of the form $ax^2 + bx + c$, where $a \neq 1$, by trial and error,
1. Look for a monomial GCF in all of the terms. If there is one, factor it out.
2. Write a pair of *first* terms whose product is ax^2.
3. Write a pair of *last* terms whose product is c.
4. Verify that the sum of the *inner* and *outer* products is bx (the middle term of the trinomial). If the sum of the inner and outer products is not bx, try the following:
   a. Exchange the last terms of the binomials from Step 3; then repeat Step 4.
   b. For each additional pair of last terms, repeat Steps 3 and 4.
   c. For each additional pair of first terms, repeat Steps 2–4.

Example 1: Factor $4x^2 - 9x + 5$.
Factors of 4 are 4 and 1, or 2 and 2.
Factors of 5 are 1 and 5.
Because the middle term $-9x$ is negative and the last term 5 is positive, we know that the second terms in each binomial factor will be negative.

$$(x - 1)(4x - 5) = 4x^2 - 9x + 5$$

To factor a trinomial of the form $ax^2 + bx + c$, where $a \neq 1$, by grouping,
1. Look for a monomial GCF in all of the terms. If there is one, factor it out.
2. Multiply a and c.
3. Find two factors of this product whose sum is b.
4. Write a four-term polynomial in which bx is written as the sum of two like terms whose coefficients are the two factors you found in Step 3.
5. Factor by grouping.

Example 2: Factor $18y^3 + 12y^2 - 48y$.
Factor out the monomial GCF, $6y$.

$$18y^3 + 12y^2 - 48y = 6y(3y^2 + 2y - 8)$$

To factor the trinomial, multiply $3(-8) = -24$; then find two factors of -24 whose sum is 2. Note that -4 and 6 work. Write the middle term, $2y$, as $-4y + 6y$; then factor by grouping.

$$\begin{aligned} &= 6y(3y^2 - 4y + 6y - 8) \\ &= 6y[y(3y - 4) + 2(3y - 4)] \\ &= 6y(3y - 4)(y + 2) \end{aligned}$$

(*continued*)

Procedures/Rules	Key Examples

SECTION 6.4 Factoring Special Products

Rules for factoring special products:

Perfect square trinomials: $a^2 + 2ab + b^2 = (a + b)^2$
$a^2 - 2ab + b^2 = (a - b)^2$

Difference of squares: $a^2 - b^2 = (a + b)(a - b)$

Note: A sum of squares, $a^2 + b^2$, cannot be factored.

Difference of cubes: $a^3 - b^3 = (a - b)(a^2 + ab + b^2)$

Sum of cubes: $a^3 + b^3 = (a + b)(a^2 - ab + b^2)$

Example 1: Factor.

a. $9x^2 + 30x + 25 = (3x + 5)^2$

b. $36m^2 - 60mn + 25n^2 = (6m - 5n)^2$

c. $4y^2 - 81 = (2y + 9)(2y - 9)$

d. $16p^2 + 1$ is prime.

e. $8x^3 - 27 = (2x - 3)(4x^2 + 6x + 9)$

f. $n^3 + 64 = (n + 4)(n^2 - 4n + 16)$

SECTION 6.5 Strategies for Factoring

To factor a polynomial, first factor out any monomial GCF; then consider the number of terms in the polynomial. If the polynomial has

I. **Four terms**, try to factor by grouping.

II. **Three terms**, determine whether the trinomial is a perfect square.
   A. If the trinomial is a perfect square, consider its form.
      1. If in form $a^2 + 2ab + b^2$, the factored form is $(a + b)^2$.
      2. If in form $a^2 - 2ab + b^2$, the factored form is $(a - b)^2$.
   B. If the trinomial is not a perfect square, consider its form.
      1. If in form $x^2 + bx + c$, find two factors of c whose sum is b and write the factored form as $(x + \text{first number})(x + \text{second number})$.
      2. If in form $ax^2 + bx + c$, where $a \neq 1$, use trial and error. Or find two factors of ac whose sum is b; write these factors as coefficients of two like terms that, when combined, equal bx; and factor by grouping.

III. **Two terms**, determine whether the binomial is a difference of squares, sum of cubes, or difference of cubes.
   A. If given a binomial that is a difference of squares $a^2 - b^2$, the factors are conjugates and the factored form is $(a + b)(a - b)$. Note that a sum of squares cannot be factored.
   B. If given a binomial that is a sum of cubes, $a^3 + b^3$, the factored form is $(a + b)(a^2 - ab + b^2)$.
   C. If given a binomial that is a difference of cubes, $a^3 - b^3$, the factored form is $(a - b)(a^2 + ab + b^2)$.

Note: Always check to see if any of the factors can be factored further.

The examples shown do not represent every possible type of problem. Each example contains a monomial GCF other than 1.

Example 1: Factoring by grouping:
$$8x^3 + 24x^2 - 2x^2y - 6xy$$
$$= 2x(4x^2 + 12x - xy - 3y)$$
$$= 2x[4x(x + 3) - y(x + 3)]$$
$$= 2x(x + 3)(4x - y)$$

Example 2: Factoring a perfect square trinomial:
$$75x^4 + 60x^3y + 12x^2y^2$$
$$= 3x^2(25x^2 + 20xy + 4y^2)$$
$$= 3x^2(5x + 2y)^2$$

Example 3: Factoring a difference of squares:
$$81y^5 - 16y = y(81y^4 - 16)$$
$$= y(9y^2 + 4)(9y^2 - 4)$$
$$= y(9y^2 + 4)(3y + 2)(3y - 2)$$

Example 4: Factoring a sum of cubes:
$$54n^4 + 16nm^3 = 2n(27n^3 + 8m^3)$$
$$= 2n(3n + 2m)(9n^2 - 6mn + 4m^2)$$

(continued)

Procedures/Rules	Key Examples

SECTION 6.6 Solving Quadratic Equations by Factoring

Zero-Factor Theorem

If a and b are real numbers and $ab = 0$, then $a = 0$ or $b = 0$.

To solve an equation in which two or more factors are equal to 0, use the zero-factor theorem.

1. Set each factor equal to zero.
2. Solve each of those equations.

To solve a quadratic equation using factoring,

1. Write the equation in standard form ($ax^2 + bx + c = 0$).
2. Write the variable expression in factored form.
3. Use the zero-factor theorem to solve.

Example 1: Solve $(2x - 5)(x + 4) = 0$.

$$2x - 5 = 0 \quad \text{or} \quad x + 4 = 0$$
$$2x = 5 \qquad\qquad x = -4$$
$$x = \frac{5}{2}$$

Example 2: Solve $2x^2 = 3 - x$.

$2x^2 = 3 - x$	
$2x^2 + x = 3$	Add x to both sides.
$2x^2 + x - 3 = 0$	Subtract 3 from both sides.
$(2x + 3)(x - 1) = 0$	Factor.
$2x + 3 = 0 \quad \text{or} \quad x - 1 = 0$	Use the zero-factor
$2x = -3 \qquad\qquad x = 1$	theorem.
$x = -\dfrac{3}{2}$	

Example 3: The length of a small rectangular building is to be 15 feet more than the width. The area of the base is to be 700 square feet. What must the dimensions of the base be?

Let w represent width. The length will be $w + 15$. Writing an equation for area, we have

$w(w + 15) = 700$	
$w^2 + 15w = 700$	Distribute w.
$w^2 + 15w - 700 = 0$	Subtract 700 from both sides.
$(w - 20)(w + 35) = 0$	Factor.
$w - 20 = 0 \quad \text{or} \quad w + 35 = 0$	Use the zero-factor theorem.
$w = 20 \qquad\qquad w = -35$	

Answer: The width is 20 ft., and the length is 35 ft.

Pythagorean Theorem

Given a right triangle, where a and b represent the lengths of the legs and c represents the length of the hypotenuse, $a^2 + b^2 = c^2$. Further, if a, b, and c, represent the lengths of the sides of a triangle and $a^2 + b^2 = c^2$, then the triangle is a right triangle with hypotenuse c.

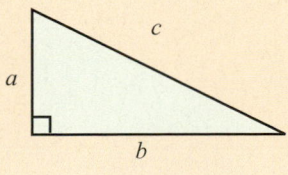

Example 4: Find the unknown length in the following right triangle.

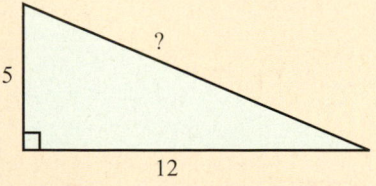

(continued)

Procedures/Rules	Key Examples

SECTION 6.6 Solving Quadratic Equations by Factoring (*continued*)

Solution: Use the Pythagorean theorem,
$a^2 + b^2 = c^2$.

$5^2 + 12^2 = c^2$	$a = 5, b = 12$
$25 + 144 = c^2$	Simplify exponential forms.
$169 = c^2$	Add.
$0 = c^2 - 169$	Subtract 169 from both sides.
$0 = (c - 13)(c + 13)$	Factor.
$c - 13 = 0, c + 13 = 0$	Use the zero-factor theorem.
$c = 13 \qquad c = -13$	

Answer: The unknown length is 13.

SECTION 6.7 Graphs of Quadratic Equations and Functions

To graph a quadratic equation,
1. Find ordered pair solutions and plot them in the coordinate plane. Continue finding and plotting solutions until the shape of the parabola can be clearly seen.
2. Connect the points to form a parabola.

Given an equation in the form $y = ax^2 + bx + c$, if $a > 0$, then the parabola opens upward; if $a < 0$, then the parabola opens downward.

Example 1: Graph $y = x^2 - 3$.

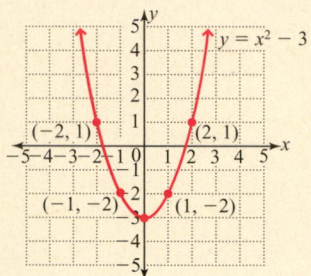

To graph a quadratic function,
1. Find enough ordered pairs by evaluating the function for various values of x so that when those ordered pairs are plotted, the shape of the parabola can be clearly seen.
2. Connect the points to form the parabola.

Example 2: Graph $f(x) = -x^2 + 4x - 1$.

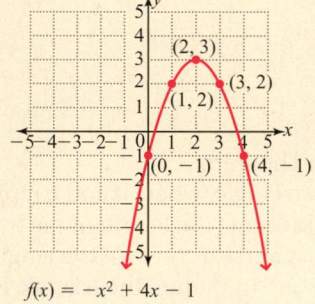

$$f(x) = -x^2 + 4x - 1$$

Chapter 6 Review Exercises

For Exercises 1–5, answer true or false.

[6.1] 1. The GCF of $24x^5$ and $36x^3$ is $6x^3$.

[6.1] 2. To factor a binomial, we use grouping.

[6.4] 3. $x^2 + 16 = (x + 4)(x + 4)$

[6.4] 4. $x^2 + 8x + 16 = (x + 4)^2$

[6.6] 5. If a and b are real numbers and $ab = 0$, then $a = 0$ or $b = 0$.

For Exercises 6–10, complete the rule.

[6.1] **6.** To factor a monomial out of a given polynomial,
   a. Find the _____ of the terms in the polynomial.
   b. Rewrite the given polynomial as a product of the GCF and the quotient of the _____ and the _____.

[6.2] **7.** Procedure: To factor a trinomial of the form $x^2 + bx + c$,
   a. Find two numbers with a product equal to _____ and a sum equal to _____.
   b. The factored trinomial will have the form $(x + \text{first number})(x + \text{second number})$.

[6.4] **8.** $a^2 - b^2 = (\quad)(\quad)$ **[6.4]** **9.** $a^3 - b^3 = (\quad)(\quad)$ **[6.4]** **10.** $a^2 - 2ab + b^2 = (\quad)^2$

Exercises 11–18   CONSTANTS AND VARIABLES

[6.1] *For Exercises 11–18, list all of the natural number factors.*

11. 60   **12.** 9   **13.** 81   **14.** 44

15. 45   **16.** 49   **17.** 30   **18.** 33

Exercises 19–76   EXPRESSIONS

[6.1] *For Exercises 19–26, find the GCF.*

19. 21, 30   **20.** 34, 51   **21.** 12, 42, 60   **22.** 10, 15, 45

23. $10y^4, 25y^2$ **24.** $4x, 20xy$ **25.** $15m^2n, 3mn^5$ **26.** $8k, 22k^9$

[6.1] *For Exercises 27–34, factor out the GCF.*

27. $4x - 2$ **28.** $5m - 35m^3$ **29.** $y^3 - y$

30. $abc + a^2b^2c^2 - a^3b^3c^3$ **31.** $x^2y + xy^2 + x^3y^3$ **32.** $105a^3b^2 - 63a^2b^3 + 84a^6b^4$

33. $100k^4 + 120k^5 - 10k + 40k^3$ **34.** $18ab^3c - 36a^2b^2c$

[6.1] *For Exercises 35–42, factor by grouping.*

35. $ax + ay + bx + by$ **36.** $ax + 2a + bx + 2b$ **37.** $y^3 + 2y^2 + 3y + 6$ **38.** $y^4 + 4y^3 - by - 4b$

39. $xy + y + x + 1$ **40.** $ax^2 - 5y^2 + ay^2 - 5x^2$ **41.** $2b^3 - 2b^2 + b - 1$ **42.** $u^2 - 3u + 4uv - 12v$

[6.2] *For Exercises 43–50, factor.*

43. $x^2 - x - 12$ **44.** $x^2 + 14x + 45$ **45.** $n^2 - 6n + 8$ **46.** $a^2 + 3a - 20$

47. $h^2 + 51h + 144$ **48.** $y^2 - 10y - 24$ **49.** $4x^2 - 24x + 36$ **50.** $3m^2 - 33m + 54$

[6.3] *For Exercises 51–58, factor completely.*

51. $6x^2 + 3x - 7$ **52.** $2u^2 + 5u + 2$ **53.** $3m^2 - 10m + 3$ **54.** $5k^2 - 7kh - 12h^2$

55. $6a^2 - 20a + 16$ **56.** $8x^2y - 4xy - y$ **57.** $3p^2 - 13pq + 4q^2$ **58.** $24x^2 - 64xy - 24y^2$

[6.4] *For Exercises 59–66, factor.*

59. $v^2 - 8v + 16$ **60.** $u^2 + 6u + 9$ **61.** $4x^2 + 20x + 25$ **62.** $9y^2 - 12y + 4$

63. $x^2 - 4$ **64.** $25 - y^2$ **65.** $x^3 - 1$ **66.** $x^3 + 27$

[6.5] *For Exercises 67–76, factor completely.*

67. $6b^2 + b - 2$ **68.** $4ab - 24ab^2$ **69.** $y^2 + 25$ **70.** $3x^2 - 3y^2$

71. $x^4 - 81$ **72.** $7u^2 - 14u - 105$ **73.** $8x^3 - 27y^3$

74. $2 - 50y^2$ **75.** $3am - 6an - 8bn + 4bm$ **76.** $3m^2 + 9m + 27$

Exercises 77–94 **EQUATIONS AND INEQUALITIES**

[6.6] *For Exercises 77–84, solve.*

77. $(m + 3)(m - 4) = 0$ **78.** $y^2 - 4 = 0$ **79.** $x^2 - 5x + 6 = 0$ **80.** $x^2 - 4x = 21$

81. $m^2 = 9$ **82.** $x^2 + 3x = 18$ **83.** $y(y + 6) = -9$ **84.** $3n^2 - 11n = 4$

[6.6] *For Exercises 85–90, translate to an equation; then solve.*

85. Find every number such that five times the square of the number is equal to twice the number.

86. The sum of the squares of two consecutive natural numbers is 61. Find the numbers.

87. The product of a number and four times that same number is 100. Find all numbers.

88. Find the dimensions of a rectangle whose length is 6 more than its width and whose area is 91 square inches.

89. The product of two consecutive natural numbers is 110. Find the numbers.

90. Find the length of the missing side of the right triangle shown.

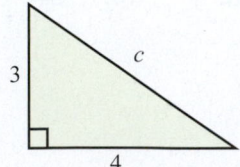

[6.7] *For Exercises 91–94, graph.*

91. $y = -5x^2$ **92.** $y = x^2 - 3$ **93.** $f(x) = x^2 + 2$ **94.** $f(x) = -3x^2 + 6x + 2$

Chapter 6 Practice Test

For Extra Help

Step-by-step test solutions are found on the Chapter Test Prep Videos available via the Video Resources on DVD, in **MyMathLab**, and on **You Tube** (search "CarsonCombinedAlg" and click on "Channels").

1. List all possible natural number factors of 80.

2. Find the GCF of $25m^2$ and $40m$.

Factor completely.

3. $5y - 30$ **4.** $6x^2y - 2y^2$ **5.** $ax - ay - bx + by$ **6.** $m^2 + 7m + 12$

7. $r^2 - 8r + 16$ **8.** $4y^2 + 20y + 25$ **9.** $3q^2 - 10q + 8$ **10.** $6x^2 - 23x + 20$

11. $ax^2 - 5ax - 24a$ **12.** $10n^3 + 38n^2 - 8n$ **13.** $c^2 - 25$ **14.** $x^2 + 4$

15. $2 - 50u^2$ **16.** $-4x^2 + 16$ **17.** $m^3 + 125$ **18.** $x^3 - 8$

Solve.

19. $a(a + 3) = 0$ **20.** $x^2 - 4x - 12 = 0$ **21.** $2n^2 + 7n = 15$

22. The product of two consecutive natural numbers is 72. Find the numbers.

23. The width of a rectangle is 5 feet less than the length. If the area is 36 square feet, find the length and width.

Graph.

24. $y = -2x^2 + 4$ **25.** $f(x) = x^2 - 6x + 5$

Chapters 1–6 Cumulative Review Exercises

For Exercises 1–4, answer true or false.

[5.4] **1.** $x^3 \cdot x^4 = x^{12}$

[6.1] **2.** The GCF of 12 and 5 is 1.

[1.7] **3.** The expression $5 + x$ can be simplified to equal $5x$.

[6.1] **4.** $4x + 12 = 2(2x + 6)$ is factored completely.

For Exercises 5 and 6, fill in the blank.

[6.4] **5.** To factor a difference of squares, we use the rule $a^2 - b^2 = $ _____.

[6.6] **6.** To solve a quadratic equation using factoring,
   a. Write the equation in _____ form $(ax^2 + bx + c = 0)$.
   b. Write the expression in _____ form.
   c. Use the zero-factor theorem to solve.

Exercises 7–19   **EXPRESSIONS**

For Exercises 7–14, simplify.

[1.5] **7.** $14 - 2 \cdot 6 \cdot 3 + 8^2$

[1.5] **8.** $-6|5 + 3^2| - \sqrt{14 + 11}$

[5.3] **9.** $(5x^2 + 6x - 1) - (2x - 3)$

[5.4] **10.** $(6x^2)(3x^2 y)$

[5.4] **11.** $(3x)(4x^3)^2$

[5.5] **12.** $(y - 8)(2y + 1)$

[5.6] **13.** $\dfrac{16h^3 + 4h^2 - 8h}{4h^2}$

[5.6] **14.** $(x^2 + 7x + 12) \div (x + 3)$

[6.1–6.5] *For Exercises 15–19, factor completely.*

15. $x^2 - 121$

16. $10x^2 + 40$

17. $x^2 - 8x + 16$

18. $2x^2 + 5x - 12$

19. $x^3 - 5x^2 + 5x - 25$

Exercises 20–30   **EQUATIONS AND INEQUALITIES**

For Exercises 20–22, solve.

[2.3] **20.** $\dfrac{3}{4} = x - \dfrac{2}{3}$

[2.3] **21.** $2.6 + 7a + 5 = 8a - 5.6$

[6.6] **22.** $x^2 - 3x = 28$

For Exercises 23–25, solve.

[2.4] **23.** Solve for x in the equation $2x - 3y = 7$.

[4.3] **24.** Find the x- and y-intercepts for $y = 2x - 6$.

[4.4] **25.** Find the slope of the line containing the points (6, 2) and (5, 2).

[4.2] **26.** Graph $y = 3x + 2$.

[2.6] *For Exercise 27 solve.* **a.** *Write the solution set using set-builder notation.*
b. *Write the solution set using interval notation.*
c. *Graph the solution set.*

27. $7x - 5 \leq 3x + 11$

For Exercises 28–30, solve.

[3.3] **28.** Find three consecutive even integers whose sum is 24.
[5.4]

[3.5] **29.** Janice invests a total of $8000 in two different accounts. The first account re-
[5.4] turns 5%, while the second account returns 8%. If the total interest earned
after one year is $565, what principal did she invest in each account?

[3.5] **30.** How much of a 35% HCl solution must be added to 50 milliliters of a 10%
[5.4] HCl solution to obtain a 20% solution?

Rational Expressions and Equations

7

> What is the most rigorous law of our being? Growth. No smallest atom of our moral, mental, or physical structure can stand still a year. It grows, it must grow . . . nothing can prevent it.
>
> —Mark Twain (1835–1910)
> American author and humorist

Throughout this book, topics are developed in a specific pattern. We first define a new type of expression, then learn how to perform arithmetic operations on those expressions. Finally, we solve equations that contain those expressions. For example, in Chapter 5, we defined polynomials, then learned how to perform operations of arithmetic with the polynomials. In Chapter 6, we solved equations containing the polynomial expressions. We follow a similar pattern of development in Chapter 7. We first define rational expressions, then learn to perform arithmetic operations on those rational expressions, and finally solve equations that contain rational expressions.

7.1 Simplifying Rational Expressions

Objectives

1. Evaluate rational expressions.
2. Find numbers that cause a rational expression to be undefined.
3. Simplify rational expressions containing only monomials.
4. Simplify rational expressions containing multiterm polynomials.

Objective 1 Evaluate rational expressions. Now that we have explored polynomials, we are ready to develop a new class of expressions called **rational expressions**.

Definition *Rational expression:* An expression that can be written in the form

$$\frac{P}{Q}, \text{ where } P \text{ and } Q \text{ are polynomials and } Q \neq 0.$$

Connection Notice how the definition of a rational expression is like the definition of a rational number. Recall from Chapter 1 that a rational number is a number that can be expressed in the form $\frac{a}{b}$, where a and b are integers and $b \neq 0$. Some texts call rational expressions algebraic fractions.

Following are some rational expressions.

$$\frac{3x^5}{18x^2} \qquad \frac{7x}{x^2 - 9} \qquad \frac{x^2 + 2x - 15}{4x + 20}$$

As with any expression, we can evaluate or rewrite rational expressions. First, let's look at evaluating rational expressions. Remember that to evaluate an algebraic expression, we replace the variables with given values and then simplify the resulting numerical expression.

Example 1 Evaluate the expression $\dfrac{3x - 5}{x + 1}$, when

a. $x = 2$

SOLUTION $\dfrac{3x - 5}{x + 1} = \dfrac{3(2) - 5}{2 + 1}$ **Replace x with 2.**

$$= \frac{6 - 5}{3}$$

$$= \frac{1}{3}$$

b. $x = -0.4$

SOLUTION $\dfrac{3x - 5}{x + 1} = \dfrac{3(-0.4) - 5}{-0.4 + 1}$ **Replace x with −0.4.**

$$= \frac{-1.2 - 5}{0.6}$$

$$= \frac{-6.2}{0.6}$$

$$= -10.\overline{3} \quad \text{or} \quad -10\frac{1}{3}$$

c. $x = -1$

SOLUTION
$$\frac{3x - 5}{x + 1} = \frac{3(-1) - 5}{-1 + 1}$$ Replace x with -1.

$$= \frac{-3 - 5}{0}$$

$$= \frac{-8}{0}, \quad \text{which is undefined.}$$

Your Turn 1 Evaluate the expression $\dfrac{x^2 - 9}{x - 2}$, when

a. $x = 4$.   **b.** $x = -3$.   **c.** $x = 2$.

Objective 2 Find numbers that cause a rational expression to be undefined. Note that in Example 1(c), the expression $\dfrac{3x - 5}{x + 1}$ is undefined when $x = -1$ because replacing x with -1 causes the denominator to be 0. We need to avoid values that cause a rational expression to be undefined, which means we need to identify and avoid values that make their denominators 0. This suggests the following procedure.

Procedure   Finding Values That Make a Rational Expression Undefined

To determine the value(s) that make a rational expression undefined,

1. Write an equation that has the denominator set equal to zero.
2. Solve the equation.

Example 2   Find every value for the variable that makes the expression undefined.

a. $\dfrac{5y}{3y - 2}$

SOLUTION

> **Note:** We do not consider the numerator because the rational expression is undefined only when the denominator is 0.

$3y - 2 = 0$ Set the denominator, $3y - 2$, equal to 0.

$3y = 2$ Add 2 to both sides.

$y = \dfrac{2}{3}$ Divide both sides by 3.

The expression $\dfrac{5y}{3y - 2}$ is undefined if y is replaced with $\dfrac{2}{3}$ because the denominator equals 0.

b. $\dfrac{7}{x^2 + x - 12}$

SOLUTION $x^2 + x - 12 = 0$ Set the denominator equal to 0.

$(x + 4)(x - 3) = 0$ Factor.

$x + 4 = 0$ or $x - 3 = 0$ Use the zero-factor theorem.

$x = -4$ $x = 3$

The original expression is undefined if x is replaced by -4 or 3.

Answers   Your Turn 1

a. $\dfrac{7}{2}$ **b.** 0   **c.** undefined

c. $\dfrac{n-6}{n^2+9}$

SOLUTION $n^2+9=0$ Set the denominator equal to 0.

For all real-number values of n, $n^2 \geq 0$. Consequently $n^2+9 \neq 0$ for all real numbers. So $\dfrac{n-6}{n^2+9}$ is defined for all real numbers.

Your Turn 2   Find every value that can replace the variable in the expression and cause the expression to be undefined.

a. $\dfrac{y+4}{3y-1}$ **b.** $\dfrac{2x}{x^2+5x}$ **c.** $\dfrac{m}{m^2+1}$

Objective ③ Simplify rational expressions containing only monomials.   Now that we have defined rational expressions, we are ready to begin simplifying them. Recall from Section 1.2 that a fraction is in lowest terms when the only common factor of its numerator and denominator is 1. Also in Section 1.2, we used the rule $\dfrac{an}{bn} = \dfrac{a \cdot 1}{b \cdot 1} = \dfrac{a}{b}$, where $b \neq 0$ and $n \neq 0$, to reduce fractions. We can rewrite this rule so that it applies to rational expressions.

Rule $\dfrac{PR}{QR} = \dfrac{P \cdot 1}{Q \cdot 1} = \dfrac{P}{Q}$, where P, Q, and R are polynomials and Q and R are not 0.

The rule indicates that a factor common to both the numerator and denominator can be divided out of a rational expression. Consider the following comparison between simplifying a fraction and simplifying a similar rational expression.

$$\dfrac{10}{15} = \dfrac{2 \cdot 5}{3 \cdot 5} = \dfrac{2 \cdot 1}{3 \cdot 1} = \dfrac{2}{3} \qquad \dfrac{2x}{3x} = \dfrac{2 \cdot x}{3 \cdot x} = \dfrac{2 \cdot 1}{3 \cdot 1} = \dfrac{2}{3}$$

5 is the common factor here.

x is the common factor here.

Note: When working with rational expressions, we assume that the variables will not be replaced by any value that would cause the expression to be undefined or indeterminate.

This suggests the following procedure for simplifying rational expressions.

Procedure   Simplifying Rational Expressions to Lowest Terms

To simplify a rational expression to lowest terms,

1. Factor the numerator and denominator completely.

2. Divide out all common factors in the numerator and denominator.

3. Multiply the remaining factors in the numerator and the remaining factors in the denominator.

We first simplify rational expressions that are a monomial over a monomial.

Answers   Your Turn 2

a. $\dfrac{1}{3}$ **b.** 0 or -5

c. defined for all real numbers

Connection In Section 5.6 and Chapter 6, we divided monomials using $\dfrac{a^m}{a^n} = a^{m-n}$ and $a^{-n} = \dfrac{1}{a^n}$. In this chapter, we divide monomials (simplify to lowest terms) by factoring the numerator and/or denominator into prime factors, as we did in Section 1.2, for consistency of approach. As we will see later in this section, this approach is the only way rational expressions with multiterm polynomials in the numerator and/or denominator can be simplified to lowest terms.

Example 3 Simplify $\dfrac{3x^5}{15x^2}$.

SOLUTION Factor the numerator and denominator completely; then divide out all common factors.

$$\frac{3x^5}{15x^2} = \frac{3 \cdot x \cdot x \cdot x \cdot x \cdot x}{3 \cdot 5 \cdot x \cdot x}$$

Note: The common factors are a single 3 and two x's. These form the GCF that we divide out, which is $3x^2$.

Note: From here on, we omit this step with the 1's and simply highlight the common factors that are eliminated.

$$= \frac{1 \cdot x \cdot x \cdot x \cdot 1}{1 \cdot 5 \cdot 1}$$

$$= \frac{x^3}{5}$$

There are different styles for showing the process of dividing out common factors. For example, we could have used cancellation marks to show that the two x's and one 3 divide out in Example 3.

$$\frac{3x^5}{15x^2} = \frac{\overset{1}{\cancel{3}} \cdot \overset{1}{\cancel{x}} \cdot \overset{1}{\cancel{x}} \cdot x \cdot x \cdot x}{3 \cdot 5 \cdot \underset{1}{\cancel{x}} \cdot \underset{1}{\cancel{x}}} = \frac{x^3}{5}$$

Develop a style that works best for you, keeping in mind that no matter the style, the bottom line is that you will be dividing out the GCF of the numerator and denominator to get a rational expression in lowest terms.

Example 4 Simplify.

a. $-\dfrac{4x^2}{28x^3}$

Connection
Using cancellation marks:

$$-\frac{4x^2}{28x^3} = -\frac{\overset{1}{\cancel{2}} \cdot \overset{1}{\cancel{2}} \cdot \overset{1}{\cancel{x}} \cdot \overset{1}{\cancel{x}}}{\underset{1}{\cancel{2}} \cdot \underset{1}{\cancel{2}} \cdot 7 \cdot \underset{1}{\cancel{x}} \cdot \underset{1}{\cancel{x}} \cdot x}$$

$$= -\frac{1}{7x}$$

SOLUTION $-\dfrac{4x^2}{28x^3} = -\dfrac{2 \cdot 2 \cdot x \cdot x}{2 \cdot 2 \cdot 7 \cdot x \cdot x \cdot x}$

Factor the numerator and denominator completely. Then eliminate the common factors, which are two 2's and two x's. Multiply the remaining factors.

$$= -\frac{1}{7x}$$

Note: Because all of the factors in the numerator divided out, the numerator is 1.

b. $\dfrac{36ab^3c}{54a^2b^3}$

SOLUTION $\dfrac{36ab^3c}{54a^2b^3} = \dfrac{2 \cdot 2 \cdot 3 \cdot 3 \cdot a \cdot b \cdot b \cdot b \cdot c}{2 \cdot 3 \cdot 3 \cdot 3 \cdot a \cdot a \cdot b \cdot b \cdot b}$

Factor the numerator and denominator completely. Then divide out the common factors, which are one 2, two 3's, one a, and three b's. Multiply the remaining factors.

$$= \frac{2c}{3a}$$

Your Turn 4 Simplify.

a. $-\dfrac{12t^6}{30t^2}$
b. $\dfrac{9n}{18n^3}$
c. $-\dfrac{72x^2y}{60x^2y^3z}$

Answers Your Turn 4

a. $-\dfrac{2t^4}{5}$ b. $\dfrac{1}{2n^2}$ c. $-\dfrac{6}{5y^2z}$

Connection Dividing out common factors when simplifying rational expressions with monomial numerators and/or denominators to lowest terms can be viewed as dividing monomials using the rules of exponents, as we did in Section 5.6.

Dividing out common factors:

$$\frac{12x^4y^2}{6x^2y} = \frac{2 \cdot 2 \cdot 3 \cdot x \cdot x \cdot x \cdot x \cdot y \cdot y}{2 \cdot 3 \cdot x \cdot x \cdot y}$$

$$= 2 \cdot x \cdot x \cdot y = 2x^2y$$

Using rules of exponents:

$$\frac{12x^4y^2}{6x^2y} = \frac{12}{6} \cdot \frac{x^4}{x^2} \cdot \frac{y^2}{y}$$

$$= 2x^{4-2}y^{2-1} = 2x^2y$$

The coefficients can be simplified as we do with ordinary fractions. Because we are dividing out two of the four x's and one of the two y's in the numerator, we are left with two x's and one y in the numerator. Subtracting the exponents accounts for eliminating the two common factors of x and the one common factor of y. This connection also applies when the greater exponent is in the denominator.

Dividing out common factors:

$$\frac{15a^2}{10a^5} = \frac{3 \cdot 5 \cdot a \cdot a}{2 \cdot 5 \cdot a \cdot a \cdot a \cdot a \cdot a}$$

$$= \frac{3}{2 \cdot a \cdot a \cdot a} = \frac{3}{2a^3}$$

Using rules of exponents:

$$\frac{15a^2}{10a^5} = \frac{15}{10} \cdot \frac{a^2}{a^5} = \frac{3}{2} \cdot a^{2-5} = \frac{3}{2}a^{-3}$$

$$= \frac{3}{2} \cdot \frac{1}{a^3} = \frac{3}{2a^3}$$

Notice that when dividing out like bases, the position of the base with the greater exponent determines whether we place the result in the numerator or denominator.

Objective **4** **Simplify rational expressions containing multiterm polynomials.** The rational expressions we have considered so far have had monomials in both the numerator and denominator. We now consider rational expressions that contain multiterm polynomials. Remember that we can divide out only common factors, so we must first factor polynomials completely.

Example 5 Simplify $\dfrac{9ab}{3a + 6}$.

SOLUTION $\dfrac{9ab}{3a + 6} = \dfrac{3 \cdot 3 \cdot a \cdot b}{3 \cdot (a + 2)}$ **Factor the numerator and denominator completely. Then divide out the common factor, which is 3. Multiply the remaining factors.**

$$= \frac{3ab}{a + 2}$$

Your Turn 5 Simplify.

a. $\dfrac{6xy}{3x + 12}$

b. $\dfrac{2a + 8}{10a^2}$

We now consider rational expressions with multiterm polynomials in both the numerator and denominator. Again, because we can divide out only common factors, we must first factor the polynomials.

Answers **Your Turn 5**

a. $\dfrac{2xy}{x + 4}$ b. $\dfrac{a + 4}{5a^2}$

Example 6 Simplify.

a. $\dfrac{2x^2 + 8x}{3x^2 + 12x}$

SOLUTION $\dfrac{2x^2 + 8x}{3x^2 + 12x} = \dfrac{2 \cdot x \cdot (x + 4)}{3 \cdot x \cdot (x + 4)}$ Factor the numerator and denominator completely. Then divide out the common factors, x and $x + 4$.

$$= \dfrac{2}{3}$$

b. $\dfrac{y^2 - 16}{y^2 + 4y - 32}$

SOLUTION $\dfrac{y^2 - 16}{y^2 + 4y - 32} = \dfrac{(y + 4)(y - 4)}{(y + 8)(y - 4)}$ Factor the numerator and denominator completely. Then divide out the common factor, $y - 4$.

$$= \dfrac{y + 4}{y + 8}$$

c. $\dfrac{2x^2 - 3x - 20}{2x^2 + x - 10} = \dfrac{(2x + 5)(x - 4)}{(2x + 5)(x - 2)}$ Factor the numerator and denominator completely. Then divide out the common factor, $2x + 5$.

$$= \dfrac{x - 4}{x - 2}$$

Warning: A common error is to divide out terms instead of factors. Recall that the terms of a polynomial are connected by plus and minus signs. You have factors only when multiplying. Below are examples of both the correct and incorrect way of simplifying a rational expression to lowest terms.

Correct: $\dfrac{x^2 + x - 6}{x^2 + 5x + 6} = \dfrac{(x + 3)(x - 2)}{(x + 3)(x + 2)} = \dfrac{x - 2}{x + 2}$ Divide out factors.

Incorrect: $\dfrac{x^2 + x - 6}{x^2 + 5x - 6} = \dfrac{\overset{1}{x^2} + \overset{1}{x} - \overset{1}{6}}{\underset{1}{x^2} + 5\underset{1}{x} - \underset{1}{6}} = \dfrac{1}{5}$ Do not divide out terms.

Your Turn 6 Simplify.

a. $\dfrac{5x^2 + 10x}{7x^2 + 14x}$ **b.** $\dfrac{y^2 - 36}{y^2 - 2y - 24}$ **c.** $\dfrac{3n^2 - 23n - 8}{3n^2 + 13n + 4}$

Recall from Chapter 5 that to factor multiterm polynomials completely, we must sometimes factor the first factorization we get as in Example 7.

Example 7 Simplify.

a. $\dfrac{6y^4 + 2y^3 - 4y^2}{36y^3 - 42y^2 + 12y}$

SOLUTION $\dfrac{6y^4 + 2y^3 - 4y^2}{36y^3 - 42y^2 + 12y} = \dfrac{2y^2(3y^2 + y - 2)}{6y(6y^2 - 7y + 2)}$ Factor out the monomial GCFs.

$$= \dfrac{2 \cdot y \cdot y \cdot (y + 1) \cdot (3y - 2)}{2 \cdot 3 \cdot y \cdot (2y - 1) \cdot (3y - 2)}$$ Factor the polynomial factors.

$$= \dfrac{y(y + 1)}{3(2y - 1)} \quad \text{or} \quad \dfrac{y^2 + y}{6y - 3}$$ Divide out the common factors 2, y, and $3y - 2$.

Answers Your Turn 6

a. $\dfrac{5}{7}$ **b.** $\dfrac{y + 6}{y + 4}$ **c.** $\dfrac{n - 8}{n + 4}$

b. $\dfrac{3x^2 + 6x - 24}{x^4 - 16}$

SOLUTION Factor the numerator and denominator completely. Then divide out all common factors.

$$\frac{3x^2 + 6x - 24}{x^4 - 16} = \frac{3(x^2 + 2x - 8)}{(x^2 + 4)(x^2 - 4)}$$

> Factor 3 out of $3x^2 + 6x - 24$ and factor $x^4 - 16$, which is a difference of squares.

$$= \frac{3 \cdot (x + 4) \cdot (x - 2)}{(x^2 + 4) \cdot (x + 2) \cdot (x - 2)}$$

> Factor the polynomial factors. Notice that $x^2 - 4$ is a difference of squares.

$$= \frac{3(x + 4)}{(x^2 + 4)(x + 2)} \quad \text{or} \quad \frac{3x + 12}{x^3 + 2x^2 + 4x + 8}$$

> Divide out the common factor, $x - 2$.

Your Turn 7 Simplify.

a. $\dfrac{14x^3 + 70x^2 + 84x}{7x^4 - 14x^3 - 105x^2}$

b. $\dfrac{5x^2 - 5x - 60}{x^4 - 81}$

Binomial factors that are additive inverses can be tricky.

Example 8 Simplify $\dfrac{3x^2 - 7x - 20}{8 - 2x}$.

SOLUTION Factor the numerator and denominator completely. Then divide out all common factors.

$$\frac{3x^2 - 7x - 20}{8 - 2x} = \frac{(3x + 5)(x - 4)}{2(4 - x)}$$

It appears that there are no common factors. However, $x - 4$ and $4 - x$ are additive inverses, which is more apparent when $4 - x$ is written in descending order as $-x + 4$. We can get matching binomial factors by factoring -1 out of $x - 4$ or $4 - x$.

$$x - 4 = -1(-x + 4) = -1(4 - x) \quad \text{or} \quad 4 - x = -1(-4 + x) = -1(x - 4)$$

We can use either form. We will use $x - 4 = -1(4 - x)$.

> **Note:** We could have factored the -1 out of the denominator:
>
> $$\frac{(3x + 5)(x - 4)}{2(-1)(x - 4)} = \frac{3x + 5}{-2}$$

$$= \frac{(3x + 5)(-1)(4 - x)}{2(4 - x)}$$

$$= \frac{-3x - 5}{2} \quad \text{or} \quad -\frac{3x + 5}{2}$$

> Divide out the common factor, $4 - x$; then multiply the remaining factors.

Example 8 also illustrates the rule of sign placement in a fraction or rational expression. In a negative fraction (or rational expression), the minus sign can be placed in the numerator or denominator or aligned with the fraction line. Note, however, that it is generally considered unsightly to write the minus sign in the denominator.

Answers Your Turn 7

a. $\dfrac{2(x + 2)}{x(x - 5)}$ or $\dfrac{2x + 4}{x^2 - 5x}$

b. $\dfrac{5(x - 4)}{(x^2 + 9)(x - 3)}$ or $\dfrac{5x - 20}{x^3 - 3x^2 + 9x - 27}$

$$-\frac{P}{Q} = \frac{-P}{Q} = \frac{P}{-Q}, \text{ where } P \text{ and } Q \text{ are polynomials and } Q \neq 0.$$

Answer   Your Turn 8

$-\dfrac{x}{x + 6}$

Your Turn 8 Simplify $\dfrac{2x - x^2}{x^2 + 4x - 12}$.

7.1 Exercises   For Extra Help   *MyMathLab*

PRACTICE   WATCH   DOWNLOAD

1. Explain how to evaluate an expression.

2. In general, what causes a rational expression to be undefined?

3. How do you determine what values cause a rational expression to be undefined?

4. When simplifying a rational expression, we divide out all common _____ in the numerator and denominator.

5. Explain why you cannot divide out the y's in $\dfrac{x + y}{2y}$.

6. What is the first step when simplifying rational expressions?

For Exercises 7–14, evaluate the rational expression. See Example 1.

7. $\dfrac{4x^2}{9y}$
      a. when $x = -2, y = 5$ **b.** when $x = 3, y = -4$ **c.** when $x = 0, y = 7$

8. $\dfrac{3x}{7y^2}$
      a. when $x = 1, y = 2$ **b.** when $x = -2, y = 3$ **c.** when $x = -4, y = -2$

9. $\dfrac{2x}{x + 4}$
      a. when $x = 4$ **b.** when $x = -4$ **c.** when $x = -2.4$

10. $\dfrac{4x}{x - 3}$
      a. when $x = 3$ **b.** when $x = -3$ **c.** when $x = 2.5$

11. $\dfrac{2x + 5}{3x - 1}$
      a. when $x = 3$ **b.** when $x = -2$ **c.** when $x = -1.3$

12. $\dfrac{3x + 2}{4x - 5}$
      a. when $x = 2$ **b.** when $x = -1$ **c.** when $x = 1.6$

13. $\dfrac{x^2 - 4}{x + 3}$ **a.** when $x = 0$ **b.** when $x = -1$ **c.** when $x = -4.2$

14. $\dfrac{x^2 + 3}{2x + 1}$ **a.** when $x = 1$ **b.** when $x = -2$ **c.** when $x = -3.5$

For Exercises 15–26, find every value for the variable that makes the expression undefined. See Example 2.

15. $\dfrac{2x}{x - 3}$ **16.** $\dfrac{3y}{y + 5}$ **17.** $\dfrac{3}{x^2 - 9}$ **18.** $\dfrac{4r}{r^2 - 4}$

19. $\dfrac{6}{x^2 - 14x + 45}$ **20.** $\dfrac{2x + 3}{x^2 + 5x + 6}$ **21.** $\dfrac{2m}{m^2 - 3m}$ **22.** $\dfrac{2p + 3}{p^2 + 6p}$

23. $\dfrac{4a}{2a^2 - 3a - 5}$ **24.** $\dfrac{t}{2t^2 + 7t + 3}$ **25.** $\dfrac{x - 5}{x^2 + 16}$ **26.** $\dfrac{3x - 2}{y^2 + 36}$

For Exercises 27–78, simplify. See Examples 3–8.

27. $\dfrac{9x^2}{12xy}$ **28.** $\dfrac{14h^2k}{21h^3}$ **29.** $-\dfrac{4m^3n}{mn^2}$ **30.** $-\dfrac{5a^3b}{a^2b^4}$ **31.** $\dfrac{54x^4y}{36x^6yz^2}$

32. $-\dfrac{48t^5uv^4}{32\,tuv^2}$ **33.** $-\dfrac{15m^2n}{5m + 10n}$ **34.** $-\dfrac{28r^3t}{7r + 14t}$ **35.** $\dfrac{7}{21(x + 5)}$ **36.** $\dfrac{9}{18(a - 1)}$

37. $\dfrac{3(m - 2)}{5(m - 2)}$ **38.** $\dfrac{7(k + 2)}{15(k + 2)}$ **39.** $\dfrac{10x - 20}{15x - 30}$ **40.** $\dfrac{12y + 2}{18y + 3}$ **41.** $\dfrac{30a + 15b}{18a + 9b}$

42. $\dfrac{8m - 24n}{12m - 36n}$ **43.** $\dfrac{x^2 + 5x}{4x + 20}$ **44.** $\dfrac{ab - b^2}{2a - 2b}$ **45.** $\dfrac{m^2 - 6m}{m - 6}$ **46.** $\dfrac{x^2 + 5x}{x + 5}$

47. $\dfrac{5y^2 - 10y - 15}{xy^2 - 2xy - 3x}$ **48.** $\dfrac{3m^3 - 2m^2 + m}{6m^2 - 4m + 2}$ **49.** $\dfrac{x^2 - y^2}{x^2 + 2xy + y^2}$ **50.** $\dfrac{a^2 - b^2}{a^2 + 2ab + b^2}$ **51.** $\dfrac{x^2 - 5x}{x^2 - 7x + 10}$

52. $\dfrac{x^2 - 2x}{x^2 - 6x + 8}$ **53.** $\dfrac{t^2 - 9}{t^2 + 5t + 6}$ **54.** $\dfrac{k^2 - 4}{k^2 + k - 2}$ **55.** $\dfrac{4a^2 - 4a - 3}{6a^2 - a - 2}$ **56.** $\dfrac{3x^2 + 16x - 35}{5x^2 + 33x - 14}$

57. $\dfrac{6b^2 + b - 12}{10b^2 + 13b - 3}$ **58.** $\dfrac{8y^2 - 6y - 9}{12y^2 + y - 6}$ **59.** $\dfrac{x^3 - 16x}{x^3 + 6x^2 + 8x}$ **60.** $\dfrac{m^3 - 4m}{m^3 - 4m^2 + 4m}$

61. $\dfrac{px + py + qx + qy}{mx - nx + my - ny}$ **62.** $\dfrac{by - ay + bx - ax}{x^2 + ax + xy + ay}$ **63.** $\dfrac{6u^3 - 4u^2 - 3u + 2}{3u^2 + u - 2}$ **64.** $\dfrac{8n^3 + 20n^2 - 2n - 5}{4n^2 + 12n + 5}$

65. $\dfrac{x - 3}{3 - x}$ **66.** $\dfrac{5 - b}{b - 5}$ **67.** $\dfrac{4x - 4}{1 - x}$ **68.** $\dfrac{12 - 4m}{m - 3}$ **69.** $\dfrac{y^2 - 4}{2 - y}$,

70. $\dfrac{a^2 - b^2}{b - a}$ **71.** $\dfrac{3 - w}{w^2 - 2w - 3}$ **72.** $\dfrac{1 - u}{u^2 - 3u + 2}$ **73.** $\dfrac{x^2 - 4}{x^3 + 8}$ **74.** $\dfrac{r^2 - s^2}{r^3 + s^3}$

75. $\dfrac{x^3 + y^3}{x^3 - x^2y + xy^2}$ **76.** $\dfrac{x^3 - 8}{x^3 + 2x^2 + 4x}$ **77.** $\dfrac{3 - m}{m^3 - 27}$ **78.** $\dfrac{4 - x}{x^3 - 64}$

Find
✕ the mistake

For Exercises 79–84, find and explain the mistake; then correct the mistake.

79. $-\dfrac{5x^2y}{10x^3y} = -\dfrac{x}{2}$

80. $\dfrac{8x^2y}{4x^3} = 2xy$

81. $\dfrac{x - \overset{3}{\cancel{6}}}{\underset{1}{\cancel{2}}y} = \dfrac{x - 3}{y}$

82. $\dfrac{y - \overset{3}{\cancel{12}}}{\underset{1}{\cancel{4}}y - 3} = \dfrac{y - 3}{y - 3} = 1$

83. $\dfrac{yx - x^2}{x - y} = \dfrac{x(y - x)}{x - y} = x$

84. $\dfrac{x^2 - x - 6}{x^3 + 2x^2 - 9x - 18} = \dfrac{(x + 2)(x - 3)}{(x + 2)(x^2 - 9)} = \dfrac{x - 3}{x^2 - 9}$

For Exercises 85–90, evaluate. See Example 1.

85. A person's body mass index (BMI) is a measure of the amount of body fat based on height, h, in inches, and weight, w, in pounds. The formula for BMI is

$$\text{BMI} = \dfrac{704.5w}{h^2}.$$

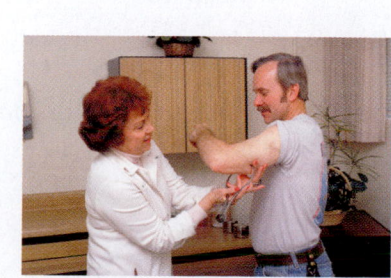

a. Use the formula to find the BMI of each person listed here.

Height	Weight	BMI
5' 4"	138	
5' 9"	155	
6' 2"	220	

b. Use the following table from the National Institutes of Health to classify each person in part a as underweight, normal weight, overweight, or obese.

Underweight:	BMI $\leq$ 18.5
Normal Weight:	BMI $= 18.5 - 24.9$
Overweight:	BMI $= 25 - 29.9$
Obese:	BMI $\geq$ 30

Connection In Section 2.5, Exercise 68, you translated the BMI formula from words to symbols.

86. Recall the following formula for slope from Chapter 4:

$$m = \frac{y_2 - y_1}{x_2 - x_1}$$

a. Find the slope of a line passing through the points $(3, 5)$ and $(-1, -4)$.

b. Find the slope of a line passing through the points $(-4, 7)$ and $(-2, -1)$.

Connection Although we hadn't formally defined rational expressions in Chapter 4, we were evaluating the rational expression $\frac{y_2 - y_1}{x_2 - x_1}$ to find slope.

87. Given the diameter in inches, d, and the number of revolutions per minute, N, of a revolving tool such as a circular saw or drill, the cutting speed, C, in feet per minute, of the tool can be found by the formula

$$C = \frac{\pi d N}{12}.$$

(*Source:* Robert D. Smith, *Mathematics for Machine Technology*, 5th ed., Cengage Learning, 2003)

Use the formula to complete the following table for a table saw. Leave answers in terms of π.

Blade Diameter	rpm	Cutting Speed
8 inches	800 rpm	
8 inches	1000 rpm	
10 inches	800 rpm	
10 inches	1000 rpm	

88. The cutting time, T, in minutes, that it takes a drill to cut through something depends on its speed of revolution, N, in rpm; the thickness of the material, L, in inches; and the tool feed, F, in inches per revolution. The formula for calculating cutting time is

$$T = \frac{L}{FN}.$$

(*Source:* Robert D. Smith, *Mathematics for Machine Technology*, 5th ed., Cengage Learning, 2003)

Use the formula to complete the following table for a drill operating at a speed of 480 rpm and a feed of 0.04 inches per revolution.

Material Thickness (inches)	Cut Time (minutes)
0.25	
0.5	
0.75	
1	

89. The formula for the volume of a cone is $V = \frac{1}{3}\pi r^2 h$.

   a. Solve this formula for h so that we have a formula for finding the height of a cone given its volume and radius. (The formula for height of a cone will contain a rational expression.)

   b. Suppose a cone has a volume of approximately 150.8 cubic centimeters with a radius of 4 centimeters. Find the height of the cone.

90. The formula for the volume of a pyramid is $V = \frac{1}{3}lwh$.

   a. Solve this formula for h so that we have a formula for finding the height of a pyramid given its volume, length, and width.

   b. Suppose a decorative onyx pyramid has a volume of 126 cubic inches, a width of 3.5 inches, and a length of 4.5 inches. Find the height of the pyramid.

For Exercises 91 and 92, simplify. See Examples 5–7.

91. In analyzing circuits, electrical engineers often have to simplify rational expressions. Suppose the following rational expression describes an electrical circuit.

$$\frac{s^2 - 9}{s^2 + 7s + 12}$$

Simplify the expression to lowest terms.

92. An engineer derives the following rational expression, which describes a certain circuit.

$$\frac{4s^2 - 6s}{2s^2 + 7s - 15}$$

Simplify the expression to lowest terms.

Collaborative Exercises Graphs of Rational Expressions

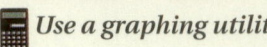 *Use a graphing utility.*

1. Graph $f(x) = \dfrac{x}{x - 3}$. The vertical line you see is called an *asymptote*.

   a. What is the equation of the asymptote?

   b. What do you think the asymptote indicates?

2. Graph $f(x) = \dfrac{5}{(x + 2)(x - 4)}$.

   a. Why are there two asymptotes?

   b. What are the equations of the two asymptotes?

   Note: The suggested window settings in Exercise 3 are for a TI-83 or TI-84. If a different graphing utility is used, the settings may need to be changed in order to see the hole.

3. Graph $f(x) = \dfrac{x^2 - x - 6}{x^2 + 3x + 2}$ with these window settings: $x_{min} = -4.7$, $x_{max} = 4.7$, $y_{min} = -10$, and $y_{max} = 10$.

   a. Are there any asymptotes? If so, write the equation for each.

   b. Rewrite $f(x) = \dfrac{x^2 - x - 6}{x^2 + 3x + 2}$ so that the numerator and denominator are in factored form.

   c. Find $f(-2)$.

   d. Look at the graph of $f(x) = \dfrac{x^2 - x - 6}{x^2 + 3x + 2}$ in the second quadrant. The blank space you see in the graph where $x = -2$ is a *removable discontinuity* and is often called a *hole*. How is a hole different from an asymptote?

4. Writing $\dfrac{x^2 - x - 6}{x^2 + 3x + 2}$ in simplest form, we have $\dfrac{x - 3}{x + 1}$. Graph $f(x) = \dfrac{x - 3}{x + 1}$. Compare the graph of $f(x) = \dfrac{x - 3}{x + 1}$ with the graph of $f(x) = \dfrac{x^2 - x - 6}{x^2 + 3x + 2}$. What do you notice about the graphs?

5. Given a rational function, how can you identify its asymptotes and holes?

Review Exercises

Exercises 1–6 **EXPRESSIONS**

[1.4] *For Exercises 1 and 2, multiply.*

1. $\dfrac{2}{3} \cdot \dfrac{6}{4}$

2. $-\dfrac{15}{16} \cdot \dfrac{4}{5}$

[1.4] *For Exercises 3 and 4, divide.*

3. $\dfrac{3}{8} \div \dfrac{5}{6}$

4. $-\dfrac{7}{9} \div \left(-\dfrac{5}{12}\right)$

[6.4] *For Exercises 5 and 6, factor completely.*

5. $4y^2 - 25$

[6.3]   **6.** $3x^2 + 13x - 10$

7.2 Multiplying and Dividing Rational Expressions

Objectives

1. Multiply rational expressions.
2. Divide rational expressions.
3. Convert units of measurement using dimensional analysis.

Objective 1   Multiply rational expressions.   In Section 1.4, we used the rule $\dfrac{a}{b} \cdot \dfrac{c}{d} = \dfrac{ac}{bd}$, where $b \neq 0$ and $d \neq 0$, to multiply fractions. We can rewrite this rule so that it applies to multiplying rational expressions.

Rule   **Multiplying Rational Expressions**

$\dfrac{P}{Q} \cdot \dfrac{R}{S} = \dfrac{PR}{QS}$, where P, Q, R, and S are polynomials and $Q \neq 0$ and $S \neq 0$.

Also remember that when multiplying fractions, we can simplify after multiplying or we can simplify before multiplying by dividing out factors common to both the numerator and denominator. For example,

Multiply, then eliminate common factors.   or   Eliminate common factors; then multiply.

$$\frac{8}{15} \cdot \frac{3}{4} = \frac{24}{60} = \frac{2 \cdot 2 \cdot 2 \cdot 3}{2 \cdot 2 \cdot 3 \cdot 5} = \frac{2}{5}$$

$$\frac{8}{15} \cdot \frac{3}{4} = \frac{2 \cdot 2 \cdot 2}{3 \cdot 5} \cdot \frac{3}{2 \cdot 2} = \frac{2}{5}$$

Multiplying rational expressions works the same way.

Procedure   **Multiplying Rational Expressions:** To multiply rational expressions,

1. Factor each numerator and denominator completely.
2. Divide out any numerator factor with any matching denominator factor.
3. Multiply numerator by numerator and denominator by denominator.
4. Simplify as needed.

Example 1   Multiply $\dfrac{9x}{5y} \cdot \dfrac{15xy^2}{12}$.

SOLUTION $\dfrac{9x}{5y} \cdot \dfrac{15xy^2}{12} = \dfrac{3 \cdot 3 \cdot x}{5 \cdot y} \cdot \dfrac{3 \cdot 5 \cdot x \cdot y \cdot y}{2 \cdot 2 \cdot 3}$ **Factor the numerators and denominators completely.**

$= \dfrac{3 \cdot 3 \cdot x}{1} \cdot \dfrac{x \cdot y}{2 \cdot 2}$ **Divide out the common factors, which are 3, 5, and y.**

$= \dfrac{9x^2y}{4}$ **Multiply the remaining numerator factors and denominator factors.**

Multiply.

a. $-\dfrac{36m}{8n^4} \cdot \dfrac{14mn^2}{28n}$

b. $-\dfrac{9}{4xy^2} \cdot \dfrac{20xz}{15y}$

Now consider rational expressions that contain multiterm polynomials.

Example 2 Multiply.

a. $-\dfrac{2x}{3x-9} \cdot \dfrac{5x-15}{20x^3}$

SOLUTION $-\dfrac{2x}{3x-9} \cdot \dfrac{5x-15}{20x^3}$

$= -\dfrac{2 \cdot x}{3 \cdot (x-3)} \cdot \dfrac{5 \cdot (x-3)}{2 \cdot 2 \cdot 5 \cdot x \cdot x \cdot x}$ Factor the numerators and denominators completely.

$= -\dfrac{1}{3} \cdot \dfrac{1}{2 \cdot x \cdot x}$ Divide out the common factors, which are 2, 5, x, and $(x-3)$.

$= -\dfrac{1}{6x^2}$ Multiply the remaining numerator factors and denominator factors.

b. $\dfrac{12-4x}{15x} \cdot \dfrac{5x^2}{2x-6}$

SOLUTION $\dfrac{12-4x}{15x} \cdot \dfrac{5x^2}{2x-6} = \dfrac{4 \cdot (3-x)}{3 \cdot 5 \cdot x} \cdot \dfrac{5 \cdot x \cdot x}{2 \cdot (x-3)}$

$= \dfrac{2 \cdot 2 \cdot (-1)(x-3)}{3 \cdot 5 \cdot x} \cdot \dfrac{5 \cdot x \cdot x}{2 \cdot (x-3)}$

$= \dfrac{2 \cdot (-1)}{3} \cdot \dfrac{x}{1}$ Divide out the common factors, which are 2, 5, x, and $(x-3)$.

$= \dfrac{-2x}{3}$ Multiply the remaining numerator factors and denominator factors.

> **Note:** We can factor 4 further. Also, because $3-x$ and $x-3$ are additive inverses, we can factor -1 out of one of them.

c. $\dfrac{x^2+x-6}{9x^3} \cdot \dfrac{12x^2-6x}{2x^2-5x+2}$

SOLUTION $\dfrac{x^2+x-6}{9x^3} \cdot \dfrac{12x^2-6x}{2x^2-5x+2}$

$= \dfrac{(x-2)(x+3)}{3 \cdot 3 \cdot x \cdot x \cdot x} \cdot \dfrac{2 \cdot 3 \cdot x \cdot (2x-1)}{(x-2)(2x-1)}$ Factor the numerators and denominators completely.

$= \dfrac{x+3}{3 \cdot x \cdot x} \cdot \dfrac{2}{1}$ Divide out the common factors, which are $(x-2)$, 3, x, and $(2x-1)$.

$= \dfrac{2(x+3)}{3x^2}$ or $\dfrac{2x+6}{3x^2}$ Multiply the remaining numerator factors and denominator factors.

Answers Your Turn 1

a. $-\dfrac{9m^2}{4n^3}$ **b.** $\dfrac{3z}{y^3}$

Answers Your Turn 2

a. $\dfrac{4}{3h^2}$ **b.** $\dfrac{3}{y}$ **c.** $\dfrac{m-5}{2m}$

Your Turn 2 Multiply.

a. $\dfrac{-8h^2}{2h+10} \cdot -\dfrac{6h+30}{18h^4}$

b. $-\dfrac{9y-3}{y^2+3y} \cdot \dfrac{2y+6}{2-6y}$

c. $\dfrac{m^2-25}{20m} \cdot \dfrac{30m+10}{3m^2+16m+5}$

Objective **2** **Divide rational expressions.** In Section 1.4, we learned how to divide fractions using the rule $\dfrac{a}{b} \div \dfrac{c}{d} = \dfrac{a}{b} \cdot \dfrac{d}{c}$, where b, c, and d are not 0, which indicates to multiply by the reciprocal of the divisor. We can rewrite this rule so that it applies to dividing rational expressions.

Rule **Dividing Rational Expressions**

$\dfrac{P}{Q} \div \dfrac{R}{S} = \dfrac{P}{Q} \cdot \dfrac{S}{R}$, where P, Q, R, and S are polynomials and $Q \neq 0$, $R \neq 0$, and $S \neq 0$.

Following is a procedure for dividing rational expressions.

Procedure **Dividing Rational Expressions**

To divide rational expressions,

1. Write an equivalent multiplication statement with the reciprocal of the divisor.
2. Factor each numerator and denominator completely. (Steps 1 and 2 are interchangeable.)
3. Divide out any numerator factor with any matching denominator factor.
4. Multiply numerator by numerator and denominator by denominator.
5. Simplify as needed.

Example 3   Divide $-\dfrac{9x^2y}{28z^3} \div \dfrac{12y}{7z}$.

SOLUTION

$-\dfrac{9x^2y}{28z^3} \div \dfrac{12y}{7z} = -\dfrac{9x^2y}{28z^3} \cdot \dfrac{7z}{12y}$

Write an equivalent multiplication statement by changing the division sign to multiplication and changing the divisor to its reciprocal.

$= -\dfrac{3 \cdot 3 \cdot x \cdot x \cdot y}{2 \cdot 2 \cdot 7 \cdot z \cdot z \cdot z} \cdot \dfrac{7 \cdot z}{2 \cdot 2 \cdot 3 \cdot y}$

Factor the numerators and denominators completely.

$= -\dfrac{3 \cdot x \cdot x}{2 \cdot 2 \cdot z \cdot z} \cdot \dfrac{1}{2 \cdot 2}$

Divide out the common factors, which are 3, 7, y, and z.

$= -\dfrac{3x^2}{16z^2}$

Multiply the remaining factors.

Your Turn 3   Divide.

a. $\dfrac{15a^3}{28b^2} \div \dfrac{10ab}{7}$

b. $-\dfrac{2t}{33u^3v} \div \left(-\dfrac{8t^3v}{21u^4} \right)$

Now consider dividing rational expressions that contain multiterm polynomials.

Example 4 Divide.

a. $\dfrac{2x - 10}{15x^3} \div \dfrac{x^2 - 5x}{12x}$

SOLUTION

$$\dfrac{2x - 10}{15x^3} \div \dfrac{x^2 - 5x}{12x} = \dfrac{2x - 10}{15x^3} \cdot \dfrac{12x}{x^2 - 5x}$$

Write an equivalent multiplication statement by changing the division sign to multiplication and changing the divisor to its reciprocal.

$$= \dfrac{2 \cdot (x - 5)}{3 \cdot 5 \cdot x \cdot x \cdot x} \cdot \dfrac{2 \cdot 2 \cdot 3 \cdot x}{x \cdot (x - 5)}$$

Factor the numerators and denominators completely.

$$= \dfrac{2}{5 \cdot x \cdot x} \cdot \dfrac{2 \cdot 2}{x}$$

Divide out the common factors, which are 3, x, and $(x - 5)$.

$$= \dfrac{8}{5x^3}$$

Multiply the remaining factors.

b. $\dfrac{12x^2 + 3x}{2x^2 + 3x - 9} \div \dfrac{4x^2 + 9x + 2}{6x^3 + 18x^2}$

SOLUTION

$$\dfrac{12x^2 + 3x}{2x^2 + 3x - 9} \div \dfrac{4x^2 + 9x + 2}{6x^3 + 18x^2}$$

$$= \dfrac{12x^2 + 3x}{2x^2 + 3x - 9} \cdot \dfrac{6x^3 + 18x^2}{4x^2 + 9x + 2}$$

Write an equivalent multiplication statement by changing the division sign to multiplication and changing the divisor to its reciprocal.

$$= \dfrac{3x(4x + 1)}{(2x - 3)(x + 3)} \cdot \dfrac{6x^2(x + 3)}{(4x + 1)(x + 2)}$$

Factor the numerators and denominators completely.

$$= \dfrac{3 \cdot x \cdot (4x + 1)}{(2x - 3)(x + 3)} \cdot \dfrac{2 \cdot 3 \cdot x \cdot x \cdot (x + 3)}{(4x + 1) \cdot (x + 2)}$$

$$= \dfrac{3 \cdot x}{2x - 3} \cdot \dfrac{2 \cdot 3 \cdot x \cdot x}{x + 2}$$

Divide out the common factors, which are $(4x + 1)$ and $(x + 3)$.

$$= \dfrac{18x^3}{(2x - 3)(x + 2)}$$

Multiply the remaining factors.

Note: It is customary not to multiply out remaining binomial factors.

c. $\dfrac{8x^2 - 6x^3}{6xy} \div \dfrac{9x^2 - 16}{5y^2 - 10y}$

SOLUTION

$$\dfrac{8x^2 - 6x^3}{6xy} \div \dfrac{9x^2 - 16}{5y^2 - 10y} = \dfrac{8x^2 - 6x^3}{6xy} \cdot \dfrac{5y^2 - 10y}{9x^2 - 16}$$

Write an equivalent multiplication statement by changing the division sign to multiplication and changing the divisor to its reciprocal.

$$= \dfrac{2x^2(4 - 3x)}{6xy} \cdot \dfrac{5y(y - 2)}{(3x + 4)(3x - 4)}$$

Factor the numerators and denominators completely.

$$= \dfrac{2 \cdot x \cdot x \cdot (-1) \cdot (3x - 4)}{2 \cdot 3 \cdot x \cdot y} \cdot \dfrac{5 \cdot y \cdot (y - 2)}{(3x + 4) \cdot (3x - 4)}$$

Note: $4 - 3x$ and $3x - 4$ are additive inverses, so we factored -1 out of $4 - 3x$.

$$= \dfrac{x \cdot (-1)}{3} \cdot \dfrac{5 \cdot (y - 2)}{3x + 4}$$

Divide out the common factors, which are 2, x, y, and $(3x - 4)$.

$$= \dfrac{-5x(y - 2)}{3(3x + 4)} \text{ or } \dfrac{-5xy + 10x}{9x + 12} \text{ or } \dfrac{10x - 5xy}{9x + 12}$$

Multiply the remaining factors.

a. $\dfrac{6yz^3}{2y - y^2} \div \dfrac{8yz}{6 - 3y}$

b. $\dfrac{mn - 3m}{14m^4} \div \dfrac{n}{7n + 7}$

c. $\dfrac{4y^2 - 25}{9y^2 + 18y} \div \dfrac{2y^2 + 3y - 5}{3y^4 - 3y^3}$

d. $\dfrac{3x^2 + 10x + 8}{18 - 6x} \div \dfrac{10 + 5x}{4x^3 - 12x^2}$

Objective ❸ Convert units of measurement using dimensional analysis. We can convert from one unit of a measurement to another using a method called *dimensional analysis*, which involves writing measurement facts as conversion factors. A conversion factor is a ratio of equivalent measures.

Measurement fact:	Conversion factor:
1 foot = 12 inches	$\dfrac{1 \text{ foot}}{12 \text{ inches}}$ or $\dfrac{12 \text{ inches}}{1 \text{ foot}}$
1 pound = 16 ounces	$\dfrac{1 \text{ pound}}{16 \text{ ounces}}$ or $\dfrac{16 \text{ ounces}}{1 \text{ pound}}$

Note: Because a conversion factor is a ratio of equivalent measures, its ratio always equals 1.

To use dimensional analysis, we multiply a measurement by appropriate conversion factors so that the undesired units divide out, leaving only the desired units of measurement. For example, if we want to convert 6 feet to inches, we multiply by the conversion factor $\dfrac{12 \text{ inches}}{1 \text{ foot}}$ so that the units of feet divide out, leaving inches.

Connection The ratio of the conversion factor $\dfrac{12 \text{ in.}}{1 \text{ ft.}}$ is 1. So multiplying 6 ft. by this ratio means that the result, 72 in., is equal to 6 ft.

$$6 \text{ ft.} = \frac{6 \text{ ft.}}{1} \cdot \frac{12 \text{ in.}}{1 \text{ ft.}} = 72 \text{ in.}$$

Procedure   Using Dimensional Analysis to Convert between Units of Measurement

To convert units using dimensional analysis, multiply the given measurement by conversion factors so that the undesired units divide out, leaving the desired units.

Example 5   Convert.

a. 500 ounces to pounds

Connection Dividing out units in dimensional analysis is like dividing out common factors when multiplying rational expressions.

Solution There are 16 ounces to 1 pound. We write this fact as a unit ratio and multiply so that ounces divides out, leaving pounds.

$$500 \text{ oz.} = \frac{500 \text{ oz.}}{1} \cdot \frac{1 \text{ lb.}}{16 \text{ oz.}} = \frac{500}{16} \text{ lb.} = 31.25 \text{ lb.}$$

b. 6 miles to yards

Solution There are 5280 feet to 1 mile and 3 feet to 1 yard. Thus, we write these facts as unit fractions so that miles and feet divide out, leaving yards.

$$6 \text{ mi.} = \frac{6 \text{ mi.}}{1} \cdot \frac{5280 \text{ ft.}}{1 \text{ mi.}} \cdot \frac{1 \text{ yd.}}{3 \text{ ft.}} = \frac{31,680}{3} \text{ yd.} = 10,560 \text{ yd.}$$

Answers   Your Turn 4

a. $\dfrac{9z^2}{4y}$ **b.** $\dfrac{(n-3)(n+1)}{2m^3 n}$

c. $\dfrac{y^2(2y-5)}{3(y+2)}$ or $\dfrac{2y^3 - 5y^2}{3y + 6}$

d. $-\dfrac{2x^2(3x+4)}{15}$ or $-\dfrac{6x^3 + 8x^2}{15}$

c. 40 miles per hour to feet per second

SOLUTION First, 40 miles per hour means $\dfrac{40 \text{ miles}}{1 \text{ hour}}$. Notice that we must convert miles to feet and hours to seconds. For the distance conversion, we use the fact that there are 5280 feet to 1 mile. For the time conversion, there are 60 minutes in 1 hour and 60 seconds in 1 minute.

$$\frac{40 \text{ miles}}{1 \text{ hour}} = \frac{40 \text{ mi.}}{1 \text{ hr.}} \cdot \frac{5280 \text{ ft.}}{1 \text{ mi.}} \cdot \frac{1 \text{ hr.}}{60 \text{ min.}} \cdot \frac{1 \text{ min.}}{60 \text{ sec.}} = \frac{211{,}200 \text{ ft.}}{3600 \text{ sec.}} = 58.\overline{6} \text{ ft./sec.}$$

Your Turn 5 Convert.

 a. 192 ounces to pounds   **b.** 540 inches to yards   **c.** 44 feet per second into miles per hour

Answers Your Turn 5
a. 12 lb.   **b.** 15 yd.   **c.** 30 mph

7.2 Exercises   For Extra Help   *MyMathLab*

Note: Exercises marked with a ★ represent challenging exercises.

1. When multiplying rational expressions, why is it important to factor the numerators and denominators completely?

2. If after factoring you determine that there are two binomial factors that are additive inverses, how can you get them to be common binomial factors?

3. Why are $\dfrac{x}{y}$ and $\dfrac{y}{x}$ multiplicative inverses (or reciprocals)?

4. To divide rational expressions, we write an equivalent multiplication with the _____ _____ of the divisor.

5. What conversion factor would be used to convert 3.8 pounds to ounces?

6. Explain how to convert one unit of measurement to another using dimensional analysis.

For Exercises 7–42, multiply. See Examples 1 and 2.

7. $\dfrac{m}{n} \cdot \dfrac{2m}{5n}$

8. $\dfrac{x}{y} \cdot \dfrac{3x}{2y}$

9. $\dfrac{r^3}{s^3} \cdot \dfrac{s^5}{r^6}$

10. $\dfrac{a^5}{b^3} \cdot \dfrac{b^2}{a^3}$

11. $\dfrac{3}{y} \cdot \dfrac{y^3}{3}$

12. $\dfrac{y^4}{7} \cdot \dfrac{7}{y^2}$

13. $-\dfrac{n}{rs^2} \cdot \dfrac{rs}{mn^3}$

14. $\dfrac{ab}{xy} \cdot -\dfrac{x}{a^3b^2}$

15. $-\dfrac{6r^2s}{5r^2s^2} \cdot -\dfrac{15r^4s^2}{2r^3s^2}$

16. $-\dfrac{14mn^2}{8m^2n} \cdot -\dfrac{16mz^2}{7n^2z}$

17. $\dfrac{5n^2}{15m} \cdot \dfrac{n}{2} \cdot \dfrac{3m^2n}{n^2}$

18. $\dfrac{a}{b^2} \cdot \dfrac{3b^2}{4a} \cdot \dfrac{ab^2}{9}$

19. $\dfrac{3b - 12}{14} \cdot \dfrac{21}{5b - 20}$

20. $\dfrac{9}{2a + 4} \cdot \dfrac{3a + 6}{15}$

21. $\dfrac{2x - 6}{3y + 12} \cdot \dfrac{2y + 8}{5x - 15}$

22. $\dfrac{3a + 12}{5b - 30} \cdot \dfrac{3b - 18}{4a + 16}$

23. $\dfrac{4x^2 - 25}{3x - 4} \cdot \dfrac{9x^2 - 16}{2x + 5}$

24. $\dfrac{16x^2 - 9}{4x + 3} \cdot \dfrac{25x^2 - 1}{5x + 1}$

25. $\dfrac{r^2 + 3r}{r^2 - 36} \cdot \dfrac{r^2 - 6r}{r^2 - 9}$

26. $\dfrac{m^2 + 5m}{m^2 - 16} \cdot \dfrac{m^2 - 4m}{m^2 - 25}$

27. $\dfrac{12q}{q^2 - 2q + 1} \cdot \dfrac{q^2 - 1}{24q^2}$

28. $\dfrac{36x^3}{x^2 - 25} \cdot \dfrac{x^2 - 10x + 25}{9x}$

29. $\dfrac{w^2 - 4w - 5}{w^2 - 1} \cdot \dfrac{w^2 - w - 20}{w^2 - 10w + 25}$

30. $\dfrac{y^2 - 2y - 24}{y^2 - 16} \cdot \dfrac{y^2 + 10y + 25}{y^2 - y - 30}$

31. $\dfrac{n^2 - 6n + 9}{n^2 - 9} \cdot \dfrac{n^2 + 4n + 3}{n^2 - 3n}$

32. $\dfrac{4u^2 + 4u + 1}{2u^2 + u} \cdot \dfrac{u^2 + 3u - 4}{2u^2 - u - 1}$

33. $\dfrac{6x^2 + x - 12}{2x^2 - 5x - 12} \cdot \dfrac{3x^2 - 14x + 8}{9x^2 - 18x + 8}$

34. $\dfrac{2x^2 + x - 15}{4x^2 + 11x - 3} \cdot \dfrac{4x^2 - 9x + 2}{2x^2 - 9x + 10}$

35. $\dfrac{ac + 3a + 2c + 6}{ad + a + 2d + 2} \cdot \dfrac{ad - 5a + 2d - 10}{ac + 2c + 3a + 6}$

36. $\dfrac{mn + 2m + 4n + 8}{np - 3n + 2p - 6} \cdot \dfrac{pq + 5p - 3q - 15}{mq + 4m + 4q + 16}$

37. $\dfrac{x^3 + 27}{x^3 - 3x^2 + 9x} \cdot \dfrac{x^2 - 7x + 10}{x^2 - 2x - 15}$

38. $\dfrac{x^2 + 6x + 8}{x^3 + 5x^2 + 25x} \cdot \dfrac{x^3 - 125}{x^2 - x - 20}$

39. $\dfrac{x^2 + 7xy + 10y^2}{x^2 + 6xy + 5y^2} \cdot \dfrac{x + 2y}{y} \cdot \dfrac{x + y}{x^2 + 4xy + 4y^2}$

40. $\dfrac{x^2 + 6xy + 9y^2}{x^2 + xy - 6y^2} \cdot \dfrac{x + 4y}{x} \cdot \dfrac{x - 2y}{x^2 + 7xy + 12y^2}$

41. $\dfrac{4x^2y^3}{2x - 6} \cdot \dfrac{12 - 4x}{6x^3y}$

42. $\dfrac{6b^3c^3}{70 - 10b} \cdot \dfrac{5b - 35}{8b^2c^5}$

43. $\dfrac{2x^2 - 11x + 12}{2x^2 + 11x - 21} \cdot \dfrac{3x^2 + 20x - 7}{4 - x}$

44. $\dfrac{3x^2 - 17x + 10}{3x^2 + 7x - 6} \cdot \dfrac{2x^2 + x - 15}{5 - x}$

For Exercises 45–74, divide. See Examples 3 and 4.

45. $\dfrac{x}{y^4} \div \dfrac{x^3}{y^2}$

46. $\dfrac{a^3}{b^3} \div \dfrac{a^5}{b^2}$

47. $\dfrac{2m}{3n^2} \div \dfrac{8m^3}{15n}$

48. $\dfrac{21c^2}{7d^3} \div \dfrac{8c^4}{12d^2}$

49. $-\dfrac{12w^2y}{5z^2} \div \left(-\dfrac{12w^2}{y}\right)$

50. $-\dfrac{7a^2b}{2c^2} \div \left(-\dfrac{7a^2}{b}\right)$

51. $\dfrac{8x^7g^3}{15x^2g^4} \div \left(-\dfrac{3xg^2}{4x^2g^3}\right)$

52. $-\dfrac{10c^5g^5}{12c^2g^4} \div \dfrac{8cg^2}{4c^2g}$

53. $\dfrac{4a - 8}{15a} \div \dfrac{3a - 6}{5a^2}$

54. $\dfrac{6b + 24}{7b^2} \div \dfrac{7b + 28}{14b^3}$

55. $\dfrac{4p + 12q}{3p - 6q} \div \dfrac{5p + 15q}{6p - 12q}$

56. $\dfrac{5r + 10s}{3r - 12s} \div \dfrac{6r + 12s}{4r - 16s}$

57. $\dfrac{a^2 - b^2}{x^2 - y^2} \div \dfrac{a + b}{x - y}$

58. $\dfrac{x + 4}{y - 3} \div \dfrac{x^2 - 16}{y^2 - 9}$

59. $\dfrac{x^2 + 16x + 64}{x^2 - 16x + 64} \div \dfrac{x + 8}{x - 8}$

60. $\dfrac{w + 3}{w - 3} \div \dfrac{w^2 + 6w + 9}{w^2 - 6w + 9}$

61. $\dfrac{x^2 + 7x + 10}{x^2 + 2x} \div \dfrac{x + 5}{x^2 - 4x}$

62. $\dfrac{t^2 - 2t}{t + 1} \div \dfrac{t^2 + 3t}{t^2 + 4t + 3}$

63. $\dfrac{a^2 - b^2}{a^2 + 2ab + b^2} \div \dfrac{a^2 - 3ab + 2b^2}{a^2 + 3ab + 2b^2}$

64. $\dfrac{3w^2 - 7w - 6}{w^2 - 9} \div \dfrac{3w^2 - 10w - 8}{3w^2 + 7w - 6}$

65. $\dfrac{3k + 2}{3k^4 + 2k^3} \div \dfrac{10k^2 + 3k - 1}{5k^3 - k^2}$

66. $\dfrac{5y - 3}{4y^2 - y - 3} \div \dfrac{5y^3 - 3y^2}{4y^2 + 3y}$

67. $\dfrac{8x^3 - 27}{3x^2 + 20x - 7} \div \dfrac{2x^2 - 7x + 6}{x^2 + 5x - 14}$

68. $\dfrac{x^2 + 6x - 16}{3x^2 - 10x + 8} \div \dfrac{2x^2 + 13x - 24}{27x^3 - 64}$

69. $\dfrac{x^2 - 2x - 8}{x^2 + 6x + 8} \div (x^2 - 3x - 4)$

70. $\dfrac{u^2 + 7u + 12}{u^2 + u - 6} \div (u^2 + u - 12)$

71. $\dfrac{ab + 3a + 2b + 6}{bc + 4b + 3c + 12} \div \dfrac{ac - 3a + 2c - 6}{bc + 4b - 4c - 16}$

72. $\dfrac{xy - 3x + 4y - 12}{xy + 6x - 3y - 18} \div \dfrac{xy + 5x + 4y + 20}{xy + 5x - 3y - 15}$

73. $\dfrac{b^2 - 16}{b^2 + b - 12} \div \dfrac{4 - b}{b^2 + 2b - 15}$

74. $\dfrac{y^2 - y - 20}{y^2 - 2y - 24} \div \dfrac{5 - y}{y^2 - 3y - 18}$

★ **For Exercises 75–80, perform the indicated operations. (Remember that the order of operations is to multiply or divide from left to right.)**

75. $\dfrac{y^2 + y - 6}{2y^2 - 3y - 2} \cdot \dfrac{2y^2 + 9y + 4}{y^2 + 7y + 12} \div \dfrac{4y + y^2}{7 + y}$

76. $\dfrac{2m^2 - 7m - 15}{4m^2 - 9} \cdot \dfrac{2m^2 - 5m - 3}{2m^2 - 7m + 3} \div \dfrac{m^2 - 5m}{2m - 1}$

77. $\dfrac{6x^2 - 5x - 6}{x^4 - 81} \div \dfrac{3x - 2x^2}{x^3 + 3x^2 + 9x + 27} \div \dfrac{3x + 2}{x^4}$

78. $\dfrac{12h^2 + 11h - 5}{h^4 - 16} \div \dfrac{h - 3h^2}{h^3 + 4h - 2h^2 - 8} \div \dfrac{4h + 5}{h^3}$

79. $\dfrac{6x^2}{x^2 - 9} \cdot \dfrac{3x^2 + x - 24}{20y - 6xy} \cdot \dfrac{x - 3}{3x - 8} \div \dfrac{3x^2 + 9x}{3x^2 - x - 30}$

80. $\dfrac{8y^3}{3y^2 - 4y - 4} \div \dfrac{8xy - 20x}{2y^2 - 9y + 10} \cdot \dfrac{y + 5}{7 - 4y} \cdot \dfrac{12y^2 - 13y - 14}{2y^2 + 10y}$

For Exercises 81–84, explain the mistake; then correct the mistake.

81. $\dfrac{5y}{9} \div \dfrac{3y}{10} = \dfrac{y}{3} \cdot \dfrac{y}{2} = \dfrac{y^2}{6}$

82. $\dfrac{12x^3y}{5} \cdot \dfrac{3y}{6x^2} = \dfrac{2xy}{5} \cdot \dfrac{3y}{1} = \dfrac{6y^2}{5}$

83. $\dfrac{x^2 - 9}{15y} \cdot \dfrac{y - 2}{x + 3} = \dfrac{(x - 3)(x + 3)}{15y} \cdot \dfrac{y - 2}{x + 3}$

$\quad\quad = \dfrac{x - 3}{15} \cdot (-2)$

$\quad\quad = \dfrac{-2x + 6}{15}$

84. $\dfrac{x^2 + 3x + 2}{x^2 - 1} \cdot \dfrac{x - 1}{x + 1} = \dfrac{(x + 1)(x + 2)}{1} \cdot \dfrac{1}{x + 1}$

$\quad\quad = x + 2$

For Exercises 85–90, use dimensional analysis and the following facts to convert units of length. See Example 5.

$$1 \text{ foot} = 12 \text{ inches} \quad\quad 1 \text{ yard} = 3 \text{ feet} \quad\quad 1 \text{ mile} = 5280 \text{ feet}$$

85. 18.5 feet to inches

86. 24.2 miles to feet

87. 6 miles to yards

88. 2400 yards to miles

89. 8 yards to inches

90. 216 inches to yards

For Exercises 91–96, use dimensional analysis and the following values to convert units of weight. See Example 5.

$$1 \text{ pound} = 16 \text{ ounces} \quad\quad 1 \text{ ton} = 2000 \text{ pounds}$$

91. 5.5 pounds to ounces

92. 36 ounces to pounds

93. 25,400 pounds to tons

94. 2.5 tons to pounds

95. 3 tons to ounces

96. 48,000 ounces to tons

For Exercises 97–104, use dimensional analysis. See Example 5.

97. The fastest winning speed at the Daytona 500 was 177.602 miles per hour by Buddy Baker in 1980. How many feet per second is this? (*Source:* www.daytonainternationalspeedway.com)

98. The speed of sound is 741.8 miles per hour at 32°F at sea level. How many feet per second is this?

99. A sprinter's best time in the 40-yard dash is 4.3 seconds. What is that speed in miles per hour?

100. A runner knows that she can run 400 yards in 87.2 seconds. If she can maintain that pace, how many minutes will it take her to run a mile?

101. The mean orbital velocity of Earth is 29.78 kilometers per second. How many kilometers does Earth orbit in one year? (Hint: One year is $365\frac{1}{4}$ days.)

(*Source:* Windows to the Universe)

102. The mean orbital velocity of Pluto is 4.74 kilometers per second. How many kilometers does Pluto orbit in one year? (*Source:* Windows to the Universe)

★**103.** One light-year is the distance light travels in a year. If light travels at a speed of approximately 186,000 miles per second, how many miles does light travel in one year?

★**104.** The light from the Sun takes about 8 minutes to reach Earth. Using the fact that light travels at a speed of about 186,000 miles per second, determine Earth's distance from the Sun in miles.

Puzzle Problem

If $\frac{1}{1000}$ of a light-year is about 9.46×10^9 kilometers, find the speed of light in meters per second.

Of Interest

The nearest star to the Sun is Proxima Centauri, which is about 4.2 light-years away. This means that when we look at Proxima Centauri, we are seeing an image of that star that is 4.2 years old. Some objects we see in the night sky are much farther from Earth. For example, the Andromeda galaxy is about 2,140,000 light-years from Earth, which means that when we look at it, we are seeing it as it was 2,140,000 years ago.

For Exercises 105–110, use dimensional analysis and the following table of exchange rates as of 1/11/08 to convert. See Example 5.

	USD($)	GBP(£)	CAN($)	EUR(€)
USD($)	1	1.9580	0.981258	1.47895
GBP(£)	0.510725	1	0.500951	0.755093
CAN($)	1.01910	1.99620	1	1.50761
EUR(€)	0.676155	1.32434	0.6633	1

Source: Federal Reserve Bank of New York

Note: Each number is the ratio of the row's unit to 1 of the column's unit. For example, 1.50761 means that there are 1.50761 Can$ to 1€.

105. Convert $500 into £ (Great Britain pound).

106. Convert $250 into Can$ (Canadian dollars).

107. Convert 450 € (euros) into U.S. dollars.

108. Convert 700 £ into U.S. dollars.

★**109.** Convert 650 £ to €.

110. Convert 400 Can$ to €.

Review Exercises

Exercises 1–6 EXPRESSIONS

For Exercises 1–6, simplify.

[1.3] **1.** $\frac{1}{3} + \frac{2}{3}$

[1.3] **2.** $\frac{1}{4} - \frac{3}{4}$

[1.7] **3.** $\frac{1}{2}x + \frac{3}{2}x$

[1.7] **4.** $\frac{3}{8}y - 5 - \frac{1}{8}y + 1$

[5.3] **5.** $(6x^2 + x - 1) + (2x^2 - 7x - 3)$

[5.3] **6.** $(3n^2 - 5n + 7) - (4n^2 - 5n + 3)$

Adding and Subtracting Rational Expressions with the Same Denominator

Objective 1 Add or subtract rational expressions with the same denominator.

Recall that when adding or subtracting fractions with the same denominator, we add or subtract the numerators and keep the same denominator.

$$\frac{3}{7} + \frac{1}{7} = \frac{3+1}{7} = \frac{4}{7} \quad \text{or} \quad \frac{7}{9} - \frac{2}{9} = \frac{7-2}{9} = \frac{5}{9}$$

We add and subtract rational expressions the same way.

$$\frac{3x}{7} + \frac{x}{7} = \frac{3x+x}{7} = \frac{4x}{7} \quad \text{or} \quad \frac{7x}{9} - \frac{2x}{9} = \frac{7x-2x}{9} = \frac{5x}{9}$$

Procedure **Adding or Subtracting Rational Expressions (Same Denominator)**

To add or subtract rational expressions that have the same denominator,

1. Add or subtract the numerators and keep the same denominator.
2. Simplify to lowest terms. (Remember to factor the numerators and denominators completely in order to simplify.)

Example 1 Add $\dfrac{5x}{14} + \dfrac{x}{14}$.

SOLUTION
$$\frac{5x}{14} + \frac{x}{14} = \frac{5x+x}{14}$$

Because the rational expressions have the same denominator, we add numerators and keep the same denominator.

$$= \frac{6x}{14}$$

Combine like terms.

$$= \frac{2 \cdot 3 \cdot x}{2 \cdot 7}$$

Factor.

$$= \frac{3x}{7}$$

Divide out the common factor, 2.

Your Turn 1 Subtract $\dfrac{2a^2}{15} - \dfrac{7a^2}{15}$.

Numerators with No Like Terms

In Example 1, the numerators were like terms. If the numerators have no like terms, we write a multiterm polynomial expression in the numerator of the sum or difference.

Answer **Your Turn 1**

$-\dfrac{a^2}{3}$

Example 2 Subtract.

a. $\dfrac{y}{y+5} - \dfrac{3}{y+5}$

SOLUTION $\dfrac{y}{y+5} - \dfrac{3}{y+5} = \dfrac{y-3}{y+5}$

Subtract numerators and keep the same denominator. Because y and 3 are not like terms, we express their difference as a polynomial.

Warning: Although it may be tempting to do so, we *cannot* divide out the y's because they are terms, not factors.

b. $\dfrac{n^2}{n+3} - \dfrac{9}{n+3}$

SOLUTION $\dfrac{n^2}{n+3} - \dfrac{9}{n+3} = \dfrac{n^2-9}{n+3}$

Note: The numerator can be factored, so we may be able to simplify.

$= \dfrac{(n-3)(n+3)}{n+3}$ Factor the numerator.

$= n - 3$ Divide out the common factor, $n+3$.

c. $\dfrac{4m}{15} - \left(-\dfrac{8}{15}\right)$

SOLUTION $\dfrac{4m}{15} - \left(-\dfrac{8}{15}\right) = \dfrac{4m-(-8)}{15}$

$= \dfrac{4m+8}{15}$

Note: $4m + 8$ can be factored.

$4m + 8 = 4(m + 2)$

However, because there are no common factors in the denominator, we cannot simplify further.

Your Turn 2 Add or subtract.

a. $\dfrac{x}{x-4} + \dfrac{6}{x-4}$ **b.** $\dfrac{16n^2}{4n+5} - \dfrac{25}{4n+5}$ **c.** $\dfrac{5x^2}{7} - \left(-\dfrac{2x}{7}\right)$

Numerators That Are Multiterm Polynomials

So far, the numerators in the given rational expressions were monomials. Now look at some cases where the numerators in the given rational expressions are multiterm polynomials. Remember from Chapter 5 that to add polynomials, we combine like terms.

Example 3 Add.

a. $\dfrac{7y-13}{5} + \dfrac{y+8}{5}$

SOLUTION $\dfrac{7y-13}{5} + \dfrac{y+8}{5} = \dfrac{(7y-13)+(y+8)}{5}$

Combine $7y$ with y to get $8y$ and combine -13 with 8 to get -5.

$= \dfrac{8y-5}{5}$

Warning: It may be tempting to divide out the 5, but the 5 in $8y - 5$ is not a factor. Because $8y - 5$ cannot be factored, we cannot simplify this rational expression.

Answers Your Turn 2

a. $\dfrac{x+6}{x-4}$ **b.** $4n-5$

c. $\dfrac{5x^2+2x}{7}$

b. $\dfrac{x^2 - 5x - 13}{x - 6} + \dfrac{3x - 11}{x - 6}$

SOLUTION $\dfrac{x^2 - 5x - 13}{x - 6} + \dfrac{3x - 11}{x - 6} = \dfrac{(x^2 - 5x - 13) + (3x - 11)}{x - 6}$

Note: Because the numerator can be factored, we may be able to simplify.

$= \dfrac{x^2 - 2x - 24}{x - 6}$ Combine like terms in the numerator.

$= \dfrac{(x - 6)(x + 4)}{x - 6}$ Factor the numerator.

$= x + 4$ Divide out the common factor, $x - 6$.

Your Turn 3   Add.

a. $\dfrac{n + 7}{11n} + \dfrac{2 - 5n}{11n}$ **b.** $\dfrac{3t^2 - 5t}{t - 3} + \dfrac{t^2 - 8t + 3}{t - 3}$

Simplifying When the Denominator Is Factorable

Sometimes you also may be able to factor the denominator. If the denominator can be factored, you will need to write it in factored form in order to simplify.

Example 4   Add $\dfrac{4x^2 + 9x + 1}{2x^3 - 18x} + \dfrac{6x^2 + 23x + 5}{2x^3 - 18x}$.

SOLUTION $\dfrac{4x^2 + 9x + 1}{2x^3 - 18x} + \dfrac{6x^2 + 23x + 5}{2x^3 - 18x} = \dfrac{(4x^2 + 9x + 1) + (6x^2 + 23x + 5)}{2x^3 - 18x}$

$= \dfrac{10x^2 + 32x + 6}{2x^3 - 18x}$ Combine like terms in the numerator.

$= \dfrac{2(5x^2 + 16x + 3)}{2x(x^2 - 9)}$ Factor 2 out of the numerator and $2x$ out of the denominator.

$= \dfrac{2(5x + 1)(x + 3)}{2x(x - 3)(x + 3)}$ Factor the trinomial in the numerator and the difference of squares in the denominator.

$= \dfrac{5x + 1}{x(x - 3)}$ or $\dfrac{5x + 1}{x^2 - 3x}$ Divide out the common factors, 2 and $x + 3$.

Your Turn 4   Add $\dfrac{x^3 + 3x^2 - 7x}{3x^4 - 12x^2} + \dfrac{5x - 2x^2}{3x^4 - 12x^2}$.

Subtracting with Numerators That Are Multiterm Polynomials

Now let's look at subtraction of rational expressions that have the same denominator and multiterm polynomials in the numerators. Recall from Chapter 5 that when subtracting polynomials, we write an equivalent addition with the additive inverse of the subtrahend.

Answers   Your Turn 3

a. $\dfrac{9 - 4n}{11n}$ **b.** $4t - 1$

Answer   Your Turn 4

$\dfrac{x - 1}{3x(x - 2)}$ or $\dfrac{x - 1}{3x^2 - 6x}$

Example 5   Subtract.

a. $\dfrac{7t^2 - t + 5}{t + 2} - \dfrac{3t^2 - 6t - 2}{t + 2}$

SOLUTION $\dfrac{7t^2 - t + 5}{t + 2} - \dfrac{3t^2 - 6t - 2}{t + 2} = \dfrac{(7t^2 - t + 5) - (3t^2 - 6t - 2)}{t + 2}$

Note: To write an equivalent addition, change the operation symbol from a minus sign to a plus sign and change all of the signs in the subtrahend (second) polynomial.

$$= \dfrac{(7t^2 - t + 5) + (-3t^2 + 6t + 2)}{t + 2}$$

$$= \dfrac{4t^2 + 5t + 7}{t + 2} \qquad \text{Combine like terms.} \quad 4t^2 + 5t + 7 \text{ is prime.}$$

b. $\dfrac{x^3 + 4x - 3}{5x^2 - 10x} - \dfrac{4x + 5}{5x^2 - 10x}$

SOLUTION $\dfrac{x^3 + 4x - 3}{5x^2 - 10x} - \dfrac{4x + 5}{5x^2 - 10x} = \dfrac{(x^3 + 4x - 3) - (4x + 5)}{5x^2 - 10x}$

$$= \dfrac{(x^3 + 4x - 3) + (-4x - 5)}{5x^2 - 10x} \qquad \text{Write the equivalent addition.}$$

Note: The numerator is a difference of cubes.

$$= \dfrac{x^3 - 8}{5x^2 - 10x} \qquad \text{Combine like terms.}$$

$$= \dfrac{(x - 2)(x^2 + 2x + 4)}{5x(x - 2)} \qquad \text{Factor the numerator and denominator. Divide out the common factor, } x - 2.$$

$$= \dfrac{x^2 + 2x + 4}{5x} \qquad x^2 + 2x + 4 \text{ is prime.}$$

Your Turn 5   Subtract.

a. $\dfrac{3t^2 - 4t - 7}{t^2 + 10t + 21} - \dfrac{2t^2 - 6t - 4}{t^2 + 10t + 21}$ **b.** $\dfrac{y^3 + y - 20}{3y^2 - 7y - 6} - \dfrac{y + 7}{3y^2 - 7y - 6}$

Answers   Your Turn 5

a. $\dfrac{t - 1}{t + 7}$ **b.** $\dfrac{y^2 + 3y + 9}{3y + 2}$

7.3 Exercises   For Extra Help   *MyMathLab* 

1. Two rational expressions with monomial numerators are added to form a rational expression whose numerator is a multiterm polynomial. Why would this happen?

2. After adding or subtracting two rational expressions, what is the first step in simplifying the resulting rational expression?

3. Two rational expressions that have numerators that are multiterm polynomials and the same denominator are to be subtracted. Explain the process.

4. After adding or subtracting two rational expressions, the resulting rational expression contains a multiterm polynomial in the numerator that can be factored. Does this mean that the rational expression can be simplified? Explain.

For Exercises 5–50, add or subtract. Simplify your answers to lowest terms. See Examples 1–5.

5. $\dfrac{3x}{10} + \dfrac{2x}{10}$

6. $\dfrac{2x}{9} + \dfrac{x}{9}$

7. $\dfrac{4a^2}{9} + \left(-\dfrac{a^2}{9}\right)$

8. $\dfrac{6x^3}{11} + \left(-\dfrac{x^3}{11}\right)$

9. $\dfrac{6}{a} - \left(-\dfrac{3}{a}\right)$

10. $\dfrac{3}{w} - \left(-\dfrac{2}{w}\right)$

11. $\dfrac{6x}{7y^2} - \dfrac{2x}{7y^2} - \dfrac{5x}{7y^2}$

12. $\dfrac{9a}{11x^2} - \dfrac{8a}{11x^2} - \dfrac{3a}{11x^2}$

13. $\dfrac{n}{n+5} + \dfrac{3}{n+5}$

14. $\dfrac{n}{n+2} + \dfrac{7}{n+2}$

15. $\dfrac{2r}{r+2} + \dfrac{r}{r+2}$

16. $\dfrac{5z}{b-6} + \dfrac{7z}{b-6}$

17. $\dfrac{1-2q}{3q} + \dfrac{5q+2}{3q}$

18. $\dfrac{6x-3}{4x} + \dfrac{2x+5}{4x}$

19. $\dfrac{19u+6}{5u} - \dfrac{4u+6}{5u}$

20. $\dfrac{8y+5}{3y} - \dfrac{2y+5}{3y}$

21. $\dfrac{2-2a}{2a-3} + \dfrac{8a-11}{2a-3}$

22. $\dfrac{4-2x}{3x-2} + \dfrac{14x-12}{3x-2}$

23. $\dfrac{w^2+5}{w+1} + \dfrac{6-w^2}{w+1}$

24. $\dfrac{x^2+2}{x+1} + \dfrac{6-x^2}{x+1}$

25. $\dfrac{4m+3}{m-1} - \dfrac{m+4}{m-1}$

26. $\dfrac{2x+5}{x+1} - \dfrac{x+8}{x+1}$

27. $\dfrac{1-3t}{2t-3} - \dfrac{8t-2}{2t-3}$

28. $\dfrac{3-4a}{5a-2} - \dfrac{2a-6}{5a-2}$

29. $\dfrac{2w+3}{w+5} + \dfrac{5w-2}{w+5} - \dfrac{w+6}{w+5}$

30. $\dfrac{6q+21}{q+1} - \dfrac{2q+8}{q+1} + \dfrac{6q-5}{q+1}$

31. $\dfrac{g}{g^2-9} - \dfrac{3}{g^2-9}$

32. $\dfrac{x}{x^2-49} + \dfrac{7}{x^2-49}$

33. $\dfrac{z^2+2z}{z-7} + \dfrac{21-12z}{z-7}$

34. $\dfrac{u^2+5u}{u+1} + \dfrac{2-2u}{u+1}$

35. $\dfrac{2s}{s^2+4s+4} - \dfrac{s-2}{s^2+4s+4}$

36. $\dfrac{v+8}{v^2+8v+16} - \dfrac{4}{v^2+8v+16}$

37. $\dfrac{y^2}{y^2+3y} - \dfrac{9}{y^2+3y}$

38. $\dfrac{x^2}{x^2-5x} - \dfrac{25}{x^2-5x}$

39. $\dfrac{x^2+3x}{x^2-x-12} + \dfrac{2x+6}{x^2-x-12}$

40. $\dfrac{y^2+2y}{y^2+5y+4} + \dfrac{5y+12}{y^2+5y+4}$

41. $\dfrac{t^2-4t}{t^2+10t+21} - \dfrac{-6t+3}{t^2+10t+21}$

42. $\dfrac{v^2+6}{v^2-3v-10} - \dfrac{2v+14}{v^2-3v-10}$

43. $\dfrac{3x^3}{x^2-2x+4} + \dfrac{24}{x^2-2x+4}$

44. $\dfrac{2x^3}{x^2-4x+16} + \dfrac{128}{x^2-4x+16}$

45. $\dfrac{x-5}{x^2-25} + \dfrac{2x+10}{x^2-25} + \dfrac{-2x}{x^2-25}$

46. $\dfrac{3x-3}{x^2-1} + \dfrac{2x+2}{x^2-1} + \dfrac{-4x}{x^2-1}$

47. $\dfrac{3x^2+2x}{x^2-2x-15} + \dfrac{3x-8}{x^2-2x-15} + \dfrac{2x+2}{x^2-2x-15}$

48. $\dfrac{m^2-6m}{m^2-5m+6} + \dfrac{2m-4}{m^2-5m+6} + \dfrac{6m-4}{m^2-5m+6}$

49. $\dfrac{2x^2+x-3}{x^2+6x+5} + \dfrac{x^2-2x+4}{x^2+6x+5} - \dfrac{2x^2+x+4}{x^2+6x+5}$

50. $\dfrac{3x^2-2x+3}{x^2+x-12} - \dfrac{x^2+x-2}{x^2+x-12} - \dfrac{x^2+4x-7}{x^2+x-12}$

Find ✕ **the mistake**   *For Exercises 51–54, explain the mistake; then find the correct sum or difference.*

51. $\dfrac{4y}{5} + \dfrac{1}{5} = \dfrac{5y}{5} = y$

52. $\dfrac{7}{m} + \dfrac{n}{m} = \dfrac{7+n}{2m}$

53.
$$\dfrac{x+1}{3} - \dfrac{x-5}{3} = \dfrac{(x+1)-(x-5)}{3}$$
$$= \dfrac{(x+1)+(-x-5)}{3}$$
$$= -\dfrac{4}{3}$$

54.
$$\dfrac{x}{x-2} + \dfrac{x-4}{x-2} = \dfrac{x^2-4}{x-2}$$
$$= \dfrac{(x+2)(x-2)}{x-2}$$
$$= x+2$$

For Exercises 55–58, find the perimeter of the figure shown.

55.

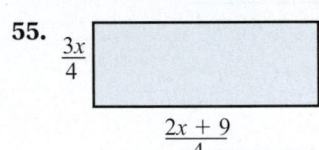

$\dfrac{3x}{4}$ $\dfrac{2x+9}{4}$

56.

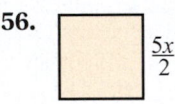

$\dfrac{5x}{2}$

57.

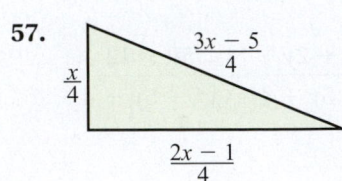

$\dfrac{x}{4}$ $\dfrac{3x-5}{4}$ $\dfrac{2x-1}{4}$

58.

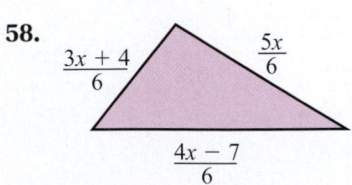

$\dfrac{3x+4}{6}$ $\dfrac{5x}{6}$ $\dfrac{4x-7}{6}$

Review Exercises

Exercises 1–6 **EXPRESSIONS**

For Exercises 1–6, simplify.

[1.3] **1.** $\dfrac{1}{2} + \dfrac{3}{5}$
[1.3] **2.** $\dfrac{5}{6} - \dfrac{3}{8}$
[1.3] **3.** $\dfrac{4}{7} - \left(-\dfrac{3}{5}\right)$
[1.7] **4.** $\dfrac{1}{3}x - \dfrac{1}{6} + \dfrac{2}{5}x + \dfrac{3}{4}$

[5.3] **5.** $\left(x^2 - 9x - \dfrac{5}{8}\right) + \left(x^2 - 3x + \dfrac{1}{6}\right)$
[5.3] **6.** $\left(3y^2 + \dfrac{3}{4}y - 8\right) - \left(y^2 + \dfrac{4}{5}y + 2\right)$

7.4 Adding and Subtracting Rational Expressions with Different Denominators

Objectives

1 Find the LCD of two or more rational expressions.

2 Given two rational expressions, write equivalent rational expressions with their LCD.

3 Add or subtract rational expressions with different denominators.

Adding or subtracting rational expressions with different denominators follows three main stages: (1) finding the LCD, (2) rewriting the expressions with the LCD, and (3) adding or subtracting the numerators. We explore the first two stages in Objectives 1 and 2, then put them back in their proper context in Objective 3.

Objective 1 **Find the LCD of two or more rational expressions.** Remember that when adding or subtracting fractions with different denominators, we must first find a common denominator. It is helpful to use the least common denominator (LCD), which is the smallest number that is evenly divisible by all of the denominators. For example, given the fractions $\dfrac{1}{8}$ and $\dfrac{7}{12}$, the LCD is 24.

To help us find the LCD when denominators are more complex, as in rational expressions, let's consider a method involving prime factorizations. Consider the prime factorizations of the denominators, 8 and 12, compared to the prime factorization of the LCD, 24, to determine a procedure.

$$8 = 2^3 \qquad 12 = 2^2 \cdot 3$$

$$\text{LCD} = 2^3 \cdot 3 = 24$$

Both 8 and 12 have factors of 2, but 8 has the greatest number of 2's, which is why 2^3 is included in the factorization for the LCD instead of 2^2. Similarly, 12 has the greatest number of factors of 3 (a single factor, whereas 8 has no factors of 3), so a single factor of 3 is included in the factorization of the LCD. This suggests the following procedure.

Finding the LCD

To find the LCD of two or more rational expressions,

1. Find the prime factorization of each denominator.
2. Write a product that contains each unique prime factor the greatest number of times that factor occurs in any factorization. Or if you prefer to use exponents, write the product that contains each unique prime factor raised to the greatest exponent that occurs on that factor in any factorization.
3. Simplify the product found in Step 2.

First, let's consider cases where the denominators are monomials.

Example 1   Find the LCD of $\dfrac{3}{4x^2}$ and $\dfrac{5}{6x^3}$.

SOLUTION   We first factor the denominators $4x^2$ and $6x^3$ by writing their prime factorizations.

$$4x^2 = 2^2 \cdot x^2$$
$$6x^3 = 2 \cdot 3 \cdot x^3$$

The unique factors are 2, 3, and x. To generate the LCD, include 2, 3, and x the greatest number of times each appears in any of the factorizations.

The greatest number of times that 2 appears is twice (in $2^2 \cdot x^2$).
The greatest number of times that 3 appears is once (in $2 \cdot 3 \cdot x^3$).
The greatest number of times that x appears is three times (in $2 \cdot 3 \cdot x^3$).

$$\text{LCD} = 2^2 \cdot 3 \cdot x^3 = 12x^3$$

Your Turn 1   Find the LCD.

a. $\dfrac{y}{12x^5}$ and $\dfrac{5}{9x^2}$

b. $\dfrac{p}{10mn}$ and $\dfrac{7}{15n^2}$

Now consider denominators that are multiterm polynomials.

Example 2   Find the LCD.

a. $\dfrac{7}{5x - 15}$ and $\dfrac{5}{2x - 6}$

SOLUTION   Factor the denominators $5x - 15$ and $2x - 6$.

$$5x - 15 = 5(x - 3)$$
$$2x - 6 = 2(x - 3)$$

The unique factors are 2, 5, and $x - 3$.
The greatest number of times that $x - 3$ appears in any of the factorizations is once.
The greatest number of times that 2 appears is once [in $2(x - 3)$].
The greatest number of times that 5 appears is once [in $5(x - 3)$].

$$\text{LCD} = 2 \cdot 5 \cdot (x - 3) = 10(x - 3) \text{ or } 10x - 30$$

Answers   Your Turn 1
a. $36x^5$ b. $30mn^2$

b. $\dfrac{7}{x^2 + 10x + 25}$ and $\dfrac{17}{6x^4 + 30x^3}$

SOLUTION Factor $x^2 + 10x + 25$ and $6x^4 + 30x^3$.

$$x^2 + 10x + 25 = (x + 5)^2$$
$$6x^4 + 30x^3 = 6x^3(x + 5) = 2 \cdot 3 \cdot x^3(x + 5)$$

The factors are 2, 3, x, and $x + 5$.

The greatest number of times that 2 appears is once [in $2 \cdot 3 \cdot x^3(x + 5)$].

The greatest number of times that 3 appears is once [in $2 \cdot 3 \cdot x^3(x + 5)$].

The greatest number of times that $x + 5$ appears is twice [in $(x + 5)^2$].

The greatest number of times that x appears is three times [in $2 \cdot 3 \cdot x^3(x + 5)$].

$$\text{LCD} = 2 \cdot 3 \cdot x^3 \cdot (x + 5)^2 = 6x^3(x + 5)^2$$

Your Turn 2 Find the LCD.

a. $\dfrac{y}{3y + 6}$ and $\dfrac{5}{4y + 8}$

b. $\dfrac{9}{x^2 - 2x + 1}$ and $\dfrac{17}{2x^3 - 2x}$

Objective ② Given two rational expressions, write equivalent rational expressions with their LCD. Writing equivalent rational expressions is much like writing equivalent fractions, which we reviewed in Section 1.2. Remember that to write an equivalent fraction, we multiply the numerator and denominator by the same nonzero number.

Example 3 Write $\dfrac{3}{4x^2}$ and $\dfrac{5}{6x^3}$ as equivalent rational expressions with their LCD.

SOLUTION In Example 1, we found the LCD to be $12x^3$. For each rational expression, we multiply both the numerator and denominator by an appropriate factor so that the denominator becomes $12x^3$.

$$\dfrac{3}{4x^2} = \dfrac{3 \cdot 3x}{4x^2 \cdot 3x} = \dfrac{9x}{12x^3}$$

Note: We can determine the appropriate factor by dividing the LCD by the original denominator. In this case, $12x^3 \div 4x^2 = 3x$.

$$\dfrac{5}{6x^3} = \dfrac{5 \cdot 2}{6x^3 \cdot 2} = \dfrac{10}{12x^3}$$

Now consider rational expressions with denominators that are multiterm polynomials. Rewriting these rational expressions with their LCD is usually easier if we leave their denominators in factored form.

Answers Your Turn 2

a. $12(y + 2)$

b. $2x(x - 1)^2(x + 1)$

Example 4 Write $\dfrac{7}{x^2 + 10x + 25}$ and $\dfrac{17}{6x^4 + 30x^3}$ as equivalent rational expressions with their LCD.

SOLUTION In Example 2(b), we found the LCD to be $6x^3(x + 5)^2$.

$$\dfrac{7}{x^2 + 10x + 25} = \dfrac{7}{(x + 5)^2}$$ Factor the denominator.

> **Note:** Another way to determine the appropriate factor is to think of it as the factor in the LCD that is missing from the original denominator. In this case, $6x^3$ is in the LCD, but not in the original denominator.

$$= \dfrac{7 \cdot 6x^3}{(x + 5)^2 \cdot 6x^3}$$ Multiply the numerator and denominator by the same factor, $6x^3$, to get the LCD, $6x^3(x + 5)^2$.

$$= \dfrac{42x^3}{6x^3(x + 5)^2}$$

> **Note:** Although we could multiply out $6x^3(x + 5)^2$, it is simpler to leave it in this form.

$$\dfrac{17}{6x^4 + 30x^3} = \dfrac{17}{6x^3(x + 5)}$$ Factor the denominator.

> **Note:** There are two factors of $x + 5$ in the LCD, whereas only one factor of $x + 5$ appears in the original denominator.

$$= \dfrac{17 \cdot (x + 5)}{6x^3(x + 5) \cdot (x + 5)}$$ Multiply the numerator and denominator by the same factor, $x + 5$, to get the LCD, $6x^3(x + 5)^2$.

$$= \dfrac{17(x + 5)}{6x^3(x + 5)^2}$$

Answers   Your Turn 4

a. $\dfrac{3np}{30mn^2}$ and $\dfrac{14m}{30mn^2}$

b. $\dfrac{18x(x + 1)}{2x(x - 1)^2(x + 1)}$ and

$\dfrac{17(x - 1)}{2x(x - 1)^2(x + 1)}$

Your Turn 4 Write equivalent rational expressions with their LCD. You found the LCD of these expressions in Your Turns 1b and 2b.

a. $\dfrac{p}{10mn}$ and $\dfrac{7}{15n^2}$ b. $\dfrac{9}{x^2 - 2x + 1}$ and $\dfrac{17}{2x^3 - 2x}$

Objective ③ Add or subtract rational expressions with different denominators. To add or subtract rational expressions with different denominators, we use the same process that we use to add or subtract fractions.

Procedure   Adding or Subtracting Rational Expressions with Different Denominators

To add or subtract rational expressions with different denominators,

1. Find the LCD.
2. Write each rational expression as an equivalent expression with the LCD.
3. Add or subtract the numerators and keep the LCD.
4. Simplify.

Example 5 Add or subtract as indicated.

a. $\dfrac{3}{4x^2} + \dfrac{5}{6x^3}$

> **Note:** Because $9x$ and 10 are not like terms, we must indicate the addition as a polynomial expression. Also, because $9x + 10$ cannot be factored, we cannot simplify this rational expression.

SOLUTION We found the LCD for these expressions in Example 1. We then wrote equivalent expressions in Example 3. Now we add the numerators and keep the LCD.

$$\dfrac{3}{4x^2} + \dfrac{5}{6x^3} = \dfrac{9x}{12x^3} + \dfrac{10}{12x^3}$$ Rewrite with the LCD.

$$= \dfrac{9x + 10}{12x^3}$$ Add numerators.

b. $\dfrac{7}{x^2 + 10x + 25} - \dfrac{17}{6x^4 + 30x^3}$

SOLUTION We found the LCD for these expressions in Example 2(b) and wrote equivalent expressions in Example 4. Now we combine the numerators and keep the LCD.

Note: Leaving the LCD in factored form helps us determine whether the final expression can be reduced. Because $42x^3 - 17x - 85$ cannot be factored, we cannot reduce this rational expression.

$$\dfrac{7}{x^2 + 10x + 25} - \dfrac{17}{6x^4 + 30x^3} = \dfrac{42x^3}{6x^3(x+5)^2} - \dfrac{17(x+5)}{6x^3(x+5)^2} \qquad \text{Rewrite with the LCD.}$$

Warning: Do not divide out the $(x + 5)$'s because $(x + 5)$ is not a factor of the numerator.

$$= \dfrac{42x^3 - 17(x+5)}{6x^3(x+5)^2} \qquad \text{Subtract numerators.}$$

$$= \dfrac{42x^3 - 17x - 85}{6x^3(x+5)^2} \qquad \text{Simplify in the numerator.}$$

Now that we have seen all of the pieces of the process spread out over several examples, let's put them together using new rational expressions.

Example 6 Add or subtract as indicated.

a. $\dfrac{2x}{9y^3} + \dfrac{3}{4xy}$

SOLUTION First, find the LCD.

$$4xy = 2^2 \cdot x \cdot y \qquad 9y^3 = 3^2 \cdot y^3 \qquad \text{LCD} = 2^2 \cdot 3^2 \cdot x \cdot y^3 = 36xy^3$$

$$\dfrac{2x}{9y^3} + \dfrac{3}{4xy} = \dfrac{2x(4x)}{9y^3(4x)} + \dfrac{3(9y^2)}{4xy(9y^2)} \qquad \text{Write equivalent rational expressions with the LCD, } 36xy^3.$$

$$= \dfrac{8x^2}{36xy^3} + \dfrac{27y^2}{36xy^3}$$

Note: The numerator cannot be factored, so we cannot reduce.

$$= \dfrac{8x^2 + 27y^2}{36xy^3} \qquad \text{Add numerators.}$$

b. $\dfrac{4}{m - 5} - \dfrac{3}{m + 2}$

SOLUTION First, find the LCD. Because both $m - 5$ and $m + 2$ are prime, the LCD is their product, $(m - 5)(m + 2)$.

$$\dfrac{4}{m-5} - \dfrac{3}{m+2} = \dfrac{4(m+2)}{(m-5)(m+2)} - \dfrac{3(m-5)}{(m+2)(m-5)} \qquad \text{Write equivalent rational expressions with the LCD, } (m-5)(m+2).$$

$$= \dfrac{4m+8}{(m-5)(m+2)} - \dfrac{3m-15}{(m-5)(m+2)} \qquad \text{Multiply numerators.}$$

$$= \dfrac{(4m+8) - (3m-15)}{(m-5)(m+2)} \qquad \text{Subtract numerators.}$$

$$= \dfrac{(4m+8) + (-3m+15)}{(m-5)(m+2)} \qquad \text{Rewrite the subtraction as addition.}$$

$$= \dfrac{m+23}{(m-5)(m+2)} \qquad \text{Combine like terms.}$$

c. $\dfrac{x}{x + 2} - \dfrac{10}{x^2 - x - 6}$

SOLUTION First, find the LCD by factoring $x + 2$ and $x^2 - x - 6$.

Factored forms: $x + 2$ cannot be factored; $x^2 - x - 6 = (x + 2)(x - 3)$
LCD $= (x + 2)(x - 3)$

$$\dfrac{x}{x + 2} - \dfrac{10}{x^2 - x - 6} = \dfrac{x}{x + 2} - \dfrac{10}{(x + 2)(x - 3)}$$

Factor the denominators.

$$= \dfrac{x(x - 3)}{(x + 2)(x - 3)} - \dfrac{10}{(x + 2)(x - 3)}$$

Write equivalent rational expressions with the LCD, $(x + 2)(x - 3)$.

$$= \dfrac{x^2 - 3x}{(x + 2)(x - 3)} - \dfrac{10}{(x + 2)(x - 3)}$$

Distribute.

$$= \dfrac{x^2 - 3x - 10}{(x + 2)(x - 3)}$$

Subtract numerators.

$$= \dfrac{(x + 2)(x - 5)}{(x + 2)(x - 3)}$$

Factor the numerator.

$$= \dfrac{x - 5}{x - 3}$$

Divide out the common factor, $x + 2$.

d. $\dfrac{x - 1}{x^2 + 4x} - \dfrac{3x + 1}{x^2 - 16}$

SOLUTION To determine the LCD, we write the factored forms of $x^2 + 4x$ and $x^2 - 16$.

Factored forms: $x^2 + 4x = x(x + 4)$; $x^2 - 16 = (x + 4)(x - 4)$
LCD $= x(x + 4)(x - 4)$

$$\dfrac{x - 1}{x^2 + 4x} - \dfrac{3x + 1}{x^2 - 16} = \dfrac{x - 1}{x(x + 4)} - \dfrac{3x + 1}{(x + 4)(x - 4)}$$

Factor the denominators.

$$= \dfrac{(x - 1)(x - 4)}{x(x + 4)(x - 4)} - \dfrac{(3x + 1)(x)}{(x + 4)(x - 4)(x)}$$

Write equivalent rational expressions with the LCD, $x(x + 4)(x - 4)$.

$$= \dfrac{x^2 - 5x + 4}{x(x + 4)(x - 4)} - \dfrac{3x^2 + x}{x(x + 4)(x - 4)}$$

Multiply in the numerator.

$$= \dfrac{(x^2 - 5x + 4) - (3x^2 + x)}{x(x + 4)(x - 4)}$$

Subtract numerators.

Note: Remember that to subtract polynomials, we write an equivalent addition.

$$= \dfrac{(x^2 - 5x + 4) + (-3x^2 - x)}{x(x + 4)(x - 4)}$$

$$= \dfrac{-2x^2 - 6x + 4}{x(x + 4)(x - 4)}$$

Note: Although the numerator can be factored to $-2(x^2 + 3x - 2)$, none of those factors match any factor in the denominator; so the result is in lowest terms.

Your Turn 6 Add or subtract as indicated.

a. $\dfrac{5}{6x^2y} + \dfrac{3}{8xy^3}$ **b.** $\dfrac{7}{x + 3} - \dfrac{3}{x - 4}$ **c.** $\dfrac{x}{x + 3} - \dfrac{-6x + 24}{x^2 - x - 12}$ **d.** $\dfrac{2}{x^2 + x} - \dfrac{x - 4}{x^2 - 1}$

If denominators are additive inverses, we can multiply the numerator and denominator of either rational expression by -1 to obtain the LCD.

Example 7 Add $\dfrac{5x}{x-3} + \dfrac{2}{3-x}$.

SOLUTION
$$\dfrac{5x}{x-3} + \dfrac{2}{3-x} = \dfrac{5x}{x-3} + \dfrac{2(-1)}{(3-x)(-1)}$$

$$= \dfrac{5x}{x-3} + \dfrac{-2}{x-3}$$

$$= \dfrac{5x-2}{x-3}$$

Because $x-3$ and $3-x$ are additive inverses, we obtain the LCD by multiplying the numerator and denominator of one of the rational expressions by -1. We chose the second rational expression.

Your Turn 7 Add or subtract as indicated.

a. $\dfrac{7y}{2y-3} - \dfrac{4}{3-2y}$

b. $\dfrac{3x}{x-2} + \dfrac{6}{2-x}$

Answers Your Turn 7
a. $\dfrac{7y+4}{2y-3}$ b. 3

7.4 Exercises For Extra Help MyMathLab

 PRACTICE   WATCH   DOWNLOAD

Note: Exercises marked with a ★ represent challenging exercises.

1. Explain how to find the LCD of two rational expressions.

2. Suppose the LCD of two fractions is $48x^3yz$. If one of the fraction's denominators is $8x^2y$, by what factor do you multiply when writing it as an equivalent fraction with the LCD? Explain how you determine this factor.

3. In general, if you are given rational expressions that contain multiterm polynomials in the denominator(s), what is your first step toward finding the LCD?

4. Instead of using the LCM as a common denominator, a student rewrites $\dfrac{3}{2x-y} + \dfrac{5}{y-2x}$ as $\dfrac{-3}{y-2x} + \dfrac{5}{y-2x}$. Explain what the student did.

5. Suppose you have found the LCD of two rational expressions and have written the equivalent rational expressions. If the problem is now at the stage $\dfrac{(7x+21)-(2x-10)}{(x-5)(x+3)}$, how do you proceed?

★6. Suppose we have found the LCD of two rational expressions and have written the equivalent rational expressions. If the problem is now at the stage $\dfrac{(7x+21)-(2x-10)}{(x-5)(x+3)}$, what might the original problem have been? Explain.

For Exercises 7–32, find the least common denominator for the rational expressions and write equivalent rational expressions with the LCD. See Examples 1–4.

7. $\dfrac{2}{n}, \dfrac{3}{m}$

8. $\dfrac{6}{r}, \dfrac{5}{t}$

9. $\dfrac{2t}{a^2b^5}, \dfrac{3z}{a^3b^3}$

10. $\dfrac{6u}{h^4k^2}, \dfrac{4v}{h^3k^7}$

11. $\dfrac{2}{3a}, \dfrac{3}{5a^2}$

12. $\dfrac{5}{4b^2}, \dfrac{3}{5b}$

13. $\dfrac{3}{7x^2y}, \dfrac{5}{14xy^2}$

14. $\dfrac{2}{9mn^3}, \dfrac{7}{27m^3n}$

15. $\dfrac{8}{15m^3n^3}, \dfrac{7}{18m^2n}$

16. $\dfrac{5}{10x^2y^3}, \dfrac{3}{14x^2y^2}$

17. $\dfrac{3}{y+2}, \dfrac{4y}{y-2}$

18. $\dfrac{3x}{x-5}, \dfrac{2}{x+1}$

19. $\dfrac{3y}{4y-20}, \dfrac{7y}{6y-30}$

20. $\dfrac{2x}{6x-12}, \dfrac{3x}{8x-16}$

21. $\dfrac{5}{6x+18}, \dfrac{2}{x^2+3x}$

22. $\dfrac{3}{4w-8}, \dfrac{5}{w^2-2w}$

23. $\dfrac{2}{p^2-4}, \dfrac{3+p}{p+2}$

24. $\dfrac{5}{y^2-36}, \dfrac{y-2}{y-6}$

25. $\dfrac{4u}{(u+2)^2}, \dfrac{2}{3(u+2)}$

26. $\dfrac{7m}{(x-3)^2}, \dfrac{4}{7(x-3)}$

27. $\dfrac{2t-4}{t^2-36}, \dfrac{3t+2}{t^2-7t+6}$

28. $\dfrac{4a+2}{a^2-9}, \dfrac{2a-1}{a^2-a-6}$

29. $\dfrac{x+2}{x^2+3x-4}, \dfrac{x-3}{x^2+6x+8}$

30. $\dfrac{b-4}{b^2-2b-35}, \dfrac{b+3}{b^2+7b+10}$

31. $\dfrac{2}{x-1}, \dfrac{3}{x^2-3x+2}, \dfrac{5x}{x-2}$

32. $\dfrac{3}{y-3}, \dfrac{2}{y^2+2y-15}, \dfrac{4y}{y+5}$

For Exercises 33–70, add or subtract as indicated. See Examples 5–7.

33. $\dfrac{3a-5}{9} - \dfrac{2a+7}{6}$

34. $\dfrac{3x-4}{6} - \dfrac{2x-1}{4}$

35. $\dfrac{5}{2c} + \dfrac{1}{6c}$

36. $\dfrac{5}{8r} + \dfrac{3}{4r}$

37. $\dfrac{5}{x} - \dfrac{2}{x^3}$

38. $\dfrac{5a}{b^2} - \dfrac{3a}{b}$

39. $\dfrac{3}{y+2} + \dfrac{5}{y-4}$

40. $\dfrac{2}{c+4} + \dfrac{3}{c+3}$

41. $\dfrac{3}{x-2} - \dfrac{2}{x+2}$

42. $\dfrac{4}{x+3} - \dfrac{3}{x-3}$

43. $\dfrac{4}{x^2-25} + \dfrac{2}{x-5}$

44. $\dfrac{5}{r^2-9} + \dfrac{4}{r+3}$

45. $\dfrac{2p}{p^2-2p+1} + \dfrac{8}{p-1}$

46. $\dfrac{u}{u-1} + \dfrac{2u}{u^2-2u+1}$

47. $\dfrac{z+5}{z^2+z-12} - \dfrac{z+2}{z-3}$

48. $\dfrac{a+6}{a^2+8a+15} - \dfrac{a-3}{a+3}$

49. $\dfrac{n+2}{n-4} + \dfrac{n-3}{n+3}$

50. $\dfrac{y-8}{y+5} + \dfrac{y-9}{y-6}$

51. $\dfrac{t+2}{t+4} - \dfrac{2t-1}{t+6}$

52. $\dfrac{3s+1}{s-2} - \dfrac{4s+1}{s-3}$

53. $\dfrac{c+5}{c^2+10c+25} + \dfrac{3}{2c+10}$

54. $\dfrac{k+3}{k^2+6k+9} - \dfrac{7}{2k+6}$

55. $\dfrac{5y+6}{y^2+3y} + \dfrac{y}{y+3}$

56. $\dfrac{2b-8}{b^2+4b} + \dfrac{b}{b+4}$

57. $\dfrac{3r}{r^2-1} - \dfrac{2r+1}{r^2-2r+1}$

58. $\dfrac{x+1}{x^2-4x+4} - \dfrac{4x}{x^2+3x-10}$

59. $\dfrac{q+5}{q^2-9} - \dfrac{1}{q^2+3q}$

60. $\dfrac{x-2}{x^2-16} - \dfrac{1}{x^2-4x}$

61. $\dfrac{w-7}{w^2+4w-5} - \dfrac{w-9}{w^2+3w-10}$

62. $\dfrac{y-1}{y^2-3y+2} - \dfrac{y+2}{y^2+y-2}$

63. $\dfrac{2v+5}{v^2-16} - \dfrac{v-9}{v^2-v-12}$

64. $\dfrac{4t}{t^2-9} - \dfrac{3t-2}{t^2-8t+15}$

65. $\dfrac{3}{t-3} + \dfrac{6}{3-t}$

66. $\dfrac{10}{2r-1} - \dfrac{6}{1-2r}$

67. $\dfrac{4}{2x-1} + \dfrac{x}{1-2x}$

68. $\dfrac{3}{t-4} + \dfrac{t}{4-t}$

69. $\dfrac{y}{3y-6} - \dfrac{2}{2-y}$

70. $\dfrac{x}{6x-2} - \dfrac{3x}{1-3x}$

7.5 Complex Rational Expressions

Objective ① **Simplify complex rational expressions.** Sometimes a rational expression may contain rational expressions in its numerator or denominator. Such an expression is called a **complex rational expression** or a **complex fraction**.

Definition *Complex rational expression:* A rational expression that contains rational expressions in the numerator or denominator.

Here are some complex rational expressions:

$$\dfrac{\dfrac{3}{4}}{\dfrac{5}{6}} \qquad \dfrac{x + \dfrac{x}{3}}{\dfrac{y}{2}} \qquad \dfrac{7}{\dfrac{t+2}{t}} \qquad \dfrac{\dfrac{2}{n} + \dfrac{1}{4}}{\dfrac{1}{n} - \dfrac{n}{3}}$$

There are two common methods for simplifying complex rational expressions. One method is to rewrite the complex rational expression as a horizontal division problem.

For example, the complex fraction $\dfrac{\dfrac{3}{4}}{\dfrac{5}{6}}$ can be expressed as the division problem $\dfrac{3}{4} \div \dfrac{5}{6}$, then simplified.

$$\frac{\dfrac{3}{4}}{\dfrac{5}{6}} = \frac{3}{4} \div \frac{5}{6} = \frac{3}{\overset{}{\underset{2}{4}}} \cdot \frac{\overset{3}{\cancel{6}}}{5} = \frac{9}{10}$$

Another method is to multiply the numerator and denominator of the complex fraction by the LCD of the numerator and denominator fractions. In this case, the denominators are 4 and 6, which means that the LCD is 12.

Connection Notice that the essence of this second method is writing an equivalent fraction by multiplying both the numerator and denominator by the same factor.

$$\frac{\dfrac{3}{4} \cdot 12}{\dfrac{5}{6} \cdot 12} = \frac{\dfrac{3}{4} \cdot \dfrac{\overset{3}{\cancel{12}}}{1}}{\dfrac{5}{6} \cdot \dfrac{\overset{2}{\cancel{12}}}{1}} = \frac{\dfrac{9}{1}}{\dfrac{10}{1}} = \frac{9}{10}$$

Procedure **Simplifying Complex Rational Expressions**

To simplify a complex rational expression, use one of the following methods.

Method 1
1. Simplify the numerator and denominator if needed.
2. Rewrite as a horizontal division problem.

Method 2
1. Multiply the numerator and denominator of the complex rational expression by their LCD.
2. Simplify.

Example 1 Simplify.

a. $\dfrac{\dfrac{2}{x}}{\dfrac{y}{3}}$

Method 1

Note: Method 1 is best used when the numerator and denominator are both a single rational expression, as in Example 1(a). Otherwise, method 2 is at least as easy as method 1.

$$\dfrac{\dfrac{2}{x}}{\dfrac{y}{3}} = \dfrac{2}{x} \div \dfrac{y}{3} \qquad \text{Write the complex fraction as a division problem.}$$

$$= \dfrac{2}{x} \cdot \dfrac{3}{y} \qquad \text{Write an equivalent multiplication problem.}$$

$$= \dfrac{6}{xy}$$

Method 2 Multiply the numerator and denominator of the rational expression by the LCD of all rational expressions involved. In this case, that LCD is $3x$.

$$\dfrac{\dfrac{2}{x}}{\dfrac{y}{3}} = \dfrac{\dfrac{2}{x} \cdot \mathbf{3x}}{\dfrac{y}{3} \cdot \mathbf{3x}} \qquad \text{Multiply the numerator and denominator by } 3x.$$

$$= \dfrac{\dfrac{2}{\overset{1}{\cancel{x}}} \cdot \dfrac{3\overset{1}{\cancel{x}}}{1}}{\dfrac{y}{\underset{1}{\cancel{3}}} \cdot \dfrac{\cancel{3}x}{1}} \qquad \text{Divide out the common factors.}$$

$$= \dfrac{6}{xy} \qquad \text{Simplify.}$$

b. $\dfrac{\dfrac{1}{4} + \dfrac{2}{3}}{\dfrac{3}{8} - \dfrac{1}{6}}$

Method 1 Simplify the numerator and denominator and then divide.

$$\dfrac{\dfrac{1}{4} + \dfrac{2}{3}}{\dfrac{3}{8} - \dfrac{1}{6}} = \dfrac{\dfrac{1(3)}{4(3)} + \dfrac{2(4)}{3(4)}}{\dfrac{3(3)}{8(3)} - \dfrac{1(4)}{6(4)}} \qquad \text{Write the numerator fractions as equivalent fractions with their LCD, 12, and write the denominator fractions with their LCD, 24.}$$

$$= \dfrac{\dfrac{3}{12} + \dfrac{8}{12}}{\dfrac{9}{24} - \dfrac{4}{24}}$$

$$= \dfrac{\dfrac{11}{12}}{\dfrac{5}{24}} \qquad \text{Add in the numerator and subtract in the denominator.}$$

$$= \frac{11}{12} \div \frac{5}{24}$$

Write the complex fraction as a division problem.

$$= \frac{11}{\underset{1}{\cancel{12}}} \cdot \frac{\overset{2}{\cancel{24}}}{5}$$

Write an equivalent multiplication problem, divide out the common factors, and then multiply.

$$= \frac{22}{5}$$

Method 2 Multiply the numerator and denominator of the rational expression by the LCD of all rational expressions involved. In this case, that LCD is 24.

$$\frac{\dfrac{1}{4} + \dfrac{2}{3}}{\dfrac{3}{8} - \dfrac{1}{6}} = \frac{\left(\dfrac{1}{4} + \dfrac{2}{3}\right) \cdot 24}{\left(\dfrac{3}{8} - \dfrac{1}{6}\right) \cdot 24}$$

Multiply the numerator and denominator by 24.

$$= \frac{\dfrac{1}{\underset{1}{\cancel{4}}} \cdot \dfrac{\overset{6}{\cancel{24}}}{1} + \dfrac{2}{\underset{1}{\cancel{3}}} \cdot \dfrac{\overset{8}{\cancel{24}}}{1}}{\dfrac{3}{\underset{1}{\cancel{8}}} \cdot \dfrac{\overset{3}{\cancel{24}}}{1} - \dfrac{1}{\underset{1}{\cancel{6}}} \cdot \dfrac{\overset{4}{\cancel{24}}}{1}}$$

Distribute the 24 to each fraction and divide out the common factors.

$$= \frac{6 + 16}{9 - 4}$$

Simplify the numerator and denominator.

$$= \frac{22}{5}$$

c. $\dfrac{5 + \dfrac{x}{3}}{\dfrac{4}{x} - \dfrac{1}{2x}}$

Method 1 Simplify the numerator and denominator and then divide.

$$\frac{5 + \dfrac{x}{3}}{\dfrac{4}{x} - \dfrac{1}{2x}} = \frac{\dfrac{5(3)}{1(3)} + \dfrac{x(1)}{3(1)}}{\dfrac{4(2)}{x(2)} - \dfrac{1(1)}{2x(1)}}$$

Write the numerator fractions as equivalent fractions with their LCD, 3, and write the denominator fractions with their LCD, 2x.

$$= \frac{\dfrac{15}{3} + \dfrac{x}{3}}{\dfrac{8}{2x} - \dfrac{1}{2x}}$$

$$= \frac{\dfrac{15 + x}{3}}{\dfrac{7}{2x}}$$

Add in the numerator and subtract in the denominator.

$$= \frac{15 + x}{3} \div \frac{7}{2x}$$

Write the complex fraction as a division problem.

$$= \frac{15 + x}{3} \cdot \frac{2x}{7}$$

Write an equivalent multiplication problem and then multiply.

$$= \frac{30x + 2x^2}{21}$$

Method 2 Multiply the numerator and denominator of the rational expression by the LCD of all rational expressions involved. In this case, that LCD is $6x$.

$$\frac{5 + \dfrac{x}{3}}{\dfrac{4}{x} - \dfrac{1}{2x}} = \frac{\left(5 + \dfrac{x}{3}\right) \cdot 6x}{\left(\dfrac{4}{x} - \dfrac{1}{2x}\right) \cdot 6x}$$

Multiply the numerator and denominator by $6x$.

$$= \frac{\dfrac{5}{1} \cdot \dfrac{6x}{1} + \dfrac{x}{\underset{1}{\cancel{3}}} \cdot \dfrac{\overset{2}{\cancel{6}x}}{1}}{\dfrac{4}{\underset{1}{\cancel{x}}} \cdot \dfrac{\overset{1}{\cancel{6}x}}{1} - \dfrac{1}{\underset{1}{\cancel{2}x}} \cdot \dfrac{\overset{3}{\cancel{6}x}}{1}}$$

Distribute the $6x$ to each fraction and divide out the common factors.

$$= \frac{30x + 2x^2}{24 - 3}$$

Simplify the numerator and denominator.

$$= \frac{30x + 2x^2}{21}$$

d. $\dfrac{1 - \dfrac{2}{x} - \dfrac{15}{x^2}}{1 + \dfrac{7}{x} + \dfrac{12}{x^2}}$

Method 1 Simplify the numerator and denominator and then divide.

$$\frac{1 - \dfrac{2}{x} - \dfrac{15}{x^2}}{1 + \dfrac{7}{x} + \dfrac{12}{x^2}} = \frac{1 \cdot \dfrac{x^2}{x^2} - \dfrac{2}{x} \cdot \dfrac{x}{x} - \dfrac{15}{x^2}}{1 \cdot \dfrac{x^2}{x^2} + \dfrac{7}{x} \cdot \dfrac{x}{x} + \dfrac{12}{x^2}}$$

Write the numerator fractions as equivalent fractions with their LCD, x^2, and write the denominator fractions with their LCD, also x^2.

$$= \frac{\dfrac{x^2}{x^2} - \dfrac{2x}{x^2} - \dfrac{15}{x^2}}{\dfrac{x^2}{x^2} + \dfrac{7x}{x^2} + \dfrac{12}{x^2}}$$

Simplify.

$$= \frac{\dfrac{x^2 - 2x - 15}{x^2}}{\dfrac{x^2 + 7x + 12}{x^2}}$$

Subtract in the numerator and add in the denominator.

$$= \frac{x^2 - 2x - 15}{x^2} \div \frac{x^2 + 7x + 12}{x^2}$$

Write the complex fraction as a division problem.

$$= \frac{x^2 - 2x - 15}{x^2} \cdot \frac{x^2}{x^2 + 7x + 12}$$

Write as an equivalent multiplication problem.

$$= \frac{(x - 5)(x + 3)}{\underset{1}{\cancel{x^2}}} \cdot \frac{\overset{1}{\cancel{x^2}}}{(x + 4)(x + 3)}$$

Factor, divide out the common factors, and then multiply the remaining factors.

$$= \frac{x - 5}{x + 4}$$

Method 2 Multiply the numerator and denominator of the rational expression by the LCD of all rational expressions involved. In this case, that LCD is x^2.

$$\frac{1 - \dfrac{2}{x} - \dfrac{15}{x^2}}{1 + \dfrac{7}{x} + \dfrac{12}{x^2}} = \frac{\left(1 - \dfrac{2}{x} - \dfrac{15}{x^2}\right) \cdot x^2}{\left(1 + \dfrac{7}{x} + \dfrac{12}{x^2}\right) \cdot x^2}$$

Multiply the numerator and denominator by x^2.

$$= \frac{1 \cdot x^2 - \dfrac{2}{\cancel{x}} \cdot \dfrac{\overset{x}{\cancel{x^2}}}{1} - \dfrac{15}{\cancel{x^2}} \cdot \dfrac{\overset{1}{\cancel{x^2}}}{1}}{1 \cdot x^2 + \dfrac{7}{\cancel{x}} \cdot \dfrac{\overset{x}{\cancel{x^2}}}{1} + \dfrac{12}{\cancel{x^2}} \cdot \dfrac{\overset{1}{\cancel{x^2}}}{1}}$$

Distribute the x^2 to each fraction and divide out the common factors.

$$= \frac{x^2 - 2x - 15}{x^2 + 7x + 12}$$

Simplify the numerator and denominator.

$$= \frac{(x - 5)\cancel{(x + 3)}}{(x + 4)\cancel{(x + 3)}}$$

Factor and divide out the common factor, $x + 3$.

$$= \frac{x - 5}{x + 4}$$

Your Turn 1 Simplify.

Answers   Your Turn 1

a. $\dfrac{ab}{15}$ **b.** $\dfrac{25}{26}$

c. $\dfrac{6y - 3}{3y^2 - y}$ **d.** $\dfrac{x - 7}{x - 5}$

a. $\dfrac{\dfrac{a}{5}}{\dfrac{3}{b}}$

b. $\dfrac{2 - \dfrac{3}{4}}{\dfrac{4}{5} + \dfrac{1}{2}}$

c. $\dfrac{2 - \dfrac{1}{y}}{y - \dfrac{1}{3}}$

d. $\dfrac{1 - \dfrac{3}{x} - \dfrac{28}{x^2}}{1 - \dfrac{1}{x} - \dfrac{20}{x^2}}$

7.5 Exercises   For Extra Help   *MyMathLab*

 PRACTICE   WATCH   DOWNLOAD

1. Identify the numerator and denominator of the complex rational expression

$$\frac{\dfrac{4}{x + 3}}{\dfrac{3}{x}}.$$

2. Given the complex rational expression $\dfrac{2x}{\dfrac{x - 3}{4}}$, is the numerator $2x$ or $\dfrac{2x}{x - 3}$?

How could the expression be written so that the numerator and denominator are easily identified?

3. Write $\dfrac{2x}{x - 3} \div \dfrac{4 - x}{2x - 6}$ as a complex rational expression.

4. Write $\dfrac{5}{x-1} \div 7x$ as a complex rational expression.

5. Which of the two methods discussed in this section would you use to simplify
$$\dfrac{\dfrac{x}{6} - \dfrac{3}{4}}{\dfrac{2}{3} + x}?\,\text{Why?}$$

6. Which of the two methods discussed in this section would you use to simplify
$$\dfrac{\dfrac{x}{7}}{\dfrac{x}{x+1}}?\,\text{Why?}$$

For Exercises 7–48, simplify. See Example 1.

7. $\dfrac{\dfrac{3}{4}}{\dfrac{3}{2}}$

8. $\dfrac{\dfrac{2}{3}}{\dfrac{3}{2}}$

9. $\dfrac{\dfrac{m}{n}}{\dfrac{q}{p}}$

10. $\dfrac{\dfrac{a}{b}}{\dfrac{c}{d}}$

11. $\dfrac{\dfrac{1}{2} + \dfrac{3}{5}}{\dfrac{3}{2} - \dfrac{1}{5}}$

12. $\dfrac{\dfrac{2}{3} + \dfrac{1}{4}}{\dfrac{4}{3} - \dfrac{3}{4}}$

13. $\dfrac{5 + \dfrac{2}{a}}{3 + \dfrac{4}{a}}$

14. $\dfrac{4 + \dfrac{1}{x}}{3 + \dfrac{5}{x}}$

15. $\dfrac{\dfrac{1}{y} - 1}{\dfrac{1}{y} + 1}$

16. $\dfrac{x - \dfrac{1}{x}}{1 + \dfrac{1}{x}}$

17. $\dfrac{r - \dfrac{3}{4}}{\dfrac{1}{2} - r}$

18. $\dfrac{\dfrac{5}{6} - x}{x - \dfrac{2}{3}}$

19. $\dfrac{\dfrac{3}{x} - 1}{\dfrac{9}{x^2} - 1}$

20. $\dfrac{1 - \dfrac{2}{p}}{1 - \dfrac{4}{p^2}}$

21. $\dfrac{x + y}{\dfrac{1}{x} + \dfrac{1}{y}}$

22. $\dfrac{\dfrac{2}{a} + \dfrac{3}{b}}{a + b}$

23. $\dfrac{x + 2}{\dfrac{x^2 - 4}{x}}$

24. $\dfrac{\dfrac{a}{b} - 1}{a^2 - b^2}$

25. $\dfrac{\dfrac{6}{3x - 2}}{\dfrac{x}{2x + 1}}$

26. $\dfrac{\dfrac{5}{2x - 1}}{\dfrac{x}{x + 1}}$

27. $\dfrac{\dfrac{a}{a^2 - b^2}}{\dfrac{b}{a + b}}$

28. $\dfrac{\dfrac{x}{4x^2 - 1}}{\dfrac{5}{2x - 1}}$

29. $\dfrac{\dfrac{y^2 - 25}{5y}}{\dfrac{y^2 + 5y}{2y^3}}$

30. $\dfrac{\dfrac{x^2 + 4x}{y^2}}{\dfrac{x^2 - 16}{3y}}$

31. $\dfrac{2 + \dfrac{2}{u - 1}}{\dfrac{2}{u - 1}}$

32. $\dfrac{\dfrac{1}{w+1}-1}{\dfrac{1}{w+1}}$

33. $\dfrac{\dfrac{a}{a-2}-1}{1+\dfrac{a}{a-2}}$

34. $\dfrac{1-\dfrac{x}{x+2}}{\dfrac{x}{x+2}+1}$

35. $\dfrac{\dfrac{x-3}{x}-1}{\dfrac{x^2-9}{x}-x}$

36. $\dfrac{q-\dfrac{q^2-1}{q}}{1-\dfrac{q-1}{q}}$

37. $\dfrac{\dfrac{1}{x-1}-\dfrac{1}{x+1}}{\dfrac{1}{x^2-1}}$

38. $\dfrac{\dfrac{16}{x^2-16}}{\dfrac{2}{x+4}-\dfrac{2}{x-4}}$

39. $\dfrac{1+\dfrac{1}{y}-\dfrac{2}{y^2}}{\dfrac{1}{y}+\dfrac{2}{y^2}}$

40. $\dfrac{1-\dfrac{3}{x}}{1-\dfrac{2}{x}-\dfrac{3}{x^2}}$

41. $\dfrac{1-\dfrac{1}{n}}{n-2+\dfrac{1}{n}}$

42. $\dfrac{y-1-\dfrac{6}{y}}{1+\dfrac{2}{y}}$

43. $\dfrac{1-\dfrac{1}{2y+1}}{\dfrac{1}{2y^2-5y-3}-\dfrac{1}{y-3}}$

44. $\dfrac{1-\dfrac{1}{f-3}}{\dfrac{1}{f^2-f-6}-\dfrac{1}{f+2}}$

45. $\dfrac{\dfrac{x^2-2x-8}{x^2-16}}{\dfrac{x^2+2x}{x^2+3x-4}}$

46. $\dfrac{\dfrac{v^2+v-2}{v^2+4v}}{\dfrac{v^2-4}{v^2+2v-8}}$

47. $\dfrac{\dfrac{2}{x+h}-\dfrac{2}{x}}{h}$

48. $\dfrac{h}{\dfrac{3}{y+h}-\dfrac{3}{y}}$

Find the mistake

For Exercises 49–56, explain the mistake; then find the correct answer.

49. $\dfrac{x-\dfrac{1}{2}}{y-\dfrac{1}{2}}=\dfrac{x}{y}$

50. $\dfrac{\dfrac{2}{3}}{\dfrac{4}{6}}=\dfrac{\overset{1}{\cancel{2}}}{3}\cdot\dfrac{4}{\underset{3}{\cancel{6}}}$

$=\dfrac{4}{9}$

51. $\dfrac{\dfrac{m}{6}}{\dfrac{3}{m}}=\dfrac{\overset{1}{\cancel{m}}}{\underset{2}{\cancel{6}}}\div\dfrac{\overset{1}{\cancel{3}}}{\cancel{m}}$

$=\dfrac{1}{2}$

52. $\dfrac{\dfrac{1}{x}+\dfrac{1}{y}}{x^2-y^2}=\dfrac{\dfrac{x}{1}\cdot\dfrac{1}{x}+\dfrac{y}{1}\cdot\dfrac{1}{y}}{x\cdot x^2-y\cdot y^2}$

$=\dfrac{1+1}{x^3-y^3}$

$=\dfrac{2}{x^3-y^3}$

53. $\dfrac{c+\dfrac{1}{c}}{\dfrac{1}{c}}=\dfrac{c+\dfrac{c}{1}\cdot\dfrac{1}{c}}{\dfrac{c}{1}\cdot\dfrac{1}{c}}$

$=\dfrac{c+1}{1}$

$=c+1$

54. $\dfrac{1+\dfrac{2}{r}}{1-\dfrac{4}{r^2}}=\dfrac{r\cdot1+\dfrac{r}{1}\cdot\dfrac{2}{r}}{r\cdot1-\dfrac{r}{1}\cdot\dfrac{4}{r^2}}$

$=\dfrac{r+2}{r-4r}$

$=\dfrac{r+2}{-3r}$

55. $\dfrac{\dfrac{3}{x}}{\dfrac{x}{6}} = \dfrac{3}{x} \div \dfrac{x}{6}$

$= \dfrac{3}{x} \cdot \dfrac{6}{x}$

$= \dfrac{18}{2x}$

$= \dfrac{9}{x}$

56. $\dfrac{\dfrac{1}{x}+\dfrac{1}{y}}{\dfrac{1}{x}-\dfrac{1}{y}} = \dfrac{\dfrac{xy}{1}\cdot\dfrac{1}{x}+\dfrac{xy}{1}\cdot\dfrac{1}{y}}{\dfrac{xy}{1}\cdot\dfrac{1}{x}-\dfrac{xy}{1}\cdot\dfrac{1}{y}}$

$= \dfrac{y+x}{y-x}$

$= -1$

For Exercises 57 and 58, average rate can be found using the following formula:

$$\text{Average rate} = \frac{\text{Total distance}}{\text{Total time}}$$

57. Suppose a person travels 10 miles in $\dfrac{1}{3}$ of an hour, then returns by traveling that same 10 miles in $\dfrac{1}{4}$ of an hour. What is that person's average rate?

58. Juan travels 50 miles in $\dfrac{3}{4}$ of an hour and then encounters a traffic jam so that the next 15 miles take $\dfrac{1}{3}$ of an hour. The last 40 miles of his trip take him $\dfrac{2}{3}$ of an hour. What was his average rate for the trip?

59. Given the area and length of a rectangle, the width can be found using the formula $w = \dfrac{A}{l}$. Suppose the area of a rectangle is described by $\dfrac{5n-2}{8}$ square feet and the length is described by $\dfrac{n-1}{6}$ feet. Find an expression for the width.

60. Given the area and base of a triangle, the height can be found using the formula $h = \dfrac{2A}{b}$. Suppose the area of a triangle is described by $\dfrac{n+1}{6}$ and its base is described by $\dfrac{5n-2}{9}$. Find an expression for the height.

61. In electrical circuits, if two resistors R_1 and R_2 are wired in parallel, the resistance of the circuit is found using the complex rational expression $\dfrac{1}{\dfrac{1}{R_1}+\dfrac{1}{R_2}}$.

  a. Simplify this complex rational expression.

  b. If a 100-ohm and 10-ohm resistor are wired in parallel, find the resistance of the circuit.

  c. Two 16-ohm speakers are wired in parallel; find the resistance of the circuit.

62. Suppose three resistors R_1, R_2, and R_3 are wired in parallel. The resistance of the circuit is found using the complex rational expression $\dfrac{1}{\dfrac{1}{R_1} + \dfrac{1}{R_2} + \dfrac{1}{R_3}}$.

    a. Simplify this complex rational expression.

    b. A 20-ohm, a 10-ohm, and a 40-ohm resistor are wired in parallel. Find the resistance of the circuit.

    c. A 300-ohm, a 500-ohm, and a 200-ohm resistor are wired in parallel. Find the resistance of the circuit.

Review Exercises

Exercises 1–6 EQUATIONS AND INEQUALITIES

[2.3] *For Exercises 1–4, solve.*

1. $-6x = 42$ **2.** $9t - 8 = 19$ **3.** $\dfrac{x}{4} = \dfrac{5}{8}$ **4.** $\dfrac{3}{4}y - \dfrac{1}{3} = \dfrac{1}{6}y + 2$

[2.1]   **5.** A runner travels 1.2 miles in 15 minutes. Find the runner's average rate in miles per hour.

[3.4]   **6.** An eastbound car and westbound car pass each other on a highway. The eastbound car is traveling 40 miles per hour, and the westbound car is traveling 60 miles per hour. How much time elapses from the time they pass each other until they are 9 miles apart?

7.6 Solving Equations Containing Rational Expressions

Objective 1 **Solve equations containing rational expressions.** In Section 2.3, we learned that we can eliminate fractions from an equation by multiplying both sides of the equation by their LCD. For example, if given the equation $\dfrac{1}{2}x - \dfrac{3}{4} = \dfrac{2}{3}$, we can eliminate the fractions by multiplying both sides by 12, which is the LCD of $\dfrac{1}{2}, \dfrac{3}{4},$ and $\dfrac{2}{3}$.

$$\overset{6}{\cancel{12}} \cdot \frac{1}{\underset{1}{\cancel{2}}}x - \overset{3}{\cancel{12}} \cdot \frac{3}{\underset{1}{\cancel{4}}} = \overset{4}{\cancel{12}} \cdot \frac{2}{\underset{1}{\cancel{3}}}$$

$$6x - 9 = 8$$

Because the rewritten equation contains only integers, it is easier to solve. Similarly, we can use the LCD to eliminate rational expressions in an equation.

Example 1 Solve $\dfrac{x}{5} = \dfrac{x}{2} - \dfrac{1}{4}$.

SOLUTION Eliminate the rational expressions by multiplying both sides by the LCD, 20.

$$20\left(\dfrac{x}{5}\right) = 20\left(\dfrac{x}{2} - \dfrac{1}{4}\right)$$ **Multiply both sides by 20.**

$$\overset{4}{\cancel{20}} \cdot \dfrac{x}{\cancel{5}_1} = \overset{10}{\cancel{20}} \cdot \dfrac{x}{\cancel{2}_1} - \overset{5}{\cancel{20}} \cdot \dfrac{1}{\cancel{4}_1}$$ **Use the distributive property to multiply every term by 20 and then divide out the common factors.**

$$4x = 10x - 5$$

$$-6x = -5$$ **Subtract 10x from both sides.**

$$\dfrac{-6x}{-6} = \dfrac{-5}{-6}$$ **Divide both sides by −6.**

$$x = \dfrac{5}{6}$$ **Simplify both sides.**

Check Using the original equation, we replace x with $\dfrac{5}{6}$ and see if the equation is true. It is helpful to write $\dfrac{x}{5}$ as $\dfrac{1}{5}x$ and $\dfrac{x}{2}$ as $\dfrac{1}{2}x$.

$$\dfrac{1}{5}x = \dfrac{1}{2}x - \dfrac{1}{4}$$

$$\dfrac{1}{\cancel{5}_1}\left(\dfrac{\overset{1}{\cancel{5}}}{6}\right) \overset{?}{=} \dfrac{1}{2}\left(\dfrac{5}{6}\right) - \dfrac{1}{4}$$ **Replace x with $\dfrac{5}{6}$ and then simplify.**

$$\dfrac{1}{6} \overset{?}{=} \dfrac{5}{12} - \dfrac{3}{12}$$

$$\dfrac{1}{6} = \dfrac{2}{12}$$ **True; therefore, $\dfrac{5}{6}$ is a solution.**

Now consider equations that contain rational expressions with variables in the denominator.

Example 2 Solve $\dfrac{1}{2} = \dfrac{3}{x} - \dfrac{1}{4}$.

SOLUTION Eliminate the rational expressions by multiplying both sides by the LCD, $4x$.

$$4x\left(\dfrac{1}{2}\right) = 4x\left(\dfrac{3}{x} - \dfrac{1}{4}\right)$$ **Multiply both sides by 4x.**

$$\overset{2}{\cancel{4}}x \cdot \dfrac{1}{\cancel{2}_1} = 4\cancel{x} \cdot \dfrac{3}{\cancel{x}_1} - \overset{1}{\cancel{4}}x \cdot \dfrac{1}{\cancel{4}_1}$$ **Use the distributive property to multiply every term by 4x and then divide out the common factors.**

$$2x = 12 - x$$

$$3x = 12$$ **Add x to both sides.**

$$x = 4$$ **Divide both sides by 3.**

Check We will leave the check to the reader.

Your Turn 2 Solve.

a. $\dfrac{x}{2} = 4 - \dfrac{x}{3}$

b. $\dfrac{1}{x} + \dfrac{3}{8} = \dfrac{5}{6}$

Extraneous Solutions

If an equation contains rational expressions with variables in the denominators, we must make sure that no solution causes any of the expressions in the equation to be undefined. In Example 2, if the solution had turned out to be 0, the expression $\dfrac{3}{x}$ would have been undefined. By inspecting the denominators of each rational expression, you can determine the values that will cause the expressions to be undefined before solving the equation. If by solving an equation, you obtain a number that causes an expression in the original equation to be undefined, we say that the number is an **extraneous solution** and we discard it.

> **Definition** *Extraneous solution:* An apparent solution that does not solve its equation.

The following procedure summarizes the process of solving equations containing rational expressions.

> **Procedure** **Solving Equations Containing Rational Expressions**
>
> To solve an equation that contains rational expressions,
>
> 1. Eliminate the rational expressions by multiplying both sides of the equation by their LCD.
> 2. Solve the equation using the methods we learned in Chapters 2 (linear equations) and 6 (quadratic equations).
> 3. Check your solution(s) in the original equation. Discard any extraneous solutions.

Connection The equation in Example 3 is a proportion. We could have solved the equation by cross multiplying, which leads to the same equation as in the second step of the solution in Example 3.

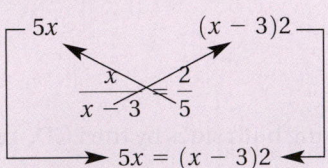

Warning: You can use this method only if the equation is a proportion.

Example 3 Solve. $\dfrac{x}{x-3} = \dfrac{2}{5}$.

SOLUTION Notice that if $x = 3$, then $\dfrac{x}{x-3}$ is undefined; so the solution cannot be 3. To solve, we first multiply both sides by the LCD, $5(x-3)$.

$$5(\overset{1}{\cancel{x-3}}) \cdot \dfrac{x}{\underset{1}{\cancel{x-3}}} = \overset{1}{\cancel{5}}(x-3) \cdot \dfrac{2}{\underset{1}{\cancel{5}}} \qquad \text{Divide out the common factors.}$$

$$5x = (x-3)2$$

$$5x = 2x - 6 \qquad \text{Distribute 2 to clear the parentheses.}$$

$$3x = -6 \qquad \text{Subtract } 2x \text{ from both sides.}$$

$$x = -2 \qquad \text{Divide both sides by 3.}$$

Check We will leave the check to the reader.

Your Turn 3 Solve $\dfrac{2}{x+5} = \dfrac{4}{x-1}$.

Answers Your Turn 2

a. $\dfrac{24}{5}$ b. $\dfrac{24}{11}$

Answer Your Turn 3

-11

In Examples 1–3, after eliminating the rational expressions, we were left with linear equations. Now let's look at some equations that transform to quadratic equations after we use the LCD to eliminate the rational expressions. Remember that when solving quadratic equations, we manipulate the equation so that one side is 0; then we factor and use the zero-factor theorem.

Example 4 Solve.

a. $\dfrac{3}{x} - 5 = x - 3$

SOLUTION Notice that because the denominator is x, the solution cannot be 0.

$$x\left(\dfrac{3}{x} - 5\right) = x(x - 3) \qquad \text{Multiply both sides by the LCD, } x.$$

$$\overset{1}{\cancel{x}} \cdot \dfrac{3}{\underset{1}{\cancel{x}}} - x \cdot 5 = x \cdot x - x \cdot 3 \qquad \text{Distribute } x; \text{ then divide out common factors.}$$

$3 - 5x = x^2 - 3x \qquad$ Because this equation is quadratic, we set it equal to 0 by subtracting 3 from and adding $5x$ to both sides.

$0 = x^2 - 3x + 5x - 3$

$0 = x^2 + 2x - 3 \qquad$ Combine like terms.

$0 = (x + 3)(x - 1) \qquad$ Factor.

$x + 3 = 0 \quad \text{or} \quad x - 1 = 0 \qquad$ Use the zero-factor theorem.

$x = -3 \qquad\qquad x = 1$

Check Verify that both -3 and 1 make the original equation true.

$$\dfrac{3}{-3} - 5 \overset{?}{=} -3 - 3 \qquad\qquad\qquad \dfrac{3}{1} - 5 \overset{?}{=} 1 - 3$$

$$-1 - 5 \overset{?}{=} -6 \qquad\qquad\qquad\qquad 3 - 5 \overset{?}{=} -2$$

$$-6 = -6 \qquad\qquad \text{True} \qquad\qquad -2 = -2 \qquad\qquad \text{True}$$

Both -3 and 1 are solutions.

b. $\dfrac{x}{x + 1} = \dfrac{6}{3x - 1}$

SOLUTION We could multiply both sides by the LCD, $(x + 1)(3x - 1)$, but because the equation is a proportion, we choose to cross multiply instead. Note that $\dfrac{1}{3}$ and -1 cannot be solutions.

$$\dfrac{x}{x + 1} = \dfrac{6}{3x - 1}$$

$$3x^2 - x \qquad\qquad 6x + 6$$

$$\dfrac{x}{x + 1} \times \dfrac{6}{3x - 1} \qquad \text{Cross multiply.}$$

$3x^2 - x = 6x + 6 \qquad$ Because this equation is quadratic, we set it equal to 0 by subtracting $6x$ and 6 from both sides.

$3x^2 - 7x - 6 = 0$

$(3x + 2)(x - 3) = 0 \qquad$ Factor.

$3x + 2 = 0, x - 3 = 0 \qquad$ Use the zero-factor theorem.

$3x = -2 \qquad x = 3 \qquad$ Solve.

$x = -\dfrac{2}{3}$

Check We will leave the check to the reader.

a. $\dfrac{1}{3} - \dfrac{2}{x} = x + 4$ **b.** $\dfrac{x}{x + 1} = \dfrac{6}{4x + 1}$

Example 5   Solve.

a. $\dfrac{x^2}{x + 2} - 3 = \dfrac{x + 6}{x + 2} - 4$

SOLUTION   By inspecting the denominators, notice that the solution cannot be -2.

$$(x + 2)\left(\dfrac{x^2}{x + 2} - 3\right) = (x + 2)\left(\dfrac{x + 6}{x + 2} - 4\right)$$
    Multiply both sides by the LCD, $x + 2$.

$$(\cancel{x + 2}) \cdot \dfrac{x^2}{\cancel{x + 2}} - (x + 2) \cdot 3 = (\cancel{x + 2}) \cdot \dfrac{x + 6}{\cancel{x + 2}} - (x + 2) \cdot 4$$
    Distribute $x + 2$ and then divide out the common factors.

$$x^2 - 3x - 6 = x + 6 - 4x - 8$$ **Combine like terms.**

$$x^2 - 3x - 6 = -3x - 2$$

$$x^2 - 4 = 0$$ **Because the equation is quadratic, we set it equal to 0 by adding $3x$ and 2 to both sides.**

$$(x + 2)(x - 2) = 0$$ **Factor.**

$$x + 2 = 0 \quad \text{or} \quad x - 2 = 0$$ **Use the zero-factor theorem.**

$$x = -2 \qquad\qquad x = 2$$

Check   In the original equation $\dfrac{x^2}{x + 2}$ and $\dfrac{x + 6}{x + 2}$ are both undefined when $x = -2$ so -2 is extraneous. Therefore, we need to check only 2.

$$\dfrac{2^2}{2 + 2} - 3 \stackrel{?}{=} \dfrac{2 + 6}{2 + 2} - 4$$ **In the original equation, replace x with 2 and then simplify.**

$$\dfrac{4}{4} - 3 \stackrel{?}{=} \dfrac{8}{4} - 4$$

$$1 - 3 \stackrel{?}{=} 2 - 4$$

$$-2 = -2$$ **True, so 2 is a solution and the only solution.**

b. $\dfrac{5}{x - 3} = \dfrac{x}{x - 2} + \dfrac{x}{x^2 - 5x + 6}$

SOLUTION   To determine the LCD of all of the denominators, we first factor quadratic form $x^2 - 5x + 6$.

Note: Inspecting the denominators after factoring, we see that neither 2 nor 3 can be a solution.

$$\dfrac{5}{x - 3} = \dfrac{x}{x - 2} + \dfrac{x}{(x - 2)(x - 3)}$$ **Factor the denominator $x^2 - 5x + 6$.**

$$(x - 2)(x - 3)\left(\dfrac{5}{x - 3}\right) = (x - 2)(x - 3)\left(\dfrac{x}{x - 2} + \dfrac{x}{(x - 2)(x - 3)}\right)$$ **Multiply both sides by the LCD, $(x - 2)(x - 3)$.**

Distribute $(x - 2)(x - 3)$; then divide out the common factors.

$$(x - 2)(\cancel{x - 3}) \cdot \dfrac{5}{\cancel{(x - 3)}} = (\cancel{x - 2})(x - 3) \cdot \dfrac{x}{\cancel{(x - 2)}} + (\cancel{x - 2})(\cancel{x - 3}) \cdot \dfrac{x}{\cancel{(x - 2)}\cancel{(x - 3)}}$$

$$(x - 2)5 = (x - 3)x + x$$

$$5x - 10 = x^2 - 3x + x$$ **Distribute to clear the parentheses.**

$$5x - 10 = x^2 - 2x$$ **Combine like terms.**

Answers   **Your Turn 4**

a. $-\dfrac{2}{3}$ and -3 **b.** $-\dfrac{3}{4}$ and 2

$$0 = x^2 - 7x + 10$$ Because the equation is quadratic, we set it equal to 0 by subtracting $5x$ from and adding 10 to both sides.

$$0 = (x - 2)(x - 5)$$ Factor.

$$x - 2 = 0 \quad \text{or} \quad x - 5 = 0$$ Use the zero-factor theorem.

$$x = 2 \qquad\qquad x = 5$$

Check In the original equation, note that $\dfrac{x}{x - 2}$ is undefined when $x = 2$; so 2 is an extraneous solution. Consequently, 5 is the only solution. We will leave the check for 5 to the reader.

Your Turn 5 Solve.

a. $\dfrac{3x + 4}{x - 4} = \dfrac{x^2}{x - 4} + 3$

b. $\dfrac{2}{x + 5} - \dfrac{6}{x^3 + 5x^2} = \dfrac{1}{x}$

Warning: Make sure you understand the difference between performing operations with rational expressions and solving equations containing rational expressions. For example,

$\dfrac{5}{x} + \dfrac{1}{3}$ is an *expression* that means two rational expressions are to be added. Remember,

we cannot solve expressions; we can only evaluate or rewrite them. To rewrite this expression, we begin by writing each rational expression with the LCD, $3x$.

$$\dfrac{5}{x} + \dfrac{1}{3}$$

$$= \dfrac{5(3)}{x(3)} + \dfrac{1(x)}{3(x)}$$

$$= \dfrac{15}{3x} + \dfrac{x}{3x}$$

$$= \dfrac{15 + x}{3x}$$

To solve an *equation* such as $\dfrac{3}{x} + \dfrac{1}{4} = \dfrac{1}{2}$, we first eliminate the rational expressions by multiplying both sides of the equation by the LCD, $4x$.

$$4x\left(\dfrac{3}{x} + \dfrac{1}{4}\right) = 4x\left(\dfrac{1}{2}\right).$$

$$12 + x = 2x$$

$$12 = x \qquad \text{Subtract x from both sides.}$$

Answers Your Turn 5

a. -4, (4 is extraneous)

b. 6 and -1

7.6 Exercises   For Extra Help   *MyMathLab*

Note: Exercises marked with a ★ represent challenging exercises.

1. How are rational expressions eliminated from an equation?

2. What is the LCD for the rational expressions in $\dfrac{x}{x^2 + 4x - 5} = \dfrac{5}{x} - 2x$?

3. What is an extraneous solution for an equation that contains rational expressions?

4. Why might an equation with rational expressions lead to one or more extraneous solutions?

5. What values might be extraneous solutions in the equation $\dfrac{3}{x-4} = \dfrac{5x}{x+2}$? Why?

6. What values might be extraneous solutions in the equation $\dfrac{x}{x^2+4x-5} = \dfrac{5}{x} - 2x$? Why?

For Exercises 7–12, check the given values to see if they are solutions to the equation.

7. $\dfrac{3}{x} + \dfrac{2}{3} = \dfrac{5}{6}$; $x = 9$

8. $\dfrac{3}{4} - \dfrac{1}{x} = \dfrac{2}{3}$; $x = 12$

9. $\dfrac{4}{m} - \dfrac{2}{5} = \dfrac{3}{4m}$; $m = \dfrac{65}{8}$

10. $\dfrac{5}{6n} + \dfrac{1}{4} = \dfrac{2}{3n}$; $n = -\dfrac{2}{3}$

11. $\dfrac{1}{t+2} + \dfrac{1}{4} = \dfrac{t}{16}$; $t = -4$

12. $\dfrac{1}{y-3} - \dfrac{5}{6} = -\dfrac{y}{12}$; $y = -7$

For Exercises 13–62, solve and check. Identify any extraneous solutions. See Examples 1–5.

13. $\dfrac{2n}{3} + \dfrac{3n}{2} = \dfrac{13}{3}$

14. $\dfrac{3x}{5} - \dfrac{x}{2} = \dfrac{19}{5}$

15. $3y - \dfrac{4y}{5} = 22$

16. $\dfrac{3t}{4} - 2 = \dfrac{t}{4}$

17. $\dfrac{r-7}{2} - 1 = \dfrac{r+9}{9} + \dfrac{1}{3}$

18. $\dfrac{d+5}{3} - 4 = \dfrac{d-8}{4} - \dfrac{1}{2}$

19. $\dfrac{4}{x} + \dfrac{3}{2x} = \dfrac{11}{6}$

20. $\dfrac{3}{2u} - \dfrac{1}{3} = \dfrac{5}{6u}$

21. $\dfrac{4-6t}{2t} + \dfrac{3}{5} = -\dfrac{2}{5t}$

22. $\dfrac{2x+3}{2x} = \dfrac{5}{12} + \dfrac{3x+2}{3x}$

23. $\dfrac{5}{20} = \dfrac{x-7}{x+2}$

24. $\dfrac{y-2}{y+3} = \dfrac{3}{8}$

25. $\dfrac{6}{y-2} = \dfrac{5}{y-3}$

26. $\dfrac{4}{3x-3} = \dfrac{7}{4x+1}$

27. $\dfrac{x+4}{x+2} - 5 = \dfrac{6}{x+2}$

28. $\dfrac{m+5}{m-3} - 5 = \dfrac{4}{m-3}$

29. $\dfrac{6f-5}{6} = \dfrac{2f-1}{2} - \dfrac{f+2}{2f+5}$

30. $\dfrac{2x-1}{8} = \dfrac{x}{4} - \dfrac{x-1}{5x-2}$

31. $\dfrac{2}{x^2-2x-3} - \dfrac{3}{x-3} = \dfrac{2}{x+1}$

32. $\dfrac{5}{t^2-9} = \dfrac{3}{t+3} - \dfrac{2}{t-3}$

33. $\dfrac{1}{u-4} + \dfrac{2}{u^2-16} = \dfrac{3}{u+4}$

34. $\dfrac{5}{h+5} - \dfrac{2}{h^2+2h-15} = \dfrac{2}{h-3}$

35. $\dfrac{x}{2-3x} = \dfrac{1}{3x+2}$

36. $\dfrac{5p}{p+4} = \dfrac{p}{p-1}$

37. $\dfrac{5}{2x} + \dfrac{15}{2x+16} = \dfrac{6}{x+8}$

38. $\dfrac{4}{3x} - \dfrac{2}{x+4} = \dfrac{2}{3x+12}$

39. $\dfrac{2}{x^2-1} = \dfrac{-3}{7x+7}$

40. $\dfrac{3}{y^2-4} = \dfrac{-2}{5y+10}$

41. $\dfrac{3}{x+5} - \dfrac{x-1}{2x} = \dfrac{-3}{2x^2+10x}$

42. $\dfrac{-1}{3x^2+9x} = \dfrac{1}{x+3} - \dfrac{x-3}{3x}$

43. $\dfrac{3m+1}{m^2-9} - \dfrac{m+3}{m-3} = \dfrac{1-5m}{m+3}$

44. $\dfrac{2n+10}{n^2-1} - \dfrac{n-1}{n+1} + \dfrac{2n+1}{n-1} = 0$

45. $\dfrac{x+3}{x-1} + \dfrac{2}{x+3} = \dfrac{5}{3}$

46. $\dfrac{m+2}{m-2} - \dfrac{2}{m+2} = -\dfrac{7}{3}$

47. $1 - \dfrac{3}{x-2} = -\dfrac{12}{x^2-4}$

48. $\dfrac{1}{x-3} + \dfrac{1}{3} = \dfrac{6}{x^2-9}$

49. $\dfrac{2r}{r-2} - \dfrac{4r}{r-3} = -\dfrac{7r}{r^2-5r+6}$

50. $\dfrac{3a}{a^2-2a-15} - \dfrac{a}{a+3} = \dfrac{2a}{a-5}$

51. $\dfrac{4u^2+3u+4}{u^2+u-2} - \dfrac{3u}{u+2} = \dfrac{-2u-1}{u-1}$

52. $\dfrac{2z}{2z-3} - \dfrac{3z}{2z+3} = \dfrac{15-32z^2}{4z^2-9}$

53. $\dfrac{2k}{k+7} - 1 = \dfrac{1}{k^2+10k+21} + \dfrac{k}{k+3}$

54. $\dfrac{p-5}{p+5} - 2 + \dfrac{p+15}{p-5} = \dfrac{50}{p^2-25}$

55. $\dfrac{6}{t^2+t-12} = \dfrac{4}{t^2-t-6} + \dfrac{1}{t^2+6t+8}$

56. $\dfrac{7}{v^2-6v+5} - \dfrac{2}{v^2-4v-5} = \dfrac{3}{v^2-1}$

57. $\dfrac{x+1}{x^2-2x-15} - \dfrac{6}{x^2-3x-10} = \dfrac{2}{x^2+5x+6}$

58. $\dfrac{x+1}{x^2+2x-24} - \dfrac{3}{x^2-x-12} = \dfrac{3}{x^2+9x+18}$

★ 59. $\dfrac{1}{d^2-5d+6} + \dfrac{d-2}{d^2-d-6} - \dfrac{d}{d^2-4} = 0$

★ 60. $\dfrac{1}{y^2-5y+6} + \dfrac{y}{y^2-2y-3} = \dfrac{y+2}{y^2-y-2}$

★ 61. $\dfrac{1}{a-3} - \dfrac{6}{a^2-1} = \dfrac{12}{a^3-3a^2-a+3}$

★ 62. $\dfrac{1}{b+2} - \dfrac{2}{b^2-1} = \dfrac{-2}{b^3+2b^2-b-2}$

For Exercises 63 and 64, explain the mistake; then find the correct solution(s).

63.
$$\frac{y}{3} + \frac{y}{y - 3} = \frac{3}{y - 3}$$

$$3(y - 3) \cdot \frac{y}{3} + 3(y - 3) \cdot \frac{y}{y - 3} = 3(y - 3) \cdot \frac{3}{y - 3}$$

$$y^2 - 3y + 3y = 9$$

$$y^2 - 9 = 0$$

$$(y + 3)(y - 3) = 0$$

$$y = -3, 3$$

64.
$$\frac{x}{x + 3} + \frac{3}{x + 3} = 0$$

$$(x + 3) \cdot \frac{x}{x + 3} + (x + 3) \cdot \frac{3}{x + 3} = (x + 3) \cdot 0$$

$$x + 3 = 0$$

$$x = -3$$

65. Explain how the LCD is used differently in $\frac{3x}{4} - \frac{5}{6}$ and in $\frac{3x}{4} - \frac{5}{6} = \frac{x}{2}$.

66. Explain how the LCD is used differently in $\frac{5x}{8} + \frac{1}{3}$ and in $\frac{5x}{8} + \frac{1}{3} = \frac{x}{6}$.

For Exercises 67–70, use the formula for the total resistance R in an electrical circuit with resistors R_1, R_2, R_3, ..., R_n that are wired in parallel.

$$\frac{1}{R} = \frac{1}{R_1} + \frac{1}{R_2} + \frac{1}{R_3} + \cdots + \frac{1}{R_n}$$

67. Two resistors are wired in parallel, one of which is 100 ohms. If the total resistance is to be 80 ohms, what must the value be of the other resistor?

★68. Three resistors are wired in parallel, two of which are equal in value. If the third resistor is 200 ohms and the total resistance is to be 20 ohms, what are the values of the other two resistors?

★69. Two resistors are wired in parallel. One resistor is to have a value that is 10 ohms more than the other resistor, and the total resistance is to be 12 ohms. Find the value of both resistors.

★70. Two resistors are wired in parallel. One resistor is to have a value that is 20 ohms less than the other resistor, and the total resistance is to be 24 ohms. Find the value of both resistors.

In optics, a lens can be used to bend light from an object through a focal point to reproduce an inverted image that is larger (or smaller) than the original object. The following formula describes how the image of an object is affected by a lens, where o represents the object's distance from the lens, i represents the image's distance from the lens, and f represents the focal length of the lens.

$$\text{Formula: } \frac{1}{o} + \frac{1}{i} = \frac{1}{f}$$

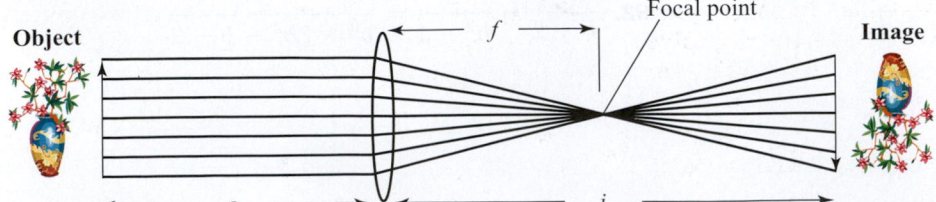

71. An object is 40 feet from a lens with a focal length of 2 feet. What is the image's distance from the lens?

72. An object is placed 50 centimeters from your eye. Your lens focuses the image of the object onto your retina, which is about 2.5 centimeters from your lens. What is the focal length of your lens?

★ **73.** A lens is being designed so that the focal length will be 5 millimeters less than the image length. Suppose an object is placed 60 millimeters from the lens. What are the focal length and image length?

★ **74.** A lens is being designed so that the image length is 9 millimeters more than the focal length. Suppose an object is placed 40 millimeters from the lens. What are the focal length and image length?

Review Exercises

Exercises 1–6 **EQUATIONS AND INEQUALITIES**

[2.1] **1.** Andrew notes that he traveled a distance of 120 miles in $1\frac{3}{4}$ hours. What was his average rate in miles per hour?

[3.1] **2.** A grocery store has a 10.5-ounce can of soup on sale for $0.88. Write a unit ratio in simplest form of the price to capacity. Interpret the answer.

[3.1] **3.** Daniel knows that he can paint 800 square feet of wall space in 4 hours. If he were able to maintain that rate, how long would it take him to paint 1200 square feet?

[3.3] **4.** The sum of two positive consecutive even integers is 146. What are the numbers?

[3.3] **5.** A rectangular eraser has a perimeter of 36 millimeters. If the length is three less than twice the width, find the length and width.

[3.4] **6.** Two joggers pass each other going in opposite directions on a path. If one jogger is traveling at a rate of 6 miles per hour and the other is traveling at a rate of 4 miles per hour, how long will it take them to be 1.5 miles apart?

7.7 Applications with Rational Expressions, Including Variation

Objectives

1. Use tables to solve problems with two unknowns involving rational expressions.
2. Solve problems involving direct variation.
3. Solve problems involving inverse variation.
4. Solve problems involving joint variation.
5. Solve problems involving combined variation.

Objective 1 Use tables to solve problems with two unknowns involving rational expressions. In Chapter 3, we used tables to organize information involving two unknown amounts. Now let's look at some similar problems that lead to equations containing rational expressions.

Problems Involving Work

Tables are helpful in problems involving two or more people (or machines) working together to complete a task. In these problems, we are given each person's rate of work and are asked to find the time for them to complete the task if they work together. For each person involved, the rate of work, time at work, and amount of the task completed are related as follows:

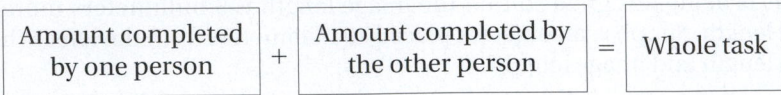

Because the people are working together, the sum of each individual's amount of the task completed equals the whole task.

| Amount completed by one person | + | Amount completed by the other person | = | Whole task |

Example 1   Karen and Jeff own a cleaning business. Karen can clean an average-size house in 2 hours. Jeff can clean a similar house in 3 hours. How long will it take them to clean an average-size house working together?

Understand   Karen cleans at a rate of 1 house in 2 hours, or $\frac{1}{2}$ of a house per hour.

Jeff cleans at a rate of 1 house in 3 hours, or $\frac{1}{3}$ of a house per hour.

Category	Rate of Work	Time at Work	Amount of Task Completed
Karen	$\frac{1}{2}$	t	$\frac{1}{2}t$ or $\frac{t}{2}$
Jeff	$\frac{1}{3}$	t	$\frac{1}{3}t$ or $\frac{t}{3}$

Note: Because they are working together on the same job for the same amount of time, we let t represent that amount of time.

Note: Multiplying the rate of work and the time at work gives an expression of the amount of work completed. For example, if Karen works 4 hours at a rate of $\frac{1}{2}$ of a house every hour, she can clean $4 \cdot \frac{1}{2} = 2$ houses.

The total job in this case is 1 house, so we can write an equation that combines the individual expressions for the task completed and set this sum equal to 1 house.

Plan and Execute   Karen's amount completed + Jeff's amount completed = 1 house

$$\frac{t}{2} + \frac{t}{3} = 1$$

$$6\left(\frac{t}{2} + \frac{t}{3}\right) = 6(1)$$ **Multiply both sides by the LCD, 6.**

$$\overset{3}{\cancel{6}} \cdot \frac{t}{\underset{1}{\cancel{2}}} + \overset{2}{\cancel{6}} \cdot \frac{t}{\underset{1}{\cancel{3}}} = 6$$ **Distribute and then divide out common factors.**

$$3t + 2t = 6$$

$$5t = 6$$ **Combine like terms.**

$$t = \frac{6}{5}$$ **Divide both sides by 5.**

Answer Working together, it takes Karen and Jeff $\frac{6}{5}$, or $1\frac{1}{5}$, hours to clean an average-size house.

Check Karen cleans $\frac{1}{2}$ of a house per hour; so if she works alone $1\frac{1}{5}$ hours, she cleans $\frac{1}{2} \cdot \frac{6}{5} = \frac{3}{5}$ of a house. Jeff cleans $\frac{1}{3}$ of a house per hour; so in $1\frac{1}{5}$ hours, he cleans $\frac{1}{3} \cdot \frac{6}{5} = \frac{2}{5}$ of a house. Combining their individual amounts, we see that in $1\frac{1}{5}$ hours, they clean $\frac{3}{5} + \frac{2}{5} = \frac{5}{5} = 1$ house.

Your Turn 1 Terrell owns a landscaping company. Every Monday, he services the same group of clients, and it takes him 5 hours to complete the service. On one occasion, he hired a student, Jarod, to service those same clients, and it took Jarod 6 hours. Working together, how much time would it take them to service the clients?

Motion Problems

Recall that the formula for calculating distance, given the rate of travel and time of travel, is $d = rt$. If we isolate r, we have $r = \frac{d}{t}$. If we isolate t, we have $t = \frac{d}{r}$. These equations suggest that we use rational expressions when describing rate or time.

Example 2 Warren runs 3 miles away from the gym, turns around, and runs an average of 2 miles per hour faster on the return trip. If the total time of his run is $1\frac{1}{4}$ hours, what is his speed in the outbound leg and the inbound leg of his run?

Understand We are to find the rates for each leg of Warren's run. Because this situation involves two different rates, we use a table to organize the distance, rate, and time of each leg of the run.

Category	Distance	Rate	Time
Outbound	3 miles	r	$\frac{3}{r}$
Inbound	3 miles	$r + 2$	$\frac{3}{r + 2}$

Note: Warren runs 2 mph faster during the inbound leg of his run; so we let r represent the outbound rate and add 2 mph to r for the inbound rate.

Note: Because $d = rt$, to describe time, we divide the distance by the rate: $t = \frac{d}{r}$.

Because the total time of the trip is $1\frac{1}{4}$ $\left(\text{or } \frac{5}{4}\right)$ hours, we can write an equation that is the sum of the outbound and inbound times.

Answer   Your Turn 1

$2\frac{8}{11}$ hr.

Plan and Execute   Outbound time + Inbound time = $\dfrac{5}{4}$ hours

$$\frac{3}{r} + \frac{3}{r+2} = \frac{5}{4}$$

$$4r(r+2)\left(\frac{3}{r} + \frac{3}{r+2}\right) = 4r(r+2)\left(\frac{5}{4}\right)$$ Multiply both sides by the LCD, $4r(r+2)$.

$$4\cancel{r}(r+2)\cdot\frac{3}{\cancel{r}} + 4r\cancel{(r+2)}\cdot\frac{3}{\cancel{r+2}} = \cancel{4}r(r+2)\cdot\frac{5}{\cancel{4}}$$ Distribute and then divide out common factors.

$$12(r+2) + 12r = 5r(r+2)$$

$$12r + 24 + 12r = 5r^2 + 10r$$ Distribute.

$$24r + 24 = 5r^2 + 10r$$ Combine like terms.

$$0 = 5r^2 - 14r - 24$$ Subtract $24r$ and 24 from both sides.

$$0 = (5r+6)(r-4)$$ Factor.

$$5r + 6 = 0 \quad \text{or} \quad r - 4 = 0$$ Use the zero-factor theorem.

$$5r = -6 \qquad\qquad r = 4$$

$$r = -\frac{6}{5}$$

Answer   Although $-\dfrac{6}{5}$ is a solution to the equation, a negative rate does not make sense in this situation; so we do not consider it as an answer. Therefore, the rate of the outbound leg is 4 miles per hour. Because the rate of the inbound leg is 2 miles per hour faster, the rate of speed must be 6 miles per hour.

Check   Verify that traveling 3 miles at 4 miles per hour, then 3 miles at 6 miles per hour takes a total of $1\dfrac{1}{4}$ hours.

$$\text{Outbound time} = \frac{3}{4}\text{ hour} \qquad \text{Inbound time} = \frac{3}{6} = \frac{1}{2}\text{ hour}$$

$$\text{Total time} = \frac{3}{4} + \frac{1}{2} = \frac{3}{4} + \frac{2}{4} = \frac{5}{4} = 1\frac{1}{4}\text{ hour}$$

Your Turn 2   A plane travels 1200 miles against the jet stream, causing its airspeed to be decreased by 20 miles per hour. On the return flight, the plane travels with the jet stream so that its airspeed is increased by 20 miles per hour. If the total flight time of the round trip is $6\dfrac{1}{3}$ hours, what is the plane's rate in still air?

Objective ➋ **Solve problems involving direct variation.**   In 2008, the Internal Revenue Service allowed a $0.505 per mile deduction for each mile driven for business purposes. If d is the amount of the deduction and n is the number of miles driven, then $d = 0.505n$. In the following table, we use that formula to determine the deduction for various values of n.

Number of Miles	Deduction ($d = 0.505n$)
1	0.505
2	1.01
3	1.515
4	2.02
5	2.525

Answer   **Your Turn 2**
380 mph

From the table, we see that as the number of miles increases, so does the amount of the deduction. Or, more formally, as values of n increase, so do values of d. In $d = 0.505n$, the two variables, d and n, are said to **vary directly**, or are *directly proportional*, and 0.505 is the constant of variation.

Definition   ***Direct variation:*** Two variables y and x vary directly if $y = kx$. If y varies directly as the nth power of x, then $y = kx^n$, where k is the constant of variation.

In words, direct variation is written as *y varies directly as x* or *y is directly proportional to x*, and these phrases translate to $y = kx$. The expression $y = kx^n$ is translated as *y varies directly as the nth power of x* or *y is directly proportional to the nth power of x*.

In all variation problems in this section, we are given one set of values of the variables, which we use to find the constant of variation, k. We then use this value of k and the equation to find other values of the variable(s).

Procedure   **Solving Variation Problems**

1. Write the equation.
2. Substitute the initial values and find k.
3. Substitute for k in the equation found in Step 1.
4. Substitute the additional data into the equation from Step 3 and solve for the unknown.

Example 3   Solve the direct variations.

a. Suppose y varies directly as x. If $y = 18$ when $x = 5$, find y when $x = 8$.

SOLUTION   Translating "y varies directly as x," we have $y = kx$. We replace y with 18 and x with 5 in $y = kx$ and solve for k.

$$18 = k \cdot 5 \qquad \text{Replace } y \text{ with 18 and } x \text{ with 5.}$$
$$3.6 = k \qquad \text{Divide both sides by 5.}$$

Now we can replace k with 3.6 in $y = kx$ so that we have $y = 3.6x$. We use this equation to find y when $x = 8$.

$$y = 3.6(8) = 28.8$$

b. Suppose u varies directly as the square of v. If $u = 54$ when $v = 3$, find u when $v = 5$.

SOLUTION   Translating "u varies directly as the square of v," we have $u = kv^2$. We replace u with 54 and v with 3 in $u = kv^2$ and solve for k.

$$54 = k \cdot 3^2 \qquad \text{Replace } u \text{ with 54 and } v \text{ with 3.}$$
$$54 = 9k \qquad \text{Square 3.}$$
$$6 = k \qquad \text{Divide both sides by 9.}$$

Now we can replace k with 6 in $u = kv^2$ so that we have $u = 6v^2$. We use this equation to find u when $v = 5$.

$$u = 6(5)^2 \qquad \text{Substitute 5 for } v.$$
$$u = 6(25) \qquad \text{Square 5.}$$
$$u = 150 \qquad \text{Multiply.}$$

Your Turn 3   Suppose m varies directly as the square of n. If $m = 24$ when $n = 2$, find m when $n = -3$.

Answer   Your Turn 3
54

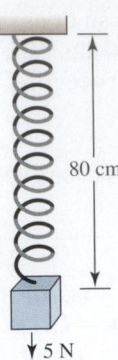

80 cm

↓5 N

Example 4 In physical science, Hooke's law states that the distance a spring of uniform material and thickness is stretched varies directly with the force applied to the spring. If a force of 5 newtons stretches a spring 80 centimeters, how much force is required to stretch the spring 128 centimeters?

Understand Translating "the distance a spring of uniform material and thickness is stretched varies directly with the force," we write $d = kF$, where d represents distance and F represents the force.

Plan Use $d = kF$, replacing d with 80 centimeters and F with 5 newtons to solve for the value of k. Then use that value in $d = kF$ to solve for the force required to stretch the spring 128 centimeters.

Execute $80 = k \cdot 5$ **Replace d with 80 and F with 5.**

$16 = k$ **Divide both sides by 5.**

Replacing k with 16 in $d = kF$, we have $d = 16F$, which we use to solve for F when d is 128 centimeters.

$$128 = 16F$$ **Substitute 128 for d.**

$$8 = F$$ **Divide both sides by 16.**

Answer To stretch the spring 128 centimeters, a force of 8 newtons is applied.

Check A force of 8 newtons stretches the spring $d = 16(8) = 128$ centimeters.

3 ft.

Your Turn 4 The pressure exerted by a liquid varies directly as the depth beneath the surface. If an object is submerged in seawater to a depth of 3 feet, the pressure is 192 pounds per square foot. Find the pressure per square foot if an object is submerged in seawater to a depth of 8 feet.

Objective ③ Solve problems involving inverse variation. Suppose a campaign worker must stuff 100 envelopes. If r is the rate at which she can stuff the envelopes and t is the time required, then $t = \dfrac{100}{r}$. For example, if she can stuff 2 envelopes per minute, it will take her $\dfrac{100}{2} = 50$ minutes to stuff the envelopes. In the following table, we use $t = \dfrac{100}{r}$ to see the relationship between the rate at which she can stuff envelopes and the amount of time required to complete the job.

Rate r	Time $\left(t = \dfrac{100}{r}\right)$
1 per minute	100 minutes
2 per minute	50 minutes
4 per minute	25 minutes

From the table, we see that as the rate at which she stuffs the envelopes increases, the time required decreases. More formally, as values of r increase, values of t decrease. In $t = \dfrac{100}{r}$, the two variables, t and r, are said to be in **inverse variation**, or are *inversely proportional*, and 100 is the constant of variation.

Definition   *Inverse variation:* Two variables y and x vary inversely if $y = \dfrac{k}{x}$. If $y = \dfrac{k}{x^n}$, then y varies inversely as the nth power of x, where k is the constant of variation.

Answer   Your Turn 4

512 lb./ft.^{2}

In words, inverse variation is written as *y varies inversely as x* or *y is inversely proportional to x*, and these phrases translate to $y = \dfrac{k}{x}$. Similarly, *y varies inversely as the nth power of x* or *y is inversely proportional to the nth power of x* translate to $y = \dfrac{k}{x^n}$.

Example 5   Boyle's law states that if the temperature is held constant, the volume of a gas in a closed container is inversely proportional to the pressure applied to it. If a gas has a volume of 4 cubic feet when the pressure is 10 pounds per square foot, find the volume when the pressure is 2.5 pounds per square foot.

Understand   Because the volume and pressure vary inversely, we can write $V = \dfrac{k}{P}$, where *V* represents volume and *p* represents pressure.

Plan   Use the fact that the volume is 4 cubic feet when the pressure is 10 pounds per square foot to find the value of the constant, *k*. Then use this value of the constant to find the volume when the pressure is 2.5 pounds per square foot.

Execute
$$4 = \frac{k}{10} \qquad \textcolor{blue}{\textbf{Substitute for } V \textbf{ and } P.}$$

$$10 \cdot 4 = 10 \cdot \frac{k}{10} \qquad \textcolor{blue}{\textbf{Multiply both sides by 10.}}$$

$$40 = k$$

Replacing *k* with 40 in $V = \dfrac{k}{P}$, we have $V = \dfrac{40}{P}$, which we use to solve for *V* when *P* is 2.5 pounds per square foot.

$$V = \frac{40}{2.5} \qquad \textcolor{blue}{\textbf{Substitute 40 for } P.}$$

$$V = 16$$

Answer   With a pressure of 2.5 pounds per square foot, the volume is 16 cubic feet.

Your Turn 5   The intensity of a light varies inversely as the square of the distance from the light source. If the intensity is 150 foot-candles (fc) when the distance is 2 meters, find the intensity if the distance is 4 meters.

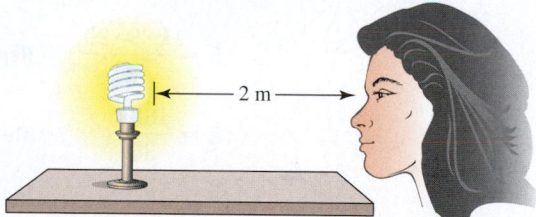

Objective ④   Solve problems involving joint variation.   Often one quantity will vary as the product of two or more quantities. This is called **joint variation**.

Definition   *Joint variation:* If *y* varies jointly as *x* and *z*, then $y = kxz$, where *k* is the constant of variation.

In words, joint variation is written as *y varies jointly as x and z* or *y is jointly proportional to x and z*, and these phrases translate to $y = kxz$.

Answer   Your Turn 5

37.5 fc

Example 6   Suppose y varies jointly with x and z. If $y = 72$ when $x = 4$ and $z = 6$, find y when $x = 2$ and $z = 5$.

SOLUTION   Translating "y varies jointly with x and z," we have $y = kxz$. We replace y with 72, x with 4, and z with 6 in $y = kxz$ and solve for k.

$$72 = k \cdot 4 \cdot 6 \qquad \text{Replace } y \text{ with 72, } x \text{ with 4, and } z \text{ with 6.}$$
$$72 = 24k \qquad \text{Multiply 4 and 6.}$$
$$3 = k \qquad \text{Divide both sides by 24.}$$

Now we can replace k with 3 in $y = kxz$ so that we have $y = 3xz$. We use this equation to find y when $x = 2$ and $z = 5$.

$$y = 3 \cdot 2 \cdot 5 \qquad \text{Substitute for } x \text{ and } z.$$
$$y = 30 \qquad \text{Multiply.}$$

Your Turn 6   Suppose n varies jointly with p and the square of q. If $n = 90$ when $p = 5$ and $q = 3$, find n when $p = 2$ and $q = 5$.

Objective ⑤ Solve problems involving combined variation.   Problems involving more than one type of variation are called *combined variation*.

Example 7   Coulomb's law states that the force, F, between two charges q_1 and q_2 in a vacuum varies jointly with the charges and inversely with the square of the distance, d, between them. If the force is 5 dynes when q_1 is 9 electrostatic units and q_2 is 20 electrostatic units and the distance between the charges is 6 centimeters, find the force between the particles if q_1 is 16 electrostatic units, q_2 is 12 electrostatic units, and the distance between the particles is 8 centimeters.

SOLUTION   Translating "the force, F, between two charges q_1 and q_2 in a vacuum varies jointly with the charges and inversely with the square of the distance, d, between them," we have $F = \dfrac{kq_1q_2}{d^2}$. Replace F with 5, q_1 with 9, q_2 with 20, and d with 6 in $F = \dfrac{kq_1q_2}{d^2}$ and solve for k.

$$5 = \frac{k \cdot 9 \cdot 20}{6^2} \qquad \text{Replace } F \text{ with 5, } q_1 \text{ with 9, } q_2 \text{ with 20, and } d \text{ with 6.}$$

$$5 = \frac{180k}{36} \qquad \text{Multiply.}$$

$$5 = 5k \qquad \text{Simplify.}$$

$$1 = k \qquad \text{Divide both sides by 5.}$$

Now we can replace k with 1 in $F = \dfrac{kq_1q_2}{d^2}$ so that we have $F = \dfrac{q_1q_2}{d^2}$. We use this equation to find F when $q_1 = 16$, $q_2 = 12$, and $d = 8$.

$$F = \frac{16 \cdot 12}{8^2} \qquad \text{Substitute for } q_1, q_2, \text{ and } d.$$

$$F = \frac{192}{64} \qquad \text{Multiply.}$$

$$F = 3 \qquad \text{Simplify.}$$

Answer   **Your Turn 6**

$n = 100$

Answer   **Answer**   The force is 3 dynes.

Your Turn 7   The maximum height, h, obtained by an object that is launched vertically upward varies directly with the square of the velocity, v, and inversely as the acceleration due to gravity, g. If an object that is launched vertically upward with a velocity of 64 feet per second and acceleration due to gravity of 32 feet per second per second obtains a maximum height of 64 feet, find the maximum height obtained by an object launched vertically upward with a velocity of 128 feet per second and acceleration due to gravity of 32 feet per second per second.

7.7 Exercises   For Extra Help   *MyMathLab*   Math XL PRACTICE   WATCH   DOWNLOAD

Note: Exercises marked with a ★ represent challenging exercises.

1. If a person can complete a task in x hours, what part of the task can he complete in 1 hour?

2. If a represents the number of hours for one person to complete a task and b represents the number of hours for a second person to complete the same task, represent the part of the task completed in 1 hour while both people work together.

3. If a vehicle travels a distance of 100 miles at a rate of r, write an expression for the time, t, it takes the vehicle to travel the 100 miles. Which type of variation is this?

4. If Fred can row a canoe x miles per hour in still water and the current in the Wekiva River is 3 miles per hour, write an expression for Fred's rate rowing upstream and an expression for his rate rowing downstream in the river.

5. If p varies directly as q, then as q decreases, what happens to the value of p? (Assume $k > 0$).

6. If m is inversely proportional to n, then as the n quantity increases, what happens to the value of m? (Assume that $k > 0$.)

For Exercises 7–22, use a table to organize the information; then solve. See Examples 1 and 2.

7. Jason can wash and wax his car in 4 hours. His younger sister can wash and wax the same car in 6 hours. Working together, how fast can they wash and wax the car?

8. Joe can paint the outside of a house in 6 days working alone. His helper, Frank, can paint the outside of the same house in 9 days working alone. How long will it take to paint the outside of the house working together?

9. Alicia and Geraldine have volunteered to make quilts for a charity auction. If Alicia can make a quilt in 25 days and Geraldine can make a quilt in 35 days, in how many days can they make a quilt working together?

10. It takes Alice 90 minutes to put a futon frame together, and it takes Maya 60 minutes to put the same type of frame together. If they worked together, how long would it take to put a frame together?

11. Working together, it takes two roofers 4 hours to put a new roof on a portable classroom. If the first roofer can do the job by himself in 6 hours, how many hours will it take the second roofer to do the job by himself?

12. Working together, Rita and Tiffany can cut and trim a lawn in 2 hours. If it takes Rita 5 hours to do the lawn by herself, how long will it take Tiffany to do the lawn by herself?

13. With both the cold water and hot water faucets open, it takes 9 minutes to fill a bathtub. The cold water faucet alone takes 15 minutes to fill the tub. How long will it take to fill the tub with just the hot water faucet?

14. The cargo hold of a ship has two loading pipes. Used together, the two pipes can fill the cargo hold in 6 hours. If the larger pipe alone can fill the hold in 8 hours, how many hours will it take the smaller pipe to fill the hold by itself?

15. A bus leaves Valdosta, Georgia, at 10 A.M. traveling north on I-75 at an average rate of 63 miles per hour. At the same time, a car leaves Valdosta traveling south on I-75 at an average rate of 72 miles per hour. At what time will they be 675 miles apart?

16. Sailing in opposite directions, an aircraft carrier and a destroyer leave their base in Hawaii at 5 A.M. If the destroyer sails at 30 miles per hour and the aircraft carrier sails at 20 miles per hour, at what time will the two ships be 300 miles apart?

17. Rapid City is 360 miles from Sioux Falls. At 6 A.M., a freight train leaves Rapid City for Sioux Falls, and at the same time, a passenger train leaves Sioux Falls for Rapid City. The two trains meet at 9 A.M. If the freight train travels $\frac{3}{5}$ of the speed of the passenger train, how fast does the passenger train travel?

18. Houston and Calgary are about 2100 miles apart. At 2 P.M., an airplane leaves Houston for Calgary, flying at 250 miles per hour. At the same time, an airplane leaves Calgary for Houston, flying 450 miles per hour. How long will it be before the two airplanes meet?

19. Jack leaves his home in Atlanta, traveling north on I-75 at an average rate of 45 miles per hour. Two hours later, his wife, Frances, leaves home and takes the same route, traveling at an average rate of 60 miles per hour. How long will it take Frances to catch Jack?

20. A ship leaves port traveling 15 miles per hour. Two hours later, a speedboat leaves the same port traveling 40 miles per hour. How long will it take the speedboat to overtake the ship?

21. An airliner flies against the wind from Washington, D.C., to San Francisco in 5.5 hours. It flies back to Washington, D.C., with the wind in 5 hours. If the average speed of the wind is 21 miles per hour, what is the speed of the airliner in still air?

22. A river has a current of 3 miles per hour. A boat goes 40 miles upstream in the same time it goes 50 miles downstream. Find the speed of the boat in still water.

For Exercises 23–32, solve the direct variations. See Examples 3 and 4.

23. Suppose a varies directly as b. If $a = 4$ when $b = 9$, find a when $b = 27$.

24. Suppose r varies directly as s. If $r = 6$ when $s = 9$, find r when $s = 15$.

25. Suppose y varies directly as the square of x. If $y = 100$ when $x = 5$, find y when $x = 3$.

26. Suppose t varies directly as the square of u. If $t = 45$ when $u = 3$, find t when $u = -2$.

27. Suppose m varies directly as n. If $m = 6$ when $n = 8$, what is the value of n when $m = 9$?

28. Suppose x varies directly as y. If $x = 12$ when $y = 15$, what is the value of y when $x = 4$?

29. The price of salmon at a fish market is constant, so the cost increases with the quantity purchased. Tamika notices that 2.5 pounds cost $16.25. If she plans to buy 6 pounds of salmon, how much will she pay?

30. At a produce stand, the cost of zucchini is constant and sells 3 for $0.89. To the nearest cent, find the cost of 5 zucchini.

31. According to Charles's law, if the pressure is held constant, the volume of a gas varies directly with the temperature measured on the Kelvin scale. If the volume of a gas is 288 cubic centimeters when the temperature is 80 K, find the volume when the temperature is 50 K.

32. According to Ohm's law, the current, I, which is measured in amperes, in an electrical circuit varies directly with the voltage, V. If the current is 0.64 ampere when the voltage is 24 volts, find the current when the voltage is 48 volts.

*In Exercises 33–36, use the fact that, ignoring air resistance, the distance
a free-falling body falls is directly proportional to the square of the time
it has been falling. See Example 4.*

33. On Earth, if an object falls 144 feet in 3 seconds, how many seconds will it take for the object to fall 400 feet?

34. On Earth, if an object falls 39.2 meters in 2 seconds, how far will it fall in 5 seconds?

35. On the Moon, an object falls 6.48 meters in 2 seconds. How far will it fall in 5 seconds?

36. On the Moon, an object falls 14.58 meters in 3 seconds. How long will it take for the object to fall 58.32 meters?

37. The circumference of a circle varies directly with its diameter. If the circumference is 12.56 feet when the diameter is 4 feet, find the diameter when the circumference is 21.98 feet. Use $\pi \approx 3.14$.

38. The pressure on an object submerged in a liquid varies directly as the depth beneath the surface. If an object is submerged in gasoline to a depth of 6 feet, the pressure is 253.8 pounds per square foot. Find the pressure if the object is submerged 9 feet.

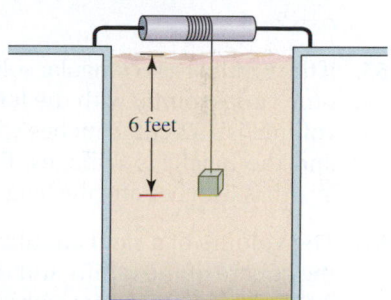

6 feet

For Exercises 39–46, solve the inverse variations. See Example 5.

39. Suppose a varies inversely as b. If $a = 3.2$ when $b = 5$, what is b when a is 8?

40. Suppose p varies inversely as q. If p is 8 when q is 3.25, what is q when p is 2?

41. Suppose m varies inversely as n. If n is 6 when m is 11, what is n when m is 8?

42. Suppose x varies inversely as y. If $y = 12$ when $x = 3$, what is y when x is 12?

43. By Charles's law, if the temperature is held constant, the pressure that a gas exerts against the walls of a container is inversely proportional to the volume of the container. A gas is inside a cylinder with a piston at one end that can vary the volume of the cylinder. When the volume of the cylinder is 20 cubic inches, the pressure inside is 40 psi (pounds per square inch). Find the pressure of the gas if the piston compresses the gas to a volume of 16 cubic inches.

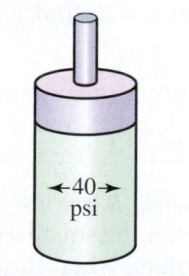

←40→
psi

40 psi at 20 cubic inches

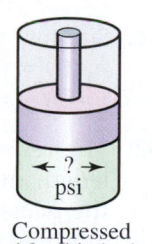

← ? →
psi

Compressed to 16 cubic inches

44. Find the volume in the cylinder from Exercise 43 if the piston compresses the gas to a pressure of 30 psi.

45. In an electrical conductor, the current, I, which is measured in amperes, varies inversely as the resistance, R, which is measured in ohms. If the current is 10 amperes when the resistance is 15 ohms, find the resistance when the current is 25 amperes.

46. In Exercise 45, find the current when the resistance is 15 ohms.

In Exercises 47 and 48, use the fact that the length of a radio wave varies inversely with its frequency. See Example 5.

47. If the length of a radio wave is 400 meters when the frequency is 900 kilohertz, find the wavelength when the frequency is 600 kilohertz.

48. If the frequency of a radio wave is 800 kilohertz when the wavelength is 200 meters, find the frequency when the radio wavelength is 500 meters.

49. The f-stop setting for a camera lens is inversely proportional to the aperture, which is the size of the opening in the lens. For a particular lens, the f-stop setting is 5.6 when the aperture is 50 mm. Find the aperture when the f-stop is 2.8.

50. The weight of an object varies inversely with the square of the distance from the center of the Earth. At Earth's surface, the weight of the space shuttle is about 4.5 million pounds. At that point, it is 4000 miles from the center of the Earth. How much will the space shuttle weigh when it is in orbit 200 miles above the surface of the Earth?

For Exercises 51–58, solve the joint variations. See Example 6.

51. Suppose a varies jointly with b and c. If $a = 96$ when $b = 6$ and $c = 4$, find a when $b = 2$ and $c = 8$.

52. Suppose m varies jointly with p and q. If $m = 70$ when $p = 5$ and $q = 2$, find m when $p = 6$ and $q = 8$.

53. Suppose a varies jointly as the square of b and c. If $a = 96$ when $b = 2$ and $c = 6$, find a when $b = 3$ and $c = 2$.

54. Suppose x varies jointly with y and the square of z. If $x = 40$ when $y = 1$ and $z = 2$, find x when $y = 2$ and $z = 1$.

55. If the width of a rectangular solid is held constant, the volume varies jointly with the length and the height. If the volume is 192 cubic inches when the length is 8 inches and the height is 6 inches, find the volume when the length is 7 inches and the height is 12 inches.

56. For a fixed amount of principal, the simple interest varies jointly with the rate and the time. If the simple interest is $1000 when the rate is 4% and the time is 5 years, find the simple interest when the rate is 6% and the time is 10 years.

57. The volume of a right circular cylinder varies jointly as the square of the radius and the height. If the volume is 301.6 cubic centimeters when the radius is 4 centimeters and the height is 6 centimeters, find the volume when the radius is 3 centimeters and the height is 6 centimeters.

58. The number of units produced varies jointly with the number of workers and the number of hours worked per worker. If 80 units can be produced by 10 workers who work 40 hours each, how many units can be produced by 15 workers who work 30 hours each?

For Exercises 59–66, solve the combined variations. See Example 7.

59. Suppose y varies directly with x and inversely as z. If $y = 8$ when $x = 4$ and $z = 6$, find y when $x = 5$ and $z = 10$.

60. Suppose m varies directly with n and inversely as p. If $m = 27$ when $n = 6$ and $p = 4$, find m when $n = 9$ and $p = 6$.

61. Suppose y varies jointly with x and z and inversely with n. If $y = 81$ when $x = 4$, $z = 9$, and $n = 8$, find n when $x = 6$, $y = 8$, and $z = 12$.

62. Suppose p varies jointly with q and r and inversely with s. If $p = 15$ when $q = 5$, $r = 3$, and $s = 6$, find p when $q = 5$, $r = 3$, and $s = 9$.

63. The resistance of a wire, R, varies directly with the length and inversely with the square of the diameter. If the resistance is 7.5 ohms when the wire is 6 meters long and the diameter is 0.02 meter, find the resistance in a wire of the same material if the length is 10 meters and the diameter is 0.04 meter.

64. The universal gas law states that the volume of a gas varies directly with the temperature and inversely with the pressure. If the volume of a gas is 1.75 cubic meters when the temperature is 70 K and the pressure is 20 grams per square centimeter, find the volume when the temperature is 80 K and the pressure is 40 grams per square centimeter.

65. Newton's Law of Universal Gravitation states that the force of attraction between two bodies is jointly proportional to their masses and inversely proportional to the square of the distance between them. If the force of attraction between two masses of 4 grams and 6 grams that are 3 centimeters apart is 48 dynes, find the force of attraction between two masses of 2 grams and 12 grams that are 6 centimeters apart.

66. The weight-carrying capacity of a rectangular beam varies jointly with its width and the square of its height and inversely as its length. If a beam is 4 inches wide, 6 inches high, and 10 feet long, it has a carrying capacity of 1400 pounds. Find the carrying capacity of a beam made of the same material if it is 3 inches wide, 5 inches high, and 12 feet long.

★ **67.** The table lists the number of miles a car can drive on the given number of gallons of gas.

Number of Miles	Number of Gallons
135	6
180	8
225	10
270	12

   a. Plot the ordered pairs of data as points in the coordinate plane. Connect the points to form a graph.

   b. Are the number of miles driven and the number of gallons of gas directly proportional or inversely proportional? Explain.

   c. Find the constant of variation. What does it represent?

   d. Does the data represent a function? Explain.

★ **68.** The table lists the time required to make a long trip at various average driving speeds.

Speed (in mph)	Driving Time (in hr.)
60	10
50	12
40	15
30	20
20	30

   a. Plot the ordered pairs of data as points in the coordinate plane. Connect the points to form a graph.

   b. Are speeds and driving times directly proportional or inversely proportional? Explain.

   c. Find the constant of variation. What does it represent?

   d. Do the data represent a function? Explain.

Review Exercises

Exercises 1 and 2 **EXPRESSIONS**

[1.1] *For Exercises 1 and 2, find the absolute values.*

1. $|-4.3|$

2. $\left|\dfrac{5}{8}\right|$

Exercises 3–6 **EQUATIONS AND INEQUALITIES**

[2.6] *For Exercises 3 and 4, graph on the number line.*

3. $x \geq 2$

4. $x < -3$

[2.6] *For Exercises 5 and 6, solve.*

5. $2x + 11 > 3$

6. $-4x - 3 \leq -11$

Chapter 7 Summary

Defined Terms

Section 7.1
Rational expression
(p. 488)

Section 7.5
Complex rational
expression (p. 528)

Section 7.6
Extraneous solution
(p. 538)

Section 7.7
Direct variation (p. 549)
Inverse variation (p. 550)
Joint variation (p. 551)

Formulas

Direct variation: $y = kx$ and $y = kx^n$ (p. 549) $d = rt$ (p. 547)

Inverse variation: $y = \dfrac{k}{x}$ and $y = \dfrac{k}{x^n}$ (p. 550) $t = \dfrac{d}{r}$ (p. 547)

Joint variation: $y = kxz$ (p. 551) $r = \dfrac{d}{t}$ (p. 547)

Procedures, Rules, and Key Examples

Procedures/Rules	Key Examples

SECTION 7.1 Simplifying Rational Expressions

To determine the value(s) that make a rational expression undefined, 1. Write an equation that has the denominator set equal to zero. 2. Solve the equation.	**Example 1:** Find every value for the variable that makes the expression undefined. a. $\dfrac{5x}{x-6}$ Solution: $x - 6 = 0$ $x = 6$ b. $\dfrac{n}{n^2 - 4n - 5}$ Solution: $n^2 - 4n - 5 = 0$ $(n - 5)(n + 1) = 0$ $n - 5 = 0$ or $n + 1 = 0$ $n = 5$ $n = -1$
To simplify a rational expression to lowest terms, 1. Factor the numerator and denominator completely. 2. Divide out all common factors in the numerator and denominator. 3. Multiply the remaining factors in the numerator and the remaining factors in the denominator.	**Example 2:** Simplify. a. $\dfrac{18x^3y}{24x^2yz} = \dfrac{2 \cdot 3 \cdot 3 \cdot x \cdot x \cdot x \cdot y}{2 \cdot 2 \cdot 2 \cdot 3 \cdot x \cdot x \cdot y \cdot z}$ $= \dfrac{3 \cdot x}{2 \cdot 2 \cdot z}$ $= \dfrac{3x}{4z}$ b. $\dfrac{5m^2 - 15m}{m^2 - 9} = \dfrac{5 \cdot m \cdot (m - 3)}{(m + 3)(m - 3)}$ $= \dfrac{5m}{m + 3}$

(continued)

Procedures/Rules	Key Examples

SECTION 7.2 Multiplying and Dividing Rational Expressions

To multiply rational expressions,
1. Factor each numerator and denominator completely.
2. Divide out any numerator factor with any matching denominator factor.
3. Multiply numerator by numerator and denominator by denominator.
4. Simplify as needed.

Example 1: Multiply.

a.
$$-\frac{8u^3}{3t} \cdot \frac{9t^2}{10u^3} =$$

$$-\frac{2 \cdot 2 \cdot 2 \cdot u \cdot u \cdot u}{3 \cdot t} \cdot \frac{3 \cdot 3 \cdot t \cdot t}{2 \cdot 5 \cdot u \cdot u \cdot u}$$

$$= -\frac{2 \cdot 2}{1} \cdot \frac{3 \cdot t}{5}$$

$$= -\frac{12t}{5}$$

b.
$$\frac{x^2 + 4x}{x^2 - 3x - 10} \cdot \frac{3x + 6}{12x}$$

$$= \frac{x \cdot (x + 4)}{(x + 2) \cdot (x - 5)} \cdot \frac{3 \cdot (x + 2)}{2 \cdot 2 \cdot 3 \cdot x}$$

$$= \frac{(x + 4)}{(x - 5)} \cdot \frac{1}{2 \cdot 2}$$

$$= \frac{x + 4}{4(x - 5)} \text{ or } \frac{x + 4}{4x - 20}$$

To divide rational expressions,
1. Write an equivalent multiplication statement with the reciprocal of the divisor.
2. Factor each numerator and denominator completely. (Steps 1 and 2 are interchangeable.)
3. Divide out any numerator factor with any matching denominator factor.
4. Multiply numerator by numerator and denominator by denominator.
5. Simplify as needed.

Example 2: Divide.

a.
$$\frac{3n^3}{14m} \div \frac{12mn^2}{7m}$$

$$= \frac{3n^3}{14m} \cdot \frac{7m}{12mn^2}$$

$$= \frac{3 \cdot n \cdot n \cdot n}{2 \cdot 7 \cdot m} \cdot \frac{7 \cdot m}{2 \cdot 2 \cdot 3 \cdot m \cdot n \cdot n}$$

$$= \frac{n}{2} \cdot \frac{1}{2 \cdot 2 \cdot m}$$

$$= \frac{n}{8m}$$

b.
$$\frac{r - 4}{3r^2 - 17r + 10} \div \frac{r^2 - 3r - 4}{3r^2 + r - 2}$$

$$= \frac{r - 4}{3r^2 - 17r + 10} \cdot \frac{3r^2 + r - 2}{r^2 - 3r - 4}$$

$$= \frac{r - 4}{(3r - 2) \cdot (r - 5)} \cdot \frac{(3r - 2) \cdot (r + 1)}{(r - 4) \cdot (r + 1)}$$

$$= \frac{1}{r - 5}$$

To convert units using dimensional analysis, multiply the given measurement by conversion factors so that the undesired units divide out, leaving the desired units.

Example 3: Convert.

a. 20 feet to inches

$$20 \text{ ft.} = \frac{20 \text{ ft.}}{1} \cdot \frac{12 \text{ in.}}{1 \text{ ft.}} = 240 \text{ in.}$$

b. 6 miles to yards

$$6 \text{ mi.} = \frac{6 \text{ mi.}}{1} \cdot \frac{5280 \text{ ft.}}{1 \text{ mi.}} \cdot \frac{1 \text{ yd.}}{3 \text{ ft.}} = 10,560 \text{ yd.}$$

(*continued*)

Procedures/Rules	Key Examples

SECTION 7.3 Adding and Subtracting Rational Expressions with the Same Denominator

To add or subtract rational expressions that have the same denominator,
1. Add or subtract the numerators and keep the same denominator.
2. Simplify to lowest terms. (Remember to factor the numerators and denominators completely in order to simplify.)

Example 1: Add or subtract.

a. $\dfrac{7y}{12} + \dfrac{y}{12} = \dfrac{7y + y}{12} = \dfrac{8y}{12} = \dfrac{2y}{3}$

b. $\dfrac{u}{u + 2} - \dfrac{3}{u + 2} = \dfrac{u - 3}{u + 2}$

c. $\dfrac{7x^2 + 5x}{x + 1} - \dfrac{x + 3}{x + 1}$

$= \dfrac{(7x^2 + 5x) - (x + 3)}{x + 1}$

$= \dfrac{(7x^2 + 5x) + (-x - 3)}{x + 1}$

$= \dfrac{7x^2 + 4x - 3}{x + 1}$

$= \dfrac{(7x - 3)(x + 1)}{x + 1}$

$= 7x - 3$

SECTION 7.4 Adding and Subtracting Rational Expressions with Different Denominators

To find the LCD of two or more rational expressions,
1. Find the prime factorization of each denominator.
2. Write a product that contains each unique prime factor the greatest number of times that factor occurs in any factorization. Or if you prefer to use exponents, write the product that contains each unique prime factor raised to the greatest exponent that occurs on that factor in any factorization.
3. Simplify the product found in Step 2.

Example 1: Find the LCD.

a. $\dfrac{5}{6t}$ and $\dfrac{1}{8t^2}$

$6t = 2 \cdot 3 \cdot t \quad 8t^2 = 2^3 \cdot t^2$
$\text{LCD} = 2^3 \cdot 3 \cdot t^2 = 24t^2$

b. $\dfrac{2}{5t - 15}$ and $\dfrac{t}{t^2 - 9}$

$5t - 15 = 5(t - 3)$
$t^2 - 9 = (t - 3)(t + 3)$
$\text{LCD} = 5(t - 3)(t + 3)$

To add or subtract rational expressions with different denominators,
1. Find the LCD.
2. Write each rational expression as an equivalent expression with the LCD.
3. Add or subtract the numerators and keep the LCD.
4. Simplify.

Example 2: Add or subtract.

a. $\dfrac{5}{6t} + \dfrac{1}{8t^2} = \dfrac{5(4t)}{6t(4t)} + \dfrac{1(3)}{8t^2(3)}$

$= \dfrac{20t}{24t^2} + \dfrac{3}{24t^2}$

$= \dfrac{20t + 3}{24t^2}$

b. $\dfrac{2}{5t - 15} - \dfrac{t}{t^2 - 9}$

$= \dfrac{2}{5(t - 3)} - \dfrac{t}{(t - 3)(t + 3)}$

$= \dfrac{2(t + 3)}{5(t - 3)(t + 3)} - \dfrac{t(5)}{(t - 3)(t + 3)(5)}$

$= \dfrac{2t + 6}{5(t - 3)(t + 3)} - \dfrac{5t}{5(t - 3)(t + 3)}$

$= \dfrac{2t + 6 - 5t}{5(t - 3)(t + 3)}$

$= \dfrac{-3t + 6}{5(t - 3)(t + 3)}$

(continued)

Procedures/Rules	Key Examples

SECTION 7.5 Complex Rational Expressions

To simplify a complex rational expression, use one of the following methods.

Method 1
1. Simplify the numerator and denominator if needed.
2. Rewrite as a horizontal division problem.

Example 1: Simplify $\dfrac{\dfrac{5t}{t+1}}{\dfrac{10}{t^2-1}}$.

$$\dfrac{\dfrac{5t}{t+1}}{\dfrac{10}{t^2-1}} = \dfrac{5t}{t+1} \div \dfrac{10}{t^2-1}$$

$$= \dfrac{5t}{t+1} \cdot \dfrac{t^2-1}{10}$$

$$= \dfrac{5\cdot t}{t+1} \cdot \dfrac{(t+1)(t-1)}{2\cdot 5}$$

$$= \dfrac{t}{1} \cdot \dfrac{t-1}{2}$$

$$= \dfrac{t^2-t}{2}$$

Method 2
1. Multiply the numerator and denominator of the complex rational expression by their LCD.
2. Simplify.

Example 2: Simplify $\dfrac{\dfrac{5}{6}-\dfrac{t}{2}}{\dfrac{t}{4}+\dfrac{1}{3}}$.

$$\dfrac{\dfrac{5}{6}-\dfrac{t}{2}}{\dfrac{t}{4}+\dfrac{1}{3}} = \dfrac{12\left(\dfrac{5}{6}-\dfrac{t}{2}\right)}{12\left(\dfrac{t}{4}+\dfrac{1}{3}\right)}$$

$$= \dfrac{\dfrac{\overset{2}{\cancel{12}}}{1}\cdot\dfrac{5}{\underset{1}{\cancel{6}}} - \dfrac{\overset{6}{\cancel{12}}}{1}\cdot\dfrac{t}{\underset{1}{\cancel{2}}}}{\dfrac{\overset{3}{\cancel{12}}}{1}\cdot\dfrac{t}{\underset{1}{\cancel{4}}} + \dfrac{\overset{4}{\cancel{12}}}{1}\cdot\dfrac{1}{\underset{1}{\cancel{3}}}}$$

$$= \dfrac{10-6t}{3t+4}$$

SECTION 7.6 Solving Equations Containing Rational Expressions

To solve an equation that contains rational expressions,
1. Eliminate the rational expressions by multiplying both sides of the equation by their LCD.
2. Solve the equation using the methods we learned in Chapters 2 (linear equations) and 6 (quadratic equations).
3. Check your solution(s) in the original equation. Discard any extraneous solutions.

Example 1: Solve

$$\dfrac{x^2-6}{x-3} + 2 = \dfrac{x^2-2x}{x-3} - x.$$

Note that x cannot be 3 because it would cause the denominators to be 0, making those expressions undefined.

$$(x-3)\left(\dfrac{x^2-6}{x-3}+2\right)$$

$$= (x-3)\left(\dfrac{x^2-2x}{x-3}-x\right)$$

(continued)

$$(x - 3) \cdot \frac{x^2 - 6}{x - 3} + (x - 3) \cdot 2$$

$$= (x - 3) \cdot \frac{x^2 - 2x}{x - 3} - (x - 3) \cdot x$$

$$x^2 - 6 + 2x - 6 = x^2 - 2x - x^2 + 3x$$
$$x^2 + 2x - 12 = x$$
$$x^2 + x - 12 = 0$$
$$(x - 3)(x + 4) = 0$$
$$x - 3 = 0 \quad \text{or} \quad x + 4 = 0$$
$$x = 3 \qquad\qquad x = -4$$

3 is extraneous.

−4 is the solution. We leave the check to the reader.

SECTION 7.7 Applications with Rational Expressions, Including Variation

If y varies directly as x, then $y = kx$. If y varies directly as the nth power of x, then $y = kx^n$, where k is the constant of variation.

Example 1: The distance a car can travel varies directly with the number of gallons of gas used. On the highway, a Toyota Corolla traveled 280 miles using 8 gallons of gas. How far can the same car travel on 12 gallons of gas?

Solution: Translating "the distance a car can travel varies directly with the number of gallons of gas used," we write $d = kn$, where d represents the distance and n is the number of gallons of gas. Use the fact that $d = 280$ when $n = 8$ to find k.

$$280 = k(8) \qquad \text{Substitute 280 for } d \text{ and 8 for } n.$$
$$\frac{280}{8} = \frac{8k}{8} \qquad \text{Divide both sides by 8.}$$
$$35 = k$$

Now use $d = kn$ with $k = 35$ to solve for d when $n = 12$.

$$d = (35)(12) = 420 \text{ miles}$$

If y varies inversely as x, then $y = \dfrac{k}{x}$. If y varies inversely as the nth power of x, then $y = \dfrac{k}{x^n}$, where k is the constant of variation.

Example 2: The intensity of light is inversely proportional to the square of the distance from the source. A light meter 6 feet from a lightbulb measures the intensity to be 16 foot-candles. What is the intensity 24 feet from the bulb?

Solution: Translating "the intensity of light is inversely proportional to the square of the distance from the source," we write $I = \dfrac{k}{d^2}$, where I is the intensity and d is the distance. Use the fact that $I = 16$ when $d = 6$ to find k.

$$16 = \frac{k}{6^2} \qquad \text{Substitute 16 for } I \text{ and 6 for } d.$$
$$36(16) = 36 \cdot \frac{k}{36} \qquad \text{Multiply both sides by 36.}$$
$$576 = k$$

(*continued*)

SECTION 7.7 Applications with Rational Expressions, Including Variation (*continued*)

	Now use $I = \dfrac{k}{d^2}$ again with $k = 576$ to solve for I when $d = 24$. $$I = \frac{576}{24^2}$$ $$I = \frac{576}{576} = 1 \text{ foot-candle}$$
If y varies jointly as x and z, then $y = kxz$, where k is the constant of variation.	**Example 3:** Suppose m varies jointly as n and p. If $m = 288$ when $n = 3$ and $p = 6$, find p when $m = 240$ and $n = 5$. Solution: Translating "m varies jointly as n and p," we write $m = knp$. Use the fact that $m = 288$ when $n = 3$ and $p = 6$ to find k. $$288 = k(3)(6)$$ $$288 = 18k$$ $$16 = k$$ Now use $m = knp$ with $k = 16$ to solve for p when $m = 240$ and $n = 5$. $$240 = 16(5)(p)$$ $$240 = 80p$$ $$3 = p$$

Chapter 7 Review Exercises

For Exercises 1–5, answer true or false.

[7.1]   **1.** The rational expression $\dfrac{2x - 1}{x}$ is undefined when $x = \dfrac{1}{2}$.

[7.1]   **2.** A fraction is in lowest terms when the GCF of the numerator and denominator is 1.

[7.1]   **3.** The expression $\dfrac{2 - x}{x - 2}$ simplifies to -1.

[7.1]   **4.** The expression $-x + 5$ can be rewritten as $-1(x - 5)$.

[7.3]   **5.** $\dfrac{2x}{3} + \dfrac{x}{3} = \dfrac{3x}{3} = x$

For Exercises 6–10, complete the rule.

[7.2]   **6.** To divide rational expressions,
   a. Write an equivalent _____ statement with the reciprocal of the divisor.
   b. Factor each numerator and denominator completely. (Steps 1 and 2 are interchangeable.)

c. _____ out any numerator factor with any matching denominator factor.

d. Multiply numerator by numerator and denominator by denominator.

e. Simplify as needed.

[7.3]   7. To add or subtract rational expressions that have the same denominator,

a. Add or subtract the _____ and keep the same _____.

b. Simplify to lowest terms. (Remember to factor the numerators and denominators completely in order to reduce.)

[7.5]   8. A _____ rational expression is a rational expression that contains rational expressions in the numerator or denominator.

[7.5]   9. One method of simplifying a complex rational expression is as follows:

a. Simplify the _____ and _____ of the complex rational expression.

b. Divide.

[7.6]   10. To solve an equation that contains rational expressions,

a. Eliminate the rational expressions by multiplying through by their _____.

b. Solve the equation using the methods we learned in Chapters 2 (linear equations) and 6 (quadratic equations).

c. Check your solution(s) in the original equation.

Exercises 11–70 **EXPRESSIONS**

[7.1] *For Exercises 11 and 12, evaluate the rational expression.*

11. $\dfrac{4x + 5}{2x}$ when

    a. $x = 3$
    b. $x = -1.3$
    c. $x = -2$

12. $\dfrac{5x}{x + 2}$ when

    a. $x = 3$
    b. $x = -3$
    c. $x = 1.2$

[7.1] *For Exercises 13–16, find every value that can replace the variable in the expression and cause the expression to be undefined.*

13. $\dfrac{9y}{5 - y}$

14. $\dfrac{2x}{x^2 - 4}$

15. $\dfrac{6}{x^2 + 5x - 6}$

16. $\dfrac{2x + 3}{x^2 + 6x}$

[7.1] *For Exercises 17–24, simplify to lowest terms.*

17. $\dfrac{3(x + 5)}{9}$

18. $\dfrac{2y + 12}{3y + 18}$

19. $\dfrac{x^2 + 6x}{4x + 24}$

20. $\dfrac{xy - y^2}{3x - 3y}$

21. $\dfrac{a^2 - b^2}{a^2 - 2ab + b^2}$

22. $\dfrac{3x + 9}{x^4 - 81}$

23. $\dfrac{x^2 - y^2}{y - x}$

24. $\dfrac{1 - w}{w^2 + 2w - 3}$

[7.2] For Exercises 25–32, multiply.

25. $\dfrac{x}{y} \cdot \dfrac{7x}{4y}$

26. $-\dfrac{2n^2}{8m^2n} \cdot \dfrac{24mp}{10n^2p^5}$

27. $\dfrac{6}{2m + 4} \cdot \dfrac{3m + 6}{15}$

28. $\dfrac{x^2 + 3x}{x^2 - 16} \cdot \dfrac{x^2 - 4x}{x^2 - 9}$

29. $\dfrac{n^2 - 6n + 9}{n^2 - 9} \cdot \dfrac{n^2 + 4n + 3}{n^2 - 3n}$

30. $\dfrac{4r^2 + 4r + 1}{r + 2r^2} \cdot \dfrac{2r}{2r^2 - r - 1}$

31. $\dfrac{4m}{m^2 - 2m + 1} \cdot \dfrac{m^2 - 1}{16m^2}$

32. $\dfrac{w^2 - 2w - 24}{w^2 - 16} \cdot \dfrac{(w + 5)(w + 4)}{w^2 - w - 30}$

[7.2] For Exercises 33–40, divide.

33. $\dfrac{y^2}{ab} \div \dfrac{x}{y}$

34. $-\dfrac{7y^2}{10b^2} \div -\dfrac{21y^2}{25b^2}$

35. $\dfrac{v + 1}{3} \div \dfrac{3v + 3}{18}$

36. $\dfrac{2a + 4}{5} \div \dfrac{4a + 8}{25a}$

37. $\dfrac{x^2 - y^2}{c^2 - d^2} \div \dfrac{x - y}{c + d}$

38. $\dfrac{b^3 - 6b^2 + 8b}{6b} \div \dfrac{2b - 4}{10b + 40}$

39. $\dfrac{3j + 2}{5j^2 - j} \div \dfrac{6j^2 + j - 2}{10j^2 + 3j - 1}$

40. $\dfrac{u^2 - 2u - 8}{u^2 + 3u + 2} \div (u^2 - 3u - 4)$

[7.3] For Exercises 41–48, add or subtract as indicated.

41. $\dfrac{5x}{18} + \dfrac{x}{18}$

42. $\dfrac{2r}{r + 3} + \dfrac{r}{r + 3}$

43. $\dfrac{8m + 5}{6m^2} - \dfrac{10m + 5}{6m^2}$

44. $\dfrac{4x + 5}{x + 1} - \dfrac{x + 4}{x + 1}$

45. $\dfrac{2x}{x + y} - \dfrac{x - y}{x + y}$

46. $\dfrac{2r - 5s}{r^2 - s^2} - \dfrac{2r - 6s}{r^2 - s^2}$

47. $\dfrac{9x^2 - 24x + 16}{2x + 1} + \dfrac{2x^2 + 3x - 9}{2x + 1}$

48. $\dfrac{q^2 + 2q}{q - 7} - \dfrac{12q - 21}{q - 7}$

[7.4] For Exercises 49–54, write equivalent rational expressions with their LCD.

49. $\dfrac{2}{4c}, \dfrac{3}{3c^3}$

50. $\dfrac{5}{xy^3}, \dfrac{3}{x^2y^2}$

51. $\dfrac{4}{m - 1}, \dfrac{4y}{m + 1}$

52. $\dfrac{3}{4h - 8}, \dfrac{5}{h^2 - 2h}$

53. $\dfrac{x}{x^2 - 1}, \dfrac{2 + x}{x + 1}$

54. $\dfrac{2}{p^2 - 4}, \dfrac{p}{p^2 + 4p + 4}$

[7.4] *For Exercises 55–62, add or subtract as indicated.*

55. $\dfrac{3a - 4}{18} - \dfrac{2a + 5}{12}$

56. $\dfrac{4}{x + 3} - \dfrac{5}{x - 3}$

57. $\dfrac{3}{y - 2} + \dfrac{5}{y - 4}$

58. $\dfrac{5}{x^2 - 4} + \dfrac{4}{x + 2}$

59. $\dfrac{6t}{(t - 3)^2} - \dfrac{3t}{2t - 6}$

60. $\dfrac{a + 6}{a^2 + 7a + 12} - \dfrac{a - 3}{a + 3}$

61. $\dfrac{10}{2y - 1} - \dfrac{5}{1 - 2y}$

62. $\dfrac{3}{3x - 12} + \dfrac{15}{x^2 - 16}$

[7.5] *For Exercises 63–70, simplify.*

63. $\dfrac{\frac{a}{b}}{\frac{x}{y}}$

64. $\dfrac{\frac{1}{3} - \frac{1}{2}}{\frac{1}{2} - \frac{1}{3}}$

65. $\dfrac{\frac{5}{3x + 5}}{\frac{x}{x - 2}}$

66. $\dfrac{\frac{x + 3}{x^2 - 9}}{x}$

67. $\dfrac{\frac{x}{4x^2 - 1}}{\frac{5}{2x + 1}}$

68. $\dfrac{r - \frac{r^2 - 1}{r}}{1 - \frac{r - 1}{r}}$

69. $\dfrac{\frac{1}{x^2} + \frac{1}{y^2}}{\frac{7}{xy}}$

70. $\dfrac{\frac{1}{h + 1} - 1}{\frac{1}{h + 1}}$

Exercises 71–87 EQUATIONS AND INEQUALITIES

[7.6] *For Exercises 71–78, solve and check. Identify any extraneous solutions.*

71. $\dfrac{3x}{2} - \dfrac{x}{5} = \dfrac{19}{5}$

72. $\dfrac{3t}{6} - 2 = \dfrac{t}{6}$

73. $\dfrac{2x + 1}{4x} = \dfrac{5}{12} + \dfrac{3x - 2}{3x}$

74. $\dfrac{y + 5}{y - 3} - 5 = \dfrac{4}{y - 3}$

75. $\dfrac{x}{x - 2} = \dfrac{6}{x - 1}$

76. $\dfrac{6g - 5}{6} = \dfrac{2g - 1}{2} - \dfrac{g + 2}{2g + 5}$

77. $\dfrac{1}{p - 4} + \dfrac{1}{4} = \dfrac{8}{p^2 - 16}$

78. $\dfrac{3}{r - 2} - \dfrac{4}{r + 3} = -\dfrac{6}{r^2 + r - 6}$

[7.7] *For Exercises 79–87, solve.*

79. Suppose p varies directly as q. If $p = 24$ when $q = 4$, find p when $q = 7$.

80. Suppose s varies directly as the square of t. If $s = 72$ when $t = 3$, find s when $t = 5$.

81. Suppose y varies inversely as x. If $y = 8$ when $x = 3$, find x when $y = 4$.

82. Suppose m varies inversely as the square of n. If $m = 16$ when $n = 3$, find m when $n = 4$.

83. Suppose y varies jointly with x and z. If $y = 40$ when $x = 2$ and $z = 5$, find y when $x = 4$ and $z = 2$.

84. Suppose m varies directly with n and inversely with p. If $m = 2$ when $n = 3$ and $p = 9$, find m when $n = 6$ and $p = 4$.

85. The distance a car can travel varies directly with the amount of gas it carries. On a trip, a Chevy Corvette travels 156 miles using 6 gallons of fuel. How many gallons are required to travel 234 miles?

86. If the wavelength of a wave remains constant, the velocity, v, of a wave is inversely proportional to its period, T. In an experiment, waves are created in a fluid so that the period is 7 seconds and the velocity is 4 centimeters per second. If the period is increased to 12 seconds, what is the velocity?

87. The volume of a right circular cylinder varies jointly with the radius squared and the height. If the volume is 62.8 cubic inches when the radius is 2 inches and the height is 5 inches, find the volume when the radius is 4 inches and the height is 2 inches.

Chapter 7 Practice Test

For Extra Help Step-by-step test solutions are found on the Chapter Test Prep Videos available via the Video Resources on DVD, in *MyMathLab*, and on You Tube (search "CarsonCombinedAlg" and click on "Channels").

For Exercises 1 and 2, find the value(s) that can replace the variable in the expression and cause the expression to be undefined.

1. $\dfrac{2x}{x - 7}$

2. $\dfrac{8 - m}{m^2 - 16}$

For Exercises 3 and 4, simplify to lowest terms.

3. $\dfrac{12 - 3x}{x^2 - 8x + 16}$

4. $\dfrac{x - y}{x^2 - y^2}$

For Exercises 5 and 6, write equivalent rational expressions with their LCD.

5. $\dfrac{6}{fg^3}, \dfrac{2}{f^2}$

6. $\dfrac{5x}{x^2 - 9}, \dfrac{2}{x^2 + 6x + 9}$

For Exercises 7–18, perform the indicated operation.

7. $\dfrac{4a^2b^3}{15x^3y} \cdot \dfrac{25x^5y}{16ab}$

8. $-\dfrac{9x^3y^4}{16ab^2} \div \dfrac{45x^5y^2}{14a^7b^9}$

9. $\dfrac{4ab - 8b}{x^2} \div \dfrac{2b - ab}{x^3}$

10. $\dfrac{12x^2 - 6x}{x^2 + 6x + 5} \cdot \dfrac{2x^2 + 10x}{4x^2 - 1}$

11. $\dfrac{2x}{2x + 3} + \dfrac{5x}{2x + 3}$

12. $\dfrac{x}{x^2 - 9} - \dfrac{3}{x^2 - 9}$

13. $\dfrac{6}{x^2 - 4} + \dfrac{x - 3}{x^2 - 2x}$

14. $\dfrac{4}{x - 1} + \dfrac{5}{x + 2}$

15. $\dfrac{5}{x - 4} - \dfrac{2}{x + 1}$

16. $\dfrac{x}{2x + 4} - \dfrac{2}{x^2 + 2x}$

17. $\dfrac{2 + \dfrac{1}{y}}{3 - \dfrac{1}{y}}$

18. $\dfrac{\dfrac{x - y}{2}}{\dfrac{x^2 - y^2}{4}}$

For Exercises 19–22, solve the equation. Identify any extraneous solutions.

19. $\dfrac{3x - 1}{4} - \dfrac{7}{6} = \dfrac{2}{3}$

20. $\dfrac{4}{5m - 1} = \dfrac{2}{2m - 1}$

21. $\dfrac{y}{y + 2} - \dfrac{y + 6}{y^2 - 4} + \dfrac{2}{y - 2} = 0$

22. $\dfrac{g}{g - 1} = \dfrac{8}{g + 2}$

For Exercises 23–25, solve.

23. If Erin can paint a room in 3 hours and her husband can paint the same room in 2 hours, how long will it take them to paint the room together?

24. Jake runs 3 miles away from the gym, turns around, and runs an average of 2 miles per hour slower on the return trip. If the total time of his run is $1\dfrac{1}{4}$ hours, what was his speed in the outbound leg and the inbound leg of his run?

25. The weight of an object is directly proportional to its mass. Suppose an object with a mass of 40 kilograms weighs 392 newtons. How much would an object with a mass of 54 kilograms weigh?

Chapter 1–7 Cumulative Review Exercises

For Exercises 1–3, answer true or false.

[2.2] 1. The equation $y = x^2 - 3$ is a linear equation.

[4.3] 2. The x-intercept for $y = 2x$ is $(0, 2)$.

[7.4] 3. The LCD of $\dfrac{2}{x^2 - 25}$ and $\dfrac{4x}{x + 5}$ is $(x^2 - 25)(x + 5)$.

For Exercises 4–7, fill in the blank.

[7.3]   4. To add or subtract rational expressions that have the same denominator,
   a. Add or subtract the numerators and keep the same _____.
   b. Simplify to lowest terms. (Remember to factor the numerators and denominators completely in order to simplify.)

[7.1]   5. Explain in your own words what it means to simplify a rational expression to lowest terms.

[5.4]   6. To multiply monomials,

   a. _____ coefficients.

   b. _____ the exponents of the like bases.
   c. Rewrite any unlike bases in the product.

[7.5]   7. A _____ is a rational expression that contains rational expressions in the numerator or denominator.

Exercises 8–19   **EXPRESSIONS**

[1.5] For Exercises 8 and 9, simplify.

8. $5^2 - (-4^2) - 20$

9. $16 \div (-8) + 2^3$

For Exercises 10 and 11, simplify.

[5.6]   10. $\dfrac{15a^4b}{-5a^{-9}b}$

[5.5]   11. $(m + 3)(m^2 - 6m + 1)$

For Exercises 12–14, factor completely.

[6.1]   12. $ax - 2x + a - 2$

[6.3]   13. $3y^2 - 3y - 60$

[6.5]   14. $-3h^2 + 75$

For Exercises 15–19, simplify.

[7.2]   15. $\dfrac{3x - 6}{4} \cdot \dfrac{12}{x - 2}$

[7.2]   16. $\dfrac{8}{m - 4} \div \dfrac{4}{m^2 - 16}$

[7.3]   17. $\dfrac{x}{x^2 - 9} - \dfrac{3}{x^2 - 9}$

[7.4]   18. $\dfrac{4}{x^2 - 49} + \dfrac{2}{x^2 - 7x}$

[7.5]   19. $\dfrac{\dfrac{y^2 - y - 12}{4y}}{\dfrac{y - 4}{8}}$

[2.3] **20.** Solve and check: $20 + 24x = 8 - 3(5 - 9x)$

[4.4] **21.** For the equation $y = \dfrac{2}{3}x + 1$,

      a. Find the slope.   **b.** Graph.

[6.6] **22.** Solve $2x^2 - 13x = 7$.   [2.4] **23.** Solve $V = \dfrac{1}{3}\pi r^2 h$ for h.

[7.6] *For Exercises 24 and 25, solve.*

24. $\dfrac{u^2}{u - 3} - \dfrac{12}{u - 3} = 8$ **25.** $\dfrac{12}{r} = \dfrac{4}{r - 4}$

For Exercises 26–30, solve.

[3.4] **26.** Two planes leaving Seattle at the same time are going in opposite directions. If the first plane travels 600 miles per hour and the second plane travels 450 miles per hour, how long will it be until the two planes are 2625 miles apart?

[3.5] **27.** Jackson invested $7000. He put some of the $7000 into an account earning 9% interest annually and the rest in an account earning 7%. If he earned $600 in interest after one year, how much did he invest at each rate?

[3.5] **28.** How many ounces of a 90% balsamic dressing need to be added to 6 ounces of a 70% balsamic dressing to get an 80% balsamic dressing?

[7.7] **29.** Jake can clean the garage in 8 hours, and his brother can do it in 6 hours. Working together, how long will it take them to clean the garage?

[7.7] **30.** If y varies directly as x and $y = 7$ when $x = 2$, find x when y is 21.

More on Inequalities, Absolute Value, and Functions

8

In this chapter, we build upon what we have learned about inequalities, absolute value, and functions. We have solved linear inequalities in which only a single inequality symbol is involved. Now, we learn to solve compound inequalities, which involve two inequality symbols. We then learn to solve absolute value equations and inequalities. We will see that compound inequalities are used in solving absolute value inequalities. Finally, after a quick review of functions and graphing, we learn to perform function operations.

8.1 Compound Inequalities

Objectives

① Solve compound inequalities involving *and*.

② Solve compound inequalities involving *or*.

In Section 2.6, we learned to solve linear inequalities and represent their solution sets with set-builder notation and interval notation and graphically on number lines. For example, the solution set for $x > 3$ is represented in those three ways here.

Set-builder notation: $\{x | x > 3\}$

Interval notation: $(3, \infty)$

Graph:

Note: Remember that the parenthesis indicates that 3 is **not** included in the solution set.

In this section, we build on this foundation and explore **compound inequalities**.

Definition *Compound inequality:* Two inequalities joined by either *and* or *or*.

Some examples of compound inequalities are as follows:

$$x > 3 \text{ and } x \leq 8 \qquad -2 \geq x \text{ or } x > 4$$

Objective ① Solve compound inequalities involving *and*.

Interpreting Compound Inequalities with *and*

First, let's consider inequalities involving *and*, such as $x > 3$ *and* $x \leq 8$. The word *and* indicates that the solution set contains only values that satisfy *both* inequalities. Therefore, the solution set is the **intersection**, or overlap, of the two inequalities' solution sets.

Definition *Intersection:* For two sets A and B, the intersection of A and B, symbolized by $A \cap B$, is a set containing only elements that are in both A and B.

For example, if $A = \{1, 2, 3, 4, 5\}$ and $B = \{3, 4, 5, 6, 7\}$, then $A \cap B = \{3, 4, 5\}$ because 3, 4, and 5 are the only elements in both A and B. Let's look at our compound inequality $x > 3$ and $x \leq 8$. First, we will graph the two inequalities separately, then consider their intersection.

$x > 3$

Note: We use the blue field to indicate the intersection and the dashed lines to indicate the intersection boundaries.

$x \leq 8$

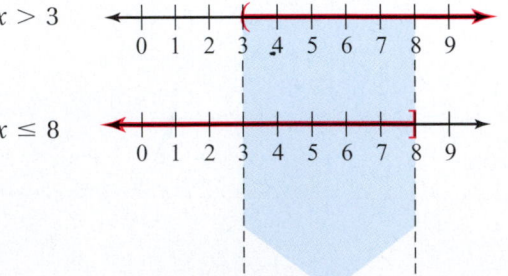

Note: The intersection of the two inequalities contains elements that are only in **both** of their solution sets. Notice that 3 is excluded in the intersection because it was excluded in one of the individual graphs.

Intersection of $x > 3$ and $x \leq 8$:

So $x > 3$ and $x \leq 8$ means that x is any number greater than 3 *and* less than or equal to 8. Because this inequality indicates that x is between two values, we can write it without the word *and*: $3 < x \leq 8$. In set-builder notation, we write $\{x | 3 < x \leq 8\}$. Using interval notation, we write $(3, 8]$.

Example 1   For the compound inequality $x > -2$ and $x < 3$, graph the solution set and write the compound inequality without *and* if possible. Then write in set-builder notation and in interval notation.

SOLUTION   The solution set is the region of intersection.

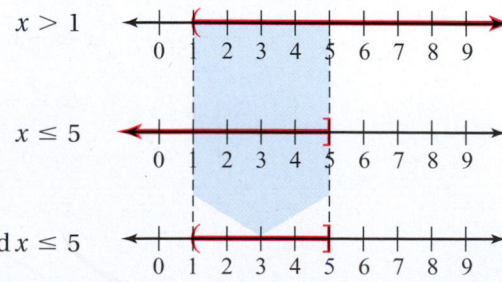

$x > -2$

$x < 3$

Solution set: $x > -2$ and $x < 3$

Without *and*: $-2 < x < 3$

Set-builder notation: $\{x \mid -2 < x < 3\}$

Interval notation: $(-2, 3)$

Warning: Be careful not to confuse the interval notation $(-2, 3)$ with an ordered pair.

Your Turn 1   For each compound inequality, graph the solution set and write the compound inequality without *and* if possible. Then write in set-builder notation and in interval notation.

   a. $x > 4$ and $x < 7$ **b.** $x \geq -8$ and $x \leq -2$

Solving Compound Inequalities with *and*

Now let's consider solution sets for more complex inequalities involving *and*.

Example 2   For the inequality $3x - 1 > 2$ and $2x + 4 \leq 14$, graph the solution set. Then write the solution set in set-builder notation and in interval notation.

SOLUTION   First, we solve each inequality in the compound inequality.

$$3x - 1 > 2 \quad \text{and} \quad 2x + 4 \leq 14$$
$$3x > 3 \quad \text{and} \qquad 2x \leq 10 \qquad \textcolor{blue}{\textbf{Use the addition principle to isolate each } x \textbf{ term.}}$$
$$x > 1 \quad \text{and} \qquad x \leq 5 \qquad \textcolor{blue}{\textbf{Divide out each coefficient.}}$$

The solution set is the intersection of the two individual solution sets.

$x > 1$

$x \leq 5$

Solution set: $x > 1$ and $x \leq 5$

Set-builder notation: $\{x \mid 1 < x \leq 5\}$

Interval notation: $(1, 5]$

 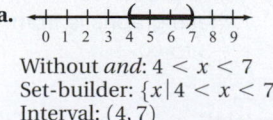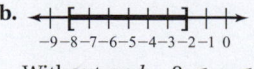

Example 2 suggests the following procedure:

Procedure   **Solving Compound Inequalities Involving *and***

To solve a compound inequality involving *and*,

1. Solve each inequality in the compound inequality.
2. The solution set will be the intersection of the individual solution sets.

Finally, we consider some special situations with compound inequalities involving *and*.

Example 3   For the compound inequality, graph the solution set. Then write the solution set in set-builder notation and in interval notation.

a. $-2 \leq -3x + 7 \leq 1$

SOLUTION   Note that $-2 \leq -3x + 7 \leq 1$ means $-2 \leq -3x + 7$ and $-3x + 7 \leq 1$. Solving the two inequalities gives us

$$-3x + 7 \geq -2 \quad \text{and} \quad -3x + 7 \leq 1 \qquad \color{blue}{-2 \leq -3x + 7 \text{ is the same as } -3x + 7 \geq -2.}$$
$$-3x \geq -9 \quad \text{and} \qquad -3x \leq -6 \qquad \color{blue}{\text{Subtract 7 from both sides of each inequality.}}$$
$$x \leq 3 \quad \text{and} \qquad x \geq 2 \qquad \color{blue}{\text{Divide both sides of each inequality by } -3.}$$

> **Note:** Remember, when we multiply or divide both sides of an inequality by a negative number, we must change the direction of the inequality symbol.

Remember, we can write $x \leq 3$ and $x \geq 2$ as $2 \leq x \leq 3$. Graphing this solution set we have

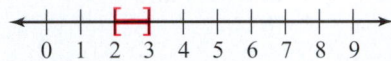

We also could have solved the compound inequality in its original form.

$$-2 \leq -3x + 7 \leq 1$$
$$-9 \leq -3x \leq -6 \qquad \color{blue}{\text{Subtract 7 from all three parts of the compound inequality.}}$$
$$3 \geq x \geq 2 \qquad \color{blue}{\text{Divide all three parts of the compound inequality by } -3.}$$

Note that $3 \geq x \geq 2$ is the same as $2 \leq x \leq 3$.

Set-builder notation: $\{x \mid 2 \leq x \leq 3\}$

Interval notation: $[2, 3]$

b. $2x + 11 \geq 5$ and $2x + 11 > 3$

SOLUTION   To graph the solution set, we first solve each inequality.

$$2x + 11 \geq 5 \quad \text{and} \quad 2x + 11 > 3$$
$$2x \geq -6 \quad \text{and} \qquad 2x > -8 \qquad \color{blue}{\text{Subtract 11 from both sides of each inequality.}}$$
$$x \geq -3 \quad \text{and} \qquad x > -4 \qquad \color{blue}{\text{Divide both sides of each inequality by 2.}}$$

$x \geq -3$

$x > -4$

Solution set:

> **Note:** The region of intersection for these graphs is from -3 to ∞.

Set-builder notation: $\{x \mid x \geq -3\}$

Interval notation: $[-3, \infty)$

c. $-4x - 3 > 1$ and $-4x - 3 < -11$

SOLUTION   First, we solve each inequality.

$-4x - 3 > 1$	and	$-4x - 3 < -11$
$-4x > 4$	and	$-4x < -8$
$x < -1$	and	$x > 2$

Add 3 to both sides of each inequality.
Divide both sides of each inequality by -4.

$x < -1$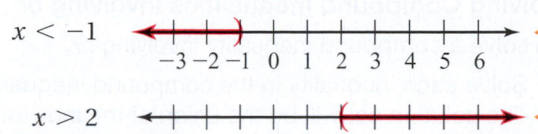

Note: There is no region of intersection for these graphs, so the solution set contains no values and is said to be empty.

$x > 2$

To graph an empty set, we draw a number line with no shading.

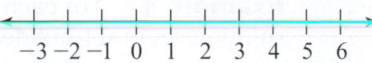

Set-builder notation: $\{ \ \}$ or $\varnothing$

Interval notation: We do not write interval notation because there are no values in the solution set.

Your Turn 3   For each compound inequality, graph the solution set. Then write the solution set in set-builder notation and in interval notation.

a. $-9 \leq -5x + 11 < 21$ **b.** $3x - 2 \leq 4$ and $5x + 7 \leq 32$

c. $-4x - 3 > 1$ and $-4x - 3 < -5$

Objective ❷ **Solve compound inequalities involving *or*.**

Interpreting Compound Inequalities with *or*

In a compound inequality such as $x > 2$ or $x \leq -1$, the word *or* indicates that the solution set contains values that satisfy *either* inequality. Therefore, the solution set is the **union**, or joining, of the two inequalities' solution sets.

Definition   ***Union:*** For two sets A and B, the union of A and B, symbolized by $A \cup B$, is a set containing every element in A or in B.

For example, if $A = \{1, 2, 3, 4, 5\}$ and $B = \{3, 4, 5, 6, 7\}$, then $A \cup B = \{1, 2, 3, 4, 5, 6, 7\}$. Consider our compound inequality $x > 2$ or $x \leq -1$. Let's graph the two inequalities separately, then consider their union.

 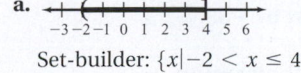
$x > 2$

$x \leq -1$

Solution set:
$x > 2$ or $x \leq -1$

Note: This graph is the union of the two individual solution sets because it contains every element in either of the individual solution sets.

Using set-builder notation, we write the solution set as $\{x | x > 2 \text{ or } x \leq -1\}$. Interval notation is a little more challenging for compound inequalities such as $x > 2$ or $x \leq -1$ because the solution set is a combination of two intervals. As we scan from left to right, the first interval is $(-\infty, -1]$. The second interval is $(2, \infty)$. To indicate the union of these two intervals, we write $(-\infty, -1] \cup (2, \infty)$.

Solving Compound Inequalities with *or*

Now let's solve more complex inequalities involving *or*. The process is similar to the process we used for solving compound inequalities involving *and*.

Procedure   **Solving Compound Inequalities Involving *or***

To solve a compound inequality involving *or*,

1. Solve each inequality in the compound inequality.
2. The solution set will be the union of the individual solution sets.

Example 4   For each compound inequality, graph the solution set. Then write the solution set in set-builder notation and in interval notation.

a. $\dfrac{2}{3}x - 5 \le -7$ or $\dfrac{2}{3}x - 5 \ge -1$

SOLUTION   First, we solve each inequality in the compound inequality.

$$\frac{2}{3}x - 5 \le -7 \quad \text{or} \quad \frac{2}{3}x - 5 \ge -1$$

$$\frac{2}{3}x \le -2 \quad \text{or} \quad \frac{2}{3}x \ge 4 \qquad \textcolor{red}{\textbf{Add 5 to both sides of each inequality.}}$$

$$x \le -3 \quad \text{or} \quad x \ge 6 \qquad \textcolor{red}{\textbf{Multiply both sides by } \frac{3}{2} \textbf{ in each inequality.}}$$

The solution set is the union of the two individual solution sets.

LEARNING Strategy

VISUAL   If you are a visual learner, imagine placing one graph on top of the other to form their union.

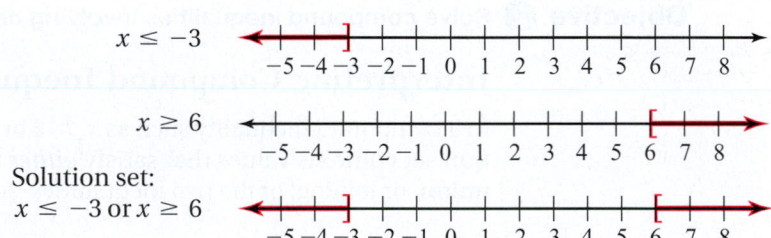

$x \le -3$

$x \ge 6$

Solution set:
$x \le -3 \text{ or } x \ge 6$

Set-builder notation: $\{x \mid x \le -3 \text{ or } x \ge 6\}$

Interval notation: $(-\infty, -3] \cup [6, \infty)$

b. $-3x + 8 < -1 \text{ or } -3x + 8 \le 20$

SOLUTION $-3x + 8 < -1 \quad \text{or} \quad -3x + 8 \le 20$

$-3x < -9 \quad \text{or} \quad -3x \le 12 \qquad \textcolor{red}{\textbf{Subtract 8 from both sides of each inequality.}}$

$x > 3 \quad \text{or} \quad x \ge -4 \qquad \textcolor{red}{\textbf{Divide both sides of each inequality by } -3.}$

Now we graph the individual solution sets and consider their union.

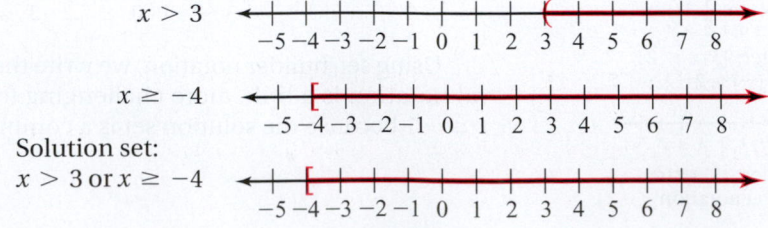

$x > 3$

$x \ge -4$

Solution set:
$x > 3 \text{ or } x \ge -4$

Note: When we join the two graphs, the graph of $x \ge -4$ covers all of the graph of $x > 3$; so the union of the two graphs is actually $x \ge -4$.

Set-builder notation: $\{x \mid x \geq -4\}$

Interval notation: $[-4, \infty)$

c. $9x - 16 > -34$ or $9x - 16 \leq 29$

SOLUTION
$$9x - 16 > -34 \quad \text{or} \quad 9x - 16 \leq 29$$
$$9x > -18 \quad \text{or} \quad 9x \leq 45 \qquad \text{Add 16 to both sides of each inequality.}$$
$$x > -2 \quad \text{or} \quad x \leq 5 \qquad \text{Divide both sides of each inequality by 9.}$$

Now we graph the individual solution sets and consider their union.

$x > -2$ ← number line from −5 to 8, open circle at −2, shaded right

$x \leq 5$ ← number line from −5 to 8, closed bracket at 5, shaded left

Solution set:
$x > -2$ or $x \leq 5$ ← entire number line from −5 to 8 shaded

Note: When we join the two graphs, the entire number line is covered; so the union of the two graphs is the set of real numbers.

Set-builder notation: $\{x \mid x \text{ is a real number}\}$, or $\mathbb{R}$.

Interval notation: $(-\infty, \infty)$

Answers Your Turn 4

a. 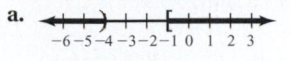 number line from −6 to 3, open circle at −4, closed at −1

Set-builder:
$\{x \mid x < -4 \text{ or } x \geq -1\}$
Interval: $(-\infty, -4) \cup [-1, \infty)$

b. 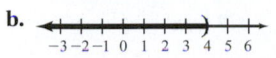 number line from −3 to 6

Set-builder: $\{x \mid x < 4\}$
Interval: $(-\infty, 4)$

c. 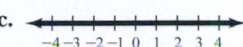 number line from −4 to 4

Set-builder:
$\{x \mid x \text{ is a real number}\}$, or $\mathbb{R}$
Interval: $(-\infty, \infty)$

Your Turn 4 For each compound inequality, graph the solution set. Then write the solution set in set-builder notation and in interval notation.

a. $-5x - 8 > 12$ or $-5x - 8 \leq -3$

b. $\dfrac{3}{4}x + 5 < -1$ or $\dfrac{3}{4}x + 5 < 8$

c. $5 - 3x \geq 11$ or $5 - 3x < 17$

8.1 Exercises   For Extra Help   *MyMathLab*

 Math XL PRACTICE   WATCH   DOWNLOAD

Note: Exercises marked with a ★ represent challenging exercises.

1. What is a compound inequality?

2. Compound inequalities involve what two key words?

3. Describe the intersection of two sets.

4. Describe the union of two sets.

5. Describe how to graph a compound inequality involving *and*.

6. Describe how to graph a compound inequality involving *or*.

For Exercises 7–14: **a.** *Find the intersection of the given sets. See Objective 1.*
b. *Find the union of the given sets. See Objective 2.*

7. $A = \{1, 3, 5\}$
$B = \{1, 3, 5, 7, 9\}$

8. $A = \{2, 4, 6\}$
$B = \{2, 4, 6, 8, 10\}$

9. $A = \{5, 6, 7, 8\}$
$B = \{7, 8, 9\}$

10. $A = \{8, 10, 12, 14, 16\}$
$B = \{14, 16, 18\}$

11. $A = \{c, a, t\}$
$B = \{d, o, g\}$

12. $A = \{l, o, v, e\}$
$B = \{m, a, t, h\}$

13. $A = \{x, y, z\}$
$B = \{w, x, y, z\}$

14. $A = \{a, b, c, d, e\}$
$B = \{c, d, e, f\}$

For Exercises 15–22, write each inequality without and. *See Example 1.*

15. $x > -4$ and $x < 5$

16. $n \geq -3$ and $n < 7$

17. $y > 0$ and $y \leq 2$

18. $m \geq 0$ and $m < 3$

19. $w > -7$ and $w < 3$

20. $r > -1$ and $r \leq 1$

21. $u \geq 0$ and $u \leq 2$

22. $t > 7$ and $t < 15$

For Exercises 23–30, graph the compound inequality. See Examples 1–3.

23. $x > 2$ and $x < 7$

24. $x > 3$ and $x < 9$

25. $x > -1$ and $x \leq 5$

26. $x \geq -4$ and $x < 0$

27. $1 \leq x \leq 10$

28. $-2 < x < -1$

29. $-3 \leq x < 4$

30. $0 < x \leq 8$

For Exercises 31–50: **a.** *Graph the solution set.*
b. *Write the solution set using set-builder notation.* $\Big\}$ *See Examples 1–3.*
c. *Write the solution set using interval notation.*

31. $x > -3$ and $x < -1$

32. $x > 4$ and $x < 7$

33. $x + 2 > 5$ and $x - 4 \leq 2$

34. $x - 3 \geq 1$ and $x + 2 < 10$

35. $-x > 1$ and $-2x \leq -10$

36. $-4x \geq 8$ and $-3x \leq -15$

37. $3x + 8 \geq -7$ and $4x - 7 < 5$

38. $3x + 5 \geq -1$ and $5x - 2 < 13$

39. $-3x - 8 > 1$ and $-4x + 5 \leq -3$

40. $-6x + 4 < -14$ and $-x - 3 \geq -2$

41. $-3 < x + 4 < 1$

42. $-2 \leq x - 2 \leq 2$

43. $-7 \leq 4x - 3 \leq 5$

44. $8 < 3x - 4 < 14$

45. $0 \leq 2 + 3x < 8$

46. $0 \leq -2 + 5x \leq 13$

47. $-1 < -2x + 5 \leq 5$

48. $-3 < 5 - 2x < 1$

49. $3 \leq 6 - x \leq 6$

50. $-4 \leq -4 - x \leq 2$

For Exercises 51–58, graph each inequality. See Objective 2.

51. $x < -2$ or $x > 6$

52. $n \leq 3$ or $n > 4$

53. $y < -3$ or $y \geq 0$

54. $m \leq 2$ or $m > 8$

55. $w > -3$ or $w > 2$

56. $r > -3$ or $r \geq -1$

57. $u \geq 0$ or $u \leq 2$

58. $t > 6$ or $t < 13$

For Exercises 59–74: **a.** *Graph the solution set.*
                             b. *Write the solution set using set-builder notation.* $\Big\}$ *See Example 4.*
                             c. *Write the solution set using interval notation.*

59. $y + 2 < -7$ or $y + 2 > 7$

60. $a - 4 \leq -2$ or $a - 4 \geq 2$

61. $4r - 3 < -11$ or $2r - 3 > -1$

62. $3t - 5 \leq -14$ or $5t + 5 \geq -5$

63. $-w + 2 \leq -5$ or $-w + 2 \geq 3$

64. $-2x - 3 \leq -1$ or $-2x - 3 \geq 5$

65. $7 - 4k \leq -5$ or $6 - 2k \geq 2$

66. $8 - 2q \leq 2$ or $8 - 2q \geq 4$

67. $\frac{2}{3}x - 4 \geq -2$ or $\frac{2}{5}x - 2 \leq -4$

68. $\frac{3}{4}x + 3 \geq 6$ or $\frac{3}{5}x - 3 \leq -6$

69. $-2(c - 1) < -4$ or $-2(c - 2) < -6$

70. $-3(y - 1) \leq 6$ or $-4(y - 2) \leq 4$

71. $2(x + 3) + 3 \leq 1$ or $2(x + 1) + 7 \leq -3$

72. $5(d - 1) + 8 < 8$ or $5(d + 1) - 2 < 18$

73. $-3(x - 1) - 1 \leq -1$ or $-3(x + 1) + 5 \geq -2$

74. $-4(k - 2) - 1 \geq -1$ or $-4(k - 2) - 1 \leq 5$

For Exercises 75–86: **a.** *Graph the solution set.*
                             b. *Write the solution set using set-builder notation.* $\Big\}$ *See Examples 2, 3, and 4.*
                             c. *Write the solution set using interval notation.*

75. $-3 < x + 4$ and $x + 4 < 7$

76. $0 < x - 1$ and $x - 1 \leq 3$

77. $4x + 3 < -5$ or $3x + 2 > 8$

78. $6x - 2 < -8$ or $4x - 3 > 9$

79. $4 < -2x < 6$

80. $9 \le -3x \le 15$

81. $7 \le 4x - 5 \le 19$

82. $-9 < 2x - 7 < -3$

83. $-5 < 1 - 2x < -1$

84. $4 \le -3x + 1 < 7$

★ **85.** $x + 3 < 2x + 1 < 3x$

★ **86.** $2x - 2 \le x + 1 \le 2x + 5$

For Exercises 87–96, solve. Then: **a.** *Graph the solution set.*
b. *Write the solution set using set-builder notation.*
c. *Write the solution set using interval notation.*

87. If Andrea's current long-distance bill is between $36 and $45 in a month, she can save money by switching to another company. Her current rate is $0.09 per minute. What is the range of minutes that Andrea must use long distance to justify switching to the other company?

88. A mail-order music club offers one bonus CD if the total of an order is from $75 to $100. Dayle decides to buy the lowest-priced CDs, which are $12.50 each. In what range would the number of CDs he orders have to be for him to receive a bonus CD?

89. Juan has taken four of the five tests in his history course. To get a B in the course, his average needs to be at least 80 and less than 90. If his scores on the first four tests are 95, 80, 82, and 88, in what range of values can his score on the fifth test be so that he has a B average?

90. Students in a chemistry course receive a C if the average of four tests is at least 70 and less than 80. Suppose a student has the following scores: 76, 72, and 84. What range of scores on the fourth test would result in the student receiving a C?

91. To conserve energy, it is recommended that a home's thermostat be set at 5° above 73°F during summer or 5° below 73° during winter. If the heat pump does not run when the temperature is at or between those values, in what range of temperatures is the heat pump off?

92. A house thermostat is set so that the heat pump/air conditioner comes on if the temperature is 2° or more above or below 70°. In what range of temperatures is the heat pump off?

93. In the maintenance of a saltwater aquarium, the temperature should be within 4° of 76°. What range of temperatures would be acceptable for saltwater fish? (*Source: The Conscientious Marine Aquarist*, Robert M. Fenner)

94. When driving in Mississippi, motorists risk a ticket if they drive more than 15 miles per hour over or under the 55 miles per hour speed limit on a state highway. In what range of speeds would a motorist not risk receiving a ticket? (*Source:* Mississippi State Code 63-3-509)

95. A building is to be designed in the shape of a trapezoid as shown. Find the range of values for the length of the back side of the building so that the square footage is from 6000 to 8000 square feet.

[Use $A = \frac{1}{2}h(a + b)$.]

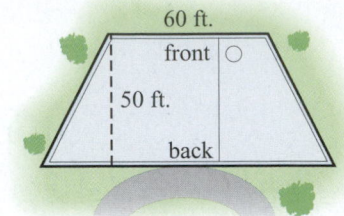

96. At 32°F and below, water is a solid (ice). At 212°F and above, water is a gas (steam). Use a compound inequality to describe the range of values in degrees Celsius for water in its liquid state. [Use $C = \frac{5}{9}(F - 32)$.]

Puzzle Problem

A cyclist is involved in a multiple-day race. She believes that she needs to complete today's 40 kilometers between 1 hour and 45 minutes and 2 hours. Find the range of values her rate can be to complete this leg of the race in her desired time frame.

Review Exercises

Exercises 1–3 **EXPRESSIONS**

[1.1] **1.** Is the absolute value of a number always positive? Explain.

[1.5] *For Exercises 2 and 3, simplify.*

    2. $-|2 - 3^2| - |-4|$
        3. $-|-|-16||$

Exercises 4–6 **EQUATIONS AND INEQUALITIES**

[2.3] *For Exercises 4 and 5, solve.*

    4. $3x - 14x = 17 - 12x - 11$
    5. $\frac{3}{5}(25 - 5x) = 15 - \frac{3}{5}$

[2.5] **6.** One-half the sum of a number and 2 is zero. Find the number.

8.2 Equations Involving Absolute Value

Objective **1** **Solve equations involving absolute value.** We learned in Section 1.1 that the absolute value of a number is its distance from zero on a number line. For example, $|4| = 4$ and $|-4| = 4$ because both 4 and -4 are 4 units from 0 on a number line.

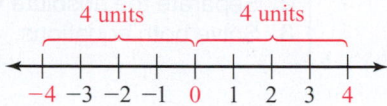

Now we will solve equations in which a variable appears within the absolute value symbols, as in $|x| = 4$. Notice that the solutions are 4 and -4, which suggests the following rule:

Rule **Absolute Value Property**

If $|x| = a$, where x is a variable or an expression and $a \geq 0$, then $x = a$ or $x = -a$.

> **Note:** The absolute value property does not apply if $a < 0$. For example, $|x| = -5$ has no solution because there is no real number whose absolute value is -5.

Example 1 Solve.

a. $|2x - 5| = 9$

SOLUTION Using the absolute value property, we separate the equation into a positive case and a negative case, then solve the two cases separately.

$$2x - 5 = 9 \quad \text{or} \quad 2x - 5 = -9$$
$$2x = 14 \quad \text{or} \quad 2x = -4$$
$$x = 7 \quad \text{or} \quad x = -2$$

The solutions are 7 and -2.

b. $|4x + 1| = -11$

SOLUTION This equation has the absolute value equal to a negative number, -11. Because the absolute value of every real number is a positive number or zero, this equation has no solution.

From Example 1(b), we can say that an absolute value equation in the form $|x| = a$, where $a < 0$, has no solution.

Your Turn 1 Solve.

a. $|3x + 4| = 8$ **b.** $|2x - 1| = 0$ **c.** $|5x - 2| = -3$

To use the absolute value property, we need the **absolute value isolated**.

Example 2 Solve: $8 - 2|4x - 8| = -16$

SOLUTION
$$-2|4x - 8| = -24 \qquad \text{Subtract 8 from both sides of the equation.}$$
$$|4x - 8| = 12 \qquad \text{Divide both sides by } -2 \text{ to isolate the absolute value.}$$
$$4x - 8 = 12 \quad \text{or} \quad 4x - 8 = -12 \qquad \text{Use the absolute value property.}$$
$$4x = 20 \quad \text{or} \quad 4x = -4$$
$$x = 5 \quad \text{or} \quad x = -1$$

The solutions are 5 and -1.

Answers **Your Turn 1**

a. $\frac{4}{3}$ and -4 **b.** $\frac{1}{2}$

c. no solution

Examples 1 and 2 suggest the following procedure:

Procedure **Solving Equations Containing a Single Absolute Value**

To solve an equation containing a single absolute value,

1. **Isolate the absolute value** so that the equation is in the form $|ax + b| = c$. If $c > 0$, proceed to Steps 2 and 3. If $c < 0$, the equation has no solution.
2. Separate the absolute value into two equations, $ax + b = c$ and $ax + b = -c$.
3. Solve both equations.

Your Turn 2   Solve.

   a. $12 - 3|3x - 6| = -6$ **b.** $|2x + 1| + 6 = 4$

More Than One Absolute Value

Some equations contain more than one absolute value expression, such as $|x + 3| = |2x - 8|$. If two absolute values are equal, they must contain expressions that are *equal* or *opposites*. Some simple examples follow.

Equal   **Opposites**

$|7| = |7|$ or $|7| = |-7|$

$|-5| = |-5|$ or $|-5| = |5|$

For the equation $|x + 3| = |2x - 8|$, therefore, the expressions $x + 3$ and $2x - 8$ must be equal or opposites.

Equal   **Opposites**

$x + 3 = 2x - 8$ or $x + 3 = -(2x - 8)$

This suggests the following procedure:

Procedure **Solving Equations in the Form $|ax + b| = |cx + d|$**

To solve an equation in the form $|ax + b| = |cx + d|$,

1. Separate the absolute value equation into two equations: $ax + b = cx + d$ and $ax + b = -(cx + d)$.
2. Solve both equations.

Example 3   Solve.

a. $|3x + 5| = |x - 9|$

SOLUTION   Separate the absolute value equation into two equations.

Equal   **Opposites**

$3x + 5 = x - 9$ or $3x + 5 = -(x - 9)$ **Note:** Think of $-(x - 9)$ as $-1(x - 9)$.

$2x + 5 = -9$ or $3x + 5 = -x + 9$

$2x = -14$ or $4x + 5 = 9$

$x = -7$ or $4x = 4$

$x = 1$

The solutions are -7 and 1.

b. $|3x - 7| = |5 - 3x|$

SOLUTION

    Equal   **Opposites**

$3x - 7 = 5 - 3x$ or $3x - 7 = -(5 - 3x)$ **Separate into two equations.**

$\quad 6x - 7 = 5$ or $3x - 7 = -5 + 3x$

$\qquad\quad 6x = 12$ or $-7 = -5$

$\qquad\qquad x = 2$

This absolute value equation has only one solution, 2.

> **Note:** Subtracting $3x$ from both sides gives us a false equation; so the "opposites" equation is a contradiction and thus has no solution.

Your Turn 3   Solve.

a. $|5x - 3| = |3x - 13|$ **b.** $|9 + 4x| = |1 - 4x|$

Answers   **Your Turn 3**
a. $-5, 2$ **b.** -1

8.2 Exercises   **For Extra Help**   *MyMathLab*   PRACTICE   WATCH   DOWNLOAD

1. What does absolute value mean in terms of a number line?

2. How do we interpret $|x| = 5$?

3. State the absolute value property in your own words.

4. If given an equation in the form $|ax + b| = c$ and $c < 0$, what can you conclude about its solution(s)?

5. Explain the first step in solving an equation containing two absolute values, such as $|2x + 1| = |3x - 5|$.

6. How can you tell that an absolute value equation containing two absolute values has only one solution?

For Exercises 7–22, solve using the absolute value property. See Example 1.

7. $|x| = 2$ **8.** $|y| = 5$ **9.** $|a| = -4$ **10.** $|r| = -1$

11. $|x + 3| = 8$ **12.** $|w - 1| = 4$ **13.** $|2m - 5| = 1$ **14.** $|3s + 6| = 12$

15. $|6 - 5x| = 1$ **16.** $|2 - 3x| = 5$ **17.** $|4 - 3w| = 6$ **18.** $|1 - 2x| = 2$

19. $|4m - 2| = -5$ **20.** $|6p + 5| = -1$ **21.** $|4w - 3| = 0$ **22.** $|3m - 5| = 0$

For Exercises 23–38, isolate the absolute value; then use the absolute value property. See Example 2.

23. $|2y| - 3 = 5$ **24.** $|3r| + 7 = 10$ **25.** $|y + 1| + 2 = 4$ **26.** $|x + 3| - 1 = 5$

27. $|b - 4| - 6 = 2$ **28.** $|v - 2| + 2 = 4$ **29.** $3 + |5x - 1| = 7$ **30.** $2 + |5t - 2| = 5$

31. $1 - |2k + 3| = -4$ **32.** $-3 = 1 - |4u - 8|$ **33.** $4 - 3|z - 2| = -8$ **34.** $3 - 2|x - 5| = -7$

35. $6 - 2|3 - 2w| = -18$ **36.** $15 - 2|5 - 2x| = -5$ **37.** $|3x - 2(x + 5)| = 10$ **38.** $|4y - 2(6 - y)| = 18$

For Exercises 39–48, solve by separating the two absolute values into equal and opposite cases. See Example 3.

39. $|2x + 1| = |x + 5|$ **40.** $|2p + 5| = |3p + 10|$ **41.** $|x + 3| = |2x - 4|$ **42.** $|p - 1| = |2p + 8|$

43. $|3v + 4| = |1 - 2v|$ **44.** $|3 - 6c| = |2c + 3|$ **45.** $|2n + 3| = |3 + 2n|$ **46.** $|2r - 1| = |1 - 2r|$

47. $|2k + 1| = |2k - 5|$ **48.** $|4 + 2q| = |2q + 8|$

For Exercises 49–56, solve. See Examples 1 and 2.

49. $|10 - (5 - h)| = 8$ **50.** $|7 - (2 - n)| = 17$ **51.** $\left|\dfrac{b}{2} - 1\right| = 4$ **52.** $\left|\dfrac{u}{3} - 2\right| = 3$

53. $\left|\dfrac{4 - 3x}{2}\right| = \dfrac{3}{4}$ **54.** $\left|\dfrac{6 - 5w}{6}\right| = \dfrac{2}{3}$ **55.** $\left|2y + \dfrac{3}{2}\right| - 2 = 5$ **56.** $\left|p + \dfrac{2}{3}\right| - 1 = 7$

Review Exercises

Exercise 1   EXPRESSIONS

[8.1]   **1.** Use interval notation to represent the solution set shown on the graph to the right.

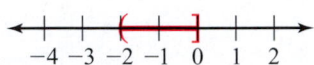

Exercises 2–6   EQUATIONS AND INEQUALITIES

[8.1]   **2.** Graph the compound inequality: $-1 < x \leq 0$

[2.6] *For Exercises 3 and 4, solve.*

     3. $6 - 4x > -5x - 1$ **4.** $-4x \leq 12$

[8.1]   **5.** Solve the compound inequality: $-1 < \dfrac{3x + 2}{4} < 4$

[2.6]   **6.** Translate to a linear inequality and solve: Two less than a number is less than five.

8.3 Inequalities Involving Absolute Value

Objectives

1. Solve absolute value inequalities involving less than.
2. Solve absolute value inequalities involving greater than.

Objective 1 **Solve absolute value inequalities involving less than.** Now let's solve inequalities that contain absolute value. Think about $|x| \leq 3$. A solution for this inequality is any number whose absolute value is less than or equal to 3. This means that the solutions are a distance of 3 units or less from 0 on a number line.

> **Note:** Solutions for the **equal to** part of $|x| \leq 3$ are 3 and −3. The **less than** part of $|x| \leq 3$ means all numbers between 3 and −3 because their absolute values are less than 3.

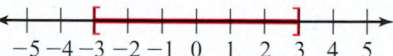

Notice that the solution region corresponds to the compound inequality $x \geq -3$ and $x \leq 3$, which we can write as $-3 \leq x \leq 3$. In set-builder notation, the solution is $\{x | -3 \leq x \leq 3\}$, and in interval notation, we write $[-3, 3]$.

Our examples suggest the following procedure:

Procedure **Solving Inequalities in the Form $|x| < a$, where $a > 0$**

To solve an inequality in the form $|x| < a$, where $a > 0$,

1. Rewrite the inequality as a compound inequality involving *and*: $x > -a$ and $x < a$ (or use $-a < x < a$).
2. Solve the compound inequality.

Similarly, to solve $|x| \leq a$, we write $x \geq -a$ and $x \leq a$ (or $-a \leq x \leq a$).

Example 1 For each inequality, solve, graph the solution set, and write the solution set in both set-builder and interval notation.

a. $|x| < 4$

SOLUTION $x > -4$ **and** $x < 4$ *Rewrite as a compound inequality.*

Recall that $x > -4$ and $x < 4$ means that $-4 < x < 4$. So our graph is as follows:

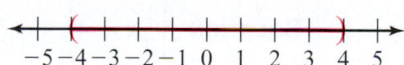

Set-builder notation: $\{x | -4 < x < 4\}$

Interval notation: $(-4, 4)$

b. $|x - 4| \leq 3$

SOLUTION $x - 4 \geq -3$ **and** $x - 4 \leq 3$ *Rewrite as a compound inequality.*

                 $x \geq 1$ **and** $x \leq 7$ *Add 4 to both sides of each inequality.*

Recall that $x \geq 1$ and $x \leq 7$ means that $1 \leq x \leq 7$. So our graph is as follows:

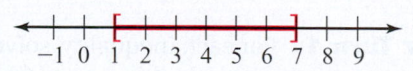

> **Note:** The solution set for $|x - 4| \leq 3$ contains every number whose distance from 4 is 3 units or less.

Set-builder notation: $\{x | 1 \leq x \leq 7\}$

Interval notation: $[1, 7]$

c. $|3x + 1| < 5$

SOLUTION Instead of $3x + 1 > -5$ and $3x + 1 < 5$, we use the more compact form.

$-5 < 3x + 1 < 5$ **Rewrite as a compound inequality.**

$-6 < 3x < 4$ **Subtract 1 from all three parts of the inequality.**

$-2 < x < \dfrac{4}{3}$ **Divide all three parts of the inequality by 3 to isolate x.**

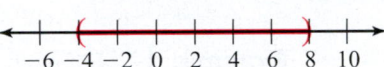

Set-builder notation: $\left\{ x \mid -2 < x < \dfrac{4}{3} \right\}$

Interval notation: $\left(-2, \dfrac{4}{3} \right)$

d. $|-0.5x + 1| - 2 < 1$

SOLUTION Notice that the equation is not in the form $|x| < a$; so our first step is to isolate the absolute value.

$|-0.5x + 1| < 3$ **Add 2 to both sides to isolate the absolute value.**

$-3 < -0.5x + 1 < 3$ **Rewrite as a compound inequality.**

$-4 < -0.5x < 2$ **Subtract 1 from all three parts of the inequality.**

$8 > x > -4$ **Divide all three parts of the inequality by -0.5 to isolate x.**

> **Note:** Because we divided by a negative, we changed the direction of the inequality symbols.

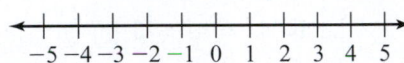

Set-builder notation: $\{ x \mid -4 < x < 8 \}$

Interval notation: $(-4, 8)$

e. $\left| \dfrac{1}{3}x - 4 \right| + 6 < -1$

> **Note:** Any inequality of the form $|x| < a$, $a \le 0$ or $|x| \le a$, $a < 0$, has no solution because the absolute value of any real number is nonnegative and, therefore, can only be zero or positive.

SOLUTION $\left| \dfrac{1}{3}x - 4 \right| < -7$ **Subtract 6 from both sides to isolate the absolute value.**

Because absolute values cannot be negative, this inequality has no solution; so the solution set is empty.

Set-builder notation: $\{ \ \}$ or $\varnothing$

Interval notation: We do not write interval notation because there are no values in the solution set.

Your Turn 1   For each inequality, solve, graph the solution set, and write the solution set in both set-builder notation and interval notation.

a. $|x + 5| < 2$ **b.** $|2x - 3| \le 7$

c. $\left| -\dfrac{1}{3}x + 1 \right| - 2 \le 1$ **d.** $\left| \dfrac{1}{2}x - 1 \right| + 3 < -2$

Objective ❷ Solve absolute value inequalities involving greater than. Now we consider inequalities with greater than, such as $|x| \geq 5$. A solution for $|x| \geq 5$ is any number whose absolute value is greater than or equal to 5. As the graph shows, the solution set contains all values that are a distance of 5 units or more from 0.

Note: Solutions for the **equal to** part of $|x| \geq 5$ are 5 and -5. The **greater than** part means all values that are farther from 0 than 5 and -5 because their absolute values are greater than 5.

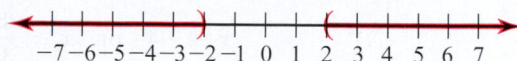

Notice that the solutions are equivalent to $x \leq -5$ or $x \geq 5$.

An inequality such as $|x| > 2$ uses parentheses.

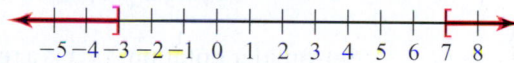

The solutions are equivalent to $x < -2$ or $x > 2$. Our examples suggest that we can split these inequalities into a compound inequality involving *or*.

Procedure

Solving Inequalities in the Form $|x| > a$, where $a > 0$

To solve an inequality in the form $|x| > a$, where $a > 0$,

1. Rewrite the inequality as a compound inequality involving *or*: $x < -a$ or $x > a$.
2. Solve the compound inequality.

Similarly, to solve $|x| \geq a$, we write $x \leq -a$ or $x \geq a$.

Warning: When solving absolute value inequalities, look carefully at the order symbol. The rules for $<$ and $>$ are very different.

Example 2 For each inequality, solve, graph the solution set, and write the solution set in both set-builder notation and interval notation.

a. $|x - 2| \geq 5$

SOLUTION $\quad x - 2 \leq -5 \quad$ **or** $\quad x - 2 \geq 5 \qquad$ Rewrite as a compound inequality.

$\qquad\qquad\qquad\quad x \leq -3 \quad$ **or** $\qquad\quad x \geq 7 \qquad$ Add 2 to both sides of each inequality.

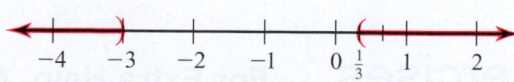

Set-builder notation: $\{x \mid x \leq -3 \text{ or } x \geq 7\}$

Interval notation: $(-\infty, -3] \cup [7, \infty)$

b. $|-3x - 4| > 5$

SOLUTION $\quad -3x - 4 < -5 \quad$ or $\quad -3x - 4 > 5 \qquad$ Rewrite as a compound inequality.

$\qquad\qquad\qquad\quad -3x < -1 \quad$ or $\qquad\quad -3x > 9 \qquad$ Add 4 to both sides of each inequality.

$\qquad\qquad\qquad\quad x > \dfrac{1}{3} \quad$ or $\qquad\qquad x < -3 \qquad$ Divide both sides of each inequality by -3.

Set-builder notation: $\left\{ x \mid x < -3 \text{ or } x > \dfrac{1}{3} \right\}$

Interval notation: $(-\infty, -3) \cup \left(\dfrac{1}{3}, \infty \right)$

c. $\left| \dfrac{3}{4}x - 2 \right| + 1 \geq 6$

SOLUTION $\left| \dfrac{3}{4}x - 2 \right| \geq 5$ Subtract 1 from both sides to isolate the absolute value.

$\dfrac{3}{4}x - 2 \leq -5$ or $\dfrac{3}{4}x - 2 \geq 5$ Rewrite as a compound inequality.

$\dfrac{3}{4}x \leq -3$ or $\dfrac{3}{4}x \geq 7$ Add 2 to both sides of each inequality.

$x \leq -4$ or $x \geq \dfrac{28}{3}$ Multiply both sides of each inequality by $\dfrac{4}{3}$.

Set-builder notation: $\left\{ x \mid x \leq -4 \text{ or } x \geq \dfrac{28}{3} \right\}$

Interval notation: $(-\infty, -4] \cup \left[\dfrac{28}{3}, \infty \right)$

d. $|0.2x + 1| - 3 > -7$

SOLUTION $|0.2x + 1| > -4$ Add 3 to both sides to isolate the absolute value.

This inequality indicates that the absolute value is greater than a negative number. Because the absolute value of every real number is either positive or 0, the solution set is $\mathbb{R}$. The graph is the entire number line.

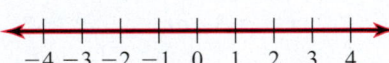

Set-builder notation: $\{x \mid x \text{ is a real number}\}$ or $\mathbb{R}$

Interval notation: $(-\infty, \infty)$

> **Note:** The solutions of any inequality of the form $|x| > a$, $a < 0$, is all real numbers because the absolute value of any real number is nonnegative and, therefore, greater than any negative number.

Your Turn 2   For each inequality, solve, graph the solution set, and write the solution set in both set-builder notation and interval notation.

a. $|x - 3| > 2$

b. $|4x + 1| \geq 9$

c. $|-0.4x + 1| + 2 \geq 3$

d. $\left| \dfrac{1}{3}x - 4 \right| - 2 > -9$

Answers   Your Turn 2

a.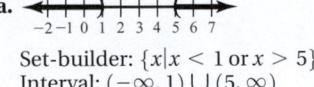

Set-builder: $\{x \mid x < 1 \text{ or } x > 5\}$
Interval: $(-\infty, 1) \cup (5, \infty)$

b.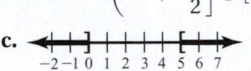

Set-builder:
$\left\{ x \mid x \leq -\dfrac{5}{2} \text{ or } x \geq 2 \right\}$
Interval: $\left(-\infty, -\dfrac{5}{2} \right] \cup [2, \infty)$

c.

Set-builder: $\{x \mid x \leq 0 \text{ or } x \geq 5\}$
Interval: $(-\infty, 0] \cup [5, \infty)$

d.

Set-builder:
$\{x \mid x \text{ is a real number}\}$
Interval: $(-\infty, \infty)$

8.3 Exercises   **For Extra Help**   *MyMathLab*

Note: Exercises marked with a ★ represent challenging exercises.

For Exercises 1–4, we assume a is a positive number.

1. What compound inequality is related to $|x| \leq a$, where $a > 0$?

2. What compound inequality is related to $|x| \geq a$, where $a > 0$?

3. How would you characterize the graph of $|x| < a$, where $a > 0$?

4. How would you characterize the graph of $|x| > a$, where $a > 0$?

5. Under what conditions does $|x| < a$ have no solution?

6. Under what conditions does the solution set for $|x| > a$ contain all real numbers?

For Exercises 7–20, solve the inequality. Then: *a.*   **Graph the solution set.**
b.   **Write the solution set using set-builder notation.**
c.   **Write the solution set using interval notation.**

     See Example 1.

7. $|x| < 5$

8. $|y| \leq 3$

9. $|x + 3| \leq 7$

10. $|m - 5| < 2$

11. $|s + 3| + 6 < 9$

12. $|p - 3| + 4 \leq 8$

13. $|2m - 5| - 3 < 6$

14. $|4x + 12| + 2 < 10$

15. $|-3k + 5| + 7 \leq 8$

16. $|-5h - 1| - 6 < 8$

17. $2|x| + 7 \leq 3$

18. $3|u| + 2 < -7$

19. $2|w - 3| + 4 < 10$

20. $4|n + 2| - 3 \leq 9$

For Exercises 21–34, solve the inequality. Then: **a.** *Graph the solution set.*
b. *Write the solution set using set-builder notation.*
c. *Write the solution set using interval notation.*
See Example 2.

21. $|c| > 12$

22. $|h| \geq 6$

23. $|y + 2| \geq 7$

24. $|a - 4| > 3$

25. $|p - 6| - 3 > 5$

26. $|x + 5| - 3 \geq 6$

27. $|3x + 6| - 3 \geq 9$

28. $|2x - 5| - 1 \geq 10$

29. $|-4n - 5| + 3 > 8$

30. $|-6h + 1| - 7 > 4$

31. $4|v| + 3 \geq 7$

32. $5|m| + 2 > 12$

33. $4|y + 2| - 1 > 3$

34. $3|x - 2| - 5 \geq 1$

For Exercises 35–54, solve the inequality. Then: **a.** *Graph the solution set.*
b. *Write the solution set using set-builder notation.*
c. *Write the solution set using interval notation.*
See Examples 1 and 2.

35. $|4m + 8| - 2 > 10$

36. $|2x + 4| - 2 \geq 10$

37. $|-3x + 6| < 6$

38. $|-2y + 3| \leq 3$

39. $|2r - 3| > -3$

40. $|3b + 7| > -2$

41. $4 - 2|x + 3| > 2$

42. $5 - 2|u + 4| \geq 1$

43. $6|2x - 1| - 3 < 3$

44. $4|3p + 6| - 2 < 22$

45. $5 - |w + 4| > 10$

46. $4 - |5 + k| > 7$

47. $\left|2 - \dfrac{3}{2}k\right| \leq 5$

48. $\left|3 - \dfrac{1}{2}x\right| \leq 7$

49. $|0.25x - 3| + 2 > 4$

50. $|0.5y - 3| + 4 > 5$

51. $\left|2.4 - \dfrac{3}{4}y\right| \leq 7.2$

52. $\left|5.3 - \dfrac{2}{3}w\right| \geq 5.3$

53. $|2p - 8| + 5 > 1$

54. $|8p + 7| + 4 > 3$

★ **For Exercises 55–64, write an inequality involving absolute value that describes the graph shown.**

55.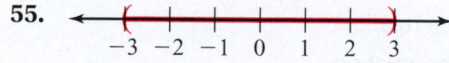

56.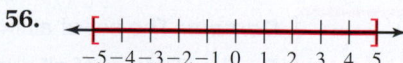

57.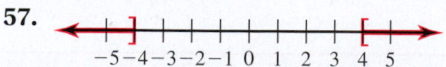

58.

59.

60.

61. [number line with shaded region from 1 to 5, brackets at 1 and 5, ticks at 0 1 2 3 4 5]

62. [number line with shaded region from -5 to 3, brackets at -5 and 3, ticks at -6 -5 -4 -3 -2 -1 0 1 2 3 4]

63. [number line with arrow extending left, ticks at -3 -2 -1 0 1]

64. [number line, ticks at -4 -3 -2 -1 0 1 2 3 4]

Review Exercises

Exercises 1–6  EQUATIONS AND INEQUALITIES

[1.7] *For Exercises 1–4, evaluate the expression using the given value of x.*

1. $-5x + 3; x = 4$

2. $0.5x - 2; x = -6$

3. $-2x^2 + 3; x = -2$

4. $|x - 2| + 1; x = -4$

[4.4] *For Exercises 5 and 6, find the slope and y-intercept.*

5. $y = -\dfrac{3}{4}x + 2$

6. $3x - 5y = 10$

8.4 Functions and Graphing

Objectives

1 Identify the domain and range of a relation and determine if the relation is a function.

2 Find the value of a function.

3 Graph functions.

In this section, we review and expand on what we have learned about functions and graphing.

Objective 1 Identify the domain and range of a relation and determine if the relation is a function. Let's review **relations**, **domain**, **range**, and **functions**, which we introduced in Section 4.7.

Definitions *Relation:* A set of ordered pairs.

Domain: The set of all input values (*x*-values) for a relation.

Range: The set of all output values (*y*-values) for a relation.

Function: A relation in which every value in the domain is paired with exactly one value in the range.

For example, the linear equation $y = 2x$ is a relation because it pairs a given *x*-value with a corresponding *y*-value. We list here some values in the domain and range for $y = 2x$ and draw its graph.

Domain (x-values)	Range (y-values)
−1	−2
0	0
1	2
2	4

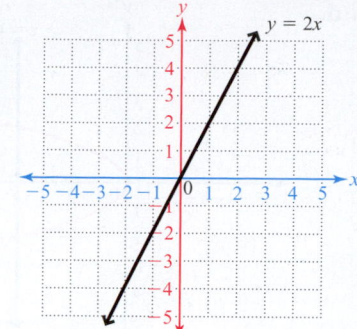

Note: The domain is the set of all x-values represented on the graph and is $\mathbb{R}$.

Note: The range is the set of all y-values represented on the graph and is $\mathbb{R}$.

We can also conclude that $y = 2x$ is a function because every value in its domain is paired with only one value in its range. In Section 4.7, we developed the vertical line test to determine whether a graph is a graph of a function. With a function, a vertical line drawn through any point of its graph intersects the graph only once.

Example 1   Identify the domain and range of each relation and determine whether it is a function.

a. The following relation pairs players on the 2008 Florida Gators National Championship team with their position(s).

Note: In this text, when given a relation in two-column-table form, assume that the domain is in the left column and the range is in the right column.

Player	Position(s)
Tim Tebow	Quarterback
Chas Henry	Punter
Brandon James	Running Back and Kick Returner
Percy Harvin	Wide Receiver and Running Back

SOLUTION   Domain: {Tim Tebow, Chas Henry, Brandon James, Percy Harvin}
Range: {Quarterback, Punter, Running Back, Kick Returner, Wide Receiver}

It is not a function because two elements in the domain, Brandon James and Percy Harvin, are each paired with two elements in the range.

b. $\{(-2, -4), (-1, -1), (0, 0), (1, -1), (2, -4)\}$

SOLUTION   Domain: $\{-2, -1, 0, 1, 2\}$
Range: $\{-4, -1, 0\}$

It is a function because every value in the domain is paired with only one value in the range.

c.

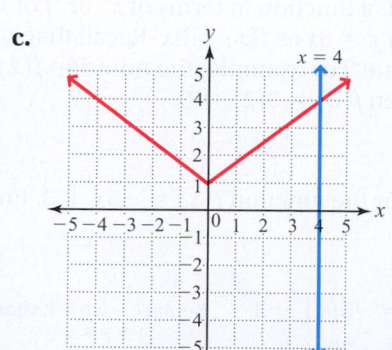

SOLUTION   Domain: all real numbers
Range: $\{y | y \geq 1\}$

It is a function because it passes the vertical line test: A vertical line drawn through any point of the graph intersects it only once. We have drawn one such line, $x = 4$, in blue.

d.

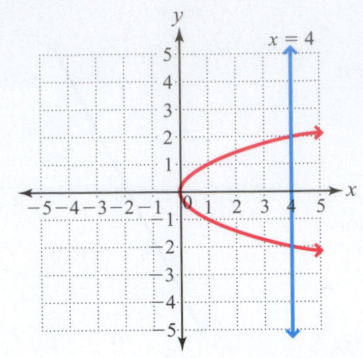

SOLUTION Domain: $\{x \mid x \geq 0\}$
Range: all real numbers

It is a not a function because there are values in the domain paired with more than one value in the range, as the vertical line test confirms. For example, the line $x = 4$ (in blue) intersects the graph twice so the value 4 in the domain is paired with two values in the range, 2 and -2.

Your Turn 1 Identify the domain and range of each relation and determine whether it is a function.

a. The following relation shows the forecasted high temperatures for Yellowstone National Park from April 24 to April 28.

Day	High Temperature
Sunday	51°
Monday	44°
Tuesday	45°
Wednesday	31°
Thursday	30°

b. $\{(0, 2), (0, -2), (3, 0), (-3, 0)\}$

Answers Your Turn 1

a. Domain: {Sunday, Monday, Tuesday, Wednesday, Thursday}
Range: {30, 31, 44, 45, 51}
It is a function.

b. Domain: {−3, 0, 3}
Range: {−2, 0, 2}
It is not a function.

c. Domain: $\{x \mid -3 \leq x \leq 3\}$
Range: $\{y \mid -2 \leq y \leq 2\}$
It is not a function.

d. Domain: $\{x \mid -3 \leq x < 5\}$
Range: {1, 2, 3}
It is a function.

c.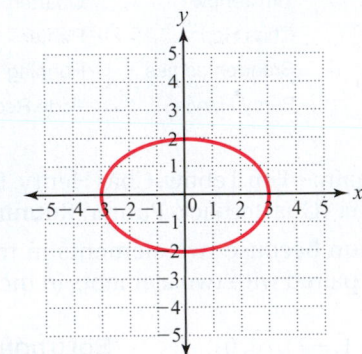

d.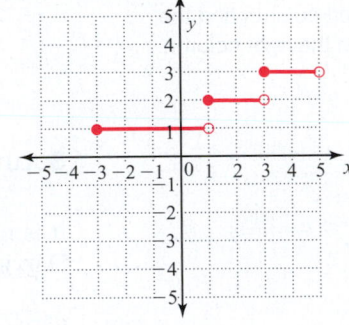

Objective ❷ **Find the value of a function.** In Section 4.7, we introduced the function notation $f(x)$, which is read "a function in terms of x" or "f of x." Using function notation, we can write the equation $y = 3x$ as $f(x) = 3x$. Recall that we evaluate functions by replacing x with the given value. For example, the notation $f(2)$ means the value of f when $x = 2$, so if $f(x) = 3x$, then $f(2) = 3(2) = 6$.

Example 2 For the function $f(x) = -5x + 3$, find the following.

a. $f(4)$

Connection $f(4) = -17$ is the same as the ordered pair $(4, -17)$.

SOLUTION $f(4) = -5(4) + 3$ **Replace x with 4, then calculate.**

$= -20 + 3$

$= -17$

b. $f(-2)$

SOLUTION $f(-2) = -5(-2) + 3$ *Replace x with −2, then calculate.*
                    $= 10 + 3$
                    $= 13$

Your Turn 2   For the function $f(x) = 0.5x - 2$, find the following.

a. $f(-6)$ **b.** $f(1)$

Objective ❸ **Graph functions.**   Now let's review how to graph functions. In general, we can graph a function by finding enough ordered pairs to determine its shape.

Linear Functions

Given a linear function in the form $f(x) = mx + b$, where m is the slope and $(0, b)$ is the y-intercept, we need to find only two ordered pairs. Note that $(0, b)$ can be one of those two ordered pairs.

Example 3   Graph: $f(x) = -\dfrac{3}{4}x + 2$

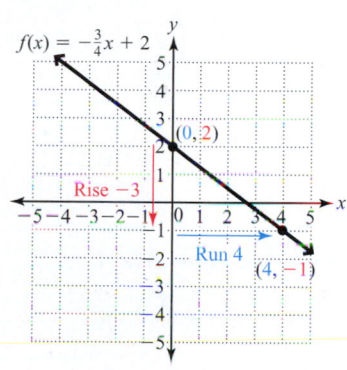

SOLUTION   We can use $(0, 2)$ as one ordered pair, then use the slope, $-\dfrac{3}{4}$, to find a second ordered pair. Recall that slope indicates the "rise" and "run" from any point on the line to another point on the line. So, $-\dfrac{3}{4}$ indicates we rise -3 and run 4 from the point $(0, 2)$ to reach a second point, $(4, -1)$.

Your Turn 3   Graph $f(x) = 2x - 3$.

Quadratic Functions

In Section 6.7, we introduced quadratic functions, which have the form $f(x) = ax^2 + bx + c$, and we saw their graphs are parabolas. To graph a parabola, we have to find many ordered pairs.

Example 4   Graph: $f(x) = -2x^2 + 3$

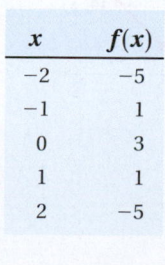
SOLUTION   We create a table of ordered pairs, plot the points, and connect them with a smooth curve.

x	$f(x)$
-2	-5
-1	1
0	3
1	1
2	-5

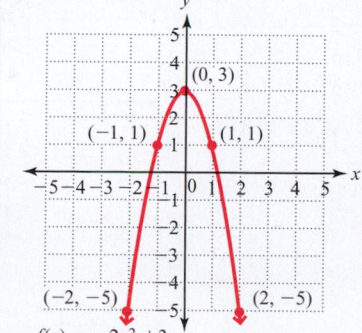

Connection   Recall from Section 6.7 that the sign of a in $f(x) = ax^2 + bx + c$ indicates whether the parabola opens up or down. If a is *positive*, the parabola opens *up*, and if a is *negative*, the parabola opens *down*.

Your Turn 4   Graph $f(x) = 3x^2 - 4$.

Absolute Value Functions

Now let's consider how to graph functions containing absolute value. In general, absolute value functions have a V shape.

Example 5   Graph each function.

a. $f(x) = |x|$

SOLUTION   We find ordered pair solutions, plot those ordered pairs, and connect the points.

x	$f(x)$
−4	4
−2	2
0	0
2	2
4	4

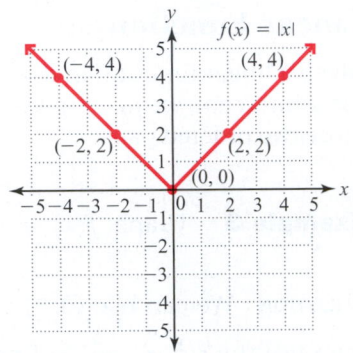

Note: $f(x) = |x|$ is equivalent to $f(x) = -x$ when $x < 0$ and $f(x) = x$ when $x \geq 0$. Notice $f(x) = -x$ and $f(x) = x$ are linear functions with opposite slopes, which explains the V shape of the graph.

b. $f(x) = |x| + 1$

SOLUTION

x	$f(x)$
−4	5
−2	3
0	1
2	3
4	5

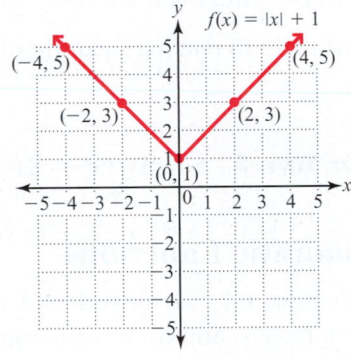

Note: Compare the graphs in Examples 5a and 5b. Notice that every point on $f(x) = |x| + 1$ is one higher than every point on $f(x) = |x|$. Adding 1 to $|x|$ shifts the graph of $f(x) = |x|$ up 1 unit.

c. $f(x) = |x - 2| + 1$

SOLUTION

x	$f(x)$
−4	7
−2	5
0	3
2	1
4	3

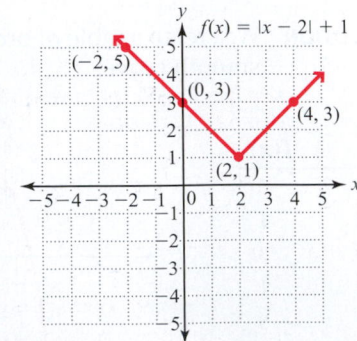

Note: As we saw in Example 5b, changing $f(x) = |x|$ to $f(x) = |x| + 1$ shifted the graph of $f(x) = |x|$ up 1 unit. Now we see that replacing x in $f(x) = |x| + 1$ with $x - 2$ shifts the graph of $f(x) = |x| + 1$ right 2 units.

Answer   Your Turn 4

d. $f(x) = -|x - 2| + 1$

a.

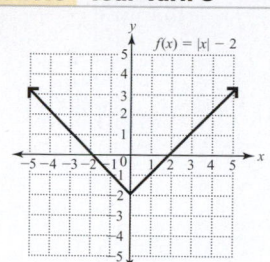

b.

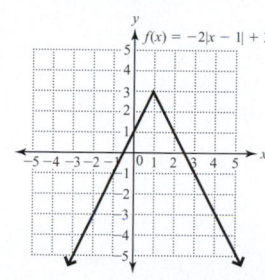

SOLUTION

x	$f(x)$
-4	-5
-2	-3
0	-1
2	1
4	-1

$f(x) = -|x - 2| + 1$

Note: Notice that the minus sign in front of the absolute value causes the graph to invert so that the **V** is "upside down."

Your Turn 5 Graph each function.

a. $f(x) = |x| - 2$ b. $f(x) = -2|x - 1| + 3$

8.4 Exercises   For Extra Help   *MyMathLab*

 PRACTICE   WATCH   DOWNLOAD

1. Explain what makes a relation a function.

2. How do you determine if a graph represents a function?

3. In your own words, define *domain* and *range*.

4. What is the shape of the graph of every linear function? How do you graph a linear function?

5. What is the shape of the graph of every quadratic function? How do you graph a quadratic function?

6. What is the shape of the graph of every absolute value function? How do you graph an absolute value function?

For Exercises 7–14, identify the domain and range of each relation and determine whether it is a function. See Example 1.

7. Most Popular USA Travel Destinations for Romantics

Rank	Destination
1	Hawaii
2	Las Vegas
3	Florida
4	California
5	Texas

Source: http://honeymoons.about.com.

8. Top Five Countries by Imports by the United States

Rank	Country
1	Canada
2	China
3	Mexico
4	Japan
5	Germany

Source: The World Almanac and Book of Facts, 2008.

9. Percent of the workforce leaving for work commute.

Percent (%)	Time of Departure
2	Work at home
3	Midnight to 4:59 A.M. *and* 10 A.M. to 11:59 A.M.
5	9 A.M. to 9:59 A.M.
14	Noon to 11:59 P.M.
15	8 A.M. to 8:59 A.M.
26	5 A.M. to 6:59 A.M.
33	7 A.M. to 7:59 A.M.

Source: U.S. Census 2000.

10. The percent of resale value to job cost for home repair/remodeling projects according to *Remodeling* magazine's 2007 Cost vs. Value Report.

Cost recouped (%)	Project
85	Deck addition
83	Siding replacement
81	Window replacement—wood
79	Window replacement—vinyl
69	Family room replacement
	Master suite addition
66	Bathroom addition

11. $\{(2, 1), (-3, 2), (0, 5), (1, 8)\}$

12. $\{(1, 1), (2, 2), (-3, -3), (4, 4)\}$

13. $\{(-3, 2), (4, 1), (-3, 5), (2, 8)\}$

14. $\{(-1, 0), (-3, 0), (4, 5), (2, 6)\}$

For Exercises 15–24, identify the domain and range of each relation and determine whether it is a function. See Example 1.

15.

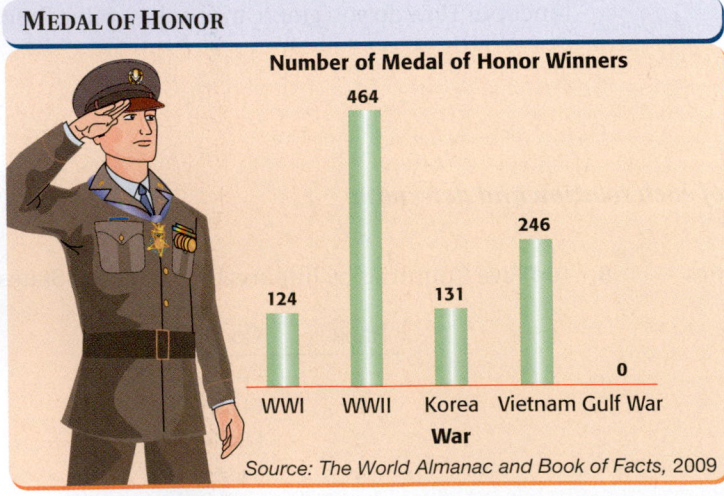

MEDAL OF HONOR

Number of Medal of Honor Winners

Source: The World Almanac and Book of Facts, 2009

16.

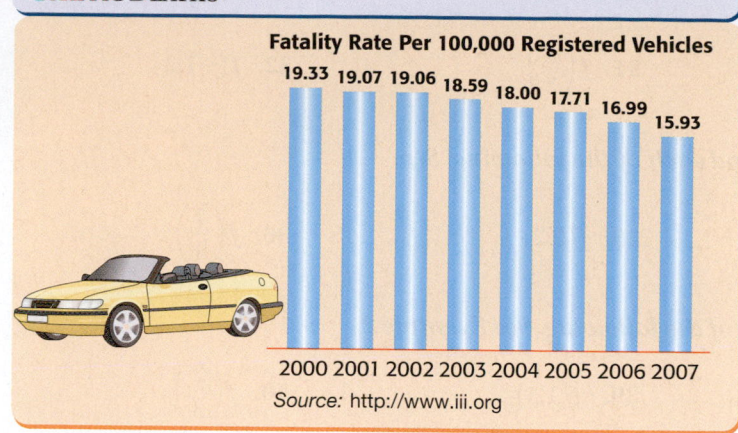

Fatality Rate Per 100,000 Registered Vehicles

19.33 19.07 19.06 18.59 18.00 17.71 16.99 15.93

2000 2001 2002 2003 2004 2005 2006 2007

Source: http://www.iii.org

17.

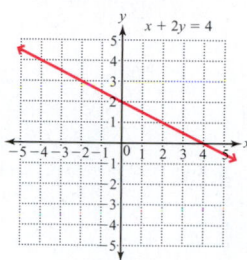

$x + 2y = 4$

18.

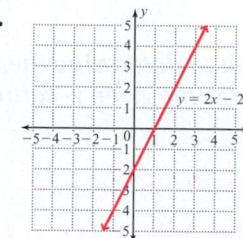

$y = 2x - 2$

19.

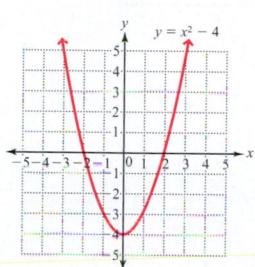

$y = x^2 - 4$

20.

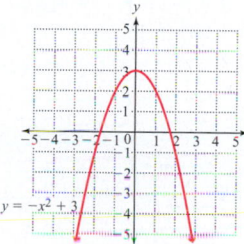

$y = -x^2 + 3$

21.

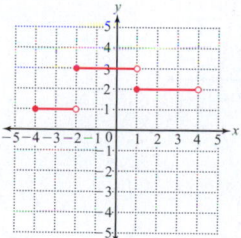

22.

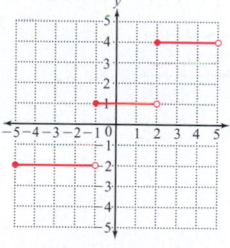

23.

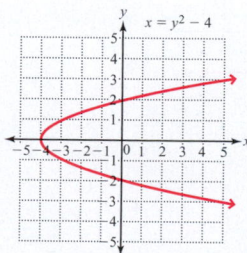

$x = y^2 - 4$

24.

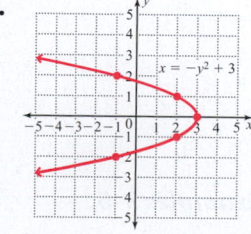

$x = -y^2 + 3$

For Exercises 25–28, given $f(x) = 2x^2 - 3x + 1$, find each of the following. See Example 2.

25. $f(-2)$

26. $f(0)$

27. $f\left(\dfrac{1}{3}\right)$

28. $f(1.5)$

For Exercises 29–32, given $f(x) = |x + 3| - 2$, find each of the following. See Example 2.

29. $f(2)$

30. $f(-5)$

31. $f\left(\dfrac{2}{3}\right)$

32. $f(-1.3)$

For Exercises 33–36, given $f(x) = x^3 - 2x^2 + 3$, find each of the following. See Example 2.

33. $f(0)$

34. $f(-2)$

35. $f(0.2)$

36. $f\left(\dfrac{1}{4}\right)$

For Exercises 37–40, given $f(x) = \dfrac{2x}{x - 1}$, find each of the following. See Example 2.

37. $f(1)$

38. $f(-2)$

39. $f(1.5)$

40. $f\left(\dfrac{2}{3}\right)$

For Exercises 41–46, use the graph to determine the value of the function. (Hint: The function value is the y-value for the given x-value.)

41. **a.** $f(2)$
    b. $f(-1)$
    c. $f(0)$

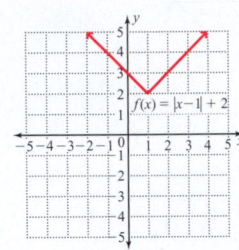

42. **a.** $f(3)$
    b. $f(0)$
    c. $f(4)$

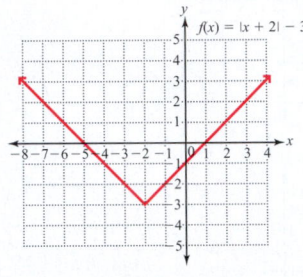

43. **a.** $f(3)$
    b. $f(-1)$
    c. $f(4)$

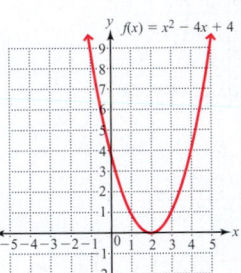

44. **a.** $f(-1)$
    b. $f(-2)$
    c. $f(3)$

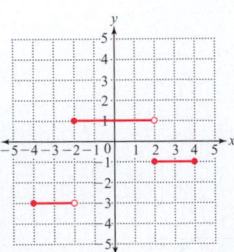

45. **a.** $f(-3)$
    b. $f(-2)$
    c. $f(2)$

46. **a.** $f(-3)$
    b. $f(0)$
    c. $f(1)$

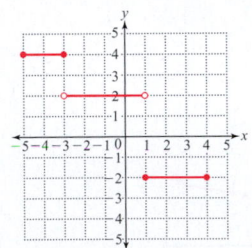

For Exercises 47–72, graph. See Examples 3–5.

47. $f(x) = 2x + 3$

48. $f(x) = 3x - 1$

49. $f(x) = -x + 4$

50. $f(x) = -2x + 5$

51. $f(x) = \frac{3}{4}x - 2$

52. $f(x) = -\frac{2}{5}x + 1$

53. $f(x) = x^2 + 4$

54. $f(x) = x^2 - 3$

55. $f(x) = -x^2 + 1$

56. $f(x) = -x^2 + 9$

57. $f(x) = \frac{1}{4}x^2 - 2$

58. $f(x) = \frac{1}{3}x^2 - 3$

59. $f(x) = x^2 - 4x + 3$

60. $f(x) = x^2 + 5x + 6$

61. $f(x) = |x| - 4$

62. $f(x) = |x| + 1$

63. $f(x) = |x + 3| - 2$

64. $f(x) = |x - 1| + 3$

65. $f(x) = 2|x - 1| + 3$

66. $f(x) = 3|x + 1| - 2$

67. $f(x) = -|x| + 3$

68. $f(x) = -|x| - 1$

69. $f(x) = -3|x + 2| - 1$

70. $f(x) = -2|x - 1| + 4$

71. $f(x) = -\frac{1}{2}|x + 1| + 1$

72. $f(x) = -\frac{1}{4}|x - 2| + 2$

Collaborative Exercises

*In the following exercises, you will use a graphing utility to explore and discuss how numbers can **shift**, **reflect**, or **stretch** a basic function. We will use the graph of $f(x) = |x|$ as our basic function.*

1. Graph $f(x) = |x + 2|$ and compare it with the graph of $f(x) = |x|$. What effect does adding 2 to x within the absolute value have on the graph?

2. Graph $f(x) = |x - 3|$ and compare it with the graph of $f(x) = |x|$. What effect does subtracting 3 from x within the absolute value have on the graph?

3. Graph $f(x) = |x| + 4$ and compare it with the graph of $f(x) = |x|$. What effect does adding 4 to $|x|$ have on the graph?

4. Graph $f(x) = |x| - 1$ and compare it with the graph of $f(x) = |x|$. What effect does subtracting 1 from $|x|$ have on the graph?

5. Notice that adding or subtracting in a function *shifts* the position of the basic function. Without graphing, how would you expect the graph of $f(x) = |x + 1| - 3$ to be shifted as compared with the basic function $f(x) = |x|$?

6. Graph $f(x) = -|x|$. How is the graph of $f(x) = -|x|$ different from that of $f(x) = |x|$?

7. Recall that $f(x) = -|x|$ is the same as $f(x) = -1 \cdot |x|$. Multiplying the basic function by a negative number causes a *reflection* of the function about the x-axis. Without graphing, how would you describe the differences between the graph of $f(x) = -|x - 2| + 1$ and the basic function $f(x) = |x|$?

8. Graph $f(x) = 2|x|$. How is the graph of $f(x) = 2|x|$ different from that of $f(x) = |x|$?

9. Graph $f(x) = \frac{1}{2}|x|$. How is the graph of $f(x) = \frac{1}{2}|x|$ different from that of $f(x) = |x|$?

10. Notice that multiplying the basic function by a number other than 1 causes the graph to be *stretched,* so that it is wider or narrower than the basic function's graph. Without graphing, describe the differences in each of the following compared to the graph of $f(x) = |x|$.

    a. $f(x) = 3|x| + 1$

    b. $f(x) = -2|x - 4|$

    c. $f(x) = \frac{1}{3}|x + 4| - 2$

For further investigation, change the basic function to $f(x) = x^2$ and work through Exercises 1–10 again. For example, Exercise 1 would read: Graph $f(x) = (x + 2)^2$ and compare it with the graph of $f(x) = x^2$. What effect does adding 2 to x have on the graph? Exercise 3 would read: Graph $f(x) = (x^2 + 4)$ and compare it with the graph of $f(x) = x^2$. What effect does adding 4 to x^2 have on the graph?

Review Exercises

Exercises 1–6   EXPRESSIONS

For Exercises 1–6, perform the indicated operation.

[5.3]   1. $(2x^2 + 3x - 1) + (3x^2 - 3x + 7)$

[5.3]   2. $(5x^2 - 3x + 1) - (-2x^2 + 8)$

[5.5]   3. $(3x - 4)(2x + 1)$

[5.5]   4. $(2x - 4)^2$

[5.6]   5. $(12x^4 + 18x^3 - 30x^2) \div 6x^2$

[5.6]   6. $(x^4 - 81) \div (x - 3)$

8.5 Function Operations

Objectives

1. Add or subtract functions.
2. Multiply functions.
3. Divide functions.

> **Note:** Letters other than f and g can be used to name functions, such as h, p, or q.

So far, we have seen how to evaluate and graph functions. Now we will consider how to add, subtract, multiply, and divide functions. Because we will be considering two different functions in the same problem, we will write one function as f and the second as g.

Objective 1 Add or subtract functions. Given two functions f and g, their sum or difference is found using the following rules.

> **Rule   Adding or Subtracting Functions**
>
> The sum of two functions, $f + g$, is found by $(f + g)(x) = f(x) + g(x)$.
> The difference of two functions, $f - g$, is found by $(f - g)(x) = f(x) - g(x)$.

Example 1   Given $f(x) = 2x + 1$ and $g(x) = 7x + 3$, find the following.

a. $f + g$

b. $f - g$

c. $(f - g)(-4)$

SOLUTION

a. $(f + g)(x) = f(x) + g(x)$

$\qquad = (2x + 1) + (7x + 3)$ Write a sum of the two functions.

$\qquad = 9x + 4$ Combine like terms.

b. $(f - g)(x) = f(x) - g(x)$

$\qquad = (2x + 1) - (7x + 3)$ Write a difference of the two functions.

$\qquad = (2x + 1) + (-7x - 3)$ Write as an equivalent addition.

$\qquad = -5x - 2$ Combine like terms.

c. In part b, we found $(f - g)(x) = -5x - 2$, so to find $(f - g)(-4)$, we replace x with -4.

$\qquad (f - g)(-4) = -5(-4) - 2$ Replace x with -4.

$\qquad\qquad\qquad = 20 - 2$

$\qquad\qquad\qquad = 18$

Answers   Your Turn 1
a. $2x^2 - 5x - 4$
b. $-x + 14$
c. 14

Your Turn 1   Given $h(x) = x^2 - 3x + 5$ and $k(x) = x^2 - 2x - 9$, find the following.

a. $h + k$ b. $h - k$ c. $(h + k)(-2)$

Objective 2 Multiply functions. Given two functions f and g, their product is found using the following rule.

Rule Multiplying Functions

The product of two functions, $f \cdot g$, is found by $(f \cdot g)(x) = f(x)\, g(x)$.

Example 2 Given $f(x) = 3x - 4$ and $g(x) = x + 5$, find $f \cdot g$.

SOLUTION $(f \cdot g)(x) = \boldsymbol{f(x)\, g(x)}$

$\qquad = (\boldsymbol{3x - 4})(\boldsymbol{x + 5})$ Write a product of the functions.

$\qquad = 3x^2 + 15x - 4x - 20$ Use FOIL to multiply the binomials.

$\qquad = 3x^2 + 11x - 20$ Combine like terms.

Your Turn 2 Given $p(x) = 7x - 2$ and $q(x) = 5x - 1$, find the following.

    a. $p \cdot q$ **b.** $(p \cdot q)(2)$

Objective ③ Divide functions. Given two functions f and g, their quotient is found using the following rule.

Rule Dividing Functions

The quotient of two functions, f/g, is found by $(f/g)(x) = \dfrac{f(x)}{g(x)}$, where $g(x) \neq 0$.

Example 3 Find f/g.

    a. $f(x) = 9x^3 - 12x^2 + 6x, \, g(x) = 3x$

Note: Because we divided by $3x$, we are assuming that $x \neq 0$ in $(f/g)(x)$.

SOLUTION $(f/g)(x) = \dfrac{\boldsymbol{f(x)}}{\boldsymbol{g(x)}} = \dfrac{\boldsymbol{9x^3 - 12x^2 + 6x}}{\boldsymbol{3x}}$ Write a quotient of the functions.

$\qquad\qquad = \dfrac{9x^3}{3x} - \dfrac{12x^2}{3x} + \dfrac{6x}{3x}$ Since the divisor is a monomial, we separate the terms.

$\qquad\qquad = 3x^2 - 4x + 2$ Divide.

    b. $f(x) = 3x^2 - 13x - 6, \, g(x) = x - 5$

SOLUTION $(f/g)(x) = \dfrac{f(x)}{g(x)} = \dfrac{3x^2 - 13x - 6}{x - 5}$ Write a quotient of the functions.

To find the quotient, we use long division.

$$
\begin{array}{r}
3x \\
x - 5\overline{)3x^2 - 13x - 6} \\
-(3x^2 - 15x)
\end{array}
$$
Change signs.

$$
\begin{array}{r}
3x \\
x - 5\overline{)3x^2 - 13x - 6} \\
-3x^2 + 15x \\
\hline
2x - 6
\end{array}
$$
Combine like terms and bring down the next term.

Answers Your Turn 2
a. $35x^2 - 17x + 2$ **b.** 108

$$
\begin{array}{r}
3x + 2 \\
x - 5\overline{)3x^2 - 13x - 6} \\
-3x^2 + 15x \\
\hline
2x - 6 \\
-(2x - 10)
\end{array}
$$
Change signs.

$$
\begin{array}{r}
3x + 2 \\
x - 5\overline{)3x^2 - 13x - 6} \\
-3x^2 - 15x \\
\hline
2x - 6 \\
-2x + 10 \\
\hline
4
\end{array}
$$
Combine like terms.

Note: The domain of the quotient function contains all real numbers except 5, so we would write the domain as $\{x \mid x \neq 5\}$ or, using interval notation, $(-\infty, 5) \cup (5, \infty)$.

Answer $3x + 2 + \dfrac{4}{x - 5}$

Your Turn 3

a. Given $f(x) = 12x^4 + 36x^3 - 21x^2$ and $g(x) = 3x^2$, find f/g and $(f/g)(-2)$.

b. Given $m(x) = 6x^2 + 19x - 21$ and $n(x) = x + 4$, find m/n and $(m/n)(1)$.

8.5 Exercises   For Extra Help   *MyMathLab*

 PRACTICE   WATCH   DOWNLOAD

1. If $p(x)$ and $q(x)$ are both polynomial functions, how do you find $p + q$?

2. If $p(x)$ and $q(x)$ are both polynomial functions, how do you find $p - q$?

3. If $p(x)$ and $q(x)$ are both binomial functions, how do you find $p \cdot q$?

4. If $p(x)$ is a trinomial function and $q(x)$ is a monomial function, how do you find p/q?

For Exercises 5–14, find $f + g$ and $f - g$. See Example 1.

5. $f(x) = 2x, g(x) = 3x - 1$

6. $f(x) = 3x, g(x) = 4x + 2$

7. $f(x) = x - 5, g(x) = 2x - 3$

8. $f(x) = 5x + 1, g(x) = x - 8$

9. $f(x) = x^2 - 4x + 7, g(x) = x - 3$

10. $f(x) = x^2 - 3x + 1, g(x) = x - 2$

11. $f(x) = 2x^2 - 5x - 3, g(x) = x^2 + 5$

12. $f(x) = -x^2 - 8x + 2, g(x) = x^2 + 2$

13. $f(x) = -5x^2 + 4x + 8, g(x) = 3x^2 - 2x - 1$

14. $f(x) = -2x^2 + 5x + 3, g(x) = 4x^2 + 2x - 9$

For Exercises 15–24, find $f \cdot g$. See Example 2.

15. $f(x) = 2x, g(x) = x - 5$

16. $f(x) = -2x, g(x) = x + 4$

17. $f(x) = x + 1, g(x) = x - 2$

18. $f(x) = x - 5, g(x) = x + 4$

19. $f(x) = 3x + 2, g(x) = x + 2$

20. $f(x) = 2x - 5, g(x) = 3x - 1$

21. $f(x) = x^2 - 3x + 2, g(x) = x^2 + 2x - 7$

22. $f(x) = x^2 - 2x - 1, g(x) = x^2 + 3x + 4$

23. $f(x) = 2x^2 - 3x + 1, g(x) = 3x^2 - x - 1$

24. $f(x) = 3x^2 - 2x - 1, g(x) = 2x^2 - 2x - 3$

For Exercises 25–34, find f/g. See Example 3.

25. $f(x) = 2x^2 - 6x, g(x) = 2x$

26. $f(x) = 3x^2 - 9x, g(x) = 3x$

27. $f(x) = 6x^2 - 3x + 6, g(x) = 3x$

28. $f(x) = 4x^2 + 4x - 8, g(x) = 2x$

29. $f(x) = x^2 - 14x + 45, g(x) = x - 9$

30. $f(x) = x^2 + 9x + 20, g(x) = x + 4$

31. $f(x) = 2x^2 - 3x - 5, g(x) = 2x - 5$

32. $f(x) = 3x^2 - 4x - 7, g(x) = 3x - 7$

33. $f(x) = 2x^2 - 5x - 7, g(x) = x + 5$

34. $f(x) = 6x^2 - 5x + 4, g(x) = 2x - 1$

For Exercises 35–38, find: a. $p + q$, b. $p - q$, c. $p \cdot q$, d. p/q. See Examples 1–3.

35. $p(x) = 5x - 1, q(x) = 2x + 3$

36. $p(x) = 3x - 4, q(x) = 5x + 2$

37. $p(x) = 2x^2 - 3x + 5, q(x) = x - 1$

38. $p(x) = 4x^2 - 4x - 1, q(x) = 2x + 3$

For Exercises 39–42, find the function values. See Example 1c.

39. Using the functions in Exercise 35, find the following.
   a. $(p + q)(-2)$ **b.** $(p - q)(5)$ **c.** $(p \cdot q)(-1)$ **d.** $(p/q)(7)$

40. Using the functions in Exercise 36, find the following.
   a. $(p + q)(-2)$ **b.** $(p - q)(3)$ **c.** $(p \cdot q)(-2)$ **d.** $(p/q)(2)$

41. Using the functions in Exercise 37, find the following.
   a. $(p + q)(3)$ **b.** $(p - q)(-4)$ **c.** $(p \cdot q)(2)$ **d.** $(p/q)(-3)$

42. Using the functions in Exercise 38, find the following.
   a. $(p + q)(-2)$ **b.** $(p - q)(3)$ **c.** $(p \cdot q)(-2)$ **d.** $(p/q)(4)$

For Exercises 43–44: **a.** *Find $f + g$.*
   b. *Complete the table shown.*
   c. *Graph f, g, and $f + g$ on the same graph.*
   d. *What do the table and graphs suggest about the sum of
        two functions?*

43. $f(x) = 3x + 1, g(x) = x + 5$

44. $f(x) = -x + 3, g(x) = 4x + 3$

x	$f(x)$	$g(x)$	$(f + g)(x)$
-4			
-2			
0			
2			
4			

x	$f(x)$	$g(x)$	$(f + g)(x)$
-4			
-2			
0			
2			
4			

For Exercises 45–46: **a.** *Find h − k.*
  b. *Complete the table shown.*
  c. *Graph h, k, and h − k on the same graph.*
  d. *What do the table and graphs suggest about the difference of two functions?*

45. $h(x) = 2x - 9, k(x) = 3x + 5$

x	$h(x)$	$k(x)$	$(h - k)(x)$
−4			
−2			
0			
2			
4			

46. $h(x) = 4x + 1, k(x) = 2x - 7$

x	$h(x)$	$k(x)$	$(h - k)(x)$
−4			
−2			
0			
2			
4			

For Exercises 47 and 48, the function w(x) describes a company's monthly costs in wages and p(x) describes the cost of production, where x represents the number of units produced.

47. $w(x) = 3x + 2, p(x) = x + 2$
  a. Find the function, $t(x)$, that describes the total cost.

  b. Calculate the total cost if $x = 200$.
  c. Graph all three functions.

48. $w(x) = 2x + 1, p(x) = x - 1$
  a. Find the function, $t(x)$, that describes the total cost.
  b. Calculate the total cost if $x = 175$.
  c. Graph all three functions.

For Exercises 49 and 50, the function r(x) describes a company's monthly revenue and c(x) describes the monthly costs where x represents the number of units produced and then sold. Use the formula Net Profit = Revenue − Cost.

49. $r(x) = x^2 - 3x - 4, c(x) = x + 8$
  a. Find the function, $p(x)$, that describes the net profit or loss.

  b. Calculate the net profit or loss if $x = 100$. Is this value a profit or a loss?

  c. Graph all three functions.
  d. How many units must be sold for the company to break even (which means the net = 0)?

50. $r(x) = x^2 - 4x - 5, c(x) = x + 1$
  a. Find the function, $p(x)$, that describes the net profit or loss.

  b. Calculate the net profit or loss if $x = 300$. Is this value a profit or a loss?

  c. Graph all three functions.
  d. How many units must be sold for the company to break even (which means the net = 0)?

*For Exercises 51 and 52, the function l(x) describes the length of a box and h(x)
describes the height, where x represents the width.*

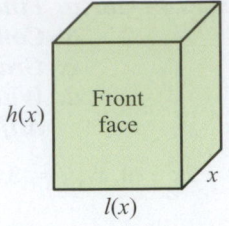

51. $l(x) = 3x + 2, h(x) = 2x$
    a. Find the function that describes the area of the front face of the box.
    b. Calculate the area of the front face if $x = 3$ feet.

52. $l(x) = 5x - 1, h(x) = 3x$
    a. Find the function that describes the area of the front face of the box.
    b. Calculate the area of the front face if $x = 2$ meters.

★ *For Exercises 53 and 54, the function V(x) describes the volume of a box, l(x)
describes its length, and w(x) describes its width.*

53. $V(x) = 4x^3 - 7x^2 - 14x - 3, l(x) = x - 3, w(x) = x + 1$
    a. Find the function, $h(x)$, that describes the height of the box.
    b. Calculate the height of the box if $x = 4$ centimeters.

54. $V(x) = 6x^3 + 11x^2 - 47x + 20, l(x) = x + 4, w(x) = 2x - 1$
    a. Find the function, $h(x)$, that describes the height of the box.
    b. Calculate the height of the box if $x = 5$ inches.

Review Exercises

Exercises 1–6   EQUATIONS AND INEQUALITIES

[2.5]   1. Let x represent the first number and y represent
           the second number, then translate to an equa-
           tion. Twice the first number plus three times the
           second number is 3.

[2.1]   2. Determine if $x = -4$ is a solution of
           $2x + 3 = -5$.

For Exercises 3 and 4, solve.

[6.6]   3. $x^2 - 16 = 0$

[8.2]   4. $|x + 5| = -8$

For Exercises 5 and 6, graph.

[4.5]   5. $y = 2x - 3$

[4.3]   6. $2x - y = 4$

Chapter 8 Summary

Defined Terms

Section 8.1
Compound inequality (p. 572)
Intersection (p. 572)
Union (p. 575)

Section 8.4
Relation (p. 594)
Domain (p. 594)
Range (p. 594)
Function (p. 594)

Procedures, Rules, and Key Examples

Procedures/Rules	Key Examples

SECTION 8.1 Compound Inequalities

To solve a compound inequality involving *and*,
1. Solve each inequality in the compound inequality.
2. The solution set will be the intersection of the individual solution sets.

For each example, solve the compound inequality; then
a. Graph the solution set.
b. Write the solution set in set-builder notation.
c. Write the solution set in interval notation.

Example 1: $-3 < -2x - 5 \leq 1$

Solution:

a.
$$-2x - 5 > -3 \quad \text{and} \quad -2x - 5 \leq 1$$
$$-2x > 2 \quad \text{and} \quad -2x \leq 6$$
$$x < -1 \quad \text{and} \quad x \geq -3$$

Graph:

b. Set-builder notation: $\{x | -3 \leq x < -1\}$
c. Interval notation: $[-3, -1)$

To solve a compound inequality involving *or*,
1. Solve each inequality in the compound inequality.
2. The solution set will be the union of the individual solution sets.

Example 2: $5x - 1 \leq -16$ or $5x - 1 > 9$

Solution:

a.
$$5x - 1 \leq -16 \quad \text{or} \quad 5x - 1 > 9$$
$$5x \leq -15 \quad \text{or} \quad 5x > 10$$
$$x \leq -3 \quad \text{or} \quad x > 2$$

Graph:

b. Set-builder notation: $\{x | x \leq -3 \text{ or } x > 2\}$
c. Interval notation: $(-\infty, -3] \cup (2, \infty)$

SECTION 8.2 Equations Involving Absolute Value

Absolute Value Property
If $|x| = a$, where x is a variable or an expression and $a \geq 0$, then $x = a$ or $x = -a$.

To solve an equation containing a single absolute value,
1. **Isolate the absolute value** so that the equation is in the form $|ax + b| = c$. If $c > 0$, proceed to steps 2 and 3. If $c < 0$, the equation has no solution.
2. Separate the absolute value into two equations, $ax + b = c$ and $ax + b = -c$.
3. Solve both equations.

Example 1: Solve $|4x + 1| - 3 = 2$.

Solution:

$|4x + 1| = 5$ **Isolate the absolute value.**

$4x + 1 = 5$ or $4x + 1 = -5$ **Use the absolute value property.**

$4x = 4$ or $4x = -6$

$x = 1$ or $x = -\dfrac{3}{2}$

(continued)

Procedures/Rules	Key Examples

SECTION 8.2 Equations Involving Absolute Value (*continued*)

To solve an equation in the form $|ax + b| = |cx + d|$,
1. Separate the absolute value equation into two equations:
   $ax + b = cx + d$ and $ax + b = -(cx + d)$.
2. Solve both equations.

Example 2: Solve $|5x - 7| = |3x + 1|$.

Solution: Separate into two equations.

Equal		Opposites
$5x - 7 = 3x + 1$	or	$5x - 7 = -(3x + 1)$
$2x - 7 = 1$	or	$5x - 7 = -3x - 1$
$2x = 8$	or	$8x = 6$
$x = 4$	or	$x = \dfrac{3}{4}$

SECTION 8.3 Inequalities Involving Absolute Value

To solve an inequality in the form $|x| < a$, where $a > 0$,
1. Rewrite the inequality as a compound inequality involving *and*: $x > -a$ and $x < a$ (or $-a < x < a$).
2. Solve the compound inequality.
   Similarly, to solve $|x| \leq a$, we write $x \geq -a$ and $x \leq a$ (or $-a \leq x \leq a$).

For each example, solve the inequality; then
a. Graph the solution set.
b. Write the solution set in set-builder notation.
c. Write the solution set in interval notation.

Example 1: $|2x - 3| - 1 < 4$

Solution:

a. $|2x - 3| < 5$ **Isolate the absolute value.**

$2x - 3 > -5$ and $2x - 3 < 5$ **Separate**
$2x > -2$ and $2x < 8$ **using *and*.**
$x > -1$ and $x < 4$

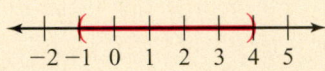

b. Set-builder notation. $\{x | -1 < x < 4\}$
c. Interval notation: $(-1, 4)$

To solve an inequality in the form $|x| > a$, where $a > 0$,
1. Rewrite the inequality as a compound inequality involving *or*: $x < -a$ or $x > a$
2. Solve the compound inequality.
   Similarly, to solve $|x| \geq a$, we write $x \leq -a$ or $x \geq a$.

Example 2: $|3x - 5| - 7 \geq 4$

Solution:

a. $|3x - 5| \geq 11$ **Isolate the absolute value.**

$3x - 5 \leq -11$ or $3x - 5 \geq 11$ **Separate using *or*.**

$3x \leq -6$ or $3x \geq 16$
$x \leq -2$ or $x \geq \dfrac{16}{3}$

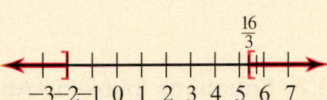

b. Set-builder notation: $\left\{ x | x \leq -2 \text{ or } x \geq \dfrac{16}{3} \right\}$

c. Interval notation: $(-\infty, -2] \cup \left[\dfrac{16}{3}, \infty \right)$

(continued)

Procedures/Rules	Key Examples

SECTION 8.4　Functions and Graphing

Relation: A set of ordered pairs.
Domain: The set of all input values for a relation.
Range: The set of all output values for a relation.
Function: A relation in which every value in the domain is paired with exactly one value in the range.

Example 1: For the relation shown, determine the domain, range, and if it is a function.

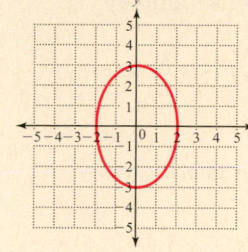

Domain: $\{x \mid -2 \leq x \leq 2\}$
Range: $\{y \mid -3 \leq y \leq 3\}$
It is not a function because it does not pass the vertical line test.

To find the value of a function replace x with the given value.

Example 2: Given $f(x) = 3x^2 - 4$, find $f(2)$.
$$f(2) = 3(2)^2 - 4$$
$$= 12 - 4$$
$$= 8$$

The graph of a linear function is a line.

A linear function has the form $f(x) = mx + b$, where m is the slope and $(0, b)$ is the y-intercept. To graph, we need to find only two ordered pairs.

Example 3: Graph.
a. $f(x) = -2x + 3$

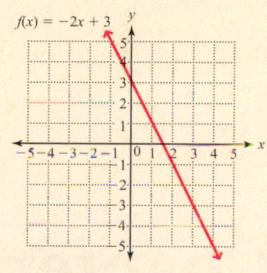

The graph of a quadratic function is a parabola.

Quadratic functions have the form $f(x) = ax^2 + bx + c$, and we saw their graphs are parabolas. To graph a parabola we have to find many ordered pairs.

b. $f(x) = x^2 - 2$

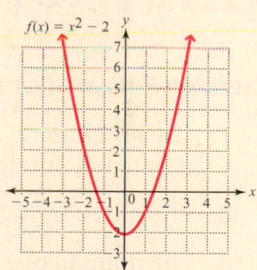

The graph of an absolute value function is a V shape.

To graph, find enough ordered pairs to generate the V shape.

c. $f(x) = |x + 2|$

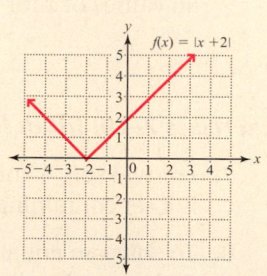

(*continued*)

Procedures/Rules	Key Examples

SECTION 8.5 Function Operations

The sum of two functions, **f + g**, is found by
$(f + g)(x) = f(x) + g(x)$.

The difference of two functions, **f − g**, is found by
$(f - g)(x) = f(x) - g(x)$.

The product of two functions, **f · g**, is found by
$(f \cdot g)(x) = f(x)\, g(x)$.

The quotient of two functions, **f/g**, is found by $(f/g)(x) = \dfrac{f(x)}{g(x)}$, where $g(x) \neq 0$.

Example 1: Given $f(x) = 2x - 5$ and $g(x) = 3x + 4$, find **a.** $f + g$ **b.** $f - g$ **c.** $f \cdot g$

a. $(f + g)(x) = (2x - 5) + (3x + 4)$
$= 5x - 1$

b. $(f - g)(x) = (2x - 5) - (3x + 4)$
$= (2x - 5) + (-3x - 4)$
$= -x - 9$

c. $(f \cdot g)(x) = (2x - 5)(3x + 4)$
$= 6x^2 + 8x - 15x - 20$
$= 6x^2 - 7x - 20$

Example 2: Given $f(x) = 12x^3 - 20x^2 + 8x$ and $g(x) = 4x$, find f/g.
Solution:

$$(f/g)(x) = \frac{f(x)}{g(x)} = \frac{12x^3 - 20x^2 + 8x}{4x}$$
$$= \frac{12x^3}{4x} - \frac{20x^2}{4x} + \frac{8x}{4x}$$
$$= 3x^2 - 5x + 2, \, x \neq 0$$

Chapter 8 Review Exercises

For Exercises 1–6, answer true or false.

[8.1] **1.** The notation $(2, 5)$ can represent an ordered pair or interval notation.

[8.4] **3.** The notation $f(x)$ means $f \cdot x$.

[8.4] **5.** The domain of a function $f(x)$ is a set containing all values of x for which the function is defined.

[8.2] **2.** To solve absolute value equations, the absolute value must be isolated.

[8.4] **4.** All functions are also relations.

[8.4] **6.** Functions always pass the vertical line test.

For Exercises 7–10, fill in the blank.

[8.1] **7.** For two sets A and B, the intersection of A and B, symbolized by _____, is a set containing only those elements that are in both A and B.

[8.2] **8.** To solve an equation containing a single absolute value,
1. Isolate the absolute value so that the equation is in the form $|ax + b| = c$.
2. Separate the absolute value into two equations, _____ and _____.
3. Solve both equations.

[8.4] **9.** The set of all output values for a relation is _____.

[8.5] **10.** The sum of two functions $f + g$ is shown by $(f + g)(x) =$ _____.

[8.1] *For Exercises 11–12, find the intersection and union of the given sets.*

11. $A = \{1, 5, 9\}$
$\quad B = \{1, 5, 7, 9\}$

12. $A = \{1, 2, 3\}$
$\quad\ \ B = \{4, 5, 6, 7\}$

Exercises 13–44 ▲ EQUATIONS AND INEQUALITIES

[8.1] *For Exercises 13–24, solve the compound inequality. Then:*
a. Graph the solution set.
b. Write the solution set using set-builder notation.
c. Write the solution set using interval notation.

13. $4 < -2x < 6$

14. $9 \leq -3x \leq 15$

15. $-3 < x + 4 < 7$

16. $0 < x - 1 \leq 3$

17. $2x + 3 \geq -1$ and $2x + 3 < 3$

18. $3x - 2 > 1$ and $3x - 2 < -8$

19. $w + 4 \leq -2$ or $w + 4 \geq 2$

20. $4w - 3 < 1$ or $4w - 3 > 0$

21. $2m - 5 < 0$ or $2m - 5 > 5$

22. $3x + 2 \leq -2$ or $3x + 2 \geq 8$

23. $-x - 6 \leq -2$ or $-x - 6 \geq 3$

24. $-4w + 1 \leq -3$ or $-4w + 1 \geq 5$

[8.2] *For Exercises 25–26, solve. Then:* **a. Graph the solution set.**
b. Write the solution set using set-builder
notation.
c. Write the solution set using interval
notation.

25. Students in a biology course receive a B if the average of four tests is 80 or higher and less than 90. Suppose a student has the following scores: 80, 89, and 83. What range of scores on the fourth test would cause the student to receive a B?

26. The width of a rectangular building is to be 80 feet. What range of values can the length have so that the area of the base of the building is between 12,000 and 16,000 square feet?

[8.2] *For Exercises 27–34, solve.*

27. $|x| = 4$ **28.** $|x - 4| = 7$ **29.** $|2w - 1| = 3$ **30.** $|5r + 8| = -3$

31. $|q - 4| - 3 = 8$ **32.** $|5w| - 2 = 13$ **33.** $2|3x - 4| = 8$ **34.** $4 - 2|r - 5| = -8$

[8.3] *For Exercises 35–44, solve the inequality. Then:* ***a.*** *Graph the solution set.*
                                                    b. *Write the solution set using set-builder notation.*
                                                    c. *Write the solution set using interval notation.*

35. $|x| < 5$ **36.** $|p| \geq 4$

37. $|x - 3| > 7$ **38.** $|2m + 6| < 4$

39. $|3s - 1| \leq -2$ **40.** $5|b| - 2 > 3$

41. $7|m + 3| \leq 21$ **42.** $-2|t - 5| < -10$

43. $5 - 2|2k - 3| \leq -15$ **44.** $3 - 7|2p + 4| \leq 24$

[8.4] *For Exercises 45–50, identify the domain and range of the relation, then determine whether it is a function.*

45. Top-Selling Comic Books

Rank	Title
1	Secret Invasion #8
2	Final Crisis #5
3	Secret Invasion Dark Reign
4	New Avengers #47
5	New Avengers #48
6	Batman #682

Source: Diamond Comics Distributors. Rankings are based on orders placed by retailers and reflect total comics ordered as of 12/08.

46. Top Business Schools for Graduate Studies as Ranked by *U.S. News and World Report* 2008

Rank	Business School
1	Harvard University (MA)
	Stanford University (CA)
3	University of Pennsylvania (Wharton)
4	Massachusetts Institute of Technology (Sloan)
	Northwestern University (Kellogg) (IL)
	University of Chicago (IL)

47. $\{(2, 1), (4, -3), (2, 4), (3, -3)\}$

48. $\{(0, 3), (-2, 3), (1, 4)\}$

Exercises 49–64 EQUATIONS AND INEQUALITIES

49.

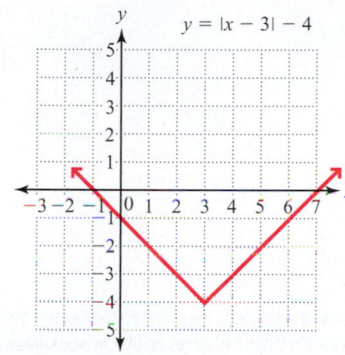

$y = |x - 3| - 4$

50.

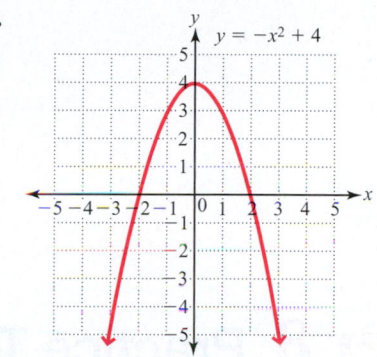

$y = -x^2 + 4$

[8.4] *For Exercises 51 and 52, find the specified value of the function.*

51. Find $f(3)$ given $f(x) = 3x^2 - 4x + 2$.

52. Find $f(-2)$ given $f(x) = -|2x - 5|$.

[8.4] *For Exercises 53–56, graph.*

53. $f(x) = -3x + 2$

54. $f(x) = 2x^2 - 3$

55. $f(x) = |x - 2|$

56. $f(x) = -2|x + 1| + 3$

[8.5] *For Exercises 57–60, find* **a.** $f + g$
b. $f - g$
c. $f \cdot g$

57. $f(x) = -3x + 9, g(x) = 4x + 1$

58. $f(x) = -x - 2, g(x) = 8x + 2$

59. $f(x) = 3x^2 - x + 5, g(x) = x + 2$

60. $f(x) = x^2 - 2x - 1, g(x) = 3x + 1$

For Exercises, 61–62, find f/g.

61. $f(x) = 6x^4 - 15x^3 + 12x^2, g(x) = 3x^2$

62. $f(x) = 2x^3 - 17x^2 + 36x - 17, g(x) = 2x - 5$

[8.5] *For Exercises 63–64, solve.*

63. The function $w(x) = 3x + 5$ describes a company's monthly expenditure on wages. The function $p(x) = x - 1$ describes the total cost of production. Note that x represents the number of units produced.
  a. Find the function, $c(x)$, that describes the total cost.
  b. If $x = 225$, then calculate the total cost.
  c. Graph all three functions.

64. The function $V(x) = 6x^3 + 41x^2 + 26x - 24$ describes the volume of a box, $l(x) = 3x + 4$ describes its length, and $w(x) = x + 6$ describes its width.
  a. Find the function, $h(x)$, that describes the height of the box.
  b. Calculate the height of the box if $x = 9$ inches.

Chapter 8 Practice Test

For Extra Help CHAPTER Test Prep VIDEOS

Step-by-step test solutions are found on the Chapter Test Prep Videos available via the Video Resources on DVD, in **MyMathLab**, and on You Tube (search "CarsonCombinedAlg" and click on "Channels").

1. Find the intersection and union of the given sets. $A = \{h, o, m, e\}$
$$B = \{h, o, u, s, e\}$$

For 2–5, solve.

2. $|x + 3| = 5$

3. $3 - |2x - 3| = -6$

4. $|2x + 3| = |x - 5|$

5. $|5x - 4| = -3$

For 6–15, solve. Then: *a.*   *Graph the solution set.*
                     b.   *Write the solution set using set-builder notation.*
                     c.   *Write the solution set using interval notation.*

6. $-3 < x + 4 \leq 7$

7. $4 < -2x \leq 6$

8. $5x + 2 \leq -3$ or $5x + 2 \geq 12$

9. $6 - 2n < 2$ or $6 - 2n < 4$

10. $|x + 4| < 9$

11. $2|x - 1| > 4$

12. $3 - 2|x + 4| > -3$

13. $|3y - 2| < -2$

14. $|8t + 4| \geq -12$

15. $2|3x - 4| \leq 10$

For Exercise 16, solve. Then: a. Graph the solution set.
b. Write the solutions set using set-builder notation.
c. Write the solution set using interval notation.

16. A mail-order book club offers one bonus book if the total of an order is from \$40 to \$50. If a person were to order only books that cost \$5, what range would the number of books ordered have to be in order to receive a bonus book?

17. Identify the domain and range of the relation shown in the graph to the right, then determine whether it is a function.

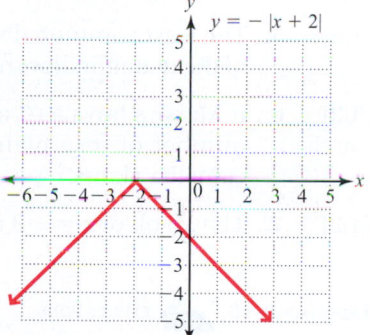

18. Given $f(x) = x^2 - 5x + 4$, find $f(-2)$.

For Exercises 19–21, graph.

19. $f(x) = -\dfrac{3}{4}x + 1$

20. $f(x) = x^2 - 3$

21. $f(x) = |x + 1| - 2$

For Exercises 22 and 23, find a. f + g
b. f − g
c. f · g

22. $f(x) = 6x - 9, g(x) = x + 2$

23. $f(x) = x^2, g(x) = 3x^2 - 2$

24. Given $f(x) = 20x^4 + 36x^3 - 28x^2$ and $g(x) = 4x^2$, find f/g.

25. The function $A(x) = 36x^2 - x - 2$ describes the area of the front face of a box and $l(x) = 4x - 1$ describes its length, where x represents its width.
   a. Find the function, $h(x)$, that describes the height of the box.
   b. Calculate the height of the box if $x = 15$ cm.

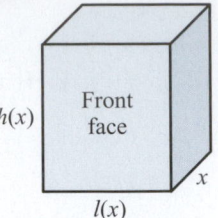

Chapters 1–8 Cumulative Review Exercises

For Exercises 1–4, answer true or false.

[5.2] **1.** All terms are monomials.

[4.6] **2.** To graph $2x - 3y > -6$, you would first graph $2x - 3y = -6$ as a dashed line.

[7.6] **3.** An extraneous solution of $\dfrac{-2}{3} = \dfrac{5a + 10}{a^2 - 4}$ is -2.

[8.2] **4.** The solution set of $|4x - 3| = -8$ is $\varnothing$.

For Exercises 5–7, fill in the blank.

[1.2] **5.** The least common denominator (LCD) is the _____ of the denominators of a given set of fractions.

[3.3] **6.** If Marsha buys a total of 24 rose and camellia plants and x represents the number of rose plants, then _____ represents the number of camellia plants.

[4.2] **7.** The graph of $x = -3$ is a _____.

Exercises 8–16   EXPRESSIONS

For Exercises 8–14, simplify.

[1.5] **8.** $10 - 2(6 - 3^2) \div 2 \cdot 3$

[5.5] **9.** $(3x - y)^2$

[5.6] **10.** $(4xyz^{-3})^2(-2x^3y^{-4})^3$

[5.6] **11.** $\dfrac{x^3 - 5x^2 + 10x}{5x^2}$

[7.2] **12.** $\dfrac{a^2 - b^2}{x^2 - y^2} \div \dfrac{a + b}{x - y}$

[7.5] **13.** $\dfrac{\dfrac{10x}{2x^2 + 7x - 4}}{\dfrac{5}{2x - 1}}$

[7.4] **14.** $\dfrac{x^2 + 4x}{x^2 + 8x + 16} + \dfrac{3}{x + 4}$

For Exercises 15 and 16, factor completely.

[6.1] **15.** $12bc - 9b + 20c - 15$

[6.5] **16.** $2x^3 - 14x^2 + 20x$

For Exercises 17–22, solve.

[2.3] **17.** $2x - 3(2x - 4) = 4 + 4x$

[3.1] **18.** $\dfrac{4}{3} = \dfrac{2x + 2}{2x - 1}$

[6.6] **19.** $3x^2 - 7x - 20 = 0$

[7.6] **20.** $\dfrac{x}{x - 2} - \dfrac{4}{x - 1} = \dfrac{2}{x^2 - 3x + 2}$

[8.1] **21.** $0 \le -1.5x - 3 \le 4.5$

[8.3] **22.** $|5x - 2| \ge 8$

For Exercises 23 and 24, graph.

[4.7] **23.** $f(x) = -\dfrac{2}{3}x + 4$

[4.6] **24.** $x - 2y \le 4$

[3.2] **25.** What percent of 90 is 32.4?

[4.7] **26.** Given $f(x) = \dfrac{x + 4}{x - 2}$, find

    a. $f(0)$ **b.** $f(5)$ **c.** $f(2)$

For Exercises 27–30, solve.

[3.5] **27.** How much of a 40% saline solution must be added to 50 ml of a 15% saline solution so that a 30% saline solution is created?

[3.4] **28.** A ship leaves port traveling at 15 mph. Two hours later a speedboat leaves the same port traveling in the same direction at 40 mph. How long will it take for the speedboat to overtake the ship?

[7.7] **29.** If y varies jointly as x and the square of z, and $y = 48$ when $x = 4$ and $z = 2$, find y when $x = 3$ and $z = 1$.

[6.6] **30.** A rock is thrown vertically upward from ground level with an initial velocity of 96 feet per second. The equation giving its height above the ground is $h = 96t - 16t^2$ where h is the height in feet and t is the number of seconds after it was thrown. Find the number of seconds when the rock is 80 feet above the ground.

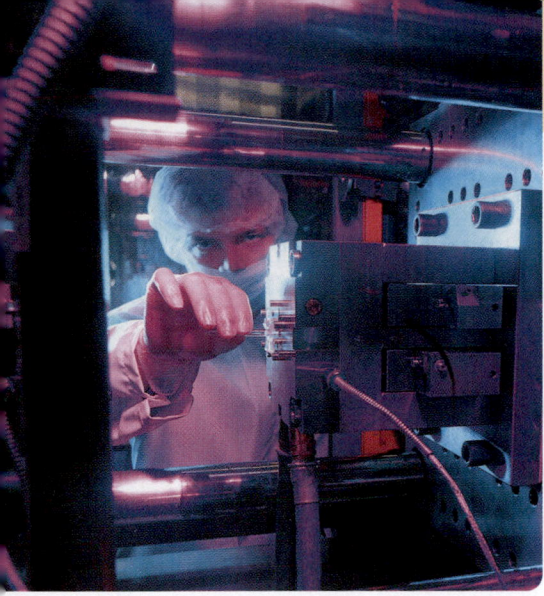

Systems of Linear Equations and Inequalities

9

> "Everybody's a mad scientist, and life is their lab. We're all trying to experiment to find a way to live, to solve problems, to fend off madness and chaos."
>
> —David Cronenberg (1943–)
> Canadian filmmaker

> "The problem is that we attempt to solve the simplest questions cleverly, thereby rendering them unusually complex. One should seek the simple solution."
>
> —Anton Chekhov (1717–1783)
> Russian author

In Chapter 3, we encountered problems involving two unknown amounts, such as finding two consecutive integers whose sum is 63. At that time, we only knew how to solve equations with a *single* variable, so we had to let one of the unknown amounts be represented by a variable, then describe the other unknown amount in terms of that variable. For example, in the consecutive integer problem just mentioned, we would let x represent the first integer and then describe the consecutive integer as $x + 1$.

In this chapter, we again solve problems involving two or more unknowns; however, we now assign a different variable to each unknown and translate the problem to several equations, called a *system of equations*. As we shall see, it is easier to translate a problem involving two or more unknowns to a system of equations than trying to translate the problem to a single equation in terms of a single variable. The challenge with systems of equations is in solving them. We will learn several techniques for solving a system of equations: graphing (Section 9.1), substitution (Section 9.2), elimination (Section 9.3), matrices (Section 9.5), and Cramer's Rule (Section 9.6). Finally, in Section 9.7, we solve systems of linear inequalities.

9.1 Solving Systems of Linear Equations Graphically

Objectives

1. Determine whether an ordered pair is a solution for a system of equations.
2. Solve a system of linear equations graphically.
3. Classify systems of linear equations in two unknowns.

Objective 1 **Determine whether an ordered pair is a solution for a system of equations.** Problems involving multiple unknowns can be represented by a group of equations called a **system of equations**.

Definition   *System of equations:* A group of two or more equations.

For example, look at the following problem: The sum of two numbers is 3. Twice the first number plus three times the second number is 8. What are the two numbers? If x represents the first number and y represents the second number, we can translate each sentence to an equation.

Sentence in the problem:	**Translation:**	
The sum of two numbers is 3.	$x + y = 3$	(Equation 1)
Twice the first number plus three times the second number is 8.	$2x + 3y = 8$	(Equation 2)

The two equations together form a system of equations that describes the problem.

$$\text{System of equations} \quad \begin{cases} x + y = 3 & \text{(Equation 1)} \\ 2x + 3y = 8 & \text{(Equation 2)} \end{cases}$$

A **solution for a system of equations** is an ordered set of numbers that satisfies *all* equations in the system.

Definition   *Solution for a system of equations:* An ordered set of numbers that makes all equations in the system true.

For example, in the preceding system of equations, $x = 1$ and $y = 2$ make both equations true; so they form a solution to the system of equations. We can check by substituting the values in place of the corresponding variables.

Equation 1:	**Equation 2:**
$x + y = 3$	$2x + 3y = 8$
$1 + 2 \overset{?}{=} 3$	$2(1) + 3(2) \overset{?}{=} 8$
$3 = 3$ True	$2 + 6 = 8$ True

We can write a solution to a system of equations in two variables as an ordered pair, in this case, (1, 2). This suggests the following procedure for checking solutions.

Procedure   **Checking a Solution to a System of Equations**

To verify or check a solution to a system of equations,

1. Replace each variable in each equation with its corresponding value.
2. Verify that each equation is true.

Example 1   Determine whether each ordered pair is a solution to the system of equations.

$$\begin{cases} x + y = 5 & \text{(Equation 1)} \\ y = 2x - 4 & \text{(Equation 2)} \end{cases}$$

a. $(-1, 6)$

SOLUTION

$x + y = 5$ (Equation 1) $y = 2x - 4$ (Equation 2)

$-1 + 6 \overset{?}{=} 5$ $6 \overset{?}{=} 2(-1) - 4$ **In both equations,**

$\quad\quad 5 = 5$ True $6 \overset{?}{=} -2 - 4$ **replace x with -1**
                                                                                    and y with 6.

$\quad\quad\quad\quad\quad\quad\quad\quad\quad\quad 6 = -6$ False

Because $(-1, 6)$ does not satisfy both equations, it is not a solution for the system.

b. $(3, 2)$

SOLUTION

$x + y = 5$ (Equation 1) $y = 2x - 4$ (Equation 2)

$3 + 2 \overset{?}{=} 5$ $2 \overset{?}{=} 2(3) - 4$ **In both equations,**

$\quad\quad 5 = 5$ True $2 \overset{?}{=} 6 - 4$ **replace x with 3 and y**
                                                                                    with 2.

$\quad\quad\quad\quad\quad\quad\quad\quad\quad\quad 2 = 2$ True

Because $(3, 2)$ satisfies both equations, it is a solution for the system.

Your Turn 1   Determine whether each ordered pair is a solution to the system of equations.

$$\begin{cases} 2x + y = 5 & \text{(Equation 1)} \\ y = 1 - x & \text{(Equation 2)} \end{cases}$$

a. $(3, -1)$ **b.** $(4, -3)$

Objective ② Solve a system of linear equations graphically.   A system of two linear equations in two variables can have one solution, no solution, or an infinite number of solutions. To see why, let's look at the graphs of the equations in three different systems. First, let's graph the equations in the system from Example 1.

$$\begin{cases} x + y = 5 \\ y = 2x - 4 \end{cases}$$

Note: In Chapter 4, we learned that the graph of an equation represents every ordered pair solution for the equation; so when two graphs intersect, the point of intersection is a solution for both equations.

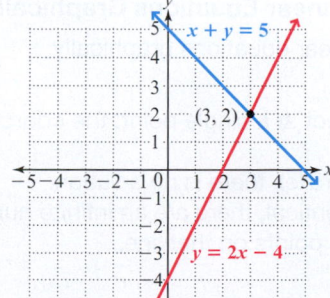

Notice that the graphs of $x + y = 5$ and $y = 2x - 4$ intersect at the point $(3, 2)$, which is the solution for the system. If two linear graphs intersect at a single point, the system has a *single solution* at the point of intersection.

Note: These lines also have different slopes, which is always the case when a system of two linear equations has a single solution.

Now let's look at a system with no solution. Look at the graphs of the equations in the system $\begin{cases} y = 3x + 1 \\ y = 3x - 2 \end{cases}$.

Answers   Your Turn 1

a. $(3, -1)$ is not a solution.

b. $(4, -3)$ is a solution.

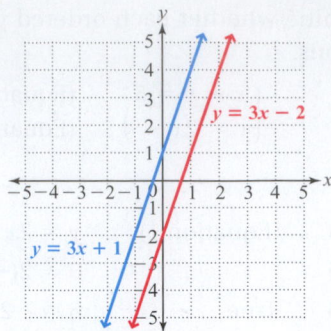

Notice that the equations $y = 3x + 1$ and $y = 3x - 2$ have the same slope, 3, and different y-intercepts; so their graphs are parallel lines. Because the graphs of the equations in this system have no point of intersection, the system has no solution.

Finally, let's examine a system with an infinite number of solutions, such as

$$\begin{cases} x + 2y = 4 & \text{(Equation 1)} \\ 2x + 4y = 8 & \text{(Equation 2)} \end{cases}$$

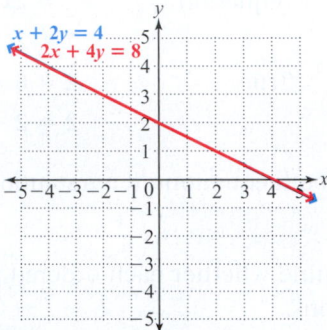

Notice that the graphs of the two equations are identical. Consequently, an infinite number of solutions to this system of equations lie on that line.

Note: Having an infinite number of solutions is not the same as having all real numbers as solutions, as in the solutions of identities.

By writing the equations in slope–intercept form, we can see without graphing that the two equations describe the same line.

$x + 2y = 4$ (Equation 1)

$2y = -x + 4$ **Subtract x from both sides.**

$y = -\dfrac{1}{2}x + \dfrac{4}{2}$ **Divide both sides by 2 to isolate y.**

$y = -\dfrac{1}{2}x + 2$ **Simplify the fraction.**

$2x + 4y = 8$ (Equation 2)

$4y = -2x + 8$ **Subtract $2x$ from both sides.**

$y = -\dfrac{2}{4}x + \dfrac{8}{4}$ **Divide both sides by 4 to isolate y.**

$y = -\dfrac{1}{2}x + 2$ **Simplify the fraction.**

Our work suggests the following graphical method for solving systems of equations.

Procedure   **Solving Systems of Linear Equations Graphically**

To solve a system of linear equations graphically,

1. Graph each equation.
   a. If the lines intersect at a single point, the coordinates of that point form the solution.
   b. If the lines are parallel, there is no solution.
   c. If the lines are identical, there are an infinite number of solutions, which are the coordinates of all points on that line.
2. Check your solution.

Example 2   Solve the system of equations graphically.

a. $\begin{cases} y = -3x + 1 & \text{(Equation 1)} \\ 2x - y = 4 & \text{(Equation 2)} \end{cases}$

SOLUTION Graph each equation.

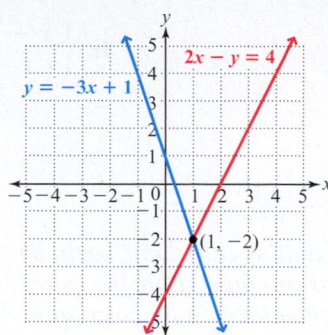

The lines intersect at a single point, which appears to be $(1, -2)$. We can verify that it is the solution by substituting into the equations.

$$
\begin{array}{ll}
y = -3x + 1 & 2x - y = 4 \\
-2 \stackrel{?}{=} -3(1) + 1 & 2(1) - (-2) \stackrel{?}{=} 4 \\
-2 \stackrel{?}{=} -3 + 1 & 2 + 2 \stackrel{?}{=} 4 \\
-2 = -2 \quad \text{True} & 4 = 4 \quad \text{True}
\end{array}
$$

Warning: Graphing by hand can be imprecise, and not all solutions have integer coordinates. So you should always check your solutions by substituting them into the original equations.

Answer   Because $(1, -2)$ makes both equations true, it is the solution.

b. $\begin{cases} 2x - 3y = 6 & \text{(Equation 1)} \\ y = \dfrac{2}{3}x + 1 & \text{(Equation 2)} \end{cases}$

SOLUTION Graph each equation.

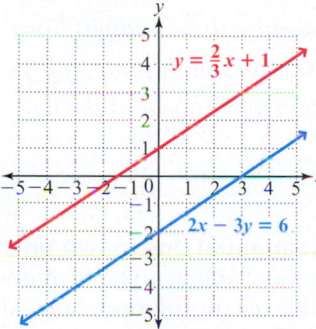

The lines appear to be parallel, which we can verify by comparing the slopes and y-intercepts. The slope of equation 2 is $\dfrac{2}{3}$, and the y-intercept is $(0, 1)$. To determine the slope and y-intercept of equation 1, we rewrite it in slope–intercept form.

$$
\begin{array}{ll}
2x - 3y = 6 & \\
-3y = -2x + 6 & \text{\textcolor{blue}{Subtract $2x$ from both sides.}} \\
y = \dfrac{2}{3}x - 2 & \text{\textcolor{blue}{Divide both sides by -3 to isolate y.}}
\end{array}
$$

The slope of equation 1 also is $\dfrac{2}{3}$, and the y-intercept is $(0, -2)$. Because the slopes are the same and the y-intercepts are different, the lines are indeed parallel.

Answer   The system has no solution.

c. $\begin{cases} 6x - 2y = 2 & \text{(Equation 1)} \\ y = 3x - 1 & \text{(Equation 2)} \end{cases}$

SOLUTION Graph each equation.

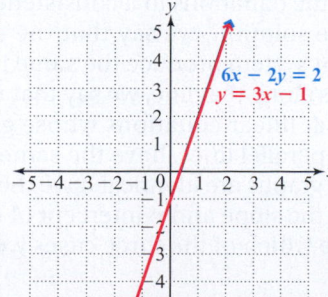

The lines appear to be identical, which we can verify by rewriting $6x - 2y = 2$ in slope–intercept form.

$$
\begin{array}{ll}
6x - 2y = 2 & \\
-2y = -6x + 2 & \text{\textcolor{blue}{Subtract $6x$ from both sides.}} \\
y = 3x - 1 & \text{\textcolor{blue}{Divide both sides by -2 to isolate y.}}
\end{array}
$$

The equations are identical.

Answer   All ordered pairs along the line $6x - 2y = 2$ (or $y = 3x - 1$).

Your Turn 2 Solve the system of equations graphically.

a. $\begin{cases} y = \dfrac{3}{4}x + 1 \\ 2x - y = -6 \end{cases}$

b. $\begin{cases} x + y = -3 \\ y = -x + 2 \end{cases}$

Calculator Tips

To solve a system of equations using a graphing calculator, begin by graphing each equation. Remember that you must write the equations in slope–intercept form in order to enter them in the calculator. Using the [Y=] *key, enter one equation in Y1 and the second equation in Y2; then press* [GRAPH] *to see both graphs.*

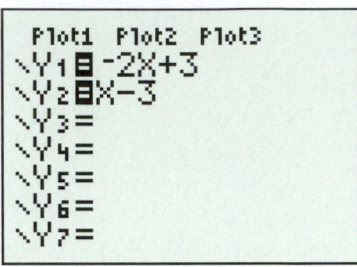

You can find the coordinates of the point of intersection by using the CALC feature, which you find by pressing [2nd] [TRACE]. *Select INTERSECT from the menu. The calculator will then prompt you to verify which lines you want it to find the intersection of. So press* [ENTER] *for each line. The calculator will then prompt you to indicate which point of intersection you want. Because these straight lines will have only one point of intersection, you can simply press* [ENTER]. *The coordinates of the point of intersection will appear at the bottom of the screen.*

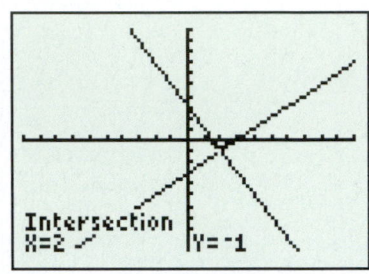

Objective ③ Classify systems of linear equations in two unknowns. As we have seen, a system of two equations with two variables can have one solution, an infinite number of solutions along a common line, or no solution. Special terms are used to indicate each of these three classifications. The first terms describe whether or not a system has a solution.

Definitions ***Consistent system of equations:*** A system of equations that has at least one solution.

Inconsistent system of equations: A system of equations that has no solution.

Answers   Your Turn 2

a. The solution is $(-4, -2)$.

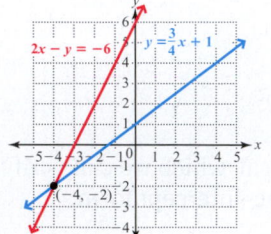

b. no solution

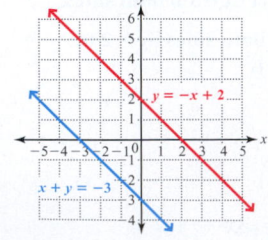

We have seen two types of consistent systems of equations. One type has a single solution because the two equations produce different lines that intersect at a single point. The other type has an infinite number of solutions because the two equations are identical and therefore produce the same line. To distinguish the two consistent cases, we use additional terms. If the equations in a consistent system produce different lines so that the system has one solution, we say that the equations are *independent*. If the equations in a consistent system produce the same line so that this system has an infinite number of solutions along that line, we say that the equations are *dependent*.

We have learned that linear equations whose graphs are different have different slopes or, in the case of parallel lines, have the same slopes and different y-intercepts. Linear equations whose graphs are identical have the same slope and same y-intercept. Therefore, by observing the slope and y-intercept of the equations in a system of equations, we can determine which of the three cases we are dealing with, which suggests the following procedure.

Procedure Classifying Systems of Equations

To classify a system of two linear equations in two unknowns, write the equations in slope–intercept form and compare the slopes and y-intercepts.

Consistent system with independent equations:	Consistent system with dependent equations:	Inconsistent system:
The system has a single solution at the point of intersection. The graphs are different. They have different slopes.	The system has an infinite number of solutions. The graphs are identical. They have the same slope and same y-intercept.	The system has no solution. The graphs are parallel lines. They have the same slope, but different y-intercepts.
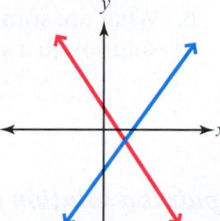	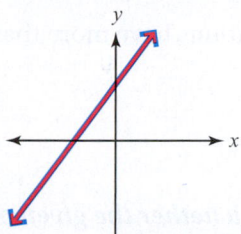	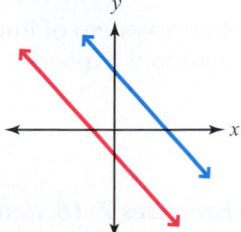

Example 3   For each of the systems of equations in Example 2, determine whether the system is consistent with independent equations, consistent with dependent equations, or inconsistent. How many solutions does the system have?

SOLUTION   When we solved the system $\begin{cases} y = -3x + 1 \\ 2x - y = 4 \end{cases}$ in Example 2(a), we found that the graphs intersected at a single point. Therefore, the system is consistent (it has a solution) with independent equations (different graphs) and has one solution: $(1, -2)$.

When we solved the system $\begin{cases} 2x - 3y = 6 \\ y = \dfrac{2}{3}x + 1 \end{cases}$ in Example 2(b), we found the graphs to be parallel lines. Therefore, the system is inconsistent and has no solution:

When we solved the system $\begin{cases} 6x - 2y = 2 \\ y = 3x - 1 \end{cases}$ in Example 2(c), we found the graphs to coincide. Therefore, the system is consistent (it has a solution) with dependent equations (same graph) and has an infinite number of solutions.

Your Turn 3

a. Determine whether the following system is consistent with independent equations, consistent with dependent equations, or inconsistent.

b. How many solutions does the system have?

$$\begin{cases} 5x + y = 1 \\ y = -5x - 2 \end{cases}$$

1. How do you check a solution for a system of equations?

2. If a system of linear equations has at least one solution (consistent), what does this indicate about the graphs of the equations?

3. If a system of linear equations has no solution (inconsistent), what does this indicate about the graphs of the equations?

4. If two equations are dependent, what does this mean about their graphs?

5. Can a system of linear equations have more than one solution? Explain.

6. What are some weaknesses in using graphing to find a solution to a system of equations?

For Exercises 7–16, determine whether the given ordered pair is a solution to the given system of equations. See Example 1.

7. $(1, 1)$; $\begin{cases} x = y \\ 9y - 13 = -4x \end{cases}$

8. $(3, -2)$; $\begin{cases} 2x = -3y \\ x - 2y = 7 \end{cases}$

9. $(2, -3)$; $\begin{cases} 4x + 3y = -1 \\ 2x - 5y = -11 \end{cases}$

10. $(4, 3)$; $\begin{cases} 3x + 4y = 24 \\ 3x + 2y = 18 \end{cases}$

11. $\left(\dfrac{2}{3}, \dfrac{4}{3}\right)$; $\begin{cases} x + y = 2 \\ 2x - y = 0 \end{cases}$

12. $\left(\dfrac{3}{2}, -\dfrac{3}{2}\right)$; $\begin{cases} x - y = 3 \\ x + y = 0 \end{cases}$

13. $(2, -4)$; $\begin{cases} 0.5x + 1.25y = 4 \\ x + y = -2 \end{cases}$

14. $(1, 5)$; $\begin{cases} 0.25x + 0.75y = 4 \\ x + y = -7 \end{cases}$

15. $(5, -2)$; $\begin{cases} x + 0.5y = 4 \\ \dfrac{1}{5}x - \dfrac{1}{2}y = 2 \end{cases}$

16. $(-1, -2)$; $\begin{cases} \dfrac{2}{3}x - y = 7 \\ 0.25x + y = -1 \end{cases}$

For Exercises 17–42, solve the system of linear equations graphically. See Example 2.

17. $\begin{cases} x + y = 5 \\ 2x - y = 7 \end{cases}$

18. $\begin{cases} x - y = 4 \\ x + y = 8 \end{cases}$

19. $\begin{cases} x + y = 1 \\ y = 7 + x \end{cases}$

20. $\begin{cases} x + y = 4 \\ y - x = 6 \end{cases}$

21. $\begin{cases} y = x - 5 \\ 2x - y = 8 \end{cases}$

22. $\begin{cases} y = x - 1 \\ 3x + y = 11 \end{cases}$

23. $\begin{cases} x = y - 4 \\ x + y = 2 \end{cases}$

24. $\begin{cases} x = y + 4 \\ x + y = 2 \end{cases}$

25. $\begin{cases} 4x + y = 8 \\ 2x - 3y = 18 \end{cases}$

26. $\begin{cases} 2x + 3y = 6 \\ x + 4 = 2y \end{cases}$

27. $\begin{cases} 2x + y = 5 \\ x + 2y = -2 \end{cases}$

28. $\begin{cases} 2x + y = 1 \\ 2x + 3y = 3 \end{cases}$

29. $\begin{cases} 2x + 2y = -2 \\ 3x - 2y = 12 \end{cases}$

30. $\begin{cases} 2x - 4y = 2 \\ 6x - 2y = -4 \end{cases}$

31. $\begin{cases} 3y = -2x + 6 \\ 4x + 6y = 18 \end{cases}$

32. $\begin{cases} 2x + y = 0 \\ 2y = 3 - 4x \end{cases}$

33. $\begin{cases} 2x + 3y = 2 \\ 6x + 9y = 6 \end{cases}$

34. $\begin{cases} x + y = 4 \\ -2x - 2y = -8 \end{cases}$

35. $\begin{cases} 3x - 2y = -6 \\ x = 2 \end{cases}$

36. $\begin{cases} 5x + 2y = -10 \\ x = -4 \end{cases}$

37. $\begin{cases} x - 2y = -6 \\ y = 2 \end{cases}$

38. $\begin{cases} 3x - y = 6 \\ y = -3 \end{cases}$

39. $\begin{cases} y = 3 \\ x = -2 \end{cases}$

40. $\begin{cases} x = 1 \\ y = -4 \end{cases}$

41. $\begin{cases} y = -\dfrac{2}{5}x \\ x - y = 7 \end{cases}$

42. $\begin{cases} y = \dfrac{1}{4}x \\ x - y = 3 \end{cases}$

For Exercises 43–48, (a) determine whether the graph shows a consistent system with independent equations, a consistent system with dependent equations, or an inconsistent system and (b) determine how many solutions the system has. See Example 3.

43.

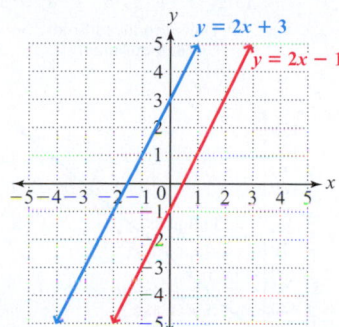

44.

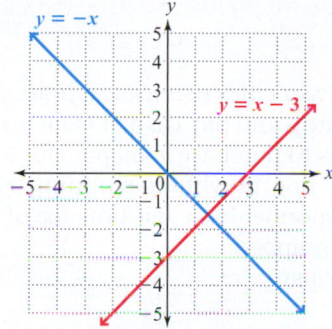

45.

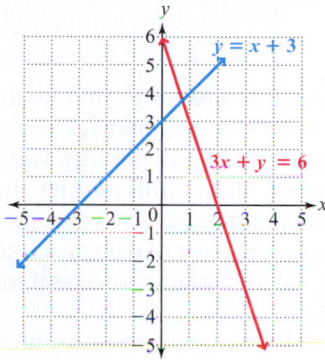

46.

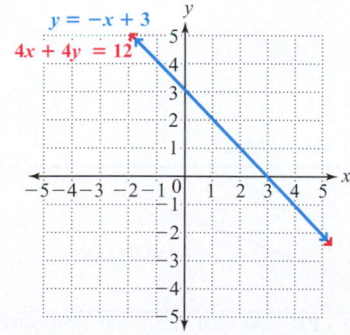

47.

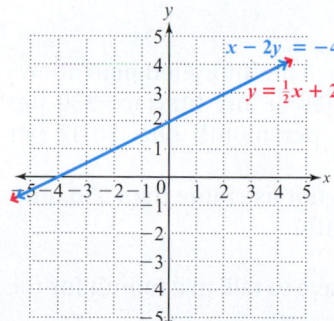

48.

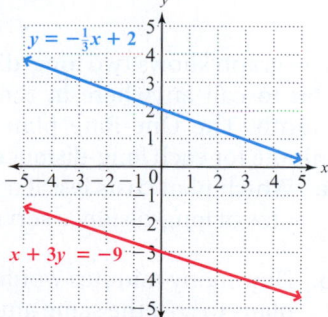

For Exercises 49–56, (a) determine whether the system of equations is consistent with independent equations, consistent with dependent equations, or inconsistent and (b) determine how many solutions the system has. See Example 3.

49. $\begin{cases} x - 4y = 9 \\ 2x - 3y = 8 \end{cases}$

50. $\begin{cases} -4y = 2x - 12 \\ x - 2y = -4 \end{cases}$

51. $\begin{cases} 3x + 2y = 5 \\ -6x - 4y = 1 \end{cases}$

52. $\begin{cases} 3x + 4y = 15 \\ 9x + 12y = 8 \end{cases}$

53. $\begin{cases} x - y = 1 \\ 2x - 2y = 2 \end{cases}$

54. $\begin{cases} 2x - y = 1 \\ -4x + 2y = -2 \end{cases}$

55. $\begin{cases} 2x - y = 1 \\ x + y = 4 \end{cases}$

56. $\begin{cases} x - y = 5 \\ x + 5y = 11 \end{cases}$

57. A business breaks even when its costs and revenue are equal. To the right is a graph of the cost of a product based on the number of units produced and the revenue based on the number of units sold.
   a. How many units must be produced and sold if the business is to break even?

   b. What amount of revenue is needed for the business to break even?

   c. Write an inequality that describes the number of units that must be sold for the business to make a profit.

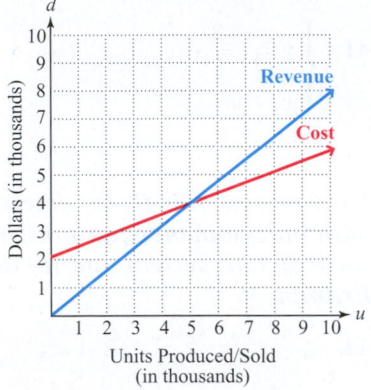

Units Produced/Sold
(in thousands)

58. John's job requires that he drive between two cities using a toll road. He has two choices for toll plans. The first plan is to stop and pay cash at each booth, which amounts to $15 per trip. The second plan is to purchase a toll pass for $12 per trip, which allows him to drive through the toll booths without stopping. However, to use the pass he also must purchase a transponder for a one-time fee of $24. To the right, the cost of using each plan has been graphed.
   a. Which plan costs less if John makes 6 trips?

   b. How many trips must he make for the costs to be equal?

   c. What other factors should he consider when choosing a plan?

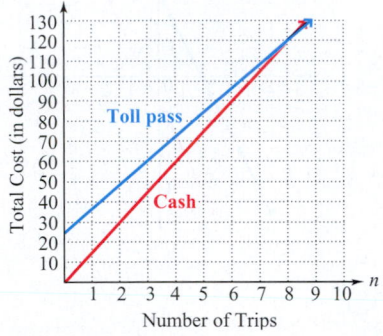

Number of Trips

59. The graph shows two long-distance plans. The Unlimited plan allows you to call anywhere in the United States at any time for $30 per month. The One-Rate plan costs $3.75 per month plus $0.07 per minute for each long-distance call.
   a. Which plan costs less if a person does not use more than 350 minutes of long-distance service each month?

   b. How many minutes would a person have to talk in a month for the plans to cost the same amount?

   c. How many minutes would a person need to talk per month for the Unlimited plan to cost less?

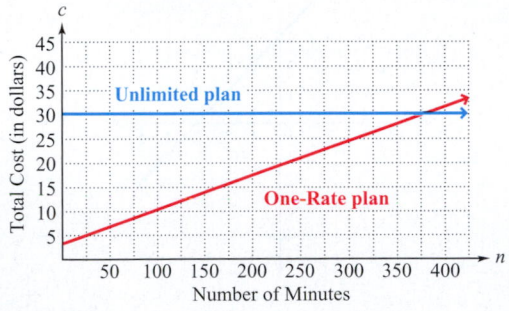

Number of Minutes

60. Deciding whether to purchase or rent a home is a difficult decision based on many factors. The graph shows the monthly mortgage payment ($850 per month) compared to the monthly rent payment (initially $600 per month and raised $50 every six months) for a 1200-square-foot house.

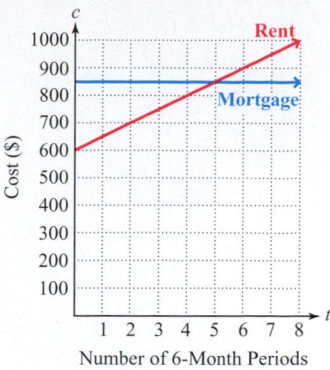

a. If a customer plans to stay only two years, ignoring any tax benefits, which costs less per month, renting or buying?

b. After how many months will the monthly rent be the same as the monthly mortgage payment?

c. In the fourth year, which costs less per month?

d. What other factors should a person consider when deciding whether to purchase or rent a home?

Review Exercises

Exercise 1 **EXPRESSIONS**

[1.7]   **1.** Simplify: $2y - 5(y + 7)$

Exercises 2–6 **EQUATIONS AND INEQUALITIES**

[2.4]   **2.** Solve for y: $x + 3y = 10$

[2.4]   **3.** Solve for x: $\frac{1}{3}x - 2y = 10$

[2.1]   **4.** A cone made of marble is used as a decorative accent piece. It is 10 inches tall, and the radius of its base is 2 inches. Find its volume. (Use $V = \frac{1}{3}\pi r^2 h$.)

[3.3]   **5.** One number exceeds another number by 62. If five times the smaller number is subtracted from two times the larger number, the difference is 91. What are the numbers?

[4.7]   **6.** Given $f(x) = 2x + 9$, find $f(x + 4)$.

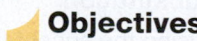

9.2 Solving Systems of Linear Equations by Substitution; Applications

Objectives

1 Solve systems of linear equations using substitution.

2 Solve applications involving two unknowns using a system of equations.

In Section 5.1, we solved systems of equations by graphing. However, if the solution to a system of equations contains fractions or decimal numbers, it could be difficult to determine those values using the graphing method. For example, look at the following system of equations:

$$\begin{cases} x + 3y = 10 & \text{(Equation 1)} \\ y = x + 4 & \text{(Equation 2)} \end{cases}$$

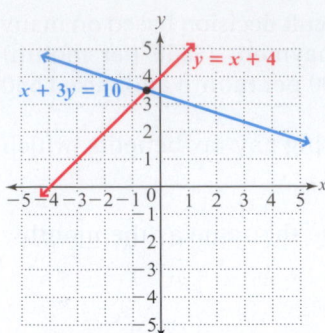

In the graph, notice that the point of intersection is between grid lines, which means the solution contains fractions. We could make a guess at the solution and check our guess in the equations, but it would be better to use a method that does not require guessing. One such method is the *substitution* method.

Objective ① **Solve systems of linear equations using substitution.** Remember that in a system's solution, the same x and y values satisfy both equations. In the preceding system, notice that Equation 2 indicates that y is equal to $x + 4$; so y must be equal to $x + 4$ in Equation 1 as well. Therefore, we can substitute $x + 4$ for y in Equation 1.

$$x + 3y = 10$$

$$x + 3\overbrace{(x + 4)} = 10$$

Now we have an equation in terms of a single variable, x, which allows us to solve for x.

$$
\begin{aligned}
x + 3(x + 4) &= 10 & \\
x + 3x + 12 &= 10 & \text{\textcolor{blue}{Distribute 3.}} \\
4x + 12 &= 10 & \text{\textcolor{blue}{Combine like terms.}} \\
4x &= -2 & \text{\textcolor{blue}{Subtract 12 from both sides.}} \\
x &= -\frac{1}{2} & \text{\textcolor{blue}{Divide both sides by 4 and simplify.}}
\end{aligned}
$$

We can now find the y-value by substituting $-\dfrac{1}{2}$ for x in either of the original equations. Equation 2 is easier because y is isolated.

$$
\begin{aligned}
y &= x + 4 & \\
y &= -\frac{1}{2} + 4 & \text{\textcolor{blue}{Substitute $-\frac{1}{2}$ for x.}} \\
y &= 3\frac{1}{2} &
\end{aligned}
$$

The solution to the system of equations is $\left(-\dfrac{1}{2}, 3\dfrac{1}{2}\right)$. Notice in our graph that the lines do appear to intersect at $\left(-\dfrac{1}{2}, 3\dfrac{1}{2}\right)$, which supports our solution graphically.

Procedure

Solving Systems of Two Linear Equations Using Substitution

To find the solution of a system of two linear equations using the substitution method,

1. Isolate one of the variables in one of the equations.
2. In the other equation, substitute the expression you found in Step 1 for that variable.
3. Solve this new equation. (It will now have only one variable.)
4. Using one of the equations containing both variables, substitute the value you found in Step 3 for that variable and solve for the value of the other variable.
5. Check the solution in the original equations.

Example 1   Solve the system of equations using substitution.

$$\begin{cases} x = 3 - y \\ 5x + 3y = 5 \end{cases}$$

SOLUTION

Step 1: We isolate a variable in one of the equations. Because x is already isolated in the first equation, we proceed to Step 2.

Step 2: Substitute the expression $3 - y$ for x in the second equation.

$$5x + 3y = 5$$
$$5(3 - y) + 3y = 5 \qquad \text{Substitute } 3 - y \text{ for } x.$$

Step 3: We now have a linear equation in terms of a single variable, y; so we solve for y.

$$15 - 5y + 3y = 5 \qquad \textbf{Distribute 5.}$$
$$15 - 2y = 5 \qquad \textbf{Combine like terms.}$$
$$-2y = -10 \qquad \textbf{Subtract 15 from both sides.}$$
$$y = 5 \qquad \textbf{Divide both sides by } -2 \text{ to isolate } y.$$

Step 4: Now we solve for x by substituting 5 for y in one of the equations containing both variables. We will use $x = 3 - y$.

> **Note:** We chose $x = 3 - y$ because x is isolated.

$$x = 3 - y$$
$$x = 3 - 5 \qquad \textbf{Substitute 5 for } y.$$
$$x = -2$$

Connection The graphs of $x = 3 - y$ and $5x + 3y = 5$ are lines that intersect at the point whose coordinates are $(-2, 5)$.

The solution is $(-2, 5)$. We will let the reader check that $(-2, 5)$ makes both equations true.

Your Turn 1   Solve the system of equations using substitution.

$$\begin{cases} x - 3y = 11 \\ y = 2x + 3 \end{cases}$$

In Step 1, if no variable is isolated in any of the given equations, we select an equation and isolate one of its variables. It is easiest to isolate a variable that has a coefficient of 1, as this will avoid fractions.

Example 2   Solve the system of equations using substitution.

$$\begin{cases} x + y = 6 \\ 5x + 2y = 8 \end{cases}$$

SOLUTION

Step 1: Isolate a variable in one of the equations. Because both x and y have a coefficient of 1 in $x + y = 6$, isolating either variable is easy. We will isolate y.

$$x + y = 6$$
$$y = 6 - x \qquad \textbf{Subtract } x \text{ from both sides to isolate } y.$$

Answer   Your Turn 1
$(-4, -5)$

Step 2: Substitute $6 - x$ for y in the second equation.

$$5x + 2y = 8$$

$$5x + 2(6 - x) = 8 \qquad \text{Substitute } 6 - x \text{ for } y.$$

Step 3: Solve for x using the equation we found in Step 2.

$$5x + 12 - 2x = 8 \qquad \text{Distribute 2.}$$

$$3x + 12 = 8 \qquad \text{Combine like terms.}$$

$$3x = -4 \qquad \text{Subtract 12 from both sides.}$$

$$x = -\frac{4}{3} \qquad \text{Divide both sides by 3 to isolate } x.$$

Step 4: Find the value of y by substituting $-\frac{4}{3}$ for x in one of the equations containing both variables. Because we isolated y in $y = 6 - x$, we will use this equation.

$$y = 6 - x$$

$$y = 6 - \left(-\frac{4}{3}\right) \qquad \text{Substitute } -\frac{4}{3} \text{ for } x.$$

$$y = \frac{18}{3} + \frac{4}{3} \qquad \text{Rewrite as addition of equivalent fractions with a common denominator.}$$

$$y = \frac{22}{3}$$

The solution is $\left(-\frac{4}{3}, \frac{22}{3}\right)$.

Step 5: We will leave the check to the reader.

Your Turn 2 Solve the system of equations by substitution.

$$\begin{cases} x - y = 1 \\ 3x + y = 11 \end{cases}$$

If none of the variables in the system of equations has a coefficient of 1 or -1, for Step 1, select the equation that seems easiest to work with and the variable in that equation that seems easiest to isolate. When selecting that variable, recognize that you'll eventually divide out its coefficient; so choose the variable whose coefficient will divide evenly into most, if not all, of the other numbers in its equation. In Section 5.3, we will discuss another method of solving systems that reduces the amount of work with fractions.

Example 3 Solve the system of equations using substitution.

$$\begin{cases} 3x + 4y = 6 \\ 2x + 6y = -1 \end{cases}$$

Note: We chose to isolate x because its coefficient, 3, divides evenly into one of the other numbers in the equation, 6, whereas y's coefficient, 4, does not divide evenly into any of the other numbers.

SOLUTION

Step 1: We will isolate x in $3x + 4y = 6$.

$$3x + 4y = 6$$

$$3x = 6 - 4y \qquad \text{Subtract } 4y \text{ from both sides.}$$

$$x = 2 - \frac{4}{3}y \qquad \text{Divide both sides by 3 to isolate } x.$$

Answer   Your Turn 2

$(3, 2)$

Step 2: Substitute $2 - \frac{4}{3}y$ for x in $2x + 6y = -1$.

$$2x + 6y = -1$$

$$2\left(2 - \frac{4}{3}y\right) + 6y = -1 \qquad \text{Substitute } 2 - \frac{4}{3}y \text{ for } x.$$

Step 3: Solve for y using the equation we found in Step 2.

$$4 - \frac{8}{3}y + 6y = -1 \qquad \text{Distribute to clear the parentheses.}$$

$$3 \cdot 4 - 3 \cdot \frac{8}{3}y + 3 \cdot 6y = 3 \cdot (-1) \qquad \text{Multiply both sides by 3 to clear the fraction.}$$

$$12 - 8y + 18y = -3$$

$$12 + 10y = -3 \qquad \text{Combine like terms.}$$

$$10y = -15 \qquad \text{Subtract 12 from both sides.}$$

$$y = -\frac{15}{10} \qquad \text{Divide both sides by 10 to isolate } y.$$

$$y = -\frac{3}{2} \qquad \text{Simplify.}$$

Step 4: Find the value of x by substituting $-\frac{3}{2}$ for y in one of the equations containing both variables. Because x is isolated in $x = 2 - \frac{4}{3}y$, we will use that equation.

$$x = 2 - \frac{4}{3}y$$

$$x = 2 - \frac{\overset{2}{\cancel{4}}}{\underset{1}{\cancel{3}}} \cdot \left(-\frac{\overset{1}{\cancel{3}}}{\underset{1}{\cancel{2}}}\right) \qquad \text{Substitute } -\frac{3}{2} \text{ for } y \text{ and simplify.}$$

$$x = 2 + 2$$

$$x = 4$$

The solution is $\left(4, -\frac{3}{2}\right)$.

Step 5: We will leave the check to the reader.

Your Turn 3 Solve the system of equations using substitution.

$$\begin{cases} 5x - 2y = 10 \\ 3x - 6y = 2 \end{cases}$$

Inconsistent Systems of Equations

In Section 5.1, we learned that inconsistent systems of linear equations have no solution because the graphs of the equations are parallel lines. Let's see what happens when we solve an inconsistent system of equations using the substitution method.

Example 4   Solve the system of equations using substitution.

$$\begin{cases} x + 2y = 4 \\ y = -\dfrac{1}{2}x + 1 \end{cases}$$

SOLUTION   Because y is isolated in the second equation, we substitute $-\dfrac{1}{2}x + 1$ in place of y in the first equation.

$$x + 2y = 4$$

$$x + 2\left(-\dfrac{1}{2}x + 1\right) = 4 \qquad \text{Substitute } -\dfrac{1}{2}x + 1 \text{ for } y.$$

Now solve for x.

$$x - x + 2 = 4 \qquad \textbf{\textcolor{blue}{Distribute to clear the parentheses.}}$$
$$2 = 4 \qquad \textbf{\textcolor{blue}{Combine like terms.}}$$

Notice that $2 = 4$ no longer has a variable and is false. This false equation with no variable indicates that the system is inconsistent and has no solution.

> **Connection** If we were to solve the system in Example 4 using graphing, we would see that the lines are parallel.

Consistent Systems with Dependent Equations

Recall from Section 5.1 that consistent systems with dependent equations have an infinite number of solutions. Let's see what happens when we use the substitution method to solve a system of dependent equations.

Example 5   Solve the system of equations using substitution.

$$\begin{cases} x = 3y + 4 \\ 2x - 6y = 8 \end{cases}$$

SOLUTION   Because x is isolated in the first equation, we substitute $3y + 4$ in place of x in the second equation.

$$2x - 6y = 8$$
$$2(3y + 4) - 6y = 8 \qquad \textbf{\textcolor{blue}{Substitute } 3y + 4 \textbf{ for } x.}$$

Now solve for y.

$$6y + 8 - 6y = 8 \qquad \textbf{\textcolor{blue}{Distribute to clear the parentheses.}}$$
$$8 = 8 \qquad \textbf{\textcolor{blue}{Combine like terms.}}$$

Notice that $8 = 8$ no longer has a variable and is true. This true equation with no variable indicates that the equations in the system are dependent; so there are an infinite number of solutions that are all of the ordered pairs along $x = 3y + 4$ (or $2x - 6y = 8$).

> **Connection** If we were to solve the system in Example 5 using graphing, we would see that the lines are identical.

Your Turn 5   Solve the systems of equations using substitution.

a. $\begin{cases} y = 2x - 7 \\ 4x - 2y = -3 \end{cases}$
b. $\begin{cases} 3x - y = 4 \\ 2y - 6x = -8 \end{cases}$

Answers   Your Turn 5

a. no solution

b. all ordered pairs along $3x - y = 4$ (or $2y - 6x = -8$)

Objective **2** **Solve applications involving two unknowns using a system of equations.** To solve a problem involving two unknowns, we must be given two relationships about those unknowns. In Chapter 3, we learned to solve these problems by selecting a variable for one of the unknowns and using one relationship to write the other unknown in terms of that variable. We then translated the second relationship to a single equation in terms of the variable.

We now solve the same types of problems using systems of equations. Using a system of equations is easier because we select two variables to represent the two unknowns, then translate each of the two given relationships to an equation in terms of those variables. We can then solve the system of equations by substitution (or graphing).

Procedure **Solving Applications Using a System of Equations**

To solve a problem with two unknowns using a system of equations,

1. Select a variable for each of the unknowns.
2. Translate each relationship to an equation.
3. Solve the system.

We repeat Example 3 of Section 3.3 using a system of equations.

Example 6 A rectangular frame is to be built so that the length is 4 inches more than the width. The perimeter of the frame is to be 72 inches. Find the dimensions of the frame.

Understand We are given two relationships about the frame, and we are to find the length and width.

Plan Select a variable for the length and another variable for the width, translate the relationships to a system of equations, and then solve the system.

Execute Let l represent the length and w represent the width.

Relationship 1: The length is 4 more than the width.

$$\text{Translation:} \quad l = w + 4$$

Relationship 2: The perimeter is 72 inches.

$$\text{Translation:} \quad 2l + 2w = 72$$

$$\text{Our system:} \quad \begin{cases} l = w + 4 \\ 2l + 2w = 72 \end{cases}$$

Because the first equation has l isolated, we substitute $w + 4$ for l in the second equation.

$$2\mathbf{l} + 2w = 72$$
$$2(\mathbf{w + 4}) + 2w = 72 \qquad \text{Substitute } w + 4 \text{ for } l.$$
$$2w + 8 + 2w = 72$$
$$4w + 8 = 72 \qquad \text{Combine like terms.}$$
$$4w = 64 \qquad \text{Subtract 8 from both sides of the equation.}$$
$$w = 16 \qquad \text{Divide both sides by 4.}$$

Now we can find the value of l using $l = w + 4$.

$$l = 16 + 4 \qquad \text{Substitute 16 for } w.$$
$$l = 20$$

Answer The length should be 20 inches, and the width should be 16 inches.

Check Verify that the length is 4 more than the width: $20 = 16 + 4$. Also verify that the perimeter is 72 inches: $2(20) + 2(16) = 40 + 32 = 72$.

Connection Prior to learning about systems of equations, when we solved problems involving two unknowns, we actually used the concepts of substitution. For example, in Section 3.3, we set up this example as follows. We selected a variable, w, for the width and translated the first relationship like this:

$$\text{Length} = w + 4$$

Then we translated the perimeter relationship like this:

$$2 \cdot \text{Length} + 2 \cdot \text{Width} = 72$$
$$2(w + 4) + 2w = 72$$

Notice that we substituted $w + 4$ in place of the word *length* in the perimeter equation.

Your Turn 6 The width of a tennis court (along the doubles lines) is 42 feet less than the length. If the perimeter is 228 feet, what are the dimensions?

As in Section 3.3, we will use a table to organize the information in problems. But we will select a variable for each unknown and write a system of equations using the relationships in the problem. We repeat Example 6 of Section 3.3 using a system of equations.

Example 7 A home improvement store sells two sizes of cans of wood stain. A small can sells for $8.95, and a large can sells for $15.95. One day the store sold twice as many of the large cans as the small cans. If the total revenue that day for these cans of stain was $694.45, how many cans of each size were sold?

Understand The two unknowns are the number of each size of cans of stain sold. There are also two relationships, one involving the number sold and the other involving the total revenue.

Plan and Execute Let x represent the number of small cans of stain sold and let y represent the number of large cans. We can use a table to organize the information.

Category	Price	Number	Revenue
Small can	8.95	x	$8.95x$
Large can	15.95	y	$15.95y$

Note: Sales revenue is found by multiplying an item's price by the number sold.

Relationship 1:   The store sold twice as many of the small cans as the large.

    Translation: $y = 2x$

Relationship 2:   The total revenue was $694.45.

    Translation: $8.95x + 15.95y = 694.45$

$$\text{Our system:} \begin{cases} y = 2x \\ 8.95x + 15.95y = 694.45 \end{cases}$$

To solve this system of equations, substitution is the best method because y is isolated in the first equation. We replace y in the second equation with $2x$ from the first equation.

$$8.95x + 15.95y = 694.45$$
$$8.95x + 15.95(2x) = 694.45 \qquad \text{Substitute } 2x \text{ for } y.$$
$$8.95x + 31.9x = 694.45$$
$$40.85x = 694.45 \qquad \text{Combine like terms.}$$
$$x = 17 \qquad \text{Divide both sides by 40.85 to isolate } x.$$

Answer **Your Turn 6**

78 ft. by 36 ft.

Now we can find the value of y using $y = 2x$ by substituting 17 for x.

$$y = 2(17) = 34$$

Answer   The home improvement store sold 17 small cans of stain and 34 large cans.

Check   Verify both given relationships. The number of large cans of stain sold, 34, is twice the number of small cans sold, 17. The total revenue, $8.95(17) + 15.95(34) = 694.45$, is also correct.

Your Turn 7   A cookware consultant sells two sizes of a rectangular baking dish. The small size sells for \$24, and the large size sells for \$35. In one month, she sold 6 more of the small-size dish than the large-size dish. If she made a total of \$439 from the sale of both dishes, how many of each size baking dish did she sell?

Example 8   Jasmine and Darius are traveling east in separate cars on the same highway. Jasmine is traveling at 60 miles per hour; Darius, at 70 miles per hour. Jasmine passes Exit 82 at 3:15 P.M. Darius passes the same exit at 3:30 P.M. At what time will Darius catch up with Jasmine?

Note: We first solved rate problems in Section 3.4.

Understand   To determine the time at which Darius catches up with Jasmine, we must calculate the amount of time it will take him to catch up with her. We can then add that amount of time to 3:30 P.M.

Plan and Execute   We will let x represent Jasmine's travel time after passing Exit 82 and y represent Darius's travel time after passing Exit 82.

Category	Rate	Time	Distance
Jasmine	60	x	$60x$
Darius	70	y	$70y$

Note: Distance is the product of rate and time.

Relationship 1:   Darius passes the exit 15 minutes after Jasmine; so when he catches up with her, she will have traveled 15 minutes longer than he has.

Translation: $x = y + \dfrac{1}{4}$

Note: We use $\frac{1}{4}$ of an hour instead of 15 minutes because the rates are in miles per hour and the time units must be consistent.

Relationship 2:   When Darius catches up, they will have traveled the same distance from Exit 82; so we set the expressions for their individual distances equal to each other.

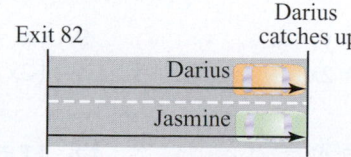

Exit 82   Darius catches up

Darius

Jasmine

Translation: $60x = 70y$

Our system: $\begin{cases} x = y + \dfrac{1}{4} \\ 60x = 70y \end{cases}$

Because we have an isolated variable, x, we will solve the system using the substitution method.

$$60x = 70y$$

$$60\left(y + \frac{1}{4}\right) = 70y \qquad \text{Substitute } y + \tfrac{1}{4} \text{ for } x.$$

$$60y + 15 = 70y \qquad \text{Distribute 60.}$$

$$15 = 10y \qquad \text{Subtract } 60y \text{ from both sides.}$$

$$1.5 = y \qquad \text{Divide both sides by 10 to isolate } y.$$

Answer   **Your Turn 7**

11 small, 5 large

Answer   It will take Darius 1.5 hours to catch up with Jasmine. So he catches up at 5 P.M.

Check   At 60 miles per hour, in $\frac{1}{4}$ of an hour, Jasmine travels 15 miles. Because Darius is traveling 70 miles per hour, he is gaining on her at a rate of 10 miles per hour (the difference of their rates). So he must make up the 15 miles that separate them at a rate of 10 miles per hour, which will indeed take 1.5 hours.

Your Turn 8   At 10:30 A.M., Celeste begins walking along a trail at an average rate of 3 miles per hour. At 10:40 A.M., Chris begins bicycling along the same trail at 8 miles per hour. At what time will Chris catch up with Celeste?

Answer   Your Turn 8

10:46 A.M.

9.2 Exercises   For Extra Help   *MyMathLab*

1. What advantages does the method of substitution have over graphing for solving systems of equations?

2. Suppose you are given the system of equations
$$\begin{cases} y = x - 6 \\ 4x + 3y = 1 \end{cases}$$
   a. Which variable would you replace in the substitution process?

   b. What expression would replace that variable?

3. Suppose you are given the system of equations
$$\begin{cases} 5x + y = 4 \\ 3x - 2y = 9 \end{cases}$$. Which variable in which equation would you isolate first? Why?

4. Suppose you are given the system of equations
$$\begin{cases} 3x + 4y = 1 \\ x - 5y = 12 \end{cases}$$. Which variable would you isolate first? Why?

5. How do you know that a system has no solution (inconsistent system) when using substitution?

6. How do you know that a system has an infinite number of solutions (consistent with dependent equations) when using substitution?

For Exercises 7–34, solve the system of equations using substitution. Note that some systems may be inconsistent or consistent with dependent equations. See Examples 1–5.

7. $\begin{cases} x + y = 6 \\ x = 2y \end{cases}$

8. $\begin{cases} x - y = 8 \\ y = 2x \end{cases}$

9. $\begin{cases} x = -3y \\ 2x - 5y = 44 \end{cases}$

10. $\begin{cases} x + 2y = 9 \\ y = -2x \end{cases}$

11. $\begin{cases} 2x - y = 4 \\ y = 3x + 6 \end{cases}$

12. $\begin{cases} x = 3y - 2 \\ 2x - 3y = 2 \end{cases}$

13. $\begin{cases} y = -2x \\ 4x + 2y = 0 \end{cases}$

14. $\begin{cases} x = 2y - 6 \\ -2x + 4y = 12 \end{cases}$

15. $\begin{cases} 2x + y = -4 \\ 3x + 2y = -5 \end{cases}$

16. $\begin{cases} 5x + y = -4 \\ 3x + 2y = -1 \end{cases}$

17. $\begin{cases} 5x - y = 1 \\ 3x + 2y = 24 \end{cases}$

18. $\begin{cases} x + 3y = 5 \\ 2y - x = 10 \end{cases}$

19. $\begin{cases} 5x - 3y = 4 \\ y - x = -2 \end{cases}$ **20.** $\begin{cases} 2x - y = 39 \\ 14 + 3x = 74y \end{cases}$ **21.** $\begin{cases} 2x = 10 - y \\ 3x - y = 5 \end{cases}$ **22.** $\begin{cases} 3y = 7 - x \\ 2x - y = 0 \end{cases}$

23. $\begin{cases} -x + 2y = 1 \\ 3x - 2y = 1 \end{cases}$ **24.** $\begin{cases} 2x + 3y = -1 \\ 5x - y = 6 \end{cases}$ **25.** $\begin{cases} 3x + 2y = 8 \\ x + 4y = 1 \end{cases}$ **26.** $\begin{cases} 4x + y = -2 \\ 8x - y = 11 \end{cases}$

27. $\begin{cases} x - 3y = -4 \\ -5x + 15y = 6 \end{cases}$ **28.** $\begin{cases} 2x - y = -1 \\ 6x - 3y = 3 \end{cases}$ **29.** $\begin{cases} 4x + 3y = -2 \\ 8x - 2y = 12 \end{cases}$ **30.** $\begin{cases} 2x + 3y = -1 \\ 6x + 3y = -9 \end{cases}$

31. $\begin{cases} 5x + 4y = 12 \\ 7x - 6y = 40 \end{cases}$ **32.** $\begin{cases} 4x - 3y = 2 \\ 3x + 5y = 16 \end{cases}$ **33.** $\begin{cases} 5x + 6y = 2 \\ 10x + 3y = -2 \end{cases}$ **34.** $\begin{cases} 7x - 6y = 1 \\ 4x - 6y = 3 \end{cases}$

For Exercises 35–38, solve using substitution; then verify the solution by graphing the equations on a graphing utility. See Examples 1–5.

35. $\begin{cases} y = 2x + 1 \\ 3x + y = 11 \end{cases}$ **36.** $\begin{cases} y = -4x + 3 \\ y - 2x = -3 \end{cases}$ **37.** $\begin{cases} 2x + y = 4 \\ x + 2y = 5 \end{cases}$ **38.** $\begin{cases} 4x - y = 12 \\ x - 3y = 14 \end{cases}$

Find

(X) the mistake   **For Exercises 39 and 40, find the mistake.**

39. $\begin{cases} 3x + 2y = 5 \\ x - 3y = 6 \end{cases}$

Rewrite: $x = 6 + 3y$

Substitute: $3(6 + 3y) + 2y = 5$

$18 + 3y + 2y = 5$

$18 + 5y = 5$

$5y = -13$

$y = -\dfrac{13}{5}$

$x = 6 + 3y$

$x = 6 + 3\left(-\dfrac{13}{5}\right)$

$x = 6 - \dfrac{39}{5}$

$x = -\dfrac{9}{5}$

Solution: $\left(-\dfrac{9}{5}, -\dfrac{13}{5}\right)$

40. $\begin{cases} 2x + 3y = 8 \\ x + 2y = 6 \end{cases}$

Rewrite: $x = 6 - 2y$

Substitute: $x + 2y = 6$

$6 - 2y + 2y = 6$

$6 - 6 - 2y + 2y = 6 - 6$

$0 = 0$

Solution: Infinite solutions along $x + 2y = 6$

For Exercises 41–60, translate the problem to a system of equations and then solve. See Examples 6–8.

41. The greater of two integers is 10 more than twice the smaller integer. The sum of the two integers is −8. Find the integers.

42. One integer is six less than three times another integer. The sum of the integers is −42. Find the integers.

43. Jon is 13 years older than Tony. The sum of their ages is 27. How old are they?

44. Patrick is twice as old as his sister Sherry. The sum of their ages is 12. How old are they?

45. The perimeter of a college basketball court is 288 feet. If the length of the court is 44 feet longer than it is wide, what are the dimensions?

46. The length of a volleyball court is twice the width. If the perimeter of the court along the doubles lines is 54 meters, what are the dimensions? (*Source: USA Volleyball Rule Book*)

47. The length of the Reflecting Pool at the National Mall in Washington, D.C., is 800 feet more than ten times the width. What are the dimensions if the perimeter is 4900 feet? (*Source:* www.answers.com)

48. The perimeter of the base of the Castillo at Chichen Itza (a pyramid in Mexico) is 300 feet. If the length is equal to the width, what are the dimensions?

49. A support beam is attached to a wall and to the bottom of a ceiling truss. The angle made with the truss on one side of the support beam is three times the angle on the other side of the support beam. Find the two angles.

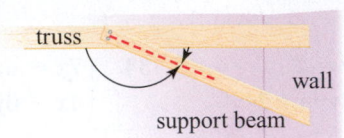

50. A laser beam is aimed at a flat photocell at an angle. The angle on one side of the beam is 25 degrees more than the angle on the other side of the beam. Find the two angles.

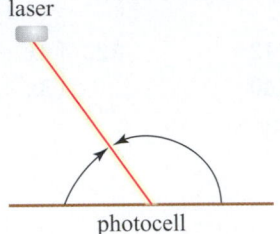

51. Troy has a checking account and a savings account with a combined balance of $4200. The balance in his savings account is four times the balance in his checking account. What are the balances of both accounts?

52. Susan has two credit cards with a combined balance of $2700. If she owes three times more on one of the cards than the other, how much does she owe on each card?

53. The Rolling Stones and Madonna were the top two moneymakers from June 2006 to June 2007 in the music industry. The Rolling Stones earned $16 million more than Madonna, and together they earned $160 million. What were the earnings of each of these performers? (*Source: Forbes*)

54. The all-time top-grossing American movie is *Titanic*, which grossed $139.8 million more than the second place *Star Wars Episode IV: A New Hope*. Together the two movies grossed $1061.8 million. How much revenue was brought in by each movie? (*Source: The World Almanac and Book of Facts,* 2008)

55. The cashier's office at a college has collected $7020 from part-time students for a total of 42 courses. If each credit hour costs $45, how many 3-hour courses and how many 5-hour courses were paid for?

56. One day a college bookstore collected $6260 for the sale of 44 beginning algebra and calculus books. If beginning algebra books sold for $130 each and calculus books sold for $160, how many of each type of book did the bookstore sell?

57. John and Karen are traveling south in separate cars on the same interstate. Karen is traveling at 65 miles per hour; John, at 70 miles per hour. Karen passes Exit 38 at 1:15 P.M. John passes the same exit at 1:30 P.M. At what time will John catch up with Karen?

58. Diane and Curtis are running in the Cooper River Bridge Run. Diane passes the first checkpoint at 10:30 A.M; Curtis, at 11:00 A.M. While Curtis is pushing himself to run at 5 miles per hour, Diane is pacing herself at 4 miles per hour. When will Curtis catch up with Diane?

59. Sharon and Rae are riding bicycles on the same trail in the same direction. Sharon crosses a bridge at 12:18 P.M. Rae crosses the same bridge at 12:33 P.M. If Sharon is going 8 miles per hour and Rae is going 10 miles per hour, when will Rae catch up with Sharon?

60. Holly and Rebecca are jogging on the same road in the same direction. Holly passes a McDonald's at 8:10 A.M. Rebecca passes the same McDonald's at 8:20 A.M. If Holly is going 4 miles per hour and Rebecca is going 6 miles per hour, when will Rebecca catch up with Holly?

Review Exercises

Exercises 1–4 **EXPRESSIONS**

[1.7] For Exercises 1 and 2, multiply.

1. $\frac{1}{4}(20x - 16y)$

2. $-3(x - 7y)$

[1.7] For Exercises 3 and 4, add or subtract.

3. $6x + 2y + 3x - 2y$

4. $5x + 3y - 5x + 10y$

Exercises 5 and 6  **EQUATIONS AND INEQUALITIES**

[2.3] For Exercises 5 and 6, solve.

5. $\frac{1}{4}x = 20$

6. $-8y = 12$

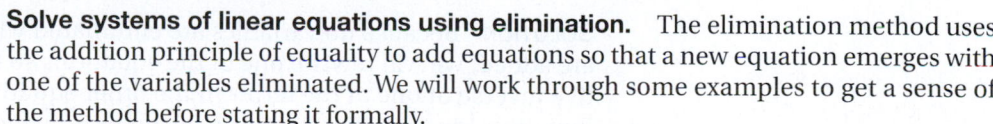

9.3 Solving Systems of Linear Equations by Elimination; Applications

Objectives

1. Solve systems of linear equations using elimination.
2. Solve applications using elimination.

We have seen two methods for solving a system of equations: graphing and substitution. Each method has advantages and disadvantages. The substitution method is advantageous over the graphing method when solutions involve fractions. Also, substitution is easy when a variable's coefficient is 1. However, substitution can be tedious when no coefficients are 1. For this reason, we turn to yet a third method, the *elimination* method.

Objective 1 **Solve systems of linear equations using elimination.** The elimination method uses the addition principle of equality to add equations so that a new equation emerges with one of the variables eliminated. We will work through some examples to get a sense of the method before stating it formally.

Example 1 Solve the system of equations using elimination.

> **Note:** This method is also called the elimination by addition method, or simply the addition method.

$$\begin{cases} x + y = 5 & \text{(Equation 1)} \\ 2x - y = 7 & \text{(Equation 2)} \end{cases}$$

SOLUTION In Chapter 2, we learned that the addition principle of equality says that adding the same amount to both sides of an equation will not affect its solution(s). Because Equation 1 indicates that $x + y$ and 5 are the same amount, if we add $x + y$ to the left side of Equation 2 and 5 to the right side of Equation 2, we are applying the addition principle of equality. Most people prefer to stack the equations and combine like terms vertically like this:

$$
\begin{array}{ll}
x + y = 5 & \text{(Equation 1)} \\
\underline{2x - y = 7} & \text{(Equation 2)} \\
3x + 0 = 12 & \text{\color{blue}{Add Equation 1 to Equation 2.}}
\end{array}
$$

> **Connection** The terms y and $-y$ are additive inverses, or opposites, which means that their sum is 0.

Notice that y is eliminated in this new equation; so we can easily solve for the value of x.

$$3x = 12$$
$$x = 4 \qquad \text{Divide both sides by 3 to isolate } x.$$

Now that we have the value of x, we can find y by substituting 4 for x in one of the original equations. We will use $x + y = 5$:

$$x + y = 5$$
$$4 + y = 5 \qquad \text{Substitute 4 for } x.$$
$$y = 1 \qquad \text{Subtract 4 from both sides to isolate } y.$$

The solution is $(4, 1)$. We can check by verifying that $(4, 1)$ makes both of the original equations true. We will leave this to the reader.

Your Turn 1 Solve the system of equations using elimination.

$$\begin{cases} 4x + 3y = 8 \\ x - 3y = 7 \end{cases}$$

Multiplying One Equation by a Number to Create Additive Inverses

Connection In Chapter 2, we learned that the multiplication principle of equality says that we can multiply both sides of an equation by the same amount without affecting its solution(s).

You may have noted that the expressions $x + y$ and $2x - y$ conveniently contained the additive inverses y and $-y$, which eliminated the y's when the expressions were added. If no such pairs of additive inverses appear in a system of equations, we will use the multiplication principle of equality to multiply both sides of one equation by a number in order to create additive inverse pairs.

Example 2 Solve the system of equations using elimination.

$$\begin{cases} x + y = 6 & \text{(Equation 1)} \\ 2x - 5y = -16 & \text{(Equation 2)} \end{cases}$$

SOLUTION Because no variables are eliminated when $x + y = 6$ and $2x - 5y = -16$ are added, we will rewrite one of the equations so that it has a term that is the additive inverse of one of the terms in the other equation. We will multiply both sides of Equation 1 by 5 so that its y term becomes $5y$, which is the opposite of the $-5y$ term in Equation 2.

$$x + y = 6$$
$$5 \cdot x + 5 \cdot y = 5 \cdot 6 \qquad \text{Multiply both sides by 5.}$$
$$5x + 5y = 30 \qquad \text{(Equation 1 rewritten)}$$

Note: We could have multiplied both sides of Equation 1 by -2. We would get a $-2x$ term in Equation 1 that is the opposite of the $2x$ term in Equation 2.

Now we add the rewritten Equation 1 to Equation 2 to eliminate a variable, just as we did in Example 1.

Note: Multiplying Equation 1 by 5 made the elimination of the y terms possible.

$$\begin{array}{ll} 5x + 5y = 30 & \text{(Equation 1 rewritten)} \\ 2x - 5y = -16 & \text{(Equation 2)} \\ \hline 7x + 0 \ = 14 & \text{Add rewritten Equation 1 to} \\ & \text{Equation 2 to eliminate } y. \end{array}$$

Answer   Your Turn 1

$\left(3, -\dfrac{4}{3}\right)$

We can now solve for x.

$$7x = 14$$
$$x = 2 \qquad \text{Divide both sides by 7 to isolate } x.$$

To finish, we substitute 2 for x in one of the equations containing both variables. We will use $x + y = 6$.

$$x + y = 6$$
$$2 + y = 6 \qquad \text{Substitute 2 for } x.$$
$$y = 4 \qquad \text{Subtract 2 from both sides to isolate } y.$$

The solution is $(2, 4)$. We will leave the check to the reader.

Your Turn 2   Solve the system of equations using elimination.

$$\begin{cases} -3x - 5y = 6 \\ x + 2y = -1 \end{cases}$$

Multiplying Each Equation by a Number to Create Additive Inverses

If every coefficient in a system of equations is other than 1, we may have to multiply each equation by a number to generate a pair of additive inverses.

Example 3   Solve the system of equations using elimination.

$$\begin{cases} 4x - 3y = -2 & \text{(Equation 1)} \\ 6x - 7y = 7 & \text{(Equation 2)} \end{cases}$$

SOLUTION   We will choose to eliminate x; so we must multiply both equations by numbers that make the x terms additive inverses. This is like finding the least common multiple (or denominator) of the coefficients of x, but with opposite signs. We will multiply Equation 1 by 3 and Equation 2 by -2.

$$4x - 3y = -2 \xrightarrow{\text{Multiply by 3.}} 12x - 9y = -6$$
$$6x - 7y = 7 \xrightarrow{\text{Multiply by } -2.} -12x + 14y = -14$$

Note: The x terms are now additive inverses, $12x$ and $-12x$.

Now we can add the rewritten equations to eliminate the x term.

$$\begin{array}{r} 12x - 9y = -6 \\ -12x + 14y = -14 \\ \hline 0 + 5y = -20 \\ y = -4 \end{array}$$

$0 + 5y = -20$ Add the rewritten equations to eliminate x.
$y = -4$ Divide both sides by 5 to isolate y.

Note: It doesn't matter which equation we use because x should be the same value in any equation in the system.

To finish, we substitute -4 for y in one of the equations and solve for x. We will use $4x - 3y = -2$.

$$4x - 3y = -2$$
$$4x - 3(-4) = -2 \qquad \text{Substitute } -4 \text{ for } y.$$
$$4x + 12 = -2$$
$$4x = -14 \qquad \text{Subtract 12 from both sides.}$$

Answer   **Your Turn 2**

$(-7, 3)$

$$x = -\frac{14}{4} \qquad \text{\color{blue}\textbf{Divide both sides by 4 to isolate } y.}$$

$$x = -\frac{7}{2} \qquad \text{\color{blue}\textbf{Simplify.}}$$

The solution is $\left(-\frac{7}{2}, -4 \right)$. We will leave the check to the reader.

> **Your Turn 3** Solve the system of equations using elimination.
>
> $$\begin{cases} 3x - 2y = 7 \\ 4x - 3y = 10 \end{cases}$$

Fractions or Decimals in a System

If any of the equations in a system of equations contains fractions or decimals, it is helpful to use the multiplication principle of equality to clear those fractions or decimals so that the equations contain only integers.

Example 4 Solve the system of equations using elimination.

Note: Since this system contains both fractions and decimals, the answers can be left as either a fraction or a decimal.

$$\begin{cases} \frac{1}{2}x - y = \frac{3}{4} & \text{(Equation 1)} \\ 0.4x - 0.3y = 1 & \text{(Equation 2)} \end{cases}$$

SOLUTION To clear the fractions in Equation 1, we can multiply both sides by the LCD, which is 4. To clear the decimals in Equation 2, we can multiply both sides by 10.

$$\frac{1}{2}x - y = \frac{3}{4} \quad \xrightarrow{\text{\color{blue}\textbf{Multiply by 4.}}} \quad 2x - 4y = 3$$

$$0.4x - 0.3y = 1 \quad \xrightarrow{\text{\color{blue}\textbf{Multiply by 10.}}} \quad 4x - 3y = 10$$

Now that both equations contain only integers, it will be easier to solve the system. We will choose to eliminate x. So we multiply the first equation by -2, then combine the equations.

$$2x - 4y = 3 \quad \xrightarrow{\text{\color{blue}\textbf{Multiply by } -2.}} \quad -4x + 8y = -6$$

$$4x - 3y = 10 \qquad\qquad\qquad\qquad \underline{4x - 3y = 10}$$

$$0 + 5y = 4 \qquad \text{\color{blue}\textbf{Add the rewritten equations to eliminate } x.}$$

$$y = \frac{4}{5} \text{ or } 0.8 \qquad \text{\color{blue}\textbf{Divide both sides by 5 to isolate } y.}$$

To finish, we substitute $\frac{4}{5}$ for y in one of the equations and solve for x. We will use $\frac{1}{2}x - y = \frac{3}{4}$.

Note: We could have multiplied both sides of the equation by 20 to clear the fraction.

$$\frac{1}{2}x - \frac{4}{5} = \frac{3}{4} \qquad \text{\color{blue}\textbf{Substitute } \frac{4}{5} \text{ for } y.}$$

$$\frac{1}{2}x = \frac{3}{4} + \frac{4}{5} \qquad \text{\color{blue}\textbf{Add } \frac{4}{5} \text{ to both sides.}}$$

Answer Your Turn 3
$(1, -2)$

$$\frac{1}{2}x = \frac{15}{20} + \frac{16}{20} \qquad \text{\color{blue}\textbf{Write the fractions with their LCD, 20.}}$$

$$\frac{1}{2}x = \frac{31}{20}$$ **Add the fractions.**

$$x = \frac{31}{\underset{10}{\cancel{20}}} \cdot \frac{\overset{1}{\cancel{2}}}{1}$$ **Multiply both sides by 2 to isolate x.**

$$x = \frac{31}{10} \text{ or } 3.1$$

The solution is $\left(\dfrac{31}{10}, \dfrac{4}{5}\right)$ or $(3.1, 0.8)$. We will leave the check to the reader.

Your Turn 4 Solve the system of equations using elimination.

$$\begin{cases} 0.6x + y = 3 \\ \dfrac{2}{5}x + \dfrac{1}{4}y = 1 \end{cases}$$

Rewriting the Equations in the Form $Ax + By = C$

Notice in Examples 1 through 4 that all of the equations are in standard form, which we learned in Chapter 4 to be $Ax + By = C$. When the elimination method is used, the equations need to be written in standard form. For example, in the system that we solved in Example 1, the equations could have been given in a nonstandard form, as follows:

$$\begin{cases} y = 5 - x & \text{(Equation 1)} \\ 2x - y = 7 & \text{(Equation 2)} \end{cases}$$

Adding x to both sides of Equation 1 puts it in standard form so that we can use elimination as we did in Example 1.

$$\begin{cases} x + y = 5 & \text{(Equation 1)} \\ 2x - y = 7 & \text{(Equation 2)} \end{cases}$$

We can now summarize the elimination method with the following procedure.

Procedure **Solving Systems of Two Linear Equations Using Elimination**

To solve a system of two linear equations using the elimination method,

1. Write the equations in standard form ($Ax + By = C$).
2. Use the multiplication principle to clear fractions or decimals (optional).
3. If necessary, multiply one or both equations by a number (or numbers) so that they have a pair of terms that are additive inverses.
4. Add the equations. The result should be an equation in terms of one variable.
5. Solve the equation from Step 4 for the value of that variable.
6. Using an equation containing both variables, substitute the value you found in Step 5 for the corresponding variable and solve for the value of the other variable.
7. Check your solution in the original equations.

Inconsistent Systems and Dependent Equations

How would we recognize an inconsistent system or a consistent system with dependent equations using the elimination method?

Answer **Your Turn 4**

$\left(1, \dfrac{12}{5}\right)$ or $(1, 2.4)$

Example 5 Solve the system of equations.

a. $\begin{cases} 2x - y = 1 \\ 2x - y = -3 \end{cases}$

SOLUTION Notice that the left sides of the equations match. Multiplying one of the equations by -1 and then adding the equations will eliminate both variables.

$$\begin{array}{l} 2x - y = 1 \\ 2x - y = -3 \end{array} \xrightarrow{\text{Multiply by } -1.} \begin{array}{l} -2x + y = -1 \\ \underline{2x - y = -3} \\ 0 = -4 \end{array}$$ **Add the rewritten equations.**

Connection If we graph the equations in Example 5a, the lines will be parallel, indicating there is no solution.

Both variables have been eliminated, and the resulting equation, $0 = -4$, is false. Therefore, there is no solution. This system of equations is inconsistent.

b. $\begin{cases} 3x + 4y = 5 \\ 9x + 12y = 15 \end{cases}$

SOLUTION To eliminate x, we could multiply the first equation by -3, then combine the equations.

$$\begin{array}{l} 3x + 4y = 5 \\ 9x + 12y = 15 \end{array} \xrightarrow{\text{Multiply by } -3.} \begin{array}{l} -9x - 12y = -15 \\ \underline{9x + 12y = 15} \\ 0 = 0 \end{array}$$ **Add the rewritten equations.**

Connection If graphed, both equations in Example 5(b) generate the same line, indicating that the equations are dependent.

Both variables have been eliminated, and the resulting equation, $0 = 0$, is true. This means that the equations are dependent. So there are an infinite number of solutions, which are all of the ordered pairs along the line $3x + 4y = 5$ (or $9x + 12y = 15$).

Answers Your Turn 5

a. all ordered pairs along
$x - 4y = 2$ or $5x - 20y = 10$
(dependent)

b. no solution (inconsistent)

Your Turn 5 Solve the system of equations.

a. $\begin{cases} x - 4y = 2 \\ 5x - 20y = 10 \end{cases}$

b. $\begin{cases} x + 2y = 3 \\ x + 2y = 1 \end{cases}$

Objective ② Solve applications using elimination. In Section 9.2, we solved problems in which the substitution method was advantageous because after translating, one of the equations had an isolated variable. In this section, we focus on problems in which elimination is advantageous. These problems translate to a system of linear equations in which every equation is in standard form ($Ax + By = C$).

Example 6 Airplanes traveling west fly against the jet stream, which decreases their average speed. When flying east, the jet stream increases average speed. Suppose a large jet flies from Atlanta to Los Angeles, a distance of approximately 2000 miles, in 5 hours. On the return trip, the plane takes only 4 hours. Find the plane's speed in still air. How much does the jet stream change the speed of the plane?

Understand We assume that without the jet stream, the plane would average the same speed going west as going east. We also assume that the amount the jet stream adds to the plane's speed is the same as the amount subtracted depending on which way the plane is traveling.

Plan and Execute Let x represent the plane's speed in still air. Let y represent the amount the jet stream increases or decreases the plane's speed depending on which way it is traveling. We can use a table to organize the information.

Note: We solved rate problems in Section 3.4.

Category	Rate	Time	Distance
West	$x - y$	5	$5(x - y)$
East	$x + y$	4	$4(x + y)$

Relationship 1: The distance from Atlanta to Los Angeles is 2000 miles.

Translation: $5(x - y) = 2000$

$$\frac{5(x - y)}{5} = \frac{2000}{5}$$ **Divide both sides by 5 to streamline the equation.**

$$x - y = 400$$

Relationship 2: The distance from Los Angeles to Atlanta is 2000 miles.

Translation: $4(x + y) = 2000$

$$\frac{4(x + y)}{4} = \frac{2000}{4}$$ **Divide both sides by 4 to streamline the equation.**

$$x + y = 500$$

Our system: $\begin{cases} x - y = 400 \\ x + y = 500 \end{cases}$

Because the equations are in standard form, we will use the elimination method.

$$
\begin{array}{l}
x - y = 400 \\
\underline{x + y = 500} \\
2x + 0 = 900 \quad \text{Add the equations to eliminate } y. \\
2x = 900 \\
x = 450 \quad \text{Divide both sides by 2 to isolate } x.
\end{array}
$$

Now we can find the value of y by substituting 450 for x in one of the equations. We will use $x + y = 500$.

$$
\begin{array}{ll}
x + y = 500 & \\
450 + y = 500 & \text{Substitute 450 for } x. \\
y = 50 & \text{Subtract 450 from both sides to isolate } y.
\end{array}
$$

Answer The plane travels 450 miles per hour in still air. The jet stream changes the plane's speed by 50 miles per hour depending whether it is traveling east or west.

Check Verify the time of travel for each part of the round trip.

Going west: $450 - 50 = 400$ miles per hour   Going east: $450 + 50 = 500$ miles per hour

Time: $t = \dfrac{d}{r} = \dfrac{2000}{400} = 5$ hours   Time: $t = \dfrac{d}{r} = \dfrac{2000}{500} = 4$ hours

Your Turn 6 To train for the swimming portion of a triathlon, an athlete swims in a river to gauge how the current will affect him. When swimming against the current, he can swim a mile in 1.25 hours. When swimming with the current, he can swim a mile in 0.5 hours. What is his rate in still water? How much did the current increase or decrease his rate?

Answer   Your Turn 6

His rate in still water is 1.4 miles per hour. The current changes his rate by 0.6 miles per hour.

Note: We first solved investment problems in Section 3.5.

Note: Recall from Chapter 3 that we write percentage rates as decimal numbers. Also, interest is the product of the APR and the principal.

Note: We could have first cleared the decimal numbers, but we chose not to because it is easy to eliminate x or y by the elimination method since both coefficients in $x + y = 5000$ are 1.

Example 7   Jeff invests $5000 in two different accounts. The first account has an annual percentage rate (APR) of 4%. The second account has an APR of 6%. If the total interest earned after one year is $252, what principal was invested in each account?

Understand   The two unknowns are the principals invested in each account. One relationship involves the total principal ($5000), and the other relationship involves the total interest earned ($252).

Plan and Execute   Let x and y represent the two amounts invested. We can use a table to organize the information.

Category	APR	Principal	Interest
Account 1	0.04	x	$0.04x$
Account 2	0.06	y	$0.06y$

Relationship 1:   The total principal is $5000.

Translation: $x + y = 5000$

Relationship 2:   The total interest is $252.

Translation: $0.04x + 0.06y = 252$

Our system: $\begin{cases} x + y = 5000 \\ 0.04x + 0.06y = 252 \end{cases}$

Because the equations are in standard form, elimination is the better method. We will eliminate x by multiplying both sides of the first equation by -0.04.

$$x + y = 5000 \xrightarrow{\text{Multiply by } -0.04.} -0.04x - 0.04y = -200$$
$$0.04x + 0.06y = 252 \qquad\qquad \underline{0.04x + 0.06y = 252} \quad \text{Add the equations to eliminate } x.$$
$$0.02y = 52 \quad$$
$$y = 2600 \quad \text{Divide both sides by 0.02 to isolate } y.$$

Now we can find x by substituting 2600 for y in one of the equations. We will use $x + y = 5000$.

$$x + y = 5000$$
$$x + 2600 = 5000 \quad \text{Substitute 2600 for } y.$$
$$x = 2400 \quad \text{Subtract 2600 from both sides to isolate } x.$$

Answer   Jeff invested $2400 in the 4% account and $2600 in the 6% account.

Check   The total investment is $2400 + $2600 = $5000, and the total interest is $0.04(2400) + 0.06(2600) = \$96 + \$156 = \$252$.

Your Turn 7   Monique invests a total of $8000 in two different accounts. The first account returns 5%, while the second account returns 8%. If the total interest earned after one year is $532, what principal was invested in each account?

Note: We first solved mixture problems in Section 3.5.

Answer   Your Turn 7

$3600 at 5%, $4400 at 8%

Example 8   How much 10% HCl solution and 30% HCl solution must be mixed together to make 500 milliliters of 15% HCl solution?

Understand   The two unknowns are the volumes of 10% solution and 30% solution that are mixed. One relationship involves concentrations of each solution in the mixture, and the other relationship involves the total volume of the final mixture (500 ml).

Plan and Execute Let x and y represent the two amounts to be mixed.

Solution	Concentration	Volume	Amount of HCl
10% solution	0.10	x	$0.10x$
30% solution	0.30	y	$0.30y$
15% solution	0.15	500	$0.15(500)$

Relationship 1:   The total volume is 500 ml.

> **Note:** The amount of HCl is the product of the concentration and the volume. Notice that we were given the volume of the 15% solution.

Translation: $x + y = 500$

Relationship 2:   The combined amount of HCl in the two mixed solutions is to be 15% of the total amount of the mixture.

Translation: $0.10x + 0.30y = 0.15(500)$

Our system: $\begin{cases} x + y = 500 \\ 0.10x + 0.30y = 0.15(500) \end{cases}$

Because the equations are in standard form, elimination is the better method. We will eliminate x by multiplying both sides of the first equation by -0.10.

$$x + y = 500 \xrightarrow{\text{Multiply by } -0.10} -0.10x - 0.10y = -50$$

$$0.10x + 0.30y = 75 \qquad \underline{0.10x + 0.30y = 75} \qquad \text{Add the equations}$$
$$0.20y = 25 \qquad \text{to eliminate } x.$$
$$y = 125 \qquad \text{Divide both sides by 0.20 to isolate } y.$$

Now we can find x by substituting 125 for y in one of the equations. We will use $x + y = 500$.

$$x + y = 500$$
$$x + 125 = 500 \qquad \text{Substitute 125 for } y.$$
$$x = 375 \qquad \text{Subtract 125 on both sides to isolate } x.$$

Answer   Mixing 375 ml of 10% solution with 125 ml of 30% solution gives 500 ml of 15% solution.

Check   The mixture volume is $125 + 375 = 500$. The amount of HCl is $0.10(375) + 0.30(125) = 37.5 + 37.5 = 75$, which means that 75 ml is HCl out of the 500 ml mixture. We can verify that 75 ml is 15% of 500 ml by multiplying: $0.15(500) = 75$.

Your Turn 8   How much 20% HCl solution and 40% HCl solution must be mixed together to make 600 milliliters of 35% HCl solution?

Answer   Your Turn 8
450 ml of 40%, 150 ml of 20%

9.3 Exercises   For Extra Help   *MyMathLab*

Math XL
PRACTICE   WATCH   DOWNLOAD

1. What advantages does the method of elimination have over graphing and substitution?

2. Suppose you are given the system of equations
$\begin{cases} x - y = 2 \\ 3x + y = 5 \end{cases}$.
Which variable would you eliminate? Why?

3. Suppose you are given the system of equations
$$\begin{cases} 6x - 2y = 1 \\ 3x + 5y = 7 \end{cases}.$$
   a. Which variable is easier to eliminate? Why?

   b. How would you eliminate the variable you chose?

4. Suppose you are given the system of equations
$$\begin{cases} 5x + 4y = 8 \\ 3x - 6y = 2 \end{cases}.$$
   a. Which variable is easier to eliminate? Why?

   b. How would you eliminate the variable you chose?

5. When using elimination, how do you know if a system of equations has no solution (inconsistent system)?

6. When using elimination, how do you know if a system of equations has an infinite number of solutions (dependent equations)?

For Exercises 7–42, solve the system of equations using the elimination method. Note that some systems may be inconsistent or consistent with dependent equations. See Examples 1–5.

7. $\begin{cases} x - y = 14 \\ x + y = -2 \end{cases}$

8. $\begin{cases} x + y = 9 \\ 5x - y = 3 \end{cases}$

9. $\begin{cases} 3x + y = 9 \\ 2x - y = 1 \end{cases}$

10. $\begin{cases} 3x + y = 10 \\ -2x - y = -7 \end{cases}$

11. $\begin{cases} 3x + 2y = -7 \\ 5x - 2y = -1 \end{cases}$

12. $\begin{cases} 2x - 3y = 8 \\ 4x + 3y = 16 \end{cases}$

13. $\begin{cases} 4x + 3y = 17 \\ 2x + 3y = 13 \end{cases}$

14. $\begin{cases} x - 3y = 7 \\ x - 5y = 13 \end{cases}$

15. $\begin{cases} 5x + y = 14 \\ 2x + y = 5 \end{cases}$

16. $\begin{cases} 3x - y = 7 \\ 5x - y = 15 \end{cases}$

17. $\begin{cases} 3x - 5y = -17 \\ 4x + y = -15 \end{cases}$

18. $\begin{cases} 7x - 4y = 4 \\ 5x + y = 26 \end{cases}$

19. $\begin{cases} 12x - 2y = -54 \\ 13x + 4y = -40 \end{cases}$

20. $\begin{cases} 5x + 2y = 3 \\ 7x - 6y = 13 \end{cases}$

21. $\begin{cases} 2x + y = -2 \\ 8x + 4y = -8 \end{cases}$

22. $\begin{cases} 4x - 2y = 3 \\ -8x + 4y = -6 \end{cases}$

23. $\begin{cases} 2x - 5y = 4 \\ -8x + 20y = -20 \end{cases}$

24. $\begin{cases} 5x + 5y = 50 \\ x + y = 2.5 \end{cases}$

25. $\begin{cases} 5x + 6y = 11 \\ 2x - 4y = -2 \end{cases}$

26. $\begin{cases} -6x + 7y = -2 \\ 9x - 5y = -8 \end{cases}$

27. $\begin{cases} 4x + 5y = -4 \\ 3x + 8y = -20 \end{cases}$

28. $\begin{cases} 3x - 4y = -14 \\ 5x + 7y = 45 \end{cases}$

29. $\begin{cases} 5x + 6y = 2 \\ 10x + 3y = -2 \end{cases}$

30. $\begin{cases} 8x + 6y = 5 \\ 2x + y = 1 \end{cases}$

31. $\begin{cases} 10x + 5y = 3.5 \\ 3x - 4y = -0.6 \end{cases}$

32. $\begin{cases} 2x - y = -0.8 \\ 3x - 2y = -1 \end{cases}$

33. $\begin{cases} \dfrac{3}{2}x - \dfrac{4}{3}y = \dfrac{11}{3} \\ \dfrac{1}{4}x - \dfrac{2}{3}y = -\dfrac{7}{6} \end{cases}$

34. $\begin{cases} \dfrac{3}{4}x - \dfrac{2}{7}y = -5 \\ \dfrac{5}{8}x + \dfrac{1}{4}y = -\dfrac{3}{4} \end{cases}$

35. $\begin{cases} 0.7x - \dfrac{1}{2}y = 9.4 \\ 0.9x + \dfrac{7}{10}y = 0 \end{cases}$

36. $\begin{cases} \dfrac{1}{3}x + 0.2y = -0.2 \\ \dfrac{2}{3}x - 0.75y = -5 \end{cases}$

37. $\begin{cases} y = 2x - 5 \\ x - y = 9 \end{cases}$

38. $\begin{cases} x + 3y = 8 \\ 2y = x + 6 \end{cases}$

39. $\begin{cases} y = \dfrac{3}{4}x + 1 \\ 4x = 2y + 3 \end{cases}$ **40.** $\begin{cases} y = -\dfrac{2}{3}x - 5 \\ 3y = 2x - 27 \end{cases}$ **41.** $\begin{cases} 0.2x - y = -3.2 \\ y = \dfrac{3}{4}x + 1 \end{cases}$ **42.** $\begin{cases} y = \dfrac{1}{5}x - 2 \\ 6y = 0.3x - 1.2 \end{cases}$

Find

 (X) the mistake   *For Exercises 43 and 44, find the mistake.*

43. $\begin{cases} x - y = 1 \\ 9x + 8y = 77 \end{cases}$

$-9(x - y) = 1 \quad \longrightarrow \quad -9x + 9y = 1$
$9x + 8y = 77 \qquad\qquad\quad \underline{9x + 8y = 77}$
$\qquad\qquad\qquad\qquad\qquad 17y = 78$
$\qquad\qquad\qquad\qquad\qquad y = \dfrac{78}{17}$

$x - y = 1$
$x - \left(\dfrac{78}{17}\right) = 1$
$x = \dfrac{95}{17}$

Solution: $\left(\dfrac{95}{17}, \dfrac{78}{17}\right)$

44. $\begin{cases} x + y = 1 \\ x + 2y = 2 \end{cases}$

$x + y = 1 \qquad\qquad\qquad x + y = 1$
$-1(x + 2y = 2) \quad \longrightarrow \quad \underline{-x - 2y = 2}$
$\qquad\qquad\qquad\qquad\qquad -y = 3$
$\qquad\qquad\qquad\qquad\qquad y = -3$

$x + y = 1$
$x + (-3) = 1$
$x = 4$

Solution: $(4, -3)$

For Exercises 45–62, translate the problem to a system of equations; then solve using the elimination method. See Examples 6–8.

45. What are the length and width of a rectangle if the length exceeds the width by 14 inches and the perimeter is 60 inches?

46. What are the length and width of a rectangle if the width is 27 feet less than its length and the perimeter is 398 feet?

47. A boat travels 24 miles upstream in 3 hours. Going downstream, it can travel 36 miles in the same amount of time. Find the speed of the current and the speed of the boat in still water.

48. A boat travels 12 miles downstream in $1\frac{1}{2}$ hours. Going upstream, the boat takes 6 hours to return. Find the speed of the current and the speed of the boat in still water.

49. If a plane can travel 500 miles per hour with the wind and only 440 miles per hour against the wind, find the speed of the wind and the speed of the plane in still air.

50. If a plane can travel 200 miles per hour with the wind and only 120 miles per hour against the wind, find the speed of the wind and the speed of the plane in still air.

51. Adam inherited $15,500. He invested part of the money at an APR of 4% and put the rest into a savings account at an APR of 3%. If his annual income from these investments is $585, how much did he invest at each rate?

52. Petunia invests some money in a plan that pays a 5% APR and invests twice as much money in a plan that pays an 8% APR. If the total interest from the two plans is $1029, how much was invested in each?

53. Ellyn owes $13,250 on two student loans. One loan has an APR of 5%; the other, an APR of 8%. If the total interest after one year is $727, what is the amount of each loan?

54. Enrique has two credit cards. One is a Visa at an APR of 14%, and the other is a department store card with an APR of 18%. If the total amount he owes is $1536 and the interest for one year is $245.36, what is the amount owed on each card?

55. Rhonda and Mike have two car loans totaling $16,250. One loan has an APR of 2.9%, and the other has an APR of 6%. If the total interest they paid after one year is $657.16, what is the amount owed on each car?

56. Corey borrowed a total of $250,000 in two loans. One loan is at an APR of 8%, and the other is at an APR of 18%. If he paid $23,000 in interest during the first year, how much was loaned at each rate?

57. How much of a 20% acid solution is to be mixed with 80 milliliters of a 15% acid solution to get an 18% solution?

58. How many gallons of milk containing 1% butterfat is to be mixed with 50 gallons of milk that is 6% butterfat to get milk that is 2% butterfat?

59. A pharmacist is preparing a 300-milliliter solution of 25% alcohol. If she has 15% and 45% alcohol solutions in stock, how much of each must she use?

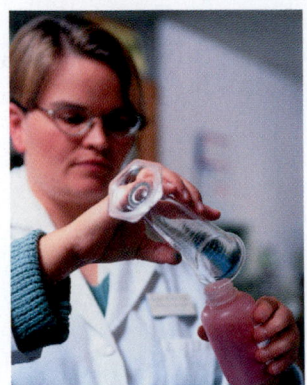

60. During chemistry class, Dr. Ellis needs a solution of 20% HCl, but all he has is 10% and 25% HCl solutions. How many milliliters of each does he need to mix to get 150 milliliters of the 20% HCl solution?

61. A coffee shop is considering a new mixture of coffee beans. It will be created from Italian Roast beans costing $9.80 per pound and the Gold Coast Blend (GCB) beans costing $11.20 per pound. The two types of beans will be mixed to create 20 pounds of coffee to be sold for $10.43 per pound. How much of each type of bean should be used?

62. A store is selling two types of jelly beans. One kind is $1.20 per pound, and the other is $0.90 per pound. If the merchant wants to sell a mixture at $1.11 per pound, how many pounds of each must be used to make 30 pounds?

Puzzle Problem

Use algebraic methods to prove that 0.999 . . . is equal to 1. (Hint: Let $x = 0.\overline{9}$. Then write a second equation using the multiplication principle so that when the two equations are added, the repeated 9 is eliminated.)

Review Exercises

Exercises 1 and 2 **EXPRESSIONS**

[1.7]   **1.** Use the distributive property: $4(5x + 7y - z)$ **[1.7]**   **2.** Simplify: $3x - 4y + z - 2x + y - z$

Exercises 3–6 **EQUATIONS AND INEQUALITIES**

[2.4]   **3.** Solve for y: $x + 3y = 6$ **[2.4]**   **4.** Solve for x: $\dfrac{1}{3}x - 2y = 10$

[2.2]   **5.** Given $x + y + 2z = 7$, find x if $y = -1$ and   **[2.2]**   **6.** Given $x - 3y + 2z = 6$, find y if $x = -2$ and
          $z = 3$. $z = 4$.

9.4 Solving Systems of Linear Equations in Three Variables; Applications

Objectives

1. Determine whether an ordered triple is a solution for a system of equations.
2. Understand the types of solution sets for systems of three equations.
3. Solve a system of three linear equations using the elimination method.
4. Solve application problems that translate to a system of three linear equations.

Objective 1 Determine whether an ordered triple is a solution for a system of equations. In this section, we solve systems of three linear equations with three unknowns. Solutions of these systems are *ordered triples* with the form (x, y, z). We check solutions for these systems just as we did in Section 9.1: Replace each variable with its corresponding value and verify that each equation is true.

Example 1 Determine whether each ordered triple is a solution to the system of equations.

$$\begin{cases} x + y + z = 3 & \text{(Equation 1)} \\ 2x + 3y + 2z = 7 & \text{(Equation 2)} \\ 3x - 4y + z = 4 & \text{(Equation 3)} \end{cases}$$

a. $(2, 1, 0)$

SOLUTION In all three equations, replace x with **2**, y with **1**, and z with **0**.

Equation 1:

$x + y + z = 3$

$2 + 1 + 0 \stackrel{?}{=} 3$

$3 = 3$ True

Equation 2:

$2x + 3y + 2z = 7$

$2(2) + 3(1) + 2(0) \stackrel{?}{=} 7$

$4 + 3 + 0 \stackrel{?}{=} 7$

$7 = 7$ True

Equation 3:

$3x - 4y + z = 4$

$3(2) - 4(1) + 0 \stackrel{?}{=} 4$

$6 - 4 + 0 \stackrel{?}{=} 4$

$2 = 4$ False

Because $(2, 1, 0)$ does not satisfy all three equations in the system, it is not a solution for the system.

b. $(3, 1, -1)$

SOLUTION

Equation 1:

$x + y + z = 3$

$3 + 1 + (-1) \stackrel{?}{=} 3$

$3 = 3$ True

Equation 2:

$2x + 3y + 2z = 7$

$2(3) + 3(1) + 2(-1) \stackrel{?}{=} 7$

$6 + 3 - 2 \stackrel{?}{=} 7$

$7 = 7$ True

Equation 3:

$3x - 4y + z = 4$

$3(3) - 4(1) + (-1) \stackrel{?}{=} 4$

$9 - 4 - 1 \stackrel{?}{=} 4$

$4 = 4$ True

Because $(3, 1, -1)$ satisfies all three equations in the system, it is a solution for the system.

Your Turn 1 Determine whether the following ordered triples are solutions to the system of equations.

a. $(4, 1, 1)$ **b.** $(-1, 5, 2)$

$$\begin{cases} x + y + z = 6 & \text{(Equation 1)} \\ 3x - 2y + 3z = -7 & \text{(Equation 2)} \\ 4x - 2y + z = -12 & \text{(Equation 3)} \end{cases}$$

Answers **Your Turn 1**

a. not a solution

b. solution

**Understand the types of solution sets for systems of three equations.** Recall that we plot an ordered pair using two axes: x and y. Similarly, we plot an ordered triple such as $(3, 4, 5)$ using three axes—x, y, and z—each of which is perpendicular to the other two, as shown.

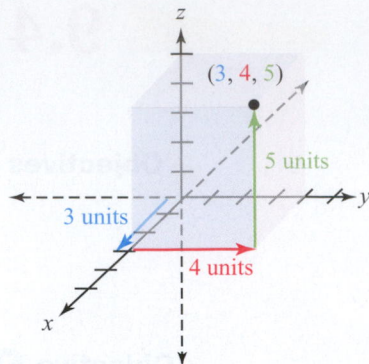

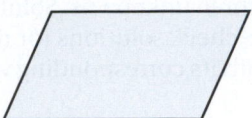

The graph of a linear equation in three variables is a *plane*, which is a flat surface much like a sheet of paper with infinite length and width. We draw a plane as a parallelogram (shown in the margin), but remember that the length and width are infinite.

In Section 9.1, we found that two lines could intersect in one point (consistent system), no points (inconsistent system), or an infinite number of points (consistent system with dependent equations). Similarly, the intersection of the planes of a three-variable system can tell us about the number of solutions for the system.

A Single Solution

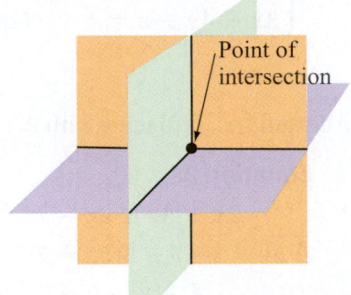

Point of intersection

If the planes intersect at a single point, that ordered triple is the solution to the system.

Infinite Number of Solutions

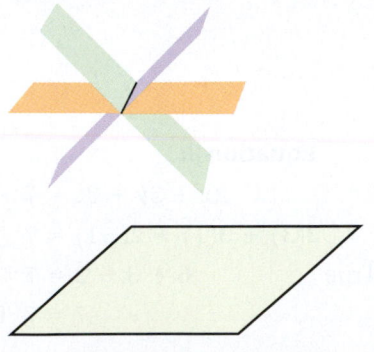

Connection Recall from Section 9.1 that a system of equations that has a solution (single or infinite) is consistent and a system that has no solution is inconsistent. This is true for systems of three equations as well. We also discussed dependent and independent equations. However, for systems of three equations, defining dependent and independent equations and discussing their graphical representations is too complicated for this course. Here, we simply say that systems with an infinite number of solutions have dependent equations. As we will see, we can identify these systems when solving them using the elimination method.

If the three planes intersect along a line, the system has an infinite number of solutions, which are the coordinates of any point along that line.

If all three graphs are the same plane, the system has an infinite number of solutions, which are the coordinates of any point in the plane.

No Solution

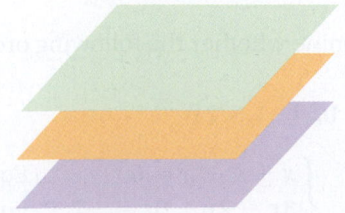

If all of the planes are parallel, the system has no solution.

Pairs of planes also can intersect, as shown. However, because all three planes do not have a common intersection, the system has no solution.

Note: Planes can be configured in still more ways so that they do not have a common intersection.

Objective ③ Solve a system of three linear equations using the elimination method. Because graphing systems of three linear equations is usually impractical, we solve them using elimination. The process is much like what we learned in Section 9.3.

Example 2 Solve the system of equations using elimination.

a.
$$\begin{cases} x + y + z = 2 & \text{(Equation 1)} \\ 2x + y + 2z = 1 & \text{(Equation 2)} \\ 3x + 2y + z = 1 & \text{(Equation 3)} \end{cases}$$

SOLUTION First, we choose a variable and eliminate that variable from two pairs of equations. Let's eliminate y using equations 1 and 2. We multiply equation 1 by -1, then add the equations.

Note: Our initial goal is to generate two equations with the same two variables.

$$\begin{array}{ll} \text{(Eq. 1)} & x + y + z = 2 \quad \xrightarrow{\text{Multiply by } -1} \quad -x - y - z = -2 \\ \text{(Eq. 2)} & 2x + y + 2z = 1 \qquad\qquad\qquad\quad \underline{2x + y + 2z = 1} \quad \text{Add the equations.} \\ & \qquad\qquad\qquad\qquad\qquad\qquad\qquad\quad x \quad\;\; + z = -1 \quad \text{(Equation 4)} \end{array}$$

Now we need to eliminate y from another pair of equations. We will choose equations 1 and 3.

$$\begin{array}{ll} \text{(Eq. 1)} & x + y + z = 2 \quad \xrightarrow{\text{Multiply by } -2} \quad -2x - 2y - 2z = -4 \\ \text{(Eq. 3)} & 3x + 2y + z = 1 \qquad\qquad\qquad\quad \underline{3x + 2y + \;\; z = 1} \quad \text{Add the equations.} \\ & \qquad\qquad\qquad\qquad\qquad\qquad\qquad\quad x \quad\;\; - z = -3 \quad \text{(Equation 5)} \end{array}$$

Equations 4 and 5 form a system of two equations with two variables x and z. We can eliminate z by adding the equations.

$$\begin{array}{ll} \text{(Eq. 4)} & x + z = -1 \\ \text{(Eq. 5)} & \underline{x - z = -3} \quad \text{Add the equations.} \\ & 2x = -4 \\ & x = -2 \quad \text{Divide both sides by 2 to isolate } x. \end{array}$$

To find z, we substitute -2 for x in equation 4 or 5. We will use equation 4.

$$\begin{array}{ll} x + z = -1 & \\ -2 + z = -1 & \text{Substitute } -2 \text{ for } x. \\ z = 1 & \text{Add 2 to both sides.} \end{array}$$

To find y, substitute -2 for x and 1 for z in any of the original equations. We will use equation 1.

$$\begin{array}{ll} x + y + z = 2 & \\ -2 + y + 1 = 2 & \text{Substitute } -2 \text{ for } x \text{ and } 1 \text{ for } z. \\ y - 1 = 2 & \text{Simplify the left side.} \\ y = 3 & \text{Add 1 to both sides to isolate } y. \end{array}$$

Note: Recall that ordered triples are written in the form (x, y, z).

The solution is $(-2, 3, 1)$. We can check the solution by verifying that the ordered triple satisfies each of the three original equations. We will leave the check to the reader.

b. $\begin{cases} 2x + 3y \quad\ \ = 3 & \text{(Equation 1)} \\ \quad\ \ 2y - 3z = -8 & \text{(Equation 2)} \\ 4x \quad\quad - z = 10 & \text{(Equation 3)} \end{cases}$

SOLUTION Notice that a variable is missing in each of the equations, which can simplify solving the system. We will eliminate z from equations 2 and 3.

Note: We chose Equation 3 because z has a coefficient of -1.

(Eq. 2) $2y - 3z = -8$ $2y - 3z = -8$

(Eq. 3) $4x - z = 10 \xrightarrow{\text{Multiply by } -3} -12x \quad\quad + 3z = -30$ **Add the equations.**

$$\overline{ -12x + 2y \quad\quad\ = -38} \quad \text{(Equation 4)}$$

Because equations 1 and 4 do not have z terms, we do not need to eliminate z from another pair of equations. We will use equations 1 and 4 to eliminate x.

(Eq. 1) $\quad\ 2x + 3y = 3 \xrightarrow{\text{Multiply by } 6} 12x + 18y = 18$

(Eq. 4) $-12x + 2y = -38$ $-12x + \ 2y = -38$ **Add the equations.**

$$\overline{ 20y = -20}$$

$$y = -1 \quad \text{**Divide both sides by 20 to isolate } y\text{.**}$$

To find x, substitute -1 for y in equation 1 or 4. We will use equation 1.

$$2x + 3(-1) = 3 \quad \text{**Substitute } -1 \text{ for } y \text{ in equation 1.**}$$

$$2x - 3 = 3 \quad \text{**Simplify.**}$$

$$2x = 6 \quad \text{**Add 3 to both sides.**}$$

$$x = 3 \quad \text{**Divide both sides by 2 to isolate } x\text{.**}$$

To find z, substitute in equation 2 or 3. (Equation 1 has no z term.) We will use equation 2.

$$2(-1) - 3z = -8 \quad \text{**Substitute } -1 \text{ for } y \text{ in equation 2.**}$$

$$-2 - 3z = -8 \quad \text{**Simplify.**}$$

$$-3z = -6 \quad \text{**Add 2 to both sides.**}$$

$$z = 2 \quad \text{**Divide both sides by } -3 \text{ to isolate } z\text{.**}$$

The solution is $(3, -1, 2)$. We will leave the check to the reader.

Following is a summary of the process of solving a system of three linear equations.

Procedure   **Solving Systems of Three Linear Equations Using Elimination**

To solve a system of three linear equations with three unknowns using elimination,

1. Write each equation in the form $Ax + By + Cz = D$.
2. Eliminate one variable from one pair of equations using the elimination method.
3. If necessary, eliminate the same variable from another pair of equations.
4. Steps 2 and 3 result in two equations with the same two variables. Solve these equations using the elimination method.
5. To find the third variable, substitute the values of the variables found in Step 4 into any of the three original equations that contain the third variable.
6. Check the ordered triple in all three original equations.

Your Turn 2 Solve the system using elimination.

a. $\begin{cases} x + 3y + 2z = 6 \\ 2x - 3y + z = -18 \\ -3x + 2y + z = 12 \end{cases}$

b. $\begin{cases} 2x - 3z = -6 \\ x + 3y = -3 \\ 2y - 3z = -16 \end{cases}$

Let's see what happens when a system of three linear equations has no solution (inconsistent) or has an infinite number of solutions (dependent).

Note: We mentioned in the Connection box on page 658 that we can identify systems of dependent equations when using the elimination method. A system of three equations is dependent if during the solution process we add two equations and get $0 = 0$. (We saw this with two equations in Example 5(b) of Section 9.3.) For example, if equation 4 of Example 3 were $-7x - z = -18$, the sum of equations 4 and 5 would be $0 = 0$.

(Eq. 4) $\quad -7x - z = -18$
(Eq. 5) $\quad \underline{7x + z = 18}$
$\qquad\qquad\qquad 0 = 0$

In such a case, we say that the equations in the system are dependent and that the system has an infinite number of solutions.

Example 3 Solve the system using elimination.

$$\begin{cases} 2x + y + 2z = -1 & \text{(Equation 1)} \\ -3x + 2y + 3z = -13 & \text{(Equation 2)} \\ 4x + 2y + 4z = 5 & \text{(Equation 3)} \end{cases}$$

Solution We need to eliminate a variable from two pairs of equations. Let's eliminate y using equations 1 and 2.

(Eq. 1) $\quad 2x + y + 2z = -1$ $\xrightarrow{\text{Multiply by } -2}$ $-4x - 2y - 4z = 2$
(Eq. 2) $-3x + 2y + 3z = -13$ $\qquad\qquad\qquad \underline{-3x + 2y + 3z = -13}$ **Add the equations.**
$\qquad\qquad\qquad\qquad\qquad\qquad\qquad\qquad\quad -7x \qquad - z = -11$ (Equation 4)

Now we eliminate y from equations 2 and 3.

(Eq. 2) $-3x + 2y + 3z = -13$ $\xrightarrow{\text{Multiply by } -1}$ $3x - 2y - 3z = 13$
(Eq. 3) $\quad 4x + 2y + 4z = 5$ $\qquad\qquad\qquad \underline{4x + 2y + 4z = 5}$ **Add the equations.**
$\qquad\qquad\qquad\qquad\qquad\qquad\qquad\qquad\quad 7x \qquad + z = 18$ (Equation 5)

Equations 4 and 5 form a system in x and z.

(Eq. 4) $\quad -7x - z = -11$
(Eq. 5) $\quad \underline{7x + z = 18}$ **Add the equations.**
$\qquad\qquad\qquad 0 = 7$

All variables are eliminated and the resulting equation is false, which means that this system has no solution; it is inconsistent.

Your Turn 3 Solve the following systems of equations using elimination.

a. $\begin{cases} 2x + y + 2z = 1 \\ x + y + z = 2 \\ 4x + 2y + 4z = 6 \end{cases}$

b. $\begin{cases} x + 3y - 6z = 9 \\ -7y + 6z = -3 \\ -2x + y + 6z = -15 \end{cases}$

Objective 4 **Solve application problems that translate to a system of three linear equations.**
We follow the same procedure to solve applications involving three unknowns as we did when solving applications with two unknowns.

Example 4 At a movie theater, John buys one popcorn, one soft drink, and one candy bar, all for $7. Fred buys two popcorns, three soft drinks, and two candy bars for $16. Carla buys one popcorn, two soft drinks, and three candy bars for $12. Find the price of one popcorn, one soft drink, and one candy bar.

Understand We have three unknowns and three relationships, and we are to find the cost of each.

Plan Select a variable for each unknown, translate the relationships to a system of three equations, and solve the system.

Execute Let x represent the cost of one popcorn, y the cost of one soft drink, and z the cost of one candy bar.

Relationship 1: One popcorn, one soft drink, and one candy bar cost $7.

$$\textbf{Translation: } x + y + z = 7$$

Relationship 2: Two popcorns, three soft drinks, and two candy bars cost $16.

$$\textbf{Translation: } 2x + 3y + 2z = 16$$

Relationship 3: One popcorn, two soft drinks, and three candy bars cost $12.

$$\textbf{Translation: } x + 2y + 3z = 12$$

$$\text{Our system:} \begin{cases} x + y + z = 7 & \text{(Equation 1)} \\ 2x + 3y + 2z = 16 & \text{(Equation 2)} \\ x + 2y + 3z = 12 & \text{(Equation 3)} \end{cases}$$

We will choose to eliminate z from two pairs of equations. We will start with equations 1 and 2.

(Eq. 1) $x + y + z = 7$ $\xrightarrow{\text{Multiply by } -2}$ $-2x - 2y - 2z = -14$

(Eq. 2) $2x + 3y + 2z = 16$ $\underline{2x + 3y + 2z = 16}$ **Add the equations.**

$y = 2$ (Equation 4)

> **Note:** Although we were trying to eliminate z, we ended up eliminating both x and z.

Equation 4 gives us the value of y, indicating that a soft drink costs $2. Now we choose another pair of equations and eliminate z again. We will use equations 1 and 3.

(Eq. 1) $x + y + z = 7$ $\xrightarrow{\text{Multiply by } -3}$ $-3x - 3y - 3z = -21$

(Eq. 3) $x + 2y + 3z = 12$ $\underline{x + 2y + 3z = 12}$ **Add the equations.**

$-2x - y = -9$ (Equation 5)

Because we already know that $y = 2$, we can substitute for y in $-2x - y = -9$.

$-2x - \mathbf{2} = -9$ **Substitute 2 for y.**

$-2x = -7$ **Add 2 to both sides.**

$x = 3.5$ **Divide both sides by -2 to isolate x.**

Because x represents the cost of one popcorn, a popcorn costs $3.50. To find z, substitute for x and y into one of the original equations. We will use equation 1.

$\mathbf{3.5} + \mathbf{2} + z = 7$ **Substitute 3.5 for x and 2 for y.**

$5.5 + z = 7$ **Simplify.**

$z = 1.5$ **Subtract 5.5 from both sides to isolate z.**

Because z represents the cost of one candy bar, a candy bar costs $1.50.

Answer Popcorn costs $3.50, a soft drink costs $2.00, and a candy bar costs $1.50.

Check Verify that at these prices—the amounts of money spent by John, Fred, and Carla—are correct. We will leave the check to the reader.

Your Turn 4   A small aquarium is in the shape of a rectangular solid. The sum of the length, width, and height is 46 inches. The sum of twice the length, three times the width, and the height is 90 inches. The sum of the length, twice the width, and three times the height is 86 inches. Find the dimensions of the aquarium.

9.4 Exercises   For Extra Help   *MyMathLab*

1. When solving a system of three equations with three variables, you eliminate a variable from one pair of equations, then eliminate the same variable from a second pair of equations. Does it matter which equations are chosen?

2. When solving a system of three equations with three variables, we eliminate a variable from one pair of equations, then eliminate the same variable from a second pair of equations. Why do we eliminate the same variable from two pairs of equations?

3. Suppose you are given this system of equations:
$$\begin{cases} x + y + z = 0 & \text{(Equation 1)} \\ 2x + 4y + 3z = 5 & \text{(Equation 2)} \\ 4x - 2y + 3z = -13 & \text{(Equation 3)} \end{cases}$$
Which variable would you choose to eliminate? Which pairs of equations would you use? Why?

4. When using elimination to solve a system of three equations with three variables, how do you know if it has no solution?

5. Two drawings were given in the text of inconsistent systems of equations with three unknowns. Make another drawing of an inconsistent system other than the ones given.

6. Where will a point be located if it solves two equations of a system of three equations in three variables but does not solve the third equation?

For Exercises 7–12, determine whether the ordered triple is a solution of the system. See Example 1.

7. $(3, -1, 1)$
$$\begin{cases} x + y + z = 3 \\ 2x - 2y - z = 7 \\ 2x + y - 2z = 3 \end{cases}$$

8. $(2, -2, 1)$
$$\begin{cases} 3x + 2y + z = 3 \\ 2x - 3y - 2z = 8 \\ -2x + 4y + 3z = -9 \end{cases}$$

9. $(1, 0, 2)$
$$\begin{cases} 2x + 3y - 3z = -4 \\ -2x + 4y - z = -4 \\ 3x - 4y + 2z = 5 \end{cases}$$

10. $(3, 4, 0)$
$$\begin{cases} x + 2y + 5z = 11 \\ 3x - 2y - 4z = 1 \\ 2x + 2y + 3z = 12 \end{cases}$$

11. $(2, -2, 4)$
$$\begin{cases} x + 2y - z = -6 \\ 2x - 3y + 4z = 26 \\ -x + 2y - 3z = -18 \end{cases}$$

12. $(0, -2, 4)$
$$\begin{cases} 3x + 2y - 3z = -16 \\ 2x - 4y - z = 4 \\ -3x + 4y - 2z = -16 \end{cases}$$

For Exercises 13–34, solve the systems of equations. See Examples 2 and 3.

13. $\begin{cases} x + y + z = 5 \\ 2x + y - 2z = -5 \\ x - 2y + z = 8 \end{cases}$

14. $\begin{cases} x + y + z = 2 \\ 3x + y - z = -2 \\ 2x - 2y + 3z = 15 \end{cases}$

15. $\begin{cases} x + y + z = 2 \\ 4x - 3y + 2z = 2 \\ 2x + 3y - 2z = -8 \end{cases}$

16. $\begin{cases} x + y - z = 7 \\ -2x + 2y - z = -5 \\ 3x - 3y + 2z = 6 \end{cases}$

17. $\begin{cases} 2x + y + 2z = 5 \\ 3x - 2y + 3z = 4 \\ -2x + 3y + z = 8 \end{cases}$

18. $\begin{cases} 2x + 3y - 2z = -4 \\ 4x - 3y + z = 25 \\ x + 2y - 4z = -12 \end{cases}$

19. $\begin{cases} x + 2y - z = 1 \\ 2x + 4y - 2z = -8 \\ 3x + y - 4z = 6 \end{cases}$

20. $\begin{cases} 3x - 2y + z = 5 \\ 4x - 5y + 2z = 7 \\ 9x - 6y + 3z = 7 \end{cases}$

21. $\begin{cases} 4x - 2y + 3z = 6 \\ 6x - 3y + 4.5z = 9 \\ 12x - 6y + 9z = 18 \end{cases}$

22. $\begin{cases} -8x + 4y + 6z = -18 \\ 2x - y - 1.5z = 4.5 \\ 4x - 2y - 3z = 9 \end{cases}$

23. $\begin{cases} x = 2y + z + 7 \\ y = -3x + 2z + 1 \\ 2x + y - z = 0 \end{cases}$

24. $\begin{cases} z = -3x + y - 10 \\ 2x + 3y - 2z = 5 \\ x = 3y - 3z - 14 \end{cases}$

25. $\begin{cases} x = 4y - z + 1 \\ 3x + 2y - z = -8 \\ x + 6y + 2z = -3 \end{cases}$

26. $\begin{cases} 2x - 3y - 4z = 3 \\ y = -4x + 8z - 1 \\ x = -5y - 2z - 4 \end{cases}$

27. $\begin{cases} 4x + 2y + 3z = 9 \\ 2x - 4y - z = 7 \\ 3x - 2z = 4 \end{cases}$

28. $\begin{cases} 4x + 3y - 2z = 19 \\ 2x + 5z = -4 \\ 3x + 2y + 3z = 5 \end{cases}$

29. $\begin{cases} 3x - 2z = -1 \\ 4x + 5y = 23 \\ y + 2z = -1 \end{cases}$

30. $\begin{cases} 4y + 3z = 2 \\ 3x + 4y = 2 \\ 2x - 5z = -6 \end{cases}$

31. $\begin{cases} 3x + 2y = -2 \\ 2x - 3z = 1 \\ 0.4y - 0.5z = -2.1 \end{cases}$

32. $\begin{cases} 0.2x - 0.3z = -1.8 \\ 3x + 2y = 5 \\ 3x + 2z = -1 \end{cases}$

33. $\begin{cases} \dfrac{3}{2}x + y - z = 0 \\ 4y - 3z = -22 \\ -0.2x + 0.3y = -2 \end{cases}$

34. $\begin{cases} -0.2x - 0.3y + 0.1z = -0.3 \\ -3x + 2y = 13 \\ \dfrac{1}{4}y - \dfrac{1}{2}z = 2 \end{cases}$

For Exercises 35–46, translate to a system of three equations; then solve.
See Example 4.

35. The sum of the measures of the angles in every triangle is 180°. The measure of one angle of a triangle is three times that of a second angle, and the measure of the second angle is 5° less than the measure of the third. Find the measure of each angle of the triangle.

36. The perimeter of a triangle is 33 inches. The sum of the length of the longest side and twice the length of the shortest side is 31 inches. Twice the length of the longest side minus both of the other side lengths is 12 inches. Find the side lengths.

37. At a fast-food restaurant, one burger, one order of fries, and one drink cost $5.00; three burgers, two orders of fries, and two drinks cost $12.50; and two burgers, four orders of fries, and three drinks cost $14.00. Find the individual cost of one burger, one order of fries, and one drink.

38. James went to the college bookstore and purchased two pens, two erasers, and one pack of paper for $6. Tamika purchased four pens, three erasers, and two packs of paper for $11.50. Jermaine purchased three pens, one eraser, and three packs of paper for $9.50. Find the individual costs of one pen, one eraser, and one pack of paper.

39. In a basketball game, John scored 14 points with a combination of 3-point field goals, 2-point field goals, and free throws (1 point each). If he scored a total of 7 times and made two more free throws than 2-point field goals, find the number of each that he made.

40. At a track meet, 10 points are awarded for each first-place finish, 5 points for each second place, and 1 point for each third place. Suppose a track team scored a total of 71 points and had two more first-place finishes than seconds and one less first-place finish than third. Find the number of first-, second-, and third-place finishes for the team.

41. A delicatessen sells Black Forest ham for $11.96 per pound, turkey breast for $8.76 per pound, and roast beef for $9.16 per pound. Suppose the deli makes a 10-pound party tray of these three meats such that the average cost is $9.80 per pound. Find the number of pounds of each meat if the number of pounds of turkey breast is equal to the sum of the number of pounds of Black Forest ham and roast beef.

42. A coffee shop sells Jamaican Blue Mountain coffee for $45.99 per pound, Hawaiian Kona for $36.99 per pound, and Sulawesi Kalossi for $12.99 per pound. A 25-pound mixture of these three coffees sells for $30.27 per pound. Find the number of pounds of each coffee if the sum of the number of pounds of Jamaica Blue Mountain and Hawaiian Kona is 5 pounds more than the number of pounds of Sulawesi Kalossi.

43. A total of $8000 is invested in three stocks, which paid 4%, 6%, and 7% dividends, respectively, in one year. The amount invested in the stock that returned 7% dividends is $1500 more than the amount that returned 4% dividends. If the total dividends in one year were $475, find the amount invested in each stock.

44. A total of $5000 is invested in three funds. The money market fund pays 5% annually, the income fund pays 6% annually, and the growth fund pays 8% annually. The total earnings for one year from the three funds are $340. If the amount invested in the growth fund is $500 less than twice the amount invested in the money market fund, find the amount in each fund.

45. The number of calories burned per hour bicycling is 120 more than the number burned from brisk walking. The number of calories burned per hour in climbing stairs is 180 more than from bicycling and is twice that from brisk walking. Find the number of calories burned per hour for each of the three activities. (*Source: Numbers: How Many, How Long, How Far, How Much*)

46. The intensity of sound is measured in decibels. The decibel reading of a rock concert is 60 less than three times the reading of normal conversation and 10 less than that of a jet at takeoff. The decibel reading of a jet takeoff is 10 more than twice that of normal conversation. Find the decibel reading of each. (*Source: Numbers: How Many, How Long, How Far, How Much*)

Review Exercises

Exercise 1 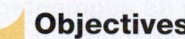 EXPRESSIONS

[1.7]   **1.** Use the distributive property to rewrite $-3(2x - 4y - z + 9)$.

Exercises 2–6 EQUATIONS AND INEQUALITIES

[2.2]   **2.** Given the equation $x - 0.5y = 8$, let $x = 6$; then solve for y.

[4.7] *For Exercises 3–6, use the function $f(x) = -2x + 1$.*

3. Find $f(1)$.

4. Find the y-intercept.

5. Find the slope.

6. Graph the function.

9.5 Solving Systems of Linear Equations Using Matrices

Objectives

1 Write a system of equations as an augmented matrix.

2 Solve a system of linear equations by transforming its augmented matrix to row echelon form.

Objective 1 Write a system of equations as an augmented matrix. Although the elimination method that we learned in Section 9.3 is effective for solving systems of linear equations, we can streamline the method by manipulating just the coefficients and constants in a **matrix**.

Definition   *Matrix:* A rectangular array of numbers.

Following are some examples of matrices (plural of *matrix*).

$$\begin{bmatrix} 1 & -2 \\ -3 & 4 \end{bmatrix} \quad \begin{bmatrix} 1 & -5 & 6 \\ -2 & 4 & 0 \end{bmatrix} \quad \begin{bmatrix} -3 & 0 & 9 \\ -2 & 4 & 7 \\ 9 & -2 & 0 \end{bmatrix}$$

Matrices are made up of horizontal rows and vertical columns.

Note: One way to remember that columns are vertical is to imagine the columns on a building, which are vertical.

Column 1   Column 2   Column 3   Column 1   Column 2

Row 1 → $\begin{bmatrix} 1 & 4 & -2 \\ 5 & -1 & 3 \end{bmatrix}$ ← Row 2   Row 1 → Row 2 → Row 3 → $\begin{bmatrix} 3 & -4 \\ -1 & 2 \\ 5 & -3 \end{bmatrix}$

The number of rows followed by the number of columns gives the *dimensions* of the matrix. So $\begin{bmatrix} 2 & 5 & -1 \\ 5 & 4 & 3 \end{bmatrix}$ is a 2×3 (read "2 by 3") matrix because it has two rows and three columns; $\begin{bmatrix} 2 & 5 \\ -4 & 6 \\ 3 & -2 \end{bmatrix}$ is a 3×2 matrix because it has three rows and two columns.

Each number in a matrix is called an *element*. To solve a system of equations using matrices, we first rewrite the system as an **augmented matrix**.

Definition   *Augmented matrix:* A matrix made up of the coefficients and the constant terms of a system. The constant terms are separated from the coefficients by a dashed vertical line.

For example, to write the system $\begin{cases} 3x - 2y = 7 \\ 4x - 3y = 10 \end{cases}$ as an augmented matrix, we omit the variables and write $\begin{bmatrix} 3 & -2 & | & 7 \\ 4 & -3 & | & 10 \end{bmatrix}$.

Example 1   Write $\begin{cases} 2x + 3y - 4z = 1 \\ 3x - 5y + z = -4 \\ 2x - 6y + 3z = 2 \end{cases}$ as an augmented matrix.

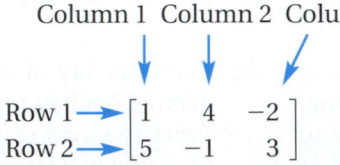

SOLUTION $\begin{bmatrix} 2 & 3 & -4 & | & 1 \\ 3 & -5 & 1 & | & -4 \\ 2 & -6 & 3 & | & 2 \end{bmatrix}$ **Note:** It is helpful to think of the dashed line as the equal sign of each equation.

Answer   **Your Turn 1**

$\begin{bmatrix} 1 & 3 & | & -2 \\ -4 & -5 & | & 7 \end{bmatrix}$

Your Turn 1   Write the augmented matrix for $\begin{cases} x + 3y = -2 \\ -4x - 5y = 7 \end{cases}$.

Objective 2 Solve a system of linear equations by transforming its augmented matrix to row echelon form. In using the elimination method, we can interchange equations, add equations, or multiply equations by a number and then add the rewritten equations. Because each row of an augmented matrix contains the constants and coefficients of each equation in the system, we can interchange rows, add rows, or multiply rows by a number and then add the rewritten rows.

Rule   **Row Operations**

The solution of a system is not affected by the following row operations in its augmented matrix.

1. Any two rows may be interchanged.
2. The elements of any row may be multiplied (or divided) by any nonzero real number.
3. Any row may be replaced by a row resulting from adding the elements of that row (or multiples of that row) to a multiple of the elements of any other row.

We use these row operations to solve a system of linear equations by transforming the augmented matrix of the system into an equivalent matrix that is in **row echelon form**.

Definition   *Row echelon form:* An augmented matrix whose coefficient portion has 1's on the diagonal from upper left to lower right and 0s below the 1's.

For example, $\begin{bmatrix} 1 & 3 & | & 3 \\ 0 & 1 & | & 4 \end{bmatrix}$ and $\begin{bmatrix} 1 & 2 & -4 & | & 5 \\ 0 & 1 & -5 & | & 7 \\ 0 & 0 & 1 & | & 5 \end{bmatrix}$ are in row echelon form.

In the following examples we let R_1 mean row 1, R_2 mean row 2, and so on.

Example 2   Solve the following linear systems by transforming their augmented matrices into row echelon form.

a. $\begin{cases} x - 2y = -4 & \text{(Equation 1)} \\ -2x + 5y = 9 & \text{(Equation 2)} \end{cases}$

SOLUTION   First, we write the augmented matrix: $\begin{bmatrix} 1 & -2 & | & -4 \\ -2 & 5 & | & 9 \end{bmatrix}$.

Now we perform row operations to transform the matrix into row echelon form. The element in the first row, first column is already 1, which is what we want. Therefore, we need to rewrite the matrix so that -2 in the second row, first column becomes 0. To do this, we multiply the first row by 2 and add it to the second row.

Note: The result of the row operations replaces the affected row in the matrix. Other rows are rewritten unchanged.

$2R_1 + R_2 \longrightarrow \begin{bmatrix} 1 & -2 & | & -4 \\ 0 & 1 & | & 1 \end{bmatrix}$

$2(1 \ -2 \ | \ -4) = (2 \ -4 \ | \ -8)$ and $(2 \ -4 \ | \ -8) + (-2 \ 5 \ | \ 9) = (0 \ 1 \ | \ 1)$ New $R_2 = 2R_1 + R_2$

The resulting matrix represents the system $\begin{cases} x - 2y = -4 \\ y = 1 \end{cases}$.

Because $y = 1$, we can solve for x using substitution.

$x - 2(1) = -4$ **Substitute 1 for y in $x - 2y = -4$.**

$x - 2 = -4$ **Simplify.**

$x = -2$ **Add 2 to both sides.**

The solution is $(-2, 1)$. The check is left to the reader.

b. $\begin{cases} 2x + 6y = 1 & \text{(Equation 1)} \\ x + 8y = 3 & \text{(Equation 2)} \end{cases}$

SOLUTION   First, we write the augmented matrix: $\begin{bmatrix} 2 & 6 & | & 1 \\ 1 & 8 & | & 3 \end{bmatrix}$.

Now we perform row operations to get row echelon form.

$\frac{1}{2}R_1 \longrightarrow \begin{bmatrix} 1 & 3 & | & \frac{1}{2} \\ 1 & 8 & | & 3 \end{bmatrix}$ We need the 2 in row 1, column 1 to be a 1; so multiply each number in row 1 by $\frac{1}{2}$.

$-R_1 + R_2 \longrightarrow \begin{bmatrix} 1 & 3 & | & \frac{1}{2} \\ 0 & 5 & | & \frac{5}{2} \end{bmatrix}$ We need the 1 in row 2, column 1 to be 0; so multiply row 1 by -1 and add it to row 2.

$\frac{1}{5}R_2 \longrightarrow \begin{bmatrix} 1 & 3 & | & \frac{1}{2} \\ 0 & 1 & | & \frac{1}{2} \end{bmatrix}$ We need the 5 in row 2, column 2 to be 1; so multiply row 2 by $\frac{1}{5}$.

This matrix represents the system $\begin{cases} x + 3y = \dfrac{1}{2} \\ y = \dfrac{1}{2} \end{cases}$.

Because $y = \dfrac{1}{2}$, we can solve for x using substitution.

$x + 3\left(\dfrac{1}{2}\right) = \dfrac{1}{2}$ Substitute $\frac{1}{2}$ for y in $x + 3y = \frac{1}{2}$.

$x + \dfrac{3}{2} = \dfrac{1}{2}$ Simplify.

$x = -1$ Subtract $\frac{3}{2}$ from both sides.

The solution is $\left(-1, \dfrac{1}{2}\right)$. The check is left to the reader.

Your Turn 2 Solve the following system by transforming its augmented matrix into row echelon form.

$$\begin{cases} 3x + 4y = 6 \\ x - 3y = -11 \end{cases}$$

Now let's use row operations to solve systems of three equations.

Example 3 Use the row echelon method to solve the following system.

$$\begin{cases} 2x + y - 2z = -3 & \text{(Equation 1)} \\ x + y + 2z = 7 & \text{(Equation 2)} \\ 4y + 3z = 5 & \text{(Equation 3)} \end{cases}$$

SOLUTION Write the augmented matrix: $\begin{bmatrix} 2 & 1 & -2 & | & -3 \\ 1 & 1 & 2 & | & 7 \\ 0 & 4 & 3 & | & 5 \end{bmatrix}$.

Now we perform row operations to get row echelon form.

$\begin{matrix} R_2 \longrightarrow \\ R_1 \longrightarrow \\ \end{matrix} \begin{bmatrix} 1 & 1 & 2 & | & 7 \\ 2 & 1 & -2 & | & -3 \\ 0 & 4 & 3 & | & 5 \end{bmatrix}$ Because we need a 1 in row 1, column 1, we interchange rows 1 and 2.

Answer Your Turn 2

$(-2, 3)$

$$-2R_1 + R_2 \longrightarrow \begin{bmatrix} 1 & 1 & 2 & | & 7 \\ 0 & -1 & -6 & | & -17 \\ 0 & 4 & 3 & | & 5 \end{bmatrix}$$

To get a 0 in row 2, column 1, we multiply the new row 1 by -2 and add it to row 2.

$$4R_2 + R_3 \longrightarrow \begin{bmatrix} 1 & 1 & 2 & | & 7 \\ 0 & -1 & -6 & | & -17 \\ 0 & 0 & -21 & | & -63 \end{bmatrix}$$

Row 3 needs only a 0 in the second column, so we multiply row 2 by 4 and add it to row 3.

$$\begin{matrix} -R_2 \longrightarrow \\ -\dfrac{1}{21}R_3 \longrightarrow \end{matrix} \begin{bmatrix} 1 & 1 & 2 & | & 7 \\ 0 & 1 & 6 & | & 17 \\ 0 & 0 & 1 & | & 3 \end{bmatrix}$$

To get a 1 in row 2, column 2, we multiply row 2 by -1.
To get a 1 in row 3, column 3, we multiply row 3 by $-\dfrac{1}{21}$.

The resulting matrix represents the system $\begin{cases} x + y + 2z = 7 \\ \quad\quad y + 6z = 17. \\ \quad\quad\quad\quad z = 3 \end{cases}$

To find y, substitute 3 for z in $y + 6z = 17$.

$$y + 6(3) = 17 \quad \text{Substitute 3 for } z.$$
$$y + 18 = 17 \quad \text{Multiply 6 and 3.}$$
$$y = -1 \quad \text{Subtract 18 from both sides.}$$

To find x, substitute -1 for y and 3 for z in $x + y + 2z = 7$.

$$x + (-1) + 2(3) = 7 \quad \text{Substitute } -1 \text{ for } y \text{ and 3 for } z.$$
$$x + 5 = 7 \quad \text{Simplify the left side of the equation.}$$
$$x = 2 \quad \text{Subtract 5 from both sides of the equation.}$$

The solution is $(2, -1, 3)$. We can check by verifying that the ordered triple satisfies all three of the original equations. This is left to the reader.

Your Turn 3 Use the row echelon method to solve the following system.

$$\begin{cases} 3x + y + z = 2 \\ x + 2y - z = -4 \\ 2x - 2y + 3z = 9 \end{cases}$$

▦ Calculator Tips

To use a graphing calculator to solve the system in Example 3, press the MATRX *key (on the TI-84, press* 2nd x^{-1} *); then use the arrow keys to highlight **EDIT**. You will see a list of matrices and their current sizes. Press 1 to select the first matrix, named [A]. We need our matrix to be 3×4; so press* 3 ENTER *, then* 4 ENTER *. The cursor will be in row 1, column 1 of a 3×4 matrix. You can now enter each number. Use the arrow keys to move the cursor to any position in the matrix. Note that the cursor position's row number, column number, and current value are displayed at the bottom of the screen. After you enter each number for Example 3, your screen should look like this:*

$$\text{MATRX }[A]\ 3 \times 4$$
$$\begin{bmatrix} 2 & 1 & -2 & -3 \\ 1 & 1 & 2 & 7 \\ 0 & 4 & 3 & 5 \end{bmatrix}$$

When you finish entering all of the numbers, press 2nd MODE *to quit editing.*

(continued)

Answer **Your Turn 3**
$(1, -2, 1)$

Now to put the matrix into row echelon form, press MATRX *again and move the cursor to highlight MATH at the top of the screen. From the menu, select A: ref (and press* ENTER *. You will be prompted to enter the name of the desired matrix. Press* MATRX *again and select 1:[A] 3 × 4. You will now see ref ([A]. Use the*) *key to close the parentheses and press* ENTER *. You now have the row echelon form, and your screen should look like this:*

Note: The row echelon form that the calculator got is different from the row echelon form that we got in Example 3. That is because there is more than one way of getting the row echelon form; so it is not unique.

ref ([A])
$$\begin{bmatrix} 1 & .5 & -1 & -1.5 \\ 0 & 1 & .75 & 1.25 \\ 0 & 0 & 1 & 3 \end{bmatrix}$$

Note: The function **ref** stands for *row echelon form*.

Notice that the last row indicates that z = 3, and we can use substitution to find the other values.

Your calculator can also show the complete solution. Instead of selecting A: ref (, select B: rref (from the math menu. Your screen will look like this:

rref ([A])
$$\begin{bmatrix} 1 & 0 & 0 & 2 \\ 0 & 1 & 0 & -1 \\ 0 & 0 & 1 & 3 \end{bmatrix}$$

Note: This form is called *reduced row echelon form*, which is why the function is **rref**.

Note: The reduced row echelon form *is* unique.

Note that the first row of this form indicates that x = 2, the second row indicates that y = −1, and the third row indicates that z = 3, which is the solution for our system.

9.5 Exercises   For Extra Help   *MyMathLab*   Math XL PRACTICE    WATCH   DOWNLOAD

Note: Exercises marked with a ★ represent challenging exercises.

1. How many rows and columns are in a 4 × 2 matrix?

2. How do you write a system of equations as an augmented matrix?

3. What does the dashed line in an augmented matrix correspond to in a system of equations?

4. Explain how solving a system of equations using row operations is similar to solving the system using elimination.

5. What is row echelon form?

6. Once an augmented matrix is in row echelon form, how do you find the values of the variables?

For Exercises 7–14, write the augmented matrix for the system of equations. See Example 1.

7. $\begin{cases} 14x + 7y = 6 \\ 7x + 6y = 8 \end{cases}$

8. $\begin{cases} 5x + 6y = 2 \\ 10x + 3y = -2 \end{cases}$

9. $\begin{cases} 7x - 6y = 1 \\ -2y = 5 \end{cases}$

10. $\begin{cases} 3x = 6 \\ 9x + 2y = -2 \end{cases}$

11. $\begin{cases} x - 3y + z = 4 \\ 2x - 4y + 2z = -4 \\ 6x - 2y + 5z = -4 \end{cases}$

12. $\begin{cases} 3x + 2y - 3z = 2 \\ 2x - 4y + 5z = -10 \\ 5x - 4y + z = 0 \end{cases}$

13. $\begin{cases} 4x + 6y - 2z = -1 \\ 8x + 3y = -12 \\ -y + 2z = 4 \end{cases}$

14. $\begin{cases} x - 3y = 3 \\ -3x + 4y + 9z = 3 \\ 2x + 7z = -9 \end{cases}$

For Exercises 15–18, given the matrices in row echelon form, find the solution for the system. See Objective 2 and Example 2.

15. $\begin{bmatrix} 1 & -3 & | & -2 \\ 0 & 1 & | & 2 \end{bmatrix}$

16. $\begin{bmatrix} 1 & -3 & | & 7 \\ 0 & 1 & | & -5 \end{bmatrix}$

17. $\begin{bmatrix} 1 & -4 & -8 & | & 6 \\ 0 & 1 & -2 & | & -7 \\ 0 & 0 & 1 & | & 1 \end{bmatrix}$

18. $\begin{bmatrix} 1 & -2 & 6 & | & 16 \\ 0 & 1 & -4 & | & -13 \\ 0 & 0 & 1 & | & 3 \end{bmatrix}$

For Exercises 19–24, complete the indicated row operation. See Objective 2 and Example 2.

19. Replace R_2 in $\begin{bmatrix} 1 & 3 & | & -1 \\ -2 & 5 & | & 6 \end{bmatrix}$ with $2R_1 + R_2$.

20. Replace R_2 in $\begin{bmatrix} 1 & 2 & | & -2 \\ 3 & 8 & | & -4 \end{bmatrix}$ with $-3R_1 + R_2$.

21. Replace R_3 in $\begin{bmatrix} 1 & -2 & 4 & | & 6 \\ 0 & 2 & -1 & | & -5 \\ 0 & 8 & -6 & | & -3 \end{bmatrix}$ with $-4R_2 + R_3$.

22. Replace R_2 in $\begin{bmatrix} 1 & 5 & -3 & | & 8 \\ -3 & 2 & -4 & | & -6 \\ 0 & 1 & -2 & | & 9 \end{bmatrix}$ with $3R_1 + R_2$.

23. Replace R_1 in $\begin{bmatrix} 4 & 8 & | & -10 \\ -1 & 3 & | & 2 \end{bmatrix}$ with $\frac{1}{4}R_1$.

24. Replace R_3 in $\begin{bmatrix} 1 & 3 & 4 & | & 8 \\ 0 & 1 & -2 & | & -6 \\ 0 & 0 & -6 & | & 24 \end{bmatrix}$ with $-\frac{1}{6}R_3$.

For Exercises 25–28, describe the next row operation that should be performed to make the matrix closer to row echelon form.

25. $\begin{bmatrix} 1 & 2 & | & -1 \\ -3 & 4 & | & 5 \end{bmatrix}$

26. $\begin{bmatrix} 1 & -3 & | & 2 \\ 5 & 6 & | & -4 \end{bmatrix}$

27. $\begin{bmatrix} 1 & -4 & 3 & | & 6 \\ 0 & 1 & -2 & | & 4 \\ 0 & 2 & -5 & | & 1 \end{bmatrix}$

28. $\begin{bmatrix} 1 & -2 & 4 & | & 7 \\ 0 & 1 & -2 & | & 3 \\ 0 & -5 & 4 & | & -9 \end{bmatrix}$

For Exercises 29–52, solve by transforming the augmented matrix into row echelon form. See Examples 2 and 3.

29. $\begin{cases} x - y = 5 \\ x + y = -1 \end{cases}$

30. $\begin{cases} x + y = 2 \\ x - y = -8 \end{cases}$

31. $\begin{cases} x + y = 3 \\ 3x - y = 1 \end{cases}$

32. $\begin{cases} x + 3y = -9 \\ -x + 2y = -11 \end{cases}$

33. $\begin{cases} x + 2y = -7 \\ 2x - 4y = 2 \end{cases}$

34. $\begin{cases} x - 3y = -13 \\ 3x + 2y = 5 \end{cases}$

35. $\begin{cases} -2x + 5y = 4 \\ x - 2y = -2 \end{cases}$

36. $\begin{cases} 2x + y = 12 \\ x - 3y = 6 \end{cases}$

37. $\begin{cases} 4x - 3y = -2 \\ 2x - 3y = -10 \end{cases}$

38. $\begin{cases} 6x - 3y = 0 \\ 2x + 4y = 0 \end{cases}$

39. $\begin{cases} 5x + 2y = 12 \\ 2x + 3y = -4 \end{cases}$

40. $\begin{cases} 4x + 3y = 14 \\ 3x - 2y = 2 \end{cases}$

41. $\begin{cases} x + y + z = 3 \\ 2x + y - 3z = -10 \\ 2x + 2y + z = 3 \end{cases}$

42. $\begin{cases} x + 2y + z = 2 \\ 3x + y - z = 3 \\ 2x + y + 2z = 7 \end{cases}$

43. $\begin{cases} x - y + z = -1 \\ 2x - 2y + z = 0 \\ x + 3y + 2z = 1 \end{cases}$

44. $\begin{cases} x + 3y - 2z = 4 \\ 2x - 3y + 2z = -7 \\ 3x + 2y - 2z = -1 \end{cases}$

45. $\begin{cases} 2x - y + z = 8 \\ x - 2y + 3z = 11 \\ 2x + 3y - z = -6 \end{cases}$ **46.** $\begin{cases} 2x + 3y - 2z = -21 \\ 2x - 4y + 3z = 15 \\ 3x + 2y - 3z = -24 \end{cases}$ **47.** $\begin{cases} 3x + 2y - 3z = 1 \\ -2x + 3y - 4z = 7 \\ 5x - 2y + z = -5 \end{cases}$ **48.** $\begin{cases} 2x - 2y + z = 6 \\ -2x + 4y + 3z = 4 \\ 4x - 3y - 2z = -7 \end{cases}$

49. $\begin{cases} 3x - 6y + z = -10 \\ 7y - z = 2 \\ 2x + 4z = 14 \end{cases}$ **50.** $\begin{cases} 6x - y + 5z = -28 \\ 4x + 2y = 10 \\ 5x + 6z = -23 \end{cases}$ **51.** $\begin{cases} 2x + 5y - z = 10 \\ x - y - z = 14 \\ x - 6y = 20 \end{cases}$ **52.** $\begin{cases} 2x + 4y - 5z = 16 \\ x - 2y - 11z = 1 \\ 4x + 5y = -6 \end{cases}$

For Exercises 53–60, solve using a matrix on a graphing calculator.

★ **53.** $\begin{cases} 2x + 5y = -14 \\ 6x + 7y = -10 \end{cases}$ **54.** $\begin{cases} 4x - 3y = 11 \\ 5x + 4y = 68 \end{cases}$ **55.** $\begin{cases} 4x - 7y = -80 \\ 9x + 5y = -14 \end{cases}$ **56.** $\begin{cases} 7x - 3y = 23 \\ 4x - 9y = -67 \end{cases}$

57. $\begin{cases} 3x + 2y - 5z = -19 \\ 4x - 7y + 6z = 67 \\ 5x - 6y - 4z = 18 \end{cases}$ **58.** $\begin{cases} 6x - 7y + 2z = -52 \\ 8x + 5y - 6z = 132 \\ 12x + 4y - 7z = 150 \end{cases}$ **59.** $\begin{cases} 8x - 7y + 2z = 108 \\ 6x + 11y - 10z = -74 \\ -2y + 13z = 77 \end{cases}$ **60.** $\begin{cases} 8x - 11y + 14z = 40 \\ 5x + 15y - 5z = 35 \\ 17x - 9y = -157 \end{cases}$

Find   (X) *the mistake*

For Exercises 61 and 62, explain the mistake; then find the correct solution.

61. $\begin{cases} x + 3y = 13 \\ -4x - y = -26 \end{cases}$

$$\begin{bmatrix} 1 & 3 & | & 13 \\ -4 & -1 & | & -26 \end{bmatrix}$$

$4R_1 + R_2 \longrightarrow \begin{bmatrix} 1 & 3 & | & 13 \\ 0 & 11 & | & -13 \end{bmatrix}$

$\dfrac{1}{11}R_2 \longrightarrow \begin{bmatrix} 1 & 3 & | & 13 \\ 0 & 1 & | & -\dfrac{13}{11} \end{bmatrix}$

Solution: $\left(\dfrac{182}{11}, -\dfrac{13}{11} \right)$

62. $\begin{cases} x - y + z = 8 \\ 3x - z = -9 \\ 4y + z = -6 \end{cases}$

$$\begin{bmatrix} 1 & -1 & 1 & | & 8 \\ 3 & 0 & -1 & | & -9 \\ 4 & 0 & 1 & | & -6 \end{bmatrix}$$

$-3R_1 + R_2 \longrightarrow \begin{bmatrix} 1 & -1 & 1 & | & 8 \\ 0 & 3 & -4 & | & -33 \\ 4 & 0 & 1 & | & -6 \end{bmatrix}$

$\begin{array}{l} \dfrac{1}{3}R_2 \longrightarrow \\ -4R_1 + R_3 \longrightarrow \end{array} \begin{bmatrix} 1 & -1 & 1 & | & 8 \\ 0 & 1 & -\dfrac{4}{3} & | & -11 \\ 0 & 4 & -3 & | & -38 \end{bmatrix}$

$-4R_2 + R_3 \longrightarrow \begin{bmatrix} 1 & -1 & 1 & | & 8 \\ 0 & 1 & -\dfrac{4}{3} & | & -11 \\ 0 & 0 & \dfrac{7}{3} & | & 6 \end{bmatrix}$

$\dfrac{3}{7}R_3 \longrightarrow \begin{bmatrix} 1 & -1 & 1 & | & 8 \\ 0 & 1 & -\dfrac{4}{3} & | & -11 \\ 0 & 0 & 1 & | & \dfrac{18}{7} \end{bmatrix}$

Solution: $\left(-\dfrac{15}{7}, -\dfrac{53}{7}, \dfrac{18}{7} \right)$

For Exercises 63–74, translate the problem to a system of equations; then solve using matrices.

63. Brad purchased three grilled chicken sandwiches and two drinks for $12.90, and Angel purchased seven grilled chicken sandwiches and four drinks for $29.30. Find the price of one grilled chicken sandwich and one drink.

64. Sharika purchased three general admission tickets and two student tickets to a college play for $55, and Yo Chen purchased two general admission tickets and four student tickets for $50. Find the cost of one general admission and one student ticket.

65. The two longest rivers in the world, the Nile and the Amazon, have a combined length of 8050 miles. The Nile is 250 miles longer than the Amazon. Find the length of both rivers.

66. The longest vehicular tunnel in the world is the Saint Gotthard Tunnel in Switzerland, and the second longest is the Arlberg Tunnel in Austria. The total length of the two tunnels is 18.8 miles. The Saint Gotthard Tunnel is 1.4 miles longer than the Arlberg. Find the length of each tunnel. (*Source: Webster's New World Book of Facts*)

67. Nikita invested a total of $10,000 in certificates of deposit that pay 5% annually and in a money market account that pays 6% annually. If the total interest earned in one year from the two investments is $536, find the principal that was invested in each.

68. An athlete received a $60,000 signing bonus, which he invested in two funds— a money market fund paying 5% annually and a growth fund paying 7% annually. The total annual interest received from the two investments is $3440. Find the principal invested in each fund.

69. John bought two CDs, four books, and three DVDs for $164, and Tanelle bought five CDs, two books, and two DVDs for $160. If the sum of the cost of one CD and one book equals the cost of one DVD, find the cost of one of each.

70. Angel bought 3 pounds of salmon, 2 pounds of tuna, and 5 pounds of cod for $63. Sara bought 6 pounds of salmon, 3 pounds of tuna, and 5 pounds of cod for $94. The cost of 1 pound of salmon and 1 pound of tuna is the same as the cost of 3 pounds of cod. Find the cost of 1 pound of each.

71. In New York's 17 to 14 victory over New England in Super Bowl XLII, New York scored touchdowns (6 points), extra points (1 point), and field goals (3 points). The number of touchdowns equaled the number of extra points. Also, the number of touchdowns was one more than the number of field goals. Find how many of each type of score New York had. (*Source:* www.nfl.com)

72. In the 2008, NCAA women's basketball championship game, Tennessee player Candace Parker scored twelve times for a total of 17 points using a combination of 3-point field goals, 2-point field goals, and free throws (1 point). If the sum of the number of 3-point and 2-point field goals was two less than the number of free throws, find the number of each type.

73. An electrical circuit has three points of connection with different voltage measurements at each of the three connections. An engineer has written the following equations to describe the voltages. Find each voltage.

$$4v_1 - v_2 = 30$$
$$-2v_1 + 5v_2 - v_3 = 10$$
$$-v_2 + 5v_3 = 4$$

74. An engineer has written the following system of equations to describe the forces in pounds acting on a steel structure. Find the forces.

$$6F_1 - F_2 = 350$$
$$9F_1 - 2F_2 - F_3 = -100$$
$$3F_2 - F_3 = 250$$

Review Exercises

Exercises 1–6 **EXPRESSIONS**

[1.5] *For Exercises 1–6, simplify.*

1. $(-3)(4) - (-4)(5)$

2. $2(2 - 8) + 3[-1 - (-6)] - 4(4 - 6)$

3. $\dfrac{(-5)(-1) - (9)(3)}{(2)(-1) - (3)(3)}$

4. $\dfrac{2\left(\frac{5}{4}\right) - 8\left(\frac{1}{4}\right)}{\left(\frac{1}{2}\right)\left(\frac{5}{4}\right) - \left(\frac{3}{2}\right)\left(\frac{1}{4}\right)}$

5. $\dfrac{(1.1)(-0.2) - (1.7)(-0.5)}{(0.2)(-0.2) - (0.5)(-0.5)}$

6. $\dfrac{1[-45 - (-9)] + 7(-9 - 2) - 2(-27 - 30)}{1(6 - 3) - 2(-9 - 2) - 2[9 - (-4)]}$

9.6 Solving Systems of Linear Equations Using Cramer's Rule

Objectives

1. Evaluate determinants of 2×2 matrices.

2. Evaluate determinants of 3×3 matrices.

3. Solve systems of equations using Cramer's rule.

Objective 1 Evaluate determinants of 2×2 matrices. In this section, we explore another way to solve systems of linear equations using a special type of matrix called a **square matrix**.

Definition *Square matrix:* A matrix that has the same number of rows and columns.

For example, $\begin{bmatrix} 2 & 3 \\ 1 & -4 \end{bmatrix}$ and $\begin{bmatrix} 1 & -3 & 5 \\ -7 & 2 & 9 \\ 0 & 4 & -6 \end{bmatrix}$ are square matrices.

Every square matrix has a *determinant*. We write the determinant of a matrix, A, as $\det(A)$ or $|A|$. The method used to find the determinant of a matrix depends on its size.

Rule **Determinant of a 2×2 Matrix**

If $A = \begin{bmatrix} a_1 & b_1 \\ a_2 & b_2 \end{bmatrix}$, then $\det(A) = \begin{vmatrix} a_1 & b_1 \\ a_2 & b_2 \end{vmatrix} = a_1 b_2 - a_2 b_1$.

Warning: Be careful to note the difference between [ ] and | |. The notation [ ] denotes a matrix, whereas | | denotes the determinant of a matrix. Also, from the context, we know that | | does not mean absolute value.

Notice that the determinant contains diagonal products, as illustrated by the following diagram.

Note: Because subtraction is not commutative, make sure you note the order of the two products.

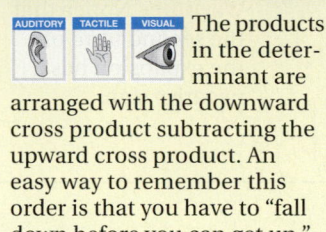
Note: An alternative notation for det(A) is $|A|$.

Example 1 Find the determinants of the following matrices.

a. $A = \begin{bmatrix} 3 & -2 \\ 2 & 4 \end{bmatrix}$

SOLUTION $\det(A) = \begin{vmatrix} 3 & -2 \\ 2 & 4 \end{vmatrix} = (3)(4) - (2)(-2) = 12 + 4 = 16$

b. $B = \begin{bmatrix} -3 & 5 \\ 4 & -2 \end{bmatrix}$

SOLUTION $\det(B) = (-3)(-2) - (4)(5) = 6 - 20 = -14$

c. $M = \begin{bmatrix} 1 & 3 \\ 3 & 9 \end{bmatrix}$

SOLUTION $\det(M) = (1)(9) - (3)(3) = 9 - 9 = 0$

Your Turn 1 Find the determinant of $\begin{bmatrix} 1 & -3 \\ -4 & 2 \end{bmatrix}$.

Objective 2 Evaluate determinants of 3 × 3 matrices. There are various methods of evaluating the determinant of a 3 × 3 matrix. One of the most common methods is *expanding by minors*. Each element of a square matrix has a number called the **minor** for that element.

Definition *Minor of an element of a matrix:* The determinant of the remaining matrix when the row and column in which the element is located are ignored.

Example 2 Find the minor of 2 in $\begin{bmatrix} 2 & -3 & -6 \\ -1 & 5 & -2 \\ 3 & -4 & 1 \end{bmatrix}$.

SOLUTION To find the minor of 2, we ignore its row and column (shown in blue) and evaluate the determinant of the remaining matrix (shown in red).

$$\begin{bmatrix} 2 & -3 & -6 \\ -1 & 5 & -2 \\ 3 & -4 & 1 \end{bmatrix}$$

$$\begin{vmatrix} 5 & -2 \\ -4 & 1 \end{vmatrix} = (5)(1) - (-4)(-2) = 5 - 8 = -3$$

Your Turn 2 Find the minor of 6 in $\begin{bmatrix} 1 & -3 & 6 \\ -2 & 2 & 0 \\ 4 & -1 & 5 \end{bmatrix}$.

To evaluate the determinant of a 3 × 3 matrix, we will expand by minors along the first column.

Answer Your Turn 1
-10

Answer Your Turn 2
$\begin{vmatrix} -2 & 2 \\ 4 & -1 \end{vmatrix} = -6$

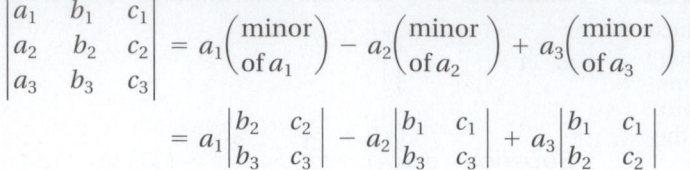

Rule Evaluating the Determinant of a 3×3 Matrix

$$\begin{vmatrix} a_1 & b_1 & c_1 \\ a_2 & b_2 & c_2 \\ a_3 & b_3 & c_3 \end{vmatrix} = a_1 \begin{pmatrix} \text{minor} \\ \text{of } a_1 \end{pmatrix} - a_2 \begin{pmatrix} \text{minor} \\ \text{of } a_2 \end{pmatrix} + a_3 \begin{pmatrix} \text{minor} \\ \text{of } a_3 \end{pmatrix}$$

Warning: Notice that the second term of the expansion has a negative sign.

$$= a_1 \begin{vmatrix} b_2 & c_2 \\ b_3 & c_3 \end{vmatrix} - a_2 \begin{vmatrix} b_1 & c_1 \\ b_3 & c_3 \end{vmatrix} + a_3 \begin{vmatrix} b_1 & c_1 \\ b_2 & c_2 \end{vmatrix}$$

Example 3 Find the determinant of $\begin{bmatrix} 2 & -3 & -4 \\ -1 & 2 & -2 \\ 3 & -4 & 1 \end{bmatrix}$.

SOLUTION Using the rule for expanding by minors along the first column, we have the following:

$$\begin{vmatrix} 2 & -3 & -4 \\ -1 & 2 & -2 \\ 3 & -4 & 1 \end{vmatrix} = 2\begin{pmatrix} \text{minor} \\ \text{of } 2 \end{pmatrix} - (-1)\begin{pmatrix} \text{minor} \\ \text{of } -1 \end{pmatrix} + 3\begin{pmatrix} \text{minor} \\ \text{of } 3 \end{pmatrix}$$

$$= 2\begin{vmatrix} 2 & -2 \\ -4 & 1 \end{vmatrix} - (-1)\begin{vmatrix} -3 & -4 \\ -4 & 1 \end{vmatrix} + 3\begin{vmatrix} -3 & -4 \\ 2 & -2 \end{vmatrix}$$

$$= 2(2 - 8) + 1(-3 - 16) + 3(6 + 8)$$

$$= -12 + (-19) + 42$$

$$= 11$$

Your Turn 3 Find the determinant of $\begin{bmatrix} 3 & -2 & 4 \\ -2 & 3 & 1 \\ 2 & -4 & 2 \end{bmatrix}$.

Objective ③ Solve systems of equations using Cramer's rule. Now we can use **Cramer's rule**, which uses determinants to solve systems of equations. To derive Cramer's rule, we solve a general system of equations using the elimination method. We show the derivation for a system of two equations in two unknowns.

$$\begin{cases} a_1x + b_1y = c_1 & \text{(Equation 1)} \\ a_2x + b_2y = c_2 & \text{(Equation 2)} \end{cases}$$

We eliminate y by multiplying equation 1 by b_2 and equation 2 by $-b_1$.

$$a_1b_2x + b_1b_2y = b_2c_1 \qquad \text{Multiply Equation 1 by } b_2.$$
$$\underline{-a_2b_1x - b_1b_2y = -b_1c_2} \qquad \text{Multiply Equation 2 by } -b_1.$$
$$a_1b_2x - a_2b_1x \qquad\quad = b_2c_1 - b_1c_2 \qquad \text{Add the equations.}$$
$$(a_1b_2 - a_2b_1)x = b_2c_1 - b_1c_2 \qquad \text{Factor out } x \text{ from the left side.}$$
$$x = \frac{b_2c_1 - b_1c_2}{a_1b_2 - a_2b_1} \qquad \text{Divide by } a_1b_2 - a_2b_1.$$

Notice that the numerator is $\begin{vmatrix} c_1 & b_1 \\ c_2 & b_2 \end{vmatrix}$ and the denominator is $\begin{vmatrix} a_1 & b_1 \\ a_2 & b_2 \end{vmatrix}$; so $x = \dfrac{\begin{vmatrix} c_1 & b_1 \\ c_2 & b_2 \end{vmatrix}}{\begin{vmatrix} a_1 & b_1 \\ a_2 & b_2 \end{vmatrix}}$.

Answer   **Your Turn 3**
26

If we repeat the same process and solve for y, we get $y = \dfrac{\begin{vmatrix} a_1 & c_1 \\ a_2 & c_2 \end{vmatrix}}{\begin{vmatrix} a_1 & b_1 \\ a_2 & b_2 \end{vmatrix}}$.

A similar approach is used to derive the rule for a system of three equations in three unknowns.

Rule   **Cramer's rule**

The solution to the system of linear equations $\begin{cases} a_1 x + b_1 y = c_1 \\ a_2 x + b_2 y = c_2 \end{cases}$ is

$$x = \dfrac{\begin{vmatrix} c_1 & b_1 \\ c_2 & b_2 \end{vmatrix}}{\begin{vmatrix} a_1 & b_1 \\ a_2 & b_2 \end{vmatrix}} = \dfrac{D_x}{D} \quad \text{and} \quad y = \dfrac{\begin{vmatrix} a_1 & c_1 \\ a_2 & c_2 \end{vmatrix}}{\begin{vmatrix} a_1 & b_1 \\ a_2 & b_2 \end{vmatrix}} = \dfrac{D_y}{D}.$$

The solution to the system of linear equations $\begin{cases} a_1 x + b_1 y + c_1 z = d_1 \\ a_2 x + b_2 y + c_2 z = d_2 \\ a_3 x + b_3 y + c_3 z = d_3 \end{cases}$ is

$$x = \dfrac{\begin{vmatrix} d_1 & b_1 & c_1 \\ d_2 & b_2 & c_2 \\ d_3 & b_3 & c_3 \end{vmatrix}}{\begin{vmatrix} a_1 & b_1 & c_1 \\ a_2 & b_2 & c_2 \\ a_3 & b_3 & c_3 \end{vmatrix}} = \dfrac{D_x}{D}, \quad y = \dfrac{\begin{vmatrix} a_1 & d_1 & c_1 \\ a_2 & d_2 & c_2 \\ a_3 & d_3 & c_3 \end{vmatrix}}{\begin{vmatrix} a_1 & b_1 & c_1 \\ a_2 & b_2 & c_2 \\ a_3 & b_3 & c_3 \end{vmatrix}} = \dfrac{D_y}{D}, \quad \text{and} \quad z = \dfrac{\begin{vmatrix} a_1 & b_1 & d_1 \\ a_2 & b_2 & d_2 \\ a_3 & b_3 & d_3 \end{vmatrix}}{\begin{vmatrix} a_1 & b_1 & c_1 \\ a_2 & b_2 & c_2 \\ a_3 & b_3 & c_3 \end{vmatrix}} = \dfrac{D_z}{D}.$$

Note: Each denominator, D, is the determinant of a matrix containing only the coefficients in the system. To find D_x, we replace the column of x-coefficients in the coefficient matrix with the constants from the system. To find D_y, we replace the column of y-coefficients in the coefficient matrix with the constant terms. We do likewise to find D_z.

Example 4   Use Cramer's rule to solve $\begin{cases} 2x + 3y = -5 \\ 3x - y = 9 \end{cases}$.

SOLUTION   First, we find D, D_x, and D_y.

Note: These are the coefficients of the system.

$$D = \begin{vmatrix} 2 & 3 \\ 3 & -1 \end{vmatrix} = (2)(-1) - (3)(3) = -2 - 9 = -11$$

Note: Replace the x-coefficients with the constants.

$$D_x = \begin{vmatrix} -5 & 3 \\ 9 & -1 \end{vmatrix} = (-5)(-1) - (9)(3) = 5 - 27 = -22$$

Note: Replace the y-coefficients with the constants.

$$D_y = \begin{vmatrix} 2 & -5 \\ 3 & 9 \end{vmatrix} = (2)(9) - (3)(-5) = 18 + 15 = 33$$

Note: If $D = 0$ and D_x and $D_y \neq 0$, the system is inconsistent (no solution). If D, D_x and $D_y = 0$, the system is dependent (infinite number of solutions).

Now we can find x and y.

$$x = \frac{D_x}{D} = \frac{-22}{-11} = 2 \qquad y = \frac{D_y}{D} = \frac{33}{-11} = -3$$

The solution is $(2, -3)$, which we can check by verifying that it satisfies both equations in the system. We will leave the check to the reader.

Your Turn 4   Use Cramer's rule to solve $\begin{cases} 3x - 2y = -16 \\ x + 3y = 2 \end{cases}$.

Answer   **Your Turn 4**

$(-4, 2)$

Note: Recall that we find the determinant of a 3×3 matrix by expanding by minors along the first column.

Example 5 Use Cramer's rule to solve $\begin{cases} x + 2y - 2x = -7 \\ 3x - 2y + z = 15 \\ 2x + 3y - 3z = -9 \end{cases}$.

SOLUTION We need to find D, D_x, D_y, and D_z.

$$D = \begin{vmatrix} 1 & 2 & -2 \\ 3 & -2 & 1 \\ 2 & 3 & -3 \end{vmatrix} = (1)\begin{vmatrix} -2 & 1 \\ 3 & -3 \end{vmatrix} - (3)\begin{vmatrix} 2 & -2 \\ 3 & -3 \end{vmatrix} + (2)\begin{vmatrix} 2 & -2 \\ -2 & 1 \end{vmatrix}$$

$$= 1(6 - 3) - 3(-6 + 6) + 2(2 - 4)$$

$$= 3 - 0 + (-4)$$

$$= -1$$

$$D_x = \begin{vmatrix} -7 & 2 & -2 \\ 15 & -2 & 1 \\ -9 & 3 & -3 \end{vmatrix} = (-7)\begin{vmatrix} -2 & 1 \\ 3 & -3 \end{vmatrix} - (15)\begin{vmatrix} 2 & -2 \\ 3 & -3 \end{vmatrix} + (-9)\begin{vmatrix} 2 & -2 \\ -2 & 1 \end{vmatrix}$$

$$= -7(6 - 3) - 15(-6 + 6) + (-9)(2 - 4)$$

$$= -21 - 0 + 18$$

$$= -3$$

$$D_y = \begin{vmatrix} 1 & -7 & -2 \\ 3 & 15 & 1 \\ 2 & -9 & -3 \end{vmatrix} = (1)\begin{vmatrix} 15 & 1 \\ -9 & -3 \end{vmatrix} - (3)\begin{vmatrix} -7 & -2 \\ -9 & -3 \end{vmatrix} + (2)\begin{vmatrix} -7 & -2 \\ 15 & 1 \end{vmatrix}$$

$$= 1(-45 + 9) - 3(21 - 18) + 2(-7 + 30)$$

$$= -36 - 9 + 46$$

$$= 1$$

$$D_z = \begin{vmatrix} 1 & 2 & -7 \\ 3 & -2 & 15 \\ 2 & 3 & -9 \end{vmatrix} = (1)\begin{vmatrix} -2 & 15 \\ 3 & -9 \end{vmatrix} - (3)\begin{vmatrix} 2 & -7 \\ 3 & -9 \end{vmatrix} + (2)\begin{vmatrix} 2 & -7 \\ -2 & 15 \end{vmatrix}$$

$$= 1(18 - 45) - 3(-18 + 21) + 2(30 - 14)$$

$$= -27 - 9 + 32$$

$$= -4$$

Note: After finding the values of two of the variables, you could find the value of the third variable by substituting these values back into any one of the original equations.

$$x = \frac{D_x}{D} = \frac{-3}{-1} = 3, \qquad y = \frac{D_y}{D} = \frac{1}{-1} = -1, \qquad z = \frac{D_z}{D} = \frac{-4}{-1} = 4$$

The solution is $(3, -1, 4)$. We will leave the check to the reader.

Your Turn 5 Use Cramer's rule to solve $\begin{cases} 2x + 3y - 2z = 11 \\ 3x + y + 4z = -8 \\ x - 3y - 2z = 4 \end{cases}$.

Answer Your Turn 5

$(1, 1, -3)$

9.6 Exercises   For Extra Help   *MyMathLab*

Note: Exercises marked with a ★ represent challenging exercises.

1. Is it possible to find the determinant of $\begin{bmatrix} 1 & 2 & 5 \\ 3 & 6 & 2 \end{bmatrix}$? Why or why not?

2. Explain the difference between a matrix and a determinant.

3. How do you find the minor of an element of a 3×3 matrix?

4. How do you find D_y when solving a system of equations using Cramer's rule?

For Exercises 5–34, find the determinant. See Examples 1–3.

5. $\begin{bmatrix} 3 & 2 \\ 1 & 5 \end{bmatrix}$

6. $\begin{bmatrix} 4 & 1 \\ 3 & 7 \end{bmatrix}$

7. $\begin{bmatrix} -3 & 5 \\ 2 & 4 \end{bmatrix}$

8. $\begin{bmatrix} 5 & 4 \\ 2 & -6 \end{bmatrix}$

9. $\begin{bmatrix} 3 & -6 \\ 2 & 4 \end{bmatrix}$

10. $\begin{bmatrix} 2 & 8 \\ -3 & 2 \end{bmatrix}$

11. $\begin{bmatrix} -3 & 4 \\ -2 & 5 \end{bmatrix}$

12. $\begin{bmatrix} -3 & -5 \\ 4 & 6 \end{bmatrix}$

13. $\begin{bmatrix} -2 & -3 \\ -4 & 5 \end{bmatrix}$

14. $\begin{bmatrix} -6 & -2 \\ 3 & -4 \end{bmatrix}$

15. $\begin{bmatrix} 0 & 3 \\ -5 & 7 \end{bmatrix}$

16. $\begin{bmatrix} -5 & 0 \\ -4 & 2 \end{bmatrix}$

17. $\begin{bmatrix} 1 & 2 & 1 \\ 3 & 1 & 4 \\ 2 & 3 & 2 \end{bmatrix}$

18. $\begin{bmatrix} 3 & 1 & 4 \\ 2 & 2 & 3 \\ 1 & 4 & 3 \end{bmatrix}$

19. $\begin{bmatrix} -1 & 2 & 0 \\ -3 & 2 & 4 \\ -4 & 2 & 3 \end{bmatrix}$

20. $\begin{bmatrix} -2 & 0 & -1 \\ 3 & -2 & 4 \\ -3 & 2 & 1 \end{bmatrix}$

21. $\begin{bmatrix} 2 & 1 & -3 \\ 0 & -3 & 2 \\ 4 & 1 & -3 \end{bmatrix}$

22. $\begin{bmatrix} 3 & -2 & 4 \\ 3 & 0 & 2 \\ -4 & -2 & 2 \end{bmatrix}$

23. $\begin{bmatrix} 0 & 4 & -2 \\ 3 & 2 & 0 \\ -1 & 4 & 3 \end{bmatrix}$

24. $\begin{bmatrix} 3 & -5 & 0 \\ 2 & -4 & 1 \\ -2 & 0 & -3 \end{bmatrix}$

25. $\begin{bmatrix} 0.3 & -0.5 \\ 1.3 & -0.6 \end{bmatrix}$

26. $\begin{bmatrix} -0.4 & 1.6 \\ -4.7 & 3.1 \end{bmatrix}$

27. $\begin{bmatrix} -0.4 & 0.7 & -1.2 \\ 3.1 & 1.5 & -3.2 \\ 1.6 & -2.2 & -1.5 \end{bmatrix}$

28. $\begin{bmatrix} 1.7 & -3.2 & 4.1 \\ 5.3 & -6.2 & -1.1 \\ -1.3 & 2.3 & -4.5 \end{bmatrix}$

29. $\begin{bmatrix} \dfrac{1}{2} & -\dfrac{1}{3} \\ \dfrac{2}{5} & \dfrac{3}{5} \end{bmatrix}$

30. $\begin{bmatrix} -\dfrac{3}{4} & -\dfrac{3}{5} \\ \dfrac{3}{2} & \dfrac{2}{5} \end{bmatrix}$

★ 31. $\begin{bmatrix} \dfrac{1}{2} & -\dfrac{3}{4} & \dfrac{2}{5} \\ \dfrac{1}{3} & \dfrac{1}{5} & -\dfrac{3}{2} \\ -\dfrac{3}{4} & \dfrac{1}{2} & \dfrac{3}{5} \end{bmatrix}$

★ 32. $\begin{bmatrix} \dfrac{1}{4} & \dfrac{3}{2} & \dfrac{4}{3} \\ \dfrac{1}{5} & -\dfrac{5}{4} & \dfrac{1}{2} \\ -\dfrac{5}{3} & \dfrac{1}{4} & -\dfrac{4}{5} \end{bmatrix}$

★ 33. $\begin{bmatrix} x & y & 1 \\ 2 & -1 & 3 \\ -2 & 0 & 1 \end{bmatrix}$

★ 34. $\begin{bmatrix} x & y & 1 \\ -3 & -2 & 4 \\ 3 & -2 & 2 \end{bmatrix}$

For Exercises 35–48, solve using Cramer's rule. See Example 4.

35. $\begin{cases} x + y = -5 \\ x - 2y = -2 \end{cases}$

36. $\begin{cases} x + 3y = 1 \\ x + y = -3 \end{cases}$

37. $\begin{cases} 2x - 3y = -6 \\ x - y = -1 \end{cases}$

38. $\begin{cases} 2x + 5y = -7 \\ x - y = -7 \end{cases}$

39. $\begin{cases} -x + 2y = -12 \\ 2x - 3y = 20 \end{cases}$

40. $\begin{cases} -x + 2y = -9 \\ 4x + 5y = -3 \end{cases}$

41. $\begin{cases} 4x - 6y = 7 \\ 6x - 9y = 8 \end{cases}$

42. $\begin{cases} 6x - 9y = 17 \\ 8x - 12y = 7 \end{cases}$

43. $\begin{cases} 8x - 3y = 10 \\ 4x + 3y = 14 \end{cases}$

44. $\begin{cases} 2x - y = 10 \\ 4x - 5y = 32 \end{cases}$

45. $\begin{cases} \dfrac{1}{2}x - \dfrac{1}{4}y = 0 \\ \dfrac{3}{4}x + \dfrac{5}{2}y = \dfrac{23}{2} \end{cases}$

46. $\begin{cases} \dfrac{2}{3}x + \dfrac{1}{4}y = \dfrac{1}{2} \\ \dfrac{3}{4}x + \dfrac{4}{3}y = -\dfrac{23}{4} \end{cases}$

47. $\begin{cases} 0.2x + 0.5y = 3.4 \\ 0.7x - 0.3y = -0.4 \end{cases}$

48. $\begin{cases} 1.2x - 0.6y = -2.4 \\ 3.1x + 1.3y = -11.9 \end{cases}$

For Exercises 49–60, solve using Cramer's rule. See Example 5.

49. $\begin{cases} x + y + z = 6 \\ 2x - 4y + 2z = 6 \\ 3x + 2y + z = 11 \end{cases}$

50. $\begin{cases} 2x + y - 3z = -1 \\ x + 2y - 2z = -3 \\ -3x - 4y + z = -3 \end{cases}$

51. $\begin{cases} 3x + y - z = -4 \\ 2x - y + 2z = -7 \\ x - 3y + z = -6 \end{cases}$

52. $\begin{cases} x - y + 3z = -10 \\ 5x + 4y - z = -7 \\ 2x + y - z = -4 \end{cases}$

53. $\begin{cases} 4x + 2y + 3z = 9 \\ 2x - 4y - z = 7 \\ 3x - 2z = 4 \end{cases}$

54. $\begin{cases} 2x - y = -1 \\ 5x - 3y + 2z = 0 \\ 3x + 2y - 3z = -8 \end{cases}$

55. $\begin{cases} 3x + 2y = -12 \\ 3y + 10z = -16 \\ 6x - 2z = 3 \end{cases}$

56. $\begin{cases} 4x + 2z = 7 \\ 8x - 2y = -7 \\ 10y - 2z = -5 \end{cases}$

57. $\begin{cases} \dfrac{1}{2}x + \dfrac{1}{3}y + \dfrac{3}{4}z = \dfrac{25}{12} \\[2mm] \dfrac{1}{3}x + \dfrac{2}{9}y + \dfrac{1}{2}z = \dfrac{25}{18} \\[2mm] \dfrac{3}{4}x + \dfrac{1}{4}y - \dfrac{2}{3}z = -\dfrac{7}{4} \end{cases}$

58. $\begin{cases} \dfrac{3}{4}x - \dfrac{1}{2}y + \dfrac{1}{2}z = \dfrac{1}{4} \\[2mm] \dfrac{7}{6}x + \dfrac{10}{9}y - \dfrac{2}{5}z = -\dfrac{623}{90} \\[2mm] \dfrac{7}{4}x + \dfrac{5}{3}y - \dfrac{3}{5}z = -\dfrac{623}{60} \end{cases}$

59. $\begin{cases} 0.3x + 0.4y - 0.6z = 2.6 \\ 0.5x - 0.2y + 0.7z = -0.8 \\ 1.4x + 1.3y - 2.2z = 9.8 \end{cases}$

60. $\begin{cases} 1.2x + 2.1y - 0.5z = 7.3 \\ 3.2x - 2.4y + 1.3z = 6.1 \\ 2.5x + 1.3y - 1.7z = 8.4 \end{cases}$

★ *For Exercises 61–64, find x.*

61. $\begin{vmatrix} 9 & x \\ -6 & 5 \end{vmatrix} = 21$

62. $\begin{vmatrix} 12 & 2 \\ x & -3 \end{vmatrix} = -22$

63. $\begin{vmatrix} 2 & -1 & 0 \\ 0 & x & 1 \\ -2 & 0 & 4 \end{vmatrix} = -38$

64. $\begin{vmatrix} 1 & -2 & 4 \\ -1 & 0 & 2 \\ 0 & x & 5 \end{vmatrix} = -28$

★ *For Exercises 65–70, use the following information. Suppose a triangle has vertices of (x_1, y_1), (x_2, y_2), and (x_3, y_3), as shown in the graph to the right. The area of the*

triangle is given by $A = \dfrac{1}{2}\left| \det \begin{bmatrix} x_1 & y_1 & 1 \\ x_2 & y_2 & 1 \\ x_3 & y_3 & 1 \end{bmatrix} \right|$. *For example, the area of the triangle*

Note: This notation indicates half of the absolute value of the determinant.

whose vertices are $(1, 2)$, $(3, -2)$, and $(-2, 5)$ is $A = \dfrac{1}{2}\left| \det \begin{bmatrix} 1 & 2 & 1 \\ 3 & -2 & 1 \\ -2 & 5 & 1 \end{bmatrix} \right|$.

Expanding about the first column we have

$$A = \frac{1}{2}\left| 1\det\begin{bmatrix} -2 & 1 \\ 5 & 1 \end{bmatrix} - 3\det\begin{bmatrix} 2 & 1 \\ 5 & 1 \end{bmatrix} - 2\det\begin{bmatrix} 2 & 1 \\ -2 & 1 \end{bmatrix} \right|$$

$$= \frac{1}{2}|(-2 - 5) - 3(2 - 5) - 2(2 + 2)|$$

$$= \frac{1}{2}|-7 + 9 - 8| = \frac{1}{2}|-6| = 3 \text{ square units.}$$

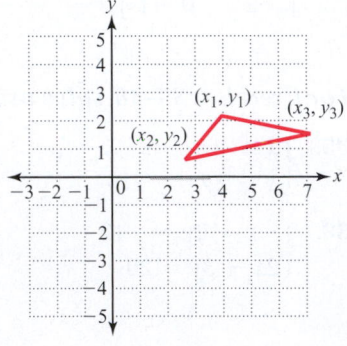

Find the area of the triangles with vertices at the given points.

65. $(2, 4)\,(4, 0)\,(6, 5)$

66. $(0, 2)\,(3, -2)\,(5, 5)$

67. $(-3, 1)\,(2, -3)\,(4, 4)$

68. $(-3, -2)\,(-1, 4)\,(3, -4)$

69. $(-4, -1)\,(1, 3)\,(3, -3)$

70. $(-3, -3)\,(2, 2)\,(4, -1)$

For Exercises 71–76, translate the problem to a system of equations; then solve using Cramer's rule.

71. The two heaviest known meteorites to be found on Earth's surface are the Hoba West, which was found in Namibia, and the Ahnighito, which was found in Greenland. The total weight of the two meteorites is 90 tons. The Hoba West is twice as heavy as the Ahnighito. Find the weight of each. (*Source: Webster's New World Book of Facts*)

72. The two largest expenses for the average American family are federal taxes and housing, including household expenses. Together these two items total 43.3% of the average family's income. The amount spent on taxes is 12.1% more than the amount spent on housing. Find the percent spent on each. (*Source: Numbers: How Many, How Long, How Far, How Much*)

73. A restaurant makes a soup that includes garbanzo and black turtle beans. The manager purchased 10 pounds of beans at a cost of $8.80. If the garbanzo beans cost $1.00 per pound and the black turtle beans cost $0.70 per pound, how many pounds of each did he purchase?

74. The perimeter of a triangle is 21 inches, and two sides are of equal length. The length of the third side is 3 inches less than the length of the two equal sides. Find the length of each side of the triangle.

75. Coinage bronze is made up of zinc, tin, and copper. The percent of tin is four times the percent of zinc. The percent of copper is nineteen times the sum of the percents of zinc and tin. Find the percent of zinc, tin, and copper in coinage bronze. (*Source: Webster's New World Book of Facts*)

76. In winning the 2009 men's NCAA basketball championship, North Carolina scored a total of 56 times in their 89 to 72 victory over Michigan State. The sum of the number of 3-point field goals and 2-point field goals was equal to the number of free throws (1 point each). How many of each did North Carolina score? (*Source: espn.go.com*)

Review Exercises

Exercises 1–6 **EQUATIONS AND INEQUALITIES**

[2.6] *For Exercises 1 and 2, solve and graph the solution set.*

1. $x > 2x - 3$

2. $2x + 3 \geq 10 - 5x$

[3.3] 3. Find two consecutive integers whose sum is 39.

[4.6] *For Exercises 4 and 5, graph.*

4. $y > 3x + 2$

5. $x + y \leq -2$

[9.2] 6. Use substitution to solve $\begin{cases} x = 3 - y \\ 5x + 3y = 5 \end{cases}$.

9.7 Solving Systems of Linear Inequalities

Objectives

1. Graph the solution set of a system of linear inequalities.
2. Solve applications involving a system of linear inequalities.

Objective 1 Graph the solution set of a system of linear inequalities. In Section 2.6, we learned to solve linear inequalities in one variable. In Section 4.6, we learned how to graph linear inequalities in two variables. In this section, we develop a graphical approach to solving *systems of linear inequalities* in two variables.

Consider the system of linear inequalities $\begin{cases} x + 2y < 6 \\ 2x - y \geq 2 \end{cases}$. First, let's graph each inequality separately.

Note: Recall from Section 4.6 that we use a dashed line with $<$ or $>$ and that points on a dashed line are not in the solution set.

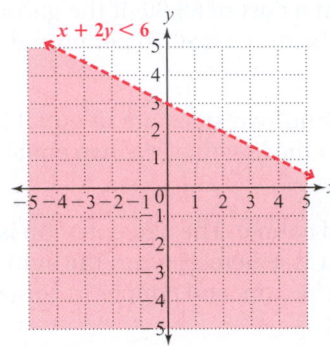

In the graph of $x + 2y < 6$, the shaded region contains all ordered pairs that make $x + 2y < 6$ true.

Note: Recall from Section 4.6 that we determine which side of the line to shade by choosing an ordered pair not on the line and checking to see if it makes the inequality true. If it does, we shade on the side of the line containing that ordered pair. If it does not, we shade the other side of the line.

Note: Recall from Section 4.6 that we use a solid line with $\leq$ or $\geq$ and that points on a solid line are part of the solution set.

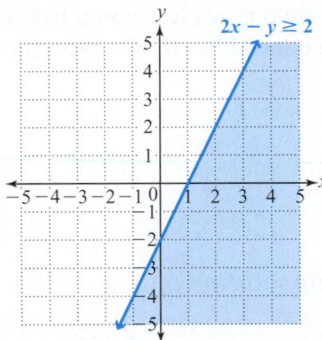

In the graph of $2x - y \geq 2$, ordered pairs in the shaded region and on the line itself make $2x - y \geq 2$ true.

Now we put the two graphs together on the same grid to determine the solution set for the system. A solution for a system of inequalities is an ordered pair that makes every inequality in the system true.

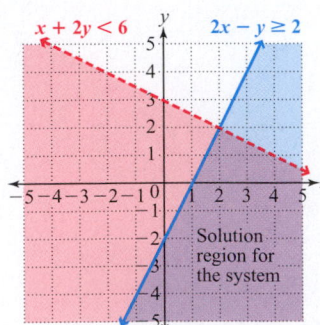

Solution region for the system

On our graph containing both $x + 2y < 6$ and $2x - y \geq 2$, the region where the shading overlaps contains ordered pairs that make both inequalities true. So all ordered pairs in this region are in the solution set for the system. Also, ordered pairs on the solid line for $2x - y \geq 2$ where it touches the region of overlap are in the solution set for the system, whereas ordered pairs on the dashed line for $x + 2y < 6$ are not. This implies the following procedure.

Procedure   **Solving a System of Linear Inequalities in Two Variables**

To solve a system of linear inequalities in two variables, graph all of the inequalities on the same grid. The solution set for the system contains all ordered pairs in the region where the inequalities' solution sets overlap along with ordered pairs on the portion of any solid line that touches the region of overlap.

To check, we can select a point in the solution region such as $(3, 0)$ and verify that it makes both inequalities true.

First inequality: **Second inequality:**

$$x + 2y < 6 \qquad\qquad 2x - y \geq 2$$

$$3 + 2(0) \overset{?}{<} 6 \qquad\qquad 2(3) - 0 \overset{?}{\geq} 2 \qquad \text{\small Replace } x \text{ with 3 and } y$$
$$\text{\small with 0 in both inequalities.}$$

$$3 < 6 \quad \text{True.} \qquad\qquad 6 \geq 2 \quad \text{True.}$$

Because $(3, 0)$ makes both inequalities true, it is a solution to the system. Although we selected only one ordered pair in the solution region, remember that *every* ordered pair in that region is a solution.

Example 1   Graph the solution set for the system of inequalities.

a. $\begin{cases} x + 3y > 6 \\ y < 2x - 1 \end{cases}$

SOLUTION Graph the inequalities on the same grid. Because both lines are dashed, the solution set for the system contains only those ordered pairs in the region of overlap (the purple shaded region).

> **Note:** Ordered pairs on dashed lines are not part of the solution region.

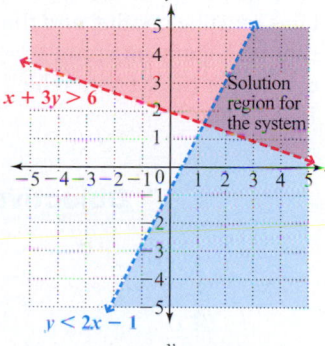

b. $\begin{cases} y < 2 \\ 5x - y \leq 4 \end{cases}$

SOLUTION Graph the inequalities on the same grid. The solution set for this system contains all ordered pairs in the region of overlap (purple shaded region) together with all ordered pairs on the portion of the solid blue line that touches the purple shaded region.

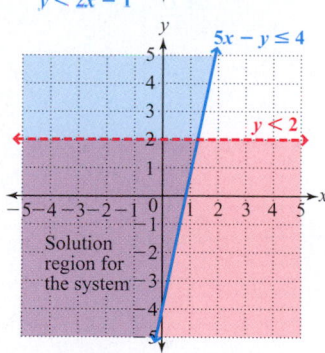

Connection Remember that for a system of two linear equations, a solution is a point of intersection of the graphs of the two equations. Similarly, for a system of linear inequalities, the solution set is a region of intersection of the graphs of the two inequalities.

Answers   **Your Turn 1**

a.

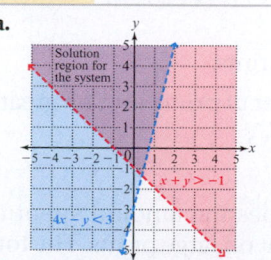

b. Solution region for the system (including the portion of the red line touching this region)

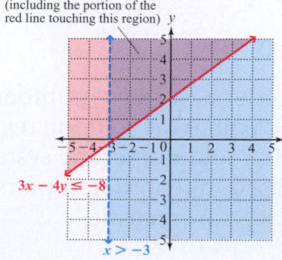

Your Turn 1   Graph the solution set for the system of inequalities.

a. $\begin{cases} x + y > -1 \\ 4x - y < 3 \end{cases}$

b. $\begin{cases} x > -3 \\ 3x - 4y \leq -8 \end{cases}$

Inconsistent Systems

Some systems of linear inequalities have no solution. We say that these systems are inconsistent.

> ### Example 2
> Graph the solution set for the system of inequalities.
>
> $$\begin{cases} x - 4y \geq 8 \\ y \geq \dfrac{1}{4}x + 3 \end{cases}$$
>
> **SOLUTION** Graph the inequalities on the same grid.

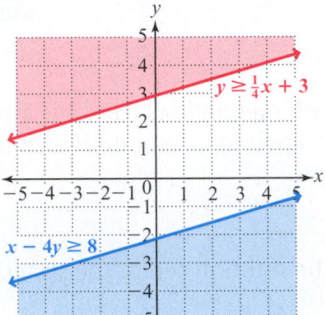

The lines appear to be parallel. We can verify by writing $x - 4y \geq 8$ in slope–intercept form and comparing the slopes and y-intercepts.

$$x - 4y \geq 8$$
$$-4y \geq -x + 8 \qquad \text{Subtract } x \text{ from both sides.}$$
$$y \leq \frac{1}{4}x - 2 \qquad \begin{array}{l}\text{Divide both sides by } -4,\\ \text{which changes the direction}\\ \text{of the inequality symbol.}\end{array}$$

The slopes are equal, and the y-intercepts differ; so the lines are in fact parallel. Because the lines are parallel and the shaded regions do not overlap, there is no solution region for this system. The system is inconsistent.

Connection Notice that the lines in Example 2 are parallel. Recall that in a system of linear equations, if the lines are parallel, the system is inconsistent because there is no point of intersection. With linear inequalities, the lines must be parallel *and* the shaded regions must not overlap for the system to be inconsistent.

Objective ② Solve applications involving a system of linear inequalities.

> ### Example 3
> A home interiors store stocks two different-size prints from an artist. The manager wants to purchase at least 15 prints for the store. The artist sells the smaller print for $20 and the larger print for $30. The manager cannot spend more than $800 for the prints. Write a system of inequalities that describes the manager's order; then solve the system by graphing.
>
> *Understand* We must translate to a system of inequalities, then solve the system.
>
> *Plan and Execute* Let x represent the number of small prints and y represent the number of large prints ordered.
>
> Relationship 1: The manager wants to purchase at least 15 prints.
>
> > The words *at least* indicate that the combined number of prints is to be greater than or equal to 15; so $x + y \geq 15$.
>
> Relationship 2: The manager cannot spend more than $800.
>
> > Because the small prints cost $20 each, $20x$ describes the amount spent on small prints. Similarly, $30y$ describes the amount spent on large prints. The total cannot exceed $800; so $20x + 30y \leq 800$.

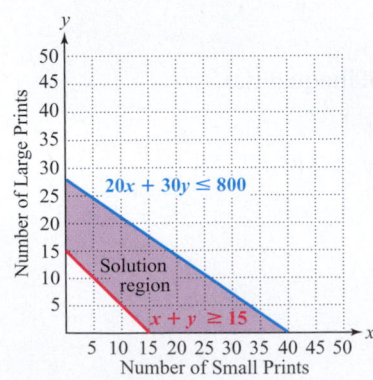

Our system: $\begin{cases} x + y \geq 15 \\ 20x + 30y \leq 800 \end{cases}$

Answer (See the graph.) Because the manager cannot order a negative number of prints, the solution set is confined to quadrant I. Any ordered pair in the solution region or on a portion of either line touching the solution region is a solution for the system. However, assuming that only whole prints can be purchased, only ordered pairs of whole numbers in the solution set, such as (10, 15) and (25, 10), are realistic.

Answer   Your Turn 3

System: $\begin{cases} 2l + 2w \le 200 \\ l \ge w + 10 \end{cases}$

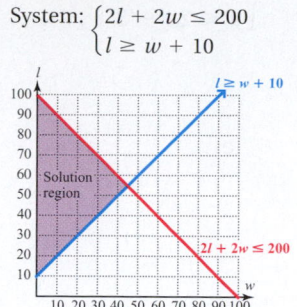

Check We will check one ordered pair in the solution region. The ordered pair (10, 15) indicates an order of 10 small prints and 15 large prints so that 25 prints are ordered (which is more than 15) that cost a total of $20(10) + 30(15) = \$650$ (which is less than \$800).

Your Turn 3 In designing a building with a rectangular base, an architect decides that the perimeter cannot exceed 200 feet. Also, the length will be at least 10 feet more than the width. Write a system of inequalities that describes the design requirements; then solve the system graphically.

Note: Because length and width must be positive numbers, the solution region is confined to quadrant 1.

9.7 Exercises   For Extra Help   *MyMathLab*

Note: Exercises marked with a ★ represent challenging exercises.

1. How would you check a possible solution to a system of inequalities?

2. How do you determine the region that is the solution set for a system of linear inequalities?

3. What circumstances would cause a system of inequalities to have no solution?

4. Write a system of linear inequalities whose solution set is the entire first quadrant.

5. Write a system of linear inequalities whose solution set is the entire third quadrant.

6. Write a system of linear inequalities whose solution set is the entire fourth quadrant.

For Exercises 7–30, graph the solution set for the system of inequalities. See Examples 1 and 2.

7. $\begin{cases} x + y > 4 \\ x - y < 6 \end{cases}$

8. $\begin{cases} x - y > 2 \\ x + y > 4 \end{cases}$

9. $\begin{cases} x + y < -2 \\ x - y > 6 \end{cases}$

10. $\begin{cases} x - y < -5 \\ x + y < 3 \end{cases}$

11. $\begin{cases} 2x + y \ge -1 \\ x - y \le 5 \end{cases}$

12. $\begin{cases} x - y < -5 \\ 2x - y < -7 \end{cases}$

13. $\begin{cases} x + y < 3 \\ x - 2y \ge 2 \end{cases}$

14. $\begin{cases} x + 3y \le 6 \\ x - y > 5 \end{cases}$

15. $\begin{cases} 2x + y > 9 \\ y > \dfrac{1}{4}x \end{cases}$

16. $\begin{cases} 3x + 4y \leq -9 \\ y < 3x \end{cases}$

17. $\begin{cases} y < x \\ y > -x + 1 \end{cases}$

18. $\begin{cases} y < x \\ y < 2x - 3 \end{cases}$

19. $\begin{cases} 3x > -4y \\ 3x + 4y \leq -8 \end{cases}$

20. $\begin{cases} 2y - x \geq 6 \\ 2x - 4y \geq 5 \end{cases}$

21. $\begin{cases} y < 3x + 1 \\ 3x - y \leq 4 \end{cases}$

22. $\begin{cases} x - 2y > 3 \\ 2x - 4y \leq 20 \end{cases}$

23. $\begin{cases} x > 2 \\ 3x - 2y > 6 \end{cases}$

24. $\begin{cases} y \leq -1 \\ x + 2y > 3 \end{cases}$

25. $\begin{cases} x \geq -3y \\ y \geq 2x \end{cases}$

26. $\begin{cases} x > 2y \\ x + y > 6 \end{cases}$

27. $\begin{cases} y > 2 \\ x < -1 \end{cases}$

28. $\begin{cases} y \geq -1 \\ x < 2 \end{cases}$

29. $\begin{cases} 2x + y \geq 1 \\ x - y > -1 \\ x > 2 \end{cases}$

30. $\begin{cases} x + y \geq 1 \\ x - y \geq -5 \\ y > -2 \end{cases}$

31. $\begin{cases} x + y > 3 \\ x - y \le 2 \end{cases}$

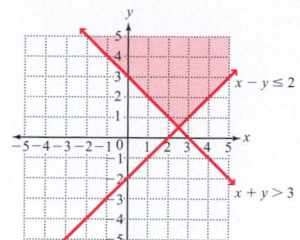

32. $\begin{cases} y < \dfrac{1}{2}x \\ y \le -3x + 1 \end{cases}$

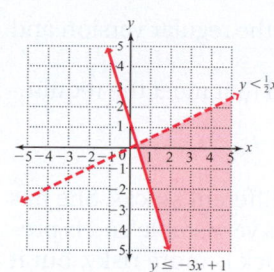

33. $\begin{cases} x < 3 \\ y > 4 \end{cases}$

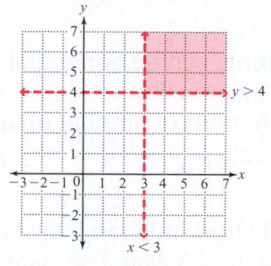

34. $\begin{cases} x > 3y \\ x + y \ge 2 \end{cases}$

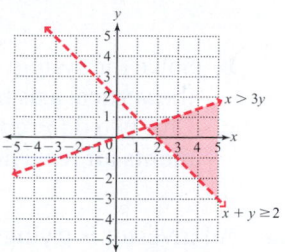

For Exercises 35–38, solve. see Example 3.

35. To be admitted to a certain college, prospective students must have a combined verbal and math score of at least 1100 on the SAT. To be admitted to the engineering program at this college, students must also have a math score of at least 500.

   a. Write two inequalities describing the requirements to be admitted to the college of engineering.

   b. The maximum score on each part of the SAT is 800. Write two inequalities describing these maximum scores.

   c. Solve the system of inequalities described by the four inequalities in parts a and b by graphing. Let the horizontal axis be verbal and the vertical axis be math.

   d. Give two different combinations of verbal and math scores that satisfy the admission requirements to the college of engineering.

36. An architect is designing a rectangular platform for an auditorium. The client wants the length of the platform to be at least twice the width, and the perimeter must not exceed 200 feet.

   a. Write a system of inequalities that describes the specifications for the platform.

   b. Solve the system by graphing. Let the horizontal axis be the width and the vertical axis be the length.

   c. Give two different combinations of length and width that satisfy the requirements of the client.

★**37.** A company sells two versions of software: the regular version and the deluxe version. The regular version sells for $15.95, and the deluxe version sells for $20.95. The company gives a bonus to any salesperson who sells at least 100 units and has at least $1700 in total sales.

   a. Write a system of inequalities that describes the requirements a salesperson must meet to receive the bonus.

   b. Solve the system by graphing. Let the horizontal axis be the regular version and the vertical axis be the deluxe version.

   c. Give two different combinations of number of units sold that meet the requirements for receiving the bonus.

★**38.** An investor is trying to decide how much to invest in two different stocks. She has decided that she will invest at most $5000. The "safe" stock is very stable and is projected to return about 5% over the next year. The other stock is more risky, but it could return about 9% if the company grows as expected. She wants her total dividend at the end of the year to be at least $300.

   a. Write a system of inequalities that describes the amount invested in the two stocks and the return on that investment.

   b. Solve the system by graphing. Let the horizontal axis be the "safe" stock and the vertical be the "risky" stock.

   c. Give two different combinations of investment amounts that return at least $300.

Collaborative Exercises   Maximizing the Profit

Linear programming is an area of mathematics that solves problems like the one described below. One of the fundamental elements of a linear programming model is the constraint. A constraint is simply an inequality that describes some limited resource required in the problem. For example, suppose you make two types of bicycles, style A and style B. Also suppose that style A requires 40 minutes of welding for each bicycle and style B requires 25 minutes of welding. If you have only 600 minutes of welding time available, the welding-time constraint would be $40A + 25B \leq 600$, where A represents the number of style A produced and B represents the number of style B produced.

An aspiring artists' group is preparing for a fund-raising sale. The artists design and construct two types of antique-reproduction tables: a semicircular foyer table and a side table. The materials cost $100 for each foyer table and $200 for each side table. Both types of table require 2 hours of cutting and carving. The foyer table requires 3.5 hours to assemble, sand, and finish, and the side table requires 1.75 hours to assemble, sand, and finish. The group has $3300 to purchase all materials; 40 hours available for cutting and carving; and 63 hours available for assembling, sanding, and finishing. For each foyer table they sell, their profit will be $350. For each side table, their profit will be $215. How many of each table should they make to maximize their profit?

   1. For the artists' fund-raising problem, there are three basic constraints: (1) cost of materials, (2) cutting and carving, and (3) assembling/sanding/finishing. For each of these constraints, write an inequality similar to the one in the bicycle example. Let x represent the number of foyer tables and y the number of side tables.

   2. Because x and y represent numbers of tables, they cannot represent negative values. Write inequalities for these two additional constraints.

   3. Now graph the system of inequalities described by these five constraints.

4. The region that is the solution of the system of inequalities is called the *feasible region*, and it includes all points that satisfy the system. The goal is to find the optimal solution, which, according to linear programming, is one of the corner points of the feasible region. There should be five corner points: one at the origin, one on each axis, and two more in quadrant I. Determine the coordinates of each corner point. Note that a corner point lying on an axis is the *x*- or *y*-intercept for the line passing through that point. Because two lines intersect to form a corner point that is not on an axis, make a system out of their two equations; then solve the system to find their point of intersection.

5. Next, develop an algebraic expression describing the profit gained for selling *x* of the foyer tables and *y* of the side tables. This is called the *objective function*.

6. Using the objective function, test each of the points found in Step 4 to see which one yields the maximum profit.

7. Using your solution, answer the following questions.
  a. How many of each type of table should the group produce?

  b. How many hours will be spent in each phase of production?

  c. How much money will they need to purchase the materials?

  d. What amount of profit will the group receive?

Review Exercises

Exercises 1–4  EXPRESSIONS

[1.5] *For Exercises 1–4, evaluate.*

1. 3^2

2. -10^2

3. $\sqrt{25}$

4. $\sqrt{\dfrac{16}{49}}$

Exercises 5 and 6 EQUATIONS AND INEQUALITIES

[4.7]   **5.** Given $f(x) = 3x - 5$, find $f(3)$.

[8.4]   **6.** Given $f(x) = 3x^2 - 2x + 5$, find $f(-2)$.

Chapter 9 Summary

Defined Terms

Section 9.1
System of equations (p. 624)
Solution for a system of equations (p. 624)

Consistent system of equations (p. 628)
Inconsistent system of equations (p. 628)

Section 9.5
Matrix (p. 666)
Augmented matrix (p. 666)
Row echelon form (p. 667)

Section 9.6
Square matrix (p. 674)
Minor of an element of a matrix (p. 675)

Procedures, Rules, and Key Examples

Procedures/Rules	Key Examples

SECTION 9.1   Solving Systems of Linear Equations Graphically

Procedures/Rules

To verify or check a solution to a system of equations,
1. Replace each variable in each equation with its corresponding value.
2. Verify that each equation is true.

Key Examples

Example 1: Determine whether the ordered pair is a solution to the system.

$$\begin{cases} 2x + 3y = 8 \\ y = 5x - 3 \end{cases}$$

a. $(-2, 4)$

$$2x + 3y = 8 \qquad\qquad y = 5x - 3$$
$$2(-2) + 3(4) \overset{?}{=} 8 \qquad 4 \overset{?}{=} 5(-2) - 3$$
$$-4 + 12 \overset{?}{=} 8 \qquad\quad 4 \overset{?}{=} -10 - 3$$
$$8 = 8 \qquad\qquad\quad 4 \neq -13$$

$(-2, 4)$ is not a solution to the system.

b. $(1, 2)$

$$2x + 3y = 8 \qquad\qquad y = 5x - 3$$
$$2(1) + 3(2) \overset{?}{=} 8 \qquad 2 \overset{?}{=} 5(1) - 3$$
$$2 + 6 \overset{?}{=} 8 \qquad\qquad 2 \overset{?}{=} 5 - 3$$
$$8 = 8 \qquad\qquad\qquad 2 = 2$$

$(1, 2)$ is a solution to the system.

Procedures/Rules

To solve a system of linear equations graphically,
1. Graph each equation.
   a. If the lines intersect at a single point, the coordinates of that point form the solution.
   b. If the lines are parallel, there is no solution.
   c. If the lines are identical, there are an infinite number of solutions, which are the coordinates of all points on that line.
2. Check your solution.

Key Examples

Example 2: Solve the system of equations graphically.

$$\begin{cases} 2x + 3y = 8 \\ y = 5x - 3 \end{cases}$$

Graph the two equations; then find the point of intersection.

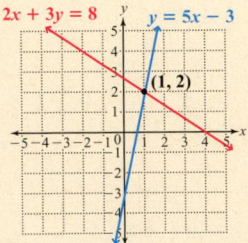

The two lines intersect at the point $(1, 2)$, so $(1, 2)$ is the solution to the system.

(continued)

Procedures/Rules	Key Examples

SECTION 9.1 Solving Systems of Linear Equations Graphically (*continued*)

To classify a system of two linear equations in two unknowns, write the equations in slope–intercept form and compare the slopes and y-intercepts.
1. **Consistent system with independent equations:** The system has a single solution at the point of intersection. The graphs are different. They have different slopes.
2. **Consistent system with dependent equations:** The system has an infinite number of solutions. The graphs are identical. They have the same slope and same y-intercept.
3. **Inconsistent system:** The system has no solution. The graphs are parallel lines. They have the same slope but different y-intercepts.

Example 3: Determine whether the system is consistent with independent equations, consistent with dependent equations, or inconsistent and discuss the number of solutions for the system.

$$\begin{cases} x + 3y = 6 \\ 2x + 6y = -5 \end{cases}$$

Solution: Write the equations in slope–intercept form.

$$x + 3y = 6 \qquad\qquad 2x + 6y = -5$$
$$3y = -x + 6 \qquad\qquad 6y = -2x - 5$$
$$y = -\frac{1}{3}x + 2 \qquad\qquad y = -\frac{1}{3}x - \frac{5}{6}$$

Because the slopes are the same $\left(\text{both } -\frac{1}{3}\right)$ but the y-intercepts are different, this system is inconsistent. This means there is no solution to the system.

SECTION 9.2 Solving Systems of Linear Equations by Substitution; Applications

To find the solution of a system of two linear equations using the substitution method,
1. Isolate one of the variables in one of the equations.
2. In the other equation, substitute the expression you found in Step 1 for that variable.
3. Solve this new equation. (It will now have only one variable.)
4. Using one of the equations containing both variables, substitute the value you found in Step 3 for that variable and solve for the value of the other variable.
5. Check the solution in the original equations.

Example 1: Solve the system using the substitution method.

$$\begin{cases} y = 4x - 1 \\ 5x - y = 2 \end{cases}$$

Because y is already isolated in the first equation, we substitute $4x - 1$ in place of y in the second equation.

$5x - y = 2$	
$5x - (4x - 1) = 2$	Substitute $4x - 1$ for y.
$5x - 4x + 1 = 2$	Distribute.
$x + 1 = 2$	Combine like terms.
$x = 1$	Subtract 1 from both sides.

Now solve for the value of y by substituting 1 in place of x in one of the equations. We will use $y = 4x - 1$.

$y = 4x - 1$	
$y = 4(1) - 1$	Substitute 1 for x.
$y = 4 - 1$	
$y = 3$	

Solution: $(1, 3)$

SECTION 9.3 Solving Systems of Linear Equations by Elimination; Applications

To solve a system of two linear equations using the elimination method,
1. Write the equations in standard form ($Ax + By = C$).
2. Use the multiplication principle of equality to clear fractions or decimals (optional).

Example 1: Solve the system using the elimination method.

$$\begin{cases} \dfrac{1}{4}x + \dfrac{1}{2}y = -1 \\ 4x + y = 5 \end{cases}$$

(continued)

SECTION 9.3 Solving Systems of Linear Equations by Elimination; Applications (*continued*)

3. If necessary, multiply one or both equations by a number (or numbers) so that they have a pair of terms that are additive inverses.
4. Add the equations. The result should be an equation in terms of one variable.
5. Solve the equation from Step 4 for the value of that variable.
6. Using an equation containing both variables, substitute the value you found in Step 5 for the corresponding variable and solve for the value of the other variable.
7. Check your solution in the original equations.

Solution: First, we clear the fractions in the first equation.

$$4\left(\frac{1}{4}x + \frac{1}{2}y\right) = 4(-1)$$
$$x + 2y = -4$$

Now we will eliminate y by multiplying the second equation by -2, then add the two equations.

$$
\begin{array}{l}
x + 2y = -4 \\
4x + y = 5
\end{array}
\xrightarrow{\text{Multiply by } -2.}
\begin{array}{l}
x + 2y = -4 \\
\underline{-8x - 2y = -10} \\
-7x + 0 = -14 \\
-7x = -14 \\
x = 2
\end{array}
$$

Add equations.
Divide both sides by −7.

Now solve for the value of y by substituting 2 for x in one of the equations. We will use $4x + y = 5$.

$$
\begin{array}{ll}
4x + y = 5 & \\
4(2) + y = 5 & \text{Substitute 2 for } x. \\
8 + y = 5 & \text{Multiply.} \\
y = -3 & \text{Subtract 8 from both sides.}
\end{array}
$$

Solution: $(2, -3)$

To solve problems using systems of linear equations,
1. Select a variable to represent each unknown.
2. Write a system of equations.
3. Solve the system.

Note: The substitution method is simpler than elimination if one of the equations has an isolated variable. If all equations in the system are in standard form, elimination is preferred.

Example 2: A vendor sells small and large drinks for $2 and $3, respectively. If the vendor sold 2400 drinks for a total of $5400, how many of each size were sold?

Let $x =$ the number of small drinks sold.
Let $y =$ the number of large drinks sold.

The total number of drinks sold, 2400, translates to $x + y = 2400$.
The total revenue is $5400, which translates to $2x + 3y = 5400$.

$$\text{System } \begin{cases} x + y = 2400 \\ 2x + 3y = 5400 \end{cases}$$

To solve the system, we use the elimination method because there are no isolated variables.

$$
\begin{array}{l}
x + y = 2400 \\
2x + 3y = 5400
\end{array}
\xrightarrow{\text{Multiply by } -3}
\begin{array}{l}
-3x - 3y = -7200 \\
\underline{2x + 3y = 5400} \\
-x + 0 = -1800 \\
x = 1800
\end{array}
$$

We eliminated y.
Solve for x.

Substitute 1800 for x in an equation.

$$
\begin{array}{ll}
1800 + y = 2400 & \text{We chose } x + y = 2400. \\
y = 600 & \text{Isolate } y.
\end{array}
$$

The vendor sold 1800 small and 600 large drinks.

(*continued*)

Procedures/Rules	Key Examples

SECTION 9.4   Solving Systems of Linear Equations in Three Variables; Applications

To solve a system of three linear equations with three unknowns using elimination,
1. Write each equation in the form $Ax + By + Cz = D$.
2. Eliminate one variable from one pair of equations using the elimination method.
3. If necessary, eliminate the same variable from another pair of equations.
4. Steps 2 and 3 result in two equations with the same two variables. Solve these equations using the elimination method.
5. To find the third variable, substitute the values of the variables found in Step 4 into any of the three original equations that contain the third variable.
6. Check the ordered triple in all three original equations.

Example 1: Solve the following system of equations.

$$\begin{cases} x + 2y - z = -6 & \text{(Equation 1)} \\ 2x - 3y + 4z = 26 & \text{(Equation 2)} \\ -x + 2y - 3z = -18 & \text{(Equation 3)} \end{cases}$$

Eliminate x by adding equation 1 and equation 3.

$$\begin{aligned} x + 2y - z &= -6 \\ \underline{-x + 2y - 3z} &= \underline{-18} \\ 4y - 4z &= -24 \qquad \text{Divide both sides by 4.} \\ y - z &= -6 \qquad \text{(Eq. 4)} \end{aligned}$$

Eliminate x again using equations 2 and 3.

$$\begin{aligned} 2x - 3y + 4z &= 26 & 2x - 3y + 4z &= 26 \\ -x + 2y - 3z &= -18 \longrightarrow & \underline{-2x + 4y - 6z} &= \underline{-36} \\ & \text{Multiply by 2.} & y - 2z &= -10 \quad \text{(Eq. 5)} \end{aligned}$$

Use equations 4 and 5 to solve for z.

$$\begin{aligned} y - z &= -6 & \text{Multiply by } -1 & & y - z &= -6 \\ y - 2z &= -10 \xrightarrow{} & & & \underline{-y + 2z} &= \underline{10} \\ & & & & z &= 4 \end{aligned}$$

Substitute 4 for z in equation 4 and solve for y.

$$\begin{aligned} y - 4 &= -6 \\ y &= -2 \end{aligned}$$

Substitute -2 for y and 4 for z in equation 1 and solve for x.

$$\begin{aligned} x + 2(-2) - 4 &= -6 \\ x - 4 - 4 &= -6 \\ x &= 2 \end{aligned}$$

The solution is $(2, -2, 4)$.

SECTION 9.5   Solving Systems of Linear Equations Using Matrices

Row operations:
The solutions of a system is not affected by the following row operations in its augmented matrix.
1. Any two rows may be interchanged.
2. The elements of any row may be multiplied (or divided) by any nonzero real number.
3. Any row may be replaced by a row resulting from adding the elements of that row (or multiples of that row) to a multiple of the elements of any other row.

Row operations can be used to solve a system of linear equations by transforming the augmented matrix of the system into an equivalent matrix that is in row echelon form. A matrix is in row echelon form if the coefficient portion of the augmented matrix has 1's on the diagonal from upper left to lower right and 0's below the 1's.

Note: Systems of three equations with three unknowns are solved in a similar manner.

Example 1: Solve $\begin{cases} 2x + 3y = -5 \\ x + 2y = -4 \end{cases}$ using the row echelon method.

The augmented matrix is $\begin{bmatrix} 2 & 3 & | & -5 \\ 1 & 2 & | & -4 \end{bmatrix}$.

$$\begin{bmatrix} 1 & 2 & | & -4 \\ 2 & 3 & | & -5 \end{bmatrix} \quad \begin{array}{l} \text{We need row 1, column 1 to be a 1;} \\ \text{so interchange the two rows.} \end{array}$$

$$-2R_1 + R_2 \longrightarrow \begin{bmatrix} 1 & 2 & | & -4 \\ 0 & -1 & | & 3 \end{bmatrix} \quad \begin{array}{l} \text{We need row 2, column 1 to} \\ \text{be 0; so multiply row 1 by} \\ -2 \text{ and add it to row 2.} \end{array}$$

$$-R_2 \longrightarrow \begin{bmatrix} 1 & 2 & | & -4 \\ 0 & 1 & | & -3 \end{bmatrix} \quad \begin{array}{l} \text{Multiply row 2 by } -1 \text{ to get} \\ \text{row echelon form.} \end{array}$$

Row 2 means $y = -3$, and row 1 means $x + 2y = -4$.

$$\begin{aligned} x + 2(-3) &= -4 \qquad \text{Substitute } -3 \text{ for } y \text{ and solve for } x \\ x &= 2 \end{aligned}$$

The solution is $(2, -3)$.

(continued)

Procedures/Rules	Key Examples

SECTION 9.6 Solving Systems of Linear Equations Using Cramer's Rule

If $A = \begin{bmatrix} a_1 & b_1 \\ a_2 & b_2 \end{bmatrix}$, then $\det(A) = \begin{vmatrix} a_1 & b_1 \\ a_2 & b_2 \end{vmatrix} = a_1b_2 - a_2b_1$.

The determinant of a 3×3 matrix is found as follows:

$$\begin{vmatrix} a_1 & b_1 & c_1 \\ a_2 & b_2 & c_2 \\ a_3 & b_3 & c_3 \end{vmatrix} = a_1\left(\begin{matrix} \text{minor} \\ \text{of } a_1 \end{matrix}\right) - a_2\left(\begin{matrix} \text{minor} \\ \text{of } a_2 \end{matrix}\right) + a_3\left(\begin{matrix} \text{minor} \\ \text{of } a_3 \end{matrix}\right)$$

$$= a_1\begin{vmatrix} b_2 & c_2 \\ b_3 & c_3 \end{vmatrix} - a_2\begin{vmatrix} b_1 & c_1 \\ b_3 & c_3 \end{vmatrix} + a_3\begin{vmatrix} b_1 & c_1 \\ b_2 & c_2 \end{vmatrix}$$

Cramer's rule:

The solution to $\begin{cases} a_1x + b_1y = c_1 \\ a_2x + b_2y = c_2 \end{cases}$ is

$$x = \frac{\begin{vmatrix} c_1 & b_1 \\ c_2 & b_2 \end{vmatrix}}{\begin{vmatrix} a_1 & b_1 \\ a_2 & b_2 \end{vmatrix}} = \frac{D_x}{D} \quad \text{and} \quad y = \frac{\begin{vmatrix} a_1 & c_1 \\ a_2 & c_2 \end{vmatrix}}{\begin{vmatrix} a_1 & b_1 \\ a_2 & b_2 \end{vmatrix}} = \frac{D_y}{D}.$$

The solution to $\begin{cases} a_1x + b_1y + c_1z = d_1 \\ a_2x + b_2y + c_2z = d_2 \\ a_3x + b_3y + c_3z = d_3 \end{cases}$ is

$$x = \frac{\begin{vmatrix} d_1 & b_1 & c_1 \\ d_2 & b_2 & c_2 \\ d_3 & b_3 & c_3 \end{vmatrix}}{\begin{vmatrix} a_1 & b_1 & c_1 \\ a_2 & b_2 & c_2 \\ a_3 & b_3 & c_3 \end{vmatrix}} = \frac{D_x}{D}, y = \frac{\begin{vmatrix} a_1 & d_1 & c_1 \\ a_2 & d_2 & c_2 \\ a_3 & d_3 & c_3 \end{vmatrix}}{\begin{vmatrix} a_1 & b_1 & c_1 \\ a_2 & b_2 & c_2 \\ a_3 & b_3 & c_3 \end{vmatrix}} = \frac{D_y}{D},$$

and $z = \dfrac{\begin{vmatrix} a_1 & b_1 & d_1 \\ a_2 & b_2 & d_2 \\ a_3 & b_3 & d_3 \end{vmatrix}}{\begin{vmatrix} a_1 & b_1 & c_1 \\ a_2 & b_2 & c_2 \\ a_3 & b_3 & c_3 \end{vmatrix}} = \dfrac{D_z}{D}.$

Example 1: Find the determinant of $\begin{bmatrix} 3 & -4 \\ 2 & -5 \end{bmatrix}$.

Solution: $\begin{vmatrix} 3 & -4 \\ 2 & -5 \end{vmatrix} = 3(-5) - 2(-4) = -15 + 8 = -7$

Example 2: Find the determinant of $\begin{bmatrix} -1 & -2 & 3 \\ 1 & -3 & 4 \\ -2 & 3 & -1 \end{bmatrix}$.

Solution:

$$\begin{vmatrix} -1 & -2 & 3 \\ 1 & -3 & 4 \\ -2 & 3 & -1 \end{vmatrix} = -1\begin{vmatrix} -3 & 4 \\ 3 & -1 \end{vmatrix} - 1\begin{vmatrix} -2 & 3 \\ 3 & -1 \end{vmatrix} + (-2)\begin{vmatrix} -2 & 3 \\ -3 & 4 \end{vmatrix}$$

$$= -1(3 - 12) - 1(2 - 9) + (-2)(-8 + 9)$$

$$= 9 + 7 - 2 = 14$$

Example 3: Solve $\begin{cases} 3x + 2y = -2 \\ 5x + 2y = 2 \end{cases}$.

Solution: $D = \begin{vmatrix} 3 & 2 \\ 5 & 2 \end{vmatrix} = -4$

$$x = \frac{\begin{vmatrix} -2 & 2 \\ 2 & 2 \end{vmatrix}}{-4} = \frac{-4 - 4}{-4} = \frac{-8}{-4} = 2$$

$$y = \frac{\begin{vmatrix} 3 & -2 \\ 5 & 2 \end{vmatrix}}{-4} = \frac{6 + 10}{-4} = \frac{16}{-4} = -4$$

The solution is $(2, -4)$.

Example 4: Solve $\begin{cases} x + 2y - z = -4 \\ 2x - 3y - 2z = 13 \\ 2x - 2y + 3z = 20 \end{cases}$.

Solution: $D = \begin{vmatrix} 1 & 2 & -1 \\ 2 & -3 & -2 \\ 2 & -2 & 3 \end{vmatrix} = -35$

$$x = \frac{\begin{vmatrix} -4 & 2 & -1 \\ 13 & -3 & -2 \\ 20 & -2 & 3 \end{vmatrix}}{-35} = \frac{-140}{-35} = 4$$

$$y = \frac{\begin{vmatrix} 1 & -4 & -1 \\ 2 & 13 & -2 \\ 2 & 20 & 3 \end{vmatrix}}{-35} = \frac{105}{-35} = -3$$

$$z = \frac{\begin{vmatrix} 1 & 2 & -4 \\ 2 & -3 & 13 \\ 2 & -2 & 20 \end{vmatrix}}{-35} = \frac{-70}{-35} = 2$$

The solution is $(4, -3, 2)$.

(continued)

Procedures/Rules	Key Examples

SECTION 9.7 Solving Systems of Linear Inequalities

To solve a system of linear inequalities, graph all of the inequalities on the same grid. The solution set for the system contains all ordered pairs in the region where the inequalities' solution sets overlap along with ordered pairs on the portion of any solid line that touches the region of overlap.

Example 1: Solve $\begin{cases} x + y > -2 \\ y \le 3x - 1 \end{cases}$.

Solution:

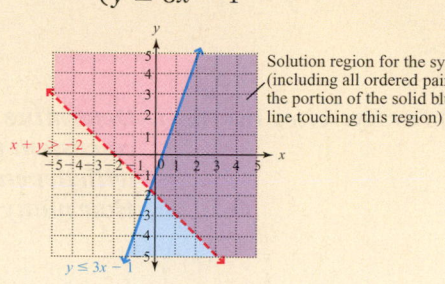

Solution region for the system (including all ordered pairs on the portion of the solid blue line touching this region)

Chapter 9 Review Exercises

For Exercises 1–5, answer true or false.

[9.1] **1.** The solution for a system of two linear equations is always exactly one ordered pair that satisfies all equations of the system.

[9.1] **2.** Given a system of two linear equations with two variables, if there is no solution, the graphs are parallel lines.

[9.2] **3.** Any solvable system of linear equations can be solved using the substitution method.

[9.5] **4.** Solving a system of linear equations using the row echelon method is like solving a system using the elimination method except that the row echelon method uses the coefficients and constant terms only.

[9.6] **5.** The determinant of a matrix is another matrix.

For Exercises 6–10, complete the rule.

[9.1] **6.** To verify or check a solution to a system of equations,

    a. _____ each variable in each equation with its corresponding value.
    b. Verify that each equation is true.

[9.1] **7.** An inconsistent system has no solution. The graphs have the same _____ but different _____.

[9.7] **8.** To solve a system of two linear inequalities,

    a. Graph each inequality.
    b. The solution of the system is _____ _____.

[9.5] **9.** A matrix is in row echelon form if _____ _____.

[9.6] **10.** When a system of linear equations is solved using Cramer's rule, the denominator is the _____ of the matrix containing only the coefficients of the system.

[9.1] *For Exercises 11 and 12, determine whether the given ordered pair is a solution to the given system of equations.*

11. $(4, 3)$; $x - y = 1$
$\qquad\qquad -x + y = -1$

12. $(1, 1)$; $2x - y = 7$
$\qquad\qquad\quad x + y = 8$

[9.1] *For Exercises 13–16:* **a.** *Determine whether the system of equations is consistent with independent equations, consistent with dependent equations, or inconsistent.*
$\qquad\qquad\qquad\qquad$ **b.** *How many solutions does the system have?*

13.

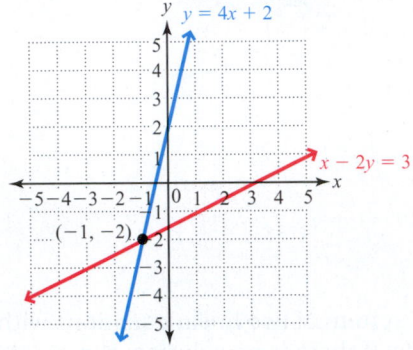

14.

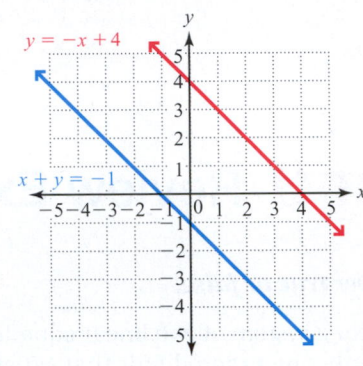

15. $\begin{cases} x - 3y = 10 \\ 2x + 3y = 5 \end{cases}$

16. $\begin{cases} x - 5y = 10 \\ 2x - 10y = 20 \end{cases}$

[9.1] *For Exercises 17–20, solve the system graphically.*

17. $\begin{cases} 4x = y \\ 3x + y = -7 \end{cases}$

18. $\begin{cases} x - y = 4 \\ 2x + 3y = 3 \end{cases}$

19. $\begin{cases} y = -2x - 3 \\ 4x + 2y = 5 \end{cases}$

20. $\begin{cases} 3x - 2y = 6 \\ y = \dfrac{3}{2}x - 3 \end{cases}$

[9.2] *For Exercises 21–24, solve the system of equations using substitution.*

21. $\begin{cases} 3x + 10y = 2 \\ x - 2y = 6 \end{cases}$

22. $\begin{cases} 2x + 5y = 8 \\ x - 10y = 9 \end{cases}$

23. $\begin{cases} 3y - 4x = 6 \\ y = \dfrac{4}{3}x + 2 \end{cases}$

24. $\begin{cases} 8x - 2y = 12 \\ y - 4x = 3 \end{cases}$

[9.3] *For Exercises 25–28, solve the system of equations using elimination.*

25. $\begin{cases} x + y = 4 \\ x - y = -2 \end{cases}$

26. $\begin{cases} 3x + 2y = 4 \\ 2x - 3y = 7 \end{cases}$

27. $\begin{cases} 0.25x + 0.75y = 4 \\ x - y = -4 \end{cases}$

28. $\begin{cases} \dfrac{1}{5}x - \dfrac{1}{3}y = 2 \\ x + y = 2 \end{cases}$

[9.4] *For Exercises 29–32, solve the system using the elimination method.*

29. $\begin{cases} x + y + z = -2 \\ 2x + 3y + 4z = -10 \\ 3x - 2y - 3z = 12 \end{cases}$
30. $\begin{cases} x + y + z = 0 \\ 2x - 4y + 3z = -12 \\ 3x - 3y + 4z = 2 \end{cases}$
31. $\begin{cases} 3x + 4y = -3 \\ -2y + 3z = 12 \\ 4x - 3z = 6 \end{cases}$
32. $\begin{cases} 2x + 2y - 3z = 5 \\ x = -3y + 4z \\ z = -x - 2y + 5 \end{cases}$

[9.5] *For Exercises 33–36, solve the system using the row echelon method.*

33. $\begin{cases} x - 4y = 8 \\ x + 2y = 2 \end{cases}$
34. $\begin{cases} 2x - y = -4 \\ 2x - 3y = 0 \end{cases}$
35. $\begin{cases} x + y + z = 2 \\ 2x + y + 2z = 1 \\ 3x + 2y + z = 1 \end{cases}$
36. $\begin{cases} x + 2y + z = -1 \\ 2x - 2y + z = -10 \\ -x + 3y - 2z = 15 \end{cases}$

[9.6] *For Exercises 37–40, solve the system of equations using Cramer's rule.*

37. $\begin{cases} 9x + 4y = 18 \\ 5x - 2y = -28 \end{cases}$
38. $\begin{cases} 5x - y = 3 \\ 10x - 2y = 5 \end{cases}$
39. $\begin{cases} x + y + z = 5 \\ 3x - 2y + z = -3 \\ x + 3y + 4z = 10 \end{cases}$
40. $\begin{cases} x + 3y + 2z = 6 \\ 2x - 3y + z = -18 \\ -3x + 2y + z = 12 \end{cases}$

[9.7] *For Exercises 41–44, graph the solution set for the system of inequalities.*

41. $\begin{cases} x + y > -5 \\ x - y \geq -1 \end{cases}$
42. $\begin{cases} 2x + y \geq -4 \\ -3x + y \geq -1 \end{cases}$
43. $\begin{cases} 2x + y > 1 \\ 3x - y \leq -1 \end{cases}$
44. $\begin{cases} -3x + 4y < 12 \\ 2x - y \leq -3 \end{cases}$

[9.2–9.6] *For Exercises 45–55, solve.*

45. The sum of two numbers is 16. Their difference is 4. What are the two numbers?

46. A movie theater charges $7.00 for adults and $5.50 for children. If 139 tickets were sold for a total of $835, how many of each ticket were sold?

47. From January through May 2008, the most visitors to the United States came from Canada and Mexico. A combined total of 3330 thousand visitors came from the two countries, with 1670 thousand more from Canada than Mexico. Find the number of visitors from each country. (*Source:* U.S. Department of Commerce)

48. Western Europe and North America contribute a combined total of 43% of the world's carbon dioxide emissions. If North America contributes 4% less than twice the percentage contributed by Western Europe, what percent does each of these regions contribute?

Of Interest

The United States alone contributes 23% of the world's carbon dioxide emissions.

49. The smaller of two complementary angles is 10 degrees less than one-third the larger angle. Find the measure of the two angles.

50. Jose invested part of $5000 in a fund that earned 7% interest and the rest in a fund that earned 9% interest. If he earned $380 interest during the year, how much did he invest in each fund?

51. Yolanda and Dee are traveling west in separate cars on the same highway. Yolanda is traveling 60 miles per hour; Dee, 70 miles per hour. Yolanda passes the I-95 exit at 4:00 P.M. Dee passes the same exit at 4:15 P.M. At what time will Dee catch up with Yolanda?

52. A plane travels 450 miles in 3 hours with the jet stream. When the plane returns, it takes 5 hours to travel the same distance against the jet stream. What is the speed of the plane in still air?

53. How much of a 10% saline solution is needed to combine with a 60% saline solution to obtain 50 milliliters of a 30% saline solution?

54. John bought vitamin supplements that cost $8 each, film that cost $4 per roll, and bags of candy that cost $3 per bag. He bought a total of eight items that cost $32. The number of rolls of film was one less than the number of bags of candy. Find the number of each type of item that he bought.

55. At a swim meet, 5 points are awarded for each first-place finish, 3 points for each second-place finish, and 1 point for each third-place finish. Shawnee Mission South High School scored a total of 38 points. The number of first-place finishes was one more than the number of second-place finishes. The number of third-place finishes was three times the number of second-place finishes. Find the number of first-, second-, and third-place finishes for the school.

[9.7] *Solve using a system of linear inequalities.*

56. A real estate course is offered in a continuing education program at a two-year college. The course is limited to 16 students. Also, to run the course, the college must receive at least $2750 in tuition. The college charges $250 per student if he or she lives in the same county as the college and $400 if he or she lives in another county (or out of state).
   a. Write a system of inequalities that describes the number of in-county students, x, enrolled in the course versus the number of out-of-county students, y, and how much they paid.
   b. Solve the system by graphing.
   c. Give a combination of students that could enroll in the course so that it runs.

Chapter 9 Practice Test

For Extra Help

CHAPTER
Test Prep
VIDEOS

Step-by-step test solutions are found on the Chapter Test Prep Videos available via the Video Resources on DVD, in **MyMathLab**, and on **YouTube** (search "CarsonCombinedAlg" and click on "Channels").

For Exercises 1 and 2, determine whether the given ordered pair is a solution to the system of equations.

1. $(-1, 3)$; $\begin{cases} 3x + 2y = 3 \\ 4x - y = -7 \end{cases}$

2. $(2, 2, 3)$; $\begin{cases} 2x - 3y + z = 1 \\ -2x + y - 3z = 11 \\ 3x + y + 3z = 14 \end{cases}$

3. Solve by graphing: $\begin{cases} x + 3y = 1 \\ -2x + y = 5 \end{cases}$

For Exercises 4–9, solve the system of equations using substitution or elimination.
Note that some systems may be inconsistent or consistent with dependent equations.

4. $\begin{cases} 2x + y = 15 \\ y = 7 - x \end{cases}$

5. $\begin{cases} x - 2y = 1 \\ 3x - 5y = 4 \end{cases}$

6. $\begin{cases} 3x - 2y = -8 \\ 2x + 3y = -14 \end{cases}$

7. $\begin{cases} 4x + 6y = 2 \\ 6x + 9y = 3 \end{cases}$

8. $\begin{cases} x + 2y + z = 2 \\ x + 4y - z = 12 \\ 3x - 3y - 2z = -11 \end{cases}$

9. $\begin{cases} x + 2y - 3z = -9 \\ 3x - y + 2z = -8 \\ 4x - 3y + 3z = -13 \end{cases}$

For Exercises 10 and 11, solve the equations using the row echelon method.

10. $\begin{cases} x + 2y = -6 \\ 3x + 4y = -10 \end{cases}$

11. $\begin{cases} x + y + z = 6 \\ 3x - 2y + 3z = -7 \\ 4x - 2y + z = -12 \end{cases}$

For Exercises 12 and 13, solve the system of equations using Cramer's rule.

12. $\begin{cases} 3x + 4y = 14 \\ 2x - 3y = -19 \end{cases}$

13. $\begin{cases} x + y + z = -1 \\ 3x - 2y + 4z = 0 \\ 2x + 5y - z = -11 \end{cases}$

14. Graph the solution set for the system of inequalities.
$\begin{cases} 2x - 3y < 1 \\ x + 2y \le -2 \end{cases}$

For Exercises 15–20, solve.

15. Excedrin surveyed workers in various professions who get headaches at least once a year on the job. Nine more accountants than waiters/waitresses in the survey got a headache. If the combined number of accountants and waiters/waitresses that got a headache was 163, how many people in each of those professions got a headache? (*Source:* the *State* newspaper)

16. When asked what the Internet most resembled, three times as many people said a library as opposed to a highway. If 240 people were polled, how many people considered the Internet to be a library? (*Source: bLINK magazine*)

17. A boat traveling with the current took 3 hours to go 30 miles. The same boat went 12 miles in 3 hours against the current. What is the rate of the boat in still water?

18. Janice invested $12,000 in two funds. One of the funds returned 6% interest, and the other returned 8% interest after one year. If the total interest for the year was $880, how much did she invest in each fund?

19. Tickets for the senior play at Apopka High School cost $3 for children, $5 for students, and $8 for adults. There were 800 tickets sold for a total of $4750. The number of adult tickets sold was 50 more than two times the number of children tickets. Find the number of each type of ticket.

20. A landscaper wants to plan a bordered rectangular garden area in a yard. Because she currently has 200 feet of border materials, the perimeter needs to be, at most, 200 feet. She thinks the garden will look best if the length is at least 10 feet more than the width.

   a. Write a system of inequalities to describe the situation.
   b. Solve the system by graphing.
   c. Give a combination of length and width that satisfy the requirements for the garden.

Chapters 1–9 Cumulative Review Exercises

For Exercises 1–3, answer true or false.

[4.4]   **1.** The graph of $y = -3x + 5$ is a line with a slope of $-3x$.

[6.4]   **2.** The expression $x^2 + 16$ cannot be factored (it is prime).

[9.1]   **3.** The ordered pair $(1, 1)$ is a solution for the system of equations

$$\begin{cases} x = y \\ 9y - 13 = -4x \end{cases}.$$

For Exercises 4–6, fill in the blank.

[1.7]   **4.** Complete using the distributive property: $4(2x + 3) =$ _____.

[3.1]   **5.** A proportion is an equation in the form $\dfrac{a}{b} =$ _____.

[5.2]   **6.** The degree of a monomial is the _____ of the exponents of all the variables

Exercises 7–16   **EXPRESSIONS**

[1.5] *For Exercises 7 and 8, simplify.*

   7. $-(6 - 2^2)^2 + 15$

   8. $-|2 - 8| + 3(2)$

For Exercises 9–11, simplify.

[5.3]   **9.** $(-3x - 4) - (3x + 2)$

[5.5]   **10.** $(x - 3)(4x^2 - 2x + 1)$

[5.6]   **11.** $(15m^2 - 22m + 14) \div (3m - 2)$

For Exercises 12 and 13, factor completely.

[6.2]   **12.** $25 + 10x + x^2$

[6.4]   **13.** $x^4 - 81$

For Exercises 14–16, simplify.

[7.2]   **14.** $\dfrac{4u^2 + 4u + 1}{u + 2u^2} \cdot \dfrac{u}{2u^2 - u - 1}$

[7.5]   **15.** $\dfrac{\dfrac{x}{4x^2 - 1}}{\dfrac{5}{2x - 1}}$

[7.4]   **16.** $\dfrac{3}{y - 2} + \dfrac{y}{y + 3}$

For Exercises 17–19, solve.

[2.3] **17.** $6x - 8 = 2(3x - 4)$

[6.6] **18.** $x^2 - 36 = 0$

[7.6] **19.** $\dfrac{5}{x-2} - \dfrac{3}{x} = \dfrac{11}{3x}$

For Exercises 20 and 21, graph.

[4.2] **20.** $y = -2x$

[8.4] **21.** $f(x) = |x + 2|$

[4.5] **22.** Write the equation of a line in standard form that passes through the point $(-2, 4)$ and is perpendicular to the line $3x - 2y = 6$.

For Exercises 23–25, solve the systems.

[9.3] **23.** $\begin{cases} 4x - 3y = -2 \\ 6x - 7y = 7 \end{cases}$

[9.4] **24.** $\begin{cases} x + y + z = 5 \\ 2x + y - 2z = -5 \\ x - 2y + z = 8 \end{cases}$

[9.7] **25.** Solve and graph the solution set: $\begin{cases} x + y < 3 \\ x - y \geq 4 \end{cases}$

For Exercises 26–30, solve.

[3.5] **26.** A chemist has a bottle containing 50 ml of 15% saline solution and a bottle of 40% saline solution. She wants a 30% solution. How much of the 40% solution must be added to the 15% solution so that a 30% concentration is created?

[5.6] **27.** If light travels about 3×10^8 meters per second, how long does it take the light from the Sun to travel to Pluto, which is a distance of about 5.9×10^{12} meters. Convert your answer to hours.

[7.7] **28.** If the wavelength of a wave remains constant, then the velocity, v, of a wave is inversely proportional to its period, T. In an experiment, waves are created in a pool of water so that the period is 8 seconds and the velocity is 6 centimeters per second. If the period is increased to 15 seconds, what is the velocity?

[7.7] **29.** LaTilla takes 3 hours to mow and edge the yard. Her husband can mow and edge in 2 hours. How long will it take them if they work on the yard together?

[9.3] **30.** In constructing a roof frame, a support beam is connected to a horizontal joist, forming two angles. If the greater angle is 15° less than twice the smaller angle, what are the measures of the two angles?

Rational Exponents, Radicals, and Complex Numbers

10

In Section 1.3, we reviewed the basic rules of square roots. In this chapter, we explore square roots and radicals in more detail. We begin in Section 10.1 with evaluating and simplifying radical expressions, much like how we began evaluating and simplifying polynomial expressions in the first part of Chapter 5. Then in Section 10.2, we see that exponents can be fractions and that those rational exponents mean roots. In Sections 10.3–10.5, we move to rewriting expressions containing radicals. Once we have explored how to rewrite radical expressions, in Section 10.6, we learn to solve equations that contain radicals. Finally, in Section 10.7, we discuss a new set of numbers called *complex numbers,* which result from square roots of negative numbers.

10.1 Radical Expressions and Functions

Objectives

1. Find the *n*th root of a number.
2. Approximate roots using a calculator.
3. Simplify radical expressions.
4. Evaluate radical functions.
5. Find the domain of radical functions.
6. Solve applications involving radical functions.

In this section through Section 10.5, we focus on the expression portion of our algebra pyramid and explore square root and radical expressions.

The Algebra Pyramid

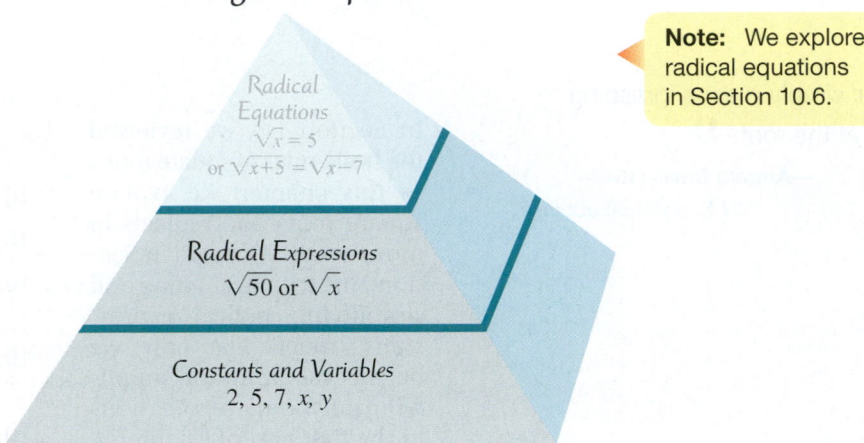

> **Note:** We explore radical equations in Section 10.6.

Objective 1 Find the *n*th root of a number. In Section 1.5, we learned that a square root of a given number is a number that, when squared, equals the given number. For example, a square root of 16 is 4 because $4^2 = 16$. However, another number can be squared to equal 16: $(-4)^2 = 16$; so -4 is also a square root of 16. Consequently, 16 has two square roots: 4 and -4. We can write them more compactly as ± 4.

Similarly, a *cube root* of a given number is a number that when cubed, equals the given number. For example, the cube root of 8 is 2 because $2^3 = 8$.

The *fourth root* of a number is a number that when raised to the fourth power, equals the given number. For example, 3 is a fourth root of 81 because $3^4 = 81$. However, $(-3)^4 = 81$ also; so ± 3 are fourth roots of 81. Our examples suggest the following definition of **nth root**.

Definition   ***nth root:*** The number b is an *n*th root of a number a if $b^n = a$.

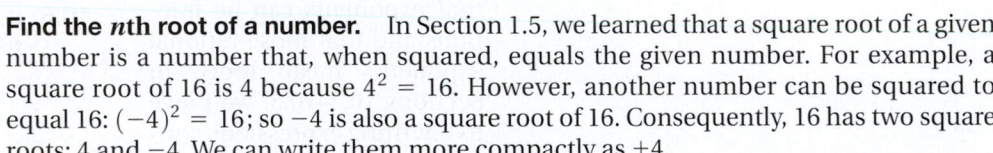

Recall from Section 1.5 that we used the symbol $\sqrt{}$, called a *radical sign,* to denote the square root of a number. To indicate roots other than square roots, we use the symbol $\sqrt[n]{a}$, read "the *n*th root of a," where n is called the *root index* and indicates which root we are to find. If no root index is given, we assume that it is 2, the square root. The number a is called the *radicand* and is the number or expression whose root we are to find. The entire expression is called a *radical,* and any expression containing a radical is called a *radical expression.*

We have seen some numbers that have more than one root, such as 16, which has two square roots, 4 and -4. To avoid confusion, the symbol $\sqrt[n]{a}$ means to find a single root. If n is even, then $\sqrt[n]{a}$ denotes the nonnegative root only and is called the *principal* root. For example, the principal square root of 16 is 4 Further, for $\sqrt[n]{a}$ to exist as a real number when n is even, a must be nonnegative because no real number can be raised to an even power to equal a negative number. For example, $\sqrt{-4}$ does not exist as a real number because there is no real number whose square is -4.

If the root index is odd, the radicand can be positive or negative because a positive number raised to an odd power is positive and a negative number raised to an odd power is negative.

Rule **Evaluating nth Roots**

When evaluating a radical expression $\sqrt[n]{a}$, the sign of a and the index n will determine possible outcomes.

If a is nonnegative, then $\sqrt[n]{a} = b$, where $b \geq 0$ and $b^n = a$.

If a is negative and n is even, then there is no real-number root.

If a is negative and n is odd, then $\sqrt[n]{a} = b$, where b is negative and $b^n = a$.

Example 1 Evaluate each root, if possible.

a. $\sqrt{225}$

SOLUTION $\sqrt{225} = 15$ Because $15^2 = 225$

b. $-\sqrt{0.81}$

SOLUTION $-\sqrt{0.81} = -0.9$ Because $(-0.9)^2 = 0.81$

> **Note:** It can be helpful to think of $-\sqrt{0.81}$ as $-1 \cdot \sqrt{0.81} = -1 \cdot 0.9 = -0.9$.

> **Note:** In Section 10.7. we define the square root of a negative number using a set of numbers called the **imaginary numbers**.

c. $\sqrt{-9}$

SOLUTION $\sqrt{-9}$ is not a real number because there is no real number whose square is -9.

d. $\pm\sqrt{121}$

SOLUTION $\pm\sqrt{121} = \pm 11$ Because $(11)^2 = 121$ and $(-11)^2 = 121$

e. $\sqrt{\dfrac{4}{9}}$

SOLUTION $\sqrt{\dfrac{4}{9}} = \dfrac{2}{3}$ Because $\left(\dfrac{2}{3}\right)^2 = \dfrac{4}{9}$

f. $\sqrt[3]{8}$

SOLUTION $\sqrt[3]{8} = 2$ Because $2^3 = 8$

g. $\sqrt[3]{-8}$

SOLUTION $\sqrt[3]{-8} = -2$ Because $(-2)^3 = -8$

h. $\sqrt[4]{16}$

SOLUTION $\sqrt[4]{16} = 2$ Because $2^4 = 16$

Your Turn 1 Evaluate each root, if possible.

a. $\sqrt{121}$ **b.** $-\sqrt{4}$ **c.** $\sqrt{-64}$ **d.** $\pm\sqrt{0.36}$

e. $\sqrt{\dfrac{25}{36}}$ **f.** $\sqrt[3]{64}$ **g.** $\sqrt[3]{-27}$ **h.** $-\sqrt[4]{81}$

Answers Your Turn 1

a. 11 **b.** −2
c. not a real number **d.** ±0.6
e. $\dfrac{5}{6}$ **f.** 4 **g.** −3 **h.** −3

Objective ② Approximate roots using a calculator. All of the roots we have considered so far have been rational. Some roots, such as $\sqrt{3}$, are called irrational. As we learned in Section 1.1, we cannot express the exact value of an irrational number using rational numbers. In fact, writing $\sqrt{3}$ with the radical sign is the only way we can express the exact value. However, we may approximate $\sqrt{3}$ using a calculator.

For example, a calculator shows that $\sqrt{2} \approx 1.414213562$, which we can round to various decimal places.

Approximating to two decimal places: $\sqrt{2} \approx 1.41$

Approximating to three decimal places: $\sqrt{2} \approx 1.414$

Note: Recall that $\approx$ means "approximately equal to."

Example 2 Approximate the roots by using a calculator. Round to three decimal places.

a. $\sqrt{12}$

Answer $\sqrt{12} \approx 3.464$

b. $-\sqrt{38}$

Answer $-\sqrt{38} \approx -6.164$

c. $\sqrt[3]{45}$

Answer $\sqrt[3]{45} \approx 3.557$

Calculator Tips

To evaluate roots higher than a square root on a scientific calculator, use the $\boxed{\sqrt[x]{}}$ *key. First, type the index; then press* $\boxed{\sqrt[x]{}}$ *and then the radicand. For example, to evaluate* $\sqrt[3]{45}$, *type* $\boxed{3}$ $\boxed{\sqrt[x]{}}$ $\boxed{4}$ $\boxed{5}$ $\boxed{\text{ENTER}}$. *On some graphing calculators,* $\sqrt[x]{}$ *is a function in a menu accessed by pressing the* $\boxed{\text{MATH}}$ *key.*

Your Turn 2 Approximate the roots using a calculator. Round to three decimal places.

a. $\sqrt{19}$ **b.** $-\sqrt{93}$ **c.** $\sqrt[3]{63}$

Objective ③ Simplify radical expressions. The definition of a root can also be used to find roots with variable radicands. Recall that with an even index, the principal root is nonnegative. So at first, we will assume that all variables represent nonnegative values. We will use $(a^m)^n = a^{mn}$ to verify the roots.

Note: Remember that with an even index, the principal root is nonnegative. Because we are assuming that the variables are nonnegative, our result accurately indicates the principal square root.

Connection To raise a power to another power using $(a^m)^n = a^{mn}$, we multiply the powers. To find a root of a power, we divide the index into the exponent.

Example 3 Find the root. Assume that variables represent nonnegative values.

a. $\sqrt{x^2}$

SOLUTION $\sqrt{x^2} = x$ Because $(x)^2 = x^2$

b. $\sqrt{a^6}$

SOLUTION $\sqrt{a^6} = a^3$ Because $(a^3)^2 = a^6$

c. $\sqrt{16x^8}$

SOLUTION $\sqrt{16x^8} = 4x^4$ Because $(4x^4)^2 = 16x^8$

d. $\sqrt{\dfrac{25x^8}{49y^2}}$

SOLUTION $\sqrt{\dfrac{25x^8}{49y^2}} = \dfrac{5x^4}{7y}$ Because $\left(\dfrac{5x^4}{7y}\right)^2 = \dfrac{25x^8}{49y^2}$

e. $\sqrt[3]{y^6}$

SOLUTION $\sqrt[3]{y^6} = y^2$ Because $(y^2)^3 = y^6$

f. $\sqrt[4]{16x^{12}}$

SOLUTION $\sqrt[4]{16x^{12}} = 2x^3$ Because $(2x^3)^4 = 16x^{12}$

Connection In parts c and f, notice the similarity between finding the root of a product and raising a product to a power. To raise a product to a power, we raise each factor to the power. To find a root of a product, we find the root of each factor.

Answers Your Turn 2

a. 4.359   **b.** −9.644   **c.** 3.979

Your Turn 3   Find the root. Assume that variables represent nonnegative values.

   a. $\sqrt{x^4}$
          b. $\sqrt{9x^{10}}$
          c. $\sqrt{36a^{12}}$

   d. $\sqrt{\dfrac{100x^4}{81y^6}}$
        e. $\sqrt[3]{27y^9}$
        f. $\sqrt[4]{b^8}$

If the variables can represent *any* real number and the index is even, we must be careful to ensure that the principal root is nonnegative by using absolute value symbols. If the root index is odd, however, we do not need absolute value symbols because the root can be positive or negative depending on the sign of the radicand.

Note: To illustrate why $\sqrt{x^2} = |x|$, suppose we were to evaluate $\sqrt{x^2}$ when $x = -3$. We would have $\sqrt{(-3)^2} = \sqrt{9} = 3$. Notice that the root, 3 is, in fact, the absolute value of -3. If we had incorrectly stated that $\sqrt{x^2} = x$, then $\sqrt{(-3)^2}$ would have to equal -3, which is not true.

Note: In part d, we do not need absolute value because y^4, with its even exponent, is always nonnegative. In parts e and f, we do not use absolute value because the indices are odd.

Example 4   Find the root. Assume that variables represent any real number.

a. $\sqrt{x^2}$

Solution $\sqrt{x^2} = |x|$

b. $\sqrt{25a^6}$

Solution $\sqrt{25a^6} = 5|a^3|$

c. $\sqrt{(n+1)^2}$

Solution $\sqrt{(n+1)^2} = |n+1|$

d. $\sqrt{25y^8}$

Solution $\sqrt{25y^8} = 5y^4$

e. $\sqrt[3]{8n^3}$

Solution $\sqrt[3]{8n^3} = 2n$

f. $\sqrt[3]{(t-2)^3}$

Solution $\sqrt[3]{(t-2)^3} = t-2$

Your Turn 4   Find the root. Assume that variables represent any real number.

   a. $\sqrt{x^4}$
        b. $\sqrt{9x^{10}}$
        c. $\sqrt{36a^{12}}$
        d. $\sqrt{1.21u^6t^2}$

   e. $\sqrt{\dfrac{100x^4}{81y^6}}$
        f. $\sqrt[3]{27y^9}$
        g. $\sqrt[4]{b^8}$

Objective ④ **Evaluate radical functions.**   Now that we have learned about radical expressions, let's examine **radical functions**.

Definition   *Radical function:* A function containing a radical expression whose radicand has a variable.

Example 5

a. Given $f(x) = \sqrt{3x-2}$, find $f(3)$.

Solution   To find $f(3)$, substitute 3 for x and simplify.
$$f(3) = \sqrt{3(3)-2} = \sqrt{9-2} = \sqrt{7}$$

b. Given $f(x) = \sqrt{2x-6}$, find $f(0)$.

Solution   To find $f(0)$, substitute 0 for x and simplify. $f(0) = \sqrt{2(0)-6} = \sqrt{-6}$, which is not a real number.

Your Turn 5   Given $f(x) = \sqrt{2x+5}$, find each of the following

   a. $f(-1)$
        b. $f(-3)$

Answers Your Turn 3
a. x^2 **b.** $3x^5$ **c.** $6a^6$
d. $\dfrac{10x^2}{9y^3}$ **e.** $3y^3$ **f.** b^2

Answers Your Turn 4
a. x^2 **b.** $3|x^5|$ **c.** $6a^6$
d. $1.1|u^3t|$ **e.** $\dfrac{10x^2}{9|y^3|}$ **f.** $3y^3$ **g.** b^2

Answers Your Turn 5
a. $f(-1) = \sqrt{3}$
b. $f(-3)$ is not a real number.

Objective ⑤ **Find the domain of radical functions.** In Example 5(b), we found that $f(0)$ did not exist because $\sqrt{-6}$ is not a real number. What does this suggest about the domains of functions involving radicals? Consider the graphs of $f(x) = \sqrt{x}$ and $f(x) = \sqrt[3]{x}$, which we can generate by choosing values for x.

x	$f(x) = \sqrt{x}$
-1	Not real
0	0
1	1
4	2

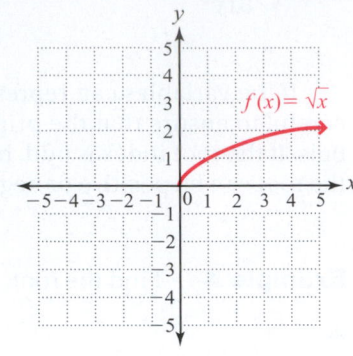

Notice that $\sqrt{x}$ is not a real number when x is negative, which means that the domain for $f(x) = \sqrt{x}$ is $\{x \mid x \geq 0\}$ or $[0, \infty)$.

x	$f(x) = \sqrt[3]{x}$
-8	-2
-1	-1
0	0
1	1
8	2

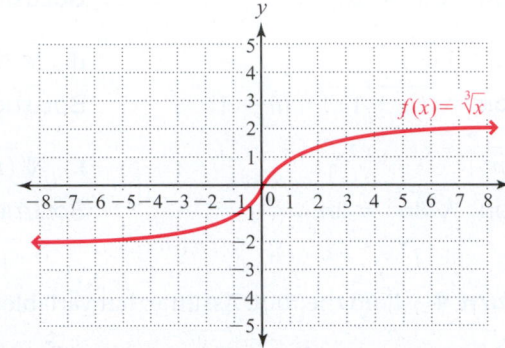

Alternatively, $\sqrt[3]{x}$, with its odd index, is a real number when x is negative; so its domain is all real numbers or $(-\infty, \infty)$. Our graphs suggest the following conclusion.

Conclusion The domain of a radical function with an even index must contain values that keep its radicand nonnegative.

Procedure **Finding the Domain of a Radical Function**

If the root index is *odd*, the domain is $\{x \mid x \text{ is a real number}\}$ or $(-\infty, \infty)$.

If the root index is *even*, the radicand must be nonnegative. Consequently, the domain is the solution set to the inequality "radicand ≥ 0."

Example 6 Find the domain of each of the following.

a. $f(x) = \sqrt{x - 4}$

SOLUTION Because the index is even, the radicand must be nonnegative. So the domain is the solution set to $x - 4 \geq 0$.

$$x - 4 \geq 0$$
$$x \geq 4 \qquad \text{Add 4 to both sides.}$$

Domain: $\{x \mid x \geq 4\}$, or $[4, \infty)$

b. $f(x) = \sqrt{-2x + 6}$

Note: Recall that when we divide *both* sides of an inequality by a negative number, we reverse the direction of the inequality.

SOLUTION The radicand must be nonnegative. So the domain is the solution set to

$$-2x + 6 \geq 0$$
$$-2x \geq -6 \qquad \text{Subtract 6 from both sides.}$$
$$x \leq 3 \qquad \text{Divide both sides by } -2.$$

Domain: $\{x \mid x \leq 3\}$, or $(-\infty, 3]$

Your Turn 6   Find the domain of each of the following.

    a. $f(x) = \sqrt{2x - 4}$ **b.** $f(x) = \sqrt{-3x - 9}$

Objective ⑥ **Solve applications involving radical functions.**   Often radical functions appear in real-world situations where one variable is a function of another.

Example 7

a. The velocity of a free-falling object is a function of the distance it has fallen. Ignoring air resistance, the velocity of an object, v, in meters per second, can be found after it has fallen h meters by using the formula $v = -\sqrt{19.6h}$. Find the velocity of a stone that has fallen 30 meters after being dropped from a cliff.

Understand   We are to find the velocity of an object after it falls 30 meters.

Plan   Use the formula $v = -\sqrt{19.6h}$, replacing h with 30.

Execute $v = -\sqrt{19.6(30)}$ **Replace h with 30.**

          $v = -\sqrt{588}$ **Multiply within the radical.**

          $v \approx -24.2$ **Evaluate the square root.**

> **Note:** A negative velocity indicates that the object is traveling downward.

Answer   After falling 30 meters, the stone is traveling at a velocity of -24.2 meters per second.

Check   We can verify the calculations, which we will leave to the reader.

b. The period of a pendulum is the amount of time it takes the pendulum to swing from the point of release to the opposite extreme and then back to the point of release. The period is a function of the length. The period, T, measured in seconds, can be found using the formula $T = 2\pi\sqrt{\dfrac{L}{9.8}}$, where L represents the length of the pendulum in meters. Find the period of a pendulum that is 0.5 meter long.

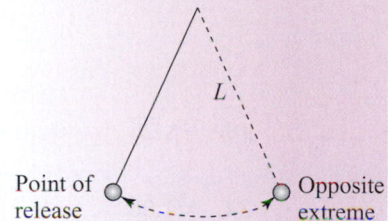

Point of release   Opposite extreme

Understand   We are to find the period of a pendulum that is 0.5 meter long.

Plan   Use the formula $T = 2\pi\sqrt{\dfrac{L}{9.8}}$, approximating π with 3.14 and replacing L with 0.5.

Execute $T \approx 2(3.14)\sqrt{\dfrac{0.5}{9.8}}$ **Substitute for π and L.**

          $T \approx 6.28\sqrt{0.051}$ **Simplify.**

          $T \approx 6.28(0.226)$ **Approximate the square root.**

          $T \approx 1.42$ **Multiply.**

Answer   The period of a pendulum that is 0.5 meter long is 1.42 seconds.

Check   We can verify the calculations, which we will leave to the reader.

Your Turn 7

a. A skydiver jumps from a plane and puts her body into a dive position so that air resistance is minimized. Find her velocity after she falls 100 meters. Use the formula in Example 7(a).

b. Find the period of a pendulum that is 0.2 meter long. Round to the nearest thousandth. Use the formula in Example 7(b).

Answers **Your Turn 6**

a. $\{x \mid x \geq 2\}$, or $[2, \infty)$

b. $\{x \mid x \leq -3\}$, or $(-\infty, -3]$

Answers **Your Turn 7**

a. -44.3 m/sec.

b. ≈ 0.898 sec.

1. Give an example of a number with rational square roots and an example of a number with irrational square roots. What makes the second number's roots irrational?

2. How can we express the exact value of a root that is irrational?

3. Why are there two square roots for every positive real number?

4. If the index is even, what is the principal root of a number?

5. Why is an even root of any negative number not a real number?

6. If x is any real number, explain why $\sqrt{x^2} = |x|$ instead of x (with no absolute value).

For Exercises 7–14, find all square roots of each number. See Objective 1.

7. 36

8. 64

9. 121

10. 81

11. 196

12. 400

13. 225

14. 289

For Exercises 15–56, evaluate each root, if possible. See Example 1.

15. $\sqrt{25}$

16. $\sqrt{49}$

17. $\sqrt{-64}$

18. $\sqrt{-25}$

19. $-\sqrt{25}$

20. $-\sqrt{100}$

21. $\pm\sqrt{25}$

22. $\pm\sqrt{100}$

23. $\sqrt{1.44}$

24. $\sqrt{1.96}$

25. $\sqrt{-10.64}$

26. $\sqrt{-2.25}$

27. $-\sqrt{0.0121}$

28. $-\sqrt{0.0169}$

29. $\sqrt{\dfrac{49}{81}}$

30. $\sqrt{\dfrac{64}{169}}$

31. $-\sqrt{\dfrac{144}{169}}$

32. $-\sqrt{\dfrac{25}{4}}$

33. $\sqrt[3]{27}$

34. $\sqrt[3]{125}$

35. $\sqrt[3]{-64}$

36. $\sqrt[3]{-216}$

37. $-\sqrt[3]{-216}$

38. $-\sqrt[3]{-125}$

39. $\sqrt[4]{256}$

40. $\sqrt[4]{81}$

41. $\sqrt[4]{-625}$

42. $\sqrt[4]{-81}$

43. $-\sqrt[4]{16}$

44. $-\sqrt[4]{625}$

45. $\sqrt[5]{32}$

46. $\sqrt[5]{243}$

47. $\sqrt[5]{-243}$

48. $\sqrt[5]{-32}$

49. $-\sqrt[5]{-32}$

50. $-\sqrt[5]{-3125}$

51. $\sqrt[6]{64}$

52. $\sqrt[6]{729}$

53. $\sqrt[3]{-\dfrac{8}{27}}$

54. $\sqrt[3]{-\dfrac{64}{125}}$

55. $\sqrt[4]{\dfrac{16}{81}}$

56. $\sqrt[5]{\dfrac{1}{32}}$

For Exercises 57–72, use a calculator to approximate each root to the nearest thousandth. See Example 2.

57. $\sqrt{7}$

58. $\sqrt{12}$

59. $-\sqrt{11}$

60. $-\sqrt{41}$

61. $\sqrt[3]{50}$

62. $\sqrt[3]{21}$

63. $\sqrt[3]{-53}$

64. $\sqrt[3]{-83}$

65. $\sqrt[4]{189}$ **66.** $\sqrt[4]{123}$ **67.** $-\sqrt[4]{85}$ **68.** $-\sqrt[4]{77}$

69. $\sqrt[5]{89}$ **70.** $\sqrt[5]{62}$ **71.** $\sqrt[6]{146}$ **72.** $\sqrt[6]{98}$

For Exercises 73–96, find the root. Assume that variables represent nonnegative values. See Example 3.

73. $\sqrt{b^4}$ **74.** $\sqrt{r^8}$ **75.** $\sqrt{16x^2}$ **76.** $\sqrt{81t^2}$ **77.** $\sqrt{100r^8s^6}$

78. $\sqrt{121x^8y^{10}}$ **79.** $\sqrt{0.25a^6b^{12}}$ **80.** $\sqrt{0.81r^4s^{14}}$ **81.** $\sqrt[3]{m^3}$ **82.** $\sqrt[3]{n^6}$

83. $\sqrt[3]{27a^9b^6}$ **84.** $\sqrt[3]{64u^{12}t^9}$ **85.** $\sqrt[3]{-64a^3b^{12}}$ **86.** $\sqrt[3]{-27r^{15}s^3}$ **87.** $\sqrt[3]{0.008x^{18}}$

88. $\sqrt[3]{0.027r^{12}}$ **89.** $\sqrt[4]{a^4}$ **90.** $\sqrt[4]{x^{12}}$ **91.** $\sqrt[4]{16x^{16}}$ **92.** $\sqrt[4]{81t^{20}}$

93. $\sqrt[5]{32x^{10}}$ **94.** $\sqrt[5]{243x^{15}}$ **95.** $\sqrt[6]{x^{12}y^6}$ **96.** $\sqrt[7]{s^7t^{21}}$

For Exercises 97–108, find the root. Assume that variables represent any real number. See Example 4.

97. $\sqrt{36m^2}$ **98.** $\sqrt{9t^6}$ **99.** $\sqrt{(r-1)^2}$ **100.** $\sqrt{(k+3)^2}$

101. $\sqrt[4]{256y^{12}}$ **102.** $\sqrt[4]{16x^{20}}$ **103.** $\sqrt[3]{27y^3}$ **104.** $\sqrt[3]{125x^6}$

105. $\sqrt{(y-3)^4}$ **106.** $\sqrt{(x+2)^4}$ **107.** $\sqrt[3]{(y-4)^6}$ **108.** $\sqrt[3]{(n+5)^9}$

For Exercises 109–112, find the indicated value of the function. See Example 5.

109. $f(x) = \sqrt{2x+4}$; find $f(0)$ **110.** $f(x) = \sqrt{3x+2}$; find $f(3)$

111. $f(x) = \sqrt{4x+3}$; find $f(3)$ **112.** $f(x) = \sqrt{-2x+3}$; find $f(-2)$

For Exercises 113–116, find the domain. See Example 6.

113. $f(x) = \sqrt{2x-8}$ **114.** $f(x) = \sqrt{3x+12}$ **115.** $f(x) = \sqrt{-4x+16}$ **116.** $f(x) = \sqrt{-2x+6}$

For Exercises 117–120:   **a.** Use a graphing calculator to graph the function. See Objective 5.
                       b. Find the domain of the function. See Objective 5 and Example 6.

117. $f(x) = \sqrt{x-2}$ **118.** $f(x) = \sqrt{x+3}$

119. $f(x) = \sqrt[3]{x + 1}$

120. $f(x) = \sqrt[4]{x - 3}$

For Exercises 121 and 122, use the following information. Ignoring air resistance, the velocity, v, in meters per second, of an object after falling h meters can be found using the formula $v = -\sqrt{19.6h}$. Round your answers to the nearest thousandth. See Example 7(a).

121. Find the velocity of a rock that has been dropped from a cliff after it falls 16 meters.

122. Find the velocity of a ball that has been dropped from a roof after it falls 9 meters.

For Exercises 123 and 124, use the following information. The period, T, of a pendulum in seconds can be found using the formula $T = 2\pi\sqrt{\dfrac{L}{9.8}}$, where L represents the length of the pendulum in meters. Round your answers to the nearest thousandth. See Example 7(b).

123. Find the period of a pendulum that is 3 meters long.

124. Find the period of a pendulum that is 6 meters long.

For Exercises 125 and 126, use the following information. The formula $S = \dfrac{7}{2}\sqrt{2D}$ can be used to approximate the speed, S, in miles per hour, that a car was traveling prior to braking and skidding a distance, D, in feet, on asphalt. (Source: Harris Technical Services Traffic Accident Reconstructionists.) Round your answers to the nearest thousandth.

125. Find the speed of a car if the skid distance is 15 feet.

126. Find the speed of a car if the skid distance is 40 feet.

For Exercises 127 and 128, use the following information. Given the lengths a and b of the sides of a right triangle, we can find the hypotenuse, c, using the formula $c = \sqrt{a^2 + b^2}$.

127. Three pieces of lumber are to be connected to form a right triangle that will be part of the frame for a roof. If the horizontal piece is to be 12 feet and the vertical piece is to be 5 feet, how long must the connecting piece be?

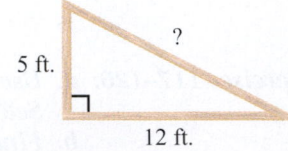

5 ft.

?

12 ft.

128. A counselor decides to create a ropes course with a zip line. She wants the line to connect from a 40-foot-tall tower to the ground at a point 100 feet from the base of the tower. Assuming that the tower and ground form a right angle, find the length of the zip line. Round your answer to the nearest tenth.

Connection In Section 6.4, we used the Pythagorean theorem, $c^2 = a^2 + b^2$, to find missing side lengths of a right triangle. The formula $c = \sqrt{a^2 + b^2}$ comes from isolating c in that theorem.

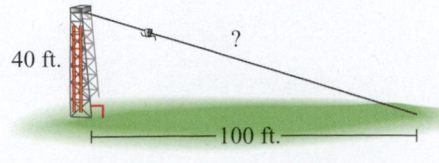

40 ft.

?

100 ft.

For Exercises 129 and 130, use the following information. Two forces, F_1 and F_2, acting on an object at a 90° angle will pull the object with a resultant force, R, at an angle between F_1 and F_2. (See the figure.) The value of the resultant force can be found using the formula $R = \sqrt{F_1^2 + F_2^2}$.

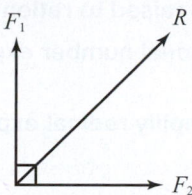

129. Find the resultant force if $F_1 = 9$ N and $F_2 = 12$ N.

> **Note:** The unit of force in the metric system is the newton (abbreviated N).

130. Find the resultant force if $F_1 = 8$ N and $F_2 = 15$ N.

131. The number of earthquakes worldwide for 2004–2008 with a magnitude of 3.0–3.9 can be approximated using the function $f(x) = 1570\sqrt{x} + 4784$, where x represents the number of years after 2004. (*Hint:* The year 2004 corresponds to $x = 0$.) (*Source:* USGS Earthquake Hazards Program)
   a. Find the approximate number of earthquakes with a magnitude of 3.0–3.9 that occurred in 2008.

   b. Find the approximate number of earthquakes with a magnitude of 3.0–3.9 that occurred in 2006.

132. The number of associate degrees, in thousands, conferred by institutions of higher education for 1970 through 2008 can be approximated using the function $f(x) = 52\sqrt{x} + 252$, where x represents the number of years after 1970. (*Hint:* The year 1970 corresponds to $x = 0$.) (*Source:* National Center for Education Statistics)
   a. How many associate degrees were awarded in 1980?
   b. How many associate degrees were awarded in 2004?

Puzzle Problem

A telemarketer calls a house and speaks to the mother of three children. The telemarketer asks the mother, "How old are you?" The woman answers, "41." The marketer then asks, "How old are your children?" The woman replies, "The product of their ages is our house number, 1296, and all of their ages are perfect squares." What are the ages of the children?

Review Exercises

Exercises 1–6 **EXPRESSIONS**

[1.5] **1.** Follow the order of operations to simplify.

   a. $\sqrt{16 \cdot 9}$ **b.** $\sqrt{16} \cdot \sqrt{9}$

For Exercises 2–6, multiply.

[5.4] 2. $x^5 \cdot x^3$ **[5.4] 3.** $(-9m^3n)(5mn^2)$ **[5.5] 4.** $4x^2(3x^2 - 5x + 1)$ **[5.5] 5.** $(7y - 4)(3y + 5)$

[5.5] 6. Write an expression for the area of the figure shown.

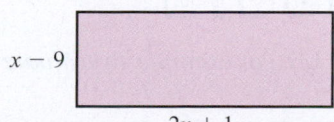

10.2 Rational Exponents

Objectives

1. Evaluate rational exponents.
2. Write radicals as expressions raised to rational exponents.
3. Simplify expressions with rational number exponents using the rules of exponents.
4. Use rational exponents to simplify radical expressions.

Objective 1 Evaluate rational exponents.

Rational Exponents with a Numerator of 1

So far we have seen only integer exponents. However, expressions can have fractional exponents also, as in $3^{1/2}$. The fractional exponent in $3^{1/2}$ is called a **rational exponent**.

Definition *Rational exponent:* An exponent that is a rational number.

To discover what rational exponents mean, consider the following.

$$\left(3^{1/2}\right)^2 = 3^{1/2 \cdot 2} = 3^1 = 3 \qquad \text{Use } (a^m)^n = a^{mn}, \text{ which we learned in Section 5.1.}$$

By the definition of square root, $\sqrt{3}$ is the number whose square is 3; so

$$\left(\sqrt{3}\right)^2 = 3.$$

Because both $3^{1/2}$ and $\sqrt{3}$ are positive, these two results suggest that a base raised to the 1/2 power and the square root of that same base are equivalent expressions, or $a^{1/2} = \sqrt{a}$. Note that the denominator, 2, of the rational exponent is the index of the root. (Remember that square roots have an unwritten index of 2.) This relationship holds for other roots as well.

Rule **Rational Exponents with a Numerator of 1**

$$a^{1/n} = \sqrt[n]{a}, \text{ where } n \text{ is a natural number other than 1.}$$

Note: If a is negative and n is odd, the root is negative. If a is negative and n is even, there is no real-number root. Such roots are imaginary and are discussed in Section 10.7.

Example 1 Rewrite using radicals; then simplify if possible. Assume that all variables represent nonnegative values.

a. $36^{1/2}$

b. $81^{1/4}$

SOLUTION $36^{1/2} = \sqrt{36} = 6$ **SOLUTION** $81^{1/4} = \sqrt[4]{81} = 3$

Calculator Tips

We use the $\boxed{\wedge}$ *key to evaluate a number raised to a rational exponent. For example, to evaluate* $81^{1/4}$, *we type*

$$\boxed{8}\ \boxed{1}\ \boxed{\wedge}\ \boxed{(}\ \boxed{1}\ \boxed{\div}\ \boxed{4}\ \boxed{)}\ \boxed{\text{ENTER}}.$$

On some scientific calculators, we use the $\boxed{y^x}$ *key. For example, to evaluate the preceding example, we type*

$$\boxed{8}\ \boxed{1}\ \boxed{y^x}\ \boxed{(}\ \boxed{1}\ \boxed{\div}\ \boxed{4}\ \boxed{)}\ \boxed{=}.$$

c. $(-125)^{1/3}$

SOLUTION $(-125)^{1/3} = \sqrt[3]{-125} = -5$

d. $(-64)^{1/4}$

SOLUTION $(-64)^{1/4} = \sqrt[4]{-64}$

There is no real-number answer.

e. $-36^{1/2}$

SOLUTION $-36^{1/2} = -\sqrt{36} = -6$

f. $x^{1/5}$

SOLUTION $x^{1/5} = \sqrt[5]{x}$

g. $(64x^6)^{1/2}$

SOLUTION $(64x^6)^{1/2} = \sqrt{64x^6} = 8x^3$

h. $12z^{1/4}$

SOLUTION $12\sqrt[4]{z}$

Note: Only the z is raised to the one-fourth power.

i. $\left(\dfrac{a^6}{36}\right)^{1/2}$

SOLUTION $\left(\dfrac{a^6}{36}\right)^{1/2} = \sqrt{\dfrac{a^6}{36}} = \dfrac{a^3}{6}$

Your Turn 1 Rewrite using radicals; then simplify if possible. Assume that all variables represent positive values.

a. $49^{1/2}$

b. $625^{1/4}$

c. $(-64)^{1/3}$

d. $-81^{1/2}$

e. $y^{1/4}$

f. $(25a^8)^{1/2}$

g. $18n^{1/5}$

h. $\left(\dfrac{49}{x^{10}}\right)^{1/2}$

Rational Exponents with a Numerator Other Than 1

Let's explore how we can rewrite expressions such as $8^{2/3}$ in which the numerator of the rational exponent is a number other than 1. Note that we could write the fraction 2/3 as a product.

$$8^{2/3} = 8^{2(1/3)} \quad \text{or} \quad 8^{2/3} = 8^{(1/3)2}$$

Using the rule of exponents $(n^a)^b = n^{ab}$ in reverse, we can write

$$8^{2/3} = 8^{2(1/3)} = (8^2)^{1/3} \quad \text{or} \quad 8^{2/3} = 8^{(1/3)2} = (8^{1/3})^2.$$

Applying the rule $a^{1/n} = \sqrt[n]{a}$, we can write

$$8^{2/3} = 8^{2(1/3)} = (8^2)^{1/3} = \sqrt[3]{8^2} \quad \text{or} \quad 8^{2/3} = 8^{(1/3)2} = (8^{1/3})^2 = (\sqrt[3]{8})^2.$$

Notice that the denominator of the rational exponent becomes the index of the radical. The numerator of the rational exponent can be written as the exponent of the radicand or as an exponent for the entire radical.

Rule **General Rule for Rational Exponents**

$a^{m/n} = \sqrt[n]{a^m} = (\sqrt[n]{a})^m$, where $a \geq 0$ and m and n are natural numbers other than 1.

Example 2 Rewrite using radicals; then simplify if possible. Assume that all variables represent nonnegative values.

a. $8^{2/3}$

SOLUTION $8^{2/3} = \sqrt[3]{8^2}$ Rewrite.

$\phantom{8^{2/3}} = \sqrt[3]{64}$ Square the radicand.

$\phantom{8^{2/3}} = 4$ Evaluate the root.

or $8^{2/3} = (\sqrt[3]{8})^2$ Rewrite.

$\phantom{8^{2/3}} = (2)^2$ Evaluate the root.

$\phantom{8^{2/3}} = 4$ Evaluate the exponential form.

b. $625^{3/4}$

SOLUTION We could rewrite $625^{3/4}$ as $\sqrt[4]{625^3}$, but calculating 625^3 is tedious without a calculator; so we use the other form of the rule.

$$625^{3/4} = (\sqrt[4]{625})^3 \qquad \text{Rewrite.}$$
$$= (5)^3 \qquad \text{Evaluate the root.}$$
$$= 125 \qquad \text{Evaluate the exponential form.}$$

c. $-9^{5/2}$

Warning: In the expression $-9^{5/2}$, the negative sign tells us to find the opposite of the value of the exponential form.

$$-9^{5/2} = -(9^{5/2}) = -(\sqrt{9})^5$$

Or some people find it helpful to think of $-9^{5/2}$ as $-1 \cdot 9^{5/2}$.

$$-9^{5/2} = -1 \cdot 9^{5/2}$$
$$= -1 \cdot (\sqrt{9})^5$$

SOLUTION $-9^{5/2} = -(\sqrt{9})^5 = -(3)^5 = -243$

d. $\left(\dfrac{1}{9}\right)^{5/2}$

Note: The simplification is usually easier if we write $a^{m/n}$ as $(\sqrt[n]{a})^m$ if a is a constant and as $\sqrt[n]{a^m}$ if a is a variable or a variable expression.

SOLUTION $\left(\dfrac{1}{9}\right)^{5/2} = \left(\sqrt{\dfrac{1}{9}}\right)^5 = \left(\dfrac{1}{3}\right)^5 = \dfrac{1}{243}$

e. $x^{2/3}$

SOLUTION $x^{2/3} = \sqrt[3]{x^2}$

Note: Because the base is a variable, we put the exponent beneath the radical sign.

f. $(3x - 5)^{3/5}$

SOLUTION $(3x - 5)^{3/5} = \sqrt[5]{(3x - 5)^3}$

Your Turn 2 Rewrite using radicals; then simplify if possible. Assume that all variables represent nonnegative values.

a. $32^{3/5}$ **b.** $-49^{3/2}$ **c.** $\left(\dfrac{1}{16}\right)^{3/2}$ **d.** $y^{5/6}$ **e.** $(3a + 4)^{4/5}$

Negative Rational Exponents

Recall from Section 5.1 that $a^{-b} = \dfrac{1}{a^b}$. For example, $2^{-3} = \dfrac{1}{2^3}$. This same rule applies to negative rational exponents.

Rule **Negative Rational Exponents**

$a^{-m/n} = \dfrac{1}{a^{m/n}}$, where $a \neq 0$ and m and n are natural numbers with $n \neq 1$.

Example 3 Rewrite using radicals; then simplify if possible.

a. $81^{-1/2}$

SOLUTION $81^{-1/2} = \dfrac{1}{81^{1/2}}$ Rewrite the exponential form with a positive exponent by inverting the base and changing the sign of the exponent.

$= \dfrac{1}{\sqrt{81}}$ Write the rational exponent in radical form.

$= \dfrac{1}{9}$ Evaluate the square root.

Answers **Your Turn 2**

a. 8 **b.** -343 **c.** $\dfrac{1}{64}$ **d.** $\sqrt[6]{y^5}$

e. $\sqrt[5]{(3a + 4)^4}$

b. $16^{-3/4}$

SOLUTION $16^{-3/4} = \dfrac{1}{16^{3/4}}$ Rewrite the exponential form with a positive exponent by inverting the base and changing the sign of the exponent.

$$= \dfrac{1}{(\sqrt[4]{16})^3}$$ Write the rational exponent in radical form.

$$= \dfrac{1}{(2)^3}$$ Evaluate the radical.

$$= \dfrac{1}{8}$$ Simplify the exponential form.

c. $\left(\dfrac{16}{25}\right)^{-1/2}$

Connection Although we used $a^{-m/n} = \dfrac{1}{a^{m/n}}$ in part c, we could have used $\left(\dfrac{a}{b}\right)^{-n} = \left(\dfrac{b}{a}\right)^{n}$, which we learned in Chapter 5.

SOLUTION $\left(\dfrac{16}{25}\right)^{-1/2} = \dfrac{1}{\left(\dfrac{16}{25}\right)^{1/2}}$ Rewrite the exponential form with a positive exponent by inverting the base and changing the sign of the exponent.

$$= \dfrac{1}{\sqrt{\dfrac{16}{25}}}$$ Write the rational exponent in radical form.

$$= \dfrac{1}{\dfrac{4}{5}}$$ Evaluate the square root.

$$= \dfrac{5}{4}$$ Simplify the complex fraction.

d. $(-64)^{-2/3}$

SOLUTION $(-64)^{-2/3} = \dfrac{1}{(-64)^{2/3}}$ Rewrite the exponential form with a positive exponent.

$$= \dfrac{1}{(\sqrt[3]{-64})^2}$$ Write the rational exponent in radical form.

$$= \dfrac{1}{(-4)^2}$$ Evaluate the cube root.

$$= \dfrac{1}{16}$$ Square -4.

Your Turn 3   Rewrite using radicals; then simplify if possible.

a. $49^{-1/2}$ **b.** $-27^{-2/3}$ **c.** $\left(\dfrac{16}{81}\right)^{-3/4}$ **d.** $(-8)^{-5/3}$

Objective ❷ Write radicals as expressions raised to rational exponents. In upper-level math courses, it is often necessary to write radical expressions in exponential form. To do so, we use the facts that $\sqrt[n]{a^m} = a^{m/n}$ and $(\sqrt[n]{a})^m = a^{m/n}$.

Answers   **Your Turn 3**

a. $\dfrac{1}{7}$ **b.** $-\dfrac{1}{9}$

c. $\dfrac{27}{8}$ **d.** $-\dfrac{1}{32}$

Example 4 Write each of the following in exponential form. Assume that all variables represent positive values.

a. $\sqrt[4]{x^3}$

SOLUTION $\sqrt[4]{x^3} = x^{3/4}$

b. $\dfrac{1}{\sqrt[3]{x^2}}$

SOLUTION $\dfrac{1}{\sqrt[3]{x^2}} = \dfrac{1}{x^{2/3}} = x^{-2/3}$

c. $(\sqrt[5]{z})^3$

SOLUTION $(\sqrt[5]{z})^3 = z^{3/5}$

d. $\sqrt[6]{(3x-5)^5}$

SOLUTION $\sqrt[6]{(3x-5)^5} = (3x-5)^{5/6}$

Your Turn 4 Write each of the following in exponential form. Assume that all variables represent positive values.

a. $\sqrt[5]{y^2}$ **b.** $\dfrac{1}{\sqrt[4]{x^3}}$ **c.** $(\sqrt[7]{a})^3$ **d.** $\sqrt[3]{(3y-2)^5}$

Objective ③ Simplify expressions with rational number exponents using the rules of exponents. Rational exponents follow the same rules that we established in Chapter 5 for integer exponents, which we review here.

Rule **Rules of Exponents Summary**

(Assume that no denominators are 0, that a and b are real numbers, and that m and n are integers.)

Zero as an exponent:	$a^0 = 1$, where $a \neq 0$
	0^0 is indeterminate.
Negative exponents:	$a^{-n} = \dfrac{1}{a^n}$ and $\dfrac{1}{a^{-n}} = a^n$
	$\left(\dfrac{a}{b}\right)^{-n} = \left(\dfrac{b}{a}\right)^n$
Product rule of exponents:	$a^m \cdot a^n = a^{m+n}$
Quotient rule for exponents:	$\dfrac{a^m}{a^n} = a^{m-n}$
Raising a power to a power:	$(a^m)^n = a^{mn}$
Raising a product to a power:	$(ab)^n = a^n b^n$
Raising a quotient to a power:	$\left(\dfrac{a}{b}\right)^n = \dfrac{a^n}{b^n}$

Example 5 Use the rules of exponents to simplify. Write the answers with positive exponents. Assume that all variables represent positive values.

a. $x^{1/5} \cdot x^{3/5}$

SOLUTION $x^{1/5} \cdot x^{3/5} = x^{1/5 + 3/5}$ Use $a^m \cdot a^n = a^{m+n}$.

$= x^{4/5}$ Add the exponents.

b. $(2a^{1/2})(4a^{1/3})$

SOLUTION $(2a^{1/2})(4a^{1/3}) = 8a^{1/2 + 1/3}$ Use $a^m \cdot a^n = a^{m+n}$.

$= 8a^{3/6 + 2/6}$ Rewrite the exponents with a common denominator of 6.

$= 8a^{5/6}$ Add the exponents.

Answers Your Turn 4

a. $y^{2/5}$ **b.** $x^{-3/4}$

c. $a^{3/7}$ **d.** $(3y-2)^{5/3}$

c. $\dfrac{5^{2/3}}{5^{3/4}}$

SOLUTION $\dfrac{5^{2/3}}{5^{3/4}} = 5^{2/3-3/4}$ Use $\dfrac{a^m}{a^n} = a^{m-n}$.

$= 5^{8/12-9/12}$ Rewrite the exponents with a common denominator of 12.

$= 5^{-1/12}$ Subtract the exponents.

$= \dfrac{1}{5^{1/12}}$ Rewrite with a positive exponent.

d. $(-4y^{-3/5})(5y^{4/5})$

SOLUTION $(-4y^{-3/5})(5y^{4/5}) = -20y^{-3/5+4/5}$ Use $a^m \cdot a^n = a^{m+n}$.

$= -20y^{1/5}$ Add the exponents.

e. $(w^{3/4})^2$

SOLUTION $(w^{3/4})^2 = w^{(3/4)2}$ Use $(a^m)^n = a^{mn}$.

$= w^{3/2}$ Multiply the exponents.

f. $(2a^{2/3}b^{3/5})^3$

SOLUTION $(2a^{2/3}b^{3/5})^3 = 2^3(a^{2/3})^3(b^{3/5})^3$ Use $(ab)^n = a^n \cdot b^n$.

$= 8a^{(2/3)3}b^{(3/5)3}$ Use $(a^m)^n = a^{mn}$.

$= 8a^2b^{9/5}$ Multiply the exponents.

g. $\dfrac{(3x^{4/3})^3}{x^3}$

SOLUTION $\dfrac{(3x^{4/3})^3}{x^3} = \dfrac{3^3(x^{4/3})^3}{x^3}$ Use $(ab)^n = a^n \cdot b^n$ in the numerator.

$= \dfrac{27x^{(4/3)3}}{x^3}$ Use $(a^m)^n = a^{mn}$.

$= \dfrac{27x^4}{x^3}$ Simplify.

$= 27x^{4-3}$ Use $\dfrac{a^m}{a^n} = a^{m-n}$.

$= 27x$ Simplify.

Your Turn 5   Use the rules of exponents to simplify. Write the answers with positive exponents. Assume that all variables represent positive values.

a. $(x^{2/3})(x^{1/2})$ **b.** $\dfrac{y^{2/7}}{y^{5/7}}$ **c.** $(b^{3/4})^2$ **d.** $(2x^{3/2}y^{2/3})^4$ **e.** $\dfrac{(2z^{5/3})^3}{z^2}$

Objective 4   **Use rational exponents to simplify radical expressions.**   Often, radical expressions can be simplified by rewriting them with rational exponents, simplifying, and then rewriting as a radical expression.

Answers   **Your Turn 5**

a. $x^{7/6}$ **b.** $\dfrac{1}{y^{3/7}}$ **c.** $b^{3/2}$

d. $16x^6y^{8/3}$ **e.** $8z^3$

Example 6   Rewrite as a radical with a smaller root index. Assume that all variables represent nonnegative values.

a. $\sqrt[4]{49}$

SOLUTION $\quad \sqrt[4]{49} = 49^{1/4}$ Rewrite in exponential form.

$\qquad\qquad\qquad = (7^2)^{1/4}$ Write 49 as 7^2.

$\qquad\qquad\qquad = 7^{2 \cdot 1/4}$ Use $(a^m)^n = a^{mn}$.

$\qquad\qquad\qquad = 7^{1/2}$ Simplify.

$\qquad\qquad\qquad = \sqrt{7}$ Write in radical form.

b. $\sqrt[6]{x^4}$

SOLUTION $\quad \sqrt[6]{x^4} = x^{4/6}$ Rewrite in exponential form.

$\qquad\qquad\qquad = x^{2/3}$ Simplify to lowest terms.

$\qquad\qquad\qquad = \sqrt[3]{x^2}$ Write in radical form.

c. $\sqrt[6]{a^4 b^2}$

SOLUTION $\quad \sqrt[6]{a^4 b^2} = (a^4 b^2)^{1/6}$ Rewrite in exponential form.

$\qquad\qquad\qquad = (a^4)^{1/6}(b^2)^{1/6}$ Use $(ab)^n = a^n b^n$.

$\qquad\qquad\qquad = a^{4 \cdot 1/6} b^{2 \cdot 1/6}$ Use $(a^m)^n = a^{mn}$.

$\qquad\qquad\qquad = a^{2/3} b^{1/3}$ Simplify.

$\qquad\qquad\qquad = (a^2 b)^{1/3}$ Use $(ab)^n = a^n b^n$.

$\qquad\qquad\qquad = \sqrt[3]{a^2 b}$ Write in radical form.

> **Your Turn 6**   Rewrite as a radical with a smaller root index. Assume that all variables represent nonnegative values.
>
> **a.** $\sqrt[4]{36}$ **b.** $\sqrt[8]{x^6}$ **c.** $\sqrt[8]{x^6 y^2}$

Writing radicals in exponential form also allows us to multiply and divide radical expressions with different root indices.

Procedure   **Multiplying and Dividing Radical Expressions with Different Root Indices**

To multiply or divide radical expressions with different root indices,

1. Change from radical to exponential form.
2. Multiply or divide using the appropriate rule(s) of exponents.
3. Write the result from Step 2 in radical form.

Example 7   Perform the indicated operations. Write the result using a radical. Assume that all variables represent positive values.

a. $\sqrt{x} \cdot \sqrt[3]{x^2}$

SOLUTION $\quad \sqrt{x} \cdot \sqrt[3]{x^2} = x^{1/2} \cdot x^{2/3}$ Write in exponential form.

$\qquad\qquad\qquad = x^{1/2 + 2/3}$ Use $a^m \cdot a^n = a^{m+n}$.

$\qquad\qquad\qquad = x^{3/6 + 4/6}$ Rewrite the exponents with their LCD.

$\qquad\qquad\qquad = x^{7/6}$ Simplify.

$\qquad\qquad\qquad = \sqrt[6]{x^7}$ Write in radical form.

> **Note:** In Section 10.3, we learn to further simplify expressions such as $\sqrt[6]{x^7}$.

b. $\dfrac{\sqrt[4]{x^3}}{\sqrt[3]{x}}$

Answers   **Your Turn 6**

a. $\sqrt{6}$ **b.** $\sqrt[4]{x^3}$ **c.** $\sqrt[4]{x^3 y}$

SOLUTION $\dfrac{\sqrt[4]{x^3}}{\sqrt[3]{x}} = \dfrac{x^{3/4}}{x^{1/3}}$ Write in exponential form.

$$= x^{3/4 - 1/3} \qquad \text{Use } \dfrac{a^m}{a^n} = a^{m-n}.$$

$$= x^{9/12 - 4/12} \qquad \text{Rewrite the exponents with their LCD.}$$

$$= x^{5/12} \qquad \text{Simplify.}$$

$$= \sqrt[12]{x^5} \qquad \text{Write in radical form.}$$

c. $\sqrt{3} \cdot \sqrt[3]{2}$

SOLUTION $\sqrt{3} \cdot \sqrt[3]{2} = 3^{1/2} \cdot 2^{1/3}$ Write in exponential form.

$$= 3^{3/6} \cdot 2^{2/6} \qquad \text{Write the exponents with their LCD.}$$

$$= (3^3 \cdot 2^2)^{1/6} \qquad \text{Use } a^n b^n = (ab)^n.$$

$$= (27 \cdot 4)^{1/6} \qquad \text{Evaluate } 3^3 \text{ and } 2^2.$$

$$= 108^{1/6} \qquad \text{Multiply.}$$

$$= \sqrt[6]{108} \qquad \text{Rewrite as a radical.}$$

Your Turn 7   Perform the indicated operations. Write the result using a radical. Assume that all variables represent positive values.

a. $\sqrt[3]{x^2} \cdot \sqrt[4]{x}$ **b.** $\dfrac{\sqrt[4]{a^3}}{\sqrt[3]{a^2}}$ **c.** $\sqrt[4]{2} \cdot \sqrt{5}$

By writing radical expressions using rational exponents, we can also find the root of a root.

Example 8   Write $\sqrt[3]{\sqrt{x}}$ as a single radical. Assume that all variables represent nonnegative values.

SOLUTION $\sqrt[3]{\sqrt{x}} = (x^{1/2})^{1/3}$ Write in exponential form.

$$= x^{(1/2)(1/3)} \qquad \text{Apply } (a^m)^n = a^{mn}.$$

$$= x^{1/6} \qquad \text{Simplify.}$$

$$= \sqrt[6]{x} \qquad \text{Write as a radical.}$$

Answers   **Your Turn 7**

a. $\sqrt[12]{x^{11}}$ **b.** $\sqrt[12]{a}$ **c.** $\sqrt[4]{50}$

Answer   **Your Turn 8**

$\sqrt[12]{y}$

Your Turn 8   Write $\sqrt[4]{\sqrt[3]{y}}$ as a single radical. Assume that all variables represent nonnegative values.

10.2 Exercises   For Extra Help   **MyMathLab**

Math XL   PRACTICE   WATCH   DOWNLOAD

1. If $4^{1/3}$ were written as a radical expression, what would be the radicand?

2. If $4^{3/5}$ were written as a radical expression, what would be the root index?

3. If $a^{m/n}$, with n even and m/n reduced to lowest terms, were written as a radical expression, what restrictions would be placed on a? Why?

4. If $a^{m/n}$, with n odd, were written as a radical expression, what restrictions would be placed on a? Why?

5. Does $\sqrt[4]{100} = \sqrt{10}$? Why or why not?

6. Does $\sqrt[3]{3} \cdot \sqrt[5]{2} = \sqrt[15]{1944}$? Why or why not?

For Exercises 7–40, rewrite each of the following using radicals; then simplify if possible. Assume that all variables represent nonnegative values. See Examples 1–3.

7. $25^{1/2}$

8. $64^{1/2}$

9. $-100^{1/2}$

10. $-64^{1/2}$

11. $27^{1/3}$

12. $216^{1/3}$

13. $(-64)^{1/3}$

14. $(-125)^{1/3}$

15. $y^{1/4}$

16. $w^{1/8}$

17. $(144x^8)^{1/2}$

18. $(121z^6)^{1/2}$

19. $18r^{1/2}$

20. $22a^{1/2}$

21. $\left(\dfrac{x^4}{81}\right)^{1/2}$

22. $\left(\dfrac{n^8}{36}\right)^{1/2}$

23. $64^{2/3}$

24. $16^{3/4}$

25. $-81^{3/4}$

26. $-16^{5/4}$

27. $(-8)^{4/3}$

28. $(-27)^{5/3}$

29. $16^{-3/4}$

30. $8^{-4/3}$

31. $x^{4/5}$

32. $m^{5/7}$

33. $8n^{2/3}$

34. $6a^{5/6}$

35. $(-32)^{-2/5}$

36. $(-216)^{-2/3}$

37. $\left(\dfrac{1}{25}\right)^{3/2}$

38. $\left(\dfrac{1}{32}\right)^{3/5}$

39. $(2a + 4)^{5/6}$

40. $(5r - 2)^{5/7}$

For Exercises 41–56, write each of the following in exponential form. Assume that all variables represent positive values. See Example 4.

41. $\sqrt[4]{25}$

42. $\sqrt[5]{42}$

43. $\sqrt[6]{z^5}$

44. $\sqrt[7]{r^5}$

45. $\dfrac{1}{\sqrt[6]{5^5}}$

46. $\dfrac{1}{\sqrt[4]{6^2}}$

47. $\dfrac{5}{\sqrt[5]{x^4}}$

48. $\dfrac{8}{\sqrt[7]{n^9}}$

49. $(\sqrt[3]{5})^7$

50. $(\sqrt[5]{8})^6$

51. $(\sqrt{x})^2$

52. $(\sqrt[3]{m})^8$

53. $\sqrt[4]{(4a - 5)^7}$

54. $\sqrt[7]{(5w + 3)^5}$

55. $(\sqrt[5]{2r - 5})^8$

56. $(\sqrt[5]{3r - 6})^9$

For Exercises 57–96, use the rules of exponents to simplify. Write the answers with positive exponents. Assume that all variables represent positive values. See Example 5.

57. $x^{1/5} \cdot x^{3/5}$

58. $n^{1/7} \cdot n^{5/7}$

59. $x^{3/2} \cdot x^{-1/3}$

60. $n^{2/3} \cdot n^{-1/2}$

61. $a^{2/3} \cdot a^{3/4}$

62. $r^{5/2} \cdot r^{5/3}$

63. $(3w^{1/7})(7w^{3/7})$

64. $(5p^{1/9})(3p^{4/9})$

65. $(-3a^{2/3})(4a^{3/4})$

66. $(8c^{4/5})(-4c^{3/2})$

67. $\dfrac{7^{7/3}}{7^{2/3}}$

68. $\dfrac{3^{7/9}}{3^{2/9}}$

69. $\dfrac{x^{1/6}}{x^{5/6}}$

70. $\dfrac{y^{2/9}}{y^{5/9}}$

71. $\dfrac{x^{3/4}}{x^{1/2}}$

72. $\dfrac{x^{5/8}}{x^{1/2}}$

73. $\dfrac{r^{3/4}}{r^{2/3}}$

74. $\dfrac{m^{3/5}}{m^{1/2}}$

75. $\dfrac{x^{-3/7}}{x^{2/7}}$

76. $\dfrac{v^{-4/5}}{v^{2/5}}$

77. $\dfrac{a^{3/4}}{a^{-3/2}}$

78. $\dfrac{b^{5/6}}{b^{-2/3}}$

79. $(5s^{-2/7})(4s^{5/7})$

80. $(6u^{8/9})(-6u^{-5/9})$

81. $(-6b^{-5/4})(4b^{3/2})$

82. $(-6y^{-5/6})(-7y^{5/3})$

83. $(x^{2/3})^3$

84. $(r^{3/4})^4$

85. $(a^{5/6})^2$

86. $(n^{3/8})^4$

87. $(b^{2/3})^{3/5}$

88. $(m^{3/2})^{2/5}$

89. $(2x^{2/3}y^{1/2})^6$

90. $(2a^{1/4}b^{3/2})^8$

91. $(8q^{3/2}t^{3/4})^{1/3}$

92. $(16x^{2/3}y^{1/3})^{3/4}$

93. $\dfrac{(3a^{5/4})^4}{a^2}$

94. $\dfrac{(5v^{5/2})^2}{v^3}$

95. $\dfrac{(9z^{7/3})^{1/2}}{z^{5/6}}$

96. $\dfrac{(36x^{3/4})^{1/2}}{x^{1/8}}$

For Exercises 97–108, represent each of the following as a radical with a smaller root index. Assume that all variables represent nonnegative values. See Example 6.

97. $\sqrt[4]{4}$

98. $\sqrt[4]{49}$

99. $\sqrt[6]{49}$

100. $\sqrt[6]{100}$

101. $\sqrt[4]{x^2}$

102. $\sqrt[6]{y^3}$

103. $\sqrt[8]{r^6}$

104. $\sqrt[10]{n^6}$

105. $\sqrt[8]{x^6y^2}$

106. $\sqrt[6]{y^2z^4}$

107. $\sqrt[10]{m^4n^6}$

108. $\sqrt[10]{a^2b^8}$

For Exercises 109–120, perform the indicated operations. Write the result using a radical. Assume that all variables represent positive values. See Example 7.

109. $\sqrt[3]{x}\cdot\sqrt{x}$

110. $\sqrt[4]{y}\cdot\sqrt[3]{y^2}$

111. $\sqrt[4]{y^2}\cdot\sqrt[3]{y^2}$

112. $\sqrt[5]{x^4}\cdot\sqrt[3]{x^2}$

113. $\dfrac{\sqrt[3]{x^4}}{\sqrt[4]{x^2}}$

114. $\dfrac{\sqrt[3]{y^5}}{\sqrt[4]{y^2}}$

115. $\dfrac{\sqrt[5]{n^4}}{\sqrt[3]{n^2}}$

116. $\dfrac{\sqrt[6]{z^4}}{\sqrt{z}}$

117. $\sqrt{5}\cdot\sqrt[3]{3}$

118. $\sqrt[3]{4}\cdot\sqrt{5}$

119. $\sqrt[4]{6}\cdot\sqrt[3]{2}$

120. $\sqrt[4]{4}\cdot\sqrt[3]{2}$

For Exercises 121–124, write each as a single radical. Assume that all variables represent nonnegative values. See Example 8.

121. $\sqrt[3]{\sqrt[3]{x}}$

122. $\sqrt[3]{\sqrt[5]{m}}$

123. $\sqrt{\sqrt[3]{n}}$

124. $\sqrt[4]{\sqrt[5]{z}}$

Review Exercises

Exercises 1–6　EXPRESSIONS

[1.5] **1.** Rewrite $2\cdot2\cdot2\cdot2\cdot x\cdot x\cdot x\cdot y\cdot y$ using exponents.

[1.5] *For Exercises 2 and 3, simplify using the order of operations agreement.*

2. $\sqrt{16}\cdot\sqrt{9}$

3. $\sqrt[3]{27}\cdot\sqrt[3]{125}$

[5.4] *For Exercises 4 and 5, simplify.*

4. $(2.5\times10^6)(3.2\times10^5)$

5. $\left(\dfrac{3}{4}x^3y\right)\left(-\dfrac{5}{6}xyz^2\right)$

[5.6] **6.** Use long division to find the quotient: $\dfrac{2x^3-2x^2-19x+18}{x+3}$

10.3 Multiplying, Dividing, and Simplifying Radicals

Objectives

① Multiply radical expressions.

② Divide radical expressions.

③ Use the product rule to simplify radical expressions.

In this section, we explore some ways to simplify expressions that involve multiplication or division of radicals.

Objective ① Multiply radical expressions. Consider the expression $\sqrt{9} \cdot \sqrt{16}$. The usual approach is to find the roots and then multiply those roots. However, we can also multiply the radicands and then find the root of the product.

Find the roots first:	Multiply the radicands first:
$\sqrt{9} \cdot \sqrt{16} = 3 \cdot 4 = 12$	$\sqrt{9} \cdot \sqrt{16} = \sqrt{9 \cdot 16} = \sqrt{144} = 12$

Both approaches give the same result, suggesting the following rule.

Rule Product Rule for Radicals

If both $\sqrt[n]{a}$ and $\sqrt[n]{b}$ are real numbers, then $\sqrt[n]{a} \cdot \sqrt[n]{b} = \sqrt[n]{a \cdot b}$.

Note: To apply this rule, the root indices must be the same.

Example 1 Find the product and simplify. Assume that all variables represent positive values.

a. $\sqrt{3} \cdot \sqrt{27}$

SOLUTION $\sqrt{3} \cdot \sqrt{27} = \sqrt{3 \cdot 27} = \sqrt{81} = 9$

> **Connection** Example 1(a) illustrates that the product of two irrational numbers can be a rational number.

b. $\sqrt{11} \cdot \sqrt{x}$

SOLUTION $\sqrt{11} \cdot \sqrt{x} = \sqrt{11x}$

c. $\sqrt[3]{4} \cdot \sqrt[3]{2}$

SOLUTION $\sqrt[3]{4} \cdot \sqrt[3]{2} = \sqrt[3]{4 \cdot 2} = \sqrt[3]{8} = 2$

d. $\sqrt[3]{3x} \cdot \sqrt[3]{4x}$

SOLUTION $\sqrt[3]{3x} \cdot \sqrt[3]{4x} = \sqrt[3]{3x \cdot 4x} = \sqrt[3]{12x^2}$

e. $\sqrt[4]{5} \cdot \sqrt[4]{7x^2}$

SOLUTION $\sqrt[4]{5} \cdot \sqrt[4]{7x^2} = \sqrt[4]{5 \cdot 7x^2} = \sqrt[4]{35x^2}$

f. $\sqrt{\dfrac{5}{x}} \cdot \sqrt{\dfrac{y}{2}}$

SOLUTION $\sqrt{\dfrac{5}{x}} \cdot \sqrt{\dfrac{y}{2}} = \sqrt{\dfrac{5}{x} \cdot \dfrac{y}{2}} = \sqrt{\dfrac{5y}{2x}}$

> **Connection** Expressions that have a fraction in a radical, such as $\sqrt{\dfrac{5y}{2x}}$, are not considered to be in simplest form. We learn how to simplify them in Section 10.5.

g. $\sqrt{x} \cdot \sqrt{x}$

SOLUTION $\sqrt{x} \cdot \sqrt{x} = \sqrt{x \cdot x} = \sqrt{x^2} = x$

Your Turn 1 Find the product and simplify. Assume that all variables represent positive values.

a. $\sqrt{2} \cdot \sqrt{32}$ b. $\sqrt{5} \cdot \sqrt{a}$ c. $\sqrt[3]{5x} \cdot \sqrt[3]{2y}$ d. $\sqrt{\dfrac{2}{a}} \cdot \sqrt{\dfrac{b}{7}}$ e. $\sqrt{a} \cdot \sqrt{a}$

Notice that in Example 1(g), $\sqrt{x} \cdot \sqrt{x} = x$. It is also true that $\sqrt{x} \cdot \sqrt{x} = (\sqrt{x})^2$. Therefore, $(\sqrt{x})^2 = x$. Similarly, $\sqrt[3]{x} \cdot \sqrt[3]{x} \cdot \sqrt[3]{x} = \sqrt[3]{x \cdot x \cdot x} = \sqrt[3]{x^3} = x$. It is also true that $\sqrt[3]{x} \cdot \sqrt[3]{x} \cdot \sqrt[3]{x} = (\sqrt[3]{x})^3$; so $(\sqrt[3]{x})^3 = x$. These examples suggest the following rule.

Rule   **Raising an *n*th Root to the *n*th Power**

For any nonnegative real number a, $(\sqrt[n]{a})^n = a$.

This means that $(\sqrt[3]{4})^3 = 4$, $(\sqrt[5]{19})^5 = 19$, and $(\sqrt[4]{3x^2})^4 = 3x^2$.

Objective ② Divide radical expressions. Earlier we developed the product rule for radicals. Now we develop a similar rule for quotients such as $\dfrac{\sqrt{100}}{\sqrt{25}}$. We can follow the order of operations and divide the roots, or we can divide the radicands and then find the square root of the quotient.

Find the roots first:
$$\frac{\sqrt{100}}{\sqrt{25}} = \frac{10}{5} = 2$$

Divide the radicands first:
$$\frac{\sqrt{100}}{\sqrt{25}} = \sqrt{\frac{100}{25}} = \sqrt{4} = 2$$

Both approaches give the same result, which suggests the following rule.

Rule   **Quotient Rule for Radicals**

If both $\sqrt[n]{a}$ and $\sqrt[n]{b}$ are real numbers, then $\dfrac{\sqrt[n]{a}}{\sqrt[n]{b}} = \sqrt[n]{\dfrac{a}{b}}$, where $b \neq 0$.

As with all equations, this rule can be used going from left to right or right to left.

Example 2 Simplify. Assume that variables represent positive values.

a. $\sqrt{\dfrac{7}{36}}$

SOLUTION $\sqrt{\dfrac{7}{36}} = \dfrac{\sqrt{7}}{\sqrt{36}} = \dfrac{\sqrt{7}}{6}$

b. $\dfrac{\sqrt{108}}{\sqrt{3}}$

SOLUTION $\dfrac{\sqrt{108}}{\sqrt{3}} = \sqrt{\dfrac{108}{3}} = \sqrt{36} = 6$

c. $\sqrt[3]{\dfrac{9}{x^3}}$

SOLUTION $\sqrt[3]{\dfrac{9}{x^3}} = \dfrac{\sqrt[3]{9}}{\sqrt[3]{x^3}} = \dfrac{\sqrt[3]{9}}{x}$

d. $\dfrac{\sqrt[3]{15}}{\sqrt[3]{5}}$

SOLUTION $\dfrac{\sqrt[3]{15}}{\sqrt[3]{5}} = \sqrt[3]{\dfrac{15}{5}} = \sqrt[3]{3}$

e. $\sqrt[4]{\dfrac{y}{81}}$

SOLUTION $\sqrt[4]{\dfrac{y}{81}} = \dfrac{\sqrt[4]{y}}{\sqrt[4]{81}} = \dfrac{\sqrt[4]{y}}{3}$

Answers   **Your Turn 1**
a. 8   b. $\sqrt{5a}$
c. $\sqrt[3]{10xy}$ d. $\sqrt{\dfrac{2b}{7a}}$
e. a

Objective ③   Use the product rule to simplify radical expressions.   Several conditions exist in which a radical is not considered to be in simplest form. One such condition is when a radicand has a factor that can be written to a power greater than or equal to the index. For example, $\sqrt[3]{81}$ is not in simplest form because the perfect cube 27 is a factor of 81. Our first step in simplifying $\sqrt[3]{81}$ is to rewrite it as $\sqrt[3]{27 \cdot 3}$ so that we can then use the product rule for radicals.

Note: We explore other conditions that require simplification in future sections.

Procedure   Simplifying nth Roots

To simplify an nth root,

1. Write the radicand as a product of the greatest possible perfect nth power and a number or an expression that has no perfect nth power factors.
2. Use the product rule $\sqrt[n]{ab} = \sqrt[n]{a} \cdot \sqrt[n]{b}$, where a is the perfect nth power.
3. Find the nth root of the perfect nth power radicand.

Connection   A list of perfect powers may be helpful.
Perfect squares:
  1, 4, 9, 16, 25, 36, 49, 64, 81, 100, ...
Perfect cubes:
  1, 8, 27, 64, 125, 216, ...
Perfect fourth powers:
  1, 16, 81, 256, 625, ...

Example 3   Simplify.

a. $\sqrt{18}$

▶ **Solution** $\sqrt{18} = \sqrt{9 \cdot 2}$ The greatest perfect square factor of 18 is 9, so we write 18 as $9 \cdot 2$.

$= \sqrt{9} \cdot \sqrt{2}$ Use the product rule of roots to separate the factors into two radicals.

$= 3\sqrt{2}$ Simplify the square root of 9.

b. $5\sqrt{72}$

Solution $5\sqrt{72} = 5 \cdot \sqrt{36 \cdot 2}$ The greatest perfect square factor of 72 is 36, so we write 72 as $36 \cdot 2$.

$= 5 \cdot \sqrt{36} \cdot \sqrt{2}$ Use the product rule of roots to separate the factors into two radicals.

$= 5 \cdot 6 \cdot \sqrt{2}$ Simplify the square root of 36.

$= 30\sqrt{2}$ Multiply $5 \cdot 6$.

Note: We can use perfect nth power factors other than the greatest perfect nth power factor. For example, in simplifying $\sqrt{72}$, instead of $\sqrt{72} = \sqrt{36 \cdot 2} = 6\sqrt{2}$, we could write

$$\sqrt{72} = \sqrt{4 \cdot 18} = 2\sqrt{18} = 2\sqrt{9 \cdot 2} = 2 \cdot 3\sqrt{2} = 6\sqrt{2}.$$

Notice that using the greatest perfect nth factor saves steps.

c. $\sqrt[3]{40}$

Solution $\sqrt[3]{40} = \sqrt[3]{8 \cdot 5}$ The greatest perfect cube factor of 40 is 8.

$= \sqrt[3]{8} \cdot \sqrt[3]{5}$ Use the product rule of roots.

$= 2\sqrt[3]{5}$ Simplify the cube root of 8.

d. $4\sqrt[4]{162}$

Solution $4\sqrt[4]{162} = 4\sqrt[4]{81 \cdot 2}$ The greatest perfect fourth power factor of 162 is 81.

$= 4\sqrt[4]{81} \cdot \sqrt[4]{2}$ Use the product rule of roots.

$= 4 \cdot 3 \cdot \sqrt[4]{2}$ Simplify the fourth root of 81.

$= 12\sqrt[4]{2}$

Your Turn 3   Simplify.

a. $\sqrt{150}$ b. $6\sqrt{80}$ c. $\sqrt[3]{108}$ d. $3\sqrt[4]{80}$

Using Prime Factorization

If the greatest perfect nth power of a particular radicand is not obvious, try using the prime factorization of the radicand. Each prime factor that appears twice will have a square root equal to one of the two factors, each prime factor that appears three times will have a cube root equal to one of the three factors, and so on. The remaining factors stay in the radical sign.

Example 4   Simplify the following radicals using prime factorizations.

a. $\sqrt{375}$

SOLUTION $\sqrt{375} = \sqrt{5 \cdot 5 \cdot 5 \cdot 3}$ Write 375 as the product of its prime factors.

$= 5\sqrt{3 \cdot 5}$ The square root of the pair of 5's is 5.

$= 5\sqrt{15}$ Multiply the prime factors in the radicand.

b. $\sqrt[3]{324}$

SOLUTION $\sqrt[3]{324} = \sqrt[3]{3 \cdot 3 \cdot 3 \cdot 3 \cdot 2 \cdot 2}$ Write 324 as the product of its prime factors.

$= 3\sqrt[3]{3 \cdot 2 \cdot 2}$ The cube root of the three 3's is 3.

$= 3\sqrt[3]{12}$ Multiply the prime factors in the radicand.

c. $\sqrt[4]{240}$

SOLUTION $\sqrt[4]{240} = \sqrt[4]{2 \cdot 2 \cdot 2 \cdot 2 \cdot 3 \cdot 5}$ Write 240 as the product of its prime factors.

$= 2\sqrt[4]{3 \cdot 5}$ The fourth root of the four 2's is 2.

$= 2\sqrt[4]{15}$ Multiply the prime factors in the radicand.

Answers   Your Turn 3
a. $5\sqrt{6}$ b. $24\sqrt{5}$
c. $3\sqrt[3]{4}$ d. $6\sqrt[4]{5}$

Answers   Your Turn 4
a. $7\sqrt{6}$ b. $3\sqrt[3]{12}$ c. $3\sqrt[4]{6}$

Your Turn 4   Simplify the following radicals using prime factorizations.

a. $\sqrt{294}$ b. $\sqrt[3]{324}$ c. $\sqrt[4]{486}$

Simplifying Radicals with Variables

We can use either procedure to simplify radicals whose radicands contain variables. Because $\sqrt[n]{a^m} = a^{m/n}$, we can find an exact root if n divides into m evenly. For example, $\sqrt{x^4} = x^{4/2} = x^2$ and $\sqrt[3]{x^{12}} = x^{12/3} = x^4$. Therefore, to find the nth root, we rewrite the radicand as a product in which one factor has the greatest possible exponent divisible by the index n.

Example 5   Simplify. Assume that variables represent nonnegative values.

a. $\sqrt{x^7}$

Connection   A list of perfect powers might be helpful.
Perfect squares:
$x^2, x^4, x^6, x^8, \ldots$
Perfect cubes:
$x^3, x^6, x^9, x^{12}, \ldots$

SOLUTION $\sqrt{x^7} = \sqrt{x^6 \cdot x}$ The greatest number smaller than 7 that is divisible by 2 is 6, so write x^7 as $x^6 \cdot x$.

$= \sqrt{x^6} \cdot \sqrt{x}$ Use the product rule of roots.

$= x^3\sqrt{x}$ Simplify $\sqrt{x^6} = x^3$.

b. $3\sqrt{24a^5b^9}$

SOLUTION $3\sqrt{24a^5b^9} = 3\sqrt{4\cdot 6\cdot a^4\cdot a\cdot b^8\cdot b}$ Write 24 as $4\cdot 6$, a^5 as $a^4\cdot a$, and b^9 as $b^8\cdot b$.

$$= 3\sqrt{4a^4b^8\cdot 6ab}$$ Regroup the factors so that perfect squares are together.

$$= 3\sqrt{4a^4b^8}\cdot\sqrt{6ab}$$ Use the product rule of roots.

$$= 3\cdot 2a^2b^4\cdot\sqrt{6ab}$$ Simplify $\sqrt{4} = 2$, $\sqrt{a^4} = a^2$, and $\sqrt{b^8} = b^4$.

$$= 6a^2b^4\sqrt{6ab}$$ Multiply $3\cdot 2 = 6$.

c. $y^2\sqrt[3]{y^8}$

SOLUTION $y^2\sqrt[3]{y^8} = y^2\sqrt[3]{y^6\cdot y^2}$ The greatest number smaller than 8 that is divisible by 3 is 6, so write y^8 as $y^6\cdot y^2$.

$$= y^2\sqrt[3]{y^6}\cdot\sqrt[3]{y^2}$$ Use the product rule of roots.

$$= y^2\cdot y^2\cdot\sqrt[3]{y^2}$$ Simplify $\sqrt[3]{y^6} = y^2$.

$$= y^4\sqrt[3]{y^2}$$ Multiply $y^2\cdot y^2 = y^4$.

d. $\sqrt[5]{64x^9y^{12}}$

SOLUTION $\sqrt[5]{64x^9y^{12}} = \sqrt[5]{32\cdot 2\cdot x^5\cdot x^4\cdot y^{10}\cdot y^2}$ Write 64 as $32\cdot 2$, x^9 as $x^5\cdot x^4$, and y^{12} as $y^{10}\cdot y^2$.

$$= \sqrt[5]{32x^5y^{10}\cdot 2x^4y^2}$$ Regroup the factors.

$$= \sqrt[5]{32x^5y^{10}}\cdot\sqrt[5]{2x^4y^2}$$ Use the product rule of roots.

$$= 2xy^2\sqrt[5]{2x^4y^2}$$ Simplify $\sqrt[5]{32} = 2$, $\sqrt[5]{x^5} = x$, and $\sqrt[5]{y^{10}} = y^2$.

Your Turn 5   Simplify. Assume that variables represent nonnegative values.

a. $\sqrt{n^{11}}$ **b.** $2\sqrt{45r^7s^3}$ **c.** $\sqrt[4]{m^{13}}$ **d.** $\sqrt[3]{a^8b^{10}}$

After using the product or quotient rules, it is often necessary to simplify the results.

Example 6   Find the product or quotient and simplify the results. Assume that variables represent positive values.

a. $\sqrt{3}\cdot\sqrt{6}$

SOLUTION $\sqrt{3}\cdot\sqrt{6} = \sqrt{18}$ Use the product rule of roots to multiply.

$$= \sqrt{9\cdot 2}$$ Write 18 as $9\cdot 2$.

$$= 3\sqrt{2}$$ Simplify $\sqrt{9} = 3$.

b. $5\sqrt{3x^3}\cdot 3\sqrt{15x^2}$

SOLUTION $5\sqrt{3x^3}\cdot 3\sqrt{15x^2} = 5\cdot 3\sqrt{3x^3}\cdot\sqrt{15x^2}$ Regroup the factors.

$$= 15\sqrt{45x^5}$$ Multiply.

$$= 15\sqrt{9\cdot 5\cdot x^4\cdot x}$$ Write 45 as $9\cdot 5$ and x^5 as $x^4\cdot x$.

$$= 15\cdot 3x^2\sqrt{5x}$$ Simplify $\sqrt{9} = 3$ and $\sqrt{x^4} = x^2$.

$$= 45x^2\sqrt{5x}$$ Multiply.

Answers   **Your Turn 5**

a. $n^5\sqrt{n}$ **b.** $6r^3s\sqrt{5rs}$

c. $m^3\sqrt[4]{m}$ **d.** $a^2b^3\sqrt[3]{a^2b}$

c. $\dfrac{\sqrt{288}}{\sqrt{6}}$

SOLUTION $\quad \dfrac{\sqrt{288}}{\sqrt{6}} = \sqrt{\dfrac{288}{6}}$ Use the quotient rule of roots.

$\qquad\qquad\qquad = \sqrt{48}$ Divide the radicand.

$\qquad\qquad\qquad = \sqrt{16 \cdot 3}$ Write 48 as $16 \cdot 3$.

$\qquad\qquad\qquad = 4\sqrt{3}$ Simplify $\sqrt{16} = 4$.

d. $\dfrac{8\sqrt{756a^8b^5}}{2\sqrt{7a^4b^2}}$

SOLUTION $\quad \dfrac{8\sqrt{756a^8b^5}}{2\sqrt{7a^4b^2}} = 4\sqrt{\dfrac{756a^8b^5}{7a^4b^2}}$ Divide coefficients and use the quotient rule of radicals.

$\qquad\qquad\qquad = 4\sqrt{108a^4b^3}$ Divide the radicand.

$\qquad\qquad\qquad = 4\sqrt{36a^4b^2 \cdot 3b}$ Rewrite the radicand with a perfect square factor.

$\qquad\qquad\qquad = 4 \cdot 6a^2b\sqrt{3b}$ Find the square roots.

$\qquad\qquad\qquad = 24a^2b\sqrt{3b}$ Multiply.

Your Turn 6   Find the product or quotient and simplify the results. Assume that variables represent positive values.

Answers   **Your Turn 6**

a. $3\sqrt{10}$ **b.** $16x^3\sqrt{21x}$

c. $6\sqrt{3}$ **d.** $21x^2y\sqrt{7xy}$

a. $\sqrt{6} \cdot \sqrt{15}$ **b.** $4\sqrt{14x^3} \cdot 2\sqrt{6x^4}$ **c.** $\dfrac{\sqrt{1296}}{\sqrt{12}}$ **d.** $\dfrac{14\sqrt{315x^{11}y^8}}{2\sqrt{5x^6y^5}}$

10.3 Exercises   For Extra Help   *MyMathLab* 

1. For $\sqrt{8} \cdot \sqrt{18}$, explain the difference between using the product rule for radicals and multiplying the approximate roots of 8 and 18.

2. Explain why $\sqrt{28}$ is not in simplest form.

3. Describe in your own words how to simplify a cube root containing a radicand with a perfect cube factor.

4. Explain why the expression $3x^2\sqrt[3]{x^5}$ is not in simplest form.

For Exercises 5–32, find the product and simplify. Assume that variables represent positive values. See Example 1.

5. $\sqrt{2} \cdot \sqrt{32}$ 6. $\sqrt{3} \cdot \sqrt{48}$ 7. $\sqrt{3x} \cdot \sqrt{27x^5}$ 8. $\sqrt{8y^3} \cdot \sqrt{2y}$

9. $\sqrt{6xy^3} \cdot \sqrt{24xy}$ 10. $\sqrt{50u^3v^2} \cdot \sqrt{2uv^4}$ 11. $\sqrt{5} \cdot \sqrt{13}$ 12. $\sqrt{6} \cdot \sqrt{11}$

13. $\sqrt{15} \cdot \sqrt{x}$ 14. $\sqrt{17} \cdot \sqrt{y}$ 15. $\sqrt[3]{3} \cdot \sqrt[3]{9}$ 16. $\sqrt[3]{4} \cdot \sqrt[3]{16}$

17. $\sqrt[3]{5y} \cdot \sqrt[3]{2y}$

18. $\sqrt[3]{6m} \cdot \sqrt[3]{2m}$

19. $\sqrt[4]{3} \cdot \sqrt[4]{7}$

20. $\sqrt[4]{7} \cdot \sqrt[4]{5}$

21. $\sqrt[4]{12w^3} \cdot \sqrt[4]{6w}$

22. $\sqrt[4]{21r^3} \cdot \sqrt[4]{7r}$

23. $\sqrt[4]{3x^2y} \cdot \sqrt[4]{5xy^2}$

24. $\sqrt[4]{2ab^2} \cdot \sqrt[4]{6ab}$

25. $\sqrt[5]{6x^3} \cdot \sqrt[5]{5x^4}$

26. $\sqrt[5]{3m^2} \cdot \sqrt[5]{8m^7}$

27. $\sqrt[6]{4x^2y^3} \cdot \sqrt[6]{2x^3y}$

28. $\sqrt[6]{ab^3} \cdot \sqrt[6]{7a^3b^2}$

29. $\sqrt{\dfrac{7}{2}} \cdot \sqrt{\dfrac{3}{5}}$

30. $\sqrt{\dfrac{5}{2}} \cdot \sqrt{\dfrac{11}{3}}$

31. $\sqrt{\dfrac{6}{x}} \cdot \sqrt{\dfrac{y}{5}}$

32. $\sqrt{\dfrac{a}{3}} \cdot \sqrt{\dfrac{7}{b}}$

For Exercises 33–48, simplify. Assume that variables represent positive values. See Example 2.

33. $\sqrt{\dfrac{25}{36}}$

34. $\sqrt{\dfrac{49}{81}}$

35. $\sqrt{\dfrac{10}{9}}$

36. $\sqrt{\dfrac{15}{81}}$

37. $\dfrac{\sqrt{180}}{\sqrt{5}}$

38. $\dfrac{\sqrt{243}}{\sqrt{3}}$

39. $\dfrac{\sqrt{15}}{\sqrt{5}}$

40. $\dfrac{\sqrt{21}}{\sqrt{3}}$

41. $\sqrt[3]{\dfrac{4}{w^6}}$

42. $\sqrt[3]{\dfrac{7}{v^9}}$

43. $\sqrt[3]{\dfrac{5y^2}{27x^9}}$

44. $\sqrt[3]{\dfrac{5a}{8r^6}}$

45. $\dfrac{\sqrt[3]{320}}{\sqrt[3]{5}}$

46. $\dfrac{\sqrt[3]{162}}{\sqrt[3]{6}}$

47. $\sqrt[4]{\dfrac{3u^3}{16x^8}}$

48. $\sqrt[4]{\dfrac{3x^2}{81y^4}}$

For Exercises 49–84, simplify. Assume that variables represent nonnegative values. See Examples 3–5.

49. $\sqrt{98}$

50. $\sqrt{96}$

51. $\sqrt{128}$

52. $\sqrt{180}$

53. $6\sqrt{80}$

54. $4\sqrt{75}$

55. $5\sqrt{112}$

56. $3\sqrt{147}$

57. $\sqrt{a^7}$

58. $\sqrt{d^5}$

59. $\sqrt{x^2y^4}$

60. $\sqrt{a^6b^2}$

61. $\sqrt{x^6y^8z^{10}}$

62. $\sqrt{p^4q^8r^8}$

63. $rs^2\sqrt{r^9s^5}$

64. $a^2b^3\sqrt{a^{11}b^3}$

65. $3\sqrt{72x^5}$

66. $6\sqrt{75d^3}$

67. $\sqrt[3]{32}$

68. $\sqrt[3]{128}$

69. $\sqrt[3]{x^7}$

70. $\sqrt[3]{b^{11}}$

71. $\sqrt[3]{x^6y^5}$

72. $\sqrt[3]{m^{13}n^9}$

73. $\sqrt[3]{128z^8}$

74. $\sqrt[3]{48h^{14}}$

75. $2\sqrt[3]{40}$

76. $4\sqrt[3]{250}$

77. $\sqrt[4]{80}$

78. $\sqrt[4]{162}$

79. $3x^2\sqrt[4]{243x^9}$

80. $3a^4\sqrt[4]{48a^7}$

81. $\sqrt[5]{486x^{16}}$

82. $\sqrt[5]{160n^{18}}$

83. $\sqrt[6]{x^8y^{14}z^{11}}$

84. $\sqrt[7]{a^{16}b^9c^{12}}$

For Exercises 85–94, find the product and write the answer in simplest form. Assume that variables represent nonnegative values. See Examples 6(a) and 6(b).

85. $\sqrt{3} \cdot \sqrt{21}$

86. $\sqrt{5} \cdot \sqrt{15}$

87. $5\sqrt{10} \cdot 3\sqrt{6}$

88. $2\sqrt{6} \cdot 5\sqrt{21}$

89. $\sqrt{y^3} \cdot \sqrt{y^2}$

90. $\sqrt{m^7} \cdot \sqrt{m^4}$

91. $x\sqrt{x^2y^3} \cdot y^2\sqrt{x^4y^4}$

92. $x\sqrt{x^5y^2} \cdot y\sqrt{xy^3}$

93. $4\sqrt{6c^3} \cdot 3\sqrt{10c^5}$

94. $6\sqrt{15c^2} \cdot 2\sqrt{10c^4}$

For Exercises 95 and 96, write an expression in simplest form for the area of the figure.

95.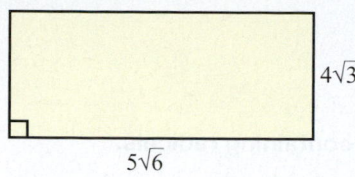

96.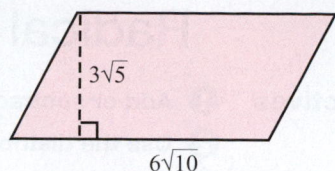

For Exercises 97–108, find the quotient and write the answer in simplest form. Assume that variables represent positive values. See Examples 6(c) and 6(d).

97. $\dfrac{\sqrt{48}}{\sqrt{6}}$

98. $\dfrac{\sqrt{54}}{\sqrt{3}}$

99. $\dfrac{9\sqrt{160}}{3\sqrt{8}}$

100. $\dfrac{10\sqrt{280}}{2\sqrt{10}}$

101. $\dfrac{\sqrt{c^5 d^6}}{\sqrt{cd^3}}$

102. $\dfrac{\sqrt{m^6 n^5}}{\sqrt{m^3 n^3}}$

103. $\dfrac{8\sqrt{45a^5}}{2\sqrt{5a}}$

104. $\dfrac{6\sqrt{48n^7}}{3\sqrt{3n^3}}$

105. $\dfrac{12\sqrt{72c^5}}{4\sqrt{6c^2}}$

106. $\dfrac{15\sqrt{48a^7}}{5\sqrt{2a^2}}$

107. $\dfrac{36\sqrt{96x^6 y^{11}}}{4\sqrt{3x^2 y^4}}$

108. $\dfrac{54\sqrt{240r^{11}s^{10}}}{9\sqrt{5r^6 s^4}}$

For Exercises 109–114, find the product and write the answer in simplest form. Assume that variables represent nonnegative values.

109. $\sqrt{\dfrac{3}{7}} \cdot \sqrt{\dfrac{8}{7}}$

110. $\sqrt{\dfrac{8}{5}} \cdot \sqrt{\dfrac{6}{5}}$

111. $\sqrt{\dfrac{a^3}{2}} \cdot \sqrt{\dfrac{a^5}{2}}$

112. $\sqrt{\dfrac{c^7}{6}} \cdot \sqrt{\dfrac{c^5}{6}}$

113. $\sqrt{\dfrac{3x^5}{2}} \cdot \sqrt{\dfrac{15x^5}{8}}$

114. $\sqrt{\dfrac{5y^3}{3}} \cdot \sqrt{\dfrac{10y^3}{27}}$

For Exercises 115 and 116, write an expression in simplest form for the area of the figure.

115.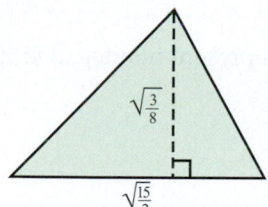

116.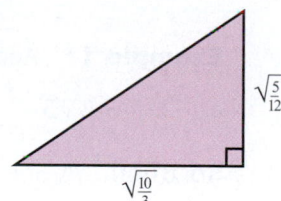

Review Exercises

Exercises 1–6  EXPRESSIONS

[1.7] For Exercises 1 and 2, use $a = 6$ and $b = 8$.

1. Does $\sqrt{a^2 + b^2} = \sqrt{a^2} + \sqrt{b^2} = a + b$?

2. Does $\sqrt{(a + b)^2} = a + b$?

[1.7] 3. Simplify: $6x^2 - 4x - 3x + 2x^2$

[5.5] For Exercises 4–6, multiply.

4. $(2a - 3b)(4a + 3b)$

5. $(3m + 5n)(3m - 5n)$

6. $(2x - 3y)^2$

10.4 Adding, Subtracting, and Multiplying Radical Expressions

Objectives

1. Add or subtract like radicals.
2. Use the distributive property in expressions containing radicals.
3. Simplify radical expressions that contain mixed operations.

Objective 1 **Add or subtract like radicals.** Recall that like terms such as $-8a^2$ and $3a^2$ have identical variables with identical exponents. Similarly, **like radicals** have identical radicands and root indices.

Definition *Like radicals:* Radical expressions with identical radicands and identical root indices.

The radicals $3\sqrt{2}$ and $5\sqrt{2}$ are like. The radicals $3\sqrt{2}$ and $5\sqrt{3}$ are unlike because their radicands are different. The radicals $3\sqrt{2}$ and $5\sqrt[3]{2}$ are unlike because their root indices are different.

Adding or subtracting like radicals is essentially the same as combining like terms. Remember that the distributive property is at work when we combine like terms because we factor out the common variable, which leaves a sum or difference of the coefficients.

Like terms:	Like radicals:
$3x + 4x = (3 + 4)x$	$3\sqrt{5} + 4\sqrt{5} = (3 + 4)\sqrt{5}$
$= 7x$	$= 7\sqrt{5}$

Our example suggests the following procedure for adding like radical expressions.

Procedure **Adding Like Radicals**

To add or subtract like radicals, add or subtract the coefficients and leave the radical parts the same.

Example 1 Add or subtract. Assume that all variables represent nonnegative values.

a. $7\sqrt{5} + 2\sqrt{5}$

SOLUTION $7\sqrt{5} + 2\sqrt{5} = (7 + 2)\sqrt{5}$
$= 9\sqrt{5}$

b. $6\sqrt[3]{4} + \sqrt[3]{4}$

SOLUTION $6\sqrt[3]{4} + \sqrt[3]{4} = (6 + 1)\sqrt[3]{4}$
$= 7\sqrt[3]{4}$

c. $6x\sqrt[4]{3x} - 2x\sqrt[4]{3x}$

SOLUTION $6x\sqrt[4]{3x} - 2x\sqrt[4]{3x} = (6x - 2x)\sqrt[4]{3x}$
$= 4x\sqrt[4]{3x}$

d. $8\sqrt{3} + 6\sqrt{2} - 5\sqrt{3} + 3\sqrt{2}$

SOLUTION $8\sqrt{3} + 6\sqrt{2} - 5\sqrt{3} + 3\sqrt{2} = 8\sqrt{3} - 5\sqrt{3} + 6\sqrt{2} + 3\sqrt{2}$ **Regroup the terms.**
$= (8 - 5)\sqrt{3} + (6 + 3)\sqrt{2}$
$= 3\sqrt{3} + 9\sqrt{2}$

Warning: Never add radicands! Consider $\sqrt{9} + \sqrt{16}$. It might be tempting to add the radicands to get $\sqrt{25}$, but $\sqrt{9} + \sqrt{16} \neq \sqrt{25}$. If we calculate each side separately, we see why they are not equivalent.

$$\sqrt{9} + \sqrt{16} \neq \sqrt{25}$$
$$3 + 4 \neq 5$$
$$7 \neq 5$$

Answers **Your Turn 1**

a. $7\sqrt{6}$ **b.** $6\sqrt[4]{5x^2}$ **c.** $-5y\sqrt[3]{7}$

d. $3x\sqrt{5x} + 5y\sqrt{3y}$

Your Turn 1   Add or subtract. Assume that all variables represent nonnegative values.

a. $5\sqrt{6} + 2\sqrt{6}$

b. $5\sqrt[4]{5x^2} + \sqrt[4]{5x^2}$

c. $7y\sqrt[3]{7} - 12y\sqrt[3]{7}$

d. $9x\sqrt{5x} - 2y\sqrt{3y} - 6x\sqrt{5x} + 7y\sqrt{3y}$

In a problem involving addition or subtraction of radicals, if the radicals are not like, it may be possible to simplify one or more of the radicals so that they are like.

Example 2   Add or subtract. Assume that variables represent nonnegative values.

a. $7\sqrt{3} + \sqrt{12}$

SOLUTION $\begin{aligned} 7\sqrt{3} + \sqrt{12} &= 7\sqrt{3} + \sqrt{4 \cdot 3} &&\text{Factor out the perfect square factor, 4.}\\ &= 7\sqrt{3} + \sqrt{4} \cdot \sqrt{3} &&\text{Use the product rule to separate the radicals.}\\ &= 7\sqrt{3} + 2\sqrt{3} &&\text{Simplify.}\\ &= 9\sqrt{3} &&\text{Combine like radicals.} \end{aligned}$

b. $3\sqrt[3]{24} - 2\sqrt[3]{3}$

SOLUTION $\begin{aligned} 3\sqrt[3]{24} - 2\sqrt[3]{3} &= 3\sqrt[3]{8 \cdot 3} - 2\sqrt[3]{3} &&\text{Rewrite 24 as } 8 \cdot 3.\\ &= 3\sqrt[3]{8} \cdot \sqrt[3]{3} - 2\sqrt[3]{3} &&\text{Use the product rule.}\\ &= 3 \cdot 2\sqrt[3]{3} - 2\sqrt[3]{3} &&\text{Simplify } \sqrt[3]{8}.\\ &= 6\sqrt[3]{3} - 2\sqrt[3]{3} &&\text{Multiply.}\\ &= 4\sqrt[3]{3} &&\text{Combine like radicals.} \end{aligned}$

c. $\sqrt{48x^3} + \sqrt{12x^3}$

SOLUTION $\begin{aligned} \sqrt{48x^3} + \sqrt{12x^3} &= \sqrt{16x^2 \cdot 3x} + \sqrt{4x^2 \cdot 3x} &&\text{Rewrite } 48x^3 \text{ as } 16x^2 \cdot 3x \text{ and } 12x^3 \text{ as } 4x^2 \cdot 3x.\\ &= \sqrt{16x^2} \cdot \sqrt{3x} + \sqrt{4x^2} \cdot \sqrt{3x} &&\text{Use the product rule.}\\ &= 4x\sqrt{3x} + 2x\sqrt{3x} &&\text{Find } \sqrt{16x^2} \text{ and } \sqrt{4x^2}.\\ &= 6x\sqrt{3x} &&\text{Combine like radicals.} \end{aligned}$

d. $3\sqrt[4]{32x^5} + \sqrt[4]{162x^9}$

SOLUTION $\begin{aligned} &3\sqrt[4]{32x^5} + \sqrt[4]{162x^9}\\ &= 3\sqrt[4]{16 \cdot 2 \cdot x^4 \cdot x} + \sqrt[4]{81 \cdot 2 \cdot x^8 \cdot x} &&\text{Write 32 as } 16 \cdot 2, x^5 \text{ as } x^4 \cdot x, 162 \text{ as } 81 \cdot 2, \text{ and } x^9 \text{ as } x^8 \cdot x.\\ &= 3 \cdot 2x\sqrt[4]{2x} + 3x^2\sqrt[4]{2x} &&\text{Find the roots.}\\ &= 6x\sqrt[4]{2x} + 3x^2\sqrt[4]{2x} &&\text{Multiply.}\\ &= (6x + 3x^2)\sqrt[4]{2x} &&\text{Factor.} \end{aligned}$

Note: Although the radicals are like, we cannot combine the coefficients further because they are not like terms.

Your Turn 2   Add or subtract. Assume that variables represent nonnegative values.

Answers **Your Turn 2**

a. $-10\sqrt{6}$ **b.** $14x^2\sqrt{2x}$

c. $10a^2\sqrt[3]{2a}$ **d.** $(6x^2 - 10x)\sqrt[3]{3x}$

a. $4\sqrt{24} - 6\sqrt{54}$

b. $4\sqrt{50x^5} - 2\sqrt{18x^5}$

c. $6a\sqrt[3]{54a^4} - 2\sqrt[3]{128a^7}$

d. $2\sqrt[3]{81x^7} - 5\sqrt[3]{24x^4}$

Objective ❷ **Use the distributive property in expressions containing radicals.**   Products involving sums and differences of radicals are found in much the same way as products of polynomials. We use the distributive property, multiply binomials using FOIL, and square a binomial (all from Section 5.5).

Example 3 Find the product. Assume that variables represent nonnegative values.

a. $\sqrt{3}(\sqrt{3} + \sqrt{15})$

SOLUTION $\sqrt{3}(\sqrt{3} + \sqrt{15}) = \sqrt{3} \cdot \sqrt{3} + \sqrt{3} \cdot \sqrt{15}$ Use the distributive property.

$= \sqrt{3 \cdot 3} + \sqrt{3 \cdot 15}$ Use the product rule.

$= \sqrt{9} + \sqrt{45}$ Multiply.

$= 3 + 3\sqrt{5}$ Find $\sqrt{9}$ and simplify $\sqrt{45}$.

b. $2\sqrt{6}(3 + 5\sqrt{5})$

SOLUTION $2\sqrt{6}(3 + 5\sqrt{5}) = 2\sqrt{6} \cdot 3 + 2\sqrt{6} \cdot 5\sqrt{5}$ Use the distributive property.

$= 2 \cdot 3\sqrt{6} + 2 \cdot 5\sqrt{6 \cdot 5}$ Use the product rule.

$= 6\sqrt{6} + 10\sqrt{30}$ Multiply.

c. $(2 + \sqrt{3})(\sqrt{5} - \sqrt{6})$

SOLUTION $(2 + \sqrt{3})(\sqrt{5} - \sqrt{6}) = 2\sqrt{5} - 2\sqrt{6} + \sqrt{3} \cdot \sqrt{5} - \sqrt{3} \cdot \sqrt{6}$ Use FOIL.

$= 2\sqrt{5} - 2\sqrt{6} + \sqrt{15} - \sqrt{18}$ Use the product rule

$= 2\sqrt{5} - 2\sqrt{6} + \sqrt{15} - 3\sqrt{2}$ Simplify $\sqrt{18}$.

d. $(3\sqrt{x} + \sqrt{y})(2\sqrt{x} - 5\sqrt{y})$

SOLUTION $(3\sqrt{x} + \sqrt{y})(2\sqrt{x} - 5\sqrt{y})$

$= 3\sqrt{x} \cdot 2\sqrt{x} - 3\sqrt{x} \cdot 5\sqrt{y} + \sqrt{y} \cdot 2\sqrt{x} - \sqrt{y} \cdot 5\sqrt{y}$ Use FOIL.

$= 6x - 15\sqrt{xy} + 2\sqrt{xy} - 5y$ Use the product rule.

$= 6x - 13\sqrt{xy} - 5y$ Combine like radicals.

e. $(5 + \sqrt{3})^2$

SOLUTION $(5 + \sqrt{3})^2 = 5^2 + 2 \cdot 5\sqrt{3} + (\sqrt{3})^2$ Use $(a + b)^2 = a^2 + 2ab + b^2$.

$= 25 + 10\sqrt{3} + 3$ Simplify.

$= 28 + 10\sqrt{3}$ Add 25 and 3.

Your Turn 3 Find the product. Assume that variables represent nonnegative values.

a. $2\sqrt{11}(3 - 3\sqrt{6})$ **b.** $(2\sqrt{a} + 3\sqrt{b})(\sqrt{a} - 3\sqrt{b})$ **c.** $(3 + \sqrt{5})^2$

Radicals in Conjugates

Radical expressions can be conjugates. Like binomial conjugates, conjugates involving radicals differ only in the sign separating the terms. For example, $7 - \sqrt{3}$ and $7 + \sqrt{3}$ are conjugates. Let's explore what happens when we multiply conjugates containing radicals.

Example 4 Find the product.

a. $(2 + \sqrt{3})(2 - \sqrt{3})$

SOLUTION $(2 + \sqrt{3})(2 - \sqrt{3}) = 2^2 - (\sqrt{3})^2$ Use $(a + b)(a - b) = a^2 - b^2$ (from Section 5.3).

$= 4 - 3$ Simplify.

$= 1$

b. $(\sqrt{5} - 3\sqrt{3})(\sqrt{5} + 3\sqrt{3})$

SOLUTION $(\sqrt{5} - 3\sqrt{3})(\sqrt{5} + 3\sqrt{3}) = (\sqrt{5})^2 - (3\sqrt{3})^2$ Use $(a - b)(a + b) = a^2 - b^2$.

$= 5 - 9 \cdot 3$ Simplify.

$= 5 - 27$ Multiply.

$= -22$ Subtract.

Your Turn 4   Find the product.

a. $(4 + \sqrt{10})(4 - \sqrt{10})$ **b.** $(3\sqrt{5} + 2\sqrt{6})(3\sqrt{5} - 2\sqrt{6})$

Objective ③ Simplify radical expressions that contain mixed operations.   Now let's use the order of operations to simplify radical expressions that have more than one operation.

Example 5   Simplify.

a. $\sqrt{2} \cdot \sqrt{10} + \sqrt{3} \cdot \sqrt{15}$

SOLUTION $\sqrt{2} \cdot \sqrt{10} + \sqrt{3} \cdot \sqrt{15} = \sqrt{2 \cdot 10} + \sqrt{3 \cdot 15}$ Use the product rule.

$= \sqrt{20} + \sqrt{45}$ Multiply.

$= \sqrt{4 \cdot 5} + \sqrt{9 \cdot 5}$ Rewrite 20 as $4 \cdot 5$ and 45 as $9 \cdot 5$.

$= \sqrt{4} \cdot \sqrt{5} + \sqrt{9} \cdot \sqrt{5}$ Use the product rule.

$= 2\sqrt{5} + 3\sqrt{5}$ Find $\sqrt{4}$ and $\sqrt{9}$.

$= 5\sqrt{5}$ Combine like radicals.

b. $\dfrac{\sqrt{54}}{\sqrt{3}} + \sqrt{32}$

SOLUTION $\dfrac{\sqrt{54}}{\sqrt{3}} + \sqrt{32} = \sqrt{\dfrac{54}{3}} + \sqrt{16 \cdot 2}$ Use the quotient rule and rewrite 32 as $16 \cdot 2$.

$= \sqrt{18} + \sqrt{16} \cdot \sqrt{2}$ Divide and use the product rule.

$= \sqrt{9 \cdot 2} + 4\sqrt{2}$ Rewrite 18 as $9 \cdot 2$ and find $\sqrt{16}$.

$= \sqrt{9} \cdot \sqrt{2} + 4\sqrt{2}$ Use the product rule.

$= 3\sqrt{2} + 4\sqrt{2}$ Find $\sqrt{9}$.

$= 7\sqrt{2}$ Combine like radicals.

🖩 Calculator Tips

We can use a graphing calculator to simplify radical expressions. For Example 5(b), we type

[2nd] [x²] [5] [4] [)] [÷] [2nd] [x²] [3] [)] [+] [2nd] [x²] [3] [2] [)] [ENTER]

The result, rounded to the nearest thousandth, is 9.899.

To see if our answer is correct, type

[7] [2nd] [x²] [2] [)] [ENTER],

which also gives 9.899 rounded to the nearest thousandth. Therefore, our answer is correct.

Answers Your Turn 4

a. 6   **b.** 21

Answers Your Turn 5

a. $34\sqrt{2}$ **b.** $5\sqrt{7}$

Your Turn 5   Simplify.

a. $2\sqrt{3} \cdot \sqrt{6} + 4\sqrt{7} \cdot \sqrt{14}$ **b.** $\sqrt{63} + \dfrac{\sqrt{140}}{\sqrt{5}}$

Note: Exercises marked with a ★ represent challenging exercises.

1. What must be identical in like radicals? What can be different?

2. Add $3x + 2x$ and then add $3\sqrt{2} + 2\sqrt{2}$. Discuss the similarities in the process.

3. Multiply $(x + 3)(x + 2)$ and then multiply $(\sqrt{5} + 3)(\sqrt{5} + 2)$. Discuss the similarities in the process.

4. Why is $\sqrt{a} + \sqrt{b} \neq \sqrt{a + b}$? Use examples if necessary.

For Exercises 5–18, add or subtract. Assume that variables represent nonnegative values. See Example 1.

5. $9\sqrt{6} - 15\sqrt{6}$

6. $4\sqrt{5} - 13\sqrt{5}$

7. $7\sqrt{a} + 2\sqrt{a}$

8. $5\sqrt{y} + 7\sqrt{y}$

9. $4\sqrt{5} - 2\sqrt{6} + 8\sqrt{5} - 6\sqrt{6}$

10. $9\sqrt{7} - 4\sqrt{11} - 3\sqrt{7} - 5\sqrt{11}$

11. $3a\sqrt{5a} - 4b\sqrt{7b} + 8a\sqrt{5a} + 2b\sqrt{7b}$

12. $12n\sqrt{2n} - 14m\sqrt{5m} - 8n\sqrt{2n} + 18m\sqrt{5m}$

13. $6x\sqrt[3]{9} - 3x\sqrt[3]{9}$

14. $4y\sqrt[3]{3} - y\sqrt[3]{3}$

15. $6x^2\sqrt[4]{5x} - 12x^2\sqrt[4]{5x}$

16. $3y^3\sqrt[4]{8y} - 9y^3\sqrt[4]{8y}$

★ 17. $3x\sqrt{5x} + 4x\sqrt[3]{5x}$

★ 18. $4z\sqrt[4]{2z} - 7z\sqrt{2z}$

For Exercises 19–38, add or subtract. Assume that variables represent nonnegative values. See Example 2.

19. $\sqrt{48} - \sqrt{75}$

20. $\sqrt{24} - \sqrt{96}$

21. $\sqrt{80y} - \sqrt{125y}$

22. $\sqrt{27x} + \sqrt{75x}$

23. $\sqrt{80} - 4\sqrt{45}$

24. $\sqrt{20} - 2\sqrt{180}$

25. $3\sqrt{96} - 2\sqrt{54}$

26. $5\sqrt{63} + 2\sqrt{28}$

27. $6\sqrt{48a^3} - 2\sqrt{75a^3}$

28. $4\sqrt{98y^5} - 7\sqrt{128y^5}$

29. $\sqrt{150} - \sqrt{54} + \sqrt{24}$

30. $\sqrt{20} + \sqrt{125} - \sqrt{80}$

31. $2\sqrt{8} - 3\sqrt{48} + 2\sqrt{98} - \sqrt{75}$

32. $3\sqrt{180} - \sqrt{192} - 4\sqrt{80} - \sqrt{48}$

33. $\sqrt[3]{128} + \sqrt[3]{54}$

34. $\sqrt[3]{24} + \sqrt[3]{81}$

35. $4\sqrt[3]{135x^5} - 6x\sqrt[3]{320x^2}$

36. $3a^2\sqrt[3]{500a^4} + 6a\sqrt[3]{108a^7}$

37. $-4\sqrt[4]{32x^9} + 2x\sqrt[4]{162x^5}$

38. $6y\sqrt[4]{243y^6} - 2y^2\sqrt[4]{48y^2}$

For Exercises 39–46 use the distributive property. Assume that variables represent nonnegative values. See Examples 3(a) and 3(b).

39. $\sqrt{2}(3 + \sqrt{2})$

40. $\sqrt{5}(4 - \sqrt{5})$

41. $\sqrt{3}(\sqrt{3} - \sqrt{15})$

42. $\sqrt{6}(\sqrt{6} + \sqrt{3})$

43. $\sqrt{5}(\sqrt{6} + 2\sqrt{10})$

44. $\sqrt{7}(\sqrt{5} - 3\sqrt{14})$

45. $4\sqrt{3x}(2\sqrt{3x} - 4\sqrt{6x})$

46. $6\sqrt{2y}(3\sqrt{2y} + 2\sqrt{10y})$

For Exercises 47–66, multiply. (Use FOIL.) Assume that variables represent nonnegative values. See Examples 3(c) and 3(d).

47. $(3 + \sqrt{5})(4 - \sqrt{2})$

48. $(2 + \sqrt{5})(6 - \sqrt{3})$

49. $(3 + \sqrt{x})(2 + \sqrt{x})$

50. $(5 - \sqrt{a})(2 + \sqrt{a})$

51. $(2 + 3\sqrt{3})(3 + 5\sqrt{2})$

52. $(7 - 3\sqrt{5})(2 - 2\sqrt{10})$

53. $(\sqrt{3} + \sqrt{5})(\sqrt{5} + \sqrt{7})$

54. $(\sqrt{5} + \sqrt{2})(\sqrt{2} + \sqrt{7})$

55. $(\sqrt{x} + 3\sqrt{y})(\sqrt{x} - 2\sqrt{y})$

56. $(2\sqrt{a} + \sqrt{b})(\sqrt{a} + 4\sqrt{b})$

57. $(4\sqrt{2} + 2\sqrt{5})(3\sqrt{7} - 3\sqrt{3})$

58. $(8\sqrt{2} - 2\sqrt{3})(2\sqrt{5} + 3\sqrt{10})$

59. $(2\sqrt{a} + 3\sqrt{b})(4\sqrt{a} - \sqrt{b})$

60. $(3\sqrt{m} - \sqrt{n})(2\sqrt{m} + 4\sqrt{n})$

61. $(\sqrt[3]{4} + 5)(\sqrt[3]{4} - 8)$

62. $(\sqrt[3]{9} + 5)(\sqrt[3]{9} - 2)$

63. $(\sqrt[3]{9} + \sqrt[3]{4})(\sqrt[3]{3} - \sqrt[3]{2})$

64. $(\sqrt[3]{5} + \sqrt[3]{9})(\sqrt[3]{25} - \sqrt[3]{3})$

★ 65. $(\sqrt[3]{x} + 2)(\sqrt[3]{x^2} - 2\sqrt[3]{x} + 4)$

★ 66. $(\sqrt[3]{r} - 3)(\sqrt[3]{r^2} + 3\sqrt[3]{r} + 9)$

For Exercises 67–74, find the product. See Example 3(e).

67. $(4 + \sqrt{6})^2$

68. $(3 + \sqrt{7})^2$

69. $(4 - \sqrt{2})^2$

70. $(5 - \sqrt{2})^2$

71. $(2 + 2\sqrt{3})^2$

72. $(3 + 2\sqrt{5})^2$

73. $(2\sqrt{3} + 3\sqrt{2})^2$

74. $(4\sqrt{2} - 5\sqrt{6})^2$

For Exercises 75–88, multiply the conjugates. Assume that variables represent nonnegative values. See Example 4.

75. $(4 + \sqrt{3})(4 - \sqrt{3})$

76. $(3 + \sqrt{5})(3 - \sqrt{5})$

77. $(\sqrt{2} + 4)(\sqrt{2} - 4)$

78. $(\sqrt{7} - 3)(\sqrt{7} + 3)$

79. $(6 + \sqrt{x})(6 - \sqrt{x})$

80. $(5 + \sqrt{y})(5 - \sqrt{y})$

81. $(\sqrt{3} + \sqrt{2})(\sqrt{3} - \sqrt{2})$

82. $(\sqrt{5} - \sqrt{3})(\sqrt{5} + \sqrt{3})$

83. $(\sqrt{x} + \sqrt{y})(\sqrt{x} - \sqrt{y})$

84. $(\sqrt{a} - \sqrt{b})(\sqrt{a} + \sqrt{b})$

85. $(4 + 2\sqrt{3})(4 - 2\sqrt{3})$

86. $(7 + 2\sqrt{3})(7 - 2\sqrt{3})$

87. $(3\sqrt{7} + \sqrt{13})(3\sqrt{7} - \sqrt{13})$

88. $(4\sqrt{5} - 3\sqrt{2})(4\sqrt{5} + 3\sqrt{2})$

For Exercises 89–96, simplify. See Example 5.

89. $\sqrt{3} \cdot \sqrt{15} + \sqrt{8} \cdot \sqrt{10}$

90. $\sqrt{18} \cdot \sqrt{6} + \sqrt{8} \cdot \sqrt{24}$

91. $3\sqrt{3} \cdot \sqrt{18} - 4\sqrt{18} \cdot \sqrt{12}$

92. $3\sqrt{2} \cdot 2\sqrt{40} - 5\sqrt{12} \cdot \sqrt{15}$

93. $\dfrac{\sqrt{54}}{\sqrt{3}} + \sqrt{72}$

94. $\dfrac{\sqrt{60}}{\sqrt{5}} + \sqrt{48}$

95. $\dfrac{\sqrt{540}}{\sqrt{3}} - 4\sqrt{125}$

96. $\dfrac{\sqrt{288}}{\sqrt{6}} - 6\sqrt{108}$

For Exercises 97 and 98, find the perimeter of the figure in simplest form.

97.

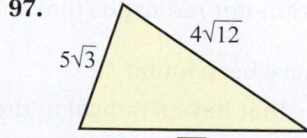

98.

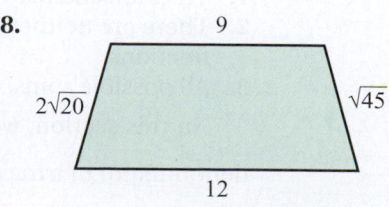

99. Crown molding, which is placed at the top of a wall, is to be installed around the perimeter of the room shown.

a. Write an expression in simplest form for the perimeter of the room.

b. Use a calculator to approximate the perimeter, rounded to the nearest tenth.

c. If crown molding costs $1.89 per foot length, how much will the crown molding cost for this room?

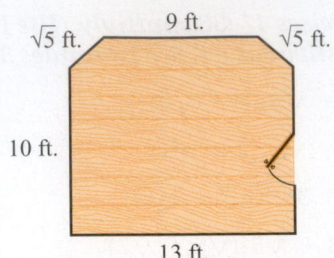

100. A tabletop is to be fitted with veneer strips along the sides.

a. Write an expression in simplest form for the perimeter of the tabletop.

b. Use a calculator to approximate the perimeter, rounded to the nearest tenth.

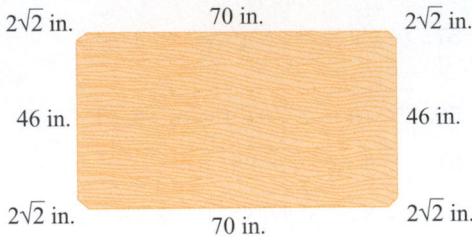

Review Exercises

Exercises 1–4   EXPRESSIONS

[5.5] **1.** What is the conjugate of $2x - 5$?

[5.5] **2.** Multiply: $(4x + 3)(4x - 3)$

[10.3] **3.** What factor could multiply $\sqrt{8}$ to equal $\sqrt{16}$?

[10.3] **4.** What factor could multiply $\sqrt[3]{2}$ to equal $\sqrt[3]{8}$?

Exercises 5 and 6  EQUATIONS AND INEQUALITIES

[4.4] **5.** For the equation $5y - 2x = 10$, find the slope and the y-intercept.

[4.2] **6.** Graph: $y = -\frac{1}{3}x + 4$

 **10.5** Rationalizing Numerators and Denominators of Radical Expressions

 **Objectives**

1 Rationalize denominators.

2 Rationalize denominators that have a sum or difference with a square root term.

3 Rationalize numerators.

We are now ready to formalize the conditions for a radical expression that is in simplest form. A radical expression is in simplest form if

1. The radicand has no factor raised to a power greater than or equal to the index.
2. There are neither radicals in the denominator of a fraction nor radicands that are fractions.
3. All possible sums, differences, products, and quotients have been found.

In this section, we explore how to simplify expressions that have a radical in the denominator of a fraction, as in $\frac{1}{\sqrt{2}}$.

Objective 1 **Rationalize denominators.** If the denominator of a fraction contains a radical, our goal is to *rationalize* the denominator, which means to rewrite the expression so that it has a rational number in the denominator. In general, we multiply the fraction by a well-chosen 1 so that the radical is eliminated. We determine that 1 by finding a factor that multiplies the *n*th root in the denominator so that its radicand is a perfect *n*th power.

Square Root Denominators

In the case of a square root in the denominator, we multiply it by a factor that makes the radicand a perfect square, which allows us to eliminate the square root.

For example, to rationalize $\dfrac{1}{\sqrt{2}}$, we could multiply by $\dfrac{\sqrt{2}}{\sqrt{2}}$ because the product's denominator is the square root of a perfect square.

$$\frac{1}{\sqrt{2}} = \frac{1}{\sqrt{2}} \cdot \frac{\sqrt{2}}{\sqrt{2}} = \frac{\sqrt{2}}{\sqrt{4}} = \frac{\sqrt{2}}{2}$$

Note: We are not changing the value of $\dfrac{1}{\sqrt{2}}$ because we are multiplying it by 1 in the form of $\dfrac{\sqrt{2}}{\sqrt{2}}$.

Any factor that produces a perfect square radicand will work. For example, we could have multiplied $\dfrac{1}{\sqrt{2}}$ by $\dfrac{\sqrt{8}}{\sqrt{8}}$.

$$\frac{1}{\sqrt{2}} = \frac{1}{\sqrt{2}} \cdot \frac{\sqrt{8}}{\sqrt{8}} = \frac{\sqrt{8}}{\sqrt{16}} = \frac{\sqrt{4 \cdot 2}}{4} = \frac{2\sqrt{2}}{4} = \frac{\sqrt{2}}{2}$$

Notice, however, that multiplying by $\dfrac{\sqrt{2}}{\sqrt{2}}$ required fewer steps to simplify than multiplying by $\dfrac{\sqrt{8}}{\sqrt{8}}$.

Example 1 Rationalize the denominator. Assume that variables represent positive values.

a. $\dfrac{2}{\sqrt{5}}$

SOLUTION $\dfrac{2}{\sqrt{5}} = \dfrac{2}{\sqrt{5}} \cdot \dfrac{\sqrt{5}}{\sqrt{5}}$ **Multiply by $\dfrac{\sqrt{5}}{\sqrt{5}}$.**

$\qquad = \dfrac{2\sqrt{5}}{\sqrt{25}}$ **Simplify.**

$\qquad = \dfrac{2\sqrt{5}}{5}$

b. $\sqrt{\dfrac{5}{8}}$

SOLUTION $\sqrt{\dfrac{5}{8}} = \dfrac{\sqrt{5}}{\sqrt{8}}$ **Use the quotient rule of square roots to separate the numerator and denominator into two radicals.**

Warning: Although it may be tempting to do so, we cannot divide out the 4 and 10 because 10 is a radicand, whereas 4 is not. We *never* divide out factors common to a radicand and a number that is not a radicand.

$\qquad = \dfrac{\sqrt{5}}{\sqrt{8}} \cdot \dfrac{\sqrt{2}}{\sqrt{2}}$ **Multiply by $\dfrac{\sqrt{2}}{\sqrt{2}}$.**

$\qquad = \dfrac{\sqrt{10}}{\sqrt{16}}$ **Multiply.**

$\qquad = \dfrac{\sqrt{10}}{4}$ **Simplify.**

Note: We chose to multiply by $\dfrac{\sqrt{2}}{\sqrt{2}}$ because it leads to a smaller perfect square than other choices such as $\dfrac{\sqrt{8}}{\sqrt{8}}$. Multiplying by $\dfrac{\sqrt{8}}{\sqrt{8}}$ produces the same final answer but requires more steps.

c. $\dfrac{3}{\sqrt{2x}}$

SOLUTION $\dfrac{3}{\sqrt{2x}} = \dfrac{3}{\sqrt{2x}} \cdot \dfrac{\sqrt{2x}}{\sqrt{2x}}$ **Multiply by $\dfrac{\sqrt{2x}}{\sqrt{2x}}$.**

$= \dfrac{3\sqrt{2x}}{\sqrt{4x^2}}$ **Multiply.**

$= \dfrac{3\sqrt{2x}}{2x}$ **Simplify.**

Your Turn 1   Rationalize the denominator. Assume that variables represent positve values.

a. $\dfrac{1}{\sqrt{7}}$ **b.** $\sqrt{\dfrac{7}{12}}$ **c.** $\dfrac{3}{\sqrt{10x}}$

nth-Root Denominators

If the denominator contains a higher-order root such as a cube root, we multiply appropriately to get a perfect cube radicand in the denominator so that we can eliminate the radical. For example, $\dfrac{2}{\sqrt[3]{5}} = \dfrac{2}{\sqrt[3]{5}} \cdot \dfrac{\sqrt[3]{25}}{\sqrt[3]{25}} = \dfrac{2\sqrt[3]{25}}{\sqrt[3]{125}} = \dfrac{2\sqrt[3]{25}}{5}$. We summarize as follows.

Procedure   **Rationalizing Denominators**

To rationalize a denominator containing a single nth root, multiply the fraction by a well-chosen 1 so that the product's denominator has a radicand that is a perfect nth power.

Example 2   Rationalize the denominator. Assume that variables represent positive values.

a. $\dfrac{3}{\sqrt[3]{2}}$

SOLUTION $\dfrac{3}{\sqrt[3]{2}} = \dfrac{3}{\sqrt[3]{2}} \cdot \dfrac{\sqrt[3]{4}}{\sqrt[3]{4}}$ **Because $\sqrt[3]{2} \cdot \sqrt[3]{4} = \sqrt[3]{8} = 2$, multiply the fraction by $\dfrac{\sqrt[3]{4}}{\sqrt[3]{4}}$.**

$= \dfrac{3\sqrt[3]{4}}{\sqrt[3]{8}}$ **Multiply.**

$= \dfrac{3\sqrt[3]{4}}{2}$ **Simplify.**

b. $\dfrac{\sqrt[3]{a}}{\sqrt[3]{b}}$

SOLUTION $\dfrac{\sqrt[3]{a}}{\sqrt[3]{b}} = \dfrac{\sqrt[3]{a}}{\sqrt[3]{b}} \cdot \dfrac{\sqrt[3]{b^2}}{\sqrt[3]{b^2}}$ **Because $\sqrt[3]{b} \cdot \sqrt[3]{b^2} = \sqrt[3]{b^3} = b$, multiply the fraction by $\dfrac{\sqrt[3]{b^2}}{\sqrt[3]{b^2}}$.**

$= \dfrac{\sqrt[3]{ab^2}}{\sqrt[3]{b^3}}$ **Multiply.**

$= \dfrac{\sqrt[3]{ab^2}}{b}$ **Simplify.**

Answers   **Your Turn 1**

a. $\dfrac{\sqrt{7}}{7}$ **b.** $\dfrac{\sqrt{21}}{6}$ **c.** $\dfrac{3\sqrt{10x}}{10x}$

c. $\sqrt[3]{\dfrac{5}{9a^2}}$

SOLUTION $\sqrt[3]{\dfrac{5}{9a^2}} = \dfrac{\sqrt[3]{5}}{\sqrt[3]{9a^2}}$ Use the quotient rule to separate the numerator and denominator.

$$= \dfrac{\sqrt[3]{5}}{\sqrt[3]{9a^2}} \cdot \dfrac{\sqrt[3]{3a}}{\sqrt[3]{3a}}$$ Because $\sqrt[3]{9a^2} \cdot \sqrt[3]{3a} = \sqrt[3]{27a^3} = 3a$, multiply the fraction by $\dfrac{\sqrt[3]{3a}}{\sqrt[3]{3a}}$.

$$= \dfrac{\sqrt[3]{15a}}{\sqrt[3]{27a^3}}$$ Multiply.

$$= \dfrac{\sqrt[3]{15a}}{3a}$$ Simplify.

d. $\dfrac{5}{\sqrt[4]{3}}$

SOLUTION $\dfrac{5}{\sqrt[4]{3}} = \dfrac{5}{\sqrt[4]{3}} \cdot \dfrac{\sqrt[4]{27}}{\sqrt[4]{27}}$ Because $\sqrt[4]{3} \cdot \sqrt[4]{27} = \sqrt[4]{81} = 3$, multiply the fraction by $\dfrac{\sqrt[4]{27}}{\sqrt[4]{27}}$.

$$= \dfrac{5\sqrt[4]{27}}{\sqrt[4]{81}}$$ Multiply.

$$= \dfrac{5\sqrt[4]{27}}{3}$$ Simplify.

Your Turn 2   Rationalize the denominator. Assume that variables represent positive values.

a. $\dfrac{6}{\sqrt[3]{3}}$ **b.** $\sqrt[3]{\dfrac{3}{x^2}}$ **c.** $\dfrac{4}{\sqrt[3]{4y}}$ **d.** $\dfrac{7}{\sqrt[4]{2}}$

Objective ➁ Rationalize denominators that have a sum or difference with a square root term.
In Example 4 of Section 10.4, we saw that the product of two conjugates containing square roots does not contain any radicals. Consequently, if the denominator of a fraction contains a sum or difference with a square root term, we can rationalize the denominator by multiplying the fraction by a 1 made up of the conjugate of the denominator. For example, to rationalize $\dfrac{5}{7 - \sqrt{3}}$, we multiply by $\dfrac{7 + \sqrt{3}}{7 + \sqrt{3}}$. Because $7 - \sqrt{3}$ and $7 + \sqrt{3}$ are conjugates, their product will not contain any radicals; so the denominator will be rationalized.

$$\dfrac{5}{7 - \sqrt{3}} = \dfrac{5}{7 - \sqrt{3}} \cdot \dfrac{7 + \sqrt{3}}{7 + \sqrt{3}} = \dfrac{5(7 + \sqrt{3})}{(7)^2 - (\sqrt{3})^2} = \dfrac{35 + 5\sqrt{3}}{49 - 3} = \dfrac{35 + 5\sqrt{3}}{46}$$

Procedure   **Rationalizing a Denominator Containing a Sum or Difference**

To rationalize a denominator containing a sum or difference with at least one square root term, multiply the fraction by a 1 whose numerator and denominator are the conjugate of the denominator.

Answers   **Your Turn 2**

a. $2\sqrt[3]{9}$ **b.** $\dfrac{\sqrt[3]{3x}}{x}$

c. $\dfrac{2\sqrt[3]{2y^2}}{y}$ **d.** $\dfrac{7\sqrt[4]{8}}{2}$

Example 3   Rationalize the denominator and simplify. Assume that variables represent positive values.

a. $\dfrac{9}{\sqrt{2} + 7}$

SOLUTION $\dfrac{9}{\sqrt{2}+7} = \dfrac{9}{\sqrt{2}+7} \cdot \dfrac{\sqrt{2}-7}{\sqrt{2}-7}$ — The conjugate of $\sqrt{2}+7$ is $\sqrt{2}-7$, so we multiply by $\dfrac{\sqrt{2}-7}{\sqrt{2}-7}$.

$= \dfrac{9(\sqrt{2}-7)}{(\sqrt{2})^2 - (7)^2}$ — Multiply. In the denominator, use the rule $(a+b)(a-b) = a^2 - b^2$

$= \dfrac{9\sqrt{2}-63}{2-49}$ — Simplify.

$= \dfrac{9\sqrt{2}-63}{-47}$ — We can simplify the negative denominator by factoring out -1 in the numerator and denominator.

$= \dfrac{-1(63 - 9\sqrt{2})}{-1(47)}$ — After factoring out the -1, the signs of the terms change. Because 63 is now positive, we write it first.

$= \dfrac{63 - 9\sqrt{2}}{47}$ — Divide out the common factor -1.

b. $\dfrac{2\sqrt{3}}{\sqrt{6}-\sqrt{2}}$

SOLUTION $\dfrac{2\sqrt{3}}{\sqrt{6}-\sqrt{2}} = \dfrac{2\sqrt{3}}{\sqrt{6}-\sqrt{2}} \cdot \dfrac{\sqrt{6}+\sqrt{2}}{\sqrt{6}+\sqrt{2}}$ — The conjugate of $\sqrt{6}-\sqrt{2}$ is $\sqrt{6}+\sqrt{2}$, so we multiply by $\dfrac{\sqrt{6}+\sqrt{2}}{\sqrt{6}+\sqrt{2}}$.

$= \dfrac{2\sqrt{3}(\sqrt{6}+\sqrt{2})}{(\sqrt{6})^2 - (\sqrt{2})^2}$

$= \dfrac{2\sqrt{18} + 2\sqrt{6}}{6-2}$ — Multiply in the numerator and evaluate the exponents in the denominator.

$= \dfrac{2\sqrt{9 \cdot 2} + 2\sqrt{6}}{4}$ — Simplify $\sqrt{18}$ by factoring out a perfect square factor in 18.

$= \dfrac{2 \cdot 3\sqrt{2} + 2\sqrt{6}}{4}$ — Simplify $\sqrt{9 \cdot 2}$ by finding the square root of 9.

$= \dfrac{2(3\sqrt{2} + \sqrt{6})}{4}$ — Factor out the common factor 2 in the numerator.

$= \dfrac{3\sqrt{2} + \sqrt{6}}{2}$ — Divide out the common factor 2.

Note: We cannot combine $3\sqrt{2}$ and $\sqrt{6}$ because they are not like radicals.

c. $\dfrac{6}{\sqrt{x}-5}$

SOLUTION $\dfrac{6}{\sqrt{x}-5} = \dfrac{6}{\sqrt{x}-5} \cdot \dfrac{\sqrt{x}+5}{\sqrt{x}+5}$ — The conjugate of $\sqrt{x}-5$ is $\sqrt{x}+5$, so we multiply by $\dfrac{\sqrt{x}+5}{\sqrt{x}+5}$.

$= \dfrac{6(\sqrt{x}+5)}{(\sqrt{x})^2 - (5)^2}$ — Multiply.

$= \dfrac{6\sqrt{x}+30}{x-25}$ — Multiply in the numerator and evaluate the exponents in the denominator.

Your Turn 3 Rationalize the denominator and simplify. Assume that variables represent positive values.

a. $\dfrac{9}{\sqrt{5}+2}$ **b.** $\dfrac{\sqrt{2}}{\sqrt{5}-\sqrt{3}}$ **c.** $\dfrac{3}{\sqrt{x}+4}$

Answers Your Turn 3
a. $9\sqrt{5} - 18$ **b.** $\dfrac{\sqrt{10}+\sqrt{6}}{2}$
c. $\dfrac{3\sqrt{x}-12}{x-16}$

Objective ③ **Rationalize numerators.** In later mathematics courses, you may need to rationalize the numerator. We use the same procedure that we use in rationalizing denominators.

Example 4 Rationalize the numerator. Assume that variables represent positive values.

a. $\dfrac{\sqrt{5x}}{4}$

SOLUTION $\dfrac{\sqrt{5x}}{4} = \dfrac{\sqrt{5x}}{4} \cdot \dfrac{\sqrt{5x}}{\sqrt{5x}}$ To create a perfect square radicand in the numerator, we multiply by $\dfrac{\sqrt{5x}}{\sqrt{5x}}$.

$= \dfrac{\sqrt{25x^2}}{4\sqrt{5x}}$ Multiply.

$= \dfrac{5x}{4\sqrt{5x}}$ Simplify.

b. $\dfrac{3 + \sqrt{2x}}{4}$

SOLUTION $\dfrac{3 + \sqrt{2x}}{4} = \dfrac{3 + \sqrt{2x}}{4} \cdot \dfrac{3 - \sqrt{2x}}{3 - \sqrt{2x}}$ The conjugate of $3 + \sqrt{2x}$ is $3 - \sqrt{2x}$, so we multiply by $\dfrac{3 - \sqrt{2x}}{3 - \sqrt{2x}}$.

$= \dfrac{3^2 - (\sqrt{2x})^2}{4(3 - \sqrt{2x})}$ Multiply.

$= \dfrac{9 - 2x}{12 - 4\sqrt{2x}}$ Simplify.

Your Turn 4 Rationalize the numerators. Assume that variables represent positive values.

a. $\dfrac{\sqrt{3a}}{7}$ b. $\dfrac{5 - \sqrt{3a}}{2}$

Answers Your Turn 4

a. $\dfrac{3a}{7\sqrt{3a}}$ b. $\dfrac{25 - 3a}{10 + 2\sqrt{3a}}$

10.5 Exercises   For Extra Help *MyMathLab*

Note: Exercises marked with a ★ represent challenging exercises.

1. Explain why each of the following expressions is not in simplest form.

   a. $\sqrt{\dfrac{3}{16}}$ b. $\dfrac{5}{\sqrt{3}}$

2. Although $\dfrac{1}{\sqrt{3}}$ and $\dfrac{\sqrt{3}}{3}$ are equal, explain why $\dfrac{\sqrt{3}}{3}$ is considered simplest form.

3. Explain in your own words how to rationalize a denominator that is the square root of a number $\left(\text{for example, } \dfrac{2}{\sqrt{3}} \text{ or } \dfrac{\sqrt{5}}{\sqrt{7}}\right)$.

4. Explain in your own words how to rationalize a denominator that is a sum or difference with a square root term $\left(\text{for example, } \dfrac{3}{5 + \sqrt{2}} \text{ or } \dfrac{2}{\sqrt{x} - \sqrt{y}}\right)$.

For Exercises 5–28, rationalize the denominator. Assume that variables represent positive values. See Example 1.

5. $\dfrac{1}{\sqrt{3}}$ 6. $\dfrac{1}{\sqrt{7}}$ 7. $\dfrac{3}{\sqrt{8}}$ 8. $\dfrac{5}{\sqrt{12}}$

9. $\sqrt{\dfrac{36}{7}}$ 10. $\sqrt{\dfrac{64}{3}}$ 11. $\sqrt{\dfrac{5}{12}}$ 12. $\sqrt{\dfrac{11}{18}}$

13. $\dfrac{\sqrt{7x^2}}{\sqrt{50}}$ 14. $\dfrac{\sqrt{3x^2}}{\sqrt{32}}$ 15. $\dfrac{\sqrt{8}}{\sqrt{56}}$ 16. $\dfrac{\sqrt{7}}{\sqrt{42}}$

17. $\dfrac{5}{\sqrt{3a}}$ 18. $\dfrac{11}{\sqrt{7b}}$ 19. $\sqrt{\dfrac{3m}{11n}}$ 20. $\sqrt{\dfrac{5r}{6s}}$

21. $\dfrac{10}{\sqrt{5x}}$ 22. $\dfrac{28}{\sqrt{7a}}$ 23. $\dfrac{\sqrt{6x}}{\sqrt{32x}}$ 24. $\dfrac{\sqrt{10a}}{\sqrt{18a}}$

25. $\dfrac{3}{\sqrt{x^3}}$ 26. $\dfrac{5}{\sqrt{b^5}}$ 27. $\dfrac{8x^2}{\sqrt{2x}}$ 28. $\dfrac{18a^4}{\sqrt{6a}}$

Find   **the mistake**   **For Exercises 29 and 30, explain the mistake; then simplify correctly.**

29. $\dfrac{\sqrt{3}}{\sqrt{2}} = \dfrac{\sqrt{3}}{\sqrt{2}} \cdot \dfrac{2}{2} = \dfrac{2\sqrt{3}}{2}$

30. $\sqrt{\dfrac{7}{3}} = \dfrac{\sqrt{7}}{\sqrt{3}} \cdot \dfrac{\sqrt{3}}{\sqrt{3}} = \dfrac{\sqrt{21}}{9}$

For Exercises 31–50, rationalize the denominators. Assume that variables represent positive values. See Example 2.

31. $\dfrac{5}{\sqrt[3]{3}}$ 32. $\dfrac{7}{\sqrt[3]{5}}$ 33. $\sqrt[3]{\dfrac{5}{2}}$ 34. $\sqrt[3]{\dfrac{3}{4}}$

35. $\dfrac{6}{\sqrt[3]{4}}$ 36. $\dfrac{9}{\sqrt[3]{9}}$ 37. $\dfrac{m}{\sqrt[3]{n}}$ 38. $\dfrac{p}{\sqrt[3]{q}}$

39. $\sqrt[3]{\dfrac{a}{b^2}}$ 40. $\sqrt[3]{\dfrac{m}{n^2}}$ 41. $\dfrac{4}{\sqrt[3]{2x}}$ 42. $\dfrac{9}{\sqrt[3]{3a}}$

43. $\sqrt[3]{\dfrac{6}{25a^2}}$ 44. $\sqrt[3]{\dfrac{5}{16b^2}}$ 45. $\dfrac{5}{\sqrt[4]{4}}$ 46. $\dfrac{7}{\sqrt[4]{9}}$

47. $\sqrt[4]{\dfrac{3}{x^2}}$ 48. $\sqrt[4]{\dfrac{5}{y^3}}$ 49. $\dfrac{9}{\sqrt[4]{3x^3}}$ 50. $\dfrac{12}{\sqrt[4]{2x^2}}$

For Exercises 51–72, rationalize the denominator and simplify. Assume that variables represent positive values. See Example 3.

51. $\dfrac{3}{\sqrt{2}+1}$ 52. $\dfrac{3}{\sqrt{5}+2}$ 53. $\dfrac{4}{2-\sqrt{3}}$ 54. $\dfrac{2}{4-\sqrt{15}}$

55. $\dfrac{5}{\sqrt{2}+\sqrt{3}}$ 56. $\dfrac{7}{\sqrt{6}+\sqrt{7}}$ 57. $\dfrac{4}{1-\sqrt{5}}$ 58. $\dfrac{6}{1-\sqrt{7}}$

59. $\dfrac{\sqrt{3}}{\sqrt{3}-1}$ 60. $\dfrac{\sqrt{5}}{\sqrt{5}-1}$ 61. $\dfrac{2\sqrt{3}}{\sqrt{3}-4}$ 62. $\dfrac{2\sqrt{5}}{\sqrt{5}-4}$

63. $\dfrac{4\sqrt{3}}{\sqrt{7}+\sqrt{2}}$

64. $\dfrac{2\sqrt{2}}{\sqrt{4}+\sqrt{3}}$

65. $\dfrac{8\sqrt{2}}{4\sqrt{2}-\sqrt{6}}$

66. $\dfrac{4\sqrt{3}}{4\sqrt{6}+\sqrt{2}}$

67. $\dfrac{6\sqrt{y}}{\sqrt{y}+1}$

68. $\dfrac{4\sqrt{x}}{\sqrt{x}+1}$

69. $\dfrac{3\sqrt{t}}{\sqrt{t}+2\sqrt{u}}$

70. $\dfrac{2\sqrt{m}}{\sqrt{n}-3\sqrt{m}}$

71. $\dfrac{\sqrt{2y}}{\sqrt{x}-\sqrt{6y}}$

72. $\dfrac{\sqrt{14h}}{\sqrt{2h}+\sqrt{k}}$

For Exercises 73–84, rationalize the numerator. Assume that variables represent positive values. See Example 4.

73. $\dfrac{\sqrt{3}}{2}$

74. $\dfrac{\sqrt{7}}{3}$

75. $\dfrac{\sqrt{2x}}{5}$

76. $\dfrac{\sqrt{7y}}{3}$

77. $\dfrac{\sqrt{8n}}{6}$

78. $\dfrac{\sqrt{20t}}{8}$

79. $\dfrac{2+\sqrt{3}}{5}$

80. $\dfrac{3+\sqrt{2}}{4}$

81. $\dfrac{\sqrt{5x}-6}{9}$

82. $\dfrac{\sqrt{2x}+7}{3}$

★**83.** $\dfrac{5\sqrt{n}+\sqrt{6n}}{2n}$

★**84.** $\dfrac{4\sqrt{k}-\sqrt{10k}}{5k}$

85. Given $f(x)=\dfrac{5\sqrt{2}}{x}$, find each of the following. Express your answer in simplest form.

    a. $f(\sqrt{6})$ **b.** $f(\sqrt{10})$ **c.** $f(\sqrt{22})$

86. Given $g(x)=\dfrac{\sqrt{2}}{x-1}$, find each of the following. Express your answer in simplest form.

    a. $g(\sqrt{5})$ **b.** $g(3\sqrt{2})$ **c.** $g(2\sqrt{6})$

87. Graph $f(x)=\dfrac{1}{\sqrt{x}}$; then graph $g(x)=\dfrac{\sqrt{x}}{x}$.

    a. What do you notice about the two graphs? What does this indicate about the two functions?

    b. Simplify $f(x)$ by rationalizing the denominator. What do you notice?

88. Graph $f(x)=-\dfrac{1}{\sqrt{x}}$; then graph $g(x)=-\dfrac{\sqrt{x}}{x}$.

    a. What do you notice about the two graphs? What does this indicate about the two functions?

    b. Simplify $f(x)$ by rationalizing the denominator. What do you notice?

89. In Section 10.1, we used the formula $T=2\pi\sqrt{\dfrac{L}{9.8}}$ to determine the period of a pendulum, where T is the period in seconds and L is the length in meters.

    a. Rewrite the formula so that the denominator is rationalized.

    b. Rewrite the formula so that the numerator is rationalized.

90. The formula $t = \sqrt{\dfrac{h}{16}}$ can be used to find the time, t, in seconds for an object to fall a distance of h feet.

    a. Rewrite the formula so that the denominator is rationalized.

    b. Rewrite the formula so that the numerator is rationalized.

91. The formula $s = \sqrt{\dfrac{3V}{h}}$ can be used to find the length, s, of each side of the base of a pyramid having a square base, volume V in cubic feet, and height h in feet.

    a. Rationalize the denominator in the formula.

    b. The volume of the Great Pyramid at Giza in Egypt is approximately 83,068,742 cubic feet, and its height is 449 feet. Find the length of each side of its base.

92. The formula $s = 2\sqrt{\dfrac{A}{6\sqrt{3}}}$ can be used to find the length, s, of each side of a regular hexagon having an area A.

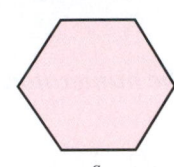

s

    a. Rationalize the denominator in the formula.

    b. If A is 100 square meters, write an expression in simplest form for the side length.

    c. Use a calculator to approximate the side lengths, rounded to three decimal places.

93. In AC circuits, voltage is often expressed as a *root-mean-square*, or *rms*, value. The formula for calculating the rms voltage, V_{rms}, given the maximum voltage, V_m, value is $V_{rms} = \dfrac{V_m}{\sqrt{2}}$.

    a. Rationalize the denominator in the formula.

    b. Given a maximum voltage of 163 V, write an expression for the rms voltage.

    c. Use a calculator to approximate the rms voltage, rounded to the nearest tenth

94. The velocity, in meters per second, of a particle can be determined by the formula $v = \sqrt{\dfrac{2E}{m}}$, where E represents the kinetic energy, in joules, of the particle and m represents the mass, in kilograms, of the particle.

    a. Rationalize the denominator in the formula.

    b. A particle with a mass of 1×10^{-6} kilograms has 2.4×10^{7} joules of kinetic energy. Write an expression for its velocity.

    c. Use a calculator to approximate the velocity rounded to the nearest tenth.

95. The resistance in a circuit is found to be $\dfrac{5\sqrt{2}}{3 + \sqrt{6}}$ ohms. Rationalize the denominator.

96. Two charged particles, q_1 and q_2, are separated by a distance of 8 centimeters. The values of the charges are $q_1 = 3 \times 10^{-6}$ coulombs and $q_2 = 1 \times 10^{-6}$ coulombs. Each charged particle exerts an electrical field. At a point between the two particles x centimeters away from q_1, the electric fields cancel each other so that the value of the fields at the point x is 0.

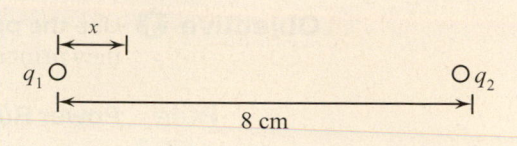

a. Use the formula $x = \dfrac{l}{1 + \sqrt{\dfrac{q_2}{q_1}}}$ to find the distance from q_1 at which the electric field is canceled, where l is the distance separating the particles. Write the distance with a rationalized denominator.

b. Use a calculator to approximate the distance, rounded to the nearest tenth.

Review Exercises

Exercises 1 and 2 **EXPRESSIONS**

[10.3] **1.** Simplify: $\pm\sqrt{28}$

[6.4] **2.** Factor: $x^2 - 6x + 9$

Exercises 3–6  **EQUATIONS AND INEQUALITIES**

[2.3] *For Exercises 3 and 4, solve.*

  3. $2x - 3 = 5$

  4. $2x - 3 = -5$

[6.6] *For Exercises 5 and 6, solve and check.*

  5. $x^2 - 36 = 0$

  6. $x^2 - 5x + 6 = 0$

 **10.6** Radical Equations and Problem Solving

Objective ① Use the power rule to solve radical equations.

We now explore how to solve **radical equations**.

Definition *Radical equation:* An equation containing at least one radical expression whose radicand has a variable.

The Algebra Pyramid

Note: We move upward on the Algebra Pyramid, making equations out of the expressions we've studied.

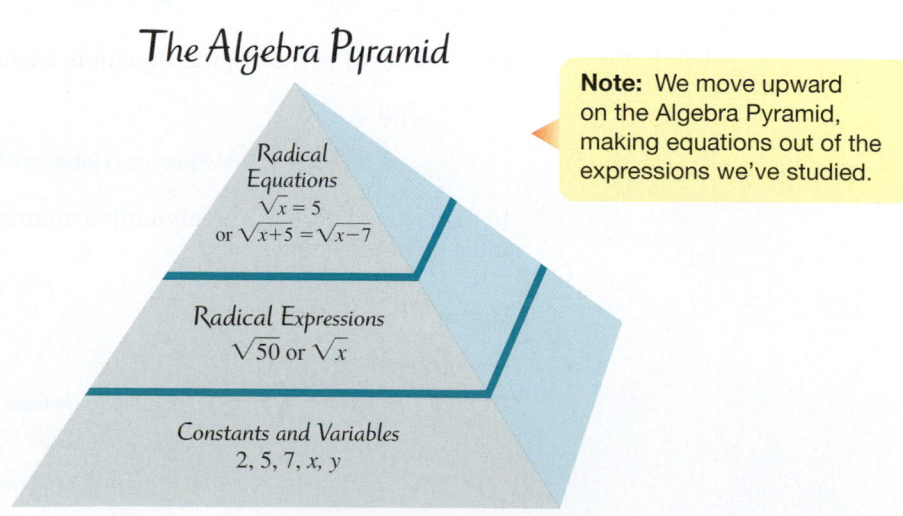

Objective ① **Use the power rule to solve radical equations.** To solve radical equations, we use a new principle of equality called the *power rule*.

Rule **Power Rule for Solving Equations**

> If both sides of an equation are raised to the same integer power, the resulting equation contains all solutions of the original equation and perhaps some solutions that do not solve the original equation. That is, the solutions of the equation $a = b$ are contained among the solutions of $a^n = b^n$, where n is an integer.

Isolated Radicals

First, we consider equations such as $\sqrt{x} = 9$ in which the radical is isolated. When we use the power rule, we raise both sides of the equation to the same integer power as the root index, then use the principle $(\sqrt[n]{x})^n = x$, which eliminates the radical, leaving its radicand.

Example 1 Solve.

a. $\sqrt{x} = 9$

SOLUTION $(\sqrt{x})^2 = (9)^2$ Because the root index is 2, we square both sides.

$$x = 81$$

Check $\sqrt{81} \overset{?}{=} 9$

$$9 = 9 \quad \text{True}$$

b. $\sqrt[3]{y} = -2$

SOLUTION $(\sqrt[3]{y})^3 = (-2)^3$ Because the root index is 3, we cube both sides.

$$y = -8$$

Check $\sqrt[3]{-8} \overset{?}{=} -2$

$$-2 = -2 \quad \text{True}$$

Extraneous Solutions

In Section 7.6, we learned that some equations have *extraneous solutions*, which are apparent solutions that do not actually make the original equation true. Using the power rule can sometimes lead to extraneous solutions, so it is important to check solutions. For example, watch what happens when we use the power rule to solve the equation $\sqrt{x} = -9$.

Warning: As we will see in the check, this result is not a solution.

$(\sqrt{x})^2 = (-9)^2$ Square both sides.

$$\longrightarrow x = 81$$

By checking 81 in the original equation, we see that it is extraneous.

$$\sqrt{81} \overset{?}{=} -9$$

$$9 = -9 \quad \text{This equation is false, so 81 is extraneous.}$$

In fact, $\sqrt{x} = -9$ has no real number solution because if x is a real number, then $\sqrt{x}$ must be nonnegative.

Example 2 Solve.

a. $\sqrt{x - 7} = 8$

SOLUTION $(\sqrt{x - 7})^2 = (8)^2$ Square both sides.

$$x - 7 = 64 \qquad \text{Simplify.}$$

$$x = 71 \qquad \text{Add 7 to both sides.}$$

Check $\quad \sqrt{71 - 7} \overset{?}{=} (8)^2$

$\qquad\qquad \sqrt{64} \overset{?}{=} 8$

$\qquad\qquad\qquad 8 = 8 \qquad$ **True. The solution is 71.**

b. $\sqrt[3]{x - 3} = -1$

SOLUTION $\quad (\sqrt[3]{x - 3})^3 = (-1)^3 \qquad$ **Cube both sides.**

$\qquad\qquad\qquad x - 3 = -1 \qquad$ **Simplify.**

$\qquad\qquad\qquad\qquad x = 2 \qquad$ **Add 3 to both sides.**

Check $\quad \sqrt[3]{2 - 3} \overset{?}{=} -1$

$\qquad\qquad \sqrt[3]{-1} \overset{?}{=} -1$

$\qquad\qquad\qquad -1 = -1 \qquad$ **True. The solution is 2.**

Note: We can see that this equation has no real-number solution because a principal square root cannot be equal to a negative number. However, we work through the steps to confirm this.

c. $\sqrt{2x + 1} = -3$

SOLUTION $\quad (\sqrt{2x + 1})^2 = (-3)^2 \qquad$ **Square both sides.**

$\qquad\qquad\qquad 2x + 1 = 9 \qquad$ **Simplify.**

$\qquad\qquad\qquad\quad 2x = 8 \qquad$ **Subtract 1 from both sides.**

$\qquad\qquad\qquad\qquad x = 4 \qquad$ **Divide both sides by 2.**

Check $\quad \sqrt{2(4) + 1} \overset{?}{=} -3$

$\qquad\qquad \sqrt{9} \overset{?}{=} -3$

$\qquad\qquad\qquad 3 = -3 \qquad$ **False, so 4 is extraneous. This equation has no real-number solution.**

Your Turn 2 Solve.

a. $\sqrt{x - 5} = 6$ $\qquad\qquad$ **b.** $\sqrt[3]{x + 2} = 3$ $\qquad\qquad$ **c.** $\sqrt{x + 3} = -7$

Radicals on Both Sides of the Equation

As we will see in Example 3, the power rule can be used to solve equations with radicals on both sides of the equal sign.

Example 3 Solve.

a. $\sqrt{6x - 1} = \sqrt{x + 2}$

SOLUTION $\quad (\sqrt{6x - 1})^2 = (\sqrt{x + 2})^2$

$\qquad\qquad\qquad 6x - 1 = x + 2 \qquad$ **Square both sides.**

$\qquad\qquad\qquad\qquad 5x = 3 \qquad$ **Subtract x from and add 1 to both sides.**

$\qquad\qquad\qquad\qquad\quad x = \dfrac{3}{5} \qquad$ **Divide both sides by 5.**

Check $\quad \sqrt{6\left(\dfrac{3}{5}\right) - 1} \overset{?}{=} \sqrt{\dfrac{3}{5} + 2}$

$\qquad\qquad \sqrt{\dfrac{18}{5} - \dfrac{5}{5}} \overset{?}{=} \sqrt{\dfrac{3}{5} + \dfrac{10}{5}}$

$\qquad\qquad \sqrt{\dfrac{13}{5}} = \sqrt{\dfrac{13}{5}} \qquad$ **True. The solution is $\dfrac{3}{5}$.**

Answers **Your Turn 2**

a. 41 $\quad$ **b.** 25

c. no real-number solution

b. $\sqrt[3]{5x - 2} = \sqrt[3]{3x + 2}$

SOLUTION $(\sqrt[3]{5x - 2})^3 = (\sqrt[3]{3x + 2})^3$ Cube both sides.

$$5x - 2 = 3x + 2 \qquad \text{Simplify.}$$
$$2x = 4 \qquad \text{Subtract } 3x \text{ and add 2 to both sides.}$$
$$x = 2 \qquad \text{Divide both sides by 2.}$$

Check $\sqrt[3]{5(2) - 2} \overset{?}{=} \sqrt[3]{3(2) + 2}$

$$\sqrt[3]{10 - 2} \overset{?}{=} \sqrt[3]{6 + 2}$$
$$\sqrt[3]{8} \overset{?}{=} \sqrt[3]{8}$$
$$2 = 2 \qquad \text{True. The solution is 2.}$$

Your Turn 3   Solve.

a. $\sqrt{8x + 5} = \sqrt{2x + 7}$ **b.** $\sqrt[3]{6x + 9} = \sqrt[3]{10x - 3}$

Multiple Solutions

Radical equations may have multiple solutions if, after using the power rule, we are left with a quadratic form.

Example 4   Solve $x + 2 = \sqrt{5x + 16}$.

SOLUTION $(x + 2)^2 = (\sqrt{5x + 16})^2$ Square both sides.

$$x^2 + 4x + 4 = 5x + 16 \qquad \text{Use FOIL on the left-hand side.}$$
$$x^2 - x - 12 = 0 \qquad \text{Because the equation is quadratic, we set it equal to 0 by subtracting } 5x \text{ and 16 from both sides.}$$
$$(x + 3)(x - 4) = 0 \qquad \text{Factor.}$$
$$x + 3 = 0 \quad \text{or} \quad x - 4 = 0 \qquad \text{Use the zero-factor theorem.}$$
$$x = -3 \qquad\qquad x = 4 \qquad \text{Solve each equation.}$$

Checks $-3 + 2 \overset{?}{=} \sqrt{5(-3) + 16}$ $4 + 2 \overset{?}{=} \sqrt{5(4) + 16}$

$$-1 \overset{?}{=} \sqrt{-15 + 16} \qquad\qquad 6 \overset{?}{=} \sqrt{20 + 16}$$
$$-1 \overset{?}{=} \sqrt{1} \qquad\qquad\qquad\quad 6 \overset{?}{=} \sqrt{36}$$
$$-1 = 1 \quad \text{False} \qquad\qquad 6 = 6 \quad \text{True. The solution is 4.}$$

Because -3 does not check, it is an extraneous solution. The only solution is 4.

Your Turn 4   Solve $\sqrt{9x + 7} = x + 3$.

Radicals Not Isolated

Now we consider radical equations in which the radical term is not isolated. In such equations, we must first isolate the radical term.

Answers   **Your Turn 3**

a. $\dfrac{1}{3}$ **b.** 3

Answer   **Your Turn 4**

2 and 1

Example 5 Solve.

a. $\sqrt{x+1} - 2x = x + 1$

SOLUTION

$$\sqrt{x+1} = 3x + 1 \qquad \text{Add } 2x \text{ to both sides to isolate the radical term.}$$

$$(\sqrt{x+1})^2 = (3x+1)^2 \qquad \text{Square both sides.}$$

$$x + 1 = 9x^2 + 6x + 1 \qquad \text{Use FOIL on the right-hand side.}$$

$$0 = 9x^2 + 5x \qquad \text{Because the equation is quadratic, we set it equal to 0 by subtracting } x \text{ and 1 from both sides.}$$

$$0 = x(9x + 5) \qquad \text{Factor.}$$

$$x = 0 \quad \text{or} \quad 9x + 5 = 0 \qquad \text{Use the zero-factor theorem.}$$

$$9x = -5$$

$$x = -\frac{5}{9}$$

Checks $\quad \sqrt{0+1} - 2(0) \overset{?}{=} 0 + 1 \qquad\qquad \sqrt{-\dfrac{5}{9}+1} - 2\left(-\dfrac{5}{9}\right) \overset{?}{=} -\dfrac{5}{9} + 1$

$$\sqrt{1} - 0 \overset{?}{=} 1 \qquad\qquad\qquad \sqrt{-\dfrac{5}{9}+\dfrac{9}{9}} - 2\left(-\dfrac{5}{9}\right) \overset{?}{=} -\dfrac{5}{9} + \dfrac{9}{9}$$

$$1 = 1 \quad \text{**True. The solution is 0.**} \qquad \sqrt{\dfrac{4}{9}} + \dfrac{10}{9} \overset{?}{=} \dfrac{4}{9}$$

$$\dfrac{2}{3} + \dfrac{10}{9} \overset{?}{=} \dfrac{4}{9}$$

> **Note:** This false equation indicates that $-\dfrac{5}{9}$ is an extraneous solution.

$$\dfrac{6}{9} + \dfrac{10}{9} \overset{?}{=} \dfrac{4}{9}$$

$$\dfrac{16}{9} = \dfrac{4}{9} \quad \text{**False**}$$

The solution is 0.

b. $\sqrt[4]{3x+4} + 5 = 7$

SOLUTION

$$\sqrt[4]{3x+4} - 2 \qquad \text{Subtract 5 from both sides to isolate the radical term.}$$

$$(\sqrt[4]{3x+4})^4 = 2^4 \qquad \text{Raise both sides to the fourth power.}$$

$$3x + 4 = 16 \qquad \text{Simplify both sides.}$$

$$3x = 12 \qquad \text{Subtract 4 from both sides.}$$

$$x = 4 \qquad \text{Divide both sides by 3.}$$

Check $\quad \sqrt[4]{3(4)+4} + 5 \overset{?}{=} 7$

$$\sqrt[4]{12+4} + 5 \overset{?}{=} 7$$

$$\sqrt[4]{16} + 5 \overset{?}{=} 7$$

$$2 + 5 \overset{?}{=} 7$$

$$7 = 7 \quad \text{**True. The solution is 4.**}$$

Your Turn 5 Solve.

a. $\sqrt{5x^2 + 6x - 7} + 3x = 5x + 1$ \qquad **b.** $\sqrt[4]{3x+6} - 7 = -4$

Using the Power Rule Twice

Some equations may require that we use the power rule twice to eliminate all radicals.

Answers **Your Turn 5**

a. 2 **b.** 25

Example 6   Solve $\sqrt{x + 21} = \sqrt{x} + 3$.

SOLUTION $(\sqrt{x + 21})^2 = (\sqrt{x} + 3)^2$ Because one of the radicals is isolated, we square both sides.

$$x + 21 = (\sqrt{x} + 3)(\sqrt{x} + 3)$$ Simplify.

$$x + 21 = x + 3\sqrt{x} + 3\sqrt{x} + 9$$ Use FOIL on the right-hand side.

$$x + 21 = x + 6\sqrt{x} + 9$$ Combine like terms.

$$12 = 6\sqrt{x}$$ Subtract x and 9 from both sides to isolate the remaining radical expression.

$$2 = \sqrt{x}$$ Divide both sides by 6.

$$(2)^2 = (\sqrt{x})^2$$ Square both sides.

$$4 = x$$

Check $\sqrt{4 + 21} \overset{?}{=} \sqrt{4} + 3$

$$\sqrt{25} \overset{?}{=} 2 + 3$$

$$5 = 5$$ True. The solution is 4.

Our examples suggest the following procedure.

Procedure   **Solving Radical Equations**

To solve a radical equation,

1. Isolate the radical if necessary. (If there is more than one radical term, isolate one of the radical terms.)
2. Raise both sides of the equation to the same power as the root index of the isolated radical.
3. If all radicals have been eliminated, solve. If a radical term remains, isolate that radical term and raise both sides to the same power as its root index.
4. Check each solution. Any apparent solution that does not check is an extraneous solution.

Answer   **Your Turn 6**

0 and 4

Your Turn 6   Solve $\sqrt{2x + 1} = \sqrt{x} + 1$.

10.6 Exercises   For Extra Help   *MyMathLab* 

PRACTICE   WATCH   DOWNLOAD

Note: Exercises marked with a ★ represent challenging exercises.

1. Explain why we must check all potential solutions to radical equations.

2. What is an extraneous solution?

3. Explain why there is no real-number solution for the radical equation $\sqrt{x} = -6$.

4. Show why $(\sqrt{a})^2 = a$, assuming that $a \geq 0$.

5. Given the radical equation $\sqrt{x + 2} + 3x = 4x - 1$, what would be the first step in solving the equation? Why?

6. Give an example of a radical equation that requires you to use the power rule twice in solving the equation. Explain why the principle must be used twice.

For Exercises 7–28, solve. See Examples 1 and 2.

7. $\sqrt{x} = 2$

8. $\sqrt{y} = 5$

9. $\sqrt{k} = -4$

10. $\sqrt{x} = -1$

11. $\sqrt[3]{y} = 3$

12. $\sqrt[3]{m} = 4$

13. $\sqrt[3]{z} = -2$

14. $\sqrt[3]{p} = -5$

15. $\sqrt{n-1} = 4$

16. $\sqrt{x-2} = 3$

17. $\sqrt{t+5} = 4$

18. $\sqrt{m+8} = 1$

19. $\sqrt{3x-2} = 4$

20. $\sqrt{2x+5} = 3$

21. $\sqrt{2x+24} = 4$

22. $\sqrt{3y+34} = 5$

23. $\sqrt{2n-8} = -3$

24. $\sqrt{5x-1} = -6$

25. $\sqrt[3]{x-3} = 2$

26. $\sqrt[3]{k+2} = 4$

27. $\sqrt[3]{3y-2} = -2$

28. $\sqrt[3]{2x+4} = -4$

For Exercises 29–38 solve. First, isolate the radical term. See Example 5.

29. $\sqrt{u-3} - 10 = 1$

30. $\sqrt{y+1} - 4 = 2$

31. $\sqrt{y-6} + 2 = 9$

32. $\sqrt{r-5} + 6 = 10$

33. $\sqrt{6x-5} - 2 = 3$

34. $\sqrt{8x+4} - 2 = 4$

35. $\sqrt[3]{n+3} - 2 = -4$

36. $\sqrt[3]{x-4} + 2 = -3$

37. $\sqrt[4]{x-2} - 2 = -4$

38. $\sqrt[4]{m+3} + 2 = 1$

For Exercises 39–64, solve. Identify any extraneous solutions. See Examples 3–6.

39. $\sqrt{3x-2} = \sqrt{8-2x}$

40. $\sqrt{m+2} = \sqrt{2m-3}$

41. $\sqrt{4x-5} = \sqrt{6x+5}$

42. $\sqrt{3x-4} = \sqrt{5x+2}$

43. $\sqrt[3]{2r+2} = \sqrt[3]{3r-1}$

44. $\sqrt[3]{3h-4} = \sqrt[3]{h+4}$

45. $\sqrt[4]{4x+4} = \sqrt[4]{5x+1}$

46. $\sqrt[4]{2x+4} = \sqrt[4]{3x-2}$

47. $\sqrt{2x+24} = x+8$

48. $\sqrt{2x+15} = x+6$

49. $y-1 = \sqrt{2y-2}$

50. $3+x = \sqrt{7+3x}$

51. $\sqrt{3x+10} - 4 = x$

52. $\sqrt{6x+1} - 1 = x$

53. $\sqrt{10n+4} - 3n = n+1$

54. $\sqrt{6x-1} - 6x = 2 - 9x$

55. $\sqrt[3]{5x+2} + 2 = 5$

56. $\sqrt[3]{4x-1} - 4 = -1$

57. $\sqrt[3]{n^2 - 2n + 5} = 2$

58. $\sqrt[3]{y^2 + y + 7} = 3$

59. $1 + \sqrt{x} = \sqrt{2x+1}$

60. $\sqrt{3x+4} - 2 = \sqrt{x}$

61. $\sqrt{3x+1} + \sqrt{3x} = 2$

62. $\sqrt{2x+5} - \sqrt{2x} = 1$

63. $\sqrt{6x+7} - 2 = \sqrt{2x+3}$

64. $\sqrt{5x-1} - 1 = \sqrt{x+2}$

Find the mistake

For Exercises 65–68, explain the mistake; then solve correctly.

65.
$$\sqrt{x} = -9$$
$$(\sqrt{x})^2 = (-9)^2$$
$$x = 81$$

66.
$$\sqrt{x} = -2$$
$$(\sqrt{x})^2 = (-2)^2$$
$$x = 4$$

67. $\sqrt{x+3} = x - 3$
$x + 3 = x^2 - 9$
$0 = x^2 - x - 12$
$0 = (x + 3)(x - 4)$
$x = -3, 4$

68. $\sqrt{x+3} = 4$
$(\sqrt{x+3})^2 = 4^2$
$x + 9 = 16$
$x + 9 - 9 = 16 - 9$
$x = 7$

For Exercises 69–72, given the period of a pendulum, find the length of the pendulum. Use the formula $T = 2\pi\sqrt{\dfrac{L}{9.8}}$, where T is the period in seconds and L is the length in meters.

69. 2π seconds

70. 6π seconds

71. $\dfrac{\pi}{2}$ seconds

72. $\dfrac{\pi}{3}$ seconds

For Exercises 73–76, find the distance an object has fallen. Use the formula $t = \sqrt{\dfrac{h}{16}}$, where t is the time in seconds it takes an object to fall a distance of h feet.

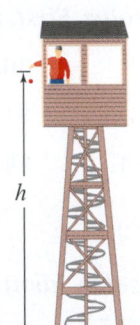

73. 0.3 second

74. 0.5 second

75. $\dfrac{1}{4}$ second

76. $\dfrac{1}{8}$ second

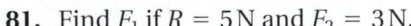

For Exercises 77–80, find the skid distance after a car has braked hard at a given speed. Use the formula $S = \dfrac{7}{2}\sqrt{2D}$, where D represents the skid distance in feet on asphalt and S represents the speed of the car in miles per hour. Round your answers to hundredths. (Source: Harris Technical Services, Traffic Accident Reconstructionists.)

77. 30 miles per hour

78. 60 miles per hour

79. 45 miles per hour

80. 75 miles per hour

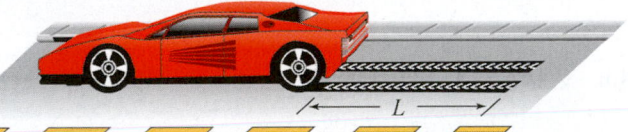

For Exercises 81–84, use the following information. Two forces, F_1 and F_2, acting on an object at a 90° angle will pull the object with a resultant force, R, at an angle between F_1 and F_2. (See the figure.) The value of the resultant force can be found by the formula $R = \sqrt{F_1^2 + F_2^2}$.

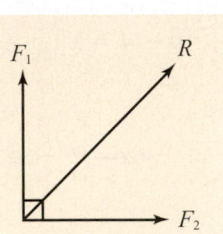

81. Find F_1 if $R = 5\,\text{N}$ and $F_2 = 3\,\text{N}$.

82. Find F_1 if $R = 10\,\text{N}$ and $F_2 = 8\,\text{N}$.

★ **83.** Find F_2 if $R = 3\sqrt{5}\,\text{N}$ and $F_1 = 3\,\text{N}$.

★ **84.** Find F_2 if $R = 2\sqrt{13}\,\text{N}$ and $F_1 = 6\,\text{N}$.

85. **a.** Complete the table of ordered pairs for the equation $y = \sqrt{x}$.

x	0	1	4	9	16
y	0	1	2		5

    b. Graph the ordered pairs in the coordinate plane; then connect the points to make a curve.

    c. Does the graph extend below or to the left of the origin? Explain.

    d. Does the curve represent a function? Explain.

86. a. Complete the table of ordered pairs for the equation $y = \sqrt{x} + 2$.

x	0	1	4	9	16
y	2	3	4		7

b. Graph the ordered pairs in the coordinate plane; then connect the points to make a curve.

c. Does the graph extend below or to the left of the origin? Explain.

d. Does the curve represent a function? Explain.

87. Use a graphing calculator to graph $y = \sqrt{x}$, $y = 2\sqrt{x}$, and $y = 3\sqrt{x}$.. Based on your observations, as you increase the size of the coefficient, what happens to the graph?

88. Use a graphing calculator to compare the graph of $y = \sqrt{x}$ to the graph of $y = -\sqrt{x}$. What effect does the negative sign have on the graph?

89. Use a graphing calculator to graph $y = \sqrt{x}$, $y = \sqrt{x} + 2$, and $y = \sqrt{x} - 2$. Based on your observations, what effect does adding or subtracting a constant have on the graph of $y = \sqrt{x}$?

90. Based on your conclusions from Exercises 87–89, without graphing, describe the graph of $y = -2\sqrt{x} + 3$. Use a graphing calculator to graph the equation to confirm your description.

Collaborative Exercises Building Time

You are constructing a grandfather clock that operates using a pendulum. The period of the pendulum is the time required to complete one full swing. The formula giving the relationship between the length, L (in meters), and the period, T (in seconds), is $T = 2\pi\sqrt{\dfrac{L}{9.8}}$.

1. Find the length of the pendulum if the period is 1 second.
2. Suppose you want the pendulum to complete one period in 2 seconds. Would you need to increase or decrease the length of the pendulum found in Exercise 1? Use the same formula to determine the length of the pendulum so that the period is 2 seconds.
3. Based on the results of Exercises 1 and 2, what can you conclude about the required length of the pendulum as the period increases?

Review Exercises

Exercises 1–6 EXPRESSIONS

[5.1] *For Exercises 1–3, evaluate.*

1. 3^4

2. $(-0.2)^3$

3. $\left(\dfrac{2}{5}\right)^{-4}$

For Exercises 4–6, use the rules of exponents to simplify.

[5.4] 4. $(x^3)(x^5)$

[5.4] 5. $(n^4)^6$

[5.6] 6. $\dfrac{y^7}{y^3}$

10.7 Complex Numbers

Objectives

1. Write imaginary numbers using *i*.
2. Perform arithmetic operations with complex numbers.
3. Raise *i* to powers.

We have said that the square root of a negative number is not a real number. In this section, we learn about the *imaginary number system,* in which square roots of negative numbers are expressed using a notation involving the letter *i*.

Objective 1 Write imaginary numbers using *i*. Using the product rule of square roots, any square root of a negative number can be rewritten as a product of a real number and the **imaginary unit**, which we express as *i*.

Definition ___*Imaginary unit:*___ The number represented by *i*, where $i = \sqrt{-1}$ and $i^2 = -1$.

A number that can be expressed as a product of a real number and the imaginary unit is called an **imaginary number**.

Definition ___*Imaginary number:*___ A number that can be expressed in the form *bi*, where *b* is a real number and *i* is the imaginary unit.

Of Interest

When the idea of finding square roots of negative numbers was first introduced, members of the established mathematics community said that this type of number existed, but only in the imagination of those finding them. From then on, the numbers have been called "imaginary numbers."

Note: In a product with the imaginary unit, it is customary to write integer factors first, then *i*, then square root factors.

Example 1 Write each imaginary number as a product of a real number and *i*.

a. $\sqrt{-9}$

SOLUTION
$$\sqrt{-9} = \sqrt{-1 \cdot 9} \qquad \text{Factor out } -1 \text{ in the radicand.}$$
$$= \sqrt{-1} \cdot \sqrt{9} \qquad \text{Use the product rule of square roots.}$$
$$= i \cdot 3 \qquad \text{Simplify.}$$
$$= 3i$$

b. $\sqrt{-54}$

SOLUTION
$$\sqrt{-54} = \sqrt{-1 \cdot 54} \qquad \text{Factor out } -1 \text{ in the radicand.}$$
$$= \sqrt{-1} \cdot \sqrt{54} \qquad \text{Use the product rule of square roots.}$$
$$= i\sqrt{9 \cdot 6} \qquad \text{Write 54 as } 9 \cdot 6.$$
$$= 3i\sqrt{6} \qquad \text{Simplify.}$$

Example 1 suggests the following procedure.

Procedure **Rewriting Imaginary Numbers**

To write an imaginary number $\sqrt{-n}$ in terms of the imaginary unit *i*,
1. Separate the radical into two factors, $\sqrt{-1} \cdot \sqrt{n}$.
2. Replace $\sqrt{-1}$ with *i*.
3. Simplify $\sqrt{n}$.

Your Turn 1 Write each imaginary number as a product of a real number and *i*.

a. $\sqrt{-64}$ **b.** $\sqrt{-18}$ **c.** $\sqrt{-48}$

Answers Your Turn 1
a. $8i$ **b.** $3i\sqrt{2}$ **c.** $4i\sqrt{3}$

Objective 2 **Perform arithmetic operations with complex numbers.** We now have two distinct sets of numbers, the set of real numbers and the set of imaginary numbers. There is yet another set of numbers, called the set of **complex numbers**, that contains both real and imaginary numbers.

Definition   **Complex number:** A number that can be expressed in the form $a + bi$, where a and b are real numbers and i is the imaginary unit.

Note: When the coefficient of i is a radical expression, as in $-2.1 - i\sqrt{5}$ or $4i\sqrt{3}$, we write i to the left of the radical sign so that i does not look like it is part of the radicand.

When written in the form $a + bi$, a complex number is said to be in *standard form.* Following are some examples of complex numbers written in standard form.

$$2 + 3i \qquad 4 - 7i \qquad -2.1 - i\sqrt{5}$$

Note that if $a = 0$, then the complex number is purely an imaginary number, such as the following:

$$-6i \qquad 4i\sqrt{3}$$

If $b = 0$, then the complex number is a real number.

The following Venn diagram shows how the set of complex numbers contains both real and imaginary numbers.

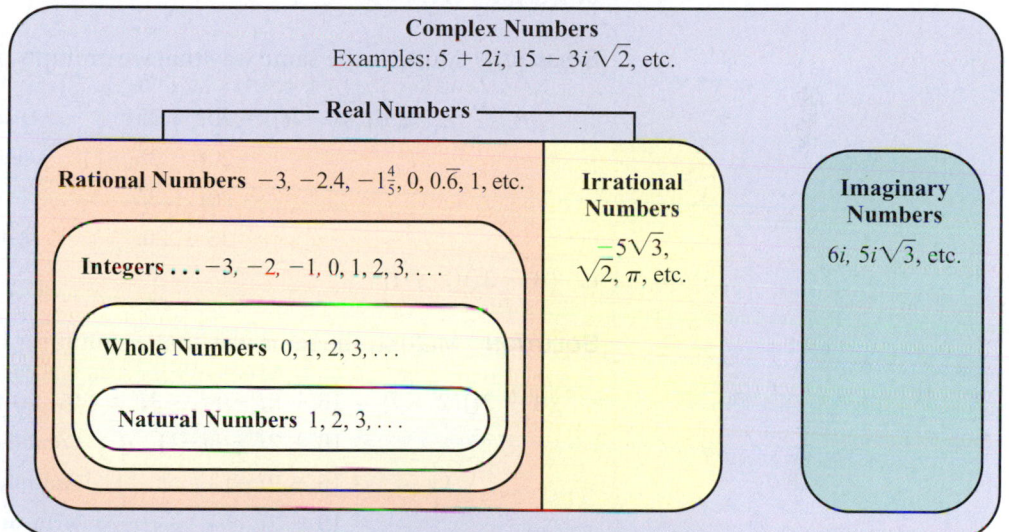

We can perform arithmetic operations with complex numbers. In general, we treat the complex numbers just like polynomials, where i is like a variable.

Adding and Subtracting Complex Numbers

Example 2   Add or subtract.

a. $(-8 + 7i) + (5 - 19i)$

SOLUTION We add complex numbers just like we add polynomials—by combining like terms.

$$(-8 + 7i) + (5 - 19i) = -3 - 12i$$

b. $(-1 + 13i) - (8 - 3i)$

SOLUTION We subtract complex numbers just like we subtract polynomials—by writing an equivalent addition and changing the signs in the second complex number.

$$(-1 + 13i) - (8 - 3i) = (-1 + 13i) + (-8 + 3i)$$
$$= -9 + 16i$$

Your Turn 2 Add or subtract.

a. $(-3 - 5i) + (4 - 6i)$ **b.** $(4 - 7i) - (-3 + 2i)$

Multiplying Complex Numbers

We multiply complex numbers the same way that we multiply monomials and binomials. However, we must be careful when simplifying because these products may contain i^2, which is equal to -1.

Example 3 Multiply.

a. $(9i)(-8i)$

SOLUTION Multiply the same way that we multiply monomials.

$$
\begin{aligned}
(9i)(-8i) &= -72i^2 && \text{Multiply.} \\
&= -72(\mathbf{-1}) && \text{Replace } i^2 \text{ with } -1. \\
&= 72
\end{aligned}
$$

b. $(5i)(4 - 7i)$

SOLUTION Multiply the same way that we multiply a binomial by a monomial.

$$
\begin{aligned}
(5i)(4 - 7i) &= 20i - 35i^2 && \text{Distribute.} \\
&= 20i - 35(-1) && \text{Replace } i^2 \text{ with } -1. \\
&= 20i + 35 \\
&= 35 + 20i && \text{Write in standard form, } a + bi.
\end{aligned}
$$

c. $(8 - 3i)(2 + i)$

SOLUTION Multiply the same way that we multiply binomials.

$$
\begin{aligned}
(8 - 3i)(2 + i) &= 16 + 8i - 6i - 3i^2 && \text{Use FOIL.} \\
&= 16 + 2i - 3(\mathbf{-1}) && \text{Combine like terms and replace } i^2 \text{ with } -1. \\
&= 16 + 2i + 3 && \text{Multiply.} \\
&= 19 + 2i && \text{Write in standard form, } a + bi.
\end{aligned}
$$

d. $(6 - 5i)(6 + 5i)$

SOLUTION Note that these complex numbers are conjugates.

$$
\begin{aligned}
(6 - 5i)(6 + 5i) &= 36 + 30i - 30i - 25i^2 && \text{Use FOIL.} \\
&= 36 - 25(\mathbf{-1}) && \text{Combine like terms and replace } i^2 \text{ with } -1. \\
&= 36 + 25 \\
&= 61
\end{aligned}
$$

Note: The product of these two complex numbers is a real number.

The complex numbers that we multiplied in Example 3(d) are called **complex conjugates**.

Definition   ***Complex conjugate:*** The complex conjugate of a complex number $a + bi$ is $a - bi$.

Other examples of complex conjugates follow.

$4 + i$ and $4 - i$

$9 - 7i$ and $9 + 7i$

Example 3(d) also illustrates the fact that the product of complex conjugates is always a real number. In general, $(a + bi)(a - bi) = a^2 - b^2i^2 = a^2 - b^2(-1) = a^2 + b^2$, which is a real number.

Your Turn 3 Multiply.

a. $(-4i)(-7i)$ **b.** $(-3i)(6 - i)$ **c.** $(8 - 3i)(5 + 7i)$ **d.** $(7 + 3i)(7 - 3i)$

Dividing Complex Numbers

Connection Recall that we rationalize denominators to clear undesired square root expressions from a denominator. The imaginary unit i represents a square root expression, $\sqrt{-1}$, which is why we rationalize denominators that contain i.

When dividing by an imaginary number, we use the fact that $i^2 = -1$, and when dividing by a complex number, we use the fact that the product of complex conjugates is a real number. The process is similar to rationalizing denominators. For example, to rationalize the denominator of $\dfrac{3}{2\sqrt{5}}$, we multiply by $\dfrac{\sqrt{5}}{\sqrt{5}}$. Similarly, to divide $\dfrac{3}{2i}$, we multiply by $\dfrac{i}{i}$.

To rationalize the denominator of $\dfrac{4}{2 + \sqrt{3}}$, we multiply by $\dfrac{2 - \sqrt{3}}{2 - \sqrt{3}}$, where $2 - \sqrt{3}$ is the conjugate of $2 + \sqrt{3}$. Similarly, to divide $\dfrac{4}{2 + i}$, we multiply by $\dfrac{2 - i}{2 - i}$, where $2 - i$ is the complex conjugate of $2 + i$.

Example 4 Divide. Write in standard form.

a. $\dfrac{9}{2i}$

SOLUTION Because $i^2 = -1$, multiplying by $\dfrac{i}{i}$ eliminates i from the denominator.

Note: We could have thought of $2i$ as $0 + 2i$; so the complex conjugate is $0 - 2i = -2i$ and is multiplied by $\dfrac{-2i}{-2i}$.

$$\frac{9}{2i} = \frac{9}{2i} \cdot \frac{-2i}{-2i} = \frac{-18i}{-4i^2}$$

$$= \frac{-18i}{-4(-1)} = \frac{-18i}{4}$$

$$= -\frac{9}{2}i$$

$$\frac{9}{2i} = \frac{9}{2i} \cdot \frac{i}{i} \qquad \text{Multiply the numerator and denominator by } i.$$

$$= \frac{9i}{2i^2} \qquad \text{Multiply.}$$

$$= \frac{9i}{2(-1)} \qquad \text{Replace } i^2 \text{ with } -1.$$

$$= \frac{9i}{-2}$$

$$= -\frac{9}{2}i$$

b. $\dfrac{6 + 5i}{7 - 2i}$

SOLUTION
$$\frac{6 + 5i}{7 - 2i} = \frac{6 + 5i}{7 - 2i} \cdot \frac{7 + 2i}{7 + 2i} \qquad \begin{array}{l}\text{Multiply numerator and denominator by} \\ \text{the complex conjugate of the denominator,} \\ \text{which is } 7 + 2i.\end{array}$$

$$= \frac{42 + 12i + 35i + 10i^2}{49 - 4i^2} \qquad \text{Multiply.}$$

$$= \frac{42 + 47i + 10(-1)}{49 - 4(-1)} \qquad \text{Simplify.}$$

$$= \frac{42 + 47i - 10}{49 + 4}$$

$$= \frac{32 + 47i}{53}$$

$$= \frac{32}{53} + \frac{47}{53}i \qquad \text{Write in standard form.}$$

63. $\dfrac{4 + 2i}{4i}$ **64.** $\dfrac{6 - 2i}{6i}$ **65.** $\dfrac{7}{2 + i}$ **66.** $\dfrac{5}{6 + i}$

67. $\dfrac{2i}{3 - 7i}$ **68.** $\dfrac{4i}{5 - 3i}$ **69.** $\dfrac{5 - 9i}{1 - i}$ **70.** $\dfrac{3 + i}{2 - i}$

71. $\dfrac{3 + i}{2 + 3i}$ **72.** $\dfrac{1 + 3i}{5 + 2i}$ **73.** $\dfrac{1 + 6i}{4 + 5i}$ **74.** $\dfrac{5 - 6i}{2 - 9i}$

For Exercises 75–90, find the powers of i. See Example 5.

75. i^{19} **76.** i^{25} **77.** i^{42} **78.** i^{51}

79. i^{38} **80.** i^{45} **81.** i^{60} **82.** i^{52}

83. i^{-20} **84.** i^{-32} **85.** i^{-30} **86.** i^{-38}

87. i^{-21} **88.** i^{-45} **89.** i^{-35} **90.** i^{-44}

Review Exercises

Exercises 1 and 2 **EXPRESSIONS**

[6.4]   **1.** Factor: $4x^2 - 12x + 9$

[10.3]   **2.** Simplify: $\pm\sqrt{48}$

Exercises 3–6 **EQUATIONS AND INEQUALITIES**

For Exercises 3 and 4, solve.

[2.3]   **3.** $3x - 2 = 4$

[6.6]   **4.** $2x^2 - x - 6 = 0$

[4.3]   **5.** Find the x- and y-intercepts for $2x + 3y = 6$.

[6.7]   **6.** Graph: $y = x^2 - 3$

Chapter 10 Summary

Defined Terms

Section 10.1
 *n*th root (p. 704)
 Radical function (p. 707)

Section 10.2
 Rational exponent (p. 714)

Section 10.4
 Like radicals (p. 732)

Section 10.6
 Radical equation (p. 747)

Section 10.7
 Imaginary unit (p. 756)
 Imaginary number (p. 756)
 Complex number (p. 757)
 Complex conjugate (p. 758)

Procedures, Rules, and Key Examples

Procedures/Rules	Key Examples

SECTION 10.1   Radical Expressions and Functions

If a is nonnegative, then $\sqrt[n]{a} = b$, where $b \geq 0$ and $b^n = a$.
If a is negative and n is even, then there is no real-number root.
If a is negative and n is odd, then $\sqrt[n]{a} = b$, where b is negative and $b^n = a$.

Example 1: Find each root, if possible.

a. $\sqrt{36} = 6$
b. $\sqrt{-25}$ is not a real number.
c. $-\sqrt{121} = -11$
d. $\sqrt{\dfrac{49}{100}} = \dfrac{7}{10}$
e. $\sqrt{16x^6} = 4|x^3|$ ◀
f. $\sqrt[3]{-27} = -3$

> **Note:** If we assume that the variables represent nonnegative numbers, the solution for part e is $\sqrt{16x^6} = 4x^3$.

To evaluate a radical function, replace the variable with the indicated value and simplify.

Example 2: If $f(x) = \sqrt{4x + 3}$, find $f(3)$.

Solution:

$$f(3) = \sqrt{4(3) + 3} = \sqrt{12 + 3} = \sqrt{15}$$

Finding the domain of a radical function: If the root index is odd, the domain is $\{x \mid x \text{ is a real number}\}$ or $(-\infty, \infty)$. If the root index is even, the radicand must be nonnegative. Consequently, the domain is the solution set to the inequality "radicand ≥ 0."

Example 3: Find the domain of

$$f(x) = \sqrt{2x - 4}.$$

Solution: Because $2x - 4$ must be nonnegative, we solve $2x - 4 \geq 0$.

$$2x - 4 \geq 0$$
$$2x \geq 4$$
$$x \geq 2$$

Domain: $\{x \mid x \geq 2\}$ or $[2, \infty)$

SECTION 10.2   Rational Exponents

$a^{1/n} = \sqrt[n]{a}$, where n is a natural number other than 1.

Note: If a is negative and n is odd, the root is negative. If a is negative and n is even, there is no real-number root.

Example 1: Evaluate.

a. $25^{1/2} = \sqrt{25} = 5$
b. $(-27)^{1/3} = \sqrt[3]{-27} = -3$
c. $(-16)^{1/4} = \sqrt[4]{-16}$, which is not a real number.

$a^{m/n} = \sqrt[n]{a^m} = (\sqrt[n]{a})^m$, where $a \geq 0$ and m and n are natural numbers other than 1.

Example 2: Simplify.

$$27^{4/3} = (\sqrt[3]{27})^4 = 3^4 = 81$$

Example 3: Write in exponential form.

a. $\sqrt[3]{x^2} = x^{2/3}$ **b.** $(\sqrt[6]{x})^5 = x^{5/6}$

(continued)

Procedures/Rules	Key Examples

SECTION 10.2 Rational Exponents (*continued*)

$a^{-m/n} = \dfrac{1}{a^{m/n}}$, where $a \neq 0$ and m and n are natural numbers with $n \neq 1$.

Example 4: Evaluate.

$$16^{-3/4} = \frac{1}{16^{3/4}} = \frac{1}{(\sqrt[4]{16})^3} = \frac{1}{2^3} = \frac{1}{8}$$

For Examples 5–7, assume that variables represent positive values.

Example 5: Write $\sqrt[8]{x^6}$ as a radical expression with a smaller root index.

Solution: $\sqrt[8]{x^6} = x^{6/8} = x^{3/4} = \sqrt[4]{x^3}$

To multiply or divide radical expressions with different root indices,
1. Change from radical to exponential form.
2. Multiply or divide using the appropriate rule(s) of exponents.
3. Write the result from Step 2 in radical form.

Example 6: Simplify.

a. $\sqrt[4]{x} \cdot \sqrt{x} = x^{1/4} \cdot x^{1/2} = x^{1/4+1/2}$
$\qquad\qquad\qquad\quad = x^{1/4+2/4}$
$\qquad\qquad\qquad\quad = x^{3/4} = \sqrt[4]{x^3}$

b. $\dfrac{\sqrt[3]{x^2}}{\sqrt[4]{x}} = \dfrac{x^{2/3}}{x^{1/4}} = x^{(2/3)-(1/4)}$
$\qquad\qquad = x^{(8/12)-(3/12)} = x^{5/12} = \sqrt[12]{x^5}$

Example 7: Write $\sqrt[3]{\sqrt[5]{x}}$ as a single radical.

Solution: $\sqrt[3]{\sqrt[5]{x}} = (x^{1/5})^{1/3} = x^{(1/5)\cdot(1/3)}$
$\qquad\qquad\qquad = x^{1/15} = \sqrt[15]{x}$

SECTION 10.3 Multiplying, Dividing, and Simplifying Radicals

Product rule for radicals:

If both $\sqrt[n]{a}$ and $\sqrt[n]{b}$ are real numbers, then

$$\sqrt[n]{a} \cdot \sqrt[n]{b} = \sqrt[n]{ab}.$$

Example 1: Simplify. Assume that variables represent non-negative values.

a. $\sqrt{2} \cdot \sqrt{50} = \sqrt{2 \cdot 50} = \sqrt{100} = 10$
b. $\sqrt{3x} \cdot \sqrt{12x^3} = \sqrt{3x \cdot 12x^3} = \sqrt{36x^4} = 6x^2$
c. $\sqrt[3]{5x} \cdot \sqrt[3]{3x} = \sqrt[3]{15x^2}$

For Examples 2–4, assume that variables represent positive values.

Example 2: Simplify $(\sqrt[4]{x^3})^4$.

Solution: $(\sqrt[4]{x^3})^4 = x^3$

Raising an *n*th root to the *n*th power:

For any nonnegative real number a, $(\sqrt[n]{a})^n = a$.

Quotient rule for radicals:

If both $\sqrt[n]{a}$ and $\sqrt[n]{b}$ are real numbers, then $\dfrac{\sqrt[n]{a}}{\sqrt[n]{b}} = \sqrt[n]{\dfrac{a}{b}}$, where $b \neq 0$.

Example 3: Simplify.

a. $\dfrac{\sqrt{45}}{\sqrt{5}} = \sqrt{\dfrac{45}{5}} = \sqrt{9} = 3$

b. $\dfrac{\sqrt[3]{128x^7}}{\sqrt[3]{2x}} = \sqrt[3]{\dfrac{128x^7}{2x}} = \sqrt[3]{64x^6} = 4x^2$

c. $\sqrt[4]{\dfrac{8}{x^4}} = \dfrac{\sqrt[4]{8}}{\sqrt[4]{x^4}} = \dfrac{\sqrt[4]{8}}{x}$

To simplify an *n*th root.
1. Write the radicand as the product of the greatest possible perfect *n*th power and a number or an expression that has no perfect *n*th power factors.
2. Use the product rule $\sqrt[n]{ab} = \sqrt[n]{a} \cdot \sqrt[n]{b}$, where a is the perfect *n*th power.
3. Find the *n*th root of the perfect *n*th power radicand.

Example 4: Simplify.

a. $\sqrt{50} = \sqrt{25 \cdot 2} = \sqrt{25} \cdot \sqrt{2} = 5\sqrt{2}$
b. $\sqrt{48x^3} = \sqrt{16x^2 \cdot 3x} = \sqrt{16x^2} \cdot \sqrt{3x} = 4x\sqrt{3x}$
c. $\sqrt[4]{48a^9} = \sqrt[4]{16a^8 \cdot 3a} = \sqrt[4]{16a^8} \cdot \sqrt[4]{3a} = 2a^2\sqrt[4]{3a}$

Example 5: Perform the operation and simplify.

$\sqrt{6} \cdot \sqrt{10} = \sqrt{60} = \sqrt{4 \cdot 15} = \sqrt{4} \cdot \sqrt{15} = 2\sqrt{15}$

(*continued*)

Procedures/Rules	Key Examples

SECTION 10.4   Adding, Subtracting, and Multiplying Radical Expressions

To add or subtract like radicals, add or subtract the coefficients and leave the radical parts the same.

Example 1: Add or subtract. Assume that variables represent nonnegative values.

a. $2\sqrt{x} - 7\sqrt{x} = (2 - 7)\sqrt{x} = -5\sqrt{x}$

b. $4\sqrt[3]{5} + 2\sqrt[3]{5} = (4 + 2)\sqrt[3]{5} = 6\sqrt[3]{5}$

Example 2: Simplify the radicals; then combine like radicals.

$$9\sqrt{3} - \sqrt{12} + 6\sqrt{75} = 9\sqrt{3} - \sqrt{4 \cdot 3} + 6\sqrt{25 \cdot 3}$$
$$= 9\sqrt{3} - 2\sqrt{3} + 6 \cdot 5\sqrt{3}$$
$$= 9\sqrt{3} - 2\sqrt{3} + 30\sqrt{3}$$
$$= 37\sqrt{3}$$

Example 3: Find the product.

$$(\sqrt{3} - 4)(\sqrt{3} + 6) = \sqrt{3} \cdot \sqrt{3} + \sqrt{3} \cdot 6 - 4 \cdot \sqrt{3} - 4 \cdot 6$$

Use FOIL.

$$= \sqrt{9} + 6\sqrt{3} - 4\sqrt{3} - 24 \qquad \textbf{Multiply.}$$
$$= 3 + 6\sqrt{3} - 4\sqrt{3} - 24 \qquad \textbf{Simplify } \sqrt{9}.$$
$$= -21 + 2\sqrt{3} \qquad \textbf{Combine like radicals.}$$

Example 4: Simplify.

a.
$$\sqrt{6} \cdot \sqrt{8} + \sqrt{5} \cdot \sqrt{15} = \sqrt{48} + \sqrt{75}$$
$$= \sqrt{16 \cdot 3} + \sqrt{25 \cdot 3}$$
$$= 4\sqrt{3} + 5\sqrt{3} = 9\sqrt{3}$$

b.
$$\frac{\sqrt{60}}{\sqrt{5}} + \sqrt{48} = \sqrt{\frac{60}{5}} + \sqrt{48}$$
$$= \sqrt{12} + \sqrt{48}$$
$$= \sqrt{4 \cdot 3} + \sqrt{16 \cdot 3}$$
$$= 2\sqrt{3} + 4\sqrt{3} = 6\sqrt{3}$$

SECTION 10.5   Rationalizing Numerators and Denominators of Radical Expressions

A radical expression is in simplest form if
1. The radicand has no factor raised to a power greater than or equal to the index.
2. There are neither radicals in the denominator of a fraction nor radicands that are fractions.
3. All possible sums, differences, products, and quotients have been found.

To rationalize a denominator containing a single *n*th root, multiply the fraction by a 1 so that the product's denominator has a radicand that is a perfect *n*th power.

To rationalize a denominator containing a sum or difference with at least one square root term, multiply the fraction by a 1 whose numerator and denominator are the conjugate of the denominator.

The numerator of a rational expression is rationalized the same way a denominator is rationalized.

Example 1: Rationalize the denominator.

a. $\dfrac{7}{\sqrt{3}} = \dfrac{7}{\sqrt{3}} \cdot \dfrac{\sqrt{3}}{\sqrt{3}} = \dfrac{7\sqrt{3}}{3}$

b. $\dfrac{4}{\sqrt[3]{3}} = \dfrac{4}{\sqrt[3]{3}} \cdot \dfrac{\sqrt[3]{9}}{\sqrt[3]{9}} = \dfrac{4\sqrt[3]{9}}{\sqrt[3]{27}} = \dfrac{4\sqrt[3]{9}}{3}$

c. $\dfrac{6}{4 - \sqrt{5}} = \dfrac{6}{4 - \sqrt{5}} \cdot \dfrac{4 + \sqrt{5}}{4 + \sqrt{5}}$

$$= \dfrac{6 \cdot 4 + 6 \cdot \sqrt{5}}{16 - 5}$$
$$= \dfrac{24 + 6\sqrt{5}}{11}$$

Example 2: Rationalize the numerator.

a. $\dfrac{\sqrt{6}}{3} = \dfrac{\sqrt{6}}{3} \cdot \dfrac{\sqrt{6}}{\sqrt{6}} = \dfrac{\sqrt{36}}{3\sqrt{6}} = \dfrac{6}{3\sqrt{6}} = \dfrac{2}{\sqrt{6}}$

b. $\dfrac{2 - \sqrt{3}}{5} = \dfrac{2 - \sqrt{3}}{5} \cdot \dfrac{2 + \sqrt{3}}{2 + \sqrt{3}}$

$$= \dfrac{4 - 3}{5(2 + \sqrt{3})} = \dfrac{1}{10 + 5\sqrt{3}}$$

(continued)

Procedures/Rules	Key Examples

SECTION 10.6 Radical Equations and Problem Solving

Power Rule for Solving Equations: If both sides of an equation are raised to the same integer power, the resulting equation contains all solutions of the original equation and perhaps some solutions that do not solve the original equation. That is, the solutions of the equation $a = b$ are contained among the solutions of $a^n = b^n$, where n is an integer.

To solve a radical equation,
1. Isolate the radical if necessary. (If there is more than one radical term, isolate one of the radical terms.)
2. Raise both sides of the equation to the same power as the root index of the isolated radical.
3. If all radicals have been eliminated, solve. If a radical term remains, isolate that radical term and raise both sides to the same power as its root index.
4. Check each solution. Any apparent solution that does not check is an extraneous solution.

Example 1: Solve $\sqrt{x - 5} = 7$.

$$(\sqrt{x - 5})^2 = 7^2 \quad \text{Square both sides.}$$
$$x - 5 = 49 \quad \text{Multiply.}$$
$$x = 54 \quad \text{Add 5 to both sides.}$$

Check: $\sqrt{54 - 5} = 7$
$\sqrt{49} = 7 \quad$ **True**

Example 2: Solve $\sqrt{n + 14} = n + 2$.

$$(\sqrt{n + 14})^2 = (n + 2)^2 \quad \text{Square both sides.}$$
$$n + 14 = n^2 + 4n + 4$$
$$0 = n^2 + 3n - 10 \quad \text{Subtract } n \text{ and 14 from both sides.}$$
$$0 = (n + 5)(n - 2) \quad \text{Factor.}$$
$$n + 5 = 0 \quad \text{or} \quad n - 2 = 0 \quad \text{Use the zero-factor theorem.}$$
$$n = -5 \qquad\qquad n = 2$$

Check: $\sqrt{-5 + 14} = -5 + 2 \qquad\qquad \sqrt{2 + 14} = 2 + 2$
$\sqrt{9} = -3 \quad$ **False** $\qquad \sqrt{16} = 4 \quad$ **True**

The only solution is 2. (-5 is extraneous.)

Example 3: Solve $3 + \sqrt{t} = \sqrt{t + 21}$.

$$(3 + \sqrt{t})^2 = (\sqrt{t + 21})^2 \quad \text{Square both sides.}$$
$$9 + 6\sqrt{t} + t = t + 21 \quad \text{Multiply.}$$
$$6\sqrt{t} = 12 \quad \text{Subtract 9 and } t \text{ from both sides.}$$
$$\sqrt{t} = 2 \quad \text{Divide both sides by 6.}$$
$$(\sqrt{t})^2 = 2^2 \quad \text{Square both sides.}$$
$$t = 4$$

We will leave the check to the reader.

SECTION 10.7 Complex Numbers

To write an imaginary number $\sqrt{-n}$ in terms of the imaginary unit i,
1. Separate the radical into two factors $\sqrt{-1} \cdot \sqrt{n}$.
2. Replace $\sqrt{-1}$ with i.
3. Simplify $\sqrt{n}$.

To add complex numbers, combine like terms.

To subtract complex numbers, write the equivalent addition and change the signs of the second complex number; then combine like terms.

To multiply complex numbers, follow the same procedures for multiplying monomials or binomials (FOIL). Remember that $i^2 = -1$.

Example 1: Write using the imaginary unit.

a. $\sqrt{-36} = \sqrt{-1} \cdot \sqrt{36} = 6i$
b. $\sqrt{-32} = \sqrt{-1} \cdot \sqrt{32}$
$\quad = i \cdot 4\sqrt{2} = 4i\sqrt{2}$

Example 2: Add $(5 - 6i) + (9 + 2i)$.

$$(5 - 6i) + (9 + 2i) = 14 - 4i$$

Example 3: Subtract $(7 - i) - (3 + 5i)$.

$$(7 - i) - (3 + 5i) = (7 - i) + (-3 - 5i)$$
$$= 4 - 6i$$

Example 4: Multiply $(6 + 5i)(2 - 3i)$.

$$(6 + 5i)(2 - 3i) = 12 - 18i + 10i - 15i^2$$
$$= 12 - 8i - 15(-1)$$
$$= 12 - 8i + 15$$
$$= 27 - 8i$$

(continued)

Procedures/Rules	Key Examples

SECTION 10.7   Complex Numbers (*continued*)

To divide by an imaginary number, multiply by $\dfrac{i}{i}$.

To divide complex numbers, rationalize the denominator using the complex conjugate.

Example 5: Divide and write $\dfrac{2-3i}{4+5i}$ in standard form.

$$\frac{2-3i}{4+5i} = \frac{2-3i}{4+5i} \cdot \frac{4-5i}{4-5i}$$

$$= \frac{8-10i-12i+15i^2}{16-25i^2}$$

$$= \frac{8-22i+15(-1)}{16-25(-1)}$$

$$= \frac{-7-22i}{41}$$

$$= -\frac{7}{41} - \frac{22}{41}i$$

Powers of i:

$$i^1 = i$$
$$i^2 = -1$$
$$i^3 = -i$$
$$i^4 = 1$$

Example 6: Find the powers of i.

a. $i^{19} = i^{16} \cdot i^3 = (i^4)^4 \cdot (-i)$
$$= 1^4(-i) = 1(-i) = -i$$

b. $i^{-21} = \dfrac{1}{i^{21}} = \dfrac{1}{i^{20} \cdot i} = \dfrac{1}{(i^4)^5 \cdot i} = \dfrac{1}{1^5 \cdot i}$

$$= \frac{1}{i} = \frac{1}{i} \cdot \frac{i}{i} = \frac{i}{i^2}$$

$$= \frac{i}{-1} = -i$$

Chapter 10 Review Exercises

For Exercises 1–5, answer true or false.

[10.1]   **1.** Every positive number has two square roots: a positive root and a negative root.

[10.1]   **2.** $\sqrt{x^2} = x$, where x is any real number.

[10.5]   **3.** The expression $\dfrac{3}{\sqrt{2}}$ is in simplest form.

[10.6]   **4.** Every radical equation has a single solution.

[10.6]   **5.** It is necessary to check all potential solutions to radical equations.

For Exercises 6–10, complete the rule.

[10.4]   **6.** Like radicals are two radical expressions with identical _____ and root indices.

[10.4]   **7.** To add or subtract like radicals, add or subtract the _____ and leave the radical parts the same.

[10.5]   **8.** To rationalize a denominator that is a single square root,
   a. _____ both the numerator and denominator of the fraction by the same square root so that the radicand in the denominator is a perfect square.
   b. Simplify.

[10.5] 9. To rationalize a denominator containing a sum or difference with a square root term, multiply the numerator and denominator by the _____ of the denominator.

[10.6] 10. An apparent solution to a radical equation that does not solve the equation is considered _____.

Exercises 11–96 ◢ EXPRESSIONS

[10.1] *For Exercises 11 and 12, find all square roots of the given number.*

11. 121

12. 49

[10.1] *For Exercises 13–16, evaluate the square root.*

13. $\sqrt{169}$

14. $-\sqrt{49}$

15. $\sqrt{-36}$

16. $\sqrt{\dfrac{1}{25}}$

[10.1] *For Exercises 17 and 18, use a calculator to approximate each root to the nearest thousandth.*

17. $\sqrt{7}$

18. $\sqrt{90}$

[10.1] *For Exercises 19–26, simplify. Assume that variables represent nonnegative values.*

19. $\sqrt{49x^8}$

20. $\sqrt{144a^6b^{12}}$

21. $\sqrt{0.16m^2n^{10}}$

22. $\sqrt[3]{x^{15}}$

23. $\sqrt[3]{-64r^9s^3}$

24. $\sqrt[4]{81x^{12}}$

25. $\sqrt[5]{32x^{15}y^{20}}$

26. $\sqrt[7]{x^{14}y^7}$

[10.1] *For Exercises 27 and 28, simplify. Assume that variables represent any real number.*

27. $\sqrt{81x^2}$

28. $\sqrt[4]{(x-1)^8}$

[10.2] *For Exercises 29–32, rewrite using radicals; then simplify if possible. Assume that variables represent nonnegative values.*

29. $(-64)^{1/3}$

30. $(24a^4)^{1/2}$

31. $\left(\dfrac{1}{32}\right)^{3/5}$

32. $(5r-2)^{5/7}$

[10.2] *For Exercises 33–38, write in exponential form.*

33. $\sqrt[8]{33}$

34. $\dfrac{8}{\sqrt[7]{n^3}}$

35. $(\sqrt[5]{8})^3$

36. $(\sqrt[3]{m})^8$

37. $(\sqrt[4]{3xw})^3$

38. $\sqrt[3]{(a+b)^4}$

[10.2] *For Exercises 39–44, use the rules of exponents to simplify. Assume that variables represent positive values.*

39. $x^{2/3} \cdot x^{4/3}$

40. $(4m^{1/4})(8m^{5/4})$

41. $\dfrac{y^{3/5}}{y^{4/5}}$

42. $\dfrac{b^{2/5}}{b^{-3/5}}$

43. $(k^{2/3})^{3/4}$

44. $(2xy^{1/5})^{3/4}$

[10.3] *For Exercises 45–56, simplify. Assume that variables represent nonnegative values.*

45. $\sqrt{3} \cdot \sqrt{27}$

46. $\sqrt{5x^5} \cdot \sqrt{20x^3}$

47. $\sqrt[3]{2} \cdot \sqrt[3]{4}$

48. $\sqrt[4]{7} \cdot \sqrt[4]{6}$

49. $\sqrt[5]{3x^2y^3} \cdot \sqrt[5]{5x^2y}$

50. $\sqrt{\dfrac{49}{121}}$

51. $\sqrt[3]{-\dfrac{27}{8}}$

52. $\sqrt{72}$

53. $4b\sqrt{27b^7}$

54. $5\sqrt[3]{108}$

55. $2\sqrt[3]{40x^{10}}$

56. $2x^4\sqrt[4]{162x^7}$

[10.3] *For Exercises 57–60, find the product and write the answer in simplified form. Assume that variables represent nonnegative values.*

57. $4\sqrt{6} \cdot 7\sqrt{15}$

58. $\sqrt{x^9} \cdot \sqrt{x^6}$

59. $4\sqrt{10c} \cdot 2\sqrt{6c^4}$

60. $a\sqrt{a^3b^2} \cdot b^2\sqrt{a^5b^3}$

[10.3] *For Exercises 61–64, find the quotient and write the answer in simplified form. Assume that variables represent positive values.*

61. $\dfrac{\sqrt{48}}{\sqrt{6}}$

62. $\dfrac{9\sqrt{160}}{3\sqrt{8}}$

63. $\dfrac{8\sqrt{45a^5}}{2\sqrt{5a}}$

64. $\dfrac{36\sqrt{96x^6y^{11}}}{4\sqrt{3x^2y^4}}$

[10.4] *For Exercises 65–72, add or subtract. Assume that variables represent nonnegative values.*

65. $-5\sqrt{n} + 2\sqrt{n}$

66. $3y^3\sqrt[4]{8y} - 9y^3\sqrt[4]{8y}$

67. $\sqrt{45} + \sqrt{20}$

68. $4\sqrt{24} - 6\sqrt{54}$

69. $\sqrt{150} - \sqrt{54} + \sqrt{24}$

70. $4\sqrt{72x^2y} - 2x\sqrt{128y} + 5\sqrt{32x^2y}$

71. $\sqrt[3]{250x^4y^5} + \sqrt[3]{128x^4y^5}$

72. $\sqrt[4]{48} + \sqrt[4]{243}$

[10.4] *For Exercises 73–82, find the product. Assume that variables represent nonnegative values.*

73. $\sqrt{5}(\sqrt{3} + \sqrt{2})$

74. $\sqrt[3]{7}(\sqrt[3]{3} + 2\sqrt[3]{7})$

75. $3\sqrt{6}(2 - 3\sqrt{6})$

76. $(\sqrt{2} - \sqrt{3})(\sqrt{5} + \sqrt{7})$

77. $(\sqrt[3]{2} - 4)(\sqrt[3]{4} + 2)$

78. $(\sqrt[4]{6x} + 2)(\sqrt[4]{2x} - 1)$

79. $(\sqrt{5a} + \sqrt{3b})(\sqrt{5a} - \sqrt{3b})$

80. $(2\sqrt{3} - \sqrt{5})(2\sqrt{3} + \sqrt{5})$

81. $(\sqrt{2} + 1)^2$

82. $(\sqrt{2} - \sqrt{5})^2$

[10.5] *For Exercises 83–92, rationalize the denominator and simplify. Assume that variables represent positive values.*

83. $\dfrac{1}{\sqrt{2}}$

84. $\dfrac{3}{\sqrt[3]{3}}$

85. $\sqrt{\dfrac{4}{7}}$

86. $\dfrac{\sqrt[4]{5x^2}}{\sqrt[4]{2}}$

87. $\sqrt[3]{\dfrac{17}{3y^2}}$

88. $\dfrac{\sqrt[4]{9}}{\sqrt[4]{27}}$

89. $\dfrac{4}{\sqrt{2} - \sqrt{3}}$

90. $\dfrac{1}{4 + \sqrt{3}}$

91. $\dfrac{1}{2 - \sqrt{n}}$

92. $\dfrac{2\sqrt{3}}{3\sqrt{2} - 2\sqrt{3}}$

[10.5] *For Exercises 93–96, rationalize the numerator. Assume that variables represent positive values.*

93. $\dfrac{\sqrt{10}}{6}$
94. $\dfrac{\sqrt{3x}}{5}$
95. $\dfrac{2 - \sqrt{3}}{8}$
★ **96.** $\dfrac{2\sqrt{t} + \sqrt{3t}}{5t}$

Exercises 97–108 EQUATIONS AND INEQUALITIES

[10.6] *For Exercises 97–108, solve. Identify any extraneous solutions.*

97. $\sqrt{x} = 9$
98. $\sqrt{y} = -3$
99. $\sqrt{w - 1} = 3$
100. $\sqrt[3]{3x - 2} = -2$

101. $\sqrt[4]{x - 2} - 3 = -1$
102. $\sqrt{y + 1} = \sqrt{2y - 4}$
103. $\sqrt{x - 6} = x + 2$
104. $\sqrt[3]{3x + 10} - 4 = 5$

105. $\sqrt[4]{x + 8} = \sqrt[4]{2x + 1}$
106. $\sqrt{5n - 1} = 4 - 2n$
107. $1 + \sqrt{x} = \sqrt{2x + 1}$
108. $\sqrt{3x + 1} = 2 - \sqrt{3x}$

Exercises 109–125 EXPRESSIONS

[10.7] *For Exercises 109 and 110, write the imaginary number using i.*

109. $\sqrt{-9}$
110. $\sqrt{-20}$

[10.7] *For Exercises 111 and 112, add or subtract.*

111. $(3 + 2i) + (5 - 8i)$
112. $(7 - 3i) - (-2 + 4i)$

[10.7] *For Exercises 113–116, multiply and simplify.*

113. $(3i)(4i)$
114. $2i(4 - i)$
115. $(6 + 2i)(4 - i)$
116. $(5 - i)^2$

[10.7] *For Exercises 117–122, divide and write in standard form.*

117. $\dfrac{5}{i}$
118. $\dfrac{3}{-i}$
119. $\dfrac{4}{3i}$

120. $\dfrac{7 + i}{5i}$
121. $\dfrac{3}{2 + i}$
122. $\dfrac{5 + i}{2 - 3i}$

[10.7] *For Exercises 123 and 124, find the powers of i.*

123. i^{20}
124. i^{15}

[10.3] **125.** Write an expression in simplest form for the area of the figure.

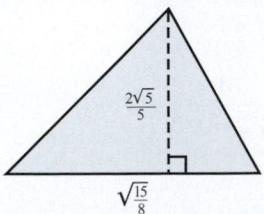

$$\frac{2\sqrt{5}}{5}$$

$$\sqrt{\frac{15}{8}}$$

For Exercises 126 and 127, use the following information. The speed of a car can be determined by the length of the skid marks using the formula $S = 2\sqrt{2L}$, where L represents the length of the skid mark in feet and S represents the speed of the car in miles per hour. (Source: Harris Technical Services, Traffic Accident Reconstructionists)

[10.1] **126.** **a.** Write an expression for the exact speed of a car if the length of the skid marks measures 40 feet.

         b. Approximate the speed to the nearest tenth.

[10.6] **127.** Find the length of skid marks if a driver brakes hard at a speed of 50 miles per hour.

For Exercises 128 and 129, use the formula $T = 2\pi\sqrt{\dfrac{L}{9.8}}$, where T is the period of a pendulum in seconds and L is the length of the pendulum in meters.

[10.1] **128.** Find the period of a pendulum with a length of 2.45 meters.

[10.6] **129.** Suppose the period of the pendulum is $\dfrac{\pi}{3}$ seconds. Find the length to the nearest thousandth.

For Exercises 130 and 131, use the formula $t = \sqrt{\dfrac{h}{16}}$, where t is the time in seconds it takes on object to fall a distance of h feet.

[10.1] **130.** **a.** Write an expression in simplest form of the time an object takes to fall 40 feet.

         b. Approximate the time to the nearest hundredth.

[10.6] **131.** Find the distance an object falls in 0.3 second.

[10.1] **132.** Three pieces of lumber are to be connected to form a right triangle that will be part of the roof frame for a small storage building. If the horizontal piece is to be 4 feet and the vertical piece is to be 3 feet, how long must the connecting piece be?

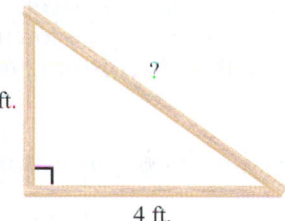
3 ft.   ?
4 ft.

Chapter 10   Practice Test

For Extra Help

CHAPTER
Test Prep
VIDEOS

Step-by-step test solutions are found on the Chapter Test Prep Videos available via the Video Resources on DVD, in **MyMathLab**, and on **You Tube** (search "CarsonCombinedAlg" and click on "Channels").

For Exercises 1 and 2, evaluate the square root.

1. $\sqrt{36}$

2. $\sqrt{-49}$

For Exercises 3–12, simplify. Assume that variables represent positive values.

3. $\sqrt{81x^2y^5}$

4. $\sqrt[3]{54}$

5. $\sqrt[4]{4x} \cdot \sqrt[4]{4x^5}$

6. $-\sqrt[3]{-27r^{15}}$

7. $\dfrac{\sqrt{5}}{\sqrt{45}}$

8. $\dfrac{\sqrt[4]{1}}{\sqrt[4]{81}}$

9. $6\sqrt{7} - \sqrt{7}$

10. $(\sqrt{3} - 1)^2$

11. $x^{2/3} \cdot x^{-4/3}$

12. $(\sqrt[3]{2} - 4)(\sqrt[3]{4} + 2)$

For Exercises 13 and 14, write in exponential form. Assume that variables represent nonnegative values.

13. $\sqrt[5]{8x^3}$

14. $\sqrt[3]{(2x + 5)^2}$

For Exercises 15 and 16, rationalize the denominator and simplify. Assume that variables represent positive values.

15. $\dfrac{1}{\sqrt[3]{4}}$

16. $\dfrac{\sqrt{x}}{\sqrt{x} + \sqrt{y}}$

For Exercises 17–19, solve the equation. Identify any extraneous solutions.

17. $\sqrt{3x - 2} = 8$

18. $\sqrt[4]{x + 8} = \sqrt[4]{2x + 1}$

19. $\sqrt{2x + 1} - \sqrt{x + 1} = 2$

For Exercises 20–22, simplify and write the answer in standard form $(a + bi)$.

20. $(2 - i) - (4 + 3i)$

21. $(4 - i)(4 + i)$

22. $\dfrac{2}{4 - 3i}$

23. Write an expression in simplest form for the area of the figure.

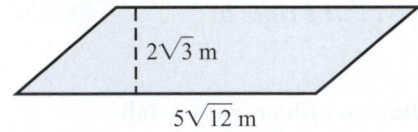

24. The formula $t = \sqrt{\dfrac{h}{16}}$ describes the amount of time t, in seconds, that an object falls a distance of h feet.
  a. Write an expression in simplest form for the exact amount of time an object falls a distance of 12 feet.

  b. Find the distance an object falls in 2 seconds.

25. The formula $S = \dfrac{7}{4}\sqrt{D}$ can be used to approximate the speed, S, in miles per hour, that a car was traveling prior to braking and skidding a distance D, in feet, on ice. (*Source:* Harris Technical Services, Traffic Accident Reconstructionists)
  a. Find the exact speed of a car if the skid length measures 40 feet long.

  b. Find the length of the skid marks that a car traveling 30 miles per hour makes if it brakes hard and skids to a halt.

Chapters 1–10 Cumulative Review Exercises

For Exercises 1–4, answer true or false

[8.3] **1.** If $|x| > 3$, then $-3 < x < 3$.

[9.6] **2.** The determinant of $\begin{bmatrix} 2 & -3 \\ 1 & 4 \end{bmatrix}$ is 11.

[10.4] **3.** The conjugate of $4 - 2\sqrt{3}$ is $4 + 2\sqrt{3}$.

[10.7] **4.** 3 is a complex number.

For Exercises 5–7, fill in the blanks.

[4.3] **5.** Given an equation of the form $ax + by = c$, to find the _____, let $y = 0$.

[9.2] **6.** A system of equations that has no solution is _____.

[7.4] **7.** The LCD for $\dfrac{5}{8x^3y^2}$ and $\dfrac{3}{12x^5y}$ is _____.

Exercises 8–18 EXPRESSIONS

[1.5] **8.** Simplify $4 - 6(5 - 3^2) + 12 \div 2 - 6$.

[5.6] **9.** Simplify $\dfrac{7.44 \times 10^{-2}}{3.1 \times 10^4}$. Write the answer in standard form.

For Exercises 10–17, simplify. Write the answers with positive exponents only.

[5.6] **10.** $\dfrac{(x^{-2})^3(x^3)^4}{(x^{-3})^3}$

[5.5] **11.** $(3x^2 - 4y)^2$

[5.6] **12.** $\dfrac{8x^3 - 22x + 8}{2x - 3}$

[7.2] **13.** $\dfrac{2x^2 + 7x + 3}{x^2 - 9} \div \dfrac{2x^2 + 11x + 5}{x^2 - 3x}$

[10.4] **14.** $2\sqrt{5a^2} \cdot 3\sqrt{10a^5}$

[10.4] **15.** $3x\sqrt[3]{24x^4} - 4x^2\sqrt[3]{81x}$

[10.2] **16.** $32^{-3/5}$

[10.7] **17.** $(4 + 5i)(2 - 3i)$

[10.5] **18.** Rationalize the denominator of $\dfrac{2 + \sqrt{2}}{4 - \sqrt{2}}$.

Exercises 19–30 EQUATIONS AND INEQUALITIES

For Exercises 19–22, solve. Identify any extraneous solutions.

[2.4] **19.** $A = \dfrac{1}{2}h(B + b)$ for b

[6.6] **20.** $(x + 2)(x + 4) = 63$

[7.6] **21.** $\dfrac{3y}{y - 4} - \dfrac{12}{y - 4} = 6$

[10.6] **22.** $\sqrt{3x + 10} = x + 2$

[4.4] **23.** Find the slope and y-intercept of the graph of $5x - 2y = -15$

[4.5] **24.** Write the equation of the line containing $(-4, 2)$ that is parallel to the graph of $y = 3x - 5$. Leave the answer in slope–intercept form.

For Exercises 25 and 26, graph.

[4.2] **25.** $y = -\dfrac{5}{3}x + 2$

[9.7] **26.** $\begin{cases} 2x - y > -4 \\ y > -\dfrac{2}{3}x - 2 \end{cases}$

For Exercises 27–30, solve.

[2.3] **27.** Hernando earns $1\frac{1}{4}$ times his normal hourly wage for all hours in excess of 40 hours per week. Last week, he worked 46 hours and earned $782.80. What is his normal hourly wage?

[7.7] **28.** A river has a current of 5 miles per hour. A boat can make a trip of 20 miles upstream in the same time it can make a trip of 30 miles downstream. Find the rate of the boat in still water.

[7.7] **29.** Suppose y varies jointly as x and the square of z. If $y = 72$ when $x = 4$ and $z = 3$, find y when $x = 3$ and $z = 4$.

[10.6] **30.** The period T, in seconds, of a pendulum whose length is L, in meters, is given by $T = 2\pi\sqrt{\dfrac{L}{9.8}}$. Find the lenght of a pendulum whose period is 4π seconds.

Quadratic Equations and Functions

11

"All our progress is unfolding like the vegetable bud."

—Ralph Waldo Emerson (1803–1882)
U.S. essayist, poet, philosopher

"For beautiful variety, no crop can be compared with this. Here is not merely the plain yellow of the grains, but nearly all the colors that we know . . ."

—Henry David Thoreau (1817–1862)
U.S. author and poet

In Section 6.6, we introduced quadratic equations and solved them by factoring. However, not all quadratic equations can be solved by factoring; so we need more powerful methods. In this chapter, we explore three other methods for solving quadratic equations, each of which uses square roots. In the first method, we solve quadratic equations using a new principle of equality called the *square root princi-ple.* Like factoring, this method has some limitations; so we develop a more powerful method called *completing the square.* Although we can complete the square to solve any quadratic equation, doing so can be rather tedious. So we develop a third method that uses a general formula, called the *quadratic for-mula,* to solve any quadratic equation. We then use these methods to solve equations that are quadratic in form and to solve quadratic inequalities. Finally, we add to what we learned in Sections 6.7 and 8.4 and learn more about the graphs of quadratic functions.

11.1 The Square Root Principle and Completing the Square

Objectives

1. Use the square root principle to solve quadratic equations.
2. Solve quadratic equations by completing the square.

Objective 1 Use the square root principle to solve quadratic equations. In Section 6.6, we solved quadratic equations such as $x^2 = 25$ by subtracting 25 from both sides, factoring, and then using the zero-factor theorem. Let's recall that process.

$$x^2 - 25 = 0$$
$$(x - 5)(x + 5) = 0 \qquad \text{Factor.}$$
$$x - 5 = 0 \quad \text{or} \quad x + 5 = 0 \qquad \text{Use the zero-factor theorem.}$$
$$x = 5 \qquad\qquad x = -5 \qquad \text{Solve for } x \text{ in each equation.}$$

Another approach to solving $x^2 = 25$ involves square roots. Notice that the solutions to this equation must be numbers that can be squared to equal 25. Those numbers are the square roots of 25, which are 5 and -5. This suggests a new rule called the *square root principle*.

Note: The expression $\pm\sqrt{a}$ is read "plus or minus the square root of a."

Rule **Square Root Principle**

If $x^2 = a$, where a is a real number, then $x = \sqrt{a}$ or $x = -\sqrt{a}$.
It is common to indicate the positive and negative solutions by writing $\pm\sqrt{a}$.

For example, if $x^2 = 64$, then $x = \sqrt{64} = 8$ or $x = -\sqrt{64} = -8$. Or we could simply write $x = \pm\sqrt{64} = \pm 8$.

Solve Equations in the Form $x^2 = a$

The square root principle is especially useful for solving equations in the form $x^2 = a$ when a is not a perfect square.

Connection In Section 10.3, we learned how to simplify square roots of numbers that have perfect square factors.

Example 1 Solve.

a. $x^2 = 40$

Solution $x^2 = 40$

$$x = \pm\sqrt{40} \qquad \text{Use the square root principle.}$$
$$x = \pm\sqrt{4 \cdot 10} \qquad \text{Simplify by factoring out a perfect square.}$$
$$x = \pm 2\sqrt{10}$$

Note: Remember that the $\pm$ symbol means that the two solutions are $2\sqrt{10}$ and $-2\sqrt{10}$.

We can check the two solutions using the original equation.

Check $2\sqrt{10}$: $(2\sqrt{10})^2 \overset{?}{=} 40$ **Check** $-2\sqrt{10}$: $(-2\sqrt{10})^2 \overset{?}{=} 40$

$\qquad\qquad\qquad 4 \cdot 10 = 40$ **True** $\qquad\qquad\qquad 4 \cdot 10 = 40$ **True**

For the remaining examples, the checks will be left to the reader.

b. $x^2 = -9$

Note: If we want to restrict the solutions of quadratic equations to real numbers only, we say, "Solve over the real numbers." In this chapter, we solve over the complex numbers, which means that complex number solutions are allowed.

Solution $x^2 = -9$

$$x = \pm\sqrt{-9} \qquad \text{Use the square root principle.}$$
$$x = \pm 3i$$

Note: The square root of a negative number is an imaginary number, which we can rewrite using i.

Answers **Your Turn 1**

a. $\pm 5\sqrt{2}$ **b.** $\pm 10i$

Your Turn 1 Solve.

a. $x^2 = 50$ **b.** $x^2 = -100$

Solve Equations in the Form $ax^2 + b = c$

If an equation is in the form $ax^2 + b = c$, we use the addition and multiplication principles of equality to isolate x^2 and then use the square root principle.

Example 2 Solve.

a. $x^2 + 2 = 100$

SOLUTION
$$x^2 + 2 = 100$$
$$x^2 = 98 \qquad \text{Subtract 2 from both sides to isolate } x^2.$$
$$x = \pm\sqrt{98} \qquad \text{Use the square root principle.}$$
$$x = \pm\sqrt{49 \cdot 2} \qquad \text{Factor out a perfect square.}$$
$$x = \pm 7\sqrt{2} \qquad \text{Simplify using } \sqrt{ab} = \sqrt{a} \cdot \sqrt{b}.$$

b. $5x^2 + 2 = 62$

SOLUTION
$$5x^2 + 2 = 62$$
$$5x^2 = 60 \qquad \text{Subtract 2 from both sides.}$$
$$x^2 = 12 \qquad \text{Divide both sides by 5.}$$
$$x = \pm\sqrt{12} \qquad \text{Use the square root principle.}$$
$$x = \pm\sqrt{4 \cdot 3} \qquad \text{Factor out a perfect square.}$$
$$x = \pm 2\sqrt{3} \qquad \text{Simplify using } \sqrt{ab} = \sqrt{a} \cdot \sqrt{b}.$$

Your Turn 2 Solve.

a. $x^2 - 6 = 42$ **b.** $2x^2 + 11 = 65$

Solve Equations in the Form $(ax + b)^2 = c$

In an equation in the form $(ax + b)^2 = c$, notice that the expression $ax + b$ is squared. We can use the square root principle to eliminate the square by thinking of this form as follows:

$$\text{If } (ax + b)^2 = c, \quad \text{then} \quad ax + b = \pm\sqrt{c}.$$

Example 3 Solve.

a. $(x + 6)^2 = 49$

SOLUTION
$$(x + 6)^2 = 49$$
$$x + 6 = \pm\sqrt{49} \qquad \text{Use the square root principle.}$$
$$x + 6 = \pm 7 \qquad \text{Simplify.}$$
$$x = -6 \pm 7 \qquad \text{Subtract 6 from both sides.}$$
$$x = -6 + 7 \quad \text{or} \quad x = -6 - 7 \qquad \text{Simplify by separating the two solutions.}$$
$$x = 1 \qquad\qquad x = -13$$

b. $(5x - 1)^2 = 10$

SOLUTION
$$(5x - 1)^2 = 10$$
$$5x - 1 = \pm\sqrt{10} \qquad \text{Use the square root principle.}$$
$$5x = 1 \pm \sqrt{10} \qquad \text{Add 1 to both sides to isolate } 5x.$$
$$x = \frac{1 \pm \sqrt{10}}{5} \qquad \text{Divide both sides by 5 to solve for } x.$$

c. $(8x + 6)^2 = -32$

SOLUTION $(8x + 6)^2 = -32$

$$8x + 6 = \pm\sqrt{-32}$$ Use the square root principle.

$$8x = -6 \pm \sqrt{-16 \cdot 2}$$ Subtract 6 from both sides and simplify the square root.

$$8x = -6 \pm 4i\sqrt{2}$$ Rewrite the imaginary number using i notation.

$$x = \frac{-6 \pm 4i\sqrt{2}}{8}$$ Divide both sides by 8 to solve for x.

$$x = -\frac{6}{8} \pm \frac{4\sqrt{2}}{8}i$$ Write the complex number in standard form $a + bi$.

$$x = -\frac{3}{4} \pm \frac{\sqrt{2}}{2}i$$ Simplify.

Your Turn 3   Solve.

a. $(x - 3)^2 = 25$ **b.** $(2x + 1)^2 = 14$ **c.** $(4x - 5)^2 = -12$

Objective ② Solve quadratic equations by completing the square.   To make use of the square root principle, we need one side of the equation to be a perfect square and the other side to be a constant, as in $(x + 2)^2 = 5$. But suppose we are given an equation such as $x^2 + 6x = 2$ whose left-hand side is an "incomplete" square. We can use the addition principle of equality to add an appropriate number (9 in this case) to both sides so that the left-hand side becomes a perfect square. We call this process *completing the square.*

$$x^2 + 6x + 9 = 2 + 9$$ Adding 9 to both sides completes the square on the left side.

We can now factor the left-hand side of the equation, then use the square root principle to finish solving the equation.

$$(x + 3)^2 = 11$$ Write the left-hand side in factored form.

$$x + 3 = \pm\sqrt{11}$$ Use the square root principle to eliminate the square.

$$x = -3 \pm \sqrt{11}$$ Subtract 3 from both sides to isolate x.

How do we know what number to add to complete the square?

Notice that squaring half of the coefficient of x gives the constant term in the completed square.

$$x^2 + 6x + 9$$

Half of 6 is 3, and
3 squared is 9.

Also notice that half of the coefficient of x is the constant in the factored form.

$$x^2 + 6x + 9 = (x + 3)^2$$

Half of 6 is 3, which is the
constant in the factored form.

Equations in the Form $x^2 + bx = c$

To solve a quadratic equation by completing the square, we need the equation in the form $x^2 + bx = c$ so that the coefficient of x^2 is 1. Once the equation is in the form $x^2 + bx = c$, we complete the square and then use the square root principle.

Connection   The product of every perfect square in the form $(x + b)^2$ is a trinomial in the form $x^2 + 2bx + b^2$. Notice that half of x's coefficient, $2b$, is b, which is then squared to equal the last term.

Answers   Your Turn 3
a. $8, -2$
b. $\dfrac{-1 \pm \sqrt{14}}{2}$ **c.** $\dfrac{5}{4} \pm \dfrac{\sqrt{3}}{2}i$

Example 4 Solve by completing the square.

a. $x^2 + 12x + 15 = 0$

SOLUTION We first write the equation in the form $x^2 + bx = c$.

Note: We found 36 by squaring half of 12.

$$\left(\frac{12}{2}\right)^2 = 6^2 = 36$$

$$x^2 + 12x = -15 \qquad \text{Subtract 15 from both sides to get the form } x^2 + bx = c.$$
$$x^2 + 12x + 36 = -15 + 36 \qquad \text{Complete the square by adding 36 to both sides.}$$
$$(x + 6)^2 = 21 \qquad \text{Factor the left side and simplify the right.}$$
$$x + 6 = \pm\sqrt{21} \qquad \text{Use the square root principle.}$$
$$x = -6 \pm \sqrt{21} \qquad \text{Subtract 6 from both sides to isolate } x.$$

b. $x^2 - 7x + 8 = 5$

SOLUTION $x^2 - 7x = -3 \qquad \text{Subtract 8 from both sides to get the form } x^2 + bx = c.$

Note: We found $\frac{49}{4}$ by squaring half of -7.

$$\left(\frac{-7}{2}\right)^2 = \frac{49}{4}$$

$$x^2 - 7x + \frac{49}{4} = -3 + \frac{49}{4} \qquad \text{Complete the square by adding } \frac{49}{4} \text{ to both sides.}$$

$$\left(x - \frac{7}{2}\right)^2 = \frac{37}{4} \qquad \text{Factor the left side and simplify the right side.}$$

$$x - \frac{7}{2} = \pm\sqrt{\frac{37}{4}} \qquad \text{Use the square root principle.}$$

$$x = \frac{7}{2} \pm \sqrt{\frac{37}{4}} \qquad \text{Add } \frac{7}{2} \text{ to both sides to isolate } x.$$

$$x = \frac{7}{2} \pm \frac{\sqrt{37}}{2} \qquad \text{Simplify the square root.}$$

$$x = \frac{7 \pm \sqrt{37}}{2} \qquad \text{Combine the fractions.}$$

Your Turn 4 Solve by completing the square.

a. $x^2 + 8x - 29 = 0$ **b.** $x^2 - 9x - 6 = 5$

Equations in the Form $ax^2 + bx = c$, Where $a \neq 1$

So far, our equations have been in the form $x^2 + bx = c$, where the coefficient of the x^2 term is already 1. To solve an equation such as $2x^2 + 12x = 3$, we need to divide both sides of the equation by 2 $\left(\text{or multiply both sides by } \frac{1}{2}\right)$ so that the coefficient of x^2 becomes 1.

Note: Our procedure for completing the square does not work unless the coefficient of the squared term is 1.

$$\frac{2x^2 + 12x}{2} = \frac{3}{2} \qquad \text{Divide both sides by 2.}$$

$$x^2 + 6x = \frac{3}{2} \qquad \text{Simplify.}$$

We can now solve by completing the square.

$$x^2 + 6x + 9 = \frac{3}{2} + 9 \qquad \text{Add 9 to both sides to complete the square.}$$

$$(x + 3)^2 = \frac{21}{2} \qquad \text{Factor the left side and simplify the right side.}$$

$$x + 3 = \pm\sqrt{\frac{21}{2}} \qquad \text{Use the square root principle.}$$

Answers **Your Turn 4**

a. $-4 \pm 3\sqrt{5}$ **b.** $\dfrac{9 \pm 5\sqrt{5}}{2}$

$$x = -3 \pm \sqrt{\frac{21}{2}} \qquad \text{Subtract 3 from both sides to isolate } x.$$

Note: The answer $-3 \pm \dfrac{\sqrt{42}}{2}$ can also be written as $\dfrac{-6 \pm \sqrt{42}}{2}$.

$$x = -3 \pm \frac{\sqrt{21}}{\sqrt{2}} \cdot \frac{\sqrt{2}}{\sqrt{2}} \qquad \text{Rationalize the denominator.}$$

$$x = -3 \pm \frac{\sqrt{42}}{2}$$

We can now write a procedure for solving any quadratic equation by completing the square.

Connection In Section 11.4, we use the process of completing the square to rewrite quadratic functions in a form that allows us to easily determine features of the graph.

Procedure **Solving Quadratic Equations by Completing the Square**

To solve a quadratic equation by completing the square,

1. Write the equation in the form $x^2 + bx = c$.
2. Complete the square by adding $\left(\dfrac{b}{2}\right)^2$ to both sides.
3. Write the completed square in factored form and simplify the right side, $c + \left(\dfrac{b}{2}\right)^2$.
4. Use the square root principle to eliminate the square.
5. Isolate the variable.

Example 5 Solve by completing the square.

a. $2x^2 - 8 = 5x$

SOLUTION
$$2x^2 - 5x = 8 \qquad \text{Rewrite in the form } ax^2 + bx = c.$$

$$\frac{2x^2 - 5x}{2} = \frac{8}{2} \qquad \text{Divide both sides by 2.}$$

$$x^2 - \frac{5}{2}x = 4 \qquad \text{Simplify.}$$

Note: To complete the square, we square half of $-\dfrac{5}{2}$.

$$\left(\frac{1}{2} \cdot -\frac{5}{2}\right)^2 = \left(-\frac{5}{4}\right)^2 = \frac{25}{16}$$

$$x^2 - \frac{5}{2}x + \frac{25}{16} = 4 + \frac{25}{16} \qquad \text{Add } \frac{25}{16} \text{ to both sides to complete the square.}$$

$$\left(x - \frac{5}{4}\right)^2 = \frac{89}{16} \qquad \text{Factor the left side and simplify the right side.}$$

$$x - \frac{5}{4} = \pm\sqrt{\frac{89}{16}} \qquad \text{Use the square root principle.}$$

$$x = \frac{5}{4} \pm \frac{\sqrt{89}}{4} \qquad \text{Add } \frac{5}{4} \text{ to both sides and simplify the square root.}$$

$$x = \frac{5 \pm \sqrt{89}}{4} \qquad \text{Combine like fractions.}$$

b. $4x^2 - 8x + 7 = 2$

SOLUTION
$$4x^2 - 8x = -5 \qquad \text{Rewrite in the form } ax^2 + bx = c.$$

$$\frac{4x^2 - 8x}{4} = -\frac{5}{4} \qquad \text{Divide both sides by 4.}$$

$$x^2 - 2x = -\frac{5}{4} \qquad \text{Simplify.}$$

$$x^2 - 2x + 1 = -\frac{5}{4} + 1 \qquad \text{Add 1 to both sides to complete the square.}$$

$$(x - 1)^2 = -\frac{1}{4} \qquad \text{Factor the left side and simplify the right side.}$$

$$x - 1 = \pm\sqrt{-\frac{1}{4}} \qquad \text{Use the square root principle.}$$

$$x = 1 \pm \frac{1}{2}i \qquad \text{Add 1 to both sides and simplify the square root.}$$

Your Turn 5 Solve by completing the square.

a. $9x^2 + 18x = -1$ **b.** $x^2 - 8x + 11 = -14$

11.1 Exercises For Extra Help *MyMathLab*

Note: Exercises marked with a ★ represent challenging exercises.

1. Explain why there are two solutions to an equation in the form $x^2 = a$.

2. Write a formula for the solutions of $ax^2 - b = c$ by solving for x.

3. Write a formula for the solutions of $(ax - b)^2 = c$ by solving for x.

4. Given an equation in the form $x^2 + bx = c$, explain how to complete the square.

5. Consider the equation $x^2 - 7x + 12 = 0$. Which is a better method for solving the equation, factoring and then using the zero-factor theorem or completing the square? Explain.

6. Consider the equation $x^2 + 4x + 5 = 0$. Which is a better method for solving the equation, factoring and then using the zero-factor theorem or completing the square? Explain.

For Exercises 7–16, solve and check. See Example 1.

7. $x^2 = 49$

8. $x^2 = 144$

9. $y^2 = \dfrac{4}{25}$

10. $t^2 = \dfrac{9}{49}$

11. $n^2 = 1.44$

12. $p^2 = 1.21$

13. $z^2 = 45$

14. $m^2 = 72$

15. $w^2 = -25$

16. $c^2 = -49$

For Exercises 17–36, solve and check. Begin by using the addition or multiplication principles of equality to isolate the squared term. See Example 2.

17. $n^2 - 7 = 42$

18. $y^2 - 5 = 59$

19. $y^2 - 16 = 65$

20. $k^2 + 5 = 30$

21. $4n^2 = 36$

22. $5y^2 = 125$

23. $25t^2 = 9$

24. $16d^2 = 49$

25. $4h^2 = -16$

26. $3k^2 = -108$

27. $\dfrac{5}{6}x^2 = \dfrac{24}{5}$

28. $-\dfrac{2}{3}m^2 = -\dfrac{27}{32}$

29. $2x^2 + 5 = 21$

30. $4x^2 - 11 = 97$

31. $5y^2 - 7 = -97$

32. $4n^2 + 20 = -76$

33. $\dfrac{3}{4}y^2 - 5 = 3$

34. $\dfrac{25}{9}m^2 - 2 = 6$

35. $0.2t^2 - 0.5 = 0.012$

36. $0.5p^2 + 1.28 = 1.6$

For Exercises 37–52, solve and check. Use the square root principle to eliminate the square. See Example 3.

37. $(x + 8)^2 = 49$

38. $(y + 3)^2 = 36$

39. $(5n - 3)^2 = 16$

40. $(6h - 5)^2 = 81$

41. $(m - 8)^2 = -16$

42. $(t - 2)^2 = -4$

43. $(4k - 1)^2 = 40$

44. $(3x - 4)^2 = 80$

45. $(m - 7)^2 = -12$

46. $(t - 5)^2 = -28$

47. $\left(y - \dfrac{3}{4}\right)^2 = \dfrac{9}{16}$

48. $\left(x + \dfrac{4}{9}\right)^2 = \dfrac{25}{81}$

49. $\left(\dfrac{5}{9}d - \dfrac{1}{2}\right)^2 = \dfrac{1}{36}$

50. $\left(\dfrac{3}{4}h + \dfrac{4}{5}\right)^2 = \dfrac{1}{100}$

51. $(0.4x + 3.8)^2 = 2.56$

52. $(0.8n - 6.8)^2 = 1.96$

Find
(X) the mistake

For Exercises 53–56, explain the mistake; then find the correct solutions.

53. $x^2 - 15 = 34$
$x^2 = 49$
$x = \sqrt{49}$
$x = 7$

54. $x^2 = 20$
$x = \sqrt{20}$
$x = 2\sqrt{5}$

55. $(x - 5)^2 = -6$
$x - 5 = \pm\sqrt{6}$
$x = 5 \pm \sqrt{6}$

56. $(x - 1)^2 = -12$
$x - 1 = \pm\sqrt{-12}$
$x = 1 \pm 2\sqrt{3}$

For Exercises 57–62, solve; then use a calculator to approximate the irrational solutions rounded to three places.

57. $x^2 = 96$

58. $t^2 = 56$

59. $y^2 - 15 = 5$

60. $x^2 - 13 = 35$

61. $(n - 6)^2 = 15$

62. $(m + 3)^2 = 10$

For Exercises 63–70: a. Add a term to the expression to make it a perfect square.
b. Factor the perfect square. See Objective 2.

63. $x^2 + 14x$

64. $c^2 + 8c$

65. $n^2 - 10n$

66. $a^2 - 12a$

67. $y^2 - 9y$

68. $m^2 - 11m$

69. $s^2 - \dfrac{2}{3}s$

70. $y^2 - \dfrac{4}{5}y$

For Exercises 71–82, solve the equation by completing the square. See Example 4.

71. $w^2 + 2w = 15$

72. $p^2 + 8p = 9$

73. $r^2 - 2r + 50 = 0$

74. $c^2 - 6c + 45 = 0$

75. $k^2 = 9k - 18$

76. $a^2 = 6a - 8$

77. $b^2 - 2b - 11 = 5$

78. $x^2 + 2x + 7 = 9$

79. $h^2 - 6h + 3 = -26$

80. $j^2 - 4j + 25 = -3$

81. $u^2 + \dfrac{1}{2}u = \dfrac{3}{2}$

82. $y^2 + \dfrac{1}{3}y = \dfrac{2}{3}$

For Exercises 83–94, solve the equation by completing the square. Begin by writing the equation in the form $x^2 + bx = c$. See Example 5.

83. $4x^2 + 16x = 9$

84. $4m^2 - 8m = 5$

85. $9x^2 - 18x + 5 = 0$

86. $16x^2 - 32x + 7 = 0$

87. $2n^2 - n - 3 = 0$ **88.** $2x^2 - 5x - 12 = 0$ **89.** $4x^2 = -16x + 7$ **90.** $4x^2 = 24x + 11$

91. $5k^2 + k - 2 = 0$ **92.** $3s^2 - 4s = 2$ **93.** $3a^2 - 8 = -4a$ **94.** $3x^2 - 4 = -8x$

Find *the mistake*   ***For Exercises 95 and 96, explain the mistake. Find the correct solution.***

95.
$$3x^2 + 4x = 2$$
$$3x^2 + 4x + 4 = 2 + 4$$
$$(3x + 2)^2 = 6$$
$$3x + 2 = \pm\sqrt{6}$$
$$3x = -2 \pm \sqrt{6}$$
$$x = \frac{-2 \pm \sqrt{6}}{3}$$

96.
$$x^2 + 6x = 7$$
$$x^2 + 6x + 9 = 7 + 9$$
$$(x + 3)^2 = 16$$
$$x + 3 = \sqrt{16}$$
$$x + 3 = 4$$
$$x = -3 + 4$$
$$x = 1$$

For Exercises 97–108, solve.

97. A square sheet of metal has an area of 196 square inches. What is the length of each side?

98. A severe thunderstorm warning is issued by the National Weather Service for a square area covering 14,400 square miles. What is the length of each side of the square area?

99. A tank is to be made for an aquarium so that the length is 8 feet, the width is twice the height, and the volume is 144 cubic feet. Find the height and width of the tank.

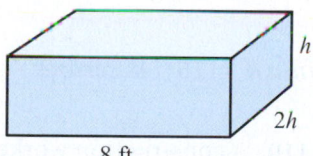

100. The length of a swimming pool is to be three times the width. If the depth is to be a constant 4 feet and the volume is to be 4800 cubic feet, find the length and width.

101. The Arecibo radio telescope is like a giant satellite dish that covers a circular area of approximately $23,256.25\pi$ square meters. Find the diameter of the dish.

102. A field is planted in a circular pattern. A watering device is to be constructed with pipe in a line extending from the center of the field to the edge of the field. The pipe is set on wheels so that it can rotate around the field and cover the entire field with water. If the area of the field is 7225π square feet, how long will the watering pipe be?

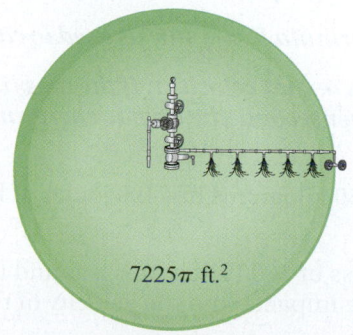

7225π ft.^{2}

103. The corner of a sheet of plywood has been cut out as shown. If the area of the piece that was removed was 256 square inches, find x.

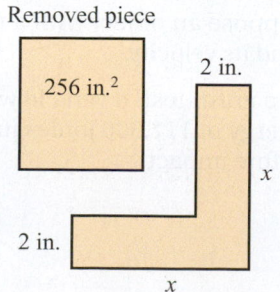

Removed piece

256 in.^{2}

2 in.

2 in.

x

x

104. The area of the hole in the washer shown is 25π square millimeters. Find r, the radius of the washer.

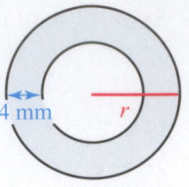

105. An LCD computer monitor measures 20 inches across its diagonal. If the width is 4 inches less than the length, find the dimensions of the monitor.

106. Two identical right triangles are placed together to form the frame of a roof. If the base of each triangle is 7 feet longer than its height and its hypotenuse is 13 feet, what are the dimensions of the base and height?

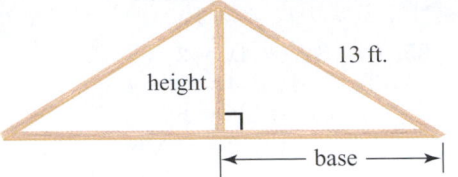

★**107.** A plastic panel is to have a rectangular hole cut as shown.

    a. Find l so that the area remaining after the hole is cut is 1230 square centimeters.

    b. Find the length and width of the plastic panel.

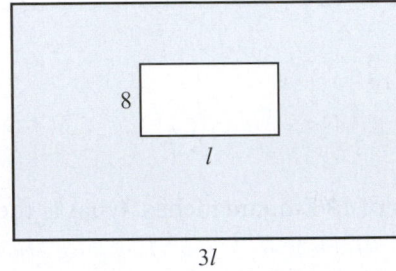

★**108.** A 6-inch-wide groove with a height of h is to be cut into a wood block as shown.

    a. Find h so that the volume remaining in the block after the groove is cut is 360 cubic inches.

    b. Find the height and width of the block.

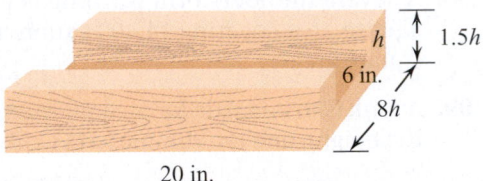

For Exercises 109 and 110, if an object is dropped, the formula $d = 16t^2$ describes the distance d in feet that the object falls in t seconds.

109. Suppose a cover for a ceiling light falls from a 9-foot ceiling. How long does the cover take to hit the floor?

110. A construction worker tosses a scrap piece of lumber from the roof of a house. How long does it take the piece of lumber to reach the ground 25 feet below?

For Exercises 111 and 112, use the following information. In physics, when an object is in motion, it has kinetic energy. The formula $E = \dfrac{1}{2}mv^2$ is used to calculate the kinetic energy E of an object with a mass m and velocity v. If the mass is measured in kilograms and the velocity is in meters per second, the kinetic energy will be in units called joules (J).

111. Suppose an object with a mass of 50 kilograms has 400 joules of kinetic energy. Find its velocity.

112. In a crash test, a vehicle with a mass of 1200 kilograms is found to have kinetic energy of 117,600 joules just before impact. Find the velocity of the vehicle just before impact.

Puzzle Problem

Without using a calculator, which of the following four numbers is a perfect square? (Hint: Write a list of smaller perfect squares and look for a pattern.)

    9,456,804,219,745,618
    2,512,339,789,576,516
    7,602,985,471,286,543
    4,682,715,204,643,182

Review Exercises

Exercises 1–6 **EXPRESSIONS**

[6.4] *For Exercises 1 and 2, factor.*

1. $x^2 - 10x + 25$

2. $x^2 + 6x + 9$

[10.1] *For Exercises 3 and 4, simplify. Assume that variables represent nonnegative values.*

3. $\sqrt{36x^2}$

4. $\sqrt{(2x + 3)^2}$

[1.5, 10.1, 10.7] *For Exercises 5 and 6, simplify.*

5. $\dfrac{-4 + \sqrt{8^2 - 4(2)(5)}}{2(2)}$

6. $\dfrac{-6 - \sqrt{6^2 - 4(5)(2)}}{2(5)}$

11.2 Solving Quadratic Equations Using the Quadratic Formula

Objectives

1. Solve quadratic equations using the quadratic formula.

2. Use the discriminant to determine the number of real solutions that a quadratic equation has.

3. Find the x- and y-intercepts of a quadratic function.

4. Solve applications using the quadratic formula.

In this section, we solve quadratic equations using a formula called the *quadratic formula*. Using this formula is much easier than completing the square.

Objective 1 Solve quadratic equations using the quadratic formula. To derive the quadratic formula, we begin with the general form of the quadratic equation, $ax^2 + bx + c = 0$, and assume that $a > 0$. We follow the procedure for solving a quadratic equation by completing the square.

$$ax^2 + bx + c = 0$$

Note: To complete the square, we square half of $\dfrac{b}{a}$.

$$\left(\frac{1}{2} \cdot \frac{b}{a}\right)^2 = \left(\frac{b}{2a}\right)^2$$

$$= \frac{b^2}{4a^2}$$

$$ax^2 + bx = -c \qquad \text{Subtract } c \text{ from both sides.}$$

$$\frac{ax^2 + bx}{a} = \frac{-c}{a} \qquad \begin{array}{l}\text{Divide both sides by } a \text{ so that the} \\ \text{coefficient of } x^2 \text{ is 1.}\end{array}$$

$$x^2 + \frac{b}{a}x = -\frac{c}{a} \qquad \text{Simplify.}$$

$$x^2 + \frac{b}{a}x + \frac{b^2}{4a^2} = -\frac{c}{a} + \frac{b^2}{4a^2} \qquad \text{Complete the square.}$$

$$\left(x + \frac{b}{2a}\right)^2 = -\frac{4ac}{4a^2} + \frac{b^2}{4a^2} \qquad \begin{array}{l}\text{On the left side, write in factored form. On the} \\ \text{right side, rewrite } -\dfrac{c}{a} \text{ with the common} \\ \text{denominator } 4a^2 \text{ so that we can combine the} \\ \text{rational expressions.}\end{array}$$

$$\left(x + \frac{b}{2a}\right)^2 = \frac{b^2 - 4ac}{4a^2} \qquad \begin{array}{l}\text{Combine the rational expressions. For simplicity,} \\ \text{we rearrange the order of the } b^2 \text{ and } -4ac \text{ terms} \\ \text{in the numerator.}\end{array}$$

$$x + \frac{b}{2a} = \pm\sqrt{\frac{b^2 - 4ac}{4a^2}} \qquad \begin{array}{l}\text{Use the square root principle to} \\ \text{eliminate the square.}\end{array}$$

$$x + \frac{b}{2a} = \pm \frac{\sqrt{b^2 - 4ac}}{2a}$$

Simplify the square root in the denominator.

$$x = -\frac{b}{2a} \pm \frac{\sqrt{b^2 - 4ac}}{2a}$$

Subtract $\frac{b}{2a}$ from both sides to isolate x.

$$x = \frac{-b \pm \sqrt{b^2 - 4ac}}{2a}$$

Combine the rational expressions.

> **Note:** We get the same result if $a < 0$.

This final equation is the *quadratic formula*. It can be used to solve any quadratic equation simply by replacing a, b, and c with the corresponding values from the given equation.

Procedure   **Using the Quadratic Formula**

> **Note:** A quadratic equation must be in the form $ax^2 + bx + c = 0$ so that a, b, and c can be identified for use in the quadratic formula.

To solve a quadratic equation in the form $ax^2 + bx + c = 0$, where $a \neq 0$, use the quadratic formula:

$$x = \frac{-b \pm \sqrt{b^2 - 4ac}}{2a}$$

Example 1   Solve.

a. $3x^2 + 10x - 8 = 0$

SOLUTION   This equation is in the form $ax^2 + bx + c = 0$, where $a = 3$, $b = 10$, and $c = -8$. So we can use the quadratic formula, $x = \frac{-b \pm \sqrt{b^2 - 4ac}}{2a}$.

$$x = \frac{-10 \pm \sqrt{10^2 - 4(3)(-8)}}{2(3)}$$

Replace a with 3, b with 10, and c with -8.

$$x = \frac{-10 \pm \sqrt{196}}{6}$$

Simplify in the radical and the denominator.

> **Connection**   Notice that the radicand 196 is a perfect square, which causes the solutions to be two rational numbers. As we consider more examples in this section, note how the radicand determines the type of solutions.

$$x = \frac{-10 \pm 14}{6}$$

Evaluate $\sqrt{196}$.

$$x = \frac{-10 + 14}{6} \quad \text{or} \quad x = \frac{-10 - 14}{6}$$

Split up the $\pm$ to calculate the two solutions.

$$x = \frac{4}{6} \qquad\qquad x = \frac{-24}{6}$$

$$x = \frac{2}{3} \qquad\qquad x = -4$$

Reduce to lowest terms.

The checks to Example 1 are left to the reader.

b. $x^2 + 8x - 15 = 14$

SOLUTION $x^2 + 8x - 29 = 0$

Subtract 14 from both sides to get the form $ax^2 + bx + c = 0$.

$$x = \frac{-8 \pm \sqrt{8^2 - 4(1)(-29)}}{2(1)}$$

Use the quadratic formula, replacing a with 1, b with 8, and c with -29.

$$x = \frac{-8 \pm \sqrt{180}}{2}$$

Simplify in the radical and the denominator.

> **Connection**   Notice that the radicand 180 is not a perfect square, which causes the two solutions to be irrational numbers.

$$x = \frac{-8 \pm \sqrt{36 \cdot 5}}{2}$$

Use the product rule of radicals.

$$x = \frac{-8 \pm 6\sqrt{5}}{2}$$

Simplify the radical.

Warning: Always make sure the radical is simplified *before* trying to reduce the fraction.

$$x = \frac{-8}{2} \pm \frac{6\sqrt{5}}{2}$$ Separate into two rational expressions to simplify.

$$x = -4 \pm 3\sqrt{5}$$ Simplify by dividing out the 2.

c. $9x^2 + 4 = -12x$

SOLUTION $9x^2 + 12x + 4 = 0$ Add $12x$ to both sides to get the form $ax^2 + bx + c = 0$.

Connection Notice that the radicand is 0, which causes this equation to have only one solution.

$$x = \frac{-12 \pm \sqrt{12^2 - 4(9)(4)}}{2(9)}$$ Use the quadratic formula, replacing a with 9, b with 12, and c with 4.

$$x = \frac{-12 \pm \sqrt{0}}{18}$$ Simplify in the radical and the denominator.

$$x = -\frac{2}{3}$$ Simplify the radical and write the fraction in lowest terms.

d. $3x^2 + 9 = -8x + 2$

SOLUTION $3x^2 + 8x + 7 = 0$ Add $8x$ to and subtract 2 from both sides to get the form $ax^2 + bx + c = 0$.

Connection Notice that the negative radicand causes the two solutions to be non-real complex numbers.

$$x = \frac{-8 \pm \sqrt{8^2 - 4(3)(7)}}{2(3)}$$ Use the quadratic formula, replacing a with 3, b with 8, and c with 7.

$$x = \frac{-8 \pm \sqrt{-20}}{6}$$ Simplify in the radical and the denominator.

Note: In standard form, the two complex solutions are

$-\frac{4}{3} + \frac{\sqrt{5}}{3}i$ and $-\frac{4}{3} - \frac{\sqrt{5}}{3}i$.

$$x = \frac{-8 \pm 2i\sqrt{5}}{6}$$ Simplify the radical.

$$x = \frac{-4 \pm i\sqrt{5}}{3}$$ Simplify to lowest terms.

Answers Your Turn 1

a. $2, -\frac{3}{4}$ **b.** $\frac{-1 \pm \sqrt{61}}{4}$

c. 4 **d.** $-2 \pm i\sqrt{2}$

Your Turn 1 Solve using the quadratic formula.

a. $4x^2 - 5x - 6 = 0$ **b.** $4x^2 = 15 - 2x$

c. $x^2 - 8x + 16 = 0$ **d.** $x^2 + 8 = 2 - 4x$

Choosing a Method for Solving Quadratic Equations

We have learned several methods for solving quadratic equations. The following table summarizes the methods and conditions that make each method the best choice.

Methods for Solving Quadratic Equations

Method	When the Method Is Beneficial
1. Factoring (Section 6.6)	Use when the quadratic equation can be easily factored.
2. Square root principle (Section 11.1)	Use when the quadratic equation can be easily written in the form $ax^2 = c$ or $(ax + b)^2 = c$.
3. Completing the square (Section 11.1)	Keep in mind that this is rarely the best method, but it is important for future topics.
4. Quadratic formula (Section 11.2)	Use when factoring is not easy, or possible, with integer coefficients.

Objective ② Use the discriminant to determine the number of real solutions that a quadratic equation has. In the Connection boxes for Example 1, we pointed out how the radicand in the quadratic formula affects the solutions to a given quadratic equation. The expression $b^2 - 4ac$, which is the radicand, is called the **discriminant**.

Definition *Discriminant:* The discriminant is the radicand, $b^2 - 4ac$, in the quadratic formula.

We use the discriminant to determine the number and type of solutions to a quadratic equation.

Procedure **Using the Discriminant**

Given a quadratic equation in the form $ax^2 + bx + c = 0$, where a, b, and c, are rational numbers and $a \neq 0$, to determine the number and type of solutions, evaluate the discriminant $b^2 - 4ac$.

Note: When the discriminant is 0, the solution is
$$\frac{-b \pm \sqrt{0}}{2a} = -\frac{b}{2a}.$$

If the **discriminant is positive**, the equation has two real-number solutions. The solutions will be rational if the discriminant is a perfect square and irrational otherwise.

If the **discriminant is 0**, the equation has one rational solution.

If the **discriminant is negative**, the equation has two nonreal complex solutions.

Example 2   Use the discriminant to determine the number and type of solutions.

a. $3x^2 - 7x = 8$

SOLUTION $3x^2 - 7x - 8 = 0$ **Write the equation in the form $ax^2 + bx + c = 0$.**

Evaluate the discriminant, $b^2 - 4ac$.

Note: Follow the order of operations when evaluating $b^2 - 4ac$ and note that $b^2 \geq 0$ if b is a real number.

$$(-7)^2 - 4(3)(-8)$$ **Replace a with 3, b with -7, and c with -8.**
$$= 49 + 96$$
$$= 145$$

Warning: 145 is the *value of the discriminant*, not a solution for the equation $3x^2 - 7x = 8$.

Because the discriminant is positive, this equation has two real-number solutions. Because 145 is not a perfect square, the solutions are irrational.

b. $x^2 = \frac{4}{5}x - \frac{4}{25}$

SOLUTION $x^2 - \frac{4}{5}x + \frac{4}{25} = 0$ **Write the equation in the form $ax^2 + bx + c = 0$.**

Note: Multiplying both sides of a quadratic equation by a constant does not change the solutions, but does change the value of determinant.

It is easier to evaluate the discriminant if the coefficients are integers, so we multiply the equation through by the LCD, which is 25.

$$25 \cdot x^2 - \frac{25}{1} \cdot \frac{4}{5}x + \frac{25}{1} \cdot \frac{4}{25} = 25 \cdot 0$$
$$25x^2 - 20x + 4 = 0$$

Now evaluate the discriminant, $b^2 - 4ac$.

Note: Because the discriminant is 0, the solution is
$$-\frac{b}{2a} = -\frac{(-20)}{2(25)} = \frac{20}{50} = \frac{2}{5},$$
which is a rational number.

$$(-20)^2 - 4(25)(4)$$ **Replace a with 25, b with -20, and c with 4.**
$$= 400 - 400$$
$$= 0$$

Because the discriminant is zero, this equation has only one real solution.

c. $0.5x^2 - 0.8x + 2.5 = 0$

Note: We could avoid calculations with decimal numbers by multiplying the original equation through by 10 so that it becomes
$$5x^2 - 8x + 25 = 0.$$

SOLUTION $(-0.8)^2 - 4(0.5)(2.5)$ **In $b^2 - 4ac$, replace a with 0.5, b with -0.8, and c with 2.5.**
$$= 0.64 - 5$$
$$= -4.36$$

Because the discriminant is negative, there are two nonreal complex solutions.

Your Turn 2 Use the discriminant to determine the number and type of solutions for the equation. If the solutions are real, state whether they are rational or irrational.

a. $5x^2 + 8x - 9 = 0$ **b.** $\dfrac{5}{2}x^2 + \dfrac{5}{6} = \dfrac{4}{3}x$ **c.** $x^2 - 0.12x = -0.0036$

Objective ③ Find the x- and y-intercepts of a quadratic function. In Chapter 4, we learned that x-intercepts are points where a graph intersects the x-axis. Because an x-intercept is on the x-axis, the y-coordinate of the point is 0. In Section 6.7, we graphed quadratic functions, which have the form $f(x) = ax^2 + bx + c$ (or $y = ax^2 + bx + c$), and learned that their graphs are parabolas. Notice that when we replace y with 0 to find the x-intercepts in $y = ax^2 + bx + c$, we have the quadratic equation $ax^2 + bx + c = 0$, which has two, one, or no real-number solutions. As a result, quadratic functions in the form $y = ax^2 + bx + c$ have two, one, or no x-intercepts. The following graphs illustrate the possibilities.

> **Note:** In Section 6.7, we noted that if $a > 0$, the parabola opens up and if $a < 0$, the parabola opens down.

Two x-intercepts: If $0 = ax^2 + bx + c$ has two real-number solutions, which occurs if $b^2 - 4ac > 0$, then the graph of $y = ax^2 + bx + c$ has two x-intercepts and looks like one of the following graphs.

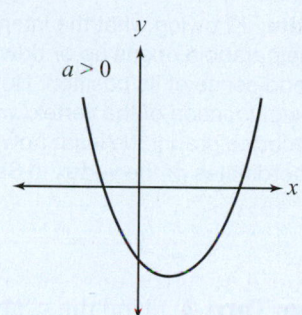 or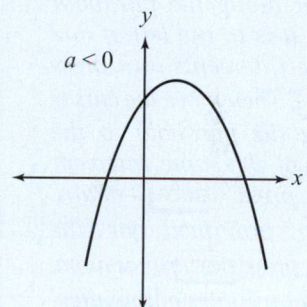

One x-intercept: If $0 = ax^2 + bx + c$ has one real-number solution, which occurs if $b^2 - 4ac = 0$, then the graph of $y = ax^2 + bx + c$ has one x-intercept and looks like one of the following graphs.

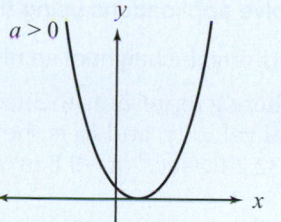 or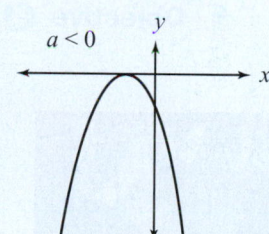

No x-intercepts: If $0 = ax^2 + bx + c$ has no real-number solution, which occurs if $b^2 - 4ac < 0$, then the graph of $y = ax^2 + bx + c$ has no x-intercepts and looks like one of the following graphs.

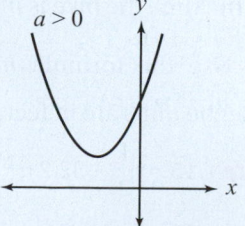 or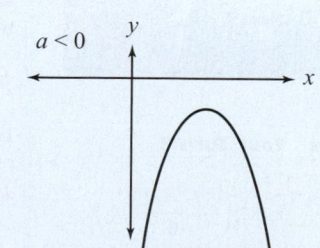

Answers Your Turn 2

a. discriminant $= 244$; two irrational solutions

b. discriminant $= -236$; two nonreal complex solutions

c. discriminant $= 0$; one rational solution

We also have learned that a y-intercept is where a graph intersects the y-axis. To find y-intercepts, we let $x = 0$ and solve for y. Notice that when we replace x with 0 in $y = ax^2 + bx + c$, we have $y = a(0)^2 + b(0) + c = c$; so the y-intercept is $(0, c)$.

Example 3 Find the x- and y-intercepts of $y = x^2 - 2x - 8$; then graph.

SOLUTION For the x-intercepts, letting $y = 0$ gives the equation $0 = x^2 - 2x - 8$, which we solve by factoring.

$$0 = x^2 - 2x - 8$$
$$0 = (x - 4)(x + 2)$$
$$x - 4 = 0 \quad \text{or} \quad x + 2 = 0$$
$$x = 4 \qquad\qquad x = -2$$

x-intercepts: $(4, 0)$ and $(-2, 0)$
Because $c = -8$, the y-intercept is $(0, -8)$.

Now we can get a rough sketch of the graph. We know that the graph opens upward because $a = 1$, which is positive.

> **Note:** Knowing what the intercepts are and whether the parabola opens up or down gives us a fairly good sense of its position. However, if we knew the exact location of the vertex, we could get a more accurate graph. We learn how to find the coordinates of the vertex in Section 11.4.

> **Connection** We could have solved $0 = x^2 - 2x - 8$ using the quadratic formula.
>
> $$x = \frac{-(-2) \pm \sqrt{(-2)^2 - 4(1)(-8)}}{2(1)}$$
> $$= \frac{2 \pm \sqrt{36}}{2} = \frac{2 \pm 6}{2}$$
> $$= 1 \pm 3$$
> $$= 4 \quad \text{or} \quad -2$$

> **Note:** We also could have calculated the y-intercept:
> $$y = (0)^2 + 2(0) - 8 = -8$$

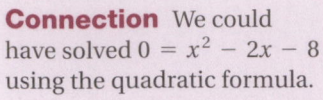

Your Turn 3 Find the x- and y-intercepts. Verify on a graphing calculator.

a. $y = 3x^2 - 5x + 1$ **b.** $y = x^2 - 8x + 16$ **c.** $y = 2x^2 - 3x + 5$

Objective ④ Solve applications using the quadratic formula. In physics, the general formula for describing the height of an object after it has been thrown upward is $h = \frac{1}{2}gt^2 + v_0t + h_0$, where g represents the acceleration due to gravity, t is the time in flight, v_0 is the initial velocity, and h_0 is the initial height. For Earth, the acceleration due to gravity is -32.2 ft./sec.^{2} or -9.8 m/sec.^{2}

Example 4 In an extreme games competition, a motorcyclist jumps with an initial velocity of 70 feet per second from a ramp height of 25 feet, landing on a ramp with a height of 15 feet. Find the time the motorcyclist is in the air.

Understand We are given the initial velocity, initial height, and final height and are to find the time the bike is in the air.

Plan Use the formula $h = \frac{1}{2}gt^2 + v_0t + h_0$ with $h = 15$, $v_0 = 70$, and $h_0 = 25$. Because the units are in feet, we use -32.2 ft./sec.^{2} for g.

Execute

$$15 = \frac{1}{2}(-32.2)t^2 + 70t + 25 \qquad \text{Substitute the values.}$$

$$15 = -16.1t^2 + 70t + 25 \qquad \text{Multiply.}$$

$$0 = -16.1t^2 + 70t + 10 \qquad \begin{array}{l}\text{Subtract 15 from both sides to get the form}\\ ax^2 + bx + c = 0.\end{array}$$

$$x = \frac{-70 \pm \sqrt{70^2 - 4(-16.1)(10)}}{2(-16.1)} \qquad \begin{array}{l}\text{Use the quadratic formula, replacing } a\\ \text{with } -16.1, b \text{ with } 70, \text{ and } c \text{ with } 10.\end{array}$$

Answers Your Turn 3

a. x-intercepts:
$\left(\frac{5 + \sqrt{13}}{6}, 0\right), \left(\frac{5 - \sqrt{13}}{6}, 0\right)$;
y-intercept: $(0, 1)$

b. x-intercept: $(4, 0)$;
y-intercept: $(0, 16)$

c. no x-intercepts;
y-intercept: $(0, 5)$

$$x = \frac{-70 \pm \sqrt{5544}}{-32.2}$$ **Simplify in the radical and the denominator.**

$$x \approx -0.138 \quad \text{or} \quad 4.486$$ **Approximate the two irrational solutions.**

Answer Because the time cannot be negative, the motorcycle is in the air approximately 4.486 seconds.

Check We can check by evaluating the original formula using $t = 4.486$ seconds to see if the motorcycle does land at a height of 15 feet. We will leave this check to the reader.

Your Turn 4 A ball is thrown from an initial height of 1.5 meters with an initial velocity of 8 meters per second. Find the time for the ball to land on the ground ($h = 0$). Approximate the time to the nearest thousandth.

Answer Your Turn 4

1.802 sec.

11.2 Exercises   For Extra Help *MyMathLab*

PRACTICE   WATCH   DOWNLOAD

Note: Exercises marked with a ★ represent challenging exercises.

1. Explain the general plan for deriving the quadratic formula.

2. Discuss the advantages and disadvantages of using the quadratic formula to solve quadratic equations.

3. Are there quadratic equations that cannot be solved using factoring? Explain.

4. What part of the quadratic formula is the discriminant?

5. Under what conditions will a quadratic equation have only one solution? Modify the quadratic formula to describe this single solution.

6. If x-intercepts for a quadratic function are imaginary numbers, what does that indicate about its graph?

For Exercises 7–14, rewrite each quadratic equation in the form $ax^2 + bx + c = 0$; then identify a, b, and c. See Objective 1.

7. $x^2 - 3x + 7 = 0$

8. $x^2 + 6x - 15 = 0$

9. $3x^2 - 9x = 4$

10. $2x^2 - 8x = -11$

11. $x = 1.5x^2 + 0.2$

12. $x = 0.8x^2 + 4.5$

13. $\frac{3}{4}x = -\frac{1}{2}x^2 + 6$

14. $\frac{1}{4}x = \frac{5}{6}x^2 + 6$

For Exercises 15–38, solve using the quadratic formula. See Example 1.

15. $x^2 + 9x + 20 = 0$

16. $x^2 + 8x + 15 = 0$

17. $4x^2 + 5x = 6$

18. $2x^2 - x = 3$

19. $x^2 - 9x = 0$

20. $x^2 - 6x = 0$

21. $x^2 - 8x = -16$

22. $x^2 + 14x = -49$

11.2 Solving Quadratic Equations Using the Quadratic Formula **791**

23. $3x^2 + 4x = 4$ | **24.** $5x^2 - 13x = 6$ | **25.** $x^2 + 2 = 2x$ | **26.** $x^2 + 5 = 4x$

27. $x^2 - x - 1 = 0$ | **28.** $x^2 + 3x - 5 = 0$ | **29.** $3x^2 + 10x + 5 = 0$ | **30.** $3x^2 - x - 3 = 0$

31. $-4x^2 = 5 - 6x$ | **32.** $-5x^2 = 3 - 4x$ | **33.** $18x^2 + 2 = -15x$ | **34.** $10x^2 - 12 = -7x$

35. $3x^2 - 4x = -3$ | **36.** $3x^2 - 6x = -8$ | **37.** $6x^2 - 4 = 3x$ | **38.** $6x^2 - 6 = 13x$

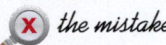 *For Exercises 39–50, solve using the quadratic formula. If the solutions are irrational, give an exact answer and an approximate answer rounded to the nearest thousandth. (Hint: You might first clear the fractions or decimals by multiplying both sides by an appropriately chosen number.)*

39. $2x^2 + 0.1x = 0.03$ | **40.** $4x^2 + 8.6x - 2.4 = 0$ | **41.** $x^2 + \frac{1}{2}x - 3 = 0$

42. $x^2 + \frac{1}{3}x - \frac{2}{3} = 0$ | **43.** $x^2 - \frac{49}{36} = 0$ | **44.** $x^2 - \frac{64}{25} = 0$

45. $\frac{1}{2}x^2 + \frac{3}{2} = x$ | **46.** $\frac{1}{3}x^2 + 2 = \frac{2}{3}x$ | **47.** $x^2 - 0.5 = -0.06x$

48. $0.5x^2 - 0.4 = 0.25x$ | **49.** $1.2x^2 - 0.6x = -0.5$ | **50.** $2.4x^2 + 4.5 = 6.3x$

Find

(X) *the mistake*

For Exercises 51–54, explain the mistake; then solve correctly.

51. Solve $3x^2 - 7x + 1 = 0$ using the quadratic formula.
$$\frac{-7 \pm \sqrt{(-7)^2 - (4)(3)(1)}}{2(3)} = \frac{-7 \pm \sqrt{49 - 12}}{6}$$
$$= \frac{-7 \pm \sqrt{37}}{6}$$

52. Solve $2x^2 - 6x - 5 = 0$ using the quadratic formula.
$$\frac{6 \pm \sqrt{(-6)^2 - (4)(2)(5)}}{2(2)} = \frac{6 \pm \sqrt{36 - 40}}{4}$$
$$= \frac{3}{2} \pm \frac{1}{2}i$$

53. Solve $x^2 - 2x + 3 = 0$ using the quadratic formula.
$$\frac{-(-2) \pm \sqrt{(-2)^2 - (4)(1)(3)}}{2(1)} = \frac{2 \pm \sqrt{4 - 12}}{2}$$
$$= \frac{2 \pm \sqrt{-8}}{2}$$

54. Solve $x^2 - 8 = 0$ using the quadratic formula.
$$\frac{-(-8) \pm \sqrt{(-8)^2 - (4)(1)(0)}}{2(1)} = \frac{8 \pm \sqrt{64 - 0}}{2}$$
$$= \frac{8 \pm \sqrt{64}}{2}$$
$$= \frac{8 \pm 8}{2}$$
$$= 8, 0$$

For Exercises 55–64, use the discriminant to determine the number and type of solutions for the equation. If the solution(s) are real, state whether they are rational or irrational. See Example 2.

55. $x^2 + 10x = -25$

56. $2x^2 - 8x + 8 = 0$

57. $\frac{1}{4}x^2 - 4x = -4$

58. $\frac{1}{2}x^2 - 5 = -3x$

59. $3x^2 + 10x - 8 = 0$

60. $2x^2 - 3x - 2 = 0$

61. $x^2 - x + 3 = 0$

62. $x^2 + 4x = -5$

63. $x^2 - 6x + 6 = 0$

64. $3x^2 = 13x - 8$

For Exercises 65–74, indicate which of the following methods is the best choice for solving the given equation: factoring, using the square root principle, or using the quadratic formula. Then solve the equation.

65. $x^2 - 81 = 0$

66. $x^2 - 64 = 0$

67. $2x^2 - 4x - 3 = 0$

68. $3x^2 - 6x - 4 = 0$

69. $x^2 + 6x = 0$

70. $x^2 + 14x + 45 = 0$

71. $(x + 7)^2 = 40$

72. $(x - 3)^2 = 48$

73. $x^2 = 8x - 19$

74. $x^2 - 4x = -10$

For Exercises 75–82, find the x- and y-intercepts. See Example 3.

75. $y = x^2 - x - 2$

76. $y = x^2 - 3x + 2$

77. $y = 4x^2 - 12x + 9$

78. $y = 9x^2 + 6x + 1$

79. $y = 2x^2 + 15x - 8$

80. $y = -15x^2 + x + 6$

81. $y = -2x^2 + 3x - 6$

82. $y = -3x^2 - 2x - 5$

For Exercises 83–90, translate to a quadratic equation; then solve using the quadratic formula.

83. A positive integer squared plus five times its consecutive integer is equal to 71. Find the integers.

84. The square of a positive integer minus twice its consecutive integer is equal to 22. Find the integers.

85. A right triangle has side lengths that are three consecutive integers. Use the Pythagorean theorem to find the lengths of those sides. (Remember that the hypotenuse in a right triangle is always the longest side.)

86. A right triangle has side lengths that are consecutive even integers. Use the Pythagorean theorem to find the lengths of those sides. (Remember that the hypotenuse in a right triangle is always the longest side.)

87. The length of a rectangular fence gate is 3.5 feet less than three times the width. Find the length and width of the gate if the area is 34 square feet.

88. A small access door for an attic storage area is designed so that the width is 9.5 feet less than two times the length. Find the length and width if the area is 52 square feet.

89. An architect is experimenting with two different shapes of a room as shown.
  a. Find x so that the rooms have the same area.

  b. Complete the dimensions for the L-shaped room.

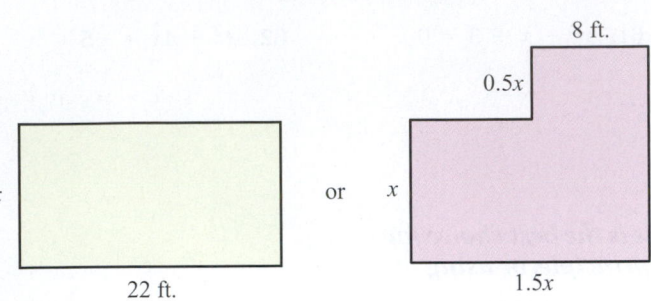

90. A cylinder is to be made so that its volume is equal to that of a sphere with a radius of 9 inches. If a cylinder is to have a height of 4 inches, find its radius.

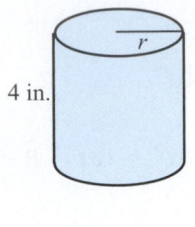 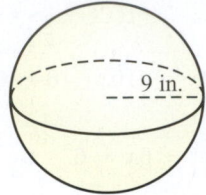

For Exercises 91–94, use the formula $h = \frac{1}{2}gt^2 + v_0t + h_0$, where g represents the acceleration due to gravity, t is the time in flight, v_0 is the initial velocity, and h_0 is the initial height. For Earth, the acceleration due to gravity is -32.2 ft./sec.^{2} or -9.8 m/sec.^{2} Approximate irrational answers to the nearest hundredth. See Example 4.

91. Robbie Knievel jumped the Grand Canyon on a motorcycle. Suppose that on take-off, his motorcycle had a vertical velocity of 48 feet per second and at the end of the launch ramp his altitude was 485 feet above sea level. If he landed on a ramp that was 460 feet above sea level, how long was he in flight?

92. In the extreme games competition, Brian Deegan successfully performed the first 360-degree flip ever attempted in a competition. Suppose his motorcycle had a vertical velocity of 12.8 meters per second from the end of a ramp that was 8 meters high and he landed on a ramp at a point 4 meters above the ground. How long was he in the air?

93. A platform diver dives from a platform that is 10 meters above the water.
  a. Write the equation that describes his height during the dive. (Assume that his initial velocity is 0.)
  b. Find the time it takes the diver to be 5 meters above the water.
  c. Find the time it takes the diver to enter the water.

94. In a cliff diving championship in Acapulco, a diver dives from a cliff at a height of 70 feet.
  a. Write the equation that describes his height during the dive. (Assume that his initial velocity is 0.)

  b. Find the time it takes the diver to be 50 feet above the water.

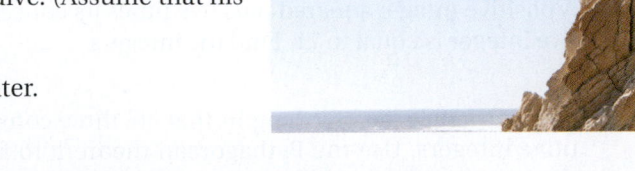

  c. Find the time for it takes the diver to enter the water.

95. The expression $0.5n^2 + 2.5n$ describes the gross income from the sale of a particular software product, where n is the number of units sold in thousands. The expression $4.5n + 16$ describes the cost of producing the n units. Find the number of units that must be produced and sold for the company to break even. (To *break even* means that the gross income and cost are the same.)

96. An economist and marketing manager discover that the expression $2n^2 + 5n$ models the price of a CD based on its demand, where n is the number of units (in millions) the market demands. The expression $-0.5n^2 + 17.1$ describes the price of the CD based on the number of units (also in millions) supplied to the market. Find the number of units that must be demanded and supplied so that the price based on demand is equal to the price based on supply.

★**97.** For the equation $2x^2 - 5x + c = 0$,
  a. Find c so that the equation has only one rational number solution.

  b. Find the range of values of c for which the equation has two real-number solutions.

  c. Find the range of values of c for which the equation has no real-number solution.

★**98.** For the equation $4x^2 + 6x + c = 0$,
  a. Find c so that the equation has only one rational number solution.

  b. Find the range of values of c for which the equation has two real-number solutions.

  c. Find the range of values of c for which the equation has no real-number solution.

★**99.** For the equation $ax^2 + 12x + 8 = 0$,
  a. Find a so that the equation has only one rational number solution.

  b. Find the range of values of a for which the equation has two real-number solutions.

  c. Find the range of values of a for which the equation has no real-number solution.

★**100.** For the equation $3x^2 + bx + 8 = 0$,
  a. Find b so that the equation has only one rational number solution.

  b. Find all positive values of b for which the equation has two real-number solutions.

  c. Find all positive values of b for which the equation has no real-number solution.

Puzzle Problem

In the equation shown, A, B, C, D, and E are five consecutive positive integers where $A < B < C < D < E$. What are the integers?

$$A^2 + B^2 + C^2 = D^2 + E^2$$

Review Exercises

Exercises 1–4 **EXPRESSIONS**

[6.2] *For Exercises 1 and 2, factor.*

  1. $u^2 - 9u + 14$

  2. $3u^2 - 2u - 16$

For Exercises 3 and 4, simplify.

[5.4]   **3.** $(x^2)^2$

[10.2]   **4.** $(x^{1/3})^2$

Exercises 5 and 6 **EQUATIONS AND INEQUALITIES**

For Exercises 5 and 6, solve.

[6.6]   **5.** $5u^2 + 13u = 6$

[7.6]   **6.** $\dfrac{7}{3u} = \dfrac{5}{u} - \dfrac{1}{u-5}$

11.3 Solving Equations That Are Quadratic in Form

Objectives

1. Solve equations by rewriting them in quadratic form.
2. Solve equations that are quadratic in form by using substitution.
3. Solve application problems using equations that are quadratic in form.

Many equations that are not quadratic equations are **quadratic in form** and can be solved using the same methods used to solve quadratic equations.

Definition An equation is **quadratic in form** if it can be rewritten as a quadratic equation $au^2 + bu + c = 0$, where $a \neq 0$ and u is a variable or an expression.

Objective 1 Solve equations by rewriting them in quadratic form.

Equations with Rational Expressions

In Section 7.6, we solved equations containing rational expressions by multiplying both sides of the equation by the LCD. Those rewritten equations are often quadratic. Remember that equations containing rational expressions sometimes have extraneous solutions.

Connection Recall that an apparent solution for an equation with rational expressions is extraneous if it causes one or more of the denominators to equal 0.

Example 1 Solve $\dfrac{3}{x+1} = 1 - \dfrac{3}{x(x+1)}$.

SOLUTION $x(x+1) \cdot \dfrac{3}{x+1} = x(x+1)\left(1 - \dfrac{3}{x(x+1)}\right)$ **Multiply both sides by the LCD $x(x+1)$.**

$x(x+1) \cdot \dfrac{3}{x+1} = x(x+1) \cdot 1 - x(x+1) \cdot \dfrac{3}{x(x+1)}$ **Distribute $x(x+1)$.**

$3x = x^2 + x - 3$ **Simplify both sides.**

$0 = x^2 - 2x - 3$ **Subtract $3x$ from both sides to get the quadratic form $ax^2 + bx + c = 0$.**

$0 = (x - 3)(x + 1)$ **Factor.**

$x - 3 = 0 \quad \text{or} \quad x + 1 = 0$ **Use the zero-factor theorem.**

$x = 3 \qquad\qquad x = -1$ **Solve each equation.**

Checks $\quad x = 3$ $\qquad\qquad\qquad\qquad x = -1$

$\dfrac{3}{3+1} \overset{?}{=} 1 - \dfrac{3}{3(3+1)} \qquad\qquad \dfrac{3}{-1+1} \overset{?}{=} 1 - \dfrac{3}{-1(-1+1)}$

$\dfrac{3}{4} \overset{?}{=} 1 - \dfrac{3}{12} \qquad\qquad\qquad \dfrac{3}{0} \overset{?}{=} 1 - \dfrac{3}{0}$

Note: The expression $\dfrac{3}{0}$ is undefined, so -1 is extraneous.

$\dfrac{3}{4} = \dfrac{3}{4}$ **True**

The solution is 3. (-1 is extraneous.)

Your Turn 1 Solve $\dfrac{x}{x-2} = \dfrac{6}{x} + \dfrac{4}{x(x-2)}$.

Equations Containing Radicals

In Section 10.6, we found that after the power rule was used on equations containing radicals, the result was often a quadratic equation. Remember that these radical equations sometimes have extraneous solutions.

Answer **Your Turn 1**
4 (2 is extraneous)

Example 2 Solve.

a. $\sqrt{x} + x = 6$

Connection Recall that an apparent solution for an equation containing radicals is extraneous if it makes the original equation false.

SOLUTION

$$\sqrt{x} = 6 - x \qquad \text{Subtract } x \text{ from both sides to isolate the radical.}$$
$$(\sqrt{x})^2 = (6 - x)^2 \qquad \text{Square both sides.}$$
$$x = 36 - 12x + x^2 \qquad \text{Simplify both sides.}$$
$$0 = x^2 - 13x + 36 \qquad \text{Subtract } x \text{ from both sides to write in quadratic form.}$$
$$0 = (x - 4)(x - 9) \qquad \text{Factor.}$$
$$x - 4 = 0 \quad \text{or} \quad x - 9 = 0 \qquad \text{Use the zero-factor theorem.}$$
$$x = 4 \qquad\qquad x = 9 \qquad \text{Solve each equation.}$$

Checks $\quad x = 4 \qquad\qquad\qquad x = 9$

$$\sqrt{4} + 4 \overset{?}{=} 6 \qquad\qquad \sqrt{9} + 9 \overset{?}{=} 6$$
$$2 + 4 = 6 \quad \text{True} \qquad 3 + 9 = 6 \quad \text{False; so 9 is extraneous.}$$

The solution is 4. (9 is extraneous.)

b. $\sqrt{x - 1} = 2x - 1$

SOLUTION $\quad (\sqrt{x - 1})^2 = (2x - 1)^2 \qquad$ Square both sides of the equation.

$$x - 1 = 4x^2 - 4x + 1 \qquad \text{Simplify both sides.}$$
$$0 = 4x^2 - 5x + 2 \qquad \text{Subtract } x \text{ from and add 1 to both sides to write in quadratic form.}$$

We cannot factor $4x^2 - 5x + 2$, so we use the quadratic formula.

Note: Written in standard form, these solutions are $\frac{5}{8} + \frac{\sqrt{7}}{8}i$ and $\frac{5}{8} - \frac{\sqrt{7}}{8}i$.

$$x = \frac{-(-5) \pm \sqrt{(-5)^2 - 4(4)(2)}}{2(4)} = \frac{5 \pm \sqrt{-7}}{8} = \frac{5 \pm i\sqrt{7}}{8}$$

Check Because the solutions are complex, we will not check them.

Answers **Your Turn 2**

a. 4 (1 is extraneous)

b. $\dfrac{2 \pm 2i\sqrt{2}}{3}$

Your Turn 2 Solve.

a. $x - \sqrt{x} - 2 = 0$

b. $\sqrt{6x - 11} = 3x - 1$

Objective ② **Solve equations that are quadratic in form by using substitution.** Recall that an equation that is quadratic in form can be written as $au^2 + bu + c = 0$, where $a \neq 0$ and u is an *expression*. We now explore a method for solving these equations where we substitute u for an expression. To use substitution, it is important to note a pattern with the exponents of the terms in a quadratic equation. Notice that the degree of the first term, au^2, is 2; the degree of the middle term, bu, is 1; and the third term is a constant. If a trinomial has one term with an expression raised to the second power, a second term with that same expression raised to the first power, and a third term that is a constant, the equation is quadratic in form and we can use substitution to solve it.

Consider the equation $x^4 - 13x^2 + 36 = 0$. Notice that we can rewrite the equation as $(x^2)^2 - 13(x^2) + 36 = 0$, so it is quadratic in form. By substituting u for each x^2, we have a "friendlier" form of quadratic equation, which we can then solve by factoring, completing the square, or using the quadratic formula.

Note: Given an equation containing a trinomial, if the degree of the first term is twice the degree of the middle term and the third term is a constant, try using substitution.

$$(x^2)^2 - 13(x^2) + 36 = 0$$
$$u^2 - 13u + 36 = 0 \qquad \text{Substitute } u \text{ for } x^2.$$
$$(u - 9)(u - 4) = 0 \qquad \text{Factor.}$$
$$u - 9 = 0 \quad \text{or} \quad u - 4 = 0 \qquad \text{Use the zero-factor theorem.}$$
$$u = 9 \qquad\qquad u = 4 \qquad \text{Solve each equation.}$$

Note that these solutions are for u, not x. We must substitute x^2 back in place of u to finish solving for x.

Note: The equation $x^4 - 13x^2 + 36 = 0$ has four solutions: 3, −3, 2, and −2.

$x^2 = 9$	$x^2 = 4$	Substitute x^2 for u.
$x = \pm\sqrt{9}$	$x = \pm\sqrt{4}$	Use the square root principle.
$x = \pm 3$	$x = \pm 2$	Simplify.

Our example suggests the following procedure.

Procedure **Using Substitution to Solve Equations That Are Quadratic in Form**

To solve equations that are quadratic in form using substitution,

1. Rewrite the equation so that it is in the form $au^2 + bu + c = 0$.
2. Solve the quadratic equation for u.
3. Substitute for u and solve.
4. Check the solutions.

Example 3   Solve.

a. $(x + 2)^2 - 2(x + 2) - 8 = 0$

SOLUTION   If we substitute u for $x + 2$, we have an equation in the form $au^2 + bu + c = 0$.

	$(x + 2)^2 - 2(x + 2) - 8 = 0$	
	$u^2 - 2u - 8 = 0$	Substitute u for $x + 2$.
	$(u - 4)(u + 2) = 0$	Factor.
$u - 4 = 0$ or	$u + 2 = 0$	Use the zero-factor theorem.
$u = 4$	$u = -2$	Solve each equation for u.
$x + 2 = 4$	$x + 2 = -2$	Substitute $x + 2$ for u.
$x = 2$	$x = -4$	Solve each equation for x.

Check   Verify that 2 and −4 make $(x + 2)^2 - 2(x + 2) - 8 = 0$ true. We will leave this check to the reader.

b. $x^{2/3} - x^{1/3} - 6 = 0$

SOLUTION   Because $x^{2/3} - x^{1/3} - 6 = 0$ can be written as $(x^{1/3})^2 - x^{1/3} - 6 = 0$, it is quadratic in form. We substitute u for $x^{1/3}$.

	$(x^{1/3})^2 - x^{1/3} - 6 = 0$	Rewrite in quadratic form.
	$u^2 - u - 6 = 0$	Substitute u for $x^{1/3}$.
	$(u - 3)(u + 2) = 0$	Factor.
$u - 3 = 0$ or	$u + 2 = 0$	Use the zero-factor theorem.
$u = 3$	$u = -2$	Solve each equation for u.
$x^{1/3} = 3$	$x^{1/3} = -2$	Substitute $x^{1/3}$ for u.
$(x^{1/3})^3 = 3^3$	$(x^{1/3})^3 = (-2)^3$	Cube both sides of the equations.
$x = 27$	$x = -8$	Simplify.

Checks

$x = 27$	$x = -8$
$27^{2/3} - 27^{1/3} - 6 \overset{?}{=} 0$	$(-8)^{2/3} - (-8)^{1/3} - 6 \overset{?}{=} 0$
$(\sqrt[3]{27})^2 - \sqrt[3]{27} - 6 \overset{?}{=} 0$	$(\sqrt[3]{-8})^2 - \sqrt[3]{-8} - 6 \overset{?}{=} 0$
$3^2 - 3 - 6 \overset{?}{=} 0$	$(-2)^2 - (-2) - 6 \overset{?}{=} 0$
$9 - 3 - 6 \overset{?}{=} 0$	$4 + 2 - 6 \overset{?}{=} 0$
$0 = 0$ **True**	$0 = 0$ **True**

Your Turn 3 Solve.

a. $x^4 - 10x^2 + 9 = 0$ **b.** $(n-2)^2 + 4(n-2) - 12 = 0$ **c.** $x^{2/3} - x^{1/3} - 2 = 0$

Objective ❸ Solve application problems using equations that are quadratic in form.

Example 4 Solve.

a. The average speed of a car is 10 miles per hour more than the average speed of a bus. The bus takes 1 hour longer than the car to travel 200 miles. Find how long it takes the car to travel 200 miles.

Understand Both the car and bus travel 200 miles, but it takes the bus 1 hour longer than it takes the car. The rate of the car is 10 miles per hour more than the rate of the bus.

Plan We use a table to organize the information, then write an equation, which we can solve.

Execute We let x represent the time for the car to travel 200 miles. Because the bus travels 1 more hour, $x + 1$ describes the time for the bus to travel 200 miles.

Vehicle	d	t	r
bus	200	$x + 1$	$\dfrac{200}{x+1}$
car	200	x	$\dfrac{200}{x}$

Note: We use $r = \dfrac{d}{t}$ to describe each rate.

Because the rate of the car is 10 miles per hour faster than the rate of the bus, we can say that (the rate of the car) = (the rate of the bus) + 10.

$$\frac{200}{x} = \frac{200}{x+1} + 10$$

$$x(x+1)\left(\frac{200}{x}\right) = x(x+1)\left(\frac{200}{x+1} + 10\right) \qquad \text{Multiply both sides by the LCD } x(x+1).$$

$$(x+1)(200) = x(x+1)\left(\frac{200}{x+1}\right) + x(x+1)(10) \qquad \text{Multiply on the left and distribute on the right.}$$

$$200x + 200 = 200x + 10x^2 + 10x \qquad \text{Simplify.}$$

$$0 = 10x^2 + 10x - 200 \qquad \text{Subtract } 200x \text{ and } 200 \text{ from both sides to get quadratic form.}$$

$$0 = x^2 + x - 20 \qquad \text{Divide both sides by 10.}$$

$$0 = (x+5)(x-4) \qquad \text{Factor.}$$

$$x + 5 = 0 \quad \text{or} \quad x - 4 = 0 \qquad \text{Use the zero-factor theorem.}$$

$$x = -5 \qquad\qquad x = 4 \qquad \text{Solve each equation.}$$

Answer Because time cannot be negative, it takes the car 4 hours to travel 200 miles.

Check The rate of the car is $\dfrac{200}{x} = \dfrac{200}{4} = 50$ miles per hour; so in 4 hours, the car travels $50(4) = 200$ miles. The bus travels 200 miles in $4 + 1 = 5$ hours at a rate of $\dfrac{200}{4+1} = \dfrac{200}{5} = 40$ miles per hour; so in 5 hours, the bus travels $5(40) = 200$ miles. Notice also that the car's rate is 10 miles per hour faster than the bus's.

b. Bobby and Pam manufacture and install blinds. Working together, they can install blinds in every window of an average-sized house in $1\dfrac{1}{3}$ hours. Working alone, Bobby takes 2 hours longer than Pam to do the same installation. Working alone, how long does each person take to install blinds in an average-sized house?

Understand   We are given the time it takes Bobby and Pam working together and are asked to find how long it takes each person individually. We also know that it takes Bobby 2 hours longer than it takes Pam.

Plan   We use a table to organize the information, write an equation, and then solve.

Execute   We let x represent the number of hours it takes Pam to install blinds working alone. Because Bobby takes 2 more hours, his time working alone is represented by $x + 2$.

Worker	Time to Complete the Job Alone	Rate of Work	Time at Work	Part of Job Completed
Pam	x	$\dfrac{1}{x}$	$\dfrac{4}{3}$	$\dfrac{4}{3x}$
Bobby	$x + 2$	$\dfrac{1}{x + 2}$	$\dfrac{4}{3}$	$\dfrac{4}{3(x + 2)}$

Note: Multiplying the rate of work and the time at work gives an expression of the part of the job completed.

The total job in this case is 1 average-sized house, so we can write an equation that combines Pam's and Bobby's individual expressions for work completed and set this sum equal to 1.

$$(\text{Part Pam does}) + (\text{Part Bobby does}) = 1 \ (\text{the entire job})$$

$$\frac{4}{3x} + \frac{4}{3(x + 2)} = 1$$

$$3x(x + 2)\left(\frac{4}{3x} + \frac{4}{3(x + 2)}\right) = 3x(x + 2)(1) \qquad \text{Multiply both sides by the LCD } 3x(x + 2).$$

$$3x(x + 2)\left(\frac{4}{3x}\right) + 3x(x + 2)\left(\frac{4}{3(x + 2)}\right) = 3x^2 + 6x \qquad \text{Distribute on the left and multiply on the right.}$$

$$4(x + 2) + 4x = 3x^2 + 6x \qquad \text{Continue simplifying.}$$

$$4x + 8 + 4x = 3x^2 + 6x \qquad \text{Distribute.}$$

$$8x + 8 = 3x^2 + 6x \qquad \text{Combine like terms.}$$

$$0 = 3x^2 - 2x - 8 \qquad \text{Subtract } 8x \text{ and } 8 \text{ from both sides.}$$

$$0 = (3x + 4)(x - 2) \qquad \text{Factor.}$$

$$3x + 4 = 0 \quad \text{or} \quad x - 2 = 0 \qquad \text{Use the zero-factor theorem.}$$

$$3x = -4 \qquad\qquad x = 2 \qquad \text{Solve each equation.}$$

$$x = -\frac{4}{3}$$

Answer   Because negative time makes no sense in the context of this problem, it takes Pam 2 hours to install blinds working alone and Bobby $x + 2 = 2 + 2 = 4$ hours working alone.

Check   Because Pam can install blinds in 2 hours, she can install them in $\dfrac{1}{2}$ of an average-sized house in 1 hour. Because Bobby takes 4 hours to do the same work, he can install blinds in $\dfrac{1}{4}$ of an average-sized house in 1 hour. Together, they can install blinds in $\dfrac{1}{2} + \dfrac{1}{4} = \dfrac{2}{4} + \dfrac{1}{4} = \dfrac{3}{4}$ of an average-sized house in 1 hour; so in $1\dfrac{1}{3}$ hours, they can install blinds in $1\dfrac{1}{3} \cdot \dfrac{3}{4} = \dfrac{4}{3} \cdot \dfrac{3}{4} = 1$ average-sized house.

Your Turn 4 Solve.

a. The average speed of a passenger train is 25 miles per hour more than the average speed of a car. The time required for the car to travel 300 miles is 2 hours more than the time required for the train to travel the same number of miles. Find the average speed of the car.

b. Terri and Tommy run a flea market. It takes Terri 2 hours longer to put out the merchandise than it does Tommy. If they can put out the merchandise together in $1\frac{7}{8}$ hours, how long does it take each working alone?

Answers   Your Turn 4

a. 50 mph

b. It takes Tommy 3 hr. and Terri 5 hr.

11.3 Exercises   For Extra Help   MyMathLab

PRACTICE   WATCH   DOWNLOAD

Note: Exercises marked with a ★ represent challenging exercises.

1. What does it mean to say that an equation is quadratic in form?

2. Is $x^{3/4} - x^{1/4} - 6 = 0$ quadratic in form? Why or why not?

3. If you solved $3\left(\dfrac{x+2}{3}\right)^2 + 13\left(\dfrac{x+2}{3}\right) - 10 = 0$ using substitution, what would your substitution be?

4. If you solved $x^4 - 7x^2 + 12 = 0$ using substitution, what would your substitution be?

5. If Alisha can clean her house in x hours, what part of her house can she clean in 1 hour?

6. If a car travels 400 miles in x hours, how many miles does it travel in 1 hour?

For Exercises 7–18, solve the equations with rational expressions. Identify any extraneous solutions. See Example 1.

7. $\dfrac{60}{x+2} = \dfrac{60}{x} - 5$

8. $\dfrac{120}{x+2} = \dfrac{120}{x} - 5$

9. $\dfrac{1}{x} + \dfrac{1}{x+2} = \dfrac{3}{4}$

10. $\dfrac{1}{x} + \dfrac{3}{x-4} = \dfrac{7}{8}$

11. $\dfrac{1}{p-4} + \dfrac{1}{4} = \dfrac{8}{p^2-16}$

12. $\dfrac{1}{y-5} - \dfrac{10}{y^2-25} = -\dfrac{1}{5}$

13. $\dfrac{6}{2y+5} = \dfrac{2}{y+5} + \dfrac{1}{5}$

14. $\dfrac{6}{2x+3} = \dfrac{2}{x-6} + \dfrac{4}{3}$

15. $1 + 2x^{-1} - 8x^{-2} = 0$

16. $1 + 5x^{-1} + 6x^{-2} = 0$

17. $3 + 13x^{-1} - 10x^{-2} = 0$

18. $3 - 5x^{-1} - 12x^{-2} = 0$

For Exercises 19–30, solve the equations with radical expressions. Identify any extraneous solutions. See Example 2.

19. $x - 8\sqrt{x} + 15 = 0$

20. $x - 3\sqrt{x} + 2 = 0$

21. $2x - 5\sqrt{x} - 7 = 0$

22. $3x + 4\sqrt{x} - 4 = 0$

23. $\sqrt{2a+5} = 3a - 3$

24. $\sqrt{3b+1} = 5b - 3$

25. $\sqrt{2m-8} - m - 1 = 0$

26. $\sqrt{8x-9} - x - 4 = 0$

27. $\sqrt{4x+1} = \sqrt{x+2} + 1$

28. $\sqrt{4x+1} = \sqrt{x+3} + 2$

29. $\sqrt{2x+1} - \sqrt{3x+4} = -1$

30. $\sqrt{2x-1} - \sqrt{4x+5} = -2$

For Exercises 31–52, solve using substitution. Identify any extraneous solutions. See Example 3.

31. $x^4 - 10x^2 + 9 = 0$

32. $x^4 - 13x^2 + 36 = 0$

33. $4x^4 - 13x^2 + 9 = 0$

34. $9x^4 - 37x^2 + 4 = 0$

35. $x^4 + 5x^2 - 36 = 0$

36. $x^4 - 12x^2 - 64 = 0$

37. $(x + 2)^2 + 6(x + 2) + 8 = 0$

38. $(x - 3)^2 + 2(x - 3) - 15 = 0$

39. $2(x + 3)^2 - 9(x + 3) - 5 = 0$

40. $3(x - 1)^2 + 4(x - 1) - 4 = 0$

41. $\left(\dfrac{x - 1}{2}\right)^2 + 8\left(\dfrac{x - 1}{2}\right) + 15 = 0$

42. $\left(\dfrac{x + 2}{3}\right)^2 - 5\left(\dfrac{x + 2}{3}\right) - 6 = 0$

43. $2\left(\dfrac{x + 2}{2}\right)^2 + \left(\dfrac{x + 2}{2}\right) - 3 = 0$

44. $3\left(\dfrac{x - 3}{3}\right)^2 + 5\left(\dfrac{x - 3}{3}\right) + 2 = 0$

45. $x^{2/3} - 5x^{1/3} + 6 = 0$

46. $x^{2/3} - 3x^{1/3} + 2 = 0$

47. $2x^{2/3} - 3x^{1/3} - 2 = 0$

48. $3x^{2/3} - 4x^{1/3} - 4 = 0$

49. $x^{1/2} - 5x^{1/4} + 6 = 0$

50. $x^{1/2} - 4x^{1/4} + 3 = 0$

51. $5x^{1/2} + 8x^{1/4} - 4 = 0$

52. $2x^{1/2} - x^{1/4} - 3 = 0$

For Exercises 53–64, solve. See Example 4.

53. The average rate of a bus is 15 miles per hour more than the average rate of a truck. The truck takes 1 hour longer than the bus to travel 180 miles. How long does the bus take to travel 180 miles?

54. A charter business has two types of small planes, a jet and a twin propeller plane. The average air speed of the prop plane is 120 miles per hour less than the jet's. The prop plane takes 1 hour longer to travel 720 miles than the jet. How long does the prop plane take to travel the 720 miles?

★**55.** The average speed of Robert Cheruiyot, who won the Boston Marathon in 2008, was 3.79 miles per hour faster than the person who finished 1 hour behind him. Find the time Robert took to run the 26-mile course.

★**56.** In the 2005 Tour de France, Lance Armstrong's average speed was 0.0378 kilometer per hour faster than the second-place finisher, Ivan Basso, who finished 4 minutes and 40 seconds behind Armstrong. Find Armstrong's time to complete the 3415-kilometer race.

57. Suppose the time required for a bus to travel 360 miles is 3 hours more than the time required for a motorcycle to travel 300 miles. If the average rate of the motorcycle is 15 miles per hour more than the average rate of the bus, find the average rate of the bus.

58. Suppose the time required for a truck to travel 400 miles is 3 hours more than the time required for a car to travel 300 miles. If the average rate of the car is 10 miles per hour more than the average rate of the truck, find the average rate of the truck.

59. After training hard for a year, a novice cyclist discovers that she has increased her average rate by 6 miles per hour and can travel 36 miles in 1 less hour than she did a year ago.
   a. What was her old average rate? What is her new average rate?
   b. What was her old time to travel 36 miles? What is her new time for 36 miles?

60. A high school track coach determines that his fastest long-distance runner runs 2 miles per hour faster than his slowest runner. Compared to the faster runner the slower runner takes a half hour longer to run 12 miles.
   a. Find the average rates of both runners.
   b. Find the time for both runners to run 12 miles.

61. Billy and Jody are commercial fishermen. Working alone, it takes Billy 2 hours longer to run the hoop nets than it takes Jody working alone. Together they can run the hoop nets in $2\frac{2}{5}$ hours. How long does it take each working alone?

62. Using a riding mower, Fran can mow the grass at the campground in 4 hours less time than it takes Donnie using a push mower. Together they can mow the grass in $2\frac{2}{3}$ hours. How long does it take each working alone?

★ **63.** A newspaper has two presses, one of which is older than the other. By itself, the newer press can print all of the copies for a typical day in a half hour less time than the older press takes. When running at the same time, the presses print all of a typical day's copies in 2 hours. How long would it take to print a day's worth if they worked alone?

★ **64.** A school's copy center prints the school newsletter in a half hour using two copiers running at the same time. If the center uses only one copier, the faster of the two copiers takes 6 minutes less time than the other copier to print all of the newsletters. How long does each copier take to print the newsletters working alone?

 On a guitar, the frequency of the vibrating string is related to the tension on the string. Suppose two strings of the same diameter and length are placed on an instrument and wound to different tensions. The formula $\dfrac{F_1^2}{F_2^2} = \dfrac{T_1}{T_2}$ describes the relationship between their frequency and tension.

★ **65.** One string is wound on a guitar to a tension of 50 pounds. A second string of the same diameter is wound to 60 pounds. If the string with the greater tension has a frequency that is 40 vibrations per second greater than the other string, what are the frequencies of the two strings? Round to the nearest ten.

★ **66.** Two strings of the same diameter are wound on a banjo to different tensions: one at 80 pounds and the other at 90 pounds. The frequency of the string under less tension is 20 vibrations per second less than the other string. What are the frequencies of the two strings? Round to the nearest ten.

Review Exercises

Exercises 1 and 2 EXPRESSIONS

[1.7] **1.** Evaluate $-\dfrac{b}{2a}$ when a is 2 and b is 8.

[1.7] **2.** Evaluate $\dfrac{4ac - b^2}{4a}$ when $a = 2$, $b = -8$, and $c = 5$.

Exercises 3–5 EQUATIONS AND INEQUALITIES

For Exercises 3 and 4, find the x- and y-intercepts.

[4.3] **3.** $2x + 3y = 6$

[4.7] **4.** If $f(x) = 2x^2 - 3x + 1$, find $f(-1)$.

[6.7] *For Exercises 5 and 6, graph.*

   5. $f(x) = x^2 - 3$

   6. $f(x) = 3x^2$

11.4 Graphing Quadratic Functions

Objectives

1. Graph quadratic functions of the form $f(x) = ax^2$.
2. Graph quadratic functions of the form $f(x) = ax^2 + k$.
3. Graph quadratic functions of the form $f(x) = a(x - h)^2$.
4. Graph quadratic functions of the form $f(x) = a(x - h)^2 + k$.
5. Graph quadratic functions of the form $f(x) = ax^2 + bx + c$.
6. Solve applications involving parabolas.

In Sections 6.7 and 8.4, we learned that quadratic functions have the form $f(x) = ax^2 + bx + c$, where a, b, and c are real numbers and $a \neq 0$. By plotting many ordered pairs, we found that the graphs of these functions are parabolas that open up if $a > 0$ and down if $a < 0$. By replacing x with 0, we found that the y-intercept is $(0, c)$.

By replacing y (or $f(x)$) with 0, we found the x-intercepts. We also learned that every parabola will have symmetry about a line called its *axis of symmetry* and the axis of symmetry always passes through a point called the *vertex*, which is the lowest point on a parabola that opens upward or the highest point on a parabola that opens downward. Look at the following graphs.

x	$f(x) = x^2$
-3	9
-2	4
-1	1
0	0
1	1
2	4
3	9

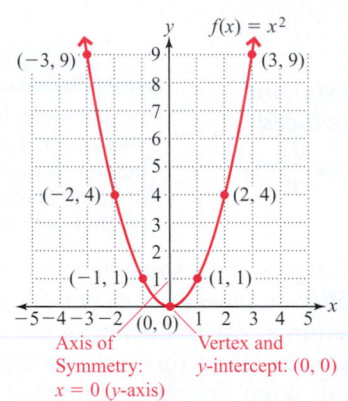

x	$f(x) = -x^2 + 4$
-3	-5
-2	0
-1	3
0	4
1	3
2	0
3	-5

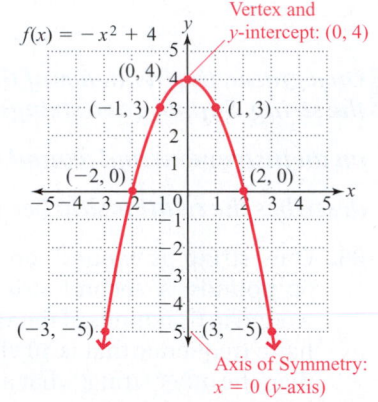

Notice that the graph of $f(x) = x^2$ opens upward because $a = 1$, which is positive, whereas the graph of $f(x) = -x^2 + 4$ opens downward because $a = -1$, which is negative. In this section, we learn more about the graphs of quadratic functions. Later in the section, we revisit the form $f(x) = ax^2 + bx + c$ and learn how to determine a parabola's vertex and axis of symmetry from the values of a and b.

Objective 1 Graph quadratic functions of the form $f(x) = ax^2$. First, we consider $f(x) = ax^2$ to discover more about how a affects the parabola. We will see that a affects not only whether the parabola opens up or down but also how wide the parabola is.

Example 1 Compare the graphs of each function.

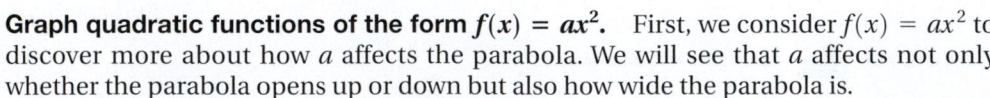

a. $f(x) = \dfrac{1}{4}x^2$, $g(x) = \dfrac{1}{3}x^2$, $h(x) = \dfrac{1}{2}x^2$, $k(x) = x^2$, and $m(x) = 2x^2$

SOLUTION We graph all five functions on the same grid.

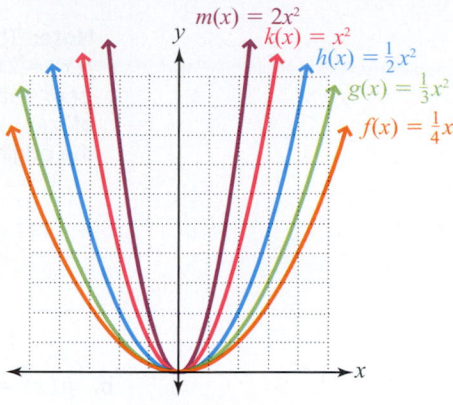

Note: All of these graphs open up because a is positive. Also notice that as a increases, the parabolas appear narrower.

b. $f(x) = -\dfrac{1}{4}x^2$, $g(x) = -\dfrac{1}{3}x^2$, $h(x) = -\dfrac{1}{2}x^2$, $k(x) = -x^2$, and $m(x) = -2x^2$

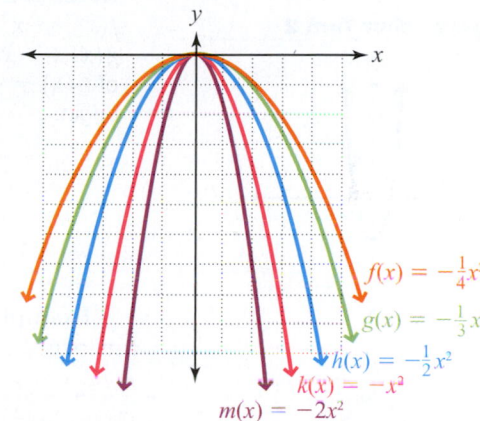

Note: All of these graphs open down because a is negative. Also notice that as $|a|$ increases, the parabolas appear narrower.

Answers   Your Turn 1

a.

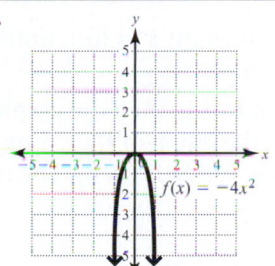

b.

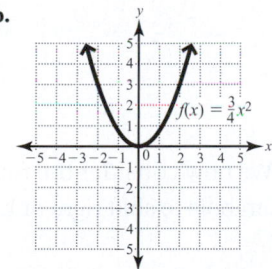

Example 1 suggests the following conclusions about functions of the form $f(x) = ax^2$.

Conclusion Given a function in the form $f(x) = ax^2$, the axis of symmetry is $x = 0$ (the y-axis) and the vertex is $(0, 0)$. Also, the greater the absolute value of a, the narrower the parabola appears (or the smaller the absolute value of a, the wider the parabola appears).

Your Turn 1   Graph the following functions.

    a. $f(x) = -4x^2$ **b.** $f(x) = \dfrac{3}{4}x^2$

Objective ② Graph quadratic functions of the form $f(x) = ax^2 + k$. We have learned that the vertex of a parabola of the form $f(x) = ax^2$ is at $(0, 0)$ and the axis of symmetry is $x = 0$ (the y-axis). Now let's consider quadratic functions of the form $f(x) = ax^2 + k$, where k is a constant. We will see that for the same value of a, the graphs of $f(x) = ax^2 + k$ and $f(x) = ax^2$ have the same width and shape but the graph of $f(x) = ax^2 + k$ is shifted up or down k units on the y-axis from the origin so that the vertex is at $(0, k)$.

Example 2   Graph the following functions.

    a. $g(x) = 2x^2 + 3$ and $h(x) = 2x^2 - 4$

Solution We graph both functions on the same grid.

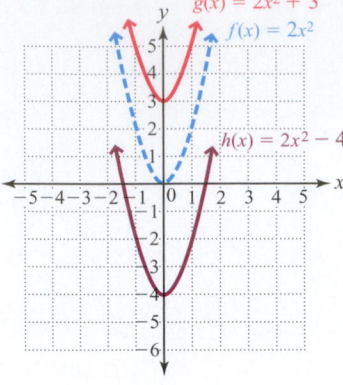

Note: The vertex of the "basic" function $f(x) = 2x^2$ is the origin $(0, 0)$. The vertex of $g(x) = 2x^2 + 3$ is shifted up 3 and the vertex of $h(x) = 2x^2 - 4$ is shifted down 4 from the origin.

b. $g(x) = -x^2 + 4$ and $h(x) = -x^2 - 2$

Solution

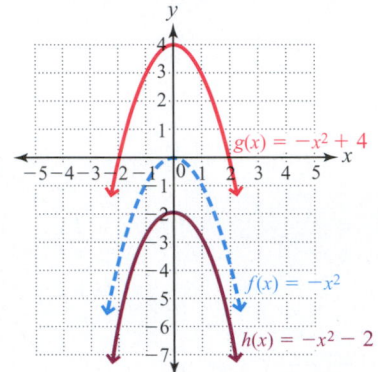

Note: The vertex of the "basic" function $f(x) = -x^2$ is the origin $(0, 0)$. The vertex of $g(x) = -x^2 + 4$ is shifted up 4 and the vertex of $h(x) = -x^2 - 2$ is shifted down 2 from the origin.

Answers Your Turn 2

a.

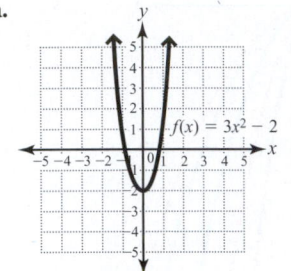

$f(x) = 3x^2 - 2$

b.

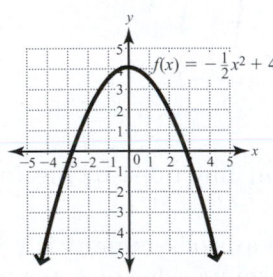

$f(x) = -\frac{1}{2}x^2 + 4$

Example 2 suggests the following conclusion about the constant k in functions of the form $f(x) = ax^2 + k$.

Conclusion Given a function in the form $f(x) = ax^2 + k$, if $k > 0$, then the graph of $f(x) = ax^2$ is shifted k units *up* from the origin. If $k < 0$, then the graph of $f(x) = ax^2$ is shifted $|k|$ units *down* from the origin. The new position of the vertex is $(0, k)$. The axis of symmetry is $x = 0$ (the y-axis).

Your Turn 2 Graph the following functions.

a. $f(x) = 3x^2 - 2$ **b.** $f(x) = -\frac{1}{2}x^2 + 4$

Objective ③ Graph quadratic functions of the form $f(x) = a(x - h)^2$. We now consider the form $f(x) = a(x - h)^2$, and we will see that the constant h causes the parabola to shift right or left.

Example 3 Graph $m(x) = 2(x - 3)^2$ and $n(x) = 2(x + 4)^2$.

Solution

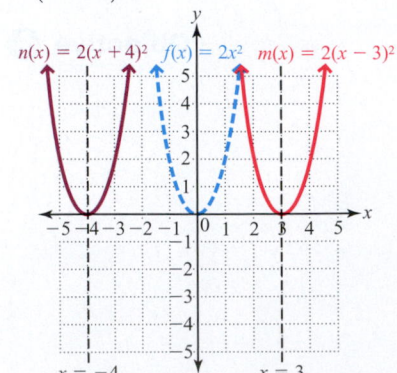

Note: The vertex of the "basic" function $f(x) = 2x^2$ is $(0, 0)$, and the axis of symmetry is $x = 0$. The graph of $m(x) = 2(x - 3)^2$ has the same shape as $f(x) = 2x^2$, only the vertex and axis of symmetry are shifted **right** 3 units from the origin along the x-axis so that they are $(3, 0)$ and $x = 3$, respectively. Similarly, the vertex and axis of symmetry of $n(x) = 2(x + 4)^2$ are shifted **left** 4 units from the origin along the x-axis so that they are $(-4, 0)$ and $x = -4$, respectively.

Example 3 suggests the following conclusion about h in $f(x) = a(x - h)^2$.

Conclusion Given a function in the form $f(x) = a(x - h)^2$, if $h > 0$, then the graph of $f(x) = ax^2$ is shifted h units *right* from the origin. If $h < 0$, then the graph of $f(x) = ax^2$ is shifted $|h|$ units *left* from the origin. The new position of the vertex is $(h, 0)$, and the axis of symmetry is $x = h$.

Your Turn 3 Graph the following functions.

  a. $f(x) = -2(x + 1)^2$ **b.** $g(x) = \dfrac{1}{3}(x - 4)^2$

Objective ④ Graph quadratic functions of the form $f(x) = a(x - h)^2 + k$. Examples 2 and 3 suggest that the graph of a function of the form $f(x) = a(x - h)^2 + k$ has the same shape as $f(x) = ax^2$ but the vertex is shifted from the origin to (h, k) and the axis of symmetry is shifted from $x = 0$ to $x = h$. We can summarize all we have learned with the following rule.

Rule **Parabola with Vertex (h, k)**

The graph of a function in the form $f(x) = a(x - h)^2 + k$ is a parabola with vertex at (h, k). The equation of the axis of symmetry is $x = h$. The parabola opens upward if $a > 0$ and downward if $a < 0$. The larger the $|a|$, the narrower the graph.

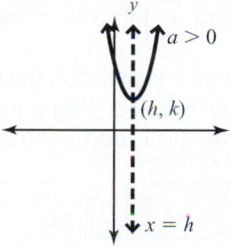 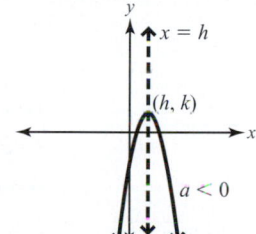

Now let's graph quadratic functions of the form $f(x) = a(x - h)^2 + k$. Recall that the domain of a function is the set of all possible x-values and the range is the set of all possible y-values.

Example 4 Given $f(x) = 2(x - 3)^2 + 1$, determine whether the graph opens up or down, find the vertex and axis of symmetry, and draw the graph. Find the domain and range.

Answers Your Turn 3

a.

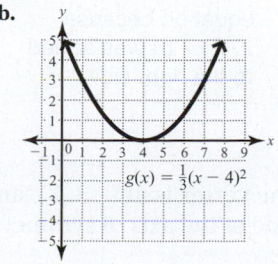

b.

SOLUTION We see that $f(x) = 2(x - 3)^2 + 1$ is in the form $f(x) = a(x - h)^2 + k$, where $a = 2$, $h = 3$, and $k = 1$. Because a is positive 2, the parabola opens upward. The vertex is at $(3, 1)$, and the axis of symmetry is $x = 3$. To complete the graph, we find a few points on either side of the axis of symmetry.

Note: We find these additional points by choosing x-values on either side of the axis of symmetry and using the equation to find the corresponding y-values.

x	y
2	3
1	9
4	3
5	9

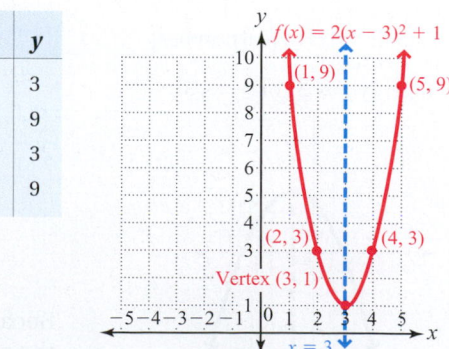

Domain: $\{x \mid x$ is a real number$\}$ or $(-\infty, \infty)$

Range: $\{y \mid y \ge 1\}$ or $[1, \infty)$

Your Turn 4 Given $f(x) = -\dfrac{1}{2}(x - 1)^2 + 3$, determine whether the graph opens upward or downward, find the vertex and axis of symmetry, and draw the graph. Find the domain and range.

Objective ⑤ **Graph quadratic functions of the form $f(x) = ax^2 + bx + c$.** The advantage of the form $f(x) = a(x - h)^2 + k$ is that we can "see" the vertex and axis of symmetry and we can "see" whether the parabola opens upward or downward by looking at the equation. If an equation is in the form $f(x) = ax^2 + bx + c$, it can be transformed into $f(x) = a(x - h)^2 + k$ by completing the square.

Example 5 Write the function in the form $f(x) = a(x - h)^2 + k$. Then determine whether the graph opens upward or downward, find the vertex and axis of symmetry, and draw the graph. Find the domain and range.

a. $f(x) = x^2 - 2x - 8$

SOLUTION

$y = x^2 - 2x - 8$	**Replace $f(x)$ with y to make manipulations more workable.**
$y + 8 = x^2 - 2x$	**Add 8 to both sides of the equation to isolate the x terms.**
$y + 8 + 1 = x^2 - 2x + 1$	**Add 1 to both sides of the equation to complete the square.**
$y + 9 = (x - 1)^2$	**Factor the right side as a perfect square.**
$y = (x - 1)^2 - 9$	**Subtract 9 from both sides to solve for y.**

Because $a = 1$ and 1 is positive, the graph opens upward. The vertex is at $(1, -9)$, and the axis of symmetry is $x = 1$. Plot a few points on either side of the axis of symmetry and graph. We include the x- and y-intercepts, which we found in Example 3 of Section 11.2.

Note: Remember, to find x-intercepts, we replace $f(x)$ (or y) with 0 and solve for x. To find y-intercepts, we replace x with 0 and solve for y.

	x	$f(x)$
x-intercept →	-2	0
	-1	-5
y-intercept →	0	-8
vertex →	1	-9
	2	-8
	3	-5
x-intercept →	4	0

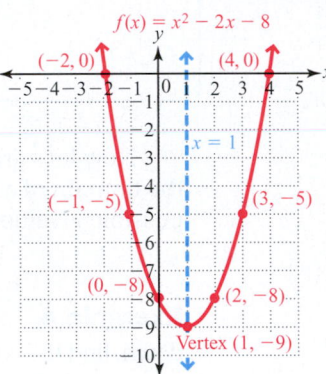

Note: The points we found are 1, 2, and 3 units to the left and right of the axis of symmetry.

Domain: $\{x \mid x$ is a real number$\}$ or $(-\infty, \infty)$

Range: $\{y \mid y \geq -9\}$ or $[-9, \infty)$

b. $f(x) = 2x^2 + 12x + 19$

SOLUTION

$y = 2x^2 + 12x + 19$	**Replace $f(x)$ with y.**
$y - 19 = 2x^2 + 12x$	**Subtract 19 from both sides of the equation.**
$y - 19 = 2(x^2 + 6x)$	**Factor 2 out of the right side of the equation.**
$y - 19 + 18 = 2(x^2 + 6x + 9)$	**Complete the square inside the parentheses.**
$y - 1 = 2(x + 3)^2$	**Combine terms on the left and factor on the right.**
$y = 2(x + 3)^2 + 1$	**Add 1 to both sides.**

Note: We added 18 to the left side of the equation because $2 \cdot 9 = 18$ was added to the right side.

Because $a = 2$ and 2 is positive, the graph opens upward. The vertex is at $(-3, 1)$, and the axis of symmetry is $x = -3$. Plot a few points on either side of the axis of symmetry and graph.

Answer **Your Turn 4**

opens downward; vertex: $(1, 3)$; axis: $x = 1$; domain: $\{x \mid x$ is a real number$\}$ or $(-\infty, \infty)$; range: $\{y \mid y \leq 3\}$ or $(-\infty, 3]$

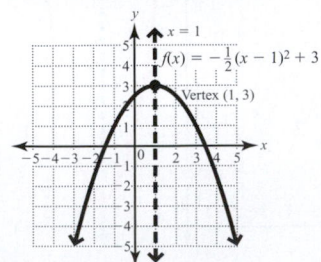

Note: The points we found are 1 and 2 units to the right and left of the axis of symmetry.

x	$f(x)$
-2	3
-1	9
-4	3
-5	9

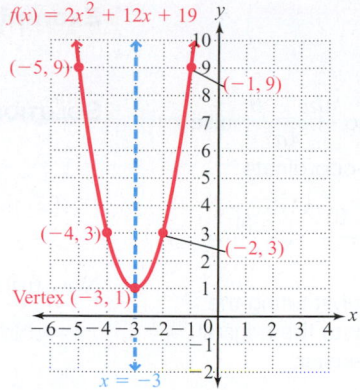

Note that there are no x-intercepts and the y-intercepts is $(0, 19)$.

Domain: $\{x | x \text{ is a real number}\}$
  or $(-\infty, \infty)$
Range: $\{y | y \geq 1\}$ or $[1, \infty)$

Answers Your Turn 5
a. x-intercepts: $(-1, 0), (3, 0)$
y-intercept: $(0, -3)$
$f(x) = (x - 1)^2 - 4$;
opens upward; vertex: $(1, -4)$;
axis: $x = 1$;

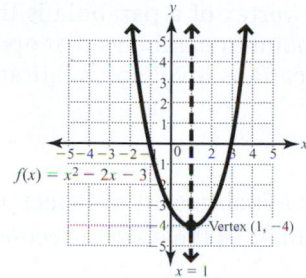

domain: $\{x | x \text{ is a real number}\}$
or $(-\infty, \infty)$;
range: $\{y | y \geq -4 \text{ or } [-4, \infty)$
b. $f(x) = -2(x - 2)^2 + 2$;
x-intercepts: $(1, 0), (3, 0)$
y-intercepts: $(0, -6)$
opens downward;
vertex: $(2, 2)$; axis: $x = 2$;

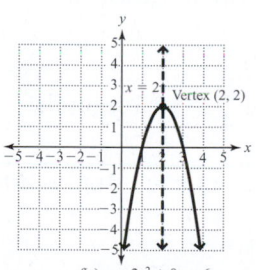

domain $\{x | x \text{ is a real number}\}$
or $(-\infty, \infty)$;
range: $\{y | y \leq 2\}$ or $(-\infty, 2]$

Your Turn 5 Write the function in the form $f(x) = a(x - h)^2 + k$; determine whether the graph opens upward or downward; find the x- and y-intercepts, vertex, and axis of symmetry; and draw the graph. Find the domain and range.

  a. $f(x) = x^2 - 2x - 3$ b. $f(x) = -2x^2 + 8x - 6$

We also can find the vertex using a formula that we derive by completing the square on the general form $f(x) = ax^2 + bx + c$.

$$y = ax^2 + bx + c \qquad \text{Replace } f(x) \text{ with } y.$$

$$y - c = ax^2 + bx \qquad \text{Isolate the } x \text{ terms.}$$

$$y - c = a\left(x^2 + \frac{b}{a}x\right) \qquad \text{Factor out } a.$$

$$y - c + \frac{b^2}{4a} = a\left(x^2 + \frac{b}{a}x + \frac{b^2}{4a^2}\right) \qquad \text{Complete the square.}$$

$$y - \frac{4ac - b^2}{4a} = a\left(x + \frac{b}{2a}\right)^2 \qquad \begin{array}{l}\text{Find the LCD of the} \\ \text{two terms on the left.} \\ \text{Factor on the right.}\end{array}$$

$$y = a\left(x + \frac{b}{2a}\right)^2 + \frac{4ac - b^2}{4a} \qquad \text{Add } \frac{4ac - b^2}{4a} \text{ to both sides.}$$

Note: Adding $\left[\frac{1}{2}\left(\frac{b}{a}\right)\right]^2$
or $\frac{b^2}{4a^2}$ completes the square inside the parentheses. Because those parentheses are multiplied by a, we add $a \cdot \frac{b^2}{4a^2} = \frac{b^2}{4a}$ to the left side to keep the equation balanced.

Notice that if we rewrite the last line above as $y = a\left(x - \left(-\frac{b}{2a}\right)\right)^2 + \frac{4ac - b^2}{4a}$,

we see that $h = -\dfrac{b}{2a}$ and $k = \dfrac{4ac - b^2}{4a}$ so that the vertex is $\left(-\dfrac{b}{2a}, \dfrac{4ac - b^2}{4a}\right)$. This also means that the axis of symmetry is $x = -\dfrac{b}{2a}$. Also, although the y-coordinate of the vertex is $\dfrac{4ac - b^2}{4a}$, it is usually easier to find the vertex by substituting the x-coordinate into the original function.

Procedure **Finding the Vertex of a Quadratic Function in the Form** $f(x) = ax^2 + bx + c$

Given an equation in the form $f(x) = ax^2 + bx + c$, to determine the vertex,

1. Find the x-coordinate using the formula $x = -\dfrac{b}{2a}$.

2. Find the y-coordinate by evaluating $f\left(-\dfrac{b}{2a}\right)$ or $\dfrac{4ac - b^2}{4a}$.

Example 6 For the function $f(x) = 2x^2 - 8x + 5$, find the coordinates of the vertex.

SOLUTION First, find the x-coordinate of the vertex using $-\dfrac{b}{2a}$.

$$x\text{-coordinate of the vertex: } -\frac{b}{2a} = -\frac{(-8)}{2(2)} = \frac{8}{4} = 2$$

Now find the y-coordinate by evaluating $f(2)$.

$$y\text{-coordinate of the vertex: } f(2) = 2(2)^2 - 8(2) + 5 = 8 - 16 + 5 = -3$$

The vertex is $(2, -3)$.

Note: Using $\dfrac{4ac - b^2}{4a}$ also gives the y-coordinate:

$$\frac{4(2)(5) - (-8)^2}{4(2)} = \frac{-24}{8} = -3$$

However, substituting the x-value, 2, into the original function is easier.

Your Turn 6 For the equation $f(x) = 3x^2 + 18x - 19$, find the vertex.

Objective ⑥ Solve applications involving parabolas. Because the vertex of a parabola is the highest point in parabolas that open down or the lowest point in parabolas that open up, we can find a minimum or maximum value in applications involving quadratic functions.

Example 7 A toy rocket is launched straight up with an initial velocity of 40 feet per second. The equation $h = -16t^2 + 40t$ describes the height, h, of the rocket t seconds after being launched.

a. Find the maximum height the rocket reaches.
b. Find the amount of time the rocket is in the air.
c. Draw the graph of the height of the rocket.

SOLUTION

a. Because the graph of $h = -16t^2 + 40t$ is a parabola that opens down ($a = -16$), the maximum height occurs at its vertex.

$$t\text{-coordinate of the vertex: } -\frac{b}{2a} = -\frac{40}{2(-16)} = -\frac{40}{-32} = \frac{5}{4} = 1.25 \text{ seconds}$$

To find the h-coordinate of the vertex, we replace t with 1.25 in $h = -16t^2 + 40t$.

$$h = -16(1.25)^2 + 40(1.25) = 25 \text{ feet}$$

The vertex is $(1.25, 25)$, so the maximum height is 25 feet, which occurs 1.25 seconds after the rocket is launched.

Note: We could have found the vertex by writing the equation in $y = a(x - h)^2 + k$ form, which would be $h = -16(t - 1.25)^2 + 25$.

b. The time the rocket is in the air is from launch until it returns to the ground. At launch and upon returning to the ground, the rocket's height is 0; so we need to find t when $h = 0$.

$$0 = -16t^2 + 40t \qquad \textcolor{blue}{\text{Replace } h \text{ with 0.}}$$
$$0 = -8t(2t - 5) \qquad \textcolor{blue}{\text{Factor out a common factor of } -8t.}$$
$$-8t = 0 \quad \text{or} \quad 2t - 5 = 0 \qquad \textcolor{blue}{\text{Use the zero-factor theorem.}}$$
$$t = 0 \qquad\qquad 2t = 5$$
$$\qquad\qquad\qquad t = 2.5$$

This means that height is 0 when $t = 0$ and when $t = 2.5$ seconds, so the rocket is in the air for 2.5 seconds.

Answer Your Turn 6

$(-3, -46)$

c. The graph of $h = -16t^2 + 40t$ shows that at 0 seconds, the rocket is at 0 feet, which is when it is launched. Its height increases to a maximum of 25 feet 1.25 seconds after launch. Then the height decreases back to 0 feet 2.5 seconds after launch.

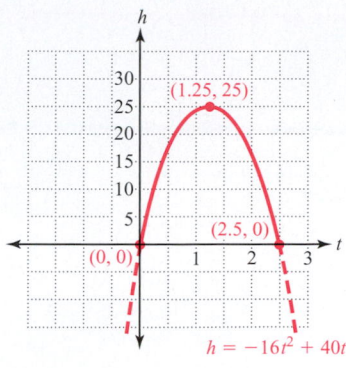

$h = -16t^2 + 40t$

Your Turn 7   A soccer player kicks a ball straight up with an initial velocity of 56 feet per second. The equation $h = -16t^2 + 56t$ describes the height, h, of the ball t seconds after being kicked.

a. After how many seconds is the ball at its maximum height?
b. What is the maximum height the ball reaches?
c. How long is the ball in the air?

If an object is thrown or shot upward and outward, gravity causes its path, or trajectory, to be in the shape of a parabola.

Note: x represents the horizontal distance the object travels, and y represents the vertical distance the object travels.

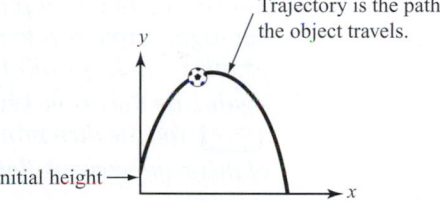

Trajectory is the path the object travels.

Initial height

Example 8   The equation $y = -0.8x^2 + 2.4x + 6$ when $x \geq 0$ and $y \geq 0$ describes the trajectory of a ball thrown upward and outward from an initial height of 6 feet.

a. What is the maximum height the ball reaches?
b. How far does the ball travel horizontally?
c. Graph the trajectory of the ball.

SOLUTION

a. The maximum height will be the y-coordinate of the vertex.

$$x\text{-coordinate of the vertex: } -\frac{b}{2a} = -\frac{(2.4)}{2(-0.8)} = 1.5$$

$$y\text{-coordinate of the vertex: } y = -0.8(1.5)^2 + 2.4(1.5) + 6 = 7.8$$

The vertex is $(1.5, 7.8)$, so the maximum height is 7.8 feet.

b. To find how far the ball travels, we need to find the x-value when it hits the ground, which is where $y = 0$.

$$0 = -0.8x^2 + 2.4x + 6$$
Substitute 0 for y in the equation.

$$x = \frac{-(2.4) \pm \sqrt{(2.4)^2 - 4(-0.8)(6)}}{2(-0.8)}$$
Substitute $a = -0.8$, $b = 2.4$, and $c = 6$ into the quadratic formula.

$$= \frac{-2.4 \pm \sqrt{24.96}}{-1.6} \approx -1.62 \text{ or } 4.62$$

The negative value does not make sense in the context of this problem, so the distance the ball travels must be approximately 4.62 feet.

Answers   Your Turn 7
a. 1.75 sec.   **b.** 49 ft.   **c.** 3.5 sec.

c.

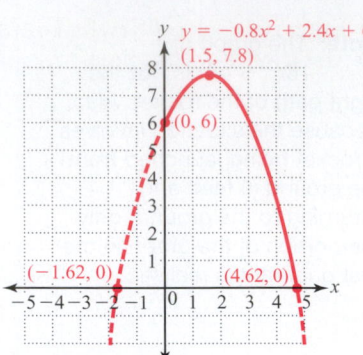

Note: Because the equation describes the trajectory when $x \geq 0$ and $y \geq 0$, we consider only the part of the graph in the first quadrant.

Calculator Tips

Besides being able to graph parabolas on the graphing calculator, you also can find the x-intercepts and the vertex.

Let's work through Example 8 using a graphing calculator. First, press the $\boxed{Y=}$ *key and enter the equation* $y = -0.8x^2 + 2.4x + 6$ *as Y1; then press GRAPH to see the graph.*

Now to find the x-intercepts, press $\boxed{2nd}$ $\boxed{TRACE}$ *, which brings up the CALC menu. Use the arrow keys to highlight ZERO in the CALC menu; then press* $\boxed{ENTER}$ *to select it. This returns you to the graph screen where you will be prompted for the left bound. Use the left–right arrow keys to move the cursor along the parabola until it is to the left of one of the x-intercepts and press* $\boxed{ENTER}$*. Next, you are prompted for the right bound. Again, use the arrow keys to move the cursor to the right of the x-intercept; then press* $\boxed{ENTER}$*. You are then prompted for a guess. Press* $\boxed{ENTER}$ *and you will see the coordinates of this x-intercept. Follow the same procedure for the other x-intercept.*

To find the vertex, we use the MAXIMUM or MINIMUM functions under the CALC menu. Because our graph opens down, the vertex will be a maximum; so we use the MAXIMUM function. After selecting the MAXIMUM function, you will be prompted for the left bound, then the right bound, then a guess. After entering the left bound and right bound and pressing $\boxed{ENTER}$ *for the guess, you will see the coordinates of the vertex.*

11.4 Exercises   For Extra Help   *MyMathLab*

Note: Exercises marked with a ★ represent challenging exercises.

1. In an equation in the form $f(x) = a(x - h)^2 + k$, what determines whether the parabola opens upward or downward?

2. In an equation in the form $f(x) = a(x - h)^2 + k$, what are the coordinates of the vertex?

3. Describe the axis of symmetry in a parabola that opens up or down. What is its equation?

4. Suppose the y-intercept for a given quadratic equation is $(0, -5)$. If the vertex is at $(3, -2)$, what can you conclude about the x-intercepts?

5. If the solutions for a quadratic equation are imaginary numbers, what does that indicate about the corresponding quadratic function?

6. Given that the x-coordinate of the vertex is $-\dfrac{b}{2a}$, what must be true about a and b if the x-coordinate of the vertex is -1?

For Exercises 7–22, find the coordinates of the vertex and write the equation of the axis of symmetry. See Examples 4–6.

7. $f(x) = 5(x - 2)^2 - 3$

8. $h(x) = 2(x - 1)^2 + 4$

9. $g(x) = -2(x + 1)^2 - 5$

10. $k(x) = -(x + 4)^2 - 3$

11. $k(x) = x^2 + 2$

12. $g(x) = x^2 + 5$

13. $h(x) = 3(x + 5)^2$

14. $f(x) = 5(x + 1)^2$

15. $f(x) = -0.5x^2$

16. $k(x) = 2.5x^2$

17. $f(x) = x^2 - 4x + 8$

18. $g(x) = x^2 + 6x + 8$

19. $k(x) = 2x^2 + 16x + 27$

20. $f(x) = 3x^2 - 6x + 1$

21. $g(x) = -3x^2 + 2x + 1$

22. $h(x) = -2x^2 - 3x + 2$

For Exercises 23–36: **a.** State whether the parabola opens upward or downward.
b. Find the coordinates of the vertex.
c. Write the equation of the axis of symmetry.
d. Graph. See Examples 1–4.

23. $h(x) = -3x^2$

24. $f(x) = 2x^2$

25. $k(x) = \frac{1}{4}x^2$

26. $g(x) = -\frac{1}{3}x^2$

27. $f(x) = 4x^2 - 3$

28. $h(x) = -2x^2 + 4$

29. $g(x) = -0.5x^2 + 2$

30. $k(x) = 0.5x^2 - 3$

31. $f(x) = (x - 3)^2 + 2$

32. $h(x) = (x - 1)^2 - 3$

33. $k(x) = -2(x + 1)^2 - 3$

34. $g(x) = -2(x + 3)^2 + 1$

35. $h(x) = \dfrac{1}{3}(x - 2)^2 - 1$

36. $f(x) = -\dfrac{1}{4}(x - 3)^2 - 2$

For Exercises 37–42: **a.** *Find the x- and y-intercepts.*
  b. *Write the equation in the form $f(x) = a(x - h)^2 + k$.*
  c. *State whether the parabola opens upward or downward.*
  d. *Find the coordinates of the vertex.*
  e. *Write the equation of the axis of symmetry.*
  f. *Graph. See Examples 5 and 6.*
  g. *Find the domain and range.*

37. $h(x) = x^2 + 6x + 9$

38. $k(x) = x^2 - 4x + 4$

39. $g(x) = -3x^2 + 6x - 5$

40. $k(x) = -x^2 - 2x + 3$

41. $f(x) = 2x^2 + 6x + 3$

42. $h(x) = -3x^2 - 6x + 4$

For Exercises 43– 54, solve. See Examples 7 and 8.

43. A toy rocket is launched with an initial velocity of 45 meters per second. The equation $h = -4.9t^2 + 45t$ describes the height, h, of the rocket in meters t seconds after being launched.
   a. After how many seconds does the rocket reach its maximum height?
   b. What is the maximum height the rocket reaches?

   c. How long is the rocket in the air?

44. A ball is dropkicked straight up with an initial velocity of 36 feet per second. The equation $h = -16t^2 + 36t$ describes the height, h, of the ball in feet t seconds after being kicked.
   a. After how many seconds does the ball reach its maximum height?
   b. What is the maximum height the ball reaches?

   c. How long is the ball in the air?

45. The equation $y = -0.8x^2 + 3.2x + 6$ models the trajectory of a ball thrown upward and outward from a height of 6 feet. (Assume that $x \geq 0$ and $y \geq 0$.)
   a. What is the maximum height the ball reaches?

   b. How far does the ball travel horizontally?

   c. Graph the trajectory of the ball.

46. The javelin toss is an event in track and field in which participants try to throw a javelin the farthest. Suppose the equation $y = -0.02x^2 + 1.3x + 8$ models the trajectory of one particular throw. (Assume that x and y represent distances in meters and that $x \geq 0$ and $y \geq 0$.)
   a. What is the maximum height the javelin reaches?
   b. How far does the javelin travel horizontally?
   c. Graph the trajectory of the javelin.

47. A record company discovers that the number of CDs sold each week after release follows a parabolic pattern. The function $n(t) = -200t^2 + 4000t$ describes the number, n, of CDs an artist sold each of t weeks after the release of the album.
   a. Which week had the greatest number of CDs sold?

   b. How many CDs sold that week?

48. The function $n(t) = -3t^2 + 42t$ describes the number, n, of tickets sold for a play each of t days after tickets went on sale.
   a. What day had the greatest number of tickets sold?

   b. How many tickets sold that day?

★ **49.** A farmer has enough materials to build a fence with a total length of 400 feet. He wants the enclosed space to be rectangular and wants to maximize the area enclosed. Find the length and width so that the area is maximized.

★ **50.** An architect wants the length and width of a rectangular building to be a total of 150 feet long. She also wants to maximize the rectangular base area that the building occupies. Find the length and width so that the base area is maximized.

51. The function $C(n) = n^2 - 110n + 5000$ describes a company's cost, C, of producing n units of its product. Find the number of units the company should produce to minimize its cost.

52. The function $P(t) = 0.001t^2 - 0.24t + 59.90$ roughly models a particular stock's closing price, P, each of t days of trading during one year.
   a. After how many days is the price at its lowest?

   b. What was the lowest price during that year?

★ **53.** One integer is 12 more than another. If their product is minimized, find the integers and their product.

★ **54.** The greater of two integers minus the smaller integer gives a result of 20. Find the two integers so that their product is minimized. Also find their product.

Collaborative Exercises Arch Span

The Gateway Arch, located in St. Louis, Missouri, was built from 1963 to 1965 and is the nation's tallest memorial. Although the arch is not a parabola, the equation $h(x) = -0.0063492063x^2 + 630$ can be used to approximate the height of the arch, where x represents the distance from the axis of symmetry to the arch and h(x) represents its height above the ground.

1. Using the equation, find the maximum height of the structure.

2. The span of the arch at a given height is the horizontal distance between the two opposing points on the arch at that height. Find the span of the arch at ground level. (*Hint:* Think of ground level as the x-axis. The height is 0 along the x-axis.)

3. How does the span at ground level relate to the maximum height of the arch?

4. The arch has foundations 60 feet below ground level. What is the span of the arch between its foundations?

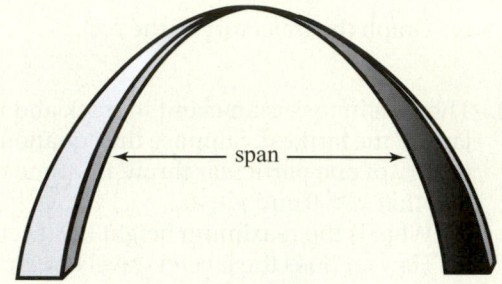

span

Review Exercises

Exercise 1  **EXPRESSIONS**

[1.5] **1.** Evaluate: $-|-2 + 3 \cdot 4| - 4^0$

Exercises 2–6 **EQUATIONS AND INEQUALITIES**

[6.6] **2.** Solve $x^2 + 2x = 15$ by factoring.

[11.1] **3.** Solve $x^2 = -20$ using the square root principle.

For Exercises 4–6, solve; then graph the solution set.

[2.6] **4.** $4x - 8 \leq 2x + 1$

[8.3] **5.** $|x - 3| \geq 8$

[9.7] **6.** $\begin{cases} x + y > 2 \\ 2x - 3y \leq 6 \end{cases}$

11.5 Solving Nonlinear Inequalities

Objectives
1. Solve quadratic and other inequalities.
2. Solve rational inequalities.

Objective 1 **Solve quadratic and other inequalities.** Now that we have solved quadratic equations, we can learn how to solve **quadratic inequalities**.

Definition   *Quadratic inequality:* An inequality that can be written in the form
$ax^2 + bx + c > 0$ or $ax^2 + bx + c < 0$, where $a \neq 0$.

Note: The symbols $<$ and $>$ can be replaced with $\leq$ and $\geq$.

For example, $2x^2 - x - 3 < 0$ and $2x^2 - x - 3 \geq 0$ are quadratic inequalities. Let's see what we can learn about these inequalities by looking at the graph of the corresponding quadratic function: $y = 2x^2 - x - 3$. Note that by letting $y = 0$, we have the corresponding equation $2x^2 - x - 3 = 0$, and solving this equation gives the x-intercepts $x = -1$ and $x = \frac{3}{2}$. Also notice that those x-intercepts divide the x-axis into three intervals: $(-\infty, -1)$, $\left(-1, \frac{3}{2}\right)$, and $\left(\frac{3}{2}, \infty\right)$.

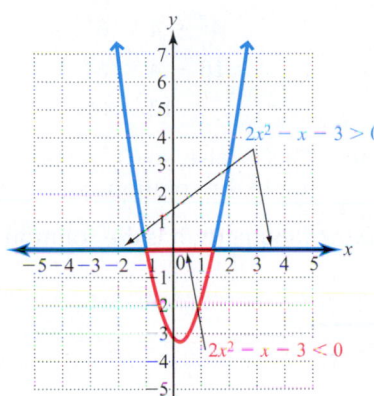

Notice that the parabola is above the x-axis in intervals $(-\infty, -1)$ and $\left(\frac{3}{2}, \infty\right)$, shown in blue. Evaluating $y = 2x^2 - x - 3$ using any x-value in these intervals produces a y-value that is greater than 0; so these intervals are the solution sets for $2x^2 - x - 3 > 0$.

The parabola is below the x-axis in the interval $\left(-1, \frac{3}{2}\right)$, shown in red. Evaluating $y = 2x^2 - x - 3$ using any x-value in this interval produces a y-value that is less than 0; so this interval is the solution set for $2x^2 - x - 3 < 0$.

The following number lines indicate various solution sets based on the inequality.

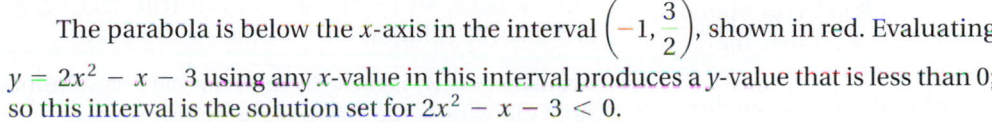

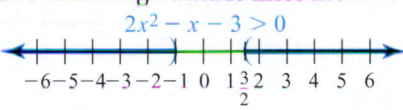

Solution set: $(-\infty, -1) \cup \left(\frac{3}{2}, \infty\right)$

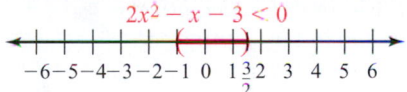

Solution set: $\left(-1, \frac{3}{2}\right)$

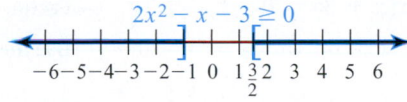

Solution set: $(-\infty, -1] \cup \left[\frac{3}{2}, \infty\right)$

Solution set: $\left[-1, \frac{3}{2}\right]$

Based on these observations, we use the following procedure to solve quadratic inequalities.

Procedure   **Solving Quadratic Inequalities**

1. Solve the related equation $ax^2 + bx + c = 0$.
2. Plot the solutions of $ax^2 + bx + c = 0$ on a number line. These solutions divide the number line into intervals.
3. Choose a test number from each interval and substitute the number into the inequality. If the test number makes the inequality *true*, then all numbers in that interval will solve the inequality. If the test number makes the inequality *false*, then no numbers in that interval will solve the inequality.
4. State the solution set of the inequality: It is the union of all intervals that solve the inequality. If the inequality symbols are $\leq$ or $\geq$, the values from Step 2 are included. If the symbols are $<$ or $>$, they are not solutions.

Example 1 Solve. Write the solution set using interval notation; then graph the solution set on a number line.

a. $x^2 - x < 6$

SOLUTION $x^2 - x = 6$ Write the related equation.

$\quad\quad\quad\quad x^2 - x - 6 = 0$ Subtract 6 from both sides to get quadratic form.

$\quad\quad\quad (x - 3)(x + 2) = 0$ Factor.

$\quad\quad x - 3 = 0 \quad \text{or} \quad x + 2 = 0$ Use the zero-factor theorem.

$\quad\quad\quad\quad x = 3 \quad\quad\quad\quad x = -2$ These are the x-intercepts of the graph of $y = x^2 - x - 6$.

Plot -2 and 3 on a number line (x-axis), which divides the number line into three intervals.

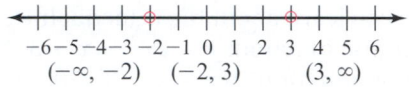

$\quad -6 \; -5 \; -4 \; -3 \; -2 \; -1 \;\; 0 \;\; 1 \;\; 2 \;\; 3 \;\; 4 \;\; 5 \;\; 6$
$\quad (-\infty, -2) \quad (-2, 3) \quad\quad\quad (3, \infty)$

> **Note:** -2 and 3 are not part of the solution set because the inequality symbol is $<$.

Choose a test number from each interval and substitute that value into $x^2 - x < 6$.

For $(-\infty, -2)$, we choose $x = -3$.

$$(-3)^2 - (-3) < 6$$
$$9 + 3 < 6$$
$$12 < 6$$

For $(-2, 3)$, we choose $x = 0$.

$$0^2 - 0 < 6$$
$$0 < 6$$

For $(3, \infty)$, we choose $x = 4$.

$$4^2 - 4 < 6$$
$$16 - 4 < 6$$
$$12 < 6$$

This is false, so $(-\infty, -2)$ is not in the solution set.

This is true, so $(-2, 3)$ is in the solution set.

This is false, so $(3, \infty)$ is not in the solution set.

Connection Notice that the solution set for $x^2 - x - 6 < 0$ corresponds to the interval where the graph of $f(x) = x^2 - x - 6$ is below the x-axis and the x-intercepts are the endpoints of the interval and are not included in the solution set.

Because $(-2, 3)$ is the only interval that has solutions to the inequality, it is the solution set. Following is the graph of the solution set.

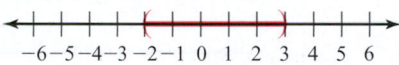

$\quad -6 \; -5 \; -4 \; -3 \; -2 \; -1 \;\; 0 \;\; 1 \;\; 2 \;\; 3 \;\; 4 \;\; 5 \;\; 6$

b. $x^2 + 3x \geq 0$

SOLUTION $x^2 + 3x = 0$ Write the related equation.

$\quad\quad\quad\quad x(x + 3) = 0$ Factor.

$\quad\quad x = 0 \quad \text{or} \quad x + 3 = 0$ Use the zero-factor theorem.

$\quad\quad\quad\quad\quad\quad\quad\quad x = -3$

Plot -3 and 0 on a number line and note the intervals.

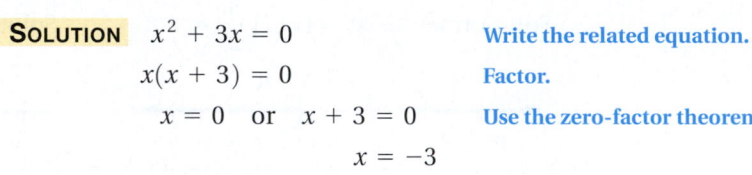

$\quad -6 \; -5 \; -4 \; -3 \; -2 \; -1 \;\; 0 \;\; 1 \;\; 2 \;\; 3 \;\; 4 \;\; 5 \;\; 6$
$\quad (-\infty, -3) \; (-3, 0) \quad\quad\quad (0, \infty)$

> **Note:** -3 and 0 are included in the solution set because the inequality symbol is $\geq$.

Note: Because we have already determined that -3 and 0 are in the solution set, we do not include them in any of our test intervals, which is why we use parentheses for each interval.

Choose a test number from each interval and test in $x^2 + 3x \geq 0$.

For $(-\infty, -3)$, we choose $x = -4$.

$$(-4)^2 + 3(-4) \geq 0$$
$$16 - 12 \geq 0$$
$$4 \geq 0$$

For $(-3, 0)$, we choose $x = -1$.

$$(-1)^2 + 3(-1) \geq 0$$
$$1 - 3 \geq 0$$
$$-2 \geq 0$$

For $(0, \infty)$, we choose $x = 1$.

$$1^2 + 3(1) \geq 0$$
$$1 + 3 \geq 0$$
$$4 \geq 0$$

This is true, so $(-\infty, -3)$ is in the solution set.

This is false, so $(-3, 0)$ is not in the solution set.

This is true, so $(0, \infty)$ is in the solution set.

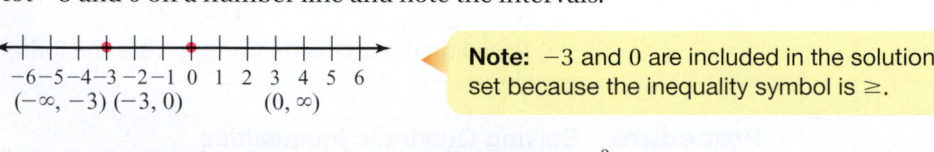

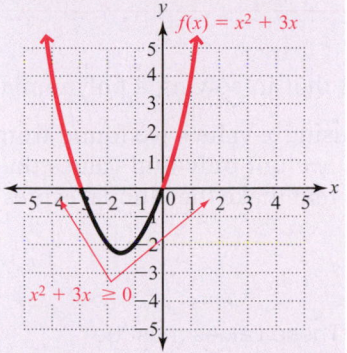
The solution set is $(-\infty, -3] \cup [0, \infty)$, which is graphed next.

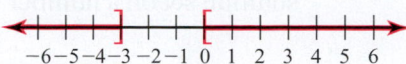

c. $(x - 1)^2 > -2$

SOLUTION Because $(x - 1)^2$ is always 0 or positive, it is always greater than -2. So every real number is a solution.

Solution set: $\mathbb{R}$ or $(-\infty, \infty)$

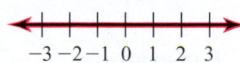

d. $(x - 1)^2 < -2$

SOLUTION Because $(x - 1)^2$ is never negative, its value can never be less than -2. So there are no real solutions for $(x - 1)^2 < -2$.

Solution set: $\varnothing$

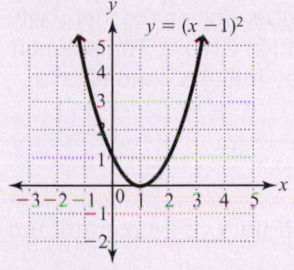
Your Turn 1 Solve. Write the solution set using interval notation; then graph the solution set on a number line.

a. $x^2 - 2x - 8 \leq 0$ **b.** $x^2 - 4x > 0$ **c.** $(3x - 8)^2 < -5$

Other Polynomial Inequalities

We use a similar procedure for expressions with more than two factors. In the following example, we condense the procedure to a table.

Example 2 Solve $(x + 4)(x + 1)(x - 3) \leq 0$. Write the solution set in interval notation; then graph the solution set on a number line.

SOLUTION $(x + 4)(x + 1)(x - 3) = 0$ **Write the related equation.**

$x + 4 = 0$ or $x + 1 = 0$ or $x - 3 = 0$ **Set each factor equal to 0.**

$\quad x = -4 \qquad\qquad x = -1 \qquad\qquad x = 3$ **Solve the equations.**

Plot -4, -1, and 3 on a number line and note the intervals.

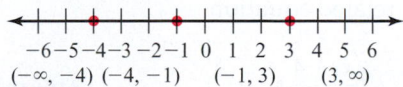

Note: -4, -1, and 3 are included in the solution set because the inequality symbol is $\leq$. Remember that we test intervals around these values, so they are not included in those intervals.

Interval	$(-\infty, -4)$	$(-4, -1)$	$(-1, 3)$	$(3, \infty)$
Test Number	-5	-2	0	4
Test Results	$-32 \leq 0$	$10 \leq 0$	$-12 \leq 0$	$40 \leq 0$
True or False	True	False	True	False

Therefore, the solution set is $(-\infty, -4] \cup [-1, 3]$, which is graphed next.

Note: Because the inequality is $\leq$, -4, -1, and 3 are included in the solution set.

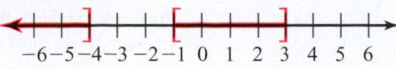

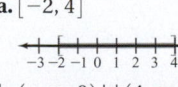
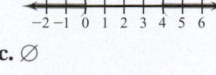
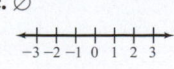

Your Turn 2 Solve. Write the solution set using interval notation; then graph the solution set on a number line.

$$(x + 3)(x - 2)(x + 5) \le 0$$

Objective 2 Solve rational inequalities. Now we consider solving **rational inequalities.**

Definition *Rational inequality:* An inequality containing a rational expression.

For example, $\dfrac{x + 2}{x - 3} > 4$ is a rational inequality. Recall that in solving a polynomial inequality, we divided the number line into intervals using x-values we found from solving its related equation. With rational inequalities, we not only find values that solve the related equation but also must consider values that make the rational expression undefined (denominator $= 0$).

Procedure **Solving Rational Inequalities**

1. Find all values that make any denominator equal to 0. These values must be excluded from the solution set.
2. Solve the related equation.
3. Plot the numbers found in Steps 1 and 2 on a number line and label the intervals.
4. Choose a test number from each interval and determine whether it solves the inequality.
5. The solution set is the union of all regions whose test number solves the inequality. If the inequality symbol is $\le$ or $\ge$, include the values found in Step 2. The solution set never includes the values found in Step 1 because they make a denominator equal to 0.

Example 3 Solve. Write the solution set using interval notation; then graph the solution set on a number line.

a. $\dfrac{x - 4}{x + 2} \ge 0$

SOLUTION First, we find the values that make the denominator equal to 0.

$$x + 2 = 0$$
$$x = -2$$

Now solve the related equation.

$$\frac{x - 4}{x + 2} = 0 \qquad \text{\color{blue}{Write the related equation.}}$$

$$(x + 2)\frac{x - 4}{x + 2} = (x + 2)(0) \qquad \text{\color{blue}{Multiply both sides by the LCD, $x + 2$.}}$$

$$x - 4 = 0 \qquad \text{\color{blue}{Simplify.}}$$

$$x = 4$$

Plot -2 and 4 on a number line and label the intervals.

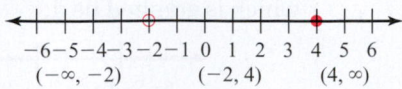

$$\begin{array}{ccc} (-\infty, -2) & (-2, 4) & (4, \infty) \end{array}$$

Answer **Your Turn 2**

$(-\infty, -5] \cup [-3, 2]$

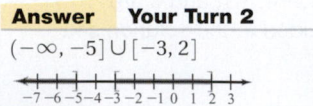

Again, we use a table.

Interval	$(-\infty, -2)$	$(-2, 4)$	$(4, \infty)$
Test Number	-3	0	5
Test Results	$7 \geq 0$	$-2 \geq 0$	$\frac{1}{7} \geq 0$
True or False	True	False	True

Note: Remember, because -2 makes $\dfrac{x-4}{x+2}$ undefined, it is not included in the solution set. Because the inequality is $\geq$, 4 is included in the solution set.

The solution set is $(-\infty, -2) \cup [4, \infty)$ and is graphed next.

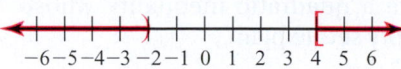

b. $\dfrac{x+5}{x-1} < 4$

SOLUTION Find the values that make the denominator equal to 0.

$$x - 1 = 0$$
$$x = 1$$

Solve the related equation.

$$\frac{x+5}{x-1} = 4 \qquad \text{\textbf{\color{blue}Write the related equation.}}$$

$$(x-1)\frac{x+5}{x-1} = (x-1)(4) \qquad \text{\textbf{\color{blue}Multiply both sides by the LCD, } } x-1.$$

$$x + 5 = 4x - 4 \qquad \text{\textbf{\color{blue}Simplify.}}$$

$$-3x = -9 \qquad \text{\textbf{\color{blue}Subtract } 4x \textbf{ and 5 from both sides.}}$$

$$x = 3$$

Plot 1 and 3 on a number line and label the intervals.

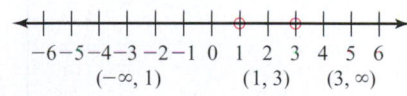

Interval	$(-\infty, 1)$	$(1, 3)$	$(3, \infty)$
Test Number	0	2	4
Test Results	$-5 < 4$	$7 < 4$	$3 < 4$
True or False	True	False	True

Note: Because 1 makes $\dfrac{x+5}{x-1}$ undefined, it is not included in the solution set. Also, 3 is not in the solution set because $\dfrac{x+5}{x-1} = 4$ when $x = 3$.

The solution set is $(-\infty, 1) \cup (3, \infty)$ and is graphed next.

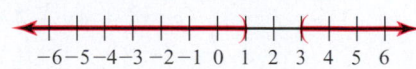

Your Turn 3 Solve. Write the solution set using interval notation; then graph the solution set on a number line.

a. $\dfrac{x+3}{x-1} < 0$

b. $\dfrac{x-2}{x+4} \leq 3$

Answers Your Turn 3

a. $(-3, 1)$

-4 -3 -2 -1 0 1 2

b. $(-\infty, -7] \cup (-4, \infty)$

-9 -8 -7 -6 -5 -4 -3 -2

Note: Exercises marked with a ★ represent challenging exercises.

1. Explain how you would find the solution set of $x^2 + 2x - 15 \geq 0$.

2. If the graph of $y = ax^2 + bx + c$ intersects the x-axis at -2 and 2, what regions would you check to solve $ax^2 + bx + c > 0$?

3. Is it possible to have a quadratic inequality whose solution set is the empty set? Explain.

4. Is it possible to have a quadratic inequality whose solution set is one number? Explain.

5. The quadratic inequality $(x - 2)(x + 5) < 0$ and the rational inequality $\dfrac{x - 2}{x + 5} < 0$ have the same solution sets. Explain how this is possible.

6. If you were to solve $\dfrac{x + 2}{(x + 5)(x - 3)} \geq 0$, what regions would you need to test?

For Exercises 7–14, the graph of a quadratic function is given. Use the graph to solve each equation and inequality. For solution sets that involve intervals, use interval notation. See Objective 1.

7. **a.** $x^2 + 6x + 5 = 0$

   b. $x^2 + 6x + 5 < 0$

   c. $x^2 + 6x + 5 > 0$

8. **a.** $x^2 + x - 2 = 0$

   b. $x^2 + x - 2 > 0$

   c. $x^2 + x - 2 < 0$

9. **a.** $-x^2 + 2x + 3 = 0$

   b. $-x^2 + 2x + 3 \leq 0$

   c. $-x^2 + 2x + 3 \geq 0$

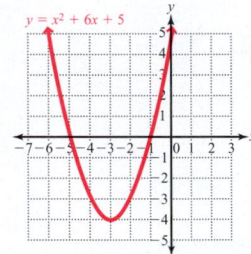 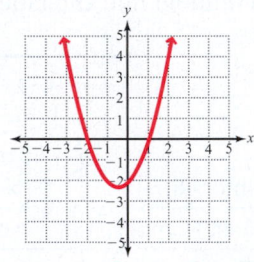 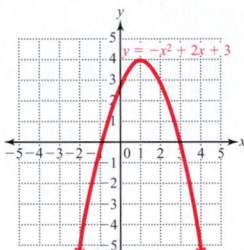

10. **a.** $-x^2 + 5x = 0$

    b. $-x^2 + 5x \leq 0$

    c. $-x^2 + 5x \geq 0$

11. **a.** $x^2 + 4x + 4 = 0$

    b. $x^2 + 4x + 4 > 0$

    c. $x^2 + 4x + 4 < 0$

12. **a.** $x^2 - 6x + 9 = 0$

    b. $x^2 - 6x + 9 < 0$

    c. $x^2 - 6x + 9 > 0$

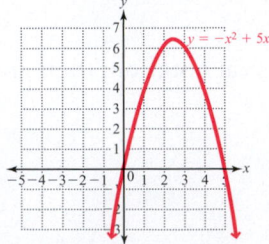 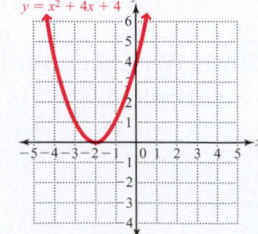 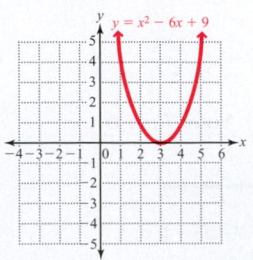

13. a. $x^2 + x + 4 = 0$

   b. $x^2 + x + 4 \leq 0$

   c. $x^2 + x + 4 \geq 0$

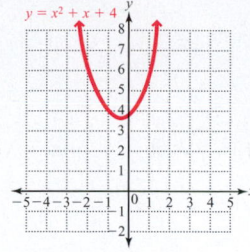

14. a. $-x^2 - x - 2 = 0$

   b. $-x^2 - x - 2 \leq 0$

   c. $-x^2 - x - 2 \geq 0$

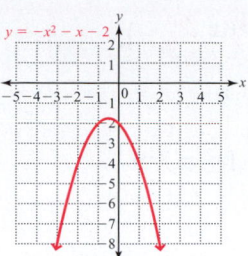

For Exercises 15–44, solve. Write the solution set using interval notation; then graph the solution set on a number line. See Examples 1 and 2.

15. $(x + 4)(x + 2) < 0$

16. $(x + 3)(x + 1) < 0$

17. $(x - 2)(x - 5) > 0$

18. $(x + 4)(x - 3) > 0$

19. $x^2 + 5x + 4 < 0$

20. $x^2 + 6x + 5 < 0$

21. $x^2 - 4x + 3 > 0$

22. $x^2 - 8x + 7 > 0$

23. $b^2 - 6b + 8 \leq 0$

24. $c^2 - 9c + 14 \leq 0$

25. $y^2 - 5 \geq 4y$

26. $z^2 - 3 \geq 2z$

27. $a^2 - 3a < 10$

28. $b^2 + 4b \leq 21$

29. $y^2 + 6y + 9 \geq 0$

30. $x^2 + 10x + 25 \geq 0$

31. $2c^2 - 4c + 7 \leq 0$

32. $2a^2 - 3a + 5 \leq 0$

33. $4r^2 + 21r + 5 > 0$

34. $4s^2 + 5s - 6 < 0$

35. $x^2 - 5x > 0$

36. $x^2 + 3x > 0$

37. $x^2 \leq 6x$

38. $x^2 \leq 3x$

39. $(x + 1)^2 \geq -9$

40. $(2x - 1)^2 > -4$

41. $(x - 4)(x + 2)(x + 4) \geq 0$

42. $(x + 5)(x - 3)(x + 1) \geq 0$

43. $(x + 2)(x + 6)(x - 1) < 0$

44. $(x - 3)(x + 3)(x + 1) < 0$

For Exercises 45–50, solve using the quadratic formula.

45. $3c^2 + 4c - 1 < 0$

46. $5a^2 - 10a + 2 < 0$

47. $4r^2 + 8r - 3 \geq 0$

48. $2y^2 - 6y - 5 \geq 0$

49. $-0.2a^2 - 1.6a - 2 \leq 0$

50. $-0.1x^2 - 1.2x + 4 > 0$

For Exercises 51–68, solve the rational inequalities. Write the solution set using interval notation; then graph the solution set on a number line. See Example 3.

51. $\dfrac{a + 4}{a - 1} > 0$

52. $\dfrac{m - 6}{m + 2} \geq 0$

53. $\dfrac{n + 1}{n + 5} \leq 0$

54. $\dfrac{b - 2}{b + 3} < 0$

55. $\dfrac{6}{x + 4} > 0$

56. $\dfrac{3}{x - 3} < 0$

57. $\dfrac{c}{c + 3} < 3$

58. $\dfrac{m}{m - 2} < 2$

59. $\dfrac{a + 5}{a - 4} > 4$

60. $\dfrac{j + 4}{j - 4} > 5$

61. $\dfrac{p + 3}{p - 3} \geq 4$ **62.** $\dfrac{c + 1}{c - 4} \geq 6$

63. $\dfrac{(k + 3)(k - 2)}{k - 5} \leq 0$ **64.** $\dfrac{(m + 1)(m - 3)}{m + 4} \geq 0$

65. $\dfrac{(2x - 1)^2}{x} \geq 0$ **66.** $\dfrac{(4x + 3)^2}{x} \geq 0$

67. $\dfrac{x^2 - 7x + 10}{x + 1} < 0$ **68.** $\dfrac{x^2 - 2x - 8}{x - 1} \geq 0$

For Exercises 69–72, solve.

69. If a ball is thrown upward with an initial velocity of 80 feet per second from the top of a building 96 feet high, the height, h, above the ground after t seconds is given by $h = -16t^2 + 80t + 96$, where h is in feet.
   a. After how many seconds will the ball hit the ground? (*Hint*: Think about the value of h when the ball is on the ground.)
   b. After how many seconds is the ball 192 feet above the ground?

   c. Find the interval of time when the ball is more than 192 feet above the ground.

   d. Use the answers to parts a, b, and c to find the intervals of time when the ball is less than 192 feet above the ground.

70. If an object is dropped from the top of a cliff that is 256 feet high, the equation giving the height, h, above the ground is $h = 256 - 16t^2$, where h is in feet and t is in seconds.
   a. After how many seconds will the object hit the ground?
   b. After how many seconds is the object 112 feet above the ground?
   c. Find the interval of time when the object is more than 112 feet above the ground.
   d. Find the interval of time when the ball is less than 112 feet above the ground.

71. In the parallelogram shown, the height is to be 2 inches less than x. The base is to be 6 inches more than x.
   a. Find the range of values for x so that the area of the parallelogram is at least 20 square inches.
   b. Find the range of values for the base and the height.

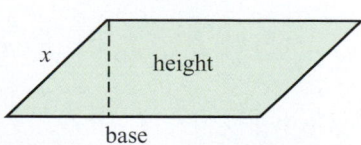

72. An auditorium is to be designed roughly in the shape of a box. The height is set to be 40 feet. It is preferred that the length of the space be 20 feet more than the width.
   a. Find the range of values for the width so that the volume of the space is at least 140,000 cubic feet.

   b. Find the range of values for the length.

For Exercises 73 and 74, use the formula for the slope of a line: $m = \dfrac{y_2 - y_1}{x_2 - x_1}$.

★ **73.** Suppose a line is to be drawn in the coordinate plane so that it passes through the point at $(2, 5)$. Find the range of values for the second point (x_2, y_2) so that $x_2 = y_2$ and the slope of the line is at most $\dfrac{1}{2}$.

★ **74.** Suppose a line is to be drawn in the coordinate plane so that it passes through the point at $(-3, 2)$. Find the range of values for the second point (x_2, y_2) so that $x_2 = y_2$ and the slope of the line is at least $\dfrac{1}{4}$.

Review Exercises

Exercises 1 and 2  **EXPRESSIONS**

[11.1] *For Exercises 1 and 2, add a term to the expression to make it a perfect square; then factor the perfect square.*

   1. $x^2 - 6x$

   2. $x^2 + 5x$

Exercises 3–6 **EQUATIONS AND INEQUALITIES**

[10.6]   **3.** Solve: $\sqrt{x - 2} = 4$

[6.7]   **4.** Does the graph of $f(x) = -x^2 + 2x - 3$ open upward or downward? Why?

[11.4]   **5.** Find the x-intercepts for $y = x^2 + 4x - 12$.

[11.4]   **6.** What are the coordinates of the vertex of the graph of $f(x) = x^2 - 6x + 5$?

Chapter 11 Summary

Defined Terms

Section 11.2
Discriminant (p. 787)

Section 11.3
Equations quadratic in form (p. 796)

Section 11.5
Quadratic inequality (p. 817)
Rational inequality (p. 820)

Formulas

The quadratic formula: Given an equation in the form $ax^2 + bx + c = 0$, $a \neq 0$,
$x = \dfrac{-b \pm \sqrt{b^2 - 4ac}}{2a}$. Given an equation in the form $f(x) = ax^2 + bx + c$, $a \neq 0$,
the x-coordinate of the vertex of a parabola is $-\dfrac{b}{2a}$ and the equation of the axis of
symmetry is $x = -\dfrac{b}{2a}$.

Procedures, Rules, and Key Examples

Procedures/Rules	Key Examples

SECTION 11.1   The Square Root Principle and Completing the Square

Square root principle:
If $x^2 = a$, where a is a real number, then $x = \sqrt{a}$ or $x = -\sqrt{a}$.

It is common to indicate the positive and negative solutions by writing $\pm\sqrt{a}$.

Example 1: Solve.

a. $x^2 = 36$

Solution: $x = \pm\sqrt{36}$
$x = \pm 6$

b. $(x - 3)^2 = 12$

Solution: $x - 3 = \pm\sqrt{12}$
$x = 3 \pm 2\sqrt{3}$

To solve a quadratic equation by completing the square,
1. Write the equation in the form $x^2 + bx = c$.
2. Complete the square by adding $\left(\dfrac{b}{2}\right)^2$ to both sides.
3. Write the completed square in factored form.
4. Use the square root principle to eliminate the square.
5. Isolate the variable.
6. Simplify as needed.

Example 2: Solve by completing the square.

$$x^2 + 6x - 7 = 8$$
$$x^2 + 6x = 15 \qquad \text{Add 7 to both sides.}$$
$$x^2 + 6x + 9 = 15 + 9 \qquad \text{Complete the square.}$$
$$(x + 3)^2 = 24 \qquad \text{Factor.}$$
$$x + 3 = \pm\sqrt{24} \qquad \text{Use the square root principle.}$$
$$x = -3 \pm 2\sqrt{6} \qquad \text{Subtract 3 from both sides and simplify the square root.}$$

SECTION 11.2   Solving Quadratic Equations Using the Quadratic Formula

To solve a quadratic equation in the form $ax^2 + bx + c = 0$, where $a \neq 0$, use the quadratic formula:

$$x = \frac{-b \pm \sqrt{b^2 - 4ac}}{2a}$$

Example 1: Solve $3x^2 - 4x + 2 = 0$.

Solution: In the quadratic formula, replace a with 3, b with -4, and c with 2.

$$x = \frac{-(-4) \pm \sqrt{(-4)^2 - 4(3)(2)}}{2(3)}$$

$$= \frac{4 \pm \sqrt{16 - 24}}{6} = \frac{4 \pm \sqrt{-8}}{6}$$

$$= \frac{4 \pm 2\sqrt{-2}}{6} = \frac{2}{3} \pm \frac{\sqrt{2}}{3}i$$

(continued)

Procedures/Rules	Key Examples

SECTION 11.2   Solving Quadratic Equations Using the Quadratic Formula (*continued*)

Given a quadratic equation in the form $ax^2 + bx + c = 0$, where a, b, and c are rational numbers and $a \neq 0$, to determine the number and type of solutions, evaluate the discriminant $b^2 - 4ac$.

If the **discriminant is positive**, the equation has two real-number solutions. They will be rational if the discriminant is a perfect square and irrational otherwise.

If the **discriminant is 0**, the equation has one rational solution.

If the **discriminant is negative,** the equation has two non-real complex solutions.

Example 2: Use the discriminant to determine the number and type of solutions for $5x^2 - 7x + 8 = 0$.

Solution: In the discriminant, replace a with 5, b with -7, and c with 8.

$$(-7)^2 - 4(5)(8) = 49 - 160$$
$$= -111$$

Because the discriminant is negative, the equation has two nonreal complex solutions.

SECTION 11.3   Solving Equations That Are Quadratic in Form

If an equation involves rational expressions, multiply both sides by the LCD. Be sure to check for extraneous solutions.

Example 1: Solve $\dfrac{6}{x-2} + \dfrac{6}{x-1} = 5$.

Solution: First, multiply both sides by the LCD, $(x-2)(x-1)$.

$$6(x-1) + 6(x-2) = 5(x-2)(x-1)$$

$$12x - 18 = 5x^2 - 15x + 10 \qquad \text{Multiply.}$$

$$0 = 5x^2 - 27x + 28 \qquad \text{Write in } ax^2 + bx + c = 0 \text{ form.}$$

$$0 = (5x - 7)(x - 4) \qquad \text{Factor.}$$

$$5x - 7 = 0 \quad \text{or} \quad x - 4 = 0 \qquad \text{Use the zero-factor theorem.}$$

$$x = \frac{7}{5} \qquad\qquad x = 4 \qquad \text{Solve each equation.}$$

Check: Verify that $\dfrac{7}{5}$ and 4 solve the original equation.

If an equation has radicals, isolate a radical and square both sides. Continue the process as needed until all radicals have been eliminated. Solve the resulting equation and check for extraneous solutions.

Example 2: Solve $\sqrt{4x + 4} = 2x - 2$.

$$4x + 4 = 4x^2 - 8x + 4 \qquad \text{Square both sides.}$$

$$0 = 4x^2 - 12x \qquad \text{Write in } ax^2 + bx + c = 0 \text{ form.}$$

$$0 = 4x(x - 3) \qquad \text{Factor.}$$

$$4x = 0 \quad \text{or} \quad x - 3 = 0 \qquad \text{Use the zero-factor theorem.}$$

$$x = 0 \qquad\qquad x = 3 \qquad \text{Solve each equation.}$$

Check:

$x = 0$
$$\sqrt{4(0) + 4} = 2(0) - 2$$
$$\sqrt{0 + 4} = 0 - 2$$
$$2 = -2 \quad \text{False}$$

$x = 3$
$$\sqrt{4(3) + 4} = 2(3) - 2$$
$$\sqrt{12 + 4} = 6 - 2$$
$$4 = 4 \quad \text{True}$$

$x = 3$ is the only solution.

To solve equations that are quadratic in form using substitution,
1. Rewrite the equation so that it is in the form $au^2 + bu + c = 0$.
2. Solve the quadratic equation for u.
3. Substitute for u and solve.
4. Check the solutions.

Example 3: Solve $(a + 2)^2 + 7(a + 2) + 12 = 0$.

$$u^2 + 7u + 12 = 0 \qquad \text{Substitute } u \text{ for } a + 2.$$

$$(u + 3)(u + 4) = 0 \qquad \text{Factor.}$$

$$u + 3 = 0 \quad \text{or} \quad u + 4 = 0 \qquad \text{Use the zero-factor}$$

$$u = -3 \qquad\qquad u = -4 \qquad \text{theorem.}$$

$$a + 2 = -3 \quad \text{or} \quad a + 2 = -4 \qquad \text{Substitute } a + 2 \text{ for } u;$$

$$a = -5 \qquad\qquad a = -6 \qquad \text{solve for } a.$$

Check: Verify that -5 and -6 solve the original equation.

(*continued*)

Procedures/Rules	Key Examples

SECTION 11.4 Graphing Quadratic Functions

The graph of a function in the form $f(x) = a(x - h)^2 + k$ is a parabola with vertex at (h, k). The equation of the axis of symmetry is $x = h$. The parabola opens upward if $a > 0$ and downward if $a < 0$. The larger the $|a|$, the narrower the graph.

Example 1: For $f(x) = -3(x + 1)^2 - 2$,

a. Determine whether the graph opens upward or downward.
b. Find the vertex.
c. Write the equation of the axis of symmetry.
d. Graph.

Answers:
a. downward
b. $(-1, -2)$
c. $x = -1$
d.

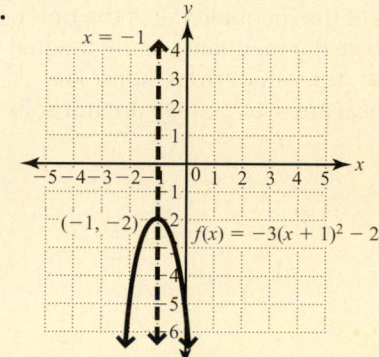

Given an equation in the form $f(x) = ax^2 + bx + c$, to determine the vertex,

1. Find the x-coordinate using the formula $x = -\dfrac{b}{2a}$.

2. Find the y-coordinate by evaluating $f\left(-\dfrac{b}{2a}\right)$ or $\dfrac{4ac - b^2}{4a}$

> **Note:** As an alternative approach to Example 2, we could have transformed $f(x) = 2x^2 - 12x + 19$ to $f(x) = 2(x - 3)^2 + 1$ by completing the square.

Example 2: For $f(x) = 2x^2 - 12x + 19$,

a. Determine whether the graph opens upward or downward.
b. Find the vertex.
c. Write the equation of the axis of symmetry.
d. Graph.

Answers:
a. upward

b. For the x-coordinate, use $-\dfrac{b}{2a}$.

$$x = -\frac{(-12)}{2(2)} = 3 \qquad a = 2, b = -12$$

For the y-coordinate, evaluate $f(3)$.

$$f(3) = 2(3)^2 - 12(3) + 19$$
$$= 18 - 36 + 19 = 1$$

Vertex: $(3, 1)$
c. $x = 3$
d.

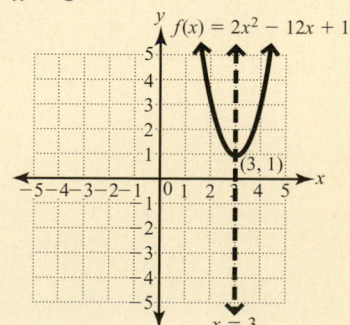

(continued)

Procedures/Rules	**Key Examples**

SECTION 11.5   Solving Nonlinear Inequalities

Solving quadratic inequalities.
1. Solve the related equation $ax^2 + bx + c = 0$.
2. Plot the solutions of $ax^2 + bx + c = 0$ on a number line. These solutions divide the number line into intervals.
3. Choose a test number from each interval and substitute the number into the inequality. If the test number makes the inequality true, then all numbers in that interval will solve the inequality. If the test number makes the inequality false, then no numbers in that interval will solve the inequality.
4. State the solution set of the inequality: It is the union of all intervals that solve the inequality. If the inequality symbols are $\leq$ or $\geq$, the values from Step 2 are included. If the symbols are $<$ or $>$, they are not solutions.

Example 1: Solve $x^2 + 2x - 15 \geq 0$.

$x^2 + 2x - 15 = 0$	Write the related equation.
$(x + 5)(x - 3) = 0$	Factor.
$x + 5 = 0$ or $x - 3 = 0$	Use the zero-factor theorem.
$x = -5$ $x = 3$	Solve each equation.

Plot -5 and 3 on a number line and label the intervals.

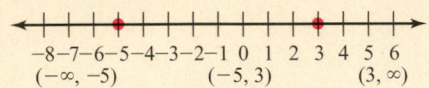

Choose a number from each interval and test it in the original inequality.

Interval	$(-\infty, -5)$	$(-5, 3)$	$(3, \infty)$
Test Value	-6	0	4
Result	$9 \geq 0$	$-15 \geq 0$	$9 \geq 0$
True/False	True	False	True

The solution set is $(-\infty, -5] \cup [3, \infty)$.

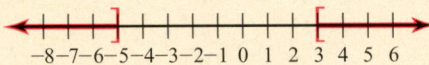

Solving rational inequalities.
1. Find all values that make any denominator equal to 0. These values must be excluded from the solution set.
2. Solve the related equation.
3. Plot the numbers found in Steps 1 and 2 on a number line and label the intervals.
4. Choose a test number from each interval and determine whether it solves the inequality.
5. The solution set is the union of all regions whose test number solves the inequality. If the inequality symbol is $\leq$ or $\geq$, include the values found in Step 2. The solution set never includes the values found in Step 1 because they make a denominator equal to 0.

Example 2: Solve $\dfrac{x + 5}{x - 1} > 4$.

Find the value(s) that make any denominator equal to 0.

$$x - 1 = 0$$
$$x = 1$$

Solve the related equation.

$\dfrac{x + 5}{x - 1} = 4$	Write the related equation.
$(x - 1)\dfrac{x + 5}{x - 1} = (x - 1)(4)$	Multiply both sides by the LCD.
$x + 5 = 4x - 4$	Simplify.
$9 = 3x$	Subtract x from and add 5 to both sides.
$3 = x$	Divide by 3.

Plot 1 and 3 on the number line and label the intervals.

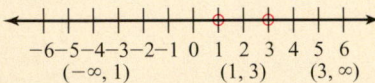

Choose a number from each interval and test it in the original inequality.

Interval	$(\infty, 1)$	$(1, 3)$	$(3, \infty)$
Test Value	0	2	4
Result	$-5 > 4$	$7 > 4$	$3 > 4$
True/False	False	True	False

The solution set is $(1, 3)$.

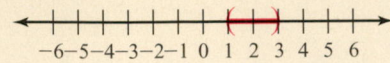

Chapter 11 Review Exercises

For Exercises 1–5, answer true or false.

[11.1] **1.** The notation $\sqrt{a}$ has the same meaning as the notation $\pm\sqrt{a}$.

$\begin{bmatrix} 11.1 \\ 11.2 \end{bmatrix}$ **3.** Every quadratic equation has a real-number solution.

[11.4] **5.** Given an equation in the form $y = ax^2 + bx + c$, if $a > 0$, the parabola opens up.

[11.3] **2.** Equations that are quadratic in form never have extraneous solutions.

[11.2] **4.** If the solutions for a quadratic equation in the form $ax^2 + bx + c = 0$ are complex, the discriminant can be used to find those solutions.

For Exercises 6–10, complete the rule.

[11.1] **6.** If $x^2 = a$ and a is a real number, then $x =$ _____ or _____.

[11.1] **7.** Given an equation in the form $x^2 + bx = c$, to complete the square, we add the constant term _____ to both sides of the equation.

[11.2] **9.** Given an equation in the form $ax^2 + bx + c = 0$, $a \neq 0$, the discriminant is _____.

[11.2] **8.** Given an equation in the form $ax^2 + bx + c = 0$, $a \neq 0$, the quadratic formula is $x =$ _____.

[11.4] **10.** Given a function in the form $f(x) = a(x - h)^2 + k$, the coordinates of the vertex are _____ and the axis of symmetry is _____.

Exercises 11–66 EQUATIONS AND INEQUALITIES

[11.1] For Exercises 11–18, solve and check.

11. $x^2 = 16$

12. $y^2 = \dfrac{1}{36}$

13. $k^2 + 2 = 30$

14. $3x^2 = 42$

15. $5h^2 + 24 = 9$

16. $(x + 7)^2 = 25$

17. $(x - 9)^2 = -16$

18. $\left(m + \dfrac{3}{5}\right)^2 = \dfrac{16}{25}$

[11.1] For Exercises 19–22, solve by completing the square.

19. $m^2 + 8m = -7$

20. $u^2 - 6u - 12 = 100$

21. $2b^2 - 6b + 7 = 0$

22. $u^2 + \dfrac{1}{4}u = \dfrac{3}{4}$

[11.2] For Exercises 23–26, solve using the quadratic formula.

23. $p^2 - 5 = -2p$

24. $3x^2 - 2x + 1 = 0$

25. $2t^2 + t - 5 = 0$

26. $2x^2 + 0.1x - 0.03 = 0$

[11.2] For Exercises 27–30, find the discriminant and determine the number and type of solutions for the equation.

27. $b^2 - 4b - 12 = 0$

28. $6z^2 - 7z + 5 = 0$

29. $k^2 + 6k + 9 = 0$

30. $0.8x^2 + 1.2x + 0.3 = 0$

[11.3] For Exercises 31–34, solve. Identify any extraneous solutions.

31. $\dfrac{1}{y} + \dfrac{1}{y + 3} = \dfrac{2}{3}$

32. $\dfrac{1}{u} + \dfrac{1}{u - 5} = \dfrac{10}{u^2 - 25}$

33. $6 - 5x^{-1} + x^{-2} = 0$

34. $2 - 3x^{-1} - x^{-2} = 0$

[11.3] *For Exercises 35–38, solve. Identify any extraneous solutions.*

35. $14\sqrt{x} + 45 = 0$ **36.** $\sqrt{4m} = 3m - 1$ **37.** $\sqrt{6r + 13} = 2r + 1$ **38.** $\sqrt{21t + 2} + t = 2 + 4t$

[11.3] *For Exercises 39–44, solve the equations using substitution. Identify any extraneous solutions.*

39. $x^4 - 5x^2 + 6 = 0$ **40.** $2m^4 - 3m^2 + 1 = 0$ **41.** $6(x + 5)^2 - 5(x + 5) + 1 = 0$

42. $\left(\dfrac{x - 1}{3}\right)^2 + 10\left(\dfrac{x - 1}{3}\right) + 9 = 0$ **43.** $p^{2/3} - 11p^{1/3} + 24 = 0$ **44.** $5a^{1/2} + 13a^{1/4} - 6 = 0$

[11.4] *For Exercises 45–48:*   *a. Find the x- and y-intercepts.*
                           b. State whether the parabola opens upward or downward.
                           c. Find the coordinates of the vertex.
                           d. Write the equation of the axis of symmetry.
                           e. Graph.

45. $f(x) = -2x^2$ **46.** $g(x) = \dfrac{1}{2}x^2 + 1$

47. $h(x) = -\dfrac{1}{3}(x - 2)^2$ **48.** $k(x) = 4(x + 3)^2 - 2$

[11.4] *For Exercises 49 and 50:*   *a. Write the equation in the form $f(x) = a(x - h)^2 + k$.*
                              b. Find the x- and y-intercepts.
                              c. State whether the parabola opens upward or downward.
                              d. Find the coordinates of the vertex.
                              e. Write the equation of the axis of symmetry.
                              f. Graph.
                              g. Find the domain and range.

49. $m(x) = x^2 + 2x - 1$ **50.** $p(x) = -0.5x^2 + 4x - 6$

51. $(x + 5)(x - 3) > 0$ **52.** $n^2 - 6n \leq -8$ **53.** $x^2 + 9x + 14 < 0$ **54.** $(x + 3)(x - 1)(x - 2) \geq 0$

[11.5] *For Exercises 55–58, solve the rational inequalities.*

55. $\dfrac{a + 3}{a - 1} \geq 0$ **56.** $\dfrac{r}{r + 2} < 2$ **57.** $\dfrac{n - 3}{n - 4} \leq 5$ **58.** $\dfrac{(k + 2)(k - 3)}{k - 5} < 0$

For Exercises 59–66, solve.

[11.1] **59.** A crop circle appeared July 7, 2003, in Windham Hill, England, and covered $22,500\pi$ square feet. Find the radius of the circle.

[11.1] **60.** Using the formula $E = \dfrac{1}{2}mv^2$, where E represents the kinetic energy in joules of an object with a mass of m kilograms and a velocity of v meters per second, find the velocity of an object with a mass of 50 kilograms and 400 joules of kinetic energy.

[11.2] **61.** The length of a small rectangular shed is 4 feet more than its width. If the area of the base is 285 square feet, what are the dimensions of the base of the shed?

[11.1] **62.** A right circular cylinder is to be constructed so that its volume is equal to that of a sphere with a radius of 9 inches. If the cylinder is to have a height of 4 inches, find the radius of the cylinder.

[11.2] **63.** A ramp is constructed so that it is a right triangle with a base that is 9 feet longer than its height. If the hypotenuse is 17 feet, find the dimensions of the base and height.

[11.4] **64.** With an initial velocity of 24 feet per second, an acrobat is launched upward from one end of a lever. The function $h = -16t^2 + 24t$ describes the height, h, of the acrobat t seconds after being launched.
   a. After how many seconds does the acrobat reach maximum height?

   b. What is the maximum height the acrobat reaches?
   c. How long is the acrobat in the air?
   d. Graph the function.

[11.4] **65.** The longest punt on record in the NFL was by Steve O'Neal in a game between the New York Jets and the Denver Broncos on September 21, 1969. The function $y = -0.03x^2 + 2.16x$ models the trajectory of the punt. (Note that x and y are distances in yards.)
   a. Find the maximum height the punt reached.

   b. Find the distance of the punt.

Of Interest

The yardage you found is how far the punt carried in the air. It was actually recorded as a 98-yard punt, which includes the amount of roll after the punt landed and was downed.

[11.5] **66.** In a triangle, the height is to be 2 inches less than x. The base is to be 4 inches more than x.
   a. Find the range of values for x so that the area of the triangle is at least 56 square inches.

   b. Find the range of values for the base and the height.

Chapter 11 Practice Test

For Extra Help

CHAPTER
Test Prep
VIDEOS

Step-by-step test solutions are found on the Chapter Test Prep Videos available via the Video Resources on DVD, in *MyMathLab*, and on YouTube (search "CarsonCombinedAlg" and click on "Channels").

For 1 and 2, use the square root principle to solve and check.

1. $x^2 = 81$

2. $(x - 3)^2 = 20$

For 3 and 4, solve by completing the square.

3. $x^2 - 8x = -4$

4. $3m^2 - 6m = 5$

For 5 and 6, solve using the quadratic formula.

5. $2x^2 + x - 6 = 0$

6. $x^2 - 8x + 15 = 0$

For 7–12, solve using any method.

7. $u^2 - 16 = -6u$

8. $x^2 = 81$

9. $4w^2 + 6w + 3 = 0$

10. $2x^2 + 4x = 0$

11. $x^2 + 16 = 0$

12. $3k^2 = -5k$

For 13–16, solve. Identify any extraneous solutions.

13. $\dfrac{1}{x + 2} + \dfrac{1}{x} = \dfrac{5}{12}$

14. $3 - x^{-1} - 2x^{-2} = 0$

15. $9\sqrt{x} + 8 = 0$

16. $\sqrt{x + 8} - x = 2$

For 17 and 18, solve the equations using substitution. Identify any extraneous solutions.

17. $9a^4 + 26a^2 - 3 = 0$

18. $(x + 1)^2 + 3(x + 1) - 4 = 0$

19. For $f(x) = -x^2 + 6x - 4$,
    a. Find the x- and y-intercepts.
    b. Write the equation in the form $f(x) = a(x - h)^2 + k$.
    c. State whether the parabola opens upward or downward.
    d. Find the coordinates of the vertex.
    e. Write the equation of the axis of symmetry.
    f. Graph.
    g. Find the domain and range.

For 20 and 21: **a.** *Solve the inequality.*

b. *Graph the solution set on a number line.*

20. $(x + 1)(x - 4) \leq 0$

21. $\dfrac{x + 2}{x - 1} > 0$

For 22–25, solve.

22. A ball is thrown downward from a window in a tall building. The distance, d, the ball traveled is given by the equation $d = 16t^2 + 32t$, where t is the time traveled in seconds. How long will it take the ball to fall 180 feet?

23. A rectangular parking space must have an area of 400 square feet. The length is to be 20 feet more than the width. Find the dimensions of the parking space.

24. An archer shoots an arrow in a field. Suppose the equation $y = -0.02x^2 + 1.3x + 8$ models the trajectory of the arrow. (Assume that x and y represent distances in meters and that $x \geq 0$ and $y \geq 0$.)

a. What is the maximum height the arrow reaches?

b. How far does the arrow travel horizontally?

c. Graph the trajectory of the arrow.

25. A zoo is planning to install a new aquarium tank. The tank is to be in the shape of a box with a height of 12 feet. It is preferred that the length be 15 feet more than the width.

a. Find the range of values for the width so that the volume of the space is, at most, 12,000 cubic feet.

b. Find the range of values for the length.

Chapters 1–11 Cumulative Review Exercises

For Exercises 1–3, answer true or false.

[4.1] **1.** The point with coordinates $(45, -12)$ is in quadrant II.

[9.1] **2.** A system of equations that has no solution is said to be consistent.

[5.2] **3.** The degree of the monomial $5x^3y$ is 4.

For Exercises 4–6, fill in the blank.

[6.7] **4.** The graph of a polynomial function of the form $f(x) = ax^2 + bx + c$ is a(n) _____ that opens up if _____.

[10.3] **5.** If $\sqrt[n]{a}$ and $\sqrt[n]{b}$ are both real numbers, then $\sqrt[n]{a} \cdot \sqrt[n]{b} =$ _____.

[11.1] **6.** If $x^2 = a$ and a is a real number, then $x =$ _____ or $x =$ _____.

Exercises 7–15 ▰ EXPRESSIONS

For Exercises 7–9, simplify.

[5.3] **7.** $(-3x - 4) - (3x + 2)$

[5.4] **8.** $(8n^2)(-7mn^3)$

[5.5] **9.** $(x - 3)(4x^2 - 2x + 1)$

[6.4] **For Exercises 10 and 11, factor completely.**

10. $m^3 + 8$

11. $x^2 + 10x + 25$

For Exercises 12–14, simplify.

[7.2] **12.** $\dfrac{4u^2 + 4u + 1}{u + 2u^2} \cdot \dfrac{u}{2u^2 - u - 1}$ [7.3] **13.** $\dfrac{3}{x - 2} - \dfrac{2}{x + 2}$ [10.2] **14.** $x^{3/4} \cdot x^{-1/4}$

[10.5] **15.** Rationalize the denominator of $\dfrac{\sqrt{n}}{\sqrt{m} - \sqrt{n}}$.

Exercises 16–30 ◣ **EQUATIONS AND INEQUALITIES**

For Exercises 16–20, solve.

[2.3] **16.** $3n - 5 = 7n + 9$ [8.2] **17.** $|3x + 4| + 3 = 7$ [6.6] **18.** $2x^2 + 7x = 15$

[7.6] **19.** $\dfrac{5}{x - 2} - \dfrac{3}{x} = \dfrac{11}{3x}$ [10.6] **20.** $\sqrt{5x - 4} = 9$ [11.2] **21.** Solve $x^2 - 6x + 11 = 0$ using the quadratic formula.

For Exercises 22 and 23: **a. Graph the solution set on a number line.**
b. Write the solution set in set-builder notation.
c. Write the solution set in interval notation.

[2.6] **22.** $-2 < x + 4 < 5$

[11.5] **23.** $x^2 + x \le 12$

[4.7] **24.** Graph $f(x) = -2x$.

[4.5] **25.** Write the equation of a line in standard form that passes through the point $(-2, 4)$ and is perpendicular to the line $3x - 2y = 6$.

[9.4] **26.** Solve the system: $\begin{cases} x + y + z = 5 \\ 2x + y - 2z = -5 \\ x - 2y + z = 8 \end{cases}$

For Exercises 27–30, solve.

$\begin{bmatrix} 3.4 \\ 9.3 \end{bmatrix}$ **27.** A trucking company finds that the average speed of experienced drivers is 10 miles per hour more than the average speed of inexperienced drivers. If inexperienced drivers take 1 hour longer than experienced drivers to drive 300 miles, find how long it takes the experienced driver to drive 300 miles.

$\begin{bmatrix} 3.5 \\ 9.3 \end{bmatrix}$ **28.** A chemist has a bottle containing 50 ml of 15% saline solution and a bottle of 40% saline solution. She wants a 30% solution. How much of the 40% solution must be added to the 15% solution so that a 30% concentration is created?

[7.7] **29.** If the wavelength of a wave remains constant, the velocity, v, of a wave is inversely proportional to its period, T. In an experiment, waves are created in a pool of water so that the period is 8 seconds and the velocity is 6 centimeters per second. If the period is increased to 15 seconds, what is the velocity?

30. The formula $t = \sqrt{\dfrac{h}{16}}$ describes the amount of time, t, in seconds, that an object falls a distance of h feet.

[10.1] **a.** Write an expression in simplest form for the exact amount of time an object falls a distance of 24 feet.

[10.6] **b.** Find the distance an object falls in 3 seconds.

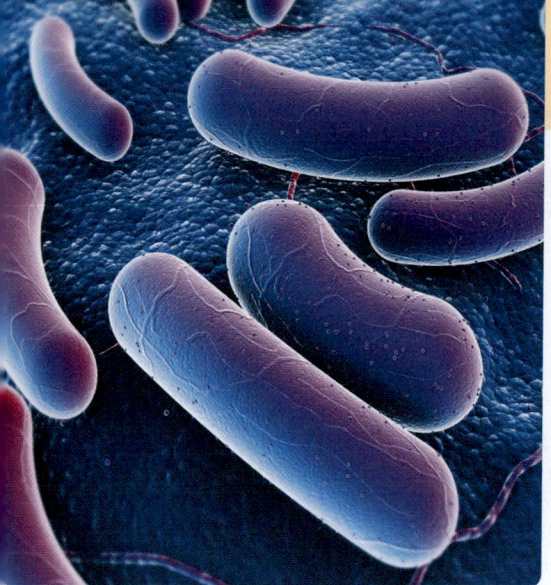

Exponential and Logarithmic Functions

12

"The mathematics of uncontrolled growth are frightening. A single cell of the bacterium *E. coli* would, under ideal circumstances, divide every twenty minutes. . . . one cell becomes two, two become four, four become eight, and so on. In this way it can be shown that in a single day, one cell of *E. coli* could produce a super-colony equal in size and weight to the entire planet Earth."

—Michael Crichton (1942–2008)
The Andromeda Strain

We begin this chapter with composite and inverse functions. Exponential and logarithmic functions, the subject of the rest of the chapter, are inverses of each other. These functions are used in modeling population growth, electrical circuits, sound decibels, pH, and the Richter scale.

12.1 Composite and Inverse Functions

Objective 1 **Find the composition of two functions.** In Section 8.5, we learned to perform the basic operations with functions. We now explore a new operation. Recall that a function pairs elements from an "input" set called the *domain* with elements in an "output" set called the *range*. If the output elements of one function are used as input elements in a second function, the resulting function is the **composition** of the two functions. Composition occurs when one quantity depends on a second quantity, which, in turn, depends on a third quantity.

We can visualize the composition of two functions as a machine with two stages, which are the two functions. For example, some coffee machines have two stages. Roasted coffee beans, which are ground in the first stage (first function) are put in the machine. The grounds are then fed into the second stage (second function), which runs hot water over the grounds to produce coffee.

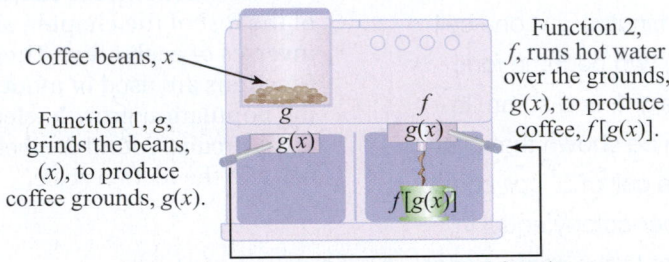

Coffee beans, x

Function 1, g, grinds the beans, (x), to produce coffee grounds, $g(x)$.

Function 2, f, runs hot water over the grounds, $g(x)$, to produce coffee, $f[g(x)]$.

Grounds, $g(x)$, are placed into second function.

Definition

Composition of functions: If f and g are functions, then the composition of f and g is defined as $(f \circ g)(x) = f[g(x)]$ for all x in the domain g for which $g(x)$ is in the domain of f. The composition of g and f is defined as $(g \circ f)(x) = g[f(x)]$ for all x in the domain of f for which $f(x)$ is in the domain of g.

Note: The notation $(g \circ f)(x)$ is read "g composed with f of x," or simply "g of f of x."

Warning: Do not confuse $(f \circ g)(x)$ with $(f \cdot g)(x)$.

For example, suppose a pebble is dropped in a calm lake, causing ripples to form in concentric circles. If the radius of the outer ripple increases at a rate of 2 feet per second, we can describe the length of the radius as a function of t by using $r(t) = 2t$, where t is in seconds. The area of the circle formed by the outer ripple is also a function described by $A(r) = \pi r^2$. If we substitute the output of $r(t)$, which is $2t$ for r, into $A(r)$, we form the composite function $(A \circ r)(t)$.

$$
\begin{aligned}
(A \circ r)(t) &= A[r(t)] \\
&= \pi(2t)^2 &&\text{Substitute } 2t \text{ for } r \text{ in } A(r) = \pi r^2. \\
&= 4\pi t^2 &&\text{Simplify.}
\end{aligned}
$$

The composite function $(A \circ r)(t)$ gives the area of the circle as a function of time.

Example 1 If $f(x) = 2x - 3$ and $g(x) = x^2 + 1$, find the following.

a. $(f \circ g)(2)$

SOLUTION $(f \circ g)(2) = f[g(2)]$ **Definition of composition**

$$= f(5) \qquad g(2) = 2^2 + 1 = 4 + 1 = 5$$

$$= 2(5) - 3 \qquad \text{In } f(x), \text{ replace } x \text{ with } 5.$$

$$= 7 \qquad \text{Simplify.}$$

> **Note:** We first find $g(2)$, which is 5, then substitute that result into f.

b. $(g \circ f)(2)$

SOLUTION $(g \circ f)(2) = g[f(2)]$ **Definition of composition**

$$= g(1) \qquad f(2) = 2(2) - 3 = 4 - 3 = 1$$

$$= 1^2 + 1 \qquad \text{In } g(x), \text{ replace } x \text{ with } 1.$$

$$= 2 \qquad \text{Simplify.}$$

c. $(f \circ g)(x)$

> **Note:** We could have found $(f \circ g)(2)$ from part a by first finding $(f \circ g)(x) = 2x^2 - 1$, then substituting 2 for x:
> $$(f \circ g)(2) = 2(2)^2 - 1$$
> $$= 8 - 1$$
> $$= 7$$

SOLUTION $(f \circ g)(x) = f[g(x)]$ **Definition of composition**

$$= f(x^2 + 1) \qquad \text{Replace } g(x) \text{ with } x^2 + 1.$$

$$= 2(x^2 + 1) - 3 \qquad \text{In } f(x), \text{ replace } x \text{ with } x^2 + 1.$$

$$= 2x^2 + 2 - 3 \qquad \text{Multiply.}$$

$$= 2x^2 - 1 \qquad \text{Add.}$$

d. $(g \circ f)(x)$

> **Note:** We could have found $(g \circ f)(2)$ from part b by first finding $(g \circ f)(x) = 4x^2 - 12x + 10$, then substituting 2 for x:
> $$(g \circ f)(2) = 4(2)^2 - 12(2) + 10$$
> $$= 16 - 24 + 10$$
> $$= 2$$

SOLUTION $(g \circ f)(x) = g[f(x)]$ **Definition of composition**

$$= g(2x - 3) \qquad \text{Replace } f(x) \text{ with } 2x - 3.$$

$$= (2x - 3)^2 + 1 \qquad \text{In } g(x), \text{ replace } x \text{ with } 2x - 3.$$

$$= 4x^2 - 12x + 9 + 1 \qquad \text{Multiply.}$$

$$= 4x^2 - 12x + 10 \qquad \text{Add.}$$

Generally, $(f \circ g)(x) \neq (g \circ f)(x)$, as in Examples 1(c) and (d), but there are special functions for which $(f \circ g)(x) = (g \circ f)(x)$.

Your Turn 1 If $f(x) = x^2 + 2$ and $g(x) = 3x + 5$, find the following.

a. $f[g(-3)]$ **b.** $g[f(2)]$ **c.** $f[g(x)]$

Objective 2 Show that two functions are inverses. Operations that undo each other, such as addition and subtraction, are called *inverse operations*. For example, if we begin with a number x and add a second number y, we have $x + y$. Now we subtract y (the number we just added) from that result, and we have $x + y - y = x$, which is the original number.

Likewise, two functions that undo each other under composition are called **inverse functions**.

Definition ***Inverse functions:*** Two functions f and g are inverses if and only if $(f \circ g)(x) = x$ for all x in the domain of g and $(g \circ f)(x) = x$ for all x in the domain of f.

Loosely speaking, f and g are inverse functions if you evaluate f for a value x in its domain, substitute that result into g and evaluate, and get x again and vice versa.

Answers Your Turn 1

a. $f[g(-3)] = 18$

b. $g[f(2)] = 23$

c. $f[g(x)] = 9x^2 + 30x + 27$

To determine whether two functions f and g are inverses of each other,

1. Show that $f[g(x)] = x$ for all x in the domain of g.
2. Show that $g[f(x)] = x$ for all x in the domain of f.

Example 2   Verify that f and g are inverses.

a. $f(x) = 3x + 2$, $g(x) = \dfrac{x - 2}{3}$

SOLUTION   We need to show that $f[g(x)] = x$ and $g[f(x)] = x$.

$$f[g(x)] = f\left(\frac{x - 2}{3}\right) \qquad \text{Substitute } \frac{x - 2}{3} \text{ for } g(x).$$

$$= 3\left(\frac{x - 2}{3}\right) + 2 \qquad \text{Substitute } \frac{x - 2}{3} \text{ for } x \text{ in } f(x).$$

$$= x - 2 + 2 \qquad \text{Multiply.}$$

$$= x \qquad \text{So } f[g(x)] = x.$$

$$g[f(x)] = g(3x + 2) \qquad \text{Substitute } 3x + 2 \text{ for } f(x).$$

$$= \frac{3x + 2 - 2}{3} \qquad \text{Replace } x \text{ with } 3x + 2 \text{ in } g(x).$$

$$= \frac{3x}{3} \qquad \text{Add.}$$

$$= x \qquad \text{So } g[f(x)] = x.$$

Because $f[g(x)] = x$ and $g[f(x)] = x$, f and g are inverses.

b. $f(x) = x^3 + 5$ and $g(x) = \sqrt[3]{x - 5}$

SOLUTION   We need to show that $f[g(x)] = x$ and $g[f(x)] = x$.

$$f[g(x)] = f(\sqrt[3]{x - 5}) \qquad \text{Replace } g(x) \text{ with } \sqrt[3]{x - 5}.$$
$$= (\sqrt[3]{x - 5})^3 + 5 \qquad \text{Replace } x \text{ with } \sqrt[3]{x - 5} \text{ in } f(x).$$
$$= x - 5 + 5 \qquad \text{Apply } (\sqrt[n]{x})^n = x.$$
$$= x \qquad \text{So } f[g(x)] = x.$$
$$g[f(x)] = g(x^3 + 5) \qquad \text{Replace } f(x) \text{ with } x^3 + 5.$$
$$= \sqrt[3]{x^3 + 5 - 5} \qquad \text{Replace } x \text{ with } x^3 + 5 \text{ in } g(x).$$
$$= \sqrt[3]{x^3} \qquad \text{Add.}$$
$$= x \qquad \text{So } g[f(x)] = x.$$

Because $f[g(x)] = x$ and $g[f(x)] = x$, f and g are inverses.

Your Turn 2   Verify that $f(x) = 5x + 6$ and $g(x) = \dfrac{x - 6}{5}$ are inverse functions.

Answer   Your Turn 2

$f[g(x)] = f\left[\dfrac{x - 6}{5}\right]$

$\quad = 5\left(\dfrac{x - 6}{5}\right) + 6$

$\quad = x - 6 + 6$

$\quad = x$

$g[f(x)] = g[5x + 6]$

$\quad = \dfrac{5x + 6 - 6}{5}$

$\quad = \dfrac{5x}{5}$

$\quad = x$

Because $f[g(x)] = x$ and $g[f(x)] = x$, f and g are inverses.

Calculator Tips

We can verify that $f[g(x)] = g[f(x)] = x$ without actually finding the composition functions by using a graphing calculator. We illustrate using $f(x) = 3x + 2$ and $g(x) = \dfrac{x - 2}{3}$, which were shown to be inverse functions in Example 2(a). Enter
(continued)

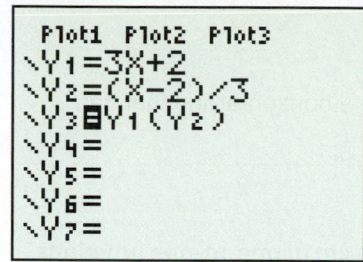

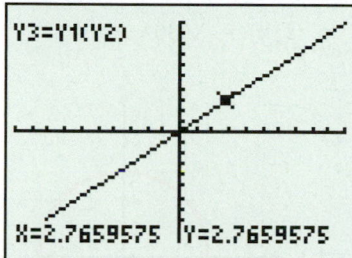

Note: $Y_1(Y_2)$ is Y_1 of Y_2, not $Y_1 \cdot Y_2$.

Warning: Do not confuse f^{-1} with raising an expression to the negative 1 power. It is usually clear from the context that f^{-1} means an inverse function.

$f(x) = Y_1 = 3x + 2$, $g(x) = Y_2 = \dfrac{x-2}{3}$. *Because* $f[g(x)] = Y_1(Y_2)$, *enter* $Y_3 = Y_1(Y_2)$.

To enter Y_1 *and* Y_2 *in the* ⌑Y=⌑ *menu on the TI-83 plus and TI-84, put the cursor at* Y_3, *press* ⌑VARS⌑, *select Y-VARS, select Function, and press* ⌑ENTER⌑. *Select your choice and press* ⌑ENTER⌑. *Graph* Y_3 *without graphing* Y_1 *or* Y_2. *To graph* Y_3 *only, press* ⌑Y=⌑, *put the cursor on the = sign of* Y_1, *and press* ⌑ENTER⌑. *Repeat for* Y_2. *Press* ⌑GRAPH⌑. *Trace to verify that the graph is* $y = x$.

Repeat for $g[f(x)]$.

Before finding inverse functions, we need to determine what types of functions have inverses. Consider the following.

$$f = \{(1,2),(3,4),(5,6),(x,y)\}; \text{domain} = \{1,3,5,x\} \text{ and range} = \{2,4,6,y\}$$
$$g = \{(2,1),(4,3),(6,5),(y,x)\}; \text{domain} = \{2,4,6,y\} \text{ and range} = \{1,3,5,x\}$$

Note that the ordered pairs in g are the ordered pairs of f with x and y interchanged. Because $(1,2)$ is in f, by definition, $f(1) = 2$. Because $(2,1)$ is in g, $g(2) = 1$. Therefore, $f[g(\mathbf{2})] = f(\mathbf{1}) = \mathbf{2}$ and $g[f(\mathbf{1})] = g(\mathbf{2}) = \mathbf{1}$. Similarly, $f[g(\mathbf{4})] = f(\mathbf{3}) = \mathbf{4}$ and $g[f(\mathbf{3})] = g(\mathbf{4}) = \mathbf{3}$. In particular, $f[g(\mathbf{y})] = f(\mathbf{x}) = \mathbf{y}$ and $g[f(\mathbf{x})] = g(\mathbf{y}) = \mathbf{x}$ for all real numbers x and y. Consequently, f and g are inverses.

In general, to find the inverse of a function, we reverse the ordered pairs by interchanging x and y. To indicate the inverse of the function f, we use a special notation f^{-1} instead of g. The following figure illustrates inverse functions.

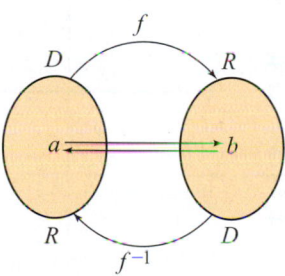

Note: Notice that f sends a to b and f^{-1} sends b back to a. Hence, $f^{-1}[f(a)] = a$. Also, f^{-1} sends b to a and f sends a back to b. Hence, $f[f^{-1}(b)] = b$. We also see that the domain of f is the range of f^{-1} and the range of f is the domain of f^{-1}.

Before we can find inverse functions, we need to explore one more concept.

Objective ③ Show that a function is one to one. If we merely reverse the ordered pairs of a function, we do not always get an inverse function. Consider the following sets in which the ordered pairs of B are the ordered pairs of A with x and y interchanged:

$$A = \{(1,2),(2,4),(3,2),(-2,5)\}$$
$$B = \{(2,1),(4,2),(2,3),(5,-2,)\}$$

Note that A represents a function but B does not because 2 in the domain of B is paired with 1 and 3. How did this happen? Because the ordered pairs of B are those of A with x and y interchanged, two ordered pairs of A having the same y-value of 2 $[(1,2)$ and $(3,2)]$ became two ordered pairs with the same x-value in B. Therefore, if the inverse of a function is to be a function, no two ordered pairs of the function may have the same y-value. Such a function is called a **one-to-one function**.

A function is one-to-one if each value in the domain corresponds to only one value in the range. In terms of x and y, two different x-values must result in two different y-values, which suggests the following formal definition.

Definition **One-to-one function:** A function f is one to one if for any two numbers a and b in its domain, when $f(a) = f(b)$, $a = b$ and when $a \neq b$, $f(a) \neq f(b)$.

It is possible to determine whether a function is one-to-one by looking at its graph. If two ordered pairs have the same y-value, the corresponding points lie on the same horizontal line. Consequently, if a function is one to one, the graph cannot be intersected by any horizontal line in more than one point.

Rule **Horizontal Line Test for One-to-One Functions**

Given a function's graph, the function is one to one if every horizontal line that can intersect the graph does so at one and only one point.

Example 3 Determine whether the following graphs are of one-to-one functions.

a. $f(x) = x^3$

b. $f(x) = x^2 + 2$

c. $f(x) = \sqrt{36 - x^2}$

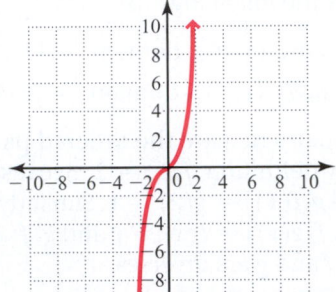

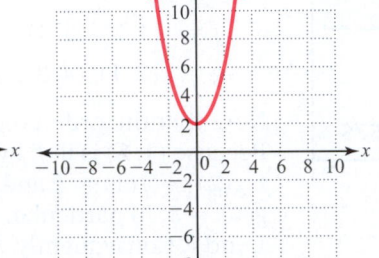

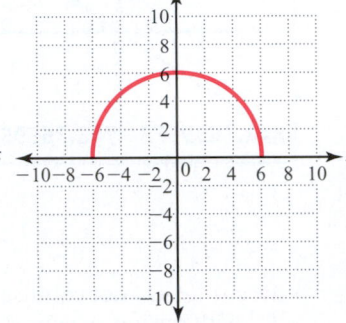

SOLUTION

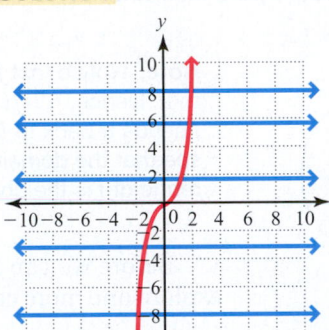

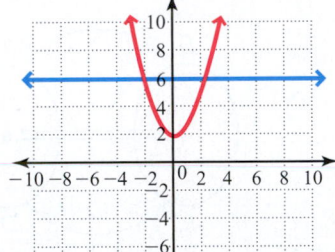

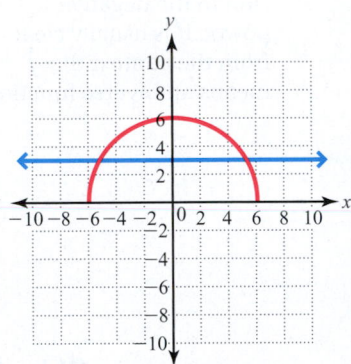

a. Every horizontal line that can intersect this graph does so at one and only one point, so the function is one to one.

b. A horizontal line can intersect this graph in more than one point, so the function is not one to one.

c. A horizontal line can intersect this graph in more than one point, so the function is not one to one.

Your Turn 3 Determine whether the following graphs are of one-to-one functions.

a.

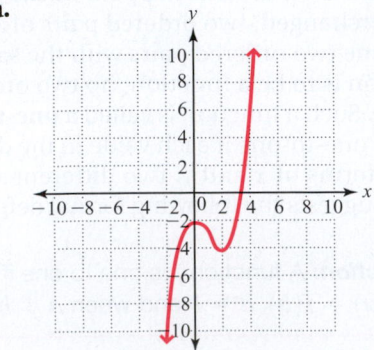

b.

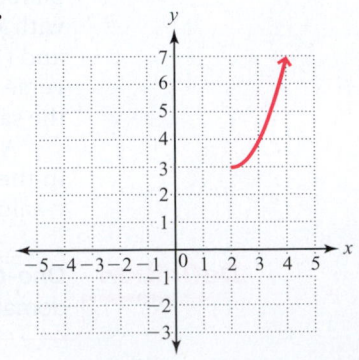

Answers **Your Turn 3**

a. no **b.** yes

Objective ④ **Find the inverse of a function.** Interchanging the ordered pairs of a function that is not one to one results in a relation that is not a function. Consequently, we have the following.

Rule **Existence of Inverse Functions**

A function has an inverse function if and only if the function is one to one.

We have already seen that to find the inverse of a function, we need to interchange the x- and y-values of all of the ordered pairs of f. That is, we replace x with y and y with x in the equation defining the function. However, we are not finished. We now have the inverse in the form $f(y) = x$, so we must solve this equation for y to have y as a function of x. The steps are summarized as follows.

Procedure **Finding the Inverse Function of a One-to-One Function**

1. If necessary, replace $f(x)$ with y.
2. Replace all x's with y's and y's with x's.
3. Solve the equation from Step 2 for y.
4. Replace y with $f^{-1}(x)$.

Example 4 Find $f^{-1}(x)$ for each of the following one-to-one functions.

a. $f(x) = 2x + 4$

The domain and range of f is the set of all real numbers, so the domain and range of f^{-1} is also the set of all real numbers.

SOLUTION

$y = 2x + 4$ Replace $f(x)$ with y.

$x = 2y + 4$ Replace x with y and y with x.

$\dfrac{x - 4}{2} = y$ Solve for y.

$f^{-1}(x) = \dfrac{x - 4}{2}$ Replace y with $f^{-1}(x)$.

To verify that we have found the inverse, we need to show that $f[f^{-1}(x)] = x$ and $f^{-1}[f(x)] = x$.

$$f[f^{-1}(x)] = f\left(\frac{x - 4}{2}\right) = 2\left(\frac{x - 4}{2}\right) + 4 = x - 4 + 4 = x$$

$$f^{-1}[f(x)] = f^{-1}(2x + 4) = \frac{(2x + 4) - 4}{2} = \frac{2x}{2} = x$$

Because $f[f^{-1}(x)] = x$ and $f^{-1}[f(x)] = x$, they are inverses.

b. $f(x) = x^3 + 2$

The domain and range of f is the set of all real numbers, so the domain and range of f^{-1} is also the set of all real numbers.

SOLUTION

$y = x^3 + 2$ Replace $f(x)$ with y.

$x = y^3 + 2$ Replace x with y and y with x.

$x - 2 = y^3$ Begin solving for y by subtracting 2 from both sides.

$\sqrt[3]{x - 2} = y$ Take the cube root of each side to solve for y.

$f^{-1}(x) = \sqrt[3]{x - 2}$ Replace y with $f^{-1}(x)$.

We can verify that f and f^{-1} are inverses, as in part a.

c. $f(x) = \sqrt{x - 3}$

The domain of f is $[3, \infty)$, and the range is $[0, \infty)$. Therefore, the domain of f^{-1} is $[0, \infty)$ and the range of f^{-1} is $[3, \infty)$.

SOLUTION

$$y = \sqrt{x - 3}$$ Replace $f(x)$ with y.

$$x = \sqrt{y - 3}$$ Replace x with y and y with x.

$$x^2 = y - 3$$ Begin solving for y by squaring both sides.

$$x^2 + 3 = y$$ Add 3 to both sides to solve for y.

The domain of $y = x^2 + 3$ is all real numbers, but the domain of f^{-1} is $[0, \infty)$. Therefore, we write $f^{-1}(x) = x^2 + 3, x \geq 0$.

To verify that these are inverses, we need to show that $f[f^{-1}(x)] = x$ and $f^{-1}[f(x)] = x$.

$$f[f^{-1}(x)] = f(x^2 + 3) = \sqrt{(x^2 + 3) - 3} = \sqrt{x^2 + 3 - 3} = \sqrt{x^2}$$

Because the domain of f^{-1} is $[0, \infty)$, $\sqrt{x^2} = x$. So $f[f^{-1}(x)] = x$.

$$f^{-1}[f(x)] = f^{-1}(\sqrt{x - 3}) = (\sqrt{x - 3})^2 + 3 = x - 3 + 3 = x$$

Because $f[f^{-1}(x)] = x$ and $f^{-1}[f(x)] = x$, they are inverses.

Your Turn 4   Find $f^{-1}(x)$ for each of the following one-to-one functions.

   a. $f(x) = 5x + 2$ **b.** $f(x) = \dfrac{x + 3}{x}$

Objective ⑤   **Graph a given function's inverse function.**   In the following figure, we have plotted pairs of points whose coordinates are interchanged and have graphed the line $y = x$.

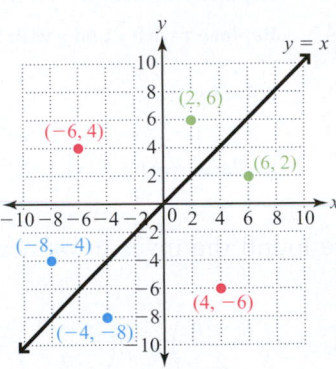

If the graph were folded along the line $y = x$, the points whose coordinates are interchanged would fall on top of each other. These points, therefore, are symmetric with respect to the line $y = x$. It can be shown that the graphs of any two points of the form (a, b) and (b, a) are symmetric with respect to the graph of $y = x$. Similarly, because the ordered pairs of f and f^{-1} have interchanged coordinates, the graphs of f and f^{-1} are symmetric with respect to the line $y = x$.

Rule   **Graphs of Inverse Functions**

The graphs of f and f^{-1} are symmetric with respect to the graph of $y = x$.

Following are the graphs of f and f^{-1} for Example 4 along with the graph of $y = x$.

   a. $f(x) = 2x + 4$ **b.** $f(x) = x^3 + 2$ **c.** $f(x) = \sqrt{x - 3}$

      $f^{-1}(x) = \dfrac{x - 4}{2}$ $f^{-1}(x) = \sqrt[3]{x - 2}$ $f^{-1}(x) = x^2 + 3, \;\; x \geq 0$

Answers   **Your Turn 4**

a. $f^{-1}(x) = \dfrac{x - 2}{5}$

b. $f^{-1}(x) = \dfrac{3}{x - 1}$

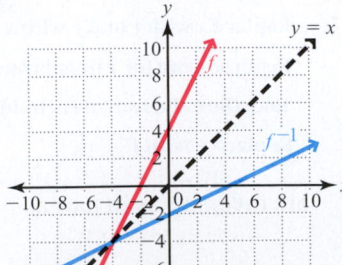

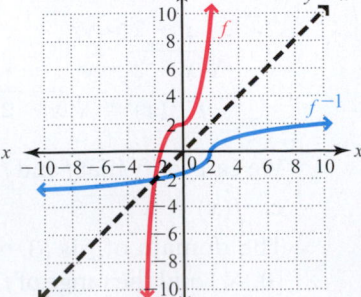

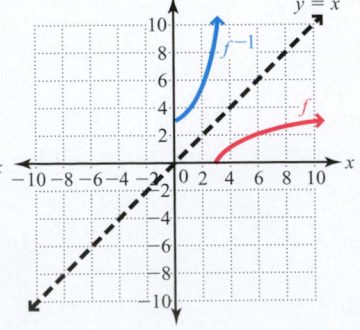

Example 5 Sketch the inverse of the functions whose graphs are shown in parts a and b.

a.

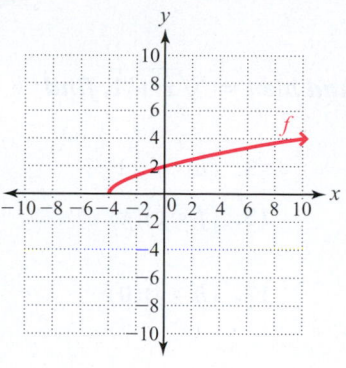

SOLUTION Draw the line $y = x$ and reflect the graph in the line.

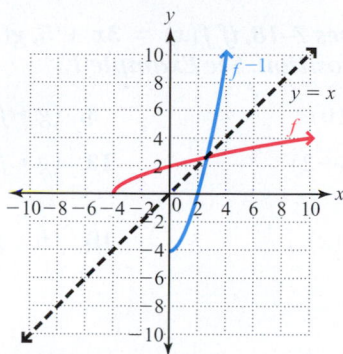

b.

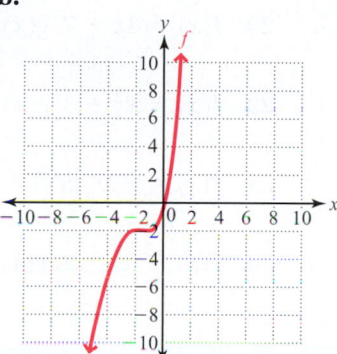

SOLUTION Draw the line $y = x$ and reflect the graph in the line.

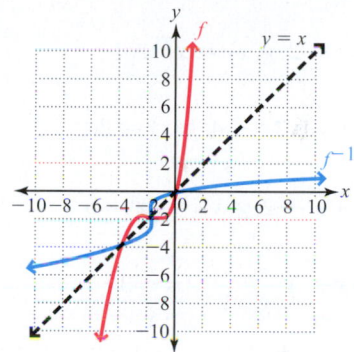

Answers Your Turn 5

a.

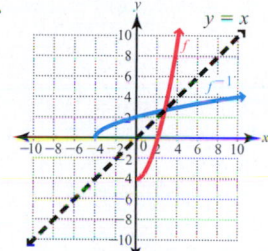

b.

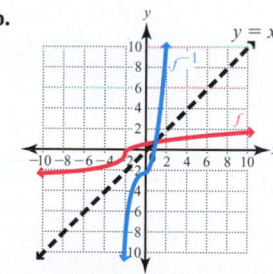

Your Turn 5 Sketch the graph of the inverse of the functions whose graphs are given.

a.

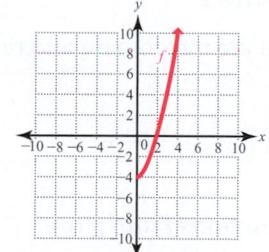

b.

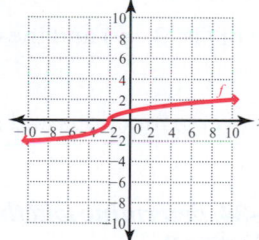

12.1 Exercises For Extra Help *MyMathLab*

PRACTICE WATCH DOWNLOAD

1. Does $f \circ g = g \circ f$ for any two functions f and g? Why or why not?

2. What must be shown to prove that the two functions f and g are inverses?

3. How are the domains and ranges of inverse functions related? Why?

4. If the coordinates of the ordered pairs of a function are interchanged, is the result always the inverse function? Why or why not?

5. How are horizontal lines used to determine whether a function is one to one?

6. How is the graph of $(-2, 3)$ related to the graph of $(3, -2)$?

For Exercises 7–18, if $f(x) = 3x + 5$, $g(x) = x^2 + 3$, and $h(x) = \sqrt{x + 1}$, find each composition. See Example 1.

7. $(f \circ g)(0)$

8. $(g \circ f)(0)$

9. $(h \circ f)(1)$

10. $(h \circ g)(3)$

11. $(f \circ g)(-2)$

12. $(g \circ f)(-1)$

13. $(f \circ g)(x)$

14. $(g \circ f)(x)$

15. $(f \circ h)(x)$

16. $(h \circ g)(x)$

17. $(h \circ f)(0)$

18. $(g \circ h)(0)$

For Exercises 19–28, find $(f \circ g)(x)$ and $(g \circ f)(x)$. See Example 1.

19. $f(x) = 2x - 2$, $g(x) = 3x + 4$

20. $f(x) = 4x + 7$, $g(x) = 3x + 4$

21. $f(x) = x + 2$, $g(x) = x^2 + 1$

22. $f(x) = x^2 - 3$, $g(x) = x + 5$

23. $f(x) = x^2 + 3x - 4$; $g(x) = 3x$

24. $f(x) = -3x$, $g(x) = x^2 + 2x + 4$

25. $f(x) = \sqrt{x + 2}$, $g(x) = 2x - 5$

26. $f(x) = 5x + 2$, $g(x) = \sqrt{x - 5}$

27. $f(x) = \dfrac{x + 1}{x}$, $g(x) = \dfrac{x - 3}{x}$

28. $f(x) = \dfrac{x + 4}{x}$, $g(x) = \dfrac{2 - x}{x}$

For Exercises 29 and 30, answer each question. See Objective 2.

29. If the domain of f is $[3, \infty)$ and the range is $[0, \infty)$, what are the domain and range of f^{-1}?

30. If f and g are inverse functions and $f(2) = 5$, then $g(5) = $ _____

For Exercises 31–34, determine whether the following functions f and g are inverses. See Objective 2.

31. $f = \{(1, 2), (-1, -3), (3, 4), (2, -5)\}$, $g = \{(2, 1), (-3, -1), (4, 3), (-5, 2)\}$

32. $f = \{(4, -3), (1, 4), (5, 2), (-3, 1), (-1, 3)\}$, $g = \{(-3, 4), (4, 1), (2, 5), (1, -3), (3, -1)\}$

33. $f = \{(-2, -2), (3, -3), (-4, 4), (-6, -6)\}$, $g = \{(-2, -2), (-3, 3), (-4, 4), (-6, -6)\}$

34. $f = \{(5, 5), (-4, 4), (2, -2), (-7, -7)\}$, $g = \{(-5, -5), (4, -4), (-2, 2), (7, 7)\}$

For Exercises 35–48, determine whether f and g are inverses by determining whether $(f \circ g)(x) = x$ and $(g \circ f)(x) = x$. See Example 2.

35. $f(x) = x + 5$, $g(x) = x - 5$

36. $f(x) = x - 8$, $g(x) = x + 8$

37. $f(x) = 6x, g(x) = \dfrac{x}{6}$

38. $f(x) = -\dfrac{x}{3}, g(x) = -3x$

39. $f(x) = 2x - 3, g(x) = \dfrac{x + 3}{2}$

40. $f(x) = \dfrac{x - 4}{5}, g(x) = 5x + 4$

41. $f(x) = x^3 - 4, g(x) = \sqrt[3]{x + 4}$

42. $f(x) = x^5 + 3, g(x) = \sqrt[5]{x - 3}$

43. $f(x) = x^2, g(x) = \sqrt{x}$

44. $f(x) = x^4, g(x) = \sqrt[4]{x}$

45. $f(x) = x^2, x \geq 0; g(x) = \sqrt{x}$

46. $f(x) = x^2 + 2, x \geq 0; g(x) = \sqrt{x - 2}$

47. $f(x) = \dfrac{3}{x + 5}, g(x) = \dfrac{3 - 5x}{x}$

48. $f(x) = \dfrac{x}{x + 4}, g(x) = \dfrac{4x}{1 - x}$

For Exercises 49 and 50, answer each question.

49. Is $f(x) = \dfrac{1}{x}$ its own inverse? Explain.

50. Is $f(x) = x$ its own inverse? Explain.

For Exercises 51–54, determine whether the function is one to one. See Example 3.

51.

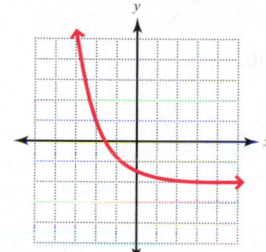

52.

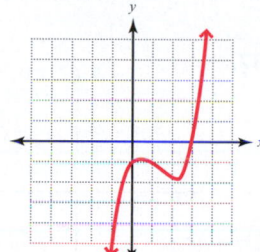

53.

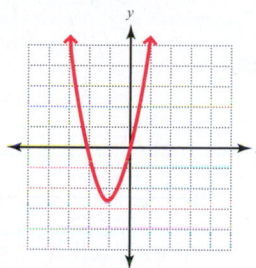

54.

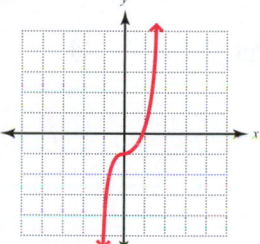

For Exercises 55–76, find $f^{-1}(x)$ for each of the following one-to-one functions f.
See Example 4.

55. $f = \{(-3, 2), (-1, -3), (0, 4), (4, 6)\}$

56. $f = \{(-4, 2), (-1, -3), (2, 3), (5, 7)\}$

57. $f = \{(7, -2), (9, 2), (-4, 1), (3, 3)\}$

58. $f = \{(1, 2), (4, 5), (-3, 3), (-9, 0)\}$

59. $f(x) = x + 6$

60. $f(x) = x - 4$

61. $f(x) = 2x + 3$

62. $f(x) = 4x - 5$

63. $f(x) = x^3 - 1$

64. $f(x) = x^3 - 3$

65. $f(x) = \dfrac{2}{x + 2}$

66. $f(x) = \dfrac{-3}{x - 3}$

67. $f(x) = \dfrac{x + 2}{x - 3}$

68. $f(x) = \dfrac{x - 4}{x + 2}$

69. $f(x) = \sqrt{x - 2}$

70. $f(x) = \sqrt{2x - 4}$

71. $f(x) = 2x^3 + 4$

72. $f(x) = 4x^3 - 5$

73. $f(x) = \sqrt[3]{x + 2}$

74. $f(x) = \sqrt[3]{x + 1}$

75. $f(x) = 2\sqrt[3]{2x + 4}$

76. $f(x) = 4\sqrt[3]{2x - 6}$

For Exercises 77–88, sketch the graph of the inverse of each of the following functions. See Example 5.

77.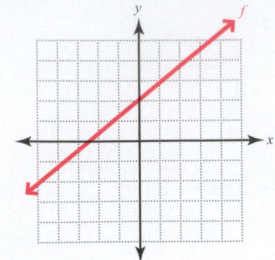

78.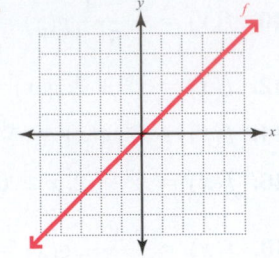

79.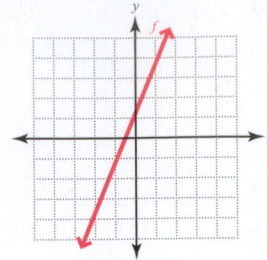

80.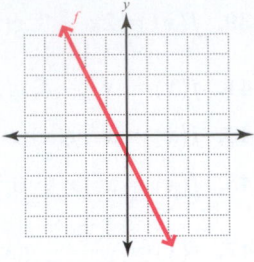

81.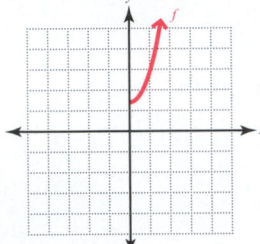

82.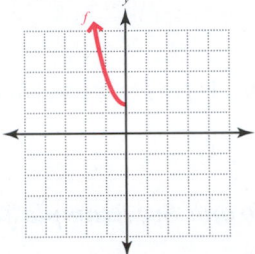

83.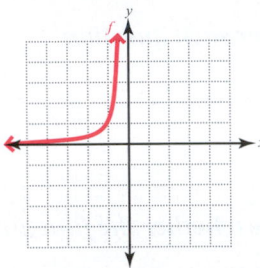

84.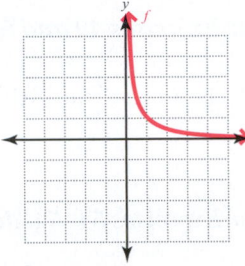

85.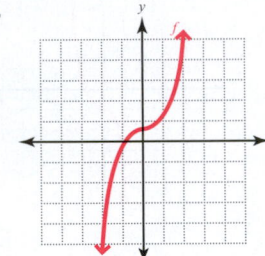

86.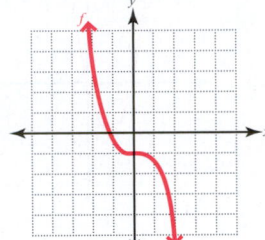

87.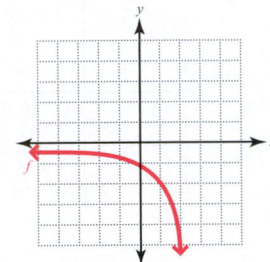

88.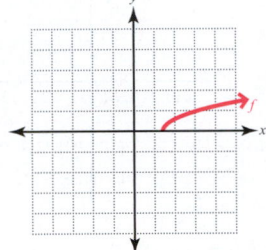

For Exercises 89–92, solve each problem.

89. If a salesperson works for $100 per week plus 5% commission on sales, the weekly salary is $y = 0.05x + 100$ where y represents the salary and x represents the sales.
   a. Find the inverse function.

   b. What does each variable of the inverse function represent?

   c. Use the inverse function to find the sales for a week in which the salary was $350.

90. If a Toyota Prius averages 50 miles per hour on a trip, then $y = 50x$, where y represents the number of miles traveled and x represents the number of hours.
   a. Find the inverse function.

   b. What does each variable of the inverse function represent?

   c. Use the inverse function to find the number of hours to travel 210 miles.

91. An office building installed 105 fans, some of which measure 36 inches with the remaining fans measuring 54 inches. If the 36-inch fans cost $45 each and the 54-inch fans cost $65 each, the cost of the fans is $y = 45x + 65(105 - x)$, where y represents the cost and x represents the number of 36-inch fans.

a. Find the inverse function.

b. What does each variable of the inverse function represent?

c. Use the inverse function to find the number of 36-inch fans if the total cost was $5225.

92. A painting contractor purchased 24 gallons of paint for the interior and exterior of a house. If he paid $18 per gallon for the interior paint and $22 per gallon for the exterior paint, the amount he paid is $y = 18x + 22(24 - x)$, where y represents the amount paid and x represents the number of gallons of interior paint.

a. Find the inverse function.

b. What does each variable of the inverse function represent?

c. Use the inverse function to find the number of gallons of interior paint if he paid a total of $488 for the 24 gallons of paint.

For Exercises 93–96, answer the question.

93. If $(-4, 5)$ is an ordered pair on the graph of g, what are the coordinates of an ordered pair on the graph of g^{-1}?

94. If $f(2) = 4$ and $f^{-1}(a) = 2$, find a.

95. A linear function is of the form $f(x) = ax + b, a \neq 0$. Find $f^{-1}(x)$.

96. The square root function is defined by $f(x) = \sqrt{x}$. Find $f^{-1}(x)$. Be careful!

Review Exercises

Exercises 1–5  **EXPRESSIONS**

[1.5] **1.** Write 32 as 2 to a power.

For Exercises 2–5, evaluate each expression.

[1.5] **2.** 2^3

[1.5] **3.** $\left(-\dfrac{1}{3}\right)^3$

[5.1] **4.** 4^{-2}

[10.2] **5.** $4^{3/2}$

Exercise 6 **EQUATIONS AND INEQUALITIES**

[11.4] **6.** What are the coordinates of the vertex of $g(x) = (x - 3)^2 + 1$?

12.2 Exponential Functions

Objectives

1 Define and graph exponential functions.

2 Solve equations of the form $b^x = b^y$ for x.

3 Use exponential functions to solve application problems.

Objective ① **Define and graph exponential functions.** Previously, we defined rational number exponents. For example, we know that $b^3 = b \cdot b \cdot b$, $b^{2/3} = \sqrt[3]{b^2}$, $b^0 = 1$, and $b^{-3} = \dfrac{1}{b^3}$.

It can be shown that irrational number exponents have meaning as well, and we can approximate expressions such as $2^{\sqrt{3}}$ and 5^{π} by using rational approximations for the exponents. Therefore, the exponential expression b^x has meaning if x is any real number (rational or irrational). So we can define the **exponential function** as follows.

Definition   *Exponential function:* If $b > 0$, $b \neq 1$, and x is any real number, then the exponential function is $f(x) = b^x$.

We graph exponential functions by plotting enough points to determine the graph's shape. We will find that the graph has one typical shape if $b > 1$ and a different typical shape if $0 < b < 1$.

Example 1   Graph.

a. $f(x) = 2^x$ and $g(x) = 4^x$

SOLUTION   Choose some values of x and find the corresponding values of $f(x)$ and $g(x)$ (which are the y-values).

x	-3	-2	-1	0	1	2	3
$f(x)$	$\frac{1}{8}$	$\frac{1}{4}$	$\frac{1}{2}$	1	2	4	8
$g(x)$	$\frac{1}{64}$	$\frac{1}{16}$	$\frac{1}{4}$	1	4	16	64

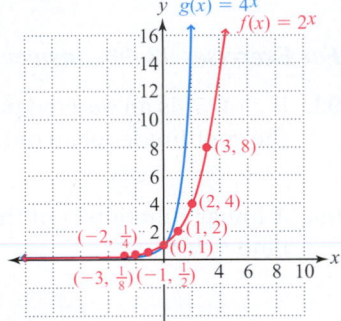

Plotting the ordered pairs for each function gives smooth curves typical of the graphs of $f(x) = b^x$ with $b > 1$. Comparing the graphs of $f(x) = 2^x$ and $g(x) = 4^x$, we can see that the greater the value of b, the steeper the graph.

b. $h(x) = \left(\dfrac{1}{2}\right)^x$ and $k(x) = \left(\dfrac{1}{4}\right)^x$

SOLUTION   Choose some values of x and find the corresponding values of $h(x)$ and $k(x)$.

x	-3	-2	-1	0	1	2	3
$h(x)$	8	4	2	1	$\frac{1}{2}$	$\frac{1}{4}$	$\frac{1}{8}$
$k(x)$	64	16	4	1	$\frac{1}{4}$	$\frac{1}{16}$	$\frac{1}{64}$

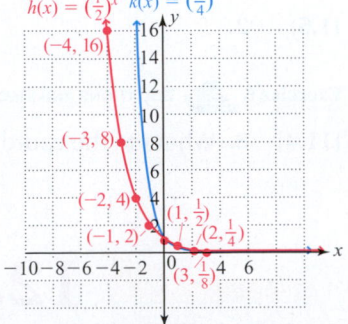

After plotting the ordered pairs for each function, we see graphs that are typical of exponential functions in the form $f(x) = b^x$ with $0 < b < 1$. Comparing the graphs of $h(x) = \left(\dfrac{1}{2}\right)^x$ and $k(x) = \left(\dfrac{1}{4}\right)^x$, we see that the smaller the value of b, the steeper the graph.

We summarize the graphs of $f(x) = b^x$ as follows.

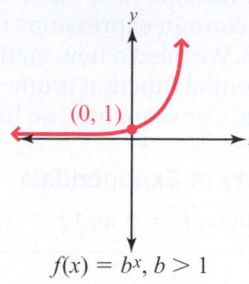

$f(x) = b^x, b > 1$ $f(x) = b^x, 0 < b < 1$

Note: Any function that increases or decreases along its entire domain is one-to-one because no horizontal line will intersect its graph at more than one point.

The function is increasing from left to right and is one to one. The graph always passes through $(0, 1)$. For negative values of x, the graph approaches the x-axis but never touches it. The larger the values of b, the steeper the graph. The domain is $(-\infty, \infty)$, and the range is $(0, \infty)$.

The function is decreasing from left to right and is one to one. The graph always passes through $(0, 1)$. For positive values of x, the graph approaches the x-axis but never touches it. The smaller the values of x, the steeper the graph. The domain is $(-\infty, \infty)$, and the range is $(0, \infty)$.

More complicated exponential functions are graphed in the same manner.

Example 2   Graph $f(x) = 3^{2x-1}$.

SOLUTION   Find some ordered pairs, plot them, and draw the graph.

Note: $f(x) = 3^{2x-1}$ is not of the form $f(x) = b^x$. Note that it does not pass through $(0, 1)$.

x	$y = f(x)$
-1	$3^{-3} = \dfrac{1}{27}$
0	$3^{-1} = \dfrac{1}{3}$
1	$3^1 = 3$
2	$3^3 = 27$

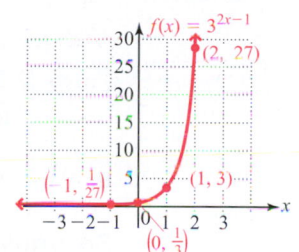

Your Turn 2   Graph $f(x) = 2^{x+1}$.

🖩 **Calculator Tips**

To graph $y = 2^x$ using a graphing calculator, enter $Y_1 = 2^x$. Set the window to $[-5, 5]$ for x and $[-10, 10]$ for y and press ⬚GRAPH⬚. We graph $y = \left(\dfrac{1}{2}\right)^x$ in the same window by entering $Y_2 = \left(\dfrac{1}{2}\right)^x$. Notice that if we folded the graph along the y-axis, the graphs would lie on top of each other, indicating that they are symmetric with respect to the y-axis.

Answer   Your Turn 2

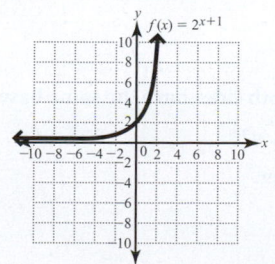

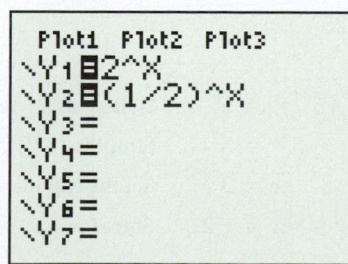

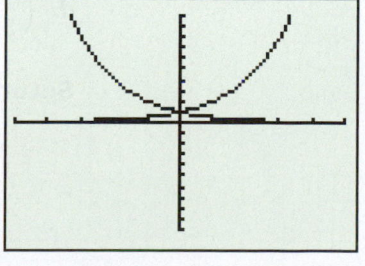

Objective ② **Solve equations of the form $b^x = b^y$ for x.** Earlier we solved equations containing expressions having a variable base and a constant exponent, such as $x^2 = 4$. Now we solve equations that contain expressions having a constant base and a variable exponent, such as $2^x = 16$. We need a new method to solve these equations. Previously, we noted that the exponential function is one to one; so for each x-value, there is a unique y-value and vice versa. Consequently, we have the following rule.

Rule **One-to-One Property of Exponentials**

Given $b > 0$ and $b \neq 1$, if $b^x = b^y$, then $x = y$.

To solve some types of exponential equations, we use the following procedure.

Procedure **Solving Exponential Equations**

1. If necessary, write both sides of the equation as a power of the same base.
2. If necessary, simplify the exponents.
3. Set the exponents equal to each other.
4. Solve the resulting equation.

Example 3 Solve.

a. $3^x = 81$

SOLUTION $3^x = 3^4$ Write 81 as 3^4 so that both sides have the same base.

$\quad\quad\quad\quad x = 4$ Set the exponents equal to each other.

Check $3^x = 81$

$\quad\quad\quad 3^4 = 81$ Replace x with 4.

$\quad\quad\quad 81 = 81$ True; so $x = 4$ is the solution.

b. $4^x = 32$

SOLUTION $(2^2)^x = 2^5$ Write 4 as 2^2 and 32 as 2^5 so that both sides have the same base.

$\quad\quad\quad 2^{2x} = 2^5$ Simplify $(2^2)^x$ by applying $(a^m)^n = a^{mn}$.

$\quad\quad\quad 2x = 5$ Set the exponents equal to each other.

$\quad\quad\quad x = \dfrac{5}{2}$ Solve for x.

Note: We leave the checks for parts b–d to the reader.

c. $8^{x+4} = 4^{2x+3}$

SOLUTION $(2^3)^{x+4} = (2^2)^{2x+3}$ Write 8 as 2^3 and 4 as 2^2 so that both sides have the same base.

$\quad\quad\quad 2^{3x+12} = 2^{4x+6}$ Simplify the exponents by applying $(a^m)^n = a^{mn}$.

$\quad\quad\quad 3x + 12 = 4x + 6$ Set the exponents equal to each other.

$\quad\quad\quad 6 = x$ Solve for x.

d. $\left(\dfrac{1}{5}\right)^x = 25$

SOLUTION $(5^{-1})^x = 5^2$ Write $\dfrac{1}{5}$ as 5^{-1} and 25 as 5^2 so that both sides have the same base.

$\quad\quad\quad 5^{-x} = 5^2$ Simplify $(5^{-1})^x$ by applying $(a^m)^n = a^{am}$.

$\quad\quad\quad -x = 2$ Set the exponents equal to each other.

$\quad\quad\quad x = -2$ Solve for x.

Your Turn 3 Solve.

    a. $2^{x+1} = 8$ **b.** $27^{x-2} = 9^x$ **c.** $\left(\dfrac{1}{2}\right)^x = 16$

Objective ③ Use exponential functions to solve application problems.

Compound Interest Formula

Exponential functions occur in many areas, especially the sciences and business. If P dollars are invested at an annual interest rate of r (written as a decimal) compounded n times per year for t years, the accumulated amount A in the account is given by the formula $A = P\left(1 + \dfrac{r}{n}\right)^{nt}$.

Example 4 Find the accumulated amount in an account if $5000 is deposited at 6% compounded quarterly for 10 years. Round the answer to the nearest cent.

Understand We are asked to find A given that **$P = \$5000$, $r = 0.06$, $n = 4$, and $t = 10$.**

Plan Use the formula $A = P\left(1 + \dfrac{r}{n}\right)^{nt}$.

Execute $A = P\left(1 + \dfrac{r}{n}\right)^{nt}$

         $A = 5000\left(1 + \dfrac{0.06}{4}\right)^{4(10)}$ Substitute 5000 for P, 0.06 for r, 4 for n, and 10 for t.

         $A = 5000(1 + 0.015)^{40}$ Simplify.

         $A = 5000(1.015)^{40}$

         $A = 9070.09$ Evaluate using a calculator and round to the nearest cent (hundredths place).

Answer After 10 years, the accumulated amount in the account is $9070.09.

Check Verify that the principal is $5000 if the accumulated amount is $9070.09 after the principal is compounded quarterly for 10 years.

$$9070.09 = P\left(1 + \frac{0.06}{4}\right)^{4(10)}$$

$$9070.09 = P(1.015)^{40}$$

$$\frac{9070.09}{(1.015)^{40}} = P$$

$$4999.998874 = P$$

Because the accumulated amount, 9070.09, was rounded, it is reasonable to expect our calculated value for the principal to be slightly different from $5000.

Your Turn 4 If $3000 is invested at 4% compounded semiannually (twice per year), how much money is in the account at the end of 8 years? Round to the nearest cent.

Half-Life

The *half-life* of a radioactive substance is the amount of time it takes until only half of the original amount of the substance remains. Suppose we begin with 100 grams of a substance that has a half-life of 10 days. After 10 days (one half-life), 50 grams will remain; after 20 days (two half-lives), 25 grams will remain; after 30 days (three half-lives),

Answers **Your Turn 3**
a. 2   **b.** 6   **c.** −4

Answer **Your Turn 4**
$4118.36

12.5 grams will remain; and so on. The formula $A = A_0\left(\dfrac{1}{2}\right)^{t/h}$ gives the amount remaining, where A_0 is the initial amount, t is the time, and h is the half-life.

Example 5   The isotope ^{45}Ca has a half-life of 165 days. How many grams of a 50-gram sample of ^{45}Ca will remain after 825 days?

Understand   Given a 50-gram sample of ^{45}Ca, we are to find the amount remaining after 825 days.

Plan   Use the formula $A = A_0\left(\dfrac{1}{2}\right)^{t/h}$.

Execute

$A = A_0\left(\dfrac{1}{2}\right)^{t/h}$

$A = 50\left(\dfrac{1}{2}\right)^{825/165}$ Substitute 50 for A_0, 825 for t, and 165 for h.

$A = 50\left(\dfrac{1}{2}\right)^{5}$ Simplify.

$A = 1.5625$ Evaluate using a calculator.

Answer   1.5625 grams of ^{45}Ca will remain after 825 days.

Check   Verify that the sample was 50 grams if 1.5625 grams remain after 825 days. We will leave this check to the reader.

Your Turn 5   The radioactive isotope ^{61}Cr has a half-life of 26 days. How much of a 10-gram sample would remain after 208 days? Give your answer to the nearest thousandth of a gram.

Aging

Example 6   The number of people in the United States aged 65 and over (in millions) is given in the following table.

Year	Number 65 and over (in millions)
1900	3.1
1910	4.0
1920	4.9
1930	6.7
1940	9.0
1950	12.4
1960	16.7
1970	20.1
1980	25.5
1990	31.4
2000	35.6

Source: U.S. Bureau of the Census

The data can be approximated by the function $y = 3.17(1.026)^x$, where x is the number of years after 1900 and y is the number of people in millions. Use the model to estimate the number of people in the United States aged 65 and over in the year 2020. Round the answer to the nearest tenth of a million.

Understand   We are given the function $y = 3.17(1.026)^x$, which approximates the data in the table. We are to find the number of people aged 65 and over in the year 2020.

Plan   Because x represents the number of years after 1900, we subtract 1900 from 2020 to find the value of x that corresponds to the year 2020. Then we can use $y = 3.17(1.026)^x$.

Execute $x = 2020 - 1900 = 120$ Subtract 1900 from 2020 to find the value of x that corresponds to the year 2020.

$y = 3.17(1.026)^{120}$ Substitute 120 for x in $y = 3.17(1.026)^x$.

$y = 69.0$ million   Evaluate using a calculator and round to the nearest tenth.

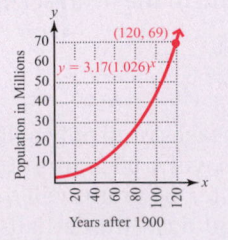

Year	Number of Computers (in millions)
1985	21.5
1988	40.8
1989	47.6
1991	62.0
1992	68.2
1993	76.5
1994	85.8
1995	96.2
2000	160.5
2005	350.6

Answer According to the function, about 69 million people will be aged 65 and over in the year 2020.

Check Verify that if $y = 69$ in the function $y = 3.17(1.026)^x$, then $x = 120$.

$$69 = 3.17(1.026)^x \qquad \text{Substitute 69 for } y \text{ in } y = 3.17(1.026)^x.$$

To solve the above equation, we need to use logarithms, which we have not learned yet. We learn about logarithms and how to solve equations like the one above in the rest of this chapter.

Your Turn 6 The table in the margin shows the number of computers (in millions) in use in the United States for selected years since 1984.

The data can be approximated by $y = 23.14(1.138)^x$, where x is the number of years after 1984 and y is the number of computers (in millions). Use the model to approximate the number of computers in use in the United States in 2015.

Answer   Your Turn 6

1272.8 million

12.2 Exercises

For Extra Help   MyMathLab  PRACTICE   WATCH   DOWNLOAD

1. Are the graphs of $f(x) = 2^x$ and $g(x) = \left(\frac{1}{2}\right)^x$ symmetric with respect to the y-axis? Explain.

2. As x gets larger, which graph is steeper: $f(x) = 3^x$ or $g(x) = 1.5^x$? Why?

3. Find the domain and range of $f(x) = 2^{x-4}$.

4. Is $f(x) = (-2)^x$ an exponential function? Why or why not?

5. For $b > 0$ and $b \neq 1$, if $b^x = b^y$, why does $x = y$?

6. If we have 10 grams of a substance that has a half-life of 50 days, how many grams will be present after 200 days?

For Exercises 7–24 graph. See Examples 1 and 2.

7. $f(x) = 3^x$

8. $f(x) = 4^x$

9. $f(x) = 4^x - 3$

10. $f(x) = 3^x + 1$

11. $f(x) = \left(\frac{1}{3}\right)^x$

12. $f(x) = \left(\frac{1}{4}\right)^x$

13. $f(x) = \left(\dfrac{2}{3}\right)^x + 2$

14. $f(x) = \left(\dfrac{3}{2}\right)^x - 1$

15. $f(x) = -3^x$

16. $f(x) = -2^x$

17. $f(x) = 2^{x-2}$

18. $f(x) = 3^{x+1}$

19. $f(x) = 2^{-x}$

20. $f(x) = 3^{-x}$

21. $f(x) = 2^{2x-3}$

22. $f(x) = 3^{2x+1}$

23. $f(x) = 3^{-x+2}$

24. $f(x) = 2^{-x-1}$

For Exercises 25–42, solve each equation. See Example 3.

25. $2^x = 8$

26. $3^x = 81$

27. $8^x = 32$

28. $27^x = 81$

29. $16^x = 4$

30. $36^x = 216$

31. $6^x = \dfrac{1}{36}$

32. $3^x = \dfrac{1}{27}$

33. $\left(\dfrac{1}{3}\right)^x = 9$

34. $\left(\dfrac{1}{5}\right)^x = 125$

35. $\left(\dfrac{2}{3}\right)^x = \dfrac{8}{27}$

36. $\left(\dfrac{3}{2}\right)^x = \dfrac{9}{4}$

37. $\left(\dfrac{1}{2}\right)^x = 16$

38. $\left(\dfrac{1}{3}\right)^x = 27$

39. $25^{x+1} = 125$

40. $9^{x+2} = 243$

41. $8^{2x-1} = 32^{x-3}$

42. $4^{2x-1} = 32^{x+2}$

For Exercises 43–46, use a graphing utility.

43. a. Graph $f(x) = 2^x$ and $g(x) = 2^{x+2}$ in the window $[-5, 5]$ for x and $[-1, 10]$ for y.
 b. How does the graph of g compare with the graph of f?

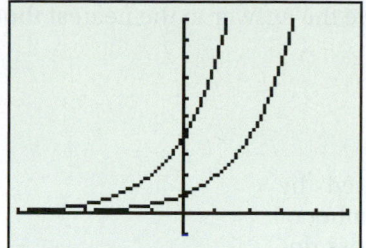

44. a. Graph $f(x) = 3^x$ and $g(x) = 3^x - 3$ in the window $[-3, 3]$ for x and $[-3, 9]$ for y.
 b. How does the graph of g compare with the graph of f?

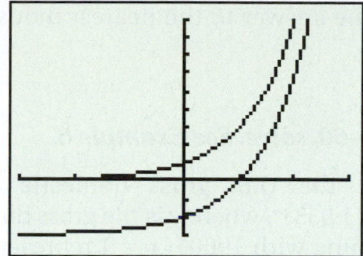

45. a. Graph $f(x) = 2^x$ and $g(x) = 2^{-x}$ in the window $[-5, 5]$ for x and $[-1, 10]$ for y.
 b. How does the graph of g compare with the graph of f?

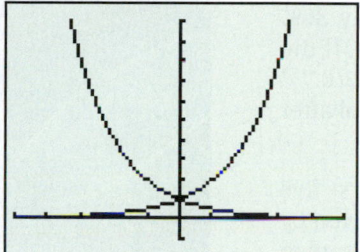

46. a. Graph $f(x) = 2^x$ and $g(x) = -2^x$ in the window $[-5, 5]$ for x and $[-5, 5]$ for y.
 b. How does the graph of g compare with the graph of f?

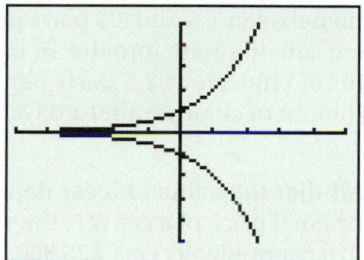

For Exercises 47 and 48, use the following information.

Under ideal conditions, a culture of *E. coli* bacteria doubles in size every 20 minutes. If A_0 is the initial amount and t is the number of minutes passed, the amount present is A, where $A = A_0 (2)^{t/20}$.

47. If a culture of *E. coli* began with 100 cells, how many cells will be present after 120 minutes?

48. If a culture of *E. coli* currently has 500 cells, how many cells were present 90 minutes earlier? (*Hint:* Let $A_0 = 500$, and the time will be negative.)

49. Under ideal conditions, human beings could double their population every 50 years. If A_0 is the initial population and t is the number of years passed, the current population is A, where $A = A_0(2)^{t/50}$. If 6 billion humans were on Earth in 2000, how many humans will there be in 2500 if their growth is uncontrolled?

50. Using the information from Exercise 49, how many human beings were on the Earth in 1900?

For Exercises 51 and 52, use the formula $A = P\left(1 + \dfrac{r}{n}\right)^{nt}$. See Example 4.

51. If $10,000 is deposited into an account paying 8% interest compounded quarterly, how much will be in the account after 12 years?

52. If $15,000 is deposited into an account paying 6% interest compounded semiannually, how much will be in the account after nine years?

For Exercises 53 and 54, use the formula $A = A_0\left(\dfrac{1}{2}\right)^{t/h}$. See Example 5.

53. Einsteinium (^{254}ES) has a half-life of 270 days. How much of a 5-gram sample would remain after 2160 days? Give the answer to the nearest thousandth of a gram.

54. Nobelium (^{257}No) has a half-life of 23 seconds. How much of a 100-gram sample would remain after 275 seconds? Give the answer to the nearest thousandth of a gram.

For Exercises 55–60, solve. See Example 6.

55. Since 1960, the U.S. gross domestic product can be approximated by $y = 2632.31(1.033)^x$, where y is the gross domestic product and x is the number of years beginning with 1960. ($x = 1$ represents 1960.) Estimate the U.S. gross domestic product in 2015.

56. Since 1970, the U.S. consumer price index can be approximated by $y = 46.56(1.045)^x$, where y is the consumer price index and x is the number of years beginning with 1970. ($x = 1$ represents 1970.) Estimate the consumer price index in 2013.

57. Chlorine is frequently used to disinfect swimming pools. The concentration should remain between 1.5 and 2.5 parts per million. On a warm, sunny day, 30% of the chlorine can dissipate into the air or combine with other chemicals. If the initial amount of chlorine is 2.5 parts per million, the function $f(x) = 2.5(0.7)^x$ models the amount of chlorine after x days. How much chlorine is in the pool after 2 days?

58. It is estimated that the value of a car depreciates 20% per year for the first five years. If the original price of a car is P, the value, A, of a car after t years is given by $A = P(0.8)^t$. If a car originally cost $25,960, find the value of the car after 3 years to the nearest dollar.

59. Between 1993 and 2008, the number of transistors that can be placed on a single chip has grown significantly, as indicated in the table. The data can be approximated by the function $T(x) = 1.41(1.54)^x$, where x is the number of years after 1993 and $T(x)$ is the number of transistors in millions. If the current trend continues, estimate the number of transistors that could be put on a single chip in 2015. (*Source:* http://en.wikipedia.org)

Year	Chip	Transistors (millions)
1993	Pentium	3.3
1995	P6	5.5
1997	Pentium II	7.5
1999	Pentium III	9.5
2000	Pentium IV	42
2004	Pentium IV Prescott	125
2006	Core 2 Duo	586
2008	Quad IV	820

60. Dave makes a cup of coffee with cream and places it on the counter to cool. The temperature of the coffee at various times is given in the table shown.

The data can be approximated by the function $T(t) = 146.9(0.989)^t$, where t is the time in seconds and T is the temperature in °F. Estimate the temperature of the coffee after 1 minute.

t (seconds)	T (°F)
0.2	155.8
8.4	133.2
16.6	117.9
24.8	107.9
33	100.7
41.1	94.9
49.3	90.5

Puzzle Problem

What is the greatest number that can be written using three numerals?

Review Exercises

Exercises 1–4 **EXPRESSIONS**

[5.1] *For Exercises 1–3, evaluate.*

1. 5^3

2. $\left(\dfrac{1}{3}\right)^{-2}$

3. 4^{-3}

[10.2] **4.** Write $5^{2/3}$ in radical notation.

Exercises 5 and 6 **EQUATIONS AND INEQUALITIES**

[12.1] **5.** If $f(x) = 3x + 4$, find $f^{-1}(x)$.

[12.1] **6.** Graph the inverse of the function whose graph is shown.

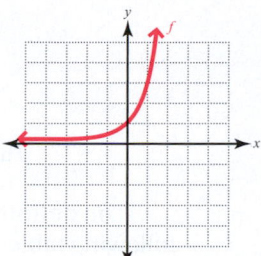

12.3 Logarithmic Functions

Objectives

1 Convert between exponential and logarithmic forms.

2 Solve logarithmic equations by changing to exponential form.

3 Graph logarithmic functions.

4 Solve applications involving logarithms.

Objective 1 **Convert between exponential and logarithmic forms.** In Section 12.2, we defined the exponential function as $f(x) = b^x$ with $b > 0$ and $b \neq 1$. Because the exponential function is a one-to-one function, it has an inverse. Following is the graph of $f(x) = 2^x$ and its inverse, which we find by reflecting the graph about the line $y = x$.

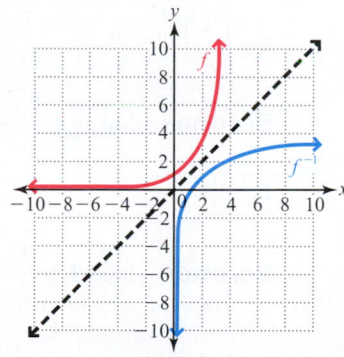

To find the inverse of $f(x) = 2^x$, we use the procedure from Section 12.1.

$$f(x) = 2^x$$
$$y = 2^x \qquad \text{Replace } f(x) \text{ with } y.$$
$$x = 2^y \qquad \text{Replace } y \text{ with } x \text{ and } x \text{ with } y.$$

The next step is to solve for y, but we haven't learned how to isolate a variable that is an exponent. To rewrite $x = 2^y$, we define a **logarithm**.

Definition **Logarithm:** If $b > 0$ and $b \neq 1$, then $y = \log_b x$ is equivalent to $x = b^y$.

If we apply this definition to $x = 2^y$ in the preceding equation, we get $y = \log_2 x$. Replacing y with $f^{-1}(x)$, we get $f^{-1}(x) = \log_2 x$; so the inverse function for $f(x) = 2^x$ is $f^{-1}(x) = \log_2 x$. To generalize, exponential functions and *logarithmic functions* are inverses. Consequently, if $y = \log_b x$, the domain is $(0, \infty)$ and the range is $(-\infty, \infty)$ with $b > 0$ and $b \neq 1$.

The expression $\log_b x$ is read "the logarithm base b of x" and *is the exponent to which b must be raised to get x.* Compare the two forms.

The exponent is the logarithm.

$$x = b^y \qquad y = \log_b x$$

The base of the exponent is the base of the logarithm.

Note: $y = \log_b x$ means that y is the power to which we raise b to get x.

The definition of a logarithm allows us to convert from one form to another. The following table contains pairs of equivalent forms.

Logarithmic Form	Exponential Form
$\log_2 8 = 3$	$2^3 = 8$
$\log_{10} \dfrac{1}{10} = -1$	$10^{-1} = \dfrac{1}{10}$
$\log_5 1 = 0$	$5^0 = 1$
$\log_{16} 4 = \dfrac{1}{2}$	$16^{1/2} = 4$
$\log_{1/2} 32 = -5$	$\left(\dfrac{1}{2}\right)^{-5} = 32$

Note: The logarithmic equations are in the form $\log_b x = y$. The values of x are positive numbers only, and the values of y are both positive and negative.

Example 1   Write in logarithmic form.

a. $3^4 = 81$

SOLUTION $\log_3 81 = 4$ The base of the exponent is the base of the logarithm, and the exponent is the logarithm.

b. $\left(\dfrac{1}{2}\right)^{-3} = 8$

SOLUTION $\log_{1/2} 8 = -3$ The base of the exponent is the base of the logarithm, and the exponent is the logarithm.

c. $9^{1/2} = 3$

SOLUTION $\log_9 3 = \dfrac{1}{2}$ The base of the exponent is the base of the logarithm, and the exponent is the logarithm.

Your Turn 1   Write in logarithmic form.

a. $4^3 = 64$ **b.** $2^{-3} = \dfrac{1}{8}$ **c.** $27^{1/3} = 3$

Answers   Your Turn 1
a. $\log_4 64 = 3$

b. $\log_2 \dfrac{1}{8} = -3$

c. $\log_{27} 3 = \dfrac{1}{3}$

Example 2   Write in exponential form.

a. $\log_6 36 = 2$

SOLUTION $6^2 = 36$ The base of the logarithm is the base of the exponent, and the logarithm is the exponent.

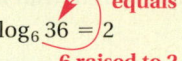
b. $\log_{16} 2 = \dfrac{1}{4}$

SOLUTION $16^{1/4} = 2$ The base of the logarithm is the base of the exponent, and the logarithm is the exponent.

c. $\log_{1/2} 16 = -4$

SOLUTION $\left(\dfrac{1}{2}\right)^{-4} = 16$ The base of the logarithm is the base of the exponent, and the logarithm is the exponent.

Your Turn 2   Write in exponential form.

   a. $\log_4 64 = 3$ **b.** $\log_8 2 = \dfrac{1}{3}$ **c.** $\log_5 \dfrac{1}{125} = -3$

Objective ②   **Solve logarithmic equations by changing to exponential form.**   A logarithmic equation in the form $\log_b x = y$ may have b, x, or y as an unknown.

Procedure   **Solving Logarithmic Equations**

To solve an equation of the form $\log_b x = y$, where b, x, or y is a variable, write the equation in exponential form, $b^y = x$, and then solve for the variable.

Example 3   Solve.

a. $\log_b 16 = 2$

SOLUTION $b^2 = 16$ Write in exponential form.

            $b = \pm 4$ Find the positive and negative square roots of 16.

            $b = 4$ The base must be positive, so $b = 4$.

b. $\log_3 \dfrac{1}{27} = y$

SOLUTION $3^y = \dfrac{1}{27}$ Write in exponential form.

            $3^y = \dfrac{1}{3^3}$ Write 27 as 3^3.

            $3^y = 3^{-3}$ Write $\dfrac{1}{3^3}$ as 3^{-3}.

            $y = -3$ Set the exponents equal to each other.

c. $\log_{25} x = \dfrac{1}{2}$

SOLUTION $25^{1/2} = x$ Write in exponential form.

            $5 = x$ Simplify. $25^{1/2} = \sqrt{25} = 5$

d. $\log_{36} \sqrt[4]{6} = y$

SOLUTION $36^y = \sqrt[4]{6}$

         $(6^2)^y = 6^{1/4}$ Write 36 as 6^2 and $\sqrt[4]{6}$ as $6^{1/4}$.

         $6^{2y} = 6^{1/4}$ $(6^2)^y = 6^{2y}$

         $2y = \dfrac{1}{4}$ Set the exponents equal to each other.

         $y = \dfrac{1}{8}$ Solve for y.

Answers   **Your Turn 2**

a. $4^3 = 64$ **b.** $8^{1/3} = 2$

c. $5^{-3} = \dfrac{1}{125}$

Your Turn 3 Solve.

a. $\log_b \dfrac{1}{9} = -2$ **b.** $\log_5 \dfrac{1}{25} = y$ **c.** $\log_{27} x = \dfrac{1}{3}$

If an equation containing a logarithm is not in the form $\log_b a = c$, try using the addition or multiplication principles of equality to rewrite the equation in that form.

Example 4   Solve $5 - 3\log_2 x = -7$.

SOLUTION   Use the addition principle of equality and multiplication principle of equality to write the equation in the form $\log_b a = c$. We can then change the equation to exponential form to solve for x.

$$-3\log_2 x = -12 \quad \text{Subtract 5 from both sides.}$$
$$\log_2 x = 4 \quad \text{Divide both sides by } -3 \text{ to isolate the logarithm.}$$
$$x = 2^4 \quad \text{Write in exponential form.}$$
$$x = 16 \quad \text{Simplify.}$$

Your Turn 4   Solve.

a. $9 + \log_3 x = 13$ **b.** $5\log_n 36 = 10$ **c.** $12 - 7\log_4 t = -9$

The definition of logarithm leads to the following two properties.

Rule   For any real number b, where $b > 0$ and $b \neq 1$,

1. $\log_b b = 1$.   **2.** $\log_b 1 = 0$.

Based on the definition of a logarithm, $\log_b b = 1$ because $b^1 = b$ and $\log_b 1 = 0$ because $b^0 = 1$.

Example 5   Find the value.

a. $\log_5 5$ **b.** $\log_e e$ **c.** $\log_{10} 1$

SOLUTION $\log_5 5 = 1$ **SOLUTION** $\log_e e = 1$ **SOLUTION** $\log_{10} 1 = 0$

Your Turn 5   Find the value.

a. $\log_{10} 10$ **b.** $\log_{\sqrt{3}} 1$

Answers Your Turn 3
a. 3   **b.** -2 **c.** 3

Answers Your Turn 4
a. 81   **b.** 6   **c.** 64

Answers Your Turn 5
a. 1   **b.** 0

Objective ③ Graph logarithmic functions.   To graph logarithmic functions, which have the form $f(x) = \log_b x$, where $b > 0$ and $b \neq 1$, we first change to exponential form so it will be easier to find ordered pairs.

Procedure   **Graphing Logarithmic Functions**

To graph a function in the form $f(x) = \log_b x$,

1. Replace $f(x)$ with y and write the logarithm in exponential form $x = b^y$.
2. Find ordered pairs that satisfy the equation by assigning values to y and finding x.
3. Plot the ordered pairs and draw a smooth curve through the points.

Example 6 Graph.

a. $f(x) = \log_2 x$

SOLUTION $y = \log_2 x$ **Replace $f(x)$ with y.**

$2^y = x$ **Write in exponential form.**

Choose values for y and find x.

y	0	1	2	3	-1	-2
x	$2^0 = 1$	$2^1 = 2$	$2^2 = 4$	$2^3 = 8$	$2^{-1} = \frac{1}{2}$	$2^{-2} = \frac{1}{4}$

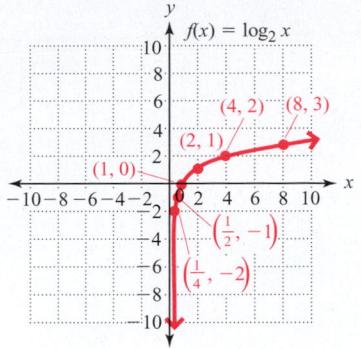

b. $f(x) = \log_{1/2} x$

SOLUTION $y = \log_{1/2} x$ **Replace $f(x)$ with x.**

$x = \left(\frac{1}{2}\right)^y$ **Write in exponential form.**

Choose values for y and find x.

y	0	1	2	3	-1	-2	-3
x	$\left(\frac{1}{2}\right)^0 = 1$	$\left(\frac{1}{2}\right)^1 = \frac{1}{2}$	$\left(\frac{1}{2}\right)^2 = \frac{1}{4}$	$\left(\frac{1}{2}\right)^3 = \frac{1}{8}$	$\left(\frac{1}{2}\right)^{-1} = 2$	$\left(\frac{1}{2}\right)^{-2} = 4$	$\left(\frac{1}{2}\right)^{-3} = 8$

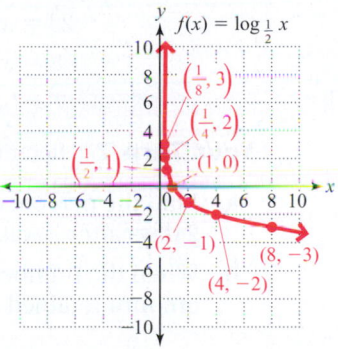

Following is a summary of the key features of the graphs of logarithmic functions.

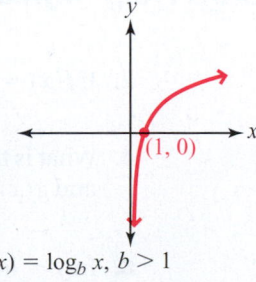

$f(x) = \log_b x,\ b > 1$

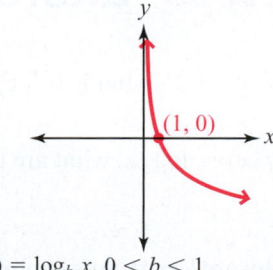

$f(x) = \log_b x,\ 0 < b < 1$

The graph passes through $(1, 0)$, approaches the y-axis, and increases. The domain is $(0, \infty)$. The range is $(-\infty, \infty)$.

The graph passes through $(1, 0)$, approaches the y-axis, and decreases. The domain is $(0, \infty)$. The range is $(-\infty, \infty)$.

Answer **Your Turn 6**

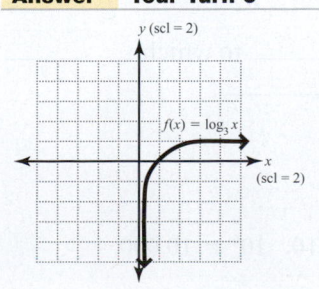

Your Turn 6 Graph $f(x) = \log_3 x$.

Example 7   The function $P = 95 - 30 \log_2 x$ models the percent, P, of students who recall the important features of a classroom lecture over time, where x is the number of days that have elapsed since the lecture was given. What percent of the students recall the important features of a lecture 8 days after it was given? (*Source: Psychology in the New Millennium*, 8th Edition, Spencer A. Rathus, Thomson Publishing Company)

Understand   We are given a function that models the percent, P, of students who recall the important features of a lecture x days after it is given. We are to find the percent of students who recall the important features of a lecture 8 days after it was given.

Plan   Evaluate $P = 95 - 30 \log_2 x$, where $x = 8$.

Execute

$P = 95 - 30 \log_2 8$	Substitute 8 for x in $P = 95 - 30 \log_2 x$.
$P = 95 - 30(3)$	$\log_2 8 = 3$ because $2^3 = 8$.
$P = 95 - 90$	Multiply.
$P = 5$	Simplify.

Answer   Five percent of the students remember the important features of a lecture 8 days after it is given.

Check   Verify that 8 days after a lecture is given, 5% of the students recall the important features of the lecture.

$5 = 95 - 30 \log_2 x$	Replace P with 5 in $P = 95 - 30 \log_2 x$.
$-90 = -30 \log_2 x$	Subtract 95 from both sides.
$3 = \log_2 x$	Divide both sides by -30.
$2^3 = x$	Write in exponential form.
$8 = x$	Simplify. It checks.

Your Turn 7   Refer to Example 7.

a. Find the percent of students who remember the important features of a lecture 2 days after it was given.

b. When the number of days was decreased by a fourth (from 8 to 2), was the amount retained also decreased by one-fourth?

Answers   Your Turn 7
a. 65%   **b.** no

12.3 Exercises   For Extra Help   *MyMathLab*

1. If $f(x) = 3^x$, what is $f^{-1}(x)$? Why?

2. If $f(x) = \log_b x$, what are the restrictions on b?

3. If $f(x) = \log_b x$, what are the restrictions on x?

4. What is the relationship between the graphs of $f(x) = 5^x$ and $g(x) = \log_5 x$? Why?

5. Why are logarithms exponents?

6. In $y = \log_b x$, y is the _____ to which _____ must be raised to get _____.

For Exercises 7–22, write in logarithmic form. See Example 1.

7. $2^5 = 32$

8. $5^4 = 625$

9. $10^3 = 1000$

10. $10^4 = 10,000$

11. $e^4 = x$

12. $e^{-2} = z$

13. $5^{-3} = \dfrac{1}{125}$

14. $7^{-4} = \dfrac{1}{2401}$

15. $10^{-2} = \dfrac{1}{100}$

16. $10^{-4} = \dfrac{1}{10,000}$

17. $625^{1/4} = 5$

18. $343^{1/3} = 7$

19. $\left(\dfrac{1}{4}\right)^2 = \dfrac{1}{16}$

20. $\left(\dfrac{3}{4}\right)^3 = \dfrac{27}{64}$

21. $7^{1/2} = \sqrt{7}$

22. $10^{1/2} = \sqrt{10}$

For Exercises 23–38, write in exponential form. See Example 2.

23. $\log_3 81 = 4$

24. $\log_4 64 = 3$

25. $\log_4 \dfrac{1}{16} = -2$

26. $\log_3 \dfrac{1}{243} = -5$

27. $\log_{10} 100 = 2$

28. $\log_{10} 1000 = 3$

29. $\log_e a = 5$

30. $\log_e y = -4$

31. $\log_e \dfrac{1}{e^4} = -4$

32. $\log_e \dfrac{1}{e^2} = -2$

33. $\log_{1/8} \dfrac{1}{64} = 2$

34. $\log_{1/4} \dfrac{1}{256} = 4$

35. $\log_{1/5} 25 = -2$

36. $\log_{1/3} 81 = -4$

37. $\log_7 \sqrt{7} = \dfrac{1}{2}$

38. $\log_6 \sqrt{6} = \dfrac{1}{2}$

For Exercises 39–58, solve. See Example 3.

39. $\log_2 x = 5$

40. $\log_3 x = 4$

41. $\log_5 x = -2$

42. $\log_4 x = -2$

43. $\log_3 81 = y$

44. $\log_2 32 = y$

45. $\log_5 \dfrac{1}{25} = y$

46. $\log_4 \dfrac{1}{16} = y$

47. $\log_b 1000 = 3$

48. $\log_b 10,000 = 4$

49. $\log_m \dfrac{1}{16} = -4$

50. $\log_n \dfrac{1}{36} = -2$

51. $\log_{1/2} x = 2$

52. $\log_{1/5} x = 3$

53. $\log_{1/3} h = -5$

54. $\log_{1/6} k = -2$

55. $\log_{1/3} \dfrac{1}{9} = y$

56. $\log_{1/4} \dfrac{1}{64} = y$

57. $\log_{1/2} 64 = t$

58. $\log_{1/5} 125 = u$

For Exercises 59–70, solve. See Example 4.

59. $\log_2 x + 4 = 8$

60. $\log_3 x + 2 = 5$

61. $\log_{1/4} h - 2 = 1$

62. $\log_{1/5} k - 4 = -1$

63. $3 \log_b 16 = 12$

64. $2 \log_b 81 = 8$

65. $\dfrac{1}{3} \log_5 c = 1$

66. $\dfrac{1}{4} \log_2 d = 2$

67. $3 \log_t 9 + 6 = 12$

68. $2 \log_u 125 + 8 = 14$

69. $\dfrac{1}{2} \log_4 m - 2 = -3$

70. $\dfrac{1}{3} \log_5 n - 4 = -5$

Exercises 71–74, graph. See Example 6.

71. $f(x) = \log_4 x$ **72.** $f(x) = \log_5 x$ **73.** $f(x) = \log_{1/3} x$ **74.** $f(x) = \log_{1/4} x$

For Exercises 75–78, solve. See Example 7.

75. The percent of adult height attained by a 5- to 15-year-old girl can be approximated by $f(x) = 62 + 35 \log_{10}(x - 4)$, where x is the age in years and $f(x)$ is the percent. At age 14, what percent of her adult height has a girl reached?

76. The percent of adult height attained by a 5- to 15-year-old boy can be approximated by $f(x) = 29 + 48.8 \log_{10}(x + 1)$, where x is the age in years and $f(x)$ is the percent. At age 9, what percent of his adult height has a boy reached?

77. In Example 7, we learned that the percent of students who recall the important features of a lecture is given by $P = 95 - 30 \log_2 x$, where P is the percent and x is the number of days that have elapsed since the lecture was given. After how many days will 35% of the students recall the important features?

78. Using the formula from Exercise 77, find the percent of students who recall the important features of a lecture 1 day after it is given.

79. Why does $\log_b b = 1$ for any value of $b > 0$ and $b \neq 1$?

80. In the definition of a logarithm, $y = \log_b x$, why must $b \neq 1$?

81. Why does $\log_b 1 = 0$?

82. If $f(x) = b^x$, the domain is $(-\infty, \infty)$ and the range is $(0, \infty)$. What are the domain and range of $f(x) = \log_b x$? Why?

Review Exercises

Exercises 1–6 **EXPRESSIONS**

[5.1]   **1.** Write $\dfrac{1}{x^6}$ as x to a negative power.

For Exercises 2–5, simplify using the rules of exponents.

[5.4]   **2.** $x^4 \cdot x^2$ **[5.4]**   **3.** $(x^3)^5$ **[5.6]**   **4.** $\dfrac{x^6}{x^3}$ **[5.6]**   **5.** $\dfrac{(x^3)^2 \cdot x^4}{x^5}$

[10.2]   **6.** Write $\sqrt[4]{x^3}$ in exponential form.

 **12.4** Properties of Logarithms

 **Objectives**   **❶** Apply the inverse properties of logarithms.

                  ❷ Apply the product, quotient, and power properties of logarithms.

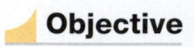 **Objective ❶ Apply the inverse properties of logarithms.**   Earlier we developed two properties of logarithms, $\log_b b = 1$ and $\log_b 1 = 0$, which were based on the definition of a logarithm. The fact that $f[f^{-1}(x)] = x$ and $f^{-1}[f(x)] = x$ gives us the following properties of logarithms.

Inverse Properties of Logarithms

For any real numbers b and x, where $b > 0$, $b \neq 1$ and $x > 0$,

1. $b^{\log_b x} = x$. **2.** $\log_b b^x = x$.

To prove that $b^{\log_b x} = x$, let $f(x) = b^x$ so $f^{-1}(x) = \log_b x$.

$$f[f^{-1}(x)] = x \qquad \text{Composition of a function with its inverse is } x.$$
$$f[\log_b x] = x \qquad \text{Replace } f^{-1}(x) \text{ with } \log_b x.$$
$$b^{\log_b x} = x \qquad \text{Replace } x \text{ with } \log_b x \text{ in } f(x).$$

To prove that $\log_b b^x = x$, use the definition of a logarithm. $\text{Log}_a\, b = c$ means that $a^c = b$; so $\log_b b^x = x$ because $b^x = b^x$.

Example 1 Find the value.

a. $3^{\log_3 8}$

b. $6^{\log_6 x}$

SOLUTION $3^{\log_3 8} = 8$ Use $b^{\log_b x} = x$.

SOLUTION $6^{\log_6 x} = x$ Use $b^{\log_b x} = x$.

c. $\log_3 3^6$

d. $\log_5 5^a$

SOLUTION $\log_3 3^6 = 6$ Use $\log_b b^x = x$.

SOLUTION $\log_5 5^a = a$ Use $\log_b b^x = x$.

Your Turn 1 Find the value.

a. $8^{\log_8 4}$ **b.** $\log_3 3^{-2}$

Objective ❷ **Apply the product, quotient, and power properties of logarithms.** Logarithms were invented to perform operations on very large or very small numbers. With the invention of handheld calculators, they are no longer used for this purpose. However, we still use the properties of logarithms, which are based on the fact that logarithms are exponents.

Further Properties of Logarithms

For real numbers x, y, and b, where $x > 0$, $y > 0$, $b > 0$, and $b \neq 1$,

Warning: There is no rule for the logarithms of sums or differences. The $\log_b(x + y) \neq \log_b x + \log_b y$.

Product Rule of Logarithms: $\log_b xy = \log_b x + \log_b y$
(The logarithm of the product of two numbers is equal to the sum of the logarithms of the numbers.)

Quotient Rule of Logarithms: $\log_b \dfrac{x}{y} = \log_b x - \log_b y$
(The logarithm of the quotient of two numbers is equal to the difference of the logarithms of the numbers.)

Note: When the product and quotient rules are used, all of the bases of all of the logarithms **must** be the same.

Power Rule of Logarithms: $\log_b x^r = r \log_b x$
(The logarithm of a number raised to a power is equal to the exponent times the logarithm of the number.)

To prove that $\log_b xy = \log_b x + \log_b y$, let $M = \log_b x$ and $N = \log_b y$.

$$x = b^M \text{ and } y = b^N \qquad \text{Write each logarithmic equation in exponential form.}$$
$$xy = b^M \cdot b^N \qquad \text{Multiply the left and right sides of the exponential forms.}$$
$$xy = b^{M+N} \qquad \text{Add the exponents.}$$
$$\log_b xy = M + N \qquad \text{Write in logarithmic form.}$$
$$\log_b xy = \log_b x + \log_b y \qquad \text{Substitute } \log_b x \text{ for } M \text{ and } \log_b y \text{ for } N, \text{ and the proof is complete.}$$

Answers **Your Turn 1**

a. 4 **b.** −2

The proofs of $\log_b \dfrac{x}{y} = \log_b x - \log_b y$ and $\log_b x^r = r \log_b x$ are similar.

Example 2 Use the product rule of logarithms to write the expression as a sum of logarithms.

a. $\log_{10} xyz$

SOLUTION $\log_{10} xyz = \log_{10} x + \log_{10} y + \log_{10} z$

b. $\log_b x(x + 3)$

SOLUTION $\log_b x(x + 3) = \log_b x + \log_b (x + 3)$ ◀ **Note:** $\log_b(x + 3) \neq \log_b x + \log_b 3$.

Your Turn 2 Use the product rule of logarithms to write the expression as a sum of logarithms.

a. $\log_a 6x$

b. $\log_2 x(3x + 2)$

Example 3 Use the product rule of logarithms in the form $\log_b x + \log_b y = \log_b xy$ to write the expression as a single logarithm.

a. $\log_3 8 + \log_3 2$

SOLUTION $\log_3 8 + \log_3 2 = \log_3 8 \cdot 2 = \log_3 16$

b. $\log_8 5 + \log_8 x + \log_8(2x - 3)$

SOLUTION $\log_8 5 + \log_8 x + \log_8(2x - 3) = \log_8 5x(2x - 3)$
$$= \log_8(10x^2 - 15x)$$

Your Turn 3 Use the product rule of logarithms to write each of the following as a single logarithm.

a. $\log_6 7 + \log_6 9$

b. $\log_7 2 + \log_7 x + \log_7(2x + 4)$

Example 4 Use the quotient rule of logarithms to write the expression as a difference of logarithms. Leave the answers in simplest form.

a. $\log_5 \dfrac{5}{11}$

SOLUTION $\log_5 \dfrac{5}{11} = \log_5 5 - \log_5 11$
$$= 1 - \log_5 11 \qquad \text{\color{blue}Remember, } \log_5 5 = 1.$$

b. $\log_4 \dfrac{x}{x - 5}$

Answers Your Turn 2
a. $\log_a 6 + \log_a x$
b. $\log_2 x + \log_2(3x + 2)$

SOLUTION $\log_4 \dfrac{x}{x - 5} = \log_4 x - \log_4(x - 5)$ ◀ **Note:** $\log_4(x - 5) \neq \log_4 x - \log_4 5$.

Answers Your Turn 3
a. $\log_6 63$ **b.** $\log_7(4x^2 + 8x)$

Answers Your Turn 4
a. $1 - \log_9 10$
b. $\log_5 x - \log_5(x + 2)$

Your Turn 4 Use the quotient rule of logarithms to write the expression as a difference of logarithms. Leave your answers in simplest form.

a. $\log_9 \dfrac{9}{10}$

b. $\log_5 \dfrac{x}{x + 2}$

Example 5   Use the quotient rule of logarithms in the form $\log_b x - \log_b y = \log_b \dfrac{x}{y}$ to write the expression as a single logarithm.

a. $\log_9 3 - \log_9 x$

SOLUTION $\log_9 3 - \log_9 x = \log_9 \dfrac{3}{x}$

b. $\log_{10} x - \log_{10}(x^2 + 4)$

SOLUTION $\log_{10} x - \log_{10}(x^2 + 4) = \log_{10} \dfrac{x}{x^2 + 4}$

Your Turn 5   Use the quotient rule of logarithms to write each of the following as a single logarithm.

a. $\log_3 15 - \log_3 5$ **b.** $\log_4(x^2 + 2) - \log_4(x + 1)$

Example 6   Use the power rule of logarithms to write the expression as a multiple of a logarithm.

a. $\log_4 a^6$

SOLUTION $\log_4 a^6 = 6 \log_4 a$

b. $\log_b \sqrt[4]{x^3}$

SOLUTION $\log_b \sqrt[4]{x^3} = \log_b x^{3/4}$ **Write $\sqrt[4]{x^3}$ as $x^{3/4}$.**

$= \dfrac{3}{4} \log_b x$

c. $\log_b \dfrac{1}{x^3}$

SOLUTION $\log_b \dfrac{1}{x^3} = \log_b x^{-3}$ **Write $\dfrac{1}{x^3}$ as x^{-3}.**

$= -3 \log_b x$

> **Note:** $\log_b \dfrac{1}{x^3}$ could be rewritten as $\log_b 1 - \log_b x^3 = 0 - 3\log_b x = -3 \log_b x.$

Your Turn 6   Use the power rule of logarithms to write the expression as a multiple of a logarithm.

a. $\log_a z^4$ **b.** $\log_a \sqrt[3]{x^2}$

Example 7   Use the power rule of logarithms in the form $r \log_b x = \log_b x^r$ to write the expression as a logarithm of a quantity to a power. Leave the answers in simplest form without negative or fractional exponents.

a. $5 \log_4 x$

SOLUTION $5 \log_4 x = \log_4 x^5$

b. $-2 \log_b 5$

SOLUTION $-2 \log_b 5 = \log_b 5^{-2}$

$= \log_b \dfrac{1}{5^2}$ **Write 5^{-2} as $\dfrac{1}{5^2}$.**

$= \log_b \dfrac{1}{25}$ **$5^2 = 25$**

Answers   Your Turn 5

a. 1   **b.** $\log_4 \dfrac{x^2 + 2}{x + 1}$

Answers   Your Turn 6

a. $4 \log_a z$ **b.** $\dfrac{2}{3} \log_a x$

c. $\dfrac{2}{3} \log_7 y$

SOLUTION $\dfrac{2}{3} \log_7 y = \log_7 y^{2/3}$

$\qquad\qquad\qquad = \log_7 \sqrt[3]{y^2}$ Write $y^{2/3}$ as $\sqrt[3]{y^2}$.

Your Turn 7   Use the power rule of logarithms to write the expression as a logarithm of a quantity to a power.

    a. $5 \log_7 x$ **b.** $-3 \log_b z$

Often it is necessary to use more than one rule to rewrite a logarithmic expression.

Example 8   Write the expression as a sum or difference of multiples of logarithms.

a. $\log_b \dfrac{z^3}{yz}$

SOLUTION $\log_b \dfrac{z^3}{yz} = \log_b z^3 - \log_b yz$ Use the quotient rule.

$\qquad\qquad\qquad\qquad = 3 \log_b z - (\log_b y + \log_b z)$ **Use the power rule and product rule. Note the use of parentheses.**

$\qquad\qquad\qquad\qquad = 3 \log_b z - \log_b y - \log_b z$ **Remove the parentheses.**

b. $\log_3 \sqrt{\dfrac{a^3}{b}}$

SOLUTION $\log_3 \sqrt{\dfrac{a^3}{b}} = \log_3 \left(\dfrac{a^3}{b}\right)^{1/2}$ Write $\sqrt{\dfrac{a^3}{b}}$ as $\left(\dfrac{a^3}{b}\right)^{1/2}$.

$\qquad\qquad\qquad\qquad = \dfrac{1}{2} \log_3 \left(\dfrac{a^3}{b}\right)$ **Use the power rule.**

$\qquad\qquad\qquad\qquad = \dfrac{1}{2}(\log_3 a^3 - \log_3 b)$ **Use the quotient rule.**

$\qquad\qquad\qquad\qquad = \dfrac{1}{2}(3 \log_3 a - \log_3 b)$ **Use the power rule.**

$\qquad\qquad\qquad\qquad = \dfrac{3}{2} \log_3 a - \dfrac{1}{2} \log_3 b$ **Distribute $\dfrac{1}{2}$.**

c. $\log_5 5^2 b^3$

SOLUTION $\log_5 5^2 b^3 = \log_5 5^2 + \log_5 b^3$ **Use the product rule.**

$\qquad\qquad\qquad\quad = 2 \log_5 5 + 3 \log_5 b$ **Use the power rule.**

$\qquad\qquad\qquad\quad = 2 + 3 \log_5 b$ $\log_5 5 = 1$

Your Turn 8   Write the expression as a sum or difference of multiples of logarithms.

    a. $\log_a \dfrac{x^2 y^3}{z}$ **b.** $\log_4 \sqrt[4]{\dfrac{x}{y^2}}$ **c.** $\log_8 8^5 m^6$

Example 9   Write the expression as a single logarithm. Leave the answers in simplest form without negative or fractional exponents.

    a. $4 \log_b 2 - 2 \log_b 3$

SOLUTION $4 \log_b 2 - 2 \log_b 3 = \log_b 2^4 - \log_b 3^2$ *Use the power rule.*

$$= \log_b \frac{2^4}{3^2} \qquad \text{\textit{Use the quotient rule.}}$$

$$= \log_b \frac{16}{9} \qquad \text{\textit{Simplify.}}$$

b. $\dfrac{1}{2}(\log_2 5 - \log_2 b)$

SOLUTION $\dfrac{1}{2}(\log_2 5 - \log_2 b) = \dfrac{1}{2} \log_2 \dfrac{5}{b}$ *Use the quotient rule.*

$$= \log_2 \left(\frac{5}{b}\right)^{1/2} \qquad \text{\textit{Use the power rule.}}$$

$$= \log_2 \sqrt{\frac{5}{b}} \qquad \text{\textit{Write} } \left(\frac{5}{b}\right)^{1/2} \text{ \textit{as} } \sqrt{\frac{5}{b}}.$$

c. $\log_a(x+2) + \log_a(x-3)$

SOLUTION $\log_a(x+2) + \log_a(x-3) = \log_a(x+2)(x-3)$ *Apply $\log_b xy = \log_b x + \log_b y$.*

$$= \log_a(x^2 - x - 6) \qquad \text{\textit{Multiply.}}$$

Your Turn 9   Write the expression as a single logarithm. Leave your answers in simplest form without negative or fractional exponents.

a. $2\log_a 4 - 3\log_a x$ **b.** $\dfrac{1}{3}(\log_a x + 2\log_a y)$ **c.** $\log_6 x + \log_6(x-2)$

Answers   **Your Turn 9**

a. $\log_a \dfrac{16}{x^3}$ **b.** $\log_a \sqrt[3]{xy^2}$

c. $\log_6(x^2 - 2x)$

12.4 Exercises   For Extra Help   *MyMathLab*

1. Why does $b^{\log_b x} = x$?

2. How can you use the facts that $\log_b x^r = r\log_b x$ and $\log_b b = 1$ to show that $\log_3 81 = 4$?

3. $\log_b(x+y) = $ _____.

4. Why does $\log_a \sqrt[3]{x^4} = \dfrac{4}{3}\log_a x$?

For Exercises 5–16, find the value. See Example 1.

5. $8^{\log_8 2}$ 6. $3^{\log_3 7}$ 7. $a^{\log_a r}$ 8. $b^{\log_b a}$ 9. $a^{\log_a 4x}$ 10. $b^{\log_b 5a}$

11. $\log_3 3^5$ 12. $\log_7 7^3$ 13. $\log_e e^y$ 14. $\log_c c^x$ 15. $\log_a a^{7x}$ 16. $\log_b b^{6y}$

For Exercises 17–24, use the product rule to write the expression as a sum of logarithms. See Example 2.

17. $\log_2 5y$ 18. $\log_3 4z$ 19. $\log_a pq$ 20. $\log_b rs$

21. $\log_4 mnp$ 22. $\log_4 pqr$ 23. $\log_a x(x-5)$ 24. $\log_b x(x+6)$

For Exercises 25–36, use the product rule to write the expression as a single logarithm. See Example 3.

25. $\log_3 5 + \log_3 8$

26. $\log_6 4 + \log_6 7$

27. $\log_4 3 + \log_4 9$

28. $\log_9 4 + \log_9 21$

29. $\log_a 7 + \log_a m$

30. $\log_b 2 + \log_b n$

31. $\log_4 a + \log_4 b$

32. $\log_6 r + \log_6 m$

33. $\log_a 2 + \log_a x + \log_a (x + 5)$

34. $\log_a 4 + \log_a y + \log_a (y - 5)$

35. $\log_4 (x + 1) + \log_4 (x + 3)$

36. $\log_6 (x - 3) + \log_6 (x + 2)$

For Exercises 37–46, use the quotient rule to write the expression as a difference of logarithms. Leave your answers in simplest form. See Example 4.

37. $\log_2 \dfrac{7}{9}$

38. $\log_4 \dfrac{5}{6}$

39. $\log_a \dfrac{x}{5}$

40. $\log_b \dfrac{y}{3}$

41. $\log_a \dfrac{a}{b}$

42. $\log_b \dfrac{a}{b}$

43. $\log_a \dfrac{x}{x - 3}$

44. $\log_b \dfrac{y}{2y + 5}$

45. $\log_4 \dfrac{2x - 3}{4x + 5}$

46. $\log_6 \dfrac{4x - 1}{3x + 2}$

For Exercises 47–58, use the quotient rule to write the expression as a single logarithm. Leave your answers in simplest form. See Example 5.

47. $\log_6 24 - \log_6 3$

48. $\log_7 18 - \log_7 3$

49. $\log_2 24 - \log_2 12$

50. $\log_3 48 - \log_3 16$

51. $\log_a x - \log_a 3$

52. $\log_a r - \log_a 5$

53. $\log_4 p - \log_4 q$

54. $\log_2 a - \log_2 b$

55. $\log_b x - \log_b(x - 4)$

56. $\log_b y - \log_b(y - 5)$

57. $\log_x(x^2 - x) - \log_x(x - 1)$

58. $\log_a(a^2 + 2a) - \log_a(a + 2)$

For Exercises 59–70, use the power rule to write the expression as a multiple of a logarithm. See Example 6.

59. $\log_4 3^6$

60. $\log_3 5^4$

61. $\log_a x^7$

62. $\log_b y^8$

63. $\log_a \sqrt{3}$

64. $\log_a \sqrt{6}$

65. $\log_3 \sqrt[3]{x^2}$

66. $\log_6 \sqrt[4]{y^3}$

67. $\log_a \dfrac{1}{6^2}$

68. $\log_a \dfrac{1}{5^4}$

69. $\log_a \dfrac{1}{y^2}$

70. $\log_a \dfrac{1}{x^5}$

For Exercises 71–82, use the power rule to write the expression as a logarithm of a quantity to a power. Leave your answers in simplest form without negative or fractional exponents. See Example 7.

71. $4 \log_3 5$

72. $5 \log_3 4$

73. $-3 \log_2 x$

74. $-4 \log_3 y$

75. $\dfrac{1}{2} \log_7 64$

76. $\dfrac{1}{3} \log_3 8$

77. $\dfrac{3}{4} \log_a x$

78. $\dfrac{5}{6} \log_b y$

79. $\dfrac{2}{3} \log_a 8$

80. $\dfrac{3}{4} \log_a 81$

81. $-\dfrac{1}{2} \log_3 x$

82. $-\dfrac{1}{3} \log_2 y$

For Exercises 83–94, write the expression as the sum or difference of multiples of logarithms. See Example 8.

83. $\log_a \dfrac{x^3}{y^4}$

84. $\log_a \dfrac{x^6}{y^5}$

85. $\log_3 a^4 b^2$

86. $\log_7 m^3 n^5$

87. $\log_a \dfrac{xy}{z}$

88. $\log_a \dfrac{pq}{r}$

89. $\log_x \dfrac{a^2}{bc^3}$

90. $\log_x \dfrac{c^2}{m^2 n}$

91. $\log_4 \sqrt[4]{\dfrac{x^3}{y}}$

92. $\log_5 \sqrt{\dfrac{a^5}{b^3}}$

93. $\log_a \sqrt[3]{\dfrac{x^2 y}{z^3}}$

94. $\log_3 \sqrt{\dfrac{c^2 d^3}{m^4}}$

For Exercises 95–108, write the expression as a single logarithm. Leave your answers in simplest form without negative or fractional exponents. See Example 9.

95. $3 \log_3 2 - 2 \log_3 4$

96. $3 \log_4 3 - 2 \log_4 9$

97. $4 \log_b x + 3 \log_b y$

98. $5 \log_b a + 4 \log_b c$

99. $\dfrac{1}{2}(\log_a 5 - \log_a 7)$

100. $\dfrac{1}{2}(\log_4 6 - \log_4 5)$

101. $\dfrac{2}{3}(\log_a x^2 + \log_a y^3)$

102. $\dfrac{3}{4}(\log_a m^3 + \log_a n^2)$

103. $\log_b x + \log_b(3x - 2)$

104. $\log_a y + \log_a(2y - 4)$

105. $3 \log_a(x - 2) - 4 \log_a(x + 1)$

106. $2 \log_b(x + 4) - 3 \log_b(2x - 1)$

107. $2 \log_a x + 4 \log_a z - 3 \log_a w - 6 \log_a u$

108. $5 \log_a b + 2 \log_a c - 4 \log_a d - 3 \log_a e$

Review Exercises

Exercises 1 and 2 EXPRESSIONS

For Exercises 1 and 2, simplify.

[5.4] **1.** $(10^{9.5})(10^{-12})$

[5.6] **2.** $\dfrac{10^{-3}}{10^{-12}}$

Exercises 3–6 EQUATIONS AND INEQUALITIES

[12.2] **3.** If \$10,000 is deposited at 6% compounded quarterly for five years, how much will be in the account?

[12.3] **4.** Write $10^{1.6990} = 50$ in logarithmic form.

[12.3] *For Exercises 5 and 6, write in exponential form.*

  5. $\log_{10} 45 = 1.6532$

  6. $\log_e 0.25 = -1.3863$

12.5 Common and Natural Logarithms

Objectives

1. Define common logarithms and evaluate them using a calculator.
2. Solve applications using common logarithms.
3. Define natural logarithms and evaluate them using a calculator.
4. Solve applications using natural logarithms.

Objective 1 Define common logarithms and evaluate them using a calculator. Of all possible bases of logarithms, two are most useful. As previously mentioned, logarithms were invented to do computations on very large and very small numbers. Because our system is a base-10 number system, base-10 logarithms were commonly used for this purpose. Consequently, base-10 logarithms are called **common logarithms** and $\log_{10} x$ is written as $\log x$, where the base 10 is understood. Base-10 logarithms are found in engineering, economics, the social sciences, and the natural sciences.

Definition *Common logarithms:* Logarithms with a base of 10. $\text{Log}_{10} x$ is written as $\log x$. Note that $\log 10 = 1$.

Connection Because logarithmic and exponential functions are inverse functions, if $f(x) = 10^x$, then $f^{-1}(x) = \log x$ and if $g(x) = \log x$, then $g^{-1}(x) = 10^x$.

Calculator Tips

On most calculators, to evaluate common logarithms, press the $\boxed{\text{LOG}}$ *key, enter the number, and then press* $\boxed{\text{ENTER}}$.

Note: Some graphing calculators automatically put a left parenthesis when the $\boxed{\text{LOG}}$ or $\boxed{\text{ln}}$ (page 876) keys are pressed. If you are evaluating the log or ln of a single number, it is not necessary to put the right parenthesis. Otherwise, it is.

To evaluate common logarithms, we use a calculator. We will round all results to four places.

Example 1 Use a calculator to approximate each common logarithm to four decimal places.

a. $\log 23$

Solution $\log 23 \approx 1.3617$

b. $\log 0.00236$

Solution $\log 0.00236 \approx -2.6271$

Connection Remember that $\log 23 \approx 1.3617$ means that $10^{1.3617} \approx 23$, which provides a way to check. Because we rounded the decimal value, evaluating $10^{1.3617}$ will not give exactly 23.

Your Turn 1 Use a calculator to approximate each common logarithm to four decimal places.

a. $\log 436$ b. $\log 0.0724$

Objective 2 Solve applications using common logarithms. Common logarithms can be used to calculate sound intensity and runway length.

Example 2

a. Sound intensity can be measured in watts per unit of area or, more commonly, in decibels. The function $d = 10 \log \dfrac{I}{I_0}$ is used to calculate sound intensity, where d represents the intensity in decibels, I represents the intensity in watts per unit of area, and I_0 represents the faintest audible sound to the average human ear (which is 10^{-12} watts per square meter). A motorcycle has a sound intensity of about $10^{-2.5}$ watts per square meter. Find the decibel reading for the motorcycle.

Answers Your Turn 1
a. 2.6395   b. −1.1403

Understand We are given the function $d = 10 \log \dfrac{I}{I_0}$ and values for I and I_0, and we are to find the decibel reading, which is d.

Plan Using $d = 10 \log \dfrac{I}{I_0}$, substitute $10^{-2.5}$ for I and 10^{-12} for I_0 and then solve for d.

Execute

$d = 10 \log \dfrac{10^{-2.5}}{10^{-12}}$ **Substitute for I and I_0 in $d = 10 \log \dfrac{I}{I_0}$.**

$d = 10 \log 10^{9.5}$ **Subtract exponents $[-2.5 - (-12) = 9.5]$.**

$d = 10(9.5)$ **Use $\log_b b^x = x$.**

$d = 95$ **Simplify.**

> **Note:** The abbreviation for decibels is dB.

Answer The motorcycle has a decibel reading of 95 dB.

Check If the sound intensity is 95 decibels, verify that I is $10^{-2.5}$.

$95 = 10 \log \dfrac{I}{10^{-12}}$ **Substitute 95 for d.**

$9.5 = \log \dfrac{I}{10^{-12}}$ **Divide both sides by 10 to isolate the logarithm.**

$10^{9.5} = \dfrac{I}{10^{-12}}$ **Write in exponential form.**

$10^{9.5} \cdot 10^{-12} = I$ **Multiply both sides by 10^{-12}.**

$10^{-2.5} = I$ **Simplify by adding the exponents. It checks.**

b. The minimum length of an airport runway needed for a plane to take off is related to the weight of the plane. For some planes, the minimum runway length may be modeled by the function $y = 3 \log x$, where x is the plane's weight in thousands of pounds and y is the length of the runway in thousands of feet. Find the minimum length of a runway needed by a Boeing 737 whose maximum takeoff weight is 174,200 pounds.

Understand We are given the function $y = 3 \log x$, where y represents the runway length in thousands of feet required for a plane with a takeoff weight of x, in thousands of pounds. We are given that a plane's takeoff weight is 174,200 pounds and must find the runway length required for that plane.

Plan Because x is the plane's weight in thousands of pounds, we first need to divide 174,200 by 1000 to find the value of x corresponding to 174,200 pounds. We can then use $y = 3 \log x$.

Execute

$x = \dfrac{174{,}200}{1000} = 174.2$

$y = 3 \log 174.2$ **Substitute 174.2 for x in $y = 3 \log x$.**

$y = 6.723$ **Evaluate using a calculator.**

Answer Because y is in thousands of feet, the minimum runway length is $(6.723)(1000) = 6723$ ft.

Check Verify that a runway length of 6723 feet corresponds to a plane whose takeoff weight is 174,200. Because y is in thousands of feet, we first divide 6723 by 1000 to find the value of y: $y = 6723/1000 = 6.723$.

$6.723 = 3 \log x$ **Substitute 6.723 for y in $y = 3 \log x$.**

$2.241 = \log x$ **Divide both sides by 3 to isolate the logarithm.**

$10^{2.241} = x$ **Write in exponential form.**

$174.2 \approx x$ **Simplify. We rounded to the nearest tenth.**

Because x is in thousands of pounds, $174.2(1000) = 174{,}200$ pounds, which checks.

Your Turn 2

a. The sound intensity of a rock band often exceeds $10^{-0.5}$ watts per square meter. Find the decibel reading of the band.

b. Find the minimum runway length needed for a B-52 Stratofortress whose maximum takeoff weight is 488,000 pounds.

Objective ③ Define natural logarithms and evaluate them using a calculator. The number e is an irrational number whose approximate value is 2.7182818285. It is a universal constant like π. In the natural sciences, base-e logarithms are more prevalent than base-10. Because base-e logarithms occur in so many "natural" situations, they are called **natural logarithms**. The notation for $\log_e x$ is $\ln x$, which is read "el en of x."

Definition   *Natural logarithms:* Base-e logarithms are called natural logarithms, and $\log_e x$ is written as $\ln x$. Note that $\ln e = 1$.

Calculator Tips

To evaluate natural logarithms, on most calculators, press the `ln` *key, enter the number, and then press* `ENTER`.

To find natural logarithms, we use a calculator.

Example 3   Use a calculator to approximate each natural logarithm to four decimal places.

a. $\ln 83$

SOLUTION $\ln 83 \approx 4.4188$

> **Connection**   Remember that $\ln 83 \approx 4.4188$ means that $e^{4.4188} \approx 83$. Similarly, $\ln 0.0055 = -5.2030$ means that $e^{-5.2030} \approx 0.0055$.

b. $\ln 0.0055$

SOLUTION $\ln 0.0055 \approx -5.2030$

Your Turn 3   Use a calculator to approximate each natural logarithm to four decimal places.

a. $\ln 102$ **b.** $\ln 0.0573$

Objective ④ Solve applications using natural logarithms. If money is deposited into an account and the interest is compounded continuously, then the time t (in years) that it will take an investment of P dollars to grow into A dollars at an interest rate r (written as a decimal) is given by $t = \dfrac{1}{r} \ln \dfrac{A}{P}$.

Example 4   An amount of $5000 is deposited into an account earning 5% annual interest compounded continuously. How many years will it take until the account has reached $10,000?

Understand   We are to find the time it takes for $5000 to grow to $10,000 if it is compounded continuously at 5%.

Plan   In $t = \dfrac{1}{r} \ln \dfrac{A}{P}$, replace P with 5000, r with 0.05, and A with 10,000; then simplify.

Answers   **Your Turn 2**

a. 115 dB   **b.** 8065 ft.

Execute $t = \dfrac{1}{0.05} \ln \dfrac{10{,}000}{5000}$ **Substitute for P, r, and A.**

$t = 20 \ln 2$ **Divide.**

$t \approx 13.86$ **Multiply using a calculator.**

Answers   **Your Turn 3**

a. 4.6250   **b.** −2.8595

Answer It will take about 13.86 years for $5000 to grow to $10,000 if it is compounded continuously at 5%.

Check We can use the formula to verify that if $5000 is compounded continuously for 13.86 years, the final amount will be $10,000.

$$13.86 = \frac{1}{0.05} \ln \frac{A}{5000}$$ **Substitute for *t*, *r*, and *P*.**

$$0.693 = \ln \frac{A}{5000}$$ **Multiply both sides by 0.05.**

$$0.693 = \ln A - \ln 5000$$ **Use $\log_b \frac{x}{y} = \log_b x - \log_b y$.**

$$0.693 + \ln 5000 = \ln A$$ **Add ln 5000 to both sides.**

$$9.21 \approx \ln A$$ **Add 0.693 and the approximate value of ln 5000.**

$$e^{9.21} \approx A$$ **Write in exponential form.**

$$9996.6 \approx A$$ **Calculate using a calculator.**

Because 13.86 is not the exact time and we rounded the sum of 0.693 and ln 5000, it is reasonable to expect that the amount will be very close to but not exactly $10,000; so $9996.60 is reasonable.

Your Turn 4 How long will it take $2000 to grow to $5000 at 8% interest if the interest is compounded continuously?

Answer   **Your Turn 4**

11.5 yr.

12.5 Exercises   **For Extra Help**   *MyMathLab*   Math XL PRACTICE    WATCH   DOWNLOAD

1. If $y = \log_b x$, what values are allowed for b? Why?

2. Is log x positive or negative for $0 < x < 1$? Why?

3. Is ln x positive or negative for $x > 1$? Why?

4. Give the symbol and meaning for **a.** common logarithms. **b.** natural logarithms.

5. If $e^{0.91629} = 2.5$, find ln 2.5.

6. Without using a calculator, find the exact value of $\log_{10} 10^{\sqrt{5}}$.

For Exercises 7–26, use a calculator to approximate each logarithm to four decimal places. See Examples 1 and 3.

7. log 64

8. log 27

9. log 0.0067

10. log 0.00087

11. log 435.6

12. log 785.4

13. $\log(1.5 \times 10^4)$

14. $\log(5.7 \times 10^7)$

15. $\log(1.6 \times 10^{-6})$

16. $\log(7.5 \times 10^{-5})$

17. ln 9.34

18. ln 5.33

19. ln 79.2

20. ln 765.4

21. ln 0.034

22. ln 0.0923

23. $\ln(5.4 \times e^4)$

24. $\ln(2.4 \times e^3)$

25. log e

26. ln 10

27. Use your calculator to find log 0. What happened? Why?

28. Use your calculator to find ln (-1). What happened? Why?

For Exercises 29–40, find the exact value of each logarithm using $\log_b b^x = x$.

29. $\log 100$

30. $\log 1000$

31. $\log \dfrac{1}{100}$

32. $\log \dfrac{1}{1000}$

33. $\log \sqrt[3]{10}$

34. $\log \sqrt[4]{10}$

35. $\log 0.001$

36. $\log 0.00001$

37. $\ln e^3$

38. $\ln e^5$

39. $\ln \sqrt{e}$

40. $\ln \sqrt[5]{e}$

For Exercises 41–44, use the formula $d = 10 \log \dfrac{I}{I_0}$, *where* $I_0 = 10^{-12}$ watts/m². *See Example 2.*

41. The sound intensity of a firecracker is 10^{-3} watts per square meter. What is the decibel reading for the firecracker?

42. The sound intensity of a race car is 10^{-1} watts per square meter. What is the decibel reading for the race car?

43. If a noisy office has a decibel reading of 60, what is the sound intensity?

44. If loud thunder has a decibel reading of 80, what is the sound intensity?

For Exercises 45–48, use the following information. In chemistry, the pH of a substance determines whether it is a base (pH > 7) or an acid (pH < 7). To find the pH of a solution, we use the formula $pH = -\log [H_3O^+]$, *where* $[H_3O^+]$ *is the hydronium ion concentration in moles per liter. Note that the pH is unitless.*

45. Find the pH of vinegar if $[H_3O^+] = 1.6 \times 10^{-3}$ moles per liter.

46. Find the pH of maple syrup if $[H_3O^+] = 2.3 \times 10^{-7}$ moles per liter.

47. Find the hydronium ion concentration of sauerkraut, which has a pH of 3.5.

48. Find the hydronium ion concentration of blood, which has a pH of 7.4.

For Exercises 49 and 50, use the formula $t = \dfrac{1}{r} \ln \dfrac{A}{P}$. *See Example 4.*

49. Find how long it will take $2000 to grow to $5000 at 4% interest if the interest is compounded continuously.

50. Find how long it will take $5000 to grow to $8000 at 5% interest if the interest is compounded continuously.

51. The magnitude of an earthquake is given by the Richter scale, whose formula is $R = \log \dfrac{I}{I_0}$, where I is the intensity of the earthquake and I_0 is the intensity of a minimal earthquake and is used for comparison purposes. The 1906 San Francisco earthquake had a magnitude of 7.8, and the 1964 Alaska earthquake had a magnitude of 8.4. Compare the intensity of the two earthquakes. (*Hint:* Express the intensity of each in terms of I_0.)

52. Using the formula from Exercise 51, compare the intensity of the 1949 Queen Charlotte Islands earthquake, whose magnitude was 8.1, with the 2004 earthquake in the Indian Ocean that killed more than 225,000 people, whose magnitude was estimated at 9.2.

53. During an earthquake, energy is released in various forms. The amount of energy radiated from the earthquake as seismic waves is given by $\log E_s = 11.8 + 1.5\,M$, where E_s is measured in ergs and M is the magnitude of the earthquake as given by the Richter scale. Vancouver Island had an earthquake whose magnitude was 7.3. How much energy was released in the form of seismic waves?

This photograph shows the destruction of San Francisco that was caused by the 1906 earthquake.

54. Using the formula from Exercise 53, find the energy released in the form of seismic waves from the Double Springs Flat earthquake of 1994, whose magnitude was 6.1 on the Richter scale.

55. The purchasing power of $1.00 in 2007 can be approximated by $y = -69.45 + 22.7 \ln x$, where y is the number of years beginning with 1970 ($y = 1$ corresponds with 1970) until 2007 and x is the purchasing power, in cents, of $1.00 in 2007. In what year was the purchasing power of $1.00 in 2007 approximately $0.40?

56. The average number of miles per gallon for light trucks can be approximated by $y = 14.27 + 2.98 \ln x$, where y is the average number of miles per gallon and x is the number of years beginning with 1975. ($y = 1$ corresponds with 1975.) Find the approximate average number of miles per gallon in 2013.

57. Using the formula from Example 2(b), $y = 3 \log x$, find the minimum runway length needed for a Boeing 717 whose maximum weight is 110,000 pounds at takeoff.

58. Walking speeds in various cities are a function of the population, because as populations increase, so does the pace of life. Average walking speeds can be modeled by the function $W = 0.35 \ln P + 2.74$, where P is the population, in thousands, and W is the walking speed in feet per second. Find the average walking speed in Chicago, whose population was approximately 2,800,000 in 2007.

Collaborative Exercises Exploring Graphs of Logarithms

Using a graphing calculator, draw the graph of $f(x) = \log x$ and $g(x) = \log(10x)$ in the window [0, 100] for x and [−2, 4] for y.

1. Press the [TRACE] key. Using up or down arrow keys, move from one graph to the other and observe the y-values of several points with the same x-values. What do you observe?

2. How is the graph of $g(x) = \log(10x)$ related to the graph of $f(x) = \log x$? Why?

3. In general, how is the graph of $g(x) = \log(kx)$ for $k > 0$ related to the graph of $f(x) = \log x$?

4. Repeat Step 1 using $f(x) = \log x$ and $g(x) = \log \dfrac{x}{10}$.

5. How is the graph of $g(x) = \log \dfrac{x}{10}$ related to the graph of $f(x) = \log x$? Why?

6. In general, how is the graph of $g(x) = \log \dfrac{x}{k}$ for $k > 0$ related to the graph of $f(x) = \log x$?

Review Exercises

Exercise 1  EXPRESSIONS

[12.4]   **1.** Write $\log_3(2x + 1) - \log_3(x - 1)$ as a single logarithm.

Exercises 2–6 EQUATIONS AND INEQUALITIES

[12.3]   **2.** Write $\log_3(2x + 5) = 2$ in exponential form.

For Exercises 3–6, solve.

[2.3]   **3.** $3x - (7x + 2) = 12 - 2(x - 4)$ [6.6]   **4.** $x^2 + 2x = 15$

[7.6]   **5.** $\dfrac{5}{x} + \dfrac{3}{x + 1} = \dfrac{23}{3x}$ [10.6]   **6.** $\sqrt{5x - 1} = 7$

12.6 Exponential and Logarithmic Equations with Applications

Objectives

① Solve equations that have variables as exponents.

② Solve equations containing logarithms.

③ Solve applications involving exponential and logarithmic functions.

④ Use the change-of-base formula.

In Section 12.2, we solved equations that could be put in the form $b^x = b^y$. For example, if $3^x = 81$, then $3^x = 3^4$; so $x = 4$. To use this form, we had to write both sides of the equation as the same base raised to a power. In this section, we solve equations like $3^x = 16$, which is the same as $3^x = 2^4$, where the bases are not the same. To solve this and other exponential and logarithmic equations, we need the following properties.

Rule **Properties for Solving Exponential and Logarithmic Equations**

For any real numbers b, x, and y, where $b > 0$ and $b \neq 1$,

1. If $b^x = b^y$, then $x = y$.
2. If $x = y$, then $b^x = b^y$.
3. For $x > 0$ and $y > 0$, if $\log_b x = \log_b y$, then $x = y$.
4. For $x > 0$ and $y > 0$, if $x = y$, then $\log_b x = \log_b y$.
5. For $x > 0$, if $\log_b x = y$, then $b^y = x$.

These properties are true because the exponential and logarithmic functions are one-to-one.

Objective ① **Solve equations that have variables as exponents.** To solve equations that have variables as exponents, we use property 4, which says that if two positive numbers are equal, then so are their logarithms.

Example 1 Solve $5^x = 16$.

SOLUTION $\log 5^x = \log 16$ Use if $x = y$, then $\log_b x = \log_b y$ (property 4).

$x \log 5 = \log 16$ Use $\log_b x^r = r \log_b x$.

$x = \dfrac{\log 16}{\log 5}$ Divide both sides of the equation by log 5.

The exact solution is $x = \dfrac{\log 16}{\log 5}$. Using a calculator, we find $x \approx 1.7227$ correct to four decimal places.

Check $5^x = 16$

$5^{1.7227} = 16$ Substitute 1.7727 for x.

$15.9998 \approx 16$ The answer is correct.

Your Turn 1 Solve $6^{2x} = 42$ for x. Round the answer to four decimal places.

In Example 1, we took the common logarithm of both sides, but we could have used natural logarithms (or logarithms of any other base) instead. If one side of the equation contains a power of e, natural logs are preferred so that we can use the fact that $\log_e e^x = x$ or, more simply, $\ln e^x = x$.

Answer **Your Turn 1**

1.0430

Example 2 Solve $e^{4x} = 23$.

SOLUTION $\ln e^{4x} = \ln 23$ Use if $x = y$; then $\log_b x = \log_b y$.

$\qquad\qquad 4x = \ln 23$ Use $\log_b b^x = x$.

$\qquad\qquad x = \dfrac{\ln 23}{4}$ Divide both sides by 4.

$\qquad\qquad x \approx 0.7839$

Note: If your calculator automatically puts the left parenthesis, this must be entered as $\ln(23) \div 4$.

We will leave the check to the reader.

Your Turn 2 Solve $e^{3x} = 5$ for x. Round the answer to four decimal places.

Objective ② **Solve equations containing logarithms.** Now that we have explored additional properties of logarithms, we can modify the procedure, learned in Section 12.3 for solving equations using logarithms.

Procedure **Solving Equations Containing Logarithms**

To solve equations containing logarithms, use the properties of logarithms to simplify each side of the equation and then use one of the following.

If the simplification results in an equation in the form $\log_b x = \log_b y$, use the fact that $x = y$ and then solve for the variable.

If the simplification results in an equation in the form $\log_b x = y$, write the equation in exponential form, $b^y = x$, and then solve for the variable (as we did in Section 12.3).

Example 3 Solve.

a. $\log_5 x + \log_5(x + 3) = \log_5 4$

SOLUTION $\log_5 x(x + 3) = \log_5 4$ Use $\log_b xy = \log_b x + \log_b y$ to simplify the left side.

$\qquad\qquad x(x + 3) = 4$ The equation is in the form $\log_b x = \log_b y$, so $x = y$.

$\qquad\qquad x^2 + 3x = 4$ Simplify.

$\qquad\qquad x^2 + 3x - 4 = 0$ Write in $ax^2 + bx + c = 0$ form.

$\qquad\qquad (x + 4)(x - 1) = 0$ Factor.

$\qquad\qquad x + 4 = 0 \quad \text{or} \quad x - 1 = 0$ Set each factor equal to 0.

$\qquad\qquad x = -4 \quad \text{or} \qquad x = 1$ Solve each equation to find possible solutions.

If -4 is substituted into the original equation, we have $\log_5(-4) + \log_5(-1) = \log_5 4$, but logarithms are defined for positive numbers only. So $x = -4$ is not a solution. A check will show that $x = 1$ is a solution.

b. $\log_3(2x + 5) = 2$

SOLUTION $3^2 = 2x + 5$ The equation is in the form $\log_b x = y$; so write it in exponential form, $b^y = x$.

$\qquad\qquad 9 = 2x + 5$ Evaluate 3^2.

$\qquad\qquad 4 = 2x$ Subtract 5 from both sides.

$\qquad\qquad 2 = x$ Solve for x.

Answer **Your Turn 2**

0.5365

We will let the reader check that $x = 2$ is a solution.

c. $\log_2(5x + 1) - \log_2(x - 1) = 3$

SOLUTION $\log_2 \dfrac{5x + 1}{x - 1} = 3$ Use $\log_b \dfrac{x}{y} = \log_b x - \log_b y$ to simplify the left side.

$\qquad\qquad \dfrac{5x + 1}{x - 1} = 2^3$ The equation is in the form $\log_b x = y$, so $b^y = x$.

$\qquad\qquad 5x + 1 = 8x - 8$ Multiply both sides by $x - 1$.

$\qquad\qquad\qquad 9 = 3x$ Isolate the x term.

$\qquad\qquad\qquad x = 3$ Solve for x.

We will let the reader check that $x = 3$ is a solution.

Your Turn 3   Solve.

    a. $\ln x + \ln(x + 2) = \ln 8$ **b.** $\log_3(4x + 2) - \log_3(x - 2) = 2$

Objective ③ Solve applications involving exponential and logarithmic functions. A wide variety of problems from business and the sciences can be solved using exponential or logarithmic functions. Earlier we used the formula for compound interest, $A = P\left(1 + \dfrac{r}{n}\right)^{nt}$, to find A when given P, r, n, and t. Using logarithms, it is also possible to find t when given A, P, r, and n.

Example 4   How long will it take $6000 invested at 6% interest compounded quarterly to grow to $10,000? Round the answer to the nearest hundredth.

Understand   We are given $A = 10,000$, $P = 6000$, $r = 0.06$, and $n = 4$.

Plan   Use $A = P\left(1 + \dfrac{r}{n}\right)^{nt}$.

Execute $10,000 = 6000\left(1 + \dfrac{0.06}{4}\right)^{4t}$ Substitute for A, P, r, and n.

$\qquad\qquad\qquad\quad \dfrac{5}{3} = (1.015)^{4t}$ Divide both sides by 6000 and simplify inside the parentheses.

$\qquad\qquad\quad \log \dfrac{5}{3} = \log 1.015^{4t}$ Use if $x = y$; then $\log_b x = \log_b y$.

$\qquad\qquad\quad \log \dfrac{5}{3} = 4t \log 1.015$ Use $\log_b x^r = r \log_b x$.

$\qquad\qquad \dfrac{\log \dfrac{5}{3}}{4 \log 1.015} = t$ Divide both sides by $4 \log 1.015$.

$\qquad\qquad\qquad 8.58 \approx t$ Evaluate using a calculator.

Answer   It will take about 8.58 years.

Check   We can use the formula to verify that if $6000 is compounded quarterly for 8.58 years, the amount will be $10,000.

$\qquad\qquad A = 6000\left(1 + \dfrac{0.06}{4}\right)^{4(8.58)}$ Substitute for P, r, n, and t.

$\qquad\qquad A = 10,001.52$ Calculate.

Because 8.58 is not the exact time, it is reasonable to expect that the amount will be very close to but not exactly, $10,000; so $10,001.52 is reasonable.

Your Turn 4   How long will it take $8000 invested at 4% annual interest compounded semiannually to grow to $12,000?

Many banks compound interest continuously. The formula for interest compounded continuously is $A = Pe^{rt}$, where A is the amount in the account, P is the amount deposited, r is the interest rate as a decimal, and t is the time in years.

Example 5   If $5000 is deposited into an account at 5% interest compounded continuously, how much will be in the account after 9 years?

Understand   We are given $P = \$5000$, $r = 0.05$, and $t = 9$ and are asked to find A.

Plan   Use $A = Pe^{rt}$.

Execute
$A = 5000e^{0.05(9)}$	**Substitute for P, r, and t.**
$A = 5000e^{0.45}$	**Apply $(a^m)^n = a^{mn}$.**
$A = \$7841.56$	**Evaluate using a calculator.**

Answer   There will be $7841.56 in the account.

Check   Use the formula $t = \dfrac{1}{r}\ln\dfrac{A}{P}$ that we learned in Example 4 of Section 12.5 to verify that it takes 9 years for $5000 to grow to $7841.56 if it is compounded continuously at 5%.

$$t = \frac{1}{0.05}\ln\frac{7841.56}{5000} \qquad \text{Substitute for } r, A, \text{ and } P.$$

$$t \approx 9 \qquad \text{Calculate. It checks.}$$

Connection Solving $A = Pe^{rt}$ for t gives the formula $t = \dfrac{1}{r}\ln\dfrac{A}{P}$.

$\dfrac{A}{P} = e^{rt}$	**Divide both sides by P to isolate e^{rt}.**
$\log_e \dfrac{A}{P} = rt$	**Write in log form.**
$\dfrac{1}{r}\ln\dfrac{A}{P} = t$	**Divide both sides by r, and by definition, $\log_e\dfrac{A}{P}$ is $\ln\dfrac{A}{P}$.**

Example 6   Since the 1980s, the greater Orlando area has been one of the fastest-growing areas in the United States. The following table shows the population of the greater Orlando area for selected years.

Year	Population in Millions
1980	0.805
1985	0.996
1990	1.225
1995	1.428
2000	1.645
2003	1.803
2007	2.033

The data can be approximated by $y = 0.823e^{0.0361t}$, where y is the population in millions and t is the number of years after 1980. (*Source:* U.S. Bureau of the Census)

a.   Assuming that the population continues to grow in the same manner, use the model to estimate the population of Orlando in 2015.

Understand   We are given the function $y = 0.823e^{0.0361t}$, where y represents the population of Orlando t years after 1980.

Plan   Because t represents the number of years after 1980, we first subtract 1980 from 2015 to determine the value of t that corresponds to 2015. We can then substitute that value into the function.

Execute $t = 2015 - 1980 = 35$

$y = 0.823e^{0.0361t}$	**Formula**
$y = 0.823e^{0.0361(35)}$	**Substitute 35 for t.**
$y = 2.912$	**Evaluate using a calculator.**

Answer   Because y is in millions, the estimated population is 2,912,000 in 2015.

Answer   Your Turn 4

10.24 yr.

Check Use $y = 0.823e^{0.0361t}$ to verify that if the population is 2,912,000 ($y = 2.912$), then the year is 2015 ($t = 35$).

$$2.912 = 0.823e^{0.0361t} \qquad \text{Substitute 2.912 for } y.$$

$$\frac{2.912}{0.823} = e^{0.0361t} \qquad \text{Divide both sides by 0.8231.}$$

$$\ln \frac{2.912}{0.823} = \ln e^{0.0361t} \qquad \text{Use if } x = y; \text{ then } \log_b x = \log_b y.$$

$$\ln \frac{2.912}{0.823} = 0.0361t \qquad \text{Use } \log_b b^x = x.$$

$$\frac{\ln \dfrac{2.912}{0.823}}{0.0361} = t \qquad \text{Divide both sides by 0.0361.}$$

$$35 \approx t \qquad \text{Calculate.}$$

Note: Our manipulations suggest the following formula for calculating t given y:

$$t = \frac{1}{0.0361} \ln \frac{y}{0.823}$$

b. Find the year in which the population will be 3,500,000.

Understand We are to find t given the population.

Plan Because y represents the population in millions, we must first divide 3,500,000 by 1,000,000 to find the value of y that corresponds to 3,500,000. We can then use $y = 0.823e^{0.0361t}$ and follow the same steps as in our check for part a. Or as we do here, we use the formula $t = \dfrac{1}{0.0361} \ln \dfrac{y}{0.823}$ that we discovered as a result of checking part a.

Execute

$$y = \frac{3{,}500{,}000}{1{,}000{,}000} = 3.5 \qquad \begin{array}{l}\text{Divide 3,500,000 by 1,000,000 to find the value of } y \\ \text{that corresponds to 3,500,000.}\end{array}$$

$$t = \frac{1}{0.0361} \ln \frac{3.5}{0.823} \approx 40.1 \qquad \begin{array}{l}\text{Substitute 3.5 for } y \text{ in } t = \dfrac{1}{0.0361} \ln \dfrac{y}{0.823} \text{ and} \\ \text{then calculate.}\end{array}$$

Answer Because t represents the number of years after 1980, the population will reach 3,500,000 in $1980 + 40.1 = 2020.1$, which means during 2020.

Check Use $y = 0.823e^{0.0361t}$ to verify that in 2020 ($t = 40.1$), the population will be 3,500,000 ($y = 3.5$). We will leave this check to the reader.

Your Turn 6

a. The bacteria *bacillus megaterium* increases at a rate of 4% per minute in a sucrose–salts medium. If 4500 bacteria were present initially, the number, A, present after t minutes is given by the equation $A = 4500e^{0.04t}$. (*Source: Todar's Online Textbook of Bacteriology*) How many bacteria are present after 26 minutes?

b. After how many minutes will 10,000 bacteria be present?

Exponential growth and decay can be represented by the equation $A = A_0 e^{kt}$, where A is the amount present, A_0 is the initial amount, t is the time, and k is a constant that is determined by the substance. Exponential growth is indicated if $k > 0$; exponential decay, if $k < 0$. Recall that the half-life of a substance is the amount of time until only one-half the original amount is present.

Plutonium-239 is frequently used as fuel in nuclear reactors to generate electricity. One of the problems with using plutonium is disposing of the radioactive waste, which is extremely dangerous for a very long period of time.

Answers Your Turn 6

a. 12,731 bacteria
b. Approximately 20 min.

Example 7 A nuclear reactor contains 10 kilograms of radioactive plutonium ^{239}P. Plutonium disintegrates according to the formula $A = A_0 e^{-0.0000284t}$.

a. How much will remain after 10,000 years?

Understand We are given $A_0 = 10$ kg and $t = 10{,}000$ years and asked to find A.

Plan   Use $A = A_0 e^{-0.0000284t}$.

Execute $A = 10 e^{-0.0000284(10,000)}$ **Substitute 10 for A_0 and 10,000 for t.**

$A = 10 e^{-0.284}$ **Simplify.**

$A \approx 7.53$ **Evaluate using a calculator.**

Answer   About 7.53 kilograms $\left(\text{or about } \dfrac{3}{4} \text{ of the original amount}\right)$ will remain after 10,000 years.

Check   Use $A = A_0 e^{-0.0000284t}$ to verify that it takes 10,000 years for 10 kilograms of ^{239}P to disintegrate to 7.53 kg.

$$7.53 = 10 e^{-0.0000284t} \qquad \text{\textbf{Substitute 7.53 for } } A \text{ \textbf{and 10 for} } A_0.$$

$$\frac{7.53}{10} = e^{-0.0000284t} \qquad \text{\textbf{Divide both sides by 10.}}$$

$$\ln \frac{7.53}{10} = \ln e^{-0.0000284t} \qquad \text{\textbf{Use if } } x = y; \text{ \textbf{then} } \log_b x = \log_b y.$$

$$\ln \frac{7.53}{10} = -0.0000284t \qquad \text{\textbf{Use} } \log_b b^x = x.$$

$$\frac{\ln \dfrac{7.53}{10}}{-0.0000284} = t \qquad \text{\textbf{Divide both sides by} } -0.0000284.$$

$$9989.1 \approx t \qquad \text{\textbf{Calculate.}}$$

Note: Our manipulations suggest the following formula for calculating t given A and A_0:

$$t = \frac{1}{-0.0000284} \ln \frac{A}{A_0}$$

Because 7.53 is not the exact amount remaining, we should expect that the corresponding time calculation will not be exactly 10,000 years. So 9989.1 years is reasonable.

b. Find the half-life of ^{239}P.

Understand   After one half-life, 5 kg will remain; so we must find t when $A = 5$.

Plan   Use $A = A_0 e^{-0.000284t}$ when $A = 5$ and $A_0 = 10$ and follow the same steps as in our check of part a. Or as we will do, use the formula $t = \dfrac{1}{-0.0000284} \ln \dfrac{A}{A_0}$, which we discovered as a result of checking part a.

Execute $t = \dfrac{1}{-0.0000284} \ln \dfrac{5}{10} \approx 24{,}406.59$ **Substitute 5 for A and 10 for A_0 and then calculate.**

Answer   The half-life of ^{239}P is about 24,400 years.

Check   Use $A = A_0 e^{-0.0000284t}$ to verify that in 24,400 years, 5 kg out of an initial 10 kg of ^{239}P will remain. We will leave this check to the reader.

Your Turn 7   Carbon-14 is a radioactive form of carbon that is present in all living things. Archaeologists and paleontologists frequently use carbon-14 dating in estimating the age of organic fossils. Carbon-14 disintegrates according to the formula $A = A_0 e^{-0.000121t}$.

    a. If a sample contains 5 grams of carbon-14, how much will be present after 1500 years?

    b. What is the half-life of carbon-14?

▸ **Objective ④** **Use the change-of-base formula.**   Sometimes applications involve logarithms other than common or natural logarithms. For example, earlier we were given the formula $P = 95 - 30 \log_2 x$, where P is the percent of students who recall the important features of a lecture after x days. To find the percent after 5 days, we need to calculate $\log_2 5$. Most calculators have only base-10 and base-e logarithms; so to calculate $\log_2 5$ using a calculator, we need to write $\log_2 5$ in terms of common or natural logarithms using the change-of-base formula. To derive the change-of-base formula, we let $y = \log_a x$.

Answers   **Your Turn 7**

a. 4.17 g   **b.** about 5728 yr.

$$a^y = x \qquad \text{\textbf{Write } } y = \log_a x \text{ \textbf{in exponential form.}}$$

$$\log_b a^y = \log_b x \qquad \text{\textbf{Take} } \log_b \text{ \textbf{of both sides.}}$$

$$y \log_b a = \log_b x \qquad \text{Use } \log_b x^r = r \log_b x.$$

$$y = \frac{\log_b x}{\log_b a} \qquad \text{Divide both sides by } \log_b a.$$

$$\log_a x = \frac{\log_b x}{\log_b a} \qquad \text{Substitute } \log_a x \text{ for } y.$$

Rule   **Change-of-Base Formula**

In general, if $a > 0$, $a \neq 1$, $b > 0$, $b \neq 1$, and $x > 0$, then $\log_a x = \dfrac{\log_b x}{\log_b a}$.

In terms of common and natural logarithms, $\log_a x = \dfrac{\log x}{\log a} = \dfrac{\ln x}{\ln a}$.

Example 8   Use the change-of-base formula to calculate $\log_5 19$. Round the answer to four decimal places.

Note: We could have used ln rather than log.

$$\log_5 19 = \frac{\ln 19}{\ln 5} \approx 1.8295$$

SOLUTION $\log_5 19 = \dfrac{\log 19}{\log 5} \approx 1.8295$ Use $\log_a x = \dfrac{\log_b x}{\log_b a}$; then evaluate using a calculator.

Check $5^{1.8295} = 19.0005 \approx 19$, so the answer is correct.

Your Turn 8   Find $\log_6 25$.

Example 9   Use $P = 95 - 30 \log_2 x$ and the change-of-base formula to find the percent of students who retained the main points of a lecture after 5 days.

SOLUTION $P = 95 - 30 \log_2 5$ Substitute 5 for x in $P = 95 - 30 \log_2 x$.

$$P = 95 - 30 \frac{\ln 5}{\ln 2} \qquad \text{Use } \log_a x = \frac{\log_b x}{\log_b a}.$$

$$P \approx 95 - 30(2.3219) \qquad \text{Evaluate } \frac{\ln 5}{\ln 2} \text{ using a calculator.}$$

$$P \approx 25.34 \qquad \text{Simplify.}$$

Answer   **Your Turn 8**

1.7965

About 25% of the students remember the main points of a lecture 5 days later.

Answer   **Your Turn 9**

5%

Your Turn 9   Use the formula from Example 9 to find the percent of students who retain the main points of a lecture 8 days later.

12.6 Exercises   For Extra Help   *MyMathLab*    PRACTICE   WATCH   DOWNLOAD

1. If $\log_a m = \log_a n$, then _____.

2. When solving $10^{x+2} = 45$, would natural or common logarithms be the better choice? Why?

3. What principle is used to solve $\log(x - 3) = \log(3x - 13)$?

4. What principle is used to solve $100 = (5)^{2n}$?

5. In solving $\log x + \log(x + 2) = \log 15$, we get possible solutions of $x = 3$ and $x = -5$. Why must $x = -5$ be rejected?

6. Suppose you were solving an equation that contained $\log(5 - x)$. You did all of the algebraic work correctly and obtained $x = -3$ as a possible solution. Would you have to reject $x = -3$ as a solution? Why or why not?

7. $2^x = 9$

8. $3^x = 20$

9. $5^{2x} = 32$

10. $6^{2x} = 48$

11. $5^{x+3} = 10$

12. $6^{x+4} = 38$

13. $8^{x-2} = 6$

14. $5^{x-3} = 12$

15. $4^{x+2} = 5^x$

16. $6^{x+4} = 10^x$

17. $2^{x+1} = 3^{x-2}$

18. $5^{x-3} = 3^{x+1}$

For Exercises 19–26, solve. Round your answers to four decimal places. See Example 2.

19. $e^{3x} = 5$

20. $e^{2x} = 7$

21. $e^{0.03x} = 25$

22. $e^{0.07x} = 32$

23. $e^{-0.022x} = 5$

24. $e^{-0.032x} = 8$

25. $\ln e^{4x} = 24$

26. $\ln e^{5x} = 35$

For Exercises 27–60, solve. Give exact answers. See Example 3.

27. $\log_4(x + 5) = 2$

28. $\log_3(x + 4) = 2$

29. $\log_4(4x - 8) = 2$

30. $\log_5(3x + 7) = 2$

31. $\log_4 x^2 = 2$

32. $\log_2 x^2 = 6$

33. $\log_6(x^2 + 5x) = 2$

34. $\log_4(x^2 + 6x) = 2$

35. $\log(4x - 3) = \log(3x + 4)$

36. $\log(4x + 1) = \log(2x + 7)$

37. $\ln(3x + 4) = \ln(x - 6)$

38. $\ln(5x + 6) = \ln(3x - 8)$

39. $\log_9(x^2 + 4x) = \log_9 12$

40. $\log_8(x^2 + x) = \log_8 30$

41. $\log_4 x + \log_4 8 = 2$

42. $\log_6 x + \log_6 4 = 2$

43. $\log_2 x - \log_2 5 = 1$

44. $\log_5 x - \log_5 3 = 2$

45. $\log_3 x + \log_3(x + 6) = 3$

46. $\log_2 x + \log_2(x - 3) = 2$

47. $\log_3(2x + 15) + \log_3 x = 3$

48. $\log_2(3x - 2) + \log_2 x = 4$

49. $\log_2(7x + 3) - \log_2(2x - 3) = 3$

50. $\log_3(3x + 3) - \log_3(x - 3) = 2$

51. $\log_2(3x + 8) - \log_2(x + 1) = 2$

52. $\log_3(2x + 1) - \log_3(x - 1) = 1$

53. $\log_8 2x + \log_8 6 = \log_8 10$

54. $\log_5 3x + \log_5 2 = \log_5 4$

55. $\ln x + \ln(2x - 1) = \ln 10$

56. $\ln x + \ln(3x - 5) = \ln 12$

57. $\log x - \log(x - 5) = \log 6$

58. $\log x - \log(x - 2) = \log 3$

59. $\log_6(3x + 4) - \log_6(x - 2) = \log_6 8$

60. $\log_7(5x + 2) - \log_7(x - 2) = \log_7 9$

For Exercises 61–80, solve. See Examples 4–7.

61. How long will it take $5000 invested at 5% compounded quarterly to grow to $8000? Round your answer to the nearest tenth of a year.

62. How long will it take $7000 invested at 6% compounded monthly to grow to $12,000? Round your answer to the nearest tenth of a year.

63. If $8000 is deposited into an account at 6% annual interest compounded continuously,
    a. How much money will be in the account after 15 years?
    b. How long will it take the $8000 to grow to $14,000?

64. If $4000 is deposited into an account at 5% annual interest compounded continuously,
    a. How much money will be in the account after 10 years?
    b. How long will it take the $4000 to grow to $10,000?

65. The probability that a person will have an accident while driving at a given blood alcohol level is approximated by $P(b) = e^{21.5b}$, where b is the blood alcohol level ($0 \le b \le 0.4$) and P is the percent probability of having an accident.
    a. What is the probability of an accident if the blood alcohol level is 0.08, which is legally drunk in many states?
    b. Estimate the blood alcohol level when the probability of an accident is 50%.

66. Atmospheric pressure (in pounds per square inch, psi) is a function of the altitude above sea level and can be modeled by $P(a) = 14.7e^{-0.21a}$, where P is the pressure and a is the altitude above sea level in miles.
    a. Find the atmospheric pressure at the peak of Mt. McKinley, Alaska, which is 3.85 miles above sea level.
    b. If the atmospheric pressure at the peak of Mt. Everest in Nepal is 4.68 pounds per square inch, find the height of Mt. Everest.

67. The population of a mosquito colony increases at a rate of 4% per day. If the initial number of mosquitoes is 500, the number present, A, at the end of t days is given by $A = 500e^{0.04t}$.
  a. How many mosquitoes are present after 2 weeks?

  b. Find the number of days until 10,000 mosquitoes are present.

68. A cake is removed from the oven at a temperature of 210°F and is left to cool on a counter where the room temperature is 70°F. The cake cools according to the function $T = 70 + 140e^{-0.0231t}$, where T is the temperature of the cake and t is in minutes.
  a. What is the temperature of the cake after 40 minutes?

  b. After how many minutes will the cake's temperature be 75°F?

69. The world population in 2005 was about 6.4 billion and is increasing at a rate of 1% per year. The world population after 2005 can be approximated by the equation $A = A_0 e^{0.01t}$, where A_0 is the world population in 2005 and t is the number of years after 2005.
  a. Assuming that the growth rate follows the same trend, find the world population in 2020.

  b. In what year will the world population reach 7 billion?

70. In 2004, Africa had a population of 885 million and a natural growth rate of 2.4% (2.4 times the world's growth rate). The population growth can be approximated by $A = A_0 e^{0.024t}$, where A_0 is the population in millions in 2004 and t is the number of years after 2004.

  a. Excluding immigration, what would the population of Africa be in 2020?

  b. Excluding immigration, in what year did Africa's population reach 1 billion?

71. The barometric pressure x miles from the eye of a hurricane is approximated by the function $P = 0.48 \ln(x + 1)$, where P is inches of mercury. (*Source:* A. Miller and R. Anthes, *Meteorology*, 4th Edition, Merrill Publishing)
  a. Find the barometric pressure 50 miles from the center of the hurricane.

  b. Find the distance from the center where the pressure is 1.5 inches of mercury.

72. The first two-year college, Joliet Junior College in Chicago, was founded in 1901. The number of two-year colleges in the United States grew rapidly, especially during the 1960s, but growth has tapered off. The total number of two-year colleges in the United States since 1960 can be approximated by the function $y = 175.6 \ln x + 513$, where x is the number of years after 1960 and y is the number of two-year colleges. Use the model to estimate the number of two-year colleges in the United States in 2015. (*Source:* American Association of Two-Year Colleges)

73. The percent, $f(x)$, of adult height attained by a boy who is x years old is modeled by $f(x) = 29 + 48.8 \log(x + 1)$, where $5 \le x \le 15$.
  a. At age 10, about what percent of his adult height has a boy reached?
  b. At what age will a boy attain 75% of his adult height?

74. The annual depreciation rate r of a car purchased for P dollars and worth A dollars after t years can be found by the formula $\log(1 - r) = \frac{1}{t} \log \frac{A}{P}$. Find the depreciation rate of a car that is purchased for $22,500 and is sold 3 years later for $10,000.

75. The following table shows the purchasing power of $1.00 in 1970 for subsequent years. For example, in 1980, $2.12 had the same purchasing power as $1.00 in 1970.

Year	Purchasing Power ($)
1970	1.00
1975	1.39
1980	2.12
1985	2.77
1990	3.37
1995	3.92
2000	4.43
2005	5.03
2007	5.34

Source: www.measuringworth.com

The data can be approximated by the exponential function $y = 1.222(1.044)^x$, where y is the purchasing power of $1.00 and x is the number of years after 1970. Assuming that the trend continues, find the approximate number of dollars in 2014 that it will take to have the same purchasing power as $1.00 in 1970.

76. The following table shows the U.S. public debt, in trillions of dollars, for selected years.

Year	Public Debt (in trillions)
1990	3.233
1992	4.065
1994	4.693
1996	5.224
1998	5.526
2000	5.874
2002	6.228
2004	7.379
2006	8.507
2007	9.008

Source: 2008 World Almanac and Book of Facts

The data can be approximated by the exponential function $y = 3.538(1.055)^x$, where y is the public debt and x is the number of years after 1990. Assuming that the trend continues, what will be the approximate public debt in 2016?

77. Since 1985, the amount spent on recreation in the United States can be approximated by the exponential equation $y = 78.62(1.092)^x$, where y is the amount spent in billions of dollars and x is the number of years after 1985. Assuming that the trend continues, what is the approximate amount that will be spent on recreation in 2014. (*Source:* Bureau of Economic Analysis, U.S. Department of Commerce)

78. Since 1980, the median income for males in the United States can be approximated by the exponential equation $y = 13192.9(1.0369)^x$, where y is the median income and x is the number of years after 1980. Assuming that the trend continues, find the approximate median income for a male in the United States in 2018. (*Source: 2008 World Almanac and Book of Facts*)

For Exercises 79–86, use the change-of-base formula to find the logarithms. Round your answers to four decimal places. See Example 8.

79. $\log_4 12$

80. $\log_5 23$

81. $\log_8 3$

82. $\log_9 5$

83. $\log_{1/2} 5$

84. $\log_{1/3} 4$

85. $\log_{1/4} \dfrac{3}{5}$

86. $\log_{1/3} \dfrac{4}{7}$

For Exercises 87 and 88, use $P = 95 - 30 \log_2 x$ and the change-of-base formula to find the percentage, P, of the lecture retained after the given number of days. Round your answer to the nearest whole percent. See Example 9.

87. 3 days

88. 7 days

Review Exercises

Exercises 1–6   **EXPRESSIONS**

For Exercises 1 and 2, simplify.

[5.3]   **1.** $(3x + 2) - (2x - 1)$

[10.4]   **2.** $(3\sqrt{5} + 2)(2\sqrt{5} - 1)$

For Exercises 3–6, use $f(x) = x^2 + 4$ and $g(x) = 2x - 1$.

[4.7]   **3.** Find $f(-5)$.

[8.5]   **4.** Find $(f + g)(x)$.

[12.1]   **5.** Find $(f \circ g)(x)$.

[12.1]   **6.** Find $(f \circ g)(2)$.

Chapter 12 Summary

Defined Terms

Section 12.1
Composition of functions (p. 838)
Inverse functions (p. 839)
One-to-one function (p. 841)

Section 12.2
Exponential function (p. 850)

Section 12.3
Logarithm (p. 859)

Section 12.5
Common logarithms (p. 874)
Natural logarithms (p. 876)

Procedures, Rules, and Key Examples

Procedures/Rules	Key Examples

SECTION 12.1   Composite and Inverse Functions

Composition of functions

$(f \circ g)(x) = f[g(x)]$ for all x in the domain of g for which $g(x)$ is in the domain of f.

$(g \circ f)(x) = g[f(x)]$ for all x in the domain of f for which $f(x)$ is in the domain of g.

Example 1: If $f(x) = x^2 - 3$ and $g(x) = 3x + 4$, find
(a) $(f \circ g)(x)$ and (b) $(g \circ f)(x)$.

a. $(f \circ g)(x) = f[g(x)]$

$= f(3x + 4)$ **Substitute $3x + 4$ for $g(x)$.**

$= (3x + 4)^2 - 3$ **Replace x with $3x + 4$ in $f(x)$.**

$= 9x^2 + 24x + 13$ **Simplify.**

b. $(g \circ f)(x) = g[f(x)]$

$= g(x^2 - 3)$ **Substitute $x^2 - 3$ for $f(x)$.**

$= 3(x^2 - 3) + 4$ **Replace x with $x^2 - 3$ in $g(x)$.**

$= 3x^2 - 5$ **Simplify.**

To determine whether two functions f and g are inverses of each other,

1. Show that $f[g(x)] = x$ for all x in the domain of g.
2. Show that $g[f(x)] = x$ for all x in the domain of f.

Example 2: Show that $f(x) = 5x - 6$ and $g(x) = \dfrac{x + 6}{5}$ are inverse functions.

$$f[g(x)] = f\left(\frac{x + 6}{5}\right)$$

$$= 5\left(\frac{x + 6}{5}\right) - 6$$

$$= x + 6 - 6 = x$$

$$g[f(x)] = g(5x - 6)$$

$$= \frac{5x - 6 + 6}{5}$$

$$= \frac{5x}{5} = x$$

Because $f[g(x)] = g[f(x)] = x$, f and g are inverse functions.

Given a function's graph, the function is one to one if every horizontal line that can intersect the graph does so at one and only one point.

A function has an inverse function if and only if the function is one to one.

Example 3: Determine whether the functions whose graphs follow are one to one.

a.

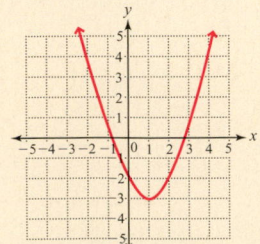

b.

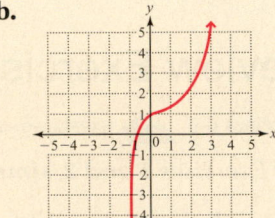

Solution: A horizontal line can intersect this graph in more than one point, so the function is not one to one.

Solution: Every horizontal line that can intersect this graph does so at one and only one point, so the function is one to one.

(continued)

Procedures/Rules	Key Examples

SECTION 12.1 Composite and Inverse Functions (*continued*)

Finding the inverse of a one-to-one function
1. If necessary, replace $f(x)$ with y.
2. Replace all x's with y's and all y's with x's.
3. Solve the equation from Step 2 for y.
4. Replace y with $f^{-1}(x)$.

Example 4: If $f(x) = 3x - 7$, find $f^{-1}(x)$.

Solution: $f(x) = 3x - 7$

$y = 3x - 7$ **Replace $f(x)$ with y.**

$x = 3y - 7$ **Interchange x and y.**

$\dfrac{x + 7}{3} = y$ **Solve for y.**

$f^{-1}(x) = \dfrac{x + 7}{3}$ **Replace y with $f^{-1}(x)$.**

Verify by showing that $f[f^{-1}(x)] = x$ and $f^{-1}[f(x)] = x$.

The graphs of f and f^{-1} are symmetric with respect to the graph of $y = x$.

Example 5: Sketch the inverse of the function whose graph follows.

Solution: Draw the line $y = x$ and reflect the graph in the line.

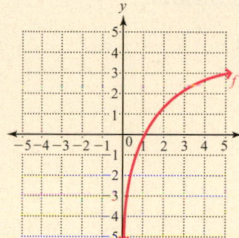 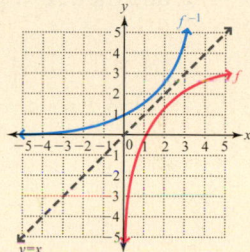

SECTION 12.2 Exponential Functions

Graphs of exponential functions

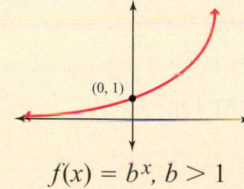 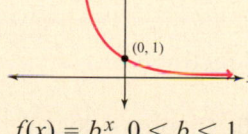

$f(x) = b^x, b > 1$ $f(x) = b^x, 0 < b < 1$

Example 1: Graph $f(x) = 2^x$.

Choose some values for x and find the corresponding values of $f(x)$, which are the y-values.

x	-2	-1	0	1	2
$f(x)$	$\frac{1}{4}$	$\frac{1}{2}$	1	2	4

Plot the points and draw the graph.

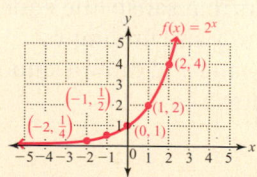

Given $b > 0$ and $b \neq 1$, if $b^x = b^y$, then $x = y$.

Solving exponential equations
1. If necessary, write both sides of the equation as a power of the same base.
2. If necessary, simplify the exponents.
3. Set the exponents equal to each other.
4. Solve the resulting equation.

Example 2: Solve $8^{x-3} = 16^{2x}$.

Solution: $8^{x-3} = 16^{2x}$

$(2^3)^{x-3} = (2^4)^{2x}$ **Rewrite 8 and 16 as powers of 2.**

$2^{3x-9} = 2^{8x}$ **Multiply exponents.**

$3x - 9 = 8x$ **Set exponents equal.**

$-\dfrac{9}{5} = x$ **Solve for x.**

(*continued*)

SECTION 12.3   Logarithmic Functions

If $b > 0$ and $b \neq 1$, then $y = \log_b x$ is equivalent to $x = b^y$.	**Example 1:** Write in logarithmic form. **a.** $3^4 = 81$ **b.** $\left(\dfrac{1}{4}\right)^{-3} = 64$ Solution: $\log_3 81 = 4$ Solution: $\log_{1/4} 64 = -3$ **Example 2:** Write in exponential form. **a.** $\log_5 125 = 3$ **b.** $\log_3 \dfrac{1}{27} = -3$ Solution: $5^3 = 125$ Solution: $3^{-3} = \dfrac{1}{27}$

To solve an equation of the form $\log_b x = y$, where b, x, or y is a variable, write the equation in exponential form, $b^y = x$, and then solve for the variable.

Example 3: Solve.

a. $\log_2 \dfrac{1}{16} = x$

Solution: $2^x = \dfrac{1}{16}$ **Change into exponential form.**

$2^x = \dfrac{1}{2^4}$ **$16 = 2^4$**

$2^x = 2^{-4}$ **$\dfrac{1}{2^4} = 2^{-4}$**

$x = -4$

b. $\log_x 49 = 2$

Solution: $x^2 = 49$ **Change into exponential form.**

$x = \pm 7$ **The square roots of 49 are ± 7.**

$x = 7$ **The base must be positive.**

For any real number b, where $b > 0$ and $b \neq 1$,
1. $\log_b b = 1$.
2. $\log_b 1 = 0$.

Example 4: Find the following logarithms.

a. $\log_8 8$ **b.** $\log_5 1$

Solution: $\log_8 8 = 1$ Solution: $\log_5 1 = 0$

To graph a function in the form $f(x) = \log_b x$,
1. Replace $f(x)$ with y and write the logarithm in exponential form $x = b^y$.
2. Find ordered pairs that satisfy the equation by assigning values to y and finding x.
3. Plot the ordered pairs and draw a smooth curve through the points.

Example 5: Graph $y = \log_3 x$.

Solution: Write as $3^y = x$ and assign values to y.

x	y
$\dfrac{1}{9}$	-2
$\dfrac{1}{3}$	-1
1	0
3	1
9	2

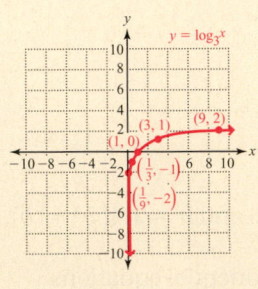

SECTION 12.4   Properties of Logarithms

For any real numbers b and x, where $b > 0$, $b \neq 1$, and $x > 0$,
1. $b^{\log_b x} = x$.
2. $\log_b b^x = x$.

Example 1: Find the value of each.

a. $6^{\log_6 5}$ **b.** $\log_8 8^a$

Solution: $6^{\log_6 5} = 5$ Solution: $\log_8 8^a = a$

(continued)

SECTION 12.4 Properties of Logarithms (*continued*)

For real numbers x, y, and b, where $x > 0$, $y > 0$, $b > 0$, and $b \neq 1$,	**Example 2:** Use the product rule.

Product Rule: $\log_b xy = \log_b x + \log_b y$

Quotient Rule: $\log_b \dfrac{x}{y} = \log_b x - \log_b y$

Power Rule: $\log_b x^r = r \log_b x$

Example 2: Use the product rule.

a. Write $\log_4 4x$ as the sum of logarithms.

Solution: $\log_4 4x = \log_4 4 + \log_4 x$

$\qquad\qquad\quad = 1 + \log_4 x \qquad$ $\log_4 4 = 1$

b. Write $\log_7 4 + \log_7 2$ as a single logarithm.

Solution: $\log_7 4 + \log_7 2 = \log_7 4 \cdot 2$

$\qquad\qquad\qquad\qquad\quad = \log_7 8$

Example 3: Use the quotient rule.

a. Write $\log_a \dfrac{w}{7}$ as the difference of logarithms.

Solution: $\log_a \dfrac{w}{7} = \log_a w - \log_a 7$

b. Write $\log_7(x + 5) - \log_7(x - 3)$ as a single logarithm.

Solution: $\log_7(x + 5) - \log_7(x - 3) = \log_7 \dfrac{x + 5}{x - 3}$

Example 4: Use the power rule.

a. Write $\log_b \sqrt[5]{x^3}$ as a multiple of a logarithm.

Solution: $\log_b \sqrt[5]{x^3} = \log_b x^{3/5} = \dfrac{3}{5} \log_b x$

b. Write $-3 \log_2 y$ as a logarithm of a quantity to a power.

Solution: $-3 \log_2 y = \log_2 y^{-3}$

$\qquad\qquad\qquad\quad = \log_2 \dfrac{1}{y^3}$

Example 5: Write $\log_a \dfrac{x^2 y}{z^4}$ as the sum or difference of multiples of logarithms.

Solution:

$$\log_a \dfrac{x^2 y}{z^4} = \log_a x^2 y - \log_a z^4 \qquad \text{\color{blue}Quotient rule}$$

$$= \log_a x^2 + \log_a y - \log_a z^4 \qquad \text{\color{blue}Product rule}$$

$$= 2 \log_a x + \log_a y - 4 \log_a z \qquad \text{\color{blue}Power rule}$$

Example 6: Write $\dfrac{1}{4}(2 \log_5 x - 3 \log_5 y)$ as a single logarithm.

Solution: $\dfrac{1}{4}(2 \log_5 x - 3 \log_5 y)$

$$= \dfrac{1}{4}(\log_5 x^2 - \log_5 y^3) \qquad \text{\color{blue}Power rule}$$

$$= \dfrac{1}{4} \log_5 \dfrac{x^2}{y^3} \qquad \text{\color{blue}Quotient rule}$$

$$= \log_5 \left(\dfrac{x^2}{y^3}\right)^{1/4} \qquad \text{\color{blue}Power rule}$$

$$= \log_5 \sqrt[4]{\dfrac{x^2}{y^3}}$$

(*continued*)

Procedures/Rules	Key Examples

SECTION 12.5   Common and Natural Logarithms

Base-10 logarithms are called common logarithms, and $\log_{10} x$ is written as $\log x$.

Common logarithms can be evaluated using the $\boxed{\text{LOG}}$ key on a calculator.

Example 1: Evaluate using a calculator and round to four decimal places.

a. $\log 356$

Solution: $\log 356 \approx 2.5514$

b. $\log 0.0059$

Solution: $\log 0.0059 \approx -2.2291$

Base-e logarithms are called natural logarithms, and $\log_e x$ is written as $\ln x$.

Natural logarithms can be evaluated using the $\boxed{\text{ln}}$ key on a calculator.

c. $\ln 72$

Solution: $\ln 72 \approx 4.2767$

d. $\ln 0.097$

Solution: $\ln 0.097 \approx -2.3330$

SECTION 12.6   Exponential and Logarithmic Equations with Applications

For any real numbers b, x, and y, where $b > 0$ and $b \neq 1$,
1. If $b^x = b^y$, then $x = y$.
2. If $x = y$, then $b^x = b^y$.
3. For $x > 0$ and $y > 0$, if $\log_b x = \log_b y$, then $x = y$.
4. For $x > 0$ and $y > 0$, if $x = y$, then $\log_b x = \log_b y$.
5. For $x > 0$, if $\log_b x = y$, then $b^y = x$.

Example 1: Solve $3^x = 4$.

Solution:
$$\log 3^x = \log 4 \qquad \text{Use property 4.}$$
$$x \log 3 = \log 4 \qquad \text{Power rule}$$
$$x = \frac{\log 4}{\log 3} \qquad \text{Divide by } \log 3.$$
$$x \approx 1.2619 \qquad \text{Evaluate.}$$

Example 2: Solve $e^{3x} = 12$.

Solution: Because the base is e, take the natural logarithm of both sides.

$$\ln e^{3x} = \ln 12 \qquad \text{Use property 4.}$$
$$3x = \ln 12 \qquad \text{Use } \log_b b^x = x.$$
$$x = \frac{\ln 12}{3} \approx 0.8283 \qquad \text{Divide by 3 and approximate.}$$

To solve equations containing logarithms, use the properties of logarithms to simplify each side of the equation and then use one of the following.

If the simplification results in an equation in the form $\log_b x = \log_b y$, use the fact that $x = y$ and then solve for the variable.

If the simplification results in an equation in the form $\log_b x = y$, write the equation in exponential form, $b^y = x$, and then solve for the variable.

Example 3: Solve $\log_3 x + \log_3(x - 3) = \log_3 10$.

Solution:
$$\log_3 x(x - 3) = \log_3 10 \qquad \text{Product rule}$$
$$x(x - 3) = 10 \qquad \text{Use property 3.}$$
$$x^2 - 3x - 10 = 0 \qquad \text{Multiply then subtract 10.}$$
$$(x - 5)(x + 2) = 0 \qquad \text{Factor.}$$
$$x - 5 = 0 \quad \text{or} \quad x + 2 = 0 \qquad \text{Use the zero-factor theorem.}$$
$$x = 5 \quad \text{or} \qquad x = -2 \qquad \text{Solve for } x$$

We must reject -2 because it results in $\log_3(-2)$ and $\log_3(-5)$ in the original equation, which do not exist.

Example 4: Solve $\log_2(2x + 2) - \log_2(x - 4) = 2$.

Solution:
$$\log_2 \frac{2x + 2}{x - 4} = 2 \qquad \text{Quotient rule}$$
$$\frac{2x + 2}{x - 4} = 2^2 \qquad \text{Use property 5.}$$
$$\frac{2x + 2}{x - 4} = 4 \qquad 2^2 = 4$$
$$2x + 2 = 4x - 16 \qquad \text{Multiply by } x - 4.$$
$$18 = 2x \qquad \text{Isolate the } x \text{ term.}$$
$$9 = x \qquad \text{Solve for } x.$$

(continued)

SECTION 12.6 Exponential and Logarithmic Equations with Applications (*continued*)

Example 5: How long will it take $12,000 invested at 4% compounded quarterly to grow to $15,000?

Solution: Use $A = P\left(1 + \dfrac{r}{n}\right)^{nt}$ with $A = 15,000$, $P = 12,000$, $r = 4\% = 0.04$, and $n = 4$. Find t.

$$15,000 = 12,000\left(1 + \frac{0.04}{4}\right)^{4t} \qquad \text{Substitute.}$$

$$\frac{5}{4} = (1.01)^{4t} \qquad \text{Divide by 12,000.}$$

$$\log \frac{5}{4} = \log 1.01^{4t} \qquad \text{Use property 4.}$$

$$\log \frac{5}{4} = 4t \log 1.01 \qquad \text{Power rule}$$

$$\frac{\log \dfrac{5}{4}}{4 \log 1.01} = t \qquad \text{Divide by 4 log 1.01.}$$

$$5.61 \approx t \qquad \text{Evaluate.}$$

It will take about 5.6 years.

Example 6: If $3000 is invested in an account paying 4.5% compounded continuously, how much will be in the account after 10 years?

Solution: Use $A = Pe^{rt}$ with $P = 3000$, $r = 0.045$, and $t = 10$.

$$A = Pe^{rt}$$
$$A = 3000e^{(0.045)(10)} \qquad \text{Substitute.}$$
$$A = 3000e^{0.45} \qquad 10(0.045) = 0.45$$
$$A = 4704.94 \qquad \text{Evaluate.}$$

Exponential growth and decay can be represented by the equation $A = A_0 e^{kt}$, where A is the amount present, A_0 is the initial amount, t is the time, and k is a constant that is determined by the substance. If $k > 0$, there is exponential growth, and if $k < 0$, there is exponential decay. The half-life of a substance is the amount of time that must pass until one half of the original substance remains.

Example 7: The element bismuth has an isotope, ^{200}Bi, that disintegrates according to the formula $A = A_0 e^{-0.0198t}$, where t is in minutes. Find the half-life of ^{200}Bi.

Solution: After one half-life, $\dfrac{1}{2} A_0$ of the original A_0 grams remain.

$$A = A_0 e^{-0.0198t}$$

$$\frac{1}{2} A_0 = A_0 e^{-0.0198t} \qquad \text{Substitute } \frac{1}{2} A_0 \text{ for } A.$$

$$\frac{1}{2} = e^{-0.0198t} \qquad \text{Divide by } A_0.$$

$$\ln \frac{1}{2} = \ln e^{-0.0198t} \qquad \text{Use if } x = y; \text{ then } \log_b x = \log_b y.$$

$$\ln \frac{1}{2} = -0.0198t \qquad \text{Use } \log_b b^x = x.$$

$$\frac{\ln \dfrac{1}{2}}{-0.0198} = t \qquad \text{Divide by } -0.0198.$$

$$35 \approx t \qquad \text{Evaluate.}$$

The half-life is approximately 35 minutes.

(*continued*)

17. $f(x) = 3x + 2, g(x) = \dfrac{x-2}{3}$

18. $f(x) = x^3 + 6, g(x) = \sqrt[3]{x-6}$

19. $f(x) = x^2 - 3, x \geq 0; g(x) = \sqrt{x+2}$

20. $f(x) = \dfrac{x}{x+4}, g(x) = \dfrac{-4x}{x+1}$

[12.1] *For Exercises 21–24, find $f^{-1}(x)$ for each of the following one-to-one functions.*

21. $f(x) = 5x + 4$

22. $f(x) = x^3 + 6$

23. $f(x) = \dfrac{4}{x+5}$

24. $f(x) = \sqrt[3]{3x+2}$

[12.1] **25.** If $(4, -6)$ is an ordered pair on the graph of f, what ordered pair is on the graph of f^{-1}?

[12.1] **26.** Fill in the blanks. If $f(a) = b$. then $f^{-1}(\_\_\_) = \_\_\_\_\_$.

[12.2] *For Exercises 27–30, graph.*

27. $f(x) = 3^x$

28. $f(x) = \left(\dfrac{1}{2}\right)^x$

29. $f(x) = 2^{x-3}$

30. $f(x) = 3^{-x+2}$

[12.2] *For Exercises 31–38, solve each equation.*

31. $5^x = 625$

32. $16^x = 64$

33. $6^x = \dfrac{1}{36}$

34. $\left(\dfrac{3}{4}\right)^x = \dfrac{16}{9}$

35. $4^{x-1} = 64$

36. $5^{x+2} = 25^x$

37. $\left(\dfrac{1}{3}\right)^{-x} = 27$

38. $8^{3x-2} = 16^{4x}$

[12.2] **39.** The median doubling time for a malignant tumor is about 100 days. If there are 500 cells initially, then the number of cells, A, after t days is given by $A = A_0 2^{t/100}$. How many cells are present after one year?

[12.2] **40.** If \$25,000 is deposited into an account paying 6% interest compounded monthly, how much will be in the account after eight years?

[12.2] **41.** The radioactive isotope ^{82}R has a half-life of 107 days. How much of a 50-gram sample remains after 300 days? $\left(\text{Use } A = A_0 \left(\dfrac{1}{2}\right)^{t/h}.\right)$

[12.2] **42.** Since 1990, the hourly minimum wage can be approximated by the exponential equation $y = 4.0144(1.028)^x$, where y is the hourly minimum wage and x is the number of years after 1990. Find the approximate minimum wage in 2015.

[12.3] *For Exercises 43–46, write in logarithmic form.*

43. $7^3 = 343$

44. $4^{-3} = \dfrac{1}{64}$

45. $\left(\dfrac{3}{2}\right)^4 = \dfrac{81}{16}$

46. $11^{1/3} = \sqrt[3]{11}$

[12.4] *For Exercises 73–76, use the power rule to write the expression as a multiple of a logarithm.*

73. $\log_3 7^4$

74. $\log_a \sqrt[3]{x}$

75. $\log_4 \dfrac{1}{a^4}$

76. $\log_a \sqrt[5]{a^4}$

[12.4] *For Exercises 77 and 78, use the power rule to write the expression as the logarithm of a quantity to a power. Simplify the answer if possible.*

77. $4 \log_a x$

78. $\dfrac{3}{5} \log_a y$

[12.4] *For Exercises 79–82, write the expression as the sum or differences of multiples of logarithms.*

79. $\log_a x^2 y^3$

80. $\log_a \dfrac{c^4}{d^3}$

81. $\log_a \dfrac{x^2 y^3}{z^4}$

82. $\log_a \sqrt{\dfrac{a^3}{b^4}}$

[12.4] *For Exercises 83–86, write the expression as a single logarithm.*

83. $3 \log_x y + 5 \log_x z$

84. $3 \log_a 4 - 2 \log_a 3$

85. $\dfrac{1}{4}(2 \log_a x + 3 \log_a y)$

86. $4 \log_a(x + 5) + 2 \log_a(x - 3)$

[12.5] *For Exercises 87–90, use a calculator to approximate each logarithm to four decimal places.*

87. $\log 326$

88. $\log 0.0035$

89. $\ln 0.043$

90. $\ln 92$

[12.5] *For Exercises 91 and 92, find the exact value without using a calculator.*

91. $\log 0.00001$

92. $\ln \sqrt[4]{e}$

Exercises 93–114 ◢◣ EQUATIONS AND INEQUALITIES

[12.5] 93. The sound intensity of a clap of thunder was $10^{-3.5}$ watts per square meter. What was the decibel reading? $\left(\text{Use } d = 10 \log \dfrac{I}{I_0}, \text{ where } I_0 = 10^{-12} \text{ watts/m}^2.\right)$

[12.5] 94. Using pH $= -\log[H_3O^+]$ (the hydronium ion concentration), find the pH of an apple whose $[H_3O^+]$ is 0.001259.

[12.5] 95. How long will it take $6000 to grow to $10,000 if it is invested at 3% annual interest compounded continuously? (Use $A = Pe^{rt}$.)

[12.5] 96. Using $R = \log \dfrac{I}{I_0}$, for Richter scale readings, compare the intensity of an, earthquake whose Richter scale reading was 7.8 with one whose reading was 6.8. What do you notice? (*Hint:* Solve for I in terms of I_0.)

[12.6] *For Exercises 97–102, solve. Round your answers to four decimal places.*

97. $9^x = 32$

98. $3^{5x} = 19$

99. $6^{2x-1} = 22$

100. $4^{2x-3} = 5^{x+1}$

101. $e^{4x} = 11$

102. $e^{-0.003x} = 5$

[12.6] *For Exercises 103–110, solve. Give exact answers.*

103. $\log_4(x + 8) = 2$

104. $\log_2(x^2 + 2x) = 3$

105. $\log(3x - 8) = \log(x - 2)$

106. $\log 5 + \log x = 2$

107. $\log_3 x - \log_3 4 = 2$

108. $\log_2 x + \log_2(x - 6) = 4$

109. $\log_4 x + \log_4(x + 2) = \log_4 8$

110. $\log_3(5x + 2) - \log_3(x - 2) = 2$

[12.6] 111. How long will it take $15,000 invested at 3% compounded monthly to grow to $18,000? Round your answer to the nearest tenth of a year. $\left(\text{Use } A = P\left(1 + \dfrac{r}{n}\right)^{nt}.\right)$

[12.6] 112. If $7000 is deposited at 7% annual interest compounded continuously,
  a. How much will be in the account after eight years? (Use $A = Pe^{rt}$.)
  b. How long will it take until $12,000 is in the account?

[12.6] 113. The population of an ant colony is 400 and is increasing at the rate of 3% per month. Use $A = A_0 e^{0.03t}$ to answer the following questions.
  a. How many ants will be in the colony after one year?
  b. After how many months will 800 ants be in the colony?

[12.6] 114. Bacteria reproduce by cell division, and the amount of time required for the cells to divide is called the generation time, G. The generation time is found using the equation $G = \dfrac{t}{3.3 \log \dfrac{b}{B}}$, where B is the number of bacteria at the beginning of the time interval, b is the number of bacteria at the end of the time interval, and t is the time interval in minutes. Find the generation time of a bacteria population that increases from 1000 to 1,000,000 cells in 4 hours. (*Source: Todar's Online Textbook of Bacteriology*)

Exercises 115–116   EXPRESSIONS

[12.6] *For Exercises 115 and 116, use the change-of-base formula to approximate each logarithm to four decimal places.*

115. $\log_4 15$

116. $\log_{1/2} 6$

Chapter 12 Practice Test

For Extra Help Step-by-step test solutions are found on the Chapter Test Prep Videos available via the Video Resources on DVD, in *MyMathLab*, and on YouTube (search "CarsonCombinedAlg" and click on "Channels").

1. If $f(x) = x^2 - 6$ and $g(x) = 3x - 5$, find $f[g(x)]$.

2. Use $f(x) = 4x - 3$ to answer a–c.
  a. Find $f^{-1}(x)$
  b. Verify that $f[f^{-1}(x)] = x$ and $f^{-1}[f(x)] = x$.
  c. What is the relationship between the graphs of $f(x)$ and $f^{-1}(x)$?

3. The graph of a function f is shown to the right. Graph f^{-1}.

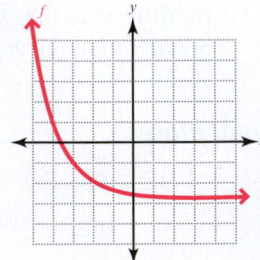

4. Graph $f(x) = 2^{x-1}$.

5. Solve $32^{x-2} = 8^{2x}$.

6. A sum of $20,000 is invested at 5% compounded quarterly.
  a. Find the amount in the account after 15 years.
  b. Find the number of years until $30,000 is in the account.

7. The isotope ^{98}Nb has a half-life of 30 minutes.
  a. How much of a 100-gram sample remains after 3.6 hours?
  b. How long will it take until only 20 grams are remaining?

8. The number of people in the United States aged 65 and over has increased rapidly since 1900 and can be approximated by $y = 3.17(1.026)^x$, where y is the number in millions aged 65 and older and x is the number of years after 1900. (*Source:* U.S. Bureau of the Census)
   a. Find the number aged 65 and older in 1960. In 2012.

   b. In what year were 31.4 million people aged 65 and older?

9. Write $\log_{1/3} 81 = -4$ in exponential form.

For Exercises 10 and 11, solve. Give exact answers.

10. $\log_6 \dfrac{1}{216} = x$

11. $\log_x 625 = 4$

12. Graph $f(x) = \log_2 x$.

For Exercises 13 and 14, write as the sum or difference of multiples of logarithms.

13. $\log_b \dfrac{x^4 y^2}{z}$

14. $\log_b \sqrt[4]{\dfrac{x^5}{y^7}}$

15. Write $\dfrac{3}{4}(2 \log_b x + 3 \log_b y)$ as a single logarithm.

16. The isotope ^{119}Sn disintegrates according to the function $A = A_0 e^{-0.0028t}$, where t is the time in days.
    a. How much of a 300-gram sample remains after 500 days?
    b. What is the half-life of ^{119}Sn?

For Exercises 17–19, solve. If necessary, use a calculator to approximate to four decimal places.

17. $6^{x-3} = 19$

18. $\log_3 x + \log_3(x + 6) = 3$

19. $\log(4x + 2) - \log(3x - 2) = \log 2$

20. The number of generations, n, of a population of bacteria cells is given by $n = 3.3 \log \dfrac{b}{B}$, where B is the number of bacteria cells at the beginning of a time interval and b is the number at the end of the time interval. (*Source: Todar's Online Textbook of Bacteriology*)
    a. Find the number of generations in a bacteria population that increases from 100 to 10,000,000 cells.
    b. Find the number of bacteria at the end of the time interval if the number at the beginning is 100 and there are 9.9 generations.

Chapters 1–12 Cumulative Review Exercises

For Exercises 1–3, answer true or false.

[12.3] 1. If $f(x) = 3^x$, then $f^{-1}(x) = \log_3 x$.

[10.2] 2. The radical expression $\sqrt[4]{x^3}$ can be written exponentially as $x^{3/4}$.

[8.4] 3. The range of $f(x) = x^2 + 4$ is $(-\infty, 4]$.

For Exercises 4–6, fill in the blank.

[5.1]   4. When 0.0000000135 is written in scientific notation, the exponent of 10 is

_____.

[6.7]   5. The solutions of $2x^2 - 9x + 10 = 0$ are the _____ of the graph of
$f(x) = 2x^2 - 9x + 10$.

[8.2]   6. If $|x - 3| = 4$, then _____ or _____.

Exercises 7–16   EXPRESSIONS

For Exercises 7–14, simplify. Write your answers with positive exponents only.

[1.5]   7. $6 - 8 \div 2 \cdot 3^2 - 4(3 - 4 \cdot 2^3)$

[5.6]   8. $\dfrac{(4a^{-4})^3(2a^2)^{-3}}{(2a^{-2})^3}$

[5.6]   9. $\dfrac{21p^2q^4 - 14p^5q^3 + 6p^3q^7}{7p^3q^5}$

[7.4]   10. $\dfrac{2x + 3}{x^2 + 6x + 9} - \dfrac{x + 4}{x + 3}$

[10.3]   11. $4\sqrt{6c^3} \cdot 3\sqrt{10c^5}$

[10.7]   12. $(4 + 3i)(2 - 4i)$

[12.4]   13. $\log_b b^{2x}$

[2.1]   14. $\dfrac{d^2 - d - 12}{2d^2} \div \dfrac{3d^2 + 13d + 12}{d}$

For Exercises 15 and 16, factor completely.

[6.4]   15. $x^4 - 9x^2 - 4x^2y^2 + 36y^2$

[6.3]   16. $24x^4y - 30x^3y^2 + 9x^2y^3$

Exercises 17–30   EQUATIONS AND INEQUALITIES

For Exercises 17–25, solve. Identify any extraneous solutions.

[8.2]   17. $|2x - 3| - 5 = -4$

[8.3]   18. $2|6x - 10| > -8$

[9.3]   19. $\begin{cases} 4x - 5y = -22 \\ 3x + 2y = -5 \end{cases}$

[11.3]   20. $(2x - 3)^2 - 2(2x - 3) = 8$

[7.6]   21. $\dfrac{x}{x - 2} - \dfrac{4}{x - 1} = \dfrac{2}{x^2 - 3x + 2}$

[10.6]   22. $\sqrt{x + 4} - \sqrt{x - 4} = 4$

[12.2]   23. $4^{x+2} = 8$

[12.3]   24. $\dfrac{1}{2}\log_2 x = 2$

[12.6]   25. $\log(3x - 5) + \log x = \log 12$

[4.4]   26. For $2x - 5y = 10$,
   a. find the slope.
   b. find the intercepts.
   c. draw the graph.

For Exercises 27–30, solve.

[6.6]   27. A frame is in the shape of a right triangle as shown. Find the length of each side.

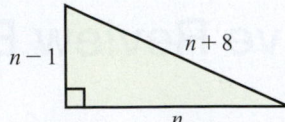

[12.2]   28. If $15,000 is deposited into an account paying 4% compounded monthly, how much will be in the account after three years?

[7.7]   29. Ethan and Pam operate a landscaping company. Ethan can mow, edge, and trim hedges on a quarter-acre lot in 2 hours. Pam takes 1.5 hours to do the same work. How much time would it take them working together?

[3.4, 9.3]   30. A car travels at an average speed that is 10 miles per hour more than the average speed of a bus. If the bus takes 1 hour longer to travel 300 miles, find how long it takes the car to travel 300 miles.

Conic Sections

13

❝The universe . . . is written in the language of mathematics, and its characters are triangles, circles, and other geometrical figures.❞

—Galileo Galilei (1564 –1642)
Italian mathematician, astronomer, physicist, and philosopher. *Il Saggiatore* (1623)

❝Mankind is not a circle with a single center but an ellipse with two focal points of which facts are one and ideas the other.❞

—Victor Hugo (1802–1885)
French poet, dramatist, novelist
Les Miserables (1862)

In this chapter, we expand our understanding of equations, inequalities, and graphs and learn about conic sections. We see that the conic sections are parabolas, circles, ellipses, and hyperbolas. These shapes are called conic sections because they are formed by the intersection of a cone with a plane. Later, we study the equations of conic sections, nonlinear systems, and nonlinear inequalities.

13.1 Parabolas and Circles

Objectives

1. Graph parabolas of the form $x = a(y - k)^2 + h$.
2. Find the distance and midpoint between two points.
3. Graph circles of the form $(x - h)^2 + (y - k)^2 = r^2$.
4. Find the equation of a circle with a given center and radius.
5. Graph circles of the form $x^2 + y^2 + dx + ey + f = 0$.

The intersection of a plane with a cone will be a circle, an ellipse, a parabola, or a hyperbola. For that reason, these curves are called **conic sections** or **conics**.

Definition   *Conic section:* A curve in a plane that is the result of intersecting the plane with a cone. More specifically, a circle, an ellipse, a parabola, or a hyperbola.

Circle

Ellipse

Parabola

Hyperbola

Recall from Section 11.4 that we graphed parabolas in the form $y = a(x - h)^2 + k$. The graph opened upward if $a > 0$, opened downward if $a < 0$, had a vertex at (h, k), and had $x = h$ as the axis of symmetry.

Example 1   For $y = 2(x - 3)^2 + 1$, determine whether the graph opens upward or downward, find the vertex and axis of symmetry, and draw the graph.

SOLUTION The graph opens upward because $a = 2$ and 2 is positive. We compare the equation with the form $y = a(x - h)^2 + k$ and observe that the vertex is at the point with coordinates $(3, 1)$ and the axis of symmetry is $x = 3$. Plot a few points on either side of the axis of symmetry by letting x have values on either side of 3 and finding y.

x	y
2	3
1	9
4	3
5	9

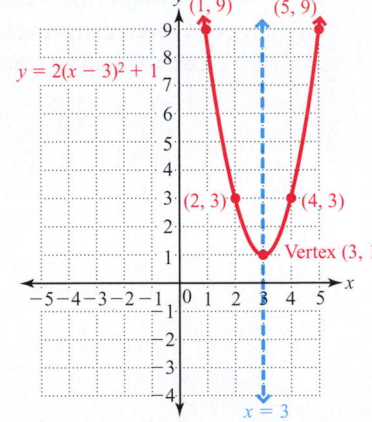

Answer   **Your Turn 1**

opens downward; vertex: $(1, 3)$;
axis of symmetry: $x = 1$

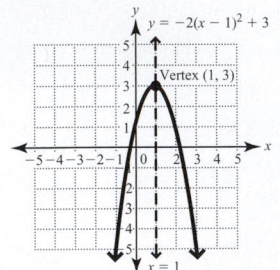

Your Turn 1   For $y = -2(x - 1)^2 + 3$, determine whether the graph opens upward or downward, find the vertex and axis of symmetry, and draw the graph.

Objective 1   **Graph parabolas of the form $x = a(y - k)^2 + h$.**   If we interchange x and y in the equations of parabolas that open upward and downward, we get the equations of parabolas that open to the left or right. To keep the vertex at (h, k), we also interchange h and k.

Rule **Equations of Parabolas Opening Left or Right**

The graph of an equation in the form $x = a(y - k)^2 + h$ is a parabola with vertex at (h, k). The parabola opens to the right if $a > 0$ and to the left if $a < 0$. The equation of the axis of symmetry is $y = k$.

> **Note:** Parabolas that open to the left or right are not functions.

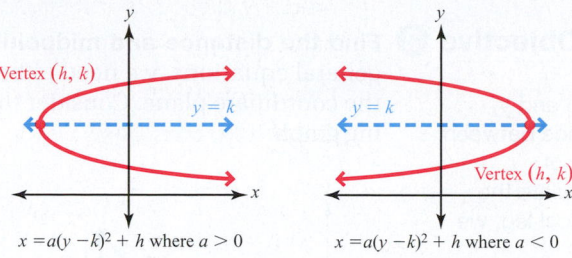

$x = a(y - k)^2 + h$ where $a > 0$ $x = a(y - k)^2 + h$ where $a < 0$

Example 2 For each equation, determine whether the graph opens left or right, find the vertex and axis of symmetry, and draw the graph.

a. $x = -2(y + 3)^2 - 2$

SOLUTION This parabola opens to the left because $a = -2$, which is negative. Rewrite the equation as $x = -2(y - (-3))^2 - 2$. Comparing this equation with $x = a(y - k)^2 + h$, we see that $h = -2$ and $k = -3$. The vertex is at the point with coordinates $(-2, -3)$, and the axis of symmetry is $y = -3$. To graph, plot a few points on either side of the axis of symmetry by letting y equal values on either side of -3 and finding x.

> **Note:** In choosing y-values we went one unit above the axis of symmetry, then two above, then one below, then two below, etc.

x	y
-4	-2
-10	-1
-4	-4
-10	-5

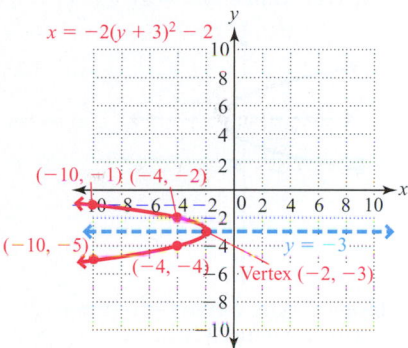

b. $x = 3y^2 + 12y + 8$

SOLUTION This parabola opens to the right because $a = 3$, which is positive. To find the vertex and axis of symmetry, we need to write the equation in the form $x = a(y - k)^2 + h$.

$x = 3y^2 + 12y + 8$	Original equation
$x - 8 = 3y^2 + 12y$	Subtract 8 from both sides.
$x - 8 = 3(y^2 + 4y)$	Factor out the common factor, 3.
$x - 8 + 12 = 3(y^2 + 4y + 4)$	Complete the square. Note that we added $3 \cdot 4 = 12$ to both sides of the equation.
$x + 4 = 3(y + 2)^2$	Simplify the left side and factor the right side.
$x = 3(y + 2)^2 - 4$	Subtract 4 from both sides of the equation.

The vertex is at the point with coordinates $(-4, -2)$, and the axis of symmetry is $y = -2$.

To complete the graph, let y equal values on either side of -2 and find x.

x	y
-1	-1
8	0
-1	-3
8	-4

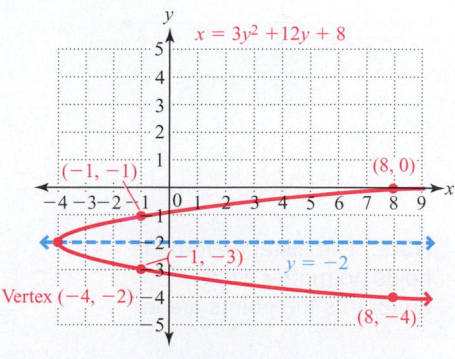

Your Turn 2 For each equation, determine whether the graph opens left or right, find the vertex and axis of symmetry, and draw the graph.

    a. $x = -(y - 2)^2 + 1$ **b.** $x = y^2 + 4y + 3$

Objective 2 **Find the distance and midpoint between two points.** To derive the other conic's general equations, we need to be able to find the distance between any two points in the coordinate plane. Consider the two points (x_1, y_1) and (x_2, y_2) shown in the following graph.

Note: If y_2 were 6 and y_1 were 2, the distance between would be $6 - 2 = 4$. Therefore, to calculate the length of the vertical leg, we calculate $y_2 - y_1$.

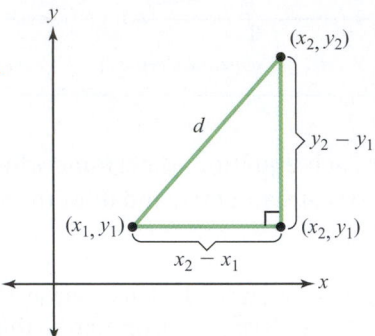

Connection Because the distance between the two points is measured along a line segment that is the hypotenuse of a right triangle, we can use the Pythagorean theorem (see below) to find the distance.

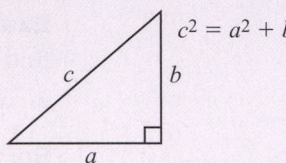

$$c^2 = a^2 + b^2$$

The lengths of the legs of the triangle are $x_2 - x_1$ and $y_2 - y_1$, as illustrated.

Answers Your Turn 2

a. opens left; vertex: $(1, 2)$; axis of symmetry: $y = 2$

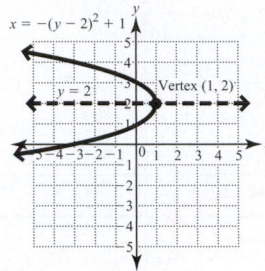

b. opens right; vertex: $(-1, -2)$; axis of symmetry: $y = -2$

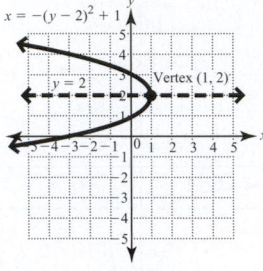

Note: If x_2 were 4 and x_1 were 1, the distance between would be $4 - 1 = 3$. Therefore, to calculate the length of the horizontal leg, we calculate $x_2 - x_1$.

    Now we can use the Pythagorean theorem, replacing a with $x_2 - x_1$, b with $y_2 - y_1$, and c with d.

$$d^2 = (x_2 - x_1)^2 + (y_2 - y_1)^2$$
$$d = \pm\sqrt{(x_2 - x_1)^2 + (y_2 - y_1)^2} \quad \text{Use the square root principle to isolate } d.$$

Because d is a distance, it must be positive; so we use only the positive value.

Rule **Distance Formula**

The distance, d, between two points with coordinates (x_1, y_1) and (x_2, y_2) can be found using the formula

$$d = \sqrt{(x_2 - x_1)^2 + (y_2 - y_1)^2}.$$

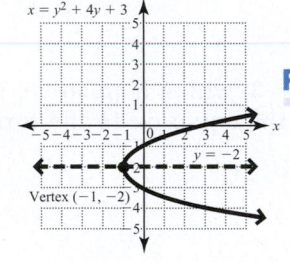

Example 3 Find the distance between $(4, 2)$ and $(-3, -1)$. If the distance is an irrational number, also give a decimal approximation rounded to three places.

SOLUTION $d = \sqrt{(x_2 - x_1)^2 + (y_2 - y_1)^2}$ **Use the distance formula.**

           $d = \sqrt{(-3 - 4)^2 + (-1 - 2)^2}$ **Let $(4, 2) = (x_1, y_1)$ and $(-3, -1) = (x_2, y_2)$.**

           $d = \sqrt{(-7)^2 + (-3)^2}$

           $d = \sqrt{49 + 9}$

           $d = \sqrt{58}$

           $d \approx 7.616$

Note: It doesn't matter which ordered pair is (x_1, y_1) and which is (x_2, y_2). Consider Example 3 again with $(-3, -1)$ as (x_1, y_1) and $(4, 2)$ as (x_2, y_2):

$$d = \sqrt{(4 - (-3))^2 + (2 - (-1))^2}$$
$$= \sqrt{(7)^2 + (3)^2} = \sqrt{49 + 9}$$
$$= \sqrt{58} \approx 7.616$$

Note: The distance formula holds no matter what quadrants the points are in.

Your Turn 3 Determine the distance between the given points. If the distance is an irrational number, also give a decimal approximation rounded to three places.

    a. $(8, 2)$ and $(3, -4)$ **b.** $(6, -5)$ and $(0, -1)$

On a line segment, the *midpoint* is located equally distant from each of the endpoints. Equivalently, the distance from either endpoint to the midpoint is half the length of the line segment.

Rule   **Midpoint Formula**

If the coordinates of the endpoints of a line segment are (x_1, y_1) and (x_2, y_2), the coordinates of the midpoint are

$$\left(\frac{x_1 + x_2}{2}, \frac{y_1 + y_2}{2} \right).$$

To find the coordinates of the midpoint of a line segment, we average the x-coordinates and the y-coordinates of the endpoints.

Example 4   Find the midpoint of the line segment whose endpoints are $(-3, 2)$ and $(5, 6)$.

SOLUTION   Midpoint $= \left(\dfrac{x_1 + x_2}{2}, \dfrac{y_1 + y_2}{2} \right)$ **Use the midpoint formula.**

$= \left(\dfrac{-3 + 5}{2}, \dfrac{2 + 6}{2} \right)$ **Let $(-3, 2) = (x_1, y_1)$ and $(5, 6) = (x_2, y_2)$.**

$= \left(\dfrac{2}{2}, \dfrac{8}{2} \right)$

$= (1, 4)$

Your Turn 4   Find the midpoint of the line segments whose endpoints are given.

    a. $(4, -5), (-6, -1)$ **b.** $(3, 1), (-2, -7)$

Objective ③ Graph circles of the form $(x - h)^2 + (y - k)^2 = r^2$. The second conic section that we consider is the **circle** with **radius r**.

Definitions   *Circle:* A set of points that are equally distant from a central point. The central point is the center.

Radius: The distance from the center of a circle to any point on the circle.

If the center of a circle is (h, k) and the radius is r, we can use the distance formula to derive the equation of the circle. If (x, y) is any point on the circle, the distance between (x, y) and (h, k) must be the radius, r.

Answers Your Turn 3
a. $\sqrt{61} \approx 7.810$
b. $2\sqrt{13} \approx 7.211$

$\sqrt{(x_2 - x_1)^2 + (y_2 - y_1)^2} = d$

$\sqrt{(x - h)^2 + (y - k)^2} = r$ **Substitute (x, y) for (x_2, y_2), (h, k) for (x_1, y_1) and r for d.**

Answers Your Turn 4
a. $(-1, -3)$ **b.** $\left(\dfrac{1}{2}, -3 \right)$

$(x - h)^2 + (y - k)^2 = r^2$ **Square both sides.**

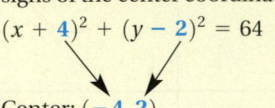

Rule   Standard Form of the Equation of a Circle

The equation of a circle with center at (h, k) and radius r is $(x - h)^2 + (y - k)^2 = r^2$.

Note: If the center of a circle is at the origin, then $(h, k) = (0, 0)$ and the equation of the circle becomes $(x - 0)^2 + (y - 0)^2 = r^2$, which simplifies to $x^2 + y^2 = r^2$.

Example 5   Find the center and radius of each circle and draw the graph.

a. $(x - 3)^2 + (y + 2)^2 = 36$

SOLUTION $(x - 3)^2 + (y - (-2))^2 = 6^2$ Write in the form $(x - h)^2 + (y - k)^2 = r^2$.

Because $h = 3$ and $k = -2$, the center is $(3, -2)$.
Because $36 = 6^2$, the radius is 6.

Note: To find the radius, we evaluate the square root of 36. Because the radius is a distance, we give only the principal square root.

$$r = \sqrt{36} = 6$$

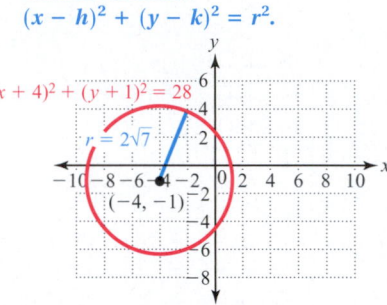

b. $(x + 4)^2 + (y + 1)^2 = 28$

SOLUTION $(x - (-4))^2 + (y - (-1))^2 = (\sqrt{28})^2$ Write in the form $(x - h)^2 + (y - k)^2 = r^2$.

Because $h = -4$ and $k = -1$, the center is $(-4, -1)$.
For this radius, $r = \sqrt{28} = 2\sqrt{7} \approx 5.292$.

Answers   Your Turn 5

a. center: $(3, -5)$; radius: 2

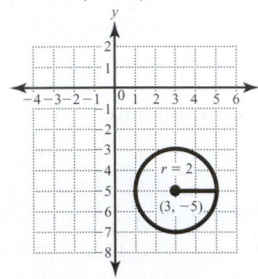

b. center: $(-1, 1)$; radius: $3\sqrt{2}$

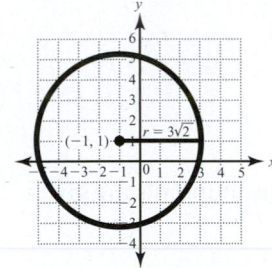

Your Turn 5   Find the center and radius of each circle and draw the graph.

a. $(x - 3)^2 + (y + 5)^2 = 4$ **b.** $(x + 1)^2 + (y - 1)^2 = 18$

Objective ④   Find the equation of a circle with a given center and radius.   We can also use the standard form of a circle to write equations of circles.

Example 6   Write the equation of each circle in standard form.

a. Center: $(-4, 2)$; radius: 8

LEARNING Strategy

VISUAL Note how the signs of the numbers in the parentheses of the general equation are opposite the signs of the center coordinates.

$(x + 4)^2 + (y - 2)^2 = 64$

Center: $(-4, 2)$

SOLUTION   Because the center is at $(-4, 2)$, $h = -4$ and $k = 2$. Also $r = 8$.

$$(x - h)^2 + (y - k)^2 = r^2 \quad \text{Standard form of a circle.}$$
$$(x - (-4))^2 + (y - 2)^2 = 8^2 \quad \text{Substitute for } h, k, \text{ and } r.$$
$$(x + 4)^2 + (y - 2)^2 = 64 \quad \text{Simplify.}$$

b. Center: $(0, 0)$; radius: 2

SOLUTION   Because the center is at $(0, 0)$, the standard form of the equation is $x^2 + y^2 = r^2$.

$$x^2 + y^2 = 2^2 \quad \text{Substitute for } r.$$
$$x^2 + y^2 = 4 \quad \text{Simplify.}$$

Answers **Your Turn 6**

a. $(x - 4)^2 + (y + 2)^2 = 25$

b. $x^2 + y^2 = 81$

Your Turn 6 Write the equation of each circle in standard form.

a. Center: $(4, -2)$; radius: 5   **b.** Center: $(0, 0)$; radius: 9

Objective ⑤ **Graph circles of the form $x^2 + y^2 + dx + ey + f = 0$.** If the equation of a circle is not given in standard form, we complete the square to write the equation in the form $(x - h)^2 + (y - k)^2 = r^2$.

Example 7 Find the center and radius of the circle whose equation is $x^2 + y^2 - 6x + 8y + 9 = 0$ and draw the graph.

SOLUTION $x^2 + y^2 - 6x + 8y = -9$ Subtract 9 from both sides to isolate the variable terms.

$(x^2 - 6x) + (y^2 + 8y) = -9$ Group x and y terms.

$(x^2 - 6x + 9) + (y^2 + 8y + 16) = -9 + 9 + 16$ Complete the square in x and y by adding 9 and 16 to both sides of the equation.

$(x - 3)^2 + (y + 4)^2 = 16$ Factor and simplify.

The center is at $(3, -4)$, and the radius is 4.

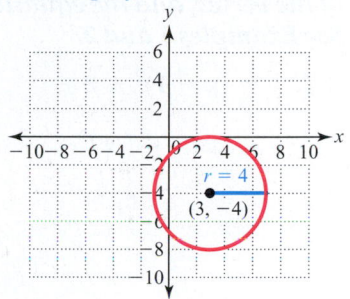

Answers **Your Turn 7**

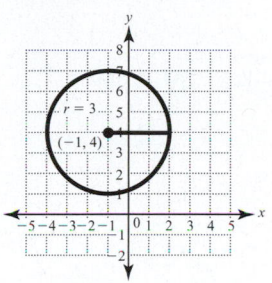

center: $(-1, 4)$; radius: 3

Your Turn 7 Find the center and radius of the circle whose equation is $x^2 + y^2 + 2x - 8y + 8 = 0$ and draw the graph.

Note: Some calculators automatically insert the left parenthesis when the root funtion is used; others do not.

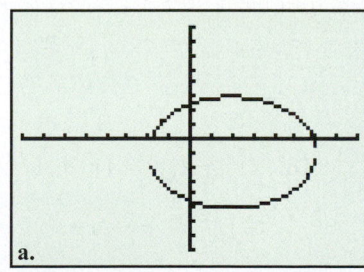

a.

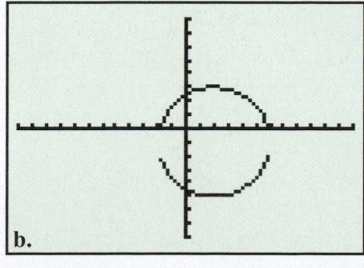

b.

🖩 Calculator Tips

The graph of a circle fails the vertical line test, so the equation of a circle is not a function. To graph equations that are not functions, we solve for y and graph the two resulting functions on the same screen.

Example Graph $(x - 2)^2 + (y + 1)^2 = 16$.

SOLUTION We solve $(x - 2)^2 + (y + 1)^2 = 16$ for y.

$(x - 2)^2 + (y + 1)^2 = 16$

$(y + 1)^2 = 16 - (x - 2)^2$ Subtract $(x - 2)^2$ from both sides.

$y + 1 = \pm\sqrt{16 - (x - 2)^2}$ Apply the square root principle.

$y = -1 \pm \sqrt{16 - (x - 2)^2}$ Subtract 1 from both sides.

*This equation defines two functions. On your graphing calculator, define $Y_1 = -1 + \sqrt{(16 - (x - 2)^2)}$ and $Y_2 = -1 - \sqrt{(16 - (x - 2)^2)}$ and graph both in the window $[-8, 8]$ for x and $[-8, 8]$ for y. The resulting graph is labeled "**a**"; note that the graph does not look like the graph of a circle. To make the graph look like a circle, select ZSquare from the ⎡ZOOM⎤ menu. This results in the graph labeled "**b**."*

For Extra Help *MyMathLab*

Note: Exercises marked with a ★ represent challenging exercises.

1. Is $x^2 + 2x - 3 + y = 0$ the equation of a circle or a parabola? Explain.

2. Find the coordinates of the vertex of the parabola $x = 3y^2 - 12y + 2$.

3. In which direction does the graph of the parabola in Exercise 2 open? Why?

4. Explain how the Pythagorean theorem is used to derive the distance formula.

5. Complete the following: The set of all points that are 4 units from the point whose coordinates are $(-3, 2)$ is a(n) _____ with center at _____ and radius of _____.

6. How is the distance formula related to the equation of a circle?

For Exercises 7–28, find the direction the parabola opens, the coordinates of the vertex, and the equation of the axis of symmetry and draw the graph. See Examples 1 and 2.

7. $y = (x - 1)^2 + 2$

8. $y = (x - 2)^2 + 3$

9. $y = -x^2 - 2x + 3$

10. $y = -x^2 + 4x - 1$

11. $x = (y + 2)^2 - 2$

12. $x = (y + 3)^2 + 2$

13. $x = -(y - 1)^2 + 3$

14. $x = -(y - 2)^2 - 1$

15. $x = 2(y + 2)^2 - 4$

16. $x = 3(y + 3)^2 - 4$

17. $x = -3(y + 2)^2 - 5$

18. $x = -2(y - 4)^2 + 1$

19. $x = y^2 + 4y + 3$ **20.** $x = y^2 - 2y - 3$ **21.** $x = -y^2 + 6y - 5$ **22.** $x = -y^2 - 4y - 3$

23. $x = 2y^2 + 8y + 3$ **24.** $x = 2y^2 - 4y + 1$ **25.** $x = 3y^2 - 6y + 1$ **26.** $x = 3y^2 - 12y + 9$

27. $x = -2y^2 + 4y + 5$ **28.** $x = -3y^2 - 6y - 2$

For Exercises 29–32, match the equation with the correct graph.

29. $x = (y + 3)^2 - 2$ **30.** $y = (x + 3)^2 - 2$ **31.** $y = 2x^2 - 8x + 5$ **32.** $x = 2y^2 - 8y + 5$

a.
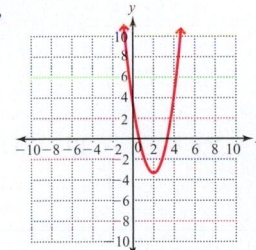

b.
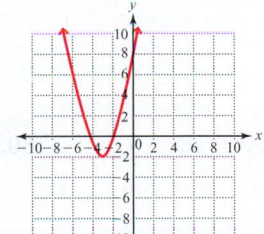

c.
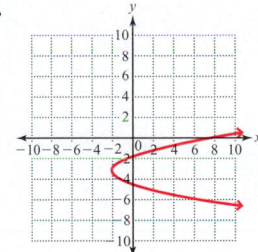

d.
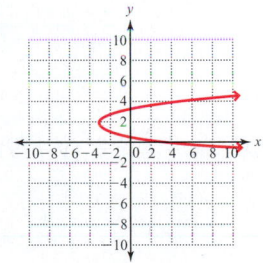

For Exercises 33–44, find the distance and midpoint between the two points.
See Examples 3 and 4.

33. $(-4, 2)$ and $(-1, 6)$ **34.** $(5, -1)$ and $(1, 2)$ **35.** $(-8, -4)$ and $(-3, 8)$ **36.** $(-3, 2)$ and $(3, -6)$

37. $(-8, -10)$ and $(4, -5)$ **38.** $(4, -6)$ and $(10, 2)$ **39.** $(2, 4)$ and $(4, 8)$ **40.** $(-3, 2)$ and $(1, -4)$

41. $(-5, 2)$ and $(3, -2)$ **42.** $(3, -4)$ and $(7, 2)$ **43.** $(6, -2)$ and $(1, -5)$ **44.** $(3, -6)$ and $(-2, 1)$

For Exercises 45–48, the coordinates of the center of a circle and a point on the circle are given. Find the radius of the circle.

45. Center: $(4, 2)$; point on the circle: $(8, -1)$

46. Center: $(-4, 6)$; point on the circle: $(2, -2)$

47. Center: $(2, -6)$; point on the circle: $(10, -1)$

48. Center: $(3, -4)$; point on the circle: $(6, 8)$

For Exercises 49–64, find the center and radius and draw the graph. See Examples 5 and 7.

49. $(x - 2)^2 + (y - 1)^2 = 4$

50. $(x - 1)^2 + (y - 3)^2 = 25$

51. $(x + 3)^2 + (y + 2)^2 = 81$

52. $(x + 5)^2 + (y + 4)^2 = 36$

53. $(x - 5)^2 + (y + 3)^2 = 49$

54. $(x + 4)^2 + (y - 2)^2 = 16$

55. $(x - 1)^2 + (y + 1)^2 = 12$

56. $(x + 1)^2 + (y - 3)^2 = 18$

57. $(x + 4)^2 + (y + 2)^2 = 32$

58. $(x - 6)^2 + (y + 5)^2 = 8$

59. $x^2 + y^2 + 8x - 6y + 16 = 0$

60. $x^2 + y^2 - 2x - 6y - 39 = 0$

61. $x^2 + y^2 + 10x - 4y - 35 = 0$

62. $x^2 + y^2 + 12x + 10y + 60 = 0$

63. $x^2 + y^2 + 14x - 4y + 49 = 0$

64. $x^2 + y^2 + 8x - 10y + 16 = 0$

For Exercises 65–68, match the equation with the correct graph.

65. $(x - 2)^2 + (y + 3)^2 = 25$

66. $(x + 2)^2 + (y - 3)^2 = 25$

67. $x^2 + y^2 + 8x + 2y - 8 = 0$

68. $x^2 + y^2 + 2x - 8y - 8 = 0$

a. 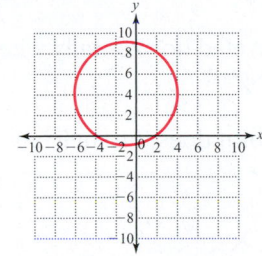   **b.** 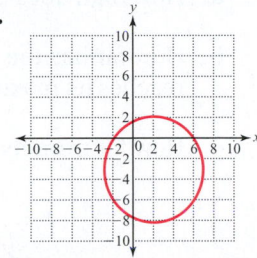   **c.** 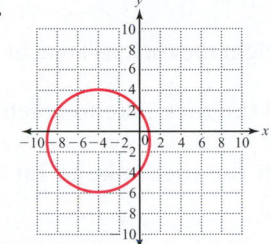   **d.**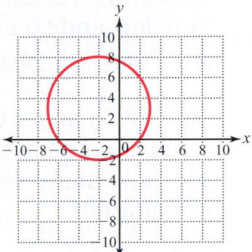

For Exercises 69–72, graph using a graphing calculator. See Calculator Tip.

69. $x^2 + y^2 = 49$

70. $x^2 + y^2 = 36$

71. $(x - 2)^2 + (y + 3)^2 = 25$

72. $(x + 4)^2 + (y - 1)^2 = 9$

For Exercises 73–80, the center and radius of a circle are given. Write the equation of the circle in standard form. See Example 6.

73. Center: $(4, 2)$; radius: 4

74. Center: $(3, 2)$; radius: 11

75. Center: $(-4, -3)$; radius: 5

76. Center: $(-6, -5)$; radius: 2

77. Center: $(6, -2)$; radius: $\sqrt{14}$

78. Center: $(-3, -3)$; radius: $\sqrt{26}$

79. Center: $(-5, 2)$; radius: $3\sqrt{5}$

80. Center: $(6, 2)$; radius: $2\sqrt{6}$

For Exercises 81–84, the center of a circle and a point on the circle are given. Write the equation of the circle in standard form.

81. Center: $(2, 4)$; point on the circle: $(5, 8)$

82. Center: $(-4, 3)$; point on the circle: $(4, 9)$

83. Center: $(2, 4)$; point on the circle: $(7, 16)$

84. Center: $(-6, 8)$; point on the circle: $(-12, 0)$

85. Write the equation of the set of all points that are a distance of 8 units from $(2, -5)$.

86. Write the equation of the set of all points that are a distance of 10 units from $(-3, -7)$.

The diameter of a circle is a line segment whose endpoints lie on the circle and contains the center of the circle. Thus, the center of a circle is the midpoint of a diameter and the diameter is twice the radius.

★ *For Exercises 87–90, the coordinates of the endpoints of a diameter are given. Find the equation of the circle.*

87. $(4, -2), (-2, 6)$

88. $(6, -6), (-2, 0)$

89. $(6, -2), (-2, 8)$

90. $(-8, 6), (-2, 4)$

91. If a rock is thrown vertically upward from the top of a building 112 feet high with an initial velocity of 96 feet per second, the height, h, above ground level after t seconds is given by $h = -16t^2 + 96t + 112$, where h is in feet and t is in seconds.
  a. What is the maximum height the rock will reach?

  b. How many seconds will the rock take to reach its maximum height?

  c. How many seconds will the rock take to hit the ground?

92. The path of a shell fired from ground level is in the shape of the parabola $y = 4x - x^2$, where x and y are given in kilometers.
  a. How high does the shell go?

  b. How far has the shell traveled horizontally when it reaches its maximum height?

  c. How far from its firing point does the shell land?

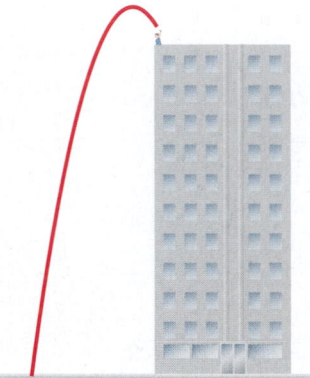

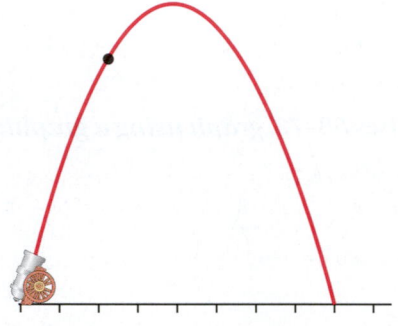

★ **93.** Arches in the shapes of parabolas are often used in construction. Find the equation of a parabolic arc that is 18 feet high at its highest point and 30 feet wide at the base, as illustrated in the following figure. Place the origin at the midpoint of the bridges.

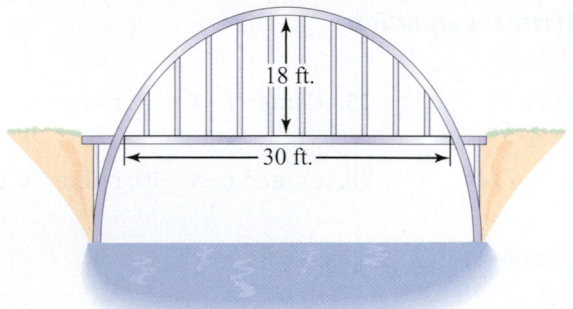

18 ft.

30 ft.

★ **94.** The cross sections of satellite dishes are in the shape of parabolas. Find the equation of a dish that is 6 feet across and 1 foot deep if the vertex is at the origin and the parabola is opening upward.

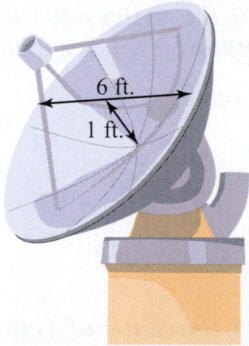

6 ft.

1 ft.

95. The percent of deaths by age per million miles driven can be approximated by the equation $y = 0.0038x^2 - 0.3475x + 8.316$, where x is the age and y is the percent. Find the percent of deaths per million miles driven for drivers 17 years old. (*Source:* Highway Traffic Safety Administration)

96. The number of drivers involved in fatal accidents for a given blood alcohol content (BAC) can be approximated by $y = -8862.5x^2 + 26622.6x + 332$, where x is the BAC and y is the number of drivers involved in fatal accidents. Find the number of drivers involved in fatal accidents who had a BAC of 0.20. (*Source:* Highway Traffic Safety Administration)

97. Bill and Don are fishing in the Gulf of Mexico in separate boats that are equipped with radios with a range of 20 miles. If we put Bill's radio at the origin of a coordinate system, what is the equation of all possible locations of Don's boat where the radios would be at their maximum range?

98. A toy plane is attached to a string pinned to the ceiling so that the plane flies in a circle. If the string is 4 feet long, write an equation that describes the path of the plane if the pin is at the origin.

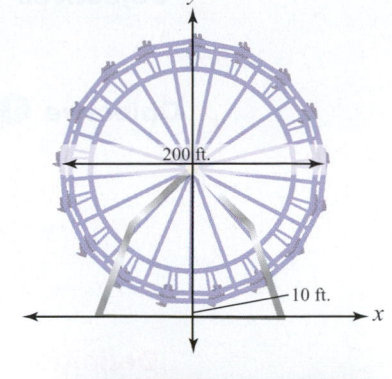

99. A Ferris wheel has a diameter of 200 feet, and the bottom of the Ferris wheel is 10 feet above the ground. Find the equation of the wheel if the origin is placed on the ground directly below the center of the wheel, as illustrated.

100. The Fermilab Tunnel houses the world's largest superconducting synchrotron. A cross section of the tunnel is a circle with radius of 1000 meters. Find the equation of a cross section of the tunnel if the origin and the center of the circle are at the center of the tunnel.

101. If a satellite is placed in a circular orbit of 230 kilometers above the Earth, what is the equation of the path of the satellite if the origin is placed at the center of the Earth (the radius of the Earth is approximately 6370 kilometers)?

102. The minute hand of Big Ben is 14 feet long. If the origin is at the center of the clock, what is the equation of the circle swept out by the tip of the hand as it makes one complete revolution?

Puzzle Problem

Using only a pencil, you can draw both a rough circle and its center without the point of the pencil losing contact with the paper, resulting in a picture like the one shown. Explain how.

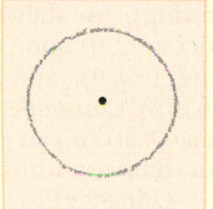

Review Exercises

Exercises 1 and 2 **EXPRESSIONS**

[4.1] **1.** Following is a coordinate system with the point $(2, 3)$ plotted. Plot the points that are 4 units to the left and right of $(2, 3)$ and give their coordinates.

[6.4] **2.** Factor: $25x^2 - 9y^2$

Exercises 3–6  **EQUATIONS AND INEQUALITIES**

[4.4] **3.** Graph $y = \frac{2}{3}x$ and $y = -\frac{2}{3}x$ on the same set of axes.

[11.1] **4.** Solve: $\frac{x^2}{16} = 1$

$\begin{bmatrix} 4.3, \\ 11.1 \end{bmatrix}$ **5.** Find the x- and y-intercepts of $9x^2 + 16y^2 = 144$

$\begin{bmatrix} 4.3, \\ 11.1 \end{bmatrix}$ **6.** Find the x- and y-intercepts of $25x^2 + 9y^2 = 225$.

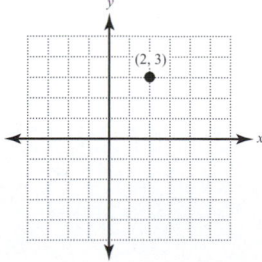

13.2 Ellipses and Hyperbolas

Objectives

1. Graph ellipses.
2. Graph hyperbolas.

Objective 1 **Graph ellipses.** Suppose you drive two nails in a board and tie a string to the two nails. Then you take a pencil, pull the string taut, and draw a figure around the two nails.

The figure you've drawn is called an **ellipse.** The locations of the two nails are the *focal points.*

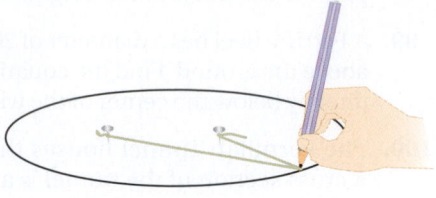

Definition | **Ellipse:** The set of all points the sum of whose distances from two fixed points is constant.

Ellipses occur in many situations. The orbits of the planets about the Sun are elliptical with the Sun at one focal point. The orbits of satellites about the Earth are also elliptical. The cams of compound bows are elliptical, which allows a decrease in the amount of effort required to hold the bow at full draw.

In the definition of an ellipse, the two fixed points are the *foci* (plural of *focus*) and the point halfway between the foci is the *center.* The following figure shows the graph of an ellipse with foci at $(c, 0)$ and $(-c, 0)$, x-intercepts at $(a, 0)$ and $(-a, 0)$, and y-intercepts at $(0, b)$ and $(0, -b)$. Consequently, the center is at the origin, $(0, 0)$. It can be shown that $c^2 = a^2 - b^2$ for ellipses in which $a > b$.

Using the distance formula, it can be shown that an ellipse with these characteristics has the following equation.

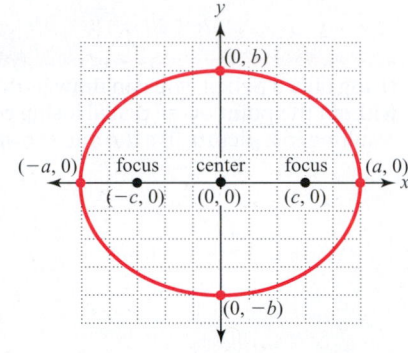

Rule | **Equation of an Ellipse Centered at $(0, 0)$**

The equation of an ellipse with center $(0, 0)$, x-intercepts $(a, 0)$ and $(-a, 0)$, and y-intercepts $(0, b)$ and $(0, -b)$ is

$$\frac{x^2}{a^2} + \frac{y^2}{b^2} = 1.$$

Example 1 Graph each ellipse and label the x- and y-intercepts.

a. $\dfrac{x^2}{16} + \dfrac{y^2}{9} = 1$

SOLUTION The equation can be rewritten as $\dfrac{x^2}{4^2} + \dfrac{y^2}{3^2} = 1$, so $a = 4$ and $b = 3$. Consequently, the x-intercepts are $(4, 0)$ and $(-4, 0)$ and the y-intercepts are $(0, 3)$ and $(0, -3)$.

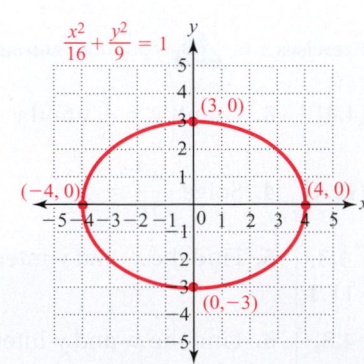

b. $25x^2 + 9y^2 = 225$

SOLUTION We first need to write the equation in standard form: $\dfrac{x^2}{a^2} + \dfrac{y^2}{b^2} = 1$.

$$\frac{25x^2}{225} + \frac{9y^2}{225} = \frac{225}{225}$$ **Divide both sides by 225.**

$$\frac{x^2}{9} + \frac{y^2}{25} = 1$$ **Simplify both sides.**

We now see that this is an equation of an ellipse with $a = 3$ and $b = 5$, so the x-intercepts are $(3, 0)$ and $(-3, 0)$ and the y-intercepts are $(0, 5)$ and $(0, -5)$.

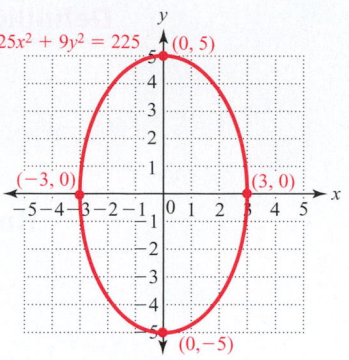

Your Turn 1 Graph the ellipse $\dfrac{x^2}{36} + \dfrac{y^2}{9} = 1$. Label the x- and y-intercepts.

If the center of an ellipse is not at the origin, it can be shown that the equation has the following form.

Rule **General Equation for an Ellipse**

The equation of an ellipse with center (h, k) is $\dfrac{(x - h)^2}{a^2} + \dfrac{(y - k)^2}{b^2} = 1$. The ellipse passes through two points that are a units to the right and left of the center and two points that are b units above and below the center.

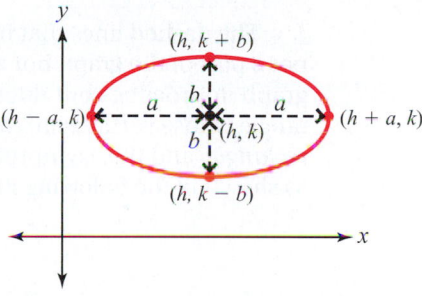

Answer Your Turn 1

$$\frac{x^2}{36} + \frac{y^2}{9} = 1$$

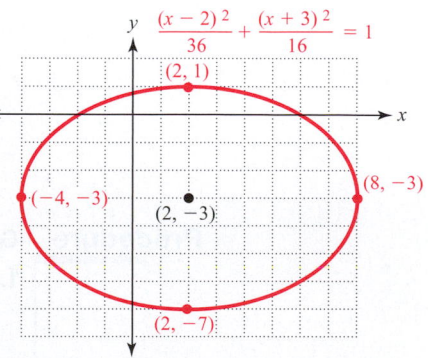

Example 2 Graph the ellipse. Label the center and the points directly above, below, to the left, and to the right of the center. $\dfrac{(x - 2)^2}{36} + \dfrac{(y + 3)^2}{16} = 1$

SOLUTION Because $h = 2$ and $k = -3$, the center of the ellipse is $(2, -3)$. Also, we see that $a = 6$, which means that the ellipse passes through two points 6 units to the right and left of $(2, -3)$. These points are $(8, -3)$ and $(-4, -3)$. Because $b = 4$, the ellipse passes through two points 4 units above and below the center. These points are $(2, 1)$ and $(2, -7)$.

Answer Your Turn 2

$$\frac{(x + 1)^2}{25} + \frac{(y - 2)^2}{36} = 1$$

Your Turn 2 Graph the ellipse $\dfrac{(x + 1)^2}{25} + \dfrac{(y - 2)^2}{36} = 1$. Label the center and the points directly above, below, to the right, and to the left of the center.

Objective 2 Graph hyperbolas. The last conic is the **hyperbola**. Applications of hyperbolas include the LORAN navigation system and the orbits of some comets.

Definition *Hyperbola:* The set of all points the difference of whose distances from two fixed points remains constant.

As with the ellipse, the two fixed points are the *foci* and the point halfway between the foci is the *center*.

Rule Equations of Hyperbolas in Standard Form

The equation of a hyperbola with center $(0, 0)$, x-intercepts $(a, 0)$ and $(-a, 0)$, and no y-intercepts is $\dfrac{x^2}{a^2} - \dfrac{y^2}{b^2} = 1$.

The equation of a hyperbola with center $(0, 0)$, y-intercepts $(0, b)$ and $(0, -b)$, and no x-intercepts is $\dfrac{y^2}{b^2} - \dfrac{x^2}{a^2} = 1$.

LEARNING Strategy

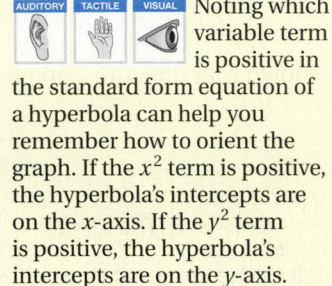 Noting which variable term is positive in the standard form equation of a hyperbola can help you remember how to orient the graph. If the x^2 term is positive, the hyperbola's intercepts are on the x-axis. If the y^2 term is positive, the hyperbola's intercepts are on the y-axis.

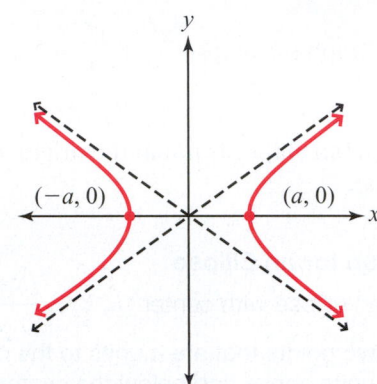

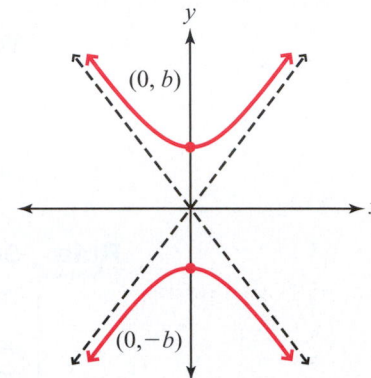

The dashed lines that intersect at the center of a hyperbola are *asymptotes*. They are not a part of the graph but are used as an aid in graphing. An asymptote is a line that the graph approaches but does not cross as the graph goes away from the origin. The rectangle whose vertices are (a, b), $(-a, b)$, $(a, -b)$, and $(-a, -b)$ is called the *fundamental rectangle,* and the asymptotes are the extended diagonals of the fundamental rectangle, as shown in the following illustration.

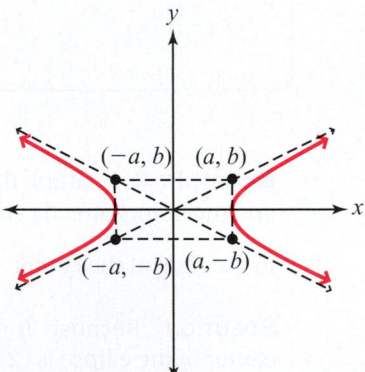

Procedure Graphing a Hyperbola in Standard Form

1. Find the intercepts. If the x^2-term is positive, the x-intercepts are $(a, 0)$ and $(-a, 0)$ and there are no y-intercepts. If the y^2-term is positive, the y-intercepts are $(0, b)$ and $(0, -b)$ and there are no x-intercepts.
2. Draw the fundamental rectangle. The vertices are (a, b), $(-a, b)$, $(a, -b)$, and $(-a, -b)$.
3. Draw the asymptotes, which are the extended diagonals of the fundamental rectangle.
4. Draw the graph so that each branch passes through an intercept and approaches the asymptotes the farther the branches are from the origin.

Example 3 Graph each hyperbola. Also show the fundamental rectangle with its corner points labeled, the asymptotes, and the intercepts.

a. $\dfrac{x^2}{16} - \dfrac{y^2}{9} = 1$

SOLUTION

1. This equation can be written as $\dfrac{x^2}{4^2} - \dfrac{y^2}{3^2} = 1$, so $a = 4$ and $b = 3$. Because the x^2-term is positive, the graph has x-intercepts at $(4, 0)$ and $(-4, 0)$.
2. The fundamental rectangle has vertices at $(4, 3)$, $(-4, 3)$, $(4, -3)$, and $(-4, -3)$.
3. Draw the asymptotes.
4. Sketch the graph so it passes through the x-intercepts and approaches the asymptotes.

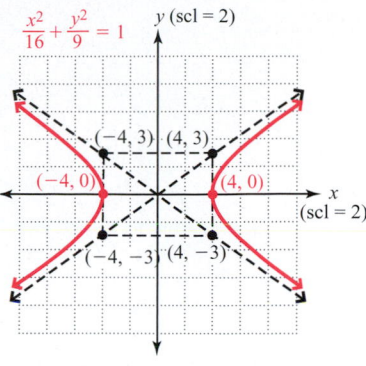

b. $\dfrac{y^2}{4} - \dfrac{x^2}{9} = 1$

SOLUTION

1. This equation can be written as $\dfrac{y^2}{2^2} - \dfrac{x^2}{3^2} = 1$, so $b = 2$ and $a = 3$. Because the y^2-term is positive, the graph has y-intercepts at $(0, 2)$ and $(0, -2)$.
2. The fundamental rectangle has vertices at $(3, 2)$, $(-3, 2)$, $(3, -2)$, and $(-3, -2)$.
3. Draw the asymptotes.
4. Sketch the graph so it passes through the y-intercepts and approaches the asymptotes.

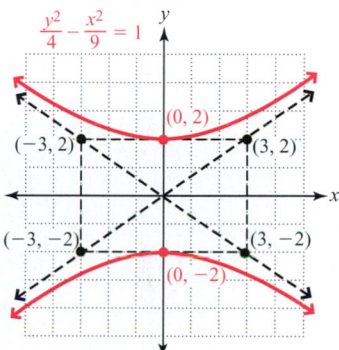

Answer   Your Turn 3

$\dfrac{x^2}{25} - \dfrac{y^2}{16} = 1$

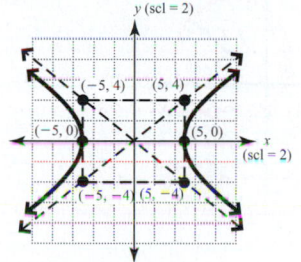

Your Turn 3 Graph the hyperbola. Also show the fundamental rectangle with its corner points labeled, the asymptotes, and the intercepts.

$$\frac{x^2}{25} - \frac{y^2}{16} = 1$$

📱 Calculator Tips

Graph the ellipse $\dfrac{x^2}{9} + \dfrac{y^2}{16} = 1$. *We solve for y as when we graph a circle.*

$$\frac{x^2}{9} + \frac{y^2}{16} = 1$$

$$\frac{y^2}{16} = 1 - \frac{x^2}{9} \qquad \text{Subtract } \frac{x^2}{9} \text{ from both sides.}$$

$$y^2 = 16\left(1 - \frac{x^2}{9}\right) \qquad \text{Multiply both sides by 16.}$$

$$y = \pm 4\sqrt{\left(1 - \frac{x^2}{9}\right)} \qquad \text{Apply the square root principle.}$$

Define $Y_1 = 4\sqrt{\left(1 - \dfrac{x^2}{9}\right)}$ *and* $Y_2 = -4\sqrt{\left(1 - \dfrac{x^2}{9}\right)}$. *Graph in a window* $[-5, 5]$ *for x and* $[-5, 5]$ *for y and square the window.*

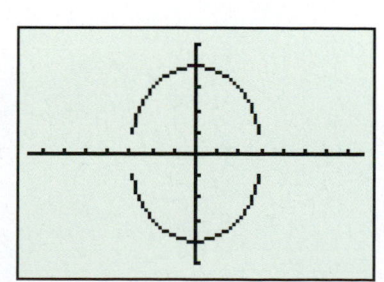

Following is a summary of the conic sections.

Standard Forms		
Parabola	$y = a(x - h)^2 + k, a > 0$ 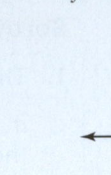	$y = a(x - h)^2 + k, a < 0$ 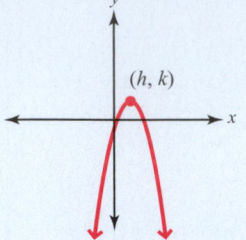
	$x = a(y - k)^2 + h, a > 0$ 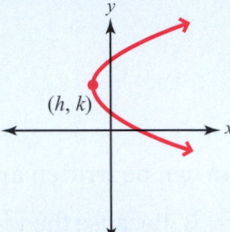	$x = a(y - k)^2 + h, a < 0$ 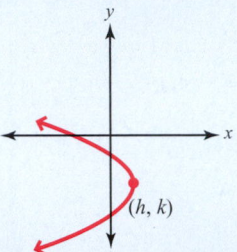
Circle	$x^2 + y^2 = r^2$ 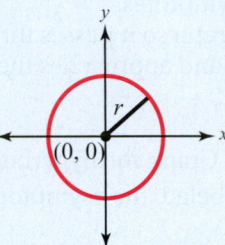	$(x - h)^2 + (y - k)^2 = r^2$ 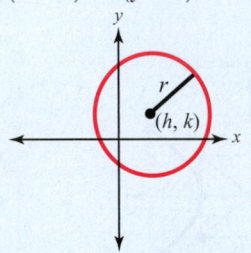
Ellipse	$\dfrac{x^2}{a^2} + \dfrac{y^2}{b^2} = 1$ 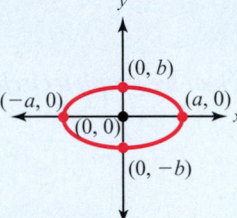	$\dfrac{(x - h)^2}{a^2} + \dfrac{(y - k)^2}{b^2} = 1$ 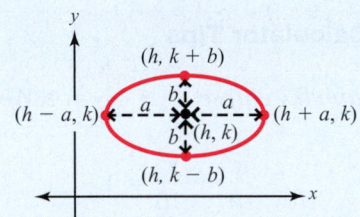
Hyperbola	$\dfrac{x^2}{a^2} - \dfrac{y^2}{b^2} = 1$ 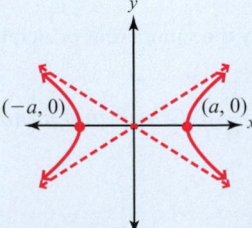	$\dfrac{y^2}{b^2} - \dfrac{x^2}{a^2} = 1$ 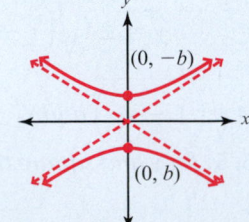

1. What is the definition of an ellipse?

2. In the definitions of the ellipse and hyperbola, two fixed points are mentioned. What are these two points called?

3. How can you tell the equation of an ellipse from the equation of a hyperbola?

4. How do you determine whether a hyperbola opens up and down or left and right?

5. Which conic has an equation that has x^2 or y^2 but not both?

6. What do the graphs of $\dfrac{x^2}{9} - \dfrac{y^2}{16} = 1$ and $\dfrac{x^2}{9} + \dfrac{y^2}{16} = 1$ have in common?

For Exercises 7–14, graph each ellipse. Label the x- and y-intercepts. See Example 1.

7. $\dfrac{x^2}{81} + \dfrac{y^2}{64} = 1$

8. $\dfrac{x^2}{25} + \dfrac{y^2}{4} = 1$

9. $\dfrac{x^2}{36} + y^2 = 1$

10. $\dfrac{x^2}{9} + \dfrac{y^2}{16} = 1$

11. $4x^2 + 9y^2 = 36$

12. $25x^2 + 4y^2 = 100$

13. $36x^2 + 4y^2 = 144$

14. $x^2 + 9y^2 = 36$

For Exercises 15–18, graph each ellipse. Label the center and the points directly above, below, to the left, and to the right of the center. See Example 2.

15. $\dfrac{(x-1)^2}{49} + \dfrac{(y+3)^2}{25} = 1$

16. $\dfrac{(x+3)^2}{25} + \dfrac{(y-2)^2}{64} = 1$

17. $\dfrac{(x-4)^2}{4} + \dfrac{(y+3)^2}{36} = 1$

18. $\dfrac{(x+5)^2}{9} + \dfrac{(y+3)^2}{25} = 1$

For Exercises 19–22, graph using a graphing calculator. See Calculator Tips.

19. $\dfrac{x^2}{25} + \dfrac{y^2}{4} = 1$

20. $\dfrac{y^2}{12} + \dfrac{x^2}{8} = 1$

21. $\dfrac{(y-2)^2}{36} + \dfrac{(x+4)^2}{25} = 1$

22. $\dfrac{(x+2)^2}{4} + \dfrac{(y+1)^2}{16} = 1$

23. If the x-intercepts of an ellipse are $(-3, 0)$ and $(3, 0)$ and the y-intercepts are $(0, 4)$ and $(0, -4)$, what is the equation of the ellipse?

24. If the x-intercepts of an ellipse are $(-6, 0)$ and $(6, 0)$ and the y-intercepts are $(0, 3)$ and $(0, -3)$, what is the equation of the ellipse?

For Exercises 25–32, graph each hyperbola. Also show the fundamental rectangle with its corner points labeled, the asymptotes, and the intercepts. See Example 3.

25. $\dfrac{x^2}{9} - \dfrac{y^2}{4} = 1$

26. $\dfrac{x^2}{16} - \dfrac{y^2}{25} = 1$

27. $\dfrac{y^2}{36} - \dfrac{x^2}{9} = 1$

28. $\dfrac{y^2}{4} - \dfrac{x^2}{25} = 1$

29. $9x^2 - y^2 = 36$

30. $x^2 - 4y^2 = 16$

31. $9y^2 - 25x^2 = 225$

32. $16y^2 - 4x^2 = 64$

For Exercises 33 and 34, graph using a graphing calculator. See Calculator Tips.

33. $\dfrac{x^2}{36} - \dfrac{y^2}{4} = 1$

34. $\dfrac{y^2}{9} - \dfrac{x^2}{25} = 1$

35. If a hyperbola opens left and right and the vertices of the fundamental rectangle are $(3, 2)$, $(3, -2)$, $(-3, 2)$, and $(-3, -2)$, what is the equation of the hyperbola?

36. If a hyperbola opens up and down and the vertices of the fundamental rectangle are $(5, 3)$, $(5, -3)$, $(-5, 3)$, and $(-5, -3)$, what is the equation of the hyperbola?

For Exercises 37–40, match the equation with the graph.

37. $\dfrac{x^2}{9} + \dfrac{y^2}{25} = 1$

38. $\dfrac{x^2}{9} - \dfrac{y^2}{25} = 1$

39. $\dfrac{x^2}{25} + \dfrac{y^2}{9} = 1$

40. $\dfrac{x^2}{25} - \dfrac{y^2}{9} = 1$

a.

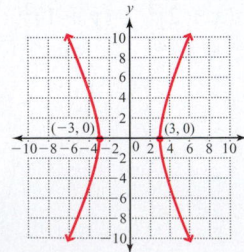

b.

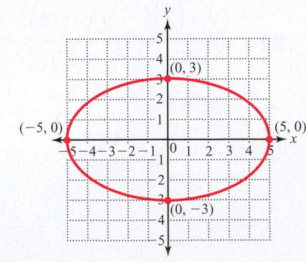

c.

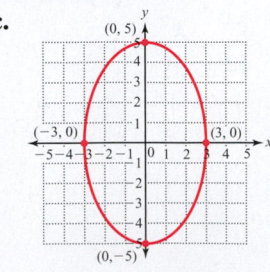

d.

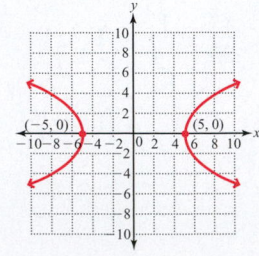

For Exercises 41–44, determine whether the graph of the equation is a circle, a parabola, an ellipse, or a hyperbola. Do not draw the graph.

41. $9x^2 + 16y^2 = 144$

42. $16y^2 - 9x^2 = 144$

43. $x^2 + y^2 - 6x + 8y - 75 = 0$

44. $2x^2 - 12x + 23 - y = 0$

For Exercises 45–58, indicate whether the graph of the given equation is a circle, a parabola, an ellipse, or a hyperbola; then draw the graph.

45. $(x - 2)^2 + (y + 2)^2 = 49$ **46.** $x^2 + y^2 = 81$

47. $y = 2(x + 1)^2 + 3$

48. $x = (y + 3)^2 - 4$

49. $\dfrac{x^2}{36} + \dfrac{y^2}{81} = 1$

50. $\dfrac{x^2}{36} + \dfrac{y^2}{16} = 1$

51. $\dfrac{x^2}{4} - \dfrac{y^2}{25} = 1$

52. $\dfrac{y^2}{49} - \dfrac{x^2}{4} = 1$

53. $y = 2x^2 + 8x + 6$

54. $x = y^2 + 6$

55. $\dfrac{x^2}{16} - \dfrac{y^2}{16} = 1$

56. $\dfrac{y^2}{25} - \dfrac{x^2}{25} = 1$

57. $\dfrac{(x + 3)^2}{9} + \dfrac{(y + 2)^2}{36} = 1$

58. $\dfrac{(x - 1)^2}{4} + (y + 3)^2 = 1$

For Exercises 59–66, solve.

59. A bridge over a waterway has an arch in the form of half an ellipse. The equation of the ellipse is $400x^2 + 256y^2 = 102,400$.

a. A sailboat, the top of whose mast is 18 feet above the water, is approaching the arch. Will the mast clear the bridge? Why or why not?

b. How wide is the base of the arch?

60. A highway passes beneath an overpass that is in the shape of half an ellipse. The overpass is 15 feet high at the center and 40 feet wide at the base.
  a. What is the equation of the ellipse?

  b. A truck that is 10 feet wide carrying a load that is 14 feet above the level of the road is approaching the bridge. If the truck goes down the middle of the road, will the load clear the bridge? Why or why not?

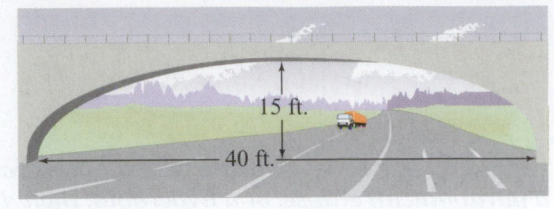

61. The comet Epoch has an orbit in the shape of an ellipse with the Sun at one of the foci. The equation is approximately $\dfrac{x^2}{3.6^2} + \dfrac{y^2}{2.88^2} = 1$, where x and y are in astronomical units. (An astronomical unit is 93,000,000 miles.) Sketch the graph of the comet Epoch. (*Source: Orbital Motion*, A. E. Roy, Institute of Physics Publishing, London)

62. The dwarf planet Pluto has an orbit in the shape of an ellipse with the Sun at one of the foci. The equation is approximately $\dfrac{x^2}{39.4^2} + \dfrac{y^2}{38.2^2} = 1$, where x and y are measured in astronomical units. Sketch the graph of the orbit of Pluto. (*Hint:* Make each unit on the x- and y-axes equal to 10.) (*Source: Orbital Motion*, A. E. Roy, Institute of Physics Publishing, London)

63. Compound bows have elliptical cams that decrease the amount of effort required to hold the bow at full draw. If a cam on a bow is 4 inches from top to bottom and 3 inches across, what is the equation of the ellipse if the origin is at the center of the cam?

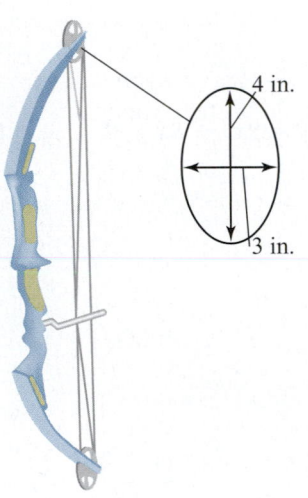

64. One of the most popular exercise machines is an elliptical trainer in which your foot moves in an elliptical path. On a typical machine, the length of the stride is about 19 inches and the height varies from 3 to 5 inches depending on the settings. Write the equation for the path that your foot takes if the total length of the stride is 19 inches and the total height is 4 inches. (*Source*: Precor National Headquarters)

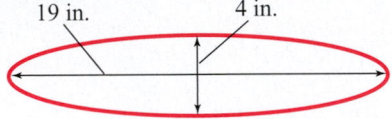

19 in.   4 in.

65. If a source of light or sound is placed at one focal point of an elliptic reflector, the light or sound is reflected through the other focal point. This principle is used in a lithotripter that uses sound waves to crush kidney stones by placing a source of sound at one focal point and the kidney stone at the other. If the elliptic reflector is based on the ellipse $\dfrac{x^2}{25} + \dfrac{y^2}{16} = 1$, how many units from the center should the kidney stone be placed? (*Hint:* $c^2 = a^2 - b^2$.)

66. The same principle used in Exercise 65 is also used in whispering rooms. The room is in the shape of an elliptic reflector where one person speaks at one focal point and the other person places his or her ear at the other focal point and can hear sounds as faint as a whisper. If a whispering room is based on the ellipse $\dfrac{x^2}{225} + \dfrac{y^2}{144} = 1$, how many units from the center of the ellipse would the two people stand? (*Hint:* $c^2 = a^2 - b^2$.)

Collaborative Exercises The Elliptical Tablecloth

Recall that an ellipse can be drawn by fixing the ends of a string to the foci and then tracing out the ellipse. By considering the following figures, we can determine an expression for the length of the string and a relationship between a, b, and c.

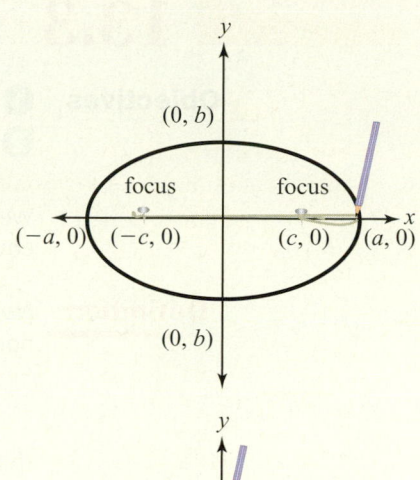

1. Use the figure to write a formula for the length of the string.

2. In the figure at right, notice that the string forms an isosceles triangle and the *y*-axis splits that triangle into two identical right triangles.
   a. Find the length of the hypotenuse of each of those right triangles.

   b. What expression describes the length of the string?

   c. Use the Pythagorean theorem to write a formula relating *a*, *b*, and *c*.

   d. Solve the formula for *c*.

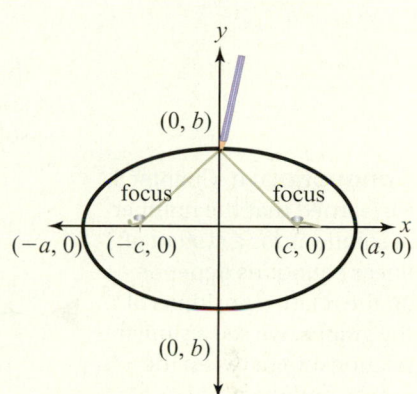

Suppose we are to make an elliptical tablecloth for an elliptical table that is 76 inches long and 58 inches wide. The tablecloth is to drape 6 inches over the edge of the table all the way around, and we need an additional inch for the hem. We have a large rectangular piece of cloth to make the tablecloth. To trace the ellipse on the cloth, we need to know the string length and the location of the foci.

3. Find the dimensions of the elliptical tablecloth, taking into account the amount it needs to drape and the hem.
4. How long must the string be so that the ellipse can be traced?
5. How far from the center are the foci located?

Puzzle Problem

Suppose an ellipse is drawn using the method described on p. 916. Assuming that the ellipse is centered at the origin, if the nails are 8 inches apart and the string is 10 inches in length, what are the *x*- and *y*-intercepts?

Review Exercises

Exercises 1–6 **EQUATIONS AND INEQUALITIES**

[2.4] **1.** Solve $3x^2 + y = 6$ for *y*.

[9.1] **2.** Solve the following system using the graphical method. $\begin{cases} x + y = 3 \\ 2x + y = 4 \end{cases}$

[9.2] **3.** Solve the following system using the substitution method. $\begin{cases} 2x + y = 1 \\ 3x + 4y = -6 \end{cases}$

[9.3] **4.** Solve the following system using the elimination method. $\begin{cases} 2x + 3y = 6 \\ 3x - 4y = -25 \end{cases}$

For Exercises 5 and 6, solve.

[6.6] **5.** $3x^2 + 10x - 8 = 0$

[11.1] **6.** $4x^2 = 36$

13.3 Nonlinear Systems of Equations

Objectives

1. Solve nonlinear systems of equations using substitution.
2. Solve nonlinear systems of equations using elimination.

We solved systems of linear equations in Chapter 9. Now we solve **nonlinear systems** of equations.

Definition   *Nonlinear system of equations:* A system of equations that contains at least one nonlinear equation.

The types of equations in a nonlinear system determine the number of solutions that are possible for the system. For example, a system containing a quadratic equation and a linear equation can have 0, 1, or 2 points of intersection and therefore 0, 1, or 2 solutions, as shown by the following figures.

Connection In Chapter 9, we learned that the number of solutions for a system of linear equations depends on the relative positions of the graphs. We see a similar relationship between the graphs and the number of solutions here.

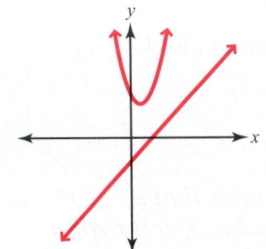

No points of intersection: no solutions

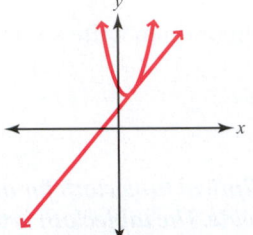

One point of intersection: one solution

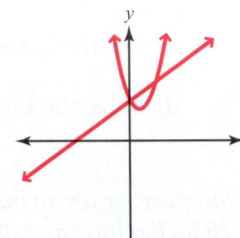

Two points of intersection: two solutions

Objective 1   Solve nonlinear systems of equations using substitution.   When solving nonlinear systems, if one equation is linear (or one of the variables in an equation is isolated), the substitution method is usually preferred.

Example 1   Solve using the substitution method.

a. $\begin{cases} y = 2(x + 2)^2 - 2 \\ 2x + y = -2 \end{cases}$

Note: The graphs are a parabola and a line, so there will be 0, 1, or 2 solutions.

SOLUTION   It is easier to solve the linear equation for one of its variables and substitute into the nonlinear equation. Because the coefficient of y is 1, we solve $2x + y = -2$ for y and get $y = -2x - 2$.

$-2x - 2 = 2(x + 2)^2 - 2$	Substitute $-2x - 2$ for y in $y = 2(x + 2)^2 - 2$.
$-2x - 2 = 2(x^2 + 4x + 4) - 2$	To solve this quadratic equation, we need to write it in the form $ax^2 + bx + c = 0$. First, we square $x + 2$.
$-2x - 2 = 2x^2 + 8x + 6$	Distribute 2; then combine like terms.
$0 = 2x^2 + 10x + 8$	Add $2x$ and 2 to both sides.
$0 = x^2 + 5x + 4$	Divide both sides by 2.
$0 = (x + 1)(x + 4)$	Factor.
$0 = x + 1 \quad \text{or} \quad 0 = x + 4$	Use the zero-factor theorem.
$-1 = x \quad\quad \text{or} \quad -4 = x$	Solve each equation.

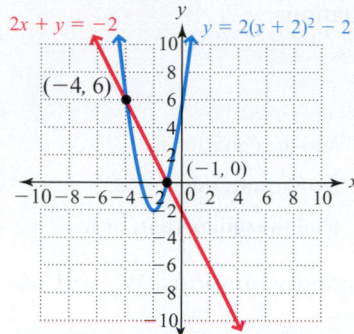

$2x + y = -2$
$y = 2(x + 2)^2 - 2$
$(-4, 6)$
$(-1, 0)$

To find y, substitute -1 and -4 for x in $y = -2x - 2$.

$$y = -2(\mathbf{-1}) - 2 \qquad\qquad y = -2(\mathbf{-4}) - 2$$
$$y = 2 - 2 \qquad\qquad\qquad y = 8 - 2$$
$$y = 0 \qquad\qquad\qquad\quad y = 6$$

The solutions are $(-1, 0)$ and $(-4, 6)$, as verified by the graph to the left.

b. $\begin{cases} y = \sqrt{x + 2} \\ x^2 + y^2 = 8 \end{cases}$

SOLUTION   Because $y = \sqrt{x + 2}$ is already solved for y, substitute $\sqrt{x + 2}$ for y in $x^2 + y^2 = 8$.

$$x^2 + (\sqrt{x + 2})^2 = 8 \qquad \text{Substitute } \sqrt{x + 2} \text{ for } y \text{ in } x^2 + y^2 = 8.$$
$$x^2 + x + 2 = 8 \qquad \text{Square } \sqrt{x + 2}.$$
$$x^2 + x - 6 = 0 \qquad \text{Because the equation is now quadratic, we subtract 8 from}$$
$$\text{both sides to get the form } ax^2 + bx + c = 0.$$
$$(x + 3)(x - 2) = 0 \qquad \text{Factor.}$$
$$x + 3 = 0 \quad \text{or} \quad x - 2 = 0 \qquad \text{Use the zero-factor theorem.}$$
$$x = -3 \quad \text{or} \qquad x = 2 \qquad \text{Solve each equation.}$$

To find y, substitute -3 and 2 for x in $y = \sqrt{x + 2}$.

$$y = \sqrt{\mathbf{-3} + 2} \qquad\qquad y = \sqrt{\mathbf{2} + 2}$$
$$y = \sqrt{-1} \qquad\qquad\qquad y = \sqrt{4}$$
$$y = 2$$

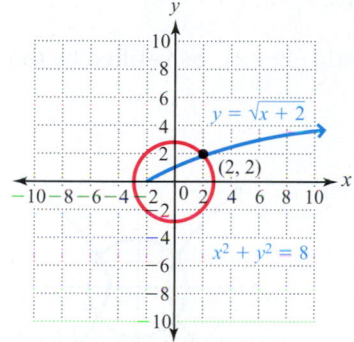

$y = \sqrt{x + 2}$
$(2, 2)$
$x^2 + y^2 = 8$

Because $\sqrt{-1}$ is imaginary, the only real solution is $(2, 2)$, as verified by the graph to the left.

Note: A system containing a root function and a circle, as in Example 1(b), can have 0, 1, or 2 solutions, as shown.

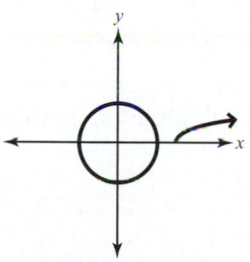

No Solutions

One Solution

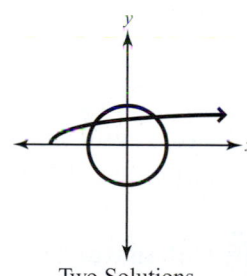

Two Solutions

Your Turn 1   Solve using the substitution method.

a. $\begin{cases} 2x - y = 1 \\ x^2 + y^2 = 1 \end{cases}$ **b.** $\begin{cases} xy = 3 \\ 4x - y = 1 \end{cases}$

Objective ②   Solve nonlinear systems of equations using elimination.   If neither equation contains a radical expression or both equations contain the same powers of the variables, the elimination method can be used.

Example 2   Solve using the elimination method: $\begin{cases} 9x^2 - 4y^2 = 20 \\ x^2 + y^2 = 8 \end{cases}$

SOLUTION $\begin{cases} 9x^2 - 4y^2 = 20 & \text{(Equation 1)} \\ x^2 + y^2 = 8 & \text{(Equation 2)} \end{cases}$

To eliminate y^2, multiply Equation 2 by 4 and add the equations.

$$9x^2 - 4y^2 = 20$$
$$x^2 + y^2 = 8 \xrightarrow{\text{Multiply by 4}}$$

$$\begin{array}{ll} 9x^2 - 4y^2 = 20 & \\ \underline{4x^2 + 4y^2 = 32} & \\ 13x^2 = 52 & \text{Add the equations.} \\ x^2 = 4 & \text{Divide both sides by 13.} \\ x = \pm 2 & \text{Find the square roots of 4.} \end{array}$$

To find y, substitute 2 and -2 for x in one of the original equations. We will use $x^2 + y^2 = 8$.

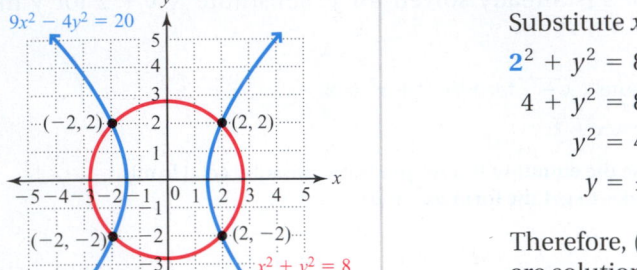

Substitute $x = 2$.

$$2^2 + y^2 = 8$$
$$4 + y^2 = 8$$
$$y^2 = 4$$
$$y = \pm 2$$

Therefore, $(2, 2)$ and $(2, -2)$ are solutions.

Substitute $x = -2$.

$$(-2)^2 + y^2 = 8$$
$$4 + y^2 = 8$$
$$y^2 = 4$$
$$y = \pm 2$$

Therefore, $(-2, 2)$ and $(-2, -2)$ are solutions.

This system has four solutions: $(2, 2)$, $(2, -2)$, $(-2, 2)$, and $(-2, -2)$, as verified by the graph to the left.

Note: If the graphs are a hyperbola and a circle both centered at the origin as in Example 2, there will be 0, 2, or 4 solutions, as shown.

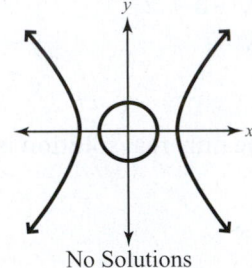

No Solutions

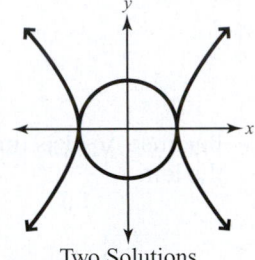

Two Solutions

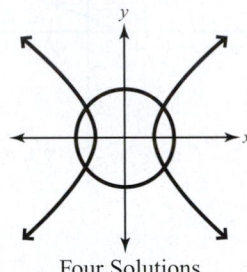

Four Solutions

Answers Your Turn 2

$(3, 2), (3, -2), (-3, 2), (-3, -2)$

Your Turn 2 Solve using the elimination method: $\begin{cases} 4x^2 + 9y^2 = 72 \\ x^2 + y^2 = 13 \end{cases}$

Calculator Tips

A system of equations can be solved using the intersect *function on a graphing calculator.*

Let's solve $\begin{cases} x^2 + y^2 = 10 \\ 3x + y = 6 \end{cases}$ *using a T1-83 or TI-84 Plus.*

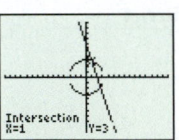

First, solve each equation for y:

$$x^2 + y^2 = 10 \qquad\qquad 3x + y = 6$$
$$y^2 = 10 - x^2 \qquad\qquad y = 6 - 3x$$
$$y = \pm\sqrt{10 - x^2}$$

Enter $Y_1 = \sqrt{(10 - x^2)}$, $Y_2 = -\sqrt{(10 - x^2)}$, *and* $Y_3 = 6 - 3x$; *then press* GRAPH .

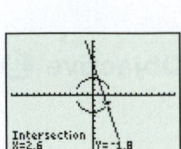

To find the intersection of y_1 and y_3 as well as y_2 and y_3, select 5: intersect from the CALC menu. You are then prompted to indicate the two graphs you want to find the intersection of and the appropriate point of intersection. We show the results of each point of intersection in the margin.

1. a. How many real solutions are possible for a system of two equations whose graphs are a circle and a parabola?

   b. Draw a figure illustrating such a system with two solutions.

2. a. How many real solutions are possible for a system of two equations whose graphs are a line and a hyperbola?

   b. Draw a figure illustrating such a system with one solution.

3. Which method (substitution or elimination) would you use to solve the system $\begin{cases} 2x + y = 8 \\ 4x^2 + 3y^2 = 24 \end{cases}$? Why? (Do not attempt to solve the system.)

4. Which method (substitution or elimination) would you use to solve the system $\begin{cases} 3x^2 - 4y^2 = 12 \\ 2x^2 + 3y^2 = 24 \end{cases}$? Why? (Do not attempt to solve the system.)

5. Without solving the system, what is the number of possible solutions of the system $\begin{cases} 3x - y = 6 \\ 4x^2 + 9y^2 = 36 \end{cases}$?

6. Is it possible for a system of two equations whose graphs are an ellipse and a hyperbola (both centered at the origin) to have three solutions? Why or why not?

For Exercises 7–42, solve. See Examples 1 and 2.

7. $\begin{cases} x^2 + 2y = 1 \\ 2x + y = 2 \end{cases}$

8. $\begin{cases} y = 2x^2 \\ 2x + y = 4 \end{cases}$

9. $\begin{cases} y = x^2 + 4x + 4 \\ 3x - y = -6 \end{cases}$

10. $\begin{cases} y = 6x - x^2 \\ 2x - y = -3 \end{cases}$

11. $\begin{cases} x^2 + y^2 = 25 \\ x - y = -1 \end{cases}$

12. $\begin{cases} x^2 + y^2 = 25 \\ x - 7y = -25 \end{cases}$

13. $\begin{cases} x^2 + y^2 = 10 \\ 3x + y = 6 \end{cases}$

14. $\begin{cases} x^2 + y^2 = 13 \\ 2x + y = 7 \end{cases}$

15. $\begin{cases} x^2 + 2y^2 = 4 \\ x + y = 5 \end{cases}$

16. $\begin{cases} 3x^2 + y^2 = 9 \\ 2x + y = 11 \end{cases}$

17. $\begin{cases} y = 2x^2 + 3 \\ 2x - y = -3 \end{cases}$

18. $\begin{cases} y = -3x^2 - 2 \\ 3x + y = -2 \end{cases}$

19. $\begin{cases} y = (x - 3)^2 + 2 \\ y = -(x - 2)^2 + 3 \end{cases}$

20. $\begin{cases} y = 2(x + 4)^2 + 2 \\ y = -2(x + 3)^2 + 4 \end{cases}$

21. $\begin{cases} x^2 + y^2 = 20 \\ x^2 - y^2 = 12 \end{cases}$

22. $\begin{cases} x^2 + y^2 = 48 \\ x^2 - y^2 = 24 \end{cases}$

23. $\begin{cases} 4x^2 + 9y^2 = 72 \\ x^2 + y^2 = 13 \end{cases}$

24. $\begin{cases} 9x^2 + 4y^2 = 145 \\ x^2 + y^2 = 25 \end{cases}$

25. $\begin{cases} 4x^2 - y^2 = 15 \\ x^2 + y^2 = 5 \end{cases}$

26. $\begin{cases} 9x^2 - 4y^2 = 32 \\ x^2 + y^2 = 5 \end{cases}$

27. $\begin{cases} x^2 + 3y^2 = 36 \\ x = y^2 - 6 \end{cases}$

28. $\begin{cases} 4x^2 + y^2 = 16 \\ y = x^2 - 4 \end{cases}$

29. $\begin{cases} 9x^2 + 4y^2 = 36 \\ 4x^2 - 9y^2 = 36 \end{cases}$

30. $\begin{cases} 9x^2 - 16y^2 = 144 \\ 4x^2 + 9y^2 = 36 \end{cases}$

31. $\begin{cases} y = x^2 \\ x^2 + y^2 = 20 \end{cases}$

32. $\begin{cases} 16x^2 + y^2 = 128 \\ y = 2x^2 \end{cases}$

33. $\begin{cases} 25x^2 - 16y^2 = 400 \\ x^2 + 4y^2 = 16 \end{cases}$

34. $\begin{cases} 9y^2 - 25x^2 = 225 \\ 4y^2 + 25x^2 = 100 \end{cases}$

35. $\begin{cases} xy = 4 \\ 2x^2 - y^2 = 4 \end{cases}$

36. $\begin{cases} xy = 2 \\ 4x^2 + y^2 = 8 \end{cases}$

37. $\begin{cases} y = x^2 - 2x - 3 \\ y = -x^2 + 6x + 7 \end{cases}$

38. $\begin{cases} y = \dfrac{1}{3}x^2 - 2 \\ 3x^2 + 9y^2 = 36 \end{cases}$

39. $\begin{cases} 4x^2 + 5y^2 = 36 \\ 4x^2 - 3y^2 = 4 \end{cases}$

40. $\begin{cases} 4x^2 + 7y^2 = 64 \\ 4x^2 - 3y^2 = 24 \end{cases}$

41. $\begin{cases} x = -y^2 + 2 \\ x^2 - 5y^2 = 4 \end{cases}$

42. $\begin{cases} x = y^2 + 2 \\ 9x^2 - 45y^2 = 36 \end{cases}$

43. Create a system of two equations whose graphs are a circle and a line for which there is no solution. Include the graphs.

44. Create a system of two equations whose graphs are a circle and a hyperbola for which there are exactly two solutions. Include the graphs.

45. The sum of the squares of two integers is 34, and the difference of their squares is 16. Find the integers. There is more than one solution.

46. The difference of the squares of two integers is 32, and their product is 12. Find the integers. There is more than one solution.

47. A computer keyboard has an area of 144 square inches and a perimeter of 52 inches. Find the length and width.

48. A rectangular living room has an area of 48 square meters and a perimeter of 28 meters. Find the dimensions.

49. If p is in dollars and x is in hundreds of units, the demand function for a certain style of chair is given by $p = -3x^2 + 120$ and the supply function is given by $p = 11x + 28$. The *market equilibrium* occurs when the number produced is equal to the number demanded. Find the number of chairs and the price per chair when market equilibrium is reached.

50. If y is in dollars and x is the number of cell phones manufactured (in thousands), a cell phone manufacturer has determined that the cost y to manufacture x cell phones is given by $y = 5x^2 + 30x + 50$ and the revenue from the sales is given by $y = 13x^2$. The *break-even point* is the point (x, y) for which the cost equals the revenue. Solve the system to find the number of units necessary to break even.

For Exercises 51–54, use a graphing calculator to verify the results of the exercise given.

51. Exercise 17

52. Exercise 18

53. Exercise 19

54. Exercise 20

Review Exercises

Exercises 1–6 ▲ EQUATIONS AND INEQUALITIES

[4.6] *For Exercises 1 and 2, determine whether the ordered pair is a solution for* $x + 3y \leq 6$.

1. $(0, 0)$

2. $(3, 4)$

[4.6] *For Exercises 3–6, graph.*

3. $y \geq 3$

4. $y < 2x + 3$

5. $2x + 3y < -6$

6. $x \geq -2$

13.4 Nonlinear Inequalities and Systems of Inequalities

Objectives

1 Graph nonlinear inequalities.

2 Graph the solution set of a system of nonlinear inequalities.

Objective 1 **Graph nonlinear inequalities.** In Section 4.6, we graphed linear inequalities like $x + 2y > 6$ by first graphing the line corresponding to $x + 2y = 6$ and then shading the appropriate region on one side of that boundary line. Recall that we used a dashed line for $<$ and $>$ and a solid line for $\leq$ and $\geq$. We determined which side to shade by using a test point on one side of the boundary.

We use a similar procedure to graph nonlinear inequalities such as $x^2 + y^2 \leq 25$. The boundary is the graph of $x^2 + y^2 = 25$, which is a circle with center at $(0, 0)$ and radius of 5. The solution set of $x^2 + y^2 \leq 25$ contains all ordered pairs on the circle (boundary) along with all ordered pairs inside it or all ordered pairs outside it. To determine which of those two regions is correct, we choose an ordered pair from one of the regions to test in the inequality. Let's test $(0, 0)$.

$0^2 + 0^2 \leq 25$ **Substitute $(0, 0)$ into $x^2 + y^2 \leq 25$.**

$0 \leq 25$ **Simplify. The inequality is true.**

Because $0 \leq 25$ is true, all ordered pairs in the region containing $(0, 0)$ are in the solution set; so we shade that region.

Ordered pairs in the region outside the circle are not in the solution set because they do not solve the inequality. To illustrate, let's test $(6, 8)$.

$6^2 + 8^2 \leq 25$ **Substitute $(6, 8)$ into $x^2 + y^2 \leq 25$.**

$100 \leq 25$ **Simplify. The inequality is false.**

Because $100 \leq 25$ is false, the region containing $(6, 8)$ is not in the solution set; so we do not shade that region.

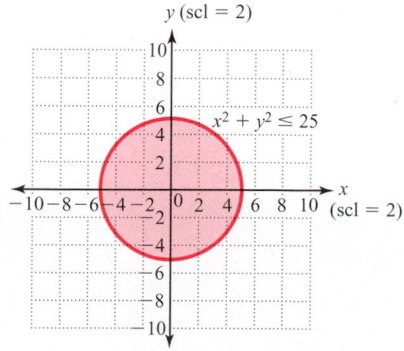

Note: If the inequality had been $x^2 + y^2 < 25$, we would have drawn a dashed circle.

Our example suggests the following procedure for graphing nonlinear inequalities.

Graphing Nonlinear Inequalities

1. Graph the related equation (the boundary curve). If the inequality symbol is $\leq$ or $\geq$, draw the graph as a solid curve. If the inequality symbol is $>$ or $<$, draw a dashed curve.
2. The graph divides the coordinate plane into at least two regions. Test an ordered pair from each region by substituting it into the inequality. If the ordered pair satisfies the inequality, shade the region containing that ordered pair.

Example 1 Graph the inequality.

a. $y > 2(x - 1)^2 + 3$

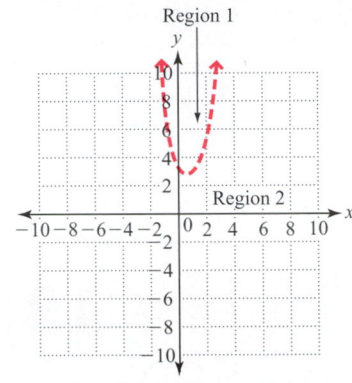

SOLUTION We first graph the related equation, $y = 2(x - 1)^2 + 3$.

> **Note:** The parabola is dashed because the inequality is $>$. Also notice that the graph divides the coordinate plane into two regions: region 1 *above* the boundary and region 2 *below*.

We now test an ordered pair from each region.

> **Note:** To make computations easier, choose ordered pairs with 0 for at least one of the coordinates.

Region 1: We choose $(0, 6)$.

$6 > 2(0 - 1)^2 + 3$ **Substitute.**

$6 > 2(-1)^2 + 3$ **Simplify.**

$6 > 5$

True; so region 1 is in the solution set.

Region 2: We choose $(0, 0)$.

$0 > 2(0 - 1)^2 + 3$ **Substitute.**

$0 > 2(-1)^2 + 3$ **Simplify.**

$0 > 5$

False; so region 2 is not in the solution set.

Because ordered pairs only in region 1 solve the inequality, we shade only that region. The solution set contains all ordered pairs in region 1. Ordered pairs on the parabola are not in the solution set.

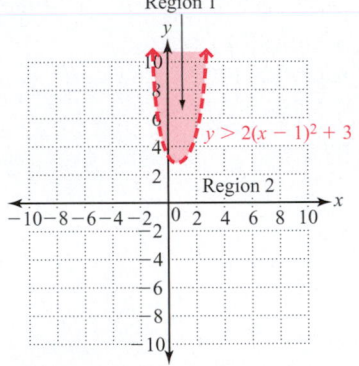

b. $\dfrac{x^2}{9} - \dfrac{y^2}{25} \leq 1$

SOLUTION Graph the related equation, $\dfrac{x^2}{9} - \dfrac{y^2}{25} = 1$.

> **Note:** The hyperbola is solid because the inequality is $\leq$. Also notice that the hyperbola divides the coordinate plane into three regions.

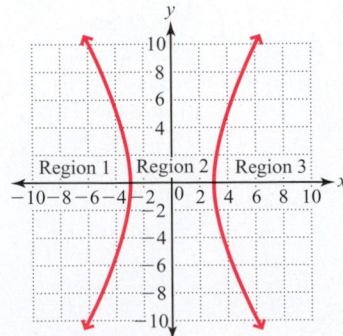

We test an ordered pair from each region.

Region 1: We choose $(-5, 0)$.

$$\frac{(-5)^2}{9} - \frac{0^2}{25} \leq 1 \qquad \textbf{Substitute.}$$

$$\frac{25}{9} \leq 1 \qquad \textbf{Simplify.}$$

False; so region 1 is not in the solution set.

Region 3: We choose $(5, 0)$.

$$\frac{(5)^2}{9} - \frac{0^2}{25} \leq 1 \qquad \textbf{Substitute.}$$

$$\frac{25}{9} \leq 1 \qquad \textbf{Simplify.}$$

False; so region 3 is not in the solution set.

Because region 2 is the only region containing ordered pairs that solve the inequality, we shade only that region. The solution set contains all ordered pairs on the hyperbola along with all ordered pairs in region 2.

Region 2: We choose $(0, 0)$.

$$\frac{0^2}{9} - \frac{0^2}{25} \leq 1 \qquad \textbf{Substitute.}$$

$$0 \leq 1 \qquad \textbf{Simplify.}$$

True; so region 2 is in the solution set.

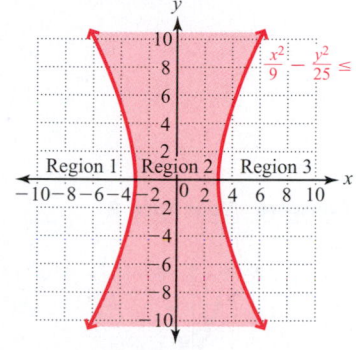

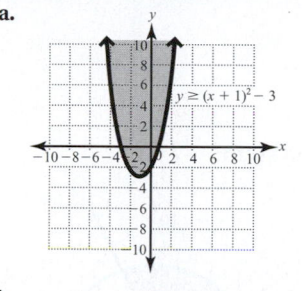
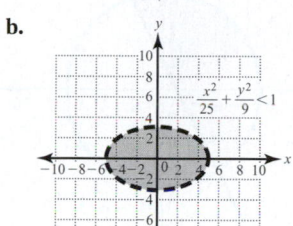
Your Turn 1   Graph the inequality.

a. $y \geq (x + 1)^2 - 3$

b. $\dfrac{x^2}{25} + \dfrac{y^2}{9} < 1$

Objective ② Graph the solution set of a system of nonlinear inequalities.   In Section 9.7, we solved systems of linear inequalities by graphing each inequality on the same grid. The solution set of the system is the intersection of the solution sets of the individual inequalities. We use a similar procedure for systems of nonlinear inequalities.

Example 2   Graph the solution set of the system of inequalities.

a. $\begin{cases} y \geq x^2 - 4 \\ 2x - y < 2 \end{cases}$

SOLUTION   We begin by graphing $y \geq x^2 - 4$ and $2x - y < 2$.

Note: The boundary graph, $y = x^2 - 4$, is a parabola opening up with vertex at $(0, -4)$. The test point $(0, 0)$ gives the true statement $0 \geq -4$, so we shade the region containing $(0, 0)$.

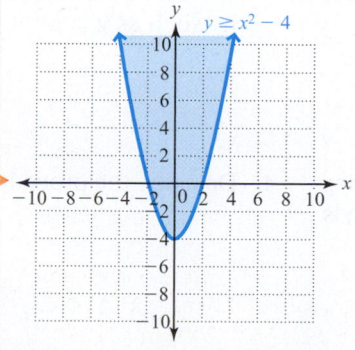

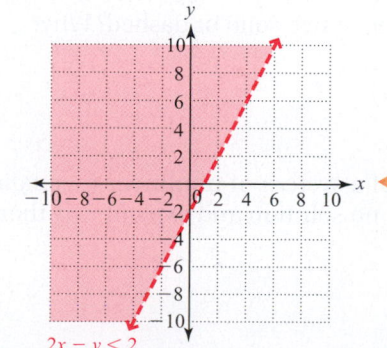

Note: The boundary graph, $2x - y = 2$, is a dashed line. The test point $(0, 0)$ gives a true statement $0 < 2$, so we shade the region containing $(0, 0)$.

If we place both graphs on the same grid, their intersection (purple shading) is the solution region for the system. In addition to ordered pairs in the purple shaded region, the solution set also contains all ordered pairs on the portions of the parabola that touch the purple shaded region.

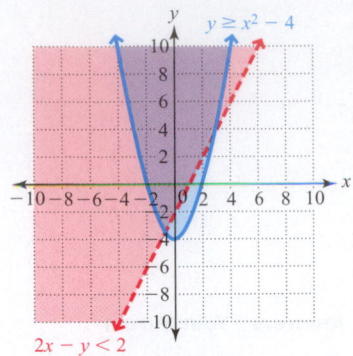

b.
$$\begin{cases} x^2 + y^2 \le 49 \\ \dfrac{x^2}{16} - \dfrac{y^2}{9} < 1 \\ y \ge 2x + 2 \end{cases}$$

SOLUTION Graph each inequality on the same coordinate system.

The graph of $x^2 + y^2 \le 49$ is a circle and its interior.

The graph of $\dfrac{x^2}{16} - \dfrac{y^2}{9} < 1$ is the region between the branches of the hyperbola with the curve dashed. The graph of $y \ge 2x + 2$ is a solid line and the region above the line. The solution set for the system contains all ordered pairs in the region where the three graphs overlap (purple shaded region) together with all ordered pairs on the portion of the circle and the line that touches the purple shaded region.

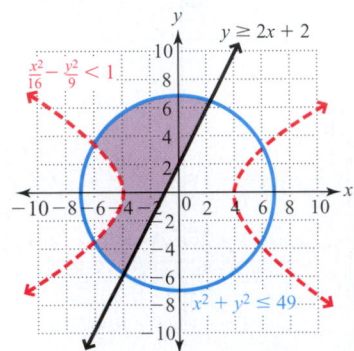

Answers Your Turn 2

a.

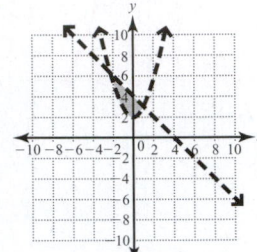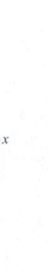

b.

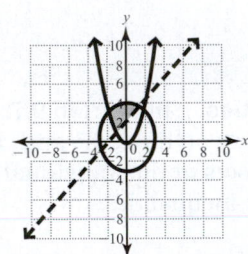

Your Turn 2 Graph the solution set of the system of inequalities.

a. $\begin{cases} y > x^2 + 2 \\ x + y < 4 \end{cases}$

b. $\begin{cases} \dfrac{x^2}{9} + \dfrac{y^2}{16} \le 1 \\ y \ge x^2 \\ -x + y > 2 \end{cases}$

13.4 Exercises   For Extra Help   *MyMathLab*

1. Do the ordered pairs that lie on the boundary curve solve the inequality $9x^2 + 4y^2 > 36$? Why or why not? Would the graph be drawn solid or dashed? Why?

2. Discuss how graphing a linear inequality such as $x + 2y \ge 4$ is similar to graphing a nonlinear inequality such as $y \ge x^2 + 3$.

3. Draw the graph of a system of nonlinear inequalities for which there is no solution and explain why there is no solution.

4. Describe the procedure used to graph $\dfrac{x^2}{9} - \dfrac{y^2}{16} > 1$.

For Exercises 5–24, graph the inequality. See Example 1.

5. $x^2 + y^2 \leq 9$

6. $x^2 + y^2 \geq 4$

7. $y > x^2$

8. $y < -x^2$

9. $y > 2(x + 2)^2 - 3$

10. $y < (x - 1)^2 + 2$

11. $\dfrac{x^2}{16} + \dfrac{y^2}{9} \leq 1$

12. $\dfrac{x^2}{4} + \dfrac{y^2}{9} \geq 1$

13. $\dfrac{x^2}{25} - \dfrac{y^2}{4} > 1$

14. $\dfrac{x^2}{36} - \dfrac{y^2}{16} < 1$

15. $x^2 + y^2 < 16$

16. $x^2 + y^2 > 25$

17. $y \geq -2(x - 3)^2 + 3$

18. $y < -(x + 3)^2 - 2$

19. $\dfrac{x^2}{16} + \dfrac{y^2}{4} \leq 1$

20. $\dfrac{x^2}{49} + \dfrac{y^2}{25} > 1$

21. $\dfrac{y^2}{25} - \dfrac{x^2}{4} \leq 1$

22. $\dfrac{y^2}{36} - \dfrac{x^2}{25} > 1$

23. $y < x^2 + 4x - 5$

24. $y > x^2 - 6x - 6$

For Exercises 25–50, graph the solution set of each system of inequalities.
See Example 2.

25. $\begin{cases} y < -x^2 \\ 2x - y < 4 \end{cases}$

26. $\begin{cases} y \geq x^2 \\ x + y \leq 3 \end{cases}$

27. $\begin{cases} 3x + 2y \geq -6 \\ x^2 + y^2 \leq 25 \end{cases}$

28. $\begin{cases} x - 2y < -4 \\ x^2 + y^2 < 16 \end{cases}$

29. $\begin{cases} x^2 + y^2 > 4 \\ x^2 + y^2 > 9 \end{cases}$

30. $\begin{cases} x^2 + y^2 \geq 9 \\ x^2 + y^2 \leq 25 \end{cases}$

31. $\begin{cases} y > x^2 + 1 \\ 2x + y < 3 \end{cases}$

32. $\begin{cases} 3x - y \leq 2 \\ y \leq -x^2 + 2 \end{cases}$

33. $\begin{cases} y < -x^2 + 3 \\ y > x^2 - 2 \end{cases}$

34. $\begin{cases} y > -x^2 + 2 \\ y < x^2 + 5 \end{cases}$

35. $\begin{cases} \dfrac{x^2}{25} + \dfrac{y^2}{9} \leq 1 \\ x^2 + y^2 \geq 4 \end{cases}$

36. $\begin{cases} \dfrac{x^2}{9} + \dfrac{y^2}{4} \leq 1 \\ x^2 + y^2 \leq 4 \end{cases}$

37. $\begin{cases} \dfrac{x^2}{25} + \dfrac{y^2}{9} < 1 \\ y > x^2 + 1 \end{cases}$

38. $\begin{cases} \dfrac{x^2}{9} + \dfrac{y^2}{4} < 1 \\ y < -x^2 + 2 \end{cases}$

39. $\begin{cases} \dfrac{x^2}{9} - \dfrac{y^2}{4} \leq 1 \\ \dfrac{x^2}{25} + \dfrac{y^2}{9} \leq 1 \end{cases}$

40. $\begin{cases} \dfrac{x^2}{16} - \dfrac{y^2}{9} \geq 1 \\ \dfrac{x^2}{36} + \dfrac{y^2}{16} \leq 1 \end{cases}$

41. $\begin{cases} \dfrac{x^2}{4} - \dfrac{y^2}{4} > 1 \\ y > 2 \end{cases}$

42. $\begin{cases} \dfrac{x^2}{9} - \dfrac{y^2}{9} > 1 \\ x > 3 \end{cases}$

43. $\begin{cases} 3x + 2y \leq 6 \\ x - y > -3 \\ x + 6y \geq 2 \end{cases}$

44. $\begin{cases} 2x + 3y < 6 \\ x - 2y > -4 \\ x + 5y \geq -4 \end{cases}$

45.
$$\begin{cases} \dfrac{x^2}{16} + \dfrac{y^2}{4} \le 1 \\ x^2 + y^2 \le 9 \\ y \le x \end{cases}$$

46.
$$\begin{cases} \dfrac{x^2}{9} + \dfrac{y^2}{16} < 1 \\ x^2 + y^2 < 9 \\ y > x + 1 \end{cases}$$

47.
$$\begin{cases} \dfrac{x^2}{49} - \dfrac{y^2}{16} \le 1 \\ \dfrac{x^2}{64} + \dfrac{y^2}{36} \le 1 \\ 2x - y \le -3 \end{cases}$$

48.
$$\begin{cases} \dfrac{x^2}{9} + \dfrac{y^2}{49} \le 1 \\ y \ge x^2 + 3 \\ x + y \le 6 \end{cases}$$

49.
$$\begin{cases} y < 2x^2 + 8 \\ 2x + y > 3 \\ x - y < 4 \end{cases}$$

50.
$$\begin{cases} y < x^2 + 2 \\ 2x + y < 4 \\ 2x - y > -5 \end{cases}$$

Review Exercises

Exercises 1–5   EXPRESSIONS

[1.5]   **1.** Simplify: $(3 \cdot 1 + 2) + (3 \cdot 2 + 2) + (3 \cdot 3 + 2) + (3 \cdot 4 + 2)$

[1.7]   **2.** Evaluate: $\dfrac{n}{2}(a_1 + a_n)$ if $n = 12$, $a_1 = -8$, and $a_n = 60$

[1.7]   **3.** Evaluate: $a_1 + (n - 1)d$ if $a_1 = -12$, $n = 25$, and $d = -3$

[1.7] *For Exercises 4 and 5, evaluate the expression for n = 1, 2, 3, and 4.*

   4. $2n^2 - 3$

   5. $\dfrac{(-1)^n}{3n - 2}$

Exercise 6   EQUATIONS AND INEQUALITIES

[3.3]   **6.** The sum of three consecutive odd integers is
207. Find the intgers.

Chapter 13 Summary

Defined Terms

Section 13.1
Conic section (p. 904)
Circle (p. 907)
Radius (p. 907)

Section 13.2
Ellipse (p. 916)
Hyperbola (p. 918)

Section 13.3
Nonlinear system of
equations (p. 926)

Procedures, Rules, and Key Examples

Procedures/Rules	Key Examples

SECTION 13.1 The Parabola and Circle

The graph of an equation in the form $x = a(y - k)^2 + h$ is a parabola with vertex at (h, k). The parabola opens to the right if $a > 0$ and to the left if $a < 0$. The equation of the axis of symmetry is $y = k$.

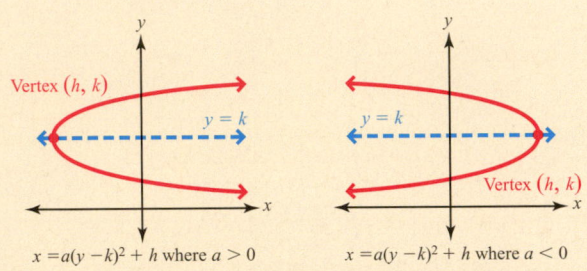

Vertex (h, k)
$y = k$
$x = a(y - k)^2 + h$ where $a > 0$

$y = k$
Vertex (h, k)
$x = a(y - k)^2 + h$ where $a < 0$

If necessary, complete the square to put the equation in the proper form.

Example 1: For $x = y^2 - 4y - 5$, determine whether the graph opens left or right, find the vertex and axis of symmetry, and draw the graph.

Solution: Because $a = 1$ and $1 > 0$, the graph opens to the right. To find the vertex and axis of symmetry, complete the square.

$$x = y^2 - 4y - 5$$
$$x + 5 = y^2 - 4y \qquad \text{Add 5 to both sides.}$$
$$x + 5 + 4 = y^2 - 4y + 4 \qquad \text{Add 4 to both sides.}$$
$$x + 9 = (y - 2)^2 \qquad \text{Simplify.}$$
$$x = (y - 2)^2 - 9 \qquad \text{Subtract 9 from both sides.}$$

The vertex is $(-9, 2)$, and the equation of the axis of symmetry is $y = 2$.

To graph, choose values of y near the axis of symmetry and plot the graph.

x	y
-5	0
-8	1
-8	3
-5	4

Vertex $(-9, 2)$
$y = 2$
$x = y^2 - 4y - 5$

The distance, d, between two points with coordinates (x_1, y_1) and (x_2, y_2) can be found using the formula

$$d = \sqrt{(x_2 - x_1)^2 + (y_2 - y_1)^2}.$$

Example 2: Find the distance and midpoint between $(-4, 3)$ and $(2, -1)$.

Solution: Let $(x_1, y_1) = (-4, 3)$ and $(x_2, y_2) = (2, -1)$.

$$d = \sqrt{(x_2 - x_1)^2 + (y_2 - y_1)^2}$$
$$d = \sqrt{(2 - (-4))^2 + (-1 - 3)^2} \qquad \text{Substitute.}$$
$$d = \sqrt{52} \qquad \text{Simplify.}$$
$$d = \sqrt{4 \cdot 13}$$
$$d = 2\sqrt{13}$$

If the coordinates of the endpoints of a line segment are (x_1, y_1) and (x_2, y_2), the coordinates of the midpoint are

$$\left(\frac{x_1 + x_2}{2}, \frac{y_1 + y_2}{2} \right).$$

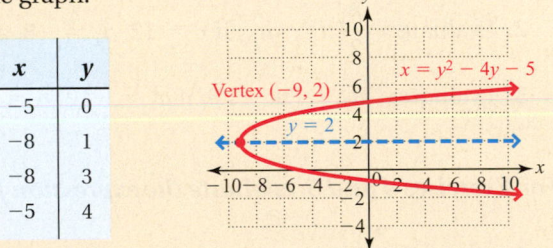

$$\text{Midpoint} = \left(\frac{x_1 + x_2}{2}, \frac{y_1 + y_2}{2} \right)$$
$$\text{Midpoint} = \left(\frac{-4 + 2}{2}, \frac{3 + (-1)}{2} \right) \qquad \text{Substitute.}$$
$$\text{Midpoint} = \left(\frac{-2}{2}, \frac{2}{2} \right) \qquad \text{Simplify.}$$
$$\text{Midpoint} = (-1, 1)$$

(*continued*)

Procedures/Rules	Key Examples

SECTION 13.1   The Parabola and Circle (*continued*)

The equation of a circle with center at (h, k) and radius r is $(x - h)^2 + (y - k)^2 = r^2$.

Example 3: Find the center and radius of the circle whose equation is $(x - 2)^2 + (y + 4)^2 = 25$ and draw the graph.

Solution: Rewrite the equation as $(x - \mathbf{2})^2 + (y - (\mathbf{-4}))^2 = \mathbf{5}^2$. Because $h = 2$ and $k = -4$, the center is $(2, -4)$ and $r = 5$.

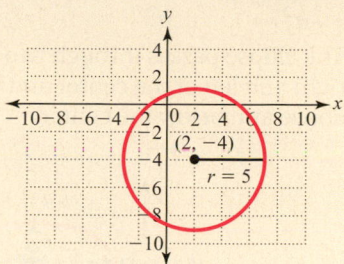

Example 4: Write the equation of the circle whose center is $(5, -6)$ and whose radius is 8.

Solution: Substitute for h, k, and r in

$$(x - h)^2 + (y - k)^2 = r^2.$$
$$(x - \mathbf{5})^2 + (y - (\mathbf{-6}))^2 = \mathbf{8}^2$$
$$(x - 5)^2 + (y + 6)^2 = 64 \qquad \textbf{Simplify.}$$

To graph circles of the form $x^2 + y^2 + dx + ey + f = 0$, complete the square in x and y to write the equation in the form $(x - h)^2 + (y - k)^2 = r^2$.

Example 5: Find the center and radius of the circle whose equation is $x^2 + y^2 + 8x - 2y + 8 = 0$ and draw the graph.

Solution: Complete the square in x and y.

$$x^2 + 8x + y^2 - 2y = -8 \qquad \textbf{Group } x \textbf{ and } y \textbf{ terms;}$$
$$\textbf{subtract 8.}$$
$$x^2 + 8x + \mathbf{16} + y^2 - 2y + \mathbf{1} = -8 + \mathbf{16} + \mathbf{1}$$
$$\textbf{Complete the square.}$$
$$(x + 4)^2 + (y - 1)^2 = 9 \qquad \textbf{Factor and simplify.}$$

The center is $(-4, 1)$, and the radius is 3.

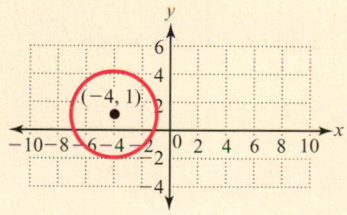

SECTION 13.2   Ellipses and Hyperbolas

The equation of an ellipse with center $(0, 0)$, x-intercepts $(a, 0)$ and $(-a, 0)$, and y-intercepts $(0, b)$ and $(0, -b)$ is $\dfrac{x^2}{a^2} + \dfrac{y^2}{b^2} = 1$.

Example 1: Graph $\dfrac{x^2}{49} + \dfrac{y^2}{36} = 1$.

Solution: The equation can be rewritten as $\dfrac{x^2}{7^2} + \dfrac{y^2}{6^2} = 1$, so $a = 7$ and $b = 6$. Consequently, the x-intercepts are $(7, 0)$ and $(-7, 0)$ and the y-intercepts are $(0, 6)$ and $(0, -6)$.

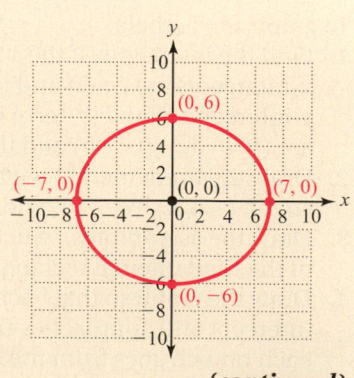

(*continued*)

Procedures/Rules	Key Examples

SECTION 13.2 Ellipses and Hyperbolas (*continued*)

The equation of an ellipse with center (h, k) is
$$\frac{(x-h)^2}{a^2} + \frac{(y-k)^2}{b^2} = 1.$$

The ellipse passes through two points that are a units to the right and left of the center and two points that are b units above and below the center.

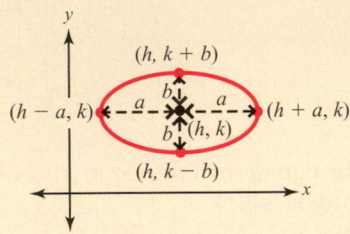

The equation of a hyperbola with center $(0, 0)$, x-intercepts $(a, 0)$ and $(-a, 0)$, and no y-intercepts is $\dfrac{x^2}{a^2} - \dfrac{y^2}{b^2} = 1$.

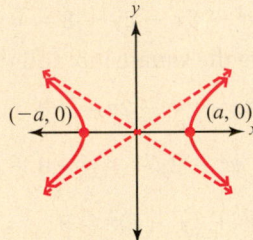

The equation of a hyperbola with center $(0, 0)$, y-intercepts $(0, b)$ and $(0, -b)$, and no x-intercepts is $\dfrac{y^2}{b^2} - \dfrac{x^2}{a^2} = 1$.

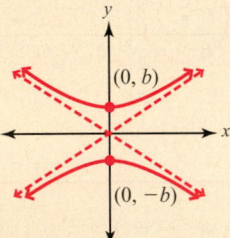

To graph a hyperbola,
1. Find the intercepts. If the x^2-term is positive, the x-intercepts are $(a, 0)$ and $(-a, 0)$ and there are no y-intercepts. If the y^2-term is positive, the y-intercepts are $(0, b)$ and $(0, -b)$ and there are no x-intercepts.
2. Draw the fundamental rectangle. The vertices are (a, b), $(-a, b)$, $(a, -b)$, and $(-a, -b)$.
3. Draw the asymptotes, which are the extended diagonals of the fundamental rectangle.
4. Draw the graph so that each branch passes through an intercept and approaches the asymptotes the farther each branch goes from the origin.

Example 2: Graph $\dfrac{(x+2)^2}{9} + \dfrac{(y-1)^2}{25} = 1$.

Solution: The equation can be rewritten as

$$\frac{(x-(-2))^2}{3^2} + \frac{(y-1)^2}{5^2} = 1,\text{ so } h = -2, k = 1, a = 3,$$

and $b = 5$. The center is $(-2, 1)$. To find other points on the ellipse, go 3 units right and left of the center and 5 units up and down from the center.

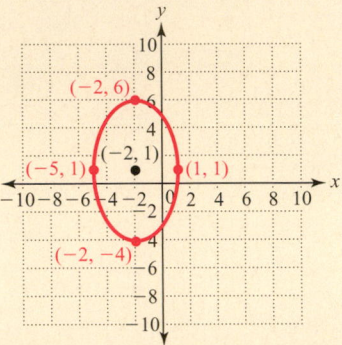

Example 3: Graph $\dfrac{x^2}{36} - \dfrac{y^2}{16} = 1$.

Solution: The equation can be rewritten as $\dfrac{x^2}{6^2} - \dfrac{y^2}{4^2} = 1$, so $a = 6$ and $b = 4$. Because the x^2-term is positive, the x-intercepts are $(6, 0)$ and $(-6, 0)$ and there are no y-intercepts. The fundamental rectangle has vertices $(6, 4)$, $(-6, 4)$, $(6, -4)$, and $(-6, -4)$.

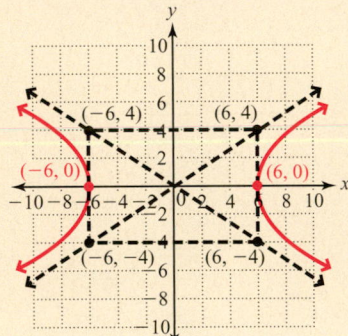

(continued)

Procedures/Rules	Key Examples

SECTION 13.3 Nonlinear Systems of Equations

If a nonlinear system has a linear equation, the substitution method is usually preferred. Solve the linear equation for one of its variables and substitute for that variable in the other equation.

Example 1: Solve the system $\begin{cases} y = (x-3)^2 + 2 \\ 2x - y = -4 \end{cases}$ by substitution.

(**Note:** This is a parabola and a line.)

Solution:

$$y = 2x + 4 \qquad \text{Solve } 2x - y = -4 \text{ for } y.$$
$$2x + 4 = (x-3)^2 + 2 \qquad \text{Substitute } 2x + 4 \text{ for } y.$$
$$2x + 4 = x^2 - 6x + 11 \qquad \text{Simplify.}$$
$$0 = x^2 - 8x + 7 \qquad \text{Write in form } ax^2 + bx + c = 0.$$
$$0 = (x-7)(x-1) \qquad \text{Factor.}$$
$$x - 7 = 0 \quad \text{or} \quad x - 1 = 0 \qquad \text{Use the zero-factor theorem.}$$
$$x = 7 \quad \text{or} \qquad x = 1 \qquad \text{Solve each equation.}$$

To find y, substitute 7 and 1 into $y = 2x + 4$.

$$y = 2(7) + 4 \qquad y = 2(1) + 4$$
$$y = 18 \qquad\qquad y = 6$$

Solutions: $(7, 18), (1, 6)$

If neither equation is linear, the elimination method is often preferred.

Example 2: Solve the system $\begin{cases} 4x^2 + 5y^2 = 36 & \text{(Equation 1)} \\ x^2 + y^2 = 8 & \text{(Equation 2)} \end{cases}$ by elimination.

(**Note:** This is an ellipse and a circle.)

Solution: To eliminate x^2, multiply equation 2 by -4 and add the equations.

$$\begin{array}{ll} 4x^2 + 5y^2 = 36 & 4x^2 + 5y^2 = 36 \\ x^2 + y^2 = 8 \longrightarrow & \underline{-4x^2 - 4y^2 = -32} \\ \quad\text{Multiply by } -4 & \qquad\quad y^2 = 4 \\ & \qquad\quad y = \pm 2 \end{array}$$

Substitute 2 and -2 for y into either equation to find x. Let's use $x^2 + y^2 = 8$.

$$\begin{array}{ll} x^2 + 2^2 = 8 & x^2 + (-2)^2 = 8 \\ x^2 + 4 = 8 & x^2 + 4 = 8 \\ x^2 = 4 & x^2 = 4 \\ x = \pm 2 & x = \pm 2 \end{array}$$

Solutions: $(2, 2), (2, -2), (-2, 2),$ and $(-2, -2)$

SECTION 13.4 Nonlinear Inequalities and Systems of Inequalities

Graphing nonlinear inequalities
1. Graph the related equation (the boundary curve). If the inequality symbol is $\leq$ or $\geq$, draw the graph as a solid curve. If the inequality symbol is $<$ or $>$, draw a dashed curve.
2. The graph divides the coordinate plane into at least two regions. Test an ordered pair from each region by substituting it into the inequality. If the ordered pair satisfies the inequality, shade the region containing that ordered pair.

Example 1: Graph $y \geq (x-2)^2 - 1$.

Solution: The related equation is a parabola with vertex $(2, -1)$ that opens upward. Because the inequality is $\geq$, all ordered pairs on the parabola are in the solution set; so we draw it with a solid curve.

We choose $(0, 0)$ as a test point.

$$0 \geq (0-2)^2 - 1$$

$0 \geq 3$, which is false; so $(0, 0)$ is not in the solution set.

Choose $(2, 0)$ as a test point.

$$0 \geq (2-2)^2 - 1$$

$0 \geq -1$, which is true; so $(2, 0)$ is in the solution set. Shade the region that contains $(2, 0)$.

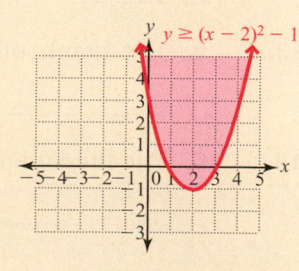

(*continued*)

SECTION 13.4 Nonlinear Inequalities and Systems of Inequalities (*continued*)

To graph a system of inequalities, graph the solution set of each individual inequality on the same grid. The solution set of the system is the intersection of the solution sets of the individual inequalities.

Example 2: Graph the solution set of

$$\begin{cases} \dfrac{x^2}{4} + \dfrac{y^2}{16} \le 1 \\ x^2 + y^2 \ge 9 \end{cases}.$$

Solution: The graph of $\dfrac{x^2}{4} + \dfrac{y^2}{16} \le 1$ is an ellipse and all points inside the ellipse. The graph of $x^2 + y^2 \ge 9$ is a circle and all points outside the circle. So the solution set of the system is the set of all points inside the ellipse and outside the circle.

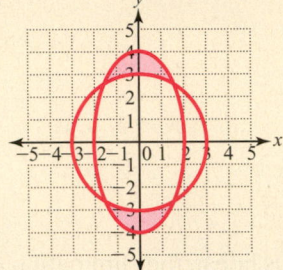

Chapter 13 Review Exercises

For Exercises 1–6, answer true or false.

[13.1] **1.** The graph of $x = -3(y - 2)^2 + 5$ opens down.

[13.1] **2.** The graph of $(x + 2)^2 + (y - 3)^2 = 36$ is a circle with center at $(-2, 3)$ and radius 6.

[13.2] **3.** A parabola has asymptotes.

[13.2] **4.** Choose any two points on an ellipse. The sum of their distances from the foci is the same.

[13.3] **5.** It is possible for a circle and a parabola to intersect in three points.

[13.4] **6.** The graph of the system $\begin{cases} x^2 + y^2 > 36 \\ 2x + y > 3 \end{cases}$ lies inside the circle and above the line.

For Exercises 7–10, fill in the blanks.

[13.1] **7.** The distance between the points whose coordinates are $(-3, 4)$ and $(3, -4)$ is _____.

[13.1] **8.** The graph of $x^2 + y^2 + 8x - 2y + 13 = 0$ is a circle with center at _____ and radius of _____.

[13.2] **9.** The graph of $\dfrac{x^2}{16} + \dfrac{y^2}{36} = 1$ is a(n) _____ with x-intercepts at _____.

[13.2] **10.** The fundamental rectangle for the graph of $\dfrac{x^2}{25} - \dfrac{y^2}{16} = 1$ has vertices at _____.

[13.1] *For Exercises 11–14, find the direction that the parabola opens, the coordinates of the vertex, and the equation of the axis of symmetry. Draw the graph.*

11. $y = 2(x - 3)^2 - 5$　　**12.** $y = -2(x + 2)^2 + 3$　　**13.** $x = -(y - 2)^2 + 4$　　**14.** $x = 2(y + 3)^2 - 2$

[13.1] *For Exercises 15–18, find the direction that the parabola opens, the coordinates of the vertex, and the equation of the axis of symmetry. Draw the graph.*

15. $x = y^2 + 6y + 8$　　**16.** $x = 2y^2 - 8y - 6$　　**17.** $x = -y^2 - 2y + 3$　　**18.** $x = -3y^2 - 12y - 9$

[13.1] *For Exercises 19 and 20, find the distance and midpoint between the two points.*

19. $(-1, -2)$ and $(-5, 1)$

20. $(2, -5)$ and $(-2, 3)$

[13.1] **21.** The center of a circle is at $(6, -4)$, and the circle passes through $(-2, 2)$. What is the radius of the circle?

[13.1] **22.** The center of a circle is at $(-6, 8)$, and the circle passes through $(3, -4)$. What is the equation of the circle?

[13.1] *For Exercises 23–26, find the center and radius and draw the graph.*

23. $(x - 3)^2 + (y + 2)^2 = 25$

24. $(x + 5)^2 + (y - 1)^2 = 4$

25. $x^2 + y^2 - 4x + 8y + 11 = 0$

26. $x^2 + y^2 + 10x + 2y + 22 = 0$

[13.1] *For Exercises 27 and 28, the center and radius of a circle are given. Write the equation of each circle in standard form.*

27. Center: $(6, -8), r = 9$

28. Center: $(-3, -5), r = 10$

[13.1]   29. If a heavy object is thrown vertically upward with an initial velocity of 32 feet per second from the top of a building 128 feet high, its height above the ground after t seconds is given by $h = -16t^2 + 32t + 128$, where h is the height in feet and t is the time in seconds.
   a. What is the maximum height the object will reach?
   b. How many seconds will the object take to reach its maximum height?
   c. How many seconds will the object take to strike the ground?

[13.1]   30. To lay out the border of a circular flower bed, sticks are tied to each end of a rope that is 20 feet long. One person holds one of the sticks stationary while another person uses the other stick to trace out a circular path while keeping the rope taut. If the center of the circle is at the stationary stick, what is the equation of the circle?

[13.2] *For Exercises 31–36, graph each ellipse. Label the center and the points directly above, below, to the left, and to the right of the center.*

31. $\dfrac{x^2}{49} + \dfrac{y^2}{25} = 1$

32. $\dfrac{x^2}{9} + \dfrac{y^2}{25} = 1$

33. $4x^2 + 9y^2 = 36$

34. $25x^2 + 9y^2 = 225$

35. $\dfrac{(x + 2)^2}{16} + \dfrac{(y - 3)^2}{4} = 1$

36. $\dfrac{(x - 1)^2}{9} + \dfrac{(y + 4)^2}{25} = 1$

[13.2] *For Exercises 37–40, graph each hyperbola. Also show the fundamental rectangle with its corner points labeled, the asymptotes, and the intercepts.*

37. $\dfrac{x^2}{25} - \dfrac{y^2}{16} = 1$

38. $\dfrac{x^2}{36} - \dfrac{y^2}{9} = 1$

39. $y^2 - 9x^2 = 36$

40. $25y^2 - 9x^2 = 225$

[13.2]   41. A bridge over a canal is in the shape of half an ellipse. If the highest point of the bridge is 15 feet above the water and the base of the bridge is 50 feet across, what is the equation of the ellipse, half of which forms the bridge?

[13.2] 42. The cam of a compound bow is elliptical and is 4 inches long and 3.25 inches wide, as shown in the figure. What is the equation of the ellipse forming the shape of the cam?

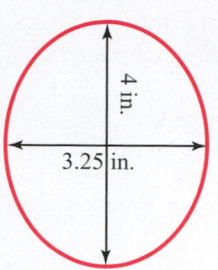

[13.3] *For Exercises 43–50, solve.*

43. $\begin{cases} y = 2x^2 - 3 \\ 2x - y = -1 \end{cases}$

44. $\begin{cases} x^2 + y^2 = 17 \\ x - y = 3 \end{cases}$

45. $\begin{cases} 25x^2 + 3y^2 = 100 \\ 2x - y = -3 \end{cases}$

46. $\begin{cases} x^2 + y^2 = 64 \\ x^2 - y^2 = 64 \end{cases}$

47. $\begin{cases} x^2 + y^2 = 25 \\ 25y^2 - 16x^2 = 256 \end{cases}$

48. $\begin{cases} x^2 + 5y^2 = 36 \\ 4x^2 - 7y^2 = 36 \end{cases}$

49. $\begin{cases} y = x^2 - 2 \\ 4x^2 + 5y^2 = 36 \end{cases}$

50. $\begin{cases} y = x^2 - 1 \\ 4y^2 - 5x^2 = 16 \end{cases}$

[13.3] 51. The sum of the squares of two integers is 89, and the difference of their squares is 39. Find the integers.

[13.3] 52. A rectangular rug has a perimeter of 36 feet and an area of 80 square feet. Find the dimensions of the rug.

[13.3] *For Exercises 53 and 54, use a graphing calculator to verify the results of the exercise given.*

53. Exercise 43

54. Exercise 49

[13.4] *For Exercises 55–58, graph the inequality.*

55. $x^2 + y^2 \le 64$

56. $y < 2(x - 3)^2 + 4$

57. $\dfrac{y^2}{25} + \dfrac{x^2}{49} > 1$

58. $\dfrac{x^2}{25} - \dfrac{y^2}{36} < 1$

[13.4] *For Exercises 59–64, graph the solution set of the system of inequalities.*

59. $\begin{cases} y \ge x^2 - 3 \\ 2x + y < 2 \end{cases}$

60. $\begin{cases} x + 2y < 4 \\ x^2 + y^2 \le 25 \end{cases}$

61. $\begin{cases} y > x^2 - 2 \\ y < -x^2 + 1 \end{cases}$

62. $\begin{cases} \dfrac{x^2}{9} + \dfrac{y^2}{25} \leq 1 \\ x^2 + y^2 \geq 9 \end{cases}$

63. $\begin{cases} \dfrac{y^2}{4} - \dfrac{x^2}{9} \leq 1 \\ \dfrac{x^2}{9} + \dfrac{y^2}{25} \leq 1 \end{cases}$

64. $\begin{cases} \dfrac{y^2}{4} + \dfrac{x^2}{16} \leq 1 \\ x^2 + y^2 \leq 9 \\ y \leq x + 1 \end{cases}$

Chapter 13 Practice Test

For Extra Help Step-by-step test solutions are found on the Chapter Test Prep Videos CHAPTER available via the Video Resources on DVD, in *MyMathLab*, and on **Test Prep** YouTube (search "CarsonCombinedAlg" and click on "Channels").

For Exercises 1–3, find the direction the parabola opens, the coordinates of the vertex, and the equation of the axis of symmetry. Draw the graph.

1. $y = 2(x + 1)^2 - 4$

2. $x = -2(y - 3)^2 + 1$

3. $x = y^2 + 4y - 3$

4. Find the distance and midpoint between the points $(2, -3)$ and $(6, -5)$.

For Exercises 5 and 6, find the center and radius and draw the graph.

5. $(x + 4)^2 + (y - 3)^2 = 36$

6. $x^2 + y^2 + 4x - 10y + 20 = 0$

7. Write the equation of the circle with center $(2, -4)$ that passes through the point $(-4, 4)$.

For Exercises 8–11, graph the equation and label relevant points. If the graph is a hyperbola, show the fundamental rectangle with its corner points labeled, the asymptotes, and the intercepts.

8. $16x^2 + 36y^2 = 576$

9. $\dfrac{(x + 2)^2}{4} + \dfrac{(y - 1)^2}{25} = 1$

10. $\dfrac{x^2}{49} - \dfrac{y^2}{25} = 1$

11. $\dfrac{y^2}{9} - \dfrac{x^2}{16} = 1$

For Exercises 12–15, solve.

12. $\begin{cases} y = (x + 1)^2 + 2 \\ 2x + y = 8 \end{cases}$

13. $\begin{cases} 3x - y = 4 \\ x^2 + y^2 = 34 \end{cases}$

14. $\begin{cases} x^2 + y^2 = 13 \\ 3x^2 + 4y^2 = 48 \end{cases}$

15. $\begin{cases} x^2 - 2y^2 = 1 \\ 4x^2 + 7y^2 = 64 \end{cases}$

For Exercises 16 and 17, graph.

16. $y \leq -2(x + 3)^2 + 2$

17. $\dfrac{x^2}{4} + \dfrac{y^2}{9} > 1$

18. Graph the solution set: $\begin{cases} y \geq x^2 - 4 \\ \dfrac{x^2}{9} + \dfrac{y^2}{16} \leq 1 \end{cases}$

19. An arch is in the shape of a parabola as shown in the figure. If we place the origin, O, as indicated, find the height of the arch and the distance across the base if the equation is $y = -\dfrac{1}{2}x^2 + 18$.

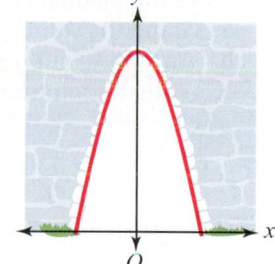

20. The path of a person's foot using an elliptical trainer on a particular setting is an ellipse. If the length of the stride is 19 inches and the height of the stride is 2 inches, what is the equation of the path of the foot?

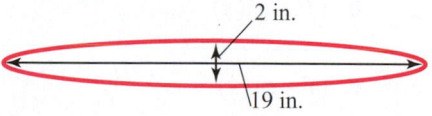

2 in.

19 in.

Chapters 1–13 Cumulative Review Exercises

For Exercises 1–3, answer true or false.

[5.1]   **1.** A positive base raised to a negative exponent simplifies to a negative number.

[3.1]   **2.** If $\dfrac{a}{b} = \dfrac{c}{d}$, then $ad = bc$ where $b \neq 0$ and $d \neq 0$.

[12.3] **3.** Logarithms are exponents.

For Exercises 4–6, fill in the blanks.

[9.6]   **4.** If $A = \begin{bmatrix} a_1 & b_1 \\ a_2 & b_2 \end{bmatrix}$, then $\det(A) = \begin{vmatrix} a_1 & b_1 \\ a_2 & b_2 \end{vmatrix} =$ _____.

[10.7]   **5.** $\sqrt{-1} =$ _____.

[13.1]   **6.** The formula for finding the distance between two points (x_1, y_1) and (x_2, y_2) is _____.

Exercises 7–11 ▲ EXPRESSIONS

For Exercises 7–9, simplify.

[1.5]   **7.** $6 - (-3) + 2^{-4}$ [5.4]   **8.** $(5m^3 n^{-2})^{-3}$ [10.7]   **9.** $\sqrt{-20}$

For Exercises 10 and 11, factor completely.

[6.1]   **10.** $ax + bx + ay + by$ [6.4]   **11.** $k^4 - 81$

Exercises 12–30 ▲ EQUATIONS AND INEQUALITIES

For Exercises 12–17, solve.

[8.2]   **12.** $|2x + 1| = |x - 3|$ [10.6]   **13.** $\sqrt{x - 3} = 7$ [11.2]   **14.** $4x^2 - 2x + 1 = 0$

[11.3]   **15.** $9 + 24x^{-1} + 16x^{-2} = 0$ [12.2]   **16.** $9^x = 27$ [12.3]   **17.** $\log_3(x + 1) - \log_3 x = 2$

[11.5]   **18.** Solve the inequality $5m^2 - 3m < 0$; then:
   a. Graph the solution set.
   b. Write the solution set in set-builder notation.
   c. Write the solution set in interval notation.

$\begin{bmatrix} 8.5, \\ 12.1 \end{bmatrix}$ **19.** Given $f(x) = 2x + 1$ and $g(x) = x^2 - 1$, find the following.
   a. $(f + g)(x)$ **b.** $(f - g)(x)$ **c.** $(f \cdot g)(x)$

   d. $(f/g)(x)$ **e.** $(f \circ g)(x)$ **f.** $(g \circ f)(x)$

[12.3]   **20.** Write $5^3 = 125$ in logarithmic form.

[12.4]   **21.** Write as a sum or difference of multiples of logarithms: $\log_5 x^5 y$

[13.1]   **22.** Write the equation of a circle in standard form with a center at $(2, 4)$ passing through the point $(7, 16)$.

948   Chapter 13   Conic Sections

For Exercises 23–25, graph.

[9.7]   23. $\begin{cases} y < -x + 2 \\ y \geq x - 4 \end{cases}$ [12.3]   24. $f(x) = \log_3 x$ [13.2]   25. $\dfrac{x^2}{4} + \dfrac{y^2}{9} = 1$

For Exercises 26–30, solve.

[2.4]   26. The volume of a cylinder can be found using the formula $V = \pi r^2 h$.

       a. Solve the formula for h.

       b. Suppose the volume of a cylinder is 15π cubic inches. Find its height.

$\begin{bmatrix} 3.3 \text{ or} \\ 9.2 \end{bmatrix}$ 27. During 2008, shrimp overtook tuna as the top-selling seafood in the United States. It is estimated that each person consumed a combined total of 6.3 pounds of these two seafoods and that each person ate $\frac{1}{2}$ pound more shrimp than tuna. How many pounds of each were consumed?

[3.5]   28. A total of $5000 is invested in three funds. The money market fund pays 5%, the income fund pays 6%, and the growth fund pays 3%. The total annual interest from the three accounts is $255, and the amount in the growth fund is $500 less than the amount invested in the money market account. Find the amount invested in each of the funds.

[12.5]   29. The magnitude of an earthquake was defined in 1935 by Charles Richter as $M = \log \dfrac{I}{I_0}$. We let I be the intensity of the earthquake measured by the amplitude of a seismograph reading taken 100 km from the epicenter of the earthquake, and we let I_0 be the intensity of a "standard earthquake" whose amplitude is 10^{-4} cm. On January 26, 2001, an earthquake in Rann of Kutch, Gujarat, was measured with an intensity of $10^{2.9}$. What was the magnitude of this earthquake?

[13.2]   30. A bridge over a canal is in the shape of half an ellipse. If the highest point of the bridge is 20 feet above the water and the base of the bridge is 150 feet across, what is the equation of the ellipse, half of which forms the bridge?

Elementary Algebra Review

R.1 Review of Expressions

Objective ❶ Evaluate an expression.   For a review of this objective, see Section 1.7, pages 70 and 71.

For Exercises 1–6, evaluate the expressions using $x = -2$, $y = 4$, and $z = -3$.

1. $3y - 2(3x + z)$
2. $-y^2 - (3z^2 - 4x)$
3. $\dfrac{2x^3 - 3y^2}{3x^3 - 2z^2}$
4. $|4x^2 - 3y(z^3 + 2y)|$
5. $\sqrt{y} - |-3x + 2z|$
6. $\sqrt{y^2 + z^2} - 2|5xy^2z|$

For additional exercises of this type, see Exercises 7–32 on pages 76 and 77.

Objective ❷ Apply the distributive property.   For a review of this objective, see Section 1.7, pages 72 and 73.

For Exercises 7–14, use the distributive property to write an equivalent expression.

7. $3(2a + 4)$
8. $4(3x + 6)$
9. $-4(r - 5)$
10. $-7(-3a + 5b)$
11. $2.3x(4.2x - 1.7)$
12. $3.5x(2.4x - 6.2)$
13. $\dfrac{3}{4}\left(2b + \dfrac{5}{6}\right)$
14. $\dfrac{5}{8}\left(6x - \dfrac{4}{3}\right)$

For additional exercises of this type, see Exercises 41–48 on page 77.

Objective ❸ Combine like terms.   For a review of this objective, see Section 1.7, pages 74–76.

For Exercises 15–22, combine like terms.

15. $8x + 3x$
16. $-13a + 4a$
17. $9xy + 5x - 3xy + x$
18. $8y - 4k + 2 - 4y + 3 - 9k$
19. $9.3a - 7.6b - 5.7a + 2.4b$
20. $6.3t - 2.8u - 4.7t - 5.7u$
21. $\dfrac{5}{8}x^2 + 4.6y - \dfrac{3}{4}x^2 - 3.8y$
22. $\dfrac{2}{3}x^2y - \dfrac{3}{5}xy^2 + \dfrac{1}{4}x^2y - \dfrac{7}{10}xy^2$

For additional exercises of this type, see Exercises 59–84 on page 78.

R.2 Review of Solving Linear Equations

Objective ❶ Verify solutions to equations.   For a review of this objective, see Section 2.1, pages 92 and 93.

For Exercises 1–8, check to see if the given number is a solution for the given equation.

1. $9y - 5 - 8y = 5y + 10 - 3y; y = -15$ 2. $2x + 6x - 8 = -23 + 5x; x = -5$

3. $5(u - 2) + 7(3 - u) = 9u; u = 1$ 4. $6(p - 4) - 3p = 18 - 3p; p = 6$

5. $x(x + 3) = 40; x = 5$ 6. $a(a + 7) = -12; a = -3$

7. $8x^2 - 15 = 14x; x = -\dfrac{3}{4}$ 8. $6x^2 - 7x = 3; x = \dfrac{1}{3}$

For additional exercises of this type, see Exercises 5–20 on pages 99 and 100.

Objective ② Solve linear equations using the addition principle. For a review of this objective, see Section 2.2, pages 106–112.

For Exercises 9–16, solve and check.

9. $7 + 5 = 6r + 8 - 5r$ 10. $3 + 6 + 6a = 7a + 4$

11. $16x - 5(3x - 2) = 12$ 12. $17x - 8 = 6 + 4(4x - 3)$

13. $-4(5x - 4) + 7(3x - 2) - 6 = 8 - 2$ 14. $3 = 2(4x + 5) - 7(x + 2)$

15. $-2(4.7y - 5.4) + 10.4y = 3.1$ 16. $-3(1.6b - 2.4) + 5.8b = -6.2$

For additional exercises of this type, see Exercises 23–74 on pages 114 and 115.

Objective ③ Solve linear equations using the multiplication principle. For a review of this objective, see Section 2.3, pages 117–119.

For Exercises 17–24, solve and check.

17. $5x = -20$ 18. $-6x = 24$ 19. $-1.2z = 4.8$

20. $3.4a = -10.2$ 21. $\dfrac{3}{4}x = 9$ 22. $\dfrac{8}{5}x = -32$

23. $\dfrac{w}{-4} = -6$ 24. $\dfrac{4}{5}r = -\dfrac{12}{25}$

For additional exercises of this type, see Exercises 7–18 on page 125.

Objective ④ Solve equations using both the addition and multiplication principles. For a review of this objective, see Section 2.3, pages 119–123.

For Exercises 25–34, solve and check.

25. $-7x - 6 = 8$ 26. $3x + 9 = -6$

27. $3x + 4 = -5x + 28$ 28. $18z - 3.8 = 5z + 4$

29. $2.3y - 1.6 - 0.8y = 4.4$ 30. $3a - 4a + 9 = -12a + 18 - 6a$

31. $14(w - 2) + 13 = 4w + 5$ 32. $5(a - 1) - 9(a - 2) = -3(2a + 1) - 2$

33. $4(b - 4) + 2(b + 2) = 2(b - 2)$ 34. $0.3n + 0.2(4n - 1) = 0.4 + 0.5n$

For additional exercises of this type, see Exercises 19–82 on pages 125 and 126.

R.3 Review of Graphing Linear Equations

Objective ① Plot points in the coordinate plane. For a review of this objective, see Section 4.1, pages 243–244.

For Exercises 1 and 2, plot the point described by the coordinates.

1. **a.** $(4, 5)$ **b.** $(0, 3)$ **c.** $(-2, -3)$

2. **a.** $(-3, 2)$ **b.** $(2, 0)$ **c.** $(0, 4)$ **d.** $(1, -5)$

For additional exercises of this type, see Exercises 9–12 on page 247.

Objective ② **Find solutions for equations in two unknowns.** For a review of this objective, see Section 4.2, pages 250–252.

For Exercises 3–6, find three solutions for each equation. (Answers may vary.)

3. $x + 4y = 8$ **4.** $3x + 2y = 6$

5. $y = -2x - 5$ **6.** $y = -\dfrac{2}{3}x + 4$

For additional exercises of this type, see the first part of Exercises 23–58 on pages 256 and 257.

Objective ③ **Graph linear equations by plotting solutions.** For a review of this objective, see Section 4.2, pages 252 and 253.

For Exercises 7–10, find three solutions for the given equations and graph.

7. $x + 2y = 4$ **8.** $3x + 2y = 6$

9. $y = -2x - 2$ **10.** $y = \dfrac{1}{3}x + 1$

For additional exercises of this type, see Exercises 23–58 on pages 256 and 257.

Objective ④ **Graph linear equations using intercepts.** For a review of this objective, see Section 4.3, pages 261–265.

For Exercises 11–16, graph using the x- and y-intercepts.

11. $x + 2y = 4$ **12.** $3x - y = 3$ **13.** $y = -2x$

14. $y = \dfrac{2}{3}x$ **15.** $y = \dfrac{1}{2}x + 2$ **16.** $y = \dfrac{2}{3}x - 2$

For additional exercises of this type, see Exercises 33–64 on pages 267 and 268.

Objective ⑤ **Graph vertical and horizontal lines.** For a review of this objective, see Section 4.2, pages 254–255 and Section 4.3, page 263.

For Exercises 17–20, graph.

17. $y = 2$ **18.** $y = -1$

19. $x = -3$ **20.** $x = 4$

For additional exercises of this type, see Exercises 53–56 on page 257 and Exercises 57–64 on page 268.

R.4 Review of Polynomials

Objective ① **Add and subtract polynomials.** For a review of this objective, see Section 5.3, pages 359–363.

For Exercises 1–4, add or subtract.

1. $(4a^2 - 7a - 4) + (2a^2 + 9a - 6)$

2. $(-3a^3 + 8a^2 - 7a - 3) + (4a^3 + 6a^2 - 9a + 7)$

3. $(6r^2 - 2r + 7) - (4r^2 - 6r + 6)$

4. $(7m^3 - m^2 + 4m - 7) - (-5m^3 + 4m^2 - 9m + 3)$

For additional exercises of this type, see Exercises 5–68 on pages 364–366.

Objective ② Multiply polynomials. For a review of this objective, see Section 5.4, pages 368 and 369, and Section 5.5, pages 378–386.

For Exercises 5–8, multiply the monomials.

5. $(3a^3b^2)(-6a^5b^6c^2)$

6. $(-2p^2q^4)(-6p^7q^3r^5)(3q^4r^5)$

7. $(-2.3m^8n^4)(0.4m^2n^3)$

8. $\left(\dfrac{2}{3}x^3y^5\right)\left(\dfrac{9}{5}x^4y^2\right)$

For additional exercises of this type, see Exercises 7–32 on page 374.

For Exercises 9–12, multiply the polynomial by the monomial.

9. $4a^3(6a^2 - 4a + 3)$

10. $-4x^2y(3x^3y^2 + 6xy^3 - 4x^5 + 5)$

11. $4x^2\left(-3x^5 - \dfrac{1}{8}x^3 + \dfrac{1}{12}x^2 - \dfrac{1}{20}x\right)$

12. $4.5a^3b^2c(2.1a^2b^2 - 3.5a^3b^5c^2 + 4)$

For additional exercises of this type, see Exercises 7–34 on pages 386 and 387.

For Exercises 13–18, multiply the polynomials.

13. $(3x - 4)(2x + 3)$

14. $(3y + 5)(y - 4)$

15. $(3p - 4q)(2p - 3q)$

16. $(4a - 5b)(3a + 2b)$

17. $(a + 2)(4a^2 + 3a - 5)$

18. $(3y - 2)(2y^2 - 4y - 5)$

For additional exercises of this type, see Exercises 39–72 on pages 387 and 388.

For Exercises 19 and 20, multiply the conjugates.

19. $(2y + 5)(2y - 5)$

20. $(3m - 7n)(3m + 7n)$

For additional exercises of this type, see Exercises 81–90 on page 388.

For Exercises 21 and 22, square the binomials.

21. $(7x + 2y)^2$

22. $(5a - 3b)^2$

For additional exercises of this type, see Exercises 91–102 on page 388.

Objective ③ Divide polynomials. For a review of this objective, see Section 5.6, pages 392–397.

For Exercises 23 and 24, divide the polynomial by the monomial.

23. $\dfrac{18x^4 - 12x^3 + 24x^2 - 30x}{6x^2}$

24. $\dfrac{45u^3v^2 - 20u^4v + 15u^2v^2}{5u^2v}$

For additional exercises of this type, see Exercises 59–76 on page 401.

For Exercises 25 and 26, divide the polynomials.

25. $\dfrac{16x - 12x^2 + 3x^3 - 12}{x - 2}$

26. $\dfrac{2x^3 - 14x + 5}{2x - 4}$

For additional exercises of this type, see Exercises 81–104 on page 402.

R.5 Review of Factoring

Objective 1 Write a polynomial as a product of a GCF and a polynomial. For a review of this objective, see Section 6.1, pages 418–422.

For Exercises 1–4, factor by factoring out the GCF.

1. $18c^3d^4 - 12c^2d^6$
2. $16c^3d^2 - 24c^2d^4 + 36cd^2$
3. $-10x^3 - 15x^2 + 25x$
4. $-24m^2n - 28mn + 16mn^2$

For additional exercises of this type, see Exercises 35–80 on page 424.

Objective 2 Factoring by grouping. For a review of this objective, see Section 6.1, pages 422 and 423.

For Exercises 5–8, factor by grouping.

5. $a^3 + 4a^2 + 3a + 12$
6. $8x^2 - 2xy + 12x - 3y$
7. $6ac + 4ad - 9bc - 6bd$
8. $15ac - 5ad - 3bc + bd$

For additional exercises of this type, see Exercises 81–104 on page 425.

Objective 3 Factor trinomials of the form $x^2 + bx + c$. For a review of this objective, see Section 6.2, pages 426–430.

For Exercises 9–12, factor the trinomial of the form $x^2 + bx + c$.

9. $z^2 + 11z + 18$
10. $y^2 + y - 42$
11. $x^2 - 12xy + 35y^2$
12. $2a^2 + 20ab - 48b^2$

For additional exercises of this type, see Exercises 13–72 on pages 430 and 431.

Objective 4 Factoring trinomials of the form $ax^2 + bx + c$ where $a \neq 1$. For a review of this objective, see Section 6.3, pages 433–436.

For Exercises 13–16, factor the trinomial of the form $ax^2 + bx + c$.

13. $9x^2 - 18x + 8$
14. $6c^2 - 13cd + 6d^2$
15. $18c^2 - 57cd + 45d^2$
16. $12a^2 + 7ab - 12b^2$

For additional exercises of this type, see Exercises 11–42 on page 439.

Objective 5 Factor special products. For a review of this objective, see Section 6.4, pages 441–445.

For Exercises 17–20, factor completely the perfect square trinomials.

17. $a^2 - 16a + 64$
18. $25x^2y - 10xy + y$
19. $4x^2 + 20xy + 25y^2$
20. $36m^2 - 84mn + 49n^2$

For additional exercises of this type, see Exercises 7–24 on page 446.

For Exercises 21–23, factor the difference of squares.

21. $4x^2 - 25y^2$
22. $y^4 - 81$
23. $36x^2 + 49y^2$

For additional exercises of this type, see Exercises 25–44 on page 446.

For Exercises 24 and 25, factor the sum or difference of cubes.

24. $r^3 - 125$
25. $27x^3 + 64$

For additional exercises of this type, see Exercises 45–64 on pages 446 and 447.

Appendix A Arithmetic Sequences and Series

 **Objectives**

1. Find the terms of a sequence when given the general term.
2. Define and write arithmetic sequences, find their common difference, and find a particular term.
3. Define and write series, find partial sums, and use summation notation.
4. Write arithmetic series and find their sums.

The word **sequence** is used in mathematics much as it is in everyday life. For example, the classes you attend on a given day occur in a particular order, or sequence. As such, a sequence is an ordered list. Suppose we have a bacteria colony that has an initial population of 10,000 and increases at a rate of 10% each day.

On the second day, we have
$10{,}000 + 0.10(10{,}000) = 10{,}000 + 1000 = 11{,}000.$
On the third day, we have $11{,}000 + 0.10(11{,}000) = 11{,}000 + 1100 = 12{,}100.$
On the fourth day, we have $12{,}100 + 0.10(12{,}100) = 12{,}100 + 1210 = 13{,}310.$
On the fifth day, we have $13{,}310 + 0.10(13{,}310) = 13{,}310 + 1331 = 14{,}641.$

The number of bacteria present after each day forms a sequence that we can summarize as follows.

Days	1	2	3	4	5
Number of Bacteria	10,000	11,000	12,100	13,310	14,641

Based on this, we have the following definition.

Definition *Sequence:* A function list whose domain is $1, 2, 3, \ldots, n$.

The *domain* of our bacteria example is the numbers of the days $\{1, 2, 3, 4, 5\}$. Each number in the *range* of a sequence is called a *term*. So the terms of our bacteria sequence are the numbers of bacteria: 10,000, 11,000, 12,100, 13,310, 14,641. Sequences can be finite or infinite depending on the number of terms. Our bacteria example is a **finite sequence** because it has a finite number of terms. The sequence $2, 4, 6, 8, 10, \ldots$ is an **infinite sequence** because it has an infinite number of terms.

Definitions *Finite sequence:* A function with a domain that is the set of natural numbers from 1 to n.

Infinite sequence: A function with a domain that is the set of natural numbers.

Objective 1 **Find the terms of a sequence when given the general term.** Because a sequence is a function, we could describe sequences using functional notation. Instead, we use a different notation that emphasizes the fact that the domain is a subset of the natural numbers. We think of the terms of a sequence as $a_1, a_2, a_3, \ldots, a_n$, where the subscript gives the number of the term. Thus, a_n is the nth or *general term* of the sequence and we represent a sequence by giving a formula for a_n. Consider the following sequence.

Term	a_1	a_2	a_3	a_4	a_5	a_n
Term of Sequence	1	4	9	16	25	n^2

We represent this sequence by writing $a_n = n^2$, which means $a_1 = 1^2$, $a_2 = 2^2$, and so on.

Example 1 Find the first three terms of the following sequences and the 25th term.

a. $a_n = 3n - 1$

SOLUTION We let $n = 1, 2, 3$, and 25 and evaluate.

$$a_1 = 3(\mathbf{1}) - 1 = 3 - 1 = 2 \qquad \text{Let } n = 1 \text{ and evaluate.}$$
$$a_2 = 3(\mathbf{2}) - 1 = 6 - 1 = 5 \qquad \text{Let } n = 2 \text{ and evaluate.}$$
$$a_3 = 3(\mathbf{3}) - 1 = 9 - 1 = 8 \qquad \text{Let } n = 3 \text{ and evaluate.}$$
$$a_{25} = 3(\mathbf{25}) - 1 = 75 - 1 = 74 \qquad \text{Let } n = 25 \text{ and evaluate.}$$

The first three terms of the sequence are 2, 5, and 8. The 25th term is 74.

b. $a_n = \dfrac{(-1)^n}{n^2 + 1}$

SOLUTION We let $n = 1, 2, 3$, and 25 and evaluate.

$$a_1 = \frac{(-1)^{\mathbf{1}}}{\mathbf{1}^2 + 1} = \frac{-1}{1 + 1} = -\frac{1}{2} \qquad \text{Let } n = 1 \text{ and evaluate.}$$

$$a_2 = \frac{(-1)^{\mathbf{2}}}{\mathbf{2}^2 + 1} = \frac{1}{4 + 1} = \frac{1}{5} \qquad \text{Let } n = 2 \text{ and evaluate.}$$

$$a_3 = \frac{(-1)^{\mathbf{3}}}{\mathbf{3}^2 + 1} = \frac{-1}{9 + 1} = -\frac{1}{10} \qquad \text{Let } n = 3 \text{ and evaluate.}$$

$$a_{25} = \frac{(-1)^{\mathbf{25}}}{\mathbf{25}^2 + 1} = \frac{-1}{625 + 1} = -\frac{1}{626} \qquad \text{Let } n = 25 \text{ and evaluate.}$$

The first three terms of the sequence are $-\dfrac{1}{2}, \dfrac{1}{5}$, and $-\dfrac{1}{10}$. The 25th term is $-\dfrac{1}{626}$.

Objective ② Define and write arithmetic sequences, find their common difference, and find a particular term. In the sequence $-3, 1, 5, 9, 13, \ldots$, notice that each term after the first is found by adding 4 to the previous term. This is an example of an **arithmetic sequence** or *arithmetic progression*. Any two successive terms of an arithmetic sequence differ by the same amount, which is called the **common difference** and is denoted by d. To find d, choose any term (except the first) and subtract the previous term.

Definition

Arithmetic sequence: A sequence in which each term after the first is found by adding the same number to the previous term.

Common difference of an arithmetic sequence: The value of d is found by $d = a_n - a_{n-1}$, where a_n is any term in the sequence (except the first) and a_{n-1} is the previous term.

Example 2 Write the first four terms of the following arithmetic sequences.

a. The first term is 2, and the common difference is 5.

SOLUTION Begin with 2 and find each successive term by adding 5 to the previous term.

$$a_1 = 2, a_2 = 2 + 5 = 7, a_3 = 7 + 5 = 12, a_4 = 12 + 5 = 17$$

The first four terms of the sequence are 2, 7, 12, 17.

b. $a_1 = -1, d = -2$

SOLUTION Begin with -1 and find each successive term by adding -2 to the previous term.

$$a_1 = -1, a_2 = -1 - 2 = -3, a_3 = -3 - 2 = -5, a_4 = -5 - 2 = -7$$

The first four terms of the sequence are $-1, -3, -5, -7$.

Example 3 Find the common difference, d, for the following arithmetic sequences.

a. $-4, -1, 2, 5, 8, \ldots$

SOLUTION Pick any term (except the first) and subtract the term before it.

$$d = -1 - (-4) = -1 + 4 = 3$$

Or we could use $d = 5 - 2 = 3$, and so on.

b. $8, 6, 4, 2, 0, \ldots$

SOLUTION Pick any term (except the first) and subtract the term before it.

$$d = 6 - 8 = -2$$

Or we could use $d = 2 - 4 = -2$, and so on.

If the first term of an arithmetic sequence is a_1 and the common difference is d, then the arithmetic sequence can be written as $a_1, a_1 + d, a_1 + 2d, a_1 + 3d, a_1 + 4d, \ldots$. Note that the coefficient of d is one less than the number of the term; so we have the following rule.

Rule ***n*th Term of an Arithmetic Sequence**

The formula for finding the nth term of an arithmetic sequence is $a_n = a_1 + (n - 1)d$, where a_1 is the first term and d is the common difference.

Example 4 Find the 23rd term and an expression for the nth term of an arithmetic sequence in which $a_1 = -8$ and $d = 3$.

SOLUTION

$$a_n = a_1 + (n - 1)d$$
$$a_{23} = -8 + (23 - 1)(3) \qquad \text{Substitute 23 for } n, -8 \text{ for } a_1, \text{ and 3 for } d.$$
$$a_{23} = 58 \qquad \text{Evaluate.}$$
$$n\text{th term: } a_n = -8 + (n - 1)3 \qquad \text{Substitute for } a_1 \text{ and } d.$$
$$a_n = -11 + 3n \qquad \text{Simplify.}$$

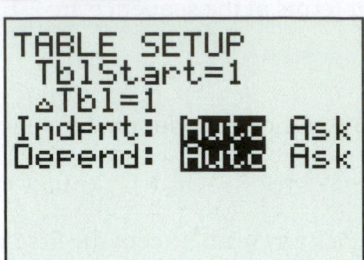
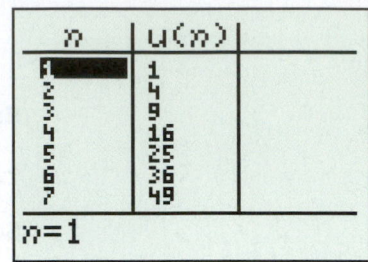
If we know the first term and one other term, we can use the formula for the nth term of an arithmetic sequence to find the common difference. Consequently, we can find the sequence.

Example 5   The first term of an arithmetic sequence is 1, and the 20th term (a_{20}) is 58. Find the common difference and the first four terms of the sequence.

SOLUTION   We use the fact that we know a_{20} and the formula for the nth term to find d.

$$a_n = a_1 + (n - 1)d$$
$$a_{20} = 1 + (20 - 1)d \qquad \text{Substitute 1 for } a_1 \text{ and 20 for } n.$$
$$58 = 1 + 19d \qquad \text{Substitute 58 for } a_{20} \text{ and solve.}$$
$$3 = d$$

The first four terms are $1, 1 + 3 = 4, 4 + 3 = 7, 7 + 3 = 10$.

Objective ③ **Define and write series, find partial sums, and use summation notation.**   When the terms of a sequence are added, the sum is called a **series**.

Definition   *Series:* The sum of the terms of a sequence.

Given a sequence $a_1, a_2, a_3, \ldots, a_n$, then $a_1 + a_2 + a_3 + \cdots + a_n$ is the corresponding series. Finite series correspond to finite sequences, and infinite series correspond to infinite sequences. If an expression for the nth term is known, we can write the series in *summation notation* using the capital Greek letter *sigma* (Σ) as follows.

$$a_1 + a_2 + a_3 + \cdots + a_n = \sum_{i=1}^{n} a_i$$

The i is called the *index of summation*; 1 is the *lower limit* of i, and n is the *upper limit* of i. If the upper limit is a natural number, the series is finite, and if the upper limit is ∞, the series is infinite. To find a finite series from summation notation, replace the index of summation (usually i) with its lower limit and evaluate that term, replace i with the lower limit plus 1 and evaluate that term, and so on, until you reach the upper limit. This gives the series. To find the sum, add the terms of the series.

Example 6

a. Write the terms of $\sum_{i=1}^{5}(3i + 2)$ and find the sum of the series.

SOLUTION Replace i with 1, 2, 3, 4, and 5. Then find the sum.

$$\sum_{i=1}^{5}(3i + 2) = (3 \cdot 1 + 2) + (3 \cdot 2 + 2) + (3 \cdot 3 + 2) + (3 \cdot 4 + 2) + (3 \cdot 5 + 2)$$

$$= 5 + 8 + 11 + 14 + 17$$

$$= 55$$

b. Write the first five terms of the infinite series $\sum_{i=1}^{\infty} 2i^2$.

SOLUTION Replace i with 1, 2, 3, 4, and 5. Then evaluate.

$$\sum_{i=1}^{\infty} 2i^2 = 2 \cdot 1^2 + 2 \cdot 2^2 + 2 \cdot 3^2 + 2 \cdot 4^2 + 2 \cdot 5^2 + \cdots$$

$$= 2 + 8 + 18 + 32 + 50 + \cdots$$

Objective **4** **Write arithmetic series and find their sums.** When the terms of an arithmetic sequence are added, it is called an **arithmetic series**.

Definition *Arithmetic series:* The sum of the terms of an arithmetic sequence.

Consequently, an arithmetic series has the form $a_1 + (a_1 + d) + (a_1 + 2d) + (a_1 + 3d) + \cdots + [a_1 + (n - 1)d]$.

Adding a finite number of terms of an infinite series gives a *partial sum.* The symbol S_n is used to indicate the sum of the first n terms. For example, S_5 means add the first five terms. Let's derive a formula for S_n for an arithmetic series as follows.

$$S_n = a_1 + (a_1 + d) + (a_1 + 2d) + (a_1 + 3d) + \cdots + a_n$$

We also need to include the terms between $(a_1 + 3d)$ and a_n. So we write another version of S_n beginning with the last term, a_n, and subtracting the common difference, d, from the previous term.

$$S_n = a_n + (a_n - d) + (a_n - 2d) + (a_n - 3d) + \cdots + a_1$$

To describe the entire sum, we add our two versions of S_n.

$$
\begin{aligned}
S_n &= a_1 && + (a_1 + d) && + (a_1 + 2d) + (a_1 + 3d) + \cdots + a_n \\
+\; S_n &= a_n && + (a_n - d) && + (a_n - 2d) + (a_n - 3d) + \cdots + a_1 \\
\hline
2S_n &= (a_1 + a_n) + (a_1 + a_n) + (a_1 + a_n) + (a_1 + a_n) + \cdots + (a_1 + a_n)
\end{aligned}
$$

Because S_n has n terms, there are n terms of $(a_1 + a_n)$; so

$$2S_n = n(a_1 + a_n), \quad \text{or} \quad S_n = \frac{n}{2}(a_1 + a_n).$$

Rule **Partial Sum, S_n, of an Arithmetic Series**

The sum of the first n terms of an arithmetic series, S_n, called the nth partial sum, is given by

$$S_n = \frac{n}{2}(a_1 + a_n),$$

where n is the number of terms, a_1 is the first term, and a_n is the nth term.

Example 7 Find the sum of the first 20 terms (S_{20}) of the arithmetic series $-6 - 2 + 2 + 6 + \cdots$.

SOLUTION To find the S_{20}, we first need to find the 20th term. Because $d = 4$,

$$a_{20} = -6 + (20 - 1)(4) = 70$$

$$S_n = \frac{n}{2}(a_1 + a_n) \qquad \text{Rule for } n\text{th partial sum}$$

$$S_{20} = \frac{20}{2}(-6 + 70) \qquad \text{Substitute 20 for } n, -6 \text{ for } a_1, \text{ and 70 for } a_n.$$

$$S_{20} = 10(64) = 640 \qquad \text{Evaluate.}$$

Appendix A

Exercises   For Extra Help   **MyMathLab**   **Math XL** PRACTICE   WATCH   DOWNLOAD

1. What set of numbers makes up the domain of a sequence?

2. A sequence with an unlimited number of terms is called a(n) _____ sequence.

3. What is a series?

4. What is an arithmetic sequence? How do you find the common difference?

5. The sequence $-10, -5, 0, 5, 10, \ldots$ is an example of a(n) _____ sequence with a common difference of _____.

6. What does S_n represent?

For Exercises 7–14, write the first four terms of the sequence and the indicated term. See Example 1.

7. $a_n = 2n + 1$, 20th term

8. $a_n = 3n - 4$, 18th term

9. $a_n = n^2 + 2$, 15th term

10. $a_n = n^2 - 3$, 12th term

11. $a_n = \dfrac{n}{n + 2}$, 22nd term

12. $a_n = \dfrac{2n}{n + 3}$, 10th term

13. $a_n = \dfrac{(-1)^n}{n^2 + 1}$, 15th term

14. $a_n = \dfrac{(-1)^n}{n^2 - 3}$, 26th term

For Exercises 15–20, find the common difference, d, for each arithmetic sequence. See Example 3.

15. $2, 7, 12, 17, \ldots$

16. $3, 11, 19, 27, \ldots$

17. $25, 22, 19, 16, \ldots$

18. $42, 36, 30, 24, \ldots$

19. $-12, -5, 2, 9, \ldots$

20. $-24, -27, -30, -33, \ldots$

For Exercises 21–26, find the indicated term and an expression for the nth term of the given arithmetic sequence. See Example 4.

21. a_{14} if $a_1 = 14$ and $d = 4$

22. a_{24} if $a_1 = 16$ and $d = 6$

23. a_{28} if $a_1 = -8$ and $d = -3$

24. a_{30} if $a_1 = -7$ and $d = -6$

25. a_{34} of $-5, -1, 3, 7, \ldots$

26. a_{21} of $8, 15, 22, 29, \ldots$

For Exercises 27–32, write the first four terms of the arithmetic sequence with the given characteristics. See Examples 2 and 5.

27. $a_1 = -6, d = 7$

28. $a_1 = -2, d = 6$

29. $a_1 = -5, d = -3$

30. $a_1 = -13, d = -5$

31. $a_1 = 7, a_{18} = 75$

32. $a_1 = 6, a_{22} = 153$

33. Find the first term of an arithmetic sequence if $a_{45} = 143$ and $d = 3$.

34. Find the first term of an arithmetic sequence if $a_{39} = 181$ and $d = 6$.

35. Find the common difference of an arithmetic sequence if the first term is -110 and the 29th term is 2.

36. Find the common difference of an arithmetic sequence if the first term is 78 and the 15th term is -76.

For Exercises 37–40, write the first four terms of the arithmetic sequence with the given d and a_n.

37. $d = 7, a_8 = 41$

38. $d = 4, a_{11} = 27$

39. $d = -3, a_7 = 9$

40. $d = -6, a_{10} = -36$

For Exercises 41–48, write the series and find the sum. See Example 6.

41. $\displaystyle\sum_{i=1}^{6} i^2$

42. $\displaystyle\sum_{i=1}^{3} 3i^2$

43. $\displaystyle\sum_{i=1}^{4} (2i - 5)$

44. $\displaystyle\sum_{i=1}^{5} (4i + 1)$

45. $\displaystyle\sum_{i=1}^{3} (3i^2 - 4)$

46. $\displaystyle\sum_{i=1}^{4} (-2i^2 + 5)$

47. $\displaystyle\sum_{i=3}^{6} (4i - 3)$

48. $\displaystyle\sum_{i=2}^{5} (4i - 2)$

For Exercises 49–56, find the given S_n for the arithmetic series. See Example 7.

49. If $a_1 = 10$ and $d = 4$, find S_{25}.

50. If $a_1 = -8$ and $d = 2$, find S_{30}.

51. If $a_1 = 12$ and $d = -3$, find S_{22}.

52. If $a_1 = 18$ and $d = -2$, find S_{30}.

53. $3 + 9 + 15 + 21 + \cdots$. Find S_{15}.

54. $5 + 9 + 13 + 17 + \cdots$. Find S_{25}.

55. $54 + 46 + 38 + 30 + \cdots$. Find S_{12}.

56. $44 + 39 + 34 + 29 + \cdots$. Find S_{18}.

For Exercises 57–64, answer each question.

57. Find the sum of the first 100 natural numbers.

58. Find the sum of the even integers 2 through 200.

59. A concert hall has 60 seats in the first row, 64 in the second, 68 in the third, and so on.
    a. How many seats are in the 22nd row?

    b. How many seats are in the concert hall if there are 35 rows?

60. Johanna bought a car for $24,000 with a special no-interest loan and made monthly payments of $400. Using 24,000 as the first term, write the first five terms of an arithmetic sequence that gives the amount she still owes at the end of each month. How much will she owe at the end of the 25th month?

61. Fence posts are arranged in a triangular stack with 25 on the bottom row, 24 on the next row, 23 on the next row, and so on until a single post is on the top.
    a. How many posts are on the 10th row from the bottom?

    b. How many posts are in the stack?

62. Carlos is doing sit-ups to strengthen his abdominal muscles. He did 25 the first night and plans to add 1 each night.
    a. Write the first five terms of an arithmetic sequence that gives the number of sit-ups he does each night.

    b. Find the number he does on the 30th night.

    c. Find the total number of sit-ups he has done after 30 nights.

63. Tanisha takes a job that pays $28,000 the first year and gives a raise of $1500 per year.
    a. Write the first five terms of an arithmetic sequence that gives her salary at the end of each year.

    b. What will her salary be for the 10th year?

    c. What are her total earnings for her first 10 years?

64. You have a choice of two jobs. Job A has a starting salary of $25,000 with raises of $900 per year, and job B has a starting salary of $28,000 with raises of $600 per year.
    a. Which job will pay the most during the 10th year?

    b. Which job will pay the biggest total amount during the first 15 years?

Appendix B Geometric Sequences and Series

Objectives

1. Write a geometric sequence and find its common ratio and a specified term.
2. Find partial sums of geometric series.
3. Find the sums of infinite geometric series.
4. Solve applications using geometric series.

Objective 1 Write a geometric sequence and find its common ratio and a specified term. In Appendix A, we generated an arithmetic sequence by adding the same number to each term to get the next term. In this section, we multiply each term by the same number to generate a **geometric sequence**. The number we multiply by is called the **common ratio** and is denoted as r. We can find r by dividing any term (except the first) by the term before it.

Definition *Geometric sequence:* A sequence in which every term after the first is found by multiplying the previous term by the same number, called the **common ratio**, r,

where $r = \dfrac{a_n}{a_{n-1}}$.

> **Note:** In geometric sequences, the first term is usually denoted as a rather than a_1.

If the first term is a and the common ratio is r, a geometric sequence has the form $a, ar, ar^2, ar^3, ar^4, \ldots, ar^{n-1}$. Thus, the general term of a geometric series is $a_n = ar^{n-1}$.

We notice that the exponent of r is one less than the number of the term, so we have the following rule.

Rule *n*th **Term of a Geometric Sequence**

The formula for finding the nth term of a geometric sequence is $a_n = ar^{n-1}$, where a is the first term and r is the common ratio.

Example 1 A geometric sequence has a first term of 3 and a common ratio of 4.

a. Write the first five terms.

SOLUTION To find each term, multiply the term before it by 4. So

$$a_1 = 3, a_2 = 3(4) = 12, a_3 = 12(4) = 48, a_4 = 48(4) = 192, a_5 = 192(4) = 768.$$

The first five terms of the sequence are 3, 12, 48, 192, 768.

b. Find the 10th term.

SOLUTION
$a_n = ar^{n-1}$ Formula for a_n
$a_{10} = 3(4)^{10-1}$ Substitute 10 for n, 3 for a, and 4 for r.
$a_{10} = 786{,}432$ Evaluate.

c. Find the general term.

SOLUTION
$a_n = ar^{n-1}$ Formula for a_n
$a_n = 3(4)^{n-1}$ Substitute for a and r.

Example 2 Given the geometric sequence $32, -16, 8, -4, \ldots$, find the common ratio r.

SOLUTION To find r, divide any term (except the first) by the term before it.

$$r = \frac{-16}{32} = -\frac{1}{2} \quad \text{or} \quad r = \frac{8}{-16} = -\frac{1}{2}, \text{ and so on.}$$

Geometric sequences often occur in populations and other applications.

Example 3 The population of rabbits in a large pen increases at a rate of 12% per month. If there are currently 50 rabbits, find the population after 15 months.

SOLUTION Let P_0 be the initial number of rabbits, P_1 the number after 1 month, P_2 the number after 2 months, and so on. The number of rabbits at the end of each month is the number at the beginning of the month plus an increase of 12% of that number. So the number at the end of each month can be found as follows.

$$P_1 = P_0 + 0.12P_0 = 1.12P_0$$
$$P_2 = P_1 + 0.12P_1 = 1.12P_1 = 1.12(1.12P_0) = (1.12)^2P_0$$
$$P_3 = P_2 + 0.12P_2 = 1.12P_2 = 1.12[(1.12)^2P_0] = (1.12)^3P_0$$

The number of rabbits at the end of each month can be represented by the sequence $P_0, P_1, P_2, P_3, \ldots P_n = P_0, 1.12P_0, (1.12)^2P_0, (1.12)^3P_0, \ldots, (1.12)^{n-1}P_0$, which is a geometric sequence whose first term is P_0 and $r = 1.12$. Consequently, the nth term is $P_0(1.12)^{n-1}$. The number of rabbits at the end of 15 months is the 16th term of the sequence; so

$$P_{16} = P_0(1.12)^{16-1}$$
$$P_{16} = 50(1.12)^{15}$$
$$P_{16} = 273.68.$$

Answer There are about 274 rabbits at the end of 15 months.

> **Note:** It can be shown using this procedure that if a population is growing at $p\%$ per unit time and the initial population is P_0, then the population at the end of each unit of time forms a geometric sequence whose first term is P_0 and whose common ratio is $(1 + p)$, where p is $p\%$ written as a decimal. So the geometric sequence is P_0, $P_0(1 + p), P_0(1 + p)^2$, $P_0(1 + p)^3, \ldots, P_0(1 + p)^{n-1}$. The population after n time periods is the $(n + 1)$st term of the sequence, which is $P_0(1 + p)^n$.

Objective ❷ Find partial sums of geometric series. If we add the terms of an arithmetic sequence, we get an arithmetic series. Likewise, if we add the terms of a geometric sequence, we get a **geometric series**.

Definition *Geometric series:* The sum of the terms of a geometric sequence.

A geometric series is of the form $a + ar + ar^2 + \cdots + ar^{n-1}$ if the series is finite and $a + ar + ar^2 + \cdots$ if the series is infinite.

In the geometric series $3 + 6 + 12 + 24 + \cdots$, we have $a = 3$ and $r = 2$. In $243 - 81 + 27 - 9 + \cdots$, we have $a = 243$ and $r = -\frac{1}{3}$. As with arithmetic series, we can find a formula for partial sum, S_n. Begin with

$$
\begin{array}{ll}
S_n = a + ar + ar^2 + \cdots + ar^{n-1} & \\
\underline{-rS_n = \quad -ar - ar^2 - \cdots - ar^{n-1} - ar^n} & \text{\color{blue}Multiply both sides of the equation by $-r$.} \\
S_n - rS_n = a - ar^n & \text{\color{blue}Add the equations.} \\
S_n(1 - r) = a(1 - r^n) & \text{\color{blue}Factor both sides.} \\
S_n = \dfrac{a(1 - r^n)}{1 - r} & \text{\color{blue}Divide both sides by $1 - r$.}
\end{array}
$$

> **Rule** | **Partial Sum, S_n, of a Geometric Series**
>
> The sum of the first n terms of a geometric series, S_n, called the nth partial sum, is given by
>
> $$S_n = \frac{a(1 - r^n)}{1 - r},$$
>
> where n is the number of terms, a is the first term, and r is the common ratio ($r \neq 1$).

Example 4 Find the sum of the first 10 terms of the geometric series: $1 - 3 + 9 - 27 + \cdots$

SOLUTION We first find r: $r = \dfrac{-3}{1} = -3$.

$$S_n = \frac{a(1 - r^n)}{1 - r} \qquad \text{Formula for } S_n$$

$$S_{10} = \frac{1(1 - (-3)^{10})}{1 - (-3)} \qquad \text{Substitute 10 for } n, \text{ 1 for } a, \text{ and } -3 \text{ for } r.$$

$$S_{10} = -14{,}762 \qquad \text{Evaluate.}$$

Objective ❸ Find the sums of infinite geometric series. If the common ratio satisfies $|r| > 1$, the partial sums become infinitely large as n becomes infinitely large. However, if $|r| < 1$, the partial sums approach a value as n becomes infinitely large. This value is called the *limit* of the partial sums and is the sum of the infinite series.

For example, for the series $2 + 1 + \dfrac{1}{2} + \dfrac{1}{4} + \cdots$, we have

$$S_5 = \frac{2\left(1 - \left(\frac{1}{2}\right)^5\right)}{1 - \frac{1}{2}} = 3.875$$

$$S_{10} = \frac{2\left(1 - \left(\frac{1}{2}\right)^{10}\right)}{1 - \frac{1}{2}} \approx 3.9960938$$

$$S_{15} = \frac{2\left(1 - \left(\frac{1}{2}\right)^{15}\right)}{1 - \frac{1}{2}} \approx 3.9998779.$$

Notice that the greater the value of n, the closer the sum is to 4. The partial sums approach 4 because as n gets larger, $\left(\frac{1}{2}\right)^n$ approaches 0. In the preceding example, $\left(\frac{1}{2}\right)^5 = 0.03125$, $\left(\frac{1}{2}\right)^{10} \approx 0.000977$, and $\left(\frac{1}{2}\right)^{15} \approx 0.0000305$. In general, if $|r| < 1$, then r^n approaches 0 as n becomes large. Consequently, if $|r| < 1$, the formula $S_n = \dfrac{a(1 - r^n)}{1 - r}$ becomes $S_\infty = \dfrac{a(1 - 0)}{1 - r} = \dfrac{a}{1 - r}$ as n becomes infinitely large. We denote the sum as S_∞ rather than S_n.

Rule **Sum of an Infinite Geometric Series**

If $|r| < 1$, the sum of an infinite geometric series, S_∞, is given by the formula

$$S_\infty = \frac{a}{1 - r},$$

where a is the first term and r is the common ratio. If $|r| \geq 1$, S_∞ does not exist.

Example 5   Find the sum of the infinite geometric series:

$$2 + 1 + \frac{1}{2} + \frac{1}{4} + \cdots$$

SOLUTION   We know that $a = 2$ and $r = \frac{1}{2}$. Because $|r| < 1$, the sum exists.

$$S_\infty = \frac{a}{1 - r}$$

$$S_\infty = \frac{2}{1 - \frac{1}{2}} \qquad \text{Substitute 2 for } a \text{ and } \frac{1}{2} \text{ for } r.$$

$$S_\infty = 4 \qquad \text{Evaluate.}$$

Infinite geometric series also provides us with a method of changing a repeating decimal into a fraction.

Example 6   Write as a fraction: $0.\overline{37}$

SOLUTION   First, write $0.\overline{37}$ as an infinite geometric series as follows:

$$0.\overline{37} = 0.37 + 0.0037 + 0.000037 + \cdots$$

$$0.\overline{37} = \frac{37}{100} + \frac{37}{10,000} + \frac{37}{1,000,000} + \cdots$$

The last line is an infinite geometric series with $a = \frac{37}{100}$ and $r = \frac{1}{100}$. Because $|r| < 1$, the sum of this series exists.

$$S_\infty = \frac{a}{1 - r}$$

$$S_\infty = \frac{\frac{37}{100}}{1 - \frac{1}{100}} \qquad \text{Substitute } \frac{37}{100} \text{ for } a \text{ and } \frac{1}{100} \text{ for } r.$$

$$S_\infty = \frac{\frac{37}{100}}{\frac{99}{100}} \qquad \text{Simplify.}$$

$$S_\infty = \frac{37}{99}$$

Answer $0.\overline{37} = \frac{37}{99}$

Example 7 To save for their child's college education, the McBride family put $1000 into a savings account the first year. Each year thereafter they deposited 10% more than the previous year.

a. Write the first five terms of the geometric sequence that gives the amount of money deposited in the account each year.

SOLUTION From the note next to Example 3, this is a geometric series in which $a = 1000$ and $r = (1 + 0.10) = 1.1$.

$$a_1 = 1000, a_2 = 1000(1.1) = 1100, a_3 = 1000(1.1)^2 = 1210,$$
$$a_4 = 1000(1.10)^3 = 1331, a_5 = 1000(1.10)^4 = 1464.10$$

The first five terms of the sequence are 1000, 1100, 1210, 1331, 1464.10.

b. How much money will the McBrides have deposited in the account at the end of 15 years?

SOLUTION The series $1000 + 1100 + 1210 + 1331 + 1464.10 + \cdots$ is geometric with $a = 1000$ and $r = 1.1$.

$$S_n = \frac{a(1 - r^n)}{1 - r}$$

$$S_{15} = \frac{1000(1 - 1.1^{15})}{1 - 1.1}$$ Substitute 15 for n, 1000 for a, and 1.1 for r.

$$S_{15} = 31{,}772.48$$ Evaluate.

Answer They will have deposited $31,772.48 in the account after 15 years.

Appendix B

Exercises For Extra Help MyMathLab

1. If you divide a term of a geometric sequence by the term before it, what do you get?

2. What is the form of a geometric sequence?

3. Can you find the sum of the first n terms of the geometric series $2 + 6 + 18 + \cdots$ using the formula $S_n = \frac{a(1 - r^n)}{1 - r}$? Why or why not?

4. Can you find the sum of the infinite geometric series $2 + 6 + 18 + \cdots$ using the formula $S_\infty = \frac{a}{1 - r}$? Why or why not?

5. The fourth partial sum, S_4, of $2 + 6 + 18 + \cdots$ is _____ + _____ + _____ + _____ = _____.

6. Write $0.\overline{23}$ as an infinite geometric series. What are a and r?

For Exercises 7–14: **a.** **Find the common ratio, r, for the given geometric sequence. See Example 2.**
   b. **Find the indicated term. See Example 1.**
   c. **Find an expression for the general term, a_n. See Example 1.**

7. $1, 3, 9, 27, \ldots$;
   9th term

8. $7, 14, 28, 56, \ldots$;
   11th term

9. $-2, 4, -8, 16, \ldots$;
   15th term

10. $-8, 24, -72, 216, \ldots$;
   9th term

11. $243, 81, 27, 9, \ldots$;
   10th term

12. $8, 4, 2, 1, \ldots$;
   8th term

13. $128, -32, 8, -2, \ldots$;
   7th term

14. $125, -25, 5, -1, \ldots$;
   9th term.

For Exercises 15–18: **a.** **Write the first five terms of the geometric sequence, satisfying the given conditions. See Example 1.**
   b. **Find the indicated term. See Example 1.**

15. $a = -4, r = -3$;
   8th term

16. $a = -6, r = -2$;
   10th term

17. $a = 243, r = \dfrac{1}{3}$;
   7th term

18. $a = 256, r = -\dfrac{1}{2}$;
   10th term

For Exercises 19–24: **a.** **Use the formula for the nth term to find r.**
   b. **Write the first four terms of the geometric sequence.**

19. $a = 1, r > 0, a_5 = 16$

20. $a = 3, a_4 = -24$

21. $a = -6, r > 0, a_5 = -96$

22. $a = -4, a_4 = -32$

23. $a = 128, a_4 = 2$

24. $a = -243, a_6 = 1$

25. Find the first term of a geometric sequence in which $r = 3$ and the fifth term is -486.

26. Find the first term of a geometric sequence in which $r = -\dfrac{1}{2}$ and the fifth term is $-\dfrac{1}{8}$.

For Exercises 27–32, find the sum of the first n terms of each geometric series for the given value of n. See Example 4.

27. $3 + 9 + 27 + 81 + \cdots, n = 11$

28. $-1 - 2 - 4 - 8 - \cdots, n = 9$

29. $32 + 16 + 8 + 4 + \cdots, n = 9$

30. $81 + 27 + 9 + 3 + \cdots, n = 7$

31. $128 - 32 + 8 - 2 + \cdots, n = 9$

32. $625 - 125 + 25 - 5 + \cdots, n = 8$

For Exercises 33–38, find the sum of the infinite geometric series if possible. If it is not possible, explain why. See Example 5.

33. $27 + 9 + 3 + \cdots$

34. $8 + 4 + 2 + \cdots$

35. $15 - 9 + \dfrac{27}{5} - \cdots$

36. $15 - 10 + \dfrac{20}{3} - \cdots$

37. $9 + 12 + 16 + \cdots$

38. $16 + 20 + 25 + \cdots$

For Exercises 39–42, write each repeating decimal as a fraction. See Example 6.

39. $0.\overline{4}$

40. $0.\overline{7}$

41. $0.\overline{17}$

42. $0.\overline{25}$

For Exercises 43 and 44, answer each question.

43. If $a_n = 500(1.04)^n$,
  a. Find the first five terms of the sequence.

  b. Find the 10th term of the sequence.

44. If $a_n = 350(1.06)^n$,
  a. Find the first five terms of the sequence.

  b. Find the 8th term of the sequence.

For Exercises 45–50, solve. See Examples 3 and 7.

45. A population of mink is increasing at a rate of 8% per month. The current mink population is 100.
  a. Using 100 as the first term, find the first four terms of the geometric sequence that gives the number of mink at the beginning of each month.

  b. Find the number of mink present at the beginning of the eighth month.

  c. Find the expression for the general term, a_n.

46. The generation time (the time required for the number present to double) for a particular bacteria is 1 hour. Suppose initially one bacteria was present.
  a. Using 1 as the first term, write the first five terms of the geometric sequence giving the number of bacteria present after each hour.

  b. Find an expression for the number present after the nth hour.

  c. How many bacteria are present after 1 day?

47. Suppose you took a job for a month (20 working days) that paid $0.01 the first day and your salary doubled each day.
  a. Write the first five terms of the geometric sequence that gives your salary each day.

  b. Find an expression for the amount earned on the nth day.

  c. How much would you earn on the 20th day?

  d. What are your total earnings for the month?

48. Damarys deposits $200 in the bank. Each month thereafter, she deposits 5% more than the month before. She does this for 1 year.
  a. Find an expression for the amount she deposits in the nth month.

  b. Write the first four terms of the geometric sequence that gives the amount of her deposit each month.

  c. How much did she deposit in the 10th month?

  d. How much does she deposit for the year?

49. A new boat costs $20,000 and depreciates by 7% each year. What will the boat be worth in 8 years?

50. The isotope $_{15}P^{33}$ has a half-life of 25 days. A sample has 400 grams.
  a. Find the first five terms of the geometric sequence that gives the amount present at the end of each half-life.

  b. Find an expression for the amount present after the nth half-life.

  c. Find the amount present after the 10th half-life.

Appendix C   The Binomial Theorem

Objectives

1. Expand a binomial using Pascal's triangle.
2. Evaluate factorial notation and binomial coefficients.
3. Expand a binomial using the binomial theorem.
4. Find a particular term of a binomial expansion.

In Section 5.5 we learned to square binomials. In this section, we will learn to raise binomials to natural-number powers.

Objective 1 **Expand a binomial using Pascal's triangle.** We begin by writing out $(a + b)^n$, where n is a natural number, and look for patterns. These products are called *binomial expansions*.

$$(a + b)^0 = 1$$
$$(a + b)^1 = a + b$$
$$(a + b)^2 = a^2 + 2ab + b^2$$
$$(a + b)^3 = a^3 + 3a^2b + 3ab^2 + b^3$$
$$(a + b)^4 = a^4 + 4a^3b + 6a^2b^2 + 4ab^3 + b^4$$
$$(a + b)^5 = a^5 + 5a^4b + 10a^3b^2 + 10a^2b^3 + 5ab^4 + b^5$$

Conclusions Several patterns can be observed from the preceding expansions.

1. The first term in the expansion, a, is raised to the same power as the binomial, and the power of a decreases by 1 in each successive term. Note that the last term contains a^0.
2. The exponent of b is 0 in the first term and increases by 1 on each successive term.
3. The sum of the exponents of the variables of each term equals the exponent of the binomial.
4. The number of terms in the expansion is one more than the exponent of the binomial.

Now consider the coefficients of the terms in these expansions.

Coefficients of Expansions

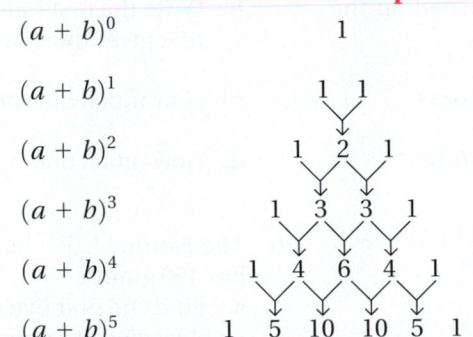

$(a + b)^0$ 1

$(a + b)^1$ 1   1

$(a + b)^2$ 1   2   1

$(a + b)^3$ 1   3   3   1

$(a + b)^4$ 1   4   6   4   1

$(a + b)^5$ 1   5   10   10   5   1

If we arrange the coefficients of each expansion in a triangular array, we see an interesting pattern. Each row begins and ends with 1. Each number inside a row is the sum of the two numbers in the row above it. For example, each 10 in the bottom row comes from adding the 4 and 6 directly above.

This triangular array of numbers is called *Pascal's triangle* in honor of the French mathematician Blaise Pascal. When these observations are used, the expansion of $(a + b)^6$ has seven terms and the variable portions of the terms are a^6, a^5b, $a^4b^2, a^3b^3, a^2b^4, ab^5, b^6$. By continuing the pattern in Pascal's triangle, we can find the coefficients for each of those terms.

$(a + b)^5$ 1   5   10   10   5   1

$(a + b)^6$ 1   6   15   20   15   6   1

Using the coefficients from the last line of Pascal's triangle and the variables previously listed gives us

$$(a + b)^6 = a^6 + 6a^5b + 15a^4b^2 + 20a^3b^3 + 15a^2b^4 + 6ab^5 + b^6.$$

Objective ② Evaluate factorial notation and binomial coefficients. Although Pascal's triangle is easy to use, it isn't practical, especially for binomials raised to large powers. Consequently, another method called the *binomial theorem* is often used. Before introducing the binomial theorem, we need **factorial notation**.

Definition   *Factorial notation:* For any natural number n, the symbol $n!$ (read "n factorial") means $n(n - 1)(n - 2)\ldots 3 \cdot 2 \cdot 1$.

$0!$ is defined to be 1, so $0! = 1$.

Example 1   Evaluate the following factorials.

a. $5!$ **b.** $7!$

SOLUTION $5! = 5 \cdot 4 \cdot 3 \cdot 2 \cdot 1 = 120$ **SOLUTION** $7! = 7 \cdot 6 \cdot 5 \cdot 4 \cdot 3 \cdot 2 \cdot 1 = 5040$

Sometimes we may not write all of the factors of a factorial. In such cases, the last desired factor is written as a factorial. Below are some alternative ways to write $7!$

$$7! = 7 \cdot 6 \cdot 5 \cdot 4 \cdot 3 \cdot 2 \cdot 1 = 7 \cdot 6! \quad \text{or} \quad 7! = 7 \cdot 6 \cdot 5! \quad \text{or} \quad 7! = 7 \cdot 6 \cdot 5 \cdot 4!$$

The coefficients of a binomial expansion can be expressed in terms of factorials using a special notation called the **binomial coefficient**. We will see exactly how the binomial coefficient is used later.

Definition   *Binomial coefficient:* A number written as $\binom{n}{r}$ and defined as $\dfrac{n!}{r!(n - r)!}$.

Example 2   Evaluate the following binomial coefficients.

a. $\binom{6}{2}$

SOLUTION $\binom{6}{2} = \dfrac{6!}{2!(6 - 2)!}$ Substitute 6 for n and 2 for r in $\dfrac{n!}{r!(n-r)!}$.

$= \dfrac{6 \cdot 5 \cdot 4 \cdot 3 \cdot 2 \cdot 1}{(2 \cdot 1)(4 \cdot 3 \cdot 2 \cdot 1)}$ Expand the factorials.

$= 15$ Simplify.

> **Note:** These factorials could have been evaluated as follows:
>
> $$\dfrac{6!}{2! \cdot 4!} = \dfrac{6 \cdot 5 \cdot 4!}{2 \cdot 1 \cdot 4!} = \dfrac{6 \cdot 5}{2}$$
>
> $$= \dfrac{30}{2} = 15$$

b. $\binom{8}{5}$

SOLUTION $\binom{8}{5} = \dfrac{8!}{5!(8 - 5)!}$ Substitute 8 for n and 5 for r in $\dfrac{n!}{r!(n-r)!}$.

$= \dfrac{8!}{5! \cdot 3!}$ Simplify.

$= \dfrac{8 \cdot 7 \cdot 6 \cdot 5!}{5! \cdot 3 \cdot 2 \cdot 1}$ Rewrite $8!$ as $8 \cdot 7 \cdot 6 \cdot 5!$ and simplify.

$= 56$ Evaluate.

Using the formula for evaluating binomial coefficients, we can prove two special cases: $\binom{n}{0} = 1$ and $\binom{n}{n} = 1$.

Calculator Tips

Factorials can be found on most graphing calculators. To find $10!$ on the TI-83 or TI-84 Plus, enter $\boxed{1}$ $\boxed{0}$ *, go to the Math menu, select PRB, and press* $\boxed{4}$ *.*

```
10!
        3628800
```

Objective ③ Expand a binomial using the binomial theorem. Let's look at the expansion of $(a + b)^6$ and make another observation.

$$(a + b)^6 = a^6 + 6a^5b + 15a^4b^2 + 20a^3b^3 + 15a^2b^4 + 6ab^5 + b^6$$

Look at the third term of the expansion, $15a^4b^2$. The coefficient is 15, and from Example 2(a), we see that $\binom{6}{2} = 15$. The coefficient of the fourth term is 20, and $\binom{6}{3} = 20$. In both cases, the n value of the binomial coefficient, $\binom{n}{r}$, is the exponent of the binomial and the r value is the exponent of b. This observation leads to the **binomial theorem**.

Definition **Binomial theorem:** For any positive integer n,

$$(a + b)^n = \binom{n}{0}a^n + \binom{n}{1}a^{n-1}b + \binom{n}{2}a^{n-2}b^2 + \binom{n}{3}a^{n-3}b^3 + \cdots + \binom{n}{n}b^n.$$

Note: This is the same result we got earlier using Pascal's triangle. Also note that

$$\binom{6}{3} = \frac{6!}{3!3!} = \frac{6 \cdot 5 \cdot 4 \cdot 3!}{3!3!}$$

$$= \frac{6 \cdot 5 \cdot 4 \cdot \cancel{3!}}{3 \cdot 2 \cdot 1 \cdot \cancel{3!}}$$

$$= \frac{\cancel{6} \cdot 5 \cdot 4}{\cancel{3} \cdot \cancel{2} \cdot 1} = 20, \text{ etc.}$$

Example 3 Expand each of the following binomials using the binomial theorem.

a. $(a + b)^6$

SOLUTION

$$(a + b)^6 = \binom{6}{0}a^6 + \binom{6}{1}a^5b + \binom{6}{2}a^4b^2 + \binom{6}{3}a^3b^3 + \binom{6}{4}a^2b^4 + \binom{6}{5}ab^5 + \binom{6}{6}b^6$$

$$= \frac{6!}{0!6!}a^6 + \frac{6!}{1!5!}a^5b + \frac{6!}{2!4!}a^4b^2 + \frac{6!}{3!3!}a^3b^3 + \frac{6!}{4!2!}a^2b^4 + \frac{6!}{5!1!}ab^5 + \frac{6!}{6!0!}b^6$$

$$= a^6 + 6a^5b + 15a^4b^2 + 20a^3b^3 + 15a^2b^4 + 6ab^5 + b^6$$

b. $(a - 3b)^5$

Note: $\binom{5}{2} = \frac{5!}{2!3!} = \frac{5 \cdot 4 \cdot 3!}{2 \cdot 1 \cdot 3!}$

$$= \frac{5 \cdot 4 \cdot \cancel{3!}}{2 \cdot 1 \cdot \cancel{3!}} = \frac{5 \cdot 4}{2}$$

$$= \frac{20}{2} = 10, \text{ etc.}$$

SOLUTION Write $(a - 3b)^5$ as $[a + (-3b)]^5$

$$[a + (-3b)]^5 = \binom{5}{0}a^5 + \binom{5}{1}a^4(-3b) + \binom{5}{2}a^3(-3b)^2 +$$

$$\binom{5}{3}a^2(-3b)^3 + \binom{5}{4}a(-3b)^4 + \binom{5}{5}(-3b)^5$$

$$= a^5 + 5a^4(-3b) + 10a^3(9b^2) + 10a^2(-27b^3) + 5a(81b^4) + (-243b^5)$$

$$= a^5 - 15a^4b + 90a^3b^2 - 270a^2b^3 + 405ab^4 - 243b^5$$

Objective ④ Find a particular term of a binomial expansion. Sometimes it is necessary to find only a specific term of a binomial expansion without writing out the entire expansion. Look again at the binomial expansion. Note that the third term (which we will call the $(2 + 1)$st term) is $\binom{n}{2}a^{n-2}b^2$ and the fourth term (which we will call the $(3 + 1)$st term) is $\binom{n}{3}a^{n-3}b^3$. Similarly, the $(m + 1)$st term is $\binom{n}{m}a^{n-m}b^m$. These observations lead to the following.

Rule **Finding the $(m + 1)$st Term of a Binomial Expansion**

The $(m + 1)$st term of the expansion $(a + b)^n$ is $\binom{n}{m}a^{n-m}b^m$.

Example 4 Find the indicated term of each of the following binomial expansions.

a. $(a + b)^{11}$, seventh term

SOLUTION Use the formula for the $(m + 1)$st term with $n = 11$ and $m = 6$ (to find the seventh term).

$$\binom{n}{m}a^{n-m}b^m = \binom{11}{6}a^{11-6}b^6 = 462a^5b^6$$

b. $(2x - 5y)^8$, fourth term

SOLUTION Write $(2x - 5y)^8$ as $[2x + (-5y)]^8$. Use the formula for the $(m + 1)$st term with $n = 8$ and $m = 3$ (to find the fourth term), $a = 2x$, and $b = -5y$.

$$\binom{n}{m}a^{n-m}b^m = \binom{8}{3}(2x)^{8-3}(-5y)^3 = 56(32x^5)(-125y^3) = -224{,}000x^5y^3$$

Appendix C

Exercises For Extra Help *MyMathLab*

PRACTICE   WATCH   DOWNLOAD

1. How many terms are in the expansion of $(5a + b)^{12}$?

2. What is the sum of the exponents on x and y for any term in the expansion of $(x + y)^9$?

3. How is the symbol 8! read?

4. What is the meaning of 8!? What is its value?

5. What is the exponent of b in the sixth term of the expansion of $(a + b)^{13}$? What is the exponent of a in that term?

6. On a term in the expansion of $(x + y)^n$, the exponent of x is 4. Find the exponent of y.

For Exercises 7–18, evaluate each expression. See Example 1.

7. $4!$

8. $7!$

9. $(4!)(3!)$

10. $(3!)(2!)$

11. $(6!)(5!)$

12. $(4!)(7!)$

13. $\dfrac{8!}{10!}$

14. $\dfrac{7!}{9!}$

15. $\dfrac{10!}{9!}$

16. $\dfrac{12!}{11!}$

17. $\dfrac{8!}{6!(8 - 6)!}$

18. $\dfrac{10!}{6!(10 - 6)!}$

For Exercises 19–26, evaluate each binomial coefficient. See Example 2.

19. $\binom{7}{3}$

20. $\binom{5}{2}$

21. $\binom{10}{4}$

22. $\binom{6}{5}$

23. $\binom{7}{7}$

24. $\binom{4}{4}$

25. $\binom{8}{0}$

26. $\binom{9}{0}$

For Exercises 27–38, use the binomial theorem to expand each of the following. See Example 3.

27. $(a + b)^5$

28. $(a + b)^7$

29. $(x - y)^4$

30. $(x - y)^3$

31. $(2a + b)^3$

32. $(a + 2b)^3$

33. $(x - 2y)^5$

34. $(x - 3y)^4$

35. $(2m + 3n)^6$

36. $(3c + 2d)^5$

37. $(3x - 4y)^4$

38. $(4a - b)^7$

For Exercises 39–46, find the indicated term of each binomial expansion.
See Example 4.

39. $(x + y)^8$, fifth term

40. $(a + b)^9$, fourth term

41. $(a - b)^{10}$, third term

42. $(m - n)^7$, second term

43. $(4x + y)^9$, sixth term

44. $(3a + b)^{11}$, seventh term

45. $(3m - 2n)^7$, fourth term

46. $(5x - 3y)^{12}$, fifth term

Appendix D Synthetic Division

Objective 1 **Use synthetic division to divide a polynomial by a binomial in the form $x - c$.** We saw in Section 5.6 that long division can be a tedious process. When a polynomial is divided by a binomial in the form $x - c$, where c is a constant, we can streamline the process by focusing on just the coefficients in the polynomials using a method called *synthetic division*.

To get a sense of how synthetic division looks, let's look again at the long division for $\dfrac{x^2 + 5x + 7}{x + 2}$, which was Example 6 in Section 5.6. To the right of that long division, we gradually remove bits so that we are left with the bare essentials for finding the quotient.

Long Division	**Variables Removed**	**Bare Essentials**
$\begin{array}{r} x + 3 \\ x + 2 \overline{)\,x^2 + 5x + 7} \\ \underline{-x^2 - 2x} \\ 3x + 7 \\ \underline{-3x - 6} \\ 1 \end{array}$	$\begin{array}{r} 1 + 3 \\ 1 + 2 \overline{)\,1 + 5 + 7} \\ \underline{-1 - 2} \\ 3 + 7 \\ \underline{-3 - 6} \\ 1 \end{array}$	$\begin{array}{r} 1 + 3 \\ 1 - (-2) \overline{)\,1 + 5 + 7} \\ \underline{-2} \\ +7 \\ \underline{-6} \\ 1 \end{array}$

Note: In polynomial long division, the terms in the positions that we've highlighted in blue are always additive inverses. Because they always cancel, we can remove them to get the "bare-essentials" version, as seen to the right.

Note: We have rewritten $x + 2$ as $x - (-2)$ so that we have the form $x - c$.

We can place these "bare-essential" values in an even more compact form, which is the form for synthetic division.

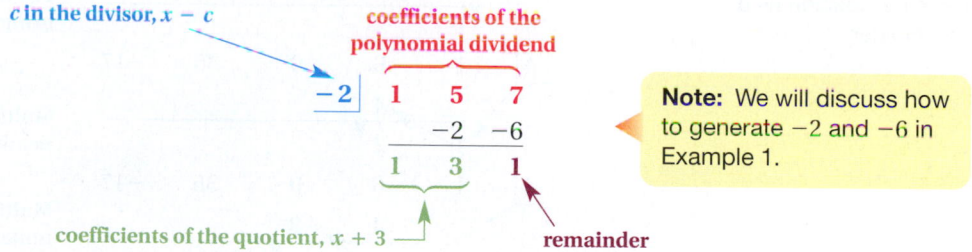

c in the divisor, $x - c$

coefficients of the polynomial dividend

coefficients of the quotient, $x + 3$

remainder

Note: We will discuss how to generate -2 and -6 in Example 1.

Now let's examine the steps of synthetic division in Example 1.

Example 1 Use synthetic division to divide.

a. $\dfrac{x^2 + 5x + 7}{x + 2}$

SOLUTION First, we need the divisor to be in the form $x - c$. So we rewrite $x + 2$ as $x - (-2)$, and we see that $c = -2$. We then set up the synthetic division with c and the coefficients of the dividend, as shown.

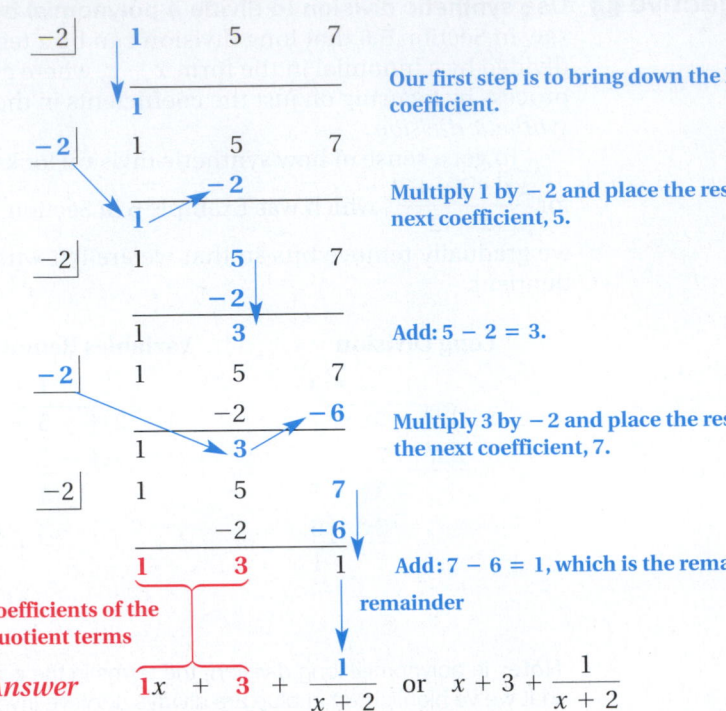

Our first step is to bring down the leading coefficient.

Multiply 1 by -2 and place the result underneath the next coefficient, 5.

Add: $5 - 2 = 3$.

Multiply 3 by -2 and place the result underneath the next coefficient, 7.

Add: $7 - 6 = 1$, which is the remainder.

coefficients of the quotient terms

remainder

Answer $1x + 3 + \dfrac{1}{x + 2}$ or $x + 3 + \dfrac{1}{x + 2}$

b. $(x^4 - 6x^3 + 35x - 17) \div (x - 4)$

SOLUTION Looking at the divisor, $x - 4$, we see that $c = 4$.

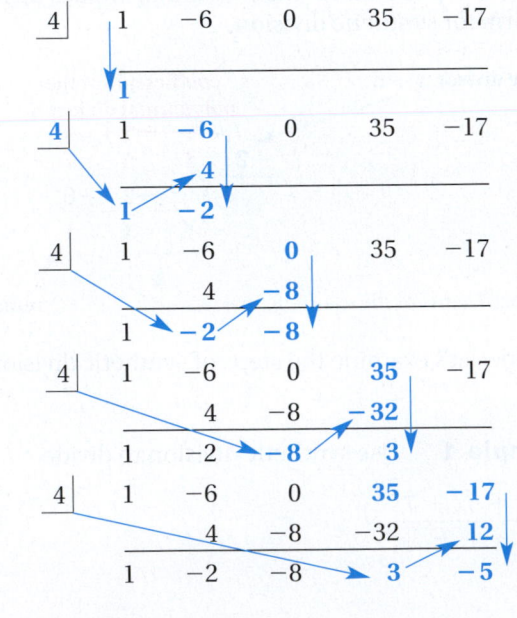

List the coefficients in the dividend and bring down the leading coefficient.

Multiply 1 by 4 and place the result underneath the next coefficient, -6. Then add.

Multiply -2 by 4 and place the result underneath the next coefficient, 0. Then add.

Multiply -8 by 4 and place the result underneath the next coefficient, 35. Then add.

Multiply 3 by 4 and place the result underneath the next coefficient, -17. Then add.

Answer $x^3 - 2x^2 - 8x + 3 - \dfrac{5}{x - 4}$

> **Note:** In synthetic division, the degree of the quotient polynomial is always one less than the degree of the dividend polynomial. Because the degree of the dividend $x^2 + 5x + 7$ is 2, the degree of the quotient, $x + 3$, is 1.

> **Note:** Because the polynomial dividend has no x^2 term, we place a 0 in the degree-2 position as a placeholder.

Appendix D

Exercises   For Extra Help

Note: Exercises marked with a ★ represent challenging exercises.

1. What form does the divisor need to have so that synthetic division can be used?

2. Suppose you are to use synthetic division to divide $\dfrac{x^3 - 8x^2 - 5x + 9}{x + 6}$. What is the value of c?

3. If you were to use synthetic division for $\dfrac{x^4 - 5x^2 - 3x + 11}{x - 2}$, what list of coefficients would you write?

4. Explain how to finish the synthetic division shown.

$$
\begin{array}{r|rrrr}
2 & 1 & 4 & -8 & -9 \\
  &   & 2 & 12 &   \\
\hline
  & 1 & 6 & 4 &
\end{array}
$$

5. Where does the remainder appear in synthetic division?

6. When using synthetic division, how do you determine the degree of the first term in the quotient expression?

For Exercises 7–12: **a.** *Write the binomial divisor.*
b. *Write the polynomial dividend.*
c. *Write the polynomial quotient with its remainder (if there is a remainder). See Objective 1.*

7.
$$
\begin{array}{r|rrr}
-3 & 1 & 7 & 12 \\
   &   & -3 & -12 \\
\hline
   & 1 & 4 & 0
\end{array}
$$

8.
$$
\begin{array}{r|rrr}
9 & 1 & -14 & 45 \\
  &   & 9 & -45 \\
\hline
  & 1 & -5 & 0
\end{array}
$$

9.
$$
\begin{array}{r|rrrr}
3 & 1 & 4 & -25 & 7 \\
  &   & 3 & 21 & -12 \\
\hline
  & 1 & 7 & -4 & -5
\end{array}
$$

10.
$$
\begin{array}{r|rrrr}
-2 & 1 & -2 & -6 & 7 \\
   &   & -2 & 8 & -4 \\
\hline
   & 1 & -4 & 2 & 3
\end{array}
$$

11.
$$
\begin{array}{r|rrrrr}
4 & 2 & -8 & -5 & 17 & 10 \\
  &   & 8 & 0 & -20 & -12 \\
\hline
  & 2 & 0 & -5 & -3 & -2
\end{array}
$$

12.
$$
\begin{array}{r|rrrrr}
-5 & 3 & 16 & 0 & -27 & -9 \\
   &   & -15 & -5 & 25 & 10 \\
\hline
   & 3 & 1 & -5 & -2 & 1
\end{array}
$$

For Exercises 13–38, divide using synthetic division. See Example 1.

13. $\dfrac{x^2 - 5x + 6}{x - 2}$

14. $\dfrac{x^2 + 8x - 9}{x - 1}$

15. $\dfrac{2x^2 - x - 5}{x - 4}$

16. $\dfrac{3x^2 - 11x - 7}{x - 5}$

17. $\dfrac{x^3 - x^2 - 5x + 6}{x - 2}$

18. $\dfrac{2x^3 - 6x^2 - 5x + 15}{x - 3}$

19. $\dfrac{3x^3 + 8x^2 + 3x - 2}{x + 2}$

20. $\dfrac{5x^3 + 32x^2 + 52x + 16}{x + 2}$

21. $(3x^3 - x^2 - 22x + 24) \div (x + 3)$

22. $(8x^3 - 10x^2 - x + 3) \div (x - 1)$

23. $\dfrac{2x^3 - 5x^2 - x + 6}{x + 2}$

24. $\dfrac{3x^3 - 5x^2 - 16x + 12}{x - 2}$

25. $(x^3 - 2x^2 + x - 3) \div (x - 2)$

26. $(x^3 + 6x^2 + 2x - 3) \div (x + 6)$

27. $\dfrac{2x^3 + x^2 - 4}{x + 3}$

28. $\dfrac{4x^3 - 2x + 5}{x - 2}$

29. $\dfrac{3x^3 - 4x^2 + 2}{x - 3}$

30. $\dfrac{x^3 + 4x^2 - 7}{x + 3}$

31. $(x^3 - 7x + 6) \div (x - 2)$

32. $(x^3 + 2x + 135) \div (x + 5)$

33. $\dfrac{x^3 + 27}{x - 3}$

34. $\dfrac{x^3 - 8}{x + 2}$

35. $\dfrac{x^4 + 16}{x + 2}$

36. $\dfrac{x^4 + 81}{x - 3}$

★ 37. $\dfrac{6x^3 + x^2 - 11x - 6}{x + \dfrac{2}{3}}$

★ 38. $\dfrac{2x^3 - 13x^2 + 27x - 18}{x - \dfrac{3}{2}}$

39. The area of the parallelogram shown is described by $9x^2 - 10x - 16$.
  a. Find an expression for the length of the base.

  b. Find the base, height, and area of the parallelogram if $x = 12$ inches.

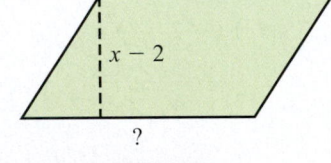

40. The area of the rectangle shown is $12n^2 + 65n - 42$.
  a. Find the length.

  b. Find the length, width, and area of the rectangle if $n = 9$ centimeters.

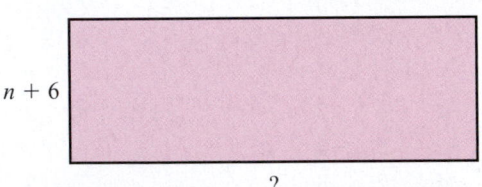

41. The volume of the walk-in freezer shown is described by $2y^3 - 7y^2 - 38y + 88$.
  a. Find an expression for the height of the room.

  b. Find the length, width, height, and volume of the room if $y = 10$ feet.

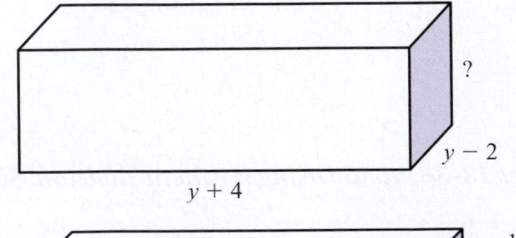

42. The volume of the storage box shown is described by $2x^3 - 168x - 320$.
  a. Find an expression for the length of the box.
  b. Find the length, width, height, and volume of the box if $x = 16$ inches.

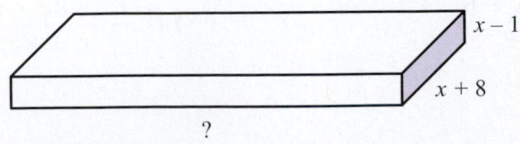

Appendix E Mean, Median, and Mode

Objectives
1. Find the mean.
2. Find the median.
3. Find the mode.

Objective 1 Find the mean.

Numbers can be used to describe characteristics of data. A number used in this way is called a **statistic**.

Definition *Statistic:* A number used to describe some characteristic of a set of data.

Oftentimes we use statistics, such as averages, which describe the *middle* or *central tendency* in a set of data. One such statistic is the **arithmetic mean**. Many people use the words *average* and *mean* interchangeably. Actually, the mean is only one type of average, called the **arithmetic average**. Other averages are the median and mode, which we will discuss later in this appendix.

Definition *Mean* or *arithmetic average:* The sum of all given numbers divided by the number of numbers.

Procedure **To find the mean, or arithmetic average, of a given set of numbers:**
1. Calculate the sum of all the given numbers.
2. Divide the sum by the number of numbers.

The variable used to indicate the mean is $\bar{x}$.

Example 1 Clerical specialist salaries within a company are listed below. What is the mean salary of a clerical specialist within the company?

$24,000	$21,900
$19,500	$19,500
$20,400	$25,200

Understand: We must calculate the arithmetic average, or mean, salary. We are given a set of salaries.

Plan: Divide the sum of the salaries by the number of salaries in the list.

Execute: $\bar{x} = \dfrac{24{,}000 + 19{,}500 + 20{,}400 + 21{,}900 + 19{,}500 + 25{,}200}{6}$

$\bar{x} = \dfrac{130{,}500}{6}$

$\bar{x} = 21{,}750$

Answer: The mean salary of the clerical specialists at this company is $21,750.

Check: We can verify the calculations by inverse operations. We will leave this to the reader.

Objective ② Find the median.

Another average or statistic that describes the *middle* or *central tendency* of a set of numbers is the **median**.

Definition   *Median:* The middle number in an ordered set of numbers.

Consider the set of numbers: 4, 5, 9, 3, 2, 10, 9

The definition of *median* indicates that the set *must be ordered*. This means that we must put the numbers in order from smallest to largest or largest to smallest. We must also write *each* repetition of a number.

$$2, 3, 4, \mathbf{5}, 9, 9, 10$$
$$\uparrow$$
$$\text{median}$$

> **Note:** Since 5 is the middle number in this ordered set, it is the median.

Notice that the median is easy to find in a set that contains an odd number of numbers. Suppose the set contains an even number of numbers as in 2, 3, 4, 5, 9, 10.

$$2, 3, 4, 5, 9, 10$$
$$\uparrow$$
$$\text{median}$$

$$\text{median} = \text{mean of } 4 \text{ and } 5 = \frac{4+5}{2} = 4.5$$

> **Note:** The middle of this set of ordered numbers is *between* 4 and 5. In fact, the number exactly halfway between 4 and 5 is the mean of 4 and 5.

Procedure   **To find the median of a set of numbers:**

1. Arrange the numbers in order from least to greatest or greatest to least.
2. Locate the middle number of the ordered set of numbers.

> **Note:** If there are an even number of numbers in the set, the median will be the mean of the two middle numbers.

Example 2   Find the median salary for clerical specialists at the company described in Example 1. The salaries were

$24,000	$21,900
$19,500	$19,500
$20,400	$25,200

Understand: We must find the median of the given salaries.

Plan: Arrange the salaries in order from least to greatest, then locate the middle salary in the ordered list. Because there is an even number of salaries, the median salary will be the mean of the middle two salaries.

Execute: 19,500   19,500   20,400   21,900   24,000   25,200

> **Note:** The median is halfway between 20,400 and 21,900, so we must find the mean of 20,400 and 21,900.

$$\text{median} = \frac{20{,}400 + 21{,}900}{2} = \frac{42{,}300}{2} = 21{,}150$$

Answer: The median income for the clerical specialists at the company is $21,150.

Check: Verify the calculations using inverse operations.

In Example 2, notice that half of the people have salaries greater than the median salary and half have salaries less than the median salary. Median does not indicate how much greater or less the other salaries might be. Also, it is possible that no one earns exactly the mean or median salary.

Objective 3 Find the mode.

Another statistic we may consider with data sets is called the **mode**.

Definition   *Mode:* The number that occurs most often in a set of numbers.

Consider the set of numbers: 2, 5, 9, 12, 9, 15, 9, 12

Since 9 is the number that occurs most often, it is the mode of this set. If no number is repeated, then there is no mode. If there is a tie between numbers that occur most often, then we list each as a mode.

Procedure   **To find the mode**

To find the mode of a set of numbers, count the number of repetitions of each number. The number with the most repetitions is the mode.

> **Note:** Ordering the data helps spot repeated numbers and the mode(s). If no numbers are repeated, then there is no mode. If there is a tie between numbers that occur most often, then list each number as a mode.

Example 3   Find the mode.

a. Twelve players in a golf tournament post the following numbers:

$$69, 66, 74, 72, 69, 70, 72, 71, 75, 65, 72, 71$$

Understand: We must find the mode in the set of numbers. The mode is the number that occurs most frequently.

Plan: Order the data and count the number of repetitions of each number. The number with the most repetitions is the mode.

Execute: 65, 66, 69, 69, 70, 71, 71, 72, 72, 72, 74, 75

                  twice   twice   three times

                           72 is the number with the most repetitions, so it is the mode.

Answer: The mode is 72.

b. Final grades for a class of 15 students:

$$85, 86, 72, 65, 80, 91, 62, 76, 80, 85, 76, 78, 80, 96, 85$$

Understand: We must find the mode in the set of numbers. The mode is the number that occurs most frequently.

Plan: Order the data and count the number of repetitions of each number. The number with the most repetitions is the mode.

Execute: 62, 65, 72, 76, 76, 78, 80, 80, 80, 85, 85, 85, 86, 91, 96

twice   three times   three times

85 and 80 both occur three times. This
means they are *both* modes.

Answer: 85 and 80 are modes.

c. Birth weights for a group of 6 siblings:

7.4 pounds   7.5 pounds   8.2 pounds   7.6 pounds   7.8 pounds   8.1 pounds

Understand: We must find the mode in the set of weights. The mode is the weight that occurs most frequently.

Plan: Order the data and count the number of repetitions of each weight. The weight with the most repetitions is the mode.

Execute: 7.4 pounds, 7.5 pounds, 7.6 pounds, 7.8 pounds, 8.1 pounds, 8.2 pounds. Because no weight is repeated, there is no mode.

Answer: There is no mode for this set of weights.

Appendix E
Exercises   For Extra Help   *MyMathLab*   Math XL
PRACTICE   WATCH   DOWNLOAD

1. What is the mean of a set of data?

2. What is the median of a set of data?

3. Explain how to find the median of a set containing an even number of numbers.

4. What is the mode of a set of data?

For Exercises 5–14, find the mean, median, and mode(s).

5. Following is a list of nurse salaries on the surgical floor of a hospital. Find the mean, median, and mode of the salaries.

$25,500	$28,700
$28,700	$27,500
$26,450	$24,200

6. Following is a list of the hourly wages for employees at a packaging plant. Find the mean, median, and mode of the wages.

$11.00	$12.00	$11.00
$15.00	$18.25	$12.50
$13.50	$16.25	$18.00

7. Find the mean, median, and mode for the test scores of students in a history class.

80	92	64	78	88
80	82	74	72	60
55	96	100	71	82
75	82	90	86	58

8. A basketball team has the following final scores. Find the mean, median, and mode of the scores.

76	82	80	78	85	75
78	80	72	70	84	88

9. A marine biologist studying leatherback sea turtles measures and records their lengths. Following is a list of lengths for the last 20 turtles. Find the mean, median, and mode of the lengths.

2.2 m	1.8 m	1.9 m	2.3 m	2.1 m
1.6 m	1.5 m	1.8 m	1.2 m	2.0 m
2.1 m	1.4 m	1.2 m	2.1 m	1.7 m
2.2 m	2.0 m	1.6 m	1.8 m	1.9 m

10. Following is a list of heights of players on a basketball team. Find the mean, median, and mode of the heights.

6.5 ft.	6.75 ft.	6.25 ft.	6.5 ft.	6.8 ft.	6.75 ft.
6.25 ft.	7 ft.	6.75 ft.	6.25 ft.	6 ft.	6.2 ft.

11. Following is a list of rainfall amounts for one month. Each amount is the total of rainfall in one 24-hour period. Calculate the mean, median, and mode.

0.4 in.	0.8 in.	1.2 in.	3.4 in.
0.6 in.	1.5 in.	0.6 in.	1.0 in.
1.4 in.	2.3 in.	1.3 in.	0.5 in.

12. The table lists daily high and low temperatures (°F) for the last two weeks. Find the mean, median, and mode of the high temperatures. Then find the mean, median, and mode of the low temperatures.

	Sun.	Mon.	Tue.	Wed.	Thu.	Fri.	Sat.
High	91	92	95	93	90	89	88
Low	72	74	78	75	72	70	71
High	90	94	96	95	92	90	91
Low	72	74	78	76	73	70	72

13. The table lists each month's electric and gas charges for a family over a two-year period. Find the mean, median, and mode of the electric and gas charges each year.

Month	Year 1	Year 2
January	$158.92	$165.98
February	$147.88	$162.85
March	$125.90	$130.45
April	$108.40	$112.55
May	$87.65	$90.45
June	$114.58	$125.91
July	$145.84	$137.70
August	$142.78	$140.19
September	$90.25	$96.15
October	$104.12	$115.75
November	$136.62	$145.21
December	$158.18	$160.25

14. The table lists water consumption, in cubic feet, by a family over a two-year period. Find the mean, median, and mode for each year.

Month	Year 1	Year 2
January	500	550
February	525	540
March	600	600
April	650	650
May	840	900
June	1100	1430
July	1000	1200
August	1400	1250
September	840	750
October	620	700
November	600	550
December	550	500

Photograph Credits

Answers

Chapter 1

1.1 Exercises 1. A collection of objects.   **3.** A rational number can be expressed as a ratio of integers, but an irrational number cannot.   **5.** true   **7.** true   **9.** False. All real numbers are either rational or irrational.   **11.** {Sunday, Monday, Tuesday, Wednesday, Thursday, Friday, Saturday}   **13.** {a, e, i, o, u}   **15.** {5, 10, 15, 20, . . . }   **17.** {9, 11, 13, 15, . . . }   **19.** {−6, −5, −4, −3}   **21.** rational   **23.** rational   **25.** irrational   **27.** rational   **29.** rational   **31.** 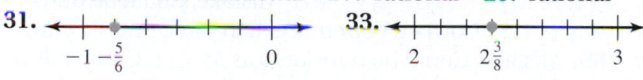   **33.**

35.   **37.**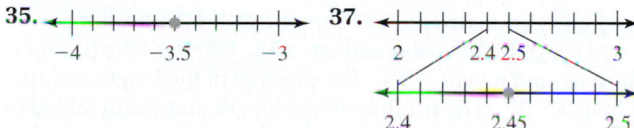

39. 23   **41.** 2   **43.** 5.7   **45.** $3\frac{1}{8}$ **47.** 0   **49.** >   **51.** >   **53.** >   **55.** <   **57.** >   **59.** >   **61.** <   **63.** =   **65.** <   **67.** =   **69.** >   **71.** >   **73.** <   **75.** <   **77.** $-4.7, -2.56, 5.4, \left|-7\frac{1}{2}\right|, |8.3|$ **79.** $-0.6, -0.44, 0, |-0.02|, 0.4, \left|1\frac{2}{3}\right|, 3\frac{1}{4}$

1.2 Exercises 1. Multiply or divide the numerator and denominator by the same nonzero number.   **3.** No, 4 is not a prime number. The correct prime factorization is $2 \cdot 2 \cdot 2 \cdot 2 \cdot 7$.   **5.** No, for example, 3 and 7 are whole numbers that are not composite. (They are prime numbers.)   **7.** True. Prime numbers are natural numbers, and every natural number is a whole number.   **9.** True. Every composite number can be written as a product of prime numbers.   **11.** $\frac{3}{10}$ **13.** $\frac{1}{3}$ **15.** $\frac{3}{4}$ **17.** $\frac{3}{8}$ **19.** $\frac{15}{16}$ **21.** 12   **23.** −35   **25.** 16   **27.** 15   **29.** $\frac{12}{15}$ and $\frac{10}{15}$ **31.** $\frac{8}{18}$ and $\frac{21}{18}$ **33.** $-\frac{44}{72}$ and $-\frac{51}{72}$ **35.** $-\frac{9}{144}$ and $-\frac{60}{144}$ **37.** $2 \cdot 2 \cdot 11$ **39.** $2 \cdot 2 \cdot 3 \cdot 3$ **41.** $2 \cdot 2 \cdot 2 \cdot 2 \cdot 2 \cdot 2$ **43.** $2 \cdot 5 \cdot 5 \cdot 5$ **45.** $\frac{4}{5}$ **47.** $\frac{7}{11}$ **49.** $-\frac{3}{4}$

51. $-\frac{8}{15}$ **53.** Incorrect. You may divide out only factors, not addends.   **55.** Incorrect. The prime factorization of 240 should be $2 \cdot 2 \cdot 2 \cdot 2 \cdot 3 \cdot 5$.   **57.** $\frac{7}{8}$ **59.** $\frac{12}{23}$ **61.** $\frac{5}{21}$ **63.** $\frac{2}{3}$ **65.** $\frac{19}{50}$ **67.** $\frac{31}{50}$ **69. a.** 1985   **b.** $\frac{77}{200}$ **71.** $\frac{37}{189}$ **73.** $\frac{6}{73}$ **75.** $\frac{2}{3}$ **77.** $\frac{202}{435}$ **79.** $\frac{4}{15}$

Review Exercises 1. {Mercury, Venus, Earth, Mars}   **2.** It is a rational number because it can be written as a ratio of the integers 8 and 10.   **3.**

4. It is an expression because it has no = sign.   **5.** 27   **6.** =

1.3 Exercises 1. The commutative property of addition changes the order, while the associative property of addition changes the grouping.   **3.** Adding 0 to a number or an expression does not change the identity of the number or expression.   **5.** When adding two numbers that have the same sign, add their absolute values and keep the same sign.   **7.** To write a subtraction statement as an equivalent addition statement, change the operation symbol from a minus sign to a plus sign and change the subtrahend to its additive inverse.   **9.** Commutative property of addition.   **11.** Additive identity.   **13.** Additive inverse.   **15.** Associative property of addition.   **17.** Commutative property of addition.   **19.** Additive inverse.   **21.** 21   **23.** −18   **25.** 9   **27.** −9   **29.** 14   **31.** −16   **33.** $\frac{3}{4}$ **35.** $-\frac{2}{3}$ **37.** $-\frac{2}{3}$ **39.** $\frac{11}{12}$ **41.** $-\frac{3}{4}$ **43.** $-\frac{9}{14}$ **45.** 0.26   **47.** 6.52   **49.** −4.38   **51.** −28   **53.** 3.18   **55.** $\frac{29}{24}$ **57.** −5   **59.** 12   **61.** 0   **63.** $-\frac{5}{6}$ **65.** 0.29   **67.** x **69.** $-\frac{m}{n}$ **71.** 2   **73.** −4   **75.** −4   **77.** −12   **79.** −9   **81.** −13   **83.** 7   **85.** −6   **87.** 0   **89.** $\frac{13}{10}$ **91.** $-\frac{18}{35}$ **93.** 0.36   **95.** −5.75   **97.** −1.6   **99.** −10   **101.** −0.9   **103.** $877,000,000   **105.** −268.2 N; the negative indicates that the beam is moving downward.   **107.** 133.40   **109.** $25.10   **111.** $-196 - (-208)$; 12   **113. a.** $18.6 - 18.8$ **b.** −0.2   **c.** The negative difference indicates that the mean composite score in 1989 was less than the score in 1986.   **115.** 2000; ≈ $16,000

Review Exercises 1. {Washington, Adams, Jefferson, Madison} 2. 7 3. $2 \cdot 2 \cdot 5 \cdot 5$ 4. $\dfrac{6}{7}$ 5. 5 6. $<$

1.4 Exercises 1. positive 3. negative 5. There is no quotient that we can multiply by 0 to get 5. 7. Their product is 1. 9. An odd number of negatives in division will result in a negative answer. 11. Distributive property. 13. Multiplicative identity. 15. Multiplicative property of 0. 17. Commutative property of multiplication. 19. Associative property of multiplication. 21. Commutative property of multiplication. 23. -18 25. -50 27. -54 29. 20 31. 42
33. $-\dfrac{1}{4}$ 35. $\dfrac{1}{2}$ 37. $-\dfrac{4}{9}$ 39. -38 41. 6.72 43. -6.536
45. 54 47. -48 49. -168 51. -144 53. 720 55. -48
57. $\dfrac{3}{2}$ 59. $-\dfrac{2}{5}$ 61. $-\dfrac{1}{5}$ 63. There is no multiplicative inverse of 0. 65. -3 67. 14 69. -2 71. 9 73. 0
75. undefined 77. indeterminate 79. 9 81. $-\dfrac{2}{3}$ 83. $\dfrac{3}{4}$
85. $-\dfrac{14}{15}$ 87. 2.1 89. 91.8 91. $-46.\overline{6}$ 93. $\dfrac{1}{2}$ 95. $-\$8680$
97. 612.8 99. -402.5 lb.; the force is downward
101. 169,202.2245 kg 103. -51.2 V 105. 3600 W

Review Exercises 1. irrational
2. [number line from -8 to -7, point between -7.2 and -7] 3. $-\dfrac{2}{3}$
4. 6.8 5. -22 6. $-\dfrac{11}{24}$

1.5 Exercises 1. Two cubed. 3. Squaring a number means to multiply the number by itself; finding its square root means to find a number whose square is the given number.
5. We can multiply and then find the square root of the product, or we can find the square root of each factor and then multiply those square roots. 7. Base: 7; exponent: 2; "seven squared." 9. Base: -5; exponent: 3; "negative five cubed."
11. Base: 2; exponent: 7; "additive inverse of two to the seventh power." 13. 81 15. 64 17. -64 19. -125
21. -125 23. 8 25. -1 27. $\dfrac{1}{25}$ 29. $-\dfrac{27}{64}$ 31. 0.008
33. 16.81 35. ± 11 37. No real-number square roots exist.
39. ± 14 41. ± 16 43. 4 45. 12 47. 0.7
49. Not a real number. 51. $\dfrac{8}{9}$ 53. 5 55. 19 57. 0
59. 10 61. -11 63. -4 65. -40 67. -20 69. 17
71. -7 73. -28 75. 26.8 77. -97.4 79. -53
81. $-10\dfrac{4}{5}$ 83. 78 85. 19 87. -29 89. $-\dfrac{253}{300}$
91. 22 93. 12 95. 0 97. 4 99. 7 101. undefined
103. Associative property of multiplication. The multiplication was not performed from left to right. 105. Distributive property. The parentheses were not simplified first.
107. Mistake: Multiplied before dividing. Correct: 1
109. Mistake: Found the square roots of the subtrahend and minuend in square root of a difference. Correct: 24 111. 77.2
113. 96 115. $\$588.41$ 117. $\$49.41$

Review Exercises 1. $\{0, 1, 2, 3, 4, 5, 6, 7, 8, 9, 10\}$
2. [number line from -1 to 0, point at $-\frac{4}{5}$] 3. $-\dfrac{1}{9}$ 4. It is an expression because it has no $=$ sign. 5. 24 6. 3

1.6 Exercises 1. Sum, plus, added. 3. Product, times, twice. 5. Less than 7. Some number plus seven, seven added to some number, the sum of some number and seven, seven more than some number, some number increased by seven. 9. Seven divided by some number, the quotient of seven and some number, the ratio of seven and some number.
11. $4x$ 13. $4x + 16$ 15. $7x - 8$ 17. $-4 \div y^3$ or $\dfrac{-4}{y^3}$
19. $8p - 4$ 21. $\dfrac{m}{14}$ 23. $x^4 + 5$ 25. $7w - \dfrac{1}{5}$
27. $-3(n - 2)$ 29. $(4 + n)^5$ 31. $mn - 5$ 33. $4 \div n - 2$ or $\dfrac{4}{n} - 2$ 35. $-27 - (a + b)$ 37. $0.6 - 4(y - 2)$
39. $(p - q) - (m + n)$ 41. $\sqrt{y} - mn$ 43. $n - 3(n - 6)$
45. Mistake: Order is incorrect. Correct: $3t - 17$
47. Mistake: Multiplied x by 9 instead of the sum of x and y.
Correct: $9(x + y)$ 49. $w + 5$ 51. $3w$ 53. $\dfrac{1}{2}d$ 55. $42 - n$
57. $t + \dfrac{1}{4}$ 59. $2w + 2l$ 61. $\dfrac{1}{3}\pi r^2 h$ 63. mc^2
65. $\sqrt{(x_2 - x_1)^2 + (y_2 - y_1)^2}$ 67. Mistake: Could be translated as $3(x + 4)$. Correct: Four more than three times a number. 69. Mistake: Could be translated as $5x - 1$. Correct: Five times the difference of x and one. 71. Mistake: Could be translated as $(n + 5)(n - 6)$. Correct: n plus the product of five and the difference of n and six. 73. One-half the product of the base and height. 75. The product of the length, width, and height. 77. The product of the length and width added to the product of the length and height added to the product of the width and height, all doubled. 79. The ratio of the difference of y_2 and y_1 to the difference of x_2 and x_1.

Review Exercises 1. $2 \cdot 5 + 2 \cdot 2 = 14$ 2. 2 3. $\dfrac{5}{12}$
4. -28 5. 12 6. -9

1.7 Exercises 1. To evaluate an expression, (1) replace the variables with their corresponding given values and (2) calculate the numerical expression using the order of operations.
3. An expression is undefined when the denominator is equal to 0. 5. A coefficient is the numerical factor in a term.
7. 9 9. 3.2 11. 23 13. $\dfrac{11}{2}$ 15. 4.6 17. -11 19. 3
21. -12 23. -21 25. -48 27. $\dfrac{3}{13}$ 29. a. 25 b. -12
31. a. 2 b. $\dfrac{2}{3}$ 33. -5 35. $-1, 3$ 37. 0 39. $\dfrac{2}{3}$
41. $6a + 12$ 43. $-32 + 24y$ 45. $\dfrac{7}{16}c - 14$
47. $0.6n - 1.6$ 49. coefficient: -6 51. coefficient: 1
53. coefficient: -1 55. coefficient: $-\dfrac{2}{3}$ 57. coefficient: $\dfrac{1}{5}$
59. $8y$ 61. $-5a$ 63. $11x$ 65. $-11r$ 67. $-1.9n$ 69. $-\dfrac{1}{3}b^2$

71. $-26c$ **73.** $5x - 1$ **75.** $-10x + 4y - 1$
77. $-9m + 5n + y - 11$ **79.** $-1.3x - 0.4$ **81.** $\frac{5}{6}c + \frac{5}{2}d + \frac{1}{7}$
83. $\frac{25}{4}a + \frac{18}{35}b^2 + 4$ **85. a.** $14 + (6n - 8n)$ **b.** $14 - 2n$
c. 20

Review Exercises
1. $-3, -2.5, 4.2, 4\frac{5}{8}, |-6|$ **2.** $\frac{6}{35}$
3. -22 **4.** 50 **5.** 1 **6.** 3

Chapter 1 Review Exercises
1. false **2.** false
3. false **4.** true **5.** true **6.** true **7.** positive
8. prime **9.** keep **10.** even **11.** {March, May}
12. {a, l, g, e, b, r} **13.** {2, 4, 6, ... }
14. { ..., -6, -4, 0, 2, 4, ... } **15.**

16.

17.

18.

19. 6.3 **20.** 8.46 **21.** $2\frac{1}{6}$

22. $4\frac{3}{8}$ **23.** $=$ **24.** $<$ **25.** $<$ **26.** $=$ **27.** $\frac{11}{15}$ **28.** $\frac{3}{8}$

29. 12 **30.** 25 **31.** 5 **32.** -10 **33.** $3 \cdot 17$ **34.** $2 \cdot 2 \cdot 5 \cdot 5$
35. $2 \cdot 2 \cdot 3 \cdot 3 \cdot 3$ **36.** $2 \cdot 2 \cdot 3 \cdot 7$ **37.** $\frac{19}{25}$ **38.** $\frac{1}{2}$ **39.** $\frac{25}{36}$
40. $\frac{2}{3}$ **41.** Associative property of addition.
42. Distributive property. **43.** Commutative property of
multiplication. **44.** Commutative property of addition.
45. Commutative property of multiplication.
46. Associative property of multiplication. **47.** -3.8
48. 15.1 **49.** $\frac{11}{15}$ **50.** $\frac{5}{12}$ **51.** 91 **52.** -48 **53.** -5
54. 10 **55.** -36 **56.** 9 **57.** 52 **58.** 9 **59.** 160 **60.** 1
61. $-\frac{28}{3}$ **62.** -8.6 **63.** -1 **64.** 27 **65.** 7 **66.** -103
67. 3 **68.** 24 **69.** ± 7 **70.** No real-number square roots
exist. **71.** $14 - 2n$ **72.** $\frac{y}{7}$ **73.** $y - 2(y + 4)$
74. $7(m - n)$ **75.** 61 **76.** 52 **77.** -8 **78.** 7 **79.** 5
80. -4 **81.** -6 **82.** $-3, 4$ **83.** $5x + 30$
84. $-15n + 24$ **85.** $2y + \frac{3}{4}$ **86.** $2.7m - 1.26$
87. $-3x - 3y - 15$ **88.** $4y^2 + 7y$ **89.** $2xy$
90. $-2x^3 - 6x^2$ **91.** $-8m$ **92.** $-14x - 6y + 6$
93. $-\$503.59$ **94. a.** -2.94 **b.** loss of $\$0.25$ **95. a.** $\$9494$
b. $\$588$ loss, or $-\$588$ **96.** $-152v$ **97.** ≈ 6.2
98. $\$54{,}158\frac{1}{3}$, or $\$54{,}158.33$

Chapter 1 Practice Test
1. [1.1]

2. 3.67 [1.1] **3.** $2 \cdot 2 \cdot 5 \cdot 5$ [1.2] **4.** $\frac{1}{3}$ [1.2] **5.** 12 [1.2]

6. Distributive property. [1.4] **7.** Commutative property of
addition. [1.3] **8.** 4 [1.3] **9.** $1\frac{17}{24}$ or $\frac{41}{24}$ [1.3] **10.** 0.6 [1.4]
11. $-\frac{5}{4}$ [1.4] **12.** -64 [1.5] **13.** -81 [1.5] **14.** -8 [1.5]
15. 2 [1.5] **16.** 5 [1.5] **17.** 10 [1.5] **18.** $2(m + n)$ [1.6]
19. $3w - 5$ [1.6] **20.** $\$250$ [1.6] **21.** 5.583 [1.6]
22. -11 [1.7] **23.** 10 [1.7] **24.** $-20y - 45$ [1.7]
25. $5.6x + 1.3$ [1.7]

Chapter 2

2.1 Exercises
1. An equation has an equal sign.
3. (1) Replace the variable in the equation with the value, and
(2) if the resulting equation is true, the value is a solution.
5. no **7.** yes **9.** no **11.** yes **13.** no **15.** yes **17.** yes
19. yes **21. a.** 66 ft. **b.** 6 **c.** $\$53.94$ **23.** ≈ 12.57 km
25. 7000 ft.^{2} **27.** $\$975$ **29. a.** 78 ft.^{2} **b.** 8 bales **c.** $\$36$
31. a. 260 in.^{2} **b.** $\$3900$ **33. a.** 3208 ft.^{2} **b.** $2138\frac{2}{3}$ ft.^{2}
c. 5 **d.** $\$530$ **35.** ≈ 6.9 ft.^{2} **37.** 82.25 ft.^{2} **39.** 47.5 m^3
41. ≈ 1005.3 cm^3 **43.** $83{,}068{,}741.\overline{6}$ ft.^{3} or $83{,}068{,}741\frac{2}{3}$ ft.^{3}
45. 64.5 mph **47.** ≈ 54.3 mph **49.** 1747.5 mi. **51.** -52 V
53. 93.2°F **55.** -297.4°F **57.** -60°C

Review Exercises
1. -51 **2.** -34 **3.** $\frac{1}{3}x + 2$
4. $-4w + 6$ **5.** $8x - 5$ **6.** $5.4y + 5.5$

2.2 Exercises
1. The addition principle of equality says
that we can add (or subtract) the same amount on both sides
of an equation without affecting its solution(s). **3.** Add 9 to
both sides of the equation. **5.** Use the addition principle of
equality to get the variable terms together on the same side of
the equal sign. **7.** yes **9.** no **11.** no **13.** yes **15.** yes
17. no **19.** yes **21.** yes **23.** 9 **25.** 5 **27.** -7 **29.** 5
31. -10 **33.** -25 **35.** $-\frac{3}{40}$ **37.** $-\frac{11}{24}$ **39.** -8.2 **41.** 4.6
43. -9 **45.** 7 **47.** 13 **49.** 6 **51.** -1 **53.** -2 **55.** 13.2
57. 36 **59.** -10 **61.** 16 **63.** 4 **65.** 9.7 **67.** All real
numbers. **69.** No solution. **71.** No solution. **73.** All real
numbers. **75.** $1947 + x = 2373$; $\$426$ **77.** $16 + x = 42$;
26 mi. **79.** $1741.62 - 1286.65 = 16.82 + 150.88 + 192.71 + x$;
$\$94.56$ **81.** $x + 12.4 + 16.3 + 27.2 = 67.2$; 11.3 cm
83. $x + 19 + 10 = 54$; 25 ft. **85.** $x + 900 + 3 \cdot 1500 +$
$2 \cdot 1245 = 8500$: $\$610$; yes, because $\$610$ is less than all of his
prior weeks. **87.** $x + \frac{1}{3} + \frac{2}{5} = 1$; $\frac{4}{15}$

Review Exercises
1. $-\frac{5}{4}$ **2.** -33.2 **3.** -9
4. $-6x + 1$ **5.** $8x^2 - 5x - 9$ **6.** $4x - 9$

2.3 Exercises
1. We can multiply (or divide) both sides
of an equation by the same amount without affecting its solu-
tion(s). **3.** Multiply both sides of the equation by the multiplicative inverse of the coefficient. **5.** Multiply both sides of
the equation by a common multiple of the denominators.

Using the LCD results in an equation with the smallest integers possible. **7.** 4 **9.** 3 **11.** 16 **13.** -20 **15.** $\dfrac{7}{6}$ **17.** $-\dfrac{4}{3}$

19. 7 **21.** 5 **23.** 5 **25.** -3 **27.** $-\dfrac{3}{2}$ **29.** 16 **31.** 0

33. -9 **35.** -5 **37.** 2 **39.** 16 **41.** 2 **43.** $\dfrac{20}{3}$ **45.** -2

47. 1 **49.** 2 **51.** No solution. **53.** 2 **55.** $-\dfrac{15}{2}$ **57.** All

real numbers. **59.** -9 **61.** No solution. **63.** $\dfrac{15}{8}$ **65.** $\dfrac{48}{13}$

67. 1 **69.** $-\dfrac{2}{3}$ **71.** -8 **73.** 9 **75.** All real numbers.

77. -9 **79.** 12 **81.** 6.5 **83.** Mistake: The minus sign was dropped in front of the 11. Correct: 2 **85.** Mistake: In the second line, the minus sign was not distributed to the 2.

Correct: $\dfrac{13}{3}$ **87.** 14 ft. **89.** 19 cm **91.** $l = 54$ ft., $w = 52$ ft.

93. 110 ft. **95.** 2 in. **97.** ≈ 6374.8 km **99.** ≈ 4.8 m
101. $w = 11$ ft., $x = 2$ ft., $w + 2 = 13$ ft., $y = 23$ ft.
103. -9.5 A **105.** 4 m/sec.² **107.** 945.5 kg
109. 413.6 m/sec. **111.** \$51.10 **113.** 15 ft./sec.

Review Exercises 1. 70 **2.** $3.8x^2 - 7.63x + 1.5$

3. $-\dfrac{9}{2}x - 6$ **4.** 20 **5.** -25 **6.** -10

2.4 Exercises 1. Add 2 to both sides. Because 2 is subtracted from w, adding 2 eliminates the 2, leaving w isolated. **3.** Subtract $2x$ from both sides. Subtracting $2x$ isolates the $3y$ term, which contains the variable we want to isolate.

5. $t = 4u + v$ **7.** $y = -\dfrac{x}{5}$ **9.** $x = \dfrac{b+3}{2}$ **11.** $m = \dfrac{y-b}{x}$

13. $y = \dfrac{8 - 3x}{4}$ **15.** $Y = \dfrac{mn}{4} - f$ **17.** $c = \dfrac{m - np - 12d}{6}$

19. $y = \dfrac{15 - 5x}{3}$ **21.** $n = c\left(\dfrac{3}{4} + 5a\right)$ **23.** $p = \dfrac{A - P}{tr}$

25. $a = P - b - c$ **27.** $l = \dfrac{A}{w}$ **29.** $d = \dfrac{C}{\pi}$ **31.** $r^2 = \dfrac{3V}{\pi h}$

33. $b = \dfrac{2A}{h}$ **35.** $w = \dfrac{P - 2l}{2}$ **37.** $l = \dfrac{2S - na}{n}$

39. $w = \dfrac{\pi r^2 h - V}{lh}$ **41.** $C = R - P$ **43.** $r = \dfrac{I}{Pt}$

45. $C = nP$ **47.** $r = \dfrac{A - P}{Pt}$ **49.** $t = \dfrac{d}{r}$ **51.** $d = \dfrac{W}{F}$

53. $t = \dfrac{W}{P}$ **55.** $t = \dfrac{v - v_0}{-32}$ or $t = \dfrac{v_0 - v}{32}$

57. $C = \dfrac{5}{9}(F - 32)$ **59.** $m = \dfrac{FR^2}{GM}$ **61.** Mistake: Subtracted

the coefficient 7 instead of dividing. Correct: $t = \dfrac{54 - 3n}{7}$

63. Mistake: Multiplied 5 by -2. Correct: $m = \dfrac{5nk + 2}{3}$

Review Exercises 1. $7n + 4$ **2.** $3(x + 2) - 9$ **3.** 0.5

4. $-\dfrac{5}{6}$ **5.** $-\dfrac{3}{4}$ **6.** -80

2.5 Exercises 1. Sum, plus, added. **3.** Of, times, twice.
5. The order of the subtraction is different. **7.** $y + 3 = -8; -11$

9. $6 - x = -3; 9$ **11.** $11m = -99; -9$ **13.** $m \div 4 = 1.6; 6.4$

15. $\dfrac{4}{5}x = \dfrac{5}{8}; \dfrac{25}{32}$ **17.** $7 + 3w = 34; 9$ **19.** $5a - 9 = 76; 17$

21. $8(8 + t) = 160; 12$ **23.** $-3(x - 2) = 12; -2$

25. $\dfrac{1}{3}(g + 2) = 1; 1$ **27.** $3m - 11 = 5 + m; 8$

29. $10 = \dfrac{r}{4} - 1; 44$ **31.** $2a - 3(a + 4) = -3; -9$

33. $(2d - 8) + (d - 12) = 13; 11$

35. $5\left(x + \dfrac{2}{3}\right) = 3x - \dfrac{2}{3}; -2$ **37.** $2x - 4 = 16 + 3x; -20$

39. $\dfrac{2}{5}x = \dfrac{1}{2}x - 2; 20$ **41.** $\dfrac{n-3}{3} = \dfrac{n-1}{4}; 9$

43. $-4(2 - x) = 2(4 - 3x) - 6; 1$ **45.** Three added to four times a number is seven. **47.** Six times the sum of a number and four is equal to the product of negative ten and the number. **49.** One-half of the difference of a number and three will result in two-thirds of the difference of the number and eight. **51.** Five-hundredths of a number added to six-hundredths of the difference of the number and eleven is twenty-two. **53.** The sum of two-thirds, three-fourths, and one-half of the same number will equal ten. **55.** Mistake: Subtraction translated in reverse order. Correct: $n - 10 = 40$ **57.** Mistake: Multiplied 5 times the unknown number instead of the difference, which requires parentheses. Correct: $5(x - 6) = -2$ **59.** Mistake: Subtracted the unknown number instead of the sum, which requires parentheses. Correct: $2t - (t + 3) = -6$ **61.** Translation: $P = 2(l + w)$

a. 84 ft. **b.** 61 cm **c.** $37\dfrac{1}{4}$ in. **63.** Translation: $P = b + 2s$

a. 27 in. **b.** 2.7 m **c.** $34\dfrac{1}{4}$ in. **65.** Translation: $I = Prt$

a. \$400 **b.** \$7.50 **c.** \$30 **67.** Translation: $t = \dfrac{e - d}{153.8}$

a. ≈ 49 sec. **b.** ≈ 52 sec. **c.** ≈ 62 sec.

Review Exercises 1. True; 19 is equal to 19. **2.** $=$

3. $>$ **4.** -5 **5.** $-\dfrac{10}{9}$ **6.** -25

2.6 Exercises 1. Any number that can replace the variable(s) in the inequality and make it true. **3.** Any number greater than -5 and -5 itself. **5. a.** $\{x | x \geq -3\}$ **b.** $[-3, \infty)$

c. ![number line from -5 to 1, closed bracket at -3] **7. a.** $\{h | h < 6\}$ **b.** $(-\infty, 6)$

c. ![number line from 0 to 8, open paren at 6]

9. a. $\left\{n | n < -\dfrac{2}{3}\right\}$

b. $\left(-\infty, -\dfrac{2}{3}\right)$ **c.** ![number line marking -4/3, -1, -2/3, -1/3]

11. a. $\{t | t \geq 2.4\}$ **b.** $[2.4, \infty)$

c. ![number line from 2 to 2.6, closed bracket at 2.4] **13. a.** $\{x | -3 < x < 6\}$

b. $(-3, 6)$ **c.** ![number line from -3 to 7, open parens at -3 and 6]

15. a. $\{n | 0 \leq n \leq 5\}$ **b.** $[0, 5]$ **c.** ![number line from -1 to 6, closed brackets at 0 and 5]

17. $\{x | x \geq -4\}, [-4, \infty)$ **19.** $\{x | x < 5), (-\infty, 5)$

21. $\{x \mid -4 \le x < 3\}, [-4, 3)$ **23. a.** $n > 5$ **b.** $\{n \mid n > 5\}$
c. $(5, \infty)$ **d.** (number line: open at 5, arrow right; ticks 2 3 4 5 6 7 8) **25. a.** $z \le -6$

b. $\{z \mid z \le -6\}$ **c.** $(-\infty, -6]$ **d.** (number line: bracket at −6, arrow left; ticks −8 −7 −6 −5 −4 −3)

27. a. $y \ge 2$ **b.** $\{y \mid y \ge 2\}$ **c.** $[2, \infty)$
d. (number line: bracket at 2, arrow right; ticks 0 1 2 3 4) **29. a.** $x \le 4$ **b.** $\{x \mid x \le 4\}$

c. $(-\infty, 4]$ **d.** (number line: bracket at 4, arrow left; ticks −1 0 1 2 3 4 5) **31. a.** $x \ge 6$

b. $\{x \mid x \ge 6\}$ **c.** $[6, \infty)$ **d.** (number line: bracket at 6, arrow right; ticks 3 4 5 6 7 8 9)

33. a. $m > 8$ **b.** $\{m \mid m > 8\}$ **c.** $(8, \infty)$
d. (number line: open at 8, arrow right; ticks 4 5 6 7 8 9 10 11) **35. a.** $y > 3$ **b.** $\{y \mid y > 3\}$

c. $(3, \infty)$ **d.** (number line: open at 3, arrow right; ticks 0 1 2 3 4 5 6) **37. a.** $x > -4$

b. $\{x \mid x > -4\}$ **c.** $(-4, \infty)$ **d.** (number line: open at −4, arrow right; ticks −5 −4 −3 −2 −1 0 1)

39. a. $a < 1$ **b.** $\{a \mid a < 1\}$ **c.** $(-\infty, 1)$
d. (number line: open at 1, arrow left; ticks 0 1 2 3 4) **41. a.** $f \le -2$ **b.** $\{f \mid f \le -2\}$

c. $(-\infty, -2]$ **d.** (number line: bracket at −2, arrow left; ticks −4 −3 −2 −1 0) **43. a.** $u \le 2$

b. $\{u \mid u \le 2\}$ **c.** $(-\infty, 2]$ **d.** (number line: bracket at 2, arrow left; ticks −2 −1 0 1 2 3)

45. a. $c > -12$ **b.** $\{c \mid c > -12\}$ **c.** $(-12, \infty)$
d. (number line: open at −12, arrow right; ticks −14 −13 −12 −11 −10 −9) **47. a.** $w \ge 3$ **b.** $\{w \mid w \ge 3\}$

c. $[3, \infty)$ **d.** (number line: bracket at 3, arrow right; ticks 0 1 2 3 4 5) **49. a.** $x \le -13$

b. $\{x \mid x \le -13\}$ **c.** $(-\infty, -13]$
d. (number line: bracket at −13, arrow left; ticks −15 −14 −13 −12 −11 −10) **51. a.** $n < 7$ **b.** $\{n \mid n < 7\}$

c. $(-\infty, 7)$ **d.** (number line: open at 7, arrow left; ticks 4 5 6 7 8 9) **53. a.** $l < 10$

b. $\{l \mid l < 10\}$ **c.** $(-\infty, 10)$ **d.** (number line: open at 10, arrow left; ticks 5 6 7 8 9 10 11)

55. a. $t \le 3.\overline{3}$ **b.** $\{t \mid t \le 3.\overline{3}\}$ **c.** $(-\infty, 3.\overline{3}]$
d. (number line: bracket at 3.3, arrow left; ticks 2, 3 3.3 3.6 4, 5)

57. $n - 4 > 24; n > 28$ **59.** $\dfrac{4}{9}x \le -8; x \le -18$

61. $8y - 36 \ge 60; y \ge 12$ **63.** $\dfrac{1}{2}a + 5 \le 2; a \le -6$

65. $4x - 8 < 2x; x < 4$ **67.** $25 \ge 6x + 7; x \le 3$
69. $i \le 10$ ft. **71.** $h \ge 15$ in. **73.** $r \le 23.89$ in.

75. $t \ge 6\dfrac{4}{13}$ hr. **77.** $R \ge \$1,475,000$ **79.** $x \ge 98$

81. $i \le 1.5$ A

Review Exercises 1. $\{1, 2, 3, 4, 5, 6, 7, 8, 9, 10\}$ **2.** $\dfrac{35}{2}$
3. $2(x - 5) + 6$ **4.** 16 **5.** $-\dfrac{135}{4}$ or $-33\dfrac{3}{4}$ **6.** 1426.208

Chapter 2 Review Exercises 1. true **2.** true
3. false **4.** false **5.** true **6.** true **7.** identify. **8.** linear
9. 1. Simplify **a.** Distribute **b.** multiplying; LCD
c. Combine **2.** addition **3.** multiplication **10.** Replace
the variable(s) in the equation with the value. If the resulting
equation is true, the value is a solution. **11.** yes **12.** no
13. no **14.** yes **15.** no **16.** yes **17.** -5 **18.** 18
19. 6 **20.** 5 **21.** -2 **22.** $\dfrac{9}{5}$ **23.** No solution. **24.** All
real numbers. **25.** 2 **26.** 18 **27.** $\dfrac{9}{4}$ **28.** $-\dfrac{1}{2}$ **29.** $\dfrac{3}{4}$

30. 5.25 **31.** $x = 1 - y$ **32.** $a^2 = c^2 - b^2$ **33.** $d = \dfrac{C}{\pi}$

34. $m = \dfrac{E}{c^2}$ **35.** $h = \dfrac{2A}{b}$ **36.** $m = \dfrac{Fr}{v^2}$ **37.** $h = \dfrac{3V}{\pi r^2}$

38. $m = \dfrac{y - b}{x}$ **39.** $w = \dfrac{P - 2l}{2}$ **40.** $h = \dfrac{2A}{a + b}$

41. $6x = -18; x = -3$ **42.** $\dfrac{1}{2}m - 3 = \dfrac{1}{4}m - 9; m = -24$

43. $2(v + 4) - 1 = 1;\ v = -3$
44. $2(y - 1) + 3y = 6y - 20; y = 18$
45. a. $x < -3$ **b.** $\{x \mid x < -3\}$ **c.** $(-\infty, -3)$
d. (number line: open at −3, arrow left; ticks −4 −3 −2 −1 0 1 2) **46. a.** $x \ge -1$ **b.** $\{x \mid x \ge -1\}$

c. $[-1, \infty)$ **d.** (number line: bracket at −1, arrow right; ticks −3 −2 −1 0 1 2 3) **47. a.** $z \le -1$

b. $\{z \mid z \le -1\}$ **c.** $(-\infty, -1]$ **d.** (number line: bracket at −1, arrow left; ticks −3 −2 −1 0 1 2 3)

48. a. $v > -4$ **b.** $\{v \mid v > -4\}$ **c.** $(-4, \infty)$
d. (number line: open at −4, arrow right; ticks −5 −4 −3 −2 −1 0 1) **49. a.** $m > 4$ **b.** $\{m \mid m > 4\}$

c. $(4, \infty)$ **d.** (number line: open at 4, arrow right; ticks 0 1 2 3 4 5 6) **50. a.** $c \ge -9$

b. $\{c \mid c \ge -9\}$ **c.** $[-9, \infty)$ **d.** (number line: bracket at −9, arrow right; ticks −10 −9 −8 −7 −6 −5 −4)

51. $13 > 3 - 10p; p > -1$ **52.** $3 + 2z < 3(z - 5); z > 18$
53. $-\dfrac{1}{2}x \ge 4; x \le -8$ **54.** $-2 \le 1 - \dfrac{1}{4}k; k \le 12$
55. 19.75 ft.^{2} **56.** 102.6 in.^{2} **57.** \$23.75 **58.** 80 in.
59. 7.1 in. **60.** 1.2 mi. **61.** 15 in. **62.** $d \le 3.18$ ft.
63. $t \ge 5\dfrac{5}{6}$ hr. **64.** $R \ge \$825,000$

Chapter 2 Practice Test 1. Equation because there
is an equal sign. [2.1] **2.** Nonlinear because there is a variable raised to an exponent other than 1. [2.2] **3.** yes [2.1]
4. no [2.1] **5.** -3 [2.3] **6.** 13.2 [2.2] **7.** 3 [2.3]

8. No solution. [2.3] **9.** $-\dfrac{2}{3}$ [2.3] **10.** 4 [2.3]

11. $x = \dfrac{7 - y}{2}$ [2.4] **12.** $r = \dfrac{I}{Pt}$ [2.4] **13.** $h = \dfrac{2A}{b}$ [2.4]

14. $F = \dfrac{9C + 160}{5}$ or $F = \dfrac{9}{5}C + 32$ [2.4] **15.** Set-builder
notation: $\{x \mid x \ge 3\}$; Interval notation: $[3, \infty)$;
Graph: (number line: bracket at 3, arrow right; ticks 0 1 2 3 4 5 6 7) [2.6]

16. Set-builder notation: $\{x|-1 \le x < 4\}$; Interval notation: $[-1, 4)$; Graph:

[2.6]

17. a. $m < 3$ **b.** $\{m|m < 3\}$ **c.** $(-\infty, 3)$

d. [2.6] **18. a.** $x \ge -8$

b. $\{x|x \ge -8\}$ **c.** $[-8, \infty)$

d. [2.6] **19. a.** $p > -\dfrac{2}{5}$

b. $\left\{p|p > -\dfrac{2}{5}\right\}$ **c.** $\left(-\dfrac{2}{5}, \infty\right)$

d. [2.6] **20. a.** $l \le -\dfrac{5}{3}$

b. $\left\{l|l \le -\dfrac{5}{3}\right\}$ **c.** $\left(-\infty, -\dfrac{5}{3}\right]$

d. [2.6]

21. $\dfrac{2}{3}n - \dfrac{1}{6} = 2n$; $n = -\dfrac{1}{8}$ [2.5] **22.** $12 + (-7x) = 5$;

$x = 1$ [2.5] **23.** $5(n - 2) - 3 = 10 - 4(n - 1)$;

$n = 3$ [2.5] **24.** $1 - n > 2n$; $n < \dfrac{1}{3}$ [2.6] **25.** 20 ft. [2.1]

26. 68.25 in.^{2} [2.1] **27.** 7 cm [2.1] **28.** $530 [2.1]
29. $l \le 3$ ft. [2.6] **30.** $x \ge 101$ [2.6]

Chapters 1 and 2 Cumulative Review

Exercises 1. true **2.** false **3.** false **4.** true
5. Understand, Plan, Execute, Answer, Check.
6. Grouping symbols, exponents/roots, multiplication/
division, addition/subtraction. **7.** Replace the variables
with their corresponding given values. Calculate the
numerical expression using the order of operations.

8. **9.** 102 **10.** -63 **11.** 12

12. -4 **13.** -5 **14.** 148 **15.** 7 **16.** $13x^2 - 13x - 16$

17. $\dfrac{31}{3}$ **18.** 3 **19.** -10 **20.** $r = \dfrac{B - P}{Pt}$

21. $w = \dfrac{3V}{lh}$ **22. a.** $h \ge -9$ **b.** $\{h|h \ge -9\}$

c. $[-9, \infty)$ **d.**

23. a. $m < -5$ **b.** $\{m|m < -5\}$ **c.** $(-\infty, -5)$
d. **24.** $3y - 40 = 7y; -10$

25. $12n - (n + 3) = -11 + 2(n - 5); -2$ **26.** 99 cm^2
27. 22 ft.^{2} **28.** 1376 ft. **29.** $x \le -5$ **30.** $x \le 8$ ft.

Chapter 3

3.1 Exercises 1.
The quantity preceding *to* (longest
side) is written in the numerator, and the quantity following
the word *to* (shortest side) is written in the denominator.
3. $ad = bc$ **5.** To solve proportion problems, (a) set up a
proportion in which the numerators and denominators of the

ratios correspond in a logical manner and (b) solve using
cross products. **7. a.** $\dfrac{4}{15}$ **b.** $\dfrac{6}{11}$ **c.** $\dfrac{2}{1}$ **9.** $\dfrac{1}{2}$ **11.** $\dfrac{2}{3}$

13. $\dfrac{63.5}{1}$; he drives an average of 63.5 miles per hour.

15. $\dfrac{0.125}{1}$; each banana costs 12.5 cents. **17.** $\approx \dfrac{16.47}{1}$; the
price of the stock is $16.47 for every $1 earned in 2009.
19. The 24-oz. container is better because it costs less per
ounce. (The unit ratio of price to quantity is less.)

21. a. $\approx \dfrac{0.80}{1}$ **b.** $\approx \dfrac{0.64}{1}$ **c.** The ratio of convictions to total

cases was greater in 1999 than in 2004. **23. a.** $\approx \dfrac{0.89}{1}$;

women in the 25–34 age group with a bachelor's degree earn
$0.89 for every $1.00 men in the corresponding group earn.

b. $\approx \dfrac{0.60}{1}$ **c.** Younger women with a bachelor's degree are

making more in relation to men with a bachelor's degree than
are older women. **25.** yes **27.** no **29.** no **31.** yes
33. no **35.** 10.5 **37.** -14 **39.** 10.5 **41.** -8 **43.** 115

45. $3\dfrac{3}{5}$ **47.** 4 **49.** 7 **51.** $183\dfrac{1}{3}$ mi. **53.** $1350.65

55. $9450 **57.** ≈ 422.25 pounds **59.** 4.875 in. **61.** $2\dfrac{1}{2}$ tsp.

63. 360 highway miles **65.** ≈ 15.8 min. **67.** 225 lb.
69. $85.91 **71.** ≈ 66 deer **73.** 617,647 **75.** 11.25 cm

77. $a = 4\dfrac{8}{19}$ ft.; $b = 5\dfrac{1}{4}$ ft.; $c = 5\dfrac{17}{19}$ ft. **79.** 300 m

81. 26.4 m

Review Exercises 1.
Yes, because it can be written as a

ratio of integers, $\dfrac{58}{100}$. **2.** $\dfrac{3}{25}$ **3.** 3.06 **4.** 0.452 **5.** 320

6. 0.4 or $\dfrac{2}{5}$

3.2 Exercises 1.
To write a percent as a decimal or frac-
tion, (a) write the percent as a ratio with 100 in the denominator and (b) simplify to the desired form. **3.** multiplication
5. In a word-for-word translation, the division yields a decimal
number that must be written as a percent. When using the
proportion method, the decimal number is multiplied

by 100, which gives the percent. **7.** $0.2, \dfrac{1}{5}$ **9.** $0.15, \dfrac{3}{20}$

11. $0.148, \dfrac{37}{250}$ **13.** $0.0375, \dfrac{3}{80}$ **15.** $0.455, \dfrac{91}{200}$ **17.** $0.\overline{3}, \dfrac{1}{3}$

19. 60% **21.** 37.5% **23.** 83.3% **25.** 66.7% **27.** 96%
29. 80% **31.** 9% **33.** 120% **35.** 2.8% **37.** 405.1%
39. 28 **41.** 3.1 **43.** 5.92 **45.** 103.2 **47.** 50 **49.** 80
51. 62.4 **53.** 20% **55.** 105% **57.** $66.\overline{6}\%$ **59.** 80%
61. 43 **63.** $4800; $400 **65.** $898.80 **67.** 5783.5 tg
69. $42.50 **71.** $300 **73.** $\approx 45.2\%$ **75.** $\approx 70\%$
77. $\approx 8.0\%$ **79.** United States $\approx 24.4\%$, China $\approx 2.7\%$,
Russia $\approx 53.6\%$; Russia **81.** 23.7% **83.** 20%
85. $6.22; $130.67 **87.** $94.90; $854.10 **89.** $2500
91. $1950; 75% **93.** $\approx 45.7\%$ **95.** $34.1\overline{6}\%$ **97.** $\approx 70.8\%$
99. 2800% **101.** 400,000 **103.** ≈ 2246.77

Review Exercises 1.
-1 **2.** $-3x - 21$
3. $-xy + 4x + 3y$ **4.** $3(n + 9) = n - 11; -19$ **5.** 2.1
6. 10.5 cm

3.3 Exercises **1.** The second number; $5n$. **3.** There is always a difference of 2 between consecutive even or odd numbers. **5. a.** $x + 3$ **b.** $6(x + 3) + 8x = 14x + 18$
7. a. $3w + 4$ **b.** $2w + 2(3w + 4) = 8w + 8$
9. a. $4x - 10$ **b.** $x + (4x - 10) = 90$ **11. a.** $-10, -8; -30$
b. $y + 2, y + 4$ **c.** $y + (y + 2) + (y + 4) = 3y + 6$
13. a. $3n$ **b.** maple: $7.99(3n)$; pecan: $11.99n$
c. $7.99(3n) + 11.99n = 35.96n$ **15. a.** $372 - x$
b. exterior: $16.99x$; interior: $12.99(372 - x)$
c. $16.99x + 12.99(372 - x) = 4x + 4832.28$ **17.** 13, 39
19. 10, 17 **21.** 11, 17 **23.** 6, 8 **25.** 12, 24 **27.** 10, 2
29. Grapefruit: 22 g; orange: 27 g **31.** Big Mac: 580 cal.; large fries: 520 cal. **33.** 25-watt: 215 lumens; 60-watt: 880 lumens
35. length: 70 ft.; width: 50 ft. **37.** length: 12 ft.; width: 4 ft.
39. length: 23 ft.; width: 13 ft. **41.** length: 11 cm; width: 7 cm
43. length: 7.5 ft.; width: 5 ft. **45.** 77°, 103° **47.** 50°, 130°
49. 26°, 64° **51.** 15°, 75° **53.** 73, 74 **55.** 12, 14
57. 92, 93, 94 **59.** 76, 78, 80 **61.** 123, 125, 127
63. 23, 24, 25 **65.** 69, 70, 71 **67.** 15, 17, 19 **69.** 24 $5 bills; 8 $50 bills **71.** 22 shares at $18; 45 shares at $32
73. 220 student tickets; 530 general public tickets
75. 10 $5 bills; 6 $10 bills

Review Exercises **1.** yes **2.** yes **3.** -16 **4.** -8
5. 3 **6.** 6.75 mi.

3.4 Exercises **1.** They will travel the same amount of time. **3.** $45t + 24t = 10$ **5.**

7. a. 165 mi. **b.** $48t$ **c.** $63(t + 2)$ **9.** Sal: $t + 2, 2t$; Jorge: t, $t - 3$ **11. a.** Their times are the same.
b.

c. Time: t, t; distance: $58t, 62t$ **d.** 500 mi. **e.** $58t + 62t = 500$ **13. a.** Polly's time is two hours more than Erick's. **b.**

c. They are the same. **d.** Time: $t + 2$; distance: $55(t + 2), 68t$
e. $55(t + 2) = 68t$ **15.** 4 hr. **17.** 3.5 hr. **19.** 2 hr.
21. 2 hr. **23.** 9:36 A.M. **25.** Car 1: 22 mph; car 2: 33 mph
27. 4.5 hr.; 247.5 mi. **29.** 5:30 P.M. **31.** 5 P.M.
33. a. 45 mph **b.** 180 mi. **35.** 6 hr. **37.** 4 hr. **39.** 79 mph
41. 3 mph, 67 mph

Review Exercises **1.** Commutative property of multiplication. **2.** 38 **3.** -43 **4.** $8x + 7y + 6$ **5.** $90.\overline{90}$
6. 700

3.5 Exercises **1. a.** Acct. 1: $3600; Acct. 2: $1500, $6800 **b.** $10,000 - $ Known amount **c.** $10,000 - P$
3. $220; $391.88; $0.05 P$; $0.065 (15,000 - P)$ **5.** Acct. 1: $P + 1200, 4000 - P$; Acct. 2: $2P, 6000 - P$ **7.** The sum of the expressions for interest is set equal to the total interest.
9. Soln. 2: 200 ml, $2n$ ml; Comb. Soln.: 120 ml, 300 ml, $3n$ ml

11. 6.5 oz.; 10.725 oz.; $0.125x$ oz.; $0.082(50 - x)$ oz.
13. $4600 at 2%; $13,800 at 4% **15.** $1600 at 2%; $2400 at 3%
17. $1250 at 8%; $1900 at 12% **19.** $1800 in plan 1; $2700 in plan 2 **21.** $940 at 5%; $660 at 4% **23.** $1000 at 5.5%; $4000 at 7% **25.** $4200 at 6%; $8400 at -2% **27.** $4000 in plan A; $6000 in plan B **29.** 6000 L **31.** 52.5 ml

33. $166\frac{2}{3}$ lb. of 26%; $133\frac{1}{3}$ lb. of 35% **35.** 50 g of 6%; 25 g of 18% **37.** 30 ml **39.** 1.2 qt. **41.** 56 lb. at $1.20; 24 lb at $0.90 **43.** 3 lb. at $1.65; 2 lb. at $1.25

45. $76\frac{2}{3}$ lb. at $0.60; $38\frac{1}{3}$ lb. at $1.20

Review Exercises **1.** $\frac{41}{108}$ **2.** -64 **3.** $12x$ **4.** $<$
5. -2 **6.** 72

Chapter 3 Review Exercises **1.** true **2.** true
3. false **4.** false **5.** false **6.** false **7.** 1. 100 **8.** congruent
9. complementary **10.** proportion **11.** $\frac{3}{8}$ **12.** $\approx \frac{\$0.085}{1 \text{ oz.}}$;
each ounce costs 8.5 cents. **13.** no **14.** yes **15.** yes
16. no **17.** 25 **18.** 22 **19.** $78\frac{3}{4}$ **20.** 9.5625 **21.** -120
22. $21.1\overline{6}$ **23.** 300 mi. **24.** 5 ft. **25.** $9.10 **26.** $2587.20
27. ≈ 220.5 mi. **28.** $270.03 **29.** 12 in. **30.** 43.2 m
31. $x = 12$ ft.; $y = 6$ ft. **32.** $a = 4\frac{5}{7}$ in., $b = 4\frac{2}{3}$ in.,
$c = 13\frac{1}{2}$ in. **33.** $0.15, \frac{3}{20}$ **34.** $0.825, \frac{33}{40}$ **35.** $0.125, \frac{1}{8}$
36. $0.0245, \frac{49}{2000}$ **37.** $0.1, \frac{1}{10}$ **38.** $0.\overline{3}, \frac{1}{3}$ **39.** 40%
40. $33\frac{1}{3}$% or $33.\overline{3}$% **41.** $36\frac{4}{11}$% or $36.\overline{36}$% **42.** 35%
43. 120% **44.** 201.6% **45.** 14.56 **46.** 289.7 **47.** 32.5%
48. 56.25% **49.** 478 **50.** $805.59 **51.** ≈ 59.3%
52. ≈ 36.1% **53.** 11, 44 **54.** 2.5, 12.5 **55.** 41, 42
56. Jack: 24; Jill: 26 **57.** 19, 21, 23 **58.** $-2, -1, 0$
59. length: 15 in.; width: 9 in. **60.** length: 45 ft.; width: 15 ft.
61. 160°, 74° **62.** 31°, 59° **63.** 3 regular; 2 control top
64. 10 gal. of interior; 14 gal. of exterior **65.** 3.5 hr.
66. 2.6 hr. **67.** 4 P.M. **68.** 75 mph **69.** $20,000 at 8%; $40,000 at 10% **70.** $10,200 at 7%; $30,600 at -2% **71.** 40 L of 90%; 60 L of 40% **72.** 60 oz.

Chapter 3 Practice Test **1.** $\frac{2}{5}$ [3.1] **2.** $\frac{0.26}{1}$;
1 ounce of cereal costs 26 cents. [3.1] **3.** $3\frac{1}{3}$ [3.1]
4. 2.25 [3.1] **5.** ≈ 49 gal. [3.1] **6.** 12.96 cm [3.1]
7. $0.22, \frac{11}{50}$ [3.2] **8.** $0.0\overline{3}, \frac{1}{30}$ [3.2] **9.** 320% [3.2]
10. 40% [3.2] **11.** 20% [3.2] **12.** 14.35 [3.2]
13. 175 [3.2] **14.** 185.5, or about 186 [3.2]
15. ≈ 67.6% [3.2] **16.** 12.5% [3.2] **17.** ostrich: 56.4 oz., albatross: 21 oz. [3.3] **18.** 54, 56 [3.3] **19.** 5 km, 11 km [3.3]
20. 102°, 78° [3.3] **21.** 15 shares of Walmart; 22 shares of Target [3.3] **22.** 0.5 hr. or 30 min. [3.4] **23.** 0.625 hr. or 37.5 min. [3.4] **24.** $3000 in plan A; $1500 in plan B [3.5]
25. 30 oz. [3.5]

Chapters 1–3 Cumulative Review Exercises

1. true **2.** false **3.** false **4.** true **5.** itself **6.** LCD
7. $2^3 \cdot 3 \cdot 5$ **8.** 3 **9.** Distributive property. **10.** Set the cross products equal to each other and solve for the variable.

11. $\dfrac{4}{9}$ **12.** 7 **13.** $\dfrac{1}{10}$ **14.** -14 **15.** $\dfrac{11}{10}$ **16.** -26.48

17. $-\dfrac{1}{2}x + 27$ **18.** 9 **19.** $\dfrac{2}{5}$ **20.** All real numbers. **21.** 5

22. $b = \dfrac{2A}{h}$ **23.** $c = P - a - b$ **24.** 6.345 **25.** $-3, -1, 1$

26. $t \geq 88$ **27.** 20 L **28.** 4 P.M. **29.** 3.75 lb. of cashews; 6.25 lb. of peanuts **30.** 300 mi.

Chapter 4

4.1 Exercises **1.** Horizontal-axis coordinate.
3.
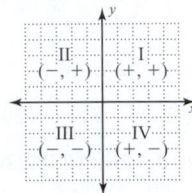

5. $A(4, 3), B(-3, 1), C(0, -2), D(2, -5)$
7. $A(-4, 0), B(2, 4), C(-2, -3), D(4, -4)$
9.
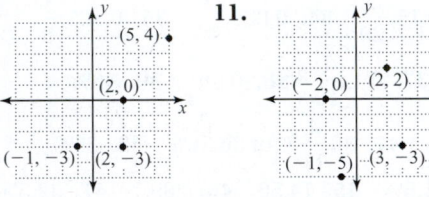
11.
13. II **15.** I

17. IV **19.** III **21.** y-axis **23.** x-axis **25.** linear
27. nonlinear **29.** linear **31.** nonlinear **33.** linear
35. Answers may vary. Some possible answers are $(-1, -5)$, $(0, -3)$, and $(2, 1)$. **37.** $A: (-6, -4); B: (-3, 2); C: (5, 2);$ $D: (8, -4)$ **39. a.** $(-3, 1), (2, 1), (1, -3), (-4, -3)$
b. $(0, 2), (5, 2), (4, -2), (-1, -2)$ **c.** $(x + 3, y + 1)$
41. a.
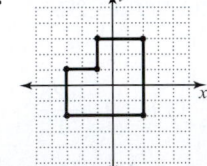
b. 20 units **c.** 21 square units

Review Exercises **1.**
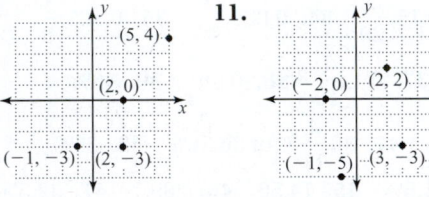
2. 2

$$-4\ -3\ -2\ -1\ \ 0\ \ 1\ \ 2$$

3. 12 **4.** 0 **5.** $\dfrac{15}{2}$ **6.** $-\dfrac{9}{10}$

4.2 Exercises **1. a.** Replace the variables in the equation with the corresponding coordinates. **b.** Verify that the equation is true. **3. a.** Choose a value for one of the variables. **b.** Replace the corresponding variable with your chosen value. **c.** Solve the equation for the value of the

other variable. **5.** A minimum of two ordered pairs are needed because two points determine a line. **7.** yes **9.** yes
11. no **13.** yes **15.** no **17.** yes **19.** no **21.** no
23. $(-1, -1), (0, 0), (1, 1)$ **25.** $(-1, -2), (0, 0), (1, 2)$

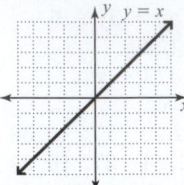 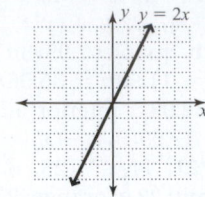

27. $(-1, 5), (0, 0), (1, -5)$ **29.** $(0, -3), (2, -1), (4, 1)$

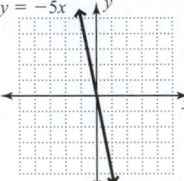 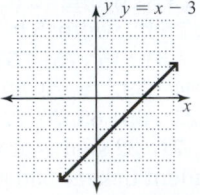

31. $(4, -6), (-4, 2)$ **33.** $(0, -5), (1, -3), (2, -1)$

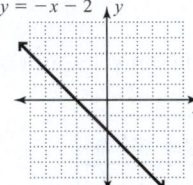 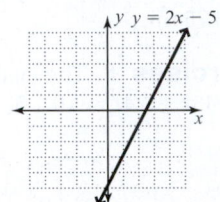

35. $(-1, 6), (0, 4), (1, 2)$ **37.** $(-2, -1), (0, 0), (2, 1)$

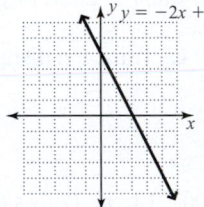 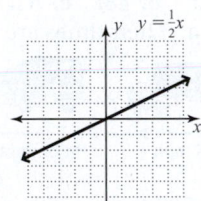

39. $(0, 0), (-3, 2), (3, -2)$ **41.** $(-3, 6), (0, 4), (3, 2)$

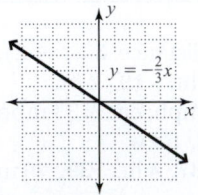 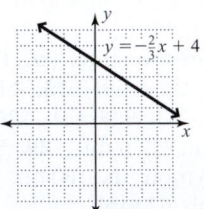

43. $(3, -5), (5, -3), (6, -2)$ **45.** $(1, 4), (3, 0), (5, -4)$

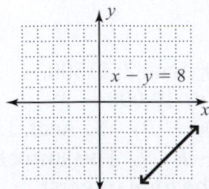 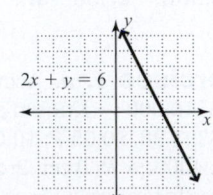

47. $(0,4), (6,0), (3,2)$

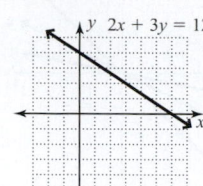
y $2x + 3y = 12$

49. $(0,4), (-3,0), (3,8)$

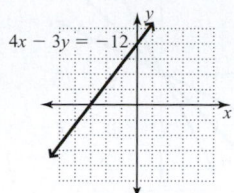
$4x - 3y = -12$

69. a. $1200 **b.** 4 yr. **c.** 8 yr. **d.**

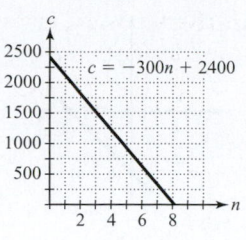
$c = -300n + 2400$

Review Exercises 1. 4 **2.** 31.5 in. **3.** 5.67 sq. in.
4. $\dfrac{5}{3}$ **5.** $x = \dfrac{v - k}{m}$ **6.**

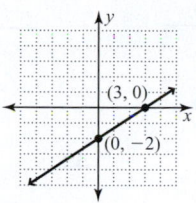
$(3, 0)$
$(0, -2)$

51. $(1, -4), (2, 0), (3, 4)$

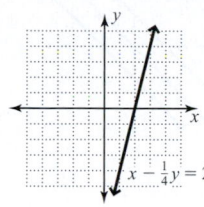
$x - \frac{1}{4}y = 2$

53. $(-1, -5), (0, -5), (1, -5)$

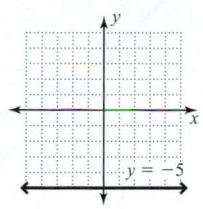
$y = -5$

4.3 Exercises 1. The point where a graph intersects the x-axis. **3. (a)** Replace y with 0 in the given equation.
(b) Solve for x. **5.** The graph of any equation in the form $y = mx$ is always a line that passes through the origin, which means the x- and y-intercepts are both at $(0, 0)$.
7. $(1, 0), (0, 3)$ **9.** $(-3, 0), (0, 4)$ **11.** $(3, 0)$; no y-intercept **13.** no x-intercept; $(0, -2)$ **15.** $(4, 0), (0, -4)$
17. $(3, 0), (0, 2)$ **19.** $\left(-\dfrac{10}{3}, 0\right), \left(0, \dfrac{5}{2}\right)$ **21.** $(2, 0), (0, -6)$
23. $\left(-\dfrac{5}{2}, 0\right), (0, 5)$ **25.** $(0, 0)$ for both **27.** $(0, 0)$ for both
29. $(5, 0)$; no y-intercept **31.** no x-intercept; $(0, -2)$

55. $(7, -1), (7, 0), (7, 1)$

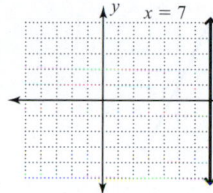
$x = 7$

57. $(-1, -2.9), (0, -2.5), (1, -2.1)$

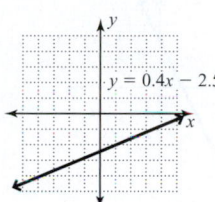
$y = 0.4x - 2.5$

59. As a gets larger, the graph gets steeper.
61. The graph will shift up b units on the y-axis.
63. a. $160 **b.** 4 hr. **c.**

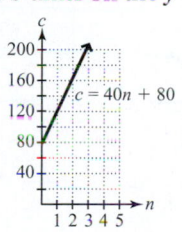
$c = 40n + 80$

d. It represents the initial charge, which is $80.
65. a. $18 **b.** 500 copies **c.**

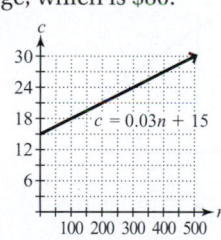
$c = 0.03n + 15$

67. a. 412.5 gal. **b.** ≈ 71.4 min. **c.** ≈ 143 min.

d.

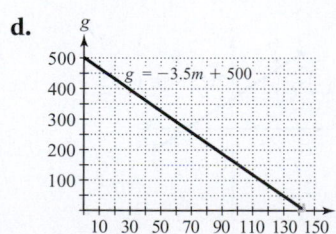
$g = -3.5m + 500$

e. It represents the original amount of water in the hot tub, which is 500 gallons. **f.** It represents the amount of time it takes for the hot tub to empty (0 gallons of water).

33.

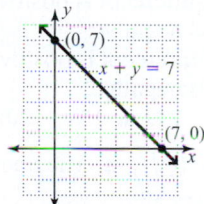
$(0, 7)$
$x + y = 7$
$(7, 0)$

35.

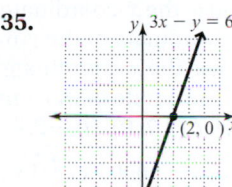
y $3x - y = 6$
$(2, 0)$
$(0, -6)$

37.

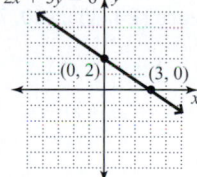
$2x + 3y = 6$
$(0, 2)$
$(3, 0)$

39.

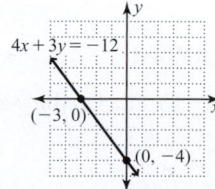
$5x + 2y = 10$
$(0, 5)$
$(2, 0)$

41.

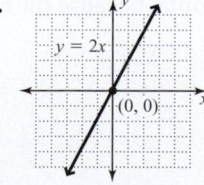
$4x + 3y = -12$
$(-3, 0)$
$(0, -4)$

43.

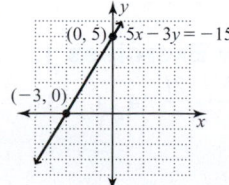
$(0, 5)$ $5x - 3y = -15$
$(-3, 0)$

45.

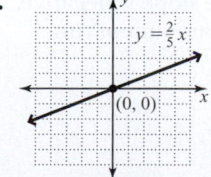
$y = 2x$
$(0, 0)$

47.

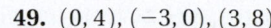
$y = \frac{2}{5}x$
$(0, 0)$

49.
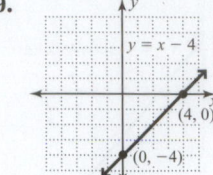

51.
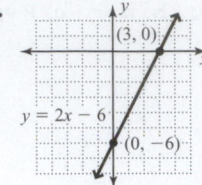

53.
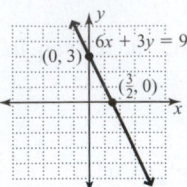

55.
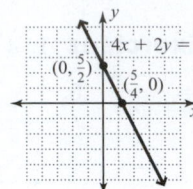

57.
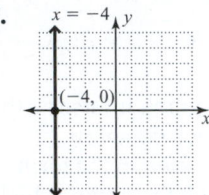

59.
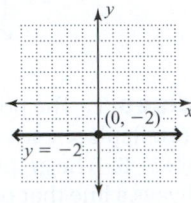

61.
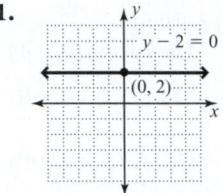

63.
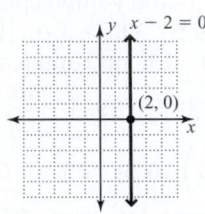

65. d because the x-coordinate of the x-intercept is positive and the y-coordinate of the y-intercept is negative.
67. c because the x-coordinate of the x-intercept is positive and the y-coordinate of the y-intercept is negative.
69. $(-2.7, 0), (2.7, 0), (0, -2.7), (0, 2.7)$ **71. a.** $15x + 20y$
b. $15x + 20y = 2000$ **c.** $\left(133\frac{1}{3}, 0\right), (0, 100)$ **d.** The number of units required to meet the goal if only one size is sold.

e. Answers may vary. Some possibilities are $(20, 85), (40, 70)$, and $(60, 55)$.   **f.**
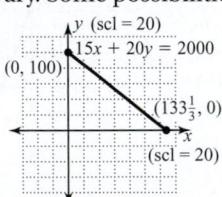

Review Exercises 1. -2 **2.** undefined
3. $w = \dfrac{P - 2l}{2}$ **4.** $4 + 3x = 5(x + 6); -13$ **5.** On the y-axis.
6. no

4.4 Exercises 1. How steep a line is.   **3.** Downhill because the slope is negative.   **5.** Horizontal because the rise is 0 for any amount of run.

7.
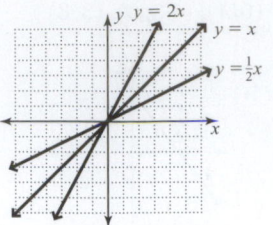

9.
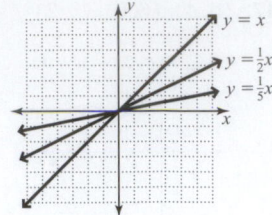

11.
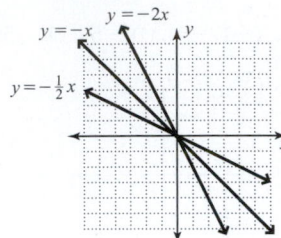

13.
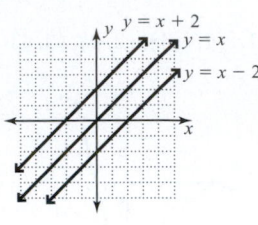

15. $m = 2; (0, 3)$ **17.** $m = -2; (0, -1)$

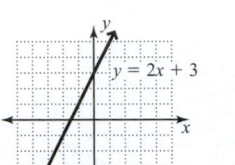
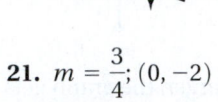

19. $m = \dfrac{1}{3}; (0, 2)$ **21.** $m = \dfrac{3}{4}; (0, -2)$

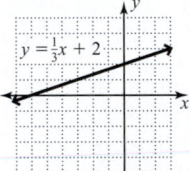
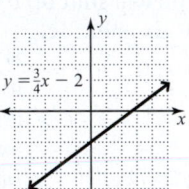

23. $m = -\dfrac{2}{3}; (0, 8)$ **25.** $m = -2; (0, 4)$

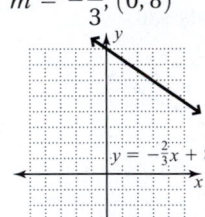
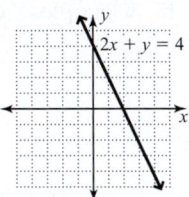

27. $m = 3; (0, 1)$ **29.** $m = -\dfrac{2}{3}; (0, 2)$

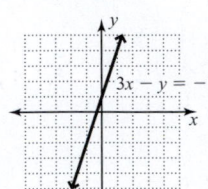
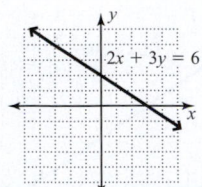

31. $m = -\frac{1}{2}; (0, -2)$ **33.** $m = \frac{3}{2}; \left(0, \frac{7}{2}\right)$

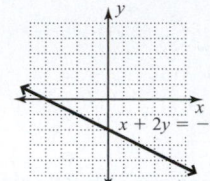
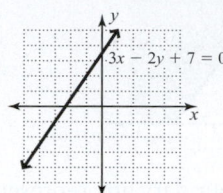

35. $m = -\frac{3}{2}; \left(0, \frac{5}{2}\right)$ **37.** $m = 3; (0, -5)$

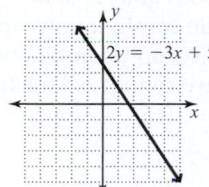
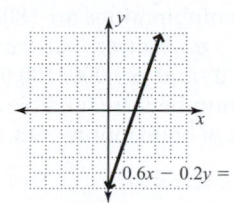

39. $y = 2x + 3$ **41.** $y = -3x - 9$ **43.** $y = \frac{3}{4}x + 5$

45. $y = -\frac{2}{5}x + \frac{7}{8}$ **47.** $y = 0.8x - 5.1$ **49.** $y = -3x$

51. $\frac{5}{2}$ **53.** $-\frac{2}{3}$ **55.** $\frac{2}{7}$ **57.** $-\frac{4}{3}$ **59.** $\frac{2}{3}$ **61.** $-\frac{5}{4}$ **63.** 0

65. undefined   **67.** d   **69.** a   **71.** f   **73.** g   **75.** h

77. $m = 2; (0, 3); y = 2x + 3$ **79.** $m = -1; (0, 2);$

$y = -x + 2$ **81.** $m = \frac{2}{3}; (0, -1); y = \frac{2}{3}x - 1$

83. $m = -\frac{1}{4}; (0, 3); y = -\frac{1}{4}x + 3$

85. m is undefined; no y-intercept; $x = 2$ **87.** $m = 0; (0, 3);$
$y = 3$ **89.** $y = -2, x = 3$ **91.** $y = 0$

93. a.   **b.** 3 and 0   **c.** They are the same.

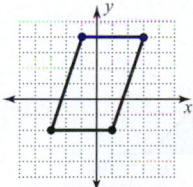

95. $0.58\overline{3}, \frac{7}{12}$ **97.** $\frac{293}{230} \approx 1.27$

Review Exercises **1.** $-\frac{3}{2}$ **2.** $11x + 7$ **3.** $-2x - 14$

4. $x = \frac{C - By}{A}$ **5.**   **6.** $\left(\frac{7}{3}, 0\right), \left(0, -\frac{7}{5}\right)$

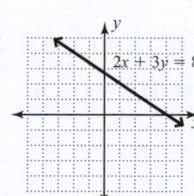

4.5 Exercises **1.** The slope–intercept form would be easiest to use because you are given the values m and b.   **3.** Find the slope.   **5.** Solve for y.   **7.** $y = 3x - 13$ **9.** $y = -x - 2$

11. $y = \frac{2}{3}x - 1$ **13.** $y = -\frac{3}{4}x - \frac{23}{4}$ **15.** $y = -\frac{4}{3}x + \frac{4}{3}$

---

17. $y = 2x$ **19.** $2x - 3y = -2$ **21.** $x + 2y = 3$

23. $y = -2x + 5$ **25.** $y = \frac{5}{2}x - \frac{13}{2}$ **27.** $y = 7x$ **29.** $x = 8$

31. $y = -1$ **33.** $y = 3x + 0.8$ **35.** $y = 2x - 2$

37. $y = -\frac{6}{5}x + 3$ **39.** $5x - 2y = 4$ **41.** $3x - 2y = 6$

43. $4x + 11y = -27$ **45.** $2x + 3y = 0$ **47.** $x = -1$

49. $y = -1$ **51. a.** $y = 4x - 14$ **b.** $4x - y = 14$

53. a. $y = -3x - 5$ **b.** $3x + y = -5$

55. a. $y = \frac{1}{3}x + \frac{1}{3}$ **b.** $x - 3y = -1$ **57. a.** $y = -\frac{3}{4}x - \frac{15}{4}$

b. $3x + 4y = -15$ **59. a.** $y = -2x + 1$ **b.** $2x + y = 1$

61. a. $y = \frac{3}{4}x + 5$ **b.** $3x - 4y = -20$ **63. a.** $y = -\frac{1}{2}x - 3$

b. $x + 2y = -6$ **65. a.** $y = -4x + 8$ **b.** $4x + y = 8$

67. a. $y = \frac{1}{2}x - 5$ **b.** $x - 2y = 10$ **69. a.** $y = -\frac{2}{3}x + \frac{5}{3}$

b. $2x + 3y = 5$ **71. a.** $y = -\frac{5}{2}x - \frac{17}{2}$ **b.** $5x + 2y = -17$

73. a. $y = \frac{3}{2}x - \frac{17}{2}$ **b.** $3x - 2y = 17$ **75.** parallel

77. perpendicular   **79.** neither   **81.** parallel   **83.** perpendicular   **85.** neither   **87.** perpendicular   **89.** parallel

91. a. $c = 0.3n + 7.75$ **b.** $19.75

c.   **93. a.**

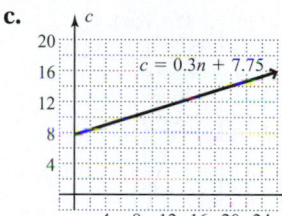
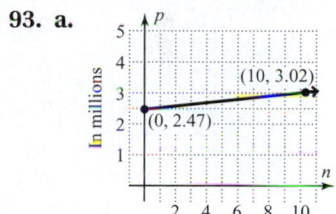

b. 0.055   **c.** $p = 0.055n + 2.47$ **d.** 2.745 million

e. 3.35 million   **95. a.**   **b.** -1.5

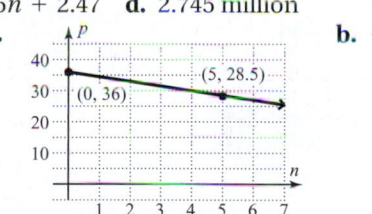

c. $p = -1.5n + 36$ **d.** $(0, 36)$; the p-intercept indicates the initial price.   **e.** $(24, 0)$; the n-intercept indicates that on the 24th day, the price will be $0.   **f.** $24

Review Exercises **1.** $<$ **2.** $\frac{36}{5}$

3. $x \leq -7$

4. $x \leq \frac{4}{3}$ **5.** $m = -\frac{3}{4}; (0, 2)$

6. $(7, 0)(0, -2)$

4.6 Exercises **1.** To determine whether an ordered pair is a solution for an inequality, replace the variables with the corresponding coordinates and see if the resulting inequality is true. If it is, the ordered pair is a solution.   **3.** A solid line indicates equality is included, $\geq$ or $\leq$. A dashed line indicates equality is excluded, $>$ or $<$.   **5.** yes   **7.** no

9. no **11.** yes

13. $y \le -x + 1$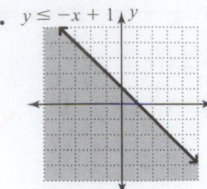

15. $y > -2x + 5$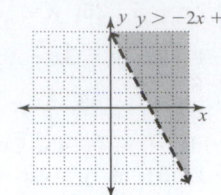

17. $y > 2x$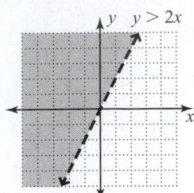

19. $y > \frac{1}{3}x$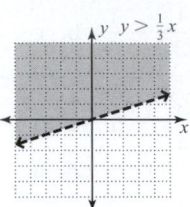

21. $y \ge \frac{3}{4}x + 2$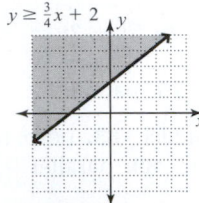

23. $3x + y \le 9$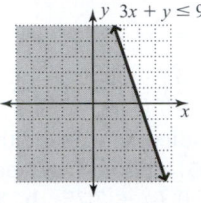

25. $5x - 2y < 10$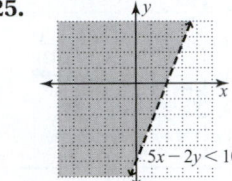

27. $3x + y \ge 8$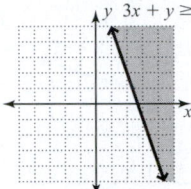

29. $2x + 3y < 9$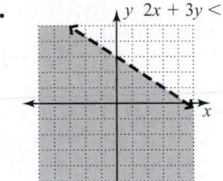

31. $3x + y \ge 0$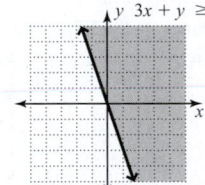

33. $x > -4$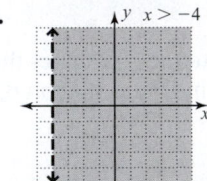

35. $y \le 2$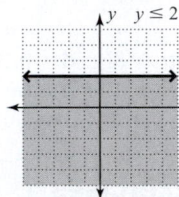

37. $x + 2 \ge 0$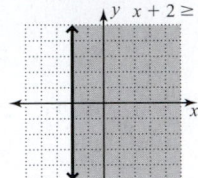

39. $y - 3 < 0$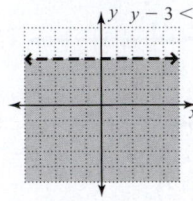

41. a. x rpresents the number of large bottles sold; y represents the number of small bottles sold.

b. $2x + 1.5y \ge 280{,}000$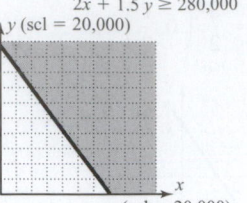

c. The boundary line represents combinations of bottle sizes sold to produce a revenue of exactly \$280,000 (breakeven). The shaded region represents combinations that produce a revenue greater than \$280,000 (profit). **d.** Answers will vary; a couple combinations are $(80{,}000, 80{,}000)$ and $(20{,}000, 160{,}000)$. **e.** Answers will vary; a couple combinations are $(150{,}000, 0)$ and $(90{,}000, 100{,}000)$. **f.** No, only combinations of whole numbers are possible because we cannot sell fractions of a bottle of hand lotion. **43. a.** $5.5x + 12.5y \le 100$

b. $5.5x + 12.5y \le 100$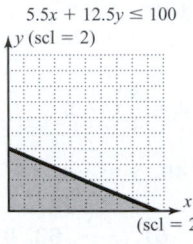

c. The combinations of plant purchases that would cost exactly \$100. **d.** The combinations of plants purchased that would cost less than \$100. **e.** Answers will vary; some combinations are $(2, 4)$, $(6, 2)$, and $(8, 4)$. **f.** Yes, $(0, 8)$. Because she cannot purchase fractional numbers of plants, the combination must be whole numbers and $(0, 8)$ is the only combination of whole numbers that costs exactly \$100.
45. a. $2l + 2w \le 180$ **b.** $2l + 2w \le 180$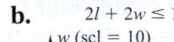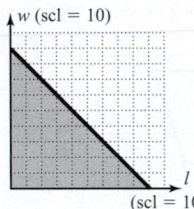

c. The combinations of lengths and widths that yield a perimeter of exactly 180 ft. **d.** The combinations of length and width that yield a perimeter that is less than 180 ft.
e. Answers will vary; some combinations are $(10, 80)$, $(20, 70)$, and $(40, 50)$. **f.** Answers will vary; some combinations are $(10, 10)$, $(20, 20)$, and $(30, 30)$.

Review Exercises 1. $\{a, e, i, o, u\}$ **2.** 1 **3.** -13
4. 20 ft. by 24 ft. **5.** $\left(\frac{8}{3}, 0\right)$, $(0, -4)$ **6.** $-x + y = 7$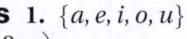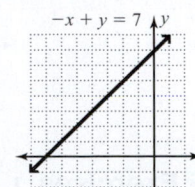

4.7 Exercises **1.** The domain is a set containing all x-values. **3.** Trail along the x-axis and determine what values of x have a corresponding y-value. **5.** Draw or imagine a vertical line through every point in the domain. If each vertical line intersects the graph in at most one point, the relation is a function. **7.** Domain: $\{2, 3, 1, -1\}$; range: $\{1, -2, 4, -1\}$. **9.** Domain: $\{4, 2, 3, -4\}$; range: $\{1, -5, 0\}$. **11.** Domain: $\{0, -3, 2, 8\}$; range: $\{0, -3, 9, -16\}$. **13.** yes **15.** no **17.** yes **19.** no **21.** Yes, every element in the domain is assigned to one element in the range. **23.** No, an element in the domain is assigned to more than one element in the range. (8 is assigned to both 2 and -1.) **25.** Yes, every element in the domain is assigned to one element in the range. **27.** No, an element in the domain is assigned to more than one element in the range. **29.** No, it fails the vertical line test. **31.** Yes, it passes the vertical line test. **33.** No, it fails the vertical line test. **35.** Yes, it passes the vertical line test. **37.** Domain: $\{x \mid 1900 \le x \le 2005\}$ or $[1900, 2005]$; range: $\{y \mid 13\,\text{million} \le y \le 66\,\text{million}\}$ or $[13\,\text{million}, 66\,\text{million}]$; function. **39.** Domain $\{1900, 1910, 1920, 1930, 1940, 1950, 1960, 1970, 1980, 1990, 2000\}$; range: $\{4.7, 5.4, 6.2, 6.9, 8.0, 8.8, 10.4, 11.6, 13.2, 13.6, 14.7\}$; function **41.** Domain: $\{x \mid -4 \le x \le 4\}$ or $[-4, 4]$; range: $\{y \mid -3 \le y \le 2\}$ or $[-3, 2]$; function. **43.** Domain: $\{x \mid x \le 0\}$ or $(-\infty, 0]$; range: all real numbers or $(-\infty, \infty)$; not a function. **45.** Domain: all real numbers or $(-\infty, \infty)$; range: $\{y \mid y \ge -1\}$ or $[-1, \infty)$; function. **47. a.** -5 **b.** -3 **c.** -9 **d.** -4 **49. a.** 7 **b.** 5 **c.** 11 **d.** 17 **51. a.** $\sqrt{3}$ **b.** Not a real number. **c.** $\sqrt{2}$ **d.** $\sqrt{a+3}$ **53. a.** 0 **b.** Not a real number.

c. 0 **d.** $\sqrt{5}$ **55. a.** -1 **b.** undefined **c.** $-\dfrac{1}{2}$

d. $\dfrac{1}{a-1}$ **57. a.** -3 **b.** -3.19 **c.** $a^2 - 2a - 3$

d. $a^2 + 2a - 3$ **59. a.** -2 **b.** $-\dfrac{5}{3}$ **c.** $-\dfrac{7}{3}$ **d.** $\dfrac{1}{3}a - 2$

61. a. $-\dfrac{3}{4}$ **b.** undefined **c.** $-\dfrac{1}{6}$ **d.** -6 **63. a.** undefined

b. $\dfrac{1}{\sqrt{3}}$ **c.** Not a real number. **d.** $\dfrac{1}{\sqrt{a+2}}$ **65. a.** 2

b. 4 **c.** 6 **d.** 10 **67. a.** 0 **b.** $-\dfrac{1}{2}$ **c.** 20 **d.** 7

69. a. $8a^2 + 6a + 1$ **b.** $12a^2 + 4a - 4$ **c.** $\dfrac{4}{3}a^2 - \dfrac{4}{3}a - 4$

d. $0.12a^2 - 0.4a - 4$ **71. a.** 3 **b.** 4 **c.** 2 **73. a.** 0 **b.** -2 **c.** undefined

75. $f(x) = 2x + 1$

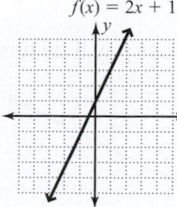

77. $f(x) = \frac{1}{3}x + 3$

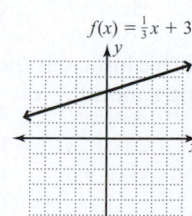

79. $f(x) = -4x + 1$

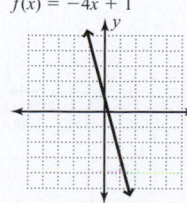

81. $f(x) = -\frac{2}{3}x$

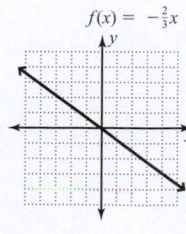

83. $f(x) = -\frac{1}{4}x - 2$

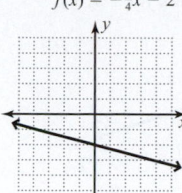

85. a. \$350 **b.** \$450

c. The cost of producing 30 items is \$750. **87. a.** \$13,500 **b.** \$9000 **c.** The value of the car after ten years is \$3000.

Review Exercises **1.** $-\dfrac{13}{30}$ **2.** $-4x - 2$

3. Commutative property of addition.

4.

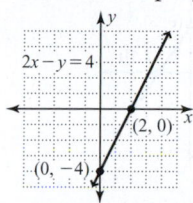

5. $y = \dfrac{3}{2}x - 3$ **6.** parallel

Chapter 4 Review Exercises **1.** true **2.** true **3.** false **4.** false **5.** true **6.** true **7. a.** y **b.** x **8. a.** x **b.** y **9.** y-intercept **10.** uphill; downhill **11.** $A(4,3), B(-2,4), C(-4,-2), D(2,-1)$ **12.** $A(3,1), B(0,4), C(-3,0), D(0,-2)$

13.

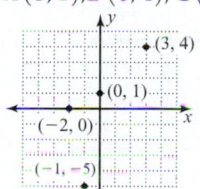

14.

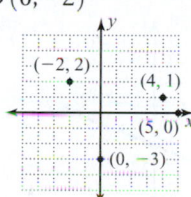

15. II **16.** III **17.** I **18.** IV **19.** yes **20.** yes **21.** no **22.** no **23.** no **24.** no **25.** $(0,0), (1,-2), (-1,2)$ **26.** $(0,0), (1,1), (-1,-1)$

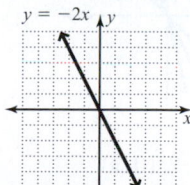

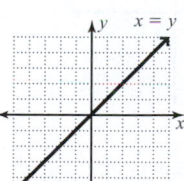

27. $(0,7), (1,6), (3,4)$ **28.** $(5,0), (1,1), (-3,2)$

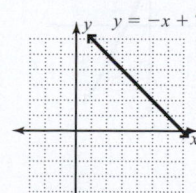

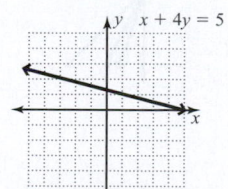

29. $(0, 2), (3, 0), (-3, 4)$ **30.** $(0, 3), (3, 2), (-3, 4)$

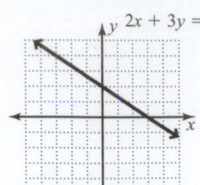

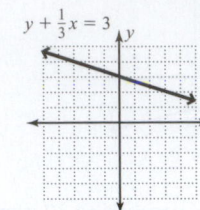

31. $(12, 0), (0, 6)$ **32.** $(0, 0)$ for both   **33.** $(3.1, 0), (0, -2)$

34. $\left(-\dfrac{15}{2}, 0\right), (0, 5)$ **35.** $(-2, 0)$, no y-intercept

36. no x-intercept, $(0, 7)$ **37.**

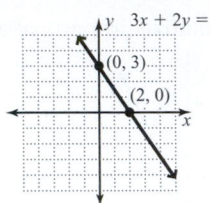

38.

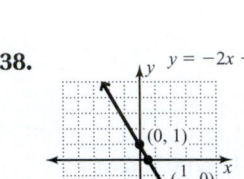

39.

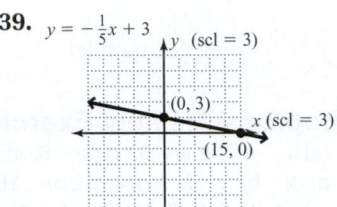

40.

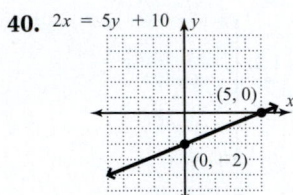

41.

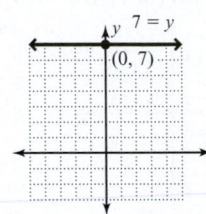

42.

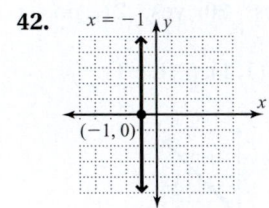

43.

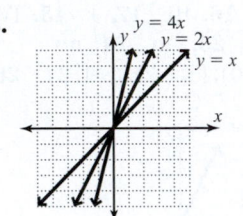

44.

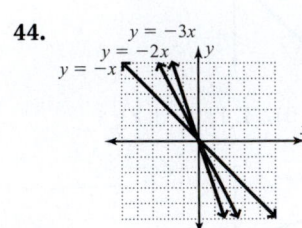

45. $m = -2; (0, 3)$

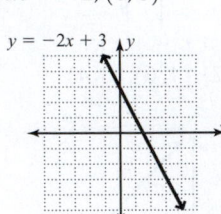

46. $m = -\dfrac{1}{4}; (0, 2)$

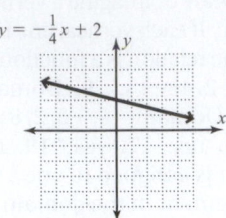

47. $m = 3; (0, 0)$

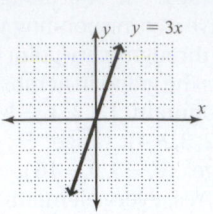

48. $m = -1; (0, 5)$

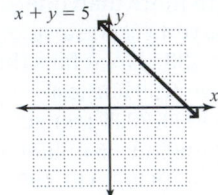

49. $m = \dfrac{1}{3}; \left(0, -\dfrac{7}{3}\right)$

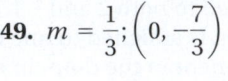

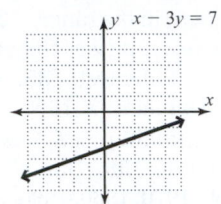

50. $m = \dfrac{2}{3}; \left(0, -\dfrac{8}{3}\right)$

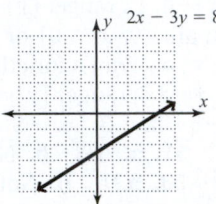

51. 3   **52.** 1   **53.** $-\dfrac{1}{3}$ **54.** $-\dfrac{3}{5}$ **55.** 0   **56.** undefined

57. $y = -x + 7$ **58.** $y = -\dfrac{1}{5}x - 8$ **59.** $y = 0.2x + 6$

60. $y = x$ **61.** $y = -2x + 9$ **62.** $y = x - 9$ **63.** $y = \dfrac{1}{3}x + \dfrac{5}{3}$

64. $y = -\dfrac{2}{5}x + \dfrac{21}{5}$ **65.** $y = 6.2x + 9.4$ **66.** $y = -0.4x + 1.6$

67. $y = -x + 4, x + y = 4$ **68.** $y = \dfrac{5}{4}x - \dfrac{37}{4}, 5x - 4y = 37$

69. $y = -\dfrac{1}{2}x + \dfrac{7}{2}, x + 2y = 7$ **70.** $y = -\dfrac{1}{5}x - \dfrac{13}{5},$

$x + 5y = -13$ **71.** $y = 2x - 2$ **72.** $y = -\dfrac{2}{3}x + \dfrac{17}{3}$

73. $y = -\dfrac{5}{3}x - \dfrac{11}{3}$ **74.** $y = \dfrac{5}{2}x + 9$ **75. a.** \$2200

b. $(0, 1000)$; This indicates the person's gross pay if he has \$0 in sales during a month.   **c.**

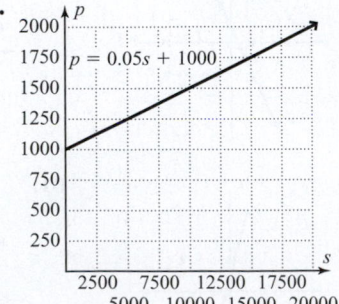

76. a.

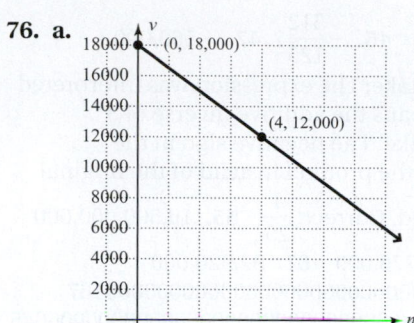

b. -1500 **c.** $v = -1500n + 18,000$ **d.** 2010 **e.** 2016
77. yes **78.** yes **79.** yes **80.** no

81.

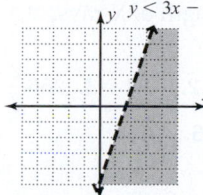

82.

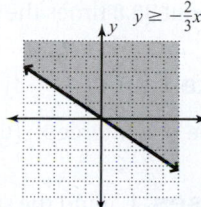

83.

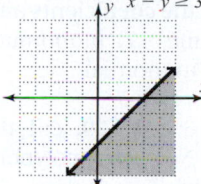

84.

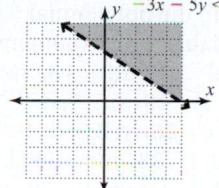

85.

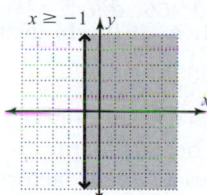

86. a. $6a + 8b \le 12,000$ **b.**

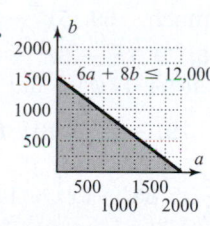

c. Answers may vary; some combinations are $(400, 1200)$, $(800, 900)$, and $(1200, 600)$. **d.** Answers may vary; some combinations are $(500, 100)$, $(1000, 200)$, and $(200, 1100)$.
87. Domain: $\{2, -1, 3, -3\}$; range: $\{3, 5, 6, -4\}$. **88.** Domain: $\{$Volkswagen, Honda, Toyota, Subaru, Nissan$\}$; range: $\{52.2, 49.7, 49.0, 47.8, 45.8\}$. **89.** no **90.** yes **91.** Domain: all real numbers or $(-\infty, \infty)$; range: all real numbers or $(-\infty, \infty)$; it is a function. **92.** Domain: $\{x \mid x \le 0\}$ or $(-\infty, 0]$; range: all real numbers or $(-\infty, \infty)$; it is not a function. **93.** Domain: $\{x \mid -4 \le x \le 5\}$ or $[-4, 5]$; range: $\{1, 2, 3\}$; it is a function. **94. a.** 1 **b.** 3 **c.** 3 **d.** 2
95. a. 0 **b.** 2 **c.** -10 **96. a.** 2 **b.** undefined **c.** $\dfrac{5}{8}$

Chapter 4 Practice Test **1.** $A(2, 4)\ B(-3, 2)\ C(0, -5)$
$D(4, -2)$ [4.1] **2.**

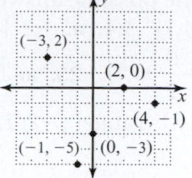

[4.1] **3.** I [4.1]

4. no [4.2] **5.** $(4, 0), (0, -2)$ [4.3] **6.** $(5, 0), (0, 2)$ [4.3]

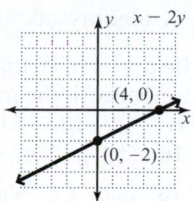 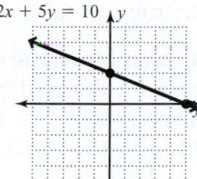

7. $(0, 0)$ [4.3] **8.** $\left(\dfrac{4}{3}, 0\right), (0, -1)$ [4.3]

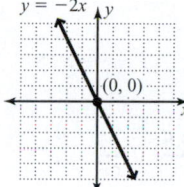 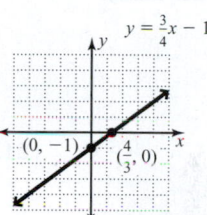

9. $m = \dfrac{3}{4}; (0, 11)$ [4.4] **10.** $m = \dfrac{5}{3}; \left(0, -\dfrac{8}{3}\right)$ [4.4]

11. $-\dfrac{10}{3}$ [4.4] **12.** 0 [4.4] **13.** $y = \dfrac{3}{5}x + 4$ [4.4]

14. $y = -\dfrac{2}{5}x + \dfrac{11}{5}$ [4.5] **15.** $2x - 3y = 1$ [4.5]

16. $x - 3y = 10$ [4.5] **17. a.**

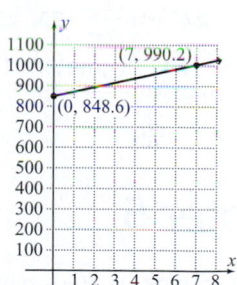

b. 20.229 [4.5] **c.** $y = 20.229x + 848.6$ [4.5]
d. 1071.1 [4.5] **18.** yes [4.6]
19.

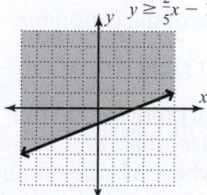

[4.6] **20.**

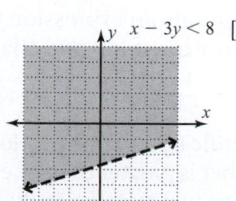

[4.6]

21. a. $10x + 8y \ge 360$ [4.6] **b.**

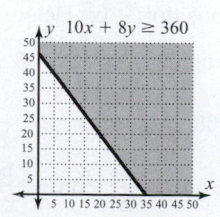

c. Answers may vary. One possible answer is $(0, 45)$. [4.6]
d. Answers may vary. One possible answer is $(10, 35)$. [4.6]
22. yes [4.7]　**23.** no [4.7]　**24. a.** -2 [4.7]

b. undefined [4.7]　**c.** $-\dfrac{9}{7}$ [4.7]　**25. a.** Domain:

$\{x \mid -3 \le x \le 3\}$ or $[-3, 3]$; range: $\{-2, 1, 2\}$ [4.7].
b. 2 [4.7]

Chapters 1–4 Cumulative Review
1. false
2. true　**3.** false　**4.** false　**5.** identity　**6.** 100　**7.** $\dfrac{33}{200}$

8. Multiply all the terms by the power of 10 that will clear the decimal from the number with the most decimal places.　**9.** -23　**10.** 15　**11.** -18　**12.** -81　**13.** -41
14. 12　**15.** $(3, 0), (0, -2)$　**16.** $(1, 1), (0, 4), (-1, 7)$

17. 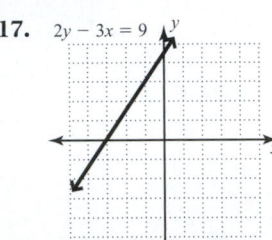 $2y - 3x = 9$　**18.** $m = -\dfrac{1}{3}$　**19.** $y = \dfrac{1}{2}x + \dfrac{1}{2}$

20. 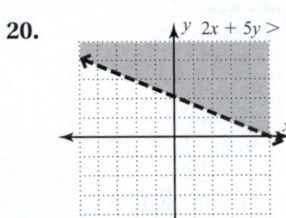 $y \; 2x + 5y > 10$　**21.** $\dfrac{85}{6}$　**22.** -12

23. -12.5　**24.** $r = \dfrac{C}{2\pi}$　**25.** 230　**26.** $\approx 1.9\%$　**27.** $36°, 54°$

28. $\approx 136\%$　**29.** 78 quarters, 112 dimes　**30.** \$20,000 at 8%, \$40,000 at 10%

Chapter 5

5.1 Exercises　**1.** If a is a real number, where $a \ne 0$ and n is a natural number, then $a^{-n} = \dfrac{1}{a^n}$.　**3.** The result is positive. We simplify an expression with a negative exponent by inverting the expression and changing the sign of the exponent: $\left(2^{-3} = \dfrac{1}{2^3} = \dfrac{1}{8}\right)$.　**5.** To write a number greater than 1 in scientific notation,　**a.** Move the decimal point so that the number is greater than or equal to 1 but less than 10.　**b.** Write the decimal number multiplied by 10^n, where n is the number of places between the new decimal position and the original decimal position.　**c.** Delete zeroes to the right of the last nonzero digit.　**7.** 1　**9.** -8　**11.** -125　**13.** -16　**15.** $\dfrac{81}{256}$

17. -0.008　**19.** 5.29　**21.** $\dfrac{1}{a^5}$　**23.** $\dfrac{1}{8}$　**25.** $-\dfrac{1}{64}$　**27.** $-\dfrac{1}{16}$

29. -125　**31.** $\dfrac{5}{m^2}$　**33.** $-\dfrac{6}{x^7}$　**35.** $\dfrac{n^5}{m^5}$　**37.** $\dfrac{49}{25}$　**39.** b^6

41. 64　**43.** $-\dfrac{1}{7,962,624}$　**45.** $-\dfrac{512}{125}$　**47.** $-160,000$

49. ≈ 0.00039　**51.** Mistake: The expression was interpreted to mean $(-3)^4$, but it means the additive inverse of 3^4. Correct: -81　**53.** Mistake: The negative sign in the exponent was passed to the product instead of the original expression being inverted. Correct: $\dfrac{1}{25}$　**55.** $16,500,000,000$

57. $299,800,000$　**59.** $6,378,000$　**61.** $92,920,000$
63. 0.0000025　**65.** $0.00000000000000000000000000167$
67. 0.00000000295　**69.** $0.0000000000000000000000006645$
71. 2.5×10^{13}　**73.** 5.3×10^9　**75.** 2.342×10^7
77. 1.3×10^9　**79.** 5.86×10^{-7}　**81.** 9.1094×10^{-31}
83. 6.25×10^{-5}　**85.** 5×10^{-8}　**87.** $8.95 \times 10^5, 7.2 \times 10^6,$
$7.5 \times 10^6, 9 \times 10^7, 1.3 \times 10^8$　**89.** The wavelength of ultraviolet light is about 38.8 times the wavelength of X-rays.

Review Exercises　**1.** $-4y + 12$　**2.** -11　**3.** $266\dfrac{2}{3}$

4. $\dfrac{1}{5}$　**5.** slope $\dfrac{2}{3}$; y-intercept: $(0, -2)$　**6.** -65

5.2 Exercises　**1.** Find the sum of the exponents on all variables in the monomial.　**3.** Find the greatest degree of any of the terms in the polynomial.　**5.** Add the coefficients and leave the variables and exponents the same.　**7.** monomial
9. monomial　**11.** not　**13.** not　**15.** monomial
17. monomial　**19.** c: -5; d: 3　**21.** c: -1; d: 5
23. c: -9; d: 0　**25.** c: 4.2; d: 4　**27.** c: 16; d: 3　**29.** c: 1; d: 1
31. trinomial, 1　**33.** monomial, 3　**35.** No special polynomial name, 3.　**37.** binomial, 3　**39.** monomial, 0
41. Not a polynomial.　**43.** 4　**45.** 5　**47.** 7　**49.** 4　**51.** 8
53. 60　**55.** -8　**57.** -4.8　**59.** 25　**61. a.** 46 ft.　**b.** 26.96 ft.
63. a. 30 units　**b.** 43 units　**65. a.** ≈ 1273.4 ft.^{3}
b. ≈ 537.2 ft.^{3}　**67. a.** 26 million　**b.** 45.95 million
c. 40.45 million　**d.** 1.3 million; answers will vary, but the prediction is most likely not reasonable because of the increase in world population and unforeseen social factors that might affect travel; it is not likely that the number will decrease this much.　**69.** $5x^7 - 3x^5 + 7x^4 - 8x + 14$
71. $-8r^6 + 18r^5 + 7r^3 - 3r^2 + 4r$
73. $-12w^4 - w^3 + 11w^2 + 5w + 20$　**75.** $5x - 3$
77. $-4a + 5$　**79.** $-2x^2 + 6x$　**81.** $2x^2 + \dfrac{5}{3}x$

83. $\dfrac{7}{6}a^2 - \dfrac{19}{15}a$　**85.** $-5x^2 + 14$　**87.** $5y^2 - 9y + 22$
89. $-2k^4 + 6k^3 - 3k^2 + 6k$　**91.** $-10a^4 + 7a^2 + a + 16$
93. $-21v^5 + 8v^3 + 35v^2 - 20$　**95.** $4a - 2b$
97. $-4y^2 + 3y$　**99.** $-9x^2 + 9y$　**101.** $\dfrac{19}{12}a^2 - \dfrac{17}{30}a$

103. $y^6 + 5yz^4 - 6yz - 3z^5 - 7z^3 - 10$
105. $-4w^2z - wz^2 + 12w^2 - 1$

Review Exercises　**1.** $-\dfrac{7}{12}$　**2.** -4.3　**3.** -1

4. 2×10^{11}　**5.** 12　**6.** 600 L

5.3 Exercises　**1.** To add polynomials, combine like terms.　**3.** Change the signs of each term in the polynomial.
5. $8x + 1$　**7.** $10y + 12$　**9.** $7x$　**11.** $3x^2 - 2x + 12$
13. $6z^2 - 11z - 2$　**15.** $5r^2 - 7r - 1$　**17.** $9y^2 + 3$

19. $3x^2 + \dfrac{1}{12}x + 1$ **21.** $2r^2 + 1.4r - 7.8$
23. $5a^3 + 3a^2 + 3a - 3$ **25.** $3p^3 - p^2 + 7p + 9$
27. $-8w^4 + 3w^3 - 7w^2 + 13w - 9$
29. $-3a^3b^2 - 2ab^2 - 8a^2b + 4ab + 3b^2 - 8$
31. $-7u^4 + u^2v^3 + 7u^3v + 7u - 13v^2 - 3$
33. $-12mnp - 16m^2n^2 - 12mn^2p + 14m^2n - 31n + 9$
35. $2a^4 - \dfrac{5}{12}a^3b + \dfrac{14}{15}ab^3 - b^4$
37. $1.8a^2bc - 1.8ab^2c^2 - 2.4a^2b^2c^2$
39. $5x - 2$ **41.** $14a - 12$ **43.** $4x + 2$ **45.** $13a^2 - 1$
47. $2x^2 - 9x + 2$ **49.** $2z^2 - 7z + 9$ **51.** $12a^2 + 10a - 15$
53. $-u^3 - 2u^2 - 6u + 8$ **55.** $-11w^5 + 3w^4 + 5w^2 + 2$
57. $-11p^4 + 3p^3 - 13p^2 + 5p - 6$
59. $11v^4 + 3v^3 + 23v^2 - 5v - 9$ **61.** $2xy$
63. $-6xy + 6xz + 2yz$ **65.** $19p^3q^2 + 2pq^2 - pq + 16q^2 + 2$
67. $1.9a^2b + 1.1ab^2 - 6.5a^2b^2$
69. $32.1x + 35.45y + 6.01z$
71. $4.22a + 1.63b + 12.13c$
73. Mistake: Only the sign of the first term in the subtrahend was changed instead of all three terms. Correct: $-4x^2 - 2x + 5$

Review Exercises 1. -17.25 **2.** 15 **3.** 0.0001
4. 10^7 **5.** $58,900,000$ **6.** 300 m^2

5.4 Exercises
1. Product rule for exponents.
3. To multiply monomials, **a.** Multiply coefficients.   **b.** Add the exponents of the like bases.   **c.** Write any unlike variable bases unchanged in the product.   **5.** Use $(a^m)^n = a^{mn}$.
So $(x^3)^8 = x^{3 \cdot 8} = x^{24}$.   **7.** a^7 **9.** 2^{13} **11.** a^5b
13. $6x^3$ **15.** $-8m^3n^2$ **17.** $-\dfrac{5}{28}s^3t^8$ **19.** $2.76a^3b^5c^6$
21. $-6r^5$ **23.** $-12x^4z^9$ **25.** $90x^6y^5z^3$ **27.** $5q^5r^{10}s^{21}$
29. $-\dfrac{1}{4}a^6b^3c^2$ **31.** $w^3x^3y^5$ **33.** Mistake: The exponents of x were multiplied instead of added. Correct: $35x^7y$
35. Mistake: The exponents of x were not added. Correct: $54x^4y^6z$ **37.** $5w^2$ **39.** $4.5w^3$ **41. a.** height $= \dfrac{1}{2}b$,
top side $= \dfrac{1}{4}b$ **b.** $A = \dfrac{5}{16}b^2$ **43.** 1.2×10^9 **45.** 2.4×10^{14}
47. 1.827×10^5 **49.** 1.1421×10^{-4} **51.** 5.612×10^{-10}
53. $4.2 \times 10^4 \text{ mi}^2$ **55.** $\approx 5.8404 \times 10^8 \text{ mi.}$
57. $\approx 1.412 \times 10^{27} \text{ m}^3$ **59.** 2.9817×10^{-19} joule
61. 2.16×10^7 joules   **63.** x^6 **65.** 3^8 **67.** h^8 **69.** $-y^6$
71. x^3y^3 **73.** $27x^6$ **75.** $-8x^9y^6$ **77.** $\dfrac{1}{64}m^6n^3$
79. $64p^{24}q^{24}r^{12}$ **81.** $-0.008x^6y^{15}z^3$ **83.** $18x^7$ **85.** r^5s^4
87. $100a^6b^4$ **89.** $\dfrac{1}{1024}a^{10}b^{15}c^5$ **91.** $360a^2b^3$ **93.** $216a^{18}b^8$
95. $324u^6v^3$ **97.** $3.24u^6v^3$ **99.** Mistake: The exponents were added instead of multiplied. Correct: $25x^6$
101. Mistake: The coefficient was multiplied by 4 instead of being raised to the 4th power. Also, the exponents were not multiplied properly. Correct: $16x^{16}y^4$

Review Exercises 1. $6x + 8$ **2.** $-1.8t^5u^2$ **3.** $-22x$
4. $-4x - 4z$ **5.** -61 **6.** 6

5.5 Exercises
1. Distributive property.   **3.** They differ only in the sign separating the terms.   **5.** The middle term

is missing.   **7.** $5x + 15$ **9.** $-12x + 24$ **11.** $28n^2 + 8n$
13. $9a^2 - 9ab$ **15.** $\dfrac{1}{4}m^2 - \dfrac{5}{8}mn$ **17.** $-0.3p^5 + 0.6p^2q^2$
19. $20x^3 + 10x^2 - 25x$ **21.** $-21x^5 + 15x^3 - 3x^2$
23. $-r^4s^4 - 3r^3s^3 + r^2s^2$
25. $-4a^5b^2 + 12a^2b^3 - 6a^3b^3 + 2a^4b^4$
27. $12a^3b^3c^3 - 6a^2b^2c^2 + 15abc$
29. $4b^6 - b^4 + \dfrac{1}{5}b^3 - \dfrac{1}{2}b^2 + 16b$
31. $-r^4t^3 + 3.2r^3t^4 - 1.64r^2t^5 + 0.8rt^2$
33. $25x^6y - 15x^5y^2 + 50x^3y^9$ **35. a.** $x + 5$ **b.** $2x$
c. $2x(x + 5)$ **d.** $2x^2 + 10x$ **e.** They describe the same area.
37. a. $x + 3$ **b.** $x + 2$ **c.** $(x + 2)(x + 3)$ **d.** $x^2 + 5x + 6$
e. They describe the same area.   **39.** $x^2 + 7x + 12$
41. $x^2 - 5x - 14$ **43.** $y^2 - 9y + 18$ **45.** $6y^2 + 19y + 10$
47. $15m^2 + 11m - 12$ **49.** $12t^2 - 26t + 10$
51. $15q^2 + 11qt - 12t^2$ **53.** $14y^2 + 34xy + 12x^2$
55. $a^4 - a^2b - 2b^2$ **57.** $x^3 - 3x + 2$
59. $3a^3 - 4a^2 - a + 2$ **61.** $6c^3 + c^2 - 11c - 6$
63. $4f^3 - 27fg^2 - 27g^3$ **65.** $27m^3 - 8n^3$ **67.** $4a^3 + 125b^3$
69. $x^4 + 3x^3y + 2x^2y^2 + xy^3 - y^4$
71. $x^4 + 14x^3 + 69x^2 + 140x + 100$ **73.** $x - 3$
75. $4x + 2y$ **77.** $4d - 3c$ **79.** $-3j + k$ **81.** $x^2 - 25$
83. $4m^2 - 25$ **85.** $x^2 - y^2$ **87.** $64r^2 - 100s^2$ **89.** $4x^2 - 9$
91. $x^2 + 6x + 9$ **93.** $16t^2 - 8t + 1$ **95.** $m^2 + 2mn + n^2$
97. $4u^2 + 12uv + 9v^2$ **99.** $81w^2 - 72wz + 16z^2$
101. $81 - 90y + 25y^2$ **103.** $10xyz + 2y^2z$ **105.** $h^2 + 4h$
107. $w^2 + 3w$ **109.** $4x^3 + 24x^2 + 20x$ **111.** $4w^3 + 6w^2$
113. $V = 3.14r^3 + 6.28r^2$, or $\pi r^3 + 2\pi r^2$

Review Exercises 1. 0.45 **2.** -5 **3.** $6,304,000$
4. x^7 **5.** $r = \dfrac{5 - d}{m}$ **6.** $9, 36$

5.6 Exercises
1. When dividing exponential forms that have the same base, subtract the divisor's exponent from the dividend's exponent and keep the same base.   **3.** The decimal factors and powers of 10 can be separated into a product of two fractions. This allows us to calculate the decimal division and divide the powers of 10 separately.   **5.** To divide a polynomial by a monomial, divide each term in the polynomial by the monomial.   **7.** $\dfrac{1}{a^3}$ **9.** $\dfrac{1}{2^5}$ **11.** y^5 **13.** 3 **15.** $\dfrac{1}{4^6}$ **17.** $\dfrac{1}{x^2}$
19. a^2 **21.** $\dfrac{1}{w^3}$ **23.** $\dfrac{1}{a^7}$ **25.** r^{10} **27.** y^8 **29.** $\dfrac{1}{p^3}$ **31.** 1
33. 2.6×10^{-1} **35.** 6×10^{-8} **37.** 7.1×10^{10} **39.** 2.7×10^4
41. 500 sec. or $8\dfrac{1}{3}$ min.   **43.** $\$36,862.75$ **45.** $\approx 9.3 \times 10^{-4}$ kg
47. $5x^3$ **49.** $\dfrac{-4}{x^3}$ **51.** $-3m^2n^4$ **53.** $-\dfrac{8}{5}q^3$ **55.** $\dfrac{4y^4}{3x^5}$ **57.** $\dfrac{4c}{3a^2}$
59. $a + 2b$ **61.** $4x^2 - 2x$ **63.** $3x + \dfrac{2}{x}$ **65.** $xy - 3y^2$
67. $-2c + 8ab$ **69.** $3x^2 + 2x - 1$ **71.** $2x - 4 + \dfrac{3}{x}$
73. $12uv^3 + \dfrac{4v^4}{u} - 5v$ **75.** $y^2 + y^4 - 3y^5 + \dfrac{1}{y^2}$ **77.** $\dfrac{5mn}{2}$
79. a. $3x - 5 + \dfrac{4}{x}$ **b.** Area $= 18$, length $= 6$, width $= 3$.
81. $x + 4$ **83.** $m - 5$ **85.** $x - 12 + \dfrac{4}{x - 5}$

87. $3x^2 + 15x + 2$ **89.** $2x^2 + 3x + 6 + \dfrac{17}{x-2}$

91. $p^2 - 5p + 25$ **93.** $3x + 2$ **95.** $2y - 2 + \dfrac{23}{7y+3}$

97. $4k^2 - 1 + \dfrac{8}{k+2}$ **99.** $7u^2 - 11u + 12 - \dfrac{30}{3u+2}$

101. $b^2 - 3b + 5 + \dfrac{3}{b-3}$ **103.** $y^2 - 2y + 3$

105. a. $x + 4$ **b.** Length = 7 ft., height = 8 ft.,
volume = 336 ft.^{3} **107.** $\dfrac{36m^4}{n^4}$ **109.** $\dfrac{x^9}{64y^6}$ **111.** $256r^9 s^7$

113. m^{16} **115.** $\dfrac{1}{x^{14}}$ **117.** $\dfrac{y^{27}}{x^9}$ **119.** x^{14} **121.** $\dfrac{27y^6}{x^3 z^6}$

123. $\dfrac{1}{x^{10}}$ **125.** $\dfrac{4x^4 z^3}{y^5}$ **127.** $\dfrac{2}{3ab^2}$ **129.** $\dfrac{1}{2x^{11}}$

Review Exercises   **1.** $2^4 \cdot 3^3 \cdot 7$ **2.** $-18x^4 y^3$

3. $3xy - 4y$ **4.** $x + \dfrac{1}{2}y$ **5.** 340 ft.^{2} **6.** -24

Chapter 5 Review Exercises   **1.** false   **2.** false
3. true   **4.** false   **5.** false   **6.** false   **7.** combine
8. a. coefficients   **b.** exponents   **9.** $(a^m)^n = a^{mn}$ **10.** $2ab$
11. $\dfrac{8}{125}$ **12.** -16 **13.** $\dfrac{1}{25}$ **14.** $\dfrac{1}{13}$ **15.** 4,500,000,000
16. 13,800,000   **17.** 0.0000000000000000000000001663
18. 0.00000000000000000000000000000002006
19. 1.661×10^{-24} **20.** 9.63×10^{-23} **21.** 3×10^8
22. 6.37×10^6 **23.** yes   **24.** yes   **25.** no   **26.** no
27. c: 6; d: 4   **28.** c: 27; d: 0   **29.** c: -2.6; d: 4   **30.** c: -1; d: 1
31. binomial   **32.** monomial   **33.** Not a polynomial.
34. No special polynomial name.
35. $9; -2x^9 + 21x^5 - 19x^3 - 3x + 15$
36. $4; -j^4 + 22j^2 + 5j - 19$
37. $5; 4u^5 + 13u^4 - 18u^3 - u - 21$
38. $8; 6v^8 + 21v^4 + 16v^3 - v^2 - 2v - 19$
39. $4y^5 + 4y^4 + 8$ **40.** $18l^5 + 5l^4 - 2l^3 + 6l^2 + 7l + 20$
41. $-2m^3 + 7m^2 + m + 3$ **42.** $-3y^2 + 7y - 15$
43. $a^2 bc - ab^2 c - 6$ **44.** $-11x^2 yz - 2xyz^2$
45. $-4c^3 - cd^2 + 11cd + 2d + 8$
46. $-8a^5 + 3abc^3 - 8abc + 8$
47. $-3j^4 + 3jk^3 + k^2 + 10jk + 12$
48. $3mn^2 - 8n^2 - 9mn + 7$ **49.** $-4ab - 18a - 19c + 8$
50. $11y - 10$ **51.** $10x^2 + 3x - 4$ **52.** $3m - 8$
53. $-2n^3 + 3n^2 + 4n + 3$ **54.** 9   **55.** $-x^3 + 4x^2 + 17$
56. $-p^3 + 10p^2 - 10p + 9$ **57.** $-16x^3 - 4x - 2$
58. $8x^2 - xy + 6y^2$ **59.** $-3m^2 + 8mn - 2$ **60.** m^5
61. $10a^9$ **62.** $-6x^6 y^6$ **63.** $25x^4 y^2$ **64.** $72u^{10}$ **65.** $32x^{10}$
66. x^8 **67.** u^7 **68.** $\dfrac{1}{s^6}$ **69.** x^6 **70.** $\dfrac{b^6}{8a^9}$ **71.** $\dfrac{4j^8}{9hk^2}$
72. $\dfrac{-5}{x^2}$ **73.** 7   **74.** 1   **75.** $2x - 42$ **76.** $-16x + 8$
77. $4a^2 - 12a$ **78.** $-12b^3 + 24b^2 + 6b$
79. $-8m^7 - 12m^5 + 4m^4 + 20m^3$
80. $-12a^2 bc^2 + 16a^2 b^4 c^2 - 8a^2 b^2 c^4 - 32abc^2$
81. $x^2 + 4x - 5$ **82.** $y^2 - 11y + 24$ **83.** $12m^2 + 28m - 5$
84. $15a^2 - 4ab - 4b^2$ **85.** $4x^2 - 1$ **86.** $x^3 - x^2 - x - 15$
87. $3y^3 - 7y^2 + 14y - 4$
88. $a^4 - a^3 b - 2a^2 b^2 - 3ab^3 - b^4$
89. $16 - x^2$ **90.** $x^2 - 12x + 36$ **91.** $9r^2 - 30r + 25$

92. $36s^2 - 4r^2$ **93. a.** $20x^2 - 28x$ **b.** 96 ft.^{2}
94. 2.25×10^8 m^2 **95.** 2.94×10^{-4} V   **96.** $x - 5$

97. $-y^2 + 2y - 3$ **98.** $1 + 10st^3 - 2t^4$ **99.** $a^2 - \dfrac{1}{c} + 2ac$

100. $2 + b$ **101.** $\dfrac{2z}{5y^2} - \dfrac{xz}{y} + \dfrac{2}{y}$ **102. a.** $9x - 10$

b. 24 in., 26 in., and 624 in.^{2} **103.** ≈ 4.2 yr.   **104.** $z + 5$

105. $3m - 2 + \dfrac{1}{m-5}$ **106.** $y - 4 - \dfrac{4}{2y+1}$

107. $2m - 2 + \dfrac{1}{3m-2}$ **108.** $x^2 - x + 1 - \dfrac{2}{x+1}$

109. $4x^2 + 6x + 9$ **110.** $2s^2 + 2s - \dfrac{3}{s-1}$

111. $5y^3 - 3y + 2$ **112.** $4x - 5$

Chapter 5 Practice Test   **1.** $\dfrac{1}{8}$ [5.1]   **2.** $\dfrac{9}{4}$ [5.1]

3. 0.006201 [5.1]   **4.** 2.75×10^8 [5.1]   **5.** 3 [5.2]
6. 4 [5.2]   **7.** 56 [5.2]   **8.** $-10x^4 - 6x^3 + 14x^2 + 12$ [5.2]
9. $8x^2 + x - 4$ [5.3]   **10.** $5x^4 - 3x^2 - x + 8$ [5.3]
11. $12x^7$ [5.4]   **12.** $-2a^6 b^4 c^7$ [5.4]   **13.** $16x^2 y^6$ [5.4]
14. $3x^3 - 12x^2 + 15x$ [5.5]   **15.** $-24t^5 u + 48t^3 u^3$ [5.5]
16. $n^2 + 3n - 4$ [5.5]   **17.** $4x^2 - 12x + 9$ [5.5]
18. $x^3 - 2x^2 - 5x + 6$ [5.5]   **19.** $6n^2 - 7n - 20$ [5.5]

20. x^5 [5.6]   **21.** $\dfrac{1}{x^5}$ [5.6]   **22.** $\dfrac{x^9 y^4}{9}$ [5.6]

23. $4x^3 + 3x$ [5.6]   **24.** $x - 4$ [5.6]

25. $5x - 4 + \dfrac{6}{3x-2}$ [5.6]

Chapters 1–5 Cumulative Review Exercises
1. false   **2.** false   **3.** true   **4.** false   **5.** coefficient,
coefficient   **6.** (0, 0) or the origin   **7.** 23   **8.** -1 **9.** 60
10. $5y^2 - 4y - 1$ **11.** $6x - 6$ **12.** $2x^2 - 11x - 21$

13. $4x^2 - 12x + 9$ **14.** $m - 2 + \dfrac{1}{3m}$ **15.** $x + 3 + \dfrac{1}{x+2}$

16. $\dfrac{x^6}{27y^9}$ **17.** $\dfrac{4x^6}{y^2}$ **18.** $-108x^{16}$

19. 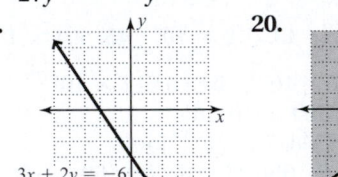   **20.**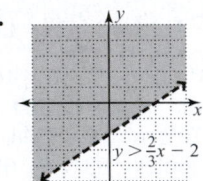

21. -12 **22.** $x < -10$ **23** 36   **24.** $-\dfrac{4}{5}$ **25.** $l = \dfrac{P - 2w}{2}$

26. 20 lb.   **27.** \$33,750 at 13%, \$11,250 at 9%   **28.** 480
29. 53.7%   **30.** 27 three-hour classes, 15 five-hour classes.

Chapter 6

6.1 Exercises   **1.** To list all of the possible factors of a
number, we can divide by 1, 2, 3, and so on, writing each divisor
and quotient pair as a product until we have all possible combinations.   **3.** The factorization for the GCF includes only
those prime factors common to all of the factorizations, each
raised to its smallest exponent.   **5.** Four-term polynomials.

7. 1, 3, 9   9. 1, 3, 11, 33   11. 1, 2, 3, 6, 9, 18   13. 1, 2, 4, 8, 16
15. 1, 2, 4, 11, 22, 44   17. 1, 2, 3, 4, 5, 6, 10, 12, 15, 20, 30, 60
19. 1, 2, 4, 7, 8, 14, 28, 56   21. 1, 2, 3, 5, 6, 9, 10, 15, 18, 30, 45, 90
23. 3   25. 8   27. 6   29. $2xy$ 31. $5h^2$ 33. $3a^4b$
35. $5(c-4)$ 37. $4(2x-3)$ 39. $x(x-1)$
41. $6z^4(3z^2-2)$ 43. $3p^3(6-5p^2)$ 45. $3ab(2a-b)$
47. $7uv(2v-1)$ 49. $25x(y-2z+4)$
51. $xy(x+y+x^2y^2)$ 53. $4ab^2c(7b-9a)$
55. $4pq(5p+6-4q)$ 57. $3mnp(n^4p+6n^2-2)$
59. $21a^2b^2(5a-3b+4a^4b^2)$
61. $3x(6x^3-3x^2+10x-4)$ 63. $-3(x-2y)$
65. $-5a(4a+3)$ 67. $-4a^3b(3a-5b^2)$
69. $-4(x^2+2x-4)$ 71. $-6x^2y^2z(4x+5yz^3-2x^2y^3z)$
73. $(a-3)(y+2)$ 75. $(2m+3n)(4a+b)$
77. $(4m-5)(5x-2)$ 79. $2(6p+5q)(r-2s)$
81. $(x+2)(b+c)$ 83. $(m-n)(a-b)$
85. $(x+2)(x^2-3)$ 87. $(1-m)(1+m^2)$
89. $(3x+5)(y+2)$ 91. $(4b-1)(b+1)$
93. $(3+b)(2-a)$ 95. $(x+2y)(3a+4b)$
97. $(y-s)(x^2-r)$ 99. $(w+3z)(w+5)$
101. $(s+y)(3t-2)$ 103. $(x^2+y^2)(a-5)$
105. $2(a+b)(a+3)$ 107. $2(2b+3)(3a+4)$
109. $3(4x-y)(2z-w)$ 111. $3a(y-4)(a+3)$
113. $2x(x+3y)(x+5)$ 115. $6x(x+6)$ 117. $3x(7x+9)$
119. $12\pi r^2(3r+2)$

Review Exercises 1. 3,740,000,000   2. 4.56×10^7
3. $x^2+7x+10$ 4. x^2-x-12 5. $x^2-8x+15$
6. x^2-7x

6.2 Exercises 1. Both are positive.   3. The number with the greater absolute value is positive, and the other number is negative.   5. 5   7. 2   9. 6   11. 2   13. $(r+3)(r+1)$
15. $(x-7)(x-1)$ 17. $(z-3)(z+1)$ 19. $(y+2)(y+3)$
21. $(u-2)(u-4)$ 23. $(u+3)(u-2)$
25. $(a+3)(a+4)$ 27. $(y-3)^2$ 29. $(w-4)(w+3)$
31. $(x-6)(x+5)$ 33. $(n-6)(n-5)$ 35. $(r-3)(r-6)$
37. $(x-8)(x+3)$ 39. $(p-9)(p+4)$ 41. prime
43. prime   45. $(p-3q)(p-7q)$ 47. $(a-9b)(a+3b)$
49. $(x-12y)(x-2y)$ 51. $(r-6s)(r+5s)$
53. $4(x-3)(x-7)$ 55. $2m(m-1)(m-6)$
57. $3b(a-8)(a+3)$ 59. $4(x-3)^2$ 61. $n^2(n+2)(n+3)$
63. $7u^2(u+5)(u+1)$ 65. $6b^2c(a-2)(a-4)$
67. $3(x-2y)(x-5y)$ 69. $2b(a-6b)(a+3b)$
71. $4x^2y(x^2-3xy-15y^2)$ 73. $(x-6)(x+1)$
75. $(x-4)(x+1)$ 77. 4, 20   79. 7, 8, 13   81. 6, 10, 12
83. 9, 16, 21, 24, 25   85. Length: $h+4$; width: $h+2$.
87. Length: $w-6$; width: $w-10$.

Review Exercises 1. 1   2. $10x^2+17x+3$
3. $15y^2-2y-24$ 4. $72n^3-60n^2+8n$
5. $5x^2y(2y^2-x-4y)$ 6. $4ab^2(2ab^2-4-3ab)$
7. $(x+2)(x-3)$ 8. $(z+5)(z-6)$ 9. $(3c-d)(5a-b)$
10. $(3c+2d)(2a-3b)$

6.3 Exercises 1. monomial GCF in all of the terms
3. Write a pair of last terms whose product is c.   5. 2, 4
7. $t, 4t$ 9. $7, 3x$ 11. $(2j+1)(j+2)$ 13. $(2y-5)(y+1)$
15. $(3m-4)(m-2)$ 17. $(6a+7)(a+1)$ 19. prime
21. $(4a-3)(a-4)$ 23. $(2x+3)(3x+5)$
25. $(2d-3)(8d+5)$ 27. $(3p+q)(p+4q)$
29. $(k+h)(5k-12h)$ 31. $(2a-5b)(6a-5b)$

33. $(8m-3n)(m-3n)$ 35. $y(8x^2-4x-1)$
37. $3(2y+1)(2y+3)$ 39. $2a(3b-7)(b-1)$
41. $2w(3w+2v)(w+2v)$ 43. $(a+1)(3a+1)$
45. $(2t-1)(t-1)$ 47. $(3x-7)(x+1)$
49. $(y-2)(2y+3)$ 51. $(2a-1)(5a-7)$
53. $(4r+3)(2r-3)$ 55. $(5x-2)(4x-3)$
57. $(2k+j)(3k+2j)$ 59. $(x-5y)(5x-y)$
61. $(5s-2t)(2s+t)$ 63. $(3u-2v)(2u+3v)$ 65. prime
67. $2(k-2)(2k-3)$ 69. $2m(2m+3)(3m-2)$
71. $8x(x-3y)(3x+y)$ 73. $4x(2x-3y)(3x+2y)$
75. Mistake: Using FOIL to check, we see that the first and last terms check but the inner and outer terms combine to give $31x$ instead of $13x$. Correct: $(2x+1)(3x+5)$
77. Mistake: Did not factor completely.
Correct: $4(n+2)(n+1)$ 79. $(5x-2)(3x-1)$
81. $2w+1$ and $3w-2$ 83. 7, 8, or 13   85. 4, 12, or 44
87. 1, 11, 19, or 41

Review Exercises 1. $9x^2-25$ 2. x^3-8
3. $4x^2-20x+25$ 4. $4cd^2(4c^2-6cd^2+9)$
5. $(b+5)(a+3)$ 6. $(m-4)(m+6)$
7. $3x(x-2)(x+5)$

6.4 Exercises 1. The first and last terms are perfect squares, and twice the product of their roots equals the middle term.   3. conjugates   5. In the trinomial factor (second term).   7. $(x+7)^2$ 9. $(b-4)^2$ 11. Not a perfect square.
13. $(5u-3)^2$ 15. $(10w+1)^2$ 17. $(y+z)^2$
19. $(2p-7q)^2$ 21. $(4g+3h)^2$ 23. $4(2t+5)^2$
25. $(x+2)(x-2)$ 27. $(4+y)(4-y)$
29. $(p+q)(p-q)$ 31. $(5u+4)(5u-4)$ 33. prime
35. $(8m+5n)(8m-5n)$ 37. $2(5x+4y)(5x-4y)$
39. $4(x^2+25y^2)$ 41. $(x^2+y^2)(x+y)(x-y)$
43. $(x^2+4)(x+2)(x-2)$ 45. $(n-3)(n^2+3n+9)$
47. $(x+3)(x^2-3x+9)$ 49. $(x-1)(x^2+x+1)$
51. $(m+n)(m^2-mn+n^2)$ 53. $(3k-2)(9k^2+6k+4)$
55. $(3k+2)(9k^2-6k+4)$ 57. $(c-4d)(c^2+4cd+16d^2)$
59. $(5x+4y)(25x^2-20xy+16y^2)$
61. $(3x-4y)(9x^2+12xy+16y^2)$
63. $(2p+qz)(4p^2-2pqz+q^2z^2)$
65. $2(x+5)(x-5)$ 67. $\left(4x+\dfrac{5}{7}\right)\left(4x-\dfrac{5}{7}\right)$
69. $2u(u+1)(u-1)$ 71. $y^3(y+4b)(y-4b)$
73. $2x(25x^2+1)$ 75. $3(y-2z)(y^2+2yz+4z^2)$
77. $\left(c-\dfrac{2}{3}\right)\left(c^2+\dfrac{2}{3}c+\dfrac{4}{9}\right)$
79. $2c(2c+d)(4c^2-2cd+d^2)$
81. $(2a-b+c)(2a-b-c)$
83. $(4-3x+3y)(4+3x-3y)$
85. $(x+3y+3z)(x^2-3xy-3xz+9y^2+18yz+9z^2)$
87. $(x+y-4d)(x^2+2xy+y^2+4dx+4dy+16d^2)$
89. 24   91. 36   93. 25   95. 16   97. $10(x+2)(x-2)$
99. $(2x-3)(4x^2+6x+9)$

Review Exercises 1. 1, 2, 3, 4, 6, 9, 12, 18, 36
2. $2\cdot2\cdot5\cdot5$ 3. $7xy^2(2x^2y-3xy^2-1)$ 4. $(x+4y)(x+3)$
5. $(a+3b)(a-6b)$ 6. $3b(a+5)(a-3)$
7. $(3n-2)(2n+3)$ 8. $2xy(2x-5y)(3x-2y)$

6.5 Exercises 1. A monomial GCF.   3. Sum of squares.
5. Try to factor by grouping when there are four terms. Grouping may also be used as a method to factor trinomials.

7. $3xy(y + 2x)$ **9.** $(x + y)(7a - b)$ **11.** $2(x + 4)(x - 4)$
13. $(a + b)(x + y)$ **15.** $abc(12a^2b + 3abc + 5c^2)$
17. $(x + 3)(x + 5)$ **19.** $(x^2 + 4)(x + 2)(x - 2)$ **21.** prime
23. $(5x - 1)(3x + 2)$ **25.** $a(x + 2)^2$ **27.** $6ab(1 - 6b)$
29. prime **31.** $(x + 7)(x - 7)$ **33.** $(p + 6)(p - 5)$
35. $u(u + 1)(u - 1)$ **37.** $2(b + 3)(b + 4)$ **39.** $3r^2(2 - 5r)$
41. $7(2u - 5)(u + 3)$ **43.** $4(h^2 + 3h + 1)$
45. $5(p + 4)(p - 4)$ **47.** $(3w + 2)(w + 1)$
49. $(2q - 3)(4q + 1)$ **51.** $(3v + 2)(4v + 5)$
53. $2(1 + 5x)(1 - 5x)$ **55.** $20k^2(2 + l)(2 - l)$ **57.** prime
59. $2(5 - t)^2$ **61.** $(x - 2y)(3a + 4b)$ **63.** $(x + 1)(x - 1)^2$
65. $(2x - 7)^2$ **67.** $2(x^2 + 9)(x + 3)(x - 3)$
69. $(a + 2b)(a + b)$ **71.** $(3 + 2m)(3 - 2m)$
73. $x(x^2 + 2y)(x^2 - 2y)$ **75.** prime
77. $(b + 5)(b^2 - 5b + 25)$ **79.** $3(y - 2)(y^2 + 2y + 4)$
81. $2x(3 - y)(9 + 3y + y^2)$ **83.** $l = 2x + 1, w = 3x - 7$
85. $l = 2x, w = 3x + 5, h = 3x + 5$ **87.** $4(5 + 2t)(5 - 2t)$
89. Current: $3i + 4$; resistance: $2r + 5$.

Review Exercises
1. $x - 4 - \dfrac{4}{2x + 1}$
2. $(6y - 1)(y + 4)$ **3.** $224\,\text{ft.}^2$ **4.** $4(x + 3) = 20; 2$
5. 17, 18, 19

6.6 Exercises
1. If a and b are real numbers and $ab = 0$,
then $a = 0$ or $b = 0$. **3.** Write the equation in standard form
$(ax^2 + bx + c = 0)$. **5.** $-5, -2$ **7.** $-3, 4$ **9.** $-\dfrac{3}{2}, \dfrac{4}{3}$

11. $0, 7$ **13.** $-2, 1, 0$ **15.** $0, 2$ **17.** $0, 4$ **19.** $0, -\dfrac{5}{3}$

21. $-2, 2$ **23.** $-3, 2$ **25.** 3 **27.** $4, -\dfrac{1}{3}$ **29.** $-\dfrac{5}{2}, \dfrac{2}{3}$

31. $-3, 7$ **33.** $1, 2$ **35.** $-3, 3$ **37.** $-7, -2$ **39.** $0, 4, -3$

41. $0, -\dfrac{5}{2}, \dfrac{2}{3}$ **43.** $\dfrac{2}{3}, 4$ **45.** $-4, \dfrac{15}{4}$ **47.** $3, 4$ **49.** $-2, 7$

51. $-\dfrac{7}{2}$ **53.** $-2, -4$ **55.** $3, -6$ **57.** $5, 11$ **59.** $11, 13$

61. 13, 14 **63.** 12 m by 21 m **65.** 172 ft. by 54 ft. **67.** 14 ft.
69. Base: 18 cm; height: 24 cm. **71.** 9 in. **73.** 10, 12, 14 or
$-16, -14, -12$ **75.** 1.25 sec. **77.** 10% **79.** 15 **81.** 25
83. 37 cm **85.** 30 blocks **87.** 45 in.

Review Exercises
1. No, it is not linear because the
variable x has an exponent other than 1. **2.** yes
3. $(-10, 0), (0, 4)$ **4.** $3x - y = -7$ **5. a.** 2 **b.** 6
6.
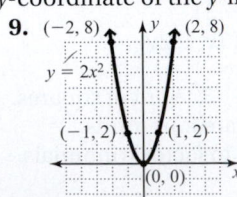

6.7 Exercises
1. If a vertical line is drawn through the
vertex, the sides of the parabola are mirror images. **3.** The
y-coordinate of the y-intercept is c. **5.** $(2, -1)$ **7.** $(3, 0)$
9. $(-2, 8), (2, 8)$
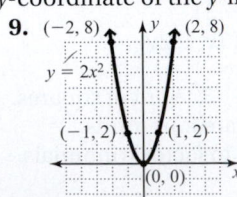
11.
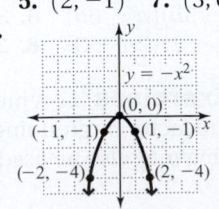

13. $y = x^2 - 3$
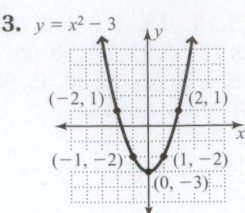

15. $y = -x^2 + 2x$
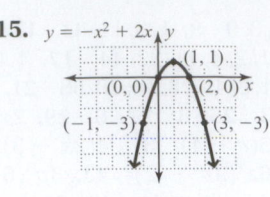

17. $y = 2x^2 - 4x + 1$
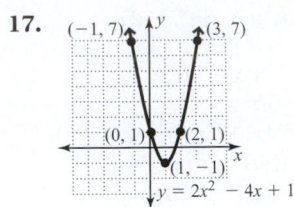

19. $y = -3x^2 + 6x + 4$
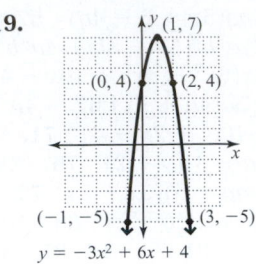

21. $y = x^2 + 1$
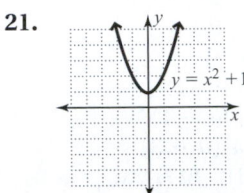

23. $y = -x^2 + 3$
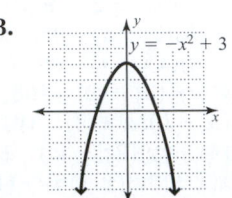

25. $y = 2x^2 - 5$
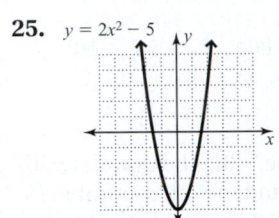

27. $y = -3x^2 + 2$
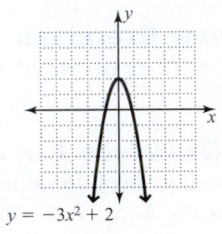

29. $y = x^2 + 4x - 1$
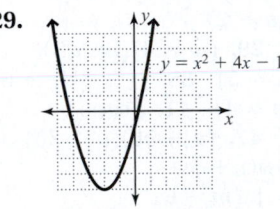

31. $y = -x^2 + 6x - 5$
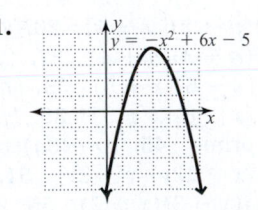

33. $y = 2x^2 - 4x - 3$
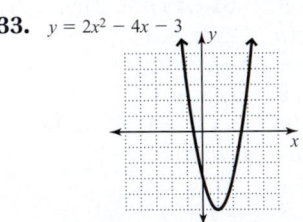

35. $y = -2x^2 + 8x - 5$
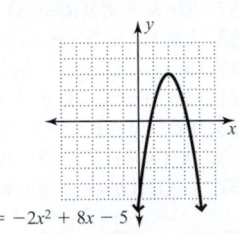

37. $f(x) = 3x^2$
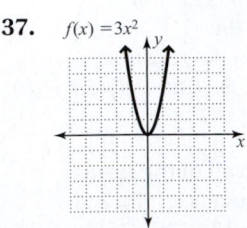

39. $f(x) = -x^2 + 1$
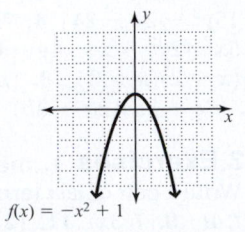

41. 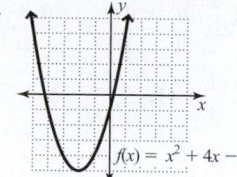 $f(x) = x^2 + 4x - 1$ **43.** 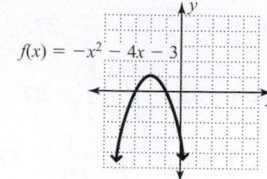 $f(x) = -x^2 - 4x - 3$

45. 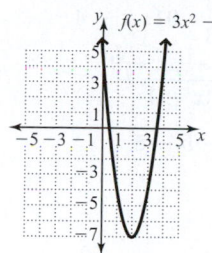 $f(x) = 3x^2 - 12x + 5$ **47.** 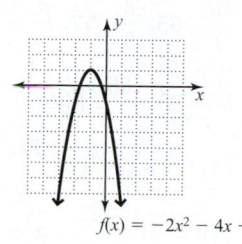 $f(x) = -2x^2 - 4x - 1$

49. A function; domain: all real numbers or $(-\infty, \infty)$; range: $\{y \mid y \geq 0\}$ or $[0, \infty)$.   **51.** Not a function; domain: $\{x \mid x \geq 0\}$ or $[0, \infty)$; range: all real numbers or $(-\infty, \infty)$.   **53.** A function; domain: all real numbers or $(-\infty, \infty)$; range: $\{y \mid y \leq 2\}$ or $(-\infty, 2]$.   **55.** Not a function; domain: $\{x \mid x \leq -1\}$ or $(-\infty, -1]$; range: all real numbers or $(-\infty, \infty)$.   **57. a.** $(0,0)$ **b.** $(0,2)$ **c.** The original graph is moved upward c units.   **d.** The original graph moved downward 3 units.   **59. a.** $y = x^2$ opens up, and $y = -x^2$ opens down. The sign of a indicates whether the parabola opens up or down.   **b.** yes   **c.** These graphs are "wider."   **d.** The smaller the absolute value of a, the wider the parabola.

Review Exercises
1. $3m^3n - 12m^2n^2 + 15mn$
2. $2x^2 - 5x - 3$ **3.** $(3n - 5)(n - 4)$ **4.** $(x + 5)(x - 5)$
5. $0, 3$ **6.** $-6, 3$

Chapter 6 Review Exercises
1. false   **2.** false
3. false   **4.** true   **5.** true   **6. a.** GCF   **b.** polynomial; GCF
7. a. c, b **8.** $(a + b)(a - b)$ **9.** $(a - b)(a^2 + ab + b^2)$
10. $(a - b)^2$ **11.** $1, 2, 3, 4, 5, 6, 10, 12, 15, 20, 30, 60$ **12.** $1,$
$3, 9$ **13.** $1, 3, 9, 27, 81$ **14.** $1, 2, 4, 11, 22, 44$ **15.** $1, 3, 5, 9,$
$15, 45$ **16.** $1, 7, 49$ **17.** $1, 2, 3, 5, 6, 10, 15, 30$ **18.** $1, 3,$
$11, 33$ **19.** 3 **20.** 17 **21.** 6 **22.** 5 **23.** $5y^2$ **24.** $4x$
25. $3mn$ **26.** $2k$ **27.** $2(2x - 1)$ **28.** $5m(1 - 7m^2)$
29. $y(y^2 - 1)$ **30.** $abc(1 + abc - a^2b^2c^2)$
31. $xy(x + y + x^2y^2)$ **32.** $21a^2b^2(5a - 3b + 4a^4b^2)$
33. $10k(10k^3 + 12k^4 - 1 + 4k^2)$ **34.** $18ab^2c(b - 2a)$
35. $(x + y)(a + b)$ **36.** $(x + 2)(a + b)$
37. $(y + 2)(y^2 + 3)$ **38.** $(y + 4)(y^3 - b)$
39. $(x + 1)(y + 1)$ **40.** $(x^2 + y^2)(a - 5)$
41. $(2b^2 + 1)(b - 1)$ **42.** $(u - 3)(u + 4v)$
43. $(x - 4)(x + 3)$ **44.** $(x + 5)(x + 9)$
45. $(n - 2)(n - 4)$ **46.** prime   **47.** $(h + 3)(h + 48)$
48. $(y - 12)(y + 2)$ **49.** $4(x - 3)^2$ **50.** $3(m - 9)(m - 2)$
51. prime   **52.** $(2u + 1)(u + 2)$ **53.** $(3m - 1)(m - 3)$
54. $(5k - 12h)(k + h)$ **55.** $2(3a - 4)(a - 2)$
56. $y(8x^2 - 4x - 1)$ **57.** $(3p - q)(p - 4q)$
58. $8(3x + y)(x - 3y)$ **59.** $(v - 4)^2$ **60.** $(u + 3)^2$
61. $(2x + 5)^2$ **62.** $(3y - 2)^2$ **63.** $(x + 2)(x - 2)$
64. $(5 + y)(5 - y)$ **65.** $(x - 1)(x^2 + x + 1)$
66. $(x + 3)(x^2 - 3x + 9)$ **67.** $(3b + 2)(2b - 1)$
68. $4ab(1 - 6b)$ **69.** prime   **70.** $3(x + y)(x - y)$
71. $(x^2 + 9)(x + 3)(x - 3)$ **72.** $7(u - 5)(u + 3)$
73. $(2x - 3y)(4x^2 + 6xy + 9y^2)$ **74.** $2(1 + 5y)(1 - 5y)$

75. $(m - 2n)(3a + 4b)$ **76.** $3(m^2 + 3m + 9)$ **77.** $-3, 4$
78. $-2, 2$ **79.** $2, 3$ **80.** $-3, 7$ **81.** $-3, 3$ **82.** $-6, 3$
83. -3 **84.** $-\dfrac{1}{3}, 4$ **85.** $5x^2 = 2x; 0, \dfrac{2}{5}$
86. $x^2 + (x + 1)^2 = 61; 5, 6$ **87.** $x \cdot 4x = 100; -5, 5$
88. $(w + 6)w = 91; 13$ in. by 7 in.   **89.** $x(x + 1) = 110; 10, 11$
90. $3^2 + 4^2 = c^2; 5$

91. 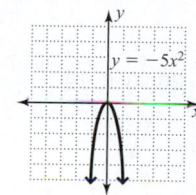 $y = -5x^2$ **92.** 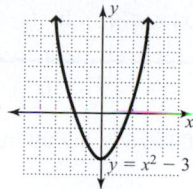 $y = x^2 - 3$

93. 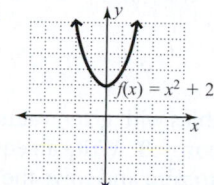 $f(x) = x^2 + 2$ **94.** 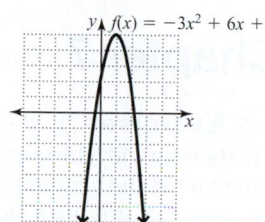 $f(x) = -3x^2 + 6x + 2$

Chapter 6 Practice Test
1. $1, 2, 4, 5, 8, 10, 16, 20, 40,$
80 [6.1]   **2.** $5m$ [6.1]   **3.** $5(y - 6)$ [6.1]
4. $2y(3x^2 - y)$ [6.1]   **5.** $(x - y)(a - b)$ [6.1]
6. $(m + 3)(m + 4)$ [6.2]   **7.** $(r - 4)^2$ [6.4]
8. $(2y + 5)^2$ [6.3]   **9.** $(q - 2)(3q - 4)$ [6.3]
10. $(3x - 4)(2x - 5)$ [6.3]   **11.** $a(x + 3)(x - 8)$ [6.2]
12. $2n(n + 4)(5n - 1)$ [6.2]   **13.** $(c + 5)(c - 5)$ [6.4]
14. prime [6.4]   **15.** $2(1 + 5u)(1 - 5u)$ [6.4]
16. $-4(x + 2)(x - 2)$ [6.4]
17. $(m + 5)(m^2 - 5m + 25)$ [6.4]
18. $(x - 2)(x^2 + 2x + 4)$ [6.4]   **19.** $-3, 0$ [6.6]
20. $6, -2$ [6.6]   **21.** $\dfrac{3}{2}, -5$ [6.6]   **22.** $8, 9$ [6.6]
23. 9ft., 4 ft. [6.6]
24. [6.7]   **25.** [6.7]

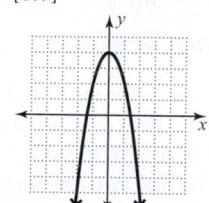 $y = -2x^2 + 4$ 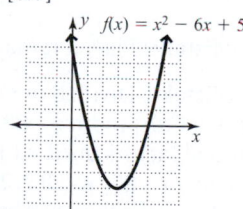 $f(x) = x^2 - 6x + 5$

Chapter 1–6 Cumulative Review Exercises
1. false   **2.** true   **3.** false   **4.** false   **5.** $(a + b)(a - b)$
6. a. standard   **b.** factored   **7.** 42 **8.** -89
9. $5x^2 + 4x + 2$ **10.** $18x^4y$ **11.** $48x^7$ **12.** $2y^2 - 15y - 8$
13. $4h + 1 - \dfrac{2}{h}$ **14.** $x + 4$ **15.** $(x + 11)(x - 11)$
16. $10(x^2 + 4)$ **17.** $(x - 4)^2$ **18.** $(2x - 3)(x + 4)$
19. $(x - 5)(x^2 + 5)$ **20.** $\dfrac{17}{12}$ **21.** 13.2 **22.** $-4, 7$
23. $x = \dfrac{7 + 3y}{2}$ **24.** $(3, 0), (0, -6)$ **25.** 0

26.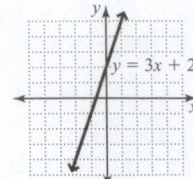

27. a. $\{x \mid x \le 4\}$ **b.** $(-\infty, 4]$

c.

$$-5\ -4\ -3\ -2\ -1\ 0\ 1\ 2\ 3\ 4\ 5$$

28. 6, 8, 10

29. \$2500 in 5%, \$5500 in 8%   **30.** $33\frac{1}{3}$ ml

Chapter 7

7.1 Exercises
1. Replace the variables with given values and then simplify the numerical expression.   **3.** Write an equation that has the denominator set equal to zero and solve the equation.   **5.** y is not a factor in the numerator, $x + y$.

7. a. $\dfrac{16}{45}$ **b.** -1 **c.** 0   **9. a.** 1 **b.** undefined **c.** -3

11. a. $\dfrac{11}{8}$ **b.** $-\dfrac{1}{7}$ **c.** $-\dfrac{24}{49}$ or ≈ -0.49 **13. a.** $-\dfrac{4}{3}$ **b.** $-\dfrac{3}{2}$

c. $-\dfrac{341}{30}$ or $-11.3\overline{6}$ **15.** 3   **17.** $-3, 3$ **19.** 5, 9   **21.** 0, 3

23. $-1, \dfrac{5}{2}$ **25.** Defined for all real numbers.   **27.** $\dfrac{3x}{4y}$

29. $-\dfrac{4m^2}{n}$ **31.** $\dfrac{3}{2x^2z^2}$ **33.** $-\dfrac{3m^2n}{m+2n}$ **35.** $\dfrac{1}{3(x+5)}$ **37.** $\dfrac{3}{5}$

39. $\dfrac{2}{3}$ **41.** $\dfrac{5}{3}$ **43.** $\dfrac{x}{4}$ **45.** m **47.** $\dfrac{5}{x}$ **49.** $\dfrac{x-y}{x+y}$ **51.** $\dfrac{x}{x-2}$

53. $\dfrac{t-3}{t+2}$ **55.** $\dfrac{2a-3}{3a-2}$ **57.** $\dfrac{3b-4}{5b-1}$ **59.** $\dfrac{x-4}{x+2}$ **61.** $\dfrac{p+q}{m-n}$

63. $\dfrac{2u^2-1}{u+1}$ **65.** -1 **67.** -4 **69.** $-y-2$ **71.** $-\dfrac{1}{w+1}$

73. $\dfrac{x-2}{x^2-2x+4}$ **75.** $\dfrac{x+y}{x}$ **77.** $-\dfrac{1}{m^2+3m+9}$

79. Mistake: Placed remaining x in numerator. Correct: $-\dfrac{1}{2x}$

81. Mistake: Divided out a part of a multiterm polynomial. Correct: Can't reduce   **83.** Mistake: $y - x$ and $x - y$ are not identical factors until -1 is factored out of one of them. Correct: $-x$ **85. a.** 23.7, 22.9, 28.3   **b.** The first two are normal weight, and the third is overweight.   **87.** $533.\overline{3}\pi, 666.\overline{6}\pi, 666.\overline{6}\pi,$ $833.\overline{3}\pi$ **89. a.** $h = \dfrac{3V}{\pi r^2}$ **b.** ≈ 9 cm   **91.** $\dfrac{s-3}{s+4}$

Review Exercises
1. 1   **2.** $-\dfrac{3}{4}$ **3.** $\dfrac{9}{20}$ **4.** $\dfrac{28}{15}$

5. $(2y+5)(2y-5)$ **6.** $(3x-2)(x+5)$

7.2 Exercises
1. To eliminate any common factors.
3. Their product is 1.   **5.** A fraction with a ratio equivalent to 1. Examples will vary.   **7.** $\dfrac{2m^2}{5n^2}$ **9.** $\dfrac{s^2}{r^3}$ **11.** y^2 **13.** $-\dfrac{1}{mn^2s}$

15. $\dfrac{9r}{s}$ **17.** $\dfrac{mn^2}{2}$ **19.** $\dfrac{9}{10}$ **21.** $\dfrac{4}{15}$ **23.** $(2x-5)(3x+4)$

25. $\dfrac{r^2}{(r+6)(r-3)}$ **27.** $\dfrac{q+1}{2q(q-1)}$ **29.** $\dfrac{w+4}{w-1}$ **31.** $\dfrac{n+1}{n}$

33. 1   **35.** $\dfrac{(d-5)}{(d+1)}$ **37.** $\dfrac{x-2}{x}$ **39.** $\dfrac{1}{y}$ **41.** $-\dfrac{4y^2}{3x}$

43. $1-3x$ **45.** $\dfrac{1}{x^2y^2}$ **47.** $\dfrac{5}{4m^2n}$ **49.** $\dfrac{y^2}{5z^2}$ **51.** $\dfrac{-32x^6}{45}$

53. $\dfrac{4a}{9}$ **55.** $\dfrac{8}{5}$ **57.** $\dfrac{a-b}{x+y}$ **59.** $\dfrac{x+8}{x-8}$ **61.** $x-4$

63. $\dfrac{a+2b}{a-2b}$ **65.** $\dfrac{1}{k(2k+1)}$ **67.** $\dfrac{4x^2+6x+9}{3x-1}$

69. $\dfrac{1}{(x+4)(x+1)}$ **71.** $\dfrac{b-4}{c-3}$ **73.** $-b-5$

75. $\dfrac{7+y}{y(y+4)}$ **77.** $-\dfrac{x^3}{x-3}$ **79.** $-\dfrac{x}{y}$

81. Mistake: Did not invert the divisor. Correct: $\dfrac{50}{27}$

83. Mistake: Divided out y, which is illegal because it is not a factor in $y - 2$. Correct: $\dfrac{(x-3)(y-2)}{15y}$ **85.** 222 in.

87. 10,560 yd.   **89.** 288 in.   **91.** 88 oz.   **93.** 12.7 T
95. 96,000 oz.   **97.** 260.48 ft./sec.   **99.** ≈ 19.03 mi./hr.
101. 939,785,328 km   **103.** $\approx 5.87 \times 10^{12}$ mi./yr.
105. 255.36£   **107.** \$665.53   **109.** 860.82€

Review Exercises
1. 1   **2.** $-\dfrac{1}{2}$ **3.** $2x$ **4.** $\dfrac{1}{4}y - 4$

5. $8x^2 - \dfrac{2}{3}x - 4$ **6.** $-\dfrac{1}{5}n^2 + 4$

7.3 Exercises
1. The numerator may not contain like terms.   **3.** Write an equivalent addition with the additive inverse of the subtrahend.   **5.** $\dfrac{x}{2}$ **7.** $\dfrac{a^2}{3}$ **9.** $\dfrac{9}{a}$ **11.** $-\dfrac{x}{7y^2}$

13. $\dfrac{n+3}{n+5}$ **15.** $\dfrac{3r}{r+2}$ **17.** $\dfrac{q+1}{q}$ **19.** 3   **21.** 3   **23.** $\dfrac{11}{w+1}$

25. $\dfrac{3m-1}{m-1}$ **27.** $\dfrac{-11t+3}{2t-3}$ **29.** $\dfrac{6w-5}{w+5}$ **31.** $\dfrac{1}{g+3}$

33. $z-3$ **35.** $\dfrac{1}{s+2}$ **37.** $\dfrac{y-3}{y}$ **39.** $\dfrac{x+2}{x-4}$ **41.** $\dfrac{t-1}{t+7}$

43. $3(x+2)$ **45.** $\dfrac{1}{x-5}$ **47.** $\dfrac{3x-2}{x-5}$ **49.** $\dfrac{x-3}{x+5}$

51. Mistake: Combined terms that are not like. Correct: $\dfrac{4y+1}{5}$

53. Mistake: Did not change sign of -5. Correct: 2

55. $\dfrac{5x+9}{2}$ **57.** $\dfrac{3x+2}{2}$

Review Exercises
1. $\dfrac{11}{10}$ **2.** $\dfrac{11}{24}$ **3.** $1\dfrac{6}{35}$ **4.** $\dfrac{11}{15}x + \dfrac{7}{12}$

5. $2x^2 - 12x - \dfrac{11}{24}$ **6.** $2y^2 - \dfrac{1}{20}y - 10$

7.4 Exercises
1. a. Find the prime factorization of each denominator.   **b.** Write a product that contains each unique prime factor the greatest number of times that factor occurs in any factorization—or if you prefer to use exponents, the product that contains each unique prime factor raised to the greatest exponent that occurs on that factor in any factorization.

c. Simplify the product found in Step 2.   **3.** Factor the denominators.   **5.** Rewrite the numerator as an equivalent addition sentence.   **7.** $mn; \dfrac{2m}{mn}, \dfrac{3n}{mn}$ **9.** $a^3b^5; \dfrac{2at}{a^3b^5}, \dfrac{3b^2z}{a^3b^5}$

11. $15a^2; \dfrac{10a}{15a^2}, \dfrac{9}{15a^2}$ **13.** $14x^2y^2; \dfrac{6y}{14x^2y^2}, \dfrac{5x}{14x^2y^2}$

15. $90m^3n^3; \dfrac{48}{90m^3n^3}, \dfrac{35mn^2}{90m^3n^3}$

17. $(y+2)(y-2); \dfrac{3y-6}{(y+2)(y-2)}, \dfrac{4y^2+8y}{(y+2)(y-2)}$

19. $12(y-5); \dfrac{9y}{12(y-5)}, \dfrac{14y}{12(y-5)}$

21. $6x(x+3); \dfrac{5x}{6x(x+3)}, \dfrac{12}{6x(x+3)}$

23. $p^2-4; \dfrac{2}{p^2-4}, \dfrac{p^2+p-6}{p^2-4}$

25. $3(u+2)^2; \dfrac{12u}{3(u-2)^2}, \dfrac{2u+4}{3(u+2)^2}$

27. $(t+6)(t-6)(t-1); \dfrac{2t^2-6t+4}{(t+6)(t-6)(t-1)}$, $\dfrac{3t^2+20t+12}{(t+6)(t-6)(t-1)}$

29. $(x+4)(x-1)(x+2); \dfrac{x^2+4x+4}{(x+4)(x-1)(x+2)}$, $\dfrac{x^2-4x+3}{(x+4)(x-1)(x+2)}$

31. $(x-1)(x-2); \dfrac{2x-4}{(x-1)(x-2)}, \dfrac{3}{(x-1)(x-2)}$, $\dfrac{5x^2-5x}{(x-1)(x-2)}$ **33.** $\dfrac{-31}{18}$ **35.** $\dfrac{8}{3c}$ **37.** $\dfrac{5x^2-2}{x^3}$

39. $\dfrac{8y-2}{(y+2)(y-4)}$ **41.** $\dfrac{x+10}{(x-2)(x+2)}$

43. $\dfrac{2x+14}{(x+5)(x-5)}$ **45.** $\dfrac{10p-8}{p^2-2p+1}$ **47.** $\dfrac{-z^2-5z-3}{(z+4)(z-3)}$

49. $\dfrac{2n^2-2n+18}{(n-4)(n+3)}$ **51.** $\dfrac{-t^2+t+16}{(t+4)(t+6)}$ **53.** $\dfrac{5}{2(c+5)}$

55. $\dfrac{y+2}{y}$ **57.** $\dfrac{r^2-6r-1}{(r+1)(r-1)^2}$ **59.** $\dfrac{q+1}{q(q-3)}$

61. $\dfrac{1}{(w-1)(w-2)}$ **63.** $\dfrac{v^2+16v+51}{(v+4)(v-4)(v+3)}$ **65.** $\dfrac{-3}{t-3}$

67. $\dfrac{4-x}{2x-1}$ **69.** $\dfrac{y+6}{3(y-2)}$ **71.** Mistake: Did not find a common denominator. Correct: $\dfrac{6-x^2}{2x}$ **73.** Mistake: Did not change the sign of 7. Correct: $\dfrac{-x-7}{x+2}$ **75.** Mistake: Added denominators. Correct: $\dfrac{7w}{x}$ **77.** $\dfrac{5t}{6}$ **79.** $\dfrac{4x+6}{x^2+3x}$

81. $\dfrac{4ah+5bh}{8}$ **83.** $\dfrac{6x^2+8x+2}{(2x-1)(2x+1)}$ **85.** xy

87. $\dfrac{3a^2+10a+20}{5a(a+2)}$ **89.** $\dfrac{2a^2+2b^2}{(a+b)^2}$ **91.** $\dfrac{x-5}{3(x+2)}$

Review Exercises **1.** $\dfrac{3}{8}$ **2.** $\dfrac{m^2}{108}$ **3.** $4(x+2)(x-2)$

4. $3y(2y+1)(y-5)$ **5.** $\dfrac{x-1}{x+2}$ **6.** $\dfrac{1}{x-3}$

7.5 Exercises **1.** $\dfrac{4}{x+3}$ is the numerator. $\dfrac{3}{x}$ is the denominator.   **3.** $\dfrac{\dfrac{2x}{x-3}}{\dfrac{4-x}{2x-6}}$ **5.** In this case, multiplying the numerator and denominator by their LCD, 12, (Method 2) is preferable because doing so requires fewer steps than simplifying $\dfrac{x}{6}-\dfrac{3}{4}$ and $\dfrac{2}{3}+x$ and then dividing the simplified expressions (Method 1). (Answers may vary.)   **7.** $\dfrac{1}{2}$ **9.** $\dfrac{mp}{nq}$

11. $\dfrac{11}{13}$ **13.** $\dfrac{5a+2}{3a+4}$ **15.** $\dfrac{1-y}{1+y}$ **17.** $\dfrac{4r-3}{2-4r}$ **19.** $\dfrac{x}{3+x}$

21. xy **23.** $\dfrac{x}{x-2}$ **25.** $\dfrac{6(2x+1)}{x(3x-2)}$ **27.** $\dfrac{a}{b(a-b)}$

29. $\dfrac{2y(y-5)}{5}$ **31.** u **33.** $\dfrac{1}{a-1}$ **35.** $\dfrac{1}{3}$ **37.** 2   **39.** $y-1$

41. $\dfrac{1}{n-1}$ **43.** $3-y$ **45.** $\dfrac{x-1}{x}$ **47.** $\dfrac{-2}{x(x+h)}$

49. Mistake: Divided out $\dfrac{1}{2}$, which is not a common factor. Correct: $\dfrac{2x-1}{2y-1}$ **51.** Mistake: Did not write as an equivalent multiplication. Correct: $\dfrac{m^2}{18}$ **53.** Mistake: Did not distribute c in the numerator. Correct: c^2+1 **55.** Mistake: Added the x's instead of multiplying them. Correct: $\dfrac{18}{x^2}$ **57.** $34\dfrac{2}{7}$ mph

59. $\dfrac{3(5n-2)}{4(n-1)}$ **61. a.** $\dfrac{R_1R_2}{R_2+R_1}$ **b.** $\approx 9.1\ \Omega$ **c.** $8\ \Omega$

Review Exercises **1.** -7 **2.** 3   **3.** 2.5   **4.** 4   **5.** 4.8 mph   **6.** 0.09 hr. or 5.4 min.

7.6 Exercises **1.** Both sides of the equation are multiplied by the LCD of the rational expressions.   **3.** An apparent solution that does not solve its equation.   **5.** 4 and -2; substituting either of those numbers for x causes an expression in the given equation to be undefined.   **7.** no   **9.** yes   **11.** yes   **13.** 2   **15.** 10   **17.** 15   **19.** 3   **21.** 1   **23.** 10   **25.** 8   **27.** -3

29. -1 **31.** 1   **33.** 9   **35.** $-2, \dfrac{1}{3}$ **37.** -5

39. $-\dfrac{11}{3}$, (-1 is extraneous)   **41.** $4, -2$ **43.** $-\dfrac{1}{4}, 5$

45. $-2, 9$ **47.** 1, (2 is extraneous)   **49.** $0, \dfrac{9}{2}$ **51.** $-3, -\dfrac{2}{3}$

53. -2 **55.** 1   **57.** $6, -1$ **59.** No solution.   **61.** 5, (1 is extraneous)   **63.** -3 is the only answer; 3 is extraneous.   **65.** In an expression, the LCD is used to combine the numerator over a single denominator. In an equation, the LCD is used to eliminate the denominator.   **67.** $400\ \Omega$ **69.** 20 and $30\ \Omega$ **71.** ≈ 2.1 ft.   **73.** $f = 15$ mm, $i = 20$ mm

Review Exercises **1.** ≈ 68.6 mph   **2.** $\approx \dfrac{0.084}{1}$; each ounce costs \$08.4   **3.** 6 hr.   **4.** 72, 74   **5.** The length is 11 mm, and the width is 7 mm.   **6.** 0.15 hr. (9 min.)

7.7 Exercises 1. $\dfrac{1}{x}$ 3. $\dfrac{100}{r}$, inverse 5. p decreases.

7. $2\dfrac{2}{5}$ hr. 9. $14\dfrac{7}{12}$ days 11. 12 hr. 13. 22.5 min.

15. 3 P.M. 17. 75 mph 19. 6 hr. 21. 441 mph 23. 12
25. 36 27. 12 29. \$39 31. $180\ \text{cm}^3$ 33. 5 sec.
35. 40.5 m 37. 7 ft. 39. 2 41. 8.25 43. 50 psi 45. $6\ \Omega$
47. 600 m 49. 100 mm 51. 64 53. 72 55. $336\ \text{in.}^3$
57. $169.65\ \text{cm}^3$ 59. 6 61. 162 63. $3.125\ \Omega$ 65. 12 dyn
67. a.

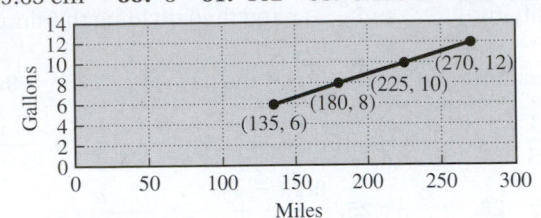

b. They are directly proportional. As the number of miles increases, so does the number of gallons. c. $k = 22.5$, which represents the miles per gallon the car gets. d. Yes, the data represent a fuction because for any given number of miles, there will be one quantity of gasoline (assuming that the miles per gallon stays constant).

Review Exercise 1. 4.3 2. $\dfrac{5}{8}$

3.
$$\begin{array}{c}\longleftarrow\!\!+\!\!+\!\!+\!\!+\!\!+\!\!\lbrack\!\!+\!\!+\!\!+\!\!+\!\!+\!\!\longrightarrow\\ -3\ -2\ -1\ \ 0\ \ 1\ \ 2\ \ 3\ \ 4\ \ 5\end{array}$$

4.
$$\begin{array}{c}\longleftarrow\!\!+\!\!+\!\!\rparen\!\!+\!\!+\!\!+\!\!+\!\!+\!\!+\!\!+\!\!\longrightarrow\\ -5\ -4\ -3\ -2\ -1\ \ 0\ \ 1\ \ 2\ \ 3\end{array}$$

5. $x > -4$ 6. $x \ge 2$

Chapter 7 Review Exercises 1. false 2. true
3. true 4. true 5. true 6. a. multiplication
b. Divide 7. a. numerator, denominator 8. complex
9. a. numerator, denominator 10. 1. LCD 11. a. $\dfrac{17}{6}$

b. ≈ 0.077 c. $\dfrac{3}{4}$ 12. a. 3 b. 15 c. 1.875 13. 5

14. $-2, 2$ 15. $-6, 1$ 16. $0, -6$ 17. $\dfrac{x + 5}{3}$ 18. $\dfrac{2}{3}$

19. $\dfrac{x}{4}$ 20. $\dfrac{y}{3}$ 21. $\dfrac{a + b}{a - b}$ 22. $\dfrac{3}{(x^2 + 9)(x - 3)}$

23. $-(x + y)$ 24. $-\dfrac{1}{w + 3}$ 25. $\dfrac{7x^2}{4y^2}$ 26. $-\dfrac{3}{5mnp^4}$

27. $\dfrac{3}{5}$ 28. $\dfrac{x^2}{(x + 4)(x - 3)}$ 29. $\dfrac{n + 1}{n}$ 30. $\dfrac{2}{r - 1}$

31. $\dfrac{m + 1}{4m(m - 1)}$ 32. $\dfrac{w + 4}{w - 4}$ 33. $\dfrac{y^3}{abx}$ 34. $\dfrac{5}{6}$ 35. 2

36. $\dfrac{5a}{2}$ 37. $\dfrac{x + y}{c - d}$ 38. $\dfrac{5(b + 4)(b - 4)}{6}$ 39. $\dfrac{2j + 1}{j(2j - 1)}$

40. $\dfrac{1}{(u + 1)^2}$ 41. $\dfrac{x}{3}$ 42. $\dfrac{3r}{r + 3}$ 43. $-\dfrac{1}{3m}$ 44. $\dfrac{3x + 1}{x + 1}$

45. 1 46. $\dfrac{s}{r^2 - s^2}$ 47. $\dfrac{11x^2 - 21x + 7}{2x + 1}$ 48. $q - 3$

49. $\dfrac{6c^2}{12c^3}, \dfrac{12}{12c^3}$ 50. $\dfrac{5x}{x^2y^3}, \dfrac{3y}{x^2y^3}$

51. $\dfrac{4m + 4}{(m - 1)(m + 1)}, \dfrac{4my - 4y}{(m - 1)(m + 1)}$

52. $\dfrac{3h}{4h(h - 2)}, \dfrac{20}{4h(h - 2)}$ 53. $\dfrac{x}{(x + 1)(x - 1)}, \dfrac{x^2 + x - 2}{(x + 1)(x - 1)}$

54. $\dfrac{2p + 4}{(p - 2)(p + 2)^2}, \dfrac{p^2 - 2p}{(p - 2)(p + 2)^2}$ 55. $-\dfrac{23}{36}$

56. $\dfrac{-x - 27}{(x + 3)(x - 3)}$ 57. $\dfrac{8y - 22}{(y - 2)(y - 4)}$

58. $\dfrac{4x - 3}{(x + 2)(x - 2)}$ 59. $\dfrac{-3t^2 + 21t}{2(t - 3)^2}$ 60. $\dfrac{18 - a^2}{(a + 3)(a + 4)}$

61. $\dfrac{15}{2y - 1}$ 62. $\dfrac{x + 19}{(x + 4)(x - 4)}$ 63. $\dfrac{ay}{bx}$ 64. -1

65. $\dfrac{5(x - 2)}{x(3x + 5)}$ 66. $\dfrac{x}{x - 3}$ 67. $\dfrac{x}{5(2x - 1)}$ 68. 1

69. $\dfrac{y^2 + x^2}{7xy}$ 70. $-h$ 71. $\dfrac{38}{13}$ 72. 6 73. 1 74. 4

75. 3, 4 76. -1 77. -8, (4 is extraneous) 78. 23
79. 42 80. 200 81. 6 82. 9 83. 32 84. 9 85. 9 gal.

86. $\dfrac{7}{3}$ cm/sec 87. $100.48\ \text{in.}^3$

Chapter 7 Practice Test 1. 7 [7.1] 2. $-4, 4$ [7.1]

3. $-\dfrac{3}{x - 4}$ [7.1] 4. $\dfrac{1}{x + y}$ [7.1] 5. $\dfrac{6f}{f^2g^3}, \dfrac{2g^3}{f^2g^3}$ [7.4]

6. $\dfrac{5x(x + 3)}{(x + 3)^2(x - 3)}, \dfrac{2(x - 3)}{(x + 3)^2(x - 3)}$ [7.4] 7. $\dfrac{5ab^2x^2}{12}$ [7.2]

8. $-\dfrac{7a^6b^7y^2}{40x^2}$ [7.2] 9. $-4x$ [7.2] 10. $\dfrac{12x^2}{(x + 1)(2x + 1)}$ [7.2]

11. $\dfrac{7x}{2x + 3}$ [7.3] 12. $\dfrac{1}{x + 3}$ [7.3] 13. $\dfrac{x^2 + 5x - 6}{x(x + 2)(x - 2)}$ [7.4]

14. $\dfrac{9x + 3}{(x - 1)(x + 2)}$ [7.4] 15. $\dfrac{3x + 13}{(x - 4)(x + 1)}$ [7.4]

16. $\dfrac{x - 2}{2x}$ [7.4] 17. $\dfrac{2y + 1}{3y - 1}$ [7.5] 18. $\dfrac{2}{x + y}$ [7.5]

19. $\dfrac{25}{9}$ [7.6] 20. -1 [7.6] 21. -1, (2 is extraneous) [7.6]

22. 2, 4 [7.6] 23. 1.2 hr. [7.7] 24. 6 mph, 4 mph [7.7]
25. 529.2 N [7.7]

Chapters 1–7 Cumulative Review Exercises
1. false 2. false 3. false 4. denominator 5. There are no common factors in the numerator and denominator besides 1. 6. multiply, add 7. complex rational expression 8. 21 9. 6 10. $-3a^{13}$
11. $m^3 - 3m^2 - 17m + 3$ 12. $(a - 2)(x + 1)$
13. $3(y - 5)(y + 4)$ 14. $-3(h + 5)(h - 5)$
15. 9 16. $2(m + 4)$ 17. $\dfrac{1}{x + 3}$ 18. $\dfrac{6x + 14}{x(x + 7)(x - 7)}$

19. $\dfrac{2(y + 3)}{y}$ 20. 9 21. a. $\dfrac{2}{3}$ b.

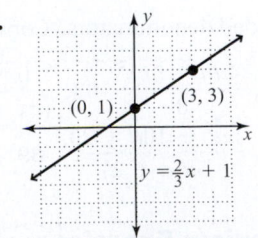

22. $-\dfrac{1}{2}, 7$ 23. $h = \dfrac{3V}{\pi r^2}$ 24. 2, 6 25. 6 26. 2.5 hr.

27. \$5500 at 9% and \$1500 at 7% 28. 6 oz. 29. $3\dfrac{3}{7}$ hr. 30. 6

Chapter 8

8.1 Exercises **1.** Two inequalities joined by either *and* or *or*. **3.** For two sets A and B, the intersection of A and B, symbolized by $A \cap B$, is a set containing only elements that are in both A and B. **5.** We graph the region of overlap of the two inequalities. **7. a.** $\{1, 3, 5\}$ **b.** $\{1, 3, 5, 7, 9\}$ **9. a.** $\{7, 8\}$ **b.** $\{5, 6, 7, 8, 9\}$ **11. a.** $\varnothing$ **b.** $\{a, c, d, g, o, t\}$ **13. a.** $\{x, y, z\}$ **b.** $\{w, x, y, z\}$ **15.** $-4 < x < 5$ **17.** $0 < y \leq 2$ **19.** $-7 < w < 3$ **21.** $0 \leq u \leq 2$

23.

25.

27.

29.

31. a.

b. $\{x | -3 < x < -1\}$ **c.** $(-3, -1)$

33. a.

b. $\{x | 3 < x \leq 6\}$ **c.** $(3, 6]$

35. a.

b. $\{ \}$ or $\varnothing$ **c.** no interval notation

37. a.

b. $\{x | -5 \leq x < 3\}$ **c.** $[-5, 3)$

39. a.

b. $\{ \}$ or $\varnothing$ **c.** no interval notation

41. a.

b. $\{x | -7 < x < -3\}$ **c.** $(-7, -3)$

43. a.

b. $\{x | -1 \leq x \leq 2\}$ **c.** $[-1, 2]$

45. a.

b. $\left\{ x \left| -\dfrac{2}{3} \leq x < 2 \right. \right\}$ **c.** $\left[-\dfrac{2}{3}, 2 \right)$

47. a.

b. $\{x | 0 \leq x < 3\}$ **c.** $[0, 3)$

49. a.

b. $\{x | 0 \leq x \leq 3\}$ **c.** $[0, 3]$

51.

53.

55.

57.

59. a.

b. $\{y | y < -9 \text{ or } y > 5\}$ **c.** $(-\infty, -9) \cup (5, \infty)$

61. a.

b. $\{r | r < -2 \text{ or } r > 1\}$ **c.** $(-\infty, -2) \cup (1, \infty)$

63. a.

b. $\{w | w \leq -1 \text{ or } w \geq 7\}$ **c.** $(-\infty, -1] \cup [7, \infty)$

65. a.

b. $\{k | k \leq 2 \text{ or } k \geq 3\}$ **c.** $(-\infty, 2] \cup [3, \infty)$

67. a.

b. $\{x | x \leq -5 \text{ or } x \geq 3\}$ **c.** $(-\infty, -5] \cup [3, \infty)$

69. a.

b. $\{c | c > 3\}$ **c.** $(3, \infty)$

71. a.

b. $\{x | x \leq -4\}$ **c.** $(-\infty, -4]$

73. a.

b. $\{x | x \text{ is a real number}\}$, or $\mathbb{R}$ **c.** $(-\infty, \infty)$

75. a.

b. $\{x | -7 < x < 3\}$ **c.** $(-7, 3)$

77. a.

b. $\{x | x < -2 \text{ or } x > 2\}$ **c.** $(-\infty, -2) \cup (2, \infty)$

79. a.

b. $\{x | -3 < x < -2\}$ **c.** $(-3, -2)$

81. a.

b. $\{x | 3 \leq x \leq 6\}$ **c.** $[3, 6]$

83. a.

b. $\{x | 1 < x < 3\}$ **c.** $(1, 3)$

85. a.

b. $\{x | x > 1\}$ **c.** $(1, \infty)$

87. a.

b. $\{x | 400 \leq x \leq 500\}$ **c.** $[400, 500]$

89. a.

b. $\{x | 55 \leq x < 105\}$ **c.** $[55, 105)$

91. a.

b. $\{x | 68° \leq x \leq 78°\}$ **c.** $[68, 78]$

93. a.

b. $\{x | 72° < x < 80°\}$ **c.** $(72, 80)$

95. a.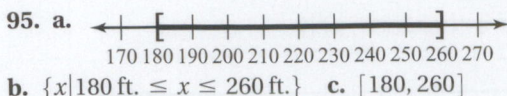
170 180 190 200 210 220 230 240 250 260 270

b. $\{x\,|\,180\,\text{ft.} \leq x \leq 260\,\text{ft.}\}$ **c.** $[180, 260]$

Review Exercises **1.** No, the absolute value of zero is zero and zero is neither negative nor positive. **2.** -11

3. -16 **4.** 6 **5.** $\dfrac{1}{5}$ **6.** -2

8.2 Exercises **1.** The distance a number is from zero.
3. If $|n| = a$, where n is a variable or an expression and $a \geq 0$, then $n = a$ or $n = -a$. **5.** Separate the absolute value equation into two equations: $ax + b = cx + d$ and $ax + b = -(cx + d)$. **7.** $-2, 2$ **9.** No solution. **11.** $-11, 5$

13. $2, 3$ **15.** $1, \dfrac{7}{5}$ **17.** $-\dfrac{2}{3}, \dfrac{10}{3}$ **19.** No solution. **21.** $\dfrac{3}{4}$

23. $-4, 4$ **25.** $1, -3$ **27.** $12, -4$ **29.** $-\dfrac{3}{5}, 1$ **31.** $-4, 1$

33. $-2, 6$ **35.** $-\dfrac{9}{2}, \dfrac{15}{2}$ **37.** $0, 20$ **39.** $-2, 4$ **41.** $\dfrac{1}{3}, 7$

43. $-5, -\dfrac{3}{5}$ **45.** All real numbers. **47.** 1 **49.** $3, -13$

51. $-6, 10$ **53.** $\dfrac{5}{6}, \dfrac{11}{6}$ **55.** $-\dfrac{17}{4}, \dfrac{11}{4}$

Review Exercises **1.** $(-2, 0]$
2.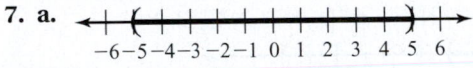
$-6\,-5\,-4\,-3\,-2\,-1\ 0\ 1\ 2\ 3\ 4\ 5\ 6$
3. $x > -7$

4. $x \geq -3$ **5.** $-2 < x < \dfrac{14}{3}$ **6.** $n - 2 < 5; n < 7$

8.3 Exercises **1.** $x \geq -a$ and $x \leq a$ or $-a \leq x \leq a$
3. Shade between the values of $-a$ and a. **5.** when $a < 0$
7. a.

$-6\,-5\,-4\,-3\,-2\,-1\ 0\ 1\ 2\ 3\ 4\ 5\ 6$

b. $\{x\,|\,-5 < x < 5\}$ **c.** $(-5, 5)$
9. a.

$-12\ -10\ -8\ -6\ -4\ -2\ 0\ 2\ 4$

b. $\{x\,|\,-10 \leq x \leq 4\}$ **c.** $[-10, 4]$
11. a.

$-6\,-5\,-4\,-3\,-2\,-1\ 0\ 1\ 2\ 3\ 4\ 5\ 6$

b. $\{s\,|\,-6 < s < 0\}$ **c.** $(-6, 0)$
13. a.

$-3\,-2\,-1\ 0\ 1\ 2\ 3\ 4\ 5\ 6\ 7\ 8$

b. $\{m\,|\,-2 < m < 7\}$ **c.** $(-2, 7)$
15. a.

$\dfrac{4}{3}$

$0 \qquad 1 \qquad 2 \qquad 3$

b. $\left\{k\,\middle|\,\dfrac{4}{3} \leq k \leq 2\right\}$ **c.** $\left[\dfrac{4}{3}, 2\right]$
17. a.

$-6\,-5\,-4\,-3\,-2\,-1\ 0\ 1\ 2\ 3\ 4\ 5\ 6$

b. $\{\ \}$ or $\varnothing$ **c.** no interval notation
19. a.

$-6\,-5\,-4\,-3\,-2\,-1\ 0\ 1\ 2\ 3\ 4\ 5\ 6$

b. $\{w\,|\,0 < w < 6\}$ **c.** $(0, 6)$
21. a.

$-12\ -10\ -8\ -6\ -4\ -2\ 0\ 2\ 4\ 6\ 8\ 10\ 12$

b. $\{c\,|\,c < -12 \text{ or } c > 12\}$ **c.** $(-\infty, -12) \cup (12, \infty)$

23. a.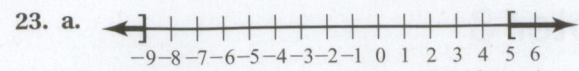
$-9\,-8\,-7\,-6\,-5\,-4\,-3\,-2\,-1\ 0\ 1\ 2\ 3\ 4\ 5\ 6$
b. $\{y\,|\,y \leq -9 \text{ or } y \geq 5\}$ **c.** $(-\infty, -9] \cup [5, \infty)$
25. a.

$-4\,-2\ 0\ 2\ 4\ 6\ 8\ 10\ 12\ 14\ 16\ 18\ 20$

b. $\{p\,|\,p < -2 \text{ or } p > 14\}$ **c.** $(-\infty, -2) \cup (14, \infty)$
27. a.

$-6\,-5\,-4\,-3\,-2\,-1\ 0\ 1\ 2\ 3\ 4$

b. $\{x\,|\,x \leq -6 \text{ or } x \geq 2\}$ **c.** $(-\infty, -6] \cup [2, \infty)$
29. a.

$-\dfrac{5}{2}$

$-4 \qquad -3 \qquad -2 \qquad -1 \qquad 0 \qquad 1 \qquad 2$

b. $\left\{n\,\middle|\,n < -\dfrac{5}{2} \text{ or } n > 0\right\}$ **c.** $\left(-\infty, -\dfrac{5}{2}\right) \cup (0, \infty)$
31. a.

$-6\,-5\,-4\,-3\,-2\,-1\ 0\ 1\ 2\ 3\ 4\ 5\ 6$

b. $\{v\,|\,v \leq -1 \text{ or } v \geq 1\}$ **c.** $(-\infty, -1] \cup [1, \infty)$
33. a.

$-6\,-5\,-4\,-3\,-2\,-1\ 0\ 1\ 2\ 3\ 4\ 5\ 6$

b. $\{y\,|\,y < -3 \text{ or } y > -1\}$ **c.** $(-\infty, -3) \cup (-1, \infty)$
35. a.

$-6\,-5\,-4\,-3\,-2\,-1\ 0\ 1\ 2\ 3\ 4\ 5\ 6$

b. $\{m\,|\,m < -5 \text{ or } m > 1\}$ **c.** $(-\infty, -5) \cup (1, \infty)$
37. a.

$-6\,-5\,-4\,-3\,-2\,-1\ 0\ 1\ 2\ 3\ 4\ 5\ 6$

b. $\{x\,|\,0 < x < 4\}$ **c.** $(0, 4)$
39. a.

$-6\,-5\,-4\,-3\,-2\,-1\ 0\ 1\ 2\ 3\ 4\ 5\ 6$

b. $\{r\,|\,r \text{ is a real number}\}$ **c.** $(-\infty, \infty)$
41. a.

$-6\,-5\,-4\,-3\,-2\,-1\ 0\ 1\ 2\ 3\ 4\ 5\ 6$

b. $\{x\,|\,-4 < x < -2\}$ **c.** $(-4, -2)$
43. a.

$-6\,-5\,-4\,-3\,-2\,-1\ 0\ 1\ 2\ 3\ 4\ 5\ 6$

b. $\{x\,|\,0 < x < 1\}$ **c.** $(0, 1)$
45. a.

$-14\,-12\,-10\,-8\,-6\,-4\,-2\ 0\ 2\ 4\ 6\ 8\ 10$

b. $\varnothing$ **c.** no interval notation
47. a.

$\dfrac{14}{3}$

$-6\,-5\,-4\,-3\,-2\,-1\ 0\ 1\ 2\ 3\ 4\ 5\ 6$

b. $\left\{k\,\middle|\,-2 \leq k \leq \dfrac{14}{3}\right\}$ **c.** $\left[-2, \dfrac{14}{3}\right]$
49. a.

$0 \quad 4 \quad 8 \quad 12 \quad 16 \quad 20 \quad 24 \quad 28$

b. $\{x\,|\,x < 4 \text{ or } x > 20\}$ **c.** $(-\infty, 4) \cup (20, \infty)$
51. a.

$-6.4 \qquad\qquad\qquad 12.8$

$-8\,-6\,-4\,-2\ 0\ 2\ 4\ 6\ 8\ 10\ 12\ 14$

b. $\{y\,|\,-6.4 \leq y \leq 12.8\}$ **c.** $[-6.4, 12.8]$
53. a.

$-5\,-4\,-3\,-2\,-1\ 0\ 1\ 2\ 3\ 4\ 5$

b. $\{p\,|\,p \text{ is a real number}\}$ **c.** $(-\infty, \infty)$
55. $|x| < 3$ **57.** $|x| \geq 4$ **59.** $|x + 1| > 1$ **61.** $|x - 3| \leq 2$
63. $|x| > $ any negative number

Review Exercises **1.** -17 **2.** -5 **3.** -5 **4.** 7
5. $m = -\dfrac{3}{4}, (0, 2)$ **6.** $m = \dfrac{3}{5}, (0, -2)$

8.4 Exercises

1. A function is a relation in which every value in the domain is paired with exactly one value in the range.   **3.** *Domain*: The set of all input values for a relation. *Range*: The set of all output values for a relation.   **5.** The graph of every quadratic function is a parabola. Find many ordered pairs, plot them, and then draw a smooth U-shaped curve through those points.   **7.** Domain: $\{1, 2, 3, 4, 5\}$; Range: $\{$Hawaii, Las Vegas, Florida, California, Texas$\}$; It is a function.   **9.** Domain: $\{2\%, 3\%, 5\%, 14\%, 15\%, 26\%, 33\%\}$;

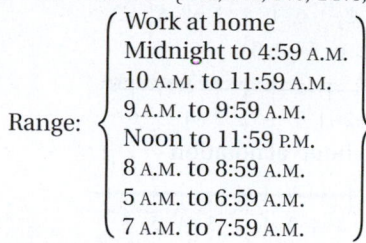

Range:
$$\left\{\begin{array}{l}\text{Work at home}\\ \text{Midnight to 4:59 A.M.}\\ \text{10 A.M. to 11:59 A.M.}\\ \text{9 A.M. to 9:59 A.M.}\\ \text{Noon to 11:59 P.M.}\\ \text{8 A.M. to 8:59 A.M.}\\ \text{5 A.M. to 6:59 A.M.}\\ \text{7 A.M. to 7:59 A.M.}\end{array}\right\};$$

It is not a function.
11. Domain: $\{-3, 0, 1, 2\}$; Range: $\{1, 2, 5, 8\}$; It is a function.
13. Domain: $\{-3, 2, 4\}$; Range: $\{1, 2, 5, 8\}$; It is not a function.
15. Domain: $\{$WWI, WWII, Korea, Vietnam, Gulf War$\}$; Range: $\{124, 464, 131, 246, 0\}$; It is a function.   **17.** Domain: all real numbers; Range: all real numbers; It is a function.
19. Domain: all real numbers; Range: $\{y|y \geq -4\}$; It is a function.   **21.** Domain: $\{x|-4 \leq x < 4\}$; Range: $\{1, 2, 3\}$; It is a function.   **23.** Domain: $\{x|x \geq -4\}$; Range: all real numbers; It is not a function.   **25.** 15   **27.** $\frac{2}{9}$ **29.** 3   **31.** $\frac{5}{3}$
33. 3   **35.** 2.928   **37.** undefined   **39.** 6   **41. a.** 3   **b.** 4   **c.** 3   **43. a.** 1   **b.** 9   **c.** 4   **45. a.** -3 **b.** 1   **c.** -1
47.

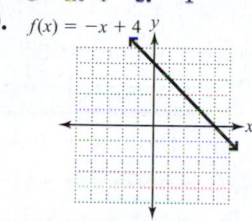

49. $f(x) = -x + 4$

51.

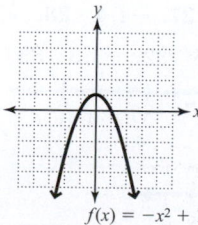

$f(x) = \frac{3}{4}x - 2$

53. $f(x) = x^2 + 4$

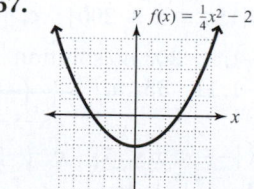

55.

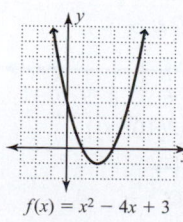

$f(x) = -x^2 + 1$

57. $f(x) = \frac{1}{4}x^2 - 2$

59.

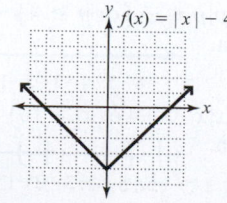

$f(x) = x^2 - 4x + 3$

61. $f(x) = |x| - 4$

63. $f(x) = |x + 3| - 2$

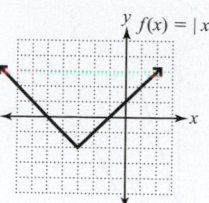

65. $f(x) = 2|x - 1| + 3$

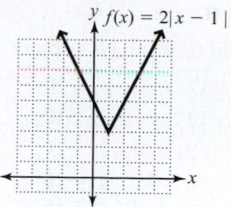

67.

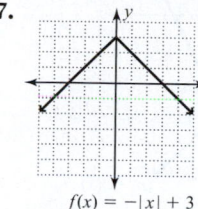

$f(x) = -|x| + 3$

69.

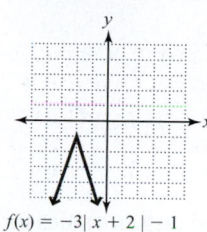

$f(x) = -3|x + 2| - 1$

71.

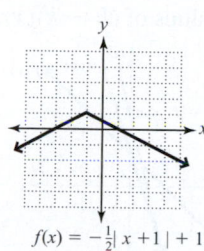

$f(x) = -\frac{1}{2}|x + 1| + 1$

Review Exercises

1. $5x^2 + 6$ **2.** $7x^2 - 3x - 7$
3. $6x^2 - 5x - 4$ **4.** $4x^2 - 16x + 16$ **5.** $2x^2 + 3x - 5$
6. $x^3 + 3x^2 + 9x + 27$

8.5 Exercises

1. $(p + q)(x) = p(x) + q(x)$
3. Use FOIL to multiply the binomials.   **5.** $5x - 1; -x + 1$
7. $3x - 8; -x - 2$ **9.** $x^2 - 3x + 4; x^2 - 5x + 10$
11. $3x^2 - 5x + 2; x^2 - 5x - 8$ **13.** $-2x^2 + 2x + 7;$
$-8x^2 + 6x + 9$ **15.** $2x^2 - 10x$ **17.** $x^2 - x - 2$
19. $3x^2 + 8x + 4$ **21.** $x^4 - x^3 - 11x^2 + 25x - 14$
23. $6x^4 - 11x^3 + 4x^2 + 2x - 1$ **25.** $x - 3, x \neq 0$
27. $2x - 1 + \frac{2}{x}, x \neq 0$ **29.** $x - 5, x \neq 9$ **31.** $x + 1, x \neq \frac{5}{2}$
33. $2x - 15 + \frac{68}{x + 5}, x \neq -5$ **35. a.** $7x + 2$ **b.** $3x - 4$
c. $10x^2 + 13x - 3$ **d.** $\frac{5x - 1}{2x + 3}, x \neq \frac{-3}{2}$
37. a. $2x^2 - 2x + 4$ **b.** $2x^2 - 4x + 6$
c. $2x^3 - 5x^2 + 8x - 5$ **d.** $2x - 1 + \frac{4}{x - 1}, x \neq 1$
39. a. -12 **b.** 11   **c.** -6 **d.** 2   **41. a.** 16   **b.** 54
c. 7   **d.** -8 **43. a.** $4x + 6$
b.

$f(x)$	$g(x)$	$(f + g)(x)$
-11	1	-10
-5	3	-2
1	5	6
7	7	14
13	9	22

c.

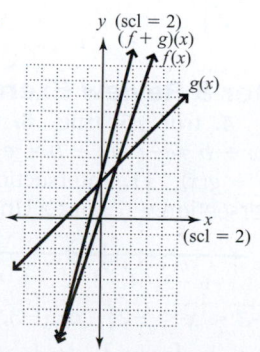

d. The output values for $f(x)$ added to the output values of $g(x)$ are equal to the output values of $(f + g)(x)$.

45. a. $-x - 14$

b.

$h(x)$	$k(x)$	$(h - k)(x)$
-17	-7	-10
-13	-1	-12
-9	5	-14
-5	11	-16
-1	17	-18

c.

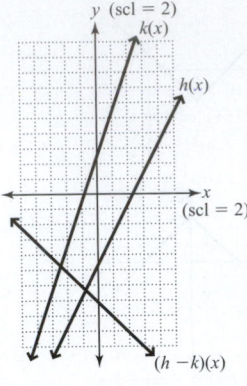

d. The output values for $k(x)$ subtracted from the output values of $h(x)$ are equal to the output values of $(h - k)(x)$.

47. a. $t(x) = 4x + 4$ **b.** \$804   **c.**

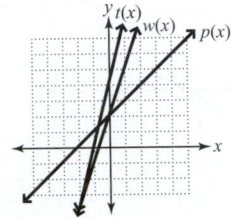

49. a. $p(x) = x^2 - 4x - 12$ **b.** \$9588 profit

c.

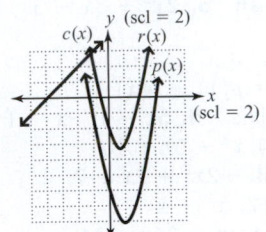

d. 6 units

51. a. $A(x) = 6x^2 + 4x$ **b.** 66 ft.^{2} **53. a.** $h(x) = 4x + 1$
b. 17 cm

Review Exercises
1. $2x + 3y = 3$ **2.** yes   **3.** $-4, 4$
4. no solution

5.

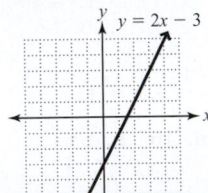

6.

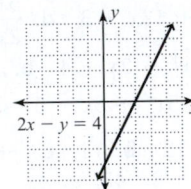

Chapter 8 Review Exercises
1. true   **2.** true
3. false   **4.** true   **5.** true   **6.** true   **7.** ∩
8. 2. $ax + b = -c, ax + b = c$ **9.** the range
10. $f(x) + g(x)$ **11.** Intersection: $\{1, 5, 9\}$; Union: $\{1, 5, 7, 9\}$
12. Intersection: $\{\ \}$ or $\varnothing$; Union: $\{1, 2, 3, 4, 5, 6, 7\}$
13. a. [number line from -6 to 6, open circles at -3 and -2]
b. $\{x \mid -3 < x < -2\}$ **c.** $(-3, -2)$
14. a. [number line from -6 to 6, brackets at -5 and -3]
b. $\{x \mid -5 \le x \le -3\}$ **c.** $[-5, -3]$

15. a. [number line from -8 to 4, parenthesis at -7 and 3]
b. $\{x \mid -7 < x < 3\}$ **c.** $(-7, 3)$
16. a. [number line from -6 to 6, parenthesis at 1 and bracket at 4]
b. $\{x \mid 1 < x \le 4\}$ **c.** $(1, 4]$
17. a. [number line from -6 to 6, bracket at -2 and parenthesis at 0]
b. $\{x \mid -2 \le x < 0\}$ **c.** $[-2, 0)$
18. a. [number line from -6 to 6]
b. $\{\ \}$ or $\varnothing$ **c.** no interval notation
19. a. [number line from -6 to 6, bracket at -6 and bracket at -2]
b. $\{w \mid w \le -6 \text{ or } w \ge -2\}$ **c.** $(-\infty, -6] \cup [-2, \infty)$
20. a. [number line from -6 to 6]
b. $\{x \mid x \text{ is a real number}\}$ **c.** $(-\infty, \infty)$
21. a. [number line from -6 to 6, parenthesis at $\frac{5}{2}$ and parenthesis at 5]
b. $\left\{m \mid m < \dfrac{5}{2} \text{ or } m > 5\right\}$ **c.** $\left(-\infty, \dfrac{5}{2}\right] \cup (5, \infty)$
22. a. [number line from -6 to 6, bracket at $-\frac{4}{3}$ and bracket at 2]
b. $\left\{x \mid x \le -\dfrac{4}{3} \text{ or } x \ge 2\right\}$ **c.** $\left(-\infty, -\dfrac{4}{3}\right] \cup [2, \infty)$
23. a. [number line from -11 to 1, brackets at -9 and -4]
b. $\{x \mid x \le -9 \text{ or } x \ge -4\}$ **c.** $(-\infty, -9] \cup [-4, \infty)$
24. a. [number line from -6 to 6, brackets at -1 and 1]
b. $\{w \mid w \le -1 \text{ or } w \ge 1\}$ **c.** $(-\infty, -1] \cup [1, \infty)$
25. a. [number line from 65 to 110, bracket at 68 and parenthesis at 108]
b. $\{x \mid 68 \le x < 108\}$ **c.** $[68, 108)$
26. a. [number line from 150 to 200, brackets at 150 and 200]
b. $\{x \mid 150 \le x \le 200\}$ **c.** $[150, 200]$ **27.** $-4, 4$ **28.** $-3, 11$
29. $-1, 2$ **30.** no solution   **31.** $-7, 15$ **32.** $-3, 3$ **33.** $0, \dfrac{8}{3}$
34. $-1, 11$ **35. a.** [number line from -6 to 6, parenthesis at -5 and 5]
b. $\{x \mid -5 < x < 5\}$ **c.** $(-5, 5)$
36. a. [number line from -6 to 6, bracket at -4 and bracket at 4]
b. $\{p \mid p \le -4 \text{ or } p \ge 4\}$ **c.** $(-\infty, -4] \cup [4, \infty)$
37. a. [number line from -6 to 11, parenthesis at -4 and 10]
b. $\{x \mid x < -4 \text{ or } x > 10\}$ **c.** $(-\infty, -4) \cup (10, \infty)$
38. a. [number line from -6 to 6, parenthesis at -5 and -1]
b. $\{m \mid -5 < m < -1\}$ **c.** $(-5, -1)$

39. a.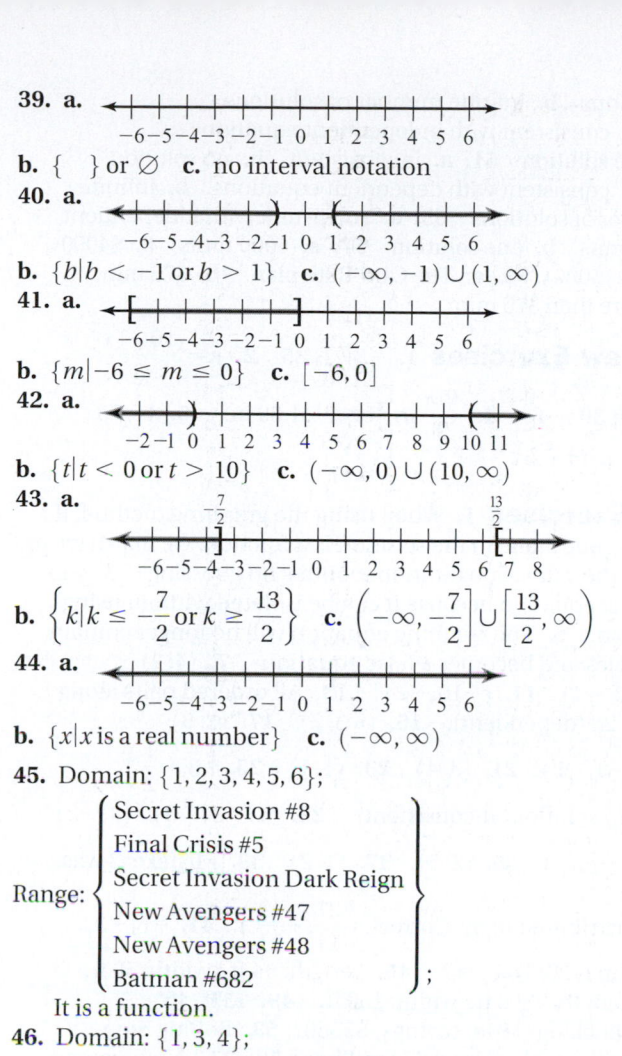

-6-5-4-3-2-1 0 1 2 3 4 5 6

b. { } or ∅   **c.** no interval notation

40. a.

-6-5-4-3-2-1 0 1 2 3 4 5 6

b. $\{b \mid b < -1 \text{ or } b > 1\}$ **c.** $(-\infty, -1) \cup (1, \infty)$

41. a.

-6-5-4-3-2-1 0 1 2 3 4 5 6

b. $\{m \mid -6 \le m \le 0\}$ **c.** $[-6, 0]$

42. a.

-2-1 0 1 2 3 4 5 6 7 8 9 10 11

b. $\{t \mid t < 0 \text{ or } t > 10\}$ **c.** $(-\infty, 0) \cup (10, \infty)$

43. a.

$-\frac{7}{2}$ $\frac{13}{2}$

-6-5-4-3-2-1 0 1 2 3 4 5 6 7 8

b. $\left\{k \mid k \le -\dfrac{7}{2} \text{ or } k \ge \dfrac{13}{2}\right\}$ **c.** $\left(-\infty, -\dfrac{7}{2}\right] \cup \left[\dfrac{13}{2}, \infty\right)$

44. a.

-6-5-4-3-2-1 0 1 2 3 4 5 6

b. $\{x \mid x \text{ is a real number}\}$ **c.** $(-\infty, \infty)$

45. Domain: $\{1, 2, 3, 4, 5, 6\}$;

Range:
$\left\{\begin{array}{l} \text{Secret Invasion \#8} \\ \text{Final Crisis \#5} \\ \text{Secret Invasion Dark Reign} \\ \text{New Avengers \#47} \\ \text{New Avengers \#48} \\ \text{Batman \#682} \end{array}\right\}$;

It is a function.

46. Domain: $\{1, 3, 4\}$;

Range:
$\left\{\begin{array}{l} \text{Harvard University (MA)} \\ \text{Stanford University (CA)} \\ \text{University of Pennsylvania (Wharton)} \\ \text{Massachusetts Institute of Technology (Sloan)} \\ \text{Northwestern University (Kellogg) (IL)} \\ \text{University of Chicago (IL)} \end{array}\right\}$;

It is not a function.

47. Domain: $\{2, 3, 4\}$; Range: $\{-3, 1, 4\}$; It is not a function.
48. Domain: $\{-2, 0, 1\}$; Range: $\{3, 4\}$; It is a function.
49. Domain: all real numbers; Range: $\{y \mid y \ge -4\}$; It is a function.   **50.** Domain: all real numbers; Range: $\{y \mid y \le 4\}$; It is a function.   **51.** 17   **52.** -9

53. $f(x) = -3x + 2$
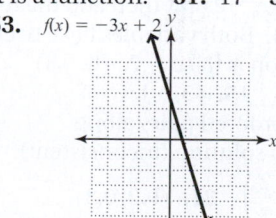

54. $f(x) = 2x^2 - 3$
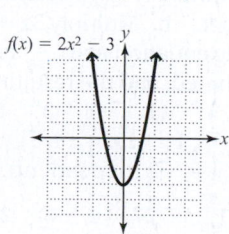

55. $f(x) = |x - 2|$
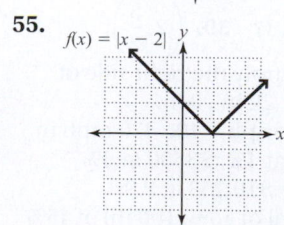

56. $f(x) = -2|x + 1| + 3$
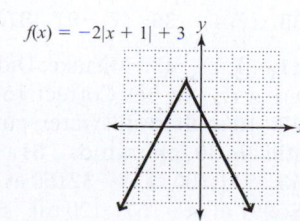

57. a. $x + 10$ **b.** $-7x + 8$ **c.** $-12x^2 + 33x + 9$

58. a. $7x$ **b.** $-9x - 4$ **c.** $-8x^2 - 18x - 4$
59. a. $3x^2 + 7$ **b.** $3x^2 - 2x + 3$ **c.** $3x^3 + 5x^2 + 3x + 10$
60. a. $x^2 + x$ **b.** $x^2 - 5x - 2$ **c.** $3x^3 - 5x^2 - 5x - 1$
61. $2x^2 - 5x + 4, x \neq 0$ **62.** $x^2 - 6x + 3 + \dfrac{-2}{2x - 5}, x \neq \dfrac{5}{2}$
63. a. $c(x) = 4x + 4$ **b.** \$904   **c.**

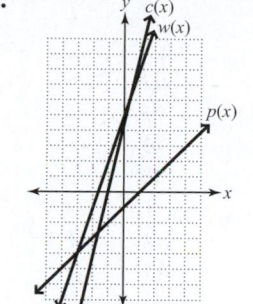

64. a. $h(x) = 2x - 1$ **b.** 17 in.

Chapter 8 Practice Test
1. $\{e, h, o\};$ $\{e, h, m, o, s, u\}$ [8.1]

2. $-8, 2$ [8.2]   **3.** $-3, 6$ [8.2]   **4.** $-8, \dfrac{2}{3}$ [8.2]

5. no solution [8.2]

6. a.

-9-8-7-6-5-4-3-2-1 0 1 2 3 4 5

b. $\{x \mid -7 < x \le 3\}$ **c.** $(-7, 3]$ [8.1]

7. a.

-6-5-4-3-2-1 0 1 2 3 4 5 6

b. $\{x \mid -3 \le x < -2\}$ **c.** $[-3, -2)$ [8.1]

8. a.

-3 -2 -1 0 1 2 3 4

b. $\{x \mid x \le -1 \text{ or } x \ge 2\}$ **c.** $(-\infty, -1] \cup [2, \infty)$ [8.1]

9. a.

-1 0 1 2 3 4 5

b. $\{n \mid n > 1\}$ **c.** $(1, \infty)$ [8.1]

10. a.

-13 5

-14-12-10-8 -6 -4 -2 0 2 4 6 8

b. $\{x \mid -13 < x < 5\}$ **c.** $(-13, 5)$ [8.3]

11. a.

-6-5-4-3-2-1 0 1 2 3 4 5 6

b. $\{x \mid x < -1 \text{ or } x > 3\}$ **c.** $(-\infty, -1) \cup (3, \infty)$ [8.3]

12. a.

-8-7-6-5-4-3-2-1 0 1 2 3 4 5 6

b. $\{x \mid -7 < x < -1\}$ **c.** $(-7, -1)$ [8.3]

13. a.

-6-5-4-3-2-1 0 1 2 3 4 5 6

b. { } or ∅   **c.** no interval notation [8.3]

14. a.

-6-5-4-3-2-1 0 1 2 3 4 5 6

b. $\{x \mid x \text{ is a real number}\}$ **c.** $(-\infty, \infty)$ [8.3]

15. a.

$-\frac{1}{3}$

-6-5-4-3-2-1 0 1 2 3 4 5 6

b. $\left\{x \mid -\dfrac{1}{3} \le x \le 3\right\}$ **c.** $\left[-\dfrac{1}{3}, 3\right]$ [8.3]

16. a.

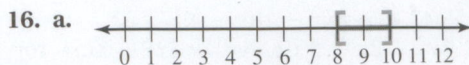

b. $\{x|8 \leq x \leq 10\}$ **c.** $[8, 10]$ [8.1]
17. Domain: all real numbers, Range: $\{y|y \leq 0\}$; It is a function. [8.4] **18.** 18 [8.4] **19.** $f(x) = -\frac{3}{4}x + 1$

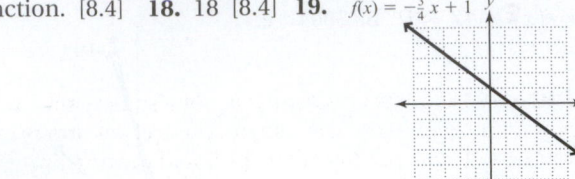

[8.4]

20.

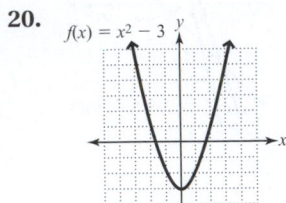

[8.4]

21.

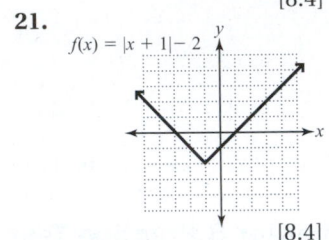

[8.4]

22. a. $7x - 7$ **b.** $5x - 11$ **c.** $6x^2 + 3x - 18$ [8.5]
23. a. $4x^2 - 2$ **b.** $-2x^2 + 2$ **c.** $3x^4 - 2x^2$ [8.5]
24. $5x^2 + 9x - 7$ [8.5] **25. a.** $h(x) = 9x + 2$ **b.** 137 cm [8.5]

Chapters 1–8 Cumulative Review Exercises

1. false **2.** true **3.** true **4.** true **5.** least common multiple (LCM) **6.** $24 - x$ **7.** vertical line **8.** 19
9. $9x^2 - 6xy + y^2$ **10.** $-\dfrac{128x^{11}}{y^{10}z^6}$ **11.** $\dfrac{1}{5}x - 1 + \dfrac{2}{x}$
12. $\dfrac{a - b}{x + y}$ **13.** $\dfrac{2x}{x + 4}$ **14.** $\dfrac{x + 3}{x + 4}$ **15.** $(4c - 3)(3b + 5)$
16. $2x(x - 5)(x - 2)$ **17.** 1 **18.** 5 **19.** $-\dfrac{5}{3}, 4$
20. 3, (2 is extraneous) **21.** $-5 < x \leq -2$
22. $x \leq -\dfrac{6}{5}$ or $x \geq 2$

23.

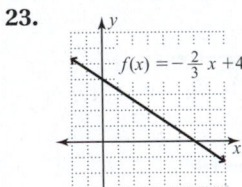

24.

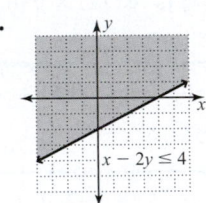

25. 36% **26. a.** -2 **b.** 3 **c.** undefined **27.** 75 ml
28. $1\dfrac{1}{5}$ hr. **29.** 9 **30.** 1 sec., 5 sec.

Chapter 9

9.1 Exercises 1. a. Replace each variable in each equation with its corresponding value. **b.** Verify that each equation is true. **3.** The graphs are parallel. **5.** Yes, when the graphs are identical. **7.** yes **9.** no **11.** yes **13.** no
15. yes **17.** $(4, 1)$ **19.** $(-3, 4)$ **21.** $(3, -2)$ **23.** $(-1, 3)$
25. $(3, -4)$ **27.** $(4, -3)$ **29.** $(2, -3)$ **31.** No solution.
33. All ordered pairs along $2x + 3y = 2$. **35.** $(2, 6)$
37. $(-2, 2)$ **39.** $(-2, 3)$ **41.** $(5, -2)$ **43. a.** inconsistent
b. no solution **45. a.** consistent with independent equations **b.** one solution **47. a.** consistent with dependent

equations **b.** infinite number of solutions
49. a. consistent with independent equations
b. one solution **51. a.** inconsistent **b.** no solution
53. a. consistent with dependent equations **b.** infinite number of solutions **55. a.** consistent with independent equations **b.** one solution **57. a.** 5000 units **b.** \$4000
c. $u > 5000$ **59. a.** The One-Rate plan. **b.** 375 min.
c. More than 375 min.

Review Exercises **1.** $-3y - 35$ **2.** $y = \dfrac{10 - x}{3}$
3. $x = 30 + 6y$ **4.** $\dfrac{40\pi}{3}$ in.^{3} or ≈ 41.89 in.^{3} **5.** 11, 73
6. $2x + 17$

9.2 Exercises 1. When using the graphing method, if either coordinate in the solution is a fraction, we may have to guess the value. Substitution requires no guessing. **3.** y in the first equation because it can be isolated without using division. **5.** The resulting equation will no longer contain variables and becomes a false equation. **7.** $(4, 2)$
9. $(12, -4)$ **11.** $(-10, -24)$ **13.** All ordered pairs along $y = -2x$ (dependent). **15.** $(-3, 2)$ **17.** $(2, 9)$
19. $(-1, -3)$ **21.** $(3, 4)$ **23.** $(1, 1)$ **25.** $\left(3, -\dfrac{1}{2}\right)$
27. No solution (inconsistent). **29.** $(1, -2)$ **31.** $(4, -2)$
33. $\left(-\dfrac{2}{5}, \dfrac{2}{3}\right)$ **35.** $(2, 5)$ **37.** $(1, 2)$ **39.** Mistake: 3 was not distributed to $3y$. Correct: $\left(\dfrac{27}{11}, -\dfrac{13}{11}\right)$ **41.** $-6, -2$
43. Jon is 20; Tony is 7. **45.** Length: 94 ft.; width: 50 ft.
47. Length: 2300 ft.; width: 150 ft. **49.** $135°, 45°$
51. checking: \$840; savings: \$3360 **53.** Rolling Stones: \$88 million; Madonna: \$72 million **55.** 27 3-hour courses, 15 5-hour courses **57.** 4:45 P.M. **59.** 1:33 P.M.

Review Exercises **1.** $5x - 4y$ **2.** $-3x + 21y$ **3.** $9x$
4. $13y$ **5.** 80 **6.** $-\dfrac{3}{2}$

9.3 Exercises 1. The elimination method is advantageous over graphing when the solution involves fractions. The elimination method is advantageous over substitution when no coefficients are 1. **3. a.** x because multiplying only the second equation by -2 allows the x to be eliminated; whereas to eliminate y, we need to multiply each equation by a number. **b.** Multiply $3x + 5y = 7$ by -2 and then add the resulting equation to $6x - 2y = 1$. **5.** Both variables have been eliminated, and the resulting equation is false. **7.** $(6, -8)$
9. $(2, 3)$ **11.** $(-1, -2)$ **13.** $(2, 3)$ **15.** $(3, -1)$
17. $(-4, 1)$ **19.** $(-4, 3)$ **21.** All ordered pairs along $2x + y = -2$ (dependent). **23.** No solution (inconsistent).
25. $(1, 1)$ **27.** $(4, -4)$ **29.** $\left(-\dfrac{2}{5}, \dfrac{2}{3}\right)$ **31.** $(0.2, 0.3)$
33. $(6, 4)$ **35.** $(7, -9)$ **37.** $(-4, -13)$ **39.** $\left(2, \dfrac{5}{2}\right)$
41. $(4, 4)$ **43.** Mistake: Did not multiply the right side of $x - y = 1$ by -9. Correct: $(5, 4)$ **45.** 22 in., 8 in.
47. 10 mph in still water; current is 2 mph **49.** 470 mph in still air, 30 mph winds **51.** \$12,000 at 4%, \$3500 at 3%
53. \$11,100 at 5%, \$2150 at 8% **55.** \$10,253 at 2.9%, \$5997 at 6% **57.** 120 ml **59.** 200 ml of 15%, 100 ml of 45%
61. 11 lb. of Italian Roast, 9 lb. of GCB

Review Exercises 1. $20x + 28y - 4z$ 2. $x - 3y$
3. $y = -\dfrac{1}{3}x + 2$ 4. $x = 6y + 30$ 5. $x = 2$ 6. $y = 0$

9.4 Exercises
1. It doesn't matter which equations are chosen as long as the second pair is different from the first pair. 3. Answers may vary. A good choice is to eliminate z using Equations 1 and 2 and 1 and 3. Multiplying Equation 1 by -3 and then adding this result to Equations 2 and 3 takes very few steps. 5. Two parallel planes intersecting a third plane.

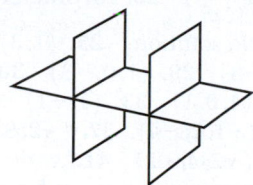

7. yes 9. no 11. yes 13. $(2, -1, 4)$ 15. $(-1, 0, 3)$
17. $(-1, 1, 3)$ 19. No solution (inconsistent). 21. Infinite number of solutions (dependent equations). 23. $(2, -3, 1)$
25. $\left(-2, -\dfrac{1}{2}, 1\right)$ 27. $(2, -1, 1)$ 29. $(-3, 7, -4)$
31. $(2, -4, 1)$ 33. $(4, -4, 2)$ 35. $105°, 35°, 40°$
37. Burger: \$2.50; fries: \$1.50; drink: \$1.00. 39. Three 3-point field goals, one 2-point field goal, three free throws.
41. Ham: 3 lb.; turkey: 5 lb.; beef: 2 lb.
43. \$2000 at 4%, \$2500 at 6%, \$3500 at 7%.
45. Bicycling: 420; walking: 300; climbing stairs: 600.

Review Exercises 1. $-6x + 12y + 3z - 27$ 2. $y = -4$
3. -1 4. $(0, 1)$ 5. $m = -2$ 6.

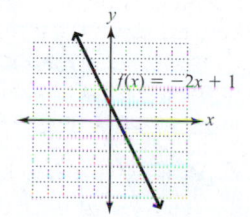
$f(x) = -2x + 1$

9.5 Exercises
1. 4 rows, 2 columns 3. The dotted line corresponds to the equal signs in the equations. 5. A matrix is in row echelon form when the coefficient portion of the augmented matrix has 1s on the diagonal from upper left to lower right and 0s below the 1s. 7. $\begin{bmatrix} 14 & 7 & | & 6 \\ 7 & 6 & | & 8 \end{bmatrix}$ 9. $\begin{bmatrix} 7 & -6 & | & 1 \\ 0 & -2 & | & 5 \end{bmatrix}$

11. $\begin{bmatrix} 1 & -3 & 1 & | & 4 \\ 2 & -4 & 2 & | & -4 \\ 6 & -2 & 5 & | & -4 \end{bmatrix}$ 13. $\begin{bmatrix} 4 & 6 & -2 & | & -1 \\ 8 & 3 & 0 & | & -12 \\ 0 & -1 & 2 & | & 4 \end{bmatrix}$

15. $(4, 2)$ 17. $(-6, -5, 1)$ 19. $\begin{bmatrix} 1 & -3 & | & -1 \\ 0 & 11 & | & 4 \end{bmatrix}$

21. $\begin{bmatrix} 1 & -2 & 4 & | & 6 \\ 0 & 2 & -1 & | & -5 \\ 0 & 0 & -2 & | & 17 \end{bmatrix}$ 23. $\begin{bmatrix} 1 & 2 & | & -2.5 \\ -1 & 3 & | & 2 \end{bmatrix}$

25. Replace R_2 with $3R_1 + R_2$. 27. Replace R_3 with $-2R_2 + R_3$. 29. $(2, -3)$ 31. $(1, 2)$ 33. $(-3, -2)$
35. $(-2, 0)$ 37. $(4, 6)$ 39. $(4, -4)$ 41. $(-1, 1, 3)$
43. $(2, 1, -2)$ 45. $(2, -3, 1)$ 47. $(-1, -1, -2)$
49. $(-3, 1, 5)$ 51. $(8, -2, -4)$ 53. $(3, -4)$ 55. $(-6, 8)$
57. $(4, -3, 5)$ 59. $(7, -6, 5)$ 61. Mistake: In the second step, $4R_1 + R_2$ is calculated incorrectly. Correct: The correct calculation is $\begin{bmatrix} 1 & 3 & | & 13 \\ 0 & 11 & | & 26 \end{bmatrix}$. The solution is $(65/11, 26/11)$.

63. Chicken sandwich: \$3.50; drink: \$1.20. 65. Nile: 4150 mi.; Amazon: 3900 mi. 67. \$6400 in CD, \$3600 in money market 69. CD: \$16; book: \$12, DVD: \$28. 71. 2 touchdowns, 2 extra points, 1 field goal 73. $v_1 = 9, v_2 = 6, v_3 = 2$

Review Exercises 1. 8 2. 11 3. 2 4. 2
5. 3 6. -1

9.6 Exercises
1. No, because it is not a square matrix. 3. Eliminate the elements in the same row and column as the element and evaluate the determinant of the remaining matrix. 5. 13 7. -22 9. 24 11. -7 13. -22
15. 15 17. 1 19. -12 21. -14 23. -64 25. 0.47
27. 14.451 29. $\dfrac{13}{30}$ 31. $-\dfrac{317}{2400}$ 33. $-x - 8y - 2$
35. $(-4, -1)$ 37. $(3, 4)$ 39. $(4, -4)$ 41. No solution.
43. $(2, 2)$ 45. $(2, 4)$ 47. $(2, 6)$ 49. $(2, 1, 3)$
51. $(-2, 1, -1)$ 53. $(2, -1, 1)$ 55. $\left(\dfrac{2}{3}, -7, \dfrac{1}{2}\right)$
57. Infinite solutions. 59. $(2, 2, -2)$ 61. $x = -4$
63. $x = -5$ 65. 9 sq. units 67. 21.5 sq. units
69. 19 sq. units 71. Hoba is 60 tons; Ahnighito is 30 tons.
73. 6 lb. of garbanzo, 4 lb. of black turtle 75. 1% zinc, 4% tin, 95% copper

Review Exercises
1. $x < 3$;

2. $x \geq 1$; 3. 19, 20
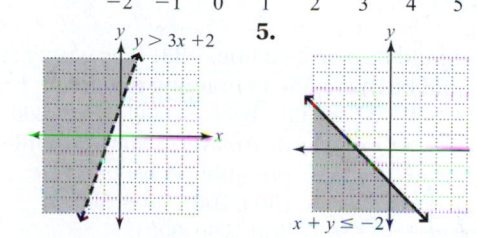
4. 5. 6. $(-2, 5)$
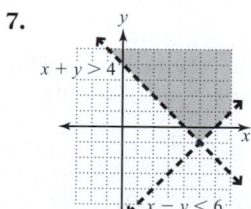
$y > 3x + 2$
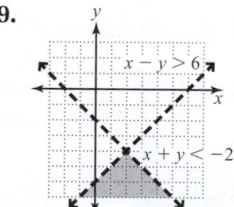
$x + y \leq -2$

9.7 Exercises
1. Select a point in the solution region and verify that it makes both inequalities true. 3. Parallel lines with shaded regions that do not overlap. 5. $\begin{cases} x < 0 \\ y < 0 \end{cases}$

7. 9.
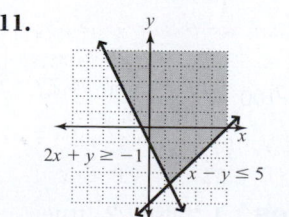
$x + y > 4$
$x - y < 6$
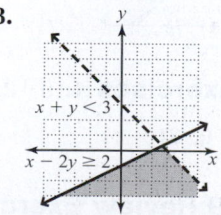
$x - y > 6$
$x + y < -2$

11. 13.

15.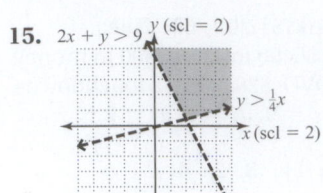

17.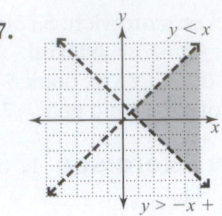

19.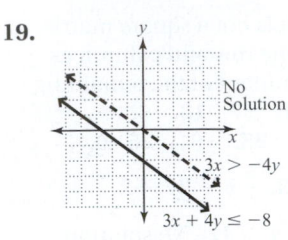

21.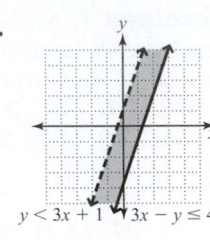

23.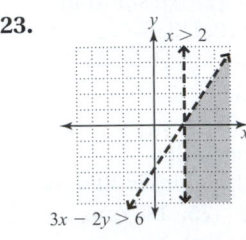

25.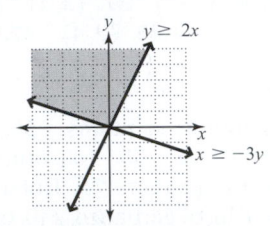

27.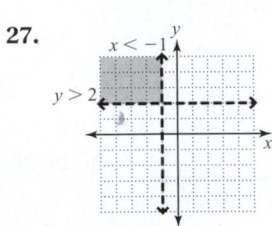

29.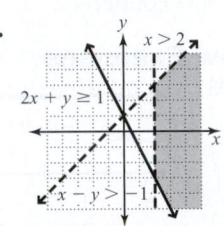

31. $x + y = 3$ should be a dashed line. **33.** The wrong area is shaded. It should be the region containing the point $(1, 5)$.

35. a. $V + M \geq 100$ $M \geq 500$ **b.** $V \leq 800$ $M \leq 800$

c.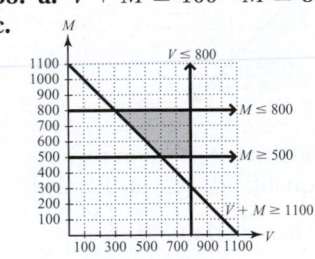

d. Answers will vary. Some possible (V, M) pairs are $(500, 700)$, $(600, 500)$, and $(700, 600)$.

37. a. $\begin{cases} R + D \geq 100 \\ 15.95R + 20.95D \geq 1700 \end{cases}$

b.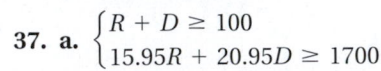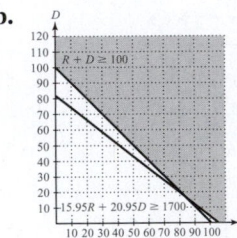

c. Answers will vary. Some possible (R, D) pairs are $(30, 90)$, $(40, 80)$, and $(50, 60)$.

Review Exercises 1. 9   **2.** -100 **3.** 5

4. $\dfrac{4}{7}$ **5.** 4   **6.** 21

Chapter 9 Review Exercises   **1.** false   **2.** true
3. true   **4.** true   **5.** false   **6.** Replace   **7.** slope, y-intercepts   **8.** the intersection (overlap) of the two

solution sets   **9.** the coefficient portion of the augmented matrix has 1s on the diagonal from upper left to lower right and 0s below the 1s   **10.** determinant   **11.** yes
12. no   **13. a.** consistent with independent equations
b. one solution   **14. a.** inconsistent   **b.** no solution
15. a. consistent with independent equations   **b.** one solution   **16. a.** consistent with dependent equations
b. infinite number of solutions   **17.** $(-1, -4)$ **18.** $(3, -1)$
19. No solution.   **20.** All ordered pairs that solve $3x - 2y = 6$.

21. $(4, -1)$ **22.** $\left(5, -\dfrac{2}{5}\right)$ **23.** All ordered pairs that solve

$3y - 4x = 6$.   **24.** No solution.   **25.** $(1, 3)$ **26.** $(2, -1)$
27. $(1, 5)$ **28.** $(5, -3)$ **29.** $(1, 0, -3)$ **30.** No solution.
31. $(3, -3, 2)$ **32.** $(4, 0, 1)$ **33.** $(4, -1)$ **34.** $(-3, -2)$
35. $(-2, 3, 1)$ **36.** $(-1, 2, -4)$ **37.** $(-2, 9)$ **38.** No solution.
39. $(2, 4, -1)$ **40.** $(-2, 4, -2)$ **41.**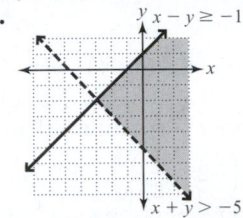

42. 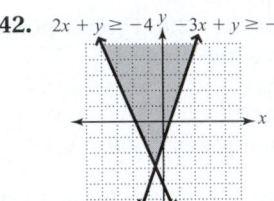   **43.**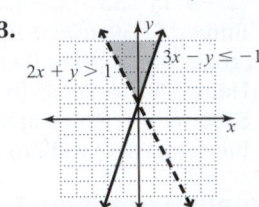

44.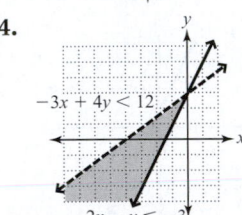

45. 10, 6
46. 47 adult, 92 children

47. Canada: 2500 thousand; Mexico: 830 thousand

48. Western Europe: $15\dfrac{2}{3}$%; North America: $27\dfrac{1}{3}$%

49. $75°, 15°$ **50.** \$3500 at 7% and \$1500 at 9%   **51.** 5:45 P.M.
52. 120 miles per hour   **53.** 30 ml   **54.** 1 vitamin supplement, 3 rolls of film, 4 bags of candy   **55.** 4 first place, 3 second place, 9 third place   **56. a.** $\begin{cases} x + y \leq 16 \\ 250x + 400y \geq 2750 \\ x \geq 0 \\ y \geq 0 \end{cases}$

b. 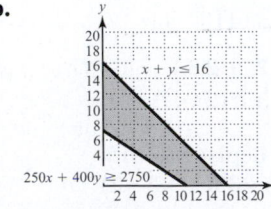   **c.** Answers may vary. One example is $(8, 4)$.

Chapter 9 Practice Test   **1.** yes [9.1]   **2.** no [9.4]
3. $(-2, 1)$ [9.1]   **4.** $(8, -1)$ [9.2]   **5.** $(3, 1)$ [9.2]
6. $(-4, -2)$ [9.3]   **7.** All ordered pairs that solve $4x + 6y = 2$.
[9.3]   **8.** $(-2, 3, -2)$ [9.4]   **9.** $(-4, 2, 3)$ [9.4]
10. $(2, -4)$ [9.5]   **11.** $(-1, 5, 2)$ [9.5]   **12.** $(-2, 5)$ [9.6]
13. $(-4, 0, 3)$ [9.6]

14.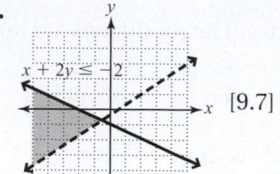

$x + 2y \le -2$ [9.7]

$2x - 3y < 1$

15. 86 accountants, 77 waiters/waitresses [9.2]

16. 180 [9.2] **17.** 7 mph [9.3] **18.** \$4000 at 6%, \$8000 at 8% [9.3] **19.** Children: 150; student: 300; adult: 350. [9.3]

20. a. $\begin{cases} 2l + 2w \le 200 \\ l \ge w + 10 \\ l > 0 \\ w > 0 \end{cases}$ **b.**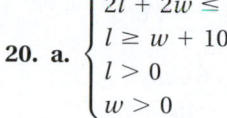

$l \ge w + 10 \qquad 2l + 2w \le 200$

c. Answers may vary. One example is $(60, 20)$. [9.7]

Chapters 1–9 Cumulative Review Exercises

1. false **2.** true **3.** true **4.** $8x + 12$ **5.** $\dfrac{c}{d}$ **6.** sum
7. 11 **8.** 0 **9.** $-6x - 6$ **10.** $4x^3 - 14x^2 + 7x - 3$
11. $5m - 4 + \dfrac{6}{3m - 2}$ **12.** $(x + 5)^2$
13. $(x^2 + 9)(x + 3)(x - 3)$ **14.** $\dfrac{1}{u - 1}$ **15.** $\dfrac{x}{10x + 5}$
16. $\dfrac{y^2 + y + 9}{(y - 2)(y + 3)}$ **17.** All real numbers **18.** ± 6 **19.** 8
20. 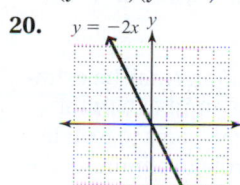 $y = -2x$

21. 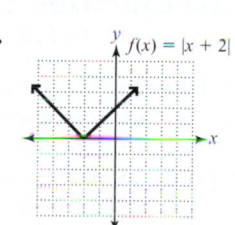 $f(x) = |x + 2|$

22. $2x + 3y = 8$ **23.** $(-3.5, -4)$ **24.** $(2, -1, 4)$
25.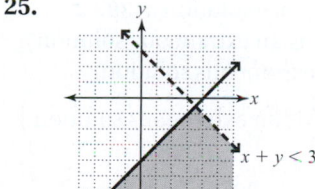

$x + y < 3$

$x - y \ge 4$

26. 75 ml **27.** ≈ 5.5 hr.

28. 3.2 cm/sec. **29.** 1.2 hr., or 1 hr. and 12 min. **30.** $115°, 65°$

Chapter 10

10.1 Exercises
1. Answers will vary. For example, 9 has rational square roots, whereas 17 has irrational square roots. The square roots of 17 are irrational because they cannot be expressed in the form $\dfrac{a}{b}$, where a and b are integers and $b \ne 0$.
3. Squaring a number or its additive inverse results in the same positive number. **5.** You cannot raise a number to an even power and get a negative value. **7.** ± 6 **9.** ± 11
11. ± 14 **13.** ± 15 **15.** 5 **17.** Not a real number.
19. -5 **21.** ± 5 **23.** 1.2 **25.** Not a real number.
27. -0.11 **29.** $\dfrac{7}{9}$ **31.** $-\dfrac{12}{13}$ **33.** 3 **35.** -4 **37.** 6 **39.** 4

41. Not a real number. **43.** -2 **45.** 2 **47.** -3 **49.** 2
51. 2 **53.** $-\dfrac{2}{3}$ **55.** $\dfrac{2}{3}$ **57.** 2.646 **59.** -3.317 **61.** 3.684
63. -3.756 **65.** 3.708 **67.** -3.036 **69.** 2.454 **71.** 2.295
73. b^2 **75.** $4x$ **77.** $10r^4s^3$ **79.** $0.5a^3b^6$ **81.** m **83.** $3a^3b^2$
85. $-4ab^4$ **87.** $0.2x^6$ **89.** a **91.** $2x^4$ **93.** $2x^2$ **95.** x^2y
97. $6|m|$ **99.** $|r - 1|$ **101.** $4|y^3|$ **103.** $3y$ **105.** $(y - 3)^2$
107. $(y - 4)^2$ **109.** 2 **111.** $\sqrt{15}$ **113.** $\{x | x \ge 4\}$ or $[4, \infty)$
115. $\{x | x \le 4\}$ or $(-\infty, 4]$
117. a. 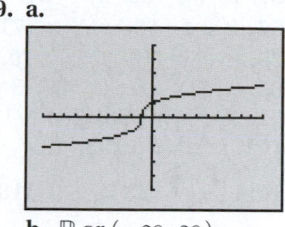 **119. a.**

b. $\{x | x \ge 2\}$ or $[2, \infty)$ **b.** $\mathbb{R}$ or $(-\infty, \infty)$
121. ≈ -17.709 m/sec. **123.** ≈ 3.476 sec.
125. ≈ 19.170 mph **127.** 13 ft. **129.** 15 N **131. a.** 7924
b. ≈ 7004

Review Exercises **1. a.** 12 **b.** 12 **2.** x^8 **3.** $-45m^4n^3$
4. $12x^4 - 20x^3 + 4x^2$ **5.** $21y^2 + 23y - 20$
6. $2x^2 - 17x - 9$

10.2 Exercises
1. 4 **3.** a would be nonnegative because with an even index, the radicand must be nonnegative.
5. Yes, because $100^{1/4} = (10^2)^{1/4} = 10^{1/2}$. **7.** $\sqrt{25} = 5$
9. $-\sqrt{100} = -10$ **11.** $\sqrt[3]{27} = 3$ **13.** $\sqrt[3]{-64} = -4$
15. $\sqrt[4]{y}$ **17.** $\sqrt{144x^8} = 12x^4$ **19.** $18\sqrt{r}$ **21.** $\sqrt{\dfrac{x^4}{81}} = \dfrac{x^2}{9}$
23. $(\sqrt[3]{64})^2 = 16$ **25.** $-(\sqrt[4]{81})^3 = -27$ **27.** $(\sqrt[3]{-8})^4 = 16$
29. $\dfrac{1}{(\sqrt[4]{16})^3} = \dfrac{1}{8}$ **31.** $\sqrt[5]{x^4}$ **33.** $8\sqrt[3]{n^2}$ **35.** $\dfrac{1}{(\sqrt[5]{-32})^2} = \dfrac{1}{4}$
37. $\left(\sqrt{\dfrac{1}{25}}\right)^3 = \dfrac{1}{125}$ **39.** $\sqrt[6]{(2a + 4)^5}$ **41.** $25^{1/4}$ **43.** $z^{5/6}$
45. $5^{-5/6}$ **47.** $5x^{-4/5}$ **49.** $5^{7/3}$ **51.** $x^{2/7}$ **53.** $(4a - 5)^{7/4}$
55. $(2r - 5)^{8/5}$ **57.** $x^{4/5}$ **59.** $x^{7/6}$ **61.** $a^{17/12}$ **63.** $21w^{4/7}$
65. $-12a^{17/12}$ **67.** $7^{5/3}$ **69.** $\dfrac{1}{x^{2/3}}$ **71.** $x^{1/4}$ **73.** $r^{1/12}$
75. $\dfrac{1}{x^{5/7}}$ **77.** $a^{9/4}$ **79.** $20s^{3/7}$ **81.** $-24b^{1/4}$ **83.** x^2
85. $a^{5/3}$ **87.** $b^{2/5}$ **89.** $64x^4y^3$ **91.** $2q^{1/2}t^{1/4}$ **93.** $81a^3$
95. $3z^{1/3}$ **97.** $\sqrt{2}$ **99.** $\sqrt[3]{7}$ **101.** $\sqrt{x}$ **103.** $\sqrt[4]{r^3}$
105. $\sqrt[4]{x^3y}$ **107.** $\sqrt[5]{m^2n^3}$ **109.** $\sqrt[6]{x^5}$ **111.** $\sqrt[6]{y^7}$
113. $\sqrt[6]{x^5}$ **115.** $\sqrt[15]{n^2}$ **117.** $\sqrt[6]{1125}$ **119.** $\sqrt[12]{3456}$
121. $\sqrt[9]{x}$ **123.** $\sqrt[6]{n}$

Review Exercises **1.** $2^4x^3y^2$ **2.** 12 **3.** 15
4. 8×10^{11} **5.** $-\dfrac{5}{8}x^4y^2z^2$ **6.** $2x^2 - 8x + 5 + \dfrac{3}{x + 3}$

10.3 Exercises
1. Multiplying the approximate roots gives $\sqrt{8} \cdot \sqrt{18} \approx 2.828 \cdot 4.243 = 11.999204$, which is tedious and inexact. Using the product rule for radicals gives $\sqrt{8} \cdot \sqrt{18} = \sqrt{8 \cdot 18} = \sqrt{144} = 12$, which is fast and exact.
3. Rewrite the expression as a product of two radicals, the first containing the perfect cube and the second containing no perfect cubes. Then simplify the first radical. For example, $\sqrt[3]{54x^8}$ can be rewritten as $\sqrt[3]{27x^6} \cdot \sqrt[3]{2x^2}$, then simplified to $3x^2\sqrt[3]{2x^2}$.

5. 8 7. $9x^3$ 9. $12xy^2$ 11. $\sqrt{65}$ 13. $\sqrt{15x}$ 15. 3
17. $\sqrt[3]{10y^2}$ 19. $\sqrt[4]{21}$ 21. $w\sqrt[4]{72}$ 23. $\sqrt[4]{15x^3y^3}$
25. $x\sqrt[5]{30x^2}$ 27. $\sqrt[6]{8x^5y^4}$ 29. $\sqrt{\dfrac{21}{10}}$ 31. $\sqrt{\dfrac{6y}{5x}}$ 33. $\dfrac{5}{6}$
35. $\dfrac{\sqrt{10}}{3}$ 37. 6 39. $\sqrt{3}$ 41. $\dfrac{\sqrt[3]{4}}{w^2}$ 43. $\dfrac{\sqrt[3]{5y^2}}{3x^3}$ 45. 4
47. $\dfrac{\sqrt[4]{3u^3}}{2x^2}$ 49. $7\sqrt{2}$ 51. $8\sqrt{2}$ 53. $24\sqrt{5}$ 55. $20\sqrt{7}$
57. $a^3\sqrt{a}$ 59. xy^2 61. $x^3y^4z^5$ 63. $r^5s^4\sqrt{rs}$ 65. $18x^2\sqrt{2x}$
67. $2\sqrt[3]{4}$ 69. $x^2\sqrt[3]{x}$ 71. $x^2y\sqrt[3]{y^2}$ 73. $4z^2\sqrt[3]{2z^2}$
75. $4\sqrt[3]{5}$ 77. $2\sqrt[4]{5}$ 79. $9x^4\sqrt[4]{3x}$ 81. $3x^3\sqrt[5]{2x}$
83. $xy^2z\sqrt[6]{x^2y^2z^5}$ 85. $3\sqrt{7}$ 87. $30\sqrt{15}$ 89. $y^2\sqrt{y}$
91. $x^4y^5\sqrt{y}$ 93. $24c^4\sqrt{15}$ 95. $60\sqrt{2}$ 97. $2\sqrt{2}$ 99. $6\sqrt{5}$
101. $c^2d\sqrt{d}$ 103. $12a^2$ 105. $6c\sqrt{3c}$ 107. $36x^2y^3\sqrt{2y}$
109. $\dfrac{2\sqrt{6}}{7}$ 111. $\dfrac{a^4}{2}$ 113. $\dfrac{3x^5\sqrt{5}}{4}$ 115. $\dfrac{3}{8}\sqrt{5}$

Review Exercises 1. no 2. yes 3. $8x^2 - 7x$
4. $8a^2 - 6ab - 9b^2$ 5. $9m^2 - 25n^2$ 6. $4x^2 - 12xy + 9y^2$

10.4 Exercises 1. Like radicals have the same index and the same radicand, but their coefficients may be different.
3. $(x + 3)(x + 2) = x^2 + 2x + 3x + 6 = x^2 + 5x + 6$ and $(\sqrt{5} + 3)(\sqrt{5} + 2) = \sqrt{5}^2 + 2\sqrt{5} + 3\sqrt{5} + 6 = 5 + 5\sqrt{5} + 6 = 11 + 5\sqrt{5}$. In both cases, the FOIL method was used and then the expansion was simplified. 5. $-6\sqrt{6}$
7. $9\sqrt{a}$ 9. $12\sqrt{5} - 8\sqrt{6}$ 11. $11a\sqrt{5a} - 2b\sqrt{7b}$
13. $3x\sqrt[3]{9}$ 15. $-6x^2\sqrt[4]{5x}$ 17. Cannot combine because the radicals are not like. 19. $-\sqrt{3}$ 21. $-\sqrt{5y}$ 23. $-8\sqrt{5}$
25. $6\sqrt{6}$ 27. $14a\sqrt{3a}$ 29. $4\sqrt{6}$ 31. $18\sqrt{2} - 17\sqrt{3}$
33. $7\sqrt[3]{2}$ 35. $-12x\sqrt[3]{5x^2}$ 37. $-2x^2\sqrt[4]{2x}$ 39. $3\sqrt{2} + 2$
41. $3 - 3\sqrt{5}$ 43. $\sqrt{30} + 10\sqrt{2}$ 45. $24x - 48x\sqrt{2}$
47. $12 - 3\sqrt{2} + 4\sqrt{5} - \sqrt{10}$ 49. $6 + 5\sqrt{x} + x$
51. $6 + 10\sqrt{2} + 9\sqrt{3} + 15\sqrt{6}$
53. $\sqrt{15} + \sqrt{21} + 5 + \sqrt{35}$ 55. $x + \sqrt{xy} - 6y$
57. $12\sqrt{14} - 12\sqrt{6} + 6\sqrt{35} - 6\sqrt{15}$
59. $8a + 10\sqrt{ab} - 3b$ 61. $2\sqrt[3]{2} - 3\sqrt[3]{4} - 40$
63. $1 - \sqrt[3]{18} + \sqrt[3]{12}$ 65. $x + 8$ 67. $22 + 8\sqrt{6}$
69. $18 - 8\sqrt{2}$ 71. $16 + 8\sqrt{3}$ 73. $30 + 12\sqrt{6}$ 75. 13
77. -14 79. $36 - x$ 81. 1 83. $x - y$ 85. 4 87. 50
89. $7\sqrt{5}$ 91. $-15\sqrt{6}$ 93. $9\sqrt{2}$ 95. $-14\sqrt{5}$ 97. $25\sqrt{3}$
99. a. $42 + 2\sqrt{5}$ ft. b. 46.5 ft. c. $87.89

Review Exercises 1. $2x + 5$ 2. $16x^2 - 9$ 3. $\sqrt{2}$
4. $\sqrt[3]{4}$ 5. Slope is $\dfrac{2}{5}$; y-intercept is $(0, 2)$.
6.

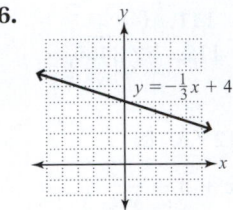

$y = -\frac{1}{3}x + 4$

10.5 Exercises 1. a. $\sqrt{16}$ is rational. b. A radical, $\sqrt{3}$, is in the denominator. 3. Multiply the fraction by a 1 so that the product's denominator has a radicand that is a perfect square. 5. $\dfrac{\sqrt{3}}{3}$ 7. $\dfrac{3\sqrt{2}}{4}$ 9. $\dfrac{6\sqrt{7}}{7}$ 11. $\dfrac{\sqrt{15}}{6}$ 13. $\dfrac{x\sqrt{14}}{10}$
15. $\dfrac{\sqrt{7}}{7}$ 17. $\dfrac{5\sqrt{3a}}{3a}$ 19. $\dfrac{\sqrt{33mn}}{11n}$ 21. $\dfrac{2\sqrt{5x}}{x}$ 23. $\dfrac{\sqrt{3}}{4}$

25. $\dfrac{3\sqrt{x}}{x^2}$ 27. $4x\sqrt{2x}$ 29. Mistake: The product of $\sqrt{2}$ and 2 is not 2. Correct: $\dfrac{\sqrt{6}}{2}$ 31. $\dfrac{5\sqrt[3]{9}}{3}$ 33. $\dfrac{\sqrt[3]{20}}{2}$ 35. $3\sqrt[3]{2}$
37. $\dfrac{m\sqrt[3]{n^2}}{n}$ 39. $\dfrac{\sqrt[3]{ab}}{b}$ 41. $\dfrac{2\sqrt[3]{4x^2}}{x}$ 43. $\dfrac{\sqrt[3]{30a}}{5a}$
45. $\dfrac{5\sqrt[4]{4}}{2}$ or $\dfrac{5\sqrt{2}}{2}$ 47. $\dfrac{\sqrt[4]{3x^2}}{x}$ 49. $\dfrac{3\sqrt[4]{27x}}{x}$ 51. $3\sqrt{2} - 3$
53. $8 + 4\sqrt{3}$ 55. $5\sqrt{3} - 5\sqrt{2}$ 57. $-1 - \sqrt{5}$
59. $\dfrac{3 + \sqrt{3}}{2}$ 61. $\dfrac{-6 - 8\sqrt{3}}{13}$ 63. $\dfrac{4\sqrt{21} - 4\sqrt{6}}{5}$
65. $\dfrac{32 + 8\sqrt{3}}{13}$ 67. $\dfrac{6y - 6\sqrt{y}}{y - 1}$ 69. $\dfrac{3t - 6\sqrt{tu}}{t - 4u}$
71. $\dfrac{\sqrt{2xy} + 2y\sqrt{3}}{x - 6y}$ 73. $\dfrac{3}{2\sqrt{3}}$ 75. $\dfrac{2x}{5\sqrt{2x}}$ 77. $\dfrac{2n}{3\sqrt{2n}}$
79. $\dfrac{1}{10 - 5\sqrt{3}}$ 81. $\dfrac{5x - 36}{9\sqrt{5x} + 54}$ 83. $\dfrac{19}{10\sqrt{n} - 2\sqrt{6n}}$
85. a. $\dfrac{5\sqrt{3}}{3}$ b. $\sqrt{5}$ c. $\dfrac{5\sqrt{11}}{11}$ 87. a. The graphs are identical. The functions are identical. b. $f(x) = g(x)$
89. a. $T = \pi\sqrt{9.8L}/4.9$ b. $T = 2\pi L/\sqrt{9.8L}$
91. a. $s = \dfrac{\sqrt{3Vh}}{h}$ b. 745 ft. 93. a. $V_{\text{rms}} = \dfrac{\sqrt{2}}{2}V_{\text{m}}$ or $\dfrac{\sqrt{2}V_{\text{m}}}{2}$
b. $V_{\text{rms}} = \dfrac{163\sqrt{2}}{2}$ c. ≈ 115.3 95. $\dfrac{15\sqrt{2} - 10\sqrt{3}}{3}\,\Omega$

Review Exercises 1. $\pm 2\sqrt{7}$ 2. $(x - 3)^2$ 3. 4
4. -1 5. $-6, 6$ 6. $2, 3$

10.6 Exercises 1. Some of the solutions may be extraneous. 3. The principal square root of a number cannot equal a negative. 5. Subtract $3x$ from both sides to isolate the radical. This allows use of the power rule to eliminate the radical. 7. 4 9. No real-number solution. 11. 27
13. -8 15. 17 17. 11 19. 6 21. -4 23. No real-number solution. 25. 11 27. -2 29. 124 31. 55
33. 5 35. -11 37. No real-number solution. 39. 2
41. No real-number solution (-5 is an extraneous solution).
43. 3 45. 3 47. -4 (-10 is an extraneous solution)
49. 1, 3 51. $-3, -2$ 53. $\dfrac{1}{2}\left(-\dfrac{3}{8}\text{ is an extraneous solution}\right)$
55. 5 57. $-1, 3$ 59. 0, 4 61. $\dfrac{3}{16}$ 63. 3 (-1 is an extraneous solution) 65. Mistake: You cannot take the principal square root of a number and get a negative. Correct: no real-number solution 67. Mistake: The binomial $x - 3$ was not squared correctly. Correct: $x = 6$ with $x = 1$ an extraneous solution 69. 9.8 m 71. 0.6125 m 73. 1.44 ft.
75. 1 ft. 77. 36.73 ft. 79. 82.65 ft. 81. 4 N 83. 6 N
85. a. 3, 4, 25 b.

$y = \sqrt{x}$, $(1, 1)$, $(4, 2)$, $(9, 3)$, $(16, 4)$, $(25, 5)$, $(0, 0)$

c. No, the x-values must be 0 or positive because real square roots exist only when $x \geq 0$. The y-values must be 0 or positive because, by definition, the principal square root is 0 or positive. d. Yes, because it passes the vertical line test.
87. The graph becomes steeper from left to right. 89. The graph rises or lowers according to the value of the constant.

Review Exercises 1. 81 2. −0.008 3. 39.0625 or $\frac{625}{16}$ 4. x^8 5. n^{24} 6. y^4

10.7 Exercises 1. $\sqrt{-1}$ 3. No, the set of complex numbers contains both the real and imaginary numbers.
5. We subtract complex numbers just like we subtract polynomials—by writing an equivalent addition and changing the signs in the second complex number. 7. $6i$ 9. $i\sqrt{5}$
11. $2i\sqrt{2}$ 13. $3i\sqrt{2}$ 15. $3i\sqrt{3}$ 17. $5i\sqrt{5}$ 19. $3i\sqrt{7}$
21. $7i\sqrt{5}$ 23. $6 + 7i$ 25. $11 - 6i$ 27. $-7 + i$
29. $8 - 2i$ 31. $-3 - 10i$ 33. 0 35. $14 + 9i$
37. $-4 + 28i$ 39. -24 41. 40 43. $14 + 12i$
45. $-72 - 32i$ 47. $19 - 3i$ 49. $50 + 9i$ 51. 65
53. $-7 - 24i$ 55. $-2i$ 57. $-\frac{4i}{5}$ 59. $-3i$ 61. $\frac{1}{2} - i$
63. $\frac{1}{2} - i$ 65. $\frac{14}{5} - \frac{7}{5}i$ 67. $-\frac{7}{29} + \frac{3}{29}i$ 69. $7 - 2i$
71. $\frac{9}{13} - \frac{7}{13}i$ 73. $\frac{34}{41} + \frac{19}{41}i$ 75. $-i$ 77. -1 79. -1
81. 1 83. 1 85. -1 87. $-i$ 89. i

Review Exercises 1. $(2x - 3)^2$ 2. $\pm 4\sqrt{3}$ 3. 2
4. $-\frac{3}{2}, 2$ 5. $(0, 2), (3, 0)$ 6.

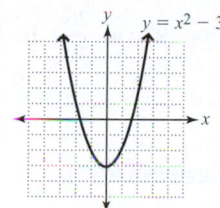

Chapter 10 Review Exercises 1. true 2. false
3. false 4. false 5. true 6. radicands 7. coefficients
8. **a.** Multiply 9. conjugate 10. extraneous 11. ± 11
12. ± 7 13. 13 14. -7 15. Not a real number.
16. $\frac{1}{5}$ 17. 2.646 18. 9.487 19. $7x^4$ 20. $12a^3b^6$
21. $0.4mn^5$ 22. x^5 23. $-4r^3s$ 24. $3x^3$ 25. $2x^3y^4$
26. x^2y 27. $9|x|$ 28. $(x - 1)^2$ 29. -4 30. $2a^2\sqrt{6}$
31. $\frac{1}{8}$ 32. $\sqrt[7]{(5r - 2)^5}$ 33. $33^{1/8}$ 34. $8n^{-3/7}$ 35. $8^{3/5}$
36. $m^{8/3}$ 37. $(3xw)^{3/4}$ 38. $(a + b)^{4/3}$ 39. x^2 40. $32m^{3/2}$
41. $\frac{1}{y^{1/5}}$ 42. b 43. $k^{1/2}$ 44. $2^{3/4}x^{3/4}y^{3/20}$ 45. 9
46. $10x^4$ 47. 2 48. $\sqrt[4]{42}$ 49. $\sqrt[5]{15x^4y^4}$ 50. $\frac{7}{11}$
51. $-\frac{3}{2}$ 52. $6\sqrt{2}$ 53. $12b^4\sqrt{3b}$ 54. $15\sqrt[3]{4}$
55. $4x^3\sqrt[3]{5x}$ 56. $6x^5\sqrt[4]{2x^3}$ 57. $84\sqrt{10}$ 58. $x^7\sqrt{x}$
59. $16c^2\sqrt{15c}$ 60. $a^5b^4\sqrt{b}$ 61. $2\sqrt{2}$ 62. $6\sqrt{5}$ 63. $12a^2$
64. $36x^2y^3\sqrt{2y}$ 65. $-3\sqrt{n}$ 66. $-6y^3\sqrt[4]{8y}$ 67. $5\sqrt{5}$
68. $-10\sqrt{6}$ 69. $4\sqrt{6}$ 70. $28x\sqrt{2y}$ 71. $9xy\sqrt[3]{2xy^2}$
72. $5\sqrt[4]{3}$ 73. $\sqrt{15} + \sqrt{10}$ 74. $\sqrt[3]{21} + 2\sqrt[3]{49}$
75. $6\sqrt{6} - 54$ 76. $\sqrt{10} + \sqrt{14} - \sqrt{15} - \sqrt{21}$
77. $-6 + 2\sqrt[3]{2} - 4\sqrt[3]{4}$ 78. $\sqrt[4]{12x^2} - \sqrt[4]{6x} + 2\sqrt[4]{2x} - 2$
79. $5a - 3b$ 80. 7 81. $3 + 2\sqrt{2}$ 82. $7 - 2\sqrt{10}$
83. $\frac{\sqrt{2}}{2}$ 84. $\sqrt[3]{9}$ 85. $\frac{2\sqrt{7}}{7}$ 86. $\frac{\sqrt[4]{40x^2}}{2}$ 87. $\frac{\sqrt[3]{153y}}{3y}$
88. $\frac{\sqrt[4]{27}}{3}$ 89. $-4\sqrt{2} - 4\sqrt{3}$ 90. $\frac{4 - \sqrt{3}}{13}$ 91. $\frac{2 + \sqrt{n}}{4 - n}$

92. $\sqrt{6} + 2$ 93. $\frac{5}{3\sqrt{10}}$ 94. $\frac{3x}{5\sqrt{3x}}$ 95. $\frac{1}{16 + 8\sqrt{3}}$
96. $\frac{1}{10\sqrt{t} - 5\sqrt{3t}}$ 97. 81 98. No real-number solution.
99. 10 100. -2 101. 18 102. 5 103. No real-number solution. 104. $\frac{719}{3}$ 105. 7 106. 1 $\left(\frac{17}{4} \text{ is extraneous}\right)$
107. $0, 4$ 108. $\frac{3}{16}$ 109. $3i$ 110. $2i\sqrt{5}$ 111. $8 - 6i$
112. $9 - 7i$ 113. -12 114. $2 + 8i$ 115. $26 + 2i$
116. $24 - 10i$ 117. $-5i$ 118. $3i$ 119. $-\frac{4i}{3}$ 120. $\frac{1}{5} - \frac{7}{5}i$
121. $\frac{6}{5} - \frac{3}{5}i$ 122. $\frac{7}{13} + \frac{17}{13}i$ 123. 1 124. $-i$ 125. $\frac{\sqrt{6}}{4}$
126. **a.** $8\sqrt{5}$ mph **b.** 17.9 mph 127. 312.5 ft. 128. π sec.
129. 0.272 m 130. **a.** $\frac{\sqrt{10}}{2}$ sec. **b.** 1.58 sec. 131. 1.44 ft.
132. 5 ft.

Chapter 10 Practice Test 1. 6 [10.1] 2. $7i$ [10.7]
3. $9xy^2\sqrt{y}$ [10.1] 4. $3\sqrt[3]{2}$ [10.1]
5. $2x\sqrt[4]{x^2}$, or $2x\sqrt{x}$ [10.3] 6. $3r^5$ [10.1] 7. $\frac{1}{3}$ [10.3]
8. $\frac{1}{3}$ [10.3] 9. $5\sqrt{7}$ [10.4] 10. $4 - 2\sqrt{3}$ [10.4]
11. $\frac{1}{x^{2/3}}$ [10.2] 12. $-6 + 2\sqrt[3]{2} - 4\sqrt[3]{4}$ [10.4]
13. $8^{1/5}x^{3/5}$ [10.2] 14. $(2x + 5)^{2/3}$ [10.2] 15. $\frac{\sqrt[3]{2}}{2}$ [10.5]
16. $\frac{x - \sqrt{xy}}{x - y}$ [10.5] 17. 22 [10.6] 18. 7 [10.6]
19. 24 (0 is an extraneous solution) [10.6]
20. $-2 - 4i$ [10.7] 21. $17 + 0i$, or 17 [10.7]
22. $\frac{8}{25} + \frac{6}{25}i$ [10.7] 23. 60 m² [10.3] 24. **a.** $\frac{\sqrt{3}}{2}$ [10.6]
b. 64 ft. [10.6] 25. **a.** $3.5\sqrt{10}$ mph [10.6]
b. 293.88 ft. or 14,400/49 ft. [10.6]

Chapters 1–10 Cumulative Review Exercises
1. false 2. true 3. true 4. true 5. x-intercept
6. inconsistent 7. $24x^5y^2$ 8. 28 9. 0.0000024 10. x^{15}
11. $9x^4 - 24x^2y + 16y^2$ 12. $4x^2 + 6x - 2 + \frac{2}{2x - 3}$
13. $\frac{x}{x + 5}$ 14. $30a^3\sqrt{2a}$ 15. $-6x^2\sqrt[3]{3x}$ 16. $\frac{1}{8}$
17. $23 - 2i$ 18. $\frac{5 + 3\sqrt{2}}{7}$ 19. $b = \frac{2A - Bh}{h}$ or $\frac{2A}{h} - B$
20. $5, -11$ 21. No solution (4 is an extraneous solution).
22. 2 (-3 is an extraneous solution)
23. slope $= \frac{5}{2}$, y-intercept $= \frac{15}{2}$ 24. $y = 3x + 14$
25.

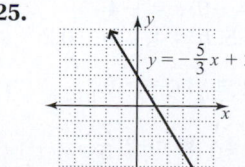

26.

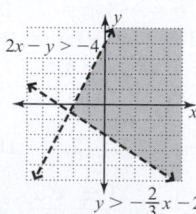

27. $16.48 28. 25 mph 29. 96 30. 39.2 m

Chapter 11

11.1 Exercises **1.** $x^2 = a$ has two solutions because squaring $\sqrt{a}$ and $-\sqrt{a}$ gives a for every real number a.
3. $x = \dfrac{b \pm \sqrt{c}}{a}$ **5.** Because $x^2 - 7x + 12$ is easy to factor, factoring is a better method as it requires fewer steps than completing the square. **7.** ± 7 **9.** $\pm \dfrac{2}{5}$ **11.** ± 1.2
13. $\pm 3\sqrt{5}$ **15.** $\pm 5i$ **17.** ± 7 **19.** ± 9 **21.** ± 3 **23.** $\pm \dfrac{3}{5}$
25. $\pm 2i$ **27.** $\pm \dfrac{12}{5}$ **29.** $\pm 2\sqrt{2}$ **31.** $\pm 3i\sqrt{2}$ **33.** $\pm \dfrac{4\sqrt{6}}{3}$
35. ± 1.6 **37.** $-15, -1$ **39.** $-\dfrac{1}{5}, \dfrac{7}{5}$ **41.** $8 \pm 4i$
43. $\dfrac{1 \pm 2\sqrt{10}}{4}$ **45.** $7 \pm 2i\sqrt{3}$ **47.** $0, \dfrac{3}{2}$ **49.** $\dfrac{3}{5}, \dfrac{6}{5}$
51. $-13.5, -5.5$ **53.** Mistake: Gave only the positive solution. Correct: ± 7 **55.** Mistake: Changed -6 to 6. Correct: $5 \pm i\sqrt{6}$ **57.** $\pm 4\sqrt{6} \approx \pm 9.798$
59. $\pm 2\sqrt{5} \approx \pm 4.472$ **61.** $6 \pm \sqrt{15} \approx 9.873, 2.127$
63. a. $x^2 + 14x + 49$ **b.** $(x+7)^2$ **65. a.** $n^2 - 10n + 25$
b. $(n-5)^2$ **67. a.** $y^2 - 9y + \dfrac{81}{4}$ **b.** $\left(y - \dfrac{9}{2}\right)^2$
69. a. $s^2 - \dfrac{2}{3}s + \dfrac{1}{9}$ **b.** $\left(s - \dfrac{1}{3}\right)^2$ **71.** $-5, 3$ **73.** $1 \pm 7i$
75. $3, 6$ **77.** $1 \pm \sqrt{17}$ **79.** $3 \pm 2i\sqrt{5}$ **81.** $-\dfrac{3}{2}, 1$
83. $-\dfrac{9}{2}, \dfrac{1}{2}$ **85.** $\dfrac{1}{3}, \dfrac{5}{3}$ **87.** $-1, \dfrac{3}{2}$ **89.** $-2 \pm \dfrac{\sqrt{23}}{2}$
91. $\dfrac{-1 \pm \sqrt{41}}{10}$ **93.** $\dfrac{-2 \pm 2\sqrt{7}}{3}$ **95.** Did not divide by 3 so that x^2 has a coefficient of 1. Then wrote an incorrect factored form. Correct: $\dfrac{-2 \pm \sqrt{10}}{3}$ **97.** 14 in. **99.** Height: 3 ft.;
width: 6 ft. **101.** 305 m **103.** 18 in. **105.** 16 ft. by 12 ft.
107. a. 15 cm **b.** Length: 45 cm; width: 30 cm. **109.** $\dfrac{3}{4}$ sec.
111. 4 m/sec.

Review Exercises **1.** $(x - 5)^2$ **2.** $(x + 3)^2$ **3.** $6x$
4. $2x + 3$ **5.** $\dfrac{-2 + \sqrt{6}}{2}$ or $-1 + \dfrac{\sqrt{6}}{2}$ **6.** $\dfrac{-3 - i}{5}$ or $-\dfrac{3}{5} - \dfrac{1}{5}i$

11.2 Exercises **1.** Follow the procedure for solving a quadratic equation by completing the square on $ax^2 + bx + c = 0$. **3.** Yes, if they have irrational or nonreal complex solutions. **5.** When the discriminant is zero. The quadratic formula becomes $x = -\dfrac{b}{2a}$.
7. $a = 1, b = -3, c = 7$ **9.** $a = 3, b = -9, c = -4$
11. $a = 1.5, b = -1, c = 0.2$ or $a = -1.5, b = 1, c = -0.2$
13. $a = -\dfrac{1}{2}, b = -\dfrac{3}{4}, c = 6$ or $a = \dfrac{1}{2}, b = \dfrac{3}{4}, c = -6$
15. $-5, -4$ **17.** $-2, \dfrac{3}{4}$ **19.** $0, 9$ **21.** 4 **23.** $-2, \dfrac{2}{3}$
25. $1 \pm i$ **27.** $\dfrac{1 \pm \sqrt{5}}{2}$ **29.** $\dfrac{-5 \pm \sqrt{10}}{3}$ **31.** $\dfrac{3 \pm i\sqrt{11}}{4}$
33. $-\dfrac{1}{6}, -\dfrac{2}{3}$ **35.** $\dfrac{2 \pm i\sqrt{5}}{3}$ **37.** $\dfrac{3 \pm \sqrt{105}}{12}$ **39.** $-0.15, 0.1$

41. $-2, \dfrac{3}{2}$ **43.** $\pm \dfrac{7}{6}$ **45.** $1 \pm i\sqrt{2}$
47. $\dfrac{-3 \pm \sqrt{5009}}{100} \approx -0.738, 0.678$ **49.** $\dfrac{3 \pm i\sqrt{51}}{12}$
51. Mistake: Did not evaluate $-b$. Correct: $\dfrac{7 \pm \sqrt{37}}{6}$
53. Mistake: The result was not completely simplified. Correct: $1 \pm i\sqrt{2}$ **55.** One rational.
57. Two irrational. **59.** Two rational. **61.** Two nonreal complex. **63.** Two irrational. **65.** Square root principle or factoring; ± 9. **67.** Quadratic formula; $\dfrac{2 \pm \sqrt{10}}{2}$.
69. Factoring; $0, -6$. **71.** Square root principle; $-7 \pm 2\sqrt{10}$.
73. Quadratic formula; $4 \pm i\sqrt{3}$. **75.** $(2, 0), (-1, 0), (0, -2)$
77. $\left(\dfrac{3}{2}, 0\right), (0, 9)$ **79.** $\left(\dfrac{1}{2}, 0\right), (-8, 0), (0, -8)$
81. No x-intercepts, $(0, -6)$. **83.** $x^2 + 5(x + 1) = 71; 6, 7$
85. $x^2 + (x + 1)^2 = (x + 2)^2; 3, 4, 5$
87. $w(3w - 3.5) = 34$; width: 4 ft.; length: 8.5 ft.
89. a. $22x = 1.5x^2 + 4x$; 12 ft. **b.** 18 ft., 12 ft., 10 ft., 6 ft., 8 ft., 18 ft. (from bottom clockwise) **91.** ≈ 3.43 sec.
93. a. $h = -4.9t^2 + 10$ **b.** ≈ 1.01 sec. **c.** ≈ 1.43 sec.
95. 8000 units **97. a.** $\dfrac{25}{8}$ **b.** $c < \dfrac{25}{8}$ **c.** $c > \dfrac{25}{8}$
99. a. $\dfrac{9}{2}$ **b.** $a < \dfrac{9}{2}$ **c.** $a > \dfrac{9}{2}$

Review Exercises **1.** $(u - 7)(u - 2)$
2. $(3u - 8)(u + 2)$ **3.** x^4 **4.** $x^{2/3}$ **5.** $\dfrac{2}{5}, -3$ **6.** 8

11.3 Exercises **1.** It can be rewritten as a quadratic equation. **3.** $u = \dfrac{x + 2}{3}$ **5.** $\dfrac{1}{x}$ **7.** $-6, 4$ **9.** $-\dfrac{4}{3}, 2$
11. -8 (4 is extraneous) **13.** $-7.5, 5$ **15.** $-4, 2$ **17.** $-5, \dfrac{2}{3}$
19. $9, 25$ **21.** $\dfrac{49}{4}$ (1 is extraneous) **23.** $2\left(\dfrac{2}{9}\text{ is extraneous}\right)$
25. $\pm 3i$ **27.** $2\left(-\dfrac{2}{9}\text{ is extraneous}\right)$ **29.** $0, 4$ **31.** $\pm 3, \pm 1$
33. $\pm 1.5, \pm 1$ **35.** $\pm 2, \pm 3i$ **37.** $-6, -4$ **39.** $-3.5, 2$
41. $-9, -5$ **43.** $-5, 0$ **45.** $8, 27$ **47.** $-\dfrac{1}{8}, 8$ **49.** $81, 16$
51. $\dfrac{16}{625}$ (16 is extraneous) **53.** 3 hr. **55.** ≈ 2.17 hr.
57. 45 mph **59. a.** Old rate = 12 mph, new rate = 18 mph.
b. Old time = 3 hr., new time = 2 hr. **61.** Jody: 4 hr.; Billy: 6 hr.
63. Older: ≈ 4.27 hr.; newer: 3.77 hr. **65.** 420 vps, 460 vps

Review Exercises **1.** -2 **2.** -3 **3.** $(3, 0), (0, 2)$
4. 6
5.
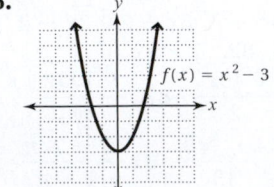
6.
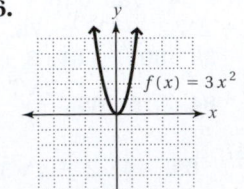

11.4 Exercises

1. The sign of a. **3.** The axis of symmetry is the vertical line through the vertex. If $y = ax^2 + bx + c$, the equation of symmetry is $x = -\dfrac{b}{2a}$. **5.** There are no x-intercepts.

7. V: $(2, -3)$; axis: $x = 2$ **9.** V: $(-1, -5)$; axis: $x = -1$
11. V: $(0, 2)$; axis: $x = 0$ **13.** V: $(-5, 0)$; axis: $x = -5$
15. V: $(0, 0)$; axis: $x = 0$ **17.** V: $(2, 4)$; axis: $x = 2$

19. V: $(-4, -5)$; axis: $x = -4$ **21.** V: $\left(\dfrac{1}{3}, \dfrac{4}{3}\right)$; axis: $x = \dfrac{1}{3}$

23. a. downward **b.** $(0, 0)$ **25. a.** upward **b.** $(0, 0)$
c. $x = 0$ **c.** $x = 0$
d. **d.** $k(x) = \frac{1}{4}x^2$
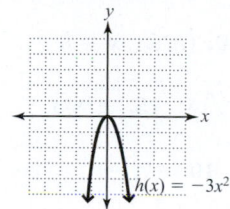 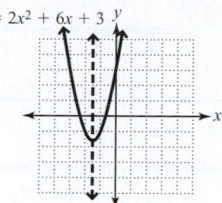

27. a. upward **b.** $(0, -3)$ **29. a.** downward **b.** $(0, 2)$
c. $x = 0$ **c.** $x = 0$
d. $f(x) = 4x^2 - 3$ **d.** $g(x) = -0.5x^2 + 2$
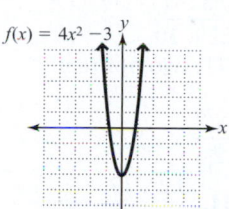

31. a. upward **b.** $(3, 2)$ **33. a.** downward
c. $x = 3$ **b.** $(-1, -3)$ **c.** $x = -1$
d. $f(x) = (x - 3)^2 + 2$ **d.** $k(x) = -2(x + 1)^2 - 3$
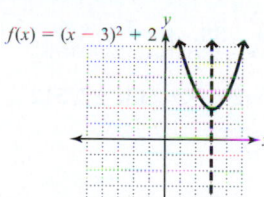

35. a. upward **d.** $h(x) = \frac{1}{3}(x - 2)^2 - 1$
b. $(2, -1)$
c. $x = 2$
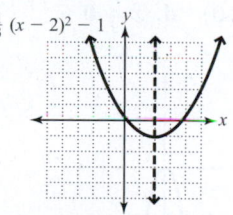

37. a. $(-3, 0), (0, 9)$ **f.** $h(x) = x^2 + 6x + 9$
b. $h(x) = (x + 3)^2$
c. upward
d. $(-3, 0)$
e. $(x = -3)$
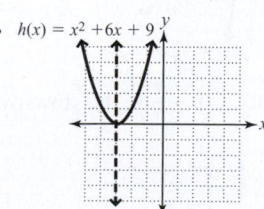

g. domain: $\{x \mid x$ is a real number or $(-\infty, \infty)$;
range: $\{y \mid y \geq 0\}$ or $[0, \infty)$.
39. a. no x-intercepts, $(0, -5)$ **f.**
b. $g(x) = -3(x - 1)^2 - 2$
c. downward
d. $(1, -2)$
e. $x = 1$
$g(x) = -3x^2 + 6x - 5$
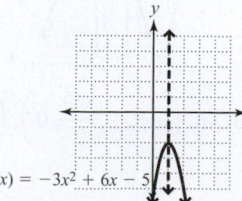

g. domain: $\{x \mid x$ is a real number$\}$ or $(-\infty, \infty)$;
range: $\{y \mid y \leq -2\}$ or $(-\infty, -2]$.

41. a. $\left(\dfrac{-3 + \sqrt{3}}{2}, 0\right), \left(\dfrac{-3 - \sqrt{3}}{2}, 0\right), (0, 3)$

b. $f(x) = 2\left(x + \dfrac{3}{2}\right)^2 - \dfrac{3}{2}$ **c.** upward **d.** $\left(-\dfrac{3}{2}, -\dfrac{3}{2}\right)$

e. $x = -\dfrac{3}{2}$ **f.** $f(x) = 2x^2 + 6x + 3$
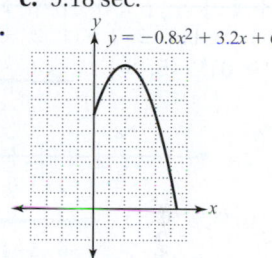

g. domain: $\{x \mid x$ is a real number$\}$ or $(-\infty, \infty)$,
range: $\left\{y \mid y \geq -\dfrac{3}{2}\right\}$ or $\left[-\dfrac{3}{2}, \infty\right)$.
43. a. 4.59 sec. **b.** 103.32 m **c.** 9.18 sec.
45. a. 9.2 ft. **b.** ≈ 5.39 ft. **c.**
$y = -0.8x^2 + 3.2x + 6$
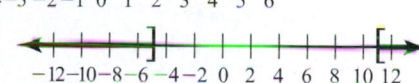

47. a. The greatest number of CDs were sold in the 10th week.
b. 20,000 **49.** The area is maximized if the length and width are 100 ft. **51.** 55 units **53.** $-6, 6$; the product is -36.

Review Exercises **1.** -11 **2.** $-5, 3$ **3.** $\pm 2i\sqrt{5}$

4. $x \leq 4.5$

$-6 \ -5 \ -4 \ -3 \ -2 \ -1 \ 0 \ 1 \ 2 \ 3 \ 4 \ 5 \ 6$

5. $x \leq -5$ or $x \geq 11$

$-12 \ -10 \ -8 \ -6 \ -4 \ -2 \ 0 \ 2 \ 4 \ 6 \ 8 \ 10 \ 12$

6. $x + y > 2$
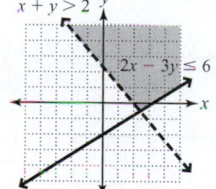
$2x - 3y \leq 6$

11.5 Exercises **1.** Solve the related equation and then determine which of the three intervals satisfy the inequality by examining one value in each interval. If this one value satisfies the inequality, then all points in the interval satisfy that inequality. **3.** Yes, for example, $x^2 + 2 \leq 0$ has an empty solution set because $x^2 + 2$ is always positive. **5.** In both cases, the intervals to check are $(-\infty, -5), (-5, 2)$, and $(2, \infty)$, and in both cases, only x-values in $(-5, 2)$ satisfy the original inequalities.
7. a. $x = -5, -1$ **b.** $(-5, -1)$ **c.** $(-\infty, -5) \cup (-1, \infty)$
9. a. $x = -1, 3$ **b.** $(-\infty, -1] \cup [3, \infty)$ **c.** $[-1, 3]$
11. a. $x = -2$ **b.** $(-\infty, -2) \cup (-2, \infty)$ **c.** $\varnothing$ **13. a.** $\varnothing$
b. $\varnothing$ **c.** $\mathbb{R}$ or $(-\infty, \infty)$
15. $(-4, -2)$

$-6 \ -5 \ -4 \ -3 \ -2 \ -1 \ 0 \ 1 \ 2 \ 3 \ 4 \ 5 \ 6$

17. $(-\infty, 2) \cup (5, \infty)$

$-6 \ -5 \ -4 \ -3 \ -2 \ -1 \ 0 \ 1 \ 2 \ 3 \ 4 \ 5 \ 6$

19. $(-4, -1)$
$\quad$ -6 -5 -4 -3 -2 -1 0 1 2 3 4 5 6

21. $(-\infty, 1) \cup (3, \infty)$
$\quad$ -6 -5 -4 -3 -2 -1 0 1 2 3 4 5 6

23. $[2, 4]$
$\quad$ -6 -5 -4 -3 -2 -1 0 1 2 3 4 5 6

25. $(-\infty, -1] \cup [5, \infty)$
$\quad$ -6 -5 -4 -3 -2 -1 0 1 2 3 4 5 6

27. $(-2, 5)$
$\quad$ -6 -5 -4 -3 -2 -1 0 1 2 3 4 5 6

29. $(-\infty, \infty)$, or $\mathbb{R}$
$\quad$ -6 -5 -4 -3 -2 -1 0 1 2 3 4 5 6

31. No solution, or $\varnothing$.
$\quad$ -6 -5 -4 -3 -2 -1 0 1 2 3 4 5 6

33. $(-\infty, -5) \cup (-0.25, \infty)$
$\quad$ -6 -5 -4 -3 -2 -1 0 1 2 3 4 5 6

35. $(-\infty, 0) \cup (5, \infty)$
$\quad$ -6 -5 -4 -3 -2 -1 0 1 2 3 4 5 6

37. $[0, 6]$
$\quad$ -6 -5 -4 -3 -2 -1 0 1 2 3 4 5 6

39. $(-\infty, \infty)$, or $\mathbb{R}$
$\quad$ -6 -5 -4 -3 -2 -1 0 1 2 3 4 5 6

41. $[-4, -2] \cup [4, \infty)$
$\quad$ -6 -5 -4 -3 -2 -1 0 1 2 3 4 5 6

43. $(-\infty, -6) \cup (-2, 1)$
$\quad$ -9 -8 -7 -6 -5 -4 -3 -2 -1 0 1 2 3

45. $\left(\dfrac{-2 - \sqrt{7}}{3}, \dfrac{-2 + \sqrt{7}}{3} \right)$

47. $\left(-\infty, \dfrac{-2 - \sqrt{7}}{2} \right] \cup \left[\dfrac{-2 + \sqrt{7}}{2}, \infty \right)$

49. $(-\infty, -4 - \sqrt{6}] \cup [-4 + \sqrt{6}, \infty)$

51. $(-\infty, -4) \cup (1, \infty)$
$\quad$ -6 -5 -4 -3 -2 -1 0 1 2 3 4 5 6

53. $(-5, -1]$
$\quad$ -6 -5 -4 -3 -2 -1 0 1 2 3 4 5 6

55. $(-4, \infty)$
$\quad$ -6 -5 -4 -3 -2 -1 0 1 2 3 4 5 6

57. $\left(-\infty, -\dfrac{9}{2} \right) \cup (-3, \infty)$
$\quad$ -6 -5 -4 -3 -2 -1 0 1 2 3 4 5 6

59. $(4, 7)$
$\quad$ -1 0 1 2 3 4 5 6 7 8 9 10 11

61. $(3, 5]$
$\quad$ -6 -5 -4 -3 -2 -1 0 1 2 3 4 5 6

63. $(-\infty, -3] \cup [2, 5)$
$\quad$ -6 -5 -4 -3 -2 -1 0 1 2 3 4 5 6

65. $(0, \infty)$
$\quad$ -6 -5 -4 -3 -2 -1 0 1 2 3 4 5 6

67. $(-\infty, -1) \cup (2, 5)$
$\quad$ -6 -5 -4 -3 -2 -1 0 1 2 3 4 5 6

69. a. 6 sec. **b.** At 2 sec. and again at 3 sec. **c.** $(2, 3)$ sec.
d. $[0, 2) \cup (3, 6]$ sec. **71. a.** $x \geq 4$ in.
b. Base ≥ 10 in., height ≥ 2 in. **73.** $2 < x \leq 8$

Review Exercises
1. $9, (x - 3)^2$ **2.** $\dfrac{25}{4}, \left(x + \dfrac{5}{2} \right)^2$

3. 18 **4.** Downward, because the coefficient of x^2 is negative. **5.** $(-6, 0)(2, 0)$ **6.** $(3, -4)$

Chapter 11 Review Exercises
1. false **2.** false
3. false **4.** false **5.** true **6.** $\sqrt{a}, -\sqrt{a}$ **7.** $\left(\dfrac{b}{2} \right)^2$

8. $\dfrac{-b \pm \sqrt{b^2 - 4ac}}{2a}$ **9.** $b^2 - 4ac$ **10.** $(h, k); x = h$

11. ± 4 **12.** $\pm \dfrac{1}{6}$ **13.** $\pm 2\sqrt{7}$ **14.** $\pm \sqrt{14}$ **15.** $\pm i\sqrt{3}$

16. $-12, -2$ **17.** $9 \pm 4i$ **18.** $-\dfrac{7}{5}, \dfrac{1}{5}$ **19.** $-7, -1$

20. $-8, 14$ **21.** $\dfrac{3 \pm i\sqrt{5}}{2}$ **22.** $-1, \dfrac{3}{4}$ **23.** $-1 \pm \sqrt{6}$

24. $\dfrac{1 \pm i\sqrt{2}}{3}$ **25.** $\dfrac{-1 \pm \sqrt{41}}{4}$ **26.** $-\dfrac{3}{20}, \dfrac{1}{10}$

27. $D = 64$; two rational **28.** $D = -71$; two nonreal complex **29.** $D = 0$; one rational

30. $D = 0.48$; two irrational **31.** $\pm \dfrac{3\sqrt{2}}{2}$

32. $-\dfrac{5}{2}$ (5 is extraneous) **33.** $\dfrac{1}{3}, \dfrac{1}{2}$ **34.** $\dfrac{3 \pm \sqrt{17}}{4}$

35. No solution. **36.** $1 \left(\dfrac{1}{9} \text{ is extraneous} \right)$

37. $2 \left(-\dfrac{3}{2} \text{ is extraneous} \right)$ **38.** $\dfrac{1}{3}, \dfrac{2}{3}$ **39.** $\pm \sqrt{2}, \pm \sqrt{3}$

40. $\pm 1, \pm \dfrac{\sqrt{2}}{2}$ **41.** $-\dfrac{9}{2}, -\dfrac{14}{3}$ **42.** $-26, -2$ **43.** $27, 512$

44. $\dfrac{16}{625}$ (81 is extraneous)

45. a. $(0, 0)$ **b.** downward
c. $(0, 0)$ **d.** $x = 0$
e.

46. a. no x-intercepts, $(0, 1)$
b. upward **c.** $(0, 1)$
d. $x = 0$
e.

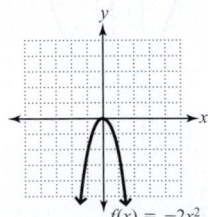

$f(x) = -2x^2$
y $g(x) = \frac{1}{2}x^2 + 1$

47. a. $(2, 0), \left(0, -\dfrac{4}{3} \right)$ **b.** downward **c.** $(2, 0)$ **d.** $x = 2$
e.

$h(x) = -\frac{1}{3}(x - 2)^2$

48. a. $\left(-3 + \dfrac{\sqrt{2}}{2}, 0 \right), \left(-3 - \dfrac{\sqrt{2}}{2}, 0 \right), (0, 34)$ **b.** upward

c. $(-3, -2)$ **d.** $x = -3$ **e.**

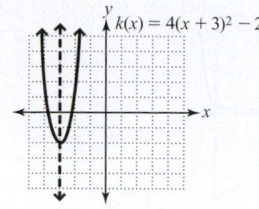

49. a. $m(x) = (x + 1)^2 - 2$
b. $(-1 + \sqrt{2}, 0), (-1 - \sqrt{2}, 0), (0, -1)$ **c.** upward
d. $(-1, -2)$ **e.** $x = -1$ **f.**

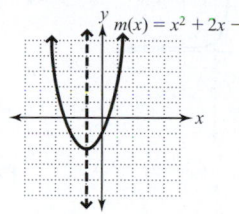

g. domain: $\{x \mid x \text{ is a real number}\}$ or $(-\infty, \infty)$;
range: $\{y \mid y \geq -2\}$ or $[-2, \infty)$.
50. a. $p(x) = -0.5(x - 4)^2 + 2$ **b.** $(2, 0)(6, 0), (0, -6)$
c. downward **d.** $(4, 2)$ **e.** $x = 4$
f.

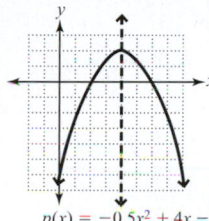

$p(x) = -0.5x^2 + 4x - 6$

g. domain: $\{x \mid x \text{ is a real number}\}$
or $(-\infty, \infty)$; range:
$\{y \mid y \leq 2\}$ or $(-\infty, 2]$.

51. $(-\infty, -5) \cup (3, \infty)$ **52.** $[2, 4]$ **53.** $(-7, -2)$
54. $[-3, 1] \cup [2, \infty)$ **55.** $(-\infty, -3] \cup (1, \infty)$
56. $(-\infty, -4) \cup (-2, \infty)$ **57.** $(-\infty, 4) \cup \left[\dfrac{17}{4}, \infty\right)$
58. $(-\infty, -2) \cup (3, 5)$ **59.** 150 ft. **60.** 4 m/sec.
61. Width: 15 ft.; length: 19 ft. **62.** $9\sqrt{3}$ in.
63. Height: 6.65 ft.; base: 15.65 ft. **64. a.** 0.75 sec. **b.** 9 ft.
c. 1.5 sec. **d.**

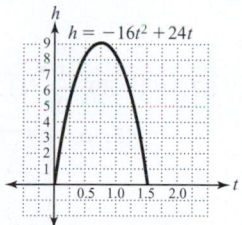

$h = -16t^2 + 24t$

65. a. 38.88 yd. **b.** 72 yd. **66. a.** $x \geq 10$ in.
b. Height ≥ 8 in., base ≥ 14 in.

Chapter 11 Practice Test **1.** ± 9 [11.1]

2. $3 \pm 2\sqrt{5}$ [11.1] **3.** $4 \pm 2\sqrt{3}$ [11.1] **4.** $\dfrac{3 \pm 2\sqrt{6}}{3}$ [11.1]

5. $-2, \dfrac{3}{2}$ [11.2] **6.** 3, 5 [11.2] **7.** $-8, 2$ [11.1, 11.2]

8. ± 9 [11.1, 11.2] **9.** $\dfrac{-3 \pm i\sqrt{3}}{4}$ [11.1, 11.2]

10. $0, -2$ [11.1, 11.2] **11.** $\pm 4i$ [11.1, 11.2]

12. $0, -\dfrac{5}{3}$ [11.1, 11.2] **13.** $-\dfrac{6}{5}, 4$ [11.3] **14.** $-\dfrac{2}{3}, 1$ [11.3]

15. No solution. [11.3] **16.** 1 (-4 is extraneous) [11.3]

17. $\pm\dfrac{1}{3}, \pm i\sqrt{3}$ [11.3] **18.** $-5, 0$ [11.3]

19. a. $(3 + \sqrt{5}, 0)(3 - \sqrt{5}, 0), (0, -4)$
b. $f(x) = -(x - 3)^2 + 5$ **c.** downward **d.** $(3, 5)$ **e.** $x = 3$
f.

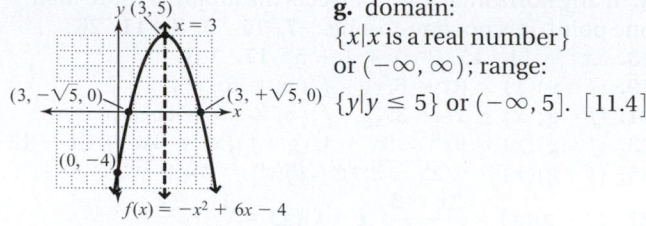

$f(x) = -x^2 + 6x - 4$

g. domain:
$\{x \mid x \text{ is a real number}\}$
or $(-\infty, \infty)$; range:
$\{y \mid y \leq 5\}$ or $(-\infty, 5]$. [11.4]

20. $[-1, 4]$

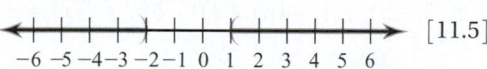

[11.5]

21. $(-\infty, -2) \cup (1, \infty)$

[11.5]

22. 2.5 sec. [11.2] **23.** Width: 12.36 ft.; length: 32.36 ft. [11.2]
24. a. 29.125 m **b.** 70.66 m **c.**

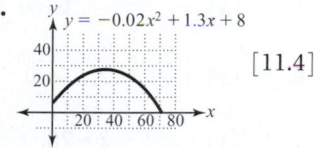

$y = -0.02x^2 + 1.3x + 8$

[11.4]

25. a. $0 < w \leq 25$ ft. **b.** $15 < l \leq 40$ ft. [11.5]

Chapters 1–11 Cumulative Review Exercises
1. false **2.** false **3.** true **4.** parabola, $a > 0$ **5.** $\sqrt[n]{ab}$
6. $\sqrt{a}, -\sqrt{a}$ **7.** $-6x - 6$ **8.** $-56mn^5$
9. $4x^3 - 14x^2 + 7x - 3$ **10.** $(m + 2)(m^2 - 2m + 4)$
11. $(x + 5)^2$ **12.** $\dfrac{1}{u - 1}$ **13.** $\dfrac{x + 10}{(x - 2)(x + 2)}$
14. $x^{1/2}$ **15.** $\dfrac{\sqrt{mn} + n}{m - n}$ **16.** $-\dfrac{7}{2}$ **17.** $-\dfrac{8}{3}, 0$ **18.** $\dfrac{3}{2}, -5$
19. 8 **20.** 17 **21.** $3 \pm i\sqrt{2}$
22. a.

b. $\{x \mid -6 < x < 1\}$ **c.** $(-6, 1)$
23. a.

b. $\{x \mid -4 \leq x \leq 3\}$ **c.** $[-4, 3]$
24.

$f(x) = -2x$

25. $2x + 3y = 8$

26. $(2, -1, 4)$ **27.** 5 hr. **28.** 75 ml **29.** 3.2 cm/sec.
30. a. $\dfrac{\sqrt{6}}{2}$ sec. **b.** 144 ft.

Chapter 12

12.1 Exercises **1.** No, $f \circ g = g \circ f$ if f and g are
inverses of each other and for a few other functions. **3.** The
domain of a function is the range of its inverse; the range of a

function is the domain of its inverse. This occurs because the ordered pairs of f and f^{-1} have each x and y interchanged.
5. If any horizontal line intersects the graph in more than one point, it is not one-to-one.   **7.** 14   **9.** 3   **11.** 26
13. $3x^2 + 14$ **15.** $3\sqrt{x+1} + 5$ **17.** $\sqrt{6}$
19. $(f \circ g)(x) = 6x + 6$; $(g \circ f)(x) = 6x - 2$
21. $(f \circ g)(x) = x^2 + 3$; $(g \circ f)(x) = x^2 + 4x + 5$
23. $(f \circ g)(x) = 9x^2 + 9x - 4$; $(g \circ f)(x) = 3x^2 + 9x - 12$
25. $(f \circ g)(x) = \sqrt{2x - 3}$; $(g \circ f)(x) = 2\sqrt{x+2} - 5$
27. $(f \circ g)(x) = \dfrac{2x - 3}{x - 3}$; $(g \circ f)(x) = \dfrac{1 - 2x}{x + 1}$
29. Domain is $[0, \infty)$ and range is $[3, \infty)$.   **31.** yes   **33.** no
35. yes   **37.** yes   **39.** yes   **41.** yes   **43.** no   **45.** yes   **47.** yes
49. Yes, because $(f \circ g)(x) = (g \circ f)(x) = x$.   **51.** yes
53. no   **55.** $f^{-1} = \{(2,-3), (-3,-1), (4,0), (6,4)\}$
57. $f^{-1} = \{(-2,7), (2,9), (1,-4), (3,3)\}$ **59.** $f^{-1}(x) = x - 6$
61. $f^{-1}(x) = \dfrac{x - 3}{2}$ **63.** $f^{-1}(x) = \sqrt[3]{x + 1}$
65. $f^{-1}(x) = \dfrac{2 - 2x}{x}$ **67.** $f^{-1}(x) = \dfrac{3x + 2}{x - 1}$
69. $f^{-1}(x) = x^2 + 2, x \geq 0$ **71.** $f^{-1}(x) = \sqrt[3]{\dfrac{x - 4}{2}}$
73. $f^{-1}(x) = x^3 - 2$ **75.** $f^{-1}(x) = \dfrac{x^3}{16} - 2$

77. 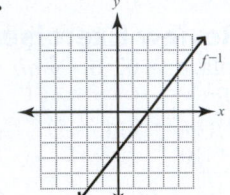   **79.**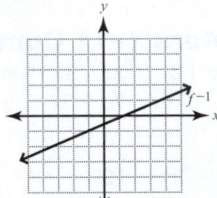

81. 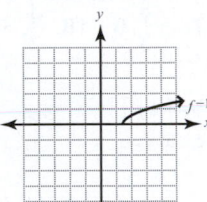   **83.**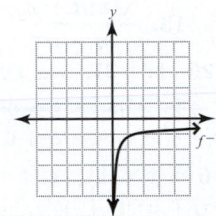

85. 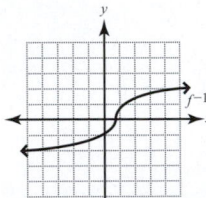   **87.**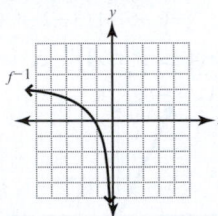

89. a. $y = \dfrac{x - 100}{0.05}$ **b.** x represents the salary; y represents the sales.   **c.** \$5000   **91. a.** $y = \dfrac{6825 - x}{20}$ **b.** x represents the cost; y represents the number of 36-in. fans.   **c.** 80
93. $(5, -4)$ **95.** $f^{-1}(x) = \dfrac{x - b}{a}$

Review Exercises 1. $32 = 2^5$ **2.** 8   **3.** $-\dfrac{1}{27}$ **4.** $\dfrac{1}{16}$
5. 8   **6.** $(3, 1)$

12.2 Exercises 1. Yes, the graphs are symmetric. The point $(1, 2)$ on $f(x)$ corresponds to the point $(-1, 2)$ on $g(x)$, and so on.   **3.** The domain is all real numbers; the range is $(0, \infty)$.
5. Because the exponential function is one-to-one.

7. 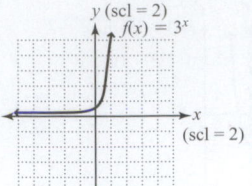   **9.**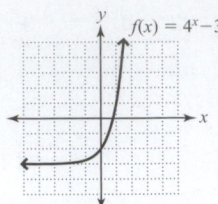

11. 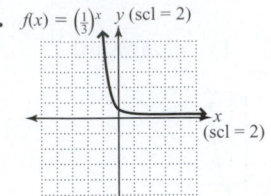   **13.**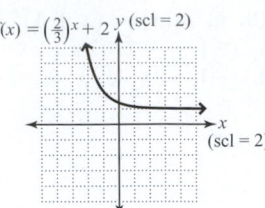

15. 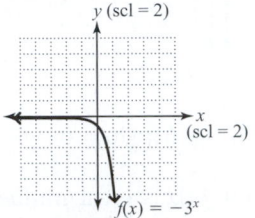   **17.**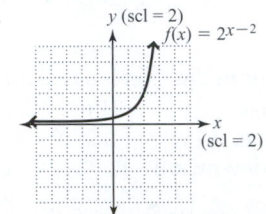

19. 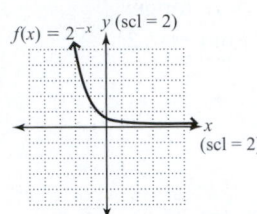   **21.**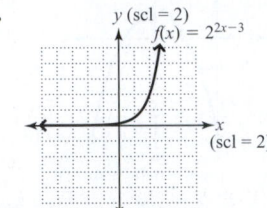

23. 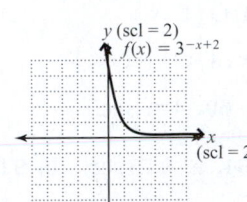   **25.** 3   **27.** $\dfrac{5}{3}$ **29.** $\dfrac{1}{2}$

31. -2 **33.** -2 **35.** 3   **37.** -4 **39.** $\dfrac{1}{2}$ **41.** -12

43. a. 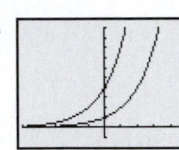   **45. a.**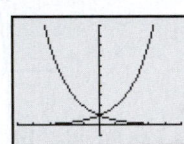

b. The graph of g is the graph of f shifted 2 units to the left.   **b.** The graph of g is the graph of f reflected about the y-axis.

47. 6400 cells   **49.** 6144 billion people   **51.** \$25,870.70
53. 0.020 g   **55.** $\approx 16{,}216.42$ **57.** 1.225 parts per million
59. $\approx 18{,}822$ million transistors

Review Exercises 1. 125   **2.** 9   **3.** $\dfrac{1}{64}$ **4.** $\sqrt[3]{5^2}$ or $(\sqrt[3]{5})^2$ **5.** $f^{-1}(x) = \dfrac{x - 4}{3}$ **6.**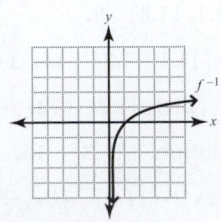

12.3 Exercises 1. $f^{-1}(x) = \log_3 x$ because logarithmic and exponential functions are inverses.   3. $x > 0$
5. Logarithms are exponents because logarithms are inverses of exponential functions.   7. $\log_2 32 = 5$ 9. $\log_{10} 1000 = 3$
11. $\log_e x = 4$ 13. $\log_5 \frac{1}{125} = -3$ 15. $\log_{10} \frac{1}{100} = -2$
17. $\log_{625} 5 = \frac{1}{4}$ 19. $\log_{1/4} \frac{1}{16} = 2$ 21. $\log_7 \sqrt{7} = \frac{1}{2}$
23. $3^4 = 81$ 25. $4^{-2} = \frac{1}{16}$ 27. $10^2 = 100$ 29. $e^5 = a$
31. $e^{-4} = \frac{1}{e^4}$ 33. $\left(\frac{1}{8}\right)^2 = \frac{1}{64}$ 35. $\left(\frac{1}{5}\right)^{-2} = 25$
37. $7^{1/2} = \sqrt{7}$ 39. 32   41. $\frac{1}{25}$ 43. 4   45. -2 47. 10
49. 2   51. $\frac{1}{4}$ 53. 243   55. 2   57. -6 59. 16   61. $\frac{1}{64}$
63. 2   65. 125   67. 3   69. $\frac{1}{16}$
71. 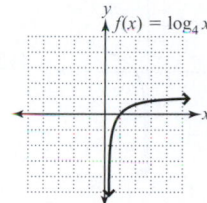   73.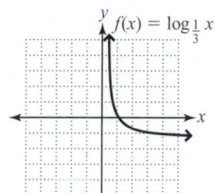
75. 97%   77. After 4 days.   79. $\log_b b = 1$ because $b^1 = b$
81. $\log_b 1 = 0$ because $b^0 = 1$.

Review Exercises 1. x^{-6} 2. x^6 3. x^{15} 4. x^3
5. x^5 6. $x^{3/4}$

12.4 Exercises 1. Because the exponential and logarithmic functions are inverses.   3. $\log_b (x + y)$; there is no rule for the logarithm of a sum.   5. 2   7. r 9. $4x$
11. 5   13. y 15. $7x$ 17. $\log_2 5 + \log_2 y$
19. $\log_a p + \log_a q$ 21. $\log_4 m + \log_4 n + \log_4 p$
23. $\log_a x + \log_a(x - 5)$ 25. $\log_3 40$ 27. $\log_4 27$
29. $\log_a 7m$ 31. $\log_4 ab$ 33. $\log_a(2x^2 + 10x)$
35. $\log_4(x^2 + 4x + 3)$ 37. $\log_2 7 - \log_2 9$
39. $\log_a x - \log_a 5$ 41. $1 - \log_a b$
43. $\log_a x - \log_a(x - 3)$ 45. $\log_4(2x - 3) - \log_4(4x + 5)$
47. $\log_6 8$ 49. 1   51. $\log_a \frac{x}{3}$ 53. $\log_4 \frac{p}{q}$ 55. $\log_b \frac{x}{x - 4}$
57. 1   59. $6\log_4 3$ 61. $7\log_a x$ 63. $\frac{1}{2}\log_a 3$ 65. $\frac{2}{3}\log_3 x$
67. $-2\log_a 6$ 69. $-2\log_a y$ 71. $\log_3 5^4$ 73. $\log_2 \frac{1}{x^3}$
75. $\log_7 8$ 77. $\log_a \sqrt[4]{x^3}$ 79. $\log_a 4$ 81. $\log_3 \frac{1}{\sqrt{x}}$
83. $3\log_a x - 4\log_a y$ 85. $4\log_3 a + 2\log_3 b$
87. $\log_a x + \log_a y - \log_a z$ 89. $2\log_x a - \log_x b - 3\log_x c$
91. $\frac{3}{4}\log_4 x - \frac{1}{4}\log_4 y$ 93. $\frac{2}{3}\log_a x + \frac{1}{3}\log_a y - \log_a z$
95. $\log_3 \frac{1}{2}$ 97. $\log_b x^4 y^3$ 99. $\log_a \sqrt{\frac{5}{7}}$ 101. $\log_a \sqrt[3]{(x^2 y^3)^2}$
103. $\log_b(3x^2 - 2x)$ 105. $\log_a \frac{(x - 2)^3}{(x + 1)^4}$ 107. $\log_a \frac{x^2 z^4}{w^3 u^6}$

Review Exercises 1. $10^{-2.5}$ 2. 10^9 3. \$13,468.55
4. $\log_{10} 50 = 1.6990$ 5. $10^{1.6532} = 45$ 6. $e^{-1.3863} = 0.25$

12.5 Exercises 1. $b > 0$, $b \neq 1$. If $y = \log_b x$, then $b^y = x$. If $b < 0$ and $y = \frac{1}{2}$ or any other fraction whose denominator is even, then x is imaginary. If $b = 1$, then $x = 1$ for all values of y and the graph is a vertical line.
3. Positive. If $y = \ln x$, then $e^y = x$. If $x > 1$, then $e^y > 1$, which is true if $y > 0$.   5. 0.91629   7. 1.8062   9. -2.1739
11. 2.6391   13. 4.1761   15. -5.7959 17. 2.2343
19. 4.3720   21. -3.3814 23. 5.6864   25. 0.4343
27. Error results because the domain of $\log_a x$ is $(0, \infty)$; so log 0 is undefined.   29. 2   31. -2 33. $\frac{1}{3}$ 35. -3 37. 3
39. $\frac{1}{2}$ 41. 90 dB   43. 10^{-6} watts/m^2 45. 2.796
47. $10^{-3.5}$ moles/L   49. 22.9, or ≈ 23 yr.   51. The 1964 Alaska earthquake was about four times as severe as the 1906 San Francisco earthquake.   53. $10^{22.75}$ ergs   55. 1985
57. 6124 ft.

Review Exercises 1. $\log_3 \frac{2x + 1}{x - 1}$ 2. $3^2 = 2x + 5$
3. -11 4. $-5, 3$ 5. 8   6. 10

12.6 Exercises 1. $m = n$ 3. If $\log_b x = \log_b y$, then $x = y$.   5. The solution $x = -5$ is rejected because substituting into the equation gives $\log(-5)$ and $\log(-3)$, which are both undefined.   7. 3.1699   9. 1.0767   11. -1.5693
13. 2.8617   15. 12.4251   17. 7.1285   19. 0.5365
21. 107.2959   23. -73.1563 25. 6   27. 11   29. 6   31. ± 4
33. 4, -9 35. 7   37. No solution.   39. $-6, 2$ 41. 2
43. 10   45. 3   47. $\frac{3}{2}$ 49. 3   51. 4   53. $\frac{5}{6}$ 55. $\frac{5}{2}$ 57. 6
59. 4   61. 9.5 yr.   63. a. \$19,676.82   b. 9.3 yr.
65. a. 5.58%   b. 0.18   67. a. 875 mosquitoes   b. 75 days
69. a. ≈ 7.44 billion   b. ≈ 9 yr. after 2005, in 2014.
71. a. 1.89 in. of mercury   b. 21.8 mi.   73. a. 79.8%
b. 7.76, so about age 8.   75. \$8.13   77. \$1009.2 billion
79. 1.7925   81. 0.5283   83. -2.3219 85. 0.3685   87. 47%

Review Exercises 1. $x + 3$ 2. $28 + \sqrt{5}$ 3. 29
4. $x^2 + 2x + 3$ 5. $4x^2 - 4x + 5$ 6. 13

Chapter 12 Review Exercises 1. true   2. false
3. false   4. false   5. true   6. false   7. 25   8. 167   9. -2
10. 14   11. $(f \circ g)(x) = 6x + 3$, $(g \circ f)(x) = 6x - 9$
12. $(f \circ g)(x) = 9x^2 - 42x + 53$, $(g \circ f)(x) = 3x^2 + 5$
13. $(f \circ g)(x) = \sqrt{2x - 4}$, $(g \circ f)(x) = 2\sqrt{x - 3} - 1$
14. $(f \circ g)(x) = \frac{4x - 4}{x - 4}$, $(g \circ f)(x) = \frac{3 - 3x}{x + 3}$ 15. no
16. yes
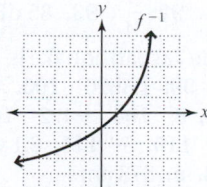
17. yes   18. yes   19. no   20. no   21. $f^{-1}(x) = \frac{x - 4}{5}$
22. $f^{-1}(x) = \sqrt[3]{x} - 6$ 23. $f^{-1}(x) = \frac{4 - 5x}{x}$
24. $f^{-1}(x) = \frac{x^3 - 2}{3}$ 25. $(-6, 4)$ 26. $f^{-1}(b) = a$

27.

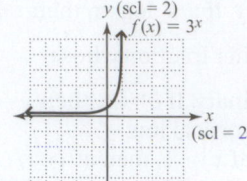

28.

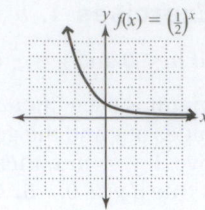

29.

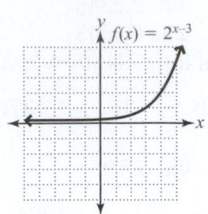

30.

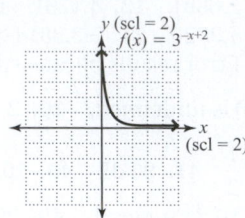

b. $f[f^{-1}(x)] = f\left[\dfrac{x+3}{4}\right] = 4\left(\dfrac{x+3}{4}\right) - 3 = x + 3 - 3 = x$

$f^{1}[f(x)] = f^{-1}(4x - 3) = \dfrac{4x - 3 + 3}{4} = \dfrac{4x}{4} = x$ [12.1]

c. The graphs are symmetric about the graph of $y = x$. [12.1]

3.

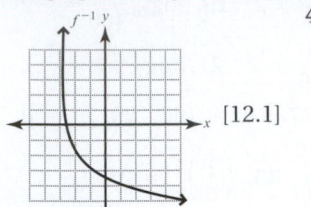

[12.1]

4.

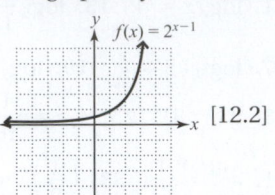

[12.2]

31. 4 **32.** $\dfrac{3}{2}$ **33.** -2 **34.** -2 **35.** 4 **36.** 2 **37.** 3

38. $-\dfrac{6}{7}$ **39.** 6277 cells **40.** \$40,353.57 **41.** 7.16 g

42. \$8.01 **43.** $\log_7 343 = 3$ **44.** $\log_4 \dfrac{1}{64} = -3$

45. $\log_{3/2} \dfrac{81}{16} = 4$ **46.** $\log_{11} \sqrt[3]{11} = \dfrac{1}{3}$ **47.** $9^2 = 81$

48. $\left(\dfrac{1}{5}\right)^{-3} = 125$ **49.** $a^4 = 16$ **50.** $e^b = c$ **51.** $\dfrac{1}{81}$

52. 4 **53.** 5 **54.** -2 **55.** 3 **56.** $\dfrac{1}{10}$ **57.** 2 **58.** 3

59.

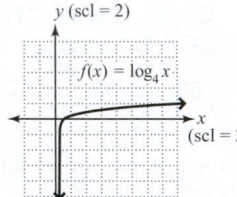

60.

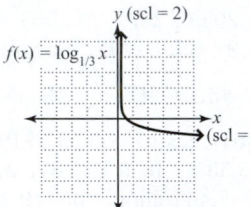

5. -10 [12.2] **6. a.** \$42,143.63 **b.** ≈ 8.2 yr. [12.2]
7. a. 0.68 g **b.** ≈ 1.16 hr. [12.2] **8. a.** 14.8 million people,
56.2 million people **b.** 1989 [12.2] **9.** $\left(\dfrac{1}{3}\right)^{-4} = 81$ [12.3]

10. -3 [12.3] **11.** 5 [12.3] **12.**

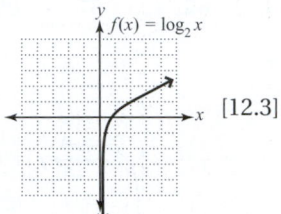

[12.3]

13. $4 \log_b x + 2 \log_b y - \log_b z$ [12.4]
14. $\dfrac{5}{4} \log_b x - \dfrac{7}{4} \log_b y$ [12.4] **15.** $\log_b \sqrt[4]{x^6 y^9}$ [12.4]
16. a. ≈ 74 g **b.** ≈ 247.6 days. [12.5] **17.** 4.6433 [12.5]
18. 3 [12.5] **19.** 3 [12.5] **20. a.** 16.5 generations
b. 100,000 [12.5]

Chapters 1–12 Cumulative Review Exercises

1. true **2.** true **3.** false **4.** -8 **5.** x-intercepts
6. $x - 3 = 4, x - 3 = -4$ **7.** 86 **8.** $\dfrac{1}{a^{12}}$ **9.** $\dfrac{3}{pq} - \dfrac{2p^2}{q^2} + \dfrac{6q^2}{7}$
10. $\dfrac{-x^2 - 5x - 9}{(x + 3)^2}$ **11.** $24c^4 \sqrt{15}$ **12.** $20 - 10i$ **13.** $2x$
14. $\dfrac{d - 4}{2d(3d + 4)}$ **15.** $(x + 2y)(x - 2y)(x + 3)(x - 3)$
16. $3x^2 y(4x - 3y)(2x - y)$ **17.** 1, 2 **18.** All real numbers.
19. $(-3, 2)$ **20.** $\dfrac{1}{2}, \dfrac{7}{2}$ **21.** 3, (2 is extraneous) **22.** No solution (5 is extraneous). **23.** $-\dfrac{1}{2}$ **24.** 16 **25.** 3 **26. a.** $\dfrac{2}{5}$
b. $(5, 0), (0, -2)$ **c.** **27.** 20, 21, 29

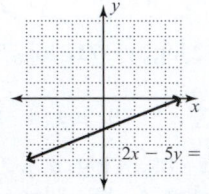

28. \$16,909.08 **29.** 0.857 hr. or ≈ 51 min. **30.** 5 hr.

61. Domain is $(-\infty, \infty)$ and range is $(0, \infty)$ because
$f(x) = \log_b x$ and $g(x) = b^x$ are inverses. **62.** 30 dB
63. 8 **64.** 6 **65.** $1 + \log_6 x$ **66.** $\log_4 x + \log_4(2x - 5)$
67. $\log_3(32)$ **68.** $\log_5(3x^2 - 6x)$ **69.** $\log_b x - \log_b 5$
70. $\log_a(3x - 2) - \log_a(4x + 3)$ **71.** $\log_8 2$

72. $\log_2 \dfrac{x + 5}{2x - 3}$ **73.** $4 \log_3 7$ **74.** $\dfrac{1}{3} \log_a x$ **75.** $-4 \log_4 a$

76. $\dfrac{4}{5}$ **77.** $\log_a x^4$ **78.** $\log_a \sqrt[5]{y^3}$ **79.** $2 \log_a x + 3 \log_a y$

80. $4 \log_a c - 3 \log_a d$ **81.** $2 \log_a x + 3 \log_a y - 4 \log_a z$

82. $\dfrac{3}{2} - 2 \log_a b$ **83.** $\log_x y^3 z^5$ **84.** $\log_a \dfrac{4^3}{3^2}$ **85.** $\log_a \sqrt[4]{x^2 y^3}$

86. $\log_a(x + 5)^4 (x - 3)^2$ **87.** 2.5132 **88.** -2.4559

89. -3.1466 **90.** 4.5218 **91.** -5 **92.** $\dfrac{1}{4}$ **93.** 85 dB

94. pH $= 2.9$ **95.** ≈ 17 yr. **96.** The 7.8 earthquake is 10 times
as severe. **97.** 1.5773 **98.** 0.5360 **99.** 1.3626 **100.** 4.9592
101. 0.5995 **102.** -536.4793 **103.** 8 **104.** $-4, 2$ **105.** 3
106. 20 **107.** 36 **108.** 8 **109.** 2 **110.** 5 **111.** 6.1 yr.
112. a. \$12,254.71 **b.** 7.7 yr. **113. a.** 573 ants
b. ≈ 23 months **114.** 24 min. **115.** 1.9534 **116.** -2.5850

Chapter 12 Practice Test
1. $f[g(x)] = 9x^2 - 30x + 19$ [12.1]

2. a. $f^{-1}(x) = \dfrac{x + 3}{4}$ [12.1]

Chapter 13

13.1 Exercises **1.** It is a parabola because only one variable is squared. **3.** The parabola opens to the right because
the variable y is squared and $a = 3$, which is positive.
5. circle; $(-3, 2)$; 4

7. Opens upward;
vertex: $(1, 2)$;
axis of symmetry: $x = 1$.

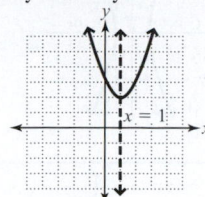

9. Opens downward;
vertex: $(-1, 4)$;
axis of symmetry: $x = -1$.

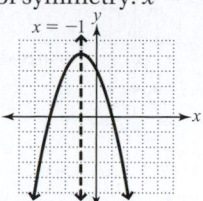

11. Opens right;
vertex: $(-2, -2)$;
axis of symmetry: $y = -2$.

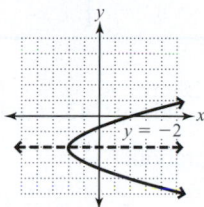

13. Opens left;
vertex: $(3, 1)$;
axis of symmetry: $y = 1$.

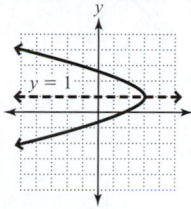

15. Opens right;
vertex: $(-4, -2)$;
axis of symmetry: $y = -2$.

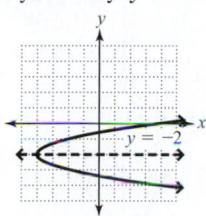

17. Opens left;
vertex: $(-5, -2)$;
axis of symmetry: $y = -2$.

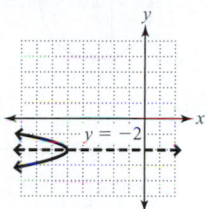

19. Opens right;
vertex: $(-1, -2)$;
axis of symmetry: $y = -2$.

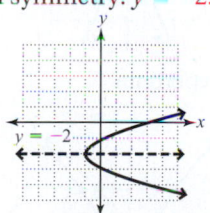

21. Opens left;
vertex: $(4, 3)$;
axis of symmetry: $y = 3$.

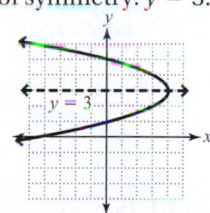

23. Opens right; vertex:
$(-5, -2)$; axis of symmetry:
$y = -2$.

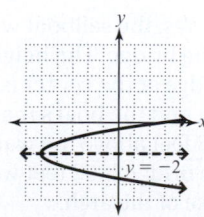

25. Opens right; vertex:
$(-2, 1)$; axis of symmetry:
$y = 1$.

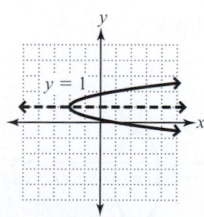

27. Opens left; vertex: $(7, 1)$; axis of symmetry: $y = 1$.

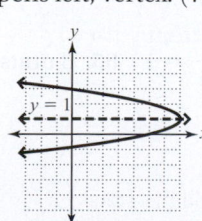

29. c   **31.** a   **33.** $5, \left(-\dfrac{5}{2}, 4\right)$ **35.** $13, \left(-\dfrac{11}{2}, 2\right)$

37. $13, \left(-2, -\dfrac{15}{2}\right)$ **39.** $2\sqrt{5}, (3, 6)$ **41.** $4\sqrt{5}, (-1, 0)$

43. $\sqrt{34}, \left(\dfrac{7}{2}, -\dfrac{7}{2}\right)$ **45.** 5   **47.** $\sqrt{89}$

49. Center: $(2, 1)$;
radius: 2.

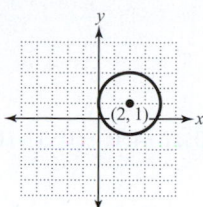

51. Center: $(-3, -2)$;
radius: 9.

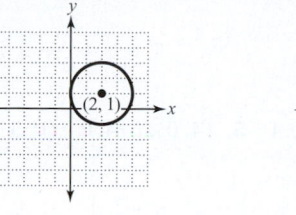

53. Center: $(5, -3)$; radius: 7.

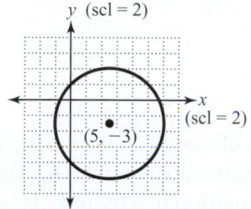

55. Center: $(1, -1)$;
radius: $2\sqrt{3}$.

57. Center: $(-4, -2)$;
radius: $4\sqrt{2}$.

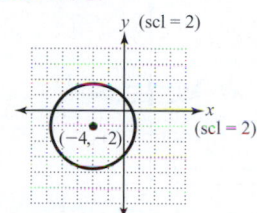

59. Center: $(-4, 3)$;
radius: 3.

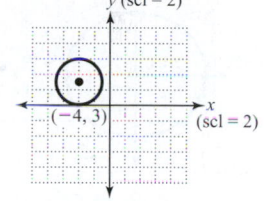

61. Center: $(-5, 2)$;
radius: 8.

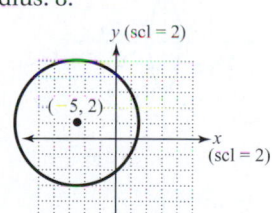

63. Center: $(-7, 2)$;
radius: 2.

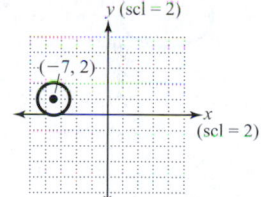

65. b   **67.** c

69.

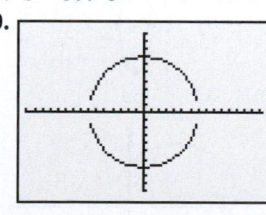

71.

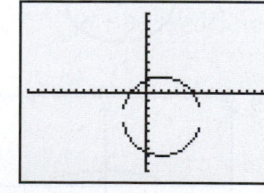

73. $(x - 4)^2 + (y - 2)^2 = 16$ **75.** $(x + 4)^2 + (y + 3)^2 = 25$
77. $(x - 6)^2 + (y + 2)^2 = 14$ **79.** $(x + 5)^2 + (y - 2)^2 = 45$
81. $(x - 2)^2 + (y - 4)^2 = 25$ **83.** $(x - 2)^2 + (y - 4)^2 = 169$
85. $(x - 2)^2 + (y + 5)^2 = 64$ **87.** $(x - 1)^2 + (y - 2)^2 = 25$
89. $(x - 2)^2 + (y - 3)^2 = 41$
91. a. 256 ft.   **b.** 3 sec.   **c.** 7 sec.
93. $y = -\dfrac{2}{25}x^2 + 18$ or $y = -0.08x^2 + 18$

95. 3.5%   **97.** $x^2 + y^2 = 400$
99. $x^2 + (y - 110)^2 = 10,000$
101. $x^2 + y^2 = 43,560,000$

Review Exercises

1.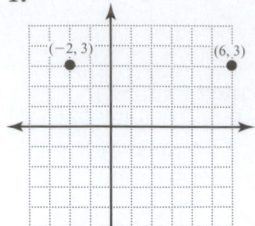

2. $(5x + 3y)(5x - 3y)$

3.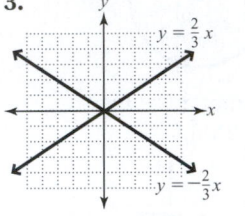

4. ± 4 **5.** $(4, 0), (-4, 0), (0, 3), (0, -3)$

6. $(3, 0), (-3, 0), (0, 5), (0, -5)$

13.2 Exercises

1. An ellipse is the set of all points the sum of whose distances from two fixed points is constant. **3.** The equation of an ellipse is a sum; the equation of a hyperbola is a difference. **5.** The parabola has only one squared term.

7.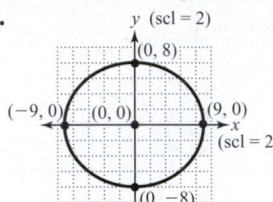

9.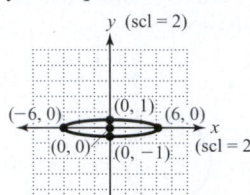

11.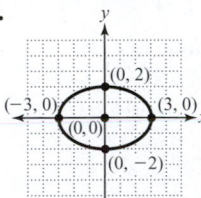

13.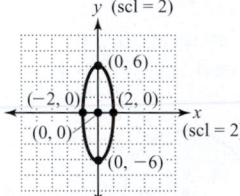

15.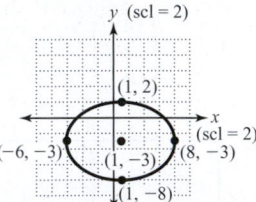

17.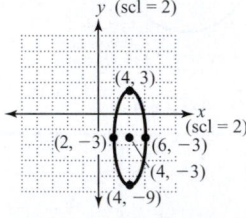

19.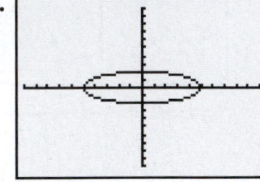

21.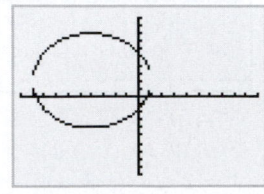

23. $\dfrac{x^2}{9} + \dfrac{y^2}{16} = 1$

25.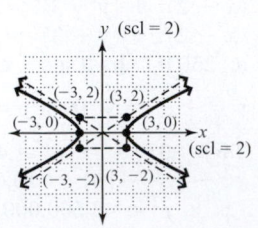

27.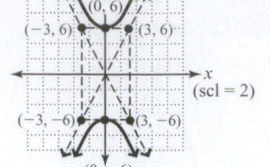

29.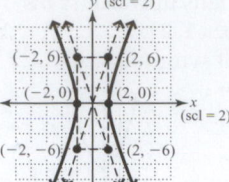

31.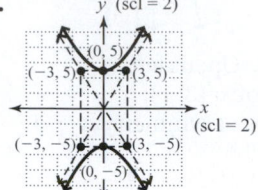

33.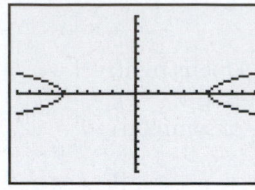

35. $\dfrac{x^2}{9} - \dfrac{y^2}{4} = 1$ **37.** c **39.** b **41.** ellipse **43.** circle

45. circle

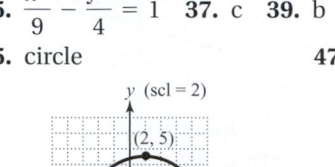

47. parabola

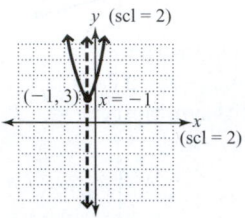

49. ellipse

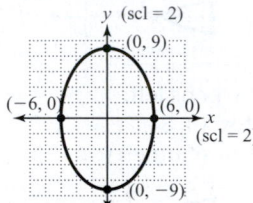

51. hyperbola

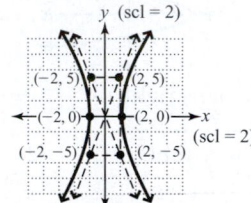

53. parabola

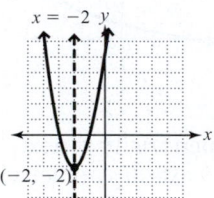

55. hyperbola

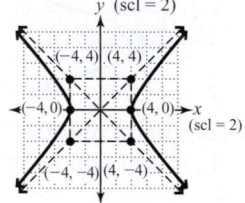

57. ellipse

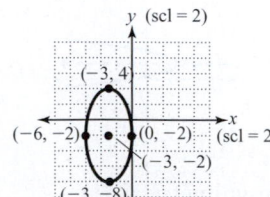

59. a. Yes, the sailboat will clear the bridge. The height of the bridge at the center is 20 feet, and the boat's mast is only 18 feet above the water. **b.** The bridge is 32 feet wide at the base of the arch.

61.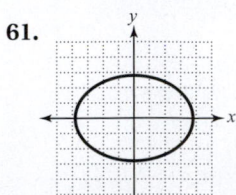

63. $\dfrac{x^2}{2.25} + \dfrac{y^2}{4} = 1$ **65.** 3 units

Review Exercises 1. $y = -3x^2 + 6$ 2. $(1, 2)$
3. $(2, -3)$ 4. $(-3, 4)$ 5. $-4, \dfrac{2}{3}$ 6. ± 3

13.3 Exercises 1. a. 0, 1, 2, 3, or 4 solutions are possible. b. Answers may vary.

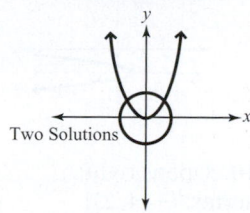
Two Solutions

3. Substitution because the first equation has a linear term and is easy to solve for y. 5. The graph of the first equation is a line, and the second is an ellipse; so the system could have 0, 1, or 2 solutions. 7. $(3, -4), (1, 0)$ 9. $(-2, 0), (1, 9)$
11. $(-4, -3), (3, 4)$ 13. $(2.6, -1.8), (1, 3)$
15. No solution. 17. $(0, 3), (1, 5)$ 19. $(3, 2), (2, 3)$
21. $(4, 2), (4, -2), (-4, -2), (-4, 2)$
23. $(3, 2), (3, -2), (-3, -2), (-3, 2)$
25. $(2, 1), (2, -1), (-2, -1), (-2, 1)$
27. $(-6, 0), (3, 3), (3, -3)$ 29. No solution.
31. $(2, 4), (-2, 4)$ 33. $(4, 0), (-4, 0)$ 35. $(2, 2), (-2, -2)$
37. $(5, 12), (-1, 0)$ 39. $(2, 2), (2, -2), (-2, -2), (-2, 2)$
41. $(-7, 3), (-7, -3), (2, 0)$

43. Answers may vary, but one possible system is $\begin{cases} x + y = 4 \\ x^2 + y^2 = 1 \end{cases}$.

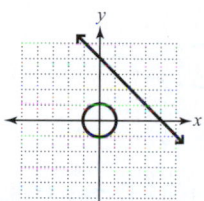

45. $(5, 3), (5, -3), (-5, -3), (-5, 3)$ 47. Length 18 in.; width: 8 in. 49. At equilibrium, there should be 400 chairs at $72 per chair.

51.
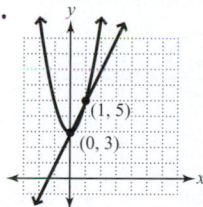

53.
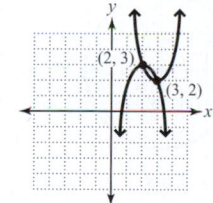

Review Exercises 1. yes 2. no

3.
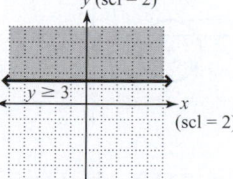
$y \geq 3$

4.
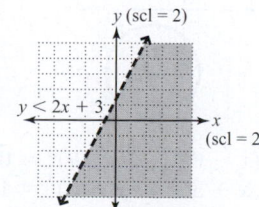
$y < 2x + 3$

5.
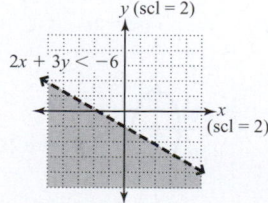
$2x + 3y < -6$

6.
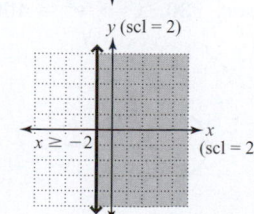
$x \geq -2$

13.4 Exercises 1. No, the coordinates of the points on the curve make $9x^2 + 4y^2 = 36$. The curve would be a dashed curve because the points are not included in the solution.

3. One example is the system $\begin{cases} x^2 + y^2 < 1 \\ x^2 + y^2 > 4 \end{cases}$. The graph of $x^2 + y^2 < 1$ is the region inside the circle with radius 1. The graph of $x^2 + y^2 > 4$ is the region outside the circle with radius 2. Thus, there is no common region of solution.

5.
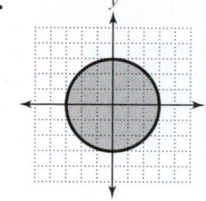

7.
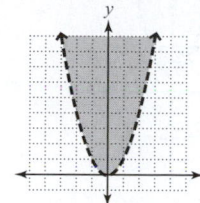

9.
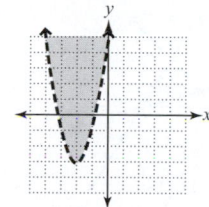

11.
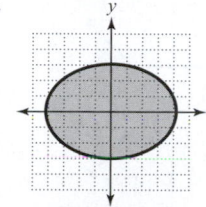

13.
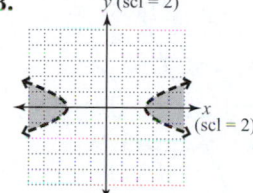

15.
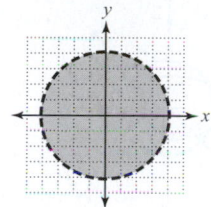

17.
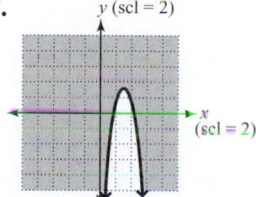

19.
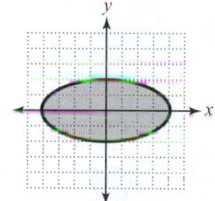

21.
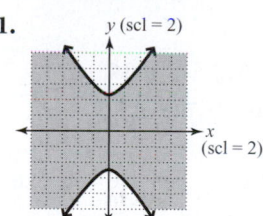

23.
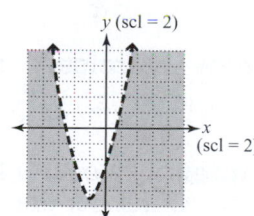

25.
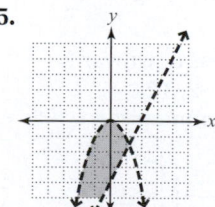

27.
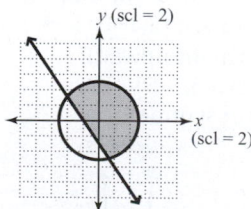

29.
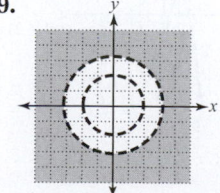

31.
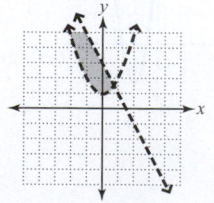

33.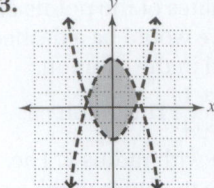

35.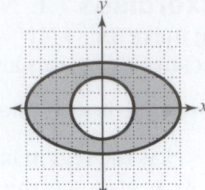

37.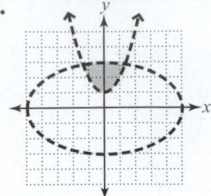

39.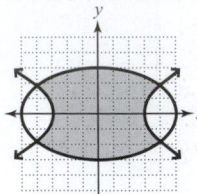

41.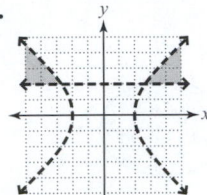

43.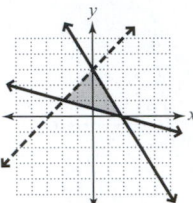

45.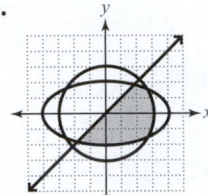

47.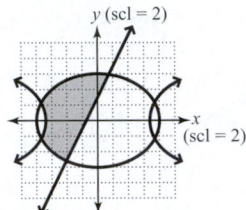

49.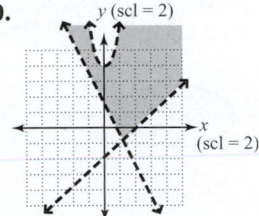

Review Exercises
1. 38 **2.** 312 **3.** −84
4. −1, 5, 15, 29 **5.** $-1, \frac{1}{4}, -\frac{1}{7}, \frac{1}{10}$ **6.** 67, 69, 71

Chapter 13 Review Exercises
1. false **2.** true
3. false **4.** true **5.** true **6.** false **7.** 10
8. (−4, 1); 2 **9.** ellipse; (4, 0) and (−4, 0)
10. (5, 4), (5, −4), (−5, −4), (−5, 4)

11. Opens upward;
vertex: (3, −5);
axis of symmetry: $x = 3$.

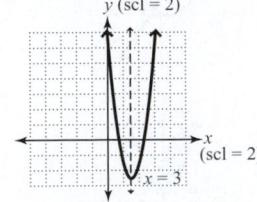

12. Opens downward;
vertex: (−2, 3);
axis of symmetry: $x = -2$.

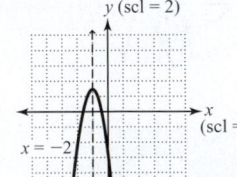

13. Opens left;
vertex: (4, 2);
axis of symmetry: $y = 2$.

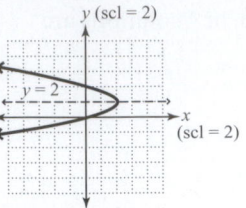

14. Opens right;
vertex: (−2, −3);
axis of symmetry: $y = -3$.

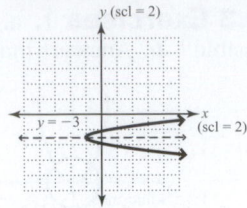

15. Opens right;
vertex: (−1, −3);
axis of symmetry: $y = -3$.

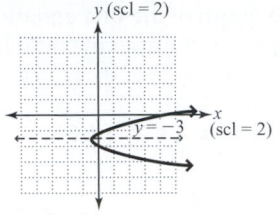

16. Opens right;
vertex: (−14, 2);
axis of symmetry: $y = 2$.

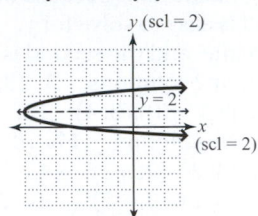

17. Opens left;
vertex: (4, −1);
axis of symmetry: $y = -1$.

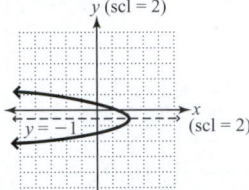

18. Opens left;
vertex: (3, −2);
axis of symmetry: $y = -2$.

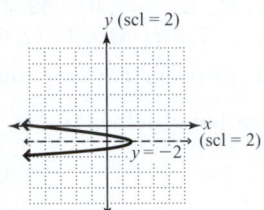

19. $5, \left(-3, -\frac{1}{2}\right)$ **20.** $4\sqrt{5}, (0, -1)$ **21.** 10
22. $(x + 6)^2 + (y - 8)^2 = 225$
23. Center: (3, −2); radius: 5. **24.** Center: (−5, 1); radius: 2.

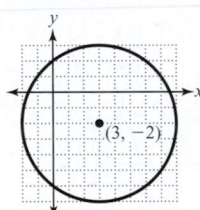

25. Center: (2, −4); radius: 3. **26.** Center: (−5, −1); radius: 2.

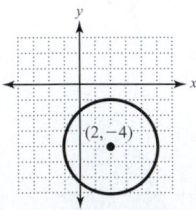

27. $(x - 6)^2 + (y + 8)^2 = 81$
28. $(x + 3)^2 + (y + 5)^2 = 100$ **29. a.** 144 ft. **b.** 1 sec.
c. 4 sec. **30.** $x^2 + y^2 = 400$
31.

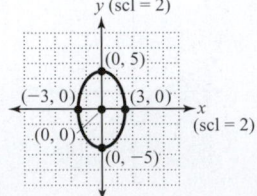

32.

33.
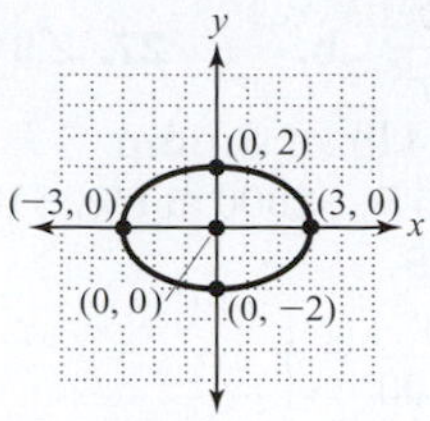

34.
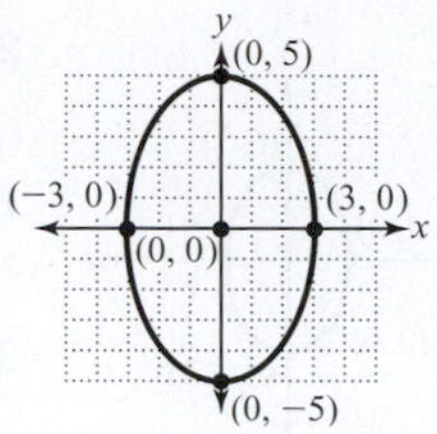

35.
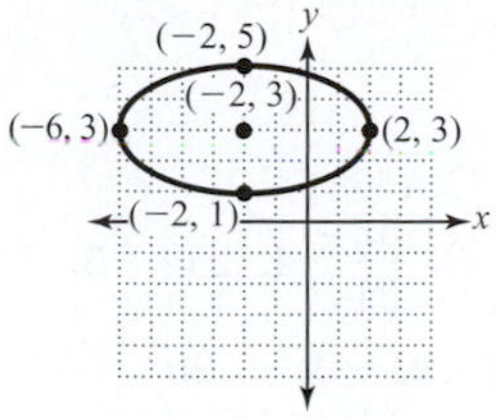

36.
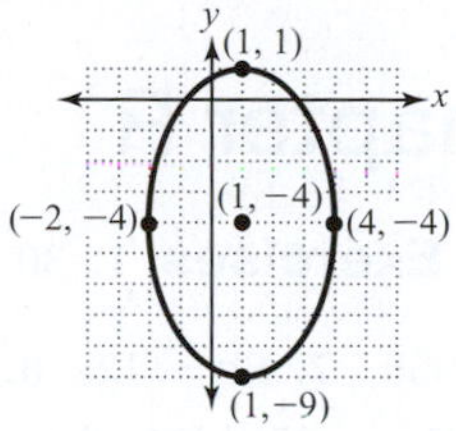

37.
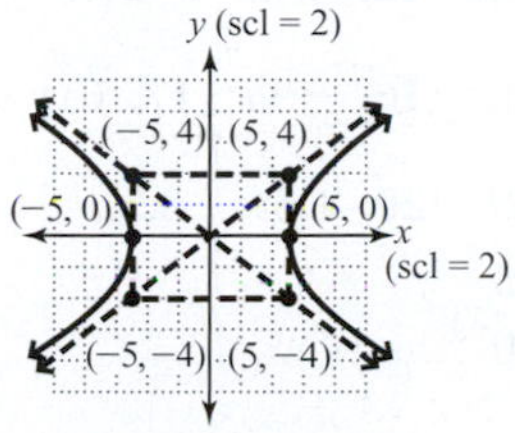

38.
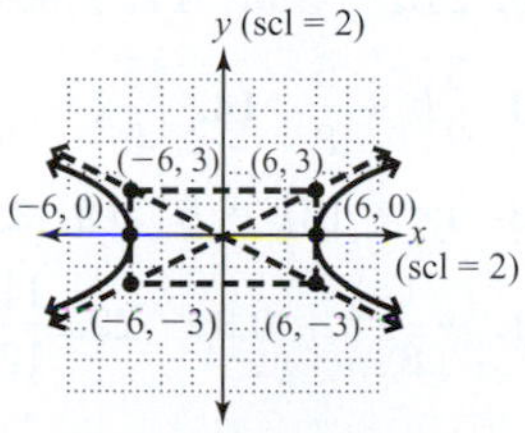

39.
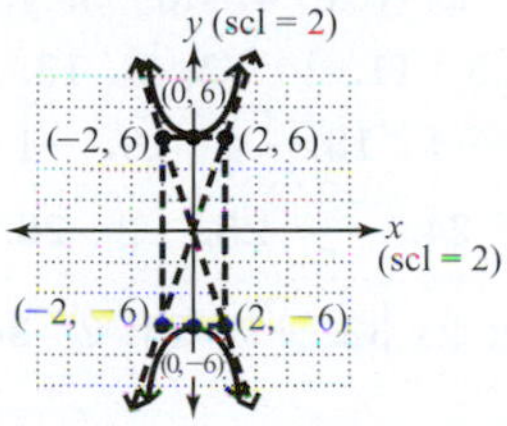

40.
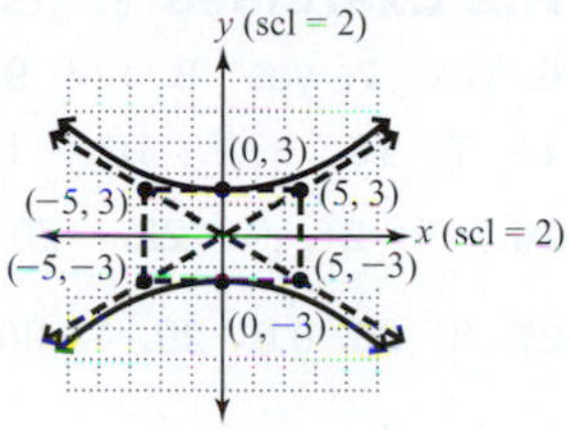

41. $\frac{x^2}{625} + \frac{y^2}{225} = 1$ **42.** $\frac{x^2}{1.625^2} + \frac{y^2}{4} = 1$
43. (2, 5), (−1, −1) **44.** (−1, −4), (4, 1)
45. (−73/37, −35/37), (1, 5) **46.** (8, 0), (−8, 0)
47. (3, 4), (3, −4), (−3, −4), (−3, 4)
48. (4, 2), (4, −2), (−4, −2), (−4, 2) **49.** (2, 2), (−2, 2)
50. (2, 3), (−2, 3) **51.** (8, 5), (8, −5), (−8, −5), (−8, 5)
52. 8 ft. by 10 ft.

53.
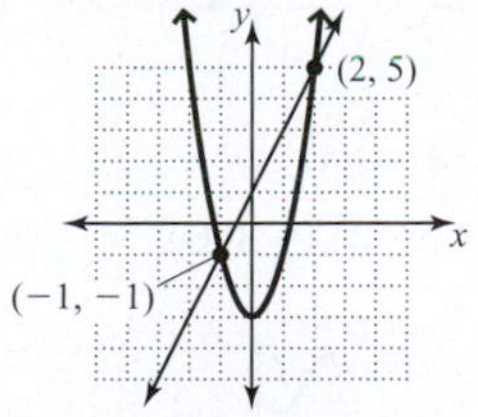

54.
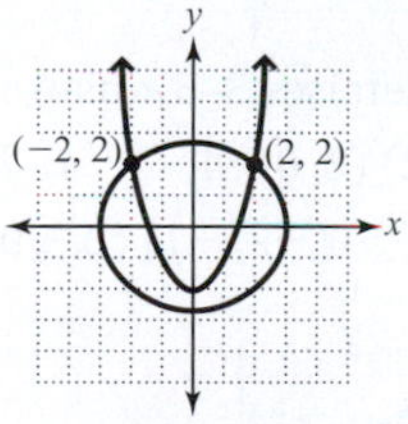

55.
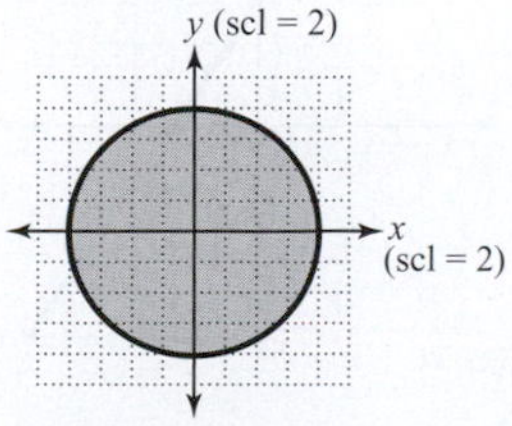

56.
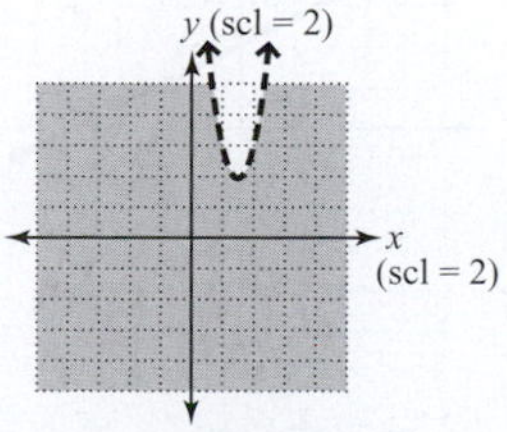

57.
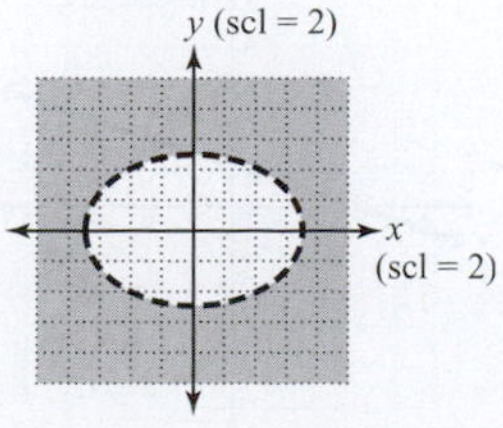

58.
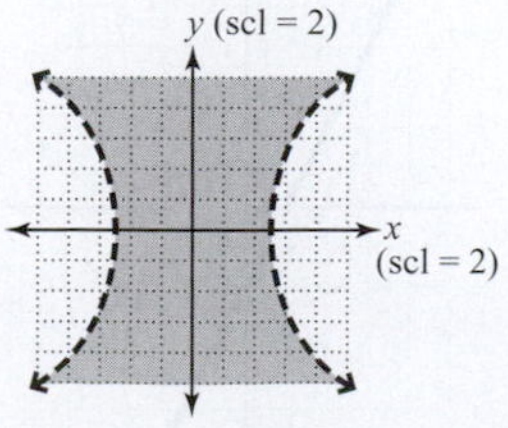

59.
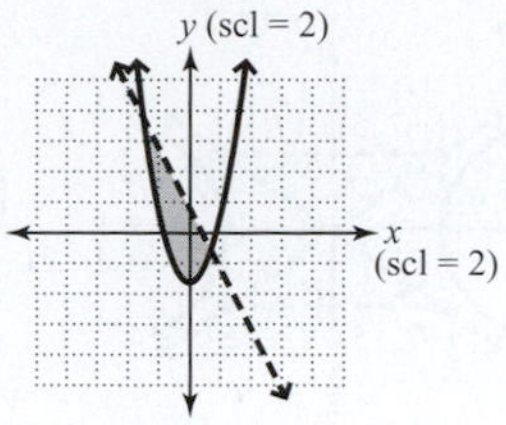

60.
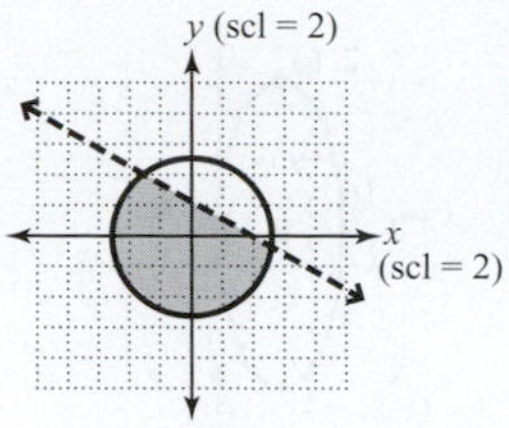

61.
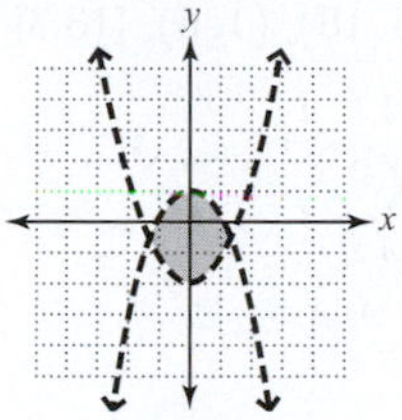

62.
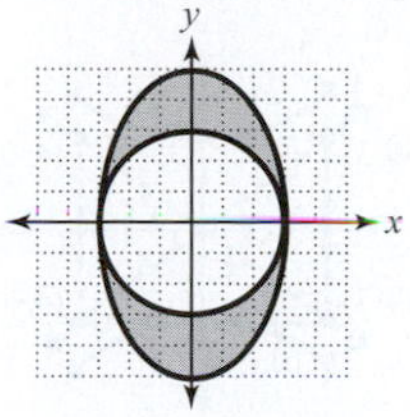

63.
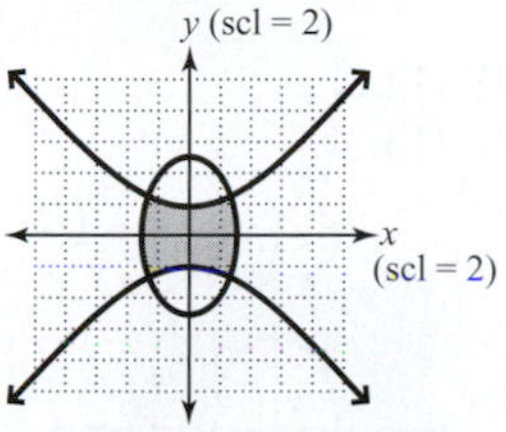

64.
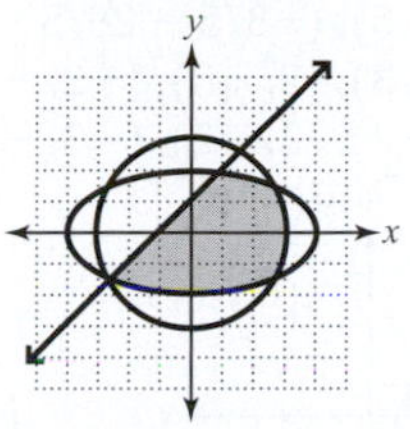

Chapter 13 Practice Test

1. Opens upward; vertex: (−1, −4); axis of symmetry: $x = -1$. [13.1]

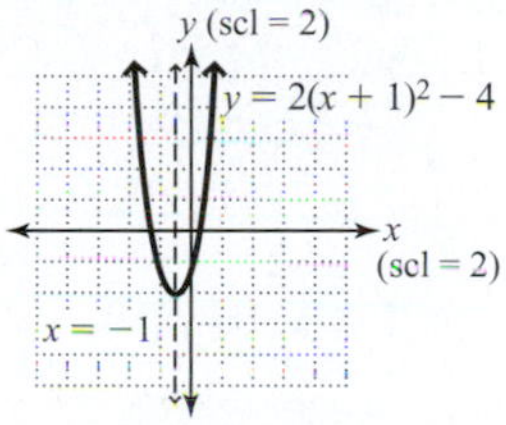

2. Opens left; vertex: (1, 3); axis of symmetry: $y = 3$. [13.1]

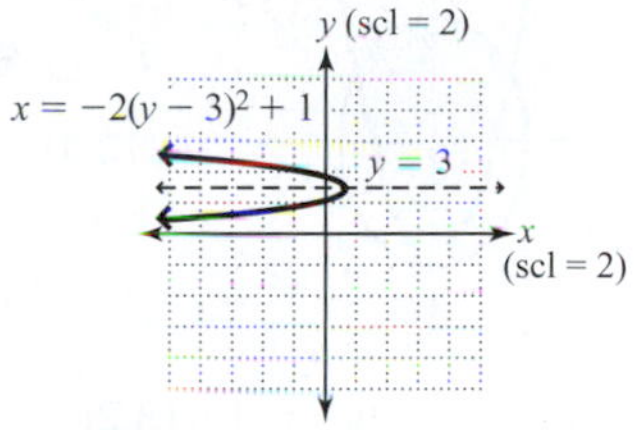

3. Opens right; vertex: (−7, −2); axis of symmetry: $y = -2$. [13.1]

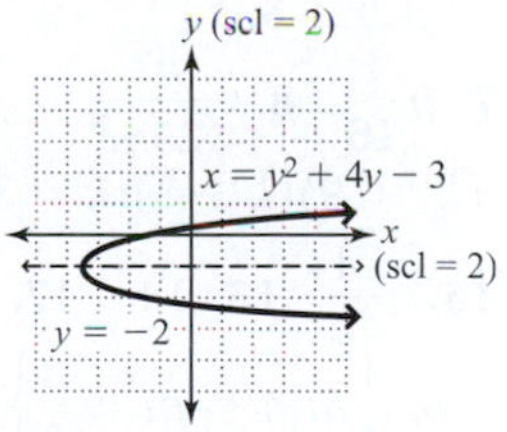

4. $2\sqrt{5}$, (4, −4) [13.1]

5. Center: (−4, 3); radius: 6. [13.1]

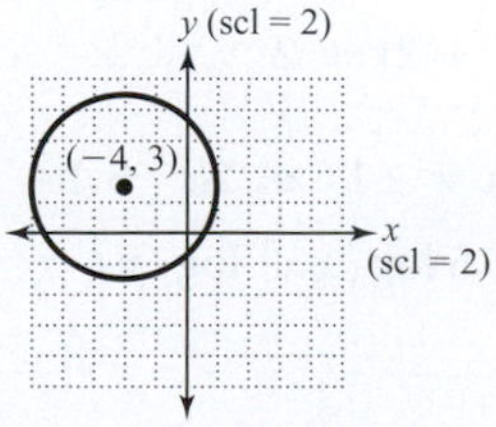

6. Center: (−2, 5), radius: 3. [13.1]

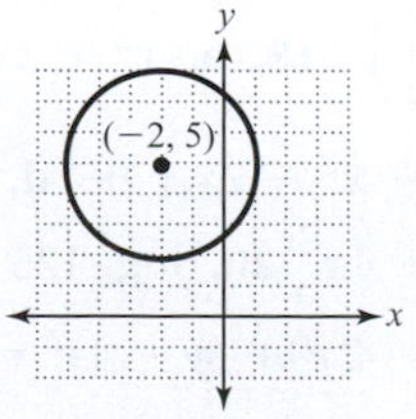

7. $(x - 2)^2 + (y + 4)^2 = 100$ [13.1]

8.
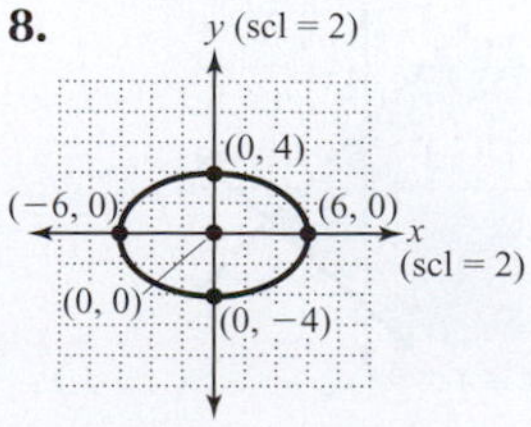

[13.2]

9.

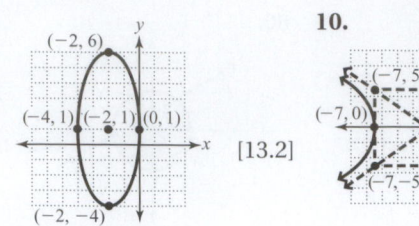

[13.2]

10.

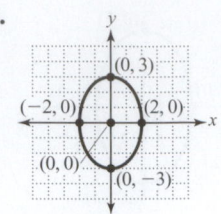

[13.2]

11.

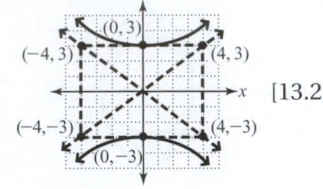

[13.2]

12. $(-5, 18), (1, 6)$ [13.3]

13. $(3, 5), (-3/5, -29/5)$ [13.3]
14. $(2, 3), (2, -3), (-2, -3), (-2, 3)$ [13.3]
15. $(3, 2), (3, -2), (-3, -2), (-3, 2)$ [13.3]

16.

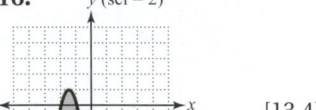

[13.4]

17.

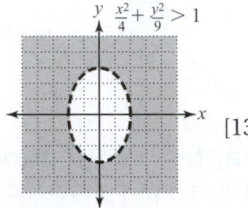

[13.4]

18.

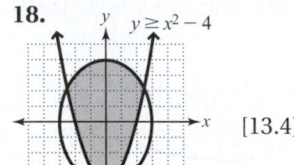

[13.4]

19. Height is 18; distance across the base is 12.   [13.1]

20. $\dfrac{x^2}{9.5^2} + y^2 = 1$ [13.2]

Chapters 1–13 Cumulative Review Exercises

1. false   **2.** true   **3.** true   **4.** $a_1 b_2 - a_2 b_1$ **5.** i

6. $d = \sqrt{(x_2 - x_1)^2 + (y_2 - y_1)^2}$ **7.** $9\dfrac{1}{16}$ **8.** $\dfrac{n^6}{125m^9}$

9. $2i\sqrt{5}$ **10.** $(x + y)(a + b)$ **11.** $(k^2 + 9)(k + 3)(k - 3)$

12. $-4, \dfrac{2}{3}$ **13.** 52   **14.** $\dfrac{1 \pm i\sqrt{3}}{4}$ **15.** $-\dfrac{4}{3}$ **16.** 1.5   **17.** $\dfrac{1}{8}$

18. a.

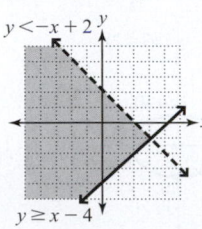

b. $\left\{ m \,\middle|\, 0 < m < \dfrac{3}{5} \right\}$

c. $\left(0, \dfrac{3}{5}\right)$ **19. a.** $x^2 + 2x$ **b.** $-x^2 + 2x + 2$

c. $2x^3 + x^2 - 2x - 1$ **d.** $\dfrac{2x + 1}{x^2 - 1}; x \neq \pm 1$ **e.** $2x^2 - 1$

f. $4x^2 + 4x$ **20.** $\log_5 125 = 3$ **21.** $5\log_5 x + \log_5 y$

22. $(x - 2)^2 + (y - 4)^2 = 169$

23.

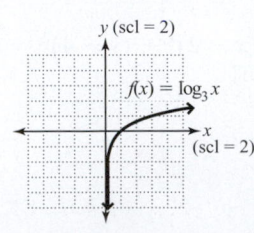

24.

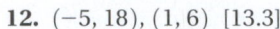

25.

26. a. $h = \dfrac{V}{\pi r^2}$ **b.** $\dfrac{15}{r^2}$ **27.** 2.9 lb. of tuna and 3.4 lb. of shrimp
28. $1500 at 5%, $2500 at 6%, $1000 at 3%   **29.** 6.9

30. $\dfrac{x^2}{5625} + \dfrac{y^2}{400} = 1$

Chapter R

R.1 Exercises
1. 30   **2.** -51 **3.** $\dfrac{32}{21}$ **4.** 244   **5.** 2

6. -955 **7.** $6a + 12$ **8.** $12x + 24$ **9.** $-4r + 20$

10. $21a - 35b$ **11.** $9.66x^2 - 3.91x$ **12.** $8.4x^2 - 21.7$

13. $\dfrac{3}{2}b + \dfrac{5}{8}$ **14.** $\dfrac{15}{4}x - \dfrac{5}{6}$ **15.** $11x$ **16.** $-9a$ **17.** $6xy + 6x$

18. $4y - 13k + 5$ **19.** $3.6a - 5.2b$ **20.** $1.6t - 8.5u$

21. $-\dfrac{1}{8}x^2 + 0.8y$ **22.** $\dfrac{11}{12}x^2 y - \dfrac{13}{10}xy^2$

R.2 Exercises
1. yes   **2.** yes   **3.** yes   **4.** no   **5.** yes

6. yes   **7.** yes   **8.** no   **9.** 4   **10.** 5   **11.** 2   **12.** 2   **13.** 10

14. 7   **15.** -7.7 **16.** -13.4 **17.** -4 **18.** -4 **19.** -4

20. -3 **21.** 12   **22.** -20 **23.** 24   **24.** $-\dfrac{3}{5}$ **25.** -2 **26.** -5

27. 3   **28.** 0.6   **29.** 4   **30.** $\dfrac{9}{17}$ **31.** 2   **32.** -9 **33.** 2   **34.** 1

R.3 Exercises
1.

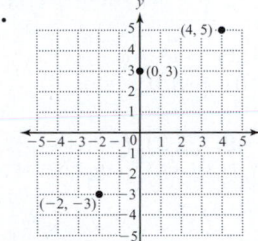

2.

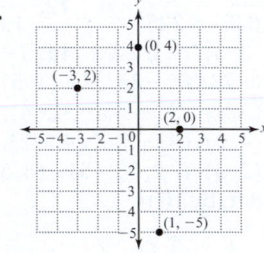

(For Exercises 3–6, answers may vary.)

3. $(0, 2)\ (8, 0)\ (4, 1)$ **4.** $(0, 3)\ (2, 0)\ (4, -3)$

5. $(0, -5)\ (-2, -1)\ (2, -9)$ **6.** $(0, 4)\ (3, 2)\ (-3, 6)$

7.

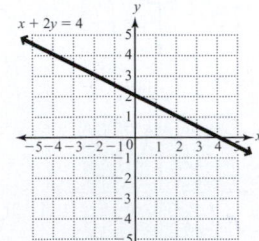

8.

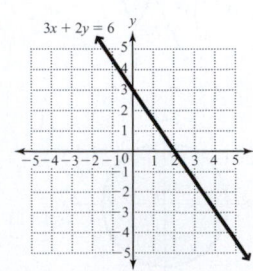

9.

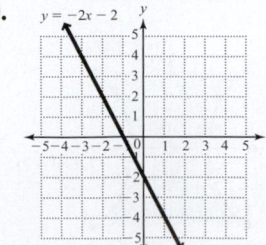

10.

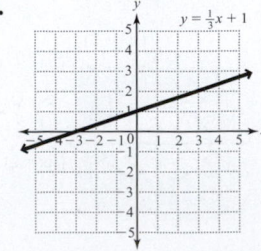

11.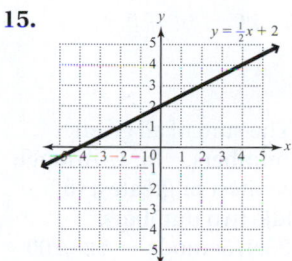

12.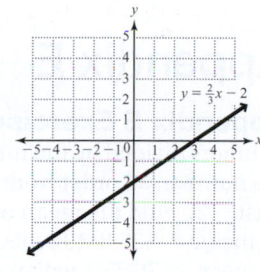

13.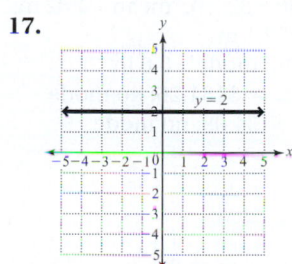

14.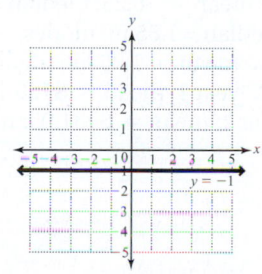

15.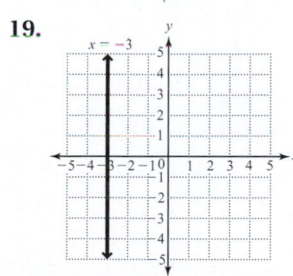

16.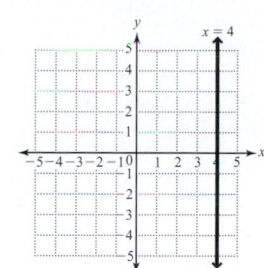

17.

18.

19.

20.

R.4 Exercises **1.** $6a^2 + 2a - 10$
2. $a^3 + 14a^2 - 16a + 4$ **3.** $2r^2 + 4r + 1$
4. $12m^3 - 5m^2 + 13m - 10$ **5.** $-18a^8b^8c^2$ **6.** $36p^9q^{11}r^{10}$
7. $-0.92m^{10}n^7$ **8.** $\frac{6}{5}x^7y^7$ **9.** $24a^5 - 16a^4 + 12a^3$
10. $-12x^5y^3 - 24x^3y^4 + 16x^7y - 20x^2y$
11. $-12x^7 - \frac{1}{2}x^5 + \frac{1}{3}x^4 - \frac{1}{5}x^3$
12. $9.45a^5b^4c - 15.75a^6b^7c^3 + 18a^3b^2c$ **13.** $6x^2 + x - 12$
14. $3y^2 - 7y - 20$ **15.** $6p^2 - 17pq + 12q^2$
16. $12a^2 - 7ab - 10b^2$ **17.** $4a^3 + 11a^2 + a - 10$
18. $6y^3 - 16y^2 - 7y + 10$ **19.** $4y^2 - 25$ **20.** $9m^2 - 49n^2$
21. $49x^2 + 28xy + 4y^2$ **22.** $25a^2 - 30ab + 9b^2$

23. $3x^2 - 2x + 4 - \frac{5}{x}$ **24.** $9uv - 4u^2 + 3v$

25. $3x^2 - 6x + 4 + \frac{-4}{x-2}$ **26.** $x^2 + 2x - 3 + \frac{-7}{2x-4}$

R.5 Exercises **1.** $6c^2d^4(3c - 2d^2)$
2. $4cd^2(4c^2 - 6cd^2 + 9)$ **3.** $-5x(2x^2 + 3x - 5)$
4. $-4mn(6m + 7 - 4n)$ **5.** $(a + 4)(a^2 + 3)$
6. $(4x - y)(2x + 3)$ **7.** $(3c + 2d)(2a - 3b)$
8. $(3c - d)(5a - b)$ **9.** $(x + 2)(z + 9)$
10. $(y + 7)(y - 6)$ **11.** $(x - 5y)(x - 7y)$
12. $2(a + 12b)(a - 2b)$ **13.** $(3x - 2)(3x - 4)$
14. $(3c - 2d)(2c - 3d)$ **15.** $3(3c - 5d)(2c - 3d)$
16. $(3a + 4b)(4a - 3b)$ **17.** $(a - 8)^2$ **18.** $y(5x - 1)^2$
19. $(2x + 5y)^2$ **20.** $(6m - 7n)^2$ **21.** $(2x + 5y)(2x - 5y)$
22. $(y^2 + 9)(y + 3)(y - 3)$ **23.** prime
24. $(r - 5)(r^2 + 5r + 25)$ **25.** $(3x + 4)(9x^2 - 12x + 16)$

Appendix A

Appendix A Exercises **1.** The domain of a sequence is the set of natural numbers 1 to n. **3.** A series is the sum of the terms of a sequence. **5.** arithmetic; 5 **7.** 3, 5, 7, 9, 41
9. 3, 6, 11, 18, 227 **11.** $\frac{1}{3}, \frac{1}{2}, \frac{3}{5}, \frac{2}{3}, \frac{11}{12}$
13. $-\frac{1}{2}, \frac{1}{5}, -\frac{1}{10}, \frac{1}{17}, -\frac{1}{226}$ **15.** 5 **17.** -3 **19.** 7
21. $a_n = 4n + 10, a_{14} = 66$ **23.** $a_n = -3n - 5, a_{28} = -89$
25. $a_n = 4n - 9, a_{34} = 127$ **27.** $-6, 1, 8, 15$
29. $-5, -8, -11, -14$ **31.** 7, 11, 15, 19 **33.** $a_1 = 11$
35. $d = 4$ **37.** $-8, -1, 6, 13$ **39.** 27, 24, 21, 18
41. $1 + 4 + 9 + 16 + 25 + 36 = 91$
43. $-3 - 1 + 1 + 3 = 0$ **45.** $-1 + 8 + 23 = 30$
47. $9 + 13 + 17 + 21 = 60$ **49.** 1450 **51.** -429 **53.** 675
55. 120 **57.** 5050 **59.** **a.** 144 seats **b.** 4480 seats
61. **a.** 16 posts **b.** 325 posts **63.** **a.** $28,000, $29,500, $31,000, $32,500, $34,000 **b.** $41,500 **c.** $347,500

Appendix B

Appendix B Exercises **1.** The common ratio, r.
3. Yes, the first n terms form a finite geometric series.
5. $2 + 6 + 18 + 54 = 80$ **7.** **a.** 3 **b.** $a_9 = 6561$
c. $a_n = 1(3)^{n-1}$ **9.** **a.** -2 **b.** $a_{15} = -32,768$
c. $a_n = -2(-2)^{n-1}$ **11.** **a.** $\frac{1}{3}$ **b.** $a_{10} = \frac{1}{81}$
c. $a_n = 243\left(\frac{1}{3}\right)^{n-1}$ **13.** **a.** $-\frac{1}{4}$ **b.** $a_7 = \frac{1}{32}$
c. $a_n = 128\left(-\frac{1}{4}\right)^{n-1}$ **15.** **a.** $-4, 12, -36, 108, -324$
b. 8748 **17.** **a.** $243, 81, 27, 9, 3$ **b.** $\frac{1}{3}$ **19.** **a.** 2 **b.** 1, 2, 4, 8
21. **a.** 2 **b.** $-6, -12, -24, -48$ **23.** **a.** $\frac{1}{4}$ **b.** 128, 32, 8, 2
25. -6 **27.** 265,719 **29.** 63.875 **31.** 102.4 **33.** 40.5

35. 9.375   **37.** The sum does not exist because $|r| \geq 1$.

39. $\dfrac{4}{9}$ **41.** $\dfrac{17}{99}$ **43. a.** 520, 540.8, 562.432, 584.92928, 608.3264512   **b.** 740.122142459   **45. a.** 100, 108, 117, 126   **b.** 171 mink   **c.** $a_n = 100(1.08)^{n-1}$

47. a. $0.01, $0.02, $0.04, $0.08, $0.16   **b.** $a_n = 0.01(2)^{n-1}$ **c.** $5242.88   **d.** $10,485.75   **49.** $11,191.64

Appendix C

Appendix C Exercises   **1.** 13   **3.** 8 factorial   **5.** 5, 8
7. 24   **9.** 144   **11.** 86,400   **13.** $\dfrac{1}{90}$, or $0.0\overline{1}$ **15.** 10   **17.** 28

19. 35   **21.** 210   **23.** 1   **25.** 1
27. $a^5 + 5a^4b + 10a^3b^2 + 10a^2b^3 + 5ab^4 + b^5$
29. $x^4 - 4x^3y + 6x^2y^2 - 4xy^3 + y^4$
31. $8a^3 + 12a^2b + 6ab^2 + b^3$
33. $x^5 - 10x^4y + 40x^3y^2 - 80x^2y^3 + 80xy^4 - 32y^5$
35. $64m^6 + 576m^5n + 2160m^4n^2 + 4320m^3n^3 +$
$4860m^2n^4 + 2916mn^5 + 729n^6$
37. $81x^4 - 432x^3y + 864x^2y^2 - 768xy^3 + 256y^4$ **39.** $70x^4y^4$
41. $45a^8b^2$ **43.** $32,256x^4y^5$ **45.** $-22,680m^4n^3$

Appendix D

Appendix D Exercises   **1.** $x - c$ **3.** $1, 0, -5, -3, 11$
5. It is the last number in the last row.   **7. a.** $x + 3$
b. $x^3 + 7x + 12$ **c.** $x + 4$ **9. a.** $x - 3$

b. $x^3 + 4x^2 - 25x + 7$ **c.** $x^2 + 7x - 4 - \dfrac{5}{x - 3}$
11. a. $x - 4$ **b.** $2x^4 - 8x^3 - 5x^2 + 17x + 10$
c. $2x^3 - 5x - 3 - \dfrac{2}{x - 4}$ **13.** $x - 3$ **15.** $2x + 7 + \dfrac{23}{x - 4}$
17. $x^2 + x - 3$ **19.** $3x^2 + 2x - 1$ **21.** $3x^2 - 10x + 8$
23. $2x^2 - 9x + 17 - \dfrac{28}{x + 2}$ **25.** $x^2 + 1 - \dfrac{1}{x - 2}$
27. $2x^2 - 5x + 15 - \dfrac{49}{x + 3}$ **29.** $3x^2 + 5x + 15 + \dfrac{47}{x - 3}$
31. $x^2 + 2x - 3$ **33.** $x^2 + 3x + 9 + \dfrac{54}{x - 3}$
35. $x^3 - 2x^2 + 4x - 8 + \dfrac{32}{x + 2}$ **37.** $6x^2 - 3x - 9$
39. a. $9x + 8$ **b.** Base: 116 in.; height: 10 in.; area: 1160 in.^{2}.
41. a. $2y - 11$ **b.** Length: 14 ft.; width: 8 ft.; height: 9 ft.;
volume: 1008 ft.^{3}.

Appendix E

Appendix E Exercises.   **1.** The sum of all given numbers divided by the number of numbers.   **3.** 1. Arrange the numbers in order from least to greatest or greatest to least.   2. Find the mean of the middle two numbers.
5. mean = $26,841.\overline{6}$; median = $26,975; mode = $28,700
7. mean = 78.25; median = 80; mode = 82   **9.** mean = 1.82 m; median = 1.85 m; modes = 1.8 m and 2.1 m
11. mean = 1.25 in.; median = 1.1 in.; mode = 0.6 in.
13. year 1: mean = $126.76; median = $131.26; no mode
year 2: mean = $131.95: median = $134.08; no mode

Glossary

Absolute value: A given number's distance from 0 on a number line.

Additive inverses: Two numbers whose sum is 0.

Area: The total number of square units that fill a figure.

Arithmetic average: The sum of all given numbers divided by the number of numbers.

Arithmetic sequence: A sequence in which each term after the first is found by adding the same number to the previous term.

Arithmetic series: The sum of the terms of an arithmetic sequence.

Augmented matrix: A matrix made up of the coefficients and the constant terms of a system. The constant terms are separated from the coefficients by a dashed vertical line.

Axis of symmetry: A line that divides a graph into two symmetrical halves.

Base: The number that is repeatedly multiplied.

Binomial: A polynomial containing two terms.

Binomial coefficient: A number written as $\binom{n}{r}$ and defined as $\dfrac{n!}{r!(n-r)!}$.

Binomial theorem: For any positive integer n,
$$(a+b)^n = \binom{n}{0}a^n + \binom{n}{1}a^{n-1}b + \binom{n}{2}a^{n-2}b^2$$
$$+ \binom{n}{3}a^{n-3}b^3 + \cdots + \binom{n}{n}b^n.$$

Circle: A set of points that are equally distant from a central point. The central point is the center.

Circumference: The distance around a circle.

Coefficient: The numerical factor in a term.

Coefficient of a monomial: The numerical factor in a monomial.

Common difference of an arithmetic sequence: The value of d is found by $d = a_n - a_{n-1}$, where a_n is any term in the sequence (except the first) and a_{n-1} is the previous term.

Common logarithms: Logarithms with a base of 10. $\log_{10} x$ is written as $\log x$. Note that $\log 10 = 1$.

Complementary angles: Two angles are complementary if the sum of their measures is $90°$.

Complex conjugates: The complex conjugate of a complex number $a + bi$ is $a - bi$.

Complex number: A number that can be expressed in the form $a + bi$, where a and b are real numbers and i is the imaginary unit.

Complex rational expression: A rational expression that contains rational expressions in the numerator or denominator.

Compound inequality: Two inequalities joined by either *and* or *or*.

Composition of functions: If f and g are functions, then the composition of f and g is defined as $(f \circ g)(x) = f[g(x)]$ for all x in the domain g for which $g(x)$ is in the domain of f. The composition of g and f is defined as $(g \circ f)(x) = g[f(x)]$ for all x in the domain of f for which $f(x)$ is in the domain of g.

Congruent angles: Angles that have the same measure. The symbol for congruent is $\cong$.

Conjugates: Binomials that differ only in the sign separating the terms.

Conic section: A curve in a plane that is the result of intersecting the plane with a cone. More specifically, a circle, an ellipse, a parabola, or a hyperbola.

Consistent system of equations: A system of equations that has at least one solution.

Constant: A symbol that does not vary in value.

Contradiction: An equation that has no real-number solution.

Degree of a monomial: The sum of the exponents of all variables in a monomial.

Degree of a polynomial: The greatest degree of any of the terms in the polynomial.

Diameter: The distance across a circle through its center.

Direct variation: Two variables y and x vary directly if $y = kx$. If y varies directly as the nth power of x, then $y = kx^n$, where k is the constant of variation.

Discriminant: The radicand, $b^2 - 4ac$, in the quadratic formula.

Domain: The set of all input values (x-values) for a relation.

Ellipse: The set of all points the sum of whose distances from two fixed points is constant.

Equation: A mathematical relationship that contains an equal sign.

Exponent: A symbol written to the upper right of a base number that indicates how many times to use the base as a factor.

Exponential function: If $b > 0$, $b \neq 1$, and x is any real number, then the exponential function is $f(x) = b^x$.

Expression: A constant, a variable, or any combination of constants, variables, and arithmetic operations that describes a calculation.

Extraneous solution: An apparent solution that does not solve its equation.

Factored form: A number or an expression written as a product of factors.

Factorial notation: For any natural number n, the symbol $n!$ (read "n factorial") means $n(n - 1)(n - 2)\ldots 3 \cdot 2 \cdot 1$. $0!$ is defined to be 1, so $0! = 1$.

Factors: If $a \cdot b = c$, then a and b are factors of c.

Finite sequence: A function with a domain that is the set of natural numbers from 1 to n.

Formula: An equation that describes a mathematical relationship.

Fraction: A quotient of two numbers or expressions a and b having the form $\frac{a}{b}$, where $b \neq 0$.

Function: A relation in which every value in the domain is paired with (or assigned to) exactly one value in the range.

Geometric sequence: A sequence in which every term after the first is found by multiplying the previous term by the same number, called the common ratio, r, where $r = \frac{a_n}{a_{n-1}}$.

Geometric series: The sum of the terms of a geometric sequence.

Greatest common factor (GCF): The largest natural number that divides all given numbers with no remainder.

Hyperbola: The set of all points the difference of whose distances from two fixed points remains constant.

Identity: An equation that has every real number as a solution (excluding any numbers that cause an expression in the equation to be undefined).

Imaginary number: A number that can be expressed in the form bi, where b is a real number and i is the imaginary unit.

Imaginary unit: The number represented by I, where $i = \sqrt{-1}$ and $i^2 = -1$.

Inconsistent system of equations: A system of equations that has no solution.

Inequality: A mathematical relationship that contains an inequality symbol ($\neq$, $<$, $>$, $\leq$, or $\geq$).

Infinite sequence: A function with a domain that is the set of natural numbers.

Intersection: For two sets A and B, the intersection of A and B, symbolized by $A \cap B$, is a set containing only elements that are in both A and B.

Inverse functions: Two functions f and g are inverses if and only if $(f \circ g)(x) = x$ for all x in the domain of g and $(g \circ f)(x) = x$ for all x in the domain of f.

Inverse variation: Two variables y and x vary inversely if $y = \frac{k}{x}$. If $y = \frac{k}{x^n}$, then y varies inversely as the nth power of x, where k is the constant of variation.

Irrational number: Any real number that is not rational.

Joint variation: If y varies jointly as x and z, then $y = kxz$, where k is the constant of variation.

Least common denominator (LCD): The least common multiple of the denominators of a given set of fractions.

Least common multiple (LCM): The smallest number that is a multiple of each number in a given set of numbers.

Like radicals: Radical expressions with identical radicands and identical root indices.

Like terms: Constant terms or variable terms that have the same variable(s) raised to the same exponents.

Linear equation: An equation in which each variable term contains a single variable raised to an exponent of 1.

Linear equation in one variable: An equation that can be written in the form $ax + b = c$, where a, b, and c are real numbers and $a \neq 0$.

Linear inequality: An inequality containing expressions in which each variable term contains a single variable with an exponent of 1.

Logarithm: If $b > 0$ and $b \neq 1$, then $y = \log_b$ is equivalent to $x = b^y$.

Lowest terms: Given a fraction $\frac{a}{b}$ and $b \neq 0$, if the only factor common to both a and b is 1, then the fraction is in lowest terms.

Matrix: A rectangular array of numbers.

Mean: The sum of all given numbers divided by the number of numbers.

Median: The middle number in an ordered set of numbers.

Minor of an element of a matrix: The determinant of the remaining matrix when the row and column in which the element is located are ignored.

Mode: The number that occurs most often in a set of numbers.

Monomial: An expression that is a constant, a variable, or a product of a constant and variable(s) that are raised to whole number powers.

Multiple: A multiple of a given integer n is the product of n and an integer.

Multiplicative inverses: Two numbers whose product is 1.

Natural logarithm: Base-e logarithms are called natural logarithms, and $\log_b x$ is written as $\ln x$. Note that $\ln e = 1$.

Nonlinear system of equations: A system of equations that contains at least one nonlinear equation.

nth root: The number b is an nth root of a number a if $b^n = a$.

One-to-one function: A function f is one to one if for any two numbers a and b in its domain, when $f(a) = f(b)$, $a = b$ and when $a \neq b$, $f(a) \neq f(b)$.

Percent: A ratio representing some part out of 100.

Perimeter: The distance around a figure.

Polynomial: A monomial or an expression that can be written as a sum of monomials.

Polynomial in one variable: A polynomial in which every variable term has the same variable.

Prime factorization: A factorization that contains only prime factors.

Prime number: A natural number that has exactly two different factors, 1 and the number itself.

Proportion: An equation in the form $\frac{a}{b} = \frac{c}{d}$, where $b \neq 0$ and $d \neq 0$.

Quadratic equation in one variable: An equation that can be written in the form $ax^2 + bx + c = 0$, where a, b, and c are real numbers and $a \neq 0$.

Quadratic equation in two variables: An equation that can be written in the form $y = ax^2 + bx + c$, where a, b, and c are real numbers and $a \neq 0$.

Quadratic inequality: An inequality that can be written in the form $ax^2 + bx + c < 0$ or $ax^2 + bx + c > 0$, where $a \neq 0$. *Note:* The symbols $<$ and $>$ can be replaced with $\leq$ and $\geq$.

Quadratic in form: An equation is quadratic in form if it can be rewritten as a quadratic equation $au^2 + bu + c = 0$, where $a \neq 0$ and u is a variable or an expression.

Radical equation: An equation containing at least one radical expression whose radicand has a variable.

Radical function: A function containing a radical expression whose radicand has a variable.

Radius: The distance from the center of a circle to any point on the circle.

Range: The set of all output values (y-values) for a relation.

Ratio: A comparison of two quantities using a quotient.

Rational exponent: An exponent that is a rational number.

Rational expression: An expression that can be written in the form $\frac{P}{Q}$, where P and Q are polynomials and $Q \neq 0$.

Rational inequality: An inequality containing a rational expression.

Rational number: Any real number that can be expressed in the form $\frac{a}{b}$, where a and b are integers and $b \neq 0$.

Real numbers: The union of the rational and irrational numbers.

Relation: A set of ordered pairs.

Row echelon form: An augmented matrix whose coefficient portion has 1's on the diagonal from upper left to lower right and 0's below the 1's.

Scientific notation: A number expressed in the form $a \times 10^n$ where a is a decimal number with $1 \leq |a| < 10$ and n is an integer.

Sequence: A function list whose domain is $1, 2, 3, \ldots, n$.

Series: The sum of the terms of a sequence.

Set: A collection of objects.

Similar figures: Figures with congruent angles and proportional side lengths.

Slope: The ratio of the vertical change between any two points on a line to the horizontal change between these points.

Solution: A number that makes an equation true when it replaces the variable in the equation.

Solution for a system of equations: An ordered set of numbers that makes all equations in the system true.

Square matrix: A matrix that has the same number of rows and columns.

Statistic: A number used to describe some characteristic of a set of data.

Supplementary angles: Two angles are supplementary if the sum of their measures is 180°.

System of equations: A group of two or more equations.

Terms: Expressions that are the addends in an expression that is a sum.

Trinomial: A polynomial containing three terms

Unit ratio: A ratio with a denominator of 1.

Union: For two sets A and B, the union of A and B, symbolized by $A \cup B$, is a set containing every element in A or in B.

Variable: A symbol that can vary in value.

Vertex of a parabola: The lowest point on a parabola that opens up or the highest point on a parabola that opens down.

Volume: The total number of cubic units that fill a space.

x-intercept: A point where a graph intersects the x-axis.

y-intercept: A point where a graph intersects the y-axis.

Index of Applications

Index

least common, 12

*n*th-root, 740

rationalization of, 739–42, 759

    containing difference, 741–42

    containing sum, 741–42

simplifying fraction with same nonzero, 14

simplifying when factorable, 513

subtraction of rational expressions with different, 517–23

Dependent equations

    consistent systems with, 629, 638–42

    inconsistent systems and, 649–50

Descartes, René, 242, 243

Determinants

    of 2×2 matrix, 674–75

    of 3×3 matrices, 675–76

Diameter, 94

Difference, 28, 30

    rationalizing denominator containing, 741–42

    square root of, 56–57

Dimensional analysis, converting units of measurement using, 505–6

Directions

    solving problems involving two objects traveling in opposite, 211–13

    solving problems involving two objects traveling in same, 213–14

Direct variation

    defined, 549

    solving problems involving, 548–50

Discriminant

    defined, 787

    in determining number of real solutions that quadratic equation has, 787

Distance

    finding between two points, 906

    formula for, 906–7

Distributive property, 380

    of multiplication over addition, 36, 72–73

    using, in expressions containing radicals, 733–34

Division

    of complex numbers, 759–60

    of decimal numbers, 44–45

    of fractions, 44

    of function, 606–7

    involving 0, 43–44

of monomials, 392–93

of numbers in scientific notation, 391–92

of polynomials by monomials, 393–94

of radical expressions, 725–26

of rational expressions, 503–5

of rational numbers, 42–44

sign rules for, 42

synthetic, App-21–22

translating word phrases to expressions in, 63

Domain, 838

    defined, 303, 594

    determining, with graph, 303–4

    finding, of radical function, 708–9

E

Earth

    gravity of, 129

    revolution of, 20

    tilt of, 104

    travel of, 376

Eiffel Tower, 181

Einstein, Albert, 335

Element

    of matrix, 675

    of set, 3

Elimination

    solving nonlinear systems of equations using, 927–28

    solving systems of linear equations using, 645–53, 659–61

Eliot, T. S., 241

Ellipses

    defined, 916

    equation of, centered at (0, 0), 916–17

    examples of, 904

    general equation for, 917

    graphing, 916

    standard form of, 920

Ellipsis, 4

Embedded parentheses, 55

Emerson, Ralph Waldo, 571, 775

Equality

    addition principle of, 105–13

    multiplication principle of, 117–24

Equations

    clearing decimals in, 122

    clearing fractions in, 121

    containing radicals, 796–97

    defined, 2, 92

    equivalent, 106

examples of, 2

graphing, in slope-intercept form, 273–75

with infinite number of solutions, 110–11

linear, 105

nonlinear, 105, 926

with no solution, 112

of parabolas, 905–6

radical, 747

with rational expressions, 796

rewriting

    in form $Ac + By = C$, 649

    in form $Ax + By = C$, 649

solving

    containing logarithms, 881–82

    containing rational expressions, 536–41

    involving absolute value, 583–85

    that are quadratic in form, 796–801

    with variables as exponents, 880

    with variable terms on both sides, 120

    translating word sentences to, 135–40

    verifying solutions to, 92–93

Equivalent equations, 106

Equivalent fractions, writing, 10–11

Eratosthenes, 13

Exponential equations

    properties for solving, 880

    solving, 852–53

Exponential forms

    defined, 50

    division of, with same bases, 390

    evaluating, 50–51

        with negative bases, 336

    evaluating numbers in, 50–51

    multiplying, 368–69

    zero as, 337

Exponential functions, 849–55

    defined, 850, 859

    graphing, 850, 851

    solving applications involving, 882–85

Exponents

    nonpositive integer, 337

    one-to-one property of, 852

    positive integer, 336

    product rule for, 368

    quotient rule for, 390–91

    rational, 714

    solving equations that have variables as, 880

GRAPHS OF COMMON FUNCTIONS

Linear Functions

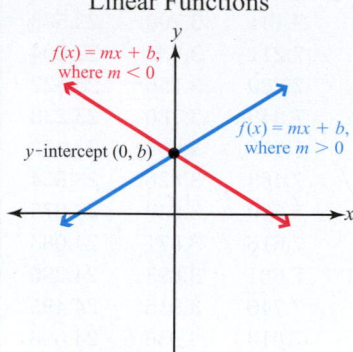

$f(x) = mx + b$, where $m < 0$

$f(x) = mx + b$, where $m > 0$

y-intercept $(0, b)$

Quadratic Functions

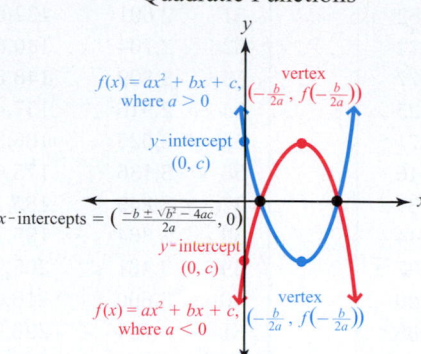

$f(x) = ax^2 + bx + c$, where $a > 0$

vertex $\left(-\frac{b}{2a}, f\left(-\frac{b}{2a}\right)\right)$

y-intercept $(0, c)$

x-intercepts $= \left(\frac{-b \pm \sqrt{b^2 - 4ac}}{2a}, 0\right)$

y-intercept $(0, c)$

vertex $\left(-\frac{b}{2a}, f\left(-\frac{b}{2a}\right)\right)$

$f(x) = ax^2 + bx + c$, where $a < 0$

Cubic Functions

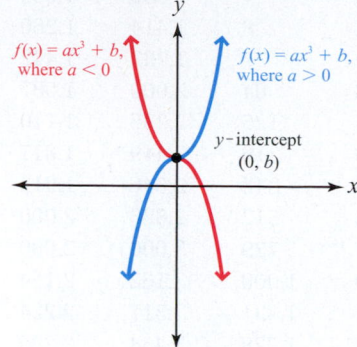

$f(x) = ax^3 + b$, where $a < 0$

$f(x) = ax^3 + b$, where $a > 0$

y-intercept $(0, b)$

Absolute Value Functions

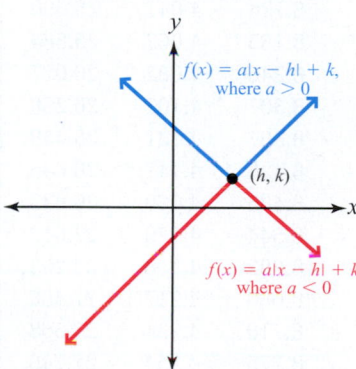

$f(x) = a|x - h| + k$, where $a > 0$

(h, k)

$f(x) = a|x - h| + k$, where $a < 0$

Radical Functions

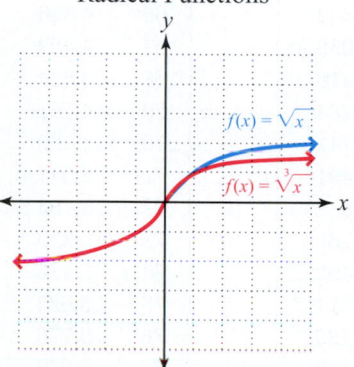

$f(x) = \sqrt{x}$

$f(x) = \sqrt[3]{x}$

Exponential and Logarithmic Functions

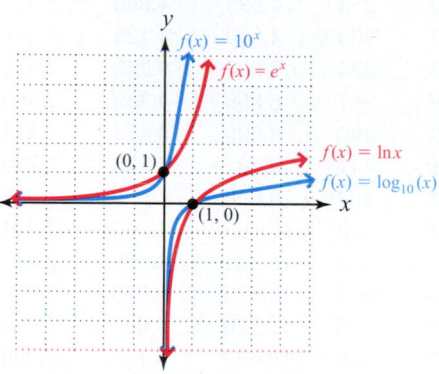

$f(x) = 10^x$

$f(x) = e^x$

$(0, 1)$

$f(x) = \ln x$

$f(x) = \log_{10}(x)$

$(1, 0)$

CONIC SECTIONS

Parabola

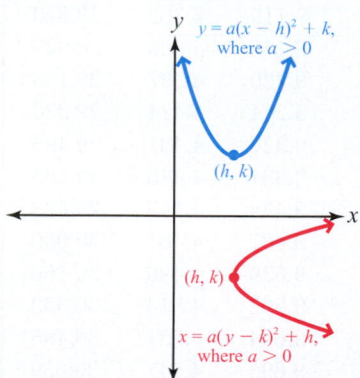

$y = a(x - h)^2 + k$, where $a > 0$

(h, k)

(h, k)

$x = a(y - k)^2 + h$, where $a > 0$

Circle

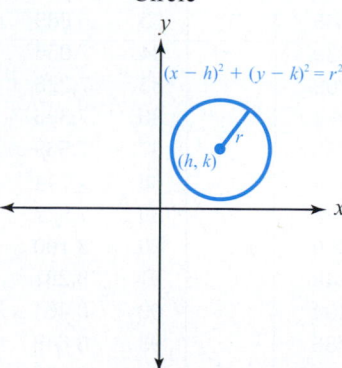

$(x - h)^2 + (y - k)^2 = r^2$

r

(h, k)

Ellipse

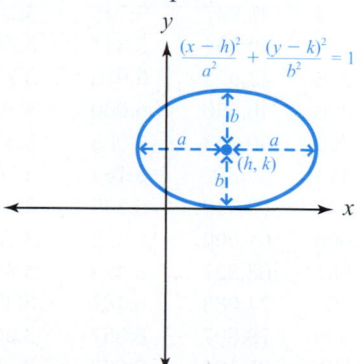

$\frac{(x - h)^2}{a^2} + \frac{(y - k)^2}{b^2} = 1$

b

a a

(h, k)

b

Hyperbola with center (0, 0)

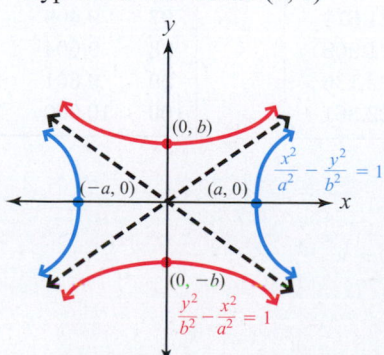

$(0, b)$

$(-a, 0)$

$(a, 0)$

$\frac{x^2}{a^2} - \frac{y^2}{b^2} = 1$

$(0, -b)$

$\frac{y^2}{b^2} - \frac{x^2}{a^2} = 1$

POWERS AND ROOTS

n	n^2	n^3	$\sqrt{n}$	$\sqrt[3]{n}$	$\sqrt{10n}$	n	n^2	n^3	$\sqrt{n}$	$\sqrt[3]{n}$	$\sqrt{10n}$
1	1	1	1.000	1.000	3.162	51	2,601	132,651	7.141	3.708	22.583
2	4	8	1.414	1.260	4.472	52	2,704	140,608	7.211	3.733	22.804
3	9	27	1.732	1.442	5.477	53	2,809	148,877	7.280	3.756	23.022
4	16	64	2.000	1.587	6.325	54	2,916	157,464	7.348	3.780	23.238
5	25	125	2.236	1.710	7.071	55	3,025	166,375	7.416	3.803	23.452
6	36	216	2.449	1.817	7.746	56	3,136	175,616	7.483	3.826	23.664
7	49	343	2.646	1.913	8.367	57	3,249	185,193	7.550	3.849	23.875
8	64	512	2.828	2.000	8.944	58	3,364	195,112	7.616	3.871	24.083
9	81	729	3.000	2.080	9.487	59	3,481	205,379	7.681	3.893	24.290
10	100	1,000	3.162	2.154	10.000	60	3,600	216,000	7.746	3.915	24.495
11	121	1,331	3.317	2.224	10.488	61	3,721	226,981	7.810	3.936	24.698
12	144	1,728	3.464	2.289	10.954	62	3,844	238,328	7.874	3.958	24.900
13	169	2,197	3.606	2.351	11.402	63	3,969	250,047	7.937	3.979	25.100
14	196	2,744	3.742	2.410	11.832	64	4,096	262,144	8.000	4.000	25.298
15	225	3,375	3.873	2.466	12.247	65	4,225	274,625	8.062	4.021	25.495
16	256	4,096	4.000	2.520	12.649	66	4,356	287,496	8.124	4.041	25.690
17	289	4,913	4.123	2.571	13.038	67	4,489	300,763	8.185	4.062	25.884
18	324	5,832	4.243	2.621	13.416	68	4,624	314,432	8.246	4.082	26.077
19	361	6,859	4.359	2.688	13.784	69	4,761	328,509	8.307	4.102	26.268
20	400	8,000	4.472	2.714	14.142	70	4,900	343,000	8.367	4.121	26.458
21	441	9,261	4.583	2.759	14.491	71	5,041	357,911	8.426	4.141	26.646
22	484	10,648	4.690	2.802	14.832	72	5,184	373,248	8.485	4.160	26.833
23	529	12,167	4.796	2.844	15.166	73	5,329	389,017	8.544	4.179	27.019
24	576	13,824	4.899	2.884	15.492	74	5,476	405,224	8.602	4.198	27.203
25	625	15,625	5.000	2.924	15.811	75	5,625	421,875	8.660	4.217	27.386
26	676	17,576	5.099	2.962	16.125	76	5,776	438,976	8.718	4.236	27.568
27	729	19,683	5.196	3.000	16.432	77	5,929	456,533	8.775	4.254	27.749
28	784	21,952	5.292	3.037	16.733	78	6,084	474,552	8.832	4.273	27.928
29	841	24,389	5.385	3.072	17.029	79	6,241	493,039	8.888	4.291	28.107
30	900	27,000	5.477	3.107	17.321	80	6,400	512,000	8.944	4.309	28.284
31	961	29,791	5.568	3.141	17.607	81	6,561	531,441	9.000	4.327	28.460
32	1,024	32,768	5.657	3.175	17.889	82	6,724	551,368	9.055	4.344	28.636
33	1,089	35,937	5.745	3.208	18.166	83	6,889	571,787	9.110	4.362	28.810
34	1,156	39,304	5.831	3.240	18.439	84	7,056	592,704	9.165	4.380	28.983
35	1,225	42,875	5.916	3.271	18.708	85	7,225	614,125	9.220	4.397	29.155
36	1,296	46,656	6.000	3.302	18.974	86	7,396	636,056	9.274	4.414	29.326
37	1,369	50,653	6.083	3.332	19.235	87	7,569	658,503	9.327	4.431	29.496
38	1,444	54,872	6.164	3.362	19.494	88	7,744	981,472	9.381	4.448	29.665
39	1,521	59,319	6.245	3.391	19.748	89	7,921	704,969	9.434	4.465	29.833
40	1,600	64,000	6.325	3.420	20.000	90	8,100	729,000	9.487	4.481	30.000
41	1,681	68,921	6.403	3.448	20.248	91	8,281	753,571	9.539	4.498	30.166
42	1,764	74,088	6.481	3.476	20.494	92	8,464	778,688	9.592	4.514	30.332
43	2,849	79,507	6.557	3.503	20.736	93	8,649	804,357	9.644	4.531	30.496
44	2,936	85,184	6.633	3.530	20.976	94	8,836	830,584	9.695	4.547	30.659
45	2,025	91,125	6.708	3.557	21.213	95	9,025	857,375	9.747	4.563	30.882
46	2,116	97,336	6.782	3.583	21.148	96	9,216	884,736	9.798	4.579	30.984
47	2,209	103,823	6.856	3.609	21.679	97	9,409	912,673	9.849	4.595	31.145
48	2,304	110,592	6.928	3.534	21.909	98	9,604	941,192	9.899	4.610	31.305
49	2,401	117,649	7.000	3.659	22.136	99	9,801	970,299	9.950	4.626	31.464
50	2,500	125,000	7.071	3.684	22.361	100	10,000	1,000,000	10.000	4.642	31.623

PROBLEM-SOLVING OUTLINE

1. **Understand** the problem.
   a. Read the question(s) (not the whole problem, just the question at the end) and write a note to yourself about what you are to find.
   b. Read the whole problem, underlining the key words.
   c. If possible or useful, make a list or table, simulate the situation, or search for a related example problem.

2. **Plan** your solution by searching for a formula or translating the key words to an equation.

3. **Execute** the plan by solving the equation/formula.

4. **Answer** the question. Look at the note about what you were to find and make sure you answer that question. Include appropriate units.

5. **Check** results.
   a. Try finding the solution in a different way, reversing the process, or estimating the answer and make sure the estimate and actual answer are reasonably close.
   b. Make sure the answer is reasonable.

USEFUL FORMULAS

Perimeter of a rectangle: $P = 2l + 2w$

Circumference of a circle: $C = \pi d$ or $C = 2\pi r$

Area of a parallelogram: $A = bh$

Area of a triangle: $A = \dfrac{1}{2}bh$

Area of a trapezoid: $A = \dfrac{1}{2}h(a + b)$

Area of a circle: $A = \pi r^2$

Surface area of a box: $SA = 2lw + 2lh + 2wh$

Volume of a box: $V = lwh$

Volume of a pyramid: $V = \dfrac{1}{3}lwh$

Volume of a cylinder: $V = \pi r^2 h$

Volume of a cone: $V = \dfrac{1}{3}\pi r^2 h$

Volume of a sphere: $V = \dfrac{4}{3}\pi r^3$

Distance, d, an object travels given its rate, r, and the time of travel, t: $d = rt$

The temperature in degrees Celsius given degrees Fahrenheit: $C = \dfrac{5}{9}(F - 32)$

The temperature in degrees Fahrenheit given degrees Celsius: $F = \dfrac{9}{5}C + 32$

The profit, P, after cost, C, is deducted from revenue, R: $P = R - C$

The Algebra Pyramid:
A Guide to the Development of Algebra Topics

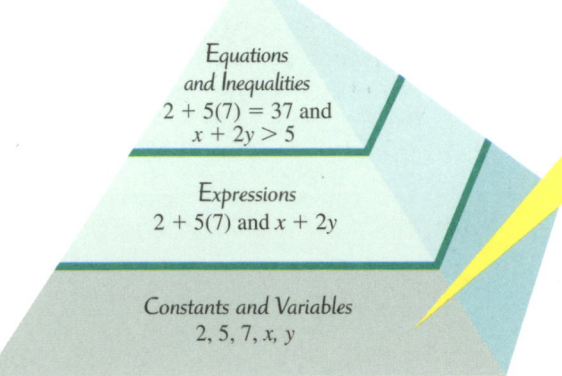

Constants and variables are the foundation of algebra. The following diagram summarizes the entire complex number system (all constants). These constants, along with variables such as x or y, are the basic building blocks that give meaning to the expressions, equations, and inequalities.

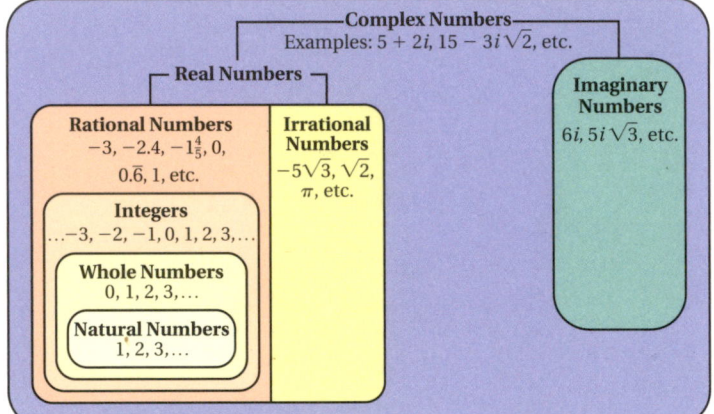

Expressions relate constants and variables using symbols such as operations of arithmetic to describe a calculation. We **evaluate** or **rewrite** expressions. To evaluate, we replace variables with numbers and perform the calculation. To rewrite, we use rules to write an alternative (and usually simpler) form.

The following list outlines the development of expressions in this text:

Chapter 1	Order of Operations
	Translating Word Phrases to Expressions
	Evaluating and Rewriting Expressions
Chapter 5	Exponent Rules and Scientific Notation
	Adding, Subtracting, Multiplying, and Dividing Polynomials
Chapter 6	Factoring Polynomials
Chapter 7	Simplifying, Adding, Subtracting, Multiplying, and Dividing Rational Expressions
	Simplifying Complex Rational Expressions
Chapter 10	Simplifying, Adding, Subtracting, Multiplying, and Dividing Radical Expressions
	Rational Exponents
	Rationalizing Numerators and Denominators of Radical Expressions
	Adding, Subtracting, Multiplying, and Dividing Complex Numbers

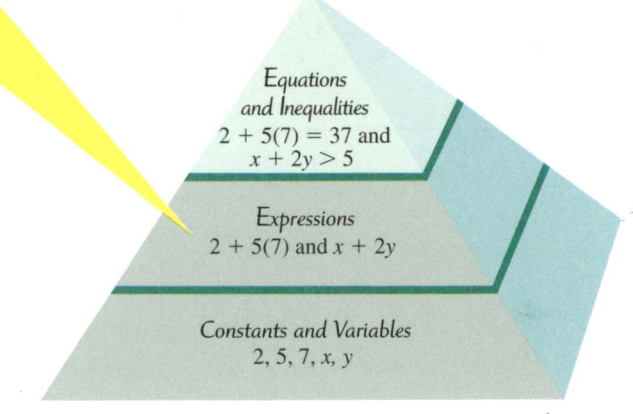